FE Civil Review

Michael R. Lindeburg, PE

PPI®
PPI2PASS.COM
A KAPLAN COMPANY

Register Your Book at ppi2pass.com

- Receive the latest exam news.
- Obtain exclusive exam tips and strategies.
- Receive special discounts.

Report Errors for This Book

PPI is grateful to every reader who notifies us of a possible error. Your feedback allows us to improve the quality and accuracy of our products. Report errata at **ppi2pass.com**.

FE CIVIL REVIEW

Current release of this edition: 2

Release History

date	edition number	revision number	update
May 2017	1	1	New book.
May 2019	1	2	Minor corrections. Minor cover updates.

Printed in the United States of America.

PPI
1250 Fifth Avenue, Belmont, CA 94002
(650) 593-9119
ppi2pass.com

ISBN: 978-1-59126-529-0

Library of Congress Control Number: 2017941424

PPI's Guarantee

This *FE Civil Review* is your best choice to prepare for the Civil Fundamentals of Engineering (FE) examination. It is the only review that

- covers every Civil FE exam knowledge area
- is based on the NCEES *FE Reference Handbook* (*NCEES Handbook*)
- provides example questions in true exam format
- provides instructional material for essentially every relevant equation, figure, and table in the *NCEES Handbook*
- can be accessed online at **ppi2pass.com**

PPI is confident that if you use this book conscientiously to prepare for the Civil FE exam, following the guidelines described in the "How to Use This Book" section, you'll pass the exam. Otherwise, regardless of where you purchased this book, with no questions asked, we will refund the purchase price of your printed book (up to PPI's published website price).

To request a refund, you must provide the following items within three months of taking the exam:

1. A summary letter listing your name, email address, and mailing address
2. Your original packing slip, store sales receipt, or online order acknowledgment showing the price you paid (price must be individually listed on receipt)
3. A dated email, notification letter, or printout of your MyNCEES webpage showing that you did not pass the FE exam
4. Your book

Mail all items to:

PPI
FE Civil Review Refund
1250 Fifth Avenue
Belmont, CA 94002

This guarantee does not extend to web access books, other products, or books with invalid registration numbers.

PPI®
PPI2PASS.COM

Topics

Topic I:	Mathematics
Topic II:	Probability and Statistics
Topic III:	Fluid Mechanics
Topic IV:	Hydraulics and Hydrologic Systems
Topic V:	Environmental Engineering
Topic VI:	Geotechnical Engineering
Topic VII:	Statics
Topic VIII:	Dynamics
Topic IX:	Mechanics of Materials
Topic X:	Materials
Topic XI:	Structural Design
Topic XII:	Transportation and Surveying
Topic XIII:	Construction
Topic XIV:	Computational Tools
Topic XV:	Engineering Economics
Topic XVI:	Ethics and Professional Practice

Mathematics
Probability/ Statistics
Fluid Mechanics
Hydrolics/ Hydrolic Sys.
Environmental Engineering
Geotechnical Engineering
Statics
Dynamics
Mechanics of Materials
Materials
Structural Design
Transportation/ Surveying
Construction
Computational Tools
Engineering Economics
Ethics/ Prof. Prac.

Table of Contents

Preface

The purpose of this book is to prepare you for the National Council of Examiners for Engineering and Surveying (NCEES) fundamentals of engineering (FE) exam.

In 2014, the NCEES adopted revised specifications for the exam. The council also transitioned from a paper-based version of the exam to a computer-based testing (CBT) version. The FE exam now requires you to sit in front of a monitor, respond to questions served up by the CBT system, access an electronic reference document, and perform your scratch calculations on a reusable notepad. You may also use an on-screen calculator with which you will likely be unfamiliar. The experience of taking the FE exam will probably be unlike anything you have ever, or will ever again, experience in your career. Similarly, preparing for the exam will be unlike preparing for any other exam.

The CBT FE exam presented three new challenges to me when I began preparing instructional material for it. (1) The subjects in the testable body of knowledge are oddly limited and do not represent a complete cross section of the traditional engineering fundamentals subjects. (2) The NCEES *FE Reference Handbook* (*NCEES Handbook*) is poorly organized, awkwardly formatted, inconsistent in presentation, and idiomatic in convention. (3) Traditional studying, doing homework while working toward a degree, and working at your own desk as a career engineer are poor preparations for the CBT exam experience.

No existing exam review book overcomes all of these challenges. But, I wanted you to have something that does. So, in order to prepare you for the CBT FE exam, this book was designed and written from the ground up. In many ways, this book is as unconventional as the exam.

This book covers all of the knowledge areas listed in the NCEES Civil FE exam specifications. For all practical purposes, this book contains the equivalent of all of the equations, tables, and figures presented in the *NCEES Handbook*, ninth edition (June 2016 revision) that you will need for the Civil FE exam. And, with the exceptions listed in the "Variables" section, for better or worse, this book duplicates the terms, variables, and formatting of the *NCEES Handbook* equations.

NCEES has selected, what it believes to be, all of the engineering fundamentals important to an early-career, minimally-qualified engineer, and has distilled them into its single reference, the *NCEES Handbook*. Personally, I cannot accept the premise that engineers learn and use so little engineering while getting their degrees and during their first few career years. However, regardless of whether you accept the NCEES subset of engineering fundamentals, one thing is certain: In serving as your sole source of formulas, theory, methods, and data during the exam, the *NCEES Handbook* severely limits the types of questions that can be included in the FE exam.

The obsolete paper-based exam required very little knowledge outside of what was presented in the previous editions of the *NCEES Handbook*. That *NCEES Handbook* supported a plug-and-chug examinee performance within a constrained body of knowledge. Based on the current FE exam specifications and the *NCEES Handbook*, the CBT FE exam is even more limited than the old paper-based exam. The number (breadth) of knowledge areas, the coverage (depth) of knowledge areas, the number of questions, and the duration of the exam are all significantly reduced. If you are only concerned about passing and/or "getting it over with" before graduation, these reductions are all in your favor. Your only deterrents will be the cost of the exam and the inconvenience of finding a time and place to take it.

Accepting that "it is what it is," I designed this book to guide you through the exam's body of knowledge.

I have several admissions to make: (1) This book contains nothing magical or illicit. (2) This book, by itself, is only one part of a complete preparation. (3) This book stops well short of being perfect. What do I mean by those admissions?

First, this book does not contain anything magical. It's called a "review" manual, and you might even learn something new from it. It will save you time in assembling review material and questions. However, it won't learn the material for you. Merely owning it is not enough. You will have to put in the time to use it.

Similarly, there is nothing clandestine or unethical about this book. It does not contain any actual exam questions. It was written in a vacuum, based entirely on the NCEES Civil FE exam specifications. This book is not based on feedback from actual examinees.

Truthfully, I expect that many exam questions will be similar to the questions I have used because NCEES and I developed content with the same set of constraints. (If anything, NCEES is even more constrained when it comes to fringe, outlier, eccentric, or original topics.)

There is a finite number of ways that questions about Ohm's law ($V = IR$) and Newton's second law of motion ($F = ma$) can be structured. Any similarity between questions in this book and questions in the exam is easily attributed to the limited number of engineering formulas and concepts, the shallowness of the coverage, and the need to keep the entire solution process (reading, researching, calculating, and responding) to less than three minutes for each question.

Let me give an example to put some flesh on the bones. As any competent engineer can attest, in order to calculate the pressure drop in a pipe network, you would normally have to (1) determine fluid density and viscosity based on the temperature, (2) convert the mass flow rate to a volumetric flow rate, (3) determine the pipe diameter from the pipe size designation (e.g., pipe schedule), (4) calculate the internal pipe area, (5) calculate the flow velocity, (6) determine the specific roughness from the conduit material, (7) calculate the relative roughness, (8) calculate the Reynolds number, (9) calculate or determine the friction factor graphically, (10) determine the equivalent length of fittings and other minor losses, (11) calculate the head loss, and finally, (12) convert the head loss to pressure drop. Length, flow quantity, and fluid property conversions typically add even more complexity. (SSU viscosity? Diameter in inches? Flow rate in SCFM?) As reasonable and conventional as that solution process is, a question of such complexity is beyond the upper time limit for an FE exam question.

To make it possible to be solved in the time allowed, any exam question you see is likely to be more limited. In fact, most or all of the information you need to answer a question will be given to you in its question statement. If only the real world were so kind!

Second, by itself, this book is inadequate. It was never intended to define the entirety of your preparation activity. While it introduces essentially all of the exam knowledge areas and content in the *NCEES Handbook*, an introduction is only an introduction. To be a thorough review, this book needs augmentation.

By design, this book has three significant inadequacies.

1. This book is "only" 736 pages long, so it cannot contain enough of everything for everyone. The number of example questions that can fit in it are limited. The number of questions needed by you, personally, to come up to speed in a particular subject may be inadequate. For example, how many questions will you have to review in order to feel comfortable about divergence, curl, differential equations, and linear algebra? (Answer: Probably more than are in all the books you will ever own!) So, additional exposure is inevitable if you want to be adequately prepared in every subject.

2. This book does not contain the *NCEES Handbook*, per se. This book is limited in helping you become familiar with the idiosyncratic sequencing, formatting, variables, omissions, and presentation of topics in the *NCEES Handbook*. The only way to remedy this is to obtain your own copy of the *NCEES Handbook* (available in printed format from PPI and as a free download from the NCEES website) and use it in conjunction with your review.

3. This book does not contain a practice examination (mock exam, sample exam, etc.). With the advent of the CBT format, any sample exam in printed format is little more than another collection of practice questions. The actual FE exam is taken sitting in front of a computer using an online reference book, so the only way to practice is to sit in front of a computer while you answer questions. Using an online reference is very different from the work environment experienced by most engineers, and it will take some getting used to.

Third, and finally, I reluctantly admit that I have never figured out how to write or publish a completely flawless first (or, even subsequent) edition. The PPI staff comes pretty close to perfection in the areas of design, editing, typography, and illustrating. Subject matter experts help immensely with calculation checking. And, beta testing before you see a book helps smooth out wrinkles. However, I still manage to muck up the content. So, I hope you will "let me have it" when you find my mistakes. PPI has established an easy way for you to report an error, as well as to review changes that resulted from errors that others have submitted. Just go to **ppi2pass.com/errata**. When you submit something, I'll receive it via email. When I answer it, you'll receive a response. We'll both benefit.

Best wishes in your examination experience. Stay in touch!

Michael R. Lindeburg, PE

Acknowledgments

Developing a book specific to the computerized Civil FE exam has been a monumental project. It involved the usual (from an author's and publisher's standpoint) activities of updating and repurposing existing content and writing new content. However, the project was made extraordinarily more difficult by two factors: (1) a new book design, and (2) the publication schedule.

Creating a definitive resource to help you prepare for the computerized FE exam was a huge team effort, and PPI's entire Publishing Services staff was heavily involved. Along the way, they had to learn new skills and competencies, solve unseen technical mysteries, and exercise professional judgment in decisions that involved publishing, resources, engineering, and user utility. They worked long hours, week after week, and month after month, often into the late evening, to publish this book for examinees taking the exam.

PPI staff members have had a lot of things to say about this book during its development. In reference to you and other examinees being unaware of what PPI staff did, one of the often-heard statements was, "They will never know."

However, I want you to know, so I'm going to tell you.

Editorial project manager Becky Morgan managed the gargantuan operation, with considerable support from Grace Wong, director of publishing services. Production services manager Cathy Schrott kept the process moving smoothly and swiftly. Steve Buehler, director of acquisitions, and Nicole Evans, acquisitions editor, arranged for all of the outside subject matter experts that were involved with this book. All of the content was eventually reviewed for consistency, PPI style, and accuracy by Jennifer Lindeburg King, editor-in-chief.

Though everyone in Publishing Services has a specialty, this project pulled everyone from his or her comfort zone. Everyone learned (with amazing speed) how to grapple with the complexities of XML and MathML while wrestling misbehaving computer code into submission. Tom Bergstrom, technical illustrator, updated existing illustrations and created new ones, and made typesetting corrections. Copy editor Tyler Hayes copy edited and proofread. Copy editors Robert Genevro and Richard Iriye proofread and corrected, and composition associate Gabby Howitz typeset, proofread, and corrected. To bring this new edition into alignment with the June 2016 revision, new content had to be added to several chapters. Two subject matter experts took on this task for PPI, producing new text, illustrations, and problems to ensure that this manual is as complete as possible. Mark Schmelter, PhD, PE, and Kenneth Wasserman, PE, took on this task, and Zane Bradley Wells, technical reviewer, helped PPI ensure the accuracy and comprehensibility of their work.

Paying customers (such as you) shouldn't have to be test pilots. So, close to the end of the process, when content was starting to coalesce out of the shapelessness of the PPI content management system, several subject matter experts became crash car dummies "for the good of engineering." They pretended to be examinees and worked through all of the content, looking for calculation errors, references that went nowhere, and logic that was incomprehensible. These engineers and their knowledge area contributions are: C. Dale Buckner, PhD, PE, SECB (Statics and Structural Design); John C. Crepeau, PhD, PE (Dynamics; Mathematics; and Mechanics of Materials); Joshua T. Frohman, PE (Computational Tools; Construction; Probability and Statistics; and Transportation Engineering); David Hurwitz, PhD (Computational Tools; Construction; Probability and Statistics; and Transportation Engineering); David Johnstone, PhD, PE (Environmental Engineering; Geotechnical Engineering; and Hydraulics and Hydrologic Systems); Liliana M. Kandic, PE (Fluid Mechanics; Statics; and Structural Design); Aparna Phadnis, PE (Engineering Economics; Environmental Engineering; Geotechnical Engineering; and Hydraulics and Hydrologic Systems); David To, PE (Dynamics; Fluid Mechanics; Mathematics; and Mechanics of Materials); and L. Adam Williamson, PE (Fluid Mechanics; Dynamics; and Materials).

Consistent with the past 40 years, I continue to thank my wife, Elizabeth, for accepting and participating in a writer's life that is full to overflowing. Even though our children have been out on their own for a long time, we seem to have even less time than we had before. As a corollary to Aristotle's "Nature abhors a vacuum," I propose: "Work expands to fill the void."

To my granddaughter, Sydney, who had to share her Gaga with his writing, I say, "Thank you for being a bright light in my life. Elmo, Grover, and I always look forward to your visits."

I also appreciate the grant of permission to reproduce materials from several other publishers. In each case, attribution is provided where the material has been included. Neither PPI nor the publishers of the reproduced material make any representations or warranties as to the accuracy of the material, nor are they liable for any damages resulting from its use.

Thank you, everyone! I'm really proud of what you've accomplished. Your efforts will be pleasing to examinees and effective in preparing them for the Civil FE exam.

Michael R. Lindeburg, PE

Codes and References Used to Prepare This Book

This book is based on the NCEES *FE Reference Handbook* (*NCEES Handbook*), ninth edition (June 2016 revision). The other documents, codes, and standards that were used to prepare this book were the most current available at the time.

NCEES does not specifically tie the FE exam to any edition (version) of any code or standard. Rather than make the FE exam subject to the vagaries of such codes and standards published by the American Concrete Institute (ACI), the American Institute of Steel Construction (AISC), the American National Standards Institute (ANSI), the American Society of Civil Engineers (ASCE), the American Society of Heating, Refrigerating and Air-Conditioning Engineers (ASHRAE), the American Society of Mechanical Engineers (ASME), ASTM International (ASTM), the International Code Council (ICC), and so on, NCEES effectively writes its own "code," the *NCEES Handbook*.

Most surely, every standard- or code-dependent concept (e.g., the design of rectangular, tied concrete columns) in the *NCEES Handbook* can be traced back to some section of some edition of a standard or code (e.g., ACI 318). So, it would be logical to conclude that you need to be familiar with everything (the limitations, surrounding sections, and commentary) in the code related to that concept. However, that does not seem to be the case. The *NCEES Handbook* is a code unto itself, and you won't need to study the parent documents. Nor will you need to know anything pertaining to related, adjacent, similar, or parallel code concepts.

Therefore, although methods and content in the *NCEES Handbook* can be ultimately traced back to some edition (version) of a relevant code, you don't need to know which. You don't need to know whether that content is current, limited in intended application, or relevant. You only need to use the content.

Introduction

PART 1: ABOUT THIS BOOK

This book is intended to guide you through the Civil Fundamentals of Engineering (FE) examination body of knowledge and the idiosyncrasies of the National Council of Examiners for Engineers and Surveyors (NCEES) *FE Reference Handbook* (*NCEES Handbook*). This book is not intended as a reference book, because you cannot use it while taking the FE examination. The only reference you may use is the *NCEES Handbook*. However, the *NCEES Handbook* is not intended as a teaching tool, nor is it an easy document to use. The *NCEES Handbook* was never intended to be something you study or learn from, or to have value as anything other than an exam-day compilation. Many of its features may distract you because they differ from what you were expecting, were exposed to, or what you currently use.

To effectively use the *NCEES Handbook*, you must become familiar with its features, no matter how odd they may seem. *FE Civil Review* will help you become familiar with the format, layout, organization, and odd conventions of the *NCEES Handbook*. This book, which displays the *NCEES Handbook* material in blue for easy identification, satisfies two important needs: it is (1) something to learn from, and (2) something to help you become familiar with the *NCEES Handbook*.

Organization

This book is organized into topics (e.g., "Structural Design") that correspond to the knowledge areas listed by NCEES in its Civil FE exam specifications. However, unlike the *NCEES Handbook*, this book arranges subtopics into chapters (e.g., "Reinforced Concrete: Beams") that build logically on one another. Each chapter contains sections (e.g., "Design of Doubly Reinforced Beams") organized around *NCEES Handbook* equations, but again, the arrangement of those equations is based on logical development, not the *NCEES Handbook*. Equations that are presented together in this book may actually be many pages apart in the *NCEES Handbook*.

The presentation of each subtopic or related group of equations uses similar components and follows a specific sequence. The components of a typical subtopic are:

- general section title
- background and developmental content
- equation name (or description) and equation number
- equation with *NCEES Handbook* formatting
- any relevant variations of the equation
- any values typically associated with the equation
- additional explanation and development
- worked quantitative example using the *NCEES Handbook* equation
- footnotes

Not all sections contain all of these features. Some features may be omitted if they are not needed. For example, "$g = 9.81\text{ m/s}^2$" would be a typical value associated with the equation $W = mg$. There would be no typical values associated with the equation $F = ma$.

Much of the information in this book and in the *NCEES Handbook* is relevant to more than one knowledge area or subtopic. For example, equations related to the Statics knowledge area also pertain to the subtopics of Structural Analysis. Many Hydraulics and Hydrologic Systems concepts correlate with Environmental Engineering subtopics. The index will help you locate all information related to any of the topics or subtopics you wish to review.

Content

This book presents equations, figures, tables, and other data equivalent to those given in the *NCEES Handbook*. For example, the *NCEES Handbook* includes tables for conversion factors, material properties, and areas and centroids of geometric shapes, so this book provides equivalent tables. Occasionally, a redundant element of the *NCEES Handbook*, or some item having no value to examinees, has been omitted.

Some elements, primarily figures and tables, that were originally published by authoritative third parties (and for whom reproduction permission has been granted) have been reprinted exactly as they appear in the *NCEES Handbook*. Other elements have been editorially and artistically reformulated, but they remain equivalent in utility to the originals.

Colors

Due to the selective nature of topics included in the *NCEES Handbook*, coverage of some topics in the *NCEES Handbook* may be incomplete. This book aims to offer more comprehensive coverage, and so, it contains material that is not covered in the *NCEES*

Handbook. This book uses color to differentiate between what is available to you during the exam, and what is supplementary content that makes a topic more interesting or easier to understand. Equations that closely parallel or duplicate the NCEES Handbook equations are printed with blue backgrounds. Headings that introduce content related to *NCEES Handbook* equations are printed in blue. Titles of figures and tables that are essentially the same as in the *NCEES Handbook* are similarly printed in blue. Headings that introduce sections, equations, figures, and tables that are NOT in the *NCEES Handbook* are printed in **black**. The **black** content is background, preliminary and supporting material, explanations, extensions to theory, and application rules that are generally missing from the *NCEES Handbook*.

Numbering

The equations, figures, and tables in the *NCEES Handbook* are unnumbered. All equations, figures, and tables in this book include unique numbers provided to help you navigate through the content.

You will find many equations in this book that have no numbers. These equations represent instructional material, often missing pieces or interim results not presented in the *NCEES Handbook*. In some cases, the material was present in the eighth edition of the *NCEES Handbook*, but is absent in the ninth edition. In some cases, I included instruction in deleted content. (This book does not contain all of the deleted *NCEES Handbook* eighth edition content, however.)

Equation and Variable Names

This book generally uses the *NCEES Handbook* terminology and naming conventions, giving standard, normal, and customary alternatives within parentheses or footnotes. For example, the *NCEES Handbook* refers to what is commonly known as Bernoulli equation as the "energy equation." This book acknowledges the *NCEES Handbook* terminology when introducing the equation, but uses the term "Bernoulli equation" thereafter.

Variables

This book makes every effort to include the *NCEES Handbook* equations exactly as they appear in the *NCEES Handbook*. While any symbol can be defined to represent any quantity, in many cases, the *NCEES Handbook*'s choice of variables will be dissimilar to what most engineers are accustomed to. For example, although there is no concept of weight in the SI system, the *NCEES Handbook* defines W as the symbol for weight with units of newtons. While engineers are comfortable with E, E_k, KE, and U representing kinetic energy, after introducing KE in its introductory pages, the *NCEES Handbook* uses T (which is used sparingly by some scientists) for kinetic energy. The *NCEES Handbook* designates power as $\dot{W}$ instead of P. The *NCEES Handbook* uses Greek omega, ω, to represent water content, rather than the industry-standard w. Because you have to be familiar with them, this book reluctantly follows all of those conventions.

This book generally follows the *NCEES Handbook* convention regarding use of italic fonts, even when doing so results in ambiguity. For example, as used by the *NCEES Handbook*, aspect ratio, AR, is indistinguishable from $A \times R$, area times radius. Occasionally, the *NCEES Handbook* is inconsistent in how it represents a particular variable, or in some sections, it drops the italic font entirely and presents all of its variables in roman font. This book maintains the publishing convention of showing all variables as italic.

There are a few important differences between the ways the *NCEES Handbook* and this book present content. These differences are intentional for the purpose of maintaining clarity and following PPI's publication policies.

- *pressure:* The *NCEES Handbook* primarily uses P for pressure, an atypical engineering convention. This book always uses p so as to differentiate it from P, which is reserved for power, momentum, and axial loading in related chapters.
- *velocity:* The *NCEES Handbook* uses v and occasionally Greek nu, ν, for velocity. This book always uses v to differentiate it from Greek upsilon, υ, which represents specific volume in some topics (e.g., thermodynamics), and Greek nu, ν, which represents absolute viscosity and Poisson's ratio.
- *specific volume:* The *NCEES Handbook* uses v for specific volume. This book always uses Greek upsilon, υ, a convention that most engineers will be familiar with.
- *units:* The *NCEES Handbook* and the FE exam generally do not emphasize the difference between pounds-mass and pounds-force. "Pounds" ("lb") can mean either force or mass. This book always distinguishes between pounds-force (lbf) and pounds-mass (lbm).

Distinction Between Mass and Weight

The *NCEES Handbook* specifies the unit weight of water, γ_w, as 9.810 N/m^3. This book follows that convention but takes every opportunity to point out that there is no concept of weight in the SI system.

Equation Formatting

The *NCEES Handbook* writes out many multilevel equations as an awkward string of characters on a single line, using a plethora of parentheses and square and curly brackets to indicate the precedence of mathematical operations. So, this book does also. However, in examples using the equations, this book reverts to normal publication style after presenting the base equation styled as it is in the *NCEES Handbook*. The change in style will show you the equations as the *NCEES Handbook* presents them, while presenting the calculations in a normal and customary typographic manner.

Footnotes

I have tried to anticipate the kinds of questions about this book and the *NCEES Handbook* that an instructor would be asked in class. Footnotes are used in this book as the preferred method of answering those questions and of drawing your attention to features in the *NCEES Handbook* that may confuse, confound, and infuriate you. Basically, *NCEES Handbook* conventions are used within the body of this book, and any inconsistencies, oddities and unconventionalities, and occasionally, even errors, are pointed out in the footnotes.

If you know the NCEES knowledge areas backward as well as forward, many of the issues pointed out in the footnotes will seem obvious. However, if you have only a superficial knowledge of the knowledge areas, the footnotes will answer many of your questions. The footnotes are intended to be factual and helpful.

Indexed Terms

The print version of this book contains an index with thousands of terms. The index will help you quickly find just what you are looking for, as well as identify related concepts and content.

PART 2: HOW YOU CAN USE THIS BOOK

IF YOU ARE A STUDENT

In reference to Isaac Asimov's *Foundation and Empire* trilogy, you'll soon experience a Seldon crisis. Given all the factors (the exam you're taking, what you learned as a student, how much time you have before the exam, and your own personality), the behaviors (strategies made evident through action) required of you will be self-evident.

Here are some of those strategies.

Get the NCEES *FE Reference Handbook*

Get a copy of the *NCEES Handbook*. Use it as you read through this book. You will want to know the sequence of the sections, what data is included, and the approximate locations of important figures and tables in the *NCEES Handbook*. You should also know the terminology (words and phrases) used in the *NCEES Handbook* to describe equations or subjects, because those are the terms you will have to look up during the exam.

The *NCEES Handbook* is available both in printed and PDF format. The index of the print version may help you locate an equation or other information you are looking for, but few terms are indexed thoroughly. The PDF version includes search functionality that is similar to what you'll have available when taking the computer-based exam. In order to find something using the PDF search function, your search term will have to match the content exactly (including punctuation).

Diagnose Yourself

Use the diagnostic exams in this book to determine how much you should study in the various knowledge areas. You can use diagnostic exams (and, other assessments) in two ways: take them before you begin studying to determine which subjects you should emphasize, or take them after you finish studying to determine if you are ready to move on.

Make a Schedule

In order to complete your review of all examination subjects, you must develop and adhere to a review schedule. If you are not taking a live review course (where the order of your preparation is determined by the lectures), you'll want to prepare your own schedule.

The amount of material in each chapter of this book, and the number of questions in the corresponding chapter of *FE Civil Practice*, were designed to fit into a practical schedule. You should be able to review one chapter in each book each day. There are 63 chapters in this book, as well as 16 diagnostic exams available for free at **ppi2pass.com** with the purchase of this book, and corresponding chapters of practice problems in the companion book *FE Civil Practice Problems*. So, you need at least 79 study days. This requires you to treat every day the same and work through weekends.

If you'd rather take all the weekends off and otherwise stick with the one-chapter-per-study-day concept, you will have to begin approximately 111 days before the exam. Use the off days to rest, review, and study questions from other books. If you are pressed for time or get behind schedule, you don't have to take the days off. That will be your choice.

Near the exam date, give yourself a week to take a realistic practice exam, to remedy any weaknesses it exposes, and to recover from the whole ordeal.

Work Through Everything

NCEES has greatly reduced the number of subjects about which you are expected to be knowledgeable and has made nothing optional. Skipping your weakest subjects is no longer a viable preparation strategy. You should study all examination knowledge areas, not just your specialty areas. That means you study every chapter in this book and skip nothing. Do not limit the number of chapters you study in hopes of finding enough questions in your areas of expertise to pass the exam.

Be Thorough

Being thorough means really doing the work. Read the material, don't skim it. Solve each numerical example using your calculator. Read through the solution, and refer back to the equations, figures, and tables it references.

Don't jump into answering questions without first reviewing the instructional text in this book. Unlike reference books that you skim or merely refer to when needed, this book requires you to read everything. That reading is going to be your only review. Reading the instructional text is a "high value" activity. There isn't much text to read in the first place, so the value per word is high. There aren't any derivations or proofs, so the text is useful. Everything in blue titled sections is in the *NCEES Handbook*, so it has a high probability of showing up on the exam.

Work Problems

You have less than an average of three minutes to answer each question on the exam. You must be able to quickly recall solution procedures, formulas, and important data. You will not have time to derive solution methods—you must know them instinctively. The best way to develop fast recall is to work as many practice problems as you can find, including those in the companion book *FE Civil Practice*.

Solve every example in this book and every problem in *FE Civil Practice*. Don't skip any of them. All of the problems were written to illustrate key points.

Finish Strong

There will be physical demands on your body during the examination. It is very difficult to remain alert, focused, and attentive for six hours or more. Unfortunately, the more time you study, the less time you have to maintain your physical condition. Thus, most examinees arrive at the examination site in high mental condition but in deteriorated physical condition. While preparing for the FE exam is not the only good reason for embarking on a physical conditioning program, it can serve as a good incentive to get in shape.

Claim Your Reward

As Hari Seldon often said in Isaac Asimov's *Foundation and Empire* trilogy, the outcome of your actions will be inevitable.

IF YOU ARE AN INSTRUCTOR

CBT Challenges

The computer-based testing (CBT) FE exam format, content, and frequent administration present several challenges to teaching a live review course. Some of the challenges are insurmountable to almost all review courses. Live review courses cannot be offered year round, a different curriculum is required for each engineering discipline, and a hard-copy, in-class mock exam taken at the end of the course no longer prepares examinees for the CBT experience. The best that instructors can do is to be honest about the limitations of their courses, and to refer examinees to any other compatible resources.

Many of the standard, tried-and-true features of live FE review courses are functionally obsolete. These obsolete features include general lectures that cover "everything," complex numerical examples with more than two or three simple steps, instructor-prepared handouts containing notes and lists of reference materials, and a hard-copy mock exam. As beneficial as those features were in the past, they are no longer best commercial practice for the CBT FE exam. However, they may still be used and provide value to examinees.

This book parallels the content of the *NCEES Handbook* and, with the exceptions listed in this Introduction, uses the same terminology and nomenclature. The figures and tables are equivalent to those in the *NCEES Handbook*. You can feel confident that I had your students and the success of your course in mind when I designed this book.

Instruction for Multiple Exams

Since this book is intended to be used by those studying for the Civil FE exam, there is no easy way to use it as the basis for more than a Civil FE exam review course.

Historically, most commercial review courses (taken primarily by engineers who already have their degrees) prepared examinees for the Other Disciplines FE examination. That is probably the only logical (practical, sustainable, etc.) course of action, even now. Few commercial review course providers have the large customer base and diverse instructors needed to offer simultaneous courses for every discipline.

University review courses frequently combine students from multiple disciplines, focusing the review course content on the core overlapping concepts and the topics covered by the Other Disciplines FE exam. The change in the FE exam scope has made it more challenging than ever to adequately prepare a diverse student group.

Lectures

Your lectures should duplicate what the examinees would be doing in a self-directed review program. That means walking through each chapter in this book in its entirety. You're basically guiding a tour through the book. By covering everything in this book, you'll cover everything on the exam.

Handouts

Everything you do in a lecture should be tied back to the *NCEES Handbook*. You will be doing your students a great disservice if you get them accustomed to using your course handouts or notes to solve problems. They can't use your notes in the exam, so train them to use the only reference they are allowed to use.

NCEES allows that the exam may require broader knowledge than the *NCEES Handbook* contains. However, there are very few areas that require formulas not present in the *NCEES Handbook*. Therefore, you shouldn't deviate too much from the subject matter of each chapter.

Homework

Students like to see and work a lot of problems. They derive great comfort from exposure to exam-like problems. They experience great reassurance in working exam-like problems and finding out how easy the problems are. However, most students are impatient. So, the repetition and reinforcement should come from working additional problems, not from more lecture.

It is unlikely that your students will be working to capacity if their work is limited to what is in this book. You will have to provide or direct your students to more problems in order to help them effectively master the concepts you will be teaching.

Schedule

I have found that a 15-week format works best for a live FE exam review course that covers everything and is intended for working engineers who already have their degrees. This schedule allows for one 2 to $2^1/_2$ hour lecture per week, with a 10-minute break each hour.

Table 1 outlines a typical format for a live commercial Civil FE review course. To some degree, the lectures build upon one another. However, a credible decision can be made to present the knowledge areas in the order they appear in the *NCEES Handbook*.

However, a 15-week course is too long for junior and senior engineering majors still working toward a degree. College students and professors don't have that much time. And, students don't need as thorough of a review as do working engineers who have forgotten more of the fundamentals. College students can get by with the most cursory of reviews in some knowledge areas, such as mathematics, fluid mechanics, and statics.

For college students, an 8-week course consisting of six weeks of lectures followed by two weeks of open questions seems appropriate. If possible, two 1-hour lectures per week are more likely to get students to attend than a single 2- or 3-hour lecture per week. The course consists of a comprehensive march through all knowledge areas except mathematics, with the major emphasis being on problem-solving rather than lecture. For current engineering majors, the main goals are to keep the students focused and to wake up their latent memories, not to teach the subjects.

Table 2 outlines a typical format for a live university review course. The sequence of the lectures is less important for a university review course than for a commercial course, because students will have recent experience in the subjects. Some may actually be enrolled in some of the related courses while you are conducting the review.

I strongly believe in the benefits of exposing all review course participants to a realistic sample examination. Unless you have made arrangements with **ppi2pass.com** for your students to take an online exam, you probably cannot provide them with an experience equivalent to the actual exam. A written take-home exam is better than nothing, but since it will not mimic the exam experience, it must be presented as little more than additional problems to solve.

I no longer recommend an in-class group final exam. Since a review course usually ends only a few days before the real FE examination, it seems inhumane to make students sit for hours into the late evening for the final exam. So, if you are going to use a written mock exam, I recommend distributing it at the first meeting of the review course and assigning it as a take-home exercise.

PART 3: ABOUT THE EXAM

EXAM STRUCTURE

The FE exam is a computer-based test that contains 110 multiple-choice questions given over two consecutive sessions (sections, parts, etc.). Each session contains approximately 55 multiple-choice questions that are grouped together by knowledge area (subject, topic, etc.). The subjects are not explicitly labeled, and the beginning and ending of the subjects are not noted. No subject spans the two exam sessions. That is, if a subject appears in the first session of the exam, it will not appear in the second.

Each question has four possible answer choices, labeled (A), (B), (C), and (D). Only one question and its answer choices is given onscreen at a time. The exam is not adaptive (i.e., your response to one question has no bearing on the next question you are given). Even if you answer the first five mathematics questions correctly, you'll still have to answer the sixth question.

In essence, the FE exam is two separate, partial exams given in sequence. During either session, you cannot view or respond to questions in the other session.

Your exam will include a limited (unknown) number of questions (known as "pretest items") that will not be scored and will not have an impact on your results. NCEES does this to determine the viability of new

Table 1 Recommended 15-Week Civil FE Exam Review Course Format for Commercial Review Courses

week	*FE Civil Review* chapter titles	*FE Civil Review* chapter numbers
1	Units; Analytic Geometry and Trigonometry; Algebra and Linear Algebra; Calculus; Differential Equations and Transforms; Probability and Statistics	1–6
2	Calculus; Differential Equations and Transforms; Probability and Statistics	4–6
3	Engineering Economics; Professional Practice; Ethics; Licensure	60–63
4	Systems of Forces and Moments; Trusses; Pulleys, Cables, and Friction; Centroids and Moments of Inertia; Indeterminate Statics	22–26
5	Kinematics; Kinetics; Kinetics of Rotational Motion; Energy and Work	27–30
6	Fluid Properties; Fluid Statics; Fluid Measurement and Similitude	7–9
7	Hydrology; Hydraulics; Groundwater	10–12
8	Water Quality; Water Supply Treatment and Distribution; Wastewater Collection and Treatment; Activated Sludge and Sludge Processing; Air Quality	13–17
9	Soil Properties and Testing; Foundations; Rigid Retaining Walls; Excavations	18–21
10	Material Properties and Testing; Engineering Materials	35–36
11	Stresses and Strains; Thermal, Hoop, and Torsional Stress; Beams; Columns	31–34
12	Structural Design: Materials and Basic Concepts; Reinforced Concrete: Beams; Reinforced Concrete: Columns; Reinforced Concrete: Slabs; Reinforced Concrete: Walls; Reinforced Concrete: Footings; Structural Steel: Beams; Structural Steel: Columns; Structural Steel: Tension Members; Structural Steel: Beam-Columns; Structural Steel: Connectors	37–47
13	Transportation Capacity and Planning; Plane Surveying; Geometric Design; Earthwork; Pavement Design; Traffic Safety	48–53
14	Construction Management, Scheduling, and Estimating; Procurement and Project Delivery Methods; Construction Documents; Construction Operations and Management; Construction Safety	54–58
15	Computer Software	59

questions for future exams. You won't know which questions are pretest items. They are not identifiable and are randomly distributed throughout the exam.

EXAM DURATION

The exam is six hours long and includes an 8-minute tutorial, a 25-minute break, and a brief survey at the conclusion of the exam. The total time you'll have to actually answer the exam questions is 5 hours and 20 minutes. The problem-solving pace works out to slightly less than 3 minutes per question. However, the exam does not pace you. You may spend as much time as you like on each question. Although the onscreen navigational interface is slightly awkward, you may work through the questions (in that session) in any sequence. If you want to go back and check your answers before you submit a session for grading, you may. However, once you submit a section you are not able to go back and review it.

You can divide your time between the two sessions any way you'd like. That is, if you want to spend 4 hours on the first section, and 1 hour and 20 minutes on the second section, you could do so. Or, if you want to spend 2 hours and 10 minutes on the first section, and 3 hours and 10 minutes on the second section, you could do that instead. Between sessions, you can take a 25-minute break. (You can take less, if you would like.) You cannot work through the break, and the break time cannot be added to the time permitted for either session. Once each session begins, you can leave your seat for personal reasons, but the "clock" does not stop for your absence. Unanswered questions are scored the same as questions answered incorrectly, so you should use the last few minutes of each session to guess at all unanswered questions.

THE NCEES NONDISCLOSURE AGREEMENT

At the beginning of your CBT experience, a nondisclosure agreement will appear on the screen. In order to begin the exam, you must accept the agreement within two minutes. If you do not accept within two minutes, your CBT experience will end, and you will forfeit your appointment and exam fees. The CBT nondisclosure agreement is discussed in the section entitled "Subversion After the Exam." The nondisclosure agreement, as stated in the December 2016 edition of the *NCEES Examinee Guide*, is as follows.

Table 2 *Recommended 8-Week Civil FE Exam Review Course Format for University Courses*

class	*FE Civil Review* chapter titles	*FE Civil Review* chapter numbers
1	Computer Software; Engineering Economics; Ethics	59, 60, 62
2	Trusses; Pulleys, Cables, and Friction; Centroids and Moments of Inertia; Indeterminate Statics; Kinematics; Kinetics; Kinetics of Rotational Motion; Energy and Work	23–30
3	Fluid Properties; Fluid Statics; Fluid Measurement and Similitude	7–9
4	Water Quality; Water Supply Treatment and Distribution; Wastewater Collection and Treatment; Activated Sludge and Sludge Processing; Air Quality; Soil Properties and Testing; Foundations; Rigid Retaining Walls; Excavations	13–21
5	Stresses and Strains; Thermal, Hoop, and Torsional Stress; Beams; Columns; Material Properties and Testing; Engineering Materials; Structural Design: Materials and Basic Concepts; Reinforced Concrete: Beams; Reinforced Concrete: Columns; Reinforced Concrete: Slabs; Reinforced Concrete: Walls; Reinforced Concrete: Footings; Structural Steel: Beams; Structural Steel: Columns; Structural Steel: Tension Members; Structural Steel: Beam-Columns; Structural Steel: Connectors	31–47
6	Transportation Capacity and Planning; Plane Surveying; Geometric Design; Earthwork; Pavement Design; Traffic Safety; Construction Management, Scheduling, and Estimating; Procurement and Project Delivery Methods; Construction Documents; Construction Operations and Management; Construction Safety	48–58
7	open questions	–
8	open questions	–

This exam is confidential and secure, owned and copyrighted by NCEES and protected by the laws of the United States and elsewhere. It is made available to you, the examinee, solely for valid assessment and licensing purposes. In order to take this exam, you must agree not to disclose, publish, reproduce, or transmit this exam, in whole or in part, in any form or by any means, oral or written, electronic or mechanical, for any purpose, without the prior express written permission of NCEES. This includes agreeing not to post or disclose any test questions or answers from this exam, in whole or in part, on any websites, online forums, or chat rooms, or in any other electronic transmissions, at any time.

YOUR EXAM IS UNIQUE

The exam that you take will not be the exam taken by the person sitting next to you. Differences between exams go beyond mere sequencing differences. NCEES says that the CBT system will randomly select different, but equivalent, questions from its database for each examinee using a linear-on-the-fly (LOFT) algorithm. Each examinee will have a unique exam of equivalent difficulty. That translates into each examinee having a slightly different minimum passing score.

So, you may conclude that either many questions are static clones of others, or that NCEES has an immense database of trusted questions with supporting econometric data.[1,2] However, there is no way to determine exactly how NCEES ensures that each examinee is given an equivalent exam. All that can be said is that looking at your neighbor's monitor would be a waste of time.

THE EXAM INTERFACE

The onscreen exam interface contains only minimal navigational tools. Onscreen navigation is limited to selecting an answer, advancing to the next question, going back to the previous question, and flagging the current question for later review. The interface also includes a timer, the current question number (e.g., 45 of 110), a pop-up scientific calculator, and access to an onscreen version of the *NCEES Handbook*.

During the exam, you can advance sequentially through the questions, but you cannot jump to any specific question, whether or not it has been flagged. After you have completed the last question in a session, however, the

[1]The FE exam draws upon a simple database of finished questions. The CBT system does not construct each examinee's questions from a set of "master" questions using randomly generated values for each question parameter constrained to predetermined ranges.

[2]Questions used in the now-obsolete paper-and-pencil exam were either 2-minute or 4-minute questions, based on the number of questions and time available in morning and afternoon sessions. Since all of the CBT exam questions are 3-minute questions, a logical conclusion is that 100% of the questions are brand new, or (more likely) that morning and afternoon questions are comingled within each subject.

navigation capabilities change, and you are permitted to review questions in any sequence and navigate to flagged questions.

THE *NCEES HANDBOOK* INTERFACE

Examinees are provided with a 24-inch computer monitor that will simultaneously display both the exam questions and a searchable PDF of the *NCEES Handbook*. The PDF's table of contents consists of live links. The search function is capable of finding anything in the *NCEES Handbook*, down to and including individual variables. However, the search function finds only precise search terms (e.g., "Hazenwilliams" will not locate "Hazen-Williams"). Like the printed version of the *NCEES Handbook*, the PDF also contains an index, but its terms and phrases are fairly limited and likely to be of little use.

WHAT IS THE REQUIRED PASSING SCORE?

Scores are based on the total number of questions answered correctly, with no deductions made for questions answered incorrectly. Raw scores may be adjusted slightly, and the adjusted scores are then scaled.

Since each question has four answer choices, the lower bound for a minimum required passing score is the performance generated by random selection, 25%. While it is inevitable that some examinees can score less than 25%, it is more likely that most examinees can score slightly more than 25% simply with judicious guessing and elimination of obvious incorrect options. So, the goal of all examinees should be to increase their scores from 25% to the minimum required passing score.

In the past, NCEES has rarely announced a minimum required passing score for the FE exam, ostensibly because the average score changed slightly with each administration of the exam. However, inside information reports that the raw percentage of questions that must be answered correctly was low—hovering around 50%. NCEES intends to release performance data on the CBT examinations approximately quarterly. That data will probably not include minimum required passing score information.

Since each state requires a passing score of 70, NCEES simply scales 50% (or whatever percentage the minimum required passing score represents) up to 70. Everyone seems happy with this practice—one of the few times that you can get something for nothing.

For the CBT examination, each examinee will have a unique exam of equivalent difficulty. This translates into a different minimum passing score for each examination. NCEES "accumulates" the passing score by summing each question's "required performance value" (RPV).[3] The RPV represents the fraction of minimally qualified examinees that it thinks will solve the question correctly. In the past, RPVs for new questions were dependent on the opinions of experts that it polled with the question, "What fraction of minimally qualified examinees do you think should be able to solve this question correctly?" For questions that have appeared in past exams, including the "pre-test" items that are used on the CBT exam, NCEES actually knows the fraction. Basically, out of all of the examinees who passed the FE exam (the "minimally qualified" part), NCEES knows how many answered a pre-test question correctly (the "fraction of examinees" part). A particularly easy question on Ohm's law might have an RPV of 0.88, while a more difficult question on Bayes' theorem might have an RPV of 0.37. Add up all of the RPVs, and bingo, you have the basis for a passing score. What could be simpler?[4]

WHAT IS THE AVERAGE PASSING RATE?

In July/August and October/November 2016, approximately 67% of first-time test takers passed the written discipline-specific Civil FE exam. The average failure rate was, accordingly, 33%. Some of those who failed the first time retook the FE exam, although the percentage of successful examinees declined precipitously with each subsequent attempt.

For the October 2013 administration, 75% of civil first-time examinees passed the FE exam; 34% of repeat test takers passed. These passing rates are higher than when the FE exam had a different format.

WHAT REFERENCE MATERIAL CAN I BRING TO THE EXAM?

Since October 1993, the FE exam has been what NCEES calls a "limited-reference exam." This means that nothing except what is supplied by NCEES may be used during the exam. Therefore, the FE exam is really an "NCEES-publication only" exam. NCEES provides its own searchable, electronic version of the *NCEES Handbook* for use during the exam. Computer screens are 24 inches wide so there is enough room to display the exam questions and the *NCEES Handbook* side-by-side. No printed books from any publisher may be used.

[3]NCEES does not actually use the term "required performance value," although it does use the method described.

[4]The flaw in this logic, of course, is that water seeks its own level. Deficient educational background and dependency on automation results in lower RPVs, which the NCEES process translates into a lower minimum passing score requirement. In the past, an "equating subtest" (a small number of questions in the exam that were associated with the gold standard of econometric data) was used to adjust the sum of RPVs based on the performance of the candidate pool. Though unmentioned in NCEES literature, that feature may still exist in the CBT exam process. However, the adjustment would still be based on the performance (good or bad) of the examinees.

WILL THE *NCEES HANDBOOK* HAVE EVERYTHING I NEED DURING THE EXAM?

In addition to not allowing examinees to be responsible for their own references, NCEES also takes no responsibility for the adequacy of coverage of its own reference. Nor does it offer any guidance or provide examples as to what else you should know, study, or memorize. The following warning statement comes from the *NCEES Handbook* preface.

> The *FE Reference Handbook* does not contain all the information required to answer every question on the exam. Basic theories, conversions, formulas, and definitions examinees are expected to know have not been included.

As open-ended as that warning statement sounds, the exam does not actually expect much knowledge outside of what is covered in the *NCEES Handbook*. For all practical purposes, the *NCEES Handbook* will have everything that you need. For example, if the *NCEES Handbook* covers only steel springs, you won't be asked to demonstrate a knowledge of copper-alloy springs. If the *NCEES Handbook* covers only ball and roller bearings, you won't be expected to know about needle or thrust bearings.

That makes it pretty simple to predict the kinds of questions that will appear on the exam. If you take your preparation seriously, the *NCEES Handbook* is pretty much a guarantee that you won't waste any time learning subjects that are not on the FE exam.

WILL THE *NCEES HANDBOOK* HAVE EVERYTHING I NEED TO STUDY FROM?

Saying that you won't need to work outside of the content published in the *NCEES Handbook* is not the same as saying the *NCEES Handbook* is adequate to study from.

From several viewpoints, the *NCEES Handbook* is marginally adequate in organization, presentation, and consistency as an examination reference. The *NCEES Handbook* was never intended to be something you study or learn from, so it is most definitely inadequate for that purpose. Background, preliminary and supporting material, explanations, extensions to the theory, and application rules are all missing from the *NCEES Handbook*. Many subtopics (e.g., contract law) listed in the exam specifications are not represented in the *NCEES Handbook*.

That is why you will notice many equations, figures, and tables in this book that are not blue. You may, for example, read several paragraphs in this book containing various **black** equations before you come across a blue equation section. While the **black** material may be less likely to appear on the exam than the blue material, it provides background information that is essential to understanding the blue material. Although memorization of the **black** material is not generally required, this material should at least make sense to you.

CIVIL FE EXAM KNOWLEDGE AREAS AND QUESTION DISTRIBUTION

The following Civil FE exam specifications have been published by NCEES. Some of the topics listed are not covered in any meaningful manner (or at all) by the *NCEES Handbook*. The only conclusion that can be drawn is that the required knowledge of these subjects is shallow, qualitative, and/or nonexistent.

1. **mathematics (7–11 questions)**: analytic geometry; calculus; roots of equations; vector analysis
2. **probability and statistics (4–6 questions)**: measures of central tendency and dispersion (e.g., mean, mode, and standard deviation); estimation for a single mean (e.g., point[5] and confidence intervals); regression and curve fitting; expected value (weighted average) in decision making
3. **computational tools (4–6 questions)**: spreadsheet computations; structured programming (e.g., if-then, loops, and macros)
4. **ethics and professional practice (4–6 questions)**: codes of ethics (professional and technical societies); professional liability; licensure; sustainability and sustainable design; professional skills (e.g., public policy, management, and business); contracts and contract law
5. **engineering economics (4–6 questions)**: discounted cash flow (e.g., equivalence, present worth (PW), equivalent annual worth, future worth (FW), and rate of return); cost (e.g., incremental, average, sunk, and estimating); analysis (e.g., breakeven, benefit-cost, and life cycle); uncertainty (e.g., expected value and risk)
6. **statics (7–11 questions)**: resultants of force systems; equivalent force systems; equilibrium of rigid bodies; frames and trusses; centroids of areas; area moments of inertia; static friction
7. **dynamics (4–6 questions)**: kinematics (e.g., particles and rigid bodies); mass moments of inertia; force acceleration (e.g., particles and rigid bodies); impulse momentum (e.g., particles and rigid bodies); work, energy, and power (e.g., particles and rigid bodies)
8. **mechanics of materials (7–11 questions)**: shear and moment diagrams; stresses and strains (e.g.,

[5]The term "point interval" is unique to NCEES. This may mean "point estimation and confidence intervals" or "confidence intervals of points," but it is ambiguous in intent.

axial, torsion, bending, shear, and thermal); deformations (e.g., axial, torsion, bending, and thermal); combined stresses; principal stresses; Mohr's circle; column analysis (e.g., buckling and boundary conditions); composite sections; elastic and plastic deformations; stress-strain diagrams

9. **materials (4–6 questions)**: mix design (e.g., concrete and asphalt); test methods and specifications (e.g., steel, concrete, aggregates, asphalt, and wood); physical and mechanical properties of concrete, ferrous and nonferrous metals, masonry, wood, engineered materials (e.g., fiber-reinforced plastic (FRP), laminated lumber, and wood/plastic composites), and asphalt
10. **fluid mechanics (4–6 questions)**: flow measurement; fluid properties; fluid statics; energy, impulse, and momentum equations
11. **hydraulics and hydrologic systems (8–12 questions)**: basic hydrology (e.g., infiltration, rainfall, runoff, detention, flood flows, and watersheds); basic hydraulics (e.g., Manning equation, Bernoulli theorem, open-channel flow, and pipe flow); pumping systems (water and wastewater); water distribution systems; reservoirs (e.g., dams, routing, and spillways); groundwater (e.g., flow, wells, and drawdown); storm sewer collection systems
12. **structural analysis (6–9 questions)**: analysis of forces in statically determinant beams, trusses, and frames; deflection of statically determinant beams, trusses, and frames; structural determinacy and stability analysis of beams, trusses, and frames; loads and load paths (e.g., dead, live, lateral, influence lines and moving loads, and tributary areas); elementary statically indeterminate structures
13. **structural design (6–9 questions)**: design of steel components (e.g., codes and design philosophies, beams, columns, beam-columns, tension members, and connections); design of reinforced concrete components (e.g., codes and design philosophies, beams, slabs, columns, walls, and footings)
14. **geotechnical engineering (9–14 questions)**: geology; index properties and soil classifications; phase relations (air-water-solid); laboratory and field tests; effective stress (buoyancy); stability of retaining walls (e.g., active pressure/passive pressure); shear strength; bearing capacity (cohesive and noncohesive); foundation types (e.g., spread footings, deep foundations, wall footings, and mats); consolidation and differential settlement; seepage/flow nets; slope stability (e.g., fills, embankments, cuts, and dams); soil stabilization (e.g., chemical additives and geosynthetics); drainage systems; erosion control
15. **transportation engineering (8–12 questions)**: geometric design of streets and highways; geometric design of intersections; pavement system design (e.g., thickness, subgrade, drainage, and rehabilitation); traffic safety; traffic capacity; traffic flow theory; traffic control devices; transportation planning (e.g., travel forecast modeling)
16. **environmental engineering (6–9 questions)**: water quality (ground and surface); basic tests (e.g., water, wastewater, and air); environmental regulations; water supply and treatment; wastewater collection and treatment
17. **construction (4–6 questions)**: construction documents; procurement methods (e.g., competitive bid and qualifications-based); project delivery methods (e.g., design-bid-build, design build, construction management, and multiple prime); construction operations and methods (e.g., lifting, rigging, dewatering and pumping, equipment production, productivity analysis and improvement, and temporary erosion control); project scheduling (e.g., critical path method (CPM) and allocation of resources); project management (e.g., owner/contractor/client relations); construction safety; construction estimating
18. **surveying (4–6 questions)**: angles, distances, and trigonometry; area computations; earthwork and volume computations; closure; coordinate systems (e.g., state plane and latitude/longitude); leveling (e.g., differential, elevations, and percent grades)

DOES THE EXAM REQUIRE LOOKING UP VALUES IN TABLES?

For some questions, you might have to look up a value, but in those cases, you must use the value in the *NCEES Handbook*. For example, you might know that the modulus of elasticity of steel is approximately 29×10^6 psi for soft steel and approximately 30×10^6 psi for hard steel. If you needed the modulus of elasticity for an elongation calculation, you would find the official *NCEES Handbook* value is "29 Mpsi." Whether or not using 30×10^6 psi will result in an (approximate) correct answer or an incorrect answer depends on whether the question writer wants to reward you for knowing something or punish you for not using the *NCEES Handbook*.

However, in order to reduce the time required to solve questions, and to reduce the variability of answers caused by examinees using different starting values, questions generally provide all required information. Unless the question is specifically determining whether you can read a table or figure, all relevant values (density, modulus of elasticity, viscosity, enthalpy, yield strength, etc.) needed to solve the question are often included in the question statement. NCEES does not want the consequences of using correct methods with ambiguous data.

DO QUESTION STATEMENTS INCLUDE SUPERFLUOUS INFORMATION?

Particularly since all relevant information is provided in the question statements, some questions end up being pretty straightforward. In order to obfuscate the solution method, some irrelevant, superfluous information will be provided in the question statement. For example, when finding the applied force from a given mass and acceleration (i.e., $F = ma$), the temperature and viscosity of the surrounding air might be given. However, if you understand the concept, this practice will be transparent to you.

Questions in this book typically do not include superfluous information. The purpose of this book is to teach you, not confuse you.

REGISTERING FOR THE EXAM

The CBT exams are administered at approved Pearson VUE testing centers. Registration is open year-round and can be completed online through your MyNCEES account.[6] Registration fees may be paid online. Once you receive notification from NCEES that you are eligible to schedule your exam, you can do so online through your MyNCEES account. Select the location where you would like to take your exam, and select from the list of available dates. You will receive a letter from Pearson VUE (via email) confirming your exam location and date.

Whether or not applying for and taking the exam is the same as applying for an FE certificate from your state depends on the state. In most cases, you might take the exam without your state board ever knowing about it. In fact, as part of the NCEES online exam application process, you will have to agree to the following statement:

> Passage of the FE exam alone does not ensure certification as an engineer intern or engineer-in-training in any U.S. state or territory. To obtain certification, you must file an application with an engineering licensing board and meet that board's requirements.

After graduation, when you are ready to obtain your FE (EIT, IE, etc.) suitable-for-wall-hanging certificate, you can apply and pay an additional fee to your state. In some cases, you will be required to take an additional nontechnical exam related to professional practice in your state. Actual procedures will vary from state to state.

WINDOWS OF OPPORTUNITY

The FE exam is administered in eight months out of the year: January, February, April, May, July, August, October, and November. There are multiple testing dates within each of those months. No exams are administered in March, June, September, or December.

WHAT TO BRING TO THE EXAM

You do not need to bring much with you to the exam. For admission, you must bring a current, signed, government-issued photographic identification. This is typically a driver's license or passport. A student ID card is not acceptable for admittance. The first and last name on the photographic ID must match the name on your appointment confirmation letter. NCEES recommends that you bring a copy of your appointment confirmation letter in order to speed up the check in process. In most cases, Pearson VUE will email this to you, or you can download it from your MyNCEES account, 2–3 weeks prior to the exam date.

Earplugs, noise-cancelling headphones, and tissues are provided at the testing center for examinees who request them. Additionally, all examinees are provided with a reusable, erasable notepad and compatible writing instrument to use for scratchwork during the exam.

Pearson VUE staff may visually examine any approved item without touching you or the item. In addition to the items provided at the testing center, the following items are permitted during the FE exam.[7]

- your ID (same one used for admittance to the exam)
- key to your test center locker
- NCEES-approved calculator without a case
- inhalers
- cough drops and prescription and nonprescription pills, including headache remedies, all unwrapped and not bottled, unless the packaging states they must remain in the packaging
- bandages, braces (for your neck, back, wrist, leg, or ankle), casts, and slings
- eyeglasses (without cases); eye patches; handheld, nonelectric magnifying glasses (without cases); and eyedrops[8]
- hearing aids
- medical/surgical face masks, medical devices attached to your body (e.g., insulin pumps and spinal cord stimulators), and medical alert bracelets (including those with USB ports)

[6]PPI is not associated with NCEES. Your MyNCEES account is not your PPI account.
[7]All items are subject to revision and reinterpretation at any time.
[8]Eyedrops can remain in their original bottle.

- pillows and cushions
- light sweaters or jackets
- canes, crutches, motorized scooters and chairs, walkers, and wheelchairs

WHAT ELSE TO BRING TO THE EXAM

Depending on your situation, any of the following items may prove useful but should be left in your test center locker.

- calculator batteries
- contact lens wetting solution
- spare calculator
- spare reading glasses
- loose shoes or slippers
- extra set of car keys
- eyeglass repair kit, including a small screwdriver for fixing glasses (or removing batteries from your calculator)

WHAT NOT TO BRING TO THE EXAM

Leave all of these items in your car or at home: pens and pencils, erasers, scratch paper, clocks and timers, unapproved calculators, cell phones, pagers, communication devices, computers, tablets, cameras, audio recorders, and video recorders.

WHAT CALCULATORS ARE PERMITTED?

To prevent unauthorized transcription and distribution of the exam questions, calculators with communicating and text editing capabilities have been banned by NCEES. You may love the reverse Polish notation of your HP 48GX, but you'll have to get used to one of the calculators NCEES has approved. If you start using one of these approved calculators at the beginning of your review, you should be familiar enough with it by the time of the exam.

Calculators permitted by NCEES are listed at **ppi2pass.com/calculators**. All of the listed calculators have sufficient engineering/scientific functionality for the exam.

At the beginning of your review program, you should purchase or borrow a spare calculator. It is preferable, but not essential, that your primary and spare calculators be identical. If your spare calculator is not identical to your primary calculator, spend some time familiarizing yourself with its functions.

Examinees found using a calculator that is not approved by NCEES will be discharged from the testing center and charged with exam subversion by their states. (See section "Exam Subversion.")

WHAT UNITS ARE USED ON THE EXAM?

You will need to learn the SI system if you are not already familiar with it. Contrary to engineering practice in the United States, the FE exam primarily uses SI units.

The *NCEES Handbook* generally presents only dimensionally consistent equations. (For example, $F = ma$ is consistent with units of newtons, kilograms, meters, and seconds. However, it is not consistent for units of pounds-force, pounds-mass, feet, and seconds.) Although pound-based data is provided parallel to the SI data in most tables, many equations cannot use the pound-based data without including the gravitational constant. After being mentioned in the first few pages, the gravitational constant ($g_c = 32.2$ ft-lbm/lbf-sec^2), which is necessary to use for equations with inconsistent U.S. units, is barely mentioned in the *NCEES Handbook* and does not appear in most equations.

Outside of the table of conversions and introductory material at its beginning, the *NCEES Handbook* does not consistently differentiate between pounds-mass and pounds-force. The labels "pound" and "lb" are used to represent both force and mass. Densities are listed in tables with units of lb/in^3.

Kips are always units of force that can be incorporated into ft-kips, units for moment, and ksi, units of stress or strength.

IS THE EXAM HARD AND/OR TRICKY?

Whether or not the exam is hard or tricky depends on who you talk to. Other than providing superfluous data (so as not to lead you too quickly to the correct formula) and anticipating common mistakes, the FE exam is not a tricky exam. The exam does not overtly try to get you to fail. The questions are difficult in their own right. NCEES does not need to provide you misleading or vague statements. Examinees manage to fail on a regular basis with perfectly straightforward questions.

Commonly made mistakes are routinely incorporated into the available answer choices. Thus, the alternative answers (known as distractors) will seem logical to many examinees. For example, if you forget to convert the pipe diameter from millimeters to meters, you'll find an answer option that is off by a factor of 1000. Perhaps, that meets your definition of "tricky."

Questions are generally practical, dealing with common and plausible situations that you might encounter on the job. In order to avoid the complications of being too practical, the ideal or perfect case is often explicitly called for in the question statement (e.g., "Assume an ideal gas."; "Disregard the effects of air friction."; or "The steam expansion is isentropic.").

You won't have to draw on any experiential knowledge or make reasonable assumptions. If a motor efficiency is required, it will be given to you. You won't have to assume a reasonable value. If a beam is to be sized to limit distressing architectural deflections, the limit will be explicitly given to you. If an allowable stress requires a factor of safety, the factor of safety will be given to you.

WHAT DOES "MOST NEARLY" REALLY MEAN?

One of the more disquieting aspects of exam questions is that answer choices generally have only two or three significant digits, and the answer choices are seldom exact. An exam question may prompt you to complete the sentence, "The value is most nearly...", or may ask "Which answer choice is closest to the correct value?" A lot of self-confidence is required to move on to the next question when you don't find an exact match for the answer you calculated, or if you have had to split the difference because no available answer choice is close.

At one time, NCEES provided this statement regarding the use of "most nearly."

> Many of the questions on NCEES exams require calculations to arrive at a numerical answer. Depending on the method of calculation used, it is very possible that examinees working correctly will arrive at a range of answers. The phrase "most nearly" is used to accommodate all these answers that have been derived correctly but which may be slightly different from the correct answer choice given on the exam. You should use good engineering judgment when selecting your choice of answer. For example, if the question asks you to calculate an electrical current or determine the load on a beam, you should literally select the answer option that is most nearly what you calculated, regardless of whether it is more or less than your calculated value. However, if the question asks you to select a fuse or circuit breaker to protect against a calculated current or to size a beam to carry a load, you should select an answer option that will safely carry the current or load. Typically, this requires selecting a value that is closest to but larger than the current or load.

The difference is significant. Suppose you were asked to calculate "most nearly" the volumetric flow rate of pure water required to dilute a contaminated stream to an acceptable concentration. Suppose, also, that you calculated 823 gpm. If the answer choices were (A) 600 gpm, (B) 800 gpm, (C) 1000 gpm, and (D) 1200 gpm, you would go with answer choice (B), because it is most nearly what you calculated. If, however, you were asked to select a pump or pipe to provide the calculated capacities, you would have to go with choice (C). Got it? If not, stop reading until you understand the distinction.

WHEN DO I FIND OUT IF I PASSED?

You will receive an email notification that your exam results are ready for viewing through your MyNCEES account 7–10 days after the exam. That email will also include instructions that you can use to proceed with your state licensing board. If you fail, you will be shown your percentage performance in each knowledge area. The diagnostic report may help you figure out what to study before taking the exam again. Because each examinee answers different questions in each knowledge area, the diagnostic report probably should not be used to compare the performance of two examinees, to determine how much smarter than another examinee you are, to rate employees, or to calculate raises and bonus.

If you fail the exam, you may take it again. NCEES's policy is that examinees may take the exam once per testing window, up to three times per 12-month period. However, you should check with your state board to see whether it imposes any restrictions on the number and frequency of retakes.

SUBVERSION DURING THE EXAM

With the CBT exam, you can no longer get kicked out of the exam room for not closing your booklet or putting down your pencil in time. However, there are still plenty of ways for you to run afoul of the rules imposed on you by NCEES, your state board, and Pearson VUE. For example, since communication devices are prohibited in the exam, occurrences as innocent as your cell phone ringing during the exam can result in the immediate invalidation of your exam.

The *NCEES Examinee Guide* gives the following statement regarding fraudulent and/or unprofessional behavior. Somewhere along the way, you will probably have to read and accept it, or something similar, before you can take the FE exam.

> Fraud, deceit, dishonesty, unprofessional behavior, and other irregular behavior in connection with taking any NCEES exam are strictly prohibited. Irregular behavior includes but is not limited to the following: failing to work independently; impersonating another individual or permitting such impersonation (surrogate testing); possessing prohibited items; communicating with other examinees or any outside parties by way of cell phone, personal computer, the Internet, or any other means during an exam; disrupting other examinees; creating safety concerns; and possessing, reproducing, or disclosing nonpublic exam questions, answers, or other information regarding the content of the exam before, during, or after the exam

administration. Evidence of an exam irregularity may be based on your performance on the exam, a report from an administrator or a third party, or other information.

The test administrator is authorized to take appropriate action to investigate, stop, or correct any observed or suspected irregular behavior, including discharging you from the test center and confiscating prohibited devices or materials. You must cooperate fully in any investigation of a suspected irregularity. NCEES reserves the right to pursue all available remedies for exam irregularities, including canceling scores and pursuing administrative, civil, and/or criminal remedies.

If you are involved in an exam irregularity, the following may occur: invalidation of results, notification to your licensing board, forfeiture of exam fees, and restrictions on future testing. Some violations may incur additional consequences, to be pursued at the discretion of NCEES.

Based on the grounds for dismissal used for the paper-and-pencil exam up through 2013, you can expect harsh treatment for

- having a cell phone in your possession
- having a device with copying, recording, or communication capabilities in your possession. These include but are not limited to cameras, pagers, personal digital assistants (PDAs), radios, headsets, tape players, calculator watches, electronic dictionaries, electronic translators, transmitting devices, digital media players (e.g., iPods), and tablets (e.g., iPads, Kindles, or Nooks)
- having papers, books, or notes
- having a calculator that is not on the NCEES-approved list
- appearing to or copying someone else's work
- talking to another examinee during the exam
- taking notes or writing on anything other than your NCEES-provided reusable, erasable notepad
- removing anything from the exam area
- leaving the exam area without authorization
- violating any other restrictions that are cause for dismissal or exam invalidation (e.g., whistling while you work, chewing gum, or being intoxicated)

If you are found to be in possession of a prohibited item (e.g., a cell phone) after the exam begins, that item will be confiscated and sent to NCEES. While you will probably eventually get your cell phone back, you won't get a refund of your exam fees.

Cheating and what is described as "subversion" are dealt with quite harshly. Proctors who observe you giving or receiving assistance, compromising the integrity of the exam, or participating in any other form of cheating during an exam will require you to surrender all exam materials and leave the test center. You won't be permitted to continue with the exam. It will be a summary execution, carried out without due process and mercy.

Of course, if you arrive with a miniature camera disguised as a pen or eyeglasses, your goose will be cooked. Talk to an adjacent examinee, and your goose will be cooked. Use a mirror to look around the room while putting on your lipstick or combing your hair, and your goose will be cooked. Bring in the wrong calculator, and your goose will be cooked. Loan your calculator to someone whose batteries have died, and your goose will be cooked. Though you get the idea, many of the ways that you might inadvertently get kicked out of the CBT exam are probably (and, unfortunately) yet to be discovered. Based on this fact, you shouldn't plan on being the first person to bring a peppermint candy in a crackly cellophane wrapper.

And, as if being escorted with your personal items out of the exam room wasn't embarrassing enough, your ordeal still won't be over. NCEES and your state will bar you from taking any exam for one or more years. Any application for licensure pending an approval for exam will be automatically rejected. You will have to reapply and pay your fees again later. By that time, you probably will have decided that the establishment's response to a minor infraction was so out of proportion that licensure as a professional engineer isn't even in the cards.

SUBVERSION AFTER THE EXAM

The NCEES testing (and financial) model is based on reusing all of its questions forever. To facilitate such reuse, the FE (and PE) exams are protected by nondisclosure agreements and a history of aggressive pursuit of actual and perceived offenses. In order to be allowed to take its exams, NCEES requires examinees to agree to its terms.

Copyright protection extends to only the exact words, phrases, and sentences, and sequences thereof, used in questions. However, the intent of the NCEES nondisclosure agreement is to grant NCEES protection beyond what is normally available through copyright protection —to prevent you from even discussing a question in general terms (e.g., "There was a question on structural bolts that stumped me. Did anyone else think the question was unsolvable?").

Most past transgressions have been fairly egregious.[9] In several prominent instances, NCEES has incurred substantial losses and expenses. In those cases, offenders have gotten what they deserved. But, even innocent public disclosures of the nature of "Hey, did anyone else have trouble solving that vertical crest curve question?" have been aggressively pursued.

A restriction against saying anything at all to anybody about any aspect of a question is probably too broad to be legally enforceable. Unfortunately, most examinees don't have the time, financial resources, or sophistication to resist what NCEES throws at them. Their only course of action is to accept whatever punishment is meted out to them by their state boards and by NCEES.

In the past, NCEES has used the U.S. courts and aggressively pursued financial redress for loss of its intellectual property and violation of its copyright. It has administratively established a standard (accounting) value of thousands of dollars for each disclosed or compromised question. You can calculate your own *pro forma* invoice from NCEES by multiplying this amount by the number of questions you discuss with others.

DOING YOUR PART, NCEES STYLE

NCEES has established a security tip line so that you can help it police the behavior of other examinees. Before, during, or after the exam, if you see any of your fellow examinees acting suspiciously, NCEES wants you to report them by phone or through the NCEES website. You'll have to identify yourself, but NCEES promises that the information you provide will be strictly confidential, and that your personal contact information will not be shared outside the NCEES compliance and security staff. Unless required by statute, rules of discovery, or a judge, of course.

PART 4: STRATEGIES FOR PASSING THE EXAM

A FEW DAYS BEFORE THE EXAM

There are a few things you should do a week or so before the examination date. For example, visit the exam site in order to find the testing center building, parking areas, examination room, and restrooms. You should also make arrangements for childcare and transportation. Since your examination may not start or end exactly at the designated times, make sure that your childcare and transportation arrangements can allow for some flexibility.

Second in importance to your scholastic preparation is the preparation of your two examination kits. (See "What to Bring to the Exam" and "What Else to Bring to the Exam" in this Introduction.) The first kit includes items that can be left in your assigned locker (e.g., your admittance letter, photo ID, and extra calculator batteries). The second kit includes items that should be left in your car in case you need them (e.g., copy of your application, warm sweater, and extra snacks or beverages).

THE DAY BEFORE THE EXAM

If possible, take the day before the examination off from work to relax. Do not cram the last night. A good prior night's sleep is the best way to start the examination. If you live far from the examination site, consider getting a hotel room in which to spend the night.

Make sure your exam kits are packed and ready to go.

THE DAY OF THE EXAM

You should arrive at least 30 minutes before your scheduled start time. This will allow time for finding a convenient parking place, bringing your items to the testing center, and checking in.

DURING THE EXAM

Once the examination has started, observe the following suggestions.

Do not spend more than four minutes working a problem. (The average time available per problem is slightly less than three minutes.) If you have not finished a question in that time, flag it for later review if you have time, and continue on.

Don't ask your proctors technical questions. Proctors are pure administrators. They don't know anything about the exam or its subjects.

Even if you do not discover them, errors in the exam (and in the *NCEES Handbook*) do occur. Rest assured that errors are almost always discovered during the scoring process, and that you will receive the performance credit for all flawed items.

[9]A candidate in Puerto Rico during the October 2006 Civil PE exam administration was found with scanning and transmitting equipment during the exam. She had recorded the entire exam, as well as the 2005 FE exam. The candidate pled guilty to two counts of fourth-degree aggravated fraud and was sentenced to six months' probation. All of the questions in both exams were compromised. NCEES obtained a civil judgment of over $1,000,000 against her.

However, NCEES has a form for reporting errors, and the test center should be able to provide it to you. If you encounter a problem with (a) missing information, (b) conflicting information, (c) no correct response from the four answer choices, or (d) more than one correct answer, use your provided reusable, erasable notepad to record the problem identification numbers. It is not necessary to tell your proctor during the exam. Wait until after the exam to ask your proctor about the procedure for reporting errors on the exam.

AFTER YOU PASS

[] Celebrate. Take someone out to dinner. Go off your diet. Get dessert.

[] Thank your family members and anyone who had to put up with your grouchiness before the exam.

[] Thank your old professors.

[] Tell everyone at the office.

[] Ask your employer for new business cards and a raise.

[] Tell your review course provider and instructors.

[] Tell the folks at PPI who were rootin' for you all along.

[] Start thinking about the PE exam.

Topic I: Mathematics

Chapter

1. Units
2. Analytic Geometry and Trigonometry
3. Algebra and Linear Algebra
4. Calculus
5. Differential Equations and Transforms

1 Units

1. INTRODUCTION

The purpose of this chapter is to eliminate some of the confusion regarding the many units available for each engineering variable. In particular, an effort has been made to clarify the use of the so-called English systems, which for years have used the *pound* unit both for force and mass—a practice that has resulted in confusion for even those familiar with it.

It is expected that most engineering problems will be stated and solved in either English engineering or SI units. Therefore, a discussion of these two systems occupies the majority of this chapter.

2. COMMON UNITS OF MASS

The choice of a mass unit is the major factor in determining which system of units will be used in solving a problem. Obviously, you will not easily end up with a force in pounds if the rest of the problem is stated in meters and kilograms. Actually, the choice of a mass unit determines more than whether a conversion factor will be necessary to convert from one system to another (e.g., between the SI and English systems). An inappropriate choice of a mass unit may actually require a conversion factor *within* the system of units.

The common units of mass are the gram, pound, kilogram, and slug. There is nothing mysterious about these units. All represent different quantities of matter, as Fig. 1.1 illustrates. In particular, note that the pound and slug do not represent the same quantity of matter. One slug is equal to 32.1740 pounds-mass.

Figure 1.1 *Common Units of Mass*

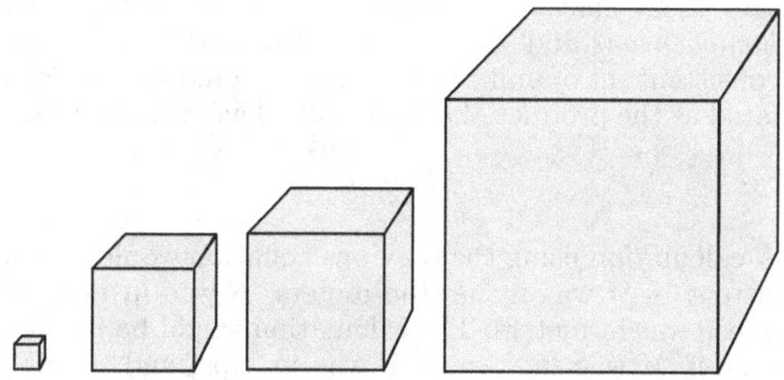

3. MASS AND WEIGHT

The SI system uses kilograms for mass and newtons for weight (force). The units are different, and there is no confusion between the variables. However, for years, the term *pound* has been used for both mass and weight. This usage has obscured the distinction between the two: mass is a constant property of an object; weight varies with the gravitational field. Even the conventional use of the abbreviations *lbm* and *lbf* (to distinguish between pounds-mass and pounds-force) has not helped eliminate the confusion.

An object with a mass of one pound will have an earthly weight of one pound, but this is true only on the earth. The weight of the same object will be much less on the moon. Therefore, care must be taken when working with mass and force in the same problem.

The relationship that converts mass to weight is familiar to every engineering student.

$$W = mg$$

This equation illustrates that an object's weight will depend on the local acceleration of gravity as well as the object's mass. The mass will be constant, but gravity will depend on location. Mass and weight are not the same.

4. ACCELERATION OF GRAVITY

Gravitational acceleration on the earth's surface is usually taken as 32.2 ft/sec^2 or 9.81 m/s^2. These values are rounded from the more exact standard values of

32.1740 ft/sec^2 and 9.8066 m/s^2. However, the need for greater accuracy must be evaluated on a problem-by-problem basis. Usually, three significant digits are adequate, since gravitational acceleration is not constant anyway, but is affected by location (primarily latitude and altitude) and major geographical features.

5. CONSISTENT SYSTEMS OF UNITS

A set of units used in a calculation is said to be *consistent* if no conversion factors are needed. (The terms *homogeneous* and *coherent* are also used to describe a consistent set of units.) For example, a moment is calculated as the product of a force and a lever arm length.

$$M = dF$$

A calculation using the previous equation would be consistent if M was in newton-meters, F was in newtons, and d was in meters. The calculation would be inconsistent if M was in ft-kips, F was in kips, and d was in inches (because a conversion factor of 1/12 would be required).

The concept of a consistent calculation can be extended to a system of units. A *consistent system of units* is one in which no conversion factors are needed for any calculation. For example, Newton's second law of motion can be written without conversion factors. Newton's second law for an object with a constant mass simply states that the force required to accelerate the object is proportional to the acceleration of the object. The constant of proportionality is the object's mass.

$$F = ma$$

Notice that this relationship is $F = ma$, not $F = Wa/g$ or $F = ma/g_c$. $F = ma$ is consistent: It requires no conversion factors. This means that in a consistent system where conversion factors are not used, once the units of m and a have been selected, the units of F are fixed. This has the effect of establishing units of work and energy, power, fluid properties, and so on.

The decision to work with a consistent set of units is desirable but unnecessary, depending often on tradition and environment. Problems in fluid flow and thermodynamics are routinely solved in the United States with inconsistent units. This causes no more of a problem than working with inches and feet when calculating moments. It is necessary only to use the proper conversion factors.

6. THE ENGLISH ENGINEERING SYSTEM

Through common and widespread use, pounds-mass (lbm) and pounds-force (lbf) have become the standard units for mass and force in the *English Engineering System.*

There are subjects in the United States where the practice of using pounds for mass is firmly entrenched. For example, most thermodynamics, fluid flow, and heat transfer problems have traditionally been solved using the units of lbm/ft^3 for density, Btu/lbm for enthalpy, and Btu/lbm-°F for specific heat. Unfortunately, some equations contain both lbm-related and lbf-related variables, as does the steady flow conservation of energy equation, which combines enthalpy in Btu/lbm with pressure in lbf/ft^2.

The units of pounds-mass and pounds-force are as different as the units of gallons and feet, and they cannot be canceled. A mass conversion factor, g_c, is needed to make the equations containing lbf and lbm dimensionally consistent. This factor is known as the *gravitational constant* and has a value of 32.1740 lbm-ft/lbf-sec^2. The numerical value is the same as the standard acceleration of gravity, but g_c is not the local gravitational acceleration, g. (It is acceptable, and recommended, that g_c be rounded to the same number of significant digits as g. Therefore, a value of 32.2 for g_c would typically be used.) g_c is a conversion constant, just as 12.0 is the conversion factor between feet and inches.

Equation 1.1: Force in Pounds-Force

$$F = ma/g_c \qquad 1.1$$

Variation

$$F \text{ in lbf} = \frac{(m \text{ in lbm})\left(a \text{ in } \dfrac{\text{ft}}{\text{sec}^2}\right)}{g_c \text{ in } \dfrac{\text{lbm-ft}}{\text{lbf-sec}^2}}$$

Description

The English Engineering System is an inconsistent system, as defined according to Newton's second law. $F = ma$ cannot be written if lbf, lbm, and ft/sec^2 are the units used. The g_c term must be included.

g_c does more than "fix the units." Since g_c has a numerical value of 32.1740, it actually changes the calculation numerically. A force of 1.0 pound will not accelerate a 1.0 pound-mass at the rate of 1.0 ft/sec^2.

In the English Engineering System, work and energy are typically measured in ft-lbf (mechanical systems) or in British thermal units, Btu (thermal and fluid systems). One Btu is equal to approximately 778 ft-lbf.

Example

What is most nearly the weight in lbf of a 1.00 lbm object in a gravitational field of 27.5 ft/sec²?

(A) 0.85 lbf

(B) 1.2 lbf

(C) 28 lbf

(D) 32 lbf

Solution

The weight is

$$F = ma/g_c = \frac{(1.00 \text{ lbm})\left(27.5 \ \frac{\text{ft}}{\text{sec}^2}\right)}{32.2 \ \frac{\text{lbm-ft}}{\text{lbf-sec}^2}} = 0.854 \text{ lbf} \quad (0.85 \text{ lbf})$$

The answer is (A).

7. OTHER FORMULAS AFFECTED BY INCONSISTENCY

It is not a significant burden to include g_c in a calculation, but it may be difficult to remember when g_c should be used. Knowing when to include the gravitational constant can be learned through repeated exposure to the formulas in which it is needed, but it is safer to carry the units along in every calculation.

Equation 1.2 Through Eq. 1.6: Equations Requiring g_c

$$\text{KE} = m\text{v}^2/2g_c \quad \text{[in ft-lbf]} \qquad 1.2$$

$$\text{PE} = mgh/g_c \quad \text{[in ft-lbf]} \qquad 1.3$$

$$p = \rho gh/g_c \quad \text{[in lbf/ft}^2\text{]} \qquad 1.4$$

$$\text{SW} = \rho g/g_c \quad \text{[in lbf/ft}^3\text{]} \qquad 1.5$$

$$\tau = (\mu/g_c)(d\text{v}/dy) \quad \text{[in lbf/ft}^2\text{]} \qquad 1.6$$

Description

Equation 1.2 through Eq. 1.6 are some representative equations that require the g_c term. Equation 1.2 calculates kinetic energy, Eq. 1.3 calculates potential energy, Eq. 1.4 calculates pressure, Eq. 1.5 calculates specific weight, and Eq. 1.6 calculates shear stress.[1] In all cases, it is assumed that the standard English Engineering System units will be used.

Example

A rocket that has a mass of 4000 lbm travels at 27,000 ft/sec. What is most nearly its kinetic energy?

(A) 1.4×10^9 ft-lbf

(B) 4.5×10^{10} ft-lbf

(C) 1.5×10^{12} ft-lbf

(D) 4.7×10^{13} ft-lbf

Solution

From Eq. 1.2, the kinetic energy is

$$\text{KE} = m\text{v}^2/2g_c = \frac{(4000 \text{ lbm})\left(27{,}000 \ \frac{\text{ft}}{\text{sec}}\right)^2}{(2)\left(32.2 \ \frac{\text{lbm-ft}}{\text{lbf-sec}^2}\right)} = 4.53 \times 10^{10} \text{ ft-lbf} \quad (4.5 \times 10^{10} \text{ ft-lbf})$$

The answer is (B).

8. WEIGHT AND SPECIFIC WEIGHT

Weight is a force exerted on an object due to its placement in a gravitational field. If a consistent set of units is used, $W = mg$ can be used to calculate the weight of a mass. In the English Engineering System, however, the following equation must be used.

$$W = \frac{mg}{g_c}$$

Both sides of this equation can be divided by the volume of an object to derive the *specific weight* (*unit weight*, *weight density*), γ, of the object. The following equation illustrates that the weight density (in lbf/ft³) can also be calculated by multiplying the mass density (in lbm/ft³) by g/g_c.

$$\frac{W}{V} = \left(\frac{m}{V}\right)\left(\frac{g}{g_c}\right)$$

[1]The NCEES *FE Reference Handbook* (*NCEES Handbook*) is not consistent in the variables it uses in Eq. 1.2 through Eq. 1.6. For example, T is used for kinetic energy and U is used for potential energy in the Dynamics section, and γ is used for specific weight in the Fluids section.

Since g and g_c usually have the same numerical values, the only effect of the following equation is to change the units of density.

$$\gamma = \frac{W}{V} = \left(\frac{m}{V}\right)\left(\frac{g}{g_c}\right) = \frac{\rho g}{g_c}$$

Weight does not occupy volume; only mass has volume. The concept of weight density has evolved to simplify certain calculations, particularly fluid calculations. For example, pressure at a depth is calculated from

$$p = \gamma h$$

Compare this to the equation for pressure at a depth.

9. THE ENGLISH GRAVITATIONAL SYSTEM

Not all English systems are inconsistent. Pounds can still be used as the unit of force as long as pounds are not used as the unit of mass. Such is the case with the consistent *English Gravitational System.*

If acceleration is given in ft/sec^2, the units of mass for a consistent system of units can be determined from Newton's second law.

$$\text{units of } m = \frac{\text{units of } F}{\text{units of } a} = \frac{\text{lbf}}{\frac{\text{ft}}{\text{sec}^2}} = \frac{\text{lbf-sec}^2}{\text{ft}}$$

The combination of units in this equation is known as a *slug*. g_c is not needed since this system is consistent. It would be needed only to convert slugs to another mass unit.

Slugs and pounds-mass are not the same, as Fig. 1.1 illustrates. However, both are units for the same quantity: mass. The following equation will convert between slugs and pounds-mass.

$$\text{no. of slugs} = \frac{\text{no. of lbm}}{g_c}$$

The number of slugs is not derived by dividing the number of pounds-mass by the local gravity. g_c is used regardless of the local gravity. The conversion between feet and inches is not dependent on local gravity; neither is the conversion between slugs and pounds-mass.

Since the English Gravitational System is consistent, the following equation can be used to calculate weight. Notice that the local gravitational acceleration is used.

$$W \text{ in lbf} = (m \text{ in slugs})\left(g \text{ in } \frac{\text{ft}}{\text{sec}^2}\right)$$

10. METRIC SYSTEMS OF UNITS

Strictly speaking, a *metric system* is any system of units that is based on meters or parts of meters. This broad definition includes *mks systems* (based on meters, kilograms, and seconds) as well as *cgs systems* (based on centimeters, grams, and seconds).

Metric systems avoid the pounds-mass versus pounds-force ambiguity in two ways. First, matter is not measured in units of force. All quantities of matter are specified as mass. Second, force and mass units do not share a common name.

The term *metric system* is not explicit enough to define which units are to be used for any given variable. For example, within the cgs system there is variation in how certain electrical and magnetic quantities are represented (resulting in the ESU and EMU systems). Also, within the mks system, it is common engineering practice today to use kilocalories as the unit of thermal energy, while the SI system requires the use of joules. Thus, there is a lack of uniformity even within the metricated engineering community.

The "metric" parts of this book are based on the SI system, which is the most developed and codified of the so-called metric systems. It is expected that there will be occasional variances with local engineering custom, but it is difficult to anticipate such variances within a book that must be consistent.

11. SI UNITS (THE MKS SYSTEM)

SI units comprise an mks system (so named because it uses the meter, kilogram, and second as dimensional units). All other units are derived from the dimensional units, which are completely listed in Table 1.1. This system is fully consistent, and there is only one recognized unit for each physical quantity (variable).

Two types of units are used: base units and derived units. The *base units* (see Table 1.1) are dependent only on accepted standards or reproducible phenomena. The previously unclassified *supplementary units*, radian and steradian, have been classified as derived units. The *derived units* (see Table 1.2 and Table 1.3) are made up of combinations of base and supplementary units.

Table 1.1 *SI Base Units*

quantity	name	symbol
length	meter	m
mass	kilogram	kg
time	second	s
electric current	ampere	A
temperature	kelvin	K
amount of substance	mole	mol
luminous intensity	candela	cd

Table 1.2 Some SI Derived Units with Special Names

quantity	name	symbol	expressed in terms of other units
frequency	hertz	Hz	1/s
force	newton	N	kg·m/s^2
pressure, stress	pascal	Pa	N/m^2
energy, work, quantity of heat	joule	J	N·m
power, radiant flux	watt	W	J/s
quantity of electricity, electric charge	coulomb	C	
electric potential, potential difference, electromotive force	volt	V	W/A
electric capacitance	farad	F	C/V
electric resistance	ohm	Ω	V/A
electric conductance	siemen	S	A/V
magnetic flux	weber	Wb	V·s
magnetic flux density	tesla	T	Wb/m^2
inductance	henry	H	Wb/A
luminous flux	lumen	lm	
illuminance	lux	lx	lm/m^2
plane angle	radian	rad	
solid angle	steradian	sr	

Table 1.3 Some SI Derived Units

quantity	description	expressed in terms of other units
area	square meter	m^2
volume	cubic meter	m^3
speed		
linear	meter per second	m/s
angular	radian per second	rad/s
acceleration		
linear	meter per second squared	m/s^2
angular	radian per second squared	rad/s^2
density, mass density	kilogram per cubic meter	kg/m^3
concentration (of amount of substance)	mole per cubic meter	mol/m^3
specific volume	cubic meter per kilogram	m^3/kg
luminance	candela per square meter	cd/m^2
absolute viscosity	pascal second	Pa·s
kinematic viscosity	square meters per second	m^2/s
moment of force	newton meter	N·m
surface tension	newton per meter	N/m
heat flux density, irradiance	watt per square meter	W/m^2
heat capacity, entropy	joule per kelvin	J/K
specific heat capacity, specific entropy	joule per kilogram kelvin	J/kg·K
specific energy	joule per kilogram	J/kg
thermal conductivity	watt per meter kelvin	W/m·K
energy density	joule per cubic meter	J/m^3
electric field strength	volt per meter	V/m
electric charge density	coulomb per cubic meter	C/m^3
surface density of charge, flux density	coulomb per square meter	C/m^2
permittivity	farad per meter	F/m
current density	ampere per square meter	A/m^2
magnetic field strength	ampere per meter	A/m
permeability	henry per meter	H/m
molar energy	joule per mole	J/mol
molar entropy, molar heat capacity	joule per mole kelvin	J/mol·K
radiant intensity	watt per steradian	W/sr

In addition, there is a set of non-SI units that may be used. This concession is primarily due to the significance and widespread acceptance of these units. Use of the non-SI units listed in Table 1.4 will usually create an inconsistent expression requiring conversion factors.

Table 1.4 *Acceptable Non-SI Units*

quantity	unit name	symbol name	relationship to SI unit
area	hectare	ha	$1\ \text{ha} = 10\,000\ \text{m}^2$
energy	kilowatt-hour	kW·h	$1\ \text{kW·h} = 3.6\ \text{MJ}$
mass	metric ton[a]	t	$1\ \text{t} = 1000\ \text{kg}$
plane angle	degree (of arc)	°	$1° = 0.017453\ \text{rad}$
speed of rotation	revolution per minute	r/min	$1\ \text{r/min} = 2\pi/60\ \text{rad/s}$
temperature interval	degree Celsius	°C	$1°\text{C} = 1\text{K}$ $(\Delta T_{°\text{C}} = \Delta T_\text{K})$
time	minute	min	$1\ \text{min} = 60\ \text{s}$
	hour	h	$1\ \text{h} = 3600\ \text{s}$
	day (mean solar)	d	$1\ \text{d} = 86\,400\ \text{s}$
	year (calendar)	a	$1\ \text{a} = 31\,536\,000\ \text{s}$
velocity	kilometer per hour	km/h	$1\ \text{km/h} = 0.278\ \text{m/s}$
volume	liter[b]	L	$1\ \text{L} = 0.001\ \text{m}^3$

[a]The international name for metric ton is *tonne*. The metric ton is equal to the *megagram* (Mg).

[b]The international symbol for liter is the lowercase l, which can be easily confused with the numeral 1. Several English-speaking countries have adopted the script ℓ and uppercase L as a symbol for liter in order to avoid any misinterpretation.

The units of force can be derived from Newton's second law.

$$\text{units of force} = (m \text{ in kg})\left(a \text{ in } \frac{\text{m}}{\text{s}^2}\right) = \frac{\text{kg·m}}{\text{s}^2}$$

This combination of units for force is known as a *newton.* Figure 1.2 illustrates common force units.

Figure 1.2 *Common Force Units and Relative Sizes*

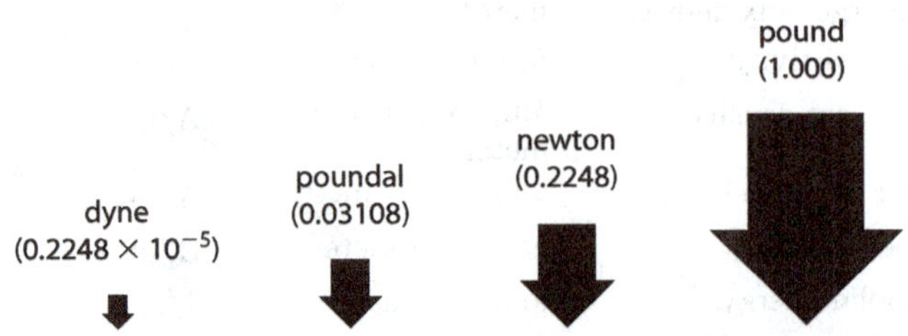

Energy variables in the SI system have units of N·m, or equivalently, kg·m^2/s^2. Both of these combinations are known as a *joule.* The units of power are joules per second, equivalent to a *watt.*

Example

A 10 kg block is raised vertically 3 m. What is most nearly the change in potential energy?

(A) 30 J

(B) 98 J

(C) 290 J

(D) 880 J

Solution

The change in potential energy is

$$\begin{aligned}\Delta\text{PE} &= mg\Delta h \\ &= (10\ \text{kg})\left(9.81\ \frac{\text{m}}{\text{s}^2}\right)(3\ \text{m}) \\ &= 294\ \text{kg·m}^2/\text{s}^2 \quad (290\ \text{J})\end{aligned}$$

The answer is (C).

12. RULES FOR USING THE SI SYSTEM

In addition to having standardized units, the SI system also has rigid syntax rules for writing the units and combinations of units. Each unit is abbreviated with a specific symbol. The following rules for writing and combining these symbols should be adhered to.

- The expressions for derived units in symbolic form are obtained by using the mathematical signs of multiplication and division; for example, units of velocity are m/s, and units of torque are N·m (not N-m or Nm).
- Scaling of most units is done in multiples of 1000.
- The symbols are always printed in roman type, regardless of the type used in the rest of the text. The only exception to this is in the use of the symbol for liter, where the use of the lowercase el (l) may be confused with the numeral one (1). In this case, "liter" should be written out in full, or the script ℓ or L should be used.
- Symbols are not pluralized: 1 kg, 45 kg (not 45 kgs).
- A period after a symbol is not used, except when the symbol occurs at the end of a sentence.
- When symbols consist of letters, there is always a full space between the quantity and the symbols: 45 kg (not 45kg). However, when the first character of a symbol is not a letter, no space is left: 32°C (not 32° C or 32 °C); or 42°12′45″ (not 42° 12′ 45″).
- All symbols are written in lowercase, except when the unit is derived from a proper name: m for meter;

s for second; A for ampere, Wb for weber, N for newton, W for watt.

- Prefixes are printed without spacing between the prefix and the unit symbol (e.g., km is the symbol for kilometer). (See Table 1.5 for a list of SI prefixes.)
- In text, symbols should be used when associated with a number. However, when no number is involved, the unit should be spelled out: The area of the carpet is 16 m^2, not 16 square meters. Carpet is sold by the square meter, not by the m^2.
- A practice in some countries is to use a comma as a decimal marker, while the practice in North America, the United Kingdom, and some other countries is to use a period (or dot) as the decimal marker. Furthermore, in some countries that use the decimal comma, a dot is frequently used to divide long numbers into groups of three. Because of these differing practices, spaces must be used instead of commas to separate long lines of digits into easily readable blocks of three digits with respect to the decimal marker: 32 453.246 072 5. A space (half-space preferred) is optional with a four-digit number: 1 234 or 1234.
- Where a decimal fraction of a unit is used, a zero should always be placed before the decimal marker: 0.45 kg (not .45 kg). This practice draws attention to the decimal marker and helps avoid errors of scale.
- Some confusion may arise with the word "tonne" (1000 kg). When this word occurs in French text of Canadian origin, the meaning may be a ton of 2000 pounds.

Table 1.5 *SI Prefixes**

prefix	symbol	value
exa	E	10^{18}
peta	P	10^{15}
tera	T	10^{12}
giga	G	10^{9}
mega	M	10^{6}
kilo	k	10^{3}
hecto	h	10^{2}
deka	da	10^{1}
deci	d	10^{-1}
centi	c	10^{-2}
milli	m	10^{-3}
micro	μ	10^{-6}
nano	n	10^{-9}
pico	p	10^{-12}
femto	f	10^{-15}
atto	a	10^{-18}

*There is no "B" (billion) prefix. In fact, the word billion means 10^9 in the United States but 10^{12} in most other countries. This unfortunate ambiguity is handled by avoiding the use of the term billion.

13. CONVERSION FACTORS AND CONSTANTS

Commonly used equivalents are given in Table 1.6. Temperature conversions are given in Table 1.7. Table 1.8 gives commonly used constants in customary U.S. and SI units, respectively. Conversion factors are given in Table 1.9.

Table 1.6 *Commonly Used Equivalents*

1 gal of water weighs	8.34 lbf
1 ft^3 of water weighs	62.4 lbf
1 in^3 of mercury weighs	0.491 lbf
The mass of 1 m^3 of water is	1000 kg
1 mg/L is*	8.34 lbf/Mgal

*mg/L is a mass concentration, and its equivalent should also be a mass concentration. However, the *NCEES Handbook* lists a weight concentration.

Table 1.7 *Temperature Conversions*

$$°F = 1.8(°C) + 32°$$

$$°C = \frac{°F - 32°}{1.8}$$

$$°R = °F + 459.69°$$

$$K = °C + 273.15°$$

Table 1.8 Fundamental Constants

quantity	symbol	customary U.S.	SI
Charge			
electron	e		-1.6022×10^{-19} C
proton	p		$+1.6021 \times 10^{-19}$ C
Density			
air [STP, 32°F, (0°C)]		0.0805 lbm/ft^3	1.29 kg/m^3
air [70°F, (20°C), 1 atm]		0.0749 lbm/ft^3	1.20 kg/m^3
earth [mean]		345 lbm/ft^3	5520 kg/m^3
mercury		849 lbm/ft^3	1.360×10^4 kg/m^3
seawater		64.0 lbm/ft^3	1025 kg/m^3
water [mean]		62.4 lbm/ft^3	1000 kg/m^3
Distance [mean]			
earth radius		2.09×10^7 ft	6.370×10^6 m
earth-moon separation		1.26×10^9 ft	3.84×10^8 m
earth-sun separation		4.89×10^{11} ft	1.49×10^{11} m
moon radius		5.71×10^6 ft	1.74×10^6 m
sun radius		2.28×10^9 ft	6.96×10^8 m
first Bohr radius	a_0	1.736×10^{-10} ft	5.292×10^{-11} m
Gravitational Acceleration			
earth [mean]	g	32.174 (32.2) ft/sec^2	9.807 (9.81) m/s^2
moon [mean]		5.47 ft/sec^2	1.67 m/s^2
Mass			
atomic mass unit	u	3.66×10^{-27} lbm	1.6606×10^{-27} kg
earth		1.32×10^{25} lbm	6.00×10^{24} kg
electron [rest]	m_e	2.008×10^{-30} lbm	9.109×10^{-31} kg
moon		1.623×10^{23} lbm	7.36×10^{22} kg
neutron [rest]	m_n	3.693×10^{-27} lbm	1.675×10^{-27} kg
proton [rest]	m_p	3.688×10^{-27} lbm	1.673×10^{-27} kg
sun		4.387×10^{30} lbm	1.99×10^{30} kg
Pressure, atmospheric		14.696 (14.7) lbf/in^2	1.0133×10^5 Pa
Temperature, standard		32°F (492°R)	0°C (273K)
Velocity			
earth escape (from surface, average)		3.67×10^4 ft/sec	1.12×10^4 m/s
light [vacuum]	c	9.84×10^8 ft/sec	$2.99792\ (3.00) \times 10^8$ m/s
sound [air, STP]	a	1090 ft/sec	331 m/s
[air, 70°F (20°C)]		1130 ft/sec	344 m/s
Volume			
molar ideal gas [STP]	V_m	359 ft^3/lbmol	22.414 m^3/kmol
			22 414 L/kmol
Fundamental Constants			
Avogadro's number	N_A		$6.0221\ (6.022) \times 10^{23}$ mol^{-1}
Bohr magneton	μ_B		9.2732×10^{-24} J/T
Boltzmann constant	k	5.65×10^{-24} ft-lbf/°R	1.3807×10^{-23} J/K
Faraday constant	F		96 485 C/mol
gravitational constant	g_c	32.174 (32.2) lbm-ft/lbf-sec^2	
gravitational constant	G	3.44×10^{-8} ft^4/lbf-sec^4	6.673×10^{-11} N·m^2/kg^2 (m^3/kg·s^2)
nuclear magneton	μ_N		5.050×10^{-27} J/T
permeability of a vacuum	μ_0		1.2566×10^{-6} N/A^2 (H/m)
permittivity of a vacuum	ϵ_0		$8.854\ (8.85) \times 10^{-12}$ C^2/N·m^2 (F/m)
Planck's constant	h		6.6256×10^{-34} J·s
Rydberg constant	R_∞		1.097×10^7 m^{-1}
specific gas constant, air	R	53.3 ft-lbf/lbm-°R	287 J/kg·K
Stefan-Boltzmann constant	σ	1.71×10^{-9} Btu/ft^2-hr-°R^4	5.67×10^{-8} W/m^2·K^4
triple point, water		32.02°F, 0.0888 psia	0.01109°C, 0.6123 kPa
universal gas constant	$\bar{R}$	1545 ft-lbf/lbmol-°R	8314 J/kmol·K
	$\bar{R}$	1.986 Btu/lbmol-°R	8.314 kPa·m^3/kmol·K
			0.08206 atm·L/mol·K

Table 1.9 Conversion Factors

multiply	by	to obtain
ac	43,560	ft^2
ampere-hr	3600	coulomb
angstrom	1×10^{-10}	m
atm	76.0	cm Hg
atm	29.92	in Hg
atm	14.70	lbf/in^2 (psia)
atm	33.90	ft water
atm	1.013×10^5	Pa
bar	1×10^5	Pa
bar	0.987	atm
barrels of oil	42	gallons of oil
Btu	1055	J
Btu	2.928×10^{-4}	kW·h
Btu	778	ft-lbf
Btu/hr	3.930×10^{-4}	hp
Btu/hr	0.293	W
Btu/hr	0.216	ft-lbf/sec
cal (g-cal)	3.968×10^{-3}	Btu
cal	1.560×10^{-6}	hp-hr
cal (g-cal)	4.186	J
cal/sec	4.184	W
cm	3.281×10^{-2}	ft
cm	0.394	in
cP	0.001	Pa·s
cP	1	g/m·s
cP	2.419	lbm/hr-ft
cSt	1×10^{-6}	m^2/s
cfs	0.646371	MGD
ft^3	7.481	gal
m^3	1000	L
eV	1.602×10^{-19}	J
ft	30.48	cm
ft	0.3048	m
ft-lbf	1.285×10^{-3}	Btu
ft-lbf	3.766×10^{-7}	kW·h
ft-lbf	0.324	g-cal
ft-lbf	1.35582	J
ft-lbf/sec	1.818×10^{-3}	hp
gal	3.785	L
gal	0.134	ft^3
gal water	8.3453	lbf water
gamma (γ, Γ)*	1×10^{-9}	T
gauss	1×10^{-4}	T
gram	2.205×10^{-3}	lbm
hectare	1×10^{-4}	m^2
hectare	2.47104	ac
hp	42.4	Btu/min
hp	745.7	W
hp	33,000	ft-lbf/min
hp	550	ft-lbf/sec
hp-hr	2545	Btu
hp-hr	1.98×10^{-4}	ft-lbf
hp-hr	2.68×10^{-4}	J
hp-hr	0.746	kW·h
in	2.54	cm
in of Hg	0.0334	atm
in of Hg	13.60	in of H_2O
in of H_2O	0.0361	lbf/in^2
in of H_2O	0.002458	atm
J	9.478×10^{-4}	Btu
J	0.7376	ft-lbf
J	1	N·m
J/s	1	W
kg	2.205	lbm
kgf	9.8066	N
km	3281	ft
km/h	0.621	mi/hr
kPa	0.145	lbf/in^2
kW	1.341	hp
kW	737.6	ft-lbf/sec
kW	3413	Btu/hr
kW·h	3413	Btu
kW·h	1.341	hp-hr
kW·h	3.6×10^6	J
kip	1000	lbf
kip	4448	N
L	61.02	in^3
L	0.264	gal
L	10×10^{-3}	m^3
L/s	2.119	ft^3/min
L/s	15.85	gal/min
m	3.281	ft
m	1.094	yd
m	196.8	ft/min
mi	5280	ft
mi	1.609	km
mph	88.0	ft/min
mph	1.609	kph
mm of Hg	1.316×10^{-3}	atm
mm of H_2O	9.678×10^{-5}	atm
N	0.225	lbf
N	1	$kg \cdot m/s^2$
N·m	0.7376	ft-lbf
N·m	1	J
Pa	9.869×10^{-6}	atm
Pa	1	N/m^2
Pa·s	10	P
lbm	0.454	kg
lbf	4.448	N
lbf-ft	1.356	N·m
lbf/in^2	0.068	atm
lbf/in^2	2.307	ft water
lbf/in^2	2.036	in Hg
lbf/in^2	6895	Pa
radian	$180/\pi$	deg
stokes	1×10^{-4}	m^2/s
therm	10^5	Btu
ton (metric)	1000	kg
ton (short)	2000	lbf
W	3.413	Btu/hr
W	1.341×10^{-3}	hp
W	1	J/s
Wb/m^2	10,000	gauss

(Atmospheres are standard; calories are gram-calories; gallons are U.S. liquid; miles are statute; pounds-mass are avoirdupois.)

*The *NCEES Handbook* is incorrect in assigning uppercase gamma (Γ) as the symbol for the magnetic field intensity unit gamma. Only the symbol γ is used.

2 Analytic Geometry and Trigonometry

1. STRAIGHT LINES

Figure 2.1 is a straight line in two-dimensional space. The *slope* of the line is m, the y-intercept is b, and the x-intercept is a. A known point on the line is represented as (x_1, y_1).

Figure 2.1 Straight Line

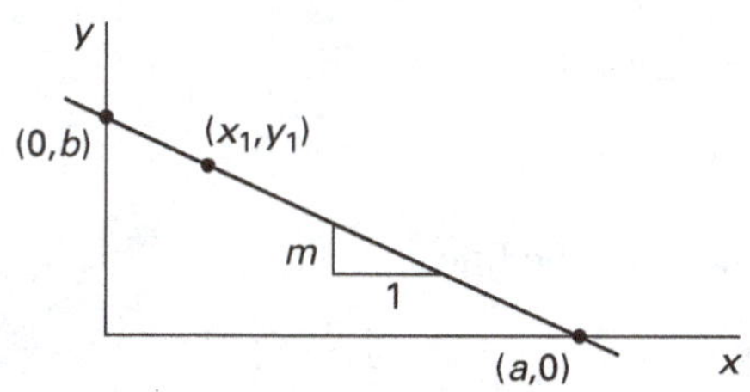

Equation 2.1: Slope

$$m = (y_2 - y_1)/(x_2 - x_1) \qquad 2.1$$

Description

Given two points on a straight line, (x_1, y_1) and (x_2, y_2), Eq. 2.1 gives the slope of the line. The slopes of two parallel lines are equal.

Example

Find the slope of the line that passes through points $(-3, 2)$ and $(5, -2)$.

(A) −2

(B) −0.5

(C) 0.5

(D) 2

Solution

Use Eq. 2.1.

$$m = (y_2 - y_1)/(x_2 - x_1) = \frac{-2-2}{5-(-3)} = -0.5$$

The answer is (B).

Equation 2.2: Slopes of Perpendicular Lines

$$m_1 = -1/m_2 \qquad 2.2$$

Description

If two lines are perpendicular to each other, then their slopes, m_1 and m_2, are negative reciprocals of each other, as shown by Eq. 2.2. For example, if the slope of a line is 5, the slope of a line perpendicular to it is $-1/5$.

Example

A line goes through the point $(4, -6)$ and is perpendicular to the line $y = 4x + 10$. What is the equation of the line?

(A) $y = -\frac{1}{4}x - 20$

(B) $y = -\frac{1}{4}x - 5$

(C) $y = \frac{1}{5}x + 5$

(D) $y = \frac{1}{4}x + 5$

Solution

The slopes of two lines that are perpendicular are related by

$$m_1 = -1/m_2$$

The slope of the line perpendicular to the line with slope $m_1 = 4$ is

$$m_2 = -1/m_1 = -1/4$$

The equation of the line is in the form $y = mx + b$. $m = -1/4$, and a known point is $(x,y) = (4,-6)$.

$$\begin{aligned} -6 &= \left(-\frac{1}{4}\right)(4) + b \\ b &= -6 - \left(-\frac{1}{4}\right)(4) \\ &= -5 \end{aligned}$$

The equation of the line is

$$y = -\tfrac{1}{4}x - 5$$

The answer is (B).

Equation 2.3: Standard Form of the Equation of a Line

$$y = mx + b \qquad 2.3$$

Description

The equation of a line can be represented in several forms. The procedure for finding the equation depends on the form chosen to represent the line. In general, the procedure involves substituting one or more known points on the line into the equation in order to determine the constants.

Equation 2.3 is the *standard form* of the equation of a line. This is also known as the *slope-intercept form* because the constants in the equation are the line's slope, m, and its y-intercept, b.

Example

What is the slope of the line defined by $y - x = 5$?

(A) -1

(B) $-1/5$

(C) $1/4$

(D) 1

Solution

The standard (or slope-intercept) form of the equation of a straight line is $y = mx + b$, where m is the slope and b is the y-intercept. Rearrange the given equation into standard form.

$$\begin{aligned} y - x &= 5 \\ y &= x + 5 \end{aligned}$$

The slope, m, is the coefficient of x, which is 1.

The answer is (D).

Equation 2.4: General Form of the Equation of a Line

$$Ax + By + C = 0 \qquad 2.4$$

Description

Equation 2.4 is the *general form* of the equation of a line.

Example

What is the general form of the equation for a line whose x-intercept is 4 and y-intercept is -6?

(A) $2x - 3y - 18 = 0$

(B) $2x + 3y + 18 = 0$

(C) $3x - 2y - 12 = 0$

(D) $3x + 2y + 12 = 0$

Solution

Find the slope of the line.

$$\begin{aligned} m &= \frac{y_2 - y_1}{x_2 - x_1} \\ &= \frac{-6 - 0}{0 - 4} \\ &= 3/2 \end{aligned}$$

Write the equation of the line in standard (or slope-intercept) form, then arrange it in the form of Eq. 2.4.

$$\begin{aligned} y &= mx + b \\ mx - y + b &= 0 \\ \tfrac{3}{2}x - y + (-6) &= 0 \\ 3x - 2y - 12 &= 0 \end{aligned}$$

The answer is (C).

Equation 2.5: Point-Slope Form of the Equation of a Line

$$y - y_1 = m(x - x_1) \qquad 2.5$$

Description

Equation 2.5 is the *point-slope form* of the equation of a line. This equation defines the line in terms of its slope, m, and one known point, (x_1, y_1).

Example

A circle with a radius of 5 is centered at the origin.

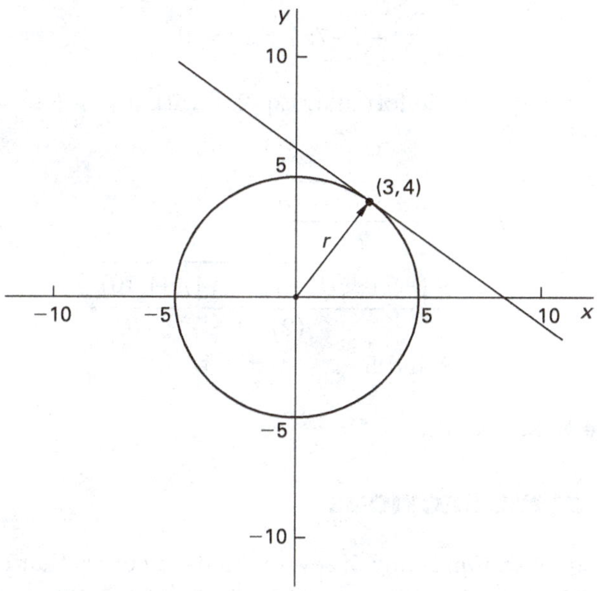

What is the standard form of the equation of the line tangent to this circle at the point (3, 4)?

(A) $x = \frac{-4}{3}y - \frac{25}{4}$

(B) $y = \frac{3}{4}x + \frac{25}{4}$

(C) $y = \frac{-3}{4}x + \frac{9}{4}$

(D) $y = \frac{-3}{4}x + \frac{25}{4}$

Solution

The slope of the radius line from point (0, 0) to point (3, 4) is 4/3. Since the radius and tangent line are perpendicular, the slope of the tangent line is −3/4.

The point-slope form of a straight line with slope $m = -3/4$ and containing point $(x_1, y_1) = (3, 4)$ is

$$y - y_1 = m(x - x_1)$$
$$y - 4 = \left(\frac{-3}{4}\right)(x - 3)$$

Rearranging this into standard form gives

$$y = \frac{-3}{4}x + \left(\frac{9}{4} + 4\right)$$
$$= \frac{-3}{4}x + \frac{25}{4}$$

The answer is (D).

Equation 2.6: Angle Between Two Lines

$$\alpha = \arctan[(m_2 - m_1)/(1 + m_2 \cdot m_1)] \qquad 2.6$$

Description

Two intersecting lines in two-dimensional space are shown in Fig. 2.2.

Figure 2.2 *Two Lines Intersecting in Two-Dimensional Space*

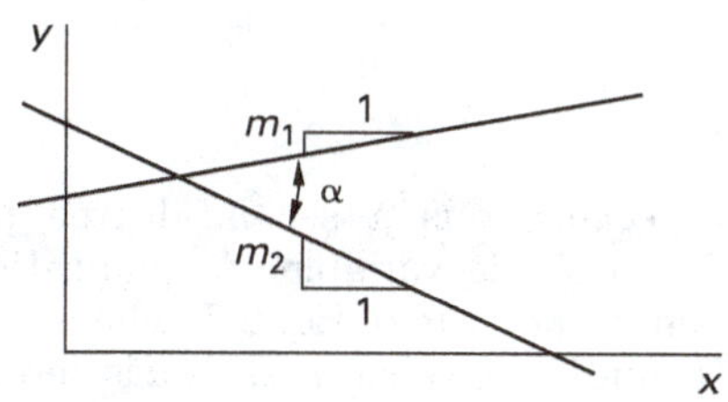

The slopes of the two lines are m_1 and m_2. The acute angle, α, between the lines is given by Eq. 2.6.[1]

[1]Since the NCEES *FE Reference Handbook* (*NCEES Handbook*) does not specify an absolute value, Eq. 2.6 is sensitive to which slopes are designated m_1 and m_2. The slopes should be intuitively selected to avoid a negative arctan argument. Alternatively, since the tangent exhibits odd symmetry, if a negative angle is calculated, the negative sign can simply be disregarded.

Example 6

The angle between the line $y = -7x + 12$ and the line $y = 3x$ is most nearly

(A) 22°

(B) 27°

(C) 33°

(D) 37°

Solution

Use Eq. 2.6.

$$\begin{aligned}\alpha &= \arctan[(m_2 - m_1)/(1 + m_2 \cdot m_1)] \\ &= \arctan\frac{-7-3}{1+(-7)(3)} = \arctan 0.5 \\ &= 26.57° \quad (27°)\end{aligned}$$

The answer is (B).

2. POLYNOMIAL FUNCTIONS

Equation 2.7 and Eq. 2.8: Quadratic Equations

$$ax^2 + bx + c = 0 \qquad 2.7$$

$$x = \frac{-b \pm \sqrt{b^2 - 4ac}}{2a} \qquad 2.8$$

Description

A *quadratic equation* is a second-degree polynomial equation with a single variable. A quadratic equation can be written in the form of Eq. 2.7, where x is the variable and a, b, and c are constants. (If a is zero, the equation is linear.)

The *roots*, x_1 and x_2, of a quadratic equation are the two values of x that satisfy the equation (i.e., make it true). These values can be found from the *quadratic formula*, Eq. 2.8.

The quantity under the radical in Eq. 2.8 is called the *discriminant*. By inspecting the discriminant, the types of roots of the equation can be determined.

- If $b^2 - 4ac > 0$, the roots are real and unequal.
- If $b^2 - 4ac = 0$, the roots are real and equal. This is known as a *double root*.
- If $b^2 - 4ac < 0$, the roots are complex and unequal.

Example 7

What are the roots of the quadratic equation $-7x + x^2 = -10$?

(A) −5 and 2

(B) −2 and 0.4

(C) 0.4 and 2

(D) 2 and 5

Solution

Rearrange the equation into the form of Eq. 2.7.

$$x^2 + (-7x) + 10 = 0$$

Use the quadratic formula, Eq. 2.8, with $a = 1$, $b = -7$, and $c = 10$.

$$\begin{aligned}x &= \frac{-b \pm \sqrt{b^2 - 4ac}}{2a} \\ &= \frac{-(-7) \pm \sqrt{(-7)^2 - (4)(1)(10)}}{(2)(1)} \\ &= 2 \text{ and } 5\end{aligned}$$

The answer is (D).

3. CONIC SECTIONS

A *conic section* is any of several kinds of curves that can be produced by passing a plane through a cone as shown in Fig. 2.3.

Equation 2.9: Eccentricity of a Cutting Plane

$$e = \cos\theta/(\cos\phi) \qquad 2.9$$

Description

If θ is the angle between the vertical axis and the cutting plane and ϕ is the *cone-generating angle*, then the *eccentricity*, e, of the conic section is given by Eq. 2.9.

Figure 2.3 *Conic Sections Produced by Cutting Planes*

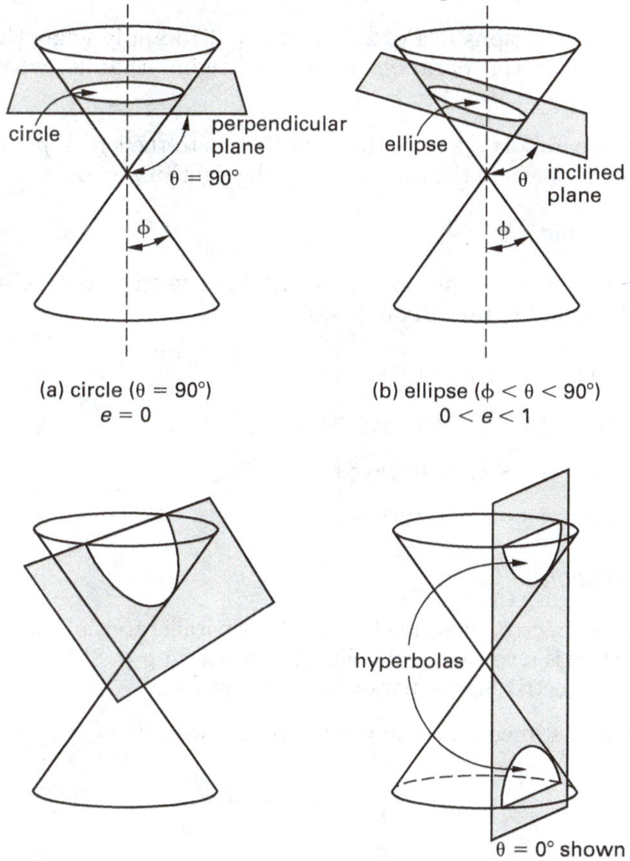

(a) circle (θ = 90°) $e = 0$

(b) ellipse (ϕ < θ < 90°) $0 < e < 1$

(c) parabola (θ = ϕ) $e = 1$

(d) hyperbolas (0° ≤ θ < ϕ) $e > 1$

Equation 2.10 Through Eq. 2.13: General Form and Normal Form of the Conic Section Equation

$$Ax^2 + Bxy + Cy^2 + Dx + Ey + F = 0 \qquad 2.10$$

$$x^2 + y^2 + 2ax + 2by + c = 0 \qquad 2.11$$

$$h = -a; \; k = -b \qquad 2.12$$

$$r = \sqrt{a^2 + b^2 - c} \qquad 2.13$$

Description

All conic sections are described by second-degree (quadratic) polynomials with two variables. The *general form* of the conic section equation is given by Eq. 2.10. x and y are variables, and A, B, C, D, E, and F are constants.

h and k are the coordinates (h, k) of the conic section's center. r is a size parameter, usually the radius of a circle or a sphere. If $r = 0$, then the conic section describes a point. If r is negative, the equation does not describe a conic section.

If $A = C$, then B must be zero for a conic section. If $A = C = 0$, the conic section is a *line*, and if $A = C \neq 0$, the conic section is a *circle*. If $A \neq C$, then if

- $B^2 - 4AC < 0$, the conic section is an *ellipse*
- $B^2 - 4AC > 0$, the conic section is a *hyperbola*
- $B^2 - 4AC = 0$, the conic section is a *parabola*

The general form of the conic section equation can be applied when the conic section is at any orientation relative to the coordinate axes. Equation 2.11 is the *normal form* of the conic section equation. It can be applied when one of the principal axes of the conic section is parallel to a coordinate axis, thereby eliminating certain terms of the general equation and reducing the number of constants needed to three: a, b, and c.

Example

What kind of conic section is described by the following equation?

$$4x^2 - y^2 + 8x + 4y = 15$$

(A) circle

(B) ellipse

(C) parabola

(D) hyperbola

Solution

The general form of a conic section is given by Eq. 2.10 as

$$Ax^2 + Bxy + Cy^2 + Dx + Ey + F = 0$$

In this case, $A = 4$, $B = 0$, and $C = -1$. Since $A \neq C$, the conic section is not a circle or line.

Calculate the discriminant.

$$B^2 - 4AC = (0)^2 - (4)(4)(-1) = 16$$

This is greater than zero, so the section is a hyperbola.

The answer is (D).

Equation 2.14: Standard Form of the Equation of a Horizontal Parabola

$$(y-k)^2 = 2p(x-h) \quad \text{[center at } (h, k)\text{]} \qquad 2.14$$

Description

A *parabola* is the locus of points equidistant from the focus (point F in Fig. 2.4) and a line called the *directrix*. The distance between the focus and the directrix is p.[2] The directrix is defined by the equation $x = h - (p/2)$. When the vertex of the parabola is at the origin, $h = k = 0$, and Eq. 2.15 and Eq. 2.16 apply.

Figure 2.4 Parabola

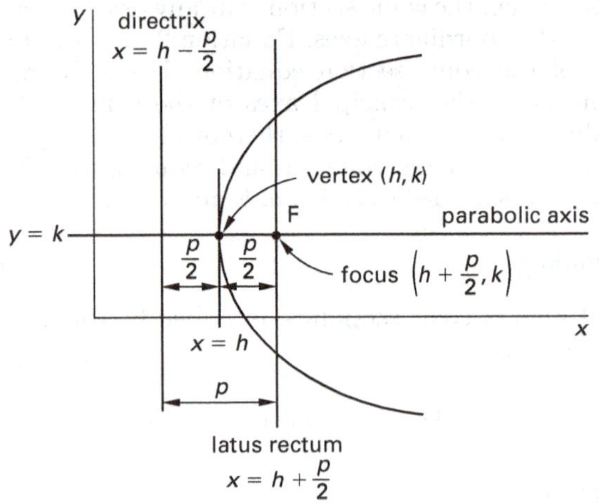

A parabola is symmetric with respect to its *parabolic axis*. The line normal to the parabolic axis and passing through the focus is known as the *latus rectum*. The eccentricity of a parabola is equal to 1.

Equation 2.14 is the *standard form* of the equation of a horizontal parabola. It can be applied when the principal axes of the parabola coincide with the coordinate axes.

The equation for a vertical parabola is similar.

$$(x-h)^2 = 2p(y-k)$$

If p is positive, the parabola opens upward. If p is negative, the parabola opens downward.

Equation 2.15 and Eq. 2.16: Parabola with Vertex at the Origin

$$\text{focus: } (p/2, 0) \qquad 2.15$$

$$x = -p/2 \qquad 2.16$$

Description

The definitions in Eq. 2.15 and Eq. 2.16 apply when the vertex of the parabola is at the origin—that is, when $(h, k) = (0, 0)$.

The parabola opens to the right (points to the left) if $p > 0$, and it opens to the left (points to the right) if $p < 0$.

Example

What is the equation of a parabola with a vertex at (4, 8) and a directrix at $y = 5$?

(A) $(x-8)^2 = 12(y-4)$

(B) $(x-4)^2 = 12(y-8)$

(C) $(x-4)^2 = 6(y-8)$

(D) $(y-8)^2 = 12(x-4)$

Solution

The directrix, described by $y = 5$, is parallel to the x-axis, so this is a vertical parabola. The vertex (at $y = 8$) is above the directrix, so the parabola opens upward.

The distance from the vertex to the directrix is

$$\frac{p}{2} = 8 - 5 = 3$$

$$p = 6$$

The focus is located a distance $p/2$ from the vertex. The focus is at $(4, 8+3)$ or $(4, 11)$.

The standard form equation for a parabola with vertex at (h, k) and opening upward is

$$(x-h)^2 = 2p(y-k)$$

$$(x-h)^2 = (2)(6)(y-8)$$

$$(x-4)^2 = 12(y-8)$$

The answer is (B).

Equation 2.17: Standard Form of the Equation of an Ellipse

$$\frac{(x-h)^2}{a^2} + \frac{(y-k)^2}{b^2} = 1 \quad \text{[center at } (h, k)\text{]} \qquad 2.17$$

[2]There are two conventions used to define the parameters of a parabola. One convention, as used in the *NCEES Handbook*, is to define p as the distance from the focus to the directrix. This results in the $2p$ term in Eq. 2.14. Another convention, arguably more prevalent, is to define p as the distance from the focus to the vertex (i.e., the distance from the focus to the directrix is $2p$), which would result in a corresponding term of $4p$ in Eq. 2.14.

Description

An *ellipse* has two foci, F_1 and F_2, separated along the *major axis* by a distance $2c$. (See Fig. 2.5.) The line perpendicular to the major axis passing through the center of the ellipse is the *minor axis.* The lines perpendicular to the major axis passing through the foci are the *latera recta.* The distance between the two vertices is $2a$. The ellipse is the locus of points such that the sum of the distances from the two foci is $2a$. The eccentricity of the ellipse is always less than one. If the eccentricity is zero, the ellipse is a circle.

Equation 2.17 is the *standard form* of the equation of an ellipse with center at (h, k), *semimajor distance a*, and *semiminor distance b*. Equation 2.17 can be applied when the principal axes of the ellipse coincide with the coordinate axes and when the major axis is oriented vertically.

Figure 2.5 *Ellipse (with horizontal major axis)*

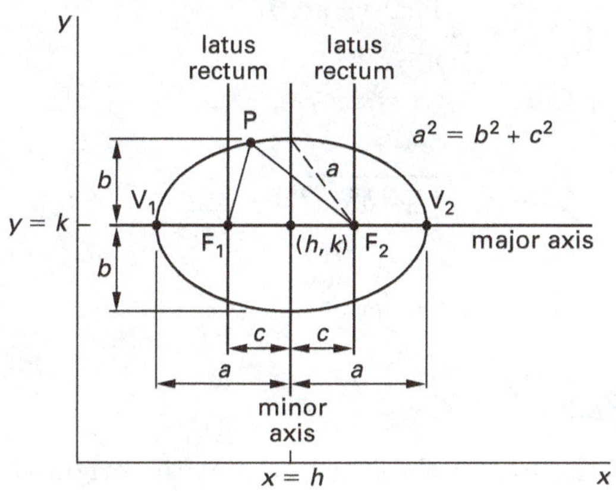

Example 10

What is the equation of the ellipse with center at $(0, 0)$, with vertical major axis, and that passes through the points $(2, 0)$, $(0, 3)$, and $(-2, 0)$?

(A) $\dfrac{x^2}{9} - \dfrac{y^2}{4} = 1$

(B) $\dfrac{x^2}{4} - \dfrac{y^2}{9} = 1$

(C) $\dfrac{x^2}{9} + \dfrac{y^2}{4} = 1$

(D) $\dfrac{x^2}{4} + \dfrac{y^2}{9} = 1$

Solution

An ellipse has the standard form

$$\frac{(x-h)^2}{a^2} + \frac{(y-k)^2}{b^2} = 1$$

The center is at $(h, k) = (0, 0)$.

$$\frac{(x-0)^2}{a^2} + \frac{(y-0)^2}{b^2} = 1$$

Substitute the known values of (x, y) to determine a and b.

For $(x, y) = (2, 0)$,

$$\frac{(2)^2}{a^2} + \frac{(0)^2}{b^2} = 1$$
$$a^2 = 4$$
$$a = 2$$

For $(x, y) = (0, 3)$,

$$\frac{(0)^2}{a^2} + \frac{(3)^2}{b^2} = 1$$
$$b^2 = 9$$
$$b = 3$$

This ellipse is oriented vertically since $b > a$.

Check: For $(x, y) = (-2, 0)$,

$$\frac{(-2)^2}{a^2} + \frac{(0)^2}{b^2} = 1$$
$$a^2 = 4$$
$$a = 2$$

[This step is not necessary as a is determined from the first point.]

The equation of the ellipse is

$$\frac{x^2}{(2)^2} + \frac{y^2}{(3)^2} = 1$$
$$\frac{x^2}{4} + \frac{y^2}{9} = 1$$

The answer is (D).

Equation 2.18 Through Eq. 2.21: Ellipse with Center at the Origin

$$\text{foci: } (\pm ae, 0) \qquad 2.18$$

$$x = \pm a/e \qquad 2.19$$

$$e = \sqrt{1 - (b^2/a^2)} = c/a \qquad 2.20$$

$$b = a\sqrt{1 - e^2} \qquad 2.21$$

Description

When the center of the ellipse is at the origin ($h = k = 0$), the foci are located at $(ae, 0)$ and $(-ae, 0)$, the directrices are located at $\pm x = a/e$, and the eccentricity and semiminor distance are given by Eq. 2.20 and Eq. 2.21, respectively. Each directrix is a vertical line located outside of the ellipse. The location of each directrix is such that the distance from a point on the ellipse to the nearest directrix is proportional to the distance from that point on the ellipse to the nearest focus.

Equation 2.22: Standard Form of the Equation of a Hyperbola

$$\frac{(x-h)^2}{a^2} - \frac{(y-k)^2}{b^2} = 1 \quad [\text{center at } (h, k)] \qquad 2.22$$

Description

As shown in Fig. 2.6, a *hyperbola* has two foci separated along the *transverse axis* by a distance $2c$. The two lines perpendicular to the transverse axis that pass through the foci are the *conjugate axes*. As the distance from the center increases, the hyperbola approaches two straight lines, called the *asymptotes*, that intersect at the hyperbola's center.

The distance from the center to either vertex is a. The distance from either vertex to either asymptote in a direction perpendicular to the transverse axis is b. The hyperbola is the locus of points such that the distances from any point to the two foci differ by $2a$. The distance from the center to either focus is c.

Equation 2.22 is the *standard form* of the equation of a hyperbola with center at (h, k) and opening horizontally.

Figure 2.6 Hyperbola

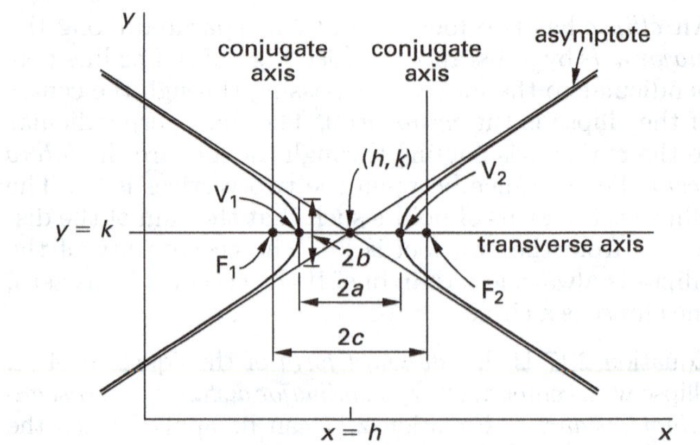

Equation 2.23 Through Eq. 2.26: Hyperbola with Center at the Origin

$$\text{foci: } (\pm ae, 0) \qquad 2.23$$

$$x = \pm a/e \qquad 2.24$$

$$e = \sqrt{1 + (b^2/a^2)} = c/a \qquad 2.25$$

$$b = a\sqrt{e^2 - 1} \qquad 2.26$$

Description

When the hyperbola is centered at the origin ($h = k = 0$), the foci are located at $(ae, 0)$ and $(-ae, 0)$, the directrices are located at $x = a/e$ and $x = -a/e$, and the eccentricity, e, and distance b are given by Eq. 2.25 and Eq. 2.26, respectively.

Example

What is most nearly the eccentricity of the hyperbola shown?

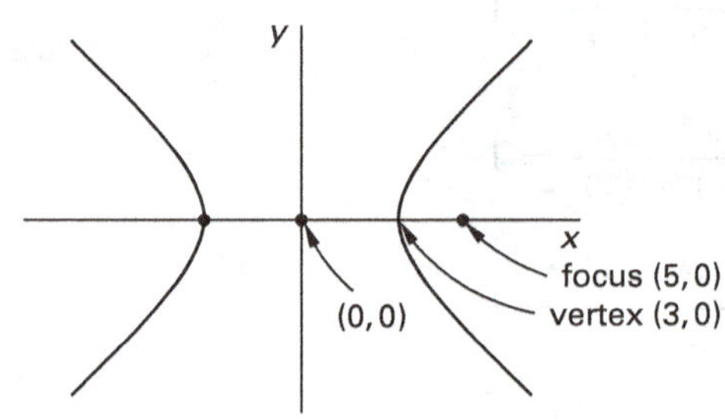

(A) 1.33

(B) 1.67

(C) 2.00

(D) 3.00

Solution

Use Eq. 2.25. a is the distance from the center to either vertex, and c is the distance from the center to either focus. The eccentricity is

$$e = c/a = \frac{5}{3} = 1.67$$

The answer is (B).

Equation 2.27 and Eq. 2.28: Standard Form of the Equation of a Circle

$$(x-h)^2 + (y-k)^2 = r^2 \quad 2.27$$

$$r = \sqrt{(x-h)^2 + (y-k)^2} \quad 2.28$$

Description

Equation 2.27 is the *standard form* (also called the *center-radius form*) of the equation of a circle with center at (h, k) and radius r. (See Fig. 2.7.) The radius is given by Eq. 2.28.

Figure 2.7 Circle

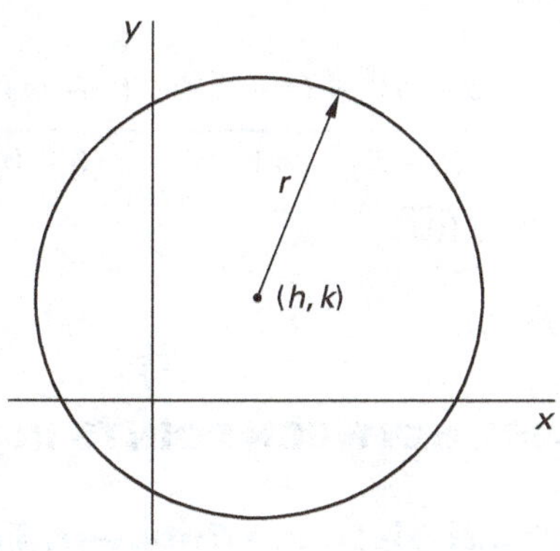

Example 12

What is the equation of the circle passing through the points $(0,0)$, $(0,4)$, and $(-4,0)$?

(A) $(x-2)^2 + (y-2)^2 = \sqrt{8}$

(B) $(x-2)^2 + (y-2)^2 = 8$

(C) $(x+2)^2 + (y-2)^2 = 8$

(D) $(x+2)^2 + (y+2)^2 = \sqrt{8}$

Solution

From Eq. 2.27, the center-radius form of the equation of a circle is

$$(x-h)^2 + (y-k)^2 = r^2$$

Substitute the first two points, $(0,0)$ and $(0,4)$.

$$(0-h)^2 + (0-k)^2 = r^2$$
$$(0-h)^2 + (4-k)^2 = r^2$$

Since both are equal to the unknown r^2, set the left-hand sides equal. Simplify and solve for k.

$$h^2 + k^2 = h^2 + (4-k)^2$$
$$k^2 = (4-k)^2$$
$$k = 2$$

Substitute the third point, $(-4,0)$, into the center-radius form.

$$(-4-h)^2 + (0-k)^2 = r^2$$

Set this third equation equal to the first equation. Simplify and solve for h.

$$(-4-h)^2 + k^2 = h^2 + k^2$$
$$(-4-h)^2 = h^2$$
$$h = -2$$

Now that h and k are known, substitute them into the first equation to determine r^2.

$$h^2 + k^2 = r^2$$
$$(-2)^2 + (2)^2 = 8$$

Substitute the known values of h, k, and r^2 into the center-radius form.

$$(x+2)^2 + (y-2)^2 = 8$$

The answer is (C).

Equation 2.29: Distance Between Two Points on a Plane

$$d = \sqrt{(y_2 - y_1)^2 + (x_2 - x_1)^2} \quad 2.29$$

Description

The distance, d, between two points (x_1, y_1) and (x_2, y_2) is given by Eq. 2.29.

Mathematics

Equation 2.30: Length of Tangent to Circle from a Point

$$t^2 = (x' - h)^2 + (y' - k)^2 - r^2 \qquad 2.30$$

Description

The length, t, of a *tangent* to a circle from a point (x', y') in two-dimensional space is illustrated in Fig. 2.8 and can be found from Eq. 2.30.

Figure 2.8 *Tangent to a Circle from a Point*

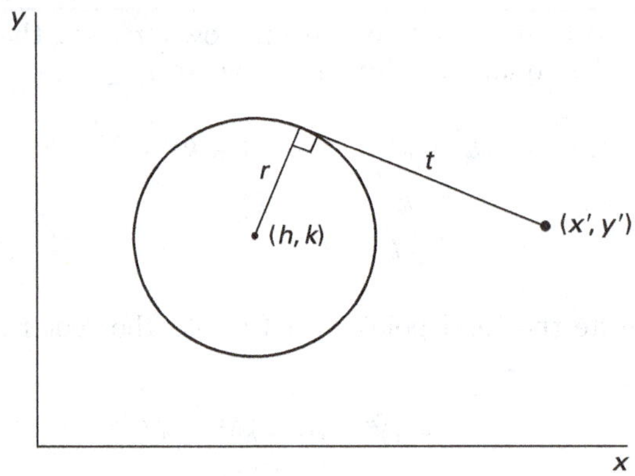

Example

What is the length of the line tangent from point (7, 1) to the circle shown?

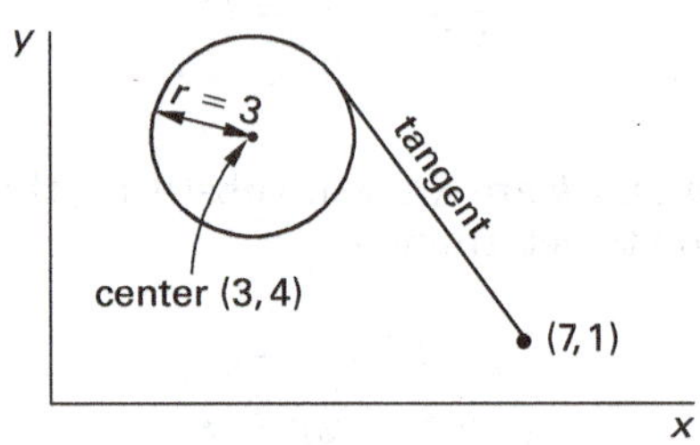

(A) 3

(B) 4

(C) 5

(D) 7

Solution

Use Eq. 2.30.

$$\begin{aligned} t^2 &= (x' - h)^2 + (y' - k)^2 - r^2 \\ &= (7-3)^2 + (1-4)^2 - 3^2 \\ &= 16 \\ t &= 4 \end{aligned}$$

The answer is (B).

4. QUADRIC SURFACE (SPHERE)

Equation 2.31: Standard Form of the Equation of a Sphere

$$(x - h)^2 + (y - k)^2 + (z - m)^2 = r^2 \qquad 2.31$$

Description

Equation 2.31 is the *standard form* of the equation of a sphere centered at (h, k, m) with radius r.

Example

Most nearly, what is the radius of a sphere with a center at the origin and that passes through the point (8, 1, 6)?

(A) 9.2

(B) $\sqrt{101}$

(C) 65

(D) 100

Solution

Use Eq. 2.31.

$$\begin{aligned} r^2 &= (x - h)^2 + (y - k)^2 + (z - m)^2 \\ r &= \sqrt{(8-0)^2 + (1-0)^2 + (6-0)^2} \\ &= \sqrt{101} \end{aligned}$$

The answer is (B).

5. DISTANCE BETWEEN POINTS IN SPACE

Equation 2.32: Distance Between Two Points in Space

$$d = \sqrt{(x_2 - x_1)^2 + (y_2 - y_1)^2 + (z_2 - z_1)^2} \qquad 2.32$$

Description

The distance between two points (x_1, y_1, z_1) and (x_2, y_2, z_2) in three-dimensional space can be found using Eq. 2.32.

Example

What is the distance between point P at $(1, -3, 5)$ and point Q at $(-3, 4, -2)$?

(A) $\sqrt{10}$

(B) $\sqrt{14}$

(C) 8

(D) $\sqrt{114}$

Solution

The distance between points P and Q is

$$\begin{aligned} d_{PQ} &= \sqrt{(x_2 - x_1)^2 + (y_2 - y_1)^2 + (z_2 - z_1)^2} \\ &= \sqrt{(-3-1)^2 + (4-(-3))^2 + (-2-5)^2} \\ &= \sqrt{114} \end{aligned}$$

The answer is (D).

6. DEGREES AND RADIANS

Degrees and *radians* are two units for measuring angles. One complete circle is divided into 360 degrees (written 360°) or 2π radians (abbreviated *rad*).[3] The conversions between degrees and radians are

multiply	by	to obtain
radians	$\frac{180}{\pi}$	degrees
degrees	$\frac{\pi}{180}$	radians

The number of radians in an angle, θ, corresponds to twice the area within a circular sector with arc length θ and a radius of one, as shown in Fig. 2.9. Alternatively, the area of a sector with central angle θ radians is $\theta/2$ for a *unit circle* (i.e., a circle with a radius of one unit).

Figure 2.9 *Radians and Area of Unit Circle*

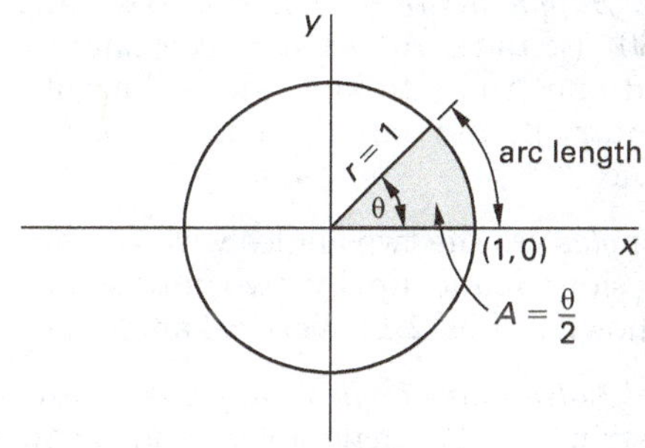

7. PLANE ANGLES

A *plane angle* (usually referred to as just an *angle*) consists of two intersecting lines and an intersection point known as the *vertex*. The angle can be referred to by a capital letter representing the vertex (e.g., B in Fig. 2.10), a letter representing the angular measure (e.g., B or β), or by three capital letters, where the middle letter is the vertex and the other two letters are two points on different lines, and either the symbol ∠ or ∢ (e.g., ∢ ABC).

Figure 2.10 *Angle*

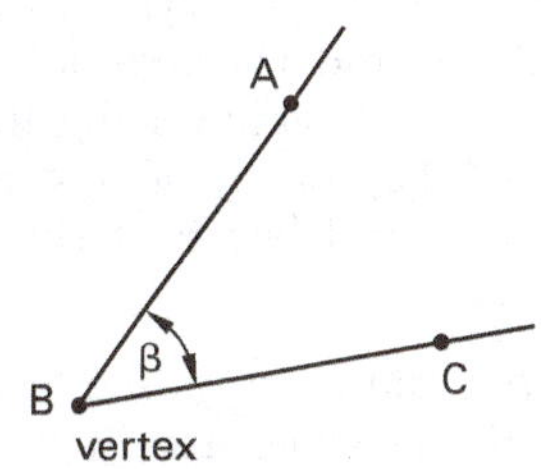

The angle between two intersecting lines generally is understood to be the smaller angle created.[4] Angles have been classified as follows.

- *acute angle:* an angle less than 90° ($\pi/2$ rad)
- *obtuse angle:* an angle more than 90° ($\pi/2$ rad) but less than 180° (π rad)
- *reflex angle:* an angle more than 180° (π rad) but less than 360° (2π rad)
- *related angle:* an angle that differs from another by some multiple of 90° ($\pi/2$ rad)
- *right angle:* an angle equal to 90° ($\pi/2$ rad)
- *straight angle:* an angle equal to 180° (π rad); that is, a straight line

[3]The abbreviation *rad* is also used to represent *radiation absorbed dose*, a measure of radiation exposure.
[4]In books on geometry, the term *ray* is used instead of *line*.

Complementary angles are two angles whose sum is 90° ($\pi/2$ rad). *Supplementary angles* are two angles whose sum is 180° (π rad). *Adjacent angles* share a common vertex and one (the interior) side. Adjacent angles are supplementary if, and only if, their exterior sides form a straight line.

Vertical angles are the two angles with a common vertex and with sides made up by two intersecting straight lines, as shown in Fig. 2.11. Vertical angles are equal.

Angle of elevation and *angle of depression* are surveying terms referring to the angle above and below the horizontal plane of the observer, respectively.

Figure 2.11 *Vertical Angles*

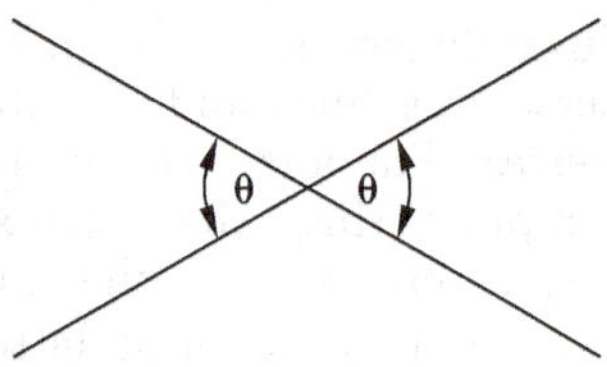

8. TRIANGLES

A *triangle* is a three-sided closed polygon with three angles whose sum is 180° (π rad). Triangles are identified by their vertices and the symbol Δ (e.g., ΔABC in Fig. 2.12). A side is designated by its two endpoints (e.g., AB in Fig. 2.12) or by a lowercase letter corresponding to the capital letter of the opposite vertex (e.g., c).

In *similar triangles*, the corresponding angles are equal and the corresponding sides are in proportion. (Since there are only two independent angles in a triangle, showing that two angles of one triangle are equal to two angles of the other triangle is sufficient to show similarity.) The symbol for similarity is $\sim$. In Fig. 2.12, ΔABC $\sim$ ΔDEF (i.e., ΔABC is similar to ΔDEF).

Figure 2.12 *Similar Triangles*

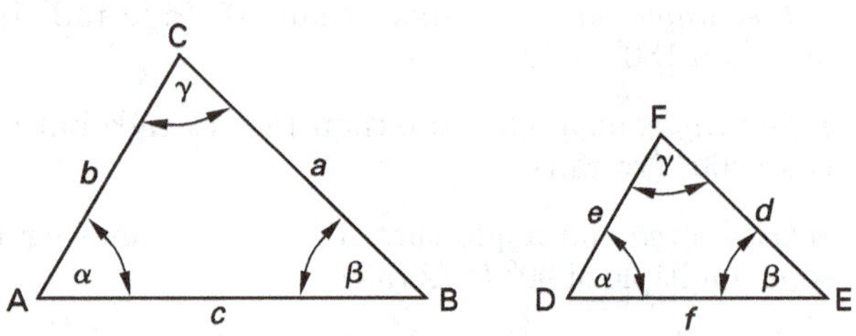

9. RIGHT TRIANGLES

A *right triangle* is a triangle in which one of the angles is 90° ($\pi/2$ rad), as shown in Fig. 2.13. Choosing one of the acute angles as a reference, the sides of the triangle are called the *adjacent side*, x, the *opposite side*, y, and the *hypotenuse*, r.

Figure 2.13 *Right Triangle*

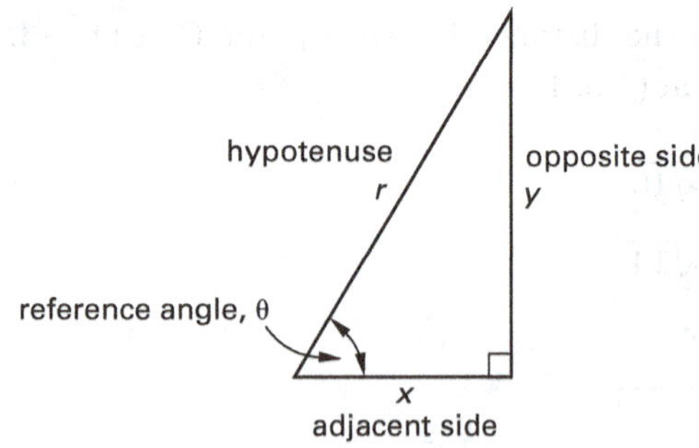

Equation 2.33 Through Eq. 2.38: Trigonometric Functions

$$\sin\theta = y/r \tag{2.33}$$

$$\cos\theta = x/r \tag{2.34}$$

$$\tan\theta = y/x \tag{2.35}$$

$$\csc\theta = r/y \tag{2.36}$$

$$\sec\theta = r/x \tag{2.37}$$

$$\cot\theta = x/y \tag{2.38}$$

Description

The trigonometric functions given in Eq. 2.33 through Eq. 2.38 are calculated from the sides of the right triangle.

The trigonometric functions correspond to the lengths of various line segments in a right triangle in a unit circle. Figure 2.14 shows such a triangle inscribed in a unit circle.

Figure 2.14 *Trigonometric Functions in a Unit Circle*

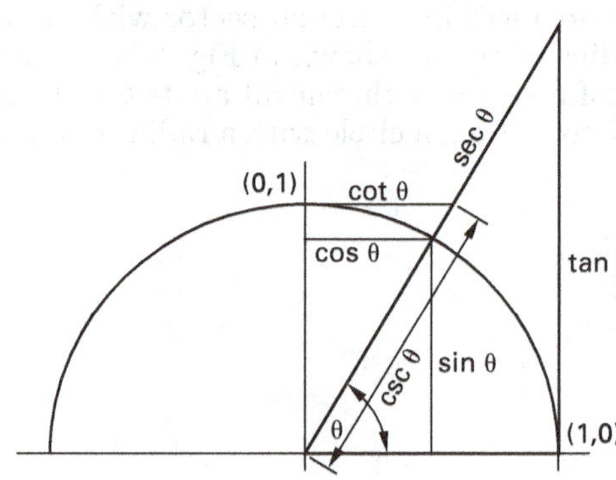

Example

The values of cos 45° and tan 45°, respectively, are

(A) 1 and $\sqrt{2}/2$

(B) 1 and $\sqrt{2}$

(C) $\sqrt{2}/2$ and 1

(D) $\sqrt{2}$ and 1

Solution

For convenience, let the adjacent side of a 45° right triangle have a length of $x = 1$. Then the opposite side has a length of $y = 1$, and the hypotenuse has a length of $r = \sqrt{2}$.

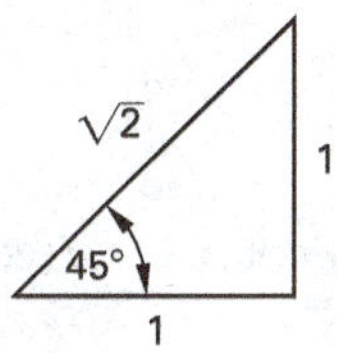

Using Eq. 2.34 and Eq. 2.35,

$$\cos 45° = x/r = \frac{1}{\sqrt{2}} = \sqrt{2}/2$$

$$\tan 45° = y/x = \frac{1}{1} = 1$$

The answer is (C).

10. TRIGONOMETRIC IDENTITIES

Equation 2.39 through Eq. 2.41 are some of the most commonly used trigonometric identities.

Equation 2.39 Through Eq. 2.41: Reciprocal Functions

$$\csc\theta = 1/\sin\theta \qquad 2.39$$

$$\sec\theta = 1/\cos\theta \qquad 2.40$$

$$\cot\theta = 1/\tan\theta \qquad 2.41$$

Description

Three pairs of the trigonometric functions are reciprocals of each other. The prefix "co-" is not a good indicator of the reciprocal functions; while the tangent and cotangent functions are reciprocals of each other, two other pairs—the sine and cosine functions and the secant and cosecant functions—are not.

Example

Simplify the expression $\cos\theta \sec\theta / \tan\theta$.

(A) 1

(B) $\cot\theta$

(C) $\csc\theta$

(D) $\sin\theta$

Solution

Use the reciprocal functions given in Eq. 2.40 and Eq. 2.41.

$$\frac{\cos\theta\sec\theta}{\tan\theta} = \frac{\cos\theta\left(\frac{1}{\cos\theta}\right)}{\tan\theta} = \frac{1}{\tan\theta} = \cot\theta$$

The answer is (B).

Equation 2.42 Through Eq. 2.47: General Identities

$$\cos\theta = \sin(\theta + \pi/2) = -\sin(\theta - \pi/2) \qquad 2.42$$

$$\sin\theta = \cos(\theta - \pi/2) = -\cos(\theta + \pi/2) \qquad 2.43$$

$$\tan\theta = \sin\theta/\cos\theta \qquad 2.44$$

$$\sin^2\theta + \cos^2\theta = 1 \qquad 2.45$$

$$\tan^2\theta + 1 = \sec^2\theta \qquad 2.46$$

$$\cot^2\theta + 1 = \csc^2\theta \qquad 2.47$$

Description

Equation 2.42 through Eq. 2.47 give some general trigonometric identities.

Example

Which of the following expressions is equivalent to the expression $\csc\theta\cos^3\theta\tan\theta$?

(A) $\sin\theta$

(B) $\cos\theta$

(C) $1 - \sin^2\theta$

(D) $1 + \sin^2\theta$

Solution

Simplify the expression using the trigonometric identities given in Eq. 2.39, Eq. 2.44, and Eq. 2.45.

$$\begin{aligned}\csc\theta\cos^3\theta\tan\theta &= \left(\frac{1}{\sin\theta}\right)\cos^3\theta\left(\frac{\sin\theta}{\cos\theta}\right)\\ &= \cos^2\theta\\ &= 1-\sin^2\theta\end{aligned}$$

The answer is (C).

Equation 2.48 Through Eq. 2.51: Double-Angle Identities

$$\sin 2\alpha = 2\sin\alpha\cos\alpha \qquad 2.48$$

$$\begin{aligned}\cos 2\alpha &= \cos^2\alpha-\sin^2\alpha\\ &= 1-2\sin^2\alpha\\ &= 2\cos^2\alpha-1\end{aligned} \qquad 2.49$$

$$\tan 2\alpha = (2\tan\alpha)/(1-\tan^2\alpha) \qquad 2.50$$

$$\cot 2\alpha = (\cot^2\alpha-1)/(2\cot\alpha) \qquad 2.51$$

Description

The identities given in Eq. 2.48 through Eq. 2.51 show equivalent expressions of trigonometric functions of double angles.

Example 17

What is an equivalent expression for $\sin 2\alpha$?

(A) $-2\sin\alpha\cos\alpha$

(B) $\frac{1}{2}\sin\alpha\cos\alpha$

(C) $\dfrac{2\sin\alpha}{\sec\alpha}$

(D) $2\sin\alpha\cos\dfrac{\alpha}{2}$

Solution

Use Eq. 2.48, the double-angle formula for the sine function.

$$\sin 2\alpha = 2\sin\alpha\cos\alpha = \frac{2\sin\alpha}{\sec\alpha}$$

The answer is (C).

Equation 2.52 Through Eq. 2.59: Two-Angle Identities

$$\sin(\alpha+\beta) = \sin\alpha\cos\beta+\cos\alpha\sin\beta \qquad 2.52$$

$$\cos(\alpha+\beta) = \cos\alpha\cos\beta-\sin\alpha\sin\beta \qquad 2.53$$

$$\tan(\alpha+\beta) = (\tan\alpha+\tan\beta)/(1-\tan\alpha\tan\beta) \qquad 2.54$$

$$\cot(\alpha+\beta) = (\cot\alpha\cot\beta-1)/(\cot\alpha+\cot\beta) \qquad 2.55$$

$$\sin(\alpha-\beta) = \sin\alpha\cos\beta-\cos\alpha\sin\beta \qquad 2.56$$

$$\cos(\alpha-\beta) = \cos\alpha\cos\beta+\sin\alpha\sin\beta \qquad 2.57$$

$$\tan(\alpha-\beta) = (\tan\alpha-\tan\beta)/(1+\tan\alpha\tan\beta) \qquad 2.58$$

$$\cot(\alpha-\beta) = (\cot\alpha\cot\beta+1)/(\cot\beta-\cot\alpha) \qquad 2.59$$

Description

The identities given in Eq. 2.52 through Eq. 2.59 show equivalent expressions of two-angle trigonometric functions.

Example 18

Simplify the following expression.

$$\frac{\cos(\alpha+\beta)+\cos(\alpha-\beta)}{\cos\beta}$$

(A) $\cos\alpha/2$

(B) $2\cos\alpha$

(C) $\sin 2\alpha$

(D) $\sin^2\alpha$

Solution

Use Eq. 2.53 and Eq. 2.57.

$$\frac{\cos(\alpha+\beta) + \cos(\alpha-\beta)}{\cos\beta} = \frac{\begin{pmatrix}\cos\alpha\cos\beta \\ -\sin\alpha\sin\beta\end{pmatrix} + \begin{pmatrix}\cos\alpha\cos\beta \\ +\sin\alpha\sin\beta\end{pmatrix}}{\cos\beta}$$

$$= \frac{2\cos\alpha\cos\beta}{\cos\beta}$$

$$= 2\cos\alpha$$

The answer is (B).

Equation 2.60 Through Eq. 2.63: Half-Angle Identities

$$\sin(\alpha/2) = \pm\sqrt{(1-\cos\alpha)/2} \qquad 2.60$$

$$\cos(\alpha/2) = \pm\sqrt{(1+\cos\alpha)/2} \qquad 2.61$$

$$\tan(\alpha/2) = \pm\sqrt{(1-\cos\alpha)/(1+\cos\alpha)} \qquad 2.62$$

$$\cot(\alpha/2) = \pm\sqrt{(1+\cos\alpha)/(1-\cos\alpha)} \qquad 2.63$$

Description

The identities given in Eq. 2.60 through Eq. 2.63 show equivalent expressions of half-angle trigonometric functions.

Equation 2.64 Through Eq. 2.70: Miscellaneous Identities

$$\sin\alpha\sin\beta = (1/2)[\cos(\alpha-\beta) - \cos(\alpha+\beta)] \qquad 2.64$$

$$\cos\alpha\cos\beta = (1/2)[\cos(\alpha-\beta) + \cos(\alpha+\beta)] \qquad 2.65$$

$$\sin\alpha\cos\beta = (1/2)[\sin(\alpha+\beta) + \sin(\alpha-\beta)] \qquad 2.66$$

$$\sin\alpha + \sin\beta = 2\sin[(1/2)(\alpha+\beta)] \times \cos[(1/2)(\alpha-\beta)] \qquad 2.67$$

$$\sin\alpha - \sin\beta = 2\cos[(1/2)(\alpha+\beta)] \times \sin[(1/2)(\alpha-\beta)] \qquad 2.68$$

$$\cos\alpha + \cos\beta = 2\cos[(1/2)(\alpha+\beta)] \times \cos[(1/2)(\alpha-\beta)] \qquad 2.69$$

$$\cos\alpha - \cos\beta = -2\sin[(1/2)(\alpha+\beta)] \times \sin[(1/2)(\alpha-\beta)] \qquad 2.70$$

Description

The identities given in Eq. 2.64 through Eq. 2.70 show equivalent expressions of other trigonometric functions.

11. GENERAL TRIANGLES

The term *general triangle* refers to any triangle, including but not limited to right triangles. Figure 2.15 shows a general triangle.

Figure 2.15 *General Triangle*

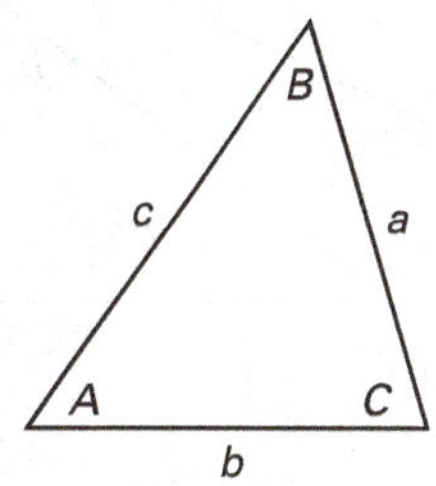

Equation 2.71: Law of Sines

$$\frac{a}{\sin A} = \frac{b}{\sin B} = \frac{c}{\sin C} \qquad 2.71$$

Description

For a general triangle, the *law of sines* relates the sines of the three angles A, B, and C and their opposite sides, a, b, and c, respectively.

Example

The vertical angle to the top of a flagpole from point A on the ground is observed to be 37°11′. The observer walks 17 m directly away from the flagpole from point A to point B and finds the new angle to be 25°43′.

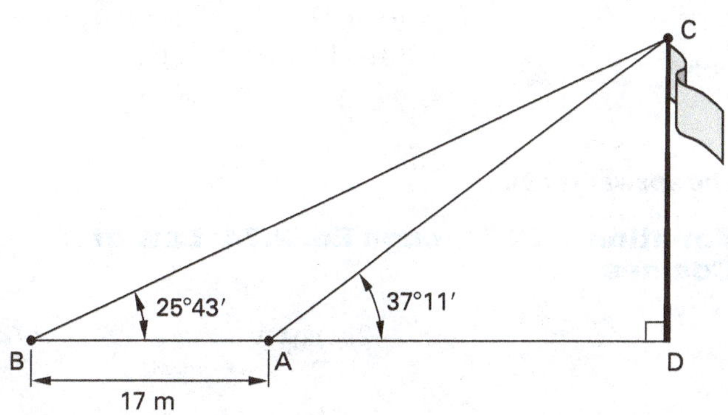

What is the approximate height of the flagpole?

(A) 10 m

(B) 22 m

(C) 82 m

(D) 300 m

Solution

The two observations lead to two triangles with a common leg, h.

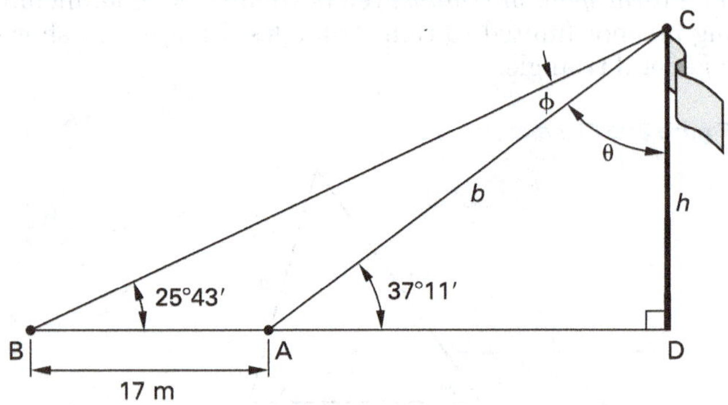

Find angle θ in triangle ADC.

$$37°11' + 90° + \theta = 180°$$
$$\theta = 52°49'$$

Find angle ϕ in triangle BDC.

$$25°43' + 90° + (52°49' + \phi) = 180°$$
$$\phi = 11°28'$$

Use the law of sines on triangle BDC to find side b.

$$\frac{\sin 11°28'}{17 \text{ m}} = \frac{\sin 25°43'}{b}$$
$$b = 37.11 \text{ m}$$

Find the flagpole height, h, using triangle ADC.

$$\sin 37°11' = \frac{h}{b}$$
$$\begin{aligned} h &= b \sin 37°11' \\ &= (37.11 \text{ m})\sin 37°11' \\ &= 22.43 \text{ m} \quad (22 \text{ m}) \end{aligned}$$

The answer is (B).

Equation 2.72 Through Eq. 2.74: Law of Cosines

$$a^2 = b^2 + c^2 - 2bc \cos A \qquad 2.72$$

$$b^2 = a^2 + c^2 - 2ac \cos B \qquad 2.73$$

$$c^2 = a^2 + b^2 - 2ab \cos C \qquad 2.74$$

Variations

$$\cos A = \frac{b^2 + c^2 - a^2}{2bc}$$

$$\cos B = \frac{a^2 + c^2 - b^2}{2ac}$$

$$\cos C = \frac{a^2 + b^2 - c^2}{2ab}$$

Description

For a general triangle, the *law of cosines* relates the cosines of the three angles A, B, and C and their opposite sides, a, b, and c, respectively.

Example

Three circles of radii 110 m, 140 m, and 220 m are tangent to one another. What are the interior angles of the triangle formed by joining the centers of the circles?

(A) 34.2°, 69.2°, and 76.6°

(B) 36.6°, 69.1°, and 74.3°

(C) 42.2°, 62.5°, and 75.3°

(D) 47.9°, 63.1°, and 69.0°

Solution

The three circles and the triangle are shown.

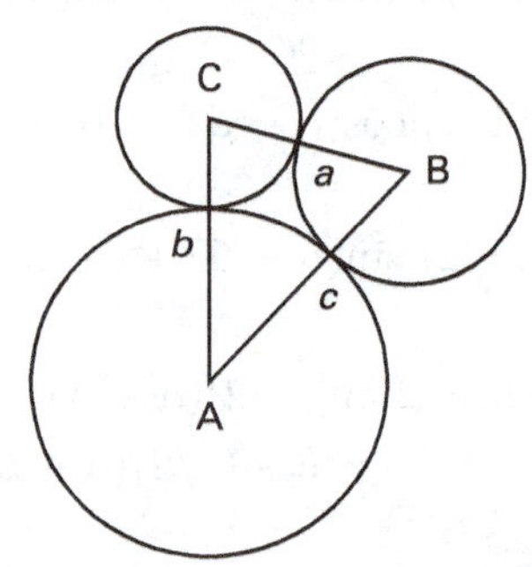

Calculate the length of each side of the triangle.

$$\begin{aligned} a &= 110 \text{ m} + 140 \text{ m} = 250 \text{ m} \\ b &= 110 \text{ m} + 220 \text{ m} = 330 \text{ m} \\ c &= 140 \text{ m} + 220 \text{ m} = 360 \text{ m} \end{aligned}$$

From Eq. 2.72,

$$a^2 = b^2 + c^2 - 2bc\cos A$$
$$\cos A = \frac{b^2 + c^2 - a^2}{2bc} = \frac{(330\text{ m})^2 + (360\text{ m})^2 - (250\text{ m})^2}{(2)(330\text{ m})(360\text{ m})} = 0.7407$$
$$A = 42.2°$$

From Eq. 2.73,

$$b^2 = a^2 + c^2 - 2ac\cos B$$
$$\cos B = \frac{a^2 + c^2 - b^2}{2ac} = \frac{(250\text{ m})^2 + (360\text{ m})^2 - (330\text{ m})^2}{(2)(250\text{ m})(360\text{ m})} = 0.4622$$
$$B = 62.5°$$

From Eq. 2.74,

$$c^2 = a^2 + b^2 - 2ab\cos C$$
$$\cos C = \frac{a^2 + b^2 - c^2}{2ab} = \frac{(250\text{ m})^2 + (330\text{ m})^2 - (360\text{ m})^2}{(2)(250\text{ m})(330\text{ m})} = 0.2533$$
$$C = 75.3°$$

The answer is (C).

12. MENSURATION OF AREAS

The dimensions, perimeter, area, and other geometric properties constitute the *mensuration* (i.e., the measurements) of a geometric shape.

Equation 2.75 and Eq. 2.76: Parabolic Segments

$$A = 2bh/3 \qquad 2.75$$

$$A = bh/3 \qquad 2.76$$

Description

Equation 2.75 and Eq. 2.76 give the area of a parabolic segment. (See Fig. 2.16.) Equation 2.75 gives the area within the curve of the parabola, and Eq. 2.76 gives the area outside the curve of the parabola.

Figure 2.16 *Parabolic Segments*

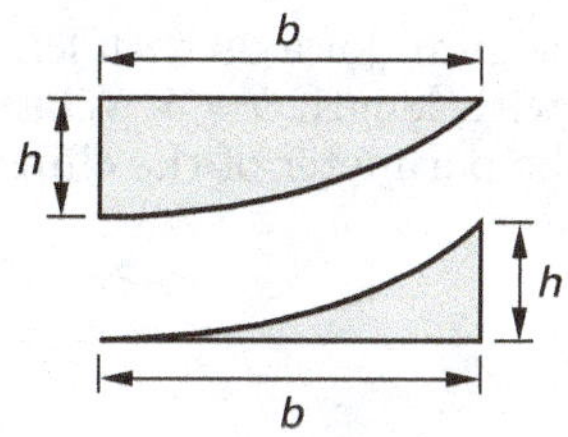

Equation 2.77 Through Eq. 2.80: Ellipses

$$A = \pi ab \qquad 2.77$$

$$P_{\text{approx}} = 2\pi\sqrt{(a^2 + b^2)/2} \qquad 2.78$$

$$P = \pi(a+b)\left|\begin{array}{l} 1 + (1/2)^2\lambda^2 + (1/2 \times 1/4)^2\lambda^4 \\ \quad + (1/2 \times 1/4 \times 3/6)^2\lambda^6 \\ \quad + (1/2 \times 1/4 \times 3/6 \times 5/8)^2\lambda^8 \\ \quad + \begin{pmatrix} 1/2 \times 1/4 \times 3/6 \\ \times 5/8 \times 7/10 \end{pmatrix}^2 \lambda^{10} \\ \quad + \cdots \end{array}\right| \qquad 2.79$$

$$\lambda = (a-b)/(a+b) \qquad 2.80$$

Description

Equation 2.77 gives the area of an ellipse. (See Fig. 2.17.) a and b are the semimajor and semiminor axes, respectively. Equation 2.78 gives an approximation of the perimeter of an ellipse. Equation 2.79 expresses the perimeter exactly, but one factor is the sum of an infinite series in which λ is defined as in Eq. 2.80.

Figure 2.17 *Ellipse*

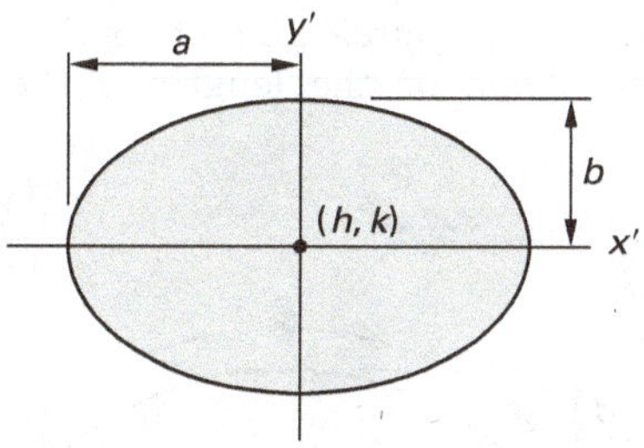

Example

An ellipse has a semimajor axis with length $a = 12$ and a semiminor axis with length $b = 3$. What is the approximate length of the perimeter of the ellipse?

(A) 24

(B) 47

(C) 55

(D) 180

Solution

Use Eq. 2.78.

$$P_{\text{approx}} = 2\pi\sqrt{(a^2 + b^2)/2} = 2\pi\sqrt{\frac{12^2 + 3^2}{2}}$$
$$= 54.96 \quad (55)$$

The answer is (C).

Equation 2.81 and Eq. 2.82: Circular Segments

$$A = [r^2(\phi - \sin\phi)]/2 \qquad 2.81$$

$$\phi = s/r = 2\{\arccos[(r - d)/r]\} \qquad 2.82$$

Description

A *circular segment* is a region bounded by a circular arc and a chord, as shown by the shaded portion in Fig. 2.18. The arc and chord are both limited by a central angle, ϕ. Use Eq. 2.81 to find the area of a circular segment when its central angle, ϕ, and the radius of the circle, r, are known; in Eq. 2.81, the central angle must be in radians. Use Eq. 2.82 to find the central angle when the radius of the circle and either the height of the circular segment, d, or the length of the arc, s, are known.

Figure 2.18 Circular Segment

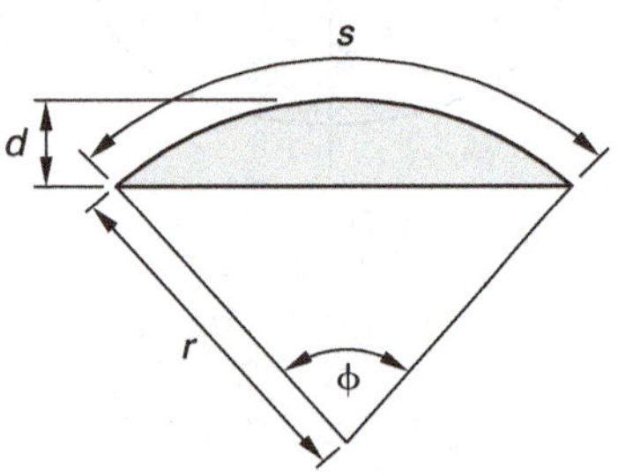

Example

Two 20 m diameter circles are placed so that the circumference of each just touches the center of the other.

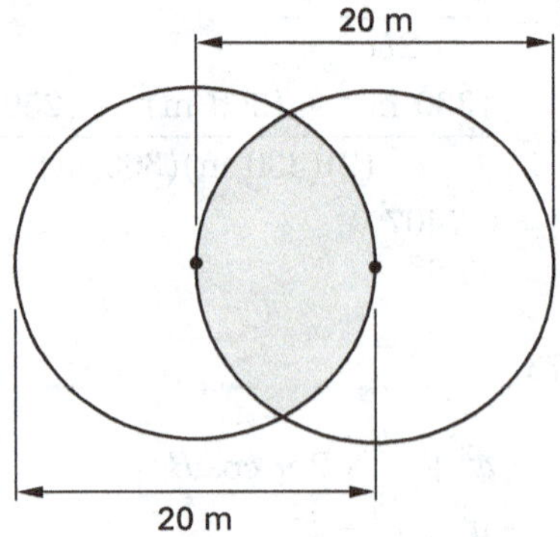

What is most nearly the area of the shared region?

(A) 62 m^2

(B) 110 m^2

(C) 120 m^2

(D) 170 m^2

Solution

The shared region can be thought of as two equal circular segments, each as shown in the illustration. The radius of each circle is $r = 10$ m. The height of each circular segment is half the radius, so $d = 5$ m.

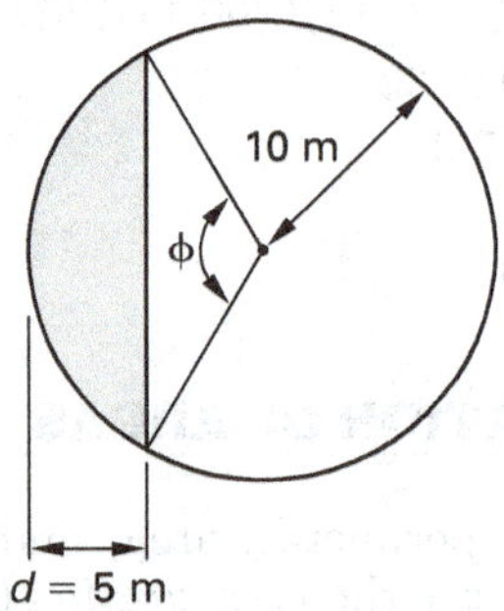

Use Eq. 2.82 to find the angle ϕ.

$$\phi = 2\{\arccos[(r - d)/r]\} = 2\arccos\left(\frac{10\text{ m} - 5\text{ m}}{10\text{ m}}\right)$$
$$= 120°$$

Convert ϕ to radians.

$$\phi = (120°)\left(\frac{2\pi}{360°}\right) = 2.094\text{ rad}$$

From Eq. 2.81, the area of a circular segment is

$$A = [r^2(\phi - \sin\phi)]/2$$
$$= \frac{(10\text{ m})^2(2.094\text{ rad} - \sin(2.094\text{ rad}))}{2}$$
$$= 61.4\text{ m}^2$$

The area of the shared region is twice this amount.

$$A_{\text{shared}} = 2A = (2)(61.4\text{ m}^2)$$
$$= 122.8\text{ m}^2 \quad (120\text{ m}^2)$$

The answer is (C).

Equation 2.83 and Eq. 2.84: Circular Sectors

$$A = \phi r^2/2 = sr/2 \qquad 2.83$$

$$\phi = s/r \qquad 2.84$$

Description

A *circular sector* is a portion of a circle bounded by two radii and an arc, as shown in Fig. 2.19. Between the two radii is the central angle, ϕ. Use Eq. 2.83 to find the area of a circular sector when its radius, r, and either its central angle or the length of its arc, s, are known; the central angle must be in radians. Use Eq. 2.84 to find the central angle in radians when the arc length and radius are known.

Figure 2.19 Circular Sector

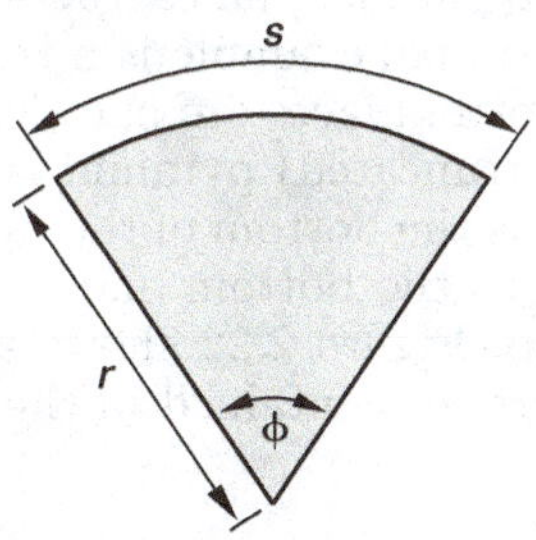

Example

A circular sector has an area of 3 m^2 and a central angle of 50°. What is most nearly the radius?

(A) 1.5 m

(B) 2.6 m

(C) 3.0 m

(D) 3.3 m

Solution

The central angle must be converted to radians.

$$\phi = (50°)\left(\frac{2\pi}{360°}\right) = 0.873\text{ rad}$$

Use Eq. 2.83.

$$A = \phi r^2/2$$
$$r = \sqrt{\frac{2A}{\phi}} = \sqrt{\frac{(2)(3\text{ m}^2)}{0.873\text{ rad}}}$$
$$= 2.62\text{ m} \quad (2.6\text{ m})$$

The answer is (B).

Equation 2.85 Through Eq. 2.89: Parallelograms

$$P = 2(a+b) \qquad 2.85$$

$$d_1 = \sqrt{a^2 + b^2 - 2ab(\cos\phi)} \qquad 2.86$$

$$d_2 = \sqrt{a^2 + b^2 + 2ab(\cos\phi)} \qquad 2.87$$

$$d_1^2 + d_2^2 = 2(a^2 + b^2) \qquad 2.88$$

$$A = ah = ab(\sin\phi) \qquad 2.89$$

Description

Equation 2.85 is the formula for the perimeter of a parallelogram. (See Fig. 2.20.) Equation 2.86 and Eq. 2.87 give the diagonals of the parallelogram when its sides and acute included angle are known. Equation 2.88 relates the sides and the diagonals, and Eq. 2.89 gives the parallelogram's area. A parallelogram with all sides of equal length is called a *rhombus*.

Figure 2.20 Parallelogram

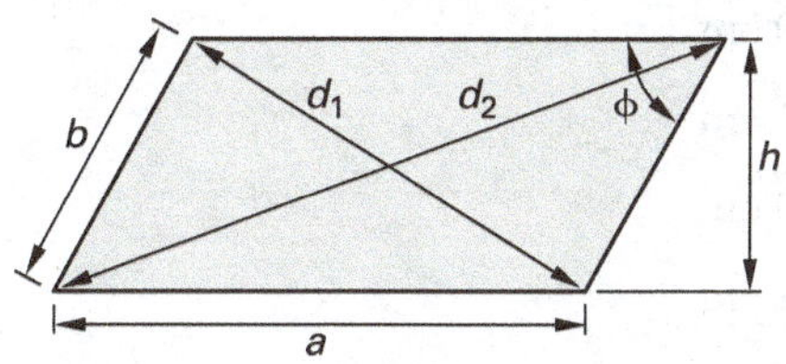

Equation 2.90 Through Eq. 2.94: Regular Polygons

$$\phi = 2\pi/n \qquad 2.90$$

$$\theta = \left[\frac{\pi(n-2)}{n}\right] = \pi\left(1-\frac{2}{n}\right) \quad 2.91$$

$$P = ns \quad 2.92$$

$$s = 2r[\tan(\phi/2)] \quad 2.93$$

$$A = (nsr)/2 \quad 2.94$$

Description

A *regular polygon* is a polygon with equal sides and equal angles. (See Fig. 2.21.) n is the number of sides. Equation 2.90 gives the central angle, ϕ, formed by two line segments drawn from the center to adjacent vertices. Equation 2.91 gives the measure of each interior angle, θ. Equation 2.92 gives the perimeter of the polygon (s is the length of one side).

Equation 2.93 can be used to find the length of one side when the polygon's central angle and apothem, r, are known. The *apothem* is a line segment drawn from the center of the polygon to the midpoint of one side; this is also the radius of a circle inscribed within the polygon. Equation 2.94 is the formula for the area of the polygon.

Figure 2.21 *Regular Polygon (n equal sides)*

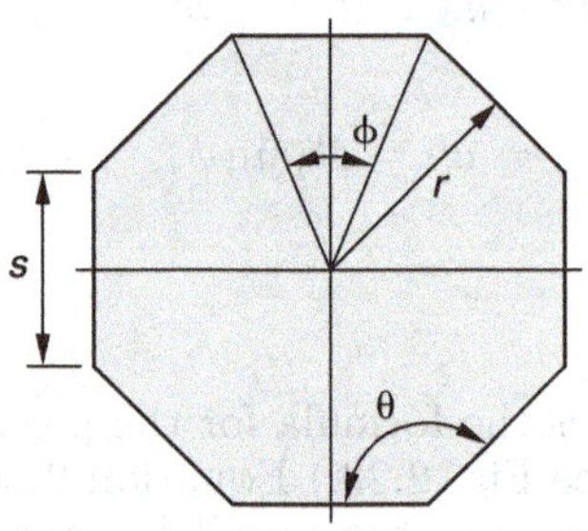

Example

A regular polygon has six sides, each with a length of 25 cm. What is most nearly the length of the apothem, r?

(A) 10 cm

(B) 15 cm

(C) 20 cm

(D) 22 cm

Solution

Use Eq. 2.90 to find the central angle.

$$\phi = 2\pi/n = \frac{2\pi \text{ rad}}{6} = 1.047 \text{ rad}$$

Convert radians to degrees.

$$\phi = (1.047 \text{ rad})\left(\frac{360°}{2\pi}\right) = 60.0°$$

Use Eq. 2.93 to find the length of the apothem.

$$\begin{aligned} s &= 2r[\tan(\phi/2)] \\ r &= \frac{s}{2\tan\frac{\phi}{2}} = \frac{25 \text{ cm}}{2\tan\frac{60°}{2}} \\ &= 21.65 \text{ cm} \quad (22 \text{ cm}) \end{aligned}$$

The answer is (D).

13. MENSURATION OF VOLUMES

Equation 2.95: Prismoids

$$V = (h/6)(A_1 + A_2 + 4A) \quad 2.95$$

Variation

$$h = \frac{6V}{A_1 + A_2 + 4A}$$

Description

A *polyhedron* is a three-dimensional solid whose faces are all flat and whose edges are all straight.

If all the vertices (corners) of the polyhedron are contained within two parallel planes, the solid is a *prismoid* (*prismatoid*). A simple example is a truncated pyramid whose top and bottom faces are parallel. Less obviously, a complete (not truncated) pyramid is also a prismoid; one plane contains the bottom of the pyramid, while the other is parallel to the bottom and contains the single vertex at the top. Figure 2.22 shows an irregular prismoid with all vertices contained in the top and bottom planes.

Figure 2.22 *Prismoid*

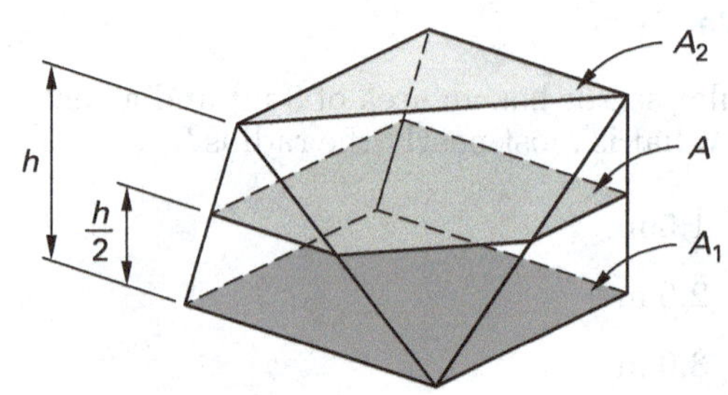

Use Eq. 2.95 to find the volume of a prismoid. h is the distance between the two parallel planes measured along a direction perpendicular to both. A_1 and A_2 are the areas of the faces contained within these two planes; if one of these planes contains only a single point (such as at the top vertex of a pyramid), the area is zero. A is the cross-sectional area of the solid halfway between the two parallel planes. Each vertex of this cross section is halfway between a vertex on the top face and another one on the bottom face, but A is not necessarily the average of A_1 and A_2.

Example

A prismoid has a volume of 100 cm^3. The area of the bottom face is 20 cm^2, the area of the top face is 5 cm^2, and the cross-sectional area halfway between the top and bottom faces is 10 cm^2. What is the approximate height of the prismoid?

(A) 8.0 cm

(B) 9.2 cm

(C) 11 cm

(D) 13 cm

Solution

Use Eq. 2.95, the formula for the volume of a prismoid.

$$\begin{aligned} V &= (h/6)(A_1 + A_2 + 4A) \\ h &= \frac{6V}{A_1 + A_2 + 4A} \\ &= \frac{(6)(100\ \text{cm}^3)}{20\ \text{cm}^2 + 5\ \text{cm}^2 + (4)(10\ \text{cm}^2)} \\ &= 9.23\ \text{cm} \quad (9.2\ \text{cm}) \end{aligned}$$

The answer is (B).

Equation 2.96 and Eq. 2.97: Spheres

$$V = 4\pi r^3/3 = \pi d^3/6 \qquad 2.96$$

$$A = 4\pi r^2 = \pi d^2 \qquad 2.97$$

Description

Equation 2.96 and Eq. 2.97 are the formulas for the volume and surface area, respectively, of a sphere whose radius, r, or diameter, d, is known. (See Fig. 2.23.)

Figure 2.23 *Sphere*

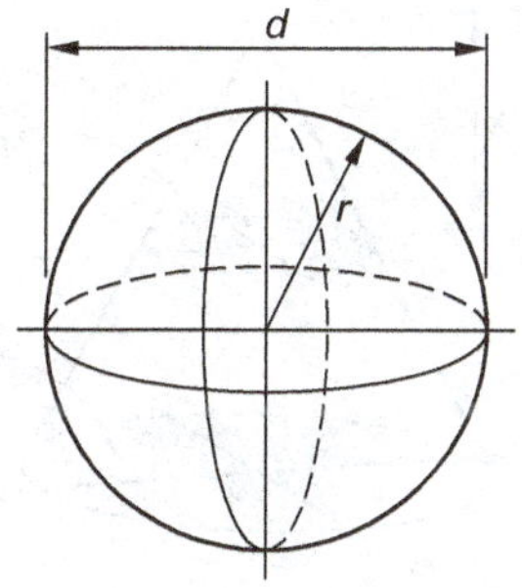

Example

A sphere has a radius of 10 cm. What is approximately the sphere's volume?

(A) 3600 cm^3

(B) 4000 cm^3

(C) 4200 cm^3

(D) 4800 cm^3

Solution

Use Eq. 2.96.

$$\begin{aligned} V &= 4\pi r^3/3 = \frac{4\pi(10\ \text{cm})^3}{3} \\ &= 4188.79\ \text{cm}^3 \quad (4200\ \text{cm}^3) \end{aligned}$$

The answer is (C).

Equation 2.98 Through Eq. 2.100: Right Circular Cones

$$V = (\pi r^2 h)/3 \qquad 2.98$$

$$A = \text{side area} + \text{base area} = \pi r\left(r + \sqrt{r^2 + h^2}\right) \qquad 2.99$$

$$A_x : A_b = x^2 : h^2 \qquad 2.100$$

Description

A *right circular cone* is a cone whose base is a circle and whose axis is perpendicular to the base. (See Fig. 2.24.) Equation 2.98 gives the volume of a right circular cone whose height, h, and base radius, r, are known. Equation 2.99 gives the cone's area. Equation 2.100 says that the cross-sectional area of the cone varies with the square of the distance from the apex.

Figure 2.24 Right Circular Cone

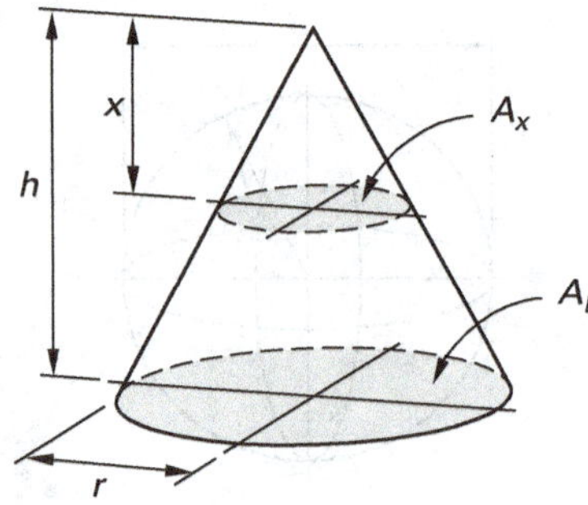

Example

A cone has a height of 100 cm. The cross section of the cone at a distance of 5 cm from the apex is a circle with an area of 20 cm^2. What is most nearly the area of the cone's base?

(A) 5000 cm^2

(B) 6000 cm^2

(C) 8000 cm^2

(D) 9000 cm^2

Solution

Use Eq. 2.100.

$$A_x{:}A_b = x^2{:}h^2$$

$$A_b = \frac{h^2 A_x}{x^2} = \frac{(100\ \text{cm})^2(20\ \text{cm}^2)}{(5\ \text{cm})^2} = 8000\ \text{cm}^2$$

The answer is (C).

Equation 2.101 and Eq. 2.102: Right Circular Cylinders

$$V = \pi r^2 h = \frac{\pi d^2 h}{4} \qquad 2.101$$

$$A = \text{side area} + \text{end areas} = 2\pi r(h + r) \qquad 2.102$$

Description

A *right circular cylinder* is a cylinder whose base is a circle and whose axis is perpendicular to the base. (See Fig. 2.25.) Equation 2.101 gives the volume of a right circular cylinder, and Eq. 2.102 gives the total surface area.

Figure 2.25 Right Circular Cylinder

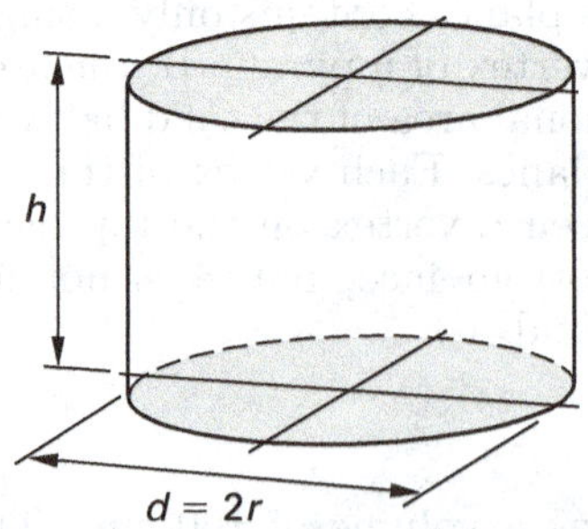

Equation 2.103: Paraboloids of Revolution

$$V = \frac{\pi d^2 h}{8} \qquad 2.103$$

Description

A *paraboloid of revolution* is the surface that is obtained by rotating a parabola around its axis. Equation 2.103 can be used to find the volume of a paraboloid of revolution if its height and diameter are known. (See Fig. 2.26.)

Figure 2.26 Paraboloid of Revolution

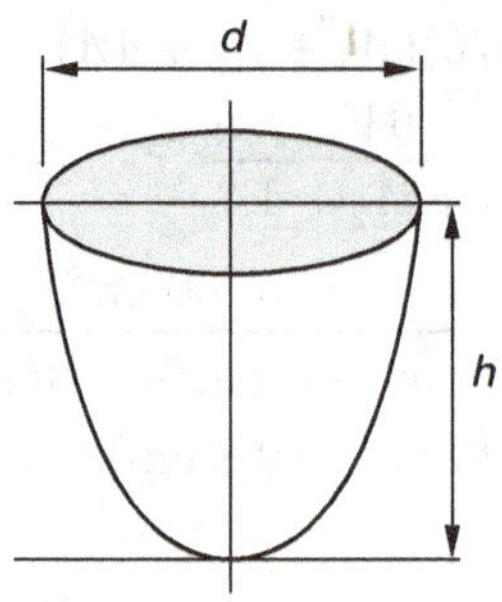

3 Algebra and Linear Algebra

1. LOGARITHMS

Logarithms can be considered to be exponents. In the equation $b^c = x$, for example, the exponent c is the logarithm of x to the base b. The two equations $\log_b x = c$ and $b^c = x$ are equivalent.

Equation 3.1 Through Eq. 3.3: Common and Natural Logarithms

$$\log_b(x) = c \quad [b^c = x] \qquad 3.1$$

$$\ln x \quad [\text{base} = e] \qquad 3.2$$

$$\log x \quad [\text{base} = 10] \qquad 3.3$$

Description

Although any number may be used as a base for logarithms, two bases are most commonly used in engineering. The base for a *common logarithm* is 10. The notation used most often for common logarithms is *log*, although *log*$_{10}$ is sometimes seen.

The base for a *natural logarithm* is 2.71828..., an irrational number that is given the symbol *e*. The most common notation for a natural logarithm is *ln*, but *log*$_e$ is sometimes seen.

Example

What is the value of $\log_{10} 1000$?

(A) 2

(B) 3

(C) 8

(D) 10

Solution

$\log_{10} 1000$ is the power of 10 that produces 1000. Use Eq. 3.1.

$$\begin{aligned} \log_b(x) &= c \quad [b^c = x] \\ \log_{10} 1000 &= c \\ 10^c &= 1000 \\ c &= 3 \end{aligned}$$

The answer is (B).

Equation 3.4 Through Eq. 3.10: Logarithmic Identities

$$\log_b b^n = n \qquad 3.4$$

$$\log x^c = c \log x \qquad 3.5$$

$$x^c = \text{antilog}(c \log x) \qquad 3.6$$

$$\log xy = \log x + \log y \qquad 3.7$$

$$\log_b b = 1 \qquad 3.8$$

$$\log 1 = 0 \qquad 3.9$$

$$\log x/y = \log x - \log y \qquad 3.10$$

Description

Logarithmic identities are useful in simplifying expressions containing exponentials and other logarithms.

Example 28

Which of the following is equal to $(0.001)^{2/3}$?

(A) $\text{antilog}\left(\frac{3}{2}\log 0.001\right)$

(B) $\frac{2}{3}\text{antilog}(\log 0.001)$

(C) $\text{antilog}\left(\log \dfrac{0.001}{\frac{2}{3}}\right)$

(D) $\text{antilog}\left(\frac{2}{3}\log 0.001\right)$

Solution

Use Eq. 3.5 and Eq. 3.6.

$$\begin{aligned}\log x^c &= c\log x\\ \log(0.001)^{2/3} &= \tfrac{2}{3}\log 0.001\\ (0.001)^{2/3} &= \text{antilog}\left(\tfrac{2}{3}\log 0.001\right)\end{aligned}$$

The answer is (D).

Equation 3.11: Changing the Base

$$\log_b x = (\log_a x)/(\log_a b) \qquad 3.11$$

Variations

$$\log_{10} x = \ln x \log_{10} e$$

$$\begin{aligned}\ln x &= \frac{\log_{10} x}{\log_{10} e}\\ &\approx 2.302585\ \log_{10} x\end{aligned}$$

Description

Equation 3.11 is often useful for calculating a logarithm with any base quickly when the available resources produce only natural or common logarithms. Equation 3.11 can also be used to convert a logarithm to a different base, such as from a common logarithm to a natural logarithm.

Example

Given that $\log_{10} 5 = 0.6990$ and $\log_{10} 9 = 0.9542$, what is the value of $\log_5 9$?

(A) 0.2550

(B) 0.7330

(C) 1.127

(D) 1.365

Solution

Use Eq. 3.11.

$$\begin{aligned}\log_b x &= (\log_a x)/(\log_a b)\\ \log_5 9 &= \frac{\log_{10} 9}{\log_{10} 5} = \frac{0.9542}{0.6990}\\ &= 1.365\end{aligned}$$

The answer is (D).

2. COMPLEX NUMBERS

A *complex number* is the sum of a *real number* and an *imaginary number*. Real numbers include the *rational numbers* and the *irrational numbers*, while imaginary numbers represent the square roots of negative numbers. Every imaginary number can be expressed in the form ib, where i represents the square root of -1 and b is a real number. Another term for i is the *imaginary unit vector*.

$$i = \sqrt{-1}$$

j is commonly used to represent the imaginary unit vector in the fields of electrical engineering and control systems engineering to avoid confusion with the variable for current, i.[1]

$$j = \sqrt{-1}$$

When a complex number is expressed in the form $a + ib$, the complex number is said to be in *rectangular* or *trigonometric form*. In the expression $a + ib$, a is the real component (or real part), and b is the imaginary component (or imaginary part). (See Fig. 3.1.)

[1]The NCEES *FE Reference Handbook* (*NCEES Handbook*) uses only j to represent the imaginary unit vector. This book uses both j and i.

Figure 3.1 Graphical Representation of a Complex Number

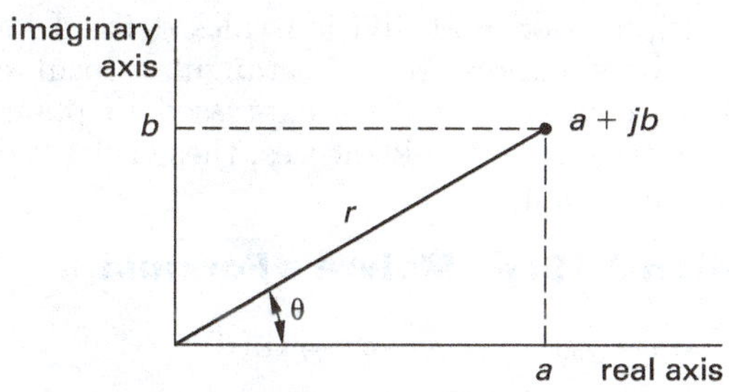

Most algebraic operations (addition, multiplication, exponentiation, etc.) work with complex numbers. When adding two complex numbers, real parts are added to real parts, and imaginary parts are added to imaginary parts.

$$(a+jb)+(c+jd)=(a+c)+j(b+d)$$

$$(a+jb)-(c+jd)=(a-c)+j(b-d)$$

Multiplication of two complex numbers in rectangular form uses the algebraic distributive law and the equivalency $j^2=-1$.

$$(a+jb)(c+jd)=(ac-bd)+j(ad+bc)$$

Division of complex numbers in rectangular form requires use of the *complex conjugate*. The complex conjugate of a complex number $a+jb$ is $a-jb$. When both the numerator and the denominator are multiplied by the complex conjugate of the denominator, the denominator becomes the real number a^2+b^2. This technique is known as *rationalizing the denominator*.

$$\begin{aligned}\frac{a+jb}{c+jd}&=\frac{(a+jb)(c-jd)}{(c+jd)(c-jd)}\\&=\frac{(ac+bd)+j(bc-ad)}{c^2+d^2}\end{aligned}$$

Example 29

Which of the following is most nearly equal to $(7+5.2j)/(3+4j)$?

(A) $-0.3+1.8j$

(B) $1.7-0.5j$

(C) $2.3-1.2j$

(D) $2.3+1.3j$

Solution

When the numerator and denominator are multiplied by the complex conjugate of the denominator, the denominator becomes a real number.

$$\begin{aligned}\frac{a+jb}{c+jd}&=\frac{(a+jb)(c-jd)}{(c+jd)(c-jd)}\\&=\frac{(ac+bd)+j(bc-ad)}{c^2+d^2}\end{aligned}$$

$$\begin{aligned}\frac{7+5.2j}{3+4j}&=\frac{\big((7)(3)+(5.2)(4)\big)+j\big((5.2)(3)-(7)(4)\big)}{(3)^2+(4)^2}\\&=1.672-0.496j\quad(1.7-0.5j)\end{aligned}$$

The answer is (B).

3. POLAR COORDINATES

Equation 3.12: Polar Form of a Complex Number

$$x+jy=r(\cos\theta+j\sin\theta)=re^{j\theta} \qquad 3.12$$

Variations

$$z\equiv r(\cos\theta+i\sin\theta)$$

$$z\equiv r\,\mathrm{cis}\,\theta$$

$$z=r\angle\theta$$

Description

A complex number can be expressed in the *polar form* $r(\cos\theta+j\sin\theta)$, where θ is the angle from the x-axis and r is the distance from the origin. r and θ are the *polar coordinates* of the complex number. Another notation for the polar form of a complex number is $re^{j\theta}$.

Equation 3.13 and Eq. 3.14: Converting from Polar Form to Rectangular Form

$$x=r\cos\theta \qquad 3.13$$

$$y=r\sin\theta \qquad 3.14$$

Description

The rectangular form of a complex number, $x+jy$, can be determined from the complex number's polar coordinates r and θ using Eq. 3.13 and Eq. 3.14.

Equation 3.15 and Eq. 3.16: Converting from Rectangular Form to Polar Form

$$r = |x + jy| = \sqrt{x^2 + y^2} \qquad 3.15$$

$$\theta = \arctan(y/x) \qquad 3.16$$

Description

The polar form of a complex number, $r(\cos\theta + j\sin\theta)$, can be determined from the complex number's rectangular coordinates x and y using Eq. 3.15 and Eq. 3.16.

Example

The rectangular coordinates of a complex number are $(4, 6)$. What are the complex number's approximate polar coordinates?

(A) $(4.0, 33°)$

(B) $(4.0, 56°)$

(C) $(7.2, 33°)$

(D) $(7.2, 56°)$

Solution

The radius and angle of the polar form can be determined from the x- and y-coordinates using Eq. 3.15 and Eq. 3.16.

$$r = \sqrt{x^2 + y^2} = \sqrt{(4)^2 + (6)^2} = 7.211 \quad (7.2)$$

$$\theta = \arctan(y/x) = \arctan\frac{6}{4}$$

$$= 56.3° \quad (56°)$$

The answer is (D).

Equation 3.17 and Eq. 3.18: Multiplication and Division with Polar Forms

$$[r_1(\cos\theta_1 + j\sin\theta_1)][r_2(\cos\theta_2 + j\sin\theta_2)] = r_1 r_2[\cos(\theta_1 + \theta_2) + j\sin(\theta_1 + \theta_2)] \qquad 3.17$$

$$\frac{r_1(\cos\theta_1 + j\sin\theta_1)}{r_2(\cos\theta_2 + j\sin\theta_2)} = \frac{r_1}{r_2}\left[\begin{matrix}\cos(\theta_1 - \theta_2) \\ +j\sin(\theta_1 - \theta_2)\end{matrix}\right] \qquad 3.18$$

Variations

$$z_1 z_2 = (r_1 r_2)\angle(\theta_1 + \theta_2)$$

$$\frac{z_1}{z_2} = \frac{r_1}{r_2}\angle(\theta_1 - \theta_2)$$

Description

The multiplication and division rules defined for complex numbers expressed in rectangular form can be applied to complex numbers expressed in polar form. Using the trigonometric identities, these rules reduce to Eq. 3.17 and Eq. 3.18.

Equation 3.19: de Moivre's Formula

$$(x + jy)^n = [r(\cos\theta + j\sin\theta)]^n = r^n(\cos n\theta + j\sin n\theta) \qquad 3.19$$

Description

Equation 3.19 is *de Moivre's formula*. This equation is valid for any real number x and integer n.

Equation 3.20 Through Eq. 3.23: Euler's Equations

$$e^{j\theta} = \cos\theta + j\sin\theta \qquad 3.20$$

$$e^{-j\theta} = \cos\theta - j\sin\theta \qquad 3.21$$

$$\cos\theta = \frac{e^{j\theta} + e^{-j\theta}}{2} \qquad 3.22$$

$$\sin\theta = \frac{e^{j\theta} - e^{-j\theta}}{2j} \qquad 3.23$$

Description

Complex numbers can also be expressed in exponential form. The relationship of the exponential form to the trigonometric form is given by *Euler's equations*, also known as *Euler's identities*.

Example

If $j = \sqrt{-1}$, which of the following is equal to j^j?

(A) j^2

(B) e^{2j}

(C) -1

(D) $e^{-\pi/2}$

Solution

j is the imaginary unit vector, so $r = 1$ and $\theta = 90°(\pi/2)$ in Fig. 3.1. From Eq. 3.19,[2]

$$(j)^n = (\cos\theta + j\sin\theta)^n$$

[2]For non-integer values of n, de Moivre's formula produces one possible solution.

From Eq. 3.20,

$$e^{j\theta} = \cos\theta + j\sin\theta$$

Since $\theta = \pi/2$,

$$j^j = \left(e^{j\pi/2}\right)^j = e^{j^2\pi/2} = e^{-\pi/2}$$

The answer is (D).

4. ROOTS

Equation 3.24: *k*th Roots of a Complex Number

$$w = \sqrt[k]{r}\left[\cos\left(\frac{\theta}{k} + n\frac{360^\circ}{k}\right) + j\sin\left(\frac{\theta}{k} + n\frac{360^\circ}{k}\right)\right] \qquad 3.24$$

Description

Use Eq. 3.24 to find the *k*th *root* of the complex number $z = r(\cos\theta + j\sin\theta)$. n can be any integer number.

Example

What is the cube root of the complex number $8e^{j60^\circ}$?

(A) $2(\cos 60^\circ + j\sin 60^\circ)$

(B) $2(j\cos 20^\circ + \sin 20^\circ)$

(C) $2.7(\cos 20^\circ + j\sin 20^\circ)$

(D) $2(\cos(20^\circ + 120^\circ n) + j\sin(20^\circ + 120^\circ n))$

Solution

From Eq. 3.24, the *k*th root of a complex number is

$$\begin{aligned} w &= \sqrt[k]{r}\left[\cos\left(\frac{\theta}{k} + n\frac{360^\circ}{k}\right) + j\sin\left(\frac{\theta}{k} + n\frac{360^\circ}{k}\right)\right] \\ &= \sqrt[3]{8}\left(\begin{array}{l} \cos\left(\frac{60^\circ}{3} + n\left(\frac{360^\circ}{3}\right)\right) \\ \quad + j\sin\left(\frac{60^\circ}{3} + n\left(\frac{360^\circ}{3}\right)\right) \end{array}\right) \\ &= 2(\cos(20^\circ + 120^\circ n) + j\sin(20^\circ + 120^\circ n)) \\ &\qquad [n = 0, 1, 2, \ldots] \end{aligned}$$

The answer is (D).

5. MATRICES

A *matrix* is an ordered set of *entries* (*elements*) arranged rectangularly and set off by brackets. The entries can be variables or numbers. A matrix by itself has no particular value; it is merely a convenient method of representing a set of numbers.

The size of a matrix is given by the number of rows and columns, and the nomenclature $m \times n$ is used for a matrix with m rows and n columns. For a square matrix, the numbers of rows and columns are the same and are equal to the *order of the matrix*.

Matrices are designated by bold uppercase letters. Matrix entries are designated by lowercase letters with subscripts, for example, a_{ij}. The term a_{23} would be the entry in the second row and third column of matrix **A**. The matrix **C** can also be designated as (c_{ij}), meaning "the matrix made up of c_{ij} entries."

Equation 3.25: Addition of Matrices

$$\begin{bmatrix} A & B & C \\ D & E & F \end{bmatrix} + \begin{bmatrix} G & H & I \\ J & K & L \end{bmatrix} = \begin{bmatrix} A+G & B+H & C+I \\ D+J & E+K & F+L \end{bmatrix} \qquad 3.25$$

Variation

$$\begin{aligned} \mathbf{C} &= \mathbf{A} + \mathbf{B} \\ &\equiv (c_{ij}) \equiv (a_{ij} + b_{ij}) \end{aligned}$$

Description

Addition and subtraction of two matrices are possible only if both matrices have the same number of rows and columns. They are accomplished by adding or subtracting the corresponding entries of the two matrices.

Equation 3.26: Multiplication of Matrices

$$\begin{aligned} \mathrm{C} &= \begin{bmatrix} A & B \\ C & D \\ E & F \end{bmatrix} \cdot \begin{bmatrix} H & I \\ J & K \end{bmatrix} \\ &= \begin{bmatrix} (A{\cdot}H + B{\cdot}J) & (A{\cdot}I + B{\cdot}K) \\ (C{\cdot}H + D{\cdot}J) & (C{\cdot}I + D{\cdot}K) \\ (E{\cdot}H + F{\cdot}J) & (E{\cdot}I + F{\cdot}K) \end{bmatrix} \end{aligned} \qquad 3.26$$

Variations

$$\mathbf{C} = \mathbf{AB}$$

$$\mathbf{C} = \mathbf{A} \times \mathbf{B}$$

$$\mathbf{C} \equiv (c_{ij}) = \left(\sum_{l=1}^{n} a_{il} b_{lj} \right)$$

Description

A matrix can be multiplied by another matrix, but only if the left-hand matrix has the same number of columns as the right-hand matrix has rows. *Matrix multiplication* occurs by multiplying the elements in each left-hand matrix row by the entries in each corresponding right-hand matrix column, adding the products, and placing the sum at the intersection point of the participating row and column.

The commutative law does not apply to matrix multiplication. That is, $\mathbf{A} \times \mathbf{B}$ is not equivalent to $\mathbf{B} \times \mathbf{A}$.

Example 33

What is the matrix product **AB** of matrices **A** and **B**?

$$\mathbf{A} = \begin{bmatrix} 2 & 1 \\ 1 & 0 \end{bmatrix} \quad \mathbf{B} = \begin{bmatrix} 4 & 3 \\ 2 & 1 \end{bmatrix}$$

(A) $\begin{bmatrix} 10 & 4 \\ 7 & 3 \end{bmatrix}$

(B) $\begin{bmatrix} 11 & 4 \\ 5 & 2 \end{bmatrix}$

(C) $\begin{bmatrix} 8 & 3 \\ 2 & 0 \end{bmatrix}$

(D) $\begin{bmatrix} 10 & 7 \\ 4 & 3 \end{bmatrix}$

Solution

Use Eq. 3.26. Multiply the elements of each row in matrix **A** by the elements of the corresponding column in matrix **B**.

$$\mathbf{C} = \begin{bmatrix} \mathrm{A} & \mathrm{B} \\ \mathrm{C} & \mathrm{D} \\ \mathrm{E} & \mathrm{F} \end{bmatrix} \cdot \begin{bmatrix} \mathrm{H} & \mathrm{I} \\ \mathrm{J} & \mathrm{K} \end{bmatrix} = \begin{bmatrix} (\mathrm{A{\cdot}H} + \mathrm{B{\cdot}J}) & (\mathrm{A{\cdot}I} + \mathrm{B{\cdot}K}) \\ (\mathrm{C{\cdot}H} + \mathrm{D{\cdot}J}) & (\mathrm{C{\cdot}I} + \mathrm{D{\cdot}K}) \\ (\mathrm{E{\cdot}H} + \mathrm{F{\cdot}J}) & (\mathrm{E{\cdot}I} + \mathrm{F{\cdot}K}) \end{bmatrix}$$

$$= \begin{bmatrix} 2 \times 4 + 1 \times 2 & 2 \times 3 + 1 \times 1 \\ 1 \times 4 + 0 \times 2 & 1 \times 3 + 0 \times 1 \end{bmatrix}$$

$$= \begin{bmatrix} 10 & 7 \\ 4 & 3 \end{bmatrix}$$

The answer is (D).

Equation 3.27: Transposes of Matrices

$$\mathbf{A} = \begin{bmatrix} \mathrm{A} & \mathrm{B} & \mathrm{C} \\ \mathrm{D} & \mathrm{E} & \mathrm{F} \end{bmatrix} \quad \mathbf{A}^T = \begin{bmatrix} \mathrm{A} & \mathrm{D} \\ \mathrm{B} & \mathrm{E} \\ \mathrm{C} & \mathrm{F} \end{bmatrix} \qquad 3.27$$

Variation

$$\boldsymbol{B} = \mathbf{A}^T$$

Description

The *transpose*, $\mathbf{A}^T$, of an $m \times n$ matrix $\mathbf{A}$ is an $n \times m$ matrix constructed by taking the ith row and making it the ith column.

Example 34

What is the transpose of matrix **A**?

$$\boldsymbol{A} = \begin{bmatrix} 5 & 8 & 5 & 8 \\ 8 & 7 & 6 & 2 \end{bmatrix}$$

(A) $\begin{bmatrix} 8 & 7 & 6 & 2 \\ 5 & 8 & 5 & 8 \end{bmatrix}$

(B) $\begin{bmatrix} 2 & 6 & 7 & 8 \\ 8 & 5 & 8 & 5 \end{bmatrix}$

(C) $\begin{bmatrix} 8 & 5 \\ 7 & 8 \\ 6 & 5 \\ 2 & 8 \end{bmatrix}$

(D) $\begin{bmatrix} 5 & 8 \\ 8 & 7 \\ 5 & 6 \\ 8 & 2 \end{bmatrix}$

Solution

The transpose of a matrix is constructed by taking the ith row and making it the ith column.

The answer is (D).

Equation 3.28: Determinants of 2×2 Matrices

$$\begin{vmatrix} a_1 & a_2 \\ b_1 & b_2 \end{vmatrix} = a_1 b_2 - a_2 b_1 \qquad 3.28$$

Variation

$$|\mathbf{A}| = \begin{vmatrix} a_1 & a_2 \\ b_1 & b_2 \end{vmatrix} = a_1 b_2 - a_2 b_1$$

Description

A *determinant* is a scalar calculated from a square matrix. The determinant of matrix **A** can be represented as D{**A**}, Det(**A**), or |**A**|. The following rules can be used to simplify the calculation of determinants.

- If **A** has a row or column of zeros, the determinant is zero.
- If **A** has two identical rows or columns, the determinant is zero.
- If **B** is obtained from **A** by adding a multiple of a row (column) to another row (column) in **A**, then $|\mathbf{B}| = |\mathbf{A}|$.
- If **A** is *triangular* (a square matrix with zeros in all positions above or below the diagonal), the determinant is equal to the product of the diagonal entries.
- If **B** is obtained from **A** by multiplying one row or column in **A** by a scalar k, then $|\mathbf{B}| = k|\mathbf{A}|$.
- If **B** is obtained from the $n \times n$ matrix **A** by multiplying by the scalar matrix k, then $|\mathbf{B}| = |\mathbf{k} \times \mathbf{A}| = k^n|\mathbf{A}|$.
- If **B** is obtained from **A** by switching two rows or columns in **A**, then $|\mathbf{B}| = -|\mathbf{A}|$.

Calculation of determinants is laborious for all but the smallest or simplest of matrices. For a 2×2 matrix, the formula used to calculate the determinant is easy to remember.

Example

What is the determinant of matrix **A**?

$$\mathbf{A} = \begin{bmatrix} 3 & 6 \\ 2 & 4 \end{bmatrix}$$

(A) 0

(B) 15

(C) 14

(D) 26

Solution

From Eq. 3.28, for a square 2×2 matrix,

$$\begin{vmatrix} a_1 & a_2 \\ b_1 & b_2 \end{vmatrix} = a_1b_2 - a_2b_1$$

$$\begin{vmatrix} 3 & 6 \\ 2 & 4 \end{vmatrix} = 3 \times 4 - 6 \times 2 = 0$$

The answer is (A).

Equation 3.29: Determinants of 3×3 Matrices

$$\begin{vmatrix} a_1 & a_2 & a_3 \\ b_1 & b_2 & b_3 \\ c_1 & c_2 & c_3 \end{vmatrix} = \begin{array}{l} a_1b_2c_3 + a_2b_3c_1 + a_3b_1c_2 - a_3b_2c_1 \\ -a_2b_1c_3 - a_1b_3c_2 \end{array} \qquad 3.29$$

Variations

$$\mathbf{A} = \begin{bmatrix} a_1 & a_2 & a_3 \\ b_1 & b_2 & b_3 \\ c_1 & c_2 & c_3 \end{bmatrix}$$

$$|\mathbf{A}| = a_1b_2c_3 + a_2b_3c_1 + a_3b_1c_2 - a_3b_2c_1 - a_2b_1c_3 - a_1b_3c_2$$

Description

In addition to the formula-based method expressed as Eq. 3.29, two methods are commonly used for calculating the determinants of 3×3 matrices by hand. The first uses an augmented matrix constructed from the original matrix and the first two columns. The determinant is calculated as the sum of the products in the left-to-right downward diagonals less the sum of the products in the left-to-right upward diagonals.

augmented **A** =

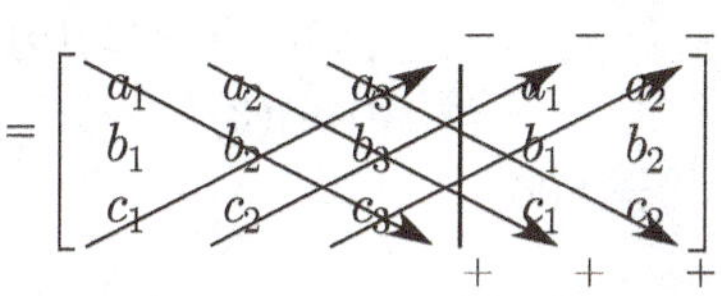

The second method of calculating the determinant is somewhat slower than the first for a 3×3 matrix but illustrates the method that must be used to calculate determinants of 4×4 and larger matrices. This method is known as *expansion by cofactors* (cofactors are explained in the following section). One row (column) is selected as the base row (column). The selection is arbitrary, but the number of calculations required to obtain the determinant can be minimized by choosing the row (column) with the most zeros. The determinant is equal to the sum of the products of the entries in the base row (column) and their corresponding cofactors.

$$\mathbf{A} = \begin{bmatrix} a_1 & a_2 & a_3 \\ b_1 & b_2 & b_3 \\ c_1 & c_2 & c_3 \end{bmatrix} \quad \begin{bmatrix} \text{first column chosen} \\ \text{as base column} \end{bmatrix}$$

$$\begin{aligned} |\mathbf{A}| &= a_1\begin{vmatrix} b_2 & b_3 \\ c_2 & c_3 \end{vmatrix} - b_1\begin{vmatrix} a_2 & a_3 \\ c_2 & c_3 \end{vmatrix} + c_1\begin{vmatrix} a_2 & a_3 \\ b_2 & b_3 \end{vmatrix} \\ &= a_1(b_2c_3 - b_3c_2) - b_1(a_2c_3 - a_3c_2) \\ &\quad + c_1(a_2b_3 - a_3b_2) \\ &= a_1b_2c_3 - a_1b_3c_2 - b_1a_2c_3 + b_1a_3c_2 \\ &\quad + c_1a_2b_3 - c_1a_3b_2 \end{aligned}$$

Example 36

For the following set of equations, what is the determinant of the coefficient matrix?

$$\begin{aligned} 10x + 3y + 10z &= 5 \\ 8x - 2y + 9z &= 5 \\ 8x + y - 10z &= 5 \end{aligned}$$

(A) 598

(B) 620

(C) 714

(D) 806

Solution

Calculate the determinant of the coefficient matrix.

$$\begin{aligned} |\mathbf{A}| &= \begin{vmatrix} a_1 & a_2 & a_3 \\ b_1 & b_2 & b_3 \\ c_1 & c_2 & c_3 \end{vmatrix} \\ &= a_1b_2c_3 + a_2b_3c_1 + a_3b_1c_2 - a_3b_2c_1 \\ &\quad - a_2b_1c_3 - a_1b_3c_2 \\ &= (10)(-2)(-10) + (3)(9)(8) + (10)(8)(1) \\ &\quad - (8)(-2)(10) - (1)(9)(10) \\ &\quad - (-10)(8)(3) \\ &= 806 \end{aligned}$$

The answer is (D).

Inverse of a Matrix

The *inverse*, $\mathbf{A}^{-1}$, of an invertible matrix, **A**, is a matrix such that the product $\mathbf{AA}^{-1}$ produces a matrix with ones along its diagonal and zeros elsewhere (i.e., above and below the diagonal). Only square matrices have inverses, but not all square matrices are invertible (i.e., have inverses). The product of a matrix and its inverse produces an identity matrix. For 3 × 3 matrices,

$$\mathbf{AA}^{-1} = \mathbf{I} = \begin{bmatrix} 1 & 0 & 0 \\ 0 & 1 & 0 \\ 0 & 0 & 1 \end{bmatrix}$$

The inverse of a 2 × 2 matrix is easily determined by the following formula.

$$\mathbf{A} = \begin{bmatrix} a_1 & a_2 \\ b_1 & b_2 \end{bmatrix}$$

$$\mathbf{A}^{-1} = \frac{\begin{bmatrix} b_2 & -a_2 \\ -b_1 & a_1 \end{bmatrix}}{|\mathbf{A}|}$$

Equation 3.30: Identity Matrix

$$[\mathbf{A}][\mathbf{A}]^{-1} = [\mathbf{A}]^{-1}[\mathbf{A}] = [\mathbf{I}] \qquad 3.30$$

Variation

$$A \times A^{-1} = A^{-1} \times A = \mathbf{I}$$

Description

The product of a matrix **A** and its *inverse*, $\mathbf{A}^{-1}$, is the *identity matrix*, **I**. A matrix has an inverse if and only if it is *nonsingular* (i.e., its determinant is nonzero).

Example 37

Using the property that $|\mathbf{AB}| = |\mathbf{A}||\mathbf{B}|$ for two square matrices, what is $|\mathbf{A}^{-1}|$ in terms of $|\mathbf{A}|$ for any invertible square matrix **A**?

(A) $\frac{1}{|\mathbf{A}|}$

(B) $\frac{1}{|\mathbf{A}^{-1}|}$

(C) $\frac{|\mathbf{A}|}{|\mathbf{A}^{-1}|}$

(D) $\frac{|\mathbf{A}^{-1}|}{|\mathbf{A}|}$

Solution

Since $|\mathbf{AB}| = |\mathbf{A}|\,|\mathbf{B}|$,

$$|\mathbf{AA}^{-1}| = |\mathbf{A}|\,|\mathbf{A}^{-1}|$$

Solving for $|\mathbf{A}^{-1}|$,

$$|\mathbf{A}^{-1}| = \frac{|\mathbf{AA}^{-1}|}{|\mathbf{A}|}$$

But $|\mathbf{AA}^{-1}| = |\mathbf{I}| = 1$. Therefore,

$$|\mathbf{A}^{-1}| = \frac{|\mathbf{AA}^{-1}|}{|\mathbf{A}|} = \frac{1}{|\mathbf{A}|}$$

The answer is (A).

Cofactors

Cofactors are determinants of submatrices associated with particular entries in the original square matrix. The *minor* of entry a_{ij} is the determinant of a submatrix resulting from the elimination of the single row i and the single column j. For example, the minor corresponding to entry a_{12} in a 3×3 matrix **A** is the determinant of the matrix created by eliminating row 1 and column 2.

$$\text{minor of } a_{12} = \begin{vmatrix} a_{21} & a_{23} \\ a_{31} & a_{33} \end{vmatrix}$$

The cofactor of entry a_{ij} is the minor of a_{ij} multiplied by either $+1$ or -1, depending on the position of the entry (i.e., the cofactor either exactly equals the minor or it differs only in sign). The sign of the cofactor of a_{ij} is positive if $(i+j)$ is even, and it is negative if $(i+j)$ is odd. For a 3×3 matrix, the multipliers in each position are

$$\begin{bmatrix} +1 & -1 & +1 \\ -1 & +1 & -1 \\ +1 & -1 & +1 \end{bmatrix}$$

For example, the cofactor of entry a_{12} in a 3×3 matrix **A** is

$$\text{cofactor of } a_{12} = -\begin{vmatrix} a_{21} & a_{23} \\ a_{31} & a_{33} \end{vmatrix}$$

Equation 3.31: Classical Adjoint

$$\mathbf{B} = \mathbf{A}^{-1} = \frac{\text{adj}(\mathbf{A})}{|\mathbf{A}|} \qquad 3.31$$

Description

The *classical adjoint*, or *adjugate*, is the transpose of the cofactor matrix. The resulting matrix can be designated as $\mathbf{A}_{\text{adj}}$, $\text{adj}\{\mathbf{A}\}$, or $\mathbf{A}^{\text{adj}}$.

For a 3×3 or larger matrix, the inverse is determined by dividing every entry in the classical adjoint by the determinant of the original matrix, as shown in Eq. 3.31.

Example

The cofactor matrix of matrix **A** is **C**.

$$\mathbf{A} = \begin{bmatrix} 4 & 2 & 3 \\ 3 & 2 & 2 \\ 2 & 1 & 4 \end{bmatrix} \qquad \mathbf{C} = \begin{bmatrix} 6 & -8 & -1 \\ -5 & 10 & 0 \\ -2 & 1 & 2 \end{bmatrix}$$

What is the inverse of matrix **A**?

(A) $\begin{bmatrix} 0.25 & 0 & 0 \\ 0 & 0.50 & 0 \\ 0 & 0 & 0.25 \end{bmatrix}$

(B) $\begin{bmatrix} 0.25 & 0.50 & 0.33 \\ 0.33 & 0.50 & 0.50 \\ 0.50 & 1.00 & 0.25 \end{bmatrix}$

(C) $\begin{bmatrix} 1.2 & -1.0 & -0.40 \\ -1.6 & 2.0 & 0.20 \\ -0.20 & 0 & 0.40 \end{bmatrix}$

(D) $\begin{bmatrix} 0.80 & 0.40 & -0.60 \\ 0.20 & -0.40 & 0.40 \\ -0.40 & 0.60 & 0.80 \end{bmatrix}$

Solution

The classical adjoint is the transpose of the cofactor matrix.

$$\text{adj}(\mathbf{A}) = \mathbf{C}^T = \begin{bmatrix} 6 & -5 & -2 \\ -8 & 10 & 1 \\ -1 & 0 & 2 \end{bmatrix}$$

Using Eq. 3.28, calculate the determinant of **A** by expanding along the top row.

$$\begin{aligned} |\mathbf{A}| &= (4)(8-2) - (2)(12-4) + (3)(3-4) \\ &= 24 - 16 - 3 \\ &= 5 \end{aligned}$$

Using Eq. 3.31, divide the classical adjoint by the determinant.

$$\begin{aligned} \mathbf{A}^{-1} &= \frac{\text{adj}(\mathbf{A})}{|\mathbf{A}|} \\ &= \frac{\begin{bmatrix} 6 & -5 & -2 \\ -8 & 10 & 1 \\ -1 & 0 & 2 \end{bmatrix}}{5} \\ &= \begin{bmatrix} 1.2 & -1.0 & -0.40 \\ -1.6 & 2.0 & 0.20 \\ -0.20 & 0 & 0.40 \end{bmatrix} \end{aligned}$$

The answer is (C).

6. WRITING SIMULTANEOUS LINEAR EQUATIONS IN MATRIX FORM

Matrices are used to simplify the presentation and solution of sets of simultaneous linear equations. For example, the following three methods of presenting simultaneous linear equations are equivalent:

$$a_{11}x_1 + a_{12}x_2 = b_1$$
$$a_{21}x_1 + a_{22}x_2 = b_2$$

$$\begin{bmatrix} a_{11} & a_{12} \\ a_{21} & a_{22} \end{bmatrix}\begin{bmatrix} x_1 \\ x_2 \end{bmatrix} = \begin{bmatrix} b_1 \\ b_2 \end{bmatrix}$$

$$\mathbf{AX} = \mathbf{B}$$

In the second and third representations, **A** is known as the *coefficient matrix*, **X** as the *variable matrix*, and **B** as the *constant matrix*.

Not all systems of simultaneous equations have solutions, and those that do may not have unique solutions. The existence of a solution can be determined by calculating the determinant of the coefficient matrix. Solution-existence rules are summarized in Table 3.1.

- If the system of linear equations is homogeneous (i.e., **B** is a zero matrix) and $|\mathbf{A}|$ is zero, there are an infinite number of solutions.
- If the system is homogeneous and $|\mathbf{A}|$ is nonzero, only the trivial solution exists.
- If the system of linear equations is nonhomogeneous (i.e., **B** is not a zero matrix) and $|\mathbf{A}|$ is nonzero, there is a unique solution to the set of simultaneous equations.
- If $|\mathbf{A}|$ is zero, a nonhomogeneous system of simultaneous equations may still have a solution. The requirement is that the determinants of all substitutional matrices (mentioned in Sec. 3.7) are zero, in which case there will be an infinite number of solutions. Otherwise, no solution exists.

Table 3.1 *Solution Existence Rules for Simultaneous Equations*

	$\mathbf{B} = 0$	$\mathbf{B} \neq 0$
$\lvert\mathbf{A}\rvert = 0$	infinite number of solutions (linearly dependent equations)	either an infinite number of solutions or no solution at all
$\lvert\mathbf{A}\rvert \neq 0$	trivial solution only ($x_i = 0$)	unique nonzero solution

7. SOLVING SIMULTANEOUS LINEAR EQUATIONS WITH CRAMER'S RULE

Gauss-Jordan elimination can be used to obtain the solution to a set of simultaneous linear equations. The coefficient matrix is augmented by the constant matrix. Then, elementary row operations are used to reduce the coefficient matrix to canonical form. All of the operations performed on the coefficient matrix are performed on the constant matrix. The variable values that satisfy the simultaneous equations will be the entries in the constant matrix when the coefficient matrix is in canonical form.

Determinants are used to calculate the solution to linear simultaneous equations through a procedure known as *Cramer's rule.*

The procedure is to calculate determinants of the original coefficient matrix **A** and of the n matrices resulting from the systematic replacement of a column in **A** by the constant matrix **B** (i.e., the *substitutional matrices*). For a system of three equations in three unknowns, there are three substitutional matrices, $\mathbf{A}_1$, $\mathbf{A}_2$, and $\mathbf{A}_3$, as well as the original coefficient matrix, for a total of four matrices whose determinants must be calculated.

The values of the unknowns that simultaneously satisfy all of the linear equations are

$$x_1 = \frac{|\mathbf{A}_1|}{|\mathbf{A}|}$$

$$x_2 = \frac{|\mathbf{A}_2|}{|\mathbf{A}|}$$

$$x_3 = \frac{|\mathbf{A}_3|}{|\mathbf{A}|}$$

Example

Using Cramer's rule, what values of x, y, and z will satisfy the following system of simultaneous equations?

$$2x + 3y - 4z = 1$$
$$3x - y - 2z = 4$$
$$4x - 7y - 6z = -7$$

(A) $x = 1,\ y = -4,\ z = -1$

(B) $x = 1,\ y = 3,\ z = 1$

(C) $x = 3,\ y = -2,\ z = 4$

(D) $x = 3,\ y = 1,\ z = 2$

Solution

The determinant of the coefficient matrix is

$$|\mathbf{A}| = \begin{vmatrix} 2 & 3 & -4 \\ 3 & -1 & -2 \\ 4 & -7 & -6 \end{vmatrix} = 82$$

The determinants of the substitutional matrices are

$$|\mathbf{A}_1| = \begin{vmatrix} 1 & 3 & -4 \\ 4 & -1 & -2 \\ -7 & -7 & -6 \end{vmatrix} = 246$$

$$|\mathbf{A}_2| = \begin{vmatrix} 2 & 1 & -4 \\ 3 & 4 & -2 \\ 4 & -7 & -6 \end{vmatrix} = 82$$

$$|\mathbf{A}_3| = \begin{vmatrix} 2 & 3 & 1 \\ 3 & -1 & 4 \\ 4 & -7 & -7 \end{vmatrix} = 164$$

The values of x, y, and z that will satisfy the linear equations are

$$x = \frac{246}{82} = 3$$

$$y = \frac{82}{82} = 1$$

$$z = \frac{164}{82} = 2$$

The answer is (D).

8. VECTORS

A physical property or quantity can be a scalar, vector, or tensor. A *scalar* has only magnitude. Knowing its value is sufficient to define a scalar. Mass, enthalpy, density, and speed are examples of scalars.

Force, momentum, displacement, and velocity are examples of *vectors*. A vector is a directed straight line with a specific magnitude. A vector is specified completely by its direction (consisting of the vector's *angular orientation* and its *sense*) and magnitude. A vector's *point of application* (*terminal point*) is not needed to define the vector. Two vectors with the same direction and magnitude are said to be equal vectors even though their *lines of action* may be different.

Unit vectors are vectors with unit magnitudes (i.e., magnitudes of one). They are represented in the same notation as other vectors. Although they can have any direction, the standard unit vectors (i.e., the *Cartesian unit vectors*, **i**, **j**, and **k**) have the directions of the x-, y-, and z-coordinate axes, respectively, and constitute the *Cartesian triad.*

A *tensor* has magnitude in a specific direction, but the direction is not unique. A tensor in three-dimensional space is defined by nine components, compared with the three that are required to define vectors. These components are written in matrix form. Stress, dielectric constant, and magnetic susceptibility are examples of tensors.

Equation 3.32: Components of a Vector

$$\mathbf{A} = a_x\mathbf{i} + a_y\mathbf{j} + a_z\mathbf{k} \qquad 3.32$$

Description

A vector **A** can be written in terms of unit vectors and its components. (See Fig. 3.2.)

If a vector is based (i.e., starts) at the origin (0, 0, 0), its magnitude (length) can be calculated as

$$|\mathbf{A}| = L_\mathbf{A}$$

$$= \sqrt{a_x^2 + a_y^2 + a_z^2}$$

Figure 3.2 *Components of a Vector*

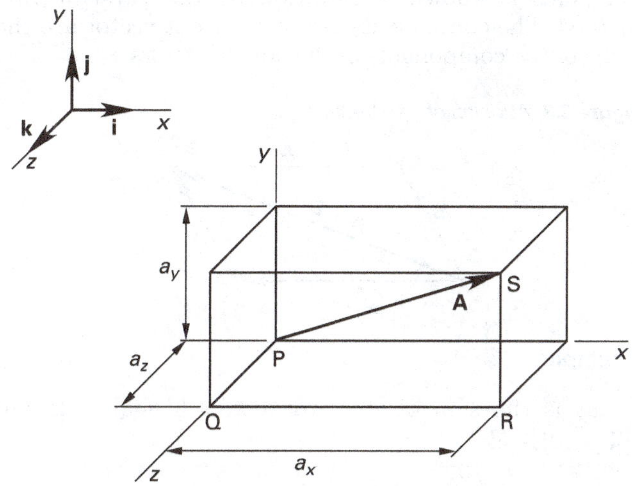

Example

Find the unit vector (i.e., the direction vector) associated with the origin-based vector $18\mathbf{i} + 3\mathbf{j} + 29\mathbf{k}$.

(A) $0.525\mathbf{i} + 0.088\mathbf{j} + 0.846\mathbf{k}$

(B) $0.892\mathbf{i} + 0.178\mathbf{j} + 0.416\mathbf{k}$

(C) $1.342\mathbf{i} + 0.868\mathbf{j} + 2.437\mathbf{k}$

(D) $6\mathbf{i} + \mathbf{j} + \frac{29}{3}\mathbf{k}$

Solution

The unit vector of a particular vector is the vector itself divided by its length.

$$\text{unit vector} = \frac{18\mathbf{i} + 3\mathbf{j} + 29\mathbf{k}}{\sqrt{(18)^2 + (3)^2 + (29)^2}} = 0.525\mathbf{i} + 0.088\mathbf{j} + 0.846\mathbf{k}$$

The answer is (A).

Equation 3.33: Vector Addition

$$\mathbf{A} + \mathbf{B} = (a_x + b_x)\mathbf{i} + (a_y + b_y)\mathbf{j} + (a_z + b_z)\mathbf{k} \qquad 3.33$$

Description

Addition of two vectors by the *polygon method* is accomplished by placing the tail of the second vector at the head (tip) of the first. The sum (i.e., the *resultant vector*) is a vector extending from the tail of the first vector to the head of the second (see Fig. 3.3). Alternatively, the two vectors can be considered as two adjacent sides of a parallelogram, while the sum represents the diagonal. This is known as addition by the *parallelogram method*. The components of the resultant vector are the sums of the components of the added vectors.

Figure 3.3 *Addition of Two Vectors*

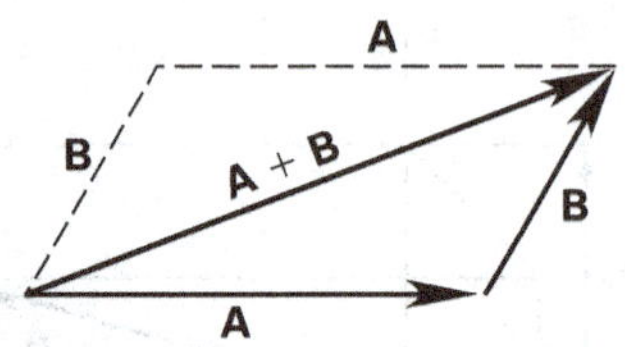

Example

What is the sum of the two vectors $5\mathbf{i} + 3\mathbf{j} - 7\mathbf{k}$ and $10\mathbf{i} - 12\mathbf{j} + 5\mathbf{k}$?

(A) $8\mathbf{i} - 7\mathbf{j} - \mathbf{k}$

(B) $10\mathbf{i} - 9\mathbf{j} + 3\mathbf{k}$

(C) $15\mathbf{i} - 9\mathbf{j} - 2\mathbf{k}$

(D) $15\mathbf{i} + 7\mathbf{j} - 3\mathbf{k}$

Solution

Use Eq. 3.33.

$$\begin{aligned}\mathbf{A} + \mathbf{B} &= (a_x + b_x)\mathbf{i} + (a_y + b_y)\mathbf{j} + (a_z + b_z)\mathbf{k} \\ &= (5 + 10)\mathbf{i} + (3 + (-12))\mathbf{j} + ((-7) + 5)\mathbf{k} \\ &= 15\mathbf{i} - 9\mathbf{j} - 2\mathbf{k}\end{aligned}$$

The answer is (C).

Equation 3.34: Vector Subtraction

$$\mathbf{A} - \mathbf{B} = (a_x - b_x)\mathbf{i} + (a_y - b_y)\mathbf{j} + (a_z - b_z)\mathbf{k} \qquad 3.34$$

Description

Vector subtraction is similar to vector addition, as shown by Eq. 3.34.

Equation 3.35 and Eq. 3.36: Vector Dot Product

$$\mathbf{A}\cdot\mathbf{B} = a_x b_x + a_y b_y + a_z b_z \qquad 3.35$$

$$\mathbf{A}\cdot\mathbf{B} = |\mathbf{A}|\,|\mathbf{B}|\cos\theta = \mathbf{B}\cdot\mathbf{A} \qquad 3.36$$

Variation

$$\theta = \arccos\left(\frac{\mathbf{A}\cdot\mathbf{B}}{|\mathbf{A}|\,|\mathbf{B}|}\right) = \arccos\left(\frac{a_x b_x + a_y b_y + a_z b_z}{|\mathbf{A}|\,|\mathbf{B}|}\right)$$

Description

The *dot product* (*scalar product*) of two vectors is a scalar that is proportional to the length of the projection of the first vector onto the second vector. (See Fig. 3.4.)

Use the variation to find the angle, θ, formed between two given vectors.

Figure 3.4 *Vector Dot Product*

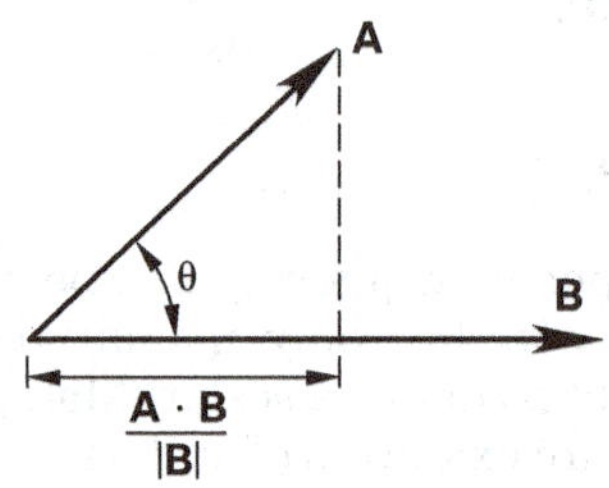

The dot product can be calculated in two ways, as Eq. 3.35 and Eq. 3.36 indicate. θ is limited to 180° and is the acute angle between the two vectors.

Example

What is the dot product, $\mathbf{A}\cdot\mathbf{B}$, of the vectors $\mathbf{A} = 2\mathbf{i} + 4\mathbf{j} + 8\mathbf{k}$ and $\mathbf{B} = -2\mathbf{i} + \mathbf{j} - 4\mathbf{k}$?

(A) $-4\mathbf{i} + 4\mathbf{j} - 32\mathbf{k}$

(B) $-4\mathbf{i} - 4\mathbf{j} - 32\mathbf{k}$

(C) -40

(D) -32

Solution

Use Eq. 3.35.

$$\begin{aligned}\mathbf{A}\cdot\mathbf{B} &= a_x b_x + a_y b_y + a_z b_z \\ &= (2)(-2) + (4)(1) + (8)(-4) \\ &= -32\end{aligned}$$

The answer is (D).

Equation 3.37 and Eq. 3.38: Vector Cross Product

$$\mathbf{A}\times\mathbf{B} = \begin{vmatrix} \mathbf{i} & \mathbf{j} & \mathbf{k} \\ a_x & a_y & a_z \\ b_x & b_y & b_z \end{vmatrix} = -\mathbf{B}\times\mathbf{A} \qquad 3.37$$

$$\mathbf{A}\times\mathbf{B} = |\mathbf{A}|\,|\mathbf{B}|\,\mathbf{n}\sin\theta \qquad 3.38$$

Description

The *cross product* (*vector product*), $\mathbf{A}\times\mathbf{B}$, of two vectors is a vector that is orthogonal (perpendicular) to the plane of the two vectors. (See Fig. 3.5.) The unit vector representation of the cross product can be calculated as a third-order determinant. $\mathbf{n}$ is the unit vector in the direction perpendicular to the plane containing $\mathbf{A}$ and $\mathbf{B}$.

Figure 3.5 *Vector Cross Product*

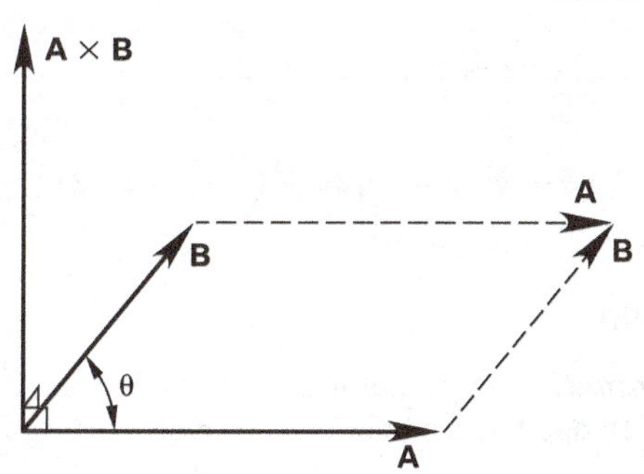

Example

What is the cross product, $\mathbf{A}\times\mathbf{B}$, of vectors $\mathbf{A}$ and $\mathbf{B}$?

$$\mathbf{A} = \mathbf{i} + 4\mathbf{j} + 6\mathbf{k}$$
$$\mathbf{B} = 2\mathbf{i} + 3\mathbf{j} + 5\mathbf{k}$$

(A) $\mathbf{i} - \mathbf{j} - \mathbf{k}$

(B) $-\mathbf{i} + \mathbf{j} + \mathbf{k}$

(C) $2\mathbf{i} + 7\mathbf{j} - 5\mathbf{k}$

(D) $2\mathbf{i} + 7\mathbf{j} + 5\mathbf{k}$

Solution

Use Eq. 3.37. The cross product of two vectors is the determinant of a third-order matrix as shown.

$$\begin{aligned}\mathbf{A}\times\mathbf{B} &= \begin{bmatrix} \mathbf{i} & \mathbf{j} & \mathbf{k} \\ a_x & a_y & a_z \\ b_x & b_y & b_z \end{bmatrix} = \begin{bmatrix} \mathbf{i} & \mathbf{j} & \mathbf{k} \\ 1 & 4 & 6 \\ 2 & 3 & 5 \end{bmatrix} \\ &= \mathbf{i}[(4)(5) - (6)(3)] - \mathbf{j}[(1)(5) - (6)(2)] \\ &\quad + \mathbf{k}[(1)(3) - (4)(2)] \\ &= 2\mathbf{i} + 7\mathbf{j} - 5\mathbf{k}\end{aligned}$$

The answer is (C).

9. VECTOR IDENTITIES

Equation 3.39 Through Eq. 3.41: Dot Product Identities

$$\mathbf{A}\cdot\mathbf{B} = \mathbf{B}\cdot\mathbf{A} \qquad 3.39$$

$$\mathbf{A}\cdot(\mathbf{B}+\mathbf{C}) = \mathbf{A}\cdot\mathbf{B} + \mathbf{A}\cdot\mathbf{C} \qquad 3.40$$

$$\mathbf{A}\cdot\mathbf{A} = |\mathbf{A}|^2 \qquad 3.41$$

Description

The dot product for vectors is commutative and distributive, as shown by Eq. 3.39 and Eq. 3.40. Equation 3.41 gives the dot product of a vector with itself, the square of its magnitude.

Equation 3.42: Dot Product of Parallel Unit Vectors

$$\mathbf{i}\cdot\mathbf{i} = \mathbf{j}\cdot\mathbf{j} = \mathbf{k}\cdot\mathbf{k} = 1 \qquad 3.42$$

Description

As indicated in Eq. 3.42, the dot product of two parallel unit vectors is one.

Example 44

What is the dot product $\mathbf{A}\cdot\mathbf{B}$ of unit vectors $\mathbf{A} = 3\mathbf{i}$ and $\mathbf{B} = 2\mathbf{i}$?

(A) –6

(B) –5

(C) 5

(D) 6

Solution

Use Eq. 3.42.

$$\mathbf{A}\cdot\mathbf{B} = 3\mathbf{i}\cdot 2\mathbf{i} = (3\cdot 2)\mathbf{i}\cdot\mathbf{i} = (6)(1) = 6$$

The answer is (D).

Equation 3.43: Dot Product of Orthogonal Vectors

$$\mathbf{i}\cdot\mathbf{j} = \mathbf{i}\cdot\mathbf{k} = \mathbf{j}\cdot\mathbf{k} = 0 \qquad 3.43$$

Description

The dot product can be used to determine whether a vector is a unit vector and to show that two vectors are orthogonal (perpendicular). As indicated in Eq. 3.43, the dot product of two non-null (nonzero) orthogonal vectors is zero.

Equation 3.44 Through Eq. 3.46: Cross Product Identities

$$\mathbf{A}\times\mathbf{B} = -\mathbf{B}\times\mathbf{A} \qquad 3.44$$

$$\mathbf{A}\times(\mathbf{B}+\mathbf{C}) = (\mathbf{A}\times\mathbf{B}) + (\mathbf{A}\times\mathbf{C}) \qquad 3.45$$

$$(\mathbf{B}+\mathbf{C})\times\mathbf{A} = (\mathbf{B}\times\mathbf{A}) + (\mathbf{C}\times\mathbf{A}) \qquad 3.46$$

Description

The vector cross product is distributive, as demonstrated in Eq. 3.45 and Eq. 3.46. However, as Eq. 3.44 shows, it is not commutative.

Equation 3.47: Cross Product of Parallel Unit Vectors

$$\mathbf{i}\times\mathbf{i} = \mathbf{j}\times\mathbf{j} = \mathbf{k}\times\mathbf{k} = 0 \qquad 3.47$$

Description

If two non-null vectors are parallel, their cross product will be zero.

Equation 3.48 and Eq. 3.49: Cross Product of Normal Unit Vectors

$$\mathbf{i}\times\mathbf{j} = \mathbf{k} = -\mathbf{j}\times\mathbf{i} \qquad 3.48$$

$$\mathbf{k}\times\mathbf{i} = \mathbf{j} = -\mathbf{i}\times\mathbf{k} \qquad 3.49$$

Description

If two non-null vectors are normal (perpendicular), their vector cross product will be perpendicular to both vectors.

10. PROGRESSIONS AND SERIES

A *progression* or *sequence*, $\{\mathbf{A}\}$, is an ordered set of numbers a_i, such as 1, 4, 9, 16, 25, ... The *terms* in a sequence can be all positive, negative, or of alternating signs. l is the last term and is also known as the *general term* of the sequence.

$$\{\mathbf{A}\} = a_1, a_2, a_3, \ldots, l$$

A sequence is said to *diverge* (i.e., be *divergent*) if the terms approach infinity, and it is said to *converge* (i.e., be *convergent*) if the terms approach any finite value (including zero).

A *series* is the sum of terms in a sequence. There are two types of series: A *finite series* has a finite number of terms. An *infinite series* has an infinite number of terms, but this does not imply that the sum is infinite. The main tasks associated with series are determining the sum of the terms and determining whether the series converges. A series is said to converge if the sum, S_n, of its terms exists. A finite series is always convergent. An infinite series may be convergent.

Equation 3.50 and Eq. 3.51: Arithmetic Progression

$$l = a + (n-1)d \qquad 3.50$$

$$S = n(a+l)/2 = n[2a + (n-1)d]/2 \qquad 3.51$$

Description

The *arithmetic progression* is a standard sequence that diverges. It has the form shown in Eq. 3.50.

In Eq. 3.50 and Eq. 3.51, a is the *first term*, d is a constant called the *common difference*, and n is the number of terms.

The difference of adjacent terms is constant in an arithmetic progression. The sum of terms in a finite arithmetic series is shown by Eq. 3.51.

Example

What is the sum of the following finite sequence of terms?

$$18, 25, 32, 39, \ldots, 67$$

(A) 181

(B) 213

(C) 234

(D) 340

Solution

Each term is 7 more than the previous term. This is an arithmetic sequence. The general mathematical representation for an arithmetic sequence is

$$l = a + (n-1)d$$

In this case, the difference term is $d = 7$. The first term is $a = 18$, and the last term is $l = 67$.

$$\begin{aligned} l &= a + (n-1)d \\ n &= \frac{l-a}{d} + 1 \\ &= \frac{67-18}{7} + 1 \\ &= 8 \end{aligned}$$

The sum of n terms is

$$\begin{aligned} S &= n[2a + (n-1)d]/2 \\ &= \frac{(8)((2)(18) + (8-1)(7))}{2} \\ &= 340 \end{aligned}$$

The answer is (D).

Equation 3.52 Through Eq. 3.55: Geometric Progression

$$l = ar^{n-1} \qquad 3.52$$

$$S = a(1-r^n)/(1-r) \quad [r \neq 1] \qquad 3.53$$

$$S = (a - rl)/(1-r) \quad [r \neq 1] \qquad 3.54$$

$$\lim_{n\to\infty} S_n = a/(1-r) \quad [r < 1] \qquad 3.55$$

Variations

$$S_n = \sum_{i=1}^{n} ar^{i-1} = \frac{a - rl}{1-r} = \frac{a(1-r^n)}{1-r}$$

$$S_n = \sum_{i=1}^{\infty} ar^{i-1} = \frac{a}{1-r}$$

Description

The *geometric progression* is another standard sequence. The quotient of adjacent terms is constant in a geometric progression. It converges for $-1 < r < 1$ and diverges otherwise.

In Eq. 3.52 through Eq. 3.55, a is the first term, and r is known as the *common ratio.*

The sum of a finite geometric series is given by Eq. 3.53 and Eq. 3.54. The sum of an infinite geometric series is given by Eq. 3.55.

Example

What is the sum of the following geometric sequence?

$$32, 80, 200, \ldots, 19{,}531.25$$

(A) 21,131.25

(B) 24,718.25

(C) 31,250.00

(D) 32,530.75

Solution

The common ratio is

$$r = \frac{80}{32} = \frac{200}{80} = 2.5$$

Since the ratio and both the initial and final terms are known, the sum can be found using Eq. 3.54.

$$\begin{aligned} S = (a - rl)/(1-r) &= \frac{32 - (2.5)(19{,}531.25)}{1 - 2.5} \\ &= 32{,}530.75 \end{aligned}$$

The answer is (D).

Equation 3.56 Through Eq. 3.59: Properties of Series

$$\sum_{i=1}^{n} c = nc \qquad 3.56$$

$$\sum_{i=1}^{n} cx_i = c\sum_{i=1}^{n} x_i \qquad 3.57$$

$$\sum_{i=1}^{n}(x_i + y_i - z_i) = \sum_{i=1}^{n} x_i + \sum_{i=1}^{n} y_i - \sum_{i=1}^{n} z_i \qquad 3.58$$

$$\sum_{x=1}^{n} x = (n + n^2)/2 \qquad 3.59$$

Description

Equation 3.56 through Eq. 3.59 list some basic properties of series. The terms x_i, y_i, and z_i represent general terms in any series. Equation 3.56 describes the obvious result of n repeated additions of a constant, c. Equation 3.57 shows that the product of a constant, c, and a serial summation of series terms is distributive. Equation 3.58 shows that addition of series is associative. Equation 3.59 gives the sum of n consecutive integers. This is not really a property of series in general; it is the property of a special kind of arithmetic sequence. It is a useful identity for use with *sum-of-the-years' depreciation.*

Equation 3.60: Power Series

$$\sum_{i=0}^{\infty} a_i(x - a)^i \qquad 3.60$$

Variation

$$\sum_{i=1}^{n} a_i x^{i-1} = a_1 + a_2 x + a_3 x^2 + \cdots + a_n x^{n-1}$$

Description

A *power series* is a series of the form shown in Eq. 3.60. The *interval of convergence* of a power series consists of the values of x for which the series is convergent. Due to the exponentiation of terms, an infinite power series can only be convergent in the interval $-1 < x < 1$.

A power series may be used to represent a function that is continuous over the interval of convergence of the series. The *power series representation* may be used to find the derivative or integral of that function.

Power series behave similarly to polynomials: They may be added together, subtracted from each other, multiplied together, or divided term by term within the interval of convergence. They may also be differentiated and integrated within their interval of convergence. If $f(x) = \sum_{i=1}^{n} a_i x^i$, then over the interval of convergence,

$$f'(x) = \sum_{i=1}^{n} \frac{d(a_i x^i)}{dx}$$

$$\int f(x)\,dx = \sum_{i=1}^{n} \int a_i x^i\,dx$$

Equation 3.61: Taylor's Series

$$f(x) = f(a) + \frac{f'(a)}{1!}(x-a) + \frac{f''(a)}{2!}(x-a)^2 + \cdots + \frac{f^{(n)}(a)}{n!}(x-a)^n + \cdots \qquad 3.61$$

Description

Taylor's series (*Taylor's formula*), Eq. 3.61, can be used to expand a function around a point (i.e., to approximate the function at one point based on the function's value at another point). The approximation consists of a series, each term composed of a derivative of the original function and a polynomial. Using Taylor's formula requires that the original function be continuous in the interval $[a, b]$. To expand a function, $f(x)$, around a point, a, in order to obtain $f(b)$, use Eq. 3.61.

If $a = 0$, Eq. 3.61 is known as a *Maclaurin series.*

To be a useful approximation, two requirements must be met: (1) Point a must be relatively close to point b, and (2) the function and its derivatives must be known or be easy to calculate.

Example

Taylor's series is used to expand the function $f(x)$ about $a = 0$ to obtain $f(b)$.

$$f(x) = \frac{1}{3x^3 + 4x + 8}$$

What are the first two terms of Taylor's series?

(A) $\frac{1}{16} + \frac{b}{8}$

(B) $\frac{1}{8} - \frac{b}{16}$

(C) $\frac{1}{8} + \frac{b}{16}$

(D) $\frac{1}{4} - \frac{b}{16}$

Solution

The first two coefficient terms of Taylor's series are

$$\begin{aligned} f(0) &= \frac{1}{(3)(0)^3 + (4)(0) + 8} \\ &= 1/8 \end{aligned}$$

$$f'(x) = \frac{-(9x^2 + 4)}{(3x^3 + 4x + 8)^2}$$

$$f'(0) = \frac{-((9)(0)^2 + 4)}{((3)(0)^3 + (4)(0) + 8)^2} = \frac{-4}{64} = -1/16$$

Using Eq. 3.61, find the first two complete terms of Taylor's series.

$$\begin{aligned} f(b) &= f(a) + \frac{f'(a)}{1!}(b - a) \\ &= \frac{1}{8} + \frac{\left(\frac{-1}{16}\right)(b-0)}{1} \\ &= \frac{1}{8} - \frac{b}{16} \end{aligned}$$

The answer is (B).

4 Calculus

1. DERIVATIVES

In most cases, it is possible to transform a continuous function, $f(x_1, x_2, x_3 \ldots)$, of one or more independent variables into a derivative function. In simple two-dimensional cases, the *derivative* can be interpreted as the slope (tangent or rate of change) of the curve described by the original function.

Equation 4.1 Through Eq. 4.3: Definitions of the Derivative

$$y' = \underset{\Delta x \to 0}{\text{limit}}[(\Delta y)/(\Delta x)] \qquad 4.1$$

$$y' = \underset{\Delta x \to 0}{\text{limit}}\{[f(x+\Delta x) - f(x)]/(\Delta x)\} \qquad 4.2$$

$$y' = \text{the slope of the curve } f(x) \qquad 4.3$$

Variation

$$f'(x) = \lim_{\Delta x \to 0}\left(\frac{f(x+\Delta x) - f(x)}{\Delta x}\right)$$

Description

Since the slope of a curve depends on x, the derivative function will also depend on x. The derivative, $f'(x)$, of a function $f(x)$ is defined mathematically by the variation given here. However, limit theory is seldom needed to actually calculate derivatives.

Equation 4.4 and Eq. 4.5: First Derivative

$$y = f(x) \qquad 4.4$$

$$D_x y = dy/dx = y' \qquad 4.5$$

Variations

$$f'(x),\ \frac{df(x)}{dx},\ \mathbf{D}f(x),\ \mathbf{D}_x f(x)$$

Description

The derivative of a function $y = f(x)$, also known as the *first derivative*, is represented in various ways, as shown by the variations.

Example

What is the slope of the curve $y = 10x^2 - 3x - 1$ when it crosses the positive part of the x-axis?

(A) 3/20

(B) 1/5

(C) 1/3

(D) 7

Solution

The curve crosses the x-axis when $y = 0$. At this point,

$$10x^2 - 3x - 1 = 0$$

Use the quadratic equation or complete the square to determine the two values of x where the curve crosses the x-axis.

$$\begin{aligned} x^2 - 0.3x &= 0.1 \\ (x - 0.15)^2 &= 0.1 + (0.15)^2 \\ x &= \pm 0.35 + 0.15 \\ &= -0.2,\ 0.5 \end{aligned}$$

Since x must be positive, $x = 0.5$. The slope of the function is the first derivative.

$$\begin{aligned} \frac{dy}{dx} &= 20x - 3 \\ &= (20)(0.5) - 3 \\ &= 7 \end{aligned}$$

The answer is (D).

Equation 4.6 Through Eq. 4.32: Derivatives

$$dc/dx = 0 \tag{4.6}$$

$$dx/dx = 1 \tag{4.7}$$

$$d(cu)/dx = c\ du/dx \tag{4.8}$$

$$d(u+v-w)/dx = du/dx + dv/dx - dw/dx \tag{4.9}$$

$$d(uv)/dx = u\ dv/dx + v\ du/dx \tag{4.10}$$

$$d(uvw)/dx = uv\ dw/dx + uw\ dv/dx + vw\ du/dx \tag{4.11}$$

$$\frac{d(u/v)}{dx} = \frac{v\ du/dx - u\ dv/dx}{v^2} \tag{4.12}$$

$$d(u^n)/dx = nu^{n-1}du/dx \tag{4.13}$$

$$d[f(u)]/dx = \{d[f(u)]/du\}\ du/dx \tag{4.14}$$

$$du/dx = 1/(dx/du) \tag{4.15}$$

$$\frac{d(\log_a u)}{dx} = (\log_a e)\frac{1}{u}\frac{du}{dx} \tag{4.16}$$

$$\frac{d(\ln u)}{dx} = \frac{1}{u}\frac{du}{dx} \tag{4.17}$$

$$\frac{d(a^u)}{dx} = (\ln a)a^u\frac{du}{dx} \tag{4.18}$$

$$d(e^u)/dx = e^u du/dx \tag{4.19}$$

$$d(u^v)/dx = vu^{v-1}du/dx + (\ln u)u^v dv/dx \tag{4.20}$$

$$d(\sin u)/dx = \cos u\ du/dx \tag{4.21}$$

$$d(\cos u)/dx = -\sin u\ du/dx \tag{4.22}$$

$$d(\tan u)/dx = \sec^2 u\ du/dx \tag{4.23}$$

$$d(\cot u)/dx = -\csc^2 u\ du/dx \tag{4.24}$$

$$d(\sec u)/dx = \sec u \tan u\ du/dx \tag{4.25}$$

$$d(\csc u)/dx = -\csc u \cot u\ du/dx \tag{4.26}$$

$$\frac{d(\sin^{-1}u)}{dx} = \frac{1}{\sqrt{1-u^2}}\frac{du}{dx} \quad [-\pi/2 \le \sin^{-1}u \le \pi/2] \tag{4.27}$$

$$\frac{d(\cos^{-1}u)}{dx} = -\frac{1}{\sqrt{1-u^2}}\frac{du}{dx} \quad [0 \le \cos^{-1}u \le \pi] \tag{4.28}$$

$$\frac{d(\tan^{-1}u)}{dx} = \frac{1}{1+u^2}\frac{du}{dx} \quad [-\pi/2 < \tan^{-1}u < \pi/2] \tag{4.29}$$

$$\frac{d(\cot^{-1}u)}{dx} = -\frac{1}{1+u^2}\frac{du}{dx} \quad [0 < \cot^{-1}u < \pi] \tag{4.30}$$

$$\frac{d(\sec^{-1}u)}{dx} = \frac{1}{u\sqrt{u^2-1}}\frac{du}{dx} \quad [0 < \sec^{-1}u < \pi/2 \text{ or } -\pi \le \sec^{-1}u < -\pi/2] \tag{4.31}$$

$$\frac{d(\csc^{-1}u)}{dx} = -\frac{1}{u\sqrt{u^2-1}}\frac{du}{dx} \quad [0 < \csc^{-1}u \le \pi/2 \text{ or } -\pi < \csc^{-1}u \le -\pi/2] \tag{4.32}$$

Description

Formulas for the derivatives of some common functional forms are listed in Eq. 4.6 through Eq. 4.32.

Example

Evaluate dy/dx for the following expression.

$$y = e^{-x}\sin 2x$$

(A) $e^{-x}(2\cos 2x - \sin 2x)$

(B) $-e^{-x}(2\sin 2x + \cos 2x)$

(C) $e^{-x}(2\sin 2x + \cos 2x)$

(D) $-e^{-x}(2\cos 2x - \sin 2x)$

Solution

Use the product rule (Eq. 4.10).

$$\begin{aligned}\frac{d}{dx}(e^{-x}\sin 2x) &= e^{-x}\frac{d}{dx}(\sin 2x) \\ &\quad + (\sin 2x)\frac{d}{dx}(e^{-x}) \\ &= e^{-x}(\cos 2x)(2) \\ &\quad + (\sin 2x)(e^{-x})(-1) \\ &= e^{-x}(2\cos 2x - \sin 2x)\end{aligned}$$

The answer is (A).

2. CRITICAL POINTS

Derivatives are used to locate the local *critical points*, that is, *extreme points* (also known as *maximum* and *minimum points*) as well as the *inflection points* (*points of contraflexure*) of functions of one variable. The plurals *extrema*, *maxima*, and *minima* are used without the word "points." These points are illustrated in Fig. 4.1. There is usually an inflection point between two adjacent local extrema.

Figure 4.1 *Critical Points*

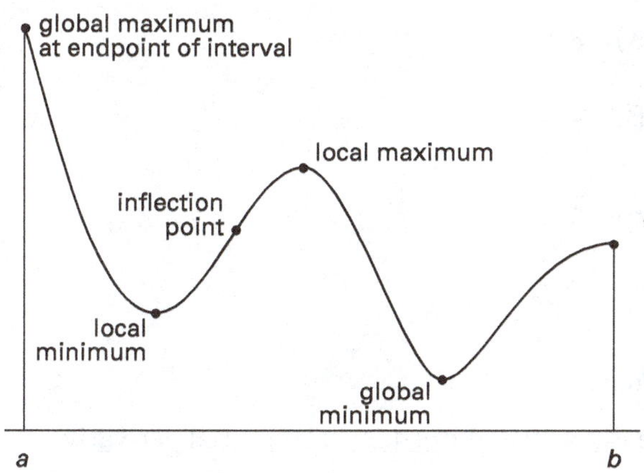

The first derivative, $f'(x)$, is calculated to determine where the critical points might be. The second derivative, $f''(x)$, is calculated to determine whether a located point is a maximum, minimum, or inflection point. With this method, no distinction is made between local and global extrema. The extrema should be compared to the function values at the endpoints of the interval.

Critical points are located where the first derivative is zero. This is a necessary, but not sufficient, requirement. That is, for a function $y = f(x)$, the point $x = a$ is a critical point if

$$f'(a) = 0$$

Equation 4.33 and Eq. 4.34: Test for a Maximum

$$f'(a) = 0 \qquad 4.33$$

$$f''(a) < 0 \qquad 4.34$$

Description

For a function $f(x)$ with an extreme point at $x = a$, if the point is a maximum, then the second derivative is negative.

Example 50

What is the maximum value of the function $f(x) = -x^2 - 8x + 1$?

(A) 1

(B) 4

(C) 8

(D) 17

Solution

Use Eq. 4.33 and Eq. 4.34.

$$\begin{aligned}f(x) &= -x^2 - 8x + 1 \\ f'(x) &= -2x - 8 \\ f''(x) &= -2\end{aligned}$$

$f'(x) = 0$ when x is equal to –4, and $f''(x)$ is less than zero, so $f(x)$ has its maximum value at $x = -4$.

$$\begin{aligned}f(x) &= -x^2 - 8x + 1 \\ &= -(-4)^2 - (8)(-4) + 1 \\ &= 17\end{aligned}$$

The answer is (D).

Equation 4.35 and Eq. 4.36: Test for a Minimum

$$f'(a) = 0 \qquad 4.35$$

$$f''(a) > 0 \qquad 4.36$$

Description

For a function $f(x)$ with a critical point at $x = a$, if the point is a minimum, then the second derivative is positive.

Mathematics

Example 51

What is the minimum value of the function $f(x) = 3x^2 + 3x - 5$?

(A) −12.0

(B) −8.0

(C) −5.75

(D) −5.00

Solution

Use Eq. 4.35 and Eq. 4.36.

$$f(x) = 3x^2 + 3x - 5$$
$$f'(x) = 6x + 3$$
$$f''(x) = 6$$

$f'(x) = 0$ when x is equal to −0.5, and $f''(x)$ is greater than zero, so $f(x)$ has its minimum value at $x = -0.5$.

$$\begin{aligned} f(x) &= 3x^2 + 3x - 5 \\ &= (3)(-0.5)^2 + (3)(-0.5) - 5 \\ &= -5.75 \end{aligned}$$

The answer is (C).

Equation 4.37: Test for a Point of Inflection

$$f''(a) = 0 \qquad 4.37$$

Description

For a function $f(x)$ with $f'(x) = 0$ at $x = a$, if the point is a point of inflection, then Eq. 4.37 is true.

3. PARTIAL DERIVATIVES

Derivatives can be taken with respect to only one independent variable at a time. For example, $f'(x)$ is the derivative of $f(x)$ and is taken with respect to the independent variable x. If a function, $f(x_1, x_2, x_3 \ldots)$, has more than one independent variable, a *partial derivative* can be found, but only with respect to one of the independent variables. All other variables are treated as constants.

Equation 4.38 and Eq. 4.39: Partial Derivative

$$z = f(x, y) \qquad 4.38$$

$$\frac{\partial z}{\partial x} = \frac{\partial f(x, y)}{\partial x} \qquad 4.39$$

Variations

Symbols for a partial derivative of $f(x, y)$ taken with respect to variable x are $\partial f/\partial x$ and $f_x(x, y)$.

Description

The geometric interpretation of a partial derivative $\partial f/\partial x$ is the slope of a line tangent to the surface (a sphere, an ellipsoid, etc.) described by the function when all variables except x are held constant. In three-dimensional space with a function described by Eq. 4.38, the partial derivative $\partial f/\partial x$ (equivalent to $\partial z/\partial x$) is the slope of the line tangent to the surface in a plane of constant y. Similarly, the partial derivative $\partial f/\partial y$ (equivalent to $\partial z/\partial y$) is the slope of the line tangent to the surface in a plane of constant x.

Example 52

What is the partial derivative with respect to x of the following function?

$$z = e^{xy}$$

(A) e^{xy}

(B) $\dfrac{e^{xy}}{x}$

(C) $\dfrac{e^{xy}}{y}$

(D) ye^{xy}

Solution

Use Eq. 4.19 and Eq. 4.39. The partial derivative is

$$\begin{aligned} d(e^u)/dx &= e^u du/dx \\ \frac{\partial z}{\partial x} = \frac{\partial e^{xy}}{\partial x} &= e^{xy}\frac{\partial (xy)}{\partial x} \\ &= ye^{xy} \end{aligned}$$

The answer is (D).

4. CURVATURE

The sharpness of a curve between two points on the curve can be defined as the rate of change of the inclination of the curve with respect to the distance traveled along the curve. As shown in Fig. 4.2, the rate of change of the inclination of the curve is the change in the angle formed by the tangents to the curve at each point and the x-axis. The distance, s, traveled along the curve is the arc length of the curve between points 1 and 2.

Figure 4.2 Curvature

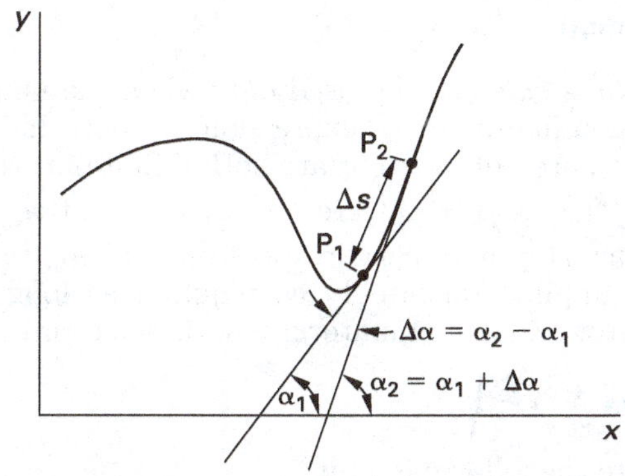

Equation 4.40: Curvature

$$K = \lim_{\Delta s \to 0} \frac{\Delta \alpha}{\Delta s} = \frac{d\alpha}{ds} \qquad 4.40$$

Description

On roadways, a "sharp" curve is one that changes direction quickly, corresponding to a small curve radius. The smaller the curve radius, the sharper the curve. Some roadway curves are circular, some are parabolic, and some are spiral. Not all curves are circular, but all curves described by polynomials have an instantaneous sharpness and radius of curvature. The sharpness, K, of a curve at a point is given by Eq. 4.40.

Equation 4.41 Through Eq. 4.43: Curvature in Rectangular Coordinates

$$K = \frac{y''}{\left[1 + (y')^2\right]^{3/2}} \qquad 4.41$$

$$x' = dx/dy \qquad 4.42$$

$$K = \frac{-x''}{\left[1 + (x')^2\right]^{3/2}} \qquad 4.43$$

Description

For an equation of a curve $f(x, y)$ given in rectangular coordinates, the curvature is defined by Eq. 4.41.

If the function $f(x, y)$ is easier to differentiate with respect to y instead of x, then Eq. 4.43 may be used.

Equation 4.44 and Eq. 4.45: Radius of Curvature

$$R = \frac{1}{|K|} \quad [K \neq 0] \qquad 4.44$$

$$R = \left| \frac{\left[1 + (y')^2\right]^{3/2}}{|y''|} \right| \quad [y'' \neq 0] \qquad 4.45$$

Description

The *radius of curvature*, R, of a curve describes the radius of a circle whose center lies on the concave side of the curve and whose tangent coincides with the tangent to the curve at that point. Radius of curvature is the absolute value of the reciprocal of the curvature.

Example

What is the approximate radius of curvature of the function $f(x)$ at the point $(x, y) = (8, 16)$?

$$f(x) = x^2 + 6x - 96$$

(A) 1.9×10^{-4}

(B) 9.8

(C) 96

(D) 5300

Solution

The first and second derivatives are

$$f'(x) = 2x + 6$$
$$f''(x) = 2$$

At $x = 8$,

$$f'(x) = (2)(8) + 6 = 22$$

From Eq. 4.45, the radius of curvature, R, is

$$\begin{aligned} R &= \left| \frac{\left[1 + f'(x)^2\right]^{3/2}}{|f''(x)|} \right| \\ &= \left| \frac{\left(1 + (22)^2\right)^{3/2}}{2} \right| \\ &= 5340.5 \quad (5300) \end{aligned}$$

The answer is (D).

5. LIMITS

A *limit* is the value a function approaches when an independent variable approaches a target value. For example, suppose the value of $y = x^2$ is desired as x approaches 5. This could be written as

$$y(5) = \operatorname*{limit}_{x \to 5} x^2$$

The power of limit theory is wasted on simple calculations such as this one, but limit theory is appreciated when the function is undefined at the target value. The object of limit theory is to determine the limit without having to evaluate the function at the target. The general case of a limit evaluated as x approaches the target value a is written as

$$\lim_{x \to a} f(x)$$

It is not necessary for the actual value, $f(a)$, to exist for the limit to be calculated. The function $f(x)$ may be undefined at point a. However, it is necessary that $f(x)$ be defined on both sides of point a for the limit to exist. If $f(x)$ is undefined on one side, or if $f(x)$ is discontinuous at $x = a$, as in Fig. 4.3(c) and Fig. 4.3(d), the limit does not exist at $x = a$.

Figure 4.3 *Existence of Limits*

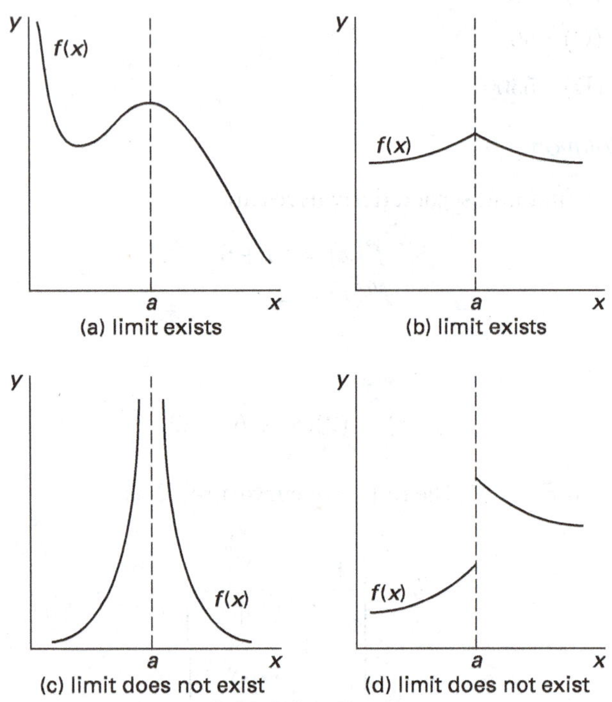

Equation 4.46: L'Hopital's Rule

$$\operatorname*{limit}_{x \to \alpha} \frac{f'(x)}{g'(x)}, \; \operatorname*{limit}_{x \to \alpha} \frac{f''(x)}{g''(x)}, \; \operatorname*{limit}_{x \to \alpha} \frac{f'''(x)}{g'''(x)} \qquad 4.46$$

Variation

$$\lim_{x \to a} \frac{f(x)}{g(x)} = \lim_{x \to a} \frac{f^k(x)}{g^k(x)}$$

Description

L'Hôpital's rule may be used only when the numerator and denominator of the expression are both indeterminate (i.e., are both zero or are both infinite) at the limit point. $f^k(x)$ and $g^k(x)$ are the kth derivatives of the functions $f(x)$ and $g(x)$, respectively. L'Hôpital's rule can be applied repeatedly as required as long as the numerator and denominator are both indeterminate.

Example

Evaluate the following limit.

$$\operatorname*{limit}_{x \to 0} \frac{1 - e^{3x}}{4x}$$

(A) $-\infty$

(B) $-3/4$

(C) 0

(D) 1/4

Solution

This limit has the indeterminate form 0/0, so use L'Hôpital's rule.

$$\operatorname*{limit}_{x \to \alpha} \frac{f(x)}{g(x)} = \operatorname*{limit}_{x \to \alpha} \frac{f'(x)}{g'(x)}$$

$$\operatorname*{limit}_{x \to 0} \frac{1 - e^{3x}}{4x} = \operatorname*{limit}_{x \to 0} \frac{-3e^{3x}}{4} = -3/4$$

The answer is (B).

6. INTEGRALS

Equation 4.47 Through Eq. 4.69: Indefinite Integrals

$$\int df(x) = f(x) \qquad 4.47$$

$$\int dx = x \qquad 4.48$$

$$\int a\, f(x)\, dx = a \int f(x)\, dx \qquad 4.49$$

$$\int[u(x)\pm v(x)]\,dx=\int u(x)\,dx\pm\int v(x)\,dx \quad 4.50$$

$$\int x^m\,dx=\frac{x^{m+1}}{m+1}\quad [m\neq -1] \quad 4.51$$

$$\int u(x)\,dv(x)=u(x)v(x)-\int v(x)\,du(x) \quad 4.52$$

$$\int\frac{dx}{ax+b}=\frac{1}{a}\ln|ax+b| \quad 4.53$$

$$\int\frac{dx}{\sqrt{x}}=2\sqrt{x} \quad 4.54$$

$$\int a^x\,dx=\frac{a^x}{\ln a} \quad 4.55$$

$$\int\sin x\,dx=-\cos x \quad 4.56$$

$$\int\cos x\,dx=\sin x \quad 4.57$$

$$\int\sin^2 x\,dx=\frac{x}{2}-\frac{\sin 2x}{4} \quad 4.58$$

$$\int\cos^2 x\,dx=\frac{x}{2}+\frac{\sin 2x}{4} \quad 4.59$$

$$\int x\sin x\,dx=\sin x-x\cos x \quad 4.60$$

$$\int x\cos x\,dx=\cos x+x\sin x \quad 4.61$$

$$\int\sin x\cos x\,dx=(\sin^2 x)/2 \quad 4.62$$

$$\int\sin ax\cos bx\,dx=-\frac{\cos(a-b)x}{2(a-b)}-\frac{\cos(a+b)x}{2(a+b)}\quad [a^2\neq b^2] \quad 4.63$$

$$\int\tan x\,dx=-\ln|\cos x|=\ln|\sec x| \quad 4.64$$

$$\int\cot x\,dx=-\ln|\csc x|=\ln|\sin x| \quad 4.65$$

$$\int\tan^2 x\,dx=\tan x-x \quad 4.66$$

$$\int\cot^2 x\,dx=-\cot x-x \quad 4.67$$

$$\int e^{ax}\,dx=(1/a)e^{ax} \quad 4.68$$

$$\int xe^{ax}\,dx=(e^{ax}/a^2)(ax-1) \quad 4.69$$

Description

Integration is the inverse operation of differentiation. There are two types of integrals: *definite integrals*, which are restricted to a specific range of the independent variable, and *indefinite integrals*, which are unrestricted. Indefinite integrals are sometimes referred to as *antiderivatives*.

Equation 4.70: Fundamental Theorem of Integral Calculus

$$\operatorname*{limit}_{n\to\infty}\sum_{i=1}^{n} f(x_i)\Delta x_i=\int_a^b f(x)\,dx \quad 4.70$$

Description

The definition of a definite integral is given by the *fundamental theorem of integral calculus.* The right-hand side of Eq. 4.70 represents the area bounded by $f(x)$ above, $y=0$ below, $x=a$ to the left, and $x=b$ to the right. This is commonly referred to as the "*area under the curve.*"

Example 55

What is the approximate total area bounded by $y=\sin x$ over the interval $0\leq x\leq 2\pi$? (x is in radians.)

(A) 0

(B) $\pi/2$

(C) 2

(D) 4

Solution

The integral of $f(x)$ represents the area under the curve $f(x)$ between the limits of integration. However, since the value of $\sin x$ is negative in the range $\pi\leq x\leq 2\pi$, the total area would be calculated as zero if the integration was carried out in one step. The integral could be

Mathematics

calculated over two ranges, but it is easier to exploit the symmetry of the sine curve.

$$\begin{aligned} A &= \int_{x_1}^{x_2} f(x)\,dx = \int_0^{2\pi} |\sin x|\,dx \\ &= 2\int_0^{\pi} \sin x\,dx \\ &= -2\cos x\Big|_0^{\pi} \\ &= (-2)(-1-1) \\ &= 4 \end{aligned}$$

The answer is (D).

7. CENTROIDS AND MOMENTS OF INERTIA

Applications of integration include the determination of the *centroid of an area* and various moments of the area, including the *area moment of inertia.*

The integration method for determining centroids and moments of inertia is not necessary for basic shapes. Formulas for basic shapes can be found in tables.

Equation 4.71 Through Eq. 4.74: Centroid of an Area

$$x_c = \frac{\int x\,dA}{A} \quad 4.71$$

$$y_c = \frac{\int y\,dA}{A} \quad 4.72$$

$$A = \int f(x)\,dx \quad 4.73$$

$$dA = f(x)\,dx = g(y)\,dy \quad 4.74$$

Description

The centroid of an area is analogous to the *center of gravity* of a homogeneous body. The location, (x_c, y_c), of the centroid of the area bounded by the x- and y-axis and the mathematical function $y = f(x)$ can be found from Eq. 4.71 through Eq. 4.74.

Example 56

What is most nearly the x-coordinate of the centroid of the area bounded by $y=0$, $f(x)$, $x=0$, and $x=20$?

$$f(x) = x^3 + 7x^2 - 5x + 6$$

(A) 7.6

(B) 9.4

(C) 14

(D) 16

Solution

Use Eq. 4.71 and Eq. 4.74.

$$\begin{aligned} \int xf(x)\,dx &= \int_0^{20} (x^4 + 7x^3 - 5x^2 + 6x)\,dx \\ &= \frac{x^5}{5} + \frac{7x^4}{4} - \frac{5x^3}{3} + \frac{6x^2}{2}\Bigg|_0^{20} \\ &= 907{,}867 \end{aligned}$$

From Eq. 4.73, the area under the curve is

$$\begin{aligned} A &= \int_a^b f(x)\,dx = \int_0^{20} (x^3 + 7x^2 - 5x + 6)\,dx \\ &= \tfrac{1}{4}x^4 + \tfrac{7}{3}x^3 - \tfrac{5}{2}x^2 + 6x\Big|_0^{20} \\ &= \left(\frac{1}{4}\right)(20)^4 + \left(\frac{7}{3}\right)(20)^3 - \left(\frac{5}{2}\right)(20)^2 + (6)(20) \\ &= 57{,}786.67 \quad (57{,}787) \end{aligned}$$

Use Eq. 4.71 to find the x-coordinate of the centroid.

$$\begin{aligned} x_c &= \frac{\int x\,dA}{A} \\ &= \frac{\int xf(x)\,dx}{A} \\ &= \frac{907{,}867}{57{,}787} \\ &= 15.71 \quad (16) \end{aligned}$$

The answer is (D).

Equation 4.75 and Eq. 4.76: First Moment of the Area

$$M_y = \int x\,dA = x_c A \quad 4.75$$

$$M_x = \int y\,dA = y_c A \quad 4.76$$

Description

The quantity $\int x dA$ is known as the *first moment of the area* or *first area moment* with respect to the y-axis. Similarly, $\int y dA$ is known as the *first moment of the area* with respect to the x-axis. Equation 4.75 and Eq. 4.76 show that the first moment of the area can be calculated from the area and centroidal distance.

Equation 4.77 and Eq. 4.78: Moment of Inertia

$$I_y = \int x^2\,dA \tag{4.77}$$

$$I_x = \int y^2\,dA \tag{4.78}$$

Description

The *second moment of the area* or *moment of inertia*, I, of the area is needed in mechanics of materials problems. The symbol I_x is used to represent a moment of inertia with respect to the x-axis. Similarly, I_y is the moment of inertia with respect to the y-axis.

Example

What is most nearly the moment of inertia about the y-axis of the area bounded by $y = 0$, $x = 0$, $x = 20$, and $f(x) = x^3 + 7x^2 - 5x + 6$?

(A) 6.3×10^5

(B) 8.2×10^6

(C) 9.9×10^6

(D) 1.5×10^7

Solution

From Eq. 4.77, the moment of inertia about the y-axis is

$$\begin{aligned} I_y &= \int x^2\,dA = \int x^2 f(x)\,dx \\ &= \int_0^{20} (x^5 + 7x^4 - 5x^3 + 6x^2)\,dx \\ &= \frac{x^6}{6} + \frac{7x^5}{5} - \frac{5x^4}{4} + \frac{6x^3}{3}\bigg|_0^{20} \\ &= 1.5 \times 10^7 \end{aligned}$$

The answer is (D).

Equation 4.79 Through Eq. 4.82: Centroidal Moment of Inertia

$$I_{\text{parallel axis}} = I_c + d^2 A \tag{4.79}$$

$$J = \int r^2 dA = I_x + I_y \tag{4.80}$$

$$I_x' = I_{x_c} + d_y^2 A \tag{4.81}$$

$$I_y' = I_{y_c} + d_x^2 A \tag{4.82}$$

Description

Moments of inertia can be calculated with respect to any axis, not just the coordinate axes. The moment of inertia taken with respect to an axis passing through the area's centroid is known as the *centroidal moment of inertia*, I_c. The centroidal moment of inertia is the smallest possible moment of inertia for the area.

If the moment of inertia is known with respect to one axis, the moment of inertia with respect to another parallel axis can be calculated from the *parallel axis theorem*, also known as the *transfer axis theorem*. (See Eq. 4.79.) This theorem is also used to evaluate the moment of inertia of areas that are composed of two or more basic shapes. In Eq. 4.79, d is the distance between the centroidal axis and the second, parallel axis.

The parallel axis theorem is generalized into two dimensions in Fig. 4.4, where the surface area of the shape $\int dA$ with the centroid located at distance d_y from the x-axis and distance d_x from the y-axis has moments of inertia I_x' and I_y' about the x- and y-axes, as defined by Eq. 4.81 and Eq. 4.82.

Figure 4.4 *Parallel Axis Theorem in Two Dimensions*

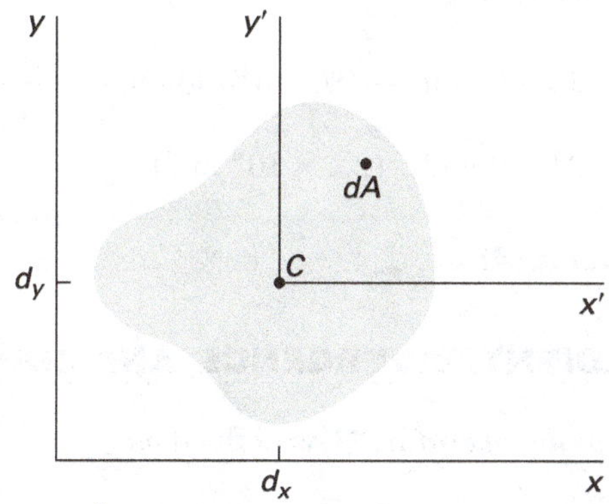

Example

The moment of inertia about the x'-axis of the cross section shown is 334 000 cm^4. The cross-sectional area is

86 cm^2, and the thicknesses of the web and the flanges are the same.

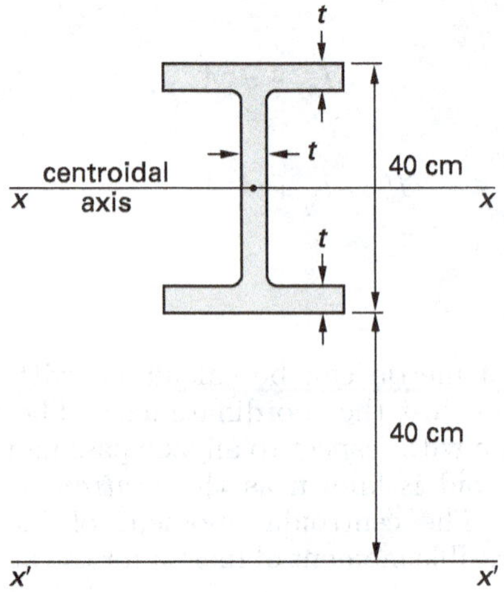

What is most nearly the moment of inertia about the centroidal axis?

(A) 2.4×10^4 cm^4

(B) 7.4×10^4 cm^4

(C) 2.0×10^5 cm^4

(D) 6.4×10^5 cm^4

Solution

Use Eq. 4.35. The moment of inertia around the centroidal axis is

$$I_x' = I_{x_c} + d_x^2 A$$

$$I_{x_c} = I_x' - d_x^2 A$$

$$= 334\,000 \text{ cm}^4 - (86 \text{ cm}^2)\left(40 \text{ cm} + \frac{40 \text{ cm}}{2}\right)^2$$

$$= 24\,400 \text{ cm}^4 \quad (2.4 \times 10^4 \text{ cm}^4)$$

The answer is (A).

8. GRADIENT, DIVERGENCE, AND CURL

The *vector del operator*, ∇, is defined as

$$\nabla = \frac{\partial}{\partial x}\mathbf{i} + \frac{\partial}{\partial y}\mathbf{j} + \frac{\partial}{\partial z}\mathbf{k}$$

Equation 4.83: Gradient of a Scalar Function

$$\nabla\phi = \left(\frac{\partial}{\partial x}\mathbf{i} + \frac{\partial}{\partial y}\mathbf{j} + \frac{\partial}{\partial z}\mathbf{k}\right)\phi \qquad 4.83$$

Variation

$$\nabla f(x,y,z) = \left(\frac{\partial f(x,y,z)}{\partial x}\right)\mathbf{i} + \left(\frac{\partial f(x,y,z)}{\partial y}\right)\mathbf{j} + \left(\frac{\partial f(x,y,z)}{\partial z}\right)\mathbf{k}$$

Description

A *scalar function* is a mathematical expression that returns a single numerical value (i.e., a *scalar*). The function may be of one or multiple variables (i.e., $f(x)$ or $f(x,y,z)$ or $f(x_1, x_2 \ldots x_n)$), but it must calculate a single number for each location. The *gradient vector field*, $\nabla\phi$, gives the maximum rate of change of the scalar function $\phi = \phi(x,y,z)$.

Equation 4.84: Divergence of a Vector Field

$$\nabla \cdot \mathbf{V} = \left(\frac{\partial}{\partial x}\mathbf{i} + \frac{\partial}{\partial y}\mathbf{j} + \frac{\partial}{\partial z}\mathbf{k}\right) \cdot (V_1\mathbf{i} + V_2\mathbf{j} + V_3\mathbf{k}) \qquad 4.84$$

Description

In three dimensions, $\mathbf{V}$ is a vector field with components V_1, V_2, and V_3. V_1, V_2, and V_3 may be specified as functions of variables, such as $P(x,y,z)$, $Q(x,y,z)$, and $R(x,y,z)$. The *divergence* of a vector field $\mathbf{V}$ is the scalar function defined by Eq. 4.84, the dot product of the del operator and the vector (i.e., the divergence is a scalar). The divergence of $\mathbf{V}$ can be interpreted as the *accumulation* of flux (i.e., a flowing substance) in a small region (i.e., at a point).

If $\mathbf{V}$ represents a flow (e.g., air moving from hot to cool regions), then $\mathbf{V}$ is incompressible if $\nabla \cdot \mathbf{V} = 0$, since the substance is not accumulating.

Example

What is the divergence of the following vector field?

$$\mathbf{V} = 2x\mathbf{i} + 2y\mathbf{j}$$

(A) 0

(B) 2

(C) 3

(D) 4

Solution

Use Eq. 4.84.

$$\begin{aligned}\nabla\cdot\mathbf{V} &= \left(\frac{\partial}{\partial x}\mathbf{i}+\frac{\partial}{\partial y}\mathbf{j}+\frac{\partial}{\partial z}\mathbf{k}\right)\cdot(V_1\mathbf{i}+V_2\mathbf{j}+V_3\mathbf{k})\\ &= \left(\frac{\partial}{\partial x}\mathbf{i}+\frac{\partial}{\partial y}\mathbf{j}+\frac{\partial}{\partial z}\mathbf{k}\right)\cdot(2x\mathbf{i}+2y\mathbf{j}+0\mathbf{k})\\ &= \frac{\partial(2x)}{\partial x}+\frac{\partial(2y)}{\partial y}+\frac{\partial(0)}{\partial z}\\ &= 2+2+0\\ &= 4\end{aligned}$$

The answer is (D).

Equation 4.85: Curl of a Vector Field

$$\nabla\times\mathbf{V}=\left(\frac{\partial}{\partial x}\mathbf{i}+\frac{\partial}{\partial y}\mathbf{j}+\frac{\partial}{\partial z}\mathbf{k}\right)\times(V_1\mathbf{i}+V_2\mathbf{j}+V_3\mathbf{k}) \qquad 4.85$$

Variation

$$\text{curl }\mathbf{V}=\nabla\times\mathbf{V}=\begin{vmatrix}\mathbf{i} & \mathbf{j} & \mathbf{k}\\ \dfrac{\partial}{\partial x} & \dfrac{\partial}{\partial y} & \dfrac{\partial}{\partial z}\\ P(x,y,z) & Q(x,y,z) & R(x,y,z)\end{vmatrix}$$

Description

The *curl*, $\nabla\times\mathbf{V}$, of a vector field $\mathbf{V}(x, y, z)$ is the vector field defined by Eq. 4.85, the cross (vector) product of the del operator and vector. For any location, the curl vector has both magnitude and direction. That is, the curl vector determines how fast the flux is rotating and in what direction the flux is going. The curl of a vector field can be interpreted as the *vorticity* per unit area of flux (i.e., a flowing substance) in a small region (i.e., at a point). One of the uses of the curl is to determine whether flow (represented in direction and magnitude by $\mathbf{V}$) is rotational. Flow is irrotational if curl $\nabla\times\mathbf{V}=0$.

Example

Determine the curl of the vector function $\mathbf{V}(x, y, z)$.

$$\mathbf{V}(x,y,z)=3x^2\mathbf{i}+7e^x y\mathbf{j}$$

(A) $7e^x y$

(B) $7e^x y\mathbf{i}$

(C) $7e^x y\mathbf{j}$

(D) $7e^x y\mathbf{k}$

Solution

Using the variation of Eq. 4.85,

$$\text{curl }\mathbf{V}=\begin{vmatrix}\mathbf{i} & \mathbf{j} & \mathbf{k}\\ \dfrac{\partial}{\partial x} & \dfrac{\partial}{\partial y} & \dfrac{\partial}{\partial z}\\ 3x^2 & 7e^x y & 0\end{vmatrix}$$

Expand the determinant across the top row.

$$\begin{aligned}&\left(\frac{\partial}{\partial y}0-\frac{\partial}{\partial z}7e^x y\right)\mathbf{i}-\left(\frac{\partial}{\partial x}0-\frac{\partial}{\partial z}3x^2\right)\mathbf{j}\\ &\quad+\left(\frac{\partial}{\partial x}7e^x y-\frac{\partial}{\partial y}3x^2\right)\mathbf{k}\\ &= (0-0)\mathbf{i}-(0-0)\mathbf{j}+(7e^x y-0)\mathbf{k}\\ &= 7e^x y\mathbf{k}\end{aligned}$$

The answer is (D).

Equation 4.86 Through Eq. 4.89: Vector Identities

$$\nabla^2\phi=\nabla\cdot(\nabla\phi)=(\nabla\cdot\nabla)\phi \qquad 4.86$$

$$\nabla\times\nabla\phi=0 \qquad 4.87$$

$$\nabla\cdot(\nabla\times\mathbf{A})=0 \qquad 4.88$$

$$\nabla\times(\nabla\times\mathbf{A})=\nabla(\nabla\cdot\mathbf{A})-\nabla^2\mathbf{A} \qquad 4.89$$

Description

Equation 4.86 through Eq. 4.89 are identities associated with gradient, divergence, and curl.

Equation 4.90: Laplacian of a Scalar Function

$$\nabla^2\phi=\frac{\partial^2\phi}{\partial x^2}+\frac{\partial^2\phi}{\partial y^2}+\frac{\partial^2\phi}{\partial z^2} \qquad 4.90$$

Description

The *Laplacian* of a scalar function, $\phi=\phi(x, y, z)$, is the divergence of the gradient function. (This is essentially the second derivative of a scalar function.) A function that satisfies Laplace's equation $\nabla^2=0$ is known as a *potential function*. Accordingly, the operator ∇^2 is commonly written as $\nabla\cdot\nabla$ or Δ. The potential function quantifies the attraction of the flux to move in a particular direction. It is used in electricity (voltage potential), mechanics (gravitational potential), mixing and diffusion (concentration gradient), hydraulics (pressure gradient), and heat transfer (thermal gradient). The term Laplacian almost always refers to three-dimensional functions, and usually functions in rectangular coordinates. The term *d'Alembertian* is used when working with four-dimensional functions. The symbol $\square^2$ (with four sides) is used in place of ∇^2. The d'Alembertian is encountered frequently when working with wave functions (including those involving relativity and quantum mechanics) of x, y, and z for location, and t for time.

Example

Determine the Laplacian of the scalar function $1/3x^3 - 9y + 5$ at the point $(3, 2, 7)$.

(A) 0

(B) 1

(C) 6

(D) 9

Solution

The Laplacian of the function is

$$\nabla^2\phi = \frac{\partial^2\phi}{\partial x^2} + \frac{\partial^2\phi}{\partial y^2} + \frac{\partial^2\phi}{\partial z^2}$$

$$\begin{aligned}\nabla^2\left(\tfrac{1}{3}x^3 - 9y + 5\right) &= \frac{\partial^2\left(\frac{1}{3}x^3 - 9y + 5\right)}{\partial x^2} \\ &\quad + \frac{\partial^2\left(\frac{1}{3}x^3 - 9y + 5\right)}{\partial y^2} \\ &\quad + \frac{\partial^2\left(\frac{1}{3}x^3 - 9y + 5\right)}{\partial z^2} \\ &= 2x + 0 + 0 \\ &= 2x\end{aligned}$$

At $(3, 2, 7)$, $2x = (2)(3) = 6$.

The answer is (C).

5 Differential Equations and Transforms

1. INTRODUCTION TO DIFFERENTIAL EQUATIONS

A *differential equation* is a mathematical expression combining a function (e.g., $y = f(x)$) and one or more of its derivatives. The *order* of a differential equation is the highest derivative in it. *First-order differential equations* contain only first derivatives of the function, *second-order differential equations* contain second derivatives (and may contain first derivatives as well), and so on.

The purpose of solving a differential equation is to derive an expression for the function in terms of the independent variable. The expression does not need to be explicit in the function, but there can be no derivatives in the expression. Since, in the simplest cases, solving a differential equation is equivalent to finding an indefinite integral, it is not surprising that *constants of integration* must be evaluated from knowledge of how the system behaves. Additional data are known as *initial values*, and any problem that includes them is known as an *initial value problem.*

Equation 5.1: Linear Differential Equation with Constant Coefficients

$$b_n \frac{d^n y(x)}{dx^n} + \cdots + b_1 \frac{dy(x)}{dx} + b_0 y(x) = f(x) \qquad 5.1$$

$[b_n, \ldots, b_i, \ldots, b_1, \text{ and } b_0 \text{ are constants}]$

Description

A *linear differential equation* can be written as a sum of multiples of the function $y(x)$ and its derivatives. If the multipliers are scalars, the differential equation is said to have *constant coefficients.* Equation 5.1 shows the general form of a linear differential equation with constant coefficients. $f(x)$ is known as the forcing function. If the forcing function is zero, the differential equation is said to be *homogeneous.*

If the function $y(x)$ or one of its derivatives is raised to some power (other than one) or is embedded in another function (e.g., y embedded in $\sin y$ or e^y), the equation is said to be *nonlinear.*

Example 62

Which of the following is NOT a linear differential equation?

(A) $5\frac{d^2y}{dt^2} - 8\frac{dy}{dt} + 16y = 4te^{-7t}$

(B) $5\frac{d^2y}{dt^2} - 8t^2\frac{dy}{dt} + 16y = 0$

(C) $5\frac{d^2y}{dt^2} - 8\frac{dy}{dt} + 16y = \frac{dy}{dy}$

(D) $5\left(\frac{dy}{dt}\right)^2 - 8\frac{dy}{dt} + 16y = 0$

Solution

A linear differential equation consists of multiples of a function, $y(t)$, and its derivatives, $d^n y/dt^n$. The multipliers may be scalar constants or functions, $g(t)$, of the independent variable, t. The forcing function, $f(t)$, (i.e., the right-hand side of the equation) may be 0, a constant, or any function of the independent variable, t. The multipliers cannot be higher powers of the function, $y(t)$.

The answer is (D).

Mathematics

2. LINEAR HOMOGENEOUS DIFFERENTIAL EQUATIONS WITH CONSTANT COEFFICIENTS

Each term of a *homogeneous differential equation* contains either the function or one of its derivatives. The forcing function is zero. That is, the sum of the function and its derivative terms is equal to zero.

$$b_n \frac{d^n y(x)}{dx^n} + \cdots + b_1 \frac{dy(x)}{dx} + b_0 y(x) = 0$$

Equation 5.2: Characteristic Equation

$$P(r) = b_n r^n + b_{n-1} r^{n-1} + \cdots + b_1 r + b_0 \quad 5.2$$

Description

A *characteristic equation* can be written for a homogeneous linear differential equation with constant coefficients, regardless of order. This characteristic equation is simply the polynomial formed by replacing all derivatives with variables raised to the power of their respective derivatives. That is, all instances of $d^n y(x)/dx^n$ are replaced with r^n, resulting in an equation of the form of Eq. 5.2.

Equation 5.3: Solving Linear Differential Equations with Constant Coefficients

$$y_h(x) = C_1 e^{r_1 x} + C_2 e^{r_2 x} + \cdots + C_i e^{r_i x} + \cdots + C_n e^{r_n x} \quad 5.3$$

Description

Homogeneous linear differential equations are most easily solved by finding the n roots of Eq. 5.2, the characteristic polynomial $P(r)$. If the roots of Eq. 5.2 are real and different, the solution is Eq. 5.3.

Equation 5.4 and Eq. 5.5: Homogeneous First-Order Linear Differential Equations

$$y' + ay = 0 \quad 5.4$$

$$y = Ce^{-at} \quad 5.5$$

Variations

$$\frac{dy}{dt} + ay = 0$$

$$f(t) = Ce^{-at}$$

Description

A homogeneous, first-order, linear differential equation with constant coefficients has the general form of Eq. 5.4.

The characteristic equation is $r + a = 0$ and has a root of $r = -a$. Equation 5.5 is the solution.

Example 63

Which of the following is the general solution to the differential equation and boundary conditions?

$$\frac{dy}{dt} - 5y = 0$$

$$y(0) = 3$$

(A) $-\frac{1}{3}e^{-5t}$

(B) $3e^{5t}$

(C) $5e^{-3t}$

(D) $\frac{1}{5}e^{-3t}$

Solution

This is a first-order, linear differential equation. The characteristic equation is $r - 5 = 0$. The root, r, is 5.

The solution is in the form of

$$y = Ce^{5t}$$

The initial condition is used to find C.

$$y(0) = Ce^{(5)(0)} = 3$$
$$C = 3$$
$$y = 3e^{5t}$$

The answer is (B).

Equation 5.6 Through Eq. 5.8: Homogeneous Second-Order Linear Differential Equations with Constant Coefficients

$$y'' + ay' + by = 0 \quad 5.6$$

$$(r^2 + ar + b)Ce^{rx} = 0 \quad 5.7$$

$$r^2 + ar + b = 0 \quad 5.8$$

Description

A second-order, homogeneous, linear differential equation has the general form given by Eq. 5.6.

The characteristic equation is Eq. 5.8.

Depending on the form of the forcing function, the solutions to most second-order differential equations will contain sinusoidal terms (corresponding to oscillatory behavior) and exponential terms (corresponding to decaying or increasing unstable behavior). Behavior of real-world systems (electrical circuits, spring-mass-dashpot, fluid flow, heat transfer, etc.) depends on the amount of system *damping* (electrical resistance, mechanical friction, pressure drop, thermal insulation, etc.).

With *underdamping* (i.e., with "light" damping) without continued energy input (i.e., a free system without a forcing function), the transient behavior will gradually decay to the steady-state equilibrium condition. Behavior in underdamped free systems will be oscillatory with diminishing magnitude. The damping is known as underdamping because the amount of damping is less than the critical damping, and the *damping ratio*, ζ, is less than 1. The characteristic equation of underdamped systems has two complex roots.

With *overdamping* ("heavy" damping), damping is greater than critical, and the damping ratio is greater than 1. Transient behavior is a sluggish gradual decrease into the steady-state equilibrium condition without oscillations. The characteristic equation of overdamped systems has two distinct real roots (zeros).

With *critical damping*, the damping ratio is equal to 1. There is no overshoot, and the behavior reaches the steady-state equilibrium condition the fastest of the three cases, without oscillations. The characteristic equation of critically damped systems has two identical real roots (zeros).

Equation 5.9 Through Eq. 5.14: Roots of the Characteristic Equation

$$r_{1,2} = \frac{-a \pm \sqrt{a^2 - 4b}}{2} \qquad 5.9$$

$$y = C_1 e^{r_1 x} + C_2 e^{r_2 x} \qquad 5.10$$

$$y = (C_1 + C_2 x) e^{r_1 x} \qquad 5.11$$

$$y = e^{\alpha x}(C_1 \cos \beta x + C_2 \sin \beta x) \qquad 5.12$$

$$\alpha = -a/2 \qquad 5.13$$

$$\beta = \frac{\sqrt{4b - a^2}}{2} \qquad 5.14$$

Description

The roots of the characteristic equation are given by the quadratic equation, Eq. 5.9.

If $a^2 > 4b$, then the two roots are real and different, and the solution is overdamped, as shown in Eq. 5.10.

If $a^2 = 4b$, then the two roots are real and the same (i.e., are *double roots*), and the solution is critically damped, as shown in Eq. 5.11.

If $a^2 < 4b$, then the two roots are imaginary and of the form $(\alpha + i\beta)$ and $(\alpha - i\beta)$, and the solution is underdamped, as shown in Eq. 5.12.

Example

What is the general solution to the following homogeneous differential equation?

$$y'' - 8y' + 16y = 0$$

(A) $y = C_1 e^{4x}$

(B) $y = (C_1 + C_2 x) e^{4x}$

(C) $y = C_1 e^{-4x} + C_2 e^{4x}$

(D) $y = C_1 e^{2x} + C_2 e^{4x}$

Solution

Find the roots of the characteristic equation.

$$r^2 - 8r + 16 = 0$$
$$a = -8$$
$$b = 16$$

From Eq. 5.9,

$$\begin{aligned} r_{1,2} &= \frac{-a \pm \sqrt{a^2 - 4b}}{2} \\ &= \frac{-(-8) \pm \sqrt{(-8)^2 - (4)(16)}}{2} \\ &= 4,4 \end{aligned}$$

Because $a^2 = 4b$, the characteristic equation has double roots. With $r = 4$, the solution takes the form

$$y = (C_1 + C_2 x) e^{rx} = (C_1 + C_2 x) e^{4x}$$

The answer is (B).

3. LINEAR NONHOMOGENEOUS DIFFERENTIAL EQUATIONS WITH CONSTANT COEFFICIENTS

In a nonhomogeneous differential equation, the sum of derivative terms is equal to a nonzero *forcing function* of the independent variable (i.e., $f(x)$ in Eq. 5.1 is nonzero). In order to solve a nonhomogeneous equation, it is often necessary to solve the homogeneous equation first. The homogeneous equation corresponding to a nonhomogeneous equation is known as the *reduced equation* or *complementary equation.*

Equation 5.15: Complete Solution to Nonhomogeneous Differential Equation

$$y(x) = y_h(x) + y_p(x) \qquad 5.15$$

Description

The complete solution to the nonhomogeneous differential equation is shown in Eq. 5.15. The term $y_h(x)$ is the *complementary solution,* which solves the complementary (i.e., homogeneous) case. The *particular solution,* $y_p(x)$, is any specific solution to the nonhomogeneous Eq. 5.1 that is known or can be found. Initial values are used to evaluate any unknown coefficients in the complementary solution after $y_h(x)$ and $y_p(x)$ have been combined. The particular solution will not have any unknown coefficients.

Table 5.1: Method of Undetermined Coefficients

Table 5.1 Method of Undetermined Coefficients

form of $f(x)$	form of $y_p(x)$
A	B
Ae^{sx}	Be^{sx}, $s \neq r_n$
$A_1 \sin \omega x + A_2 \cos \omega x$	$B_1 \sin \omega x + B_2 \cos \omega x$

Description

Two methods are available for finding a particular solution. The *method of undetermined coefficients,* as presented here, can be used only when $f(x)$ in Eq. 5.1 takes on one of the forms given in Table 5.1. $f(x)$ is known as the *forcing function.*

The particular solution can be read from Table 5.1 if the forcing function is one of the forms given. Of course, the coefficients A_i and B_i are not known—these are the *undetermined coefficients.* The exponent s is the smallest non-negative number (and will be zero, one, or two, etc.), which ensures that no term in the particular solution is also a solution to the complementary equation. α must be determined prior to proceeding with the solution procedure.

Once $y_p(x)$ (including s) is known, it is differentiated to obtain $dy_p(x)/dx$, $d^2y_p(x)/dx^2$, and all subsequent derivatives. All of these derivatives are substituted into the original nonhomogeneous equation. The resulting equation is rearranged to match the forcing function, $f(x)$, and the unknown coefficients are determined, usually by solving simultaneous equations.

The presence of an exponential of the form e^{rx} in the solution indicates that *resonance* is present to some extent.

Equation 5.16 Through Eq. 5.20: First-Order Linear Nonhomogeneous Differential Equations with Constant Coefficients, with Step Input

$$\tau \frac{dy}{dt} + y = Kx(t) \qquad 5.16$$

$$x(t) = \begin{Bmatrix} A & t < 0 \\ B & t > 0 \end{Bmatrix} \qquad 5.17$$

$$y(0) = KA \qquad 5.18$$

$$y(t) = KA + (KB - KA)\left(1 - \exp\left(\frac{-t}{\tau}\right)\right) \qquad 5.19$$

$$\frac{t}{\tau} = \ln\left[\frac{KB - KA}{KB - y}\right] \qquad 5.20$$

Variation

$$b_1 \frac{dy(t)}{dt} + b_0 y(t) = u(t) \quad [u(t) = \text{unit step function}]$$

Description

As the variation equation for Eq. 5.16 implies, a first-order, linear, nonhomogeneous differential equation with constant coefficients is an extension of Eq. 5.1. Equation 5.16 builds on the differential equation of Eq. 5.1 in the context of a specific control system scenario. It also changes the independent variable from x to t and changes the notation for the forcing function used in Eq. 5.1.

The *time constant,* τ, is the amount of time a homogeneous system (i.e., one with a zero forcing function, $x(t)$) would take to reach $(e-1)/e$, or approximately 63.2% of its final value. This could also be described as the time required to grow to within 36.8% of the final value or as the time to decay to 36.8% of the initial value. The *system gain,* K, or *amplification ratio* is a scalar constant that gives the ratio of the output response to the input response at steady state.

Equation 5.16 describes a *step function,* a special case of a generic forcing function. The forcing function is some value, typically zero ($A = 0$) until $t = 0$, at which time the forcing function immediately jumps to a constant value. Equation 5.19 gives the *step response*, the solution to Eq. 5.16.

Example

A spring-mass-dashpot system starting from a motionless state is acted upon by a step function. The response is described by the differential equation in which time, t, is given in seconds measured from the application of the ramp function.

$$\frac{dy}{dt} + 2y = 2u(0) \quad [y(0) = 0]$$

How long will it take for the system to reach 63% of its final value?

(A) 0.25 s

(B) 0.50 s

(C) 1.0 s

(D) 2.0 s

Solution

To fit this problem into the format used by Eq. 5.16, the coefficient of y must be 1. Dividing by 2,

$$0.5\frac{dy}{dt} + y = u(0)$$

$$\tau = 0.50 \text{ s}$$

The answer is (B).

4. FOURIER SERIES

Any periodic waveform can be written as the sum of an infinite number of sinusoidal terms (i.e., an infinite series), known as *harmonic terms*. Such a sum of sinusoidal terms is known as a *Fourier series*, and the process of finding the terms is *Fourier analysis*. Since most series converge rapidly, it is possible to obtain a good approximation to the original waveform with a limited number of sinusoidal terms.

Equation 5.21 and Eq. 5.22: Fourier's Theorem

$$f(t) = a_0 + \sum_{n=1}^{\infty} [a_n \cos(n\omega_0 t) + b_n \sin(n\omega_0 t)] \qquad 5.21$$

$$T = 2\pi/\omega_0 \qquad 5.22$$

Variation

$$\omega_0 = \frac{2\pi}{T} = 2\pi f$$

Description

Fourier's theorem is Eq. 5.21. The object of a Fourier analysis is to determine the *Fourier coefficients* a_n and b_n. The term a_0 can often be determined by inspection since it is the average value of the waveform.

ω_0 is the *natural* (*fundamental*) *frequency* of the waveform. It depends on the actual waveform *period*, T.

Equation 5.23 Through Eq. 5.25: Fourier Coefficients

$$a_0 = (1/T)\int_0^T f(t)\,dt \qquad 5.23$$

$$a_n = (2/T)\int_0^T f(t)\cos(n\omega_0 t)\,dt \quad [n = 1, 2, \ldots] \qquad 5.24$$

$$b_n = (2/T)\int_0^T f(t)\sin(n\omega_0 t)\,dt \quad [n = 1, 2, \ldots] \qquad 5.25$$

Description

The *Fourier coefficients* are found from the relationships shown in Eq. 5.23 through Eq. 5.25.

Example

What are the first terms in the Fourier series of the repeating function shown?

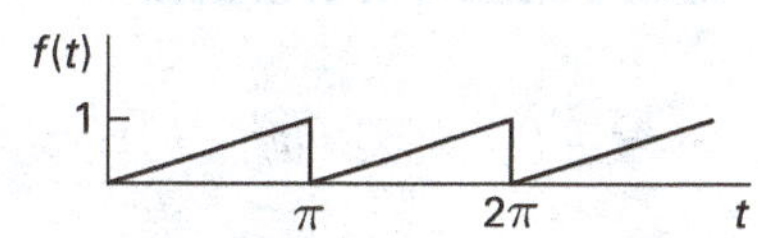

(A) $\frac{1}{2} - \cos 2t - \frac{1}{2}\cos 4t - \frac{1}{3}\cos 6t$

(B) $\frac{1}{2} - \frac{1}{\pi}\sin 2t - \frac{1}{2\pi}\sin 4t - \frac{1}{3\pi}\sin 6t$

(C) $\frac{1}{4} - \frac{1}{\pi}\begin{pmatrix} \cos 2t + \sin 2t + \cos 4t \\ + \frac{1}{2}\sin 4t + \cos 6t + \frac{1}{3}\sin 6t \end{pmatrix}$

(D) $\frac{1}{4} - \frac{1}{\pi}\begin{pmatrix} \frac{1}{\pi}\cos 2t + \sin 2t \\ + \frac{1}{2\pi}\cos 4t + \frac{1}{2}\sin 4t \\ + \frac{1}{3\pi}\cos 6t + \frac{1}{3}\sin 6t \end{pmatrix}$

Solution

A Fourier series has the form given by Eq. 5.21.

$$f(t) = a_0 + \sum_{n=1}^{\infty} [a_n \cos(n\omega_0 t) + b_n \sin(n\omega_0 t)]$$

The constant term a_0 corresponds to the average of the function. The average is seen by observation to be 1/2, so $a_0 = 1/2$.

In this problem, the triangular pulses are ramps, so $f(t)$ has the form of kt, where k is a scalar. A cycle is completed at $t = \pi$, so $T = \pi$, and $\omega_0 = 2\pi/T = 2$. Since $f(T) = 1$ (that is, $f(t) = 1$ at $t = \pi$), $f(t) = t/\pi$.

Calculate the general form of the a_n terms using Eq. 5.24.

$$\begin{aligned} a_n &= (2/T)\int_0^T f(t)\cos(n\omega_0 t)\,dt = \frac{2}{\pi^2}\int_0^{\pi} t\cos(2nt)\,dt \\ &= \frac{1}{2n^2\pi^2}(\cos(2nt) + 2nt\sin(2nt))\Big|_0^{\pi} \\ &= 0 \end{aligned}$$

There are no a_n terms in the series. From Eq. 5.21, there are no cosine terms in the expansion. There are only sine terms in the expansion.

Only choice (B) satisfies both of these requirements.

Alternatively, the values can be derived, though this would be a lengthy process.

The answer is (B).

Equation 5.26: Parseval Relation

$$F_N^2 = a_0^2 + (1/2)\sum_{n=1}^{N}(a_n^2 + b_n^2) \qquad 5.26$$

Variation

$$F_{\text{rms}} = \sqrt{a_0^2 + \frac{a_1^2 + a_2^2 + \cdots a_N^2 + b_1^2 + b_2^2 + \cdots b_N^2}{2}}$$

Description

The *Parseval relation* (also known as *Parseval's equality*) calculates the root-mean-square (rms) value of a Fourier series that has been truncated after N terms. The rms value, F_{rms}, is the square root of Eq. 5.26.

5. FOURIER TRANSFORMS

Equation 5.27 Through Eq. 5.30: Fourier Transform Pairs

$$F(\omega) = \int_{-\infty}^{\infty} f(t)e^{-j\omega t}dt \qquad 5.27$$

$$f(t) = [1/(2\pi)]\int_{-\infty}^{\infty} F(\omega)e^{j\omega t}d\omega \qquad 5.28$$

$$X(f) = \int_{-\infty}^{+\infty} x(t)e^{-j2\pi ft}dt \qquad 5.29$$

$$x(t) = \int_{-\infty}^{+\infty} X(f)e^{j2\pi ft}df \qquad 5.30$$

Description

There are several useful ways to transform a complex, general equation of one variable into the summation of one or more relatively simple terms of another variable. Functions are transformed for convenience, as when it is necessary to solve exactly for one or more of their properties, and out of necessity, as when it is necessary to approximate the behavior of a waveform that has no exact mathematical expression. In engineering, it is common to use lowercase letters for the original function (of x or t), and to use uppercase letters for the *transform*. It is also necessary to change the variable so that position or time, x or t (known as the *spatial domain*) in the original is not confused with the transform's variable, s or ω (known as the *s-domain* or *frequency domain*). The original function, $f(t)$ and its transform, $F(s)$, constitute a *transform pair*. Although transforms can be determined mathematically from their functions, working with transforms is greatly facilitated by having tables of transform pairs. Extracting $f(t)$ from $F(s)$ is often described as finding the *inverse transform*.

The *Fourier transform*, Eq. 5.27, transforms a function of time, t, into a function of frequency, ω. Essentially, the Fourier transform replaces a function with a sum of simpler sinusoidal functions of a different frequency. Equation 5.28 calculates the inverse transform. Equation 5.29 and Eq. 5.30 are variations of Eq. 5.27 and Eq. 5.28.[1] While the limited number of Fourier transform pairs listed in Table 5.2 and Table 5.3 may not appear to simplify anything, in practice, the transformation is quite useful. Fourier transforms have a wide range of applications, including waveform and image analysis, filtering, reconstruction, and compression.

[1]Table 5.2 gives additional transform pairs that apply to Eq. 5.29 and Eq. 5.30.

Equation 5.31, Eq. 5.32, Table 5.2, and Table 5.3: Additional Fourier Transform Pairs

$$f(t) = 0 \quad [t < 0] \tag{5.31}$$

$$\int_0^\infty |f(t)|\,dt < \infty \tag{5.32}$$

Table 5.2 Fourier Transform Pairs[*]

$x(t)$	$X(f)$
1	$\delta(f)$
$\delta(t)$	1
$u(t)$	$\frac{1}{2}\delta(f) + \frac{1}{j2\pi f}$
$\Pi(t/\tau)$	$\tau\,\text{sinc}(\tau f)$
$\text{sinc}(Bt)$	$\frac{1}{B}\Pi(f/B)$
$\Lambda(t/\tau)$	$\tau\,\text{sinc}^2(\tau f)$
$e^{-at}u(t)$	$\frac{1}{a + j2\pi f} \quad [a > 0]$
$te^{-at}u(t)$	$\frac{1}{(a + j2\pi f)^2} \quad [a > 0]$
$e^{-a\lvert t\rvert}$	$\frac{2a}{a^2 + (2\pi f)^2} \quad [a > 0]$
$e^{-(at)^2}$	$\frac{\sqrt{\pi}}{a} e^{-(\pi f/a)^2}$
$\cos(2\pi f_0 t + \theta)$	$\frac{1}{2}[e^{j\theta}\delta(f - f_0) + e^{-j\theta}\delta(f + f_0)]$
$\sin(2\pi f_0 t + \theta)$	$\frac{1}{2j}[e^{j\theta}\delta(f - f_0) - e^{-j\theta}\delta(f + f_0)]$
$\sum_{n=-\infty}^{n=+\infty} \delta(t - nT_s)$	$f_s \sum_{k=-\infty}^{k=+\infty} \delta(f - kf_s) \quad \left[f_s = \frac{1}{T_s}\right]$

[*]Although not explicitly defined in the NCEES *FE Reference Handbook* (*NCEES Handbook*), sinc(x) is an abbreviation for $\sin(x)/x$.

Table 5.3 Fourier Transform Pairs

$f(t)$	$F(\omega)$
$\delta(t)$	1
$u(t)$	$\pi\delta(\omega) + 1/j\omega$
$u\left(t + \frac{\tau}{2}\right) - u\left(t - \frac{\tau}{2}\right) = r_{\text{rect}}\frac{t}{\tau}$	$\tau\frac{\sin(\omega\tau/2)}{\omega\tau/2}$
$e^{j\omega_0 t}$	$2\pi\delta(\omega - \omega_0)$

Description

Table 5.3 gives some additional useful Fourier transform pairs.[2] Other pairs can be derived from the Laplace transform by replacing s in Table 5.2 with f, if the conditions given in Eq. 5.31 and Eq. 5.32 are met.

Example

The Fourier transform of an impulse $a^2\delta(t)$ of magnitude a^2 is equal to

(A) $\sqrt{a}$

(B) $a - 1$

(C) a

(D) a^2

Solution

The Fourier transform $X(f)$ of a given signal $x(t)$ is found from Eq. 5.29.

$$\begin{aligned} X(f) &= \int_{-\infty}^{+\infty} x(t)\,e^{-j2\pi ft}\,dt \\ &= \int_{-\infty}^{+\infty} a^2\delta(t)\,e^{-j2\pi ft}\,dt \\ &= a^2\int_{-\infty}^{+\infty} \delta(t)\,e^{-j2\pi ft}\,dt \end{aligned}$$

For $t = 0$, $x(t) = \delta(t) = 1$, and for all other values of t, $x(t) = 0$. This corresponds to the first line of Table 5.3.

$$\begin{aligned} X(f) &= a^2\int_{-\infty}^{+\infty} \delta(t)\,e^{-j2\pi ft}\,dt \\ &= a^2(1) \\ &= a^2 \end{aligned}$$

The answer is (D).

[2]While any variable can be used to designate any quantity, the *NCEES Handbook* uses an uncommon Fourier transform notation that may be confusing to some. A spatial or temporal function is usually described as $f(x)$ or $f(t)$, where f designates the function, and x or t is the independent variable. In that case, the Fourier transform of $f(t)$ would be designated as $F(\omega)$, where ω is an independent variable from the imaginary frequency domain. However, in Eq. 5.29 and Eq. 5.30, the *NCEES Handbook* uses x and X to designate the function and its transform, and f to designate an independent variable from the frequency domain, where $\omega = 2\pi f$. What would commonly be shown as $F(\omega)$ is shown as $X(f)$.

Table 5.4: Fourier Transform Theorems

Table 5.4 Fourier Transform Theorems

theorem	function	transform
linearity	$ax(t) + by(t)$	$aX(f) + bY(f)$
scale change	$x(at)$	$\frac{1}{\lvert a\rvert}X\left(\frac{f}{a}\right)$
time reversal	$x(-t)$	$X(-f)$
duality	$X(t)$	$x(-f)$
time shift	$x(t-t_0)$	$X(f)e^{-j2\pi ft_0}$
frequency shift	$x(t)e^{j2\pi f_0 t}$	$X(f-f_0)$
modulation	$x(t)\cos 2\pi f_0 t$	$\frac{1}{2}X(f-f_0) + \frac{1}{2}X(f+f_0)$
multiplication	$x(t)y(t)$	$X(f)^* Y(f)$
convolution	$x(t)^* y(t)$	$X(f)Y(f)$
differentiation	$\frac{d^n x(t)}{dt^n}$	$(j2\pi f)^n X(f)$
integration	$\int_{-\infty}^{t} x(\lambda)\,d\lambda$	$\frac{1}{j2\pi f}X(f) + \frac{1}{2}X(0)\delta(f)$

Description

Determining the Fourier transform of a complex mathematical function is simplified by various Fourier theorems, which are summarized in Table 5.4. While all are important, the simplest are the addition, linearity, and scale change (commonly referred to as *similarity*) theorems. In Table 5.4, the addition theorem is combined with the linearity theorem. The addition theorem states, not surprisingly, that the transform of a sum of functions is the sum of the transforms of the individual functions. In Table 5.4's nomenclature and format, this would be designated as

$$x(t) + y(t) \qquad X(f) + Y(f) \qquad \text{[addition]}$$

The asterisk symbol * is used to designate the *convolution operation*, which is not the same as multiplication. The convolution of two functions $x(t)$ and $y(t)$ is a third function defined as the integral of the product of one of the functions and the other function shifted by some given distance, x_0. The convolution essentially determines the amount of overlap between the functions when the functions are separated by x_0.

$$x(t)^* y(t) = \int_{-\infty}^{+\infty} x(t)y(t_0 - t)\,dt$$

6. LAPLACE TRANSFORMS

Traditional methods of solving nonhomogeneous differential equations by hand are usually difficult and/or time consuming. *Laplace transforms* can be used to reduce many solution procedures to simple algebra.

Equation 5.33: Laplace Transform

$$F(s) = \int_0^\infty f(t)e^{-st}dt \qquad 5.33$$

Description

Every mathematical function, $f(t)$, has a Laplace transform, written as $F(s)$ or $\mathcal{L}(s)$. The transform is written in the s-domain, regardless of the independent variable in the original function. The variable s is equivalent to a derivative operator, although it may be handled in the equations as a simple variable. Equation 5.33 converts a function into a Laplace transform.

Generally, it is unnecessary to actually obtain a function's Laplace transform by use of Eq. 5.33. Tables of these transforms are readily available (see Table 5.5).

Example

What is the Laplace transform of $f(t) = e^{-6t}$?

(A) $\frac{1}{s+6}$

(B) $\frac{1}{s-6}$

(C) e^{-6+s}

(D) e^{6+s}

Solution

The Laplace transform of a function, $F(s)$, can be calculated from the definition of a transform.

$$F(e^{-6t}) = \int_0^\infty e^{-(s+6)t}dt = -\frac{e^{-(s+6)t}}{s+6}\Bigg|_0^\infty = 0 - \left(-\frac{1}{s+6}\right)$$

$$= \frac{1}{s+6}$$

(This problem could have been solved more quickly by using a Laplace transform pair table, such as Table 5.5.)

The answer is (A).

Table 5.5: Laplace Transform Pairs

Table 5.5 Laplace Transforms

$f(t)$	$F(s)$
$\delta(t)$, impulse at $t=0$	1
$u(t)$, step at $t=0$	$1/s$
$t[u(t)]$, ramp at $t=0$	$1/s^2$
$e^{-\alpha t}$	$1/(s+\alpha)$
$te^{-\alpha t}$	$1/(s+\alpha)^2$
$e^{-\alpha t}\sin\beta t$	$\beta/[(s+\alpha)^2+\beta^2]$
$e^{-\alpha t}\cos\beta t$	$(s+\alpha)/[(s+\alpha)^2+\beta^2]$
$\dfrac{d^n f(t)}{dt^n}$	$s^n F(s)-\sum_{m=0}^{n-1} s^{n-m-1}\dfrac{d^m f(0)}{dt^m}$
$\int_0^t f(\tau)d\tau$	$(1/s)F(s)$
$\int_0^t x(t-\tau)h(t)d\tau$	$H(s)X(s)$
$f(t-\tau)u(t-\tau)$	$e^{-\tau s}F(s)$

Description

Table 5.5 gives common Laplace transforms.

Example 69

What is the Laplace transform of the step function $f(t)$?

$$f(t) = u(t-1) + u(t-2)$$

(A) $\dfrac{1}{s}+\dfrac{2}{s}$

(B) $\dfrac{e^{-s}+e^{-2s}}{s}$

(C) $1+\dfrac{e^{-2s}}{s}$

(D) $\dfrac{e^{s}}{s}+\dfrac{e^{2s}}{s}$

Solution

The notations $u(t-1)$ and $u(t-2)$ mean that a unit step input (a step of height 1) is applied at $t = 1$, and another unit step is applied at $t = 2$. (This function could be used to describe the terrain that a tracked robot would have to navigate to go up a flight of two stairs in a particular interval.) Table 5.5 contains Laplace transforms for various input functions, including steps. For steps at $t = 0$, the Laplace transform is $1/s$. However, in this example, the steps are encountered at $t = 1$ and $t = 2$. Superposition can be used to calculate the Laplace transform of the summation as the sum of the two transforms. Use the last entry in Table 5.5, with $f(t-\tau)=1$.

$$F(s) = F(u(t-1)) + F(u(t-2)) = \frac{e^{-s}}{s}+\frac{e^{-2s}}{s} = \frac{e^{-s}+e^{-2s}}{s}$$

The answer is (B).

Equation 5.34: Inverse Laplace Transform

$$f(t) = \frac{1}{2\pi j}\int_{\sigma-j\infty}^{\sigma+j\infty} F(s)e^{st}dt \qquad 5.34$$

Description

Extracting a function from its transform is the *inverse Laplace transform* operation. Although Eq. 5.34 could be used and other methods exist, this operation is almost always done using a table, such as Table 5.5.

Equation 5.35: Initial Value Theorem

$$\lim_{s\to\infty} sF(s) \qquad 5.35$$

Description

Equation 5.35 shows the *initial value theorem* (IVT).

Equation 5.36: Final Value Theorem

$$\lim_{s\to 0} sF(s) \qquad 5.36$$

Description

Equation 5.36 shows the *final value theorem* (FVT).

7. DIFFERENCE EQUATIONS

Equation 5.37: Difference Equation

$$f(t) = y' = \frac{y_{i+1}-y_i}{t_{i+1}-t_i} \qquad 5.37$$

Description

Many processes can be accurately modeled by differential equations. However, exact solutions to these models may be difficult to obtain. In such cases, discrete versions of the original differential equations can be produced. These discrete equations are known as *finite difference equations* or just *difference equations.*

Communication signal processing, heat transfer, and traffic flow are just a few of the applications of difference equations.

Difference equations are also ideal for modeling processes whose states or values are restricted to certain specified (equally spaced) points in time or space as is done with many simulation models.

A difference equation is a relationship between a function and its differences over some interval of integers. (This is analogous to a differential equation that is a relationship of functions and their derivatives over some interval of real numbers.) Any system with an input $v(t)$ and an output $y(t)$ defined only at the equally spaced intervals given by Eq. 5.37 can be described by a difference equation.

The *order* of the difference equation is the number of differences that are in the equation.

Although simple difference equations can be solved by hand, in practice, they are solved by computer using numerical analysis techniques.

Equation 5.38 and Eq. 5.39: First-Order Linear Difference Equation

$$\Delta t = t_{i+1} - t_i \qquad 5.38$$

$$y_{i+1} = y_i + y'(\Delta t) \qquad 5.39$$

Description

A *first-order difference equation* is a relationship between the values of some function at two consecutive points in time or space. The relationship can take on any form using any of the mathematical operators. For example, an additive relationship might be $y_{i+1} = y_i + 7$; a multiplicative relationship might be $y_{i+1} = 5y_i$; and an exponent relationship might be $y_{i+1} = y_i^2$. Equation 5.39 is a first-order linear difference equation that uses linear extrapolation to predict a subsequent curve point. For example, Eq. 5.39 can be interpreted as using the elevation of a projectile and slope of the path in one interval to predict the elevation reached by the projectile in the next interval.

Second-Order Difference Equation of the Fibonacci Sequence

A *second-order difference equation* is a relationship between the values of some function at three consecutive points in time or space. The relationship can take on any form using any of the mathematical operators. For example, an additive relationship might be $y_{i+1} = y_i + y_{i-1} - 2$; a multiplicative relationship might be $y_{i+1} = 2y_iy_{i-1}$; and an exponent relationship might be $y_{i+1} = y_i^2 + 2y_{i-1}$. An additive second-order difference equation that describes the Fibonacci sequence (where each term is the sum of the previous two terms) is $F_{i+1} = F_i + F_{i-1}$.

$$y(k) = y(k-1) + y(k-2)$$

$$f(k+2) = f(k+1) + f(k) \quad [f(0) = 1 \text{ and } f(1) = 1]$$

Topic II: Probability and Statistics

Chapter

6 Probability and Statistics

1. SET THEORY

A *set* (usually designated by a capital letter) is a population or collection of individual items known as *elements* or *members*. The *null set*, $\varnothing$, is empty (i.e., contains no members). If A and B are two sets, A is a *subset* of B if every member in A is also in B. A is a *proper subset* of B if B consists of more than the elements in A. These relationships are denoted as follows.[1]

$$A \subseteq B \text{ [subset]}$$

$$A \subset B \text{ [proper subset]}$$

The *universal set*, U, is one from which other sets draw their members. If A is a subset of U, then $\overline{A}$ (also designated as A', A^{-1}, $\tilde{A}$, and $-A$) is the *complement* of A and consists of all elements in U that are not in A. This is illustrated in a *Venn diagram* in Fig. 6.1(a).

The *union of two sets*, denoted by $A \cup B$ and shown in Fig. 6.1(b), is the set of all elements that are either in A or B or both. The *intersection of two sets*, denoted by $A \cap B$ and shown in Fig. 6.1(c), is the set of all elements that belong to both A and B. If $A \cap B = \varnothing$, A and B are said to be *disjoint sets*.

Figure 6.1 Venn Diagrams

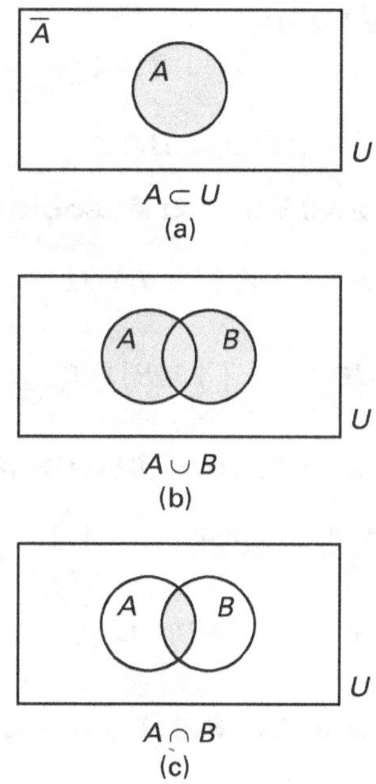

If A, B, and C are subsets of the universal set, the following laws apply.

Identity Laws

$$A \cup \varnothing = A$$

$$A \cup U = U$$

$$A \cap \varnothing = \varnothing$$

$$A \cap U = A$$

Idempotent Laws

$$A \cup A = A$$

$$A \cap A = A$$

[1]The *NCEES FE Reference Handbook* (*NCEES Handbook*) is inconsistent in its representation of sets, set members, matrices, matrix elements, and relations. Uppercase and lowercase, bold and not bold, italic, and not italic are all used interchangeably. For example, uppercase italic letters are used for set theory, while nonitalic letters are used in discrete math. In order to present these subjects in the same chapter, this book has adopted a consistent presentation style that may differ somewhat from the *NCEES Handbook*.

Complement Laws

$$A \cup \overline{A} = U$$

$$\overline{(\overline{A})} = A$$

$$A \cap \overline{A} = \varnothing$$

$$\overline{U} = \varnothing$$

Commutative Laws

$$A \cup B = B \cup A$$

$$A \cap B = B \cap A$$

Equation 6.1 and Eq. 6.2: Associative Laws

$$A \cup (B \cup C) = (A \cup B) \cup C \qquad 6.1$$

$$A \cap (B \cap C) = (A \cap B) \cap C \qquad 6.2$$

Equation 6.3 and Eq. 6.4: Distributive Laws

$$A \cup (B \cap C) = (A \cup B) \cap (A \cup C) \qquad 6.3$$

$$A \cap (B \cup C) = (A \cap B) \cup (A \cap C) \qquad 6.4$$

Equation 6.5 and Eq. 6.6: De Morgan's Laws

$$\overline{A \cup B} = \overline{A} \cap \overline{B} \qquad 6.5$$

$$\overline{A \cap B} = \overline{A} \cup \overline{B} \qquad 6.6$$

2. COMBINATIONS AND PERMUTATIONS

There are a finite number of ways in which n elements can be combined into distinctly different groups of r items. For example, suppose a farmer has a chicken, a rooster, a duck, and a cage that holds only two birds. The possible *combinations* of three birds taken two at a time are (chicken, rooster), (chicken, duck), and (rooster, duck). The birds in the cage will not remain stationary, so the combination (rooster, chicken) is not distinctly different from (chicken, rooster). That is, the combinations are not *order conscious*.

Equation 6.7: Combinations

$$C(n,r) = \frac{P(n,r)}{r!} = \frac{n!}{r!\,(n-r)!} \qquad 6.7$$

Description

The number of *combinations* of n items taken r at a time is written $C(n, r)$, C_r^n, nC_r, ${}_nC_r$, or $\binom{n}{r}$ (pronounced "n choose r"). It is sometimes referred to as the *binomial coefficient* and is given by Eq. 6.7.

Example 1

Six design engineers are eligible for promotion to pay grade G8, but only four spots are available. How many different combinations of promoted engineers are possible?

(A) 4

(B) 6

(C) 15

(D) 20

Solution

The number of combinations of $n = 6$ items taken $r = 4$ items at a time is

$$\begin{aligned} C(6,4) &= \frac{n!}{r!\,(n-r)!} = \frac{6!}{4!\,(6-4)!} \\ &= \frac{6 \times 5 \times 4 \times 3 \times 2 \times 1}{4 \times 3 \times 2 \times 1 \times 2 \times 1} \\ &= 15 \end{aligned}$$

The answer is (C).

Equation 6.8: Permutations

$$P(n, r) = \frac{n!}{(n-r)!} \qquad 6.8$$

Description

An order-conscious subset of r items taken from a set of n items is the *permutation*, $P\,(n, r)$, also written P_r^n, ${}_nP_r$, and nP_r. A permutation is order conscious because the arrangement of two items (e.g., a_i and b_i) as a_ib_i is different from the arrangement b_ia_i. The number of permutations is found from Eq. 6.8.

Example 2

An identification code begins with three letters. The possible letters are A, B, C, D, and E. If none of the letters are used more than once, how many different ways can the letters be arranged to make a code?

(A) 10

(B) 20

(C) 40

(D) 60

Solution

Since the order of the letters affects the identification code, determine the number of permutations of $n = 5$ items taken $r = 3$ items at a time using Eq. 6.8.

$$\begin{aligned} P(5,3) &= \frac{n!}{(n-r)!} = \frac{5!}{(5-3)!} \\ &= \frac{5 \times 4 \times 3 \times 2 \times 1}{2 \times 1} \\ &= 60 \end{aligned}$$

The answer is (D).

Equation 6.9: Permutations of Different Object Types

$$P(n; n_1, n_2, \ldots, n_k) = \frac{n!}{n_1!\, n_2! \ldots n_k!} \qquad 6.9$$

Description

Suppose n_1 objects of one type (e.g., color, size, shape, etc.) are combined with n_2 objects of another type and n_3 objects of yet a third type, and so on, up to k types. The collection of $n = n_1 + n_2 + \cdots + n_k$ objects forms a population from which arrangements of n items can be formed. The number of permutations of n objects taken n at a time from a collection of k types of objects is given by Eq. 6.9.

Example 3

An urn contains 13 marbles total: 4 black marbles, 2 red marbles, and 7 yellow marbles. Arrangements of 13 marbles are made. Most nearly, how many unique ways can the 13 marbles be ordered (arranged)?

(A) 800

(B) 1200

(C) 14,000

(D) 26,000

Solution

The marble colors represent different types of objects. The number of permutations of the marbles taken 13 at a time is

$$\begin{aligned} P(13; 4, 2, 7) &= \frac{n!}{n_1!\, n_2! \ldots n_k!} = \frac{13!}{4!\, 2!\, 7!} \\ &= \frac{13 \times 12 \times 11 \times 10 \times 9 \times 8 \times 7 \times 6 \times 5 \times 4 \times 3 \times 2 \times 1}{4 \times 3 \times 2 \times 1 \times 2 \times 1 \times 7 \times 6 \times 5 \times 4 \times 3 \times 2 \times 1} \\ &= 25{,}740 \quad (26{,}000) \end{aligned}$$

The answer is (D).

3. LAWS OF PROBABILITY

Probability theory determines the relative likelihood that a particular event will occur. An *event*, E, is one of the possible outcomes of a *trial.* The *probability* of E occurring is denoted as $P(E)$.

Probabilities are real numbers in the range of zero to one. If an event E is certain to occur, then the probability $P(E)$ of the event is equal to one. If the event is certain *not* to occur, then the probability $P(E)$ of the event is equal to zero. The probability of any other event is between zero and one.

The probability of an event occurring is equal to one minus the probability of the event not occurring. This is known as a *complementary probability.*

$$P(E) = 1 - P(\text{not } E)$$

Complementary probability can be used to simplify some probability calculations. For example, calculation of the probability of numerical events being "greater than" or "less than" or quantities being "at least" a certain number can often be simplified by calculating the probability of the complementary event.

Probabilities of multiple events can be calculated from the probabilities of individual events using a variety of methods. When multiple events are considered, those events can be either independent or dependent. The probability of an *independent event* does not affect (and is not affected by) other events. The assumption of independence is appropriate when sampling from infinite or very large populations, when sampling from finite populations with replacement, or when sampling from different populations (universes). For example, the outcome of a second coin toss is generally not affected by the outcome of the first coin toss. The probability of a *dependent event* is affected by what has previously happened. For example, drawing a second card from a deck of cards without replacement is affected by what was drawn as the first card.

Probability/Statistics

Events can be combined in two basic ways, according to the way the combination is described. Events can be connected by the words "and" and "or." For example, the question, "What is the probability of event A and event B occurring?" is different than the question, "What is the probability of event A or event B occurring?" The combinatorial "and" is designated in various ways: AB, $A \cdot B$, $A \times B$, $A \cap B$, and A, B, among others. In this book, the probability of A and B both occurring is designated as $P(A, B)$.

The combinatorial "or" is designated as: $A + B$ and $A \cup B$. In this book, the probability of A or B occurring is designated as $P(A + B)$.

Equation 6.10: Law of Total Probability

$$P(A+B) = P(A) + P(B) - P(A,B) \qquad 6.10$$

Description

Equation 6.10 gives the probability that either event A or B will occur. $P(A, B)$ is the probability that both A and B will occur.

Example

A deck of ten children's cards contains three fish cards, two dog cards, and five cat cards. What is the probability of drawing either a cat card or a dog card from a full deck?

(A) 1/10

(B) 2/10

(C) 5/10

(D) 7/10

Solution

The two events are mutually exclusive, so the probability of both happening, $P(A, B)$, is zero. The total probability of drawing either a cat card or a dog card is

$$\begin{aligned} P(A+B) &= P(A) + P(B) - P(A,B) \\ &= \frac{5}{10} + \frac{2}{10} - 0 \\ &= 7/10 \end{aligned}$$

The answer is (D).

Equation 6.11: Law of Compound (Joint) Probability

$$P(A, B) = P(A)P(B|A) = P(B)P(A|B) \qquad 6.11$$

Variation

$$P(A, B) = P(A)P(B) \quad \left[\begin{array}{c}\text{independent}\\ \text{events}\end{array}\right]$$

Description

Equation 6.11, the *law of compound (joint) probability*, gives the probability that events A and B will both occur. $P(B|A)$ is the *conditional probability* that B will occur given that A has already occurred. Likewise, $P(A|B)$ is the conditional probability that A will occur given that B has already occurred. It is possible that the events come from different populations (universes, sample spaces, etc.), such as when one marble is drawn from one urn and another marble is drawn from a different urn. In that case, the events will be independent and won't affect each other. If the events are independent, then $P(B|A) = P(B)$ and $P(A|B) = P(A)$. Examples of dependent events for which the probability is conditional include drawing objects from a container or cards from a deck, without replacement.

Example

A bag contains seven orange balls, eight green balls, and two white balls. Two balls are drawn from the bag without replacing either of them. Most nearly, what is the probability that the first ball drawn is white and the second ball drawn is orange?

(A) 0.036

(B) 0.052

(C) 0.10

(D) 0.53

Solution

There is a total of 17 balls. There are 2 white balls. The probability of picking a white ball as the first ball is

$$P(A) = 2/17$$

After picking a white ball first, there are 16 balls remaining, 7 of which are orange. The probability of picking an orange ball second given that a white ball was chosen first is

$$P(B|A) = 7/16$$

The probability of picking a white ball first and an orange ball second is

$$\begin{aligned} P(A,B) &= P(A)P(B|A) \\ &= \left(\frac{2}{17}\right)\left(\frac{7}{16}\right) \\ &= 0.05147 \quad (0.052) \end{aligned}$$

The answer is (B).

Equation 6.12: Bayes' Theorem

$$P(B_j|A) = \frac{P(B_j)P(A|B_j)}{\sum_{i=1}^{n} P(A|B_i)P(B_i)} \qquad 6.12$$

Variation

$$P(B_j|A) = \frac{P(B \text{ and } A)}{P(A)}$$

Description

Given two dependent sets of events, A and B, the probability that event B will occur given the fact that the dependent event A has already occurred is written as $P(B_j|A)$ and is given by *Bayes' theorem*, Eq. 6.12.

Example

A medical patient exhibits a symptom that occurs naturally 10% of the time in all people. The symptom is also exhibited by all patients who have a particular disease. The incidence of that particular disease among all people is 0.0002%. What is the probability of the patient having that particular disease?

(A) 0.002%

(B) 0.01%

(C) 0.3%

(D) 4%

Solution

This problem is asking for a conditional probability: the probability that a person has a disease, D, given that the person has a symptom, S. Use Bayes' theorem to calculate the probability. The probability that a person has the symptom S given that they have the disease D is $P(S|D)$ and is 100%. Multiply by 100% to get the answer as a percentage.

$$\begin{aligned} P(D|S) &= \frac{P(D)P(S|D)}{P(S|D)P(D) + P(S|\text{ not } D)P(\text{not } D)} \\ &= \frac{(0.000002)(1.00)}{(1.00)(0.000002) + (0.10)(0.999998)} \\ &= 0.00002 \quad (0.002\%) \end{aligned}$$

The answer is (A).

4. MEASURES OF CENTRAL TENDENCY

It is often unnecessary to present experimental data in their entirety, either in tabular or graphic form. In such cases, the data and distribution can be represented by various parameters. One type of parameter is a measure of *central tendency*. The mode, median, and mean are measures of central tendency.

Mode

The *mode* is the observed value that occurs most frequently. The mode may vary greatly between series of observations; its main use is as a quick measure of the central value, since little or no computation is required to find it. Beyond this, the usefulness of the mode is limited.

Median

The *median* is the point in the distribution that partitions the total set of observations into two parts containing equal numbers of observations. It is not influenced by the extremity of scores on either side of the distribution. The median is found by counting from either end through an ordered set of data until half of the observations have been accounted for. If the number of data points is odd, the median will be the exact middle value. If the number of data points is even, the median will be the average of the middle two values.

Equation 6.13: Arithmetic Mean

$$\overline{X} = (1/n)(X_1 + X_2 + \cdots + X_n) = (1/n)\sum_{i=1}^{n} X_i \qquad 6.13$$

Variation

$$\overline{X} = \frac{\sum f_i X_i}{\sum f_i} \quad \left[\begin{array}{c} f_i \text{ are frequencies} \\ \text{of occurrence of} \\ \text{events } i \end{array}\right]$$

Probability/ Statistics

Description

The *arithmetic mean* is the arithmetic average of the observations. The *sample mean*, $\overline{X}$, can be used as an unbiased estimator of the *population mean*, μ. The term *unbiased estimator* means that on the average, the sample mean is equal to the population mean. The mean may be found without ordering the data (as was necessary to find the mode and median) from Eq. 6.13.

Example 7

100 random samples were taken from a large population. A particular numerical characteristic of sampled items was measured. The results of the measurements were as follows.

- 45 measurements were between 0.859 and 0.900.
- 0.901 was observed once.
- 0.902 was observed three times.
- 0.903 was observed twice.
- 0.904 was observed four times.
- 45 measurements were between 0.905 and 0.958.

The smallest value was 0.859, and the largest value was 0.958. The sum of all 100 measurements was 91.170. Except those noted, no measurements occurred more than twice.

What are the (a) mean, (b) mode, and (c) median of the measurements, respectively?

(A) 0.908; 0.902; 0.902

(B) 0.908; 0.904; 0.903

(C) 0.912; 0.902; 0.902

(D) 0.912; 0.904; 0.903

Solution

(a) From Eq. 6.13, the arithmetic mean is

$$\overline{X} = (1/n)\sum_{i=1}^{n} X_i = \left(\frac{1}{100}\right)(91.170) = 0.9117 \quad (0.912)$$

(b) The mode is the value that occurs most frequently. The value of 0.904 occurred four times, and no other measurements repeated more than four times. 0.904 is the mode.

(c) The median is the value at the midpoint of an ordered (sorted) set of measurements. There were 100 measurements, so the middle of the ordered set occurs between the 50th and 51st measurements. Since these measurements are both 0.903, the average of the two is 0.903.

The answer is (D).

Equation 6.14: Weighted Arithmetic Mean

$$\overline{X}_w = \frac{\sum w_i X_i}{\sum w_i} \qquad 6.14$$

Description

If some observations are considered to be more significant than others, a *weighted mean* can be calculated. Equation 6.14 defines a *weighted arithmetic mean*, $\overline{X}_w$, where w_i is the weight assigned to observation X_i.

Example 8

A course has four exams that compose the entire grade for the course. Each exam is weighted. A student's scores on all four exams and the weight for each exam are as given.

exam	student score	weight
1	80%	1
2	95%	2
3	72%	2
4	95%	5

What is most nearly the student's final grade in the course?

(A) 82%

(B) 85%

(C) 87%

(D) 89%

Solution

The student's final grade is the weighted arithmetic mean of the individual exam scores.

$$\begin{aligned}\overline{X}_w &= \frac{\sum w_i X_i}{\sum w_i} \\ &= \frac{(1)(80\%) + (2)(95\%) + (2)(72\%) + (5)(95\%)}{1+2+2+5} \\ &= 88.9\% \quad (89\%)\end{aligned}$$

The answer is (D).

Equation 6.15: Geometric Mean

$$\text{sample geometric mean} = \sqrt[n]{X_1 X_2 X_3 \ldots X_n} \qquad 6.15$$

Description

The *geometric mean* of n nonnegative values is defined by Eq. 6.15. The geometric mean is the number that, when raised to the power of the sample size, produces the same result as the product of all samples. It is appropriate to use the geometric mean when the values being averaged are used as consecutive multipliers in other calculations. For example, the total revenue earned on an investment of C earning an effective interest rate of i_k in year k is calculated as $R = C(i_1 i_2 i_3 \ldots i_k)$. The interest rate, i, is a multiplicative element. If a \$100 investment earns 10% in year 1 (resulting in \$110 at the end of the year), then the \$110 earns 30% in year 2 (resulting in \$143), and the \$143 earns 50% in year 3 (resulting in \$215), the average interest earned each year would not be the arithmetic mean of (10% + 30%+ 50%)/3 = 30%. The average would be calculated as a geometric mean (24.66%).

Example

What is most nearly the geometric mean of the following data set?

$$0.820, 1.96, 2.22, 0.190, 1.00$$

(A) 0.79

(B) 0.81

(C) 0.93

(D) 0.96

Solution

The geometric mean of the data set is

$$\begin{aligned}\text{sample geometric mean} &= \sqrt[n]{X_1 X_2 X_3 \ldots X_n} \\ &= \sqrt[5]{\begin{matrix}(0.820)(1.96)(2.22) \\ \times (0.190)(1.00)\end{matrix}} \\ &= 0.925 \quad (0.93)\end{aligned}$$

The answer is (C).

Equation 6.16: Root-Mean-Square

$$\text{sample root-mean-square value} = \sqrt{(1/n)\sum X_i^2} \qquad 6.16$$

Description

The *root-mean-square* (rms) value of a series of observations is defined by Eq. 6.16. The variable X_{rms} is often used to represent the rms value.

Example

The water level on a tank in a chemical plant is measured every six hours. The tank has a depth of 6 m. The water levels on the tank on a certain day were found to be 2.5 m, 4.2 m, 5.6 m, and 3.3 m. What is most nearly the root-mean-square value of water level for that day?

(A) 2.0 m

(B) 3.3 m

(C) 4.1 m

(D) 5.8 m

Solution

Use Eq. 6.16 to find the root-mean-square value of water level for the day.

$$\begin{aligned}X_{\text{rms}} &= \sqrt{(1/n)\sum X_i^2} \\ &= \sqrt{\left(\frac{1}{4}\right)\left(\begin{matrix}(2.5\text{ m})^2 + (4.2\text{ m})^2 \\ +(5.6\text{ m})^2 + (3.3\text{ m})^2\end{matrix}\right)} \\ &= 4.07\text{ m} \quad (4.1\text{ m})\end{aligned}$$

The answer is (C).

5. MEASURES OF DISPERSION

Measures of dispersion describe the variability in observed data.

Equation 6.17 Through Eq. 6.21: Standard Deviation

$$\sigma_{\text{population}} = \sqrt{(1/N)\sum (X_i - \mu)^2} \qquad 6.17$$

$$\sigma_{\text{sum}} = \sqrt{\sigma_1^2 + \sigma_2^2 + \cdots + \sigma_n^2} \qquad 6.18$$

$$\sigma_{\text{series}} = \sigma\sqrt{n} \qquad 6.19$$

$$\sigma_{\text{mean}} = \frac{\sigma}{\sqrt{n}} \qquad 6.20$$

$$\sigma_{\text{product}} = \sqrt{A^2\sigma_b^2 + B^2\sigma_a^2} \qquad 6.21$$

Variation

$$\sigma = \sqrt{\frac{\sum f_i (X_i - \mu)^2}{\sum f_i}}$$

Description

One measure of dispersion is the *standard deviation*, defined in Eq. 6.17. N is the total population size, not the sample size, n. This implies that the entire population is measured.

Equation 6.17 can be used to calculate the standard deviation only when the entire population can be included in the calculation. When only a small subset is available, as when a sample is taken (see Eq. 6.22), there are two obstacles to its use. First, the population mean, μ, is not known. This obstacle is overcome by using the sample average, $\overline{X}$, which is an unbiased estimator of the population mean. Second, Eq. 6.17 is inaccurate for small samples.

When combining two or more data sets for which the standard deviations are known, the standard deviation for the combined data is found using Eq. 6.18. This equation is used even if some of the data sets are subtracted; subtracting one data set from another increases the standard deviation of the result just as adding the two data sets does.

When a series of samples is taken from the same population, the sum of the standard deviations for the series is calculated from Eq. 6.19, where σ is the population standard deviation and n is the number of samples. The standard deviation of the mean values of these samples is called the *standard deviation* (or *standard error*) *of the mean* and is found with Eq. 6.20.

The standard deviation of the product of two random variables is given by Eq. 6.21. A and B are the expected values of the two variables, and σ_a^2 and σ_b^2 are the population variances for the two variables.

Example

A cat colony living in a small town has a total population of seven cats. The ages of the cats are as shown.

age	number
7 yr	1
8 yr	1
10 yr	2
12 yr	1
13 yr	2

What is most nearly the standard deviation of the age of the cat population?

(A) 1.7 yr

(B) 2.0 yr

(C) 2.2 yr

(D) 2.4 yr

Solution

Using Eq. 6.13, the arithmetic mean of the ages is the population mean, μ.

$$\begin{aligned}\mu &= (1/n)\sum_{i=1}^{n} X_i \\ &= \left(\frac{1}{7}\right)\begin{pmatrix}(1)(7\text{ yr}) + (1)(8\text{ yr}) + (2)(10\text{ yr}) \\ +(1)(12\text{ yr}) + (2)(13\text{ yr})\end{pmatrix} \\ &= 10.4\text{ yr}\end{aligned}$$

From Eq. 6.17, the standard deviation of the ages is

$$\begin{aligned}\sigma_{\text{population}} &= \sqrt{(1/N)\sum (X_i - \mu)^2} \\ &= \sqrt{\left(\frac{1}{7}\right)\begin{pmatrix}(7\text{ yr} - 10.4\text{ yr})^2 \\ +(8\text{ yr} - 10.4\text{ yr})^2 \\ +(2)(10\text{ yr} - 10.4\text{ yr})^2 \\ +(12\text{ yr} - 10.4\text{ yr})^2 \\ +(2)(13\text{ yr} - 10.4\text{ yr})^2\end{pmatrix}} \\ &= 2.19\text{ yr} \quad (2.2\text{ yr})\end{aligned}$$

The answer is (C).

Equation 6.22: Sample Standard Deviation

$$s = \sqrt{[1/(n-1)]\sum_{i=1}^{n}(X_i - \overline{X})^2} \qquad 6.22$$

Description

The *standard deviation of a sample* (particularly a small sample) of n items calculated from Eq. 6.17 is a *biased estimator* of (i.e., on the average, it is not equal to) the population standard deviation. A different measure of dispersion, called the *sample standard deviation*, s (not the same as the standard deviation of a sample), is an unbiased estimator of the population standard deviation. The sample standard deviation can be found using Eq. 6.22.

Example 12

Samples of aluminum-alloy channels were tested for stiffness. The following distribution of results was obtained.

stiffness	frequency
2480	23
2440	35
2400	40
2360	33
2320	21

If the mean of the samples is 2400, what is the approximate standard deviation of the population from which the samples are taken?

(A) 48.2

(B) 49.7

(C) 50.6

(D) 50.8

Solution

The number of samples is

$$n = 23 + 35 + 40 + 33 + 21 = 152$$

The sample standard deviation, s, is the unbiased estimator of the population standard deviation, σ.

$$s = \sqrt{[1/(n-1)]\sum_{i=1}^{n}(X_i - \overline{X})^2}$$

$$= \sqrt{\left(\frac{1}{152-1}\right)\begin{pmatrix}(23)(2480-2400)^2 \\ +(35)(2440-2400)^2 \\ +(40)(2400-2400)^2 \\ +(33)(2360-2400)^2 \\ +(21)(2320-2400)^2\end{pmatrix}}$$

$$= 50.85 \quad (50.8)$$

The answer is (D).

Equation 6.23 Through Eq. 6.25: Variance and Sample Variance

$$\sigma^2 = (1/N)[(X_1 - \mu)^2 + (X_2 - \mu)^2 + \cdots + (X_N - \mu)^2] \quad 6.23$$

$$\sigma^2 = (1/N)\sum_{i=1}^{N}(X_i - \mu)^2 \quad 6.24$$

$$s^2 = [1/(n-1)]\sum_{i=1}^{n}(X_i - \overline{X})^2 \quad 6.25$$

Description

The *variance* is the square of the standard deviation. Since there are two standard deviations, there are two variances. The *variance of the population* (i.e., the *population variance*) is σ^2, and the *sample variance* is s^2. The population variance can be found using either Eq. 6.23 or Eq. 6.24, both derived from Eq. 6.17, and the sample variance can be found using Eq. 6.25, derived from Eq. 6.22.

Example 13

Most nearly, what is the sample variance of the following data set?

$$2, 4, 6, 8, 10, 12, 14$$

(A) 4.3

(B) 5.2

(C) 8.0

(D) 19

Solution

Find the mean using Eq. 6.13.

$$\overline{X} = (1/n)\sum_{i=1}^{n}X_i = \left(\frac{1}{7}\right)(2+4+6+8+10+12+14)$$
$$= 8$$

From Eq. 6.25, the sample variance is

$$s^2 = [1/(n-1)]\sum_{i=1}^{n}(X_i - \overline{X})^2$$

$$= \left(\frac{1}{7-1}\right)\begin{pmatrix}(2-8)^2 + (4-8)^2 + (6-8)^2 \\ +(8-8)^2 + (10-8)^2 + (12-8)^2 \\ +(14-8)^2\end{pmatrix}$$

$$= 18.67 \quad (19)$$

The answer is (D).

Probability/ Statistics

Equation 6.26: Sample Coefficient of Variation

$$CV = s/\overline{X} \qquad 6.26$$

Description

The *relative dispersion* is defined as a measure of dispersion divided by a measure of central tendency. The *sample coefficient of variation*, *CV*, is a relative dispersion calculated from the sample standard deviation and the mean.

Example

The following data were recorded from a laboratory experiment.

$$20, 25, 30, 32, 27, 22$$

The mean of the data is 26. What is most nearly the sample coefficient of variation of the data?

(A) 0.18

(B) 1.1

(C) 2.4

(D) 4.6

Solution

Find the sample standard deviation of the data using Eq. 6.22.

$$\begin{aligned} s &= \sqrt{[1/(n-1)]\sum_{i=1}^{n}(X_i - \overline{X})^2} \\ &= \sqrt{\left(\frac{1}{6-1}\right)\left(\begin{array}{c}(20-26)^2+(25-26)^2 \\ +(30-26)^2+(32-26)^2 \\ +(27-26)^2+(22-26)^2\end{array}\right)} \\ &= 4.6 \end{aligned}$$

From Eq. 6.26, the sample coefficient of variation is

$$CV = s/\overline{X} = \frac{4.6}{26} = 0.177 \quad (0.18)$$

The answer is (A).

6. NUMERICAL EVENTS

A *discrete numerical event* is an occurrence that can be described by an integer. For example, 27 cars passing through a bridge toll booth in an hour is a discrete numerical event. Most numerical events are *continuously distributed* and are not constrained to discrete or integer values. For example, the resistance of a 10% 1 Ω resistor may be any value between 0.9 Ω and 1.1 Ω.

7. PROBABILITY DENSITY FUNCTIONS (DISCRETE)

Equation 6.27: Probability Mass Function

$$f(x_k) = P(X = x_k) \quad [k = 1,2,\ldots,n] \qquad 6.27$$

Description

A *discrete random variable*, X, can take on values from a set of discrete values, x_i. The set of values can be finite or infinite, as long as each value can be expressed as an integer. The *probability mass function*, defined by Eq. 6.27, gives the probability that a discrete random variable, X, is equal to each of the set's possible values, x_k. The probabilities of all possible outcomes add up to unity.

Equation 6.28: Probability Density Function

$$P(a \le X \le b) = \int_a^b f(x)\,dx \qquad 6.28$$

Description

A *density function* is a nonnegative function whose integral taken over the entire range of the independent variable is unity. A *probability density function* (PDF) is a mathematical formula that gives the probability of a numerical event.

Various mathematical models are used to describe probability density functions. Figure 6.2 shows a graph of a continuous probability density function. The area under the probability density function is the probability that the variable will assume a value between the limits of evaluation. The total probability, or the probability that the variable will assume any value over the interval, is 1.0. The probability of an exact numerical event is zero. That is, there is no chance that a numerical event will be exactly a. It is possible to determine only the probability that a numerical event will be less than a, greater than b, or between the values of a and b.

Figure 6.2 *Probability Density Function*

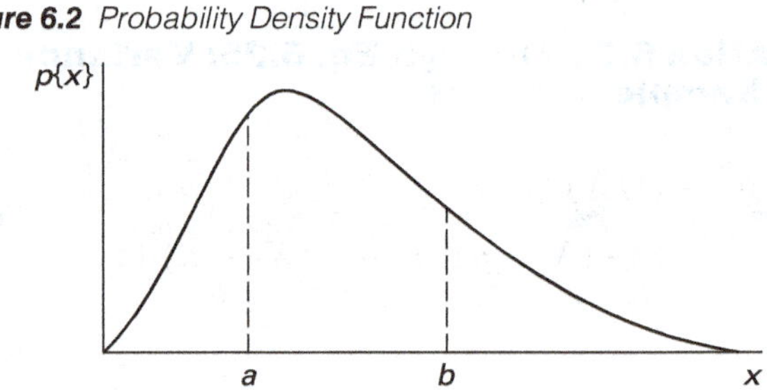

If a random variable, X, is continuous over an interval, then a nonnegative *probability density function* of that variable exists over the interval as defined by Eq. 6.28.

8. PROBABILITY DISTRIBUTION FUNCTIONS (CONTINUOUS)

A *cumulative probability distribution function*, $F(x)$, gives the probability that a numerical event will occur or the probability that the numerical event will be less than or equal to some value, x.

Equation 6.29: Cumulative Distribution Function: Discrete Random Variable

$$F(x_m) = \sum_{k=1}^{m} P(x_k) = P(X \le x_m) \quad [m = 1, 2, \ldots, n] \qquad 6.29$$

Description

For a *discrete random variable*, X, the probability distribution function is the sum of the individual probabilities of all possible events up to and including event x_m. The *cumulative distribution function* (CDF) is a function that calculates the cumulative sum of all values up to and including a particular end point. For discrete probability density functions (PDFs), $F(x_m)$, the CDF can be calculated as a summation, as shown in Eq. 6.29.

Because calculating cumulative probabilities can be cumbersome, tables of values are often used. Table 6.1 at the end of this chapter gives values for cumulative binomial probabilities, where n is the number of trials, P is the probability of success for a single trial, and x is the maximum number of successful trials.

Equation 6.30: Cumulative Distribution Function: Continuous Random Variable

$$F(x) = \int_{-\infty}^{x} f(t)\,dt \qquad 6.30$$

Description

For continuous functions, the CDF is calculated as an integral of the PDF from minus infinity to the limit of integration, as in Eq. 6.30. This integral corresponds to the area under the curve up to the limit of integration and represents the probability that the variable is less than or equal to the limit of integration. That is, $F(x) = P(x \le a)$. A CDF has a maximum value of 1.0, and for a continuous probability density function, $F(x)$ will approach 1.0 asymptotically.

Example

For the probability density function shown, what is the probability of the random variable x being less than 1/3?

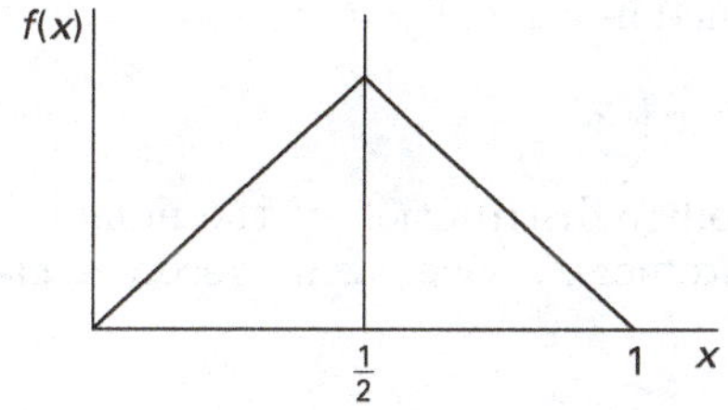

(A) 0.11

(B) 0.22

(C) 0.25

(D) 0.33

Solution

The total area under the probability density function is equal to 1. The area of two triangles is

$$A = (2)\left(\frac{1}{2}\right)bh = (2)\left(\frac{1}{2}\right)\left(\frac{1}{2}\right)h = 1$$

Therefore, the height of the curve at its peak is 2.

The equation of the line from $x = 0$ up to $x = 1/2$ is

$$f(x) = 4x \quad \left[0 \le x \le \frac{1}{2}\right]$$

The probability that $x < 1/3$ is equal to the area under the curve between 0 and 1/3. From Eq. 6.30,

$$\begin{aligned} F\left(0 < x < \frac{1}{3}\right) &= \int_0^{1/3} f(x)\,dx = \int_0^{1/3} 4x\,dx = 2x^2\Big|_0^{1/3} \\ &= (2)\left(\frac{1}{3}\right)^2 - 0 \\ &= 0.222 \quad (0.22) \end{aligned}$$

The answer is (B).

9. EXPECTED VALUES

Equation 6.31: Expected Value of a Discrete Variable

$$\mu = E[X] = \sum_{k=1}^{n} x_k f(x_k) \qquad 6.31$$

Description

The *expected value*, E, of a discrete random variable, X, is given by Eq. 6.31. $f(x_k)$ is the probability mass function as defined in Eq. 6.27.

Example

The probability distribution of the number of calls, X, that a customer service agent receives each hour is shown.

x	$f(x)$
0	0.00
2	0.04
4	0.05
6	0.10
8	0.35
10	0.46

What is most nearly the average number of phone calls that a customer service agent expects to receive in an hour?

(A) 5

(B) 7

(C) 8

(D) 9

Solution

The expected number of received calls is

$$\begin{aligned}\mu = E[X] &= \sum_{k=1}^{n} x_k f(x_k) \\ &= (0)(0.00) + (2)(0.04) + (4)(0.05) \\ &\quad +(6)(0.10) + (8)(0.35) + (10)(0.46) \\ &= 8.28 \quad (8)\end{aligned}$$

The answer is (C).

Equation 6.32: Variance of a Discrete Variable

$$\sigma^2 = V[X] = \sum_{k=1}^{n} (x_k - \mu)^2 f(x_k) \qquad 6.32$$

Description

Equation 6.32 gives the variance, σ^2, of a discrete function of variable X. To use Eq. 6.32, the population mean, μ, must be known, having been calculated from the total population of n values. The name "discrete" requires only that n be a finite number and all values of x be known. It does not limit the values of x to integers.

Equation 6.33 and Eq. 6.34: Expected Value (Mean) of a Continuous Variable

$$\mu = E[X] = \int_{-\infty}^{\infty} x f(x)\,dx \qquad 6.33$$

$$E[Y] = E[g(X)] = \int_{-\infty}^{\infty} g(x) f(x)\,dx \qquad 6.34$$

Description

Equation 6.33 calculates the population mean, μ, of a continuous variable, X, from the probability density function, $f(x)$. Equation 6.34 calculates the mean of any continuously distributed variable defined by $Y = g(x)$, whose values are observed according to the probabilities given by the probability density function (PDF) $f(x)$. Equation 6.34 is the general form of Eq. 6.33, where $g(x) = x$.

Equation 6.35: Variance of a Continuous Variable

$$\sigma^2 = V[X] = E[(X-\mu)^2] = \int_{-\infty}^{\infty} (x-\mu)^2 f(x)\,dx \qquad 6.35$$

Description

Equation 6.35 gives the variance of a continuous random variable, X. μ is the mean of X, and $f(x)$ is the density function of X.

Equation 6.36: Standard Deviation of a Continuous Variable

$$\sigma = \sqrt{V[X]} \qquad 6.36$$

Variation

$$\sigma = \sqrt{\sigma^2}$$

Description

The standard deviation is always the square root of the variance, as shown in the variation equation. Equation 6.36 gives the standard deviation for a continuous random variable, X.

Equation 6.37: Coefficient of Variation of a Continuous Variable

$$CV = \sigma/\mu \qquad 6.37$$

Description

The coefficient of variation of a continuous variable is calculated from Eq. 6.37.

10. PROBABILITY DISTRIBUTIONS

Equation 6.38 and Eq. 6.39: Binomial Distribution

$$P_n(x) = C(n,x)p^x q^{n-x} = \frac{n!}{x!\,(n-x)!}p^x q^{n-x} \qquad 6.38$$

$$q = 1 - p \qquad 6.39$$

Description

The *binomial probability function* is used when all outcomes are discrete and can be categorized as either successes or failures. The probability of success in a single trial is designated as p, and the probability of failure is the complement, q, calculated from Eq. 6.39.

Equation 6.38 gives the probability of x successes in n independent successive trials. The quantity $C(n, x)$ is the *binomial coefficient,* identical to the number of combinations of n items taken x at a time. (See Eq. 6.7.)

Example

A cat has a litter of seven kittens. If the probability that any given kitten will be female is 0.52, what is the probability that exactly two of the seven will be male?

(A) 0.07

(B) 0.18

(C) 0.23

(D) 0.29

Solution

Since the outcomes are "either-or" in nature, the outcomes (and combinations of outcomes) follow a binomial distribution. A male kitten is defined as a success. The probability of a success is

$$\begin{aligned} p &= 1 - 0.52 = 0.48 = P(\text{male kitten}) \\ q &= 0.52 = P(\text{female kitten}) \\ n &= 7 \text{ trials} \\ x &= 2 \text{ successes} \\ P_n(x) &= \frac{n!}{x!\,(n-x)!}p^x q^{n-x} \\ P_7(2) &= \left(\frac{7!}{2!\,(7-2)!}\right)(0.48)^2(0.52)^{7-2} \\ &= 0.184 \quad (0.18) \end{aligned}$$

The answer is (B).

Equation 6.40 Through Eq. 6.43: Normal Distribution

$$f(x) = \frac{1}{\sigma\sqrt{2\pi}}e^{-\frac{1}{2}\left(\frac{x-\mu}{\sigma}\right)^2} \quad [-\infty \le x \le \infty] \qquad 6.40$$

$$f(x) = \frac{1}{\sqrt{2\pi}}e^{-x^2/2} \quad [-\infty \le x \le \infty] \qquad 6.41$$

$$Z = \frac{x-\mu}{\sigma} \qquad 6.42$$

$$F(-x) = 1 - F(x) \qquad 6.43$$

Description

The *normal distribution* (*Gaussian distribution*) is a symmetrical continuous distribution, commonly referred to as the *bell-shaped curve,* which describes the distribution of outcomes of many real-world experiments, processes, and phenomena. (See Fig. 6.3.)

Figure 6.3 *Normal Distribution*

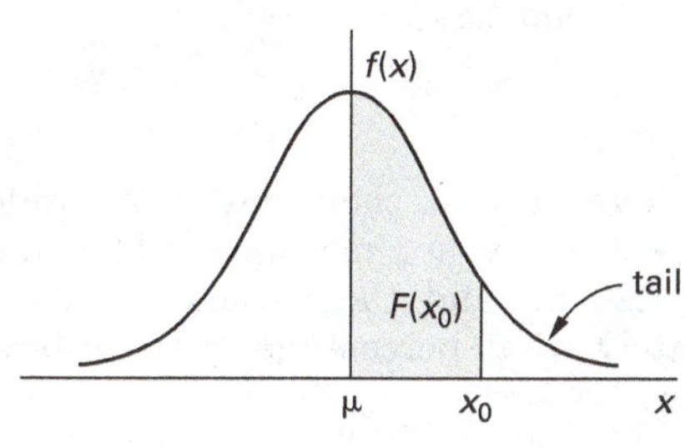

The probability density function for the normal distribution with population mean μ and population variance σ^2 is illustrated in Fig. 6.4 and is described by Eq. 6.40.

Figure 6.4 Normal Curve with Mean μ and Standard Deviation σ

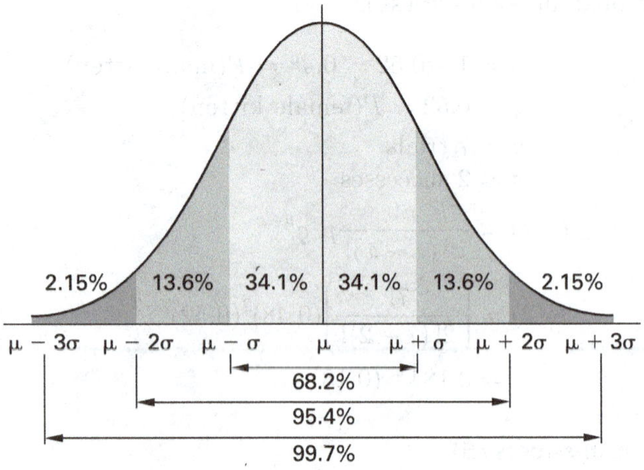

Since $f(x)$ is difficult to integrate (i.e., Eq. 6.40 is difficult to evaluate), Eq. 6.40 is seldom used directly, and a *unit normal table* is used instead. (See Table 6.2 at the end of this chapter.) The unit normal table (also called the *standard normal table*) is based on a normal distribution with a mean of zero and a standard deviation of one. The standard normal distribution is given by Eq. 6.41. In Table 6.2, $F(x)$ is the area under the curve from $-\infty$ to x, $R(x)$ is the area under the curve from x to ∞, and $W(x)$ is the area under the curve between $-x$ and x. The generic variable x used in Table 6.2 is the *standard normal variable*, Z, calculated in Eq. 6.42, not the actual measurement of the random variable, X. That is, the x used in Table 6.2 is not the x used in Eq. 6.42.

Since the range of values from an experiment or phenomenon will not generally correspond to the unit normal table, a value, x, must be converted to a standard normal variable, Z. In Eq. 6.42, μ and σ are the population mean and standard deviation, respectively, of the distribution from which x comes. The unbiased estimators for μ and σ are $\overline{X}$ and s, respectively, when a sample is used to estimate the population parameters. Both $\overline{X}$ and s approach the population values as the sample size, n, increases.

Example

The heights of several thousand fifth grade boys in Santa Clara County are measured. The mean of the heights is 1.20 m, and the variance is 25×10^{-4} m^2. Approximately what percentage of these boys is taller than 1.23 m?

(A) 27%

(B) 31%

(C) 69%

(D) 73%

Solution

To convert the normal distribution to unit normal distribution, the new variable, Z, is constructed from the height, x, mean, μ, and standard deviation, σ. The mean is known; the standard deviation is found from the variance and a variation of Eq. 6.36.

$$\sigma = \sqrt{\sigma^2} = \sqrt{25 \times 10^{-4}\ \text{m}^2} = 0.05\ \text{m}$$

For a height less than or equal to 1.23 m, from Eq. 6.42,

$$Z = \frac{x-\mu}{\sigma} = \frac{1.23\ \text{m} - 1.20\ \text{m}}{0.05\ \text{m}} = 0.6$$

From Table 6.2, the cumulative distribution function at $Z=0.6$ is $F(Z) = 0.7257$. The percentage of boys having height greater than 1.23 m is

$$\begin{aligned}\text{percentage taller than 1.23 m} &= 100\% - (0.7257)(100\%)\\ &= 27.43\% \quad (27\%)\end{aligned}$$

The answer is (A).

Equation 6.44 and Eq. 6.45: Central Limit Theorem

$$\mu_{\bar{y}} = \mu \qquad 6.44$$

$$\sigma_{\bar{y}} = \frac{\sigma}{\sqrt{n}} \qquad 6.45$$

Description

The *central limit theorem* states that the distribution of a significantly large number of sample means of n items where all items are drawn from the same (i.e., parent) population will be normal. According to the central limit theorem, the mean of sample means, $\mu_{\bar{y}}$, is equal to the population mean of the parent distribution, μ, as shown in Eq. 6.44. The standard deviation of the sample means, $\sigma_{\bar{y}}$, is equal to the standard deviation of the parent population divided by the square root of the sample size, as shown in Eq. 6.45.

Equation 6.46 and Eq. 6.47: *t*-Distribution

$$f(t) = \frac{\Gamma\left(\frac{\nu+1}{2}\right)}{\sqrt{\nu\pi}\,\Gamma\left(\frac{\nu}{2}\right)}\left(1+\frac{t^2}{\nu}\right)^{-\frac{\nu+1}{2}} \quad [-\infty \le t \le \infty] \qquad 6.46$$

$$t = \frac{\bar{x}-\mu}{s/\sqrt{n}} \quad [-\infty \le t \le \infty] \qquad 6.47$$

Description

For the *t-distribution* (commonly referred to as *Student's t-distribution*), the probability distribution function with ν *degrees of freedom* (sample size of $n+1$) is given by Eq. 6.46. The t-distribution is tabulated in Table 6.4 at the end of this chapter, with t as a function of ν and α. In Table 6.4, the column labeled "ν" lists the degrees of freedom, one less than the sample size. Degrees of freedom is sometimes given the symbol "df."

In Eq. 6.47, x is the sample mean, μ is the population mean, n is the sample size and s is the sample standard deviation.

Equation 6.48: Gamma Function

$$\Gamma(n) = \int_0^\infty t^{n-1}e^{-t}dt \quad [n > 0] \qquad 6.48$$

Description

The *gamma function*, $\Gamma(n)$, is an extension of the factorial function and is used to determine values of the factorial for complex numbers greater than zero (i.e., positive integers).

Equation 6.49: Chi-Squared Distribution

$$\chi^2 = Z_1^2 + Z_2^2 + \ldots + Z_n^2 \qquad 6.49$$

Description

The sum of the squares of n independent normal random variables will be distributed according to the *chi-squared distribution* and will have n degrees of freedom. The chi-squared distribution is often used with hypothesis testing of variances. Chi-squared values, $\chi^2_{\alpha,n}$, for selected values of α and n can be found from Table 6.5 at the end of this chapter.

11. STUDENT'S *t*-TEST

Equation 6.50: Exceedance

$$\alpha = \int_{t_{\alpha,\nu}}^\infty f(t)\,dt \qquad 6.50$$

Description

The *t-test* is a method of comparing two variables, usually to test the significance of the difference between samples. For example, the t-test can be used to test whether the populations from which two samples are drawn have the same means.

The *exceedance* (i.e., the probability of being incorrect), α, is equal to the total area under the upper tail. (See Fig. 6.5.) For a *one-tail test*, $\alpha = 1 - C$. For a *two-tail test*, $\alpha = 1 - C/2$. Since the t-distribution is symmetric about zero, $t_{1-\alpha,n} = -t_{1-\alpha,n}$. As n increases, the t-distribution approaches the normal distribution.

α can be used to find the critical values of F, as listed in Table 6.6 at the end of this chapter.

Figure 6.5 Exceedance

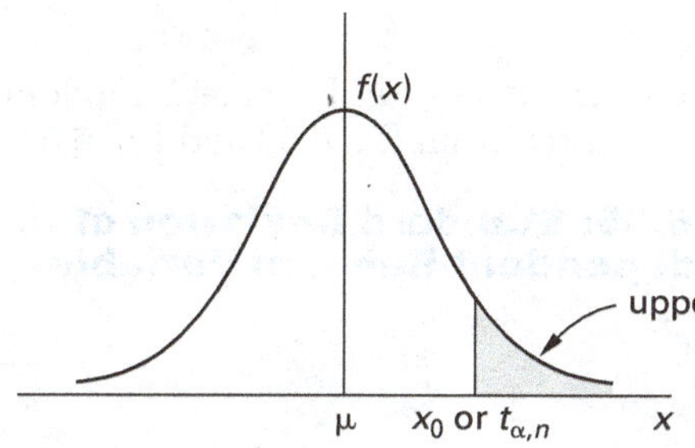

12. CONFIDENCE LEVELS

The results of experiments are seldom correct 100% of the time. Recognizing this, researchers accept a certain probability of being wrong. In order to minimize this probability, an experiment is repeated several times. The number of repetitions required depends on the level of confidence wanted in the results. For example, if the results have a 5% probability of being wrong, the *confidence level*, C, is 95% that the results are correct.

13. SUMS OF RANDOM VARIABLES

Equation 6.51: Sums of Random Variables

$$Y = a_1X_1 + a_2X_2 + \ldots + a_nX_n \qquad 6.51$$

Description

The sum of random variables, Y, is found from Eq. 6.51.

Equation 6.52: Expected Value of the Sum of Random Variables

$$\mu_y = E(Y) = a_1E(X_1) + a_2E(X_2) + \ldots + a_nE(X_n) \qquad 6.52$$

Description

The expected value of the sum of random variables, μ_y, is calculated using Eq. 6.52.

Probability/ Statistics

Equation 6.53 and Eq. 6.54: Variance of the Sum of Independent Random Variables

$$\begin{aligned}\sigma_y^2 &= V(Y)\\ &= a_1^2V(X_1) + a_2^2V(X_2) + \ldots + a_n^2V(X_n)\end{aligned} \quad 6.53$$

$$\sigma_y^2 = a_1^2\sigma_1^2 + a_2^2\sigma_2^2 + \ldots + a_n^2\sigma_n^2 \quad 6.54$$

Description

The variance of the sum of independent random variables can be calculated from Eq. 6.53 and Eq. 6.54.

Equation 6.55: Standard Deviation of the Sum of Independent Random Variables

$$\sigma_y = \sqrt{\sigma_y^2} \quad 6.55$$

Description

The standard deviation of the sum of independent random variables (see Eq. 6.51) is found from Eq. 6.55.

Equation 6.56: Standard Deviation of a Combination of Independent Random Variables

$$\sigma_Y = \sqrt{\left(\frac{\partial f}{\partial X_1}\sigma_{X_1}\right)^2 + \left(\frac{\partial f}{\partial X_2}\sigma_{X_2}\right)^2 + \ldots + \left(\frac{\partial f}{\partial X_n}\sigma_{X_n}\right)^2} \quad 6.56$$

Description

When the sum of random variables, Y, is a combination of independent random variables, the standard deviation of Y, σ_Y, is found from Eq. 6.56.

14. PROPAGATION OF ERROR

Engineering analyses frequently involve the measurement of multiple random variables that are combined mathematically to describe some dimension. Uncertainty and errors that exist in each individual variable combine and propagate from the original measurements through the mathematical combination of random variables. This results in a composite uncertainty that is directly related to the magnitudes of the uncertainty and errors in the individual variables. Quantifying the propagation of error and uncertainty requires understanding measurement errors and linear combinations of random variables.

The *true value* of the underlying phenomenon represents the value that would be measured with a "perfect" instrument (i.e., one with no errors). *Bias* is a systematic predisposition, either high or low, relative to the true value. Bias can generally be regarded as an offset of the true value. The *random error* of a measurement represents how reproducible any individual measurement is for a given state of the medium being tested. Over many measurements, the random error will by definition have a mean of zero, and will vary according to the normal distribution, with some constant variance.

Equation 6.57: Measurement Error

$$x = x_{\text{true}} + x_{\text{bias}} + x_{re} \quad 6.57$$

Description

The random variable x represents a measured value of some dimension. Any given measurement comprises the true value of the underlying phenomenon, x_{true}, the bias (or systematic error) in the measuring dimension, x_{bias}, and the random error associated with the technique, x_{re}.

Equation 6.58: Population (Arithmetic) Mean

$$\mu = \left(\frac{1}{N}\right)\sum_{i=1}^{n} x_i = \left(\frac{1}{N}\right)\sum_{i=1}^{n}(x_{\text{true}} + x_{\text{bias}})_i \quad 6.58$$

Description

The *population (arithmetic) mean* of a random variable represents the true mean value of the random variable. For example, measurements of water temperature may vary depending on the bias and random error of a certain thermometer, but at any given moment in time, there is a true value of the temperature. According to the law of large numbers, the random error of measurements of a "large" number of observations has a mean value of zero (or will approach zero as the number of observations increases). However, some constant variance will exist in any of the individual measurements. Because of this, the random error term approaches zero when calculating the mean value of a sufficiently large set of observations. Equation 6.58 shows that a population mean can be calculated by taking the sum of all the individual observations, x_i, corrected for bias where needed and divided by the number of observations, N. In this instance, bias is assumed to be a constant offset that does not change over the sample set.

Example 19

A newly compacted embankment of nearly uniform density is tested using a nuclear density gauge. Ten measurements are taken, with their densities as shown. Later, a calibration procedure determines that the gauge is over-reporting the density of samples by 100 kg/m^3.

observation	density (kg/m^3)
1	1886
2	1891
3	1918
4	1891
5	1924
6	1916
7	1893
8	1943
9	1879
10	1894

What is most nearly the true mean population density of the embankment?

(A) 1800 kg/m^3

(B) 1890 kg/m^3

(C) 1900 kg/m^3

(D) 1920 kg/m^3

Solution

Each observation consists of the true value and the bias value. The bias is constant across all measurements. Rearranging Eq. 6.58, subtract the bias from the measured values to find the true population mean density of the embankment.

$$\begin{aligned}\mu &= \left(\frac{1}{N}\right)\sum_{i=1}^{n} x_i = \left(\frac{1}{N}\right)\sum_{i=1}^{n}(x_{\text{true}} + x_{\text{bias}})_i \\ &= \left(\frac{1}{N}\right)\sum_{i=1}^{n} x_{\text{true},i} = \left(\frac{1}{N}\right)\sum_{i=1}^{n} x_i - x_{\text{bias}} \\ &= \frac{\left(\begin{array}{l} 1886\ \frac{\text{kg}}{\text{m}^3} + 1891\ \frac{\text{kg}}{\text{m}^3} + 1918\ \frac{\text{kg}}{\text{m}^3} \\ \quad +1891\ \frac{\text{kg}}{\text{m}^3} + 1924\ \frac{\text{kg}}{\text{m}^3} + 1916\ \frac{\text{kg}}{\text{m}^3} \\ \quad +1893\ \frac{\text{kg}}{\text{m}^3} + 1943\ \frac{\text{kg}}{\text{m}^3} + 1879\ \frac{\text{kg}}{\text{m}^3} \\ \quad +1894\ \frac{\text{kg}}{\text{m}^3} \end{array}\right)}{10} - 100\ \frac{\text{kg}}{\text{m}^3} \\ &= 1803\ \text{kg/m}^3 \quad (1800\ \text{kg/m}^3)\end{aligned}$$

The answer is (A).

Equation 6.59: Population Standard Deviation

$$\sigma_{\text{population}} = \sigma_{re} = \sqrt{(1/N)\sum(X_i - \mu)^2} \qquad 6.59$$

Description

The standard deviation of a population represents the random error associated with the uncertainty of the associated set of measurements. That is, the standard deviation represents the departure of observations, X_i, from a central tendency, such as the mean value of the population, μ.

Equation 6.60 and Eq. 6.61: Linear Combinations

$$\sigma_{cx} = |c|\,\sigma_x \qquad 6.60$$

$$\sigma_{c|x|+\ldots c_n x_n} = \sqrt{c_1^2\sigma_1^2 + \ldots c_n^2\sigma_{x_n}^2} \qquad 6.61$$

Description

When a random variable with standard deviation σ_x is multiplied by a constant value, c, the resulting standard deviation of the new random variable is the original standard deviation scaled by the absolute value of the constant. (See Eq. 6.60.) Generally, when any number of random variables x_1, $x_2 \ldots x_n$ with standard deviations σ_1, $\sigma_2 \ldots \sigma_n$ and constants c_1, $c_2 \ldots c_n$ are combined linearly, the resulting standard deviation is given by Eq. 6.61.

Example

A spring daily produces 1000 L of water. Water quality samples reveal copper concentrations of 0.01 mg/L, with a standard deviation of 0.001 mg/L. What is most nearly the standard deviation of the copper concentration?

(A) 0.1 mg/d

(B) 0.2 mg/d

(C) 1.0 mg/d

(D) 8.6 mg/d

Solution

From Eq. 6.60, the standard deviation of the copper concentration is

$$\begin{aligned}\sigma_{cx} &= |c|\sigma_x \\ &= \left|1000\ \frac{\text{L}}{\text{d}}\right|\left(0.001\ \frac{\text{mg}}{\text{L}}\right) \\ &= 1\ \text{mg/d} \quad (1.0\ \text{mg/d})\end{aligned}$$

The answer is (C).

Equation 6.62 Through Eq. 6.64: Measurement Uncertainties

$$\sigma_u = \sqrt{\left(\frac{\partial u}{\partial x_1}\right)^2 \sigma_{x_1}^2 + \left(\frac{\partial u}{\partial x_2}\right)^2 \sigma_{x_2}^2 + \ldots + \left(\frac{\partial u}{\partial x_n}\right)^2 \sigma_{x_n}^2} \qquad 6.62$$

$$u(x_1, x_2, \ldots x_n) = f(x_1, x_2, \ldots x_n) \qquad 6.63$$

$$\sigma_u \approx \left|\frac{du}{dx_1}\right| \sigma_{x_1} \qquad 6.64$$

Description

The propagation of measurement uncertainty from observations of individual random variables through combinations of those variables can be estimated using Eq. 6.62. If u is a function of random variables $x_1, x_2 \ldots x_n$ and each of the random variables has an associated uncertainty expressed as standard deviation $\sigma_{x1}, \sigma_{x2}, \ldots, \sigma_{x_n}$, then Eq. 6.62 quantifies the transmission of errors from the individual measurements into the function $u(x_1, x_2 \ldots x_n)$ as shown in Eq. 6.63. Eq. 6.64 is a simplified approximation in which the uncertainty of a single random variable is transformed into the function $u(x_1)$.

Example

A process is governed by the equation $u = 2x_1 + 5x_2$. The observations of random variables x_1 and x_2 have estimated uncertainties of 0.1 and 0.05, respectively. What is the total uncertainty of the process?

(A) 0.10

(B) 0.15

(C) 0.30

(D) 0.32

Solution

Use Eq. 6.62.

$$\begin{aligned}\sigma_u &= \sqrt{\left(\frac{\partial u}{\partial x_1}\right)^2 \sigma_{x_1}^2 + \left(\frac{\partial u}{\partial x_2}\right)^2 \sigma_{x_2}^2} \\ &= \sqrt{(2)^2(0.1)^2 + (5)^2(0.05)^2} \\ &= 0.320 \quad (0.32)\end{aligned}$$

The answer is (D).

15. SUMS AND DIFFERENCES OF MEANS

When two variables are sampled from two different standard normal variables (i.e., are independent), their sums will be distributed with mean $\mu_{\text{new}} = \mu_1 + \mu_2$ and variance $\sigma_{\text{new}}^2 = \sigma_1^2/n_1 + \sigma_2^2/n_2$. The sample sizes, n_1 and n_2, do not have to be the same. The relationships for confidence intervals and hypothesis testing can be used for a new variable, $x_{\text{new}} = x_1 + x_2$, if μ is replaced by μ_{new} and σ is replaced by σ_{new}.

For the difference in two standard normal variables, the mean is the difference in two population means, $\mu_{\text{new}} = \mu_1 - \mu_2$, but the variance is the sum, as it was for the sum of two standard normal variables.

16. CONFIDENCE INTERVALS

Population properties such as means and variances must usually be estimated from samples. The sample mean, $\overline{X}$, and sample standard deviation, s, are unbiased estimators, but they are not necessarily precisely equal to the true population properties. For estimated values, it is common to specify an interval expected to contain the true population properties. The interval is known as a confidence interval because a confidence level, C (e.g., 99%), is associated with it. (There is still a $1 - C$ chance that the true population property is outside of the interval.) The interval will be bounded below by its *lower confidence limit* (LCL) and above by its *upper confidence limit* (UCL).

As a consequence of the *central limit theorem*, means of samples of n items taken from a distribution that is normally distributed with mean μ and standard deviation σ will be normally distributed with mean μ and variance σ^2/n. Therefore, the probability that any given average, $\overline{X}$, exceeds some value, L, is

$$p\{\overline{X} > L\} = p\left\{x > \left|\frac{L-\mu}{\frac{\sigma}{\sqrt{n}}}\right|\right\}$$

L is the *confidence limit* for the confidence level $1 - p\{\overline{X} > L\}$ (expressed as a percentage). Values of px are read directly from the unit normal table. (See Table 6.2.) As an example, $x = 1.645$ for a 95% confidence level since only 5% of the curve is above that x in the upper tail. This is known as a *one-tail confidence limit* because all of the exceedance probability is given to one side of the variation.

With *two-tail confidence limits*, the probability is split between the two sides of variation. There will be upper and lower confidence limits: UCL and LCL, respectively. This is appropriate when it is not specifically known that the calculated parameter is too high or too low. Table 6.3 at the end of this chapter lists standard normal variables and t values for two-tail confidence limits.

$$p\{\text{LCL} < \overline{X} < \text{UCL}\} = p\left\{\frac{\text{LCL}-\mu}{\frac{\sigma}{\sqrt{n}}} < x < \frac{\text{UCL}-\mu}{\frac{\sigma}{\sqrt{n}}}\right\}$$

Equation 6.65 and Eq. 6.66: Confidence Limits and Interval for Mean of a Normal Distribution

$$\overline{X} - Z_{\alpha/2}\frac{\sigma}{\sqrt{n}} \le \mu \le \overline{X} + Z_{\alpha/2}\frac{\sigma}{\sqrt{n}} \quad [\text{known } \sigma] \qquad 6.65$$

$$\overline{X} - t_{\alpha/2}\frac{s}{\sqrt{n}} \le \mu \le \overline{X} + t_{\alpha/2}\frac{s}{\sqrt{n}} \quad [\text{unknown } \sigma] \qquad 6.66$$

Variations

$$\text{LCL} = \overline{X} - t_{\alpha/2,n-1}\left(\frac{s}{\sqrt{n}}\right)$$

$$\text{UCL} = \overline{X} + t_{\alpha/2,n-1}\left(\frac{s}{\sqrt{n}}\right)$$

Description

The *confidence limits for the mean*, μ, of a normal distribution can be calculated from Eq. 6.65 when the standard deviation, σ, is known.

If the standard deviation, σ, of the underlying distribution is not known, the confidence limits must be estimated from the sample standard deviation, s, using Eq. 6.66. Accordingly, the standard normal variable is replaced by the t-distribution parameter, $t_{\alpha/2}$, with $n-1$ degrees of freedom, where n is the sample size. $\alpha = 1 - C$, and $\alpha/2$ is the t-distribution parameter since half of the exceedance is allocated to each confidence limit.

Equation 6.67 and Eq. 6.68: Confidence Limits for the Difference Between Two Means

$$\overline{X}_1 - \overline{X}_2 - Z_{\alpha/2}\sqrt{\frac{\sigma_1^2}{n_1}+\frac{\sigma_2^2}{n_2}} \le \mu_1 - \mu_2 \le \overline{X}_1 - \overline{X}_2 + Z_{\alpha/2}\sqrt{\frac{\sigma_1^2}{n_1}+\frac{\sigma_2^2}{n_2}} \quad [\text{known } \sigma_1 \text{ and } \sigma_2] \qquad 6.67$$

$$\overline{X}_1 - \overline{X}_2 - t_{\alpha/2}\sqrt{\frac{\left(\frac{1}{n_1}+\frac{1}{n_2}\right)\left[\begin{array}{c}(n_1-1)S_1^2\\+(n_2-1)S_2^2\end{array}\right]}{n_1+n_2-2}} \le \mu_1 - \mu_2 \le \overline{X}_1 - \overline{X}_2 + t_{\alpha/2}\sqrt{\frac{\left(\frac{1}{n_1}+\frac{1}{n_2}\right)\left[\begin{array}{c}(n_1-1)S_1^2\\+(n_2-1)S_2^2\end{array}\right]}{n_1+n_2-2}} \quad [\text{unknown } \sigma_1 \text{ and } \sigma_2] \qquad 6.68$$

Description

The difference in two standard normal variables will be distributed with mean $\mu_{\text{new}} = \mu_1 - \mu_2$. Use Eq. 6.67 to calculate the confidence interval for the difference between two means, μ_1 and μ_2, if the standard deviations σ_1 and σ_2 are known. If the standard deviations σ_1 and σ_2 are unknown, use Eq. 6.68. S_1 and S_2 are the sample standard deviations for the two sample sets. The t-distribution parameter, $t_{\alpha/2}$, has $n_1 + n_2 - 2$ degrees of freedom.

Example 22

100 resistors produced by company A and 150 resistors produced by company B are tested to find their limits before burning out. The test results show that the company A resistors have a mean rating of 2 W before burning out, with a standard deviation of 0.25 W; and the company B resistors have a 3 W mean rating before burning out, with a standard deviation of 0.30 W. What are the 95% confidence limits for the difference between the two means for the company A resistors and company B resistors (i.e., A – B)?

(A) −1.1 W; −1.0 W

(B) −1.1 W; −0.93 W

(C) −1.1 W; −0.90 W

(D) −1.0 W; −0.99 W

Solution

From Table 6.3, the value of the standard normal variable for a two-tail test with 95% confidence is 1.9600.

Probability/Statistics

From Eq. 6.67, the confidence limits for the difference between the two means are

$$\begin{aligned}\text{LCL}(\mu_1 - \mu_2) &= \overline{X}_1 - \overline{X}_2 - Z_{\alpha/2}\sqrt{\frac{\sigma_1^2}{n_1} + \frac{\sigma_2^2}{n_2}} \\ &= 2\text{ W} - 3\text{ W} \\ &\quad -1.9600\sqrt{\frac{(0.25\text{ W})^2}{100} + \frac{(0.30\text{ W})^2}{150}} \\ &= -1.0686\text{ W} \quad (-1.1\text{ W})\end{aligned}$$

$$\begin{aligned}\text{UCL}(\mu_1 - \mu_2) &= \overline{X}_1 - \overline{X}_2 + Z_{\alpha/2}\sqrt{\frac{\sigma_1^2}{n_1} + \frac{\sigma_2^2}{n_2}} \\ &= 2\text{ W} - 3\text{ W} \\ &\quad +1.9600\sqrt{\frac{(0.25\text{ W})^2}{100} + \frac{(0.30\text{ W})^2}{150}} \\ &= -0.9314\text{ W} \quad (-0.93\text{ W})\end{aligned}$$

The answer is (B).

Equation 6.69: Confidence Limits and Interval for the Variance of a Normal Distribution

$$\frac{(n-1)s^2}{\chi^2_{\alpha/2,n-1}} \le \sigma^2 \le \frac{(n-1)s^2}{\chi^2_{1-\alpha/2,n-1}} \qquad 6.69$$

Description

Equation 6.69 gives the limits of a confidence interval (confidence $C = 1 - \alpha$) for an estimate of the population variance calculated as the sample variance from Eq. 6.25 with a sample size of n drawn from a normal distribution. Since the variance is a squared variable, it will be distributed as a chi-squared distribution with $n - 1$ degrees of freedom. Therefore, the denominators are the χ^2 values taken from Table 6.5 at the end of this chapter. (The values in Table 6.5 are already squared and should not be squared again.) Since the chi-squared distribution is not symmetrical, the table values for $\alpha/2$ and for $1 - (\alpha/2)$ will be different for the two confidence limits.

17. HYPOTHESIS TESTING

A *hypothesis test* is a procedure that answers the question, "Did these data come from [a particular type of] distribution?" There are many types of tests, depending on the distribution and parameter being evaluated. The most simple hypothesis test determines whether an average value obtained from n repetitions of an experiment could have come from a population with known mean μ and standard deviation σ. A practical application of this question is whether a manufacturing process has changed from what it used to be or should be. Of course, the answer (i.e., yes or no) cannot be given with absolute certainty—there will be a confidence level associated with the answer.

The following procedure is used to determine whether the average of n measurements can be assumed (with a given confidence level) to have come from a known normal population, or to determine the sample size required to make the decision with the desired confidence level.

Equation 6.70 Through Eq. 6.75: Test on Mean of Normal Distribution, Population Mean and Variance Known

step 1: Assume random sampling from a normal population.

The *null hypothesis* is

$$H_0{:}\,\mu = \mu_0 \qquad 6.70$$

The *alternative hypothesis* is

$$H_1{:}\,\mu = \mu_1 \qquad 6.71$$

A *type I error* is rejecting H_0 when it is true. The probability of a type I error is the *level of significance*.

$$\alpha = \text{probability(type I error)} \qquad 6.72$$

A *type II error* is accepting H_0 when it is false.

$$\beta = \text{probability(type II error)} \qquad 6.73$$

step 2: Choose the desired confidence level, C.

step 3: Decide on a one-tail or two-tail test. If the hypothesis being tested is that the average has or has not *increased* or has not *decreased*, use a one-tail test. If the hypothesis being tested is that the average has or has not *changed*, use a two-tail test.

step 4: Use Table 6.3 or the unit normal table to determine the x-value corresponding to the confidence level and number of tails.

step 5: Calculate the actual standard normal variable, Z, from Eq. 6.74. The relationship of the sample size, n, and the actual standard normal variable is illustrated in Eq. 6.75.

$$Z = \frac{\overline{X} - \mu_0}{\sigma/\sqrt{n}} \qquad 6.74$$

$$n = \left[\frac{Z_{\alpha/2}\sigma}{\overline{x} - \mu}\right]^2 \qquad 6.75$$

step 6: If $|Z| \geq x$, the average can be assumed (with confidence level C) to have come from a different distribution.

Equation 6.76 Through Eq. 6.83: Sample Size for Normal Distribution, α and β Known

$$H_0{:}\mu = \mu_0 \qquad 6.76$$

$$H_1{:}\mu \neq \mu_0 \qquad 6.77$$

$$\beta = \Phi\left(\frac{\mu_0 - \mu}{\sigma/\sqrt{n}} + Z_{\alpha/2}\right) - \Phi\left(\frac{\mu_0 - \mu}{\sigma/\sqrt{n}} - Z_{\alpha/2}\right) \qquad 6.78$$

$$n \simeq \frac{(Z_{\alpha/2} + Z_\beta)^2\sigma^2}{(\mu_1 - \mu_0)^2} \qquad 6.79$$

$$H_0{:}\mu = \mu_0 \qquad 6.80$$

$$H_1{:}\mu > \mu_0 \qquad 6.81$$

$$\beta = \Phi\left(\frac{\mu_0 - \mu}{\sigma/\sqrt{n}} + Z_\alpha\right) \qquad 6.82$$

$$n = \frac{(Z_\alpha + Z_\beta)^2\sigma^2}{(\mu_1 - \mu_0)^2} \qquad 6.83$$

Description

Equation 6.76 through Eq. 6.83 are used to determine the required sample size when the probabilities of type 1 and type 2 errors, α and β, respectively, are known. μ_1 is the assumed true mean. The notation $\Phi(z)$ designates the cumulative normal distribution function (i.e., the fraction of the normal curve from $-\infty$ up to z).[2] Equation 6.76 through Eq. 6.79 are used when the test is to determine if the sample mean is the same as the population mean, while Eq. 6.80 through Eq. 6.83 are used when the test is to determine if the sample mean is larger or smaller than the population mean.

Example 23

When it is operating properly, a chemical plant has a daily production rate that is normally distributed with a mean of 880 tons/day and a standard deviation of 21 tons/day. During an analysis period, the output is measured with random sampling on 50 consecutive days, and the mean output is found to be 871 tons/day. With a 95% confidence level, determine if the plant is operating properly.

(A) There is at least a 5% probability that the plant is operating properly.

(B) There is at least a 95% probability that the plant is operating properly.

(C) There is at least a 5% probability that the plant is not operating properly.

(D) There is at least a 95% probability that the plant is not operating properly.

Solution

Since a specific direction in the variation is not given (i.e., the example does not ask if the average has decreased), use a two-tail hypothesis test.

From Table 6.3, $x = 1.9600$.

Use Eq. 6.74 to calculate the actual standard normal variable.

$$Z = \frac{\overline{X} - \mu}{\sigma/\sqrt{n}} = \frac{871 - 880}{\frac{21}{\sqrt{50}}}$$

$$|Z| = -3.03$$

Since $|-3.03| > 1.9600$, the distributions are not the same. There is at least a 95% probability that the plant is not operating correctly.

The answer is (D).

18. LINEAR REGRESSION

Equation 6.84 Through Eq. 6.90: Method of Least Squares

If it is necessary to draw a straight line ($\widehat{y} = \widehat{a} + \widehat{b}x$) through n two-dimensional data points $(x_1, y_1), (x_2, y_2), \ldots, (x_n, y_n)$, the following method based on the *method of least squares* can be used.

[2]Not only is $\Phi(z)$ undefined in the *NCEES Handbook*, but it is the same as what the *NCEES Handbook* designated earlier as $F(x)$.

step 1: Calculate the following seven quantities.

$$\sum x_i \quad \sum x_i^2 \quad \left(\sum x_i\right)^2 \quad \sum x_i y_i$$
$$\sum y_i \quad \sum y_i^2 \quad \left(\sum y_i\right)^2$$

$$\overline{x} = (1/n)\left(\sum_{i=1}^{n} x_i\right) \qquad 6.84$$

$$\overline{y} = (1/n)\left(\sum_{i=1}^{n} y_i\right) \qquad 6.85$$

step 2: Calculate the slope, $\hat{b}$, of the line.

$$\hat{b} = S_{xy}/S_{xx} \qquad 6.86$$

$$S_{xy} = \sum_{i=1}^{n} x_i y_i - (1/n)\left(\sum_{i=1}^{n} x_i\right)\left(\sum_{i=1}^{n} y_i\right) \qquad 6.87$$

$$S_{xx} = \sum_{i=1}^{n} x_i^2 - (1/n)\left(\sum_{i=1}^{n} x_i\right)^2 \qquad 6.88$$

step 3: Calculate the y-intercept, $\hat{a}$.

$$\hat{a} = \overline{y} - \hat{b}\,\overline{x} \qquad 6.89$$

The equation of the straight line is

$$y = \hat{a} + \hat{b}\,x \qquad 6.90$$

Example

The least squares method is used to plot a straight line through the data points (1, 6), (2, 7), (3, 11), and (5, 13). The slope of the line is most nearly

(A) 0.87

(B) 1.7

(C) 1.9

(D) 2.0

Solution

First, calculate the following values.

$$\sum x_i = 1 + 2 + 3 + 5 = 11$$
$$\sum y_i = 6 + 7 + 11 + 13 = 37$$
$$\sum x_i^2 = (1)^2 + (2)^2 + (3)^2 + (5)^2 = 39$$
$$\sum x_i y_i = (1)(6) + (2)(7) + (3)(11) + (5)(13) = 118$$

Find the value of S_{xy} using Eq. 6.87.

$$\begin{aligned} S_{xy} &= \sum_{i=1}^{n} x_i y_i - (1/n)\left(\sum_{i=1}^{n} x_i\right)\left(\sum_{i=1}^{n} y_i\right) \\ &= 118 - \left(\frac{1}{4}\right)(11)(37) \\ &= 16.25 \end{aligned}$$

Find the value of S_{xx} from Eq. 6.88.

$$\begin{aligned} S_{xx} &= \sum_{i=1}^{n} x_i^2 - (1/n)\left(\sum_{i=1}^{n} x_i\right)^2 \\ &= 39 - \left(\frac{1}{4}\right)(11)^2 \\ &= 8.75 \end{aligned}$$

From Eq. 6.86, the slope is

$$\begin{aligned} \hat{b} &= S_{xy}/S_{xx} \\ &= \frac{16.25}{8.75} \\ &= 1.857 \quad (1.9) \end{aligned}$$

The answer is (C).

Equation 6.91 and Eq. 6.92: Standard Error of Estimate

$$S_e^2 = \frac{S_{xx}S_{yy} - S_{xy}^2}{S_{xx}(n-2)} = MSE \qquad 6.91$$

$$S_{yy} = \sum_{i=1}^{n} y_i^2 - (1/n)\left(\sum_{i=1}^{n} y_i\right)^2 \qquad 6.92$$

Description

Equation 6.91 gives the *mean squared error*, S_e^2 or *MSE*, which estimates the likelihood of a value being close to an observed value by averaging the square of the errors (i.e., the difference between the estimated value and observed value). Small *MSE* values are favorable, as they indicate a smaller likelihood of error.

Equation 6.93 and Eq. 6.94: Confidence Intervals for Slope and Intercept

$$\hat{b} \pm t_{\alpha/2,n-2}\sqrt{\frac{MSE}{S_{xx}}} \qquad 6.93$$

$$\hat{a} \pm t_{\alpha/2,n-2}\sqrt{\left(\frac{1}{n}+\frac{\bar{x}^2}{S_{xx}}\right)MSE} \qquad 6.94$$

Description

The confidence intervals for calculated slope and intercept are calculated from the mean square error using Eq. 6.93 and Eq. 6.94, respectively.

Equation 6.95 and Eq. 6.96: Sample Correlation Coefficient

$$R = \frac{S_{xy}}{\sqrt{S_{xx}S_{yy}}} \qquad 6.95$$

$$R^2 = \frac{S_{xy}^2}{S_{xx}S_{yy}} \qquad 6.96$$

Description

Once the slope of the line is calculated using the least squares method, the *goodness of fit* can be determined by calculating the *sample correlation coefficient*, R. The goodness of fit describes how well the calculated regression values, plotted as a line, match actual observed values, plotted as points.

If $\hat{b}$ is positive, R will be positive; if $\hat{b}$ is negative, R will be negative. As a general rule, if the absolute value of R exceeds 0.85, the fit is good; otherwise, the fit is poor. R equals 1.0 if the fit is a perfect straight line.

Example

The least squares method is used to plot a straight line through the data points $(5,-5)$, $(3,-2)$, $(2,3)$, and $(-1,7)$. The correlation coefficient is most nearly

(A) −0.97

(B) −0.92

(C) −0.88

(D) −0.80

Solution

First, calculate the following values.

$$\sum x_i = 5+3+2+(-1) = 9$$
$$\sum y_i = (-5)+(-2)+3+7 = 3$$
$$\sum x_i^2 = (5)^2+(3)^2+(2)^2+(-1)^2 = 39$$
$$\sum y_i^2 = (-5)^2+(-2)^2+(3)^2+(7)^2 = 87$$
$$\sum x_iy_i = (5)(-5)+(3)(-2)+(2)(3)+(-1)(7) = -32$$

From Eq. 6.95, and substituting Eq. 6.87, Eq. 6.88, and Eq. 6.92 for S_{xy}, S_{xx}, and S_{yy}, respectively, the correlation coefficient is

$$\begin{aligned} R &= \frac{S_{xy}}{\sqrt{S_{xx}S_{yy}}} \\ &= \frac{\sum x_iy_i - (1/n)\left(\sum x_i\right)\left(\sum y_i\right)}{\sqrt{\left[\sum x_i^2 - (1/n)\left(\sum x_i\right)^2\right]\left[\sum y_i^2 - (1/n)\left(\sum y_i\right)^2\right]}} \\ &= \frac{-32-\left(\frac{1}{4}\right)(9)(3)}{\sqrt{\left(39-\left(\frac{1}{4}\right)(9)^2\right)\left(87-\left(\frac{1}{4}\right)(3)^2\right)}} \\ &= -0.972 \quad (-0.97) \end{aligned}$$

The answer is (A).

Equation 6.97: Residuals

$$e_i = y_i - \hat{y}_i = y_i - (\hat{a}+\hat{b}x_i) \qquad 6.97$$

Description

Residuals are calculated by taking the observed values of the dependent variable, y_i, and subtracting the estimated response from the statistical model (e.g., linear regression), $\hat{y}_i$. Model residuals should be normally distributed with a mean value of zero and constant variance, otherwise the statistical model is not reliable.

Table 6.1 *Cumulative Binomial Probabilities $P(X \le x)$*

		P										
n	x	0.1	0.2	0.3	0.4	0.5	0.6	0.7	0.8	0.9	0.95	0.99
1	0	0.9000	0.8000	0.7000	0.6000	0.5000	0.4000	0.3000	0.2000	0.1000	0.0500	0.0100
2	0	0.8100	0.6400	0.4900	0.3600	0.2500	0.1600	0.0900	0.0400	0.0100	0.0025	0.0001
	1	0.9900	0.9600	0.9100	0.8400	0.7500	0.6400	0.5100	0.3600	0.1900	0.0975	0.0199
3	0	0.7290	0.5120	0.3430	0.2160	0.1250	0.0640	0.0270	0.0080	0.0010	0.0001	0.0000
	1	0.9720	0.8960	0.7840	0.6480	0.5000	0.3520	0.2160	0.1040	0.0280	0.0073	0.0003
	2	0.9990	0.9920	0.9730	0.9360	0.8750	0.7840	0.6570	0.4880	0.2710	0.1426	0.0297
4	0	0.6561	0.4096	0.2401	0.1296	0.0625	0.0256	0.0081	0.0016	0.0001	0.0000	0.0000
	1	0.9477	0.8192	0.6517	0.4752	0.3125	0.1792	0.0837	0.0272	0.0037	0.0005	0.0000
	2	0.9963	0.9728	0.9163	0.8208	0.6875	0.5248	0.3483	0.1808	0.0523	0.0140	0.0006
	3	0.9999	0.9984	0.9919	0.9744	0.9375	0.8704	0.7599	0.5904	0.3439	0.1855	0.0394
5	0	0.5905	0.3277	0.1681	0.0778	0.0313	0.0102	0.0024	0.0003	0.0000	0.0000	0.0000
	1	0.9185	0.7373	0.5282	0.3370	0.1875	0.0870	0.0308	0.0067	0.0005	0.0000	0.0000
	2	0.9914	0.9421	0.8369	0.6826	0.5000	0.3174	0.1631	0.0579	0.0086	0.0012	0.0000
	3	0.9995	0.9933	0.9692	0.9130	0.8125	0.6630	0.4718	0.2627	0.0815	0.0226	0.0010
	4	1.0000	0.9997	0.9976	0.9898	0.6988	0.9222	0.8319	0.6723	0.4095	0.2262	0.0490
6	0	0.5314	0.2621	0.1176	0.0467	0.0156	0.0041	0.0007	0.0001	0.0000	0.0000	0.0000
	1	0.8857	0.6554	0.4202	0.2333	0.1094	0.0410	0.0109	0.0016	0.0001	0.0000	0.0000
	2	0.9842	0.9011	0.7443	0.5443	0.3438	0.1792	0.0705	0.0170	0.0013	0.0001	0.0000
	3	0.9987	0.9830	0.9295	0.8208	0.6563	0.4557	0.2557	0.0989	0.0159	0.0022	0.0000
	4	0.9999	0.9984	0.9891	0.9590	0.9806	0.7667	0.5798	0.3446	0.1143	0.0328	0.0015
	5	1.0000	0.9999	0.9993	0.9959	0.9844	0.9533	0.8824	0.7379	0.4686	0.2649	0.0585
7	0	0.4783	0.2097	0.0824	0.0280	0.0078	0.0106	0.0002	0.0000	0.0000	0.0000	0.0000
	1	0.8503	0.5767	0.3294	0.1586	0.0625	0.0188	0.0038	0.0004	0.0000	0.0000	0.0000
	2	0.9743	0.8520	0.6471	0.4199	0.2266	0.0963	0.0288	0.0047	0.0002	0.0000	0.0000
	3	0.9973	0.9667	0.8740	0.7102	0.5000	0.2898	0.1260	0.0333	0.0027	0.0002	0.0000
	4	0.9998	0.9953	0.9712	0.9037	0.7734	0.5801	0.3529	0.1480	0.0257	0.0038	0.0000
	5	1.0000	0.9996	0.9962	0.9812	0.9375	0.8414	0.6706	0.4233	0.1497	0.0444	0.0020
	6	1.0000	1.0000	0.9998	0.9984	0.9922	0.9720	0.9176	0.7903	0.5217	0.3017	0.0679
8	0	0.4305	0.1678	0.0576	0.0168	0.0039	0.0007	0.0001	0.0000	0.0000	0.0000	0.0000
	1	0.8131	0.5033	0.2553	0.1064	0.0352	0.0085	0.0013	0.0001	0.0000	0.0000	0.0000
	2	0.9619	0.7969	0.5518	0.3154	0.1445	0.0498	0.0113	0.0012	0.0000	0.0000	0.0000
	3	0.9950	0.9437	0.8059	0.5941	0.3633	0.1737	0.0580	0.0104	0.0004	0.0000	0.0000
	4	0.9996	0.9896	0.9420	0.8263	0.6367	0.4059	0.1941	0.0563	0.0050	0.0004	0.0000
	5	1.0000	0.9988	0.9887	0.9502	0.8555	0.6846	0.4482	0.2031	0.0381	0.0058	0.0001
	6	1.0000	0.9999	0.9987	0.9915	0.9648	0.8936	0.7447	0.4967	0.1869	0.0572	0.0027
	7	1.0000	1.0000	0.9999	0.9993	0.9961	0.9832	0.9424	0.8322	0.5695	0.3366	0.0773
9	0	0.3874	0.1342	0.0404	0.0101	0.0020	0.0003	0.0000	0.0000	0.0000	0.0000	0.0000
	1	0.7748	0.4362	0.1960	0.0705	0.0195	0.0038	0.0004	0.0000	0.0000	0.0000	0.0000
	2	0.9470	0.7382	0.4628	0.2318	0.0889	0.0250	0.0043	0.0003	0.0000	0.0000	0.0000
	3	0.9917	0.9144	0.7297	0.4826	0.2539	0.0994	0.0253	0.0031	0.0001	0.0000	0.0000
	4	0.9991	0.9804	0.9012	0.7334	0.5000	0.2666	0.0988	0.0196	0.0009	0.0000	0.0000
	5	0.9999	0.9969	0.9747	0.9006	0.7461	0.5174	0.2703	0.0856	0.0083	0.0006	0.0000
	6	1.0000	0.9997	0.9957	0.9750	0.9102	0.7682	0.5372	0.2618	0.0530	0.0084	0.0001
	7	1.0000	1.0000	0.9996	0.9962	0.9805	0.9295	0.8040	0.5638	0.2252	0.0712	0.0034
	8	1.0000	1.0000	1.0000	0.9997	0.9980	0.9899	0.9596	0.8658	0.6126	0.3698	0.0865
10	0	0.3487	0.1074	0.0282	0.0060	0.0010	0.0001	0.0000	0.0000	0.0000	0.0000	0.0000
	1	0.7361	0.3758	0.1493	0.0464	0.0107	0.0017	0.0001	0.0000	0.0000	0.0000	0.0000
	2	0.9298	0.6778	0.3828	0.1673	0.0547	0.0123	0.0016	0.0001	0.0000	0.0000	0.0000
	3	0.9872	0.8791	0.6496	0.3823	0.1719	0.0548	0.0106	0.0009	0.0000	0.0000	0.0000
	4	0.9984	0.9672	0.8497	0.6331	0.3770	0.1662	0.0473	0.0064	0.0001	0.0000	0.0000
	5	0.9999	0.9936	0.9527	0.8338	0.6230	0.3669	0.1503	0.0328	0.0016	0.0001	0.0000
	6	1.0000	0.9991	0.9894	0.9452	0.8281	0.6177	0.3504	0.1209	0.0128	0.0010	0.0000
	7	1.0000	0.9999	0.9984	0.9877	0.9453	0.8327	0.6172	0.3222	0.0702	0.0115	0.0001
	8	1.0000	1.0000	0.9999	0.9983	0.9893	0.9536	0.8507	0.6242	0.2639	0.0861	0.0043
	9	1.0000	1.0000	1.0000	0.9999	0.9990	0.9940	0.9718	0.8926	0.6513	0.4013	0.0956

		P										
n	*x*	0.1	0.2	0.3	0.4	0.5	0.6	0.7	0.8	0.9	0.95	0.99
15	0	0.2059	0.0352	0.0047	0.0005	0.0000	0.0000	0.0000	0.0000	0.0000	0.0000	0.0000
	1	0.4590	0.1671	0.0353	0.0052	0.0005	0.0000	0.0000	0.0000	0.0000	0.0000	0.0000
	2	0.8159	0.3980	0.1268	0.0271	0.0037	0.0003	0.0000	0.0000	0.0000	0.0000	0.0000
	3	0.9444	0.6482	0.2969	0.0905	0.0176	0.0019	0.0001	0.0000	0.0000	0.0000	0.0000
	4	0.9873	0.8358	0.5155	0.2173	0.0592	0.0093	0.0007	0.0000	0.0000	0.0000	0.0000
	5	0.9978	0.9389	0.7216	0.4032	0.1509	0.0338	0.0037	0.0001	0.0000	0.0000	0.0000
	6	0.9997	0.9819	0.8689	0.6098	0.3036	0.0950	0.0152	0.0008	0.0000	0.0000	0.0000
	7	1.0000	0.9958	0.9500	0.7869	0.5000	0.2131	0.0500	0.0042	0.0000	0.0000	0.0000
	8	1.0000	0.9992	0.9848	0.9050	0.6964	0.3902	0.1311	0.0181	0.0003	0.0000	0.0000
	9	1.0000	0.9999	0.9963	0.9662	0.8491	0.5968	0.2784	0.0611	0.0022	0.0001	0.0000
	10	1.0000	1.0000	0.9993	0.9907	0.9408	0.7827	0.4845	0.1642	0.0127	0.0006	0.0000
	11	1.0000	1.0000	0.9999	0.9981	0.9824	0.9095	0.7031	0.3518	0.0556	0.0055	0.0000
	12	1.0000	1.0000	1.0000	0.9997	0.9963	0.9729	0.8732	0.6020	0.1841	0.0362	0.0004
	13	1.0000	1.0000	1.0000	1.0000	0.9995	0.9948	0.9647	0.8329	0.4510	0.1710	0.0096
	14	1.0000	1.0000	1.0000	1.0000	1.0000	0.9995	0.9953	0.9648	0.7941	0.5367	0.1399
20	0	0.1216	0.0115	0.0008	0.0000	0.0000	0.0000	0.0000	0.0000	0.0000	0.0000	0.0000
	1	0.3917	0.0692	0.0076	0.0005	0.0000	0.0000	0.0000	0.0000	0.0000	0.0000	0.0000
	2	0.6769	0.2061	0.0355	0.0036	0.0002	0.0000	0.0000	0.0000	0.0000	0.0000	0.0000
	3	0.8670	0.4114	0.1071	0.0160	0.0013	0.0000	0.0000	0.0000	0.0000	0.0000	0.0000
	4	0.9568	0.6296	0.2375	0.0510	0.0059	0.0003	0.0000	0.0000	0.0000	0.0000	0.0000
	5	0.9887	0.8042	0.4164	0.1256	0.0207	0.0016	0.0000	0.0000	0.0000	0.0000	0.0000
	6	0.9976	0.9133	0.6080	0.2500	0.0577	0.0065	0.0003	0.0000	0.0000	0.0000	0.0000
	7	0.9996	0.9679	0.7723	0.4159	0.1316	0.0210	0.0013	0.0000	0.0000	0.0000	0.0000
	8	0.9999	0.9900	0.8867	0.5956	0.2517	0.0565	0.0051	0.0001	0.0000	0.0000	0.0000
	9	1.0000	0.9974	0.9520	0.7553	0.4119	0.1275	0.0171	0.0006	0.0000	0.0000	0.0000
	10	1.0000	0.9994	0.9829	0.8725	0.5881	0.2447	0.0480	0.0026	0.0000	0.0000	0.0000
	11	1.0000	0.9999	0.9949	0.9435	0.7483	0.4044	0.1133	0.0100	0.0001	0.0000	0.0000
	12	1.0000	1.0000	0.9987	0.9790	0.8684	0.5841	0.2277	0.0321	0.0004	0.0000	0.0000
	13	1.0000	1.0000	0.9997	0.9935	0.9423	0.7500	0.3920	0.0867	0.0024	0.0000	0.0000
	14	1.0000	1.0000	1.0000	0.9984	0.9793	0.8744	0.5836	0.1958	0.0113	0.0003	0.0000
	15	1.0000	1.0000	1.0000	0.9997	0.9941	0.9490	0.7625	0.3704	0.0432	0.0026	0.0000
	16	1.0000	1.0000	1.0000	1.0000	0.9987	0.9840	0.8929	0.5886	0.1330	0.0159	0.0000
	17	1.0000	1.0000	1.0000	1.0000	0.9998	0.9964	0.9645	0.7939	0.3231	0.0755	0.0010
	18	1.0000	1.0000	1.0000	1.0000	1.0000	0.9995	0.9924	0.9308	0.6083	0.2642	0.0169
	19	1.0000	1.0000	1.0000	1.0000	1.0000	1.0000	0.9992	0.9885	0.8784	0.6415	0.1821

Montgomery, Douglas C., and George C. Runger, *Applied Statistics and Probability for Engineers*, 4th ed. Reproduced by permission of John Wiley & Sons, 2007.

Table 6.2 *Unit Normal Distribution*

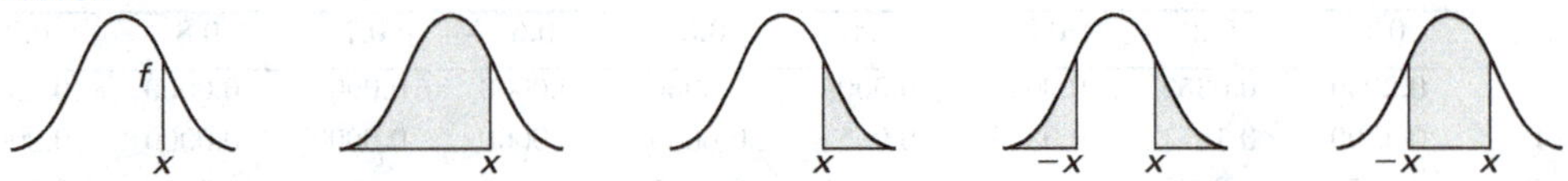

x	$f(x)$	$F(x)$	$R(x)$	$2R(x)$	$W(x)$
0.0	0.3989	0.5000	0.5000	1.0000	0.0000
0.1	0.3970	0.5398	0.4602	0.9203	0.0797
0.2	0.3910	0.5793	0.4207	0.8415	0.1585
0.3	0.3814	0.6179	0.3821	0.7642	0.2358
0.4	0.3683	0.6554	0.3446	0.6892	0.3108
0.5	0.3521	0.6915	0.3085	0.6171	0.3829
0.6	0.3332	0.7257	0.2743	0.5485	0.4515
0.7	0.3123	0.7580	0.2420	0.4839	0.5161
0.8	0.2897	0.7881	0.2119	0.4237	0.5763
0.9	0.2661	0.8159	0.1841	0.3681	0.6319
1.0	0.2420	0.8413	0.1587	0.3173	0.6827
1.1	0.2179	0.8643	0.1357	0.2713	0.7287
1.2	0.1942	0.8849	0.1151	0.2301	0.7699
1.3	0.1714	0.9032	0.0968	0.1936	0.8064
1.4	0.1497	0.9192	0.0808	0.1615	0.8385
1.5	0.1295	0.9332	0.0668	0.1336	0.8664
1.6	0.1109	0.9452	0.0548	0.1096	0.8904
1.7	0.0940	0.9554	0.0446	0.0891	0.9109
1.8	0.0790	0.9641	0.0359	0.0719	0.9281
1.9	0.0656	0.9713	0.0287	0.0574	0.9426
2.0	0.0540	0.9772	0.0228	0.0455	0.9545
2.1	0.0440	0.9821	0.0179	0.0357	0.9643
2.2	0.0355	0.9861	0.0139	0.0278	0.9722
2.3	0.0283	0.9893	0.0107	0.0214	0.9786
2.4	0.0224	0.9918	0.0082	0.0164	0.9836
2.5	0.0175	0.9938	0.0062	0.0124	0.9876
2.6	0.0136	0.9953	0.0047	0.0093	0.9907
2.7	0.0104	0.9965	0.0035	0.0069	0.9931
2.8	0.0079	0.9974	0.0026	0.0051	0.9949
2.9	0.0060	0.9981	0.0019	0.0037	0.9963
3.0	0.0044	0.9987	0.0013	0.0027	0.9973
Fractiles					
1.2816	0.1755	0.9000	0.1000	0.2000	0.8000
1.6449	0.1031	0.9500	0.0500	0.1000	0.9000
1.9600	0.0584	0.9750	0.0250	0.0500	0.9500
2.0537	0.0484	0.9800	0.0200	0.0400	0.9600
2.3263	0.0267	0.9900	0.0100	0.0200	0.9800
2.5758	0.0145	0.9950	0.0050	0.0100	0.9900

Table 6.3 *Values of x for Various Two-Tail Confidence Intervals*

confidence interval level, C	two-tail limit, x ($Z_{\alpha/2}$)
80%	1.2816
90%	1.6449
95%	1.9600
96%	2.0537
98%	2.3263
99%	2.5758

Table 6.4 *Student's t-Distribution (values of t for ν degrees of freedom (sample size n + 1); 1 − α confidence level)*

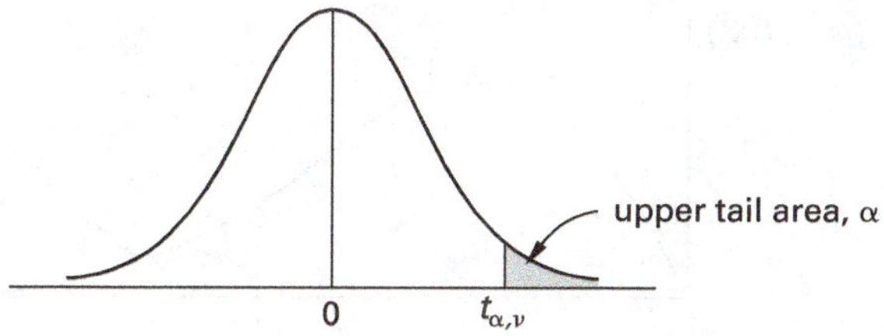

area under the upper tail

ν*	α = 0.25	α = 0.20	α = 0.15	α = 0.10	α = 0.05	α = 0.025	α = 0.01	α = 0.005	ν*
1	1.000	1.376	1.963	3.078	6.314	12.706	31.821	63.657	1
2	0.816	1.061	1.386	1.886	2.920	4.303	6.965	9.925	2
3	0.765	0.978	1.350	1.638	2.353	3.182	4.541	5.841	3
4	0.741	0.941	1.190	1.533	2.132	2.776	3.747	4.604	4
5	0.727	0.920	1.156	1.476	2.015	2.571	3.365	4.032	5
6	0.718	0.906	1.134	1.440	1.943	2.447	3.143	3.707	6
7	0.711	0.896	1.119	1.415	1.895	2.365	2.998	3.499	7
8	0.706	0.889	1.108	1.397	1.860	2.306	2.896	3.355	8
9	0.703	0.883	1.100	1.383	1.833	2.262	2.821	3.250	9
10	0.700	0.879	1.093	1.372	1.812	2.228	2.764	3.169	10
11	0.697	0.876	1.088	1.363	1.796	2.201	2.718	3.106	11
12	0.695	0.873	1.083	1.356	1.782	2.179	2.681	3.055	12
13	0.694	0.870	1.079	1.350	1.771	2.160	2.650	3.012	13
14	0.692	0.868	1.076	1.345	1.761	2.145	2.624	2.977	14
15	0.691	0.866	1.074	1.341	1.753	2.131	2.602	2.947	15
16	0.690	0.865	1.071	1.337	1.746	2.120	2.583	2.921	16
17	0.689	0.863	1.069	1.333	1.740	2.110	2.567	2.898	17
18	0.688	0.862	1.067	1.330	1.734	2.101	2.552	2.878	18
19	0.688	0.861	1.066	1.328	1.729	2.093	2.539	2.861	19
20	0.687	0.860	1.064	1.325	1.725	2.086	2.528	2.845	20
21	0.686	0.859	1.063	1.323	1.721	2.080	2.518	2.831	21
22	0.686	0.858	1.061	1.321	1.717	2.074	2.508	2.819	22
23	0.685	0.858	1.060	1.319	1.714	2.069	2.500	2.807	23
24	0.685	0.857	1.059	1.318	1.711	2.064	2.492	2.797	24
25	0.684	0.856	1.058	1.316	1.708	2.060	2.485	2.787	25
26	0.684	0.856	1.058	1.315	1.706	2.056	2.479	2.779	26
27	0.684	0.855	1.057	1.314	1.703	2.052	2.473	2.771	27
28	0.683	0.855	1.056	1.313	1.701	2.048	2.467	2.763	28
29	0.683	0.854	1.055	1.311	1.699	2.045	2.462	2.756	29
30	0.683	0.854	1.055	1.310	1.697	2.042	2.457	2.750	30
∞	0.674	0.842	1.036	1.282	1.645	1.960	2.326	2.576	∞

*The number of independent degrees of freedom, ν, is always one less than the sample size, n.

Table 6.5 *Critical Values of Chi-Squared Distribution*

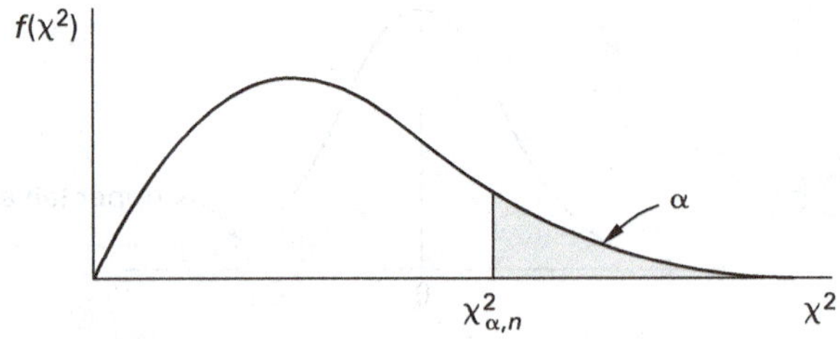

degrees of freedom, ν	$\chi^2_{0.995}$	$\chi^2_{0.990}$	$\chi^2_{0.975}$	$\chi^2_{0.950}$	$\chi^2_{0.900}$	$\chi^2_{0.100}$	$\chi^2_{0.050}$	$\chi^2_{0.025}$	$\chi^2_{0.010}$	$\chi^2_{0.005}$
1	0.0000393	0.0001571	0.0009821	0.0039321	0.0157908	2.70554	3.84146	5.02389	6.6349	7.87944
2	0.0100251	0.0201007	0.0506356	0.102587	0.21072	4.60517	5.99147	7.37776	9.21034	10.5966
3	0.0717212	0.114832	0.215795	0.351846	0.584375	6.25139	7.81473	9.3484	11.3449	12.8381
4	0.20699	0.29711	0.484419	0.710721	1.063623	7.77944	9.48773	11.1433	13.2767	14.8602
5	0.41174	0.5543	0.831211	1.145476	1.61031	9.23635	11.0705	12.8325	15.0863	16.7496
6	0.675727	0.872085	1.237347	1.63539	2.20413	10.6446	12.5916	14.4494	16.8119	18.5476
7	0.989265	1.239043	1.68987	2.16735	2.83311	12.017	14.0671	16.0128	18.4753	20.2777
8	1.344419	1.646482	2.17973	2.73264	3.48954	13.3616	15.5073	17.5346	20.0902	21.955
9	1.734926	2.087912	2.70039	3.32511	4.16816	14.6837	16.919	19.0228	21.666	23.5893
10	2.15585	2.55821	3.24697	3.9403	4.86518	15.9871	18.307	20.4831	23.2093	25.1882
11	2.60321	3.05347	3.81575	4.57481	5.57779	17.275	19.6751	21.92	24.725	26.7569
12	3.07382	3.57056	4.40379	5.22603	6.3038	18.5494	21.0261	23.3367	26.217	28.2995
13	3.56503	4.10691	5.00874	5.89186	7.0415	19.8119	22.3621	24.7356	27.6883	29.8194
14	4.07468	4.66043	5.62872	6.57063	7.78953	21.0642	23.6848	26.119	29.1413	31.3193
15	4.60094	5.22935	6.26214	7.26094	8.54675	22.3072	24.9958	27.4884	30.5779	32.8013
16	5.14224	5.81221	6.90766	7.96164	9.31223	23.5418	26.2962	28.8454	31.9999	34.2672
17	5.69724	6.40776	7.56418	8.67176	10.0852	24.769	27.5871	30.191	33.4087	35.7185
18	6.26481	7.01491	8.23075	9.39046	10.8649	25.9894	28.8693	31.5264	34.8053	37.1564
19	6.84398	7.63273	8.90655	10.117	11.6509	27.2036	30.1435	32.8523	36.1908	38.5822
20	7.43386	8.2604	9.59083	10.8508	12.4426	28.412	31.4104	34.1696	37.5662	39.9968
21	8.03366	8.8972	10.28293	11.5913	13.2396	29.6151	32.6705	35.4789	38.9321	41.401
22	8.64272	9.54249	10.9823	12.338	14.0415	30.8133	33.9244	36.7807	40.2894	42.7956
23	9.26042	10.19567	11.6885	13.0905	14.8479	32.0069	35.1725	38.0757	41.6384	44.1813
24	9.88623	10.8564	12.4011	13.8484	15.6587	33.1963	36.4151	39.3641	42.9798	45.5585
25	10.5197	11.524	13.1197	14.6114	16.4734	34.3816	37.6525	40.6465	44.3141	46.9278
26	11.1603	12.1981	13.8439	15.3791	17.2919	35.5631	38.8852	41.9232	45.6417	48.2899
27	11.8076	12.8786	14.5733	16.1513	18.1138	36.7412	40.1133	43.1944	46.963	49.6449
28	12.4613	13.5648	15.3079	16.9279	18.9392	37.9159	41.3372	44.4607	48.2782	50.9933
29	13.1211	14.2565	16.0471	17.7083	19.7677	39.0875	42.5569	45.7222	49.5879	52.3356
30	13.7867	14.9535	16.7908	18.4926	20.5992	40.256	43.7729	46.9792	50.8922	53.672
40	20.7065	22.1643	24.4331	26.5093	29.0505	51.805	55.7585	59.3417	63.6907	66.7659
50	27.9907	29.7067	32.3574	34.7642	37.6886	63.1671	67.5048	71.4202	76.1539	79.49
60	35.5346	37.4848	40.4817	43.1879	46.4589	74.397	79.0819	83.2976	88.3794	91.9517
70	43.2752	45.4418	48.7576	51.7393	55.329	85.5271	90.5312	95.0231	100.425	104.215
80	51.172	53.54	57.1532	60.3915	64.2778	96.5782	101.879	106.629	112.329	116.321
90	59.1963	61.7541	65.6466	69.126	73.2912	107.565	113.145	118.136	124.116	128.299
100	67.3276	70.0648	74.2219	77.9295	82.3581	118.498	124.342	129.561	135.807	140.169

Table 6.6 *Critical Values of F*

For a particular combination of numerator and denominator degrees of freedom, entry represents the critical values of F corresponding to a specified upper tail area (α).

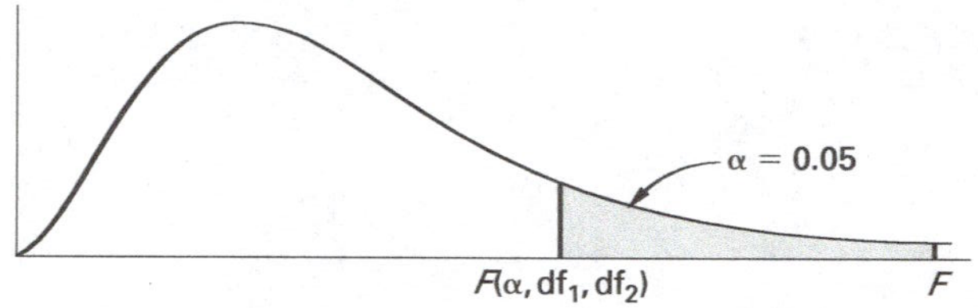

denominator df_2	numerator df_1: 1	2	3	4	5	6	7	8	9	10	12	15	20	24	30	40	60	120	∞
1	161.4	199.5	215.7	224.6	230.2	234.0	236.8	238.9	240.5	241.9	243.9	245.9	248.0	249.1	250.1	251.1	252.2	253.3	254.3
2	18.51	19.00	19.16	19.25	19.30	19.33	19.35	19.37	19.38	19.40	19.41	19.43	19.45	19.45	19.46	19.47	19.48	19.49	19.50
3	10.13	9.55	9.28	9.12	9.01	8.94	8.89	8.85	8.81	8.79	8.74	8.70	8.66	8.64	8.62	8.59	8.57	8.55	8.53
4	7.71	6.94	6.59	6.39	6.26	6.16	6.09	6.04	6.00	5.96	5.91	5.86	5.80	5.77	5.75	5.72	5.69	5.66	5.63
5	6.61	5.79	5.41	5.19	5.05	4.95	4.88	4.82	4.77	4.74	4.68	4.62	4.56	4.53	4.50	4.46	4.43	4.40	4.36
6	5.99	5.14	4.76	4.53	4.39	4.28	4.21	4.15	4.10	4.06	4.00	3.94	3.87	3.84	3.81	3.77	3.74	3.70	3.67
7	5.59	4.74	4.35	4.12	3.97	3.87	3.79	3.73	3.68	3.64	3.57	3.51	3.44	3.41	3.38	3.34	3.30	3.27	3.23
8	5.32	4.46	4.07	3.84	3.69	3.58	3.50	3.44	3.39	3.35	3.28	3.22	3.15	3.12	3.08	3.04	3.01	2.97	2.93
9	5.12	4.26	3.86	3.63	3.48	3.37	3.29	3.23	3.18	3.14	3.07	3.01	2.94	2.90	2.86	2.83	2.79	2.75	2.71
10	4.96	4.10	3.71	3.48	3.33	3.22	3.14	3.07	3.02	2.98	2.91	2.85	2.77	2.74	2.70	2.66	2.62	2.58	2.54
11	4.84	3.98	3.59	3.36	3.20	3.09	3.01	2.95	2.90	2.85	2.79	2.72	2.65	2.61	2.57	2.53	2.49	2.45	2.40
12	4.75	3.89	3.49	3.26	3.11	3.00	2.91	2.85	2.80	2.75	2.69	2.62	2.54	2.51	2.47	2.43	2.38	2.34	2.30
13	4.67	3.81	3.41	3.18	3.03	2.92	2.83	2.77	2.71	2.67	2.60	2.53	2.46	2.42	2.38	2.34	2.30	2.25	2.21
14	4.60	3.74	3.34	3.11	2.96	2.85	2.76	2.70	2.65	2.60	2.53	2.46	2.39	2.35	2.31	2.27	2.22	2.18	2.13
15	4.54	3.68	3.29	3.06	2.90	2.79	2.71	2.64	2.59	2.54	2.48	2.40	2.33	2.29	2.25	2.20	2.16	2.11	2.07
16	4.49	3.63	3.24	3.01	2.85	2.74	2.66	2.59	2.54	2.49	2.42	2.35	2.28	2.24	2.19	2.15	2.11	2.06	2.01
17	4.45	3.59	3.20	2.96	2.81	2.70	2.61	2.55	2.49	2.45	2.38	2.31	2.23	2.19	2.15	2.10	2.06	2.01	1.96
18	4.41	3.55	3.16	2.93	2.77	2.66	2.58	2.51	2.46	2.41	2.34	2.27	2.19	2.15	2.11	2.06	2.02	1.97	1.92
19	4.38	3.52	3.13	2.90	2.74	2.63	2.54	2.48	2.42	2.38	2.31	2.23	2.16	2.11	2.07	2.03	1.98	1.93	1.88
20	4.35	3.49	3.10	2.87	2.71	2.60	2.51	2.45	2.39	2.35	2.28	2.20	2.12	2.08	2.04	1.99	1.95	1.90	1.84
21	4.32	3.47	3.07	2.84	2.68	2.57	2.49	2.42	2.37	2.32	2.25	2.18	2.10	2.05	2.01	1.96	1.92	1.87	1.81
22	4.30	3.44	3.05	2.82	2.66	2.55	2.46	2.40	2.34	2.30	2.23	2.15	2.07	2.03	1.98	1.94	1.89	1.84	1.78
23	4.28	3.42	3.03	2.80	2.64	2.53	2.44	2.37	2.32	2.27	2.20	2.13	2.05	2.01	1.96	1.91	1.86	1.81	1.76
24	4.26	3.40	3.01	2.78	2.62	2.51	2.42	2.36	2.30	2.25	2.18	2.11	2.03	1.98	1.94	1.89	1.84	1.79	1.73
25	4.24	3.39	2.99	2.76	2.60	2.49	2.40	2.34	2.28	2.24	2.16	2.09	2.01	1.96	1.92	1.87	1.82	1.77	1.71
26	4.23	3.37	2.98	2.74	2.59	2.47	2.39	2.32	2.27	2.22	2.15	2.07	1.99	1.95	1.90	1.85	1.80	1.75	1.69
27	4.21	3.35	2.96	2.73	2.57	2.46	2.37	2.31	2.25	2.20	2.13	2.06	1.97	1.93	1.88	1.84	1.79	1.73	1.67
28	4.20	3.34	2.95	2.71	2.56	2.45	2.36	2.29	2.24	2.19	2.12	2.04	1.96	1.91	1.87	1.82	1.77	1.71	1.65
29	4.18	3.33	2.93	2.70	2.55	2.43	2.35	2.28	2.22	2.18	2.10	2.03	1.94	1.90	1.85	1.81	1.75	1.70	1.64
30	4.17	3.32	2.92	2.69	2.53	2.42	2.33	2.27	2.21	2.16	2.09	2.01	1.93	1.89	1.84	1.79	1.74	1.68	1.62
40	4.08	3.23	2.84	2.61	2.45	2.34	2.25	2.18	2.12	2.08	2.00	1.92	1.84	1.79	1.74	1.69	1.64	1.58	1.51
60	4.00	3.15	2.76	2.53	2.37	2.25	2.17	2.10	2.04	1.99	1.92	1.84	1.75	1.70	1.65	1.59	1.53	1.47	1.39
120	3.92	3.07	2.68	2.45	2.29	2.17	2.09	2.02	1.96	1.91	1.83	1.75	1.66	1.61	1.55	1.50	1.43	1.35	1.25
∞	3.84	3.00	2.60	2.37	2.21	2.10	2.01	1.94	1.88	1.83	1.75	1.67	1.57	1.52	1.46	1.39	1.32	1.22	1.00

Topic III: Fluid Mechanics

7 Fluid Properties

Nomenclature

A	area	m^2
d	diameter	m
F	force	N
g	gravitational acceleration, 9.81	m/s^2
h	height	m
K	power law consistency index	–
L	length	m
m	mass	kg
n	power law index	–
p	pressure	Pa
r	radius	m
SG	specific gravity	–
v	velocity	m/s
V	volume	m^3
W	weight	N

Symbols*

β	angle of contact	deg
γ	specific (unit) weight	N/m^3
δ	thickness of fluid	m
μ	absolute viscosity	Pa·s
ν	kinematic viscosity	m^2/s
ρ	density	kg/m^3
σ	surface tension	N/m
τ	stress	Pa
υ	specific volume	m^3/kg

*The NCEES *FE Reference Handbook* (*NCEES Handbook*) uses the symbol τ for both normal and shear stress. τ is almost universally interpreted in engineering practice as the symbol for shear stress. The use of τ stems from Cauchy stress tensor theory and the desire to use the same symbol for all nine stress directions. However, the stress tensor concept is not developed in the *NCEES Handbook*, and σ is used as the symbol for normal stress elsewhere, so the use of τ_n for normal stress and τ_t for tangential (shear) stress may be confusing. (This usage does avoid a symbol conflict with surface tension, σ, which is used in the *NCEES Handbook* in contexts unrelated to stress.)

Subscripts

n	normal
t	tangential (shear)
v	vapor
w	water

1. FLUIDS

A *fluid* is a substance in either the liquid or gas phase. Fluids cannot support shear, and they deform continuously to minimize applied shear forces.

In fluid mechanics, a fluid is modeled as a *continuum*—that is, a substance that can be divided into infinitesimally small volumes, with properties that are continuous functions over the entire volume. For the infinitesimally small volume ΔV, Δm is the infinitesimal mass, and ΔW is the infinitesimal weight.

Equation 7.1: Density

$$\rho = \operatorname*{limit}_{\Delta V \to 0} \Delta m / \Delta V \qquad 7.1$$

Variation

$$\rho = \frac{m}{V}$$

Description

The *density*, ρ, also called *mass density*, of a fluid is its mass per unit volume. The density of a fluid in a liquid form is usually given, known in advance, or easily obtained from tables.

If ΔV is the volume of an infinitesimally small element, the density is given as Eq. 7.1. Density is typically measured in kg/m^3.

Specific Volume

Specific volume, υ, is the volume occupied by a unit mass of fluid.

$$\upsilon = \frac{1}{\rho}$$

Specific volume is the reciprocal of density and is typically measured in m^3/kg.

Equation 7.2 Through Eq. 7.4: Specific Weight

$$\gamma = \lim_{\Delta V \to 0} \Delta W / \Delta V \qquad 7.2$$

$$\gamma = \lim_{\Delta V \to 0} g \Delta m / \Delta V = \rho g \qquad 7.3$$

$$\gamma = \rho g \qquad 7.4$$

Variation

$$\gamma = \frac{W}{V} = \frac{mg}{V}$$

Description

Specific weight, γ, also known as *unit weight*, is the weight of substance per unit volume.

The use of specific weight is most often encountered in civil engineering work in the United States, where it is commonly called *density*. The usual units of specific weight are N/m^3. Specific weight is not an absolute property of a substance since it depends on the local gravitational field.

Example

The density of a gas is 1.5 kg/m^3. The specific weight of the gas is most nearly

(A) 9.0 N/m^3

(B) 15 N/m^3

(C) 76 N/m^3

(D) 98 N/m^3

Solution

Use Eq. 7.4.

$$\gamma = \rho g = \left(1.5 \ \frac{\text{kg}}{\text{m}^3}\right)\left(9.81 \ \frac{\text{m}}{\text{s}^2}\right)$$
$$= 14.715 \ \text{kg/s}^2 \cdot \text{m}^2 \quad (15 \ \text{N/m}^3)$$

The answer is (B).

Equation 7.5: Specific Gravity

$$\text{SG} = \gamma/\gamma_w = \rho/\rho_w \qquad 7.5$$

Description

Specific gravity, SG, is the dimensionless ratio of a fluid's density to a standard reference density. For liquids and solids, the reference is the density of pure water, which is approximately 1000 kg/m^3 over the normal ambient temperature range. The temperature at which water density should be evaluated is not standardized, so some small variation in the reference density is possible. See Table 7.1 and Table 7.2 for the properties of water in SI and customary U.S. units, respectively.

Since the SI density of water is very nearly 1.000 g/cm^3 (1000 kg/m^3), the numerical values of density in g/cm^3 and specific gravity are the same.

Example

A fluid has a density of 860 kg/m^3. The specific gravity of the fluid is most nearly

(A) 0.63

(B) 0.82

(C) 0.86

(D) 0.95

Solution

Use Eq. 7.5. The specific gravity is

$$\text{SG} = \rho / \rho_w = \frac{860 \frac{\text{kg}}{\text{m}^3}}{1000 \frac{\text{kg}}{\text{m}^3}} = 0.86$$

The answer is (C).

2. PRESSURE

Fluid pressures are measured with respect to two pressure references: zero pressure and atmospheric pressure. Pressures measured with respect to a true zero pressure reference are known as *absolute pressures.* Pressures measured with respect to atmospheric pressure are known as *gage pressures.* To distinguish them, the word "gage" or "absolute" can be added to the measurement (e.g., 25.1 kPa absolute). Alternatively, the letter "g" can be added to the measurement for gage pressures (e.g., 15 kPag), and the pressure is assumed to be absolute otherwise.

Table 7.1 Properties of Water (SI units)

temperature (°C)	specific weight, γ (kN/m^3)	density, ρ (kg/m^3)	viscosity, $\mu \times 10^3$ (Pa·s)	kinematic viscosity, $\nu \times 10^6$ (m^2/s)	vapor pressure, p_v (kPa)
0	9.805	999.8	1.781	1.785	0.61
5	9.807	1000.0	1.518	1.518	0.87
10	9.804	999.7	1.307	1.306	1.23
15	9.798	999.1	1.139	1.139	1.70
20	9.789	998.2	1.002	1.003	2.34
25	9.777	997.0	0.890	0.893	3.17
30	9.764	995.7	0.798	0.800	4.24
40	9.730	992.2	0.653	0.658	7.38
50	9.689	988.0	0.547	0.553	12.33
60	9.642	983.2	0.466	0.474	19.92
70	9.589	977.8	0.404	0.413	31.16
80	9.530	971.8	0.354	0.364	47.34
90	9.466	965.3	0.315	0.326	70.10
100	9.399	958.4	0.282	0.294	101.33

Table 7.2 Properties of Water (customary U.S. units)

temperature (°F)	specific weight, γ (lbf/ft^3)	density, ρ (lbm-sec^2/ft^4)	viscosity, $\mu \times 10^{-5}$ (lbf-sec/ft^2)	kinematic viscosity, $\nu \times 10^{-5}$ (ft^2/sec)	vapor pressure, p_v (lbf/ft^2)
32	62.42	1.940	3.746	1.931	0.09
40	62.43	1.940	3.229	1.664	0.12
50	62.41	1.940	2.735	1.410	0.18
60	62.37	1.938	2.359	1.217	0.26
70	62.30	1.936	2.050	1.059	0.36
80	62.22	1.934	1.799	0.930	0.51
90	62.11	1.931	1.595	0.826	0.70
100	62.00	1.927	1.424	0.739	0.95
110	61.86	1.923	1.284	0.667	1.24
120	61.71	1.918	1.168	0.609	1.69
130	61.55	1.913	1.069	0.558	2.22
140	61.38	1.908	0.981	0.514	2.89
150	61.20	1.902	0.905	0.476	3.72
160	61.00	1.896	0.838	0.442	4.74
170	60.80	1.890	0.780	0.413	5.99
180	60.58	1.883	0.726	0.385	7.51
190	60.36	1.876	0.678	0.362	9.34
200	60.12	1.868	0.637	0.341	11.52
212	59.83	1.860	0.593	0.319	14.70

Equation 7.6 and Eq. 7.7: Absolute Pressure

$$\text{absolute pressure} = \text{atmospheric pressure} + \text{gage pressure reading} \quad 7.6$$

$$\text{absolute pressure} = \text{atmospheric pressure} - \text{vacuum gage pressure reading} \quad 7.7$$

Values

Standard atmospheric pressure is equal to 101.3 kPa or 29.921 inches of mercury.

Description

Absolute and gage pressures are related by Eq. 7.6. In this equation, "atmospheric pressure" is the actual atmospheric pressure that exists when the gage measurement is taken. It is not standard atmospheric pressure unless that pressure is implicitly or explicitly applicable. Also, since a barometer measures atmospheric pressure, *barometric pressure* is synonymous with atmospheric pressure.

A *vacuum* measurement is implicitly a pressure below atmospheric pressure (i.e., a negative gage pressure). It must be assumed that any measured quantity given as a vacuum is a quantity to be subtracted from the atmospheric pressure. (See Eq. 7.7.) When a condenser is operating with a vacuum of 4.0 inches of mercury, the absolute pressure is approximately 29.92 − 4.0 = 25.92 inches of mercury (25.92 in Hg). Vacuums are always stated as positive numbers.

Example

A vessel is initially connected to a reservoir open to the atmosphere. The connecting valve is then closed, and a vacuum of 65.5 kPa is applied to the vessel. Assume standard atmospheric pressure. What is most nearly the absolute pressure in the vessel?

(A) 36 kPa

(B) 66 kPa

(C) 86 kPa

(D) 110 kPa

Solution

From Eq. 7.7, for vacuum pressures,

$$\begin{aligned}\text{absolute pressure} &= \text{atmospheric pressure} - \text{vacuum gage pressure reading}\\ &= 101.3\ \text{kPa} - 65.5\ \text{kPa}\\ &= 35.8\ \text{kPa} \quad (36\ \text{kPa})\end{aligned}$$

The answer is (A).

3. STRESS

Stress, τ, is force per unit area. There are two primary types of stress, differing in the orientation of the loaded area: *normal stress* and *tangential* (or *shear*) *stress.* With *normal stress*, τ_n, the area is normal to the force carried. With *tangential* (or *shear*) *stress*, τ_t, the area is parallel to the force.

Ideal fluids that are inviscid and incompressible respond to normal stresses, but they cannot support shear, and they deform continuously to minimize applied shear forces.

Equation 7.8 and Eq. 7.9: Normal Stress[1]

$$\tau(1) = \lim_{\Delta A \to 0} \Delta F/\Delta A \quad 7.8$$

$$\tau_n = -p \quad 7.9$$

Description

At some arbitrary point 1, with an infinitesimal area, ΔA, subjected to a force, ΔF, the normal or shear stress is defined as in Eq. 7.8.

Normal stress is equal to the pressure of the fluid, as indicated by Eq. 7.9.

4. VISCOSITY

The *viscosity* of a fluid is a measure of that fluid's resistance to flow when acted upon by an external force, such as a pressure gradient or gravity.

The viscosity of a fluid can be determined with a *sliding plate viscometer* test. Consider two plates of area A separated by a fluid with thickness δ. The bottom plate is fixed, and the top plate is kept in motion at a constant velocity, v, by a force, F. (See Fig. 7.1.)

[1]Equation 7.8, as given in the *NCEES Handbook*, is vague. The *NCEES Handbook* calls $\tau(1)$ the "surface stress at point 1." Point 1 is undefined, and the term "surface stress" is used without explanation, although it apparently refers to both normal and shear stress. The format of using parentheses to designate the location of a stress is not used elsewhere in the *NCEES Handbook*.

Figure 7.1 *Sliding Plate Viscometer*

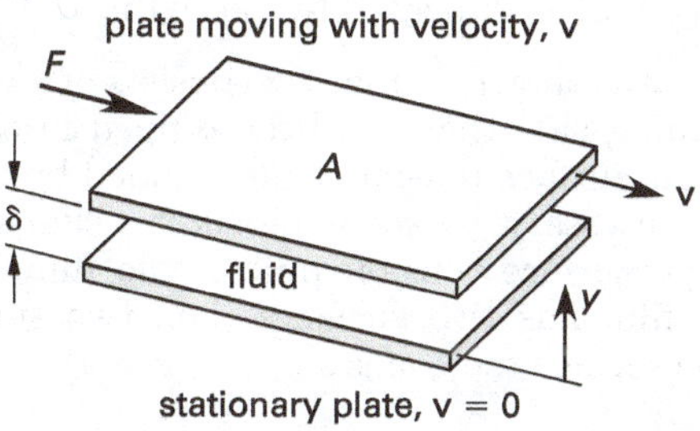

Experiments with many fluids have shown that the force, F, that is needed to maintain the velocity, v, is proportional to the velocity and the area but is inversely proportional to the separation of the plates. That is,

$$\frac{F}{A} \propto \frac{\mathrm{v}}{\delta}$$

The constant of proportionality needed to make this an equality for a particular fluid is the fluid's *absolute viscosity*, μ, also known as the *absolute dynamic viscosity*. Typical units for absolute viscosity are Pa·s (N·s/m^2).

$$\frac{F}{A} = \mu\left(\frac{\mathrm{v}}{\delta}\right)$$

F/A is the *fluid shear stress* (tangential stress), τ_t.

Equation 7.10 Through Eq. 7.12: Newton's Law of Viscosity

$$\mathrm{v}(y) = \mathrm{v}y/\delta \qquad 7.10$$

$$d\mathrm{v}/dy = \mathrm{v}/\delta \qquad 7.11$$

$$\tau_t = \mu(d\mathrm{v}/dy) \quad \text{[one-dimensional]} \qquad 7.12$$

Variation

$$\tau_t = \frac{F}{A} = \mu\left(\frac{\mathrm{v}}{\delta}\right)$$

Description

For a thin Newtonian fluid film, Eq. 7.10 and Eq. 7.11 describe the linear velocity profile. The quantity $d\mathrm{v}/dy$ is known by various names, including *rate of strain*, *shear rate*, *velocity gradient*, and *rate of shear formation*.

Equation 7.12 is known as *Newton's law of viscosity*, from which Newtonian fluids get their name. (Not all fluids are Newtonian, although many are.) For a Newtonian fluid, strains are proportional to the applied shear stress (i.e., the stress versus strain curve is a straight line with slope μ). The straight line will be closer to the τ axis if the fluid is highly viscous. For low-viscosity fluids, the straight line will be closer to the $d\mathrm{v}/dy$ axis. Equation 7.12 is applicable only to Newtonian fluids, for which the relationship is linear.

Equation 7.13: Power Law

$$\tau_t = K(d\mathrm{v}/dy)^n \qquad 7.13$$

Values

fluid	power law index, n
Newtonian	1
non-Newtonian	
pseudoplastic	<1
dilatant	>1

Description

Many fluids are not Newtonian (i.e., do not behave according to Eq. 7.12). Non-Newtonian fluids have viscosities that change with shear rate, $d\mathrm{v}/dt$. For example, *pseudoplastic fluids* exhibit a decrease in viscosity the faster they are agitated. Such fluids present no serious pumping difficulties. On the other hand, pumps for *dilatant fluids* must be designed carefully, since dilatant fluids exhibit viscosities that increase the faster they are agitated. The fluid shear stress for most non-Newtonian fluids can be predicted by the *power law*, Eq. 7.13. In Eq. 7.13, the constant K is known as the *consistency index*. The consistency index, also known as the *flow consistency index*, is actually the average fluid viscosity across the range of viscosities being modeled. For *pseudoplastic non-Newtonian fluids*, $n<1$; for *dilatant non-Newtonian fluids*, $n>1$. For Newtonian fluids, $n=1$.

Equation 7.14: Kinematic Viscosity

$$\nu = \mu/\rho \qquad 7.14$$

Description

Another quantity with the name viscosity is the ratio of absolute viscosity to mass density. This combination of variables, known as *kinematic viscosity*, ν, appears often in fluids and other problems and warrants its own symbol and name. Kinematic viscosity is merely the name given to a frequently occurring combination of variables. Typical units are m^2/s.

Example

32°C water flows at 2 m/s through a pipe that has an inside diameter of 3 cm. The viscosity of the water is 769×10^{-6} N·s/m^2, and the density of the water is

995 kg/m^3. The kinematic viscosity of the water is most nearly

(A) 0.71×10^{-6} m^2/s

(B) 0.77×10^{-6} m^2/s

(C) 0.84×10^{-6} m^2/s

(D) 0.92×10^{-6} m^2/s

Solution

The kinematic viscosity is

$$\nu = \mu/\rho = \frac{769 \times 10^{-6}\ \frac{\text{N·s}}{\text{m}^2}}{995\ \frac{\text{kg}}{\text{m}^3}}$$
$$= 0.773 \times 10^{-6}\ \text{m}^2/\text{s} \quad (0.77 \times 10^{-6}\ \text{m}^2/\text{s})$$

The answer is (B).

5. SURFACE TENSION AND CAPILLARITY

Equation 7.15: Surface Tension

$$\sigma = F/L \tag{7.15}$$

Description

The membrane or "skin" that seems to form on the free surface of a fluid is caused by intermolecular cohesive forces and is known as *surface tension*, σ. Surface tension is the reason that insects are able to sit on a pond and a needle is able to float on the surface of a glass of water, even though both are denser than the water that supports them. Surface tension also causes bubbles and droplets to form in spheres, since any other shape would have more surface area per unit volume.

Surface tension can be interpreted as the tensile force between two points a unit distance apart on the surface, or as the amount of work required to form a new unit of surface area in an apparatus similar to that shown in Fig. 7.2. Typical units of surface tension are N/m, J/m^2, and dynes/cm. (Dynes/cm are equivalent to mN/m.)

Figure 7.2 *Wire Frame for Stretching a Film*

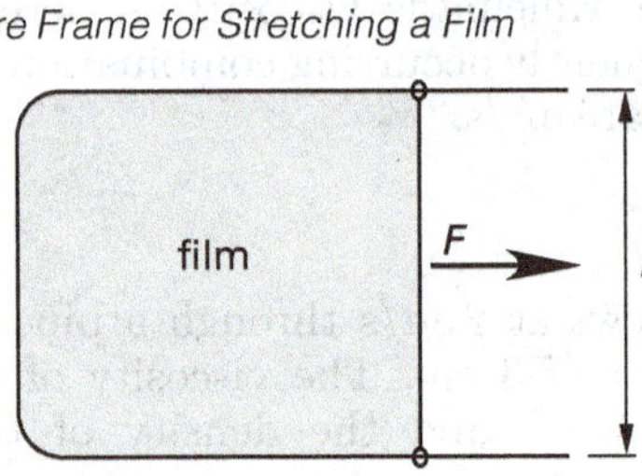

Surface tension is defined as a force, F, acting along a line of length L, as indicated by Eq. 7.15.

The apparatus shown in Fig. 7.2 consists of a wire frame with a sliding side that has been dipped in a liquid to form a film. Surface tension is determined by measuring the force necessary to keep the sliding side stationary against the surface tension pull of the film. However, since the film has two surfaces (i.e., two surface tensions), the surface tension is

$$\sigma = \frac{F}{2L} \quad \left[\begin{array}{c}\text{wire frame}\\ \text{apparatus}\end{array}\right]$$

Surface tension can also be measured by measuring the force required to pull a *Du Nouy wire ring* out of a liquid, as shown in Fig. 7.3. Because the ring's inner and outer sides are both in contact with the liquid, the wetted perimeter is twice the circumference. The surface tension is therefore

$$\sigma = \frac{F}{4\pi r} \quad \left[\begin{array}{c}\text{Du Nouy ring}\\ \text{apparatus}\end{array}\right]$$

Figure 7.3 *Du Nouy Ring Surface Tension Apparatus*

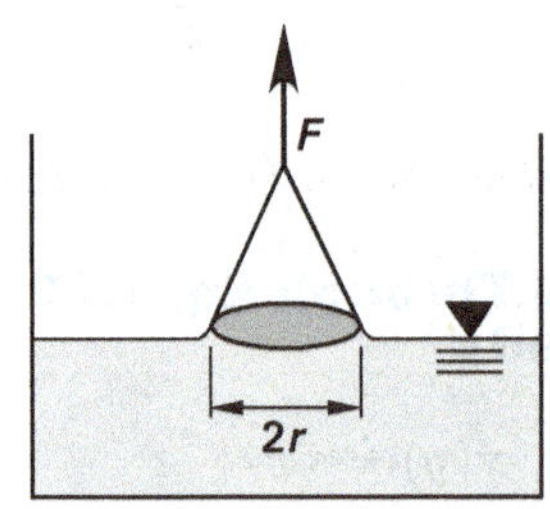

Equation 7.16: Capillary Rise or Depression

$$h = 4\sigma\cos\beta/\gamma d \tag{7.16}$$

Variation

$$h = \frac{4\sigma\cos\beta}{\rho g d_{\text{tube}}}$$

Description

Capillary action is the name given to the behavior of a liquid in a thin-bore tube. Capillary action is caused by surface tension between the liquid and a vertical solid surface. In water, the adhesive forces between the liquid molecules and the surface are greater than (i.e., dominate) the cohesive forces between the water molecules themselves. The adhesive forces cause the water to attach itself to and climb a solid vertical surface; the water rises above the general water surface level. (See

Fig. 7.4.) This is called *capillary rise*, and the curved surface of the liquid within the tube is known as a *meniscus*.

Figure 7.4 Capilarity of Liquids

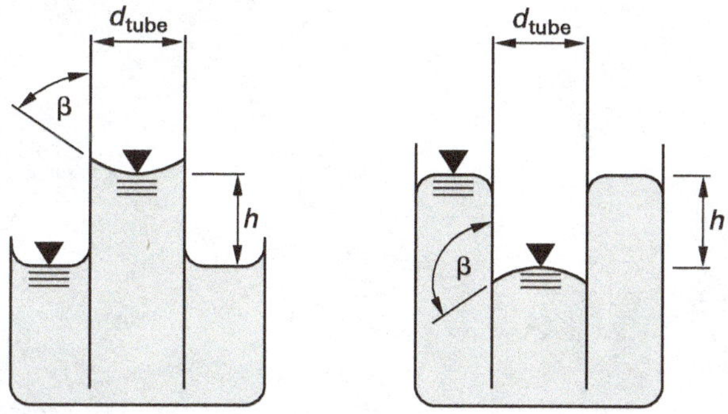

(a) adhesive force dominates (b) cohesive force dominates

For a few liquids, such as mercury, the molecules have a strong affinity for each other (i.e., the cohesive forces dominate). These liquids avoid contact with the tube surface. In such liquids, the meniscus will be below the general surface level, a state called *capillary depression*.

The *angle of contact*, β, is an indication of whether adhesive or cohesive forces dominate. For contact angles less than 90°, adhesive forces dominate. For contact angles greater than 90°, cohesive forces dominate. For water in a glass tube, the contact angle is zero; for mercury in a glass tube, the contact angle is 140°.

Equation 7.16 can be used to predict the capillary rise (if the result is positive) or capillary depression (if the result is negative) in a small-bore tube. Surface tension is a material property of a fluid, and contact angles are specific to a particular fluid-solid interface. Both may be obtained from tables.

Example

An open glass tube with a diameter of 1 mm contains mercury at 20°C. At this temperature, mercury has a surface tension of 0.519 N/m and a density of 13 600 kg/m^3. The contact angle for mercury in a glass tube is 140°. The capillary depression is most nearly

(A) 6.1 mm

(B) 8.6 mm

(C) 12 mm

(D) 17 mm

Solution

Use Eq. 7.4 and Eq. 7.16 to find the capillary depression (or negative rise).

$$h = 4\sigma\cos\beta/\gamma d = \frac{4\sigma\cos\beta}{\rho g d_{\text{tube}}}$$

$$= \frac{(4)\left(0.519\ \frac{\text{N}}{\text{m}}\right)\cos 140°\left(1000\ \frac{\text{mm}}{\text{m}}\right)}{\left(13\,600\ \frac{\text{kg}}{\text{m}^3}\right)\left(9.81\ \frac{\text{m}}{\text{s}^2}\right)(1\ \text{mm})}$$

$$= -0.0119\ \text{m} \quad (12\ \text{mm})$$

The answer is (C).

8 Fluid Statics

Nomenclature

A	area	m^2
F	force	N
g	gravitational acceleration, 9.81	m/s^2
h	vertical depth or difference in vertical depth	m
I	moment of inertia	m^4
p	pressure	Pa
R	resultant force	N
SG	specific gravity	–
V	volume	m^3
y	distance	m
z	elevation	m

Symbols

α	angle	deg
γ	specific (unit) weight	N/m^3
θ	angle	deg
ρ	density	kg/m^3

Subscripts

0	atmospheric
atm	atmospheric
B	barometer fluid
C	centroid
CP	center of pressure
f	fluid
m	manometer
R	resultant
v	vapor
x	horizontal

1. HYDROSTATIC PRESSURE

Hydrostatic pressure is the pressure a fluid exerts on an immersed object or on container walls. The term *hydrostatic* is used with all fluids, not only with water.

Pressure is equal to the force per unit area of surface.

$$p = \frac{F}{A}$$

Hydrostatic pressure in a stationary, incompressible fluid behaves according to the following characteristics.

- Pressure is a function of vertical depth and density only. If density is constant, then the pressure will be the same at two points with identical depths.
- Pressure varies linearly with vertical depth. The relationship between pressure and depth for an incompressible fluid is given by the equation

  $$p = \rho g h = \gamma h$$

  Since ρ and g are constants, this equation shows that p and h are linearly related. One determines the other.
- Pressure is independent of an object's area and size, and of the weight (or mass) of water above the object. Figure 8.1 illustrates the *hydrostatic paradox.* The pressures at depth h are the same in all four columns because pressure depends only on depth, not on volume.

Figure 8.1 *Hydrostatic Paradox*

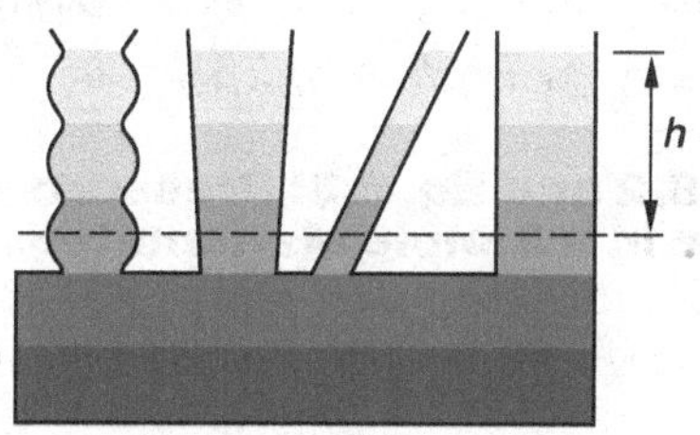

- Pressure at a point has the same magnitude in all directions (*Pascal's law*). Therefore, pressure is a scalar quantity.
- Pressure is always normal to a surface, regardless of the surface's shape or orientation. (This is a result of the fluid's inability to support shear stress.)

Equation 8.1: Pressure Difference in a Static Fluid[1]

$$p_2 - p_1 = -\gamma(z_2 - z_1) = -\gamma h = -\rho g h \qquad 8.1$$

Description

As pressure in a fluid varies linearly with depth, difference in pressure likewise varies linearly with difference in depth. This is expressed in Eq. 8.1. The variable z decreases with depth while the pressure increases with depth, so pressure and elevation have an inverse linear relationship, as indicated by the negative sign.

2. MANOMETERS

Manometers can be used to measure small pressure differences, and for this purpose, they provide good accuracy. A difference in manometer fluid surface heights indicates a pressure difference. When both ends of the manometer are connected to pressure sources, the name *differential manometer* is used. If one end of the manometer is open to the atmosphere, the name *open manometer* is used. An open manometer indicates gage pressure. It is theoretically possible, but impractical, to have a manometer indicate absolute pressure, since one end of the manometer would have to be exposed to a perfect vacuum.

Consider the simple manometer in Fig. 8.2. The pressure difference $p_2 - p_1$ causes the difference h_m in manometer fluid surface heights. Fluid column h_2 exerts a hydrostatic pressure on the manometer fluid, forcing the manometer fluid to the left. This increase must be subtracted out. Similarly, the column h_1 restricts the movement of the manometer fluid. The observed measurement must be increased to correct for this restriction. The typical way to solve for pressure differences in a manometer is to start with the pressure on one side, and then add or subtract changes in hydrostatic pressure at known points along the column until the pressure on the other side is reached.

$$\begin{aligned} p_2 &= p_1 + \rho_1 g h_1 + \rho_m g h_m - \rho_2 g h_2 \\ &= p_1 + \gamma_1 h_1 + \gamma_m h_m - \gamma_2 h_2 \end{aligned}$$

Equation 8.2 and Eq. 8.3: Pressure Difference in a Simple Manometer

$$p_0 = p_2 + \gamma_2 h_2 - \gamma_1 h_1 = p_2 + g(\rho_2 h_2 - \rho_1 h_1) \qquad 8.2$$

Figure 8.2 *Manometer Requiring Corrections*

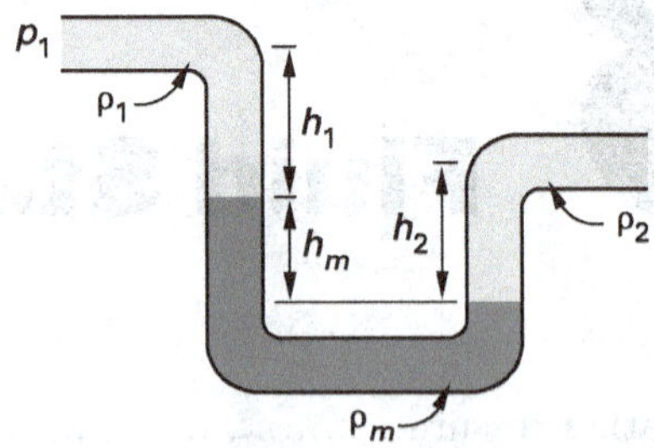

$$p_0 = p_2 + (\gamma_2 - \gamma_1)h = p_2 + (\rho_2 - \rho_1)gh \qquad 8.3$$
$$[h_1 = h_2 = h]$$

Description

Figure 8.3 illustrates an open manometer. Neglecting the air in the open end, the pressure difference is given by Eq. 8.2. $p_0 - p_2$ is the gage pressure in the vessel.

Figure 8.3 *Open Manometer*

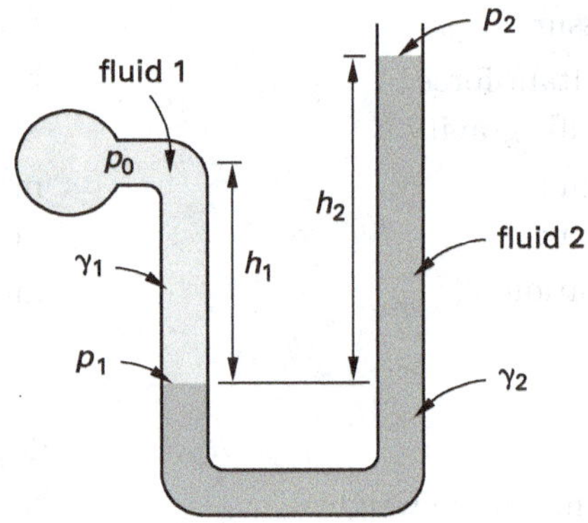

Equation 8.2 is a version of Eq. 8.1 as applied to an open manometer. Equation 8.3 is a simplified version that can be used only when h_1 is equal to h_2.[2]

Example

One leg of a mercury U-tube manometer is connected to a pipe containing water under a gage pressure of 100 kPa. The mercury in this leg stands 0.75 m below the water. The mercury in the other leg is open to the

[1]Although the variable y is used elsewhere in the NCEES *FE Reference Handbook* (*NCEES Handbook*) to represent vertical direction, the *NCEES Handbook* uses z to measure some vertical dimensions within fluid bodies. As referenced to a Cartesian coordinate system (a practice that is not continued elsewhere in the *NCEES Handbook*), Eq. 8.1 is academically correct, but it is inconsistent with normal practice, which measures z from the fluid surface, synonymous with "depth." The *NCEES Handbook* reverts to common usage of h, y, and z in its subsequent discussion of forces on submerged surfaces.

[2]h_1 and h_2 would be equal only in the most contrived situations.

air. The density of the water is 1000 kg/m^3, and the specific gravity of the mercury is 13.6.

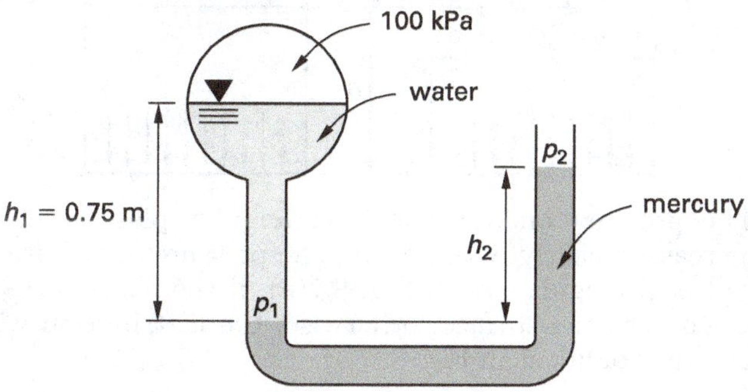

The height of the mercury in the open leg is most nearly

(A) 0.05 m

(B) 0.5 m

(C) 0.8 m

(D) 1 m

Solution

Find the specific weights of water, γ_1, and mercury, γ_2.

$$\begin{aligned}\gamma_1 &= \rho_{\text{water}} g = \left(1000\ \frac{\text{kg}}{\text{m}^3}\right)\left(9.81\ \frac{\text{m}}{\text{s}^2}\right)\\ &= 9810\ \text{N/m}^3\\ \gamma_2 &= \rho_{\text{Hg}} g = (\text{SG}_{\text{Hg}})\rho_{\text{water}} g\\ &= (13.6)\left(1000\ \frac{\text{kg}}{\text{m}^3}\right)\left(9.81\ \frac{\text{m}}{\text{s}^2}\right)\\ &= 133\,416\ \text{N/m}^3\end{aligned}$$

Use Eq. 8.2 to find the height of the mercury in the open leg. Since the 100 kPa pressure is a gage pressure, the atmospheric pressure, p_2, is zero.

$$\begin{aligned}p_0 &= p_2 + \gamma_2 h_2 - \gamma_1 h_1\\ h_2 &= \frac{p_0 + \gamma_1 h_1 - p_2}{\gamma_2}\\ &= \frac{(100\ \text{kPa})\left(1000\ \frac{\text{Pa}}{\text{kPa}}\right) + \left(9810\ \frac{\text{N}}{\text{m}^3}\right)(0.75\ \text{m}) - 0\ \text{Pa}}{133\,416\ \frac{\text{N}}{\text{m}^3}}\\ &= 0.805\ \text{m}\quad (0.8\ \text{m})\end{aligned}$$

The answer is (C).

3. BAROMETERS

The *barometer* is a common device for measuring the absolute pressure of the atmosphere. It is constructed by filling a long tube open at one end with mercury (alcohol or another liquid can also be used) and inverting the tube such that the open end is below the level of the mercury-filled container. The vapor pressure of the mercury in the tube is insignificant; if this is neglected, the fluid column is supported only by the atmospheric pressure transmitted through the container fluid at the lower, open end. (See Fig. 8.4.) In such a case, the atmospheric pressure is given by

$$p_{\text{atm}} = \rho g h = \gamma h$$

Figure 8.4 *Barometer*

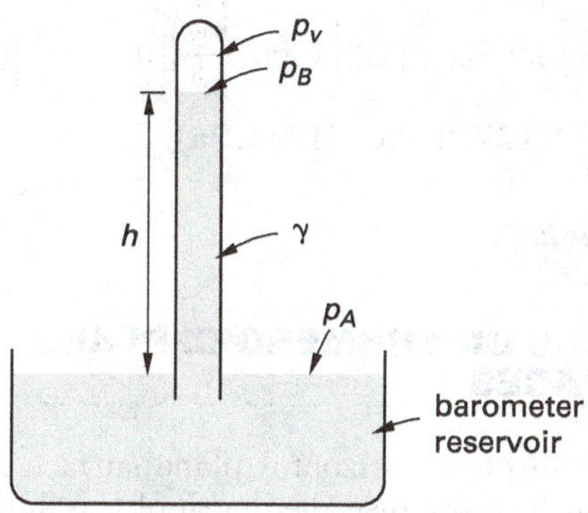

Equation 8.4: Vapor Pressure[3]

$$p_{\text{atm}} = p_A = p_v + \gamma h = p_B + \gamma h = p_B + \rho g h \qquad 8.4$$

Variation

$$p_{\text{atm}} = p_v + (\text{SG})\rho_{\text{water}} g h$$

Description

When the vapor pressure of the barometer liquid is significant, as it is with alcohol or water, the vapor pressure effectively reduces the height of the fluid column, as Eq. 8.4 indicates.

Fluid Mechanics

[3]In Eq. 8.4, the *NCEES Handbook* uses *A* as a subscript to designate location A in Fig. 8.4, not to designate "atmosphere," which is inconsistently designated by the subscripts "atm" and "0." Figure 8.4 shows both p_v and p_B, although in fact, these two pressures are the same.

Example

A fluid with a vapor pressure of 0.2 Pa and a specific gravity of 12 is used in a barometer. If the fluid's column height is 1 m, the atmospheric pressure is most nearly

(A) 9.8 kPa

(B) 12 kPa

(C) 98 kPa

(D) 120 kPa

Solution

From Eq. 8.4,

$$\begin{aligned} p_{\text{atm}} &= p_B + \rho g h = p_v + (\text{SG})\rho_{\text{water}} g h \\ &= 0.2 \text{ Pa} + (12)\left(1000 \ \frac{\text{kg}}{\text{m}^3}\right)\left(9.81 \ \frac{\text{m}}{\text{s}^2}\right)(1 \text{ m}) \\ &= 117\,720 \text{ Pa} \quad (120 \text{ kPa}) \end{aligned}$$

The answer is (D).

4. FORCES ON SUBMERGED PLANE SURFACES

The pressure on a horizontal plane surface is uniform over the surface because the depth of the fluid above is uniform. The resultant of the pressure distribution acts through the *center of pressure* of the surface, which corresponds to the centroid of the surface. (See Fig. 8.5.)

Figure 8.5 *Hydrostatic Pressure on a Horizontal Plane Surface*

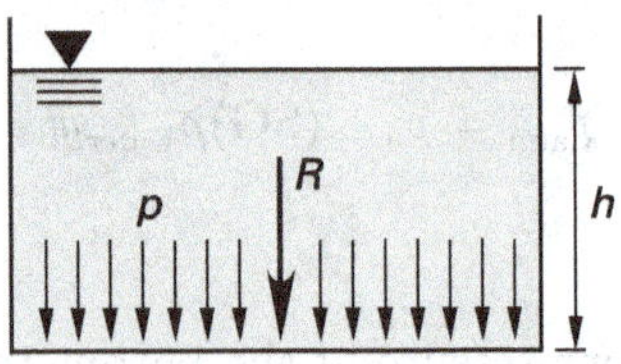

The total vertical force on the horizontal plane of area A is given by the equation

$$R = pA$$

It is not always correct to calculate the vertical force on a submerged surface as the weight of the fluid above it. Such an approach works only when there is no change in the cross-sectional area of the fluid above the surface. This is a direct result of the *hydrostatic paradox*. (See Fig. 8.1.) The two containers in Fig. 8.6 have the same distribution of pressure (or force) over their bottom surfaces.

Figure 8.6 *Two Containers with the Same Pressure Distribution*

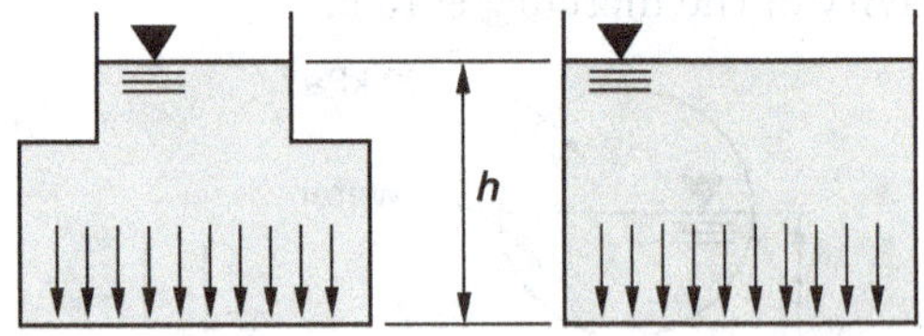

The pressure on a vertical rectangular plane surface increases linearly with depth. The pressure distribution will be triangular, as in Fig. 8.7(a), if the plane surface extends to the surface; otherwise, the distribution will be trapezoidal, as in Fig. 8.7(b).

Figure 8.7 *Hydrostatic Pressure on a Vertical Plane Surface*

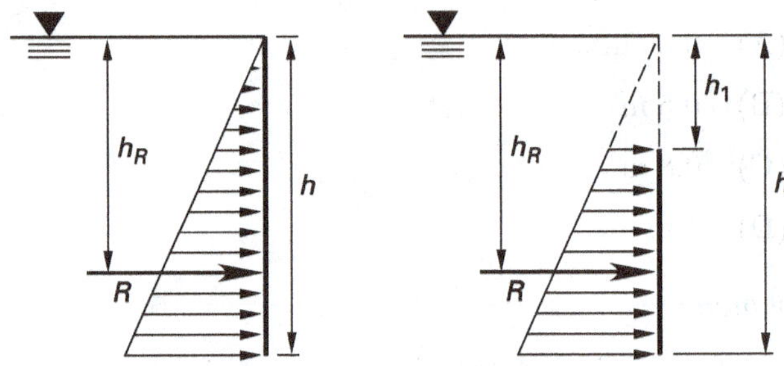

(a) full-depth wall (b) partial-depth wall

The resultant force on a vertical rectangular plane surface is

$$R = \bar{p}A$$

$\bar{p}$ is the *average pressure*, which is also equal to the pressure at the centroid of the plane area. The average pressure is

$$\bar{p} = \tfrac{1}{2}(p_1 + p_2) = \tfrac{1}{2}\rho g(h_1 + h_2) = \tfrac{1}{2}\gamma(h_1 + h_2)$$

Although the resultant is calculated from the average depth, the resultant does not act at the average depth. The resultant of the pressure distribution passes through the centroid of the pressure distribution. For the triangular distribution of Fig. 8.7(a), the resultant is located at a depth of $h_R = \tfrac{2}{3}h$. For the more general case, the center of pressure can be calculated by the method described in Sec. 8.5.

The average pressure and resultant force on an inclined rectangular plane surface are calculated in the same fashion as for the vertical plane surface. (See Fig. 8.8.) The pressure varies linearly with depth. The resultant is calculated from the average pressure, which, in turn, depends on the average depth.

Figure 8.8 *Hydrostatic Pressure on an Inclined Rectangular Plane Surface*

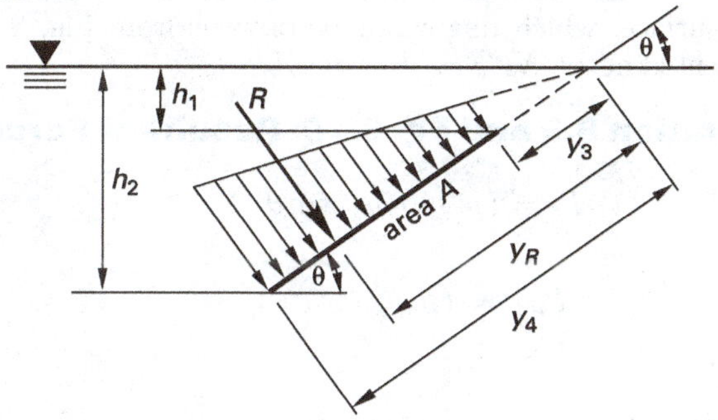

The resultant and average pressure on an inclined plane surface are given by the same equations as for a vertical plane surface.

$$R = \bar{p}A$$

$$\bar{p} = \tfrac{1}{2}(p_1 + p_2) = \tfrac{1}{2}\rho g(h_1 + h_2) = \tfrac{1}{2}\gamma(h_1 + h_2)$$

As with a vertical plane surface, the resultant acts at the centroid of the pressure distribution, not at the average depth.

5. CENTER OF PRESSURE

For the case of pressure on a general plane surface, the resultant force depends on the average pressure and acts through the *center of pressure*, CP. Figure 8.9 shows a nonrectangular plane surface of area A that may or may not extend to the liquid surface and that may or may not be inclined. The average pressure is calculated from the location of the plane surface's centroid, C, where y_C is measured parallel to the plane surface. That is, if the plane surface is inclined, y_C is an inclined distance.

The center of pressure is always at least as deep as the area's centroid. In most cases, it is deeper.

The pressure at the centroid is

$$\begin{aligned} p_C = \bar{p} &= p_{atm} + \rho g y_C \sin\theta \\ &= p_{atm} + \gamma y_C \sin\theta \end{aligned}$$

Equation 8.5: Absolute Pressure on a Point[4]

$$p = p_0 + \rho g h \quad [h \geq 0] \qquad 8.5$$

Description

Equation 8.5 gives the absolute pressure on a point at a vertical distance of h under the surface. Depth, h, must be greater than or equal to 0.[5]

Figure 8.9 *Hydrostatic Pressure on a Submerged Plane Surface**

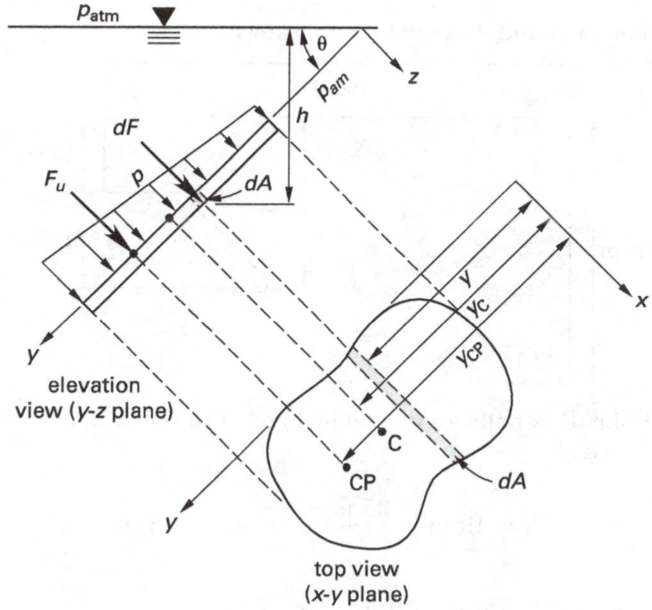

*The meaning of the inclined p_{am} in this figure as presented in the *NCEES Handbook* is unknown. Its location implies that it is not the same as p_{atm}.

Example

A closed tank with the dimensions shown contains water. The air pressure in the tank is 700 kPa. Point P is located halfway up the inclined wall.

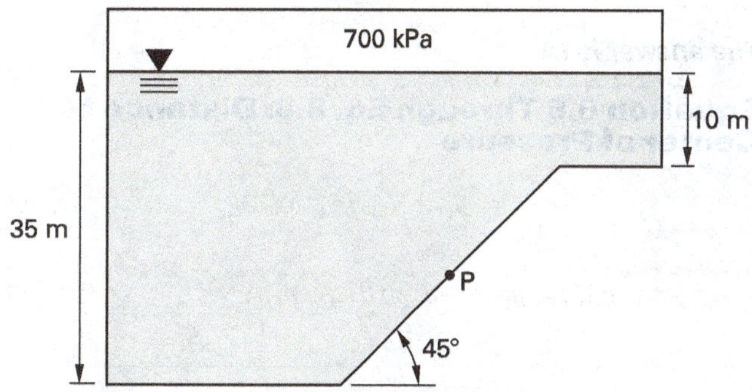

The pressure at point P is most nearly

(A) 920 kPa

(B) 1900 kPa

(C) 7200 kPa

(D) 8100 kPa

[4]The *NCEES Handbook* is inconsistent in how it designates atmospheric pressure. Although p_{atm} is used in Eq. 8.4 and in Fig. 8.9, Eq. 8.5 uses p_0.
[5]Equation 8.5 calculates the *absolute pressure* because it includes the atmospheric pressure. If the atmospheric pressure term is omitted, the *gauge pressure* (also known as *gage pressure*) will be calculated.

Solution

The tank and its geometry are shown.

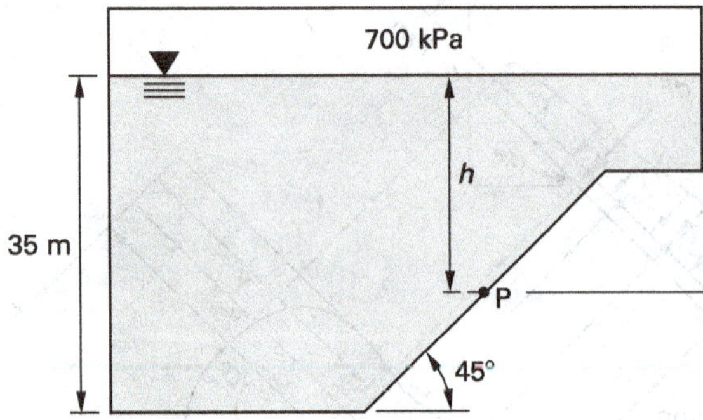

Point P is halfway up the inclined surface, so it is at a depth of

$$h = 10 \text{ m} + \frac{35 \text{ m} - 10 \text{ m}}{2} = 22.5 \text{ m}$$

The pressure at point P is

$$\begin{aligned} p &= p_0 + \rho g h \\ &= (700 \text{ kPa})\left(1000 \frac{\text{Pa}}{\text{kPa}}\right) \\ &\quad + \left(1000 \frac{\text{kg}}{\text{m}^3}\right)\left(9.81 \frac{\text{m}}{\text{s}^2}\right)(22.5 \text{ m}) \\ &= 920\,725 \text{ Pa} \quad (920 \text{ kPa}) \end{aligned}$$

The answer is (A).

Equation 8.6 Through Eq. 8.8: Distance to Center of Pressure

$$y_{CP} = y_C + I_{xC}/y_C A \qquad 8.6$$

$$y_{CP} = y_C + \rho g \sin\theta I_{xC}/p_C A \qquad 8.7$$

$$y_C = h_C/\sin\alpha \qquad 8.8$$

Description

Equation 8.6 and Eq. 8.7 apply when the atmospheric pressure acts on the liquid surface and on the dry side of the submerged surface. The distance from the surface of the liquid to the center of pressure measured along the slanted surface, y_{CP}, is found from Eq. 8.6 and Eq. 8.7. Equation 8.7 is derived from Eq. 8.6 by using the pressure-height relationship $p_C = \rho g h_C = \rho g y_C \sin\theta$.[6] y_C is the distance from the surface of the liquid to the centroid of the area, C, calculated using Eq. 8.8. In Eq. 8.6 and Eq. 8.7, the subscript x refers to a horizontal (centroidal) axis parallel to the surface, which might not be obvious from Fig. 8.9, as presented in the *NCEES Handbook.*

Equation 8.9 and Eq. 8.10: Resultant Force

$$F_R = (p_0 + \rho g y_C \sin\theta)A \qquad 8.9$$

$$F_{R_{\text{net}}} = (\rho g y_C \sin\theta)A \qquad 8.10$$

Description

The resultant force, F_R, on the wetted side of the surface is found from Eq. 8.9. Equation 8.10 calculates the net resultant force when p_0 acts on both sides of the surface.[7]

6. BUOYANCY

Buoyant force is an upward force that acts on all objects that are partially or completely submerged in a fluid. The fluid can be a liquid or a gas. There is a buoyant force on all submerged objects, not only on those that are stationary or ascending. A buoyant force caused by displaced air also exists, although it may be insignificant. Examples include the buoyant force on a rock sitting at the bottom of a pond, the buoyant force on a rock sitting exposed on the ground (since the rock is "submerged" in air), and the buoyant force on partially exposed floating objects, such as icebergs.

Buoyant force always acts to cancel the object's weight (i.e., buoyancy acts against gravity). The magnitude of the buoyant force is predicted from *Archimedes' principle* (the *buoyancy theorem*), which states that the buoyant force on a submerged or floating object is equal to the weight of the displaced fluid. An equivalent statement of Archimedes' principle is that a floating object displaces liquid equal in weight to its own weight. In the situation of an object floating at the interface between two immiscible liquids of different densities, the buoyant force equals the sum of the weights of the two displaced fluids.

In the case of stationary (i.e., not moving vertically) floating or submerged objects, the buoyant force and object weight are in equilibrium. If the forces are not in equilibrium, the object will rise or fall until equilibrium is reached—that is, the object will sink until its remaining weight is supported by the bottom, or it will rise until the weight of liquid is reduced by breaking the surface.

[6]p_C in Eq. 8.7 is defined as the "pressure at the centroid of the area," but it cannot be calculated from Eq. 8.5, because Eq. 8.5 calculates an absolute pressure. As used in Eq. 8.7, p_C is a gauge pressure, not an absolute pressure. $\sin\alpha$ in Eq. 8.8 is an error, and it should be $\sin\theta$.

[7]For all practical purposes, atmospheric pressure always acts on both sides of an object. It acts through the liquid on both sides of a submerged plate, it acts on both sides of a submerged gate, and it acts on both sides of a discharge gate/door. Except for objects in a vacuum, Eq. 8.9 is of purely academic interest.

The two forces acting on a stationary floating object are the *buoyant force* and the *object's weight.* The buoyant force acts upward through the centroid of the displaced volume (not the object's volume). This centroid is known as the *center of buoyancy.* The gravitational force on the object (i.e., the object's weight) acts downward through the entire object's center of gravity.

9 Fluid Measurement and Similitude

Nomenclature

A	area	m^2
C	coefficient	–
Ca	Cauchy number	–
d	depth	m
D	diameter	m
E	specific energy	J/kg
F	force	N
Fr	Froude number	–
g	gravitational acceleration, 9.81	m/s^2
h	head	m
h	head loss	m
h	height	m
l	characteristic length	m
p	pressure	Pa
Q	flow rate	m^3/s
Re	Reynolds number	–
v	velocity	m/s
We	Weber number	–
z	elevation	m

Symbols

γ	specific (unit) weight	N/m^3
μ	absolute viscosity	Pa·s
ρ	density	kg/m^3
σ	surface tension	N/m

Subscripts

0	stagnation (zero velocity)
c	contraction
E	elastic
G	gravitational
I	inertial
m	manometer fluid or model
p	constant pressure or prototype
s	static
T	surface tension
v	velocity
v	constant volume

1. PITOT TUBE

A *pitot tube* is simply a hollow tube that is placed longitudinally in the direction of fluid flow, allowing the flow to enter one end at the fluid's *velocity of approach.* (See Fig. 9.1.) A pitot tube is used to measure velocity of flow.

Figure 9.1 Pitot Tube

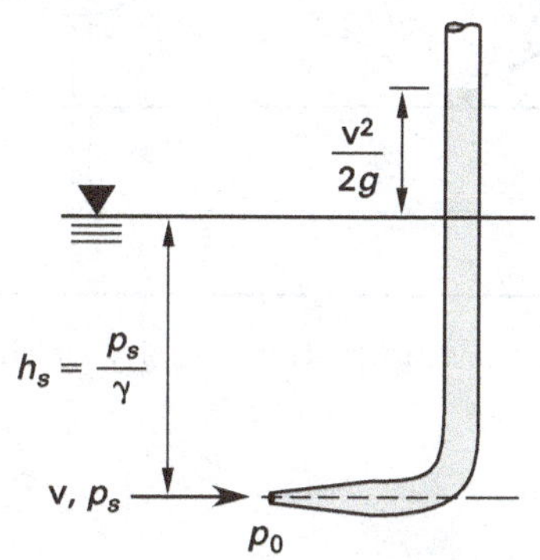

When the fluid enters the pitot tube, it is forced to come to a stop (at the *stagnation point*), and its kinetic energy is transformed into static pressure energy.

Equation 9.1: Fluid Velocity[1]

$$\mathrm{v} = \sqrt{(2/\rho)(p_0 - p_s)} = \sqrt{2g(p_0 - p_s)/\gamma} \qquad 9.1$$

[1]As used in the NCEES *FE Reference Handbook* (*NCEES Handbook*), there is no significance to the inconsistent placement of the density terms in the two forms of Eq. 9.1.

Fluid Mechanics

Description

The Bernoulli equation can be used to predict the static pressure at the stagnation point. Since the velocity of the fluid within the pitot tube is zero, the upstream velocity can be calculated if the *static*, p_s, and *stagnation*, p_0, *pressures* are known.

$$\frac{p_s}{\rho}+\frac{\mathrm{v}^2}{2}=\frac{p_0}{\rho}$$

$$\frac{p_s}{\gamma}+\frac{\mathrm{v}^2}{2g}=\frac{p_0}{\gamma}$$

In reality, the fluid may be compressible. If the Mach number is less than approximately 0.3, Eq. 9.1 for incompressible fluids may be used.

Example

The density of air flowing in a duct is 1.15 kg/m^3. A pitot tube is placed in the duct as shown. The static pressure in the duct is measured with a wall tap and pressure gage.

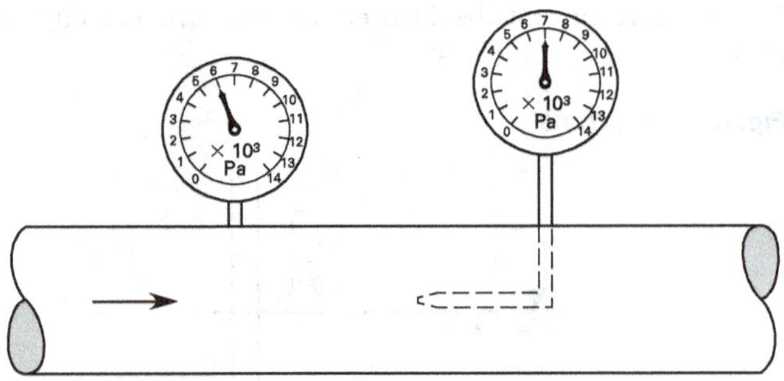

From the gage readings, the velocity of the air is most nearly

(A) 42 m/s

(B) 100 m/s

(C) 110 m/s

(D) 150 m/s

Solution

The static pressure is read from the first static pressure gage as 6000 Pa. The impact pressure is 7000 Pa. From Eq. 9.1,

$$\begin{aligned}\mathrm{v}&=\sqrt{(2/\rho)(p_0-p_s)}\\&=\sqrt{\left(\frac{2}{1.15\ \frac{\mathrm{kg}}{\mathrm{m}^3}}\right)(7000\ \mathrm{Pa}-6000\ \mathrm{Pa})}\\&=41.7\ \mathrm{m/s}\quad(42\ \mathrm{m/s})\end{aligned}$$

The answer is (A).

2. VENTURI METER

Figure 9.2 illustrates a simple *venturi meter*. This flow-measuring device can be inserted directly into a pipeline. Since the diameter changes are gradual, there is very little friction loss. Static pressure measurements are taken at the throat and upstream of the diameter change. The difference in these pressures is directly indicated by a *differential manometer*.

Figure 9.2 *Venturi Meter with Differential Manometer*

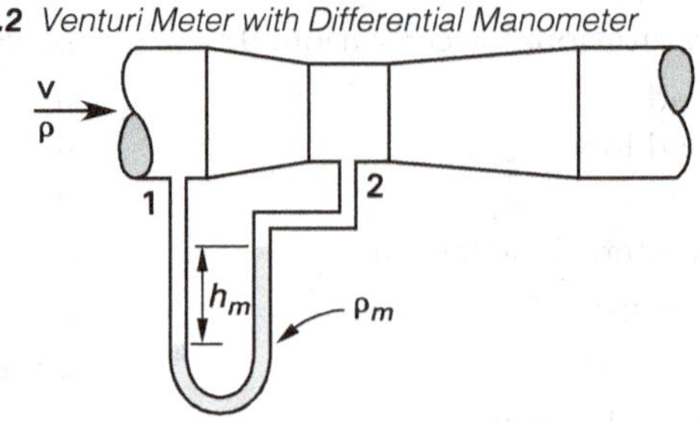

The pressure differential across the venturi meter shown can be calculated from the following equations.

$$p_1-p_2=(\rho_m-\rho)gh_m=(\gamma_m-\gamma)h_m$$

$$\frac{p_1-p_2}{\rho}=\left(\frac{\rho_m}{\rho}-1\right)gh_m$$

$$\frac{p_1-p_2}{\gamma}=\left(\frac{\gamma_m}{\gamma}-1\right)h_m$$

Equation 9.2: Flow Rate Through Venturi Meter

$$Q=\frac{C_\mathrm{v}A_2}{\sqrt{1-(A_2/A_1)^2}}\sqrt{2g\left(\frac{p_1}{\gamma}+z_1-\frac{p_2}{\gamma}-z_2\right)}\qquad 9.2$$

Variation

$$Q=\frac{C_vA_2}{\sqrt{1-\left(\frac{A_2}{A_1}\right)^2}}\sqrt{2\left(\frac{p_1}{\rho}+z_1-\frac{p_2}{\rho}-z_2\right)}$$

Values

The *coefficient of velocity*, C_v, accounts for the small effect of friction and is very close to 1.0, usually 0.98 or 0.99.

Description

The flow rate, Q, can be calculated from venturi measurements using Eq. 9.2. For a horizontal venturi meter, $z_1 = z_2$. The quotients, p/γ, in Eq. 9.2 represent the heads of the fluid flowing through a venturi meter. Therefore, the specific weight, γ, of the fluid should be used, not the specific weight of the manometer fluid.

Example

A venturi meter is installed horizontally to measure the flow of water in a pipe. The area ratio of the meter, A_2/A_1, is 0.5, the velocity through the throat of the meter is 3 m/s, and the coefficient of velocity is 0.98. The pressure differential across the venturi meter is most nearly

(A) 1.5 kPa

(B) 2.3 kPa

(C) 3.5 kPa

(D) 6.8 kPa

Solution

From Eq. 9.2, for a venturi meter,

$$Q=\frac{C_vA_2}{\sqrt{1-(A_2/A_1)^2}}\sqrt{2g\left(\frac{p_1}{\gamma}+z_1-\frac{p_2}{\gamma}-z_2\right)}$$

Dividing both sides by the area at the throat, A_2, gives

$$\frac{Q}{A_2}=\mathrm{v}_2=\frac{C_v}{\sqrt{1-\left(\frac{A_2}{A_1}\right)^2}}\sqrt{2g\left(\frac{p_1}{\gamma}+z_1-\frac{p_2}{\gamma}-z_2\right)}$$

Since the venturi meter is horizontal, $z_1 = z_2$. Reducing and solving for the pressure differential gives

$$\begin{aligned}p_1-p_2&=\frac{\mathrm{v}_2^2\left(1-\left(\frac{A_2}{A_1}\right)^2\right)\gamma}{2gC_v^2}\\&=\frac{\left(3\ \frac{\mathrm{m}}{\mathrm{s}}\right)^2\left(1-(0.5)^2\right)\left(9.81\ \frac{\mathrm{kN}}{\mathrm{m}^3}\right)\left(1000\ \frac{\mathrm{N}}{\mathrm{kN}}\right)}{(2)\left(9.81\ \frac{\mathrm{m}}{\mathrm{s}^2}\right)(0.98)^2}\\&=3514\ \mathrm{Pa}\quad(3.5\ \mathrm{kPa})\end{aligned}$$

The answer is (C).

3. ORIFICE METER

The *orifice meter* (or *orifice plate*) is used more frequently than the venturi meter to measure flow rates in small pipes. It consists of a thin or sharp-edged plate with a central, round hole through which the fluid flows.

As with the venturi meter, pressure taps are used to obtain the static pressure upstream of the orifice plate and at the *vena contracta* (i.e., at the point of minimum area and minimum pressure). A differential manometer connected to the two taps conveniently indicates the difference in static pressures. The pressure differential equations, derived for the manometer in Fig. 9.2, are also valid for the manometer configuration of the orifice shown in Fig. 9.3.

Figure 9.3 *Orifice Meter with Differential Manometer*

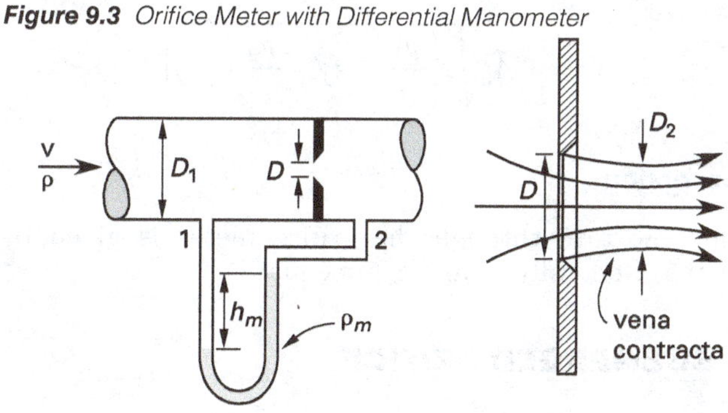

(a) orifice meter (b) orifice plate

Equation 9.3: Orifice Area

$$A_2 = C_c A \qquad 9.3$$

Description

The area of the orifice is A, and the area of the pipeline is A_1. The area at the vena contracta, A_2, can be calculated from the orifice area and the *coefficient of contraction*, C_c, using Eq. 9.3.

Equation 9.4: Coefficient of the Meter (Orifice Plate)[2]

$$C = \frac{C_v C_c}{\sqrt{1 - C_c^2 (A_0/A_1)^2}} \qquad 9.4$$

Description

The *coefficient of the meter*, C, combines the coefficients of velocity and contraction in a way that corrects the theoretical discharge of the meter for frictional flow and for contraction at the vena contracta. The coefficient of the meter is also known as the flow coefficient.[3] Approximate orifice coefficients are listed in Table 9.1.

Equation 9.5: Flow Through Orifice Plate

$$Q = CA_0 \sqrt{2g\left(\frac{p_1}{\gamma} + z_1 - \frac{p_2}{\gamma} - z_2\right)} \qquad 9.5$$

Variation

$$Q = CA_0 \sqrt{2\left(\frac{p_1}{\rho} + z_1 - \frac{p_2}{\rho} - z_2\right)}$$

Description

The flow rate through the orifice meter is given by Eq. 9.5. Generally, z_1 and z_2 are equal.

4. SUBMERGED ORIFICE

The flow rate of a jet issuing from a *submerged orifice* in a tank can be determined by modifying Eq. 9.5 in terms of the potential energy difference, or head difference, on either side of the orifice. (See Fig. 9.4.)

Table 9.1 *Approximate Orifice Coefficients for Turbulent Water*

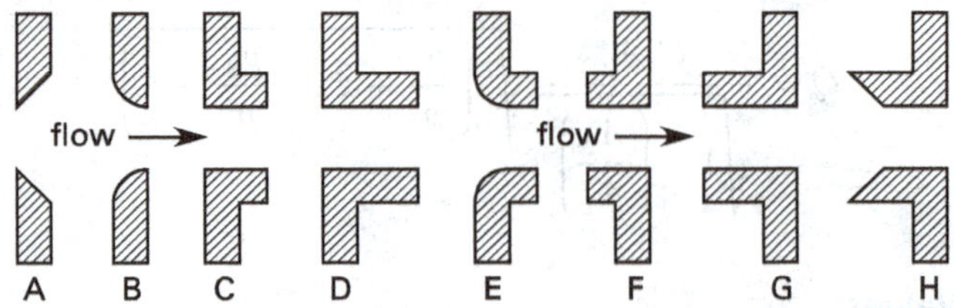

illustration	description	C	C_c	C_v
A	sharp-edged	0.61	0.62	0.98
B	round-edged	0.98	1.00	0.98
C	short tube (fluid separates from walls)	0.61	1.00	0.61
D	short tube (no separation)	0.80	1.00	0.80
E	short tube with rounded entrance	0.97	0.99	0.98
F	reentrant tube, length less than one-half of pipe diameter	0.54	0.55	0.99
G	reentrant tube, length 2–3 pipe diameters	0.72	1.00	0.72
H	Borda	0.51	0.52	0.98
(none)	smooth, well-tapered nozzle	0.98	0.99	0.99

Figure 9.4 *Submerged Orifice*

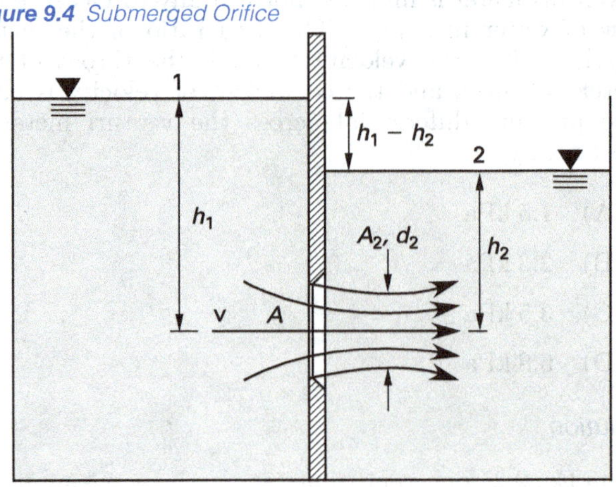

[2]The *NCEES Handbook*'s use of the symbol C for coefficient of the meter is ambiguous. In literature describing orifice plate performance, when C_d is not used, C is frequently reserved for the coefficient of discharge. The symbols C_M, C_F (for coefficient of the meter and *flow coefficient*), K, and F are typically used to avoid ambiguity.

[3]The *NCEES Handbook* lists "orifice coefficient" as a synonym for the "coefficient of the meter." However, this ambiguous usage should be avoided, as four orifice coefficients are attributed to an orifice: coefficient of contraction, coefficient of velocity, coefficient of discharge, and coefficient of resistance.

Equation 9.6 Through Eq. 9.8: Flow Through Submerged Orifice[4]

$$Q = A_2 v_2 = C_c C_v A \sqrt{2g(h_1 - h_2)} \quad 9.6$$

$$Q = CA\sqrt{2g(h_1 - h_2)} \quad 9.7$$

$$C = C_c C_v \quad 9.8$$

Description

The coefficients of velocity and contraction can be combined into the *coefficient of discharge*, C, calculated from Eq. 9.8.

5. ORIFICE DISCHARGING FREELY INTO ATMOSPHERE

If the orifice discharges from a tank into the atmosphere, Eq. 9.7 can be further simplified. (See Fig. 9.5.)

Figure 9.5 Orifice Discharging Freely into the Atmosphere

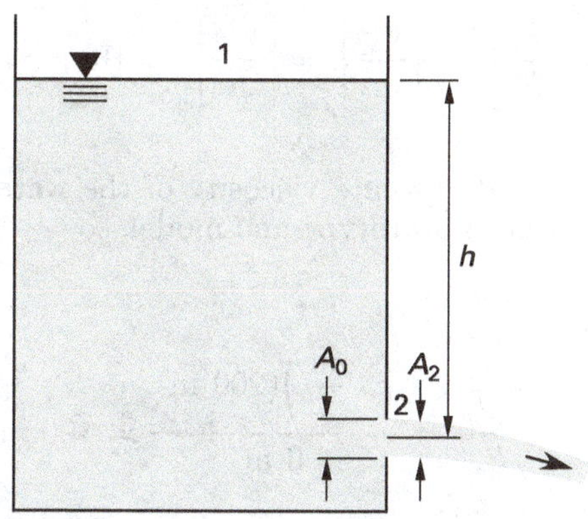

Equation 9.9: Orifice Flow with Free Discharge

$$Q = CA_0\sqrt{2gh} \quad 9.9$$

Variation

$$v_2 = \frac{Q}{A_2} = C_v\sqrt{2gh}$$

Description

A_0 is the orifice area. A_2 is the area at the vena contracta (see Eq. 9.3 and Fig. 9.5).

Example

Water under an 18 m head discharges freely into the atmosphere through a 25 mm diameter orifice. The orifice is round-edged and has a coefficient of discharge of 0.98.

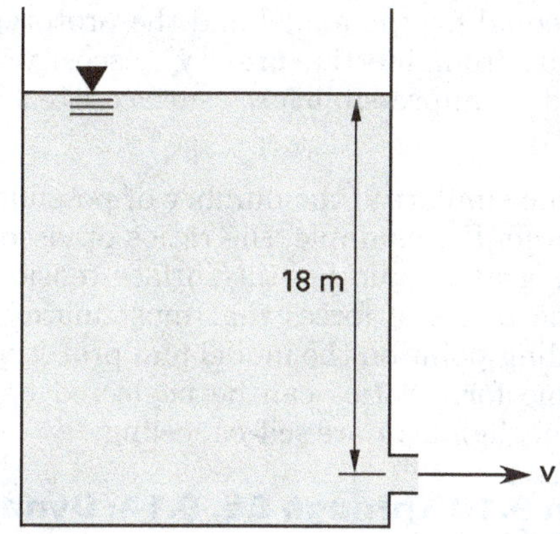

The velocity of the water as it passes through the orifice is most nearly

(A) 1.2 m/s

(B) 3.2 m/s

(C) 8.2 m/s

(D) 18 m/s

Solution

From Eq. 9.9, for an orifice discharging freely into the atmosphere,

$$Q = CA_0\sqrt{2gh}$$

Dividing both sides by A_0 gives

$$\begin{aligned} v &= C\sqrt{2gh} \\ &= 0.98\sqrt{(2)\left(9.81\ \frac{\text{m}}{\text{s}^2}\right)(18\ \text{m})} \\ &= 18.4\ \text{m/s} \quad (18\ \text{m/s}) \end{aligned}$$

The answer is (D).

6. SIMILITUDE

Similarity considerations between a *model* (subscript m) and a full-size object (subscript p, for *prototype*) imply that the model can be used to predict the performance of the prototype. Such a model is said to be *mechanically similar* to the prototype.

[4]The *NCEES Handbook*'s use of the symbol C for both coefficient of discharge (submerged orifice) and coefficient of the meter (see Eq. 9.4) makes it difficult to determine the meaning of C.

Complete mechanical similarity requires geometric, kinematic, and dynamic similarity. *Geometric similarity* means that the model is true to scale in length, area, and volume. *Kinematic similarity* requires that the flow regimes of the model and prototype be the same. *Dynamic similarity* means that the ratios of all types of forces are equal for the model and the prototype. These forces result from inertia, gravity, viscosity, elasticity (i.e., fluid compressibility), surface tension, and pressure.

For dynamic similarity, the number of possible ratios of forces is large. For example, the ratios of viscosity/inertia, inertia/gravity, and inertia/surface tension are only three of the ratios of forces that must match for every corresponding point on the model and prototype. Fortunately, some force ratios can be neglected because the forces are negligible or are self-canceling.

Equation 9.10 Through Eq. 9.14: Dynamic Similarity[5]

$$\left[\frac{F_I}{F_p}\right]_p = \left[\frac{F_I}{F_p}\right]_m = \left[\frac{\rho \text{v}^2}{p}\right]_p = \left[\frac{\rho \text{v}^2}{p}\right]_m \quad 9.10$$

$$\left[\frac{F_I}{F_V}\right]_p = \left[\frac{F_I}{F_V}\right]_m = \left[\frac{\text{v}l\rho}{\mu}\right]_p = \left[\frac{\text{v}l\rho}{\mu}\right]_m = [\text{Re}]_p = [\text{Re}]_m \quad 9.11$$

$$\left[\frac{F_I}{F_G}\right]_p = \left[\frac{F_I}{F_G}\right]_m = \left[\frac{\text{v}^2}{lg}\right]_p = \left[\frac{\text{v}^2}{lg}\right]_m = [\text{Fr}]_p = [\text{Fr}]_m \quad 9.12$$

$$\left[\frac{F_I}{F_E}\right]_p = \left[\frac{F_I}{F_E}\right]_m = \left[\frac{\rho \text{v}^2}{E_v}\right]_p = \left[\frac{\rho \text{v}^2}{E_v}\right]_m = [\text{Ca}]_p = [\text{Ca}]_m \quad 9.13$$

$$\left[\frac{F_I}{F_T}\right]_p = \left[\frac{F_I}{F_T}\right]_m = \left[\frac{\rho l \text{v}^2}{\sigma}\right]_p = \left[\frac{\rho l \text{v}^2}{\sigma}\right]_m = [\text{We}]_p = [\text{We}]_m \quad 9.14$$

Description

If Eq. 9.10 through Eq. 9.14 are satisfied for model and prototype, complete dynamic similarity will be achieved. In practice, it is rare to be able (or to even attempt) to demonstrate *complete similarity.* Usually, *partial similarity* is based on only one similarity law, and correlations, experience, and general rules of thumb are used to modify the results. For completely submerged objects (i.e., where there is no free surface), such as torpedoes in water and aircraft in the atmosphere, similarity is usually based on Reynolds numbers. For objects partially submerged and experiencing wave activity, such as surface ships, open channels, spillways, weirs, and hydraulic jumps, partial similarity is usually based on Froude numbers.

Example 12

A 200 m long submarine is being designed to travel underwater at 3 m/s. The corresponding underwater speed for a 6 m model is most nearly

(A) 0.5 m/s

(B) 2 m/s

(C) 100 m/s

(D) 200 m/s

Solution

The Reynolds numbers should be equal for model and prototype. From Eq. 9.11,

$$\left[\frac{F_I}{F_V}\right]_p = \left[\frac{F_I}{F_V}\right]_m = \left[\frac{\text{v}l\rho}{\mu}\right]_p = \left[\frac{\text{v}l\rho}{\mu}\right]_m = [\text{Re}]_p = [\text{Re}]_m$$

The density and absolute viscosity of the water will be the same for both prototype and model, so

$$\text{v}_p l_p = \text{v}_m l_m$$

$$\text{v}_m = \frac{\text{v}_p l_p}{l_m} = \frac{\left(3\ \frac{\text{m}}{\text{s}}\right)(200\ \text{m})}{6\ \text{m}} = 100\ \text{m/s}$$

The answer is (C).

[5]The *NCEES Handbook* is inconsistent in its presentation and definition of the Froude number. While some equations use y_h as the symbol for *characteristic length,* Eq. 9.12 uses l. This leads to some potentially confusing and misleading conflicts.

Topic IV: Hydraulics and Hydrologic Systems

10 Hydrology

Nomenclature

A	area	ft^2
C	rational runoff coefficient	–
CN	curve number	–
E	evaporation	in/day
F	number of years in recurrence interval	–
I	rainfall intensity	in/hr
K_p	pan coefficient	–
n	number of years or periods	–
P	precipitation	in
P	probability	–
Q	flow rate	ft^3/sec
Q	total runoff	in
S	storage capacity	in
t	time	min
t_c	time of concentration	min
V	volume	ft^3

Subscripts

ave	average
b	base
c	concentration
n	period n
p	pan
R	reservoir

1. HYDROLOGIC CYCLE

The *hydrologic cycle* is the full "life cycle" of water. The cycle begins with *precipitation*, which encompasses all of the hydrometeoric forms, including rain, snow, sleet, and hail from a storm. Precipitation can (a) fall on vegetation and structures and evaporate back into the atmosphere, (b) be absorbed into the ground and either make its way to the water table or be absorbed by plants, after which it evapotranspires back into the atmosphere, or (c) travel as surface water to a depression, watershed, or creek from which it either evaporates back into the atmosphere, infiltrates into the ground water system, or flows off in streams and rivers to an ocean or lakes. The cycle is completed when lake and ocean water evaporates into the atmosphere.

The *water balance equation* (*water budget equation*) is the application of conservation to the hydrologic cycle.

$$\begin{aligned}\text{total precipitation} &= \text{net change in surface water removed} \\ &\quad + \text{net change in ground water removed} \\ &\quad + \text{evapotranspiration} \\ &\quad + \text{interception evaporization} \\ &\quad + \text{net increase in surface water storage} \\ &\quad + \text{net increase in ground water storage} \\ P &= Q + E + \Delta S\end{aligned}$$

The total amount of water that is intercepted (and subsequently evaporates) and absorbed into ground water before runoff begins is known as the *initial abstraction*. Even after runoff begins, the soil continues to absorb some infiltrated water. Initial abstraction and infiltration do not contribute to surface runoff. The equation can be restated as

$$\begin{aligned}\text{total precipitation} &= \text{initial abstraction} + \text{infiltration} \\ &\quad + \text{surface runoff}\end{aligned}$$

2. STORM CHARACTERISTICS

Storm rainfall characteristics include the duration, total volume, intensity, and areal distribution of a storm. Storms are also characterized by their recurrence intervals.

The duration of storms is measured in hours and days. The volume of rainfall is simply the total quantity of precipitation dropping on the watershed. Average rainfall intensity is the volume divided by the duration of the storm. Average rainfall can be considered to be generated by an equivalent theoretical storm that drops the same volume of water uniformly and constantly over the entire watershed area.

A *storm hyetograph* is the instantaneous rainfall intensity measured as a function of time, as shown in Fig. 10.1(a). Hyetographs are usually bar graphs showing constant rainfall intensities over short periods of time. Hyetograph data can be reformulated as a *cumulative rainfall curve*, as shown in Fig. 10.1(b). Cumulative rainfall curves are also known as *rainfall mass curves.*

Figure 10.1 *Storm Hyetograph and Cumulative Rainfall Curves*

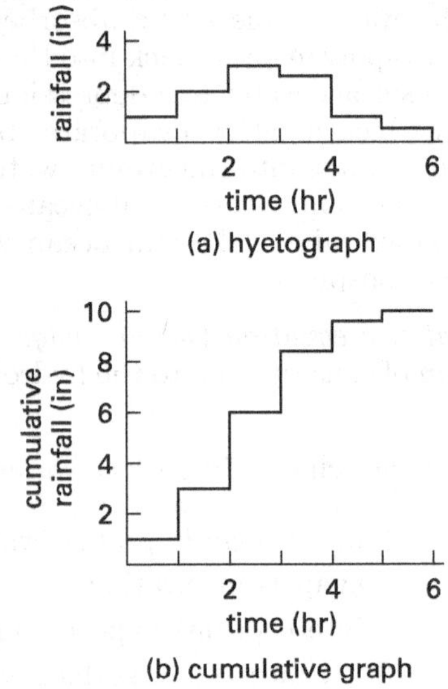

3. PRECIPITATION DATA

Precipitation data on rainfall can be collected in a number of ways, but use of an open precipitation rain gauge is the most common. This type of gauge measures only the volume of rain collected between readings, usually 24 hours.

4. FLOODS

A *flood* occurs when more water arrives than can be drained away. When a watercourse (i.e., a creek or river) is too small to contain the flow, the water overflows the banks.

The flooding may be categorized as nuisance, damaging, or devastating. *Nuisance floods* result in inconveniences such as wet feet, tire spray, and soggy lawns. *Damaging floods* soak flooring, carpeting, and first-floor furniture. *Devastating floods* wash buildings, vehicles, and livestock downstream.

Although rain causes flooding, large storms do not always cause floods. The size of a flood depends not only on the amount of rainfall, but also on the conditions within the watershed before and during the storm. Runoff will occur only when the rain falls on a very wet watershed that is unable to absorb additional water, or when a very large amount of rain falls on a dry watershed faster than it can be absorbed.

Specific terms are sometimes used to designate the degree of protection required. For example, the *probable maximum flood* (PMF) is a hypothetical flood that can be expected to occur as a result of the most severe combination of critical meteorologic and hydrologic conditions possible within a region.

Designing for the *probable maximum precipitation* (PMP) or probable maximum flood is very conservative and usually uneconomical since the recurrence interval for these events exceeds 100 years and may even approach 1000 years. Designing for 100-year floods and floods with even lower recurrence intervals is more common. (100-year floods are not necessarily caused by 100-year storms.)

The *design flood* or *design basis flood* (DBF) depends on the site. It is the flood that is adopted as the basis for design of a particular project. The DBF is usually determined from economic considerations, or it is specified as part of the contract document.

The *standard flood* or *standard project flood* (SPF) is a flood that can be selected from the most severe combinations of meteorological and hydrological conditions reasonably characteristic of the region, excluding extremely rare combinations of events. SPF volumes are commonly 40–60% of the PMF volumes.

The probability that a flooding event in any given year will equal a design basis flood with a *recurrence interval frequency* (*return interval*) of F is

$$p\{F \text{ event in one year}\} = \frac{1}{F}$$

The probability of an F event occurring in n years is

$$p\{F \text{ event in } n \text{ years}\} = 1 - \left(1 - \frac{1}{F}\right)^n$$

Planning for a 1% flood has proven to be a good compromise between not doing enough and spending too much. Although the 1% flood is a common choice for the design basis flood, shorter recurrence intervals are often used, particularly in low-value areas such as cropland. For example, a 5 year value can be used in residential areas, a 10 year value in business sections, and a 15 year value for high-value districts where flooding will result in more extensive damage. The ultimate choice of recurrence interval, however, must be made on the basis of economic considerations and trade-offs.

5. TOTAL SURFACE RUNOFF FROM STREAM HYDROGRAPH

After a rain, runoff and groundwater increases stream flow. A plot of the stream discharge versus time is known as a *hydrograph*. Hydrograph periods may be very short (e.g., hours) or very long (e.g., days, weeks, or months). A typical hydrograph is shown in Fig. 10.2. The *time base* is the length of time that the stream flow exceeds the original *base flow*. The flow rate increases on the *rising limb* (*concentration curve*) and decreases on the *falling limb* (*recession curve*).

Figure 10.2 *Stream Hydrograph*

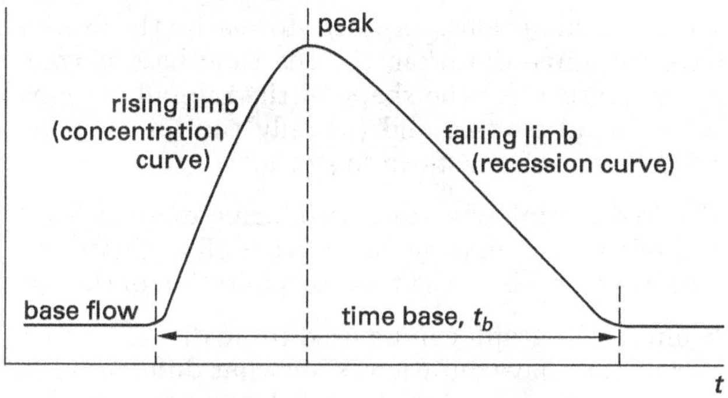

6. HYDROGRAPH SEPARATION

The stream discharge consists of both surface runoff and subsurface groundwater flows. A procedure known as *hydrograph separation* or *hydrograph analysis* is used to separate runoff (*surface flow*, *net flow*, or *overland flow*) and groundwater (*subsurface flow*, *base flow*).[1]

There are several methods of separating groundwater from runoff. Most of the methods are somewhat arbitrary. Three methods that are easily carried out manually are presented here.

Method 1: In the *straight-line method*, a horizontal line is drawn from the start of the rising limb to the falling limb. All of the flow under the horizontal line is considered base flow. This assumption is not theoretically accurate, but the error can be small. This method is illustrated in Fig. 10.3.

Method 2: In the *fixed-base method*, shown in Fig. 10.4, the base flow existing before the storm is projected graphically down to a point directly under the peak of the hydrograph. Then, a straight line is used to connect the projection to the falling limb. The duration of the recession limb is determined by inspection, or it can be calculated from correlations with the drainage area.

Figure 10.3 *Straight-Line Method*

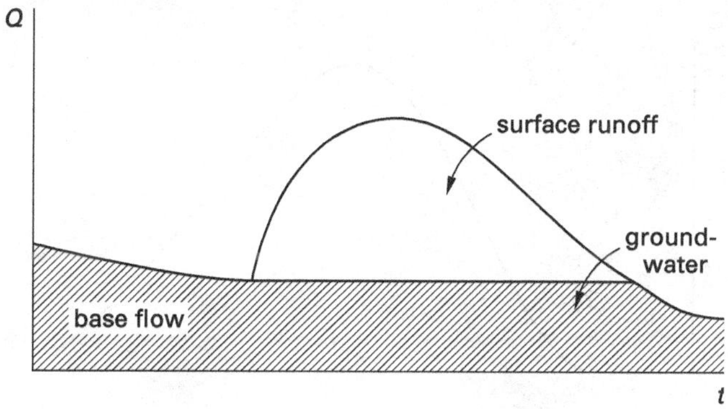

Figure 10.4 *Fixed-Base Method*

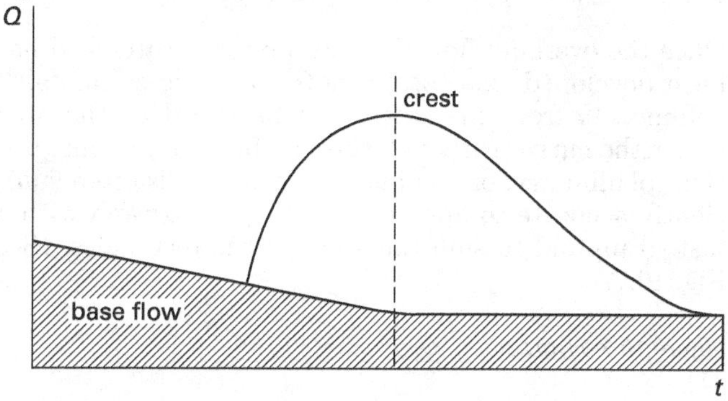

Method 3: The *variable-slope method*, as shown in Fig. 10.5, recognizes that the shape of the base flow curve before the storm will probably match the shape of the base flow curve after the storm. The groundwater curve after the storm is projected back under the hydrograph to a point under the inflection point of the falling limb. The separation line under the rising limb is drawn arbitrarily.

Figure 10.5 *Variable-Slope Method*

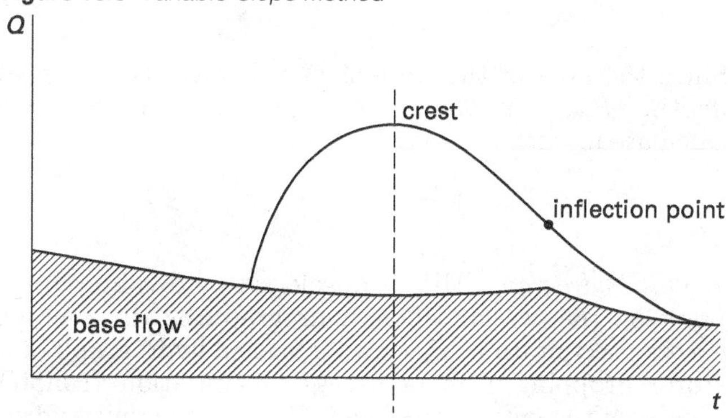

Once the base flow is separated out, the hydrograph of surface runoff will have the approximate appearance of Fig. 10.6.

[1]The total rain dropped by a storm is the *gross rain*. The rain that actually appears as immediate runoff can be called *surface runoff, overland flow, surface flow*, and *net rain*. The water that is absorbed by the soil and that does not contribute to the surface runoff can be called *base flow, groundwater, infiltration*, and *dry weather flow*.

Figure 10.6 Overland Flow Hydrograph

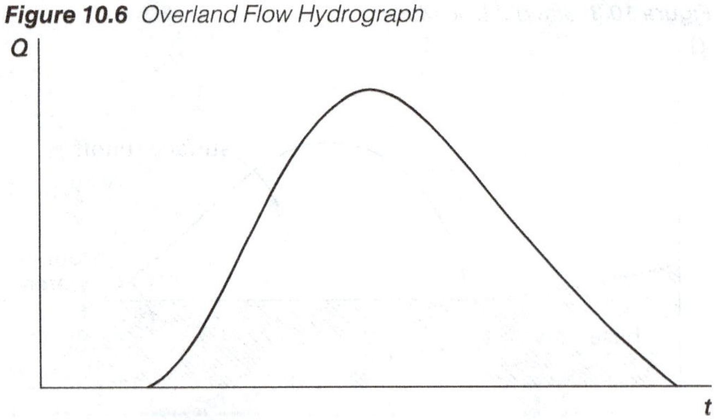

7. UNIT HYDROGRAPH

Once the overland flow hydrograph for a watershed has been developed, the total runoff (i.e., "excess rainfall") volume, V, from the storm can be found as the area under the curve. Although this can be found by integration, planimetry, or computer methods, it is often sufficiently accurate to approximate the hydrograph with a histogram and to sum the areas of the rectangles. (See Fig. 10.7.)

Figure 10.7 Hydrograph Histogram

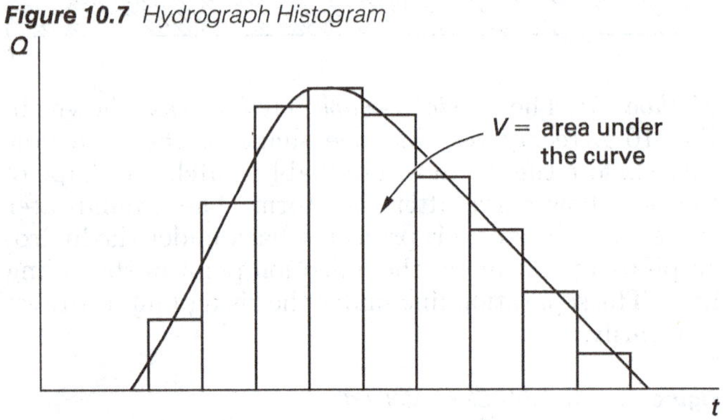

Since the area of the watershed is known, the average depth, P_{ave}, of the excess precipitation can be calculated.

$$V = A_d P_{\text{ave,excess}}$$

A *unit hydrograph* (UH) is developed by dividing every point on the overland flow hydrograph by the average excess precipitation, $P_{\text{ave,excess}}$. This is a hydrograph of a storm dropping 1 in of excess precipitation (runoff) evenly on the entire watershed. Units of the unit hydrograph are in/in. Figure 10.8 shows how a unit hydrograph compares to its surface runoff (storm) hydrograph.

Once a unit hydrograph has been developed from historical data of a particular storm volume, it can be used for other storm volumes. Such application is based on several assumptions: (a) All storms in the watershed have the same duration, (b) the time base is constant for all storms, (c) the shape of the rainfall curve is the same for all storms, and (d) only the total amount of rainfall varies from storm to storm.

Figure 10.8 Unit Hydrograph

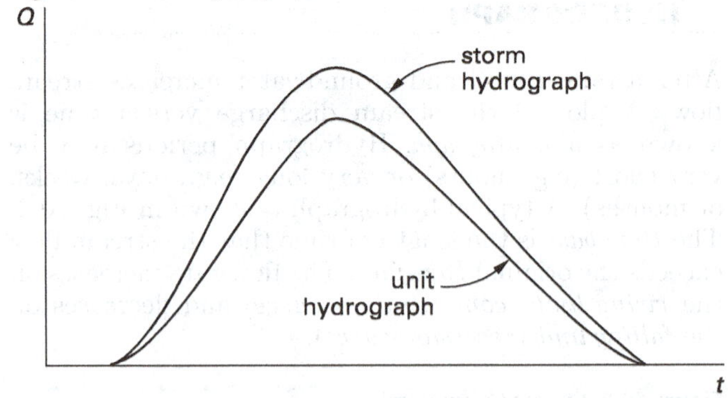

The hydrograph of a storm producing more or less than 1 in of rain is found by multiplying all ordinates on the unit hydrograph by the total precipitation of the storm.

A unit hydrograph can be used to predict the runoff for storms that have durations somewhat different than the storms used to develop the unit hydrograph. Generally, storm durations differing by up to ±25% are considered to be equivalent.

8. PEAK RUNOFF FROM THE RATIONAL METHOD

Although total runoff volume is required for reservoir and dam design, the instantaneous peak runoff is needed to size culverts and storm drains.

The rational method assumes that rainfall occurs at a constant rate. If this is true, then the peak runoff will occur when the entire drainage area is contributing to surface runoff, which will occur at the *time to concentration*, t_c. Other assumptions include (a) the recurrence interval of the peak flow is the same as for the design storm, (b) the runoff coefficient is constant, and (c) the rainfall is spatially uniform over the drainage area.

Equation 10.1: Rational Formula

$$Q = CIA \qquad 10.1$$

Description

The *rational formula* ("method," "equation," etc.), shown in Eq. 10.1, for peak discharge has been in widespread use in the United States since the early 1900s.[2] It is applicable to small areas (i.e., less than several hundred acres or so), but is seldom used for areas greater than 1–2 mi². The intensity used in Eq. 10.1 depends on

[2]In Great Britain, the rational equation is known as the *Lloyd-Davies equation*.

the time of concentration and the degree of protection desired (i.e., the recurrence interval). *Time of concentration* is defined as the time of travel from the hydraulically most remote (timewise) point in the watershed to the watershed outlet or other design point.[3]

A is in acres and I is in inches per hour, so Q should be in units of ac-in/hr. However, Q is taken as ft^3/sec because the conversion factor between these two units is 1.008.

Typical values of the runoff coefficient, C, range from 0.05 to 0.3 for farmland and parks, 0.3 to 0.7 for residential areas, and 0.5 to 0.95 for industrial and urban areas. If more than one area contributes to the runoff, the coefficient is weighted by the areas.

Example 1

A store property has a 20 ac asphalt parking lot with a runoff coefficient of 0.85, a 5 ac building roof with a runoff coefficient of 0.75, and 3.75 ac of lawn with a runoff coefficient of 0.20. A storm with an intensity of 3 in/hr occurs. The peak runoff for the store property is most nearly

(A) 65 ft^3/sec

(B) 88 ft^3/sec

(C) 110 ft^3/sec

(D) 140 ft^3/sec

Solution

The total watershed area is

$$A = 20 \text{ ac} + 5 \text{ ac} + 3.75 \text{ ac} = 28.75 \text{ ac}$$

Determine the weighted runoff coefficient for the total area.

$$C = \frac{(20 \text{ ac})(0.85) + (5 \text{ ac})(0.75) + (3.75 \text{ ac})(0.20)}{28.75 \text{ ac}} = 0.748$$

Use the rational formula to determine the peak runoff.

$$\begin{aligned} Q &= CIA \\ &= (0.748)\left(3 \ \frac{\text{in}}{\text{hr}}\right)(28.75 \text{ ac}) \\ &= 64.5 \text{ ft}^3/\text{sec} \quad (65 \text{ ft}^3/\text{sec}) \end{aligned}$$

The answer is (A).

9. NRCS CURVE NUMBER

Several methods of calculating total and peak runoff have been developed over the years by the U.S. Natural Resources Conservation Service (NRCS). These methods have generally been well correlated with actual experience, and the NRCS methods have become dominant in the United States. The NRCS methods classify the land use and soil type by a single parameter called the *curve number*, CN. This method can be used for any size homogeneous watershed with a known percentage of imperviousness. If the watershed varies in soil type or in cover, it generally should be divided into regions to be analyzed separately. A composite curve number can be calculated by weighting the curve number for each region by its area. Alternatively, the runoffs from each region can be calculated separately and added.

The *storage capacity* (*storativity* or *storage coefficient*), S, is the change (increase or decrease) in stored aquifer water volume when the aquifer thickness (i.e., piezometric head) changes (increases or decreases). Storage capacity has units of length (e.g., ft).

Equation 10.2 Through Eq. 10.4: NRCS Rainfall-Runoff

$$S = \frac{1000}{\text{CN}} - 10 \qquad 10.2$$

$$\text{CN} = \frac{1000}{S + 10} \qquad 10.3$$

$$Q = \frac{(P - 0.2S)^2}{P + 0.8S} \qquad 10.4$$

Description

The NRCS method assumes that infiltration follows an exponential decay curve with time. The storage capacity of the soil (i.e., the potential maximum retention after runoff begins) in inches, S, is calculated from the curve number by using Eq. 10.2. Similarly, the curve number, CN, can be calculated from the storage capacity in inches with Eq. 10.3.

The total *runoff* (net rain, precipitation excess, etc.) in inches, Q, is calculated from the areal rain. In Eq. 10.4, losses from interception, storm period evaporation, depression storage, and infiltration are assumed to be equal to 20% of the storage capacity, S. These losses are subtracted from the *gross rain*, P, to obtain the *net rain*. (Equation 10.4 can be derived from the water balance equation in Sec. 10.1.)

[3]When using intensity-duration-frequency (IDF) curves to size storm sewers, culverts, and other channels, it is assumed that the frequencies and probabilities of flood damage and storms are identical. This is not generally true, but the assumption is usually made anyway.

Example 2

The gross precipitation on a watershed is 12 in, and the maximum basin retention is 1 in. The runoff is most nearly

(A) 0.85 in

(B) 0.92 in

(C) 1.3 in

(D) 11 in

Solution

Use the NRCS rainfall-runoff formula, Eq. 10.4.

$$Q = \frac{(P - 0.2S)^2}{P + 0.8S} = \frac{(12\ \text{in} - (0.2)(1\ \text{in}))^2}{12\ \text{in} + (0.8)(1\ \text{in})}$$
$$= 11\ \text{in}$$

The answer is (D).

10. RESERVOIRS

An effective method of preventing flooding is to store surface runoff temporarily. After the storm is over, the stored water can be gradually released. An *impounding reservoir* (*retention watershed* or *detention watershed*) is a watershed used to store excess flow from a stream or river. The stored water is released when the stream flow drops below the minimum level that is needed to meet water demand. The *impoundment depth* is the design depth. Finding the impoundment depth is equivalent to finding the design storage capacity of the reservoir.

11. RESERVOIR SIZING: RESERVOIR ROUTING

Reservoir routing is the process by which the outflow hydrograph (i.e., the outflow over time) of a reservoir is determined from the inflow hydrograph (i.e., the inflow over time), the initial storage, and other characteristics of the reservoir. The simplest method is to keep track of increments in inflow, storage, and outflow period by period in a tabular simulation. This is the basis of the *storage indication method*, which is basically a bookkeeping process. The validity of this method is dependent on choosing time increments that are as small as possible.

step 1: Determine the starting storage volume, V_n. If the starting volume is zero or considerably different from the average steady-state storage, a large number of iterations will be required before the simulation reaches its steady-state results. A convergence criterion should be determined before the simulation begins.

step 2: For the next iteration, determine the inflow, discharge, evaporation, and seepage. The starting storage volume for the next iteration is found by solving the equation shown here.

$$V_{n+1} = V_n + (\text{inflow})_n - (\text{discharge})_n - (\text{seepage})_n - (\text{evaporation})_n$$

Repeat step 2 as many times as necessary.

Loss due to seepage is generally very small compared to inflow and discharge, and it is often neglected. Reservoir *evaporation*, E_R, can be estimated from analytical relationships or by evaluating data from *evaporation pans*. Pan data is extended to reservoir evaporation by the pan coefficient formula. In the equation for E_R, the summation is taken over the number of days in the simulation period. Units of evaporation are typically in/day. The *pan coefficient*, K_p, is typically 0.7–0.8.

$$E_R = \sum K_p E_p$$

Inflow can be taken from actual past history.

11 Hydraulics

Nomenclature

A	area	m^2
B	channel width	m
C	Hazen-Williams coefficient	$m^{1/2}/s$
d	depth	m
D	diameter	m
E	specific energy	m
f	friction factor	–
F	force	N
Fr	Froude number	–
g	acceleration of gravity, 9.81	m/s^2
h	head	m
h_c	flow depth to the cross section	m
h_f	head loss due to friction	m
H	pressure drop	lbf/ft^2
I	impulse	N·s
k_1	constant	–
K	constant	–
K	K-factor	$gpm/(psi^{1/2})$
L	length	m
m	mass	kg
$\dot{m}$	mass flow rate	kg/s
M	momentum	m^3
n	Manning's roughness coefficient	–
p	pressure	Pa
P	momentum	kg·m/s
P	pressure	lbf/ft^2
q	flow per unit width	$m^3/s{\cdot}m$
Q	discharge capacity	gpm
Q	flow rate	m^3/s
R_H	hydraulic radius	m
S	slope	–
t	time	s
T	width at surface	m
v	velocity	m/s
y	depth	m
z	height above datum	m

Symbols

α	kinetic energy correction factor	–
γ	specific weight	N/m^3
ρ	density	kg/m^3

Subscripts

c	critical
F	during test
h	hydraulic
H	hydraulic
R	rated or residual

1. INTRODUCTION

Fluid dynamics is the theoretical science of all fluids in motion. In a general sense, *hydraulics* is the study of the practical laws of incompressible fluid flow and resistance in pipes and open channels. Hydraulic formulas are often developed from experimentation, empirical factors, and curve fitting, without an attempt to justify why the fluid behaves the way it does.

2. CONSERVATION LAWS

Equation 11.1 Through Eq. 11.3: Continuity Equation

$$A_1\mathrm{v}_1 = A_2\mathrm{v}_2 \tag{11.1}$$

$$Q = A\mathrm{v} \tag{11.2}$$

$$\dot{m} = \rho Q = \rho A\mathrm{v} \tag{11.3}$$

Hydraulics/Hydrologic Sys.

Description

Fluid mass is always conserved in fluid systems, regardless of the pipeline complexity, orientation of the flow, and fluid. This single concept is often sufficient to solve simple fluid problems.

$$\dot{m}_1 = \dot{m}_2$$

When applied to fluid flow, the conservation of mass law is known as the *continuity equation.*

$$\rho_1 A_1 \mathrm{v}_1 = \rho_2 A_2 \mathrm{v}_2$$

If the fluid is incompressible, then $\rho_1 = \rho_2$. Equation 11.1 is the continuity equation for incompressible flow.

Volumetric flow rate, Q, is defined as the product of cross-sectional area and velocity, as shown in Eq. 11.2. From Eq. 11.1 and Eq. 11.2, it follows that

$$Q_1 = Q_2$$

Various units are used for volumetric flow rate. MGD (millions of gallons per day) and MGPCD (millions of gallons per capita day) are units commonly used in municipal water works problems. MMSCFD (millions of standard cubic feet per day) may be used to express gas flows.

Calculation of flow rates is often complicated by the interdependence between flow rate and friction loss. Each affects the other, so many pipe flow problems must be solved iteratively. Usually, a reasonable friction factor is assumed and used to calculate an initial flow rate. The flow rate establishes the flow velocity, from which a revised friction factor can be determined.

Example 3

An incompressible fluid flows through a pipe with an inner diameter of 10 cm at a velocity of 4 m/s. The pipe contracts to an inner diameter of 8 cm. What is most nearly the velocity of the fluid in the narrower section?

(A) 4.7 m/s

(B) 5.0 m/s

(C) 5.8 m/s

(D) 6.3 m/s

Solution

The cross-sectional areas of the two pipes are

$$A_1 = \frac{\pi D_1^2}{4} = \frac{\pi (10 \text{ cm})^2}{4} = 78.54 \text{ cm}^2$$

$$A_2 = \frac{\pi D_2^2}{4} = \frac{\pi (8 \text{ cm})^2}{4} = 50.27 \text{ cm}^2$$

Use Eq. 11.1.

$$A_1 \mathrm{v}_1 = A_2 \mathrm{v}_2$$

$$\mathrm{v}_2 = \frac{A_1 \mathrm{v}_1}{A_2} = \frac{(78.54 \text{ cm}^2)\left(4 \ \frac{\text{m}}{\text{s}}\right)}{50.27 \text{ cm}^2} = 6.25 \text{ m/s} \quad (6.3 \text{ m/s})$$

The answer is (D).

Equation 11.4 and Eq. 11.5: Bernoulli Equation

$$\frac{p_2}{\gamma} + \frac{\mathrm{v}_2^2}{2g} + z_2 = \frac{p_1}{\gamma} + \frac{\mathrm{v}_1^2}{2g} + z_1 \qquad 11.4$$

$$\frac{p_2}{\rho} + \frac{\mathrm{v}_2^2}{2} + z_2 g = \frac{p_1}{\rho} + \frac{\mathrm{v}_1^2}{2} + z_1 g \qquad 11.5$$

Description

The *Bernoulli equation*, also known as the *field equation* or the *energy equation*, is an energy conservation equation that is valid for incompressible, frictionless flow. The Bernoulli equation states that the total energy of a fluid flowing without friction losses in a pipe is constant. The total energy possessed by the fluid is the sum of its pressure, kinetic, and potential energies. In other words, the Bernoulli equation states that the total head at any two points is the same.

Example 4

The diameter of a water pipe gradually changes from 5 cm at the entrance, point A, to 15 cm at the exit, point B. The exit is 5 m higher than the entrance. The pressure is 700 kPa at the entrance and 664 kPa at the exit. Friction between the water and the pipe walls is negligible. The water density is 1000 kg/m³.

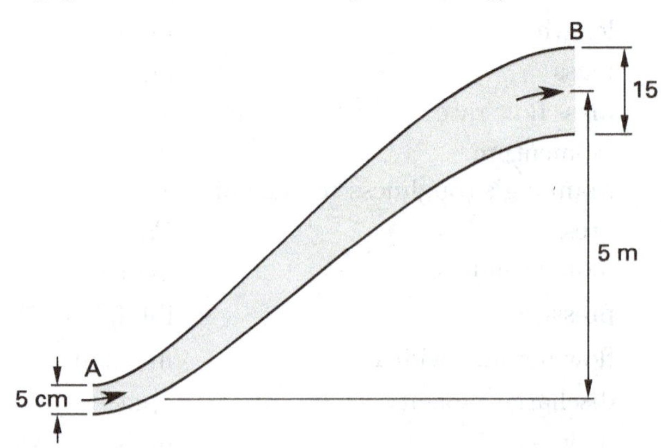

What is most nearly the rate of discharge at the exit?

(A) $0.0035 \text{ m}^3/\text{s}$

(B) $0.0064 \text{ m}^3/\text{s}$

(C) $0.010 \text{ m}^3/\text{s}$

(D) $0.018 \text{ m}^3/\text{s}$

Solution

Find the relationship between the entrance and exit velocities, v_1 and v_2, respectively. From Eq. 11.1 and substituting the pipe area equation,

$$A_1 v_1 = A_2 v_2$$

$$\left(\frac{\pi D_1^2}{4}\right) v_1 = \left(\frac{\pi D_2^2}{4}\right) v_2$$

$$v_1 = \left(\frac{D_2^2}{D_1^2}\right) v_2 = \left(\frac{(15 \text{ cm})^2 v_2}{(5 \text{ cm})^2}\right) = 9 v_2$$

Use the Bernoulli equation, Eq. 11.5, to find the velocity at the exit.

$$\frac{p_2}{\rho} + \frac{v_2^2}{2} + z_2 g = \frac{p_1}{\rho} + \frac{v_1^2}{2} + z_1 g$$

$$= \frac{p_1}{\rho} + \frac{(9v_2)^2}{2} + z_1 g$$

$$\frac{p_2 - p_1}{\rho} + g(z_2 - z_1) = \frac{81 v_2^2}{2} - \frac{v_2^2}{2}$$

$$v_2 = \sqrt{\frac{p_2 - p_1}{40\rho} + \frac{g(z_2 - z_1)}{40}}$$

$$= \sqrt{\frac{(664 \text{ kPa} - 700 \text{ kPa})\times\left(1000 \ \frac{\text{Pa}}{\text{kPa}}\right)}{(40)\left(1000 \ \frac{\text{kg}}{\text{m}^3}\right)} + \frac{\left(9.81 \ \frac{\text{m}}{\text{s}^2}\right)(5 \text{ m})}{40}}$$

$$= 0.571 \text{ m/s}$$

Multiply the velocity at the exit with the cross-sectional area to get the rate of flow.

$$Q = A_2 v_2$$

$$= \left(\frac{\pi D_2^2}{4}\right) v_2$$

$$= \left(\frac{\pi (15 \text{ cm})^2}{(4)\left(100 \ \frac{\text{cm}}{\text{m}}\right)^2}\right)\left(0.571 \ \frac{\text{m}}{\text{s}}\right)$$

$$= 0.0101 \text{ m}^3/\text{s} \quad (0.010 \text{ m}^3/\text{s})$$

The answer is (C).

3. STEADY INCOMPRESSIBLE FLOW IN PIPES AND CONDUITS

Equation 11.6 and Eq. 11.7: Energy Conservation Equation

$$\frac{p_1}{\gamma} + z_1 + \frac{v_1^2}{2g} = \frac{p_2}{\gamma} + z_2 + \frac{v_2^2}{2g} + h_f \qquad 11.6$$

$$\frac{p_1}{\rho g} + z_1 + \frac{v_1^2}{2g} = \frac{p_2}{\rho g} + z_2 + \frac{v_2^2}{2g} + h_f \qquad 11.7$$

Description

The *energy conservation* (*extended field*) *equation*, also known as the *steady-flow energy equation*, for steady incompressible flow is shown in Eq. 11.6 and Eq. 11.7. Equation 11.6 and Eq. 11.7 do not include the effects of *shaft devices* (so named because they have rotating shafts adding or extracting power) such as pumps, compressors, fans, and turbines. (Effects of shaft devices would be represented by *shaft work* terms.)

If the cross-sectional area of the pipe is the same at points 1 and 2, then $v_1 = v_2$ and $v_1^2/2g = v_2^2/2g$. If the elevation of the pipe is the same at points 1 and 2, then $z_1 = z_2$. When analyzing discharge from reservoirs and large tanks, it is common to use gauge pressures, so that $p_1 = 0$ at the surface. In addition, since the surface elevation changes slowly (or not at all) when drawing from a large tank or reservoir, $v_1 = 0$.

The *head loss due to friction* is denoted by the symbol h_f.

Example

An open reservoir with a water surface level at an elevation of 200 m drains through a 1 m diameter pipe with the outlet at an elevation of 180 m. The pipe outlet discharges to atmospheric pressure. The total head losses in the pipe and fittings are 18 m. Assume steady incompressible flow. The flow rate from the outlet is most nearly

(A) $4.9\ \mathrm{m^3/s}$

(B) $6.3\ \mathrm{m^3/s}$

(C) $31\ \mathrm{m^3/s}$

(D) $39\ \mathrm{m^3/s}$

Solution

Use the energy conservation equation, Eq. 11.7. Take point 1 at the reservoir surface and point 2 at the pipe outlet.

$$\frac{p_1}{\rho g}+z_1+\frac{\mathrm{v}_1^2}{2g}=\frac{p_2}{\rho g}+z_2+\frac{\mathrm{v}_2^2}{2g}+h_f$$

The pressure is atmospheric at the reservoir and the outlet, so $p_1 = p_2$. The velocity at the reservoir surface is $\mathrm{v}_1 \approx 0$ m/s, so the equation reduces to

$$z_1 = z_2+\frac{\mathrm{v}_2^2}{2g}+h_f$$

Solve for the velocity at the pipe outlet, v_2.

$$\begin{aligned}\mathrm{v}_2 &= \sqrt{2g(z_1-z_2-h_f)}\\ &= \sqrt{(2)\left(9.81\ \frac{\mathrm{m}}{\mathrm{s}^2}\right)(200\ \mathrm{m}-180\ \mathrm{m}-18\ \mathrm{m})}\\ &= 6.26\ \mathrm{m/s}\end{aligned}$$

The flow rate out of the pipe outlet is

$$\begin{aligned}Q &= \mathrm{v}_2 A = \mathrm{v}_2\left(\frac{\pi D^2}{4}\right)\\ &= \left(6.26\ \frac{\mathrm{m}}{\mathrm{s}}\right)\left(\frac{\pi(1\ \mathrm{m})^2}{4}\right)\\ &= 4.92\ \mathrm{m^3/s}\quad (4.9\ \mathrm{m^3/s})\end{aligned}$$

The answer is (A).

Equation 11.8: Pressure Drop

$$p_1 - p_2 = \gamma h_f = \rho g h_f \qquad 11.8$$

Description

For a pipe of constant cross-sectional area and constant elevation, the *pressure change* (*pressure drop*) from one point to another is given by Eq. 11.8.

4. FLOW IN NONCIRCULAR CONDUITS

Equation 11.9: Hydraulic Radius

$$R_H = \frac{\text{cross-sectional area}}{\text{wetted perimeter}} = \frac{D_H}{4} \qquad 11.9$$

Description

The *hydraulic radius* is defined as the cross-sectional area in flow divided by the *wetted perimeter.*

The area in flow is the cross-sectional area of the fluid flowing. When a fluid is flowing under pressure in a pipe (i.e., *pressure flow*), the area in flow will be the internal area of the pipe. However, the fluid may not completely fill the pipe and may flow simply because of a sloped surface (i.e., *gravity flow* or *open channel flow*).

The wetted perimeter is the length of the line representing the interface between the fluid and the pipe or channel. It does not include the *free surface* length (i.e., the interface between fluid and atmosphere).

For a circular pipe flowing completely full, the area in flow is πR^2. The wetted perimeter is the entire circumference, $2\pi R$. The hydraulic radius in this case is half the radius of the pipe.

$$R_H = \frac{\pi R^2}{2\pi R} = \frac{R}{2} = \frac{D}{4}$$

The hydraulic radius of a pipe flowing half full is also $R/2$, since the flow area and wetted perimeter are both halved.

Many fluid, thermodynamic, and heat transfer processes are dependent on the physical length of an object. The general name for this controlling variable is *characteristic dimension.* The characteristic dimension in evaluating fluid flow is the *hydraulic diameter,* D_H. The hydraulic diameter for a full-flowing circular pipe is simply its inside diameter. If the hydraulic radius of a noncircular duct is known, it can be used to calculate the hydraulic diameter.

$$D_H = 4R_H = 4\times\frac{\text{area in flow}}{\text{wetted perimeter}}$$

The frictional energy loss by a fluid flowing in a rectangular, annular, or other noncircular duct can be calculated from the Darcy equation by using the hydraulic diameter, D_H, in place of the diameter, D. The friction factor, f, is determined in any of the conventional manners.

Example

The 8 cm × 12 cm rectangular flume shown is filled to three-quarters of its height.

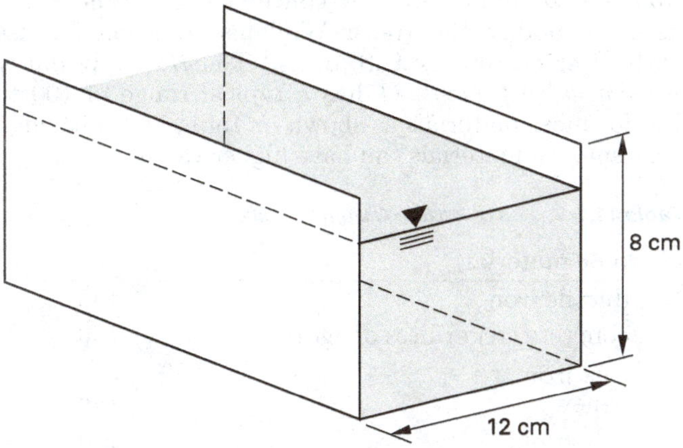

What is most nearly the hydraulic radius of the flow?

(A) 1.5 cm

(B) 2.5 cm

(C) 3.0 cm

(D) 5.0 cm

Solution

The hydraulic radius is

$$R_H = \frac{\text{cross-sectional area}}{\text{wetted perimeter}}$$
$$= \frac{(12 \text{ cm})\left(\frac{3}{4}\right)(8 \text{ cm})}{(2)\left(\frac{3}{4}\right)(8 \text{ cm}) + 12 \text{ cm}}$$
$$= 3.0 \text{ cm}$$

The answer is (C).

5. OPEN CHANNEL AND PARTIAL-AREA PIPE FLOW

An *open channel* is a fluid passageway that allows part of the fluid to be exposed to the atmosphere. This type of channel includes natural waterways, canals, culverts, flumes, and pipes flowing under the influence of gravity (as opposed to pressure conduits, which always flow full). A *reach* is a straight section of open channel with uniform shape, depth, slope, and flow quantity.

There are difficulties in evaluating open channel flow. The unlimited geometric cross sections and variations in roughness have contributed to a relatively small number of scientific observations upon which to estimate the required coefficients and exponents. Therefore, the analysis of open channel flow is more empirical and less exact than that of pressure conduit flow. This lack of precision, however, is more than offset by the percentage error in runoff calculations that generally precede the channel calculations.

Equation 11.10: Manning's Equation

$$\text{v} = (K/n)R_H^{2/3}S^{1/2} \tag{11.10}$$

Values

SI units	$K = 1$
customary U.S. units	$K = 1.486$

Description

Manning's equation has typically been used to estimate the velocity of flow in any open channel. It depends on the hydraulic radius, R_H, the slope of the energy grade line, S, and a dimensionless *Manning's roughness coefficient*, n. A conversion constant, K, modifies the equation for use with SI or customary U.S. units. The slope of the energy grade line is the terrain grade (slope) for uniform open-channel flow. The Manning roughness coefficient is typically taken as 0.013 for concrete.

Example

Water flows through the open concrete channel shown. Assume a Manning roughness coefficient of 0.013 for concrete.

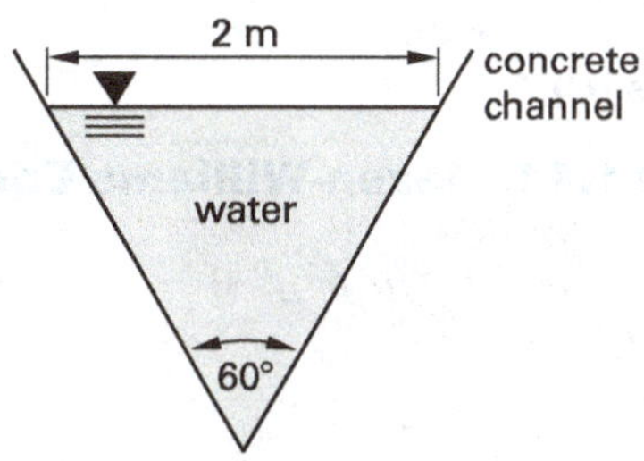

What is most nearly the minimum slope needed to maintain the flow at 3 m^3/s?

(A) 0.00015

(B) 0.00052

(C) 0.0015

(D) 0.0052

Solution

The area of flow is that of an equilateral triangle 2 m on each side, so

$$A = \frac{\sqrt{3}\,a^2}{4} = \frac{\sqrt{3}\,(2\text{ m})^2}{4} = 1.732\text{ m}^2$$

The hydraulic radius is

$$R_H = \frac{\text{cross-sectional area}}{\text{wetted perimeter}} = \frac{1.732\text{ m}^2}{2\text{ m} + 2\text{ m}} = 0.433\text{ m}$$

The velocity needed is

$$\text{v} = \frac{Q}{A} = \frac{3\ \frac{\text{m}^3}{\text{s}}}{1.732\text{ m}^2} = 1.732\text{ m/s}$$

Rearrange Manning's equation to solve for slope.

$$\text{v} = (K/n)R_H^{2/3}S^{1/2}$$

$$S = \left(\frac{\text{v}n}{KR_H^{2/3}}\right)^2 = \left(\frac{\left(1.732\ \frac{\text{m}}{\text{s}}\right)(0.013)}{(1)(0.433\text{ m})^{2/3}}\right)^2 = 0.001507 \quad (0.0015)$$

The answer is (C).

Equation 11.11: Hazen-Williams Equation

$$\text{v} = k_1 C R_H^{0.63} S^{0.54} \qquad 11.11$$

Values

SI units	$k_1 = 0.849$
customary U.S. units	$k_1 = 1.318$

Description

Although Manning's equation can be used for circular pipes flowing less than full, the *Hazen-Williams equation* is used more often. A conversion constant, k_1, is used to modify the Hazen-Williams equation for use with SI or customary U.S. units.[1] The *Hazen-Williams roughness coefficient*, C, has a typical range of 100 to 130 for most materials as shown in Table 11.1, although very smooth materials can have higher values.

Table 11.1 *Values of Hazen-Williams Coefficient, C*

pipe material	C
ductile iron	140
concrete (regardless of age)	130
cast iron:	
new	130
5 yr old	120
20 yr old	100
welded steel, new	120
wood stave (regardless of age)	120
vitrified clay	110
riveted steel, new	110
brick sewers	100
asbestos-cement	140
plastic	150

Example 8

Stormwater flows through a square concrete pipe that has a hydraulic radius of 1 m and an energy grade line slope of 0.8. Using the Hazen-Williams equation, the velocity of the water is most nearly

(A) 43 m/s

(B) 67 m/s

(C) 98 m/s

(D) 150 m/s

Solution

Use the Hazen-Williams equation, Eq. 11.11, to calculate the velocity. From Table 11.1, the Hazen-Williams roughness coefficient for concrete is 130.

$$\text{v} = k_1 C R_H^{0.63} S^{0.54} = (0.849)(130)(1\text{ m})^{0.63}(0.8)^{0.54} = 97.8\text{ m/s} \quad (98\text{ m/s})$$

The answer is (C).

[1]There is no significance to the subscript or uppercase and lowercase usage in the conversion factor. While the *NCEES Handbook* chose to use K in Eq. 11.10 and Eq. 11.11, it could just as easily have chosen to use K_1 and K_2, or k_1 and k_2.

Equation 11.12 Through Eq. 11.14: Circular Pipe Head Loss[2]

$$h_f = \frac{4.73L}{C^{1.852}D^{4.87}}Q^{1.852} \quad \text{[U.S. only]} \qquad 11.12$$

$$p = \frac{4.52Q^{1.85}}{C^{1.85}D^{4.87}} \quad \text{[U.S. only]} \qquad 11.13$$

$$p = \frac{6.05Q^{1.85}}{C^{1.85}D^{4.87}} \times 10^5 \quad \text{[SI only]} \qquad 11.14$$

Values

Multiply atmospheres by 1.013 to obtain bars.

Multiply bars by 0.987 to obtain atmospheres.

Multiply kPa by 0.01 to obtain bars.

Multiply bars by 100 to obtain kPa.

Description

Civil engineers commonly use the *Hazen-Williams equation* to calculate head loss. This method requires knowing the Hazen-Williams *roughness coefficient*, C. The advantage of using this equation is that C does not depend on the Reynolds number. The Hazen-Williams equation is empirical and is not dimensionally homogeneous.

Equation 11.12 gives the head loss expressed in feet. This equation is only valid when the variables are expressed in the correct customary U.S. units. For head loss due to friction, h_f, to be expressed in feet, the following units must be used: Hazen-Williams constant, C, dimensionless; length, L, in feet; inside diameter, D, in feet; and flow rate, Q, in ft^3/sec. Equation 11.12 cannot be used with SI units.

Equation 11.13 and Eq. 11.14 give the head loss expressed as pressure in customary U.S. and SI units, respectively.[3]

Equation 11.13 is only valid when the variables are expressed in the correct customary U.S. units. For head loss due to friction, p, to be expressed in units of psi drop per foot of pipe, the following units must be used: Hazen-Williams constant, C, dimensionless; inside diameter, D, in inches; and flow rate, Q, in ft^3/sec. Equation 11.13 cannot be used with SI units.

Equation 11.14 is only valid when the variables are expressed in the correct SI units.[4] For head loss due to friction, p, to be expressed in units of bars per meter of pipe, the following units must be used: Hazen-Williams constant, C, dimensionless; inside diameter, D, in mm; and flow rate, Q, in L/min. Equation 11.14 cannot be used with customary U.S. units. A bar is a unit approximately equal to one atmosphere.

Equation 11.15: Specific Energy

$$E = \alpha\frac{\text{v}^2}{2g} + y = \frac{\alpha Q^2}{2gA^2} + y \qquad 11.15$$

Description

Specific energy, E, is a term used primarily with open channel flow. It is the total head with respect to the channel bottom. Because the channel bottom is the reference elevation for potential energy, potential energy does not contribute to specific energy; only kinetic energy and pressure energy contribute. α is the kinetic energy correction factor, which is usually equal to 1.0. The pressure head at the channel bottom is equal to the depth of the channel, y.

In *uniform flow* (flow with constant width and depth), total head decreases due to the frictional effects, but specific energy is constant. In nonuniform flow, total head also decreases, but specific energy may increase or decrease.

Equation 11.16 Through Eq. 11.18: Critical Depth

$$y_c = \left(\frac{q^2}{g}\right)^{1/3} \qquad 11.16$$

$$q = Q/B \qquad 11.17$$

$$\frac{Q^2}{g} = \frac{A^3}{T} \qquad 11.18$$

Description

For any channel, there is some depth of flow that will minimize the energy of flow. (The depth is not minimized, however.) This depth is known as the *critical depth*, y_c. The critical depth depends on the shape of the channel, but it is independent of the channel slope.

Hydraulics/ Hydrologic Sys.

[2]The NCEES *FE Reference Handbook* (*NCEES Handbook*) uses the same symbol, D, for inside diameter in feet and in inches, and the same symbol, Q, for flow rate in gal/min and ft^3/sec.

[3]Although the proper usage is explained, the *NCEES Handbook* uses confusing nomenclature in Eq. 11.13 and Eq. 11.14. p should be $\Delta p_f/L$ or something similar. Equation 11.13 and Eq. 11.14 calculate a pressure difference, not a pressure.

[4]Equation 11.14 uses three conventional, but non-SI, units: bars, mm, and liters. A similar equation using kPa, m, and m^3 is easily derived.

For rectangular channels, the critical depth can be found with Eq. 11.16. q in this equation is *unit discharge*, the flow per unit width. q is defined in Eq. 11.17 as the ratio of the flow rate, Q, to the channel width, B.

For channels with nonrectangular shapes (including trapezoidal channels), the critical depth can be found by trial and error from Eq. 11.18. T is the surface width. To use this equation, assume trial values of the critical depth, use them to calculate the dependent quantities in the equation, and then verify the equality.

Figure 11.1 illustrates how specific energy is affected by depth, and accordingly, how specific energy relates to critical depth.

Figure 11.1 Specific Energy Diagram

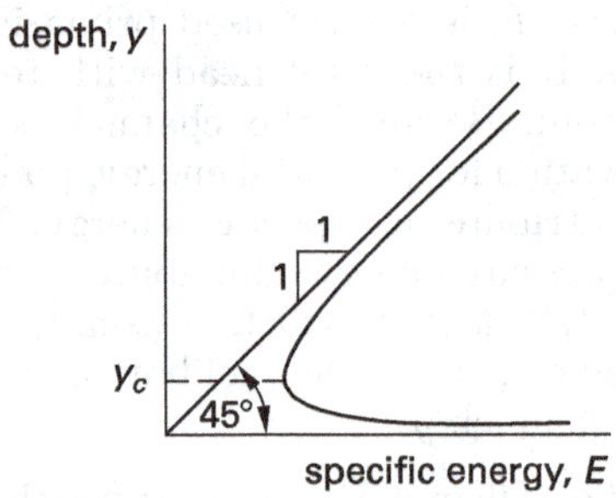

Equation 11.19: Momentum Depth Diagram

$$M = \frac{Q^2}{gA} + Ah_c \qquad 11.19$$

Description

Equation 11.19 is used to calculate momentum for any steady flow rate. Q is the discharge, g is the gravitational acceleration, A is the cross-sectional area of the flow, and h_c is the flow depth to the centroid of the cross-section.[5,6] The location of the flow area centroid depends on channel geometry.[7]

Figure 11.2 describes how momentum changes over different flow depths, with the x-axis representing momentum and the y-axis representing depth (and, by extension, discharge). The exact curve in Fig. 11.2 will apply to a single discharge, meaning the specific location and geometry of the curve will change based on flow rate but the general geometrical properties remain across all discharge values. In Fig. 11.2 there is a critical depth, y_c, where momentum is minimized and two depths are possible for other values of momentum; these are called *conjugate depths*.

Figure 11.2 Momentum-Depth Diagram

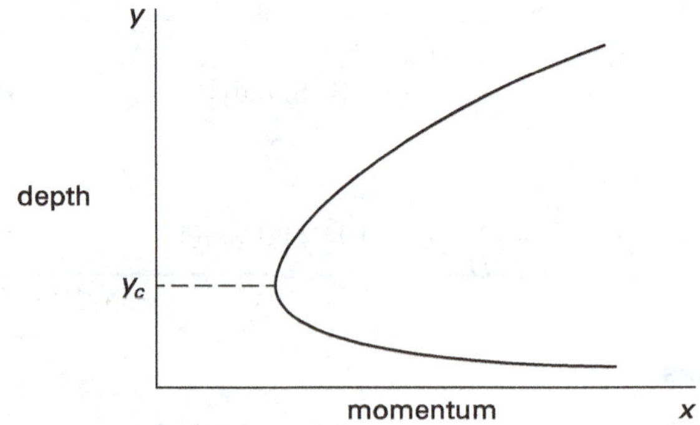

Example

A rectangular channel passes a total discharge of 7 m^3/s. The channel is 5 m wide and is flowing at a depth of 1 m. Most nearly, what is the momentum?

(A) 1.3 m^3

(B) 2.0 m^3

(C) 3.5 m^3

(D) 6.0 m^3

Solution

The cross-sectional area of the flow is

$$A = (5 \text{ m})(1 \text{ m}) = 5 \text{ m}^2$$

The height of the centroid of the cross section is half the flow depth, so the flow depth to the centroid is

$$h_c = \left(\frac{1}{2}\right)(1 \text{ m}) = 0.5 \text{ m}$$

[5]Eq. 11.19 uses the full cross-sectional discharge, Q, in units of L^3/T, rather than specific discharge, q, defined as Q/w, where w is width and therefore has units of $L^3/TL = L^2/T$. Many textbooks use specific discharge in the momentum-depth equation rather than total cross-sectional discharge.

[6]The *NCEES Handbook* uses h_c to represent the flow depth to the centroid of the cross-section and not the critical depth, y_c. Many hydraulics texts will use the variable h for depth rather than y, so caution is required when comparing the momentum-depth equation between the *NCEES Handbook* and other texts.

[7]Rectangular channels are often assumed, and so the depth to centroid is simply one-half the total flow depth. In cases where other channel geometries are used (e.g., trapezoidal, circular, etc.), the cross-sectional flow area centroid will not be 50% of the flow depth.

From Eq. 11.19, the momentum is

$$M = \frac{Q^2}{gA} + Ah_c$$

$$= \frac{\left(7\ \frac{\text{m}^3}{\text{s}}\right)^2}{\left(9.81\ \frac{\text{m}}{\text{s}^2}\right)(5\ \text{m}^2)} + (5\ \text{m})^2(0.5\ \text{m})$$

$$= 3.5\ \text{m}^3$$

The answer is (C).

Equation 11.20: Hydraulic Jump

$$y_2 = \frac{y_1}{2}\left(-1 + \sqrt{1 + 8\text{Fr}_1^2}\right) \qquad 11.20$$

Description

Hydraulic jumps occur when a supercritical flow changes rapidly to subcritical flow. The physical manifestation of this is when a shallow, high velocity flow changes to a deeper, lower velocity flow (e.g., at the bottoms of dam spillways). Equation 11.20 shows the relationship between the flow depths upstream and downstream of a hydraulic jump. Hydraulic jumps dissipate energy, so the conjugate depths from the specific energy diagram cannot be used. Instead, Eq. 11.20 uses conservation of momentum (Eq. 11.19) to relate the upstream depth to the downstream depth. The derivation of the hydraulic jump equation can be studied further in most basic hydraulic texts.

Equation 11.20 calculates the downstream subcritical flow depth, y_2, as a function of the upstream depth, y_1, and the Froude number of the upstream supercritical flow.

Example

An experiment is conducted on hydraulic jumps in a 3 m wide flume. The upstream velocity is 4 m/s, and the upstream depth is 0.3 m. What downstream depth should be expected after the hydraulic jump?

(A) 0.3 m

(B) 0.5 m

(C) 0.9 m

(D) 2.3 m

Solution

From Eq. 11.21, the Froude number at the upstream cross section is

$$\text{Fr} = \frac{\text{v}}{\sqrt{gy_h}} = \frac{4\ \frac{\text{m}}{\text{s}}}{\sqrt{\left(9.81\ \frac{\text{m}}{\text{s}^2}\right)(0.3\ \text{m})}} = 2.33$$

From Eq. 11.20,

$$y_2 = \frac{y_1}{2}\left(-1 + \sqrt{1 + 8\text{Fr}_1^2}\right) = \frac{0.3\ \text{m}}{2}\left(-1 + \sqrt{1 + (8)(2.33)^2}\right)$$
$$= 0.85\ \text{m} \quad (0.9\ \text{m})$$

The answer is (C).

Equation 11.21 and Eq. 11.22: Froude Number

$$\text{Fr} = \frac{\text{v}}{\sqrt{gy_h}} \qquad 11.21$$

$$y_h = A/T \qquad 11.22$$

Description

Equation 11.21 is the formula for the dimensionless *Froude number*, Fr, a convenient index of the flow regime. v is velocity. y_h is the *characteristic length*, also referred to as the *characteristic (length) scale*, *hydraulic depth*, *mean hydraulic depth*, and others, depending on the channel configuration.[8] For a circular channel flowing half full, $y_h = \pi D/8$. For a rectangular channel, $y_h = d$, the depth corresponding to velocity v. For trapezoidal and semicircular channels, and in general, y_h is the area in flow divided by the top width.

The Froude number can be used to determine whether the flow is subcritical or supercritical. When the Froude number is less than one, the flow is subcritical. The depth of flow is greater than the critical depth, and the flow velocity is less than the critical velocity.

When the Froude number is greater than one, the flow is supercritical. The depth of flow is less than critical depth, and the flow velocity is greater than the critical velocity.

When the Froude number is equal to one, the flow is critical.[9]

[8]The *NCEES Handbook* uses subscripts *H* and *h* inconsistently to mean "hydraulic." There is no difference between the two subscripts.

[9]The similarity of the Froude number to the Mach number used to classify gas flows is more than coincidental. Both bodies of knowledge employ parallel concepts.

Example

A 150 m long surface vessel with a speed of 40 km/h is modeled at a scale of 1:50. Similarity based on Froude numbers is appropriate for the modeling of surface vessels, so the model should travel at a speed of most nearly

(A) 0.22 m/s

(B) 1.6 m/s

(C) 2.2 m/s

(D) 16 m/s

Solution

The surface vessel and the model should have the same Froude numbers. From Eq. 11.21,

$$\begin{aligned}
\text{Fr}_{\text{vessel}} &= \text{Fr}_{\text{model}} \\
\frac{\text{v}_{\text{vessel}}}{\sqrt{gy_{h,\text{vessel}}}} &= \frac{\text{v}_{\text{model}}}{\sqrt{gy_{h,\text{model}}}} \\
\text{v}_{\text{model}} &= \text{v}_{\text{vessel}}\sqrt{\frac{y_{h,\text{model}}}{y_{h,\text{vessel}}}} \\
&= \frac{\left(40\ \frac{\text{km}}{\text{h}}\right)\left(1000\ \frac{\text{m}}{\text{km}}\right)}{3600\ \frac{\text{s}}{\text{h}}}\sqrt{\frac{1}{50}} \\
&= 1.57\ \text{m/s} \quad (1.6\ \text{m/s})
\end{aligned}$$

The answer is (B).

6. THE IMPULSE-MOMENTUM PRINCIPLE

The *momentum*, **P**, of a moving object is a vector quantity defined as the product of the object's mass and velocity.

$$\mathbf{P} = m\mathbf{v}$$

The *impulse*, **I**, of a constant force is calculated as the product of the force's magnitude and the length of time the force is applied.

$$\mathbf{I} = \mathbf{F}\Delta t$$

The *impulse-momentum principle* states that the impulse applied to a body is equal to the change in that body's momentum. This is one way of stating *Newton's second law.*

$$\mathbf{I} = \Delta\mathbf{P}$$

From this,

$$\begin{aligned}
F\Delta t &= m\Delta\text{v} \\
&= m(\text{v}_2 - \text{v}_1)
\end{aligned}$$

For fluid flow, there is a mass flow rate, $\dot{m}$, but no mass per se. Since $\dot{m} = m/\Delta t$, the impulse-momentum equation can be rewritten as

$$F = \dot{m}\Delta\text{v}$$

Substituting for the mass flow rate, $\dot{m} = \rho A\text{v}$. The quantity $Q\rho\text{v}$ is the *rate of momentum.*

$$\begin{aligned}
F &= \rho A\text{v}\Delta\text{v} \\
&= Q\rho\Delta\text{v}
\end{aligned}$$

Equation 11.23: Control Volume[10]

$$\sum F = Q_2\rho_2\text{v}_2 - Q_1\rho_1\text{v}_1 \qquad 11.23$$

Description

Eq. 11.23 results from applying the impulse-momentum principle to a control volume. $\sum F$ is the resultant of all external forces acting on the control volume. $Q_1\rho_1\text{v}_1$ and $Q_2\rho_2\text{v}_2$ represent the rate of momentum of the fluid entering and leaving the control volume, respectively, in the same direction as the force.

7. FIRE HYDRAULICS

Equation 11.24 Through Eq. 11.26: Rated Capacity of Fire Hydrants

$$Q_R = Q_F \times (H_R/H_F)^{0.54} \qquad 11.24$$

$$H_R = P_S - 20\ \text{psi} \qquad 11.25$$

$$H_F = P_S - P_R \qquad 11.26$$

Description

Fire hydrants are rated according to the amount of flow that can be delivered from them under certain conditions. A hydrant's rated capacity is the result of a field test where the *static pressure* (pressure before the valve is opened), the test flow rate, and the *residual pressure* (the pressure of the system while the test flow is discharging) are measured. Pressure and discharge are not independent variables, and in order to have a common

[10]While any symbol can be used to represent any quantity, the use of bold letters is usually reserved for vector quantities. The *NCEES Handbook* uses bold ***F*** to designate force in its fluid dynamics section. While force can indeed be a vector quantity, the fluid equations in this section are not vector equations. ***F*** in text and **F** in illustrations should be interpreted as *F*, as it is presented in this book.

flow rating at comparable pressures it is necessary to normalize the test results to a reference pressure. Equation 11.24 through Eq. 11.26 calculate the discharge of a fire hydrant at a common pressure of 20 psi, where Q_R is the rated discharge capacity at the chosen residual pressure (typically 20 psi), Q_F is the discharge observed during the test, P_S is the observed static pressure of the system (no water is flowing), and P_R is the observed residual pressure during the flow test corresponding to Q_F. The variables H_F nd H_R re the observed and target residual pressure drop parameters. Equation 11.24 through Eq. 11.26 are not unit-specific—any units can be used, so long as they are applied consistently.

Example

A fire hydrant is tested to determine its flow rating at 20 psi. The static pressure at the hydrant measures 80 psi, and the residual pressure when the hydrant is flowing at 1876 gpm is 45 psi. Most nearly, what is the rated discharge capacity at 20 psi?

(A) 1400 gpm

(B) 1900 gpm

(C) 2500 gpm

(D) 3200 gpm

Solution

The observed and target residual pressure drop parameters are

$$H_R = P_S - 20 \text{ psi} = 80 \text{ psi} - 20 \text{ psi} = 60 \text{ psi}$$

$$H_F = P_S - P_R = 80 \text{ psi} - 45 \text{ psi} = 35 \text{ psi}$$

From Eq. 11.24,

$$\begin{aligned} Q_R &= Q_F \times (H_R/H_F)^{0.54} \\ &= \left(1876 \ \frac{\text{gal}}{\text{min}}\right)\left(\frac{60 \ \frac{\text{lbf}}{\text{in}^2}}{35 \ \frac{\text{lbf}}{\text{in}^2}}\right)^{0.54} \\ &= 2509.8 \text{ gpm} \quad (2500 \text{ gpm}) \end{aligned}$$

The answer is (C).

Equation 11.27: Fire Hydrant Discharging to the Atmosphere

$$Q = 29.8D^2C_dP^{1/2} \qquad 11.27$$

Description

Fire hydrant flow tests determine the flow rate and pressure of fire hydrants to ensure they are capable of providing water at acceptable pressure and flow rates for public health and firefighting operations. *NFPA Standard 291* provides guidance on fire flow testing. During a fire hydrant flow test, a hydrant is opened to atmospheric pressure, and water is allowed to flow downstream. The rated capacity for a hydrant open to the atmosphere is determined by measuring the water discharged (i.e., the flow) using Eq. 11.27. Q is the discharge in gpm, D is the outlet orifice diameter in inches, P is the pressure measured by a pitot gage (the *pitot pressure* or *velocity head*) in psi, and C_d is the hydraulic outlet coefficient. Figure 11.3 shows three commonly found hydrant outlets and provides their corresponding hydraulic outlet coefficient.

Figure 11.3 *Hydraulic Outlet Coefficients*

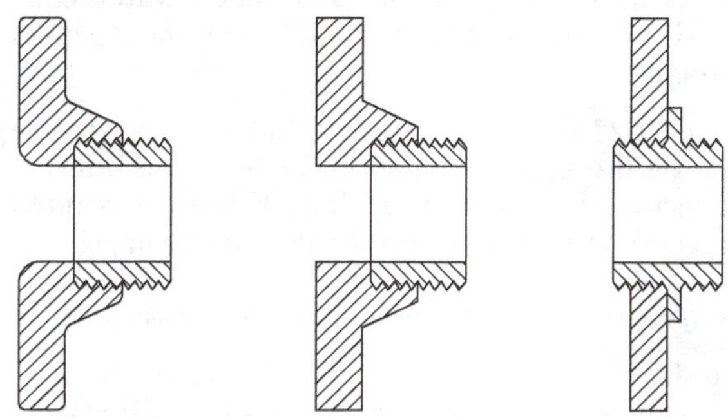

(a) outlet smooth and well-rounded coefficient 0.90

(b) outlet square and sharp coefficient 0.80

(c) outlet square projecting into barrel coefficient 0.70

Example

A smooth and rounded fire hydrant has an outlet has a diameter of 2.0 in. The pitot pressure is 26 psi. The total hydrant discharge is most nearly

(A) 270 gpm

(B) 450 gpm

(C) 550 gpm

(D) 610 gpm

Solution

From Fig. 11.3, for a smooth rounded hydrant, the hydraulic outlet coefficient is 0.90. From Eq. 11.27,

$$\begin{aligned} Q &= 29.8D^2C_dP^{1/2} \\ &= (29.8)(2.0\ \text{in})^2(0.90)\left(26\ \frac{\text{lbf}}{\text{in}^2}\right)^{1/2} \\ &= 547.0\ \text{gpm} \quad (550\ \text{gal/min}) \end{aligned}$$

The answer is (C).

Equation 11.28: Fire Sprinkler Discharge

$$Q = KP^{1/2} \qquad 11.28$$

Description

Fire suppression sprinkler systems are commonly used in many commercial, industrial and public buildings to detect and suppress fires quickly. Sprinkler systems must be properly sized so that they can discharge the correct amount of water in order to be effective; to do this, it is necessary to know how much water can be delivered by the supply line and size the sprinklers accordingly.

Equation 11.28 is used to calculate the expected discharge for a single fire suppression sprinkler based on orifice size and K-factor. (See Table 11.2). The K-factor has units of gpm/(psi$^{1/2}$). P is the pressure at the sprinkler in psi.

Table 11.2 K-Factors for Different Sized Sprinkler Orifices

orifice size (in)	name	K-factor (gpm/(psi$^{1/2}$))
1/2	standard	5.6
17/32	large	8.0
5/8	extra large	11.2

Example 14

A fire sprinkler operates at a pressure of 30 psi and has a 1/2 in orifice. Most nearly, what is the flow rate of the sprinkler?

(A) 31 gpm

(B) 62 gpm

(C) 74 gpm

(D) 170 gpm

Solution

From Table 11.2, the K-factor for a 1/2 in orifice is 5.6 gpm/(psi$^{1/2}$).

$$\begin{aligned} Q = KP^{1/2} &= \left(5.6\ \frac{\text{gpm}}{\left(\frac{\text{lbf}}{\text{in}^2}\right)^{1/2}}\right)\sqrt{30\ \frac{\text{lbf}}{\text{in}^2}} \\ &= 30.7\ \text{gpm} \quad (31\ \text{gpm}) \end{aligned}$$

The answer is (A).

Hydraulics/ Hydrologic Sys.

12 Groundwater

Nomenclature

A	area	ft^2
D	aquifer thickness	ft
D_w	well drawdown	ft
g	acceleration of gravity, 32.2	ft/sec^2
g_c	gravitational constant, 32.2	lbm-ft/lbf-sec^2
h	aquifer depth	ft
h	total hydraulic head	ft
i	hydraulic gradient	ft/ft
k	hydraulic conductivity (coefficient of permeability)	ft/sec
K	hydraulic conductivity (coefficient of permeability)	ft/sec
L	length	ft
n	porosity	–
p	pressure	lbf/ft^2
q	specific discharge	ft/sec
Q	flow quantity	ft^3/sec
r	radial distance from well	ft
S	storage capacity	ft
t	time	sec
T	transmissivity	ft^2/day
v	flow velocity	ft/sec
x	distance in x-direction	ft

Symbols

γ	specific weight	lbf/ft^3
ρ	density	lbm/ft^3

Subscripts

e	effective
o	at well
r	distance or radius

1. AQUIFERS

Underground water, also known as *subsurface water*, is contained in saturated geological formations known as *aquifers*. Aquifers are divided into two zones by the water table surface. The *vadose zone* is above the elevation of the water table. Pores in the vadose zone may be either saturated, partially saturated, or empty. The *phreatic zone* is below the elevation of the water table. Pores are always saturated in the phreatic zone.

An aquifer whose water surface is at atmospheric pressure and that can rise or fall with changes in volume is a *free aquifer*, also known as an *unconfined aquifer*. If a well is drilled into an unconfined aquifer, the water level in the well will correspond to the water table. Such a well is known as a *gravity well*.

The *storage capacity* (*storativity* or *storage coefficient*), S, is the change (increase or decrease) in stored aquifer water volume when the aquifer thickness (i.e., piezometric head) changes (increases or decreases). Storage capacity has units of length (e.g., ft).

An aquifer that is bounded on all extents is known as a *confined aquifer*. The water in confined aquifers may be under pressure. If a well is drilled into such an aquifer, the water in the well will rise to a height corresponding to the hydrostatic pressure. The *piezometric height* of the rise is

$$h = \frac{p}{\gamma} = \frac{p}{\rho} \times \frac{g_c}{g}$$

If the confining pressure is high enough, the water will be expelled from the surface, and the source is known as an *artesian well*.

Hydraulics/ Hydrologic Sys.

2. PERMEABILITY

For studies involving the flow of water through an aquifer, effects of intrinsic soil permeability and the water are combined into the *hydraulic conductivity*, also known as the *coefficient of permeability* or simply the *permeability*, K. Hydraulic conductivity can be determined from a number of water-related tests.[1] It has units of volume per unit area per unit time, which is equivalent to length divided by time (i.e., units of velocity). Volume may be expressed as cubic feet and cubic meters, or gallons and liters.

For many years in the United States, hydraulic conductivity was specified in *Meinzer units* (gallons per day per square foot). To avoid confusion related to multiple definitions and ambiguities in these definitions, hydraulic conductivity is now often specified in units of feet per day.

3. DARCY'S LAW

Equation 12.1 Through Eq. 12.3: Darcy's Law

$$Q = -KA(dh/dx) \qquad 12.1$$

$$q = -K(dh/dx) \qquad 12.2$$

$$\text{v} = q/n = -K/n(dh/dx) \qquad 12.3$$

Variations

$$Q = -KiA$$

$$q = \frac{Q}{A} = Ki$$

$$i = \frac{dh}{dx}$$

Description

Movement of groundwater through an aquifer (i.e., seepage) is given by *Darcy's law.* The negative sign in Darcy's law accounts for the fact that flow is in the direction of decreasing head. That is, the hydraulic gradient, dh/dx, is negative in the direction of flow. The hydraulic gradient is typically specified in units of in/in or ft/ft, or is sometimes treated as dimensionless.[2]

The *porosity*, n, used in Eq. 12.3 is the volumetric fraction of voids in the soil. The voids may be filled with gas or liquid (e.g., air or water) or both without affecting the value of porosity.

$$n = \frac{V_{\text{voids}}}{V_{\text{total}}} = \frac{V_{\text{total}} - V_{\text{solids}}}{V_{\text{total}}} = 1 - \frac{\gamma_{\text{bulk}}}{\gamma_{\text{solids}}}$$

The *specific discharge*, q, in Eq. 12.2 and Eq. 12.3 is also known as the *effective velocity, superficial velocity*, and *Darcy velocity*.

Darcy's law is applicable only when the Reynolds number is less than 1. Significant deviations have been noted when the Reynolds number is even as high as 2.

Example

A soil sample in a permeameter has a length of 0.5 ft and a cross-sectional area of 0.07 ft^2. The water head is 3.5 ft upstream and 0.1 ft downstream. A flow rate of 1 ft^3/day is observed. The hydraulic conductivity is most nearly

(A) 2.4×10^{-5} ft/sec

(B) 6.1×10^{-5} ft/sec

(C) 8.2×10^{-5} ft/sec

(D) 9.3×10^{-5} ft/sec

Solution

Use Darcy's law. The change in hydraulic head (which decreases in the direction of flow) over the length of the soil sample is

$$\frac{dh}{dx} = \frac{0.1\ \text{ft} - 3.5\ \text{ft}}{0.5\ \text{ft}} = -6.8\ \text{ft/ft}$$

The hydraulic conductivity is calculated from Eq. 12.1.

$$Q = -KA(dh/dx)$$

$$K = -\frac{Q}{A\dfrac{dh}{dx}}$$

$$= -\frac{1\ \dfrac{\text{ft}^3}{\text{day}}}{(0.07\ \text{ft}^2)\left(-6.8\ \dfrac{\text{ft}}{\text{ft}}\right)\left(24\ \dfrac{\text{hr}}{\text{day}}\right)\left(3600\ \dfrac{\text{sec}}{\text{hr}}\right)}$$

$$= 2.4 \times 10^{-5}\ \text{ft/sec}$$

The answer is (A).

[1]Permeability can be determined from constant-head permeability tests (sands), falling-head permeability tests (fine sands and silts), consolidation tests (clays), and field tests of wells (in situ gravels and sands).

[2]The symbol i is commonly used in practice and academia to designate the Darcy gradient, dh/dx, in groundwater seepage problems.

4. TRANSMISSIVITY

Equation 12.4: Coefficient of Transmissivity

$$T = KD \qquad 12.4$$

Description

Transmissivity (also known as the *coefficient of transmissivity*) is an index of the rate of groundwater movement. The transmissivity of flow from a saturated aquifer of thickness D and hydraulic conductivity K is given by Eq. 12.4. The thickness, D, of a confined aquifer is the difference in elevations of the bottom and top of the saturated formation. For permeable soil, the thickness, D, is the difference in elevations of the impermeable bottom and the water table. Transmissivity is given in units of ft^2/day.

Example

Given that the hydraulic conductivity of fine gravel is 100 ft/day, what is most nearly the transmissivity of a confined fine gravel aquifer 75 ft thick?

(A) 750 ft^2/day

(B) 2000 ft^2/day

(C) 5000 ft^2/day

(D) 7500 ft^2/day

Solution

Use Eq. 12.4. The transmissivity of the aquifer is

$$T = KD = \left(100\ \frac{\text{ft}}{\text{day}}\right)(75\ \text{ft}) = 7500\ \text{ft}^2/\text{day}$$

The answer is (D).

5. FLOW DIRECTION

Flow direction will be from an area with a high piezometric head (as determined from observation wells) to an area of low piezometric head. Piezometric head is assumed to vary linearly between points of known head. Similarly, all points along a line joining two points with the same piezometric head can be considered to have the same head.

6. WELLS

Water in aquifers can be extracted from *gravity wells*. However, *monitor wells* may also be used to monitor the quality and quantity of water in an aquifer. *Relief wells* are used to dewater soil.

Wells may be dug, bored, driven, jetted, or drilled in a number of ways, depending on the aquifer material and the depth of the well. Wells deeper than 100 ft are usually drilled. After construction, the well is *developed*, which includes the operations of removing any fine sand and mud. Production is *stimulated* by increasing the production rate. The fractures in the rock surrounding the well are increased in size by injecting high-pressure water or using similar operations.

Figure 12.1 illustrates a typical water-supply well. Water is removed from the well through the *riser pipe* (*eductor pipe*). Water enters the well through a perforated or slotted casing known as the *screen*. The required *open area* depends on the flow rate and is limited by the maximum permissible entrance velocity that will not lift grains larger than a certain size.

Figure 12.1 *Typical Gravity Well*

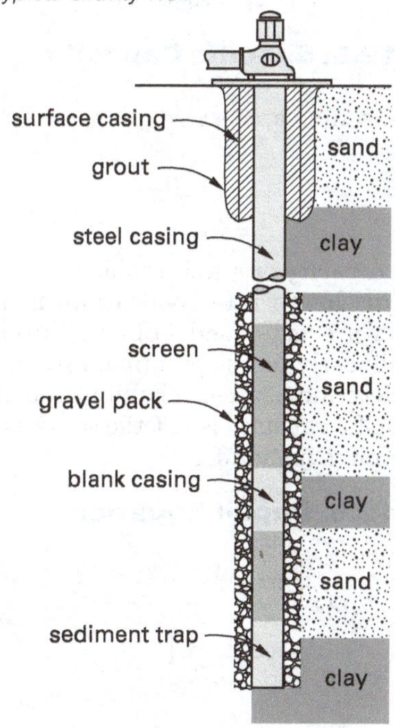

Screens do not necessarily extend the entire length of the well. The required screen length can be determined from the total amount of open area required and the open area per unit length of casing. For confined aquifers, screens are usually installed in the middle 70–80% of the well. For unconfined aquifers, screens are usually installed in the lower 30–40% of the well.

It is desirable to use screen openings as large as possible to reduce entrance friction losses. Larger openings can be tolerated if the well is surrounded by a *gravel pack* to prevent fine material from entering the well. Gravel packs are generally required in soils where the D_{90} size (i.e., the sieve size retaining 90% of the soil) is less than 0.25 mm and when the well goes through layers of sand and clay.

7. WELL DRAWDOWN IN AQUIFERS

An aquifer with a well is shown in Fig. 12.2. Once pumping begins, the water table will be lowered in the vicinity of the well. The resulting water table surface is referred to as a *cone of depression*. The decrease in water level at some distance r from the well is known as the *drawdown*, $D_{w,r}$. The drawdown at the well is denoted as $D_{w,o}$.

Figure 12.2 Well Drawdown in an Unconfined Aquifer

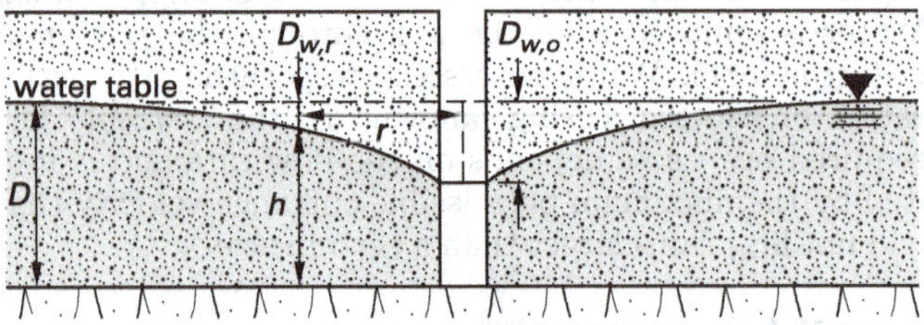

Equation 12.5: Specific Capacity

$$\text{specific capacity} = Q/D_w \qquad 12.5$$

Description

The *specific capacity* of a well is the ratio of short-term sustainable discharge rate to drawdown at the well.[3] Typical units are m^2/d and gpm/ft. Specific capacity depends nonlinearly on the pumping rate, so its primary usage is to determine when rehabilitation of the well is needed. (Specific capacity is not the same as specific discharge, q, through an aquifer.)

Equation 12.6: Dupuit Equation

$$Q = \frac{\pi k (h_2^2 - h_1^2)}{\ln\left(\frac{r_2}{r_1}\right)} \qquad 12.6$$

Description

If the drawdown, $D_{w,o}$, is small with respect to the aquifer phreatic zone thickness, D, and the well completely penetrates the aquifer, the equilibrium (steady-state) well discharge is given by the *Dupuit equation*. The Dupuit equation can only be used to determine the equilibrium flow rate a "long time" after pumping has begun.

In Eq. 12.6, h_1 and h_2 are the aquifer depths at radial distances r_1 and r_2, respectively, from the well.[4] h_2 can also be taken as the original aquifer depth, D, if r_2 is taken as the well's *radius of influence*, the distance at which the well has no effect on the water table level. k is the hydraulic conductivity (coefficient of permeability).[5]

Example

An unconfined aquifer is shown with the original water table 200 ft from its bottom. To access the water, a well with a 1 ft radius is drilled. The pumping rate is 80 gal/min, and the hydraulic conductivity is 1.0 ft/day. The well drawdown is zero at a distance of 1500 ft from the well.

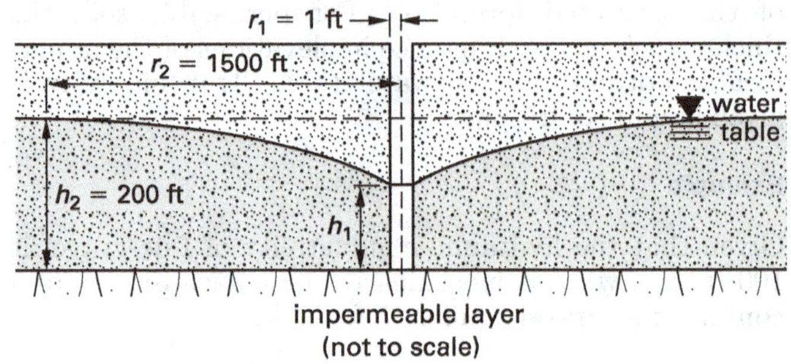

After pumping has continued long enough to reach steady-state conditions, what will be the approximate depth of water in the well?

(A) 0 ft

(B) 30 ft

(C) 65 ft

(D) 90 ft

Solution

Rearrange the Dupuit equation to solve for the depth of water in the well, h_2.

$$Q = \frac{\pi k (h_2^2 - h_1^2)}{\ln\left(\frac{r_2}{r_1}\right)}$$

[3]The NCEES *FE Reference Handbook* (*NCEES Handbook*) uses D_w to designate "well drawdown," while the symbol s is used almost exclusively in engineering practice to designate drawdown. The *NCEES Handbook* also uses D for aquifer thickness and pipe casing diameter, but the three Ds should not be confused.

[4](1) The *NCEES Handbook* unnecessarily specifies an explicit definition of r_1. Specifically, r_1 is defined as the "radius to water surface at perimeter of well, i.e., radius of well (ft)." According to the accompanying illustration, the *NCEES Handbook*'s intent is to define the inside radius of the well casing. (2) For Eq. 12.6, it is irrelevant which location is designated "1" and which is designated "2," as long as the designations are used consistently for both the radius, r, and depth, h, terms. If the designations are such that the numerator of Eq. 12.6 is negative, then the denominator of Eq. 12.6 will also be negative. The flow rate, Q, will be positive in either case. (3) Although the *NCEES Handbook* specifies that all variables incorporate units of feet, Eq. 12.6 can be used without change with SI units.

[5]The *NCEES Handbook* uses K for hydraulic conductivity in Eq. 12.1 through Eq. 12.3, but it uses k as the coefficient of permeability in Eq. 12.6. In fact, hydraulic conductivity and coefficient of permeability are the same parameter, and there is no difference between K and k.

$$h_1 = \sqrt{h_2^2 - \frac{Q \ln \frac{r_2}{r_1}}{\pi k}}$$

$$= \sqrt{(200 \text{ ft})^2 - \frac{\left(\frac{\left(80 \frac{\text{gal}}{\text{min}}\right)\left(60 \frac{\text{min}}{\text{hr}}\right)\left(24 \frac{\text{hr}}{\text{day}}\right)}{7.48 \frac{\text{gal}}{\text{ft}^3}} \right) \times \ln \frac{1500 \text{ ft}}{1 \text{ ft}}}{\pi\left(1.0 \frac{\text{ft}}{\text{day}}\right)}}$$

$$= 64.4 \text{ ft} \quad (65 \text{ ft})$$

The answer is (C).

Equation 12.7: Thiem Equation

$$Q = \frac{2\pi T(h_2 - h_1)}{\ln\left(\frac{r_2}{r_1}\right)} \qquad 12.7$$

Description

For an artesian well fed by a confined aquifer of thickness D, (see Fig. 12.3) the discharge is given by the *Thiem equation.*

Figure 12.3 *Confined Aquifer*

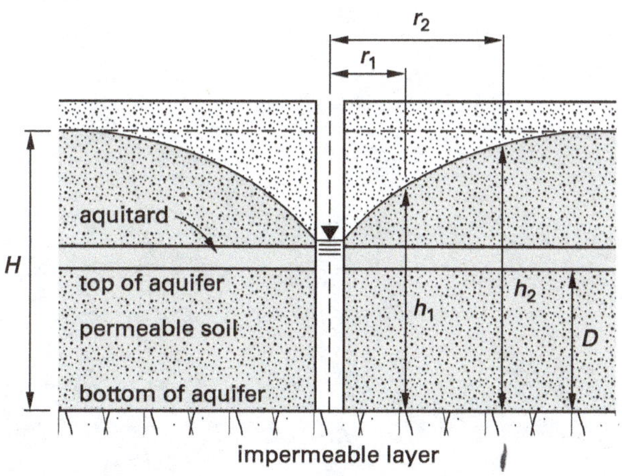

8. PUMPING POWER

Various types of pumps are used in wells. Problems with excessive suction lift are avoided by the use of submersible pumps.

Pumping power can be determined from hydraulic (water) power equations. The total head is the sum of static lift, velocity head, drawdown, pipe friction, and minor entrance losses from the casing, strainer, and screen. The Hazen-Williams equation is commonly used with a coefficient of $C = 100$ to determine the pipe friction.

Topic V: Environmental Engineering

Chapter

13. Water Quality
14. Water Supply Treatment and Distribution
15. Wastewater Collection and Treatment
16. Activated Sludge and Sludge Processing
17. Air Quality

13 Water Quality

Nomenclature

C	concentration	mg/L
EW	equivalent weight	g/mol
m	mass	g
meq	milliequivalent concentration	meq
MW	molecular weight	g/mol

1. WATER CHEMISTRY UNITS OF CONCENTRATION

There are many ways that the amount of compounds dissolved in water can be presented. For water supply and wastewater calculations, concentrations can be presented as substance or as one of several equivalents. For example, an analysis might report a sulfate concentration as 38 mg/L as substance, 790 meq/L, or 40 mg/L as $CaCO_3$.

The *as substance concentration* is the gravimetric amount of the substance in the volumetric water basis. If water contains 3.8 mg/L of CO_2, then there are 3.8 mg of CO_2 in a liter of water.

The *milliequivalent concentration* is the number of equivalent weights of the substance in the volumetric water basis times 1000 meq/eq. The *equivalent weight* is the molecular weight divided by the oxidation number (valence, or number of charges on the ion).

$$\text{meq} = \frac{m}{\text{EW}} \times 1000\ \frac{\text{meq}}{\text{eq}}$$

$$\text{EW} = \frac{\text{MW}}{\text{oxidation number}}$$

The molecular weight of $CaCO_3$ is approximately 100 g/mol and the calcium ion is Ca^{++}, so the equivalent weight is approximately 50 g/mol. If a water sample contains 25 mg/L of $CaCO_3$, concentration might be reported as 0.5 meq/L.

Water chemistry concentrations are often reported in *$CaCO_3$ equivalents.* The *as $CaCO_3$ concentration* is the amount of $CaCO_3$ in mg/L that would contribute the same number of ionic charges as the reported compound. The $CaCO_3$ equivalent is calculated from the ratio of equivalent weights.

$$C_{\text{as CaCO}_3} = C_{\text{as substance}}\left(\frac{\text{EW}_{\text{CaCO}_3}}{\text{EW}_{\text{substance}}}\right) = C_{\text{as substance}}\left(\frac{50.1\ \frac{\text{g}}{\text{mol}}}{\text{EW}_{\text{substance}}}\right)$$

Example

The concentration of magnesium in a water sample is 31 mg/L as substance. Most nearly, what is the $CaCO_3$ equivalent concentration?

(A) 12 mg/L as $CaCO_3$

(B) 64 mg/L as $CaCO_3$

(C) 130 mg/L as $CaCO_3$

(D) 250 mg/L as $CaCO_3$

Environmental Engineering

Solution

The molecular weight of magnesium, Mg^{++}, is 24.305 g/mol. The magnesium ion is doubly charged, so the equivalent weight is

$$\begin{aligned} EW_{Mg^{++}} &= \frac{MW_{Mg^{++}}}{\text{oxidation number}} = \frac{24.305\ \frac{g}{mol}}{2} \\ &= 12.153\ g/mol \end{aligned}$$

The concentration as $CaCO_3$ is

$$\begin{aligned} C_{Mg^{++},\text{as } CaCO_3} &= C_{Mg^{++},\text{as substance}}\left(\frac{EW_{CaCO_3}}{EW_{Mg^{++}}}\right) \\ &= \left(31\ \frac{mg}{L}\right)\left(\frac{50.1\ \frac{g}{mol}}{12.153\ \frac{g}{mol}}\right) \\ &= 127.8\ mg/L \quad (130\ mg/L \text{ as } CaCO_3) \end{aligned}$$

The answer is (C).

2. CATIONS AND ANIONS IN NEUTRAL SOLUTIONS

Equivalency concepts provide a useful check on the accuracy of water analyses. For the water to be electrically neutral, the sum of anion equivalents must equal the sum of cation equivalents.

Concentrations of dissolved compounds in water are usually expressed in mg/L, not equivalents. However, anionic and cationic substances can be converted to their equivalent concentrations in milliequivalents per liter (meq/L) by dividing their concentrations in mg/L by their equivalent weights.

$$C_{meq/L} = \frac{C_{mg/L}}{EW_{g/mol}}$$

Since water is an excellent solvent, it will contain the ions of the inorganic compounds to which it is exposed. A chemical analysis listing these ions does not explicitly determine the compounds from which the ions originated. Several graphical methods, such as bar graphs and Piper (Hill) trilinear diagrams, can be used for this purpose. The bar graph, also known as a *milliequivalent per liter bar chart*, is constructed by listing the cations (positive ions) in the sequence of calcium (Ca^{++}), magnesium (Mg^{++}), iron (Fe^{+++}), sodium (Na^{+}), and potassium (K^{+}), and pairing them with the anions (negative ions) in the sequence of carbonate (CO_3^{--}), bicarbonate (HCO_3^{-}), sulfate (SO_4^{--}), chloride (Cl^{-}), nitrate (NO_3^{-}), and fluoride (F^{-}). A bar chart can be used to deduce the hypothetical combinations of positive and negative ions that would have resulted in the given water analysis.

Example 2

A water analysis of lake water has the results shown, with all values reported as $CaCO_3$.

alkalinity	151.5 mg/L
sodium	120.0 mg/L
calcium	127.5 mg/L
iron (III)	0.107 mg/L
magnesium	43.5 mg/L
potassium	8.24 mg/L
chloride	39.5 mg/L
fluoride	1.05 mg/L
nitrate	1.06 mg/L
sulfate	106 mg/L

What is one of the most likely compounds dissolved in the water?

(A) $CaSO_4$

(B) $MgCl_2$

(C) $Fe_2(SO_4)_3$

(D) $NaHCO_3$

Solution

Since this is lake water, alkalinity can be assumed to be the result of dissolved CO_2 and the formation of bicarbonate, HCO_3^{-}.

Calculate the milliequivalent concentrations for each component. Since the $CaCO_3$ equivalent has been given, the molecular and equivalent weights are not needed. The milliequivalent concentration is calculated as

$$C_{meq/L} = \frac{C_{\text{as } CaCO_3,mg/L}}{EW_{CaCO_3}} = \frac{C_{\text{as } CaCO_3,mg/L}}{50.1\ \frac{g}{mol}}$$

ion/component	$CaCO_3$ equivalent (mg/L)	milliequivalents (meq/L)
Na^+	120.0	2.40
Ca^{++}	127.5	2.54
Fe^{+++}	0.107	0.00214
Mg^{++}	43.5	0.868
K^+	8.24	0.164
HCO_3^-	151.5F	3.02
Cl^-	39.5	0.788
F^-	1.05	0.0210
NO_3^-	1.06	0.0212
SO_4^{--}	106.0	2.12

Draw a milliequivalent bar chart following the Ca-Mg-Fe-Na-K and CO_3-HCO_3-SO_4-Cl-NO_3-F sequences.

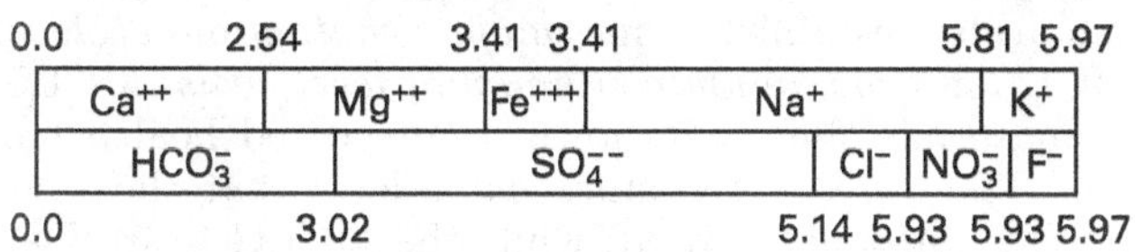

Most likely, none of the compounds listed except $Fe_2(SO_4)_3$ contributed to the lake water.

The answer is (C).

3. ACIDITY

Acidity is a measure of acids in solution. Acidity in surface water is caused by formation of carbonic acid (H_2CO_3) from carbon dioxide in the air. Carbonic acid is aggressive and must be neutralized to eliminate a cause of water pipe corrosion. If the pH of water is greater than 4.5, carbonic acid ionizes to form bicarbonate. If the pH is greater than 8.3, carbonate ions form.

Measurement of acidity is done by titration with a standard basic measuring solution. Acidity in water is typically given in terms of the $CaCO_3$ equivalent that would neutralize the acid.

4. ALKALINITY

Alkalinity is a measure of the ability of a water to neutralize acids (i.e., to absorb hydrogen ions without significant pH change). The principal alkaline ions are OH^-, CO_3^{--}, and HCO_3^-. Other radicals, such as NO_3^-, also contribute to alkalinity, but their presence is rare. The measure of alkalinity is the sum of concentrations of each of the substances measured as equivalent $CaCO_3$.

5. INDICATOR SOLUTIONS

End points for acidity and alkalinity titrations are determined by color changes in indicator dyes that are pH sensitive. Several commonly used indicators are listed in Table 13.1.

Table 13.1 *Indicator Solutions Commonly Used in Water Chemistry*

indicator	titration	end point pH	color change
bromophenol blue	acidity	3.7	yellow to blue
phenolphthalein	acidity	8.3	colorless to red-violet
phenolphthalein	alkalinity	8.3	red-violet to colorless
mixed bromocresol/ green-methyl red	alkalinity	4.5	grayish to orange-red

6. HARDNESS

Hardness in natural water is caused by the presence of polyvalent (but not singly charged) metallic cations. Principal cations causing hardness in water and the major anions associated with them are presented in Table 13.2. Because the most prevalent of these species are the divalent cations of calcium and magnesium, total hardness is typically defined as the sum of the concentration of these two elements and is expressed in terms of milligrams per liter as $CaCO_3$. (Hardness is occasionally expressed in units of *grains per gallon*, where 7000 grains are equal to a pound.)

Table 13.2 *Principal Cations and Anions Indicating Hardness*

cations	anions
Ca^{++}	HCO_3^-
Mg^{++}	SO_4^{--}
Sr^{++}	Cl^-
Fe^{++}	NO_3^-
Mn^{++}	SiO_3^{--}

Carbonate hardness is caused by cations from the dissolution of calcium or magnesium carbonate and bicarbonate in the water. Carbonate hardness is hardness that is chemically equivalent to alkalinity, where most of the alkalinity in natural water is caused by the bicarbonate and carbonate ions.

Noncarbonate hardness is caused by cations from calcium (i.e., calcium hardness) and magnesium (i.e., magnesium hardness) compounds of sulfate, chloride, or silicate that are dissolved in the water. Noncarbonate hardness is equal to the total hardness minus the carbonate hardness.

Environmental Engineering

Hardness can be classified as shown in Table 13.3. Although high values of hardness do not present a health risk, they have an impact on the aesthetic acceptability of water for domestic use. (Hardness reacts with soap to reduce its cleansing effectiveness and to form scum on the water surface.) Where feasible, carbonate hardness in potable water should be reduced to the 25–40 mg/L range and total hardness reduced to the 50–75 mg/L range.

Table 13.3 *Relationship of Hardness Concentration to Classification*

hardness (mg/L as $CaCO_3$)	classification
0 to 60	soft
61 to 120	moderately hard
121 to 180	hard
181 to 350	very hard
>350	saline; brackish

Water containing bicarbonate (HCO_3^-) can be heated to precipitate carbonate (CO_3^{--}) as a *scale*. Water used in steam-producing equipment (e.g., boilers) must be essentially hardness-free to avoid deposit of scale.

Noncarbonate hardness, also called *permanent hardness*, cannot be removed by heating. It can be removed by precipitation softening processes (typically the lime-soda ash process) or by ion exchange processes using resins selective for ions causing hardness.

Hardness is measured in the laboratory by titrating the sample using a standardized solution of ethylenediaminetetraacetic acid (EDTA) and an indicator dye such as Eriochrome Black T. The sample is titrated at a pH of approximately 10 until the dye color changes from red to blue. The standardized solution of EDTA is usually prepared such that 1 mL of EDTA is equivalent to 1 mg/L of hardness.

7. HARDNESS AND ALKALINITY

Hardness is caused by multi-positive ions. Alkalinity is caused by negative ions. Both positive and negative ions are present simultaneously. Therefore, an alkaline water can also be hard.

With some assumptions and minimal information about the water composition, it is possible to determine the ions in the water from the hardness and alkalinity. For example, Fe^{++} is an unlikely ion in most water supplies, and it is often neglected.

If hardness and alkalinity (both as $CaCO_3$) are the same and there are no monovalent cations, then there are no SO_4^{--}, Cl^-, or NO_3^- ions present. That is, there is no noncarbonate (permanent) hardness. If hardness is greater than the alkalinity, however, then noncarbonate hardness is present, and the carbonate (temporary) hardness is equal to the alkalinity. If hardness is less than the alkalinity, then all hardness is carbonate hardness, and the extra HCO_3^- comes from other sources (such as $NaHCO_3$).

8. NATIONAL PRIMARY DRINKING WATER REGULATIONS

Following passage of the Safe Drinking Water Act in the United States, the Environmental Protection Agency (EPA) established minimum primary drinking water regulations. These regulations set limits on the amount of various substances in drinking water. Every public water supply serving at least 15 service connections or 25 or more people must ensure that its water meets these minimum standards.

Accordingly, the EPA has established the National Primary Drinking Water Regulations and the National Secondary Drinking Water Regulations. The primary standards establish *maximum contaminant levels* (MCL) and *maximum contaminant level goals* (MCLG) for materials that are known or suspected health hazards. The MCL is the enforceable level that the water supplier must not exceed, while the MCLG is an unenforceable health goal equal to the maximum level of a contaminant that is not expected to cause any adverse health effects over a lifetime of exposure.

9. NATIONAL SECONDARY DRINKING WATER REGULATIONS

The national secondary drinking water regulations, outlined in Table 13.4, are not designed to protect public health. Instead, they are intended to protect "public welfare" by providing helpful guidelines regarding the taste, odor, color, and other aesthetic aspects of drinking water.

10. IRON

Even at low concentrations, iron is objectionable because it stains porcelain bathroom fixtures, causes a brown color in laundered clothing, and can be tasted. Typically, iron is a problem in groundwater pumped from anaerobic aquifers in contact with iron compounds. Soluble ferrous ions can be formed under these conditions, which, when exposed to atmospheric air at the surface or to dissolved oxygen in the water system, are oxidized to the insoluble ferric state, causing the color and staining problems mentioned. Iron determinations are made through colorimetric (i.e., wet titration) analysis.

11. MANGANESE

Manganese ions are similar in formation, effect, and measurement to iron ions.

Table 13.4 National Secondary Drinking Water Regulations (Code of Federal Regulations (CFR) Title 40, Ch. I, Part 143)

contaminant	suggested levels	effects
aluminum	0.05–0.2 mg/L	discoloration of water
chloride	250 mg/L	salty taste and pipe corrosion
color	15 color units	visible tint
copper	1.0 mg/L	metallic taste and staining
corrosivity	noncorrosive	taste, staining, and corrosion
fluoride	2.0 mg/L	dental fluorosis
foaming agents	0.5 mg/L	froth, odor, and bitter taste
iron	0.3 mg/L	taste, staining, and sediment
manganese	0.05 mg/L	taste and staining
odor	3 TON*	"rotten egg," musty, and chemical odor
pH	6.5–8.5	low pH—metallic taste and corrosion high pH—slippery feel, soda taste, and deposits
silver	0.1 mg/L	discoloration of skin and graying of eyes
sulfate	250 mg/L	salty taste and laxative effect
total dissolved solids (TDS)	500 mg/L	taste, corrosivity, and soap interference
zinc	5 mg/L	metallic taste

*threshold odor number

12. FLUORIDE

Natural fluoride is found in groundwaters as a result of dissolution from geologic formations. Surface waters generally contain much smaller concentrations of fluoride. An absence or low concentration of ingested fluoride causes the formation of tooth enamel less resistant to decay, resulting in a high incidence of dental cavities in children's teeth. Excessive concentration of fluoride causes *fluorosis*, a brownish discoloration of dental enamel. The MCL of 4.0 mg/L established by the EPA is to prevent unsightly fluorosis.

Communities with water supplies deficient in natural fluoride may chemically add fluoride during the treatment process. Since water consumption is influenced by climate, the recommended optimum concentrations listed in Table 13.5 are based on the annual average of the maximum air temperatures based on a minimum of five years of records.

Compounds commonly used as fluoride sources in water treatment are listed in Table 13.6.

Table 13.5 Recommended Optimum Concentrations of Fluoride in Drinking Water

average air temperature range (°F)	recommended optimum concentration (mg/L)
53.7 and below	1.2
53.8 to 58.3	1.1
58.4 to 63.8	1.0
63.9 to 70.6	0.9
70.7 to 79.2	0.8
79.3 to 90.5	0.7

Table 13.6 Fluoridation Chemicals

compound	formula	percentage F^- ion (%)
sodium fluoride	NaF	45
sodium silicofluoride	Na_2SiF_6	61
hydrofluosilicic acid	H_2SiF_6	79
ammonium silicofluoride*	$(NH_4)_2SiF_6$	64

*used in conjunction with chlorine disinfection where it is desired to maintain a chloramine residual in the distribution system

13. PHOSPHORUS

Phosphate content is more of a concern in wastewater treatment than in supply water, although phosphorus can enter water supplies in large amounts from runoff. Excessive phosphate discharge contributes to aquatic plant (phytoplankton, algae, and macrophytes) growth and subsequent *eutrophication*. (Eutrophication is an "over-fertilization" of receiving waters.)

Phosphorus is of considerable interest in the management of lakes and reservoirs because phosphorus is a nutrient that has a major effect on aquatic plant growth. Algae normally have a phosphorus content of 0.1–1% of dry weight. The molar N:P ratio for ideal algae growth is 16:1.

Phosphorus exists in several forms in aquatic environments. Soluble phosphorus occurs as *orthophosphate*, as condensed *polyphosphates* (from detergents), and as various organic species. Orthophosphates ($H_2PO_4^-$, HPO_4^{--}, and PO_4^{---}) and polyphosphates (such as $Na_3(PO_3)_6$) result from the use of synthetic detergents (*syndets*). A sizable fraction of the soluble phosphorus is in the organic form, originating from the decay or excretion of nucleic acids and algal storage

Environmental Engineering

products. *Particulate phosphorus* occurs in the organic form as a part of living organisms and detritus as well as in the inorganic form of minerals such as apatite.

Phosphorus is normally measured by colorimetric or digestion methods. The results are reported in terms of mg/L of phosphorus (e.g., "mg/L of P" or "mg/L of total P"), although the tests actually measure the concentration of orthophosphate. The concentration of a particular compound is found by multiplying the concentration as P by the molecular weight of the compound and dividing by the atomic weight of phosphorus (30.97).

A substantial amount of the phosphorus that enters lakes is probably not available to aquatic plants. Bioavailable compounds include orthophosphates, polyphosphates, most soluble organic phosphorus, and a portion of the particulate fraction. Studies have indicated that bioavailable phosphorus generally does not exceed 60% of the total phosphorus.

In aquatic systems, phosphorus does not enter into any redox reactions, nor are any common species volatile. Therefore, lakes retain a significant portion of the entering phosphorus. The main mechanism for retaining phosphorus is simple sedimentation of particles containing the phosphorus. Particulate phosphorus can originate from the watershed (*allochthonous material*) or can be formed within the lake (*autochthonous material*).

A large fraction of the phosphorus that enters a lake is recycled, and much of the recycling occurs at the sediment-water interface. Recycling of phosphorus is linked to iron and manganese recycling. Soluble phosphorus in the water column is removed by adsorption onto iron and manganese hydroxides, which precipitate under aerobic conditions. However, when the *hypolimnion* (i.e., the lower part of the lake that is essentially stagnant) becomes anaerobic, the iron (or manganese) is reduced, freeing up phosphorus. This is consistent with a fairly general observation that phosphorus release rates are nearly an order of magnitude higher under anaerobic conditions than under aerobic conditions.

Factors that control phosphorus recycling rates are not well understood, although it is clear that oxygen status, phosphorus speciation, temperature, and pH are important variables. Phosphorus release from sediments is usually considered to be constant (usually less than 1 mg/m^2·d under aerobic conditions).

In addition to direct regeneration from sediments, *macrophytes* (large aquatic plants) often play a significant role in phosphorus recycling. Macrophytes with highly developed root systems derive most of their phosphorus from the sediments. Regeneration to the water column can occur by excretion or through decay, effectively "pumping" phosphorus from the sediments. The internal loading generated by recycling is particularly important in shallow, eutrophic lakes. In several cases where phosphorus inputs have been reduced to control algal blooms, regeneration of phosphorus from phosphorus-rich sediments has slowed the rate of recovery.

14. NITROGEN

Compounds containing nitrogen are not abundant in virgin surface waters. However, nitrogen can reach large concentrations in groundwaters that have been contaminated with barnyard runoff or that have percolated through heavily fertilized fields. Sources of surface water contamination include agricultural runoff and discharge from sewage treatment facilities.

Of greatest interest, in order of decreasing oxidation state, are nitrates (NO_3^-), nitrites (NO_2^-), ammonia (NH_3), and organic nitrogen. These three compounds are reported as *total nitrogen*, TN, with units of "mg/L of N" or "mg/L of total N." The concentration of a particular compound is found by multiplying the concentration as N by the molecular weight of the compound and dividing by the atomic weight of nitrogen (14.01).

Excessive amounts of nitrate in water can contribute to the illness in infants known as *methemoglobinemia* ("blue baby" syndrome). As with phosphorus, nitrogen stimulates aquatic plant growth.

Un-ionized ammonia is a colorless gas at standard temperature and pressure. A pungent odor is detectable at levels above 50 mg/L. Ammonia is very soluble in water at low pH.

Ammonia levels in zero-salinity surface water increase with increasing pH and temperature. At low pH and temperature, ammonia combines with water to produce ammonium (NH_4^+) and hydroxide (OH^-) ions. The ammonium ion is nontoxic to aquatic life and not of great concern. The un-ionized ammonia (NH_3), however, can easily cross cell membranes and have a toxic effect on a wide variety of fish. The EPA has established the criteria for fresh and saltwater fish that depend on temperature, pH, species, and averaging period.

Ammonia is usually measured by a distillation and titration technique or with an ammonia-selective electrode. The results are reported as ammonia nitrogen. Nitrites are measured by a colorimetric method. The results are reported as nitrite nitrogen. Nitrates are measured by ultraviolet spectrophotometry, selective electrode, or reduction methods. The results are reported as nitrate nitrogen. Organic nitrogen is determined by a digestion process that identifies organic and ammonia nitrogen combined. Organic nitrogen is found by subtracting the ammonia nitrogen value from the digestion results.

15. COLOR

Color in water is caused by substances in solution, known as *true color*, and by substances in suspension, mostly organics, known as *apparent* or *organic color*. Iron, copper, manganese, and industrial wastes all can cause color. Color is aesthetically undesirable, and it stains fabrics and porcelain bathroom fixtures.

Water color is determined by comparison with standard platinum/cobalt solutions or by spectrophotometric methods. The standard color scales range from 0 (clear) to 70. Water samples with more intense color can be evaluated using a dilution technique.

16. TURBIDITY

Turbidity is a measure of the light-transmitting properties of water and comprises suspended and colloidal material. Turbidity is expressed in *nephelometric turbidity units* (NTU). Viruses and bacteria become attached to these particles, where they can be protected from the bactericidal and viricidal effects of chlorine, ozone, and other disinfecting agents. The organic material included in turbidity has also been identified as a potential precursor to carcinogenic disinfection by-products.

Turbidity in excess of 5 NTU is noticeable by visual observation. Turbidity in a typical clear lake is approximately 25 NTU, and muddy water exceeds 100 NTU. Turbidity is measured using an electronic instrument called a nephelometer, which detects light scattered by the particles when a focused light beam is shown through the sample.

17. SOLIDS

Solids present in a sample of water can be classified in several ways.

- *total solids* (TS): Total solids are the material residue left after the evaporation of the sample. Total solids include total suspended solids and total dissolved solids.
- *total suspended solids* (TSS): The material retained on a standard glass-fiber filter disk is defined as the suspended solids in a sample. The filter is weighed before filtration, dried, and weighed again. The gain in weight is the amount of suspended solids. Suspended solids can also be categorized into *volatile suspended solids* (VSS) and *fixed suspended solids* (FSS).
- *total dissolved solids* (TDS): These solids are in solution and pass through the pores of the standard glass-fiber filter. Dissolved solids are determined by passing the sample through a filter, collecting the filtrate in a weighed drying dish, and evaporating the liquid. The gain in weight represents the dissolved solids. Dissolved solids can be categorized into *volatile dissolved solids* (VDS) and *fixed dissolved solids* (FDS).
- *total volatile solids* (TVS): The residue from one of the previous determinations is ignited to constant weight in an electric muffle furnace. The loss in weight during the ignition process represents the volatile solids.
- *total fixed solids* (TFS): The weight of solids that remain after the ignition used to determine volatile solids represents the fixed solids.
- *settleable solids*: The volume (mL/L) of settleable solids is measured by allowing a sample to settle for one hour in a graduated conical container (*Imhoff cone*).

Waters with high concentrations of suspended solids are classified as *turbid waters*. Waters with high concentrations of dissolved solids often can be tasted. Therefore, a limit of 500 mg/L has been established in the National Secondary Drinking Water Regulations for dissolved solids.

18. CHLORINE AND CHLORAMINES

Chlorine is the most common disinfectant used in water treatment. It is a strong oxidizer that deactivates microorganisms. Its oxidizing capability also makes it useful in removing soluble iron and manganese ions.

Chlorine gas in water forms *hydrochloric* and *hypochlorous acids*. At a pH greater than 9, hypochlorous acid dissociates to hydrogen and hypochlorite ions.

Free chlorine, hypochlorous acid, and hypochlorite ions left in water after treatment are known as *free chlorine residuals*. Hypochlorous acid reacts with ammonia (if it is present) to form *chloramines*. Chloramines are known as *combined residuals*. Chloramines are more stable than free residuals, but their disinfecting ability is less.

Free and combined residual chlorine can be determined by color comparison, by titration, and with chlorine-sensitive electrodes. Color comparison is the most common field method.

19. HALOGENATED COMPOUNDS

Halogenated compounds have become a subject of concern in the treatment of water due to their potential as carcinogens. The use of chlorine as a disinfectant in water treatment generates these compounds by reacting with organic substances in the water. (For this reason, there are circumstances under which chlorine may not be the most appropriate disinfectant.)

The organic substances, called *precursors*, are not in themselves harmful, but the chlorinated end products, known by the term *disinfection by-products* (DBPs) raise serious health concerns. Typical precursors are decay by-products such as humic and fulvic acids. Several of the DBPs contain bromine, which is found in low concentrations in most surface waters and can also occur as an impurity in commercial chlorine gas.

The DBPs can take a variety of forms depending on the precursors present, the concentration of free chlorine, the contact time, the pH, and the temperature. The most common ones are the trihalomethanes (THMs), haloacetic acids (HAAs), dihaloacetonitriles (DHANs), and various trichlorophenol isomers.

Trihalomethanes are regulated by the EPA under the National Primary Drinking Water Regulations. Standards for *total haloacetic acids* (referred to as "HAA5" in consideration of the five compounds identified) have also been established.

The *haloacetic acids* (HAAs) exist in tri-, di-, and mono-forms, abbreviated as THAAs, DHAAs, and MHAAs. All of the haloacetic acids are toxic and are suspected or proven carcinogens.

The *dihaloacetonitriles* (DHANs) are formed when acetonitrile (methyl cyanide: C_2H_3N (structurally H_3CCN)) is exposed to chlorine. All are toxic and suspected carcinogens. Maximum concentration limits have not been established by the EPA for DHANs.

Trichlorophenol can exist in six isomeric forms, with varying potential toxicities. As with all halogenated organics, all six isomers are potential carcinogens.

20. AVOIDANCE OF DISINFECTION BY-PRODUCTS IN DRINKING WATER

Reduction of DBPs can best be achieved by avoiding their production in the first place. The best strategy dictates using source water with few or no organic precursors.

Often source water choices are limited, necessitating tailoring treatment processes to produce the desired result. This entails removing the precursors prior to the application of chlorine, applying chlorine at certain points in the treatment process that minimize production of DBPs, using disinfectants that do not produce significant DBPs, or a combination of these techniques.

Removal of precursors is achieved by preventing growth of vegetative material (algae, plankton, etc.) in the source water and by collecting source water at various depths to avoid concentrations of precursors. Oxidizers such as potassium permanganate and chlorine dioxide can often reduce the concentration of the precursors without forming the DBPs. Use of activated carbon, pH-adjustment processes, or dechlorination can reduce the impact of chlorination.

Chlorine application should be delayed if possible until after the flocculation, coagulation, settling, and filtration processes have been completed. In this manner, turbidity and common precursors will be reduced. If it is necessary to chlorinate early to facilitate treatment processes, chlorination can be followed by dechlorination to reduce contact time. Granular activated carbon has been used to some extent to remove DBPs after they form, but the carbon needs frequent regeneration.

Alternative disinfectants include ozone, chloramines, chlorine dioxide, iodine, bromine, potassium permanganate, hydrogen peroxide, and ultraviolet radiation. Ozone and chloramines, singularly or together, are often used for control of THMs. However, ozone creates other DBPs, including aldehydes, hydrogen peroxide, carboxylic acids, ketones, and phenols. When ozone is used as the primary disinfectant, a secondary disinfectant such as chlorine or chloramine must be used to provide an active residual that can be measured within the distribution system.

14 Water Supply Treatment and Distribution

Nomenclature

a	adsorption fraction	–	–
A	area	ft^2	m^2
b	width	ft	m
C	coefficient	–	–
C	concentration	lbm/ft^3	mg/L
D	depth	ft	m
D	diameter	ft	m
D_S	diffusivity	ft^2/sec	m^2/s
E	efficiency	–	–
E	voltage	V	V
E_1	removal efficiency	–	–
E_2	electrical efficiency	–	–
F	Faraday constant, 96 485	C/mol	A·s/mol
F	force	lbf	N
g	acceleration of gravity, 32.2 (9.81)	ft/sec^2	m/s^2
g_c	gravitational constant, 32.2	lbm-ft/lbf-sec^2	n.a.
G	mixing velocity gradient	sec^{-1}	s^{-1}
h	height	ft	m
h	Henry's law constant	atm	atm
H	Henry's law constant	atm-ft^3/lbmol	atm·m^3/kmol
H'	dimensionless Henry's law constant	–	–
H_L	head loss in mixing zone	ft	m
HTU	height of transfer unit	ft	m
I	current	A	A
J	flux	ft^3/ft^2-sec	m^3/m^2·s
k	mixing rate constant	–	–
K	coefficient	–	–
K	experimental constant	–	–
K_p	membrane solute mass transfer coefficient	–	–
K_T	impeller constant	–	–
K_La	overall transfer rate constant	sec^{-1}	s^{-1}
L	depth or length	ft	m
L	liquid molar loading rate	mol/ft^2-sec	mol/m^2·s
LRV	log removal value	–	–
m	mass	lbm	kg
M	molar density	lbmol/ft^3	kmol/m^3
n	experimental constant	–	–
n	number	–	–
n	rotational speed	rev/sec	rev/s
N	normality of solution	GEW/L	GEW/L
NTU	number of transfer units	–	–
p	pressure	lbf/ft^2	Pa
P	power	hp	W
Q	flow rate	ft^3/sec or MGD	m^3/s
r	membrane pore size	ft	m
R	resistance	Ω	Ω
R	universal gas constant	atm-ft^3/lbmol-°R	atm·m^3/kmol·K
R_S	stripping factor	–	–
t	time	sec	s
T	absolute temperature	°R	K

v	van't Hoff factor	–	–
v	velocity	ft/sec	m/s
V	specific volume	ft^3/lbmol	m^3/kmol
V	volume	ft^3	m^3
W	width	ft	m
W_p	coefficient of water permeation	lbmol/lbf-sec	kmol/m^2·s·Pa
WOR	weir overflow rate	gal/day-ft	L/d·m
x	mass fraction remaining	–	–
x	mass of solute adsorbed	lbm	kg
X	mass ratio of solid phase	–	–
y	mole fraction	–	–
Z	stripper packing height	ft	m
Z	total carbon depth	ft	m
Z_s	depth of sorption zone	ft	m

Symbols

γ	specific weight	lbf/ft^3	n.a.
δ	membrane thickness	ft	m
ε	membrane porosity	–	–
ν	van't Hoff factor	–	–
θ	hydraulic residence time	sec	s
η	porosity	–	–
μ	absolute viscosity	lbf-sec/ft^2	Pa·s
π	osmotic pressure	lbf/ft^2	Pa
Π	osmotic pressure	lbf/ft^2	Pa
ρ	density	lbm/ft^3	kg/m^2
ϕ	osmotic coefficient	–	–

Subscripts

0	influent
A	air
B	backwash or total at breakthrough
D	drag
e	equilibrium
f	flow-through, fluid, fluidized bed, or fraction
h	hydraulic
i	impeller
in	input or inside
m	membrane
mix	mixing
o	initial or overflow
out	output or outside
p	paddle or perpendicular
P	paddle
s	settling or solute
t	terminal
T	total at exhaustion or turbulent
x	cross-sectional
w	water
W	water
Z	change between exhaustion and breakthrough

1. WATER DEMAND

Normal water demand is specified in gallons per capita day (gpcd)—the average number of gallons used by each person each day. This is referred to as *average annual daily flow* (AADF) if the average is taken over a period of a year. Residential (i.e., domestic), commercial, industrial, and public uses all contribute to normal water demand, as do waste and unavoidable loss.[1]

An AADF of 165 gpcd (625 L/d) is a typical minimum for planning purposes. If large industries are present (e.g., canning, steel making, automobile production, and electronics), then their special demand requirements must be added.

Water demand varies with the time of day and season. Each community will have its own demand distribution curve.

2. PROCESS INTEGRATION

The processes and sequences used in a water treatment plant depend on the characteristics of the incoming water. However, some sequences are more appropriate than others due to economic and hydraulic considerations. *Conventional filtration*, also referred to as *complete filtration*, is a term used to describe the traditional sequence of adding coagulation chemicals, flash mixing, coagulation-flocculation, sedimentation, and subsequent filtration. Coagulants, chlorine (or an alternative disinfectant), fluoride, and other chemicals are added at various points along the path, as indicated by Fig. 14.1. Conventional filtration is still the best choice when incoming water has high color, turbidity, or other impurities.

Direct filtration refers to a modern sequence of adding coagulation chemicals, flash mixing, minimal flocculation, and subsequent filtration. In direct filtration, the physical chemical reactions of flocculation occur to some extent, but special flocculation and sedimentation facilities are eliminated. This reduces the amount of sludge that has to be treated and disposed of. Direct filtration is applicable when the incoming water is of high initial quality.

In-line filtration refers to another modern sequence that starts with adding coagulation chemicals at the filter inlet pipe. Mixing occurs during the turbulent flow

[1]"Public" use includes washing streets, flushing water and sewer mains, flushing fire hydrants, filling public fountains, and fighting fires.

Figure 14.1 Chemical Application Points

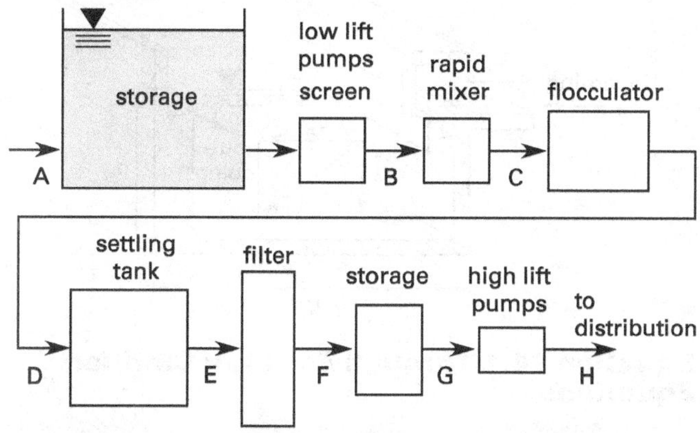

typical flow diagram of water treatment plant

category of chemicals	possible points of application							
	A	B	C	D	E	F	G	H
algicide	X				X			
disinfectant		X	X		X	X	X	X
activated carbon		X	X	X	X			
coagulants		X	X					
coagulation aids		X	X		X			
alkali								
for flocculation			X					
for corrosion control						X		
for softening			X					
acidifier			X			X		
fluoride						X		
cupric-chloramine						X		
dechlorinating agent						X		X

Note: With solids contact reactors, point C is the same as point D.

toward a filter, which is commonly of the pressure-filter variety. As with direct filtration, flocculation and sedimentation facilities are not used.

3. PRETREATMENT

Preliminary treatment is a general term that usually includes all processes prior to the first flocculation operation (i.e., pretreatment, screening, presedimentation, microstraining, aeration, and chlorination). Flow measurement is usually considered to be part of the pretreatment sequence.

4. SCREENING

Screens are used to protect pumps and mixing equipment from large objects. The degree of screening required will depend on the nature of solids expected. Screens can be either manually or automatically cleaned.

5. MICROSTRAINING

Microstrainers are effective at removing 50–95% of the algae in incoming water. Microstrainers are constructed from woven stainless steel fabric mounted on a hollow drum that rotates at 4–7 rpm. Flow is usually radially outward through the drum. The accumulated filter cake is removed by backwashing.

6. ALGAE PRETREATMENT

Biological growth in water from impounding reservoirs, lakes, storage reservoirs, and settling basins can be prevented or eliminated with an *algicide* such as copper sulfate. Such growth can produce unwanted taste and odors, clog fine-mesh filters, and contribute to the buildup of slime.

Copper sulfate is toxic and should not be used without considering and monitoring the effects on aquatic life (e.g., fish).

7. PRECHLORINATION

A prechlorination process was traditionally employed in most water treatment plants. Many plants have now eliminated all or part of the prechlorination due to the formation of trihalomethanes (THMs). In so doing, benefits such as algae control have been eliminated. Alternative disinfection chemicals (e.g., ozone and potassium permanganate) can be used. Otherwise, coagulant doses must be increased.

8. PRESEDIMENTATION

The purpose of presedimentation is to remove easily settled sand and grit. This can be accomplished by using pure sedimentation basins, sand and grit chambers, and various passive cyclone degritters. *Trash racks* may be integrated into sedimentation basins to remove leaves and other floating debris.

9. FLOW MEASUREMENT

Flow measurement is incorporated into the treatment process whenever the water is conditioned enough to be compatible with the measurement equipment. Flow measurement devices should not be exposed to scour from grit or highly corrosive chemicals (e.g., chlorine).

Flow measurement often takes place in a *Parshall flume.* Chemicals may be added in the flume to take advantage of the turbulent mixing that occurs at that point. All of the traditional fluid measurement devices are also applicable, including venturi meters, orifice plates, propeller and turbine meters, and modern passive devices such as magnetic and ultrasonic flowmeters.

Environmental Engineering

10. AERATION

Aeration is used to reduce taste- and odor-causing compounds, to lower the concentration of dissolved gases (e.g., hydrogen sulfide), to increase dissolved CO_2 (i.e., recarbonation) or decrease CO_2, to reduce iron and manganese, and to increase dissolved oxygen.

Various types of aerators are used. The best transfer efficiencies are achieved when the air-water contact area is large, the air is changed rapidly, and the aeration period is long. *Force draft air injection* is common. The release depth varies from 10 ft to 25 ft. The ideal compression power required to aerate water with a simple air injector depends on the air flow rate, Q, and head (i.e., which must be greater than the release depth), h, at the point where the air is injected. Motor-compression-distribution efficiencies are typically around 75%.

Diffused air systems with compressed air at 5–10 psig are the most efficient methods of aerating water. The air injection volume is 0.2–0.3 ft^3/gal. However, the equipment required to produce and deliver compressed air is more complex than with simple injectors. The *transfer efficiency* of a diffused air system varies with depth and bubble size. With injection depths of 5–10 ft, if coarse bubbles are produced, only 4–8% of the available oxygen will be transferred to the water. With medium-sized bubbles, the efficiency can be 6–15%, and it can approach 10–30% with fine bubble systems.

11. SEDIMENTATION PHYSICS

Water containing suspended sediment can be held in a *plain sedimentation tank* (basin) that allows the particles to settle out.[2] Settling velocity and settling time for sediment depends on the water temperature (i.e., viscosity), particle size, and particle specific gravity. (The specific gravity of sand is usually taken as 2.65.) Typical settling velocities are as follows: gravel, 1 m/s; coarse sand, 0.1 m/s; fine sand, 0.01 m/s; and silt, 0.0001 m/s. Bacteria and colloidal particles are generally considered to be nonsettleable during the detention periods available in water treatment facilities.

12. SEDIMENTATION TANKS

Clarifiers (i.e., sedimentation tanks or sedimentation basins) are usually rectangular or circular in plan, and are equipped with scrapers or raking arms to periodically remove accumulated sediment. (See Fig. 14.2.) Table 14.1 gives design criteria for sedimentation basins.

Figure 14.2 *Sedimentation Basin*

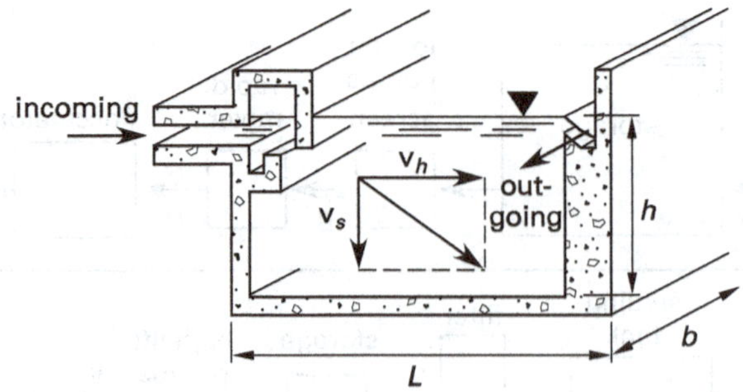

Equation 14.1 Through Eq. 14.4: Clarifier Equations

$$v_o = Q/A_{\text{surface}} \quad (14.1)$$

$$\text{WOR} = Q/\text{weir length} \quad (14.2)$$

$$v_h = Q/A_{\text{cross-section}} = Q/A_x \quad (14.3)$$

$$\text{hydraulic residence time} = V/Q = \theta \quad (14.4)$$

Description

Water flows through the tank at the average *flow-through velocity*, v_h, which should not exceed 1 ft/min. The time that water spends in the tank depends on the flow-through velocity and the tank length, L, typically 100–200 ft. The minimum settling time depends on the tank depth, h, typically 6–15 ft.

The time that water remains in the basin is known as the *hydraulic residence time* (*detention time*, *retention time*, *detention period*, etc.). Typical detention times range from 2 hr to 6 hr, although periods from 1 hr to 12 hr are used depending on the size of particles.

Basin efficiency can approach 80% for fine sediments. Virtually all of the coarse particles are removed. Theoretically, all particles with settling velocities greater than the *overflow rate*, v_o, also known as the *surface loading* or *critical velocity*, will be removed.

Weir overflow rate, WOR (also called the *weir loading*), is the flow rate divided by the total effluent weir length.

The accumulated sediment is referred to as *sludge*. It is removed either periodically or on a continual basis, when it has reached a concentration of 25 MG/L or is organic. Various methods of removing the sludge are used, including scrapers and pumps. The linear velocity of sludge scrapers should be 15 ft/min or higher.

[2]The term "plain" refers to the fact that no chemicals are used as coagulants.

Table 14.1 Design Criteria for Sedimentation Basins

	overflow rate		solids loading rate			
type of basin	average (gpd/ft² (m³/m²·d))	peak (gpd/ft² (m³/m²·d))	average (lbm/ft²-day (kg/m²·h))	peak (lbm/ft²-hr (kg/m²·h))	hydraulic residence time (hr)	depth (ft)
water treatment						
clarification following coagulation and flocculation:						
alum coagulation	350–550 (14–22)	–	–	–	4–8	12–16
ferric coagulation	550–700 (22–28)	–	–	–	4–8	12–16
upflow clarifiers						
groundwater	1500–2200 (61–90)	–	–	–	1	–
surface water	1000–1500 (41–61)	–	–	–	4	–
clarification following lime-soda softening						
conventional	550–1000 (22–41)	–	–	–	2–4	–
upflow clarifiers						
groundwater	1000–2500 (41–102)	–	–	–	1	–
surface water	1000–1800 (41–73)	–	–	–	4	–
wastewater treatment						
primary clarifiers	800–1200 (32–49)	1200–2000 (50–80)	–	–	2	10–12
settling basins following fixed film reactors	400–800 (16–33)	–	–	–	2	–
settling basins following air-activated sludge reactors						
all configurations EXCEPT extended aeration	400–700 (16–28)	–	–	–	2	12–15
extended aeration	200–400 (8–16)	1000–1200 (40–64)	19–29 (4–6)	38 (8)	2	12–15
settling basins following chemical flocculation reactors	800–1200	600–800 (24–32)	5–24 (1–5)	34 (7)	2	–

Example

A horizontal flow grit chamber is to be used to pretreat wastewater with a flow rate of 26.5 ft³/sec. The chamber is 5.9 ft deep with a width 15 times its length, and an approach velocity of 1.15 ft/sec. The width of the grit chamber is most nearly

(A) 0.43 ft

(B) 3.1 ft

(C) 3.9 ft

(D) 7.9 ft

Solution

The cross-sectional area of the grit chamber is found from Eq. 14.3.

$$\mathrm{v}_h = Q/A_x$$

$$A_x = \frac{Q}{\mathrm{v}_h} = \frac{26.5\ \frac{\text{ft}^3}{\text{sec}}}{1.15\ \frac{\text{ft}}{\text{sec}}} = 23\ \text{ft}^2$$

The width is

$$W = \frac{A_x}{D} = \frac{23\ \text{ft}^2}{5.9\ \text{ft}} = 3.9\ \text{ft}$$

The answer is (C).

13. COAGULANTS

Various chemicals can be added to remove fine solids. There are two main categories of coagulating chemicals: hydrolyzing metal ions (based on either aluminum or iron) and ionic polymers. Since the chemicals work by agglomerating particles in the water to form floc, they are known as *coagulants*. *Floc* is the precipitate that forms when the coagulant allows the colloidal particles to agglomerate.

Common *hydrolyzing metal ion* coagulants are aluminum sulfate ($Al_2(SO_4)_3 \cdot nH_2O$, commonly referred to as "alum"), ferrous sulfate ($FeSO_4 \cdot 7H_2O$, sometimes referred to as "copperas"), and chlorinated copperas (a mixture of ferrous sulfate and ferric chloride). n is the number of waters of hydration, approximately 14.3.

14. MIXERS AND MIXING KINETICS

Coagulants and other water treatment chemicals are added in *mixers*. If the mixer adds a coagulant for the removal of colloidal sediment, a downstream location (i.e., a tank or basin) with a reduced velocity gradient may be known as a *flocculator*.

There are two basic models: plug flow mixing and complete mixing. The *complete mixing* model is appropriate when the chemical is distributed throughout by impellers or paddles. If the basin volume is small, so that time for mixing is low, the tank is known as a *flash mixer*, *rapid mixer*, or *quick mixer*. The volume of flash mixers is seldom greater than 300 ft^3, and flash mixer detention time is usually 30–60 sec. Flash mixing kinetics are described by the complete mixing model.

Flash mixers are usually concrete tanks, square in horizontal cross section, and fitted with vertical shaft impellers. The size of a mixing basin can be determined from various combinations of dimensions that satisfy the volume-flow rate relationship.

The detention time required for complete mixing in a tank of volume V depends on the *mixing rate constant* and the incoming and outgoing concentrations.

The *plug flow mixing* model is appropriate when the water flows through a long narrow chamber, the chemical is added at the entrance, and there is no mechanical agitation. All of the molecules remain in the plug flow mixer for the same amount of time as they flow through. For any mixer, the maximum chemical conversion will occur with plug flow, since all of the molecules have the maximum opportunity to react.

Environmental Engineering

15. MIXING PHYSICS

Mixing time is largely a function of how much time water spends in the process. Some mixing proceeds with laminar flow over a long period of time in large tanks, and some mixing occurs in turbulent flow over a short period of time in small tanks. The power required depends on the flow regime, water properties, volume, speed, and impeller type.

When mixing power is to be calculated from basic principles (i.e., as $P = F_D\mathrm{v}$), the velocity used must be carefully considered. While the rotational speed is a reliable parameter, the term *paddle velocity*, v_p, is ambiguous, since linear speed varies with distance from the rotational axis. The tip velocity, $\mathrm{v}_{\mathrm{tip}}$, also known as the *peripheral velocity*, occurs at the paddle location most distant from the rotational axis and is the maximum speed anywhere on the paddle. This rarely coincides with the velocity to be used in calculating required power from basic principles, although some types of reel and paddle wheel mixers with long, thin rectangular blades far from the rotational axis, the average blade velocity related to the mean radius might be sufficient.

In contrast to theoretical power calculations using basic principles, practical power calculations almost always are based on the *impeller diameter*, D_i (sometimes called the *agitator diameter*), which is the maximum diameter of the impeller. Power requirement calculations then incorporate various dimensionless correction/adjustment factors correlated with the type of impeller. No effort is made to define an "effective" diameter that correlates perfectly to the power.

$$\mathrm{v}_{\mathrm{tip,ft/sec}} = \frac{\pi D_{i,\mathrm{ft}} n_{\mathrm{rpm}}}{60\ \frac{\mathrm{sec}}{\mathrm{min}}}$$

Further complicating the calculation of power from basic principles, the fluid being mixed is generally not stationary. Since it doesn't take any power at all to "go with the flow," only the relative velocity or the difference in paddle and fluid velocities ($\mathrm{v}_p - \mathrm{v}_f$) contributes to power requirement. This difference is named the *mixing velocity*, $\mathrm{v}_{\mathrm{mix}}$, also known as the *relative paddle velocity* or *effective paddle velocity*. Various methods are used to specify a useful mixing velocity from the paddle velocity, including use of a slip coefficient. The *slip coefficient* (*slip ratio*, *slip factor*, etc.) is a measure of the difference between the paddle and fluid velocities (i.e., how much of the paddle "slips by" without putting the fluid into motion). Despite appearances, the slip coefficient is not constant; it increases with increasing rotational speed. In the absence of any other information, the mixing velocity of paddle mixers is commonly assumed to be 75% of the tip velocity (i.e., the slip coefficient is 0.75).

$$\mathrm{v}_{\mathrm{mix}} = \mathrm{v}_p - \mathrm{v}_f$$

$$\text{slip coefficient} = \frac{\mathrm{v}_p - \mathrm{v}_f}{\mathrm{v}_p}$$

Equation 14.5 and Eq. 14.6: Reel and Paddle Power

$$P = \frac{C_D A_P \rho_f \mathrm{v}_r^3}{2} \qquad 14.5$$

$$\mathrm{v}_r = \mathrm{v}_p \cdot \text{slip coefficient} \qquad 14.6$$

Variations

$$P = F_D \mathrm{v}_{\mathrm{mix}}$$

$$F_D = \frac{C_D A \rho \mathrm{v}_{\mathrm{mix}}^2}{2} \quad \text{[SI only]}$$

Description

The power required is calculated by Eq. 14.5 from the drag force and the mixing velocity. The mixing velocity is approximately 0.7–0.8 times the tip speed.

The drag force on a paddle is given by the standard fluid drag force equation. For flat plates with length-to-width (L: W) ratios greater than 20, the coefficient of drag, C_D, is approximately 1.8.

Equation 14.7 and Eq. 14.8: Mixing Velocity Gradient (Rapid Mix and Flocculator)

$$G = \sqrt{\frac{P}{\mu V}} = \sqrt{\frac{\gamma H_L}{t\mu}} \qquad 14.7$$

$$Gt = 10^4 \text{ to } 10^5 \qquad 14.8$$

Description

Equation 14.7 correlates the mixing power to the mixing intensity, G, which is somewhat constant for similar types of mixers. It can be calculated from the power, P; mixing volume, V; and fluid absolute viscosity, μ (also known as the *bulk viscosity*); or from the fluid specific weight ($\gamma = g\rho$ for SI, or $\gamma = g\rho/g_c$ for U.S. units); head loss, H_L, through the mixer; and mixing time, t.

For slow-moving paddle mixers, the *mixing velocity gradient*, or *mixing intensity*, G, varies from 20 sec^{-1} to 75 sec^{-1} for a 15–30 min mixing period. Typical units are ft-lbf/sec for power (multiply hp by 550 to obtain ft-lbf/sec), lbf-sec/ft^2 for μ, and ft^3 for volume.

Equation 14.7 can also be used for rapid mixers, in which case the mean velocity gradient is much higher: approximately 500–1000 sec^{-1} for 10–30 sec mixing period, or 3000–5000 sec^{-1} for a 0.5–1.0 sec mixing period in an in-line blender configuration.

16. IMPELLER CHARACTERISTICS

Mixing equipment uses rotating impellers on rotating shafts. The blades of *radial-flow impellers* (paddle-type impellers, turbine impellers, etc.) are parallel to the drive shaft. *Axial-flow impellers* (propellers, pitched-blade impellers, etc.) have blades inclined with respect to the drive shaft. (See Fig. 14.3.) Axial-flow impellers are better at keeping materials (e.g., water softening chemicals) in suspension.

Equation 14.9 and Table 14.2: Turbulent Flow Impeller Mixer Power

$$P = K_T(n)^3(D_i)^5\rho_f \qquad 14.9$$

Figure 14.3 *Typical Axial Flow Mixing Impellers*

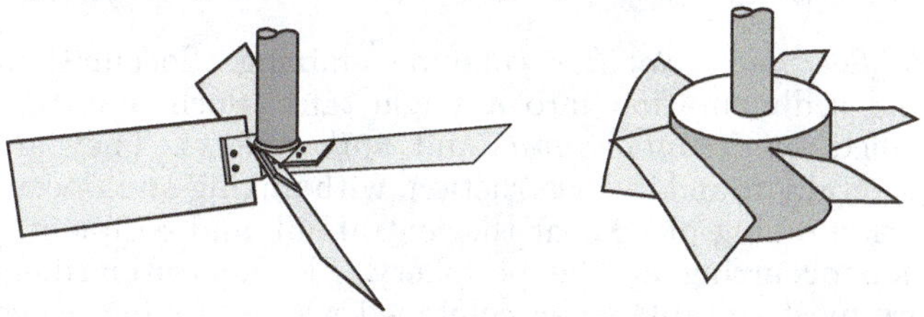

Values

Table 14.2 *Typical Impeller Constants**

impeller type	impeller constant, K_T
propeller, pitch of 1, 3 blades	0.32
propeller, pitch of 2, 3 blades	1.00
turbine, 6 flat blades, vaned disc	6.30
turbine, 6 curved blades	4.80
fan turbine, 6 blades at 45°	1.65
shrouded turbine, 6 curved blades	1.08
shrouded turbine with stator, no baffles	1.12

*Assume turbulent flow. K_T assumes baffled tanks with four baffles at the tank wall and a width equal to 10% of the tank diameter.

Description

Equation 14.9 gives the power required to drive a turbulent flow impeller mixer. Typical values of the impeller constant, K_T, are given in Table 14.2. The impeller rotational speed, n, is in revolutions per second. ρ_f is the density of the fluid mixture.

17. FLOCCULATION

After flash mixing, the floc is allowed to form during a 20–60 min period of gentle mixing. Flocculation is enhanced by the gentle agitation, but floc disintegrates with violent agitation. During this period, the flow-through velocity should be limited to 0.5–1.5 ft/min. The peripheral speed of mixing paddles should vary approximately from 0.5 ft/sec for fragile, cold-water floc to 3.0 ft/sec for warm-water floc.

Many modern designs make use of *tapered flocculation*, also known as *tapered energy*, a process in which the amount (severity) of flocculation gradually decreases as the treated water progresses through the flocculation basin.

Flocculation is followed by sedimentation for two to eight hours (four hours typical) in a low-velocity portion of the basin. (The flocculation time is determined from settling column data.) A good settling process will remove 90% of the settleable solids. Poor design, usually resulting in some form of *short-circuiting* of the flow path, will reduce the effective time in which particles have to settle.

Environmental Engineering

18. FLOCCULATOR-CLARIFIERS

A *flocculator-clarifier* combines mixing, flocculation, and sedimentation into a single tank. Such units are called *solid contact units* and *upflow tanks*. They are generally round in construction, with mixing and flocculation taking place near the central hub and sedimentation occurring at the periphery. Flocculator-clarifiers are most suitable when combined with softening, since the precipitated solids help seed the floc.

Typical operational characteristics of flocculator-clarifiers are given in Table 14.3.

Table 14.3 *Characteristics of Flocculator-Clarifiers*

typical flocculation and mixing time	20–60 min
minimum detention time	1.5–2.0 hr
maximum weir loading	10 gpm/ft
upflow rate	0.8–1.7 gpm/ft^2; 1.0 gpm/ft^2 typical
maximum sludge formation rate	5% of water flow

19. SLUDGE QUANTITIES

Sludge is the watery waste that carries off the settled floc and the water softening precipitates.

The most accurate way to calculate the mass of sludge is to extrapolate from jar or pilot test data. There is no absolute correlation between the mass generated and other water quality measurements. However, a few generalizations are possible. (a) Each unit mass (lbm, mg/L, etc.) of alum produces 0.46 unit mass of floc. (b) 100% of the reduction in suspended solids (expressed as substance) shows up as floc. (c) 100% of any supplemental flocculation aids is recovered in the sludge.

20. FILTRATION

Nonsettling floc, algae, suspended precipitates from softening, and metallic ions (iron and manganese) are removed by filtering. *Sand filters* (and in particular, rapid sand filters) are commonly used for this purpose. Sand filters are beds of gravel, sand, and other granulated materials.[3]

Filters discharge into a storage reservoir known as a *clearwell*. The *hydraulic head* (the distance between the water surfaces in the filter and clearwell) is usually 9–12 ft. This allows for a substantial decrease in available head prior to backwashing. Clearwell storage volume is 30–60% of the daily filter output, with a minimum capacity of 12 hr of the maximum daily demand so that demand can be satisfied from the clearwell while the filter is being cleaned or serviced.

The term *log removal value* refers to the logarithm of a concentration ratio, where a 1-log reduction is equivalent to 90%, a 2-log reduction is equivalent to 99%, a 3-log reduction is equivalent to 99.9%, and a 4-log reduction is equivalent to 99.99% removal. In tank processes, there are clear initial and final conditions. In continuous flow processes, it is common to refer to the influent condition as the *feed*, and the effluence condition as the *filtrate*. Microorganisms may be either removed completely (as with filtration) or inactivated (as with disinfection). Both are considered to be "removal." The *log removal value* (also inaccurately referred to as the *removal efficiency*), LRV, is calculated as

$$\text{LRV} = \log_{10} C_{\text{initial}} - \log_{10} C_{\text{final}} = \log_{10}\left(\frac{C_{\text{initial}}}{C_{\text{final}}}\right)$$

There are two ways that log removal values are commonly used.

1. Various processes may be indexed by log removal rate. Operational parameters can be selected to achieve a specific removal efficiency requirement. For example, chlorine *CT* values (product of concentration and contact time) depend on the log removal value desired, as well as temperature and pH.
2. Contractual and legislated limits on contaminants and microorganisms (i.e., bacteria and viruses) are often specified in terms of *log removal credits*. For example, in Table 14.4, *Giardia* requires a 3-log removal, and viruses require a 4-log removal. The removal and inactivation credit is different for each process and for each microorganism. The removal efficiency of existing processes may be reported in terms of log removal values. Demonstrating or achieving the desired removal efficiency is referred to as "receiving the log credit" (see Table 14.5).

Table 14.4 *Removal and Inactivation Requirements*

microorganism	required log reduction	treatment
Giardia	3-log (99.9%)	removal and/or inactivation
viruses	4-log (99.99%)	removal and/or inactivation
Cryptosporidium	2-log (99%)	removal

[3]Most early sand filter beds were designed when a turbidity level of 5 NTU was acceptable. With the U.S. federal MCL at 1 NTU, some states at 0.5 NTU, and planning "on the horizon" for 0.2 NTU, these early conventional filters are clearly inadequate.

Table 14.5 Typical Removal Credits and Inactivation Requirements for Various Treatment Technologies

	typical log removal credits		resulting disinfection log inactivation requirements	
process	*Giardia*	viruses	*Giardia*	viruses
conventional treatment	2.5	2.0	0.5	2.0
direct filtration	2.0	1.0	1.0	3.0
slow sand filtration	2.0	2.0	1.0	2.0
diatomaceous earth filtration	2.0	1.0	1.0	3.0
unfiltered	0	0	3.0	4.0

21. SAND FILTER BACKWASHING

The most common type of service needed by filters is backwashing, which is needed when the pores between the filter particles clog up. Typically, this occurs after 1–3 days of operation, when the head loss reaches 6–8 ft. There are two parameters that can trigger backwashing: head loss and turbidity. Head loss increases almost linearly with time, while turbidity remains constant for several days before suddenly increasing. The point of sudden increase is known as *breakthrough*.[4] Since head loss is more easily monitored than turbidity, it is desired to have head loss trigger the backwashing cycle.

Equation 14.10 Through Eq. 14.16: Head Loss Through a Clean Bed

$$h_f = \frac{1.067(\text{v}_s)^2 L C_D}{g\eta^4 d} \quad \text{14.10}$$

$$h_f = \frac{1.067(\text{v}_s)^2 L}{g\eta^4} \sum \frac{C_{D_{ij}} x_{ij}}{d_{ij}} \quad \text{14.11}$$

$$h_f = \frac{f' L(1-\eta)\text{v}_s^2}{\eta^3 g d_p} \quad \text{14.12}$$

$$h_f = \frac{L(1-\eta)\text{v}_s^2}{\eta^3 g} \sum \frac{f'_{ij} x_{ij}}{d_{ij}} \quad \text{14.13}$$

$$f' = 150\left(\frac{1-\eta}{\text{Re}}\right) + 1.75 \quad \text{14.14}$$

$$\text{v}_s = Q/A_{\text{plan}} \quad \text{14.15}$$

$$\text{Re} = \frac{\text{v}_s \rho d}{\mu} \quad \text{14.16}$$

Description

The head loss through clean sand filters and other granular media beds can be predicted with only fair accuracy from several theoretical models. Equation 14.10 and Eq. 14.11 are derived from the *Rose* model. Equation 14.12 and Eq. 14.13 are derived from the *Carmen-Kozeny* model.[5] Equation 14.10 and Eq. 14.12 are used for monosized media, while Eq. 14.11 and Eq. 14.13 are extensions used for filters with multiple and multisized media based on the principle of adding flow resistances in series. Equation 14.10 through Eq. 14.13 predict the friction head loss through the entire filter, not per unit bed depth. f' is the friction factor, v_s is the *superficial velocity* (also known as the *face velocity*, *approach velocity*, and the *empty bed approach velocity*), and Re is the Reynolds number.[6] f', v_s, and Re are given by Eq. 14.14, Eq. 14.15, and Eq. 14.16, respectively.[7]

Equation 14.17 Through Eq. 14.20: Bed Expansion

$$L_f = \frac{L_o(1-\eta_o)}{1-\left(\frac{\text{v}_B}{\text{v}_t}\right)^{0.22}} \quad \text{14.17}$$

$$L_f = L_o(1-\eta_o) \sum \frac{x_{ij}}{1-\left(\frac{\text{v}_B}{\text{v}_{t,ij}}\right)^{0.22}} \quad \text{14.18}$$

Environmental Engineering

[4]Technically, *breakthrough* is the point at which the turbidity rises above the MCL permitted. With low MCLs, this occurs very soon after the beginning of filter performance degradation.

[5]The difficulty with reducing theoretical models to a single formula, as has been done in the NCEES *FE Reference Handbook* (*NCEES Handbook*) with the filter bed equations, is that the formula is only as valid as the assumptions that were used to derive it. Equation 14.10, for example, is based on the more general Rose model but omits the shape factor, and therefore is valid only for spherical particles. Equation 14.12 similarly omits the shape factor, but also omits the empirical packing factor, and it combines the kinematic viscosity with the constant term, so it is valid only for spherical particles and some unspecified temperature range.

[6]The *NCEES Handbook* calls the approach velocity a "filtration rate," but it is not the flow rate through the entire filter. The superficial velocity (hence the subscript, *s*) is the flow rate per unit area, with units identical to velocity.

[7]The *NCEES Handbook* uses the symbol η to represent porosity in this section, which is inconsistent with n used in its geotechnical section. The most common symbols used for porosity outside of geotechnical subjects are ε and ϕ.

$$\eta_f = \left(\frac{v_B}{v_t}\right)^{0.22} \quad \text{14.19}$$

$$v_B = \frac{Q_B}{A_{plan}} \quad \text{14.20}$$

Description

A *fluidized bed* is a bed whose particles have been made to act like a fluid, rather than a granular plug. When the filter media are fluidized by backwashing, the media depth increases. Equation 14.17 and Eq. 14.18 are used to calculate the depth of expanded (fluidized) filter media during backwashing for monosized media and multisized media, respectively. η_f is the porosity of the fluidized bed, and can be found from Eq. 14.19.[8] v_B is the backwash velocity as determined from Eq. 14.20. Clearly, the depth of the fluidized bed depends on the backflow velocity, so Eq. 14.17 does not necessarily predict the depth after backwashing. The "initial" porosity and bed depth refer to the fouled condition before backwashing, not the initial clean condition.

Example

Sand with a particle diameter of 0.02 in is used in a sand filter. After use, the depth of the filter is 0.5 ft and the porosity is 0.35. During backwash, the porosity is 0.70. The settling velocity of sand particles in water is 0.27 ft/sec. Most nearly, what will be the depth of the filter after backwashing has doubled the porosity?

(A) 0.80 ft

(B) 1.0 ft

(C) 1.1 ft

(D) 1.3 ft

Solution

Combine Eq. 14.17 and Eq. 14.19.

$$L_f = L_o\left(\frac{1-\eta_o}{1-\eta_f}\right) = (0.5\ \text{ft})\left(\frac{1-0.35}{1-0.70}\right) = 1.08\ \text{ft} \quad (1.1\ \text{ft})$$

The answer is (C).

22. OTHER FILTRATION METHODS

Pressure filters operate similarly to rapid sand filters except that incoming water is pressurized up to 25 ft gage. Pressure filters are not used in large installations.

Biofilm filtration (*biofilm process*) uses microorganisms to remove selected contaminants (e.g., aromatics and other hydrocarbons). Operation of biofilters is similar to trickling filters used in wastewater processing.

Slow sand filters are primarily of historical interest, though there are some similarities with modern biomethods used to remediate toxic spills. Slow sand filters operate similarly to rapid sand filters except that the exposed surface (loading) area is much larger and the flow rate is much lower.

Ultrafilters are membranes that act as sieves to retain turbidity, microorganisms, and large organic molecules that are THM precursors, while allowing water, salts, and small molecules to pass through. Ultrafiltration is effective in removing particles ranging in size from 0.001 μm to 10 μm.[9] A pressure of 15–75 psig is required to drive the water through the membrane.

Equation 14.21: Ultrafiltration

$$J_w = \frac{\varepsilon r^2 \int \Delta p}{8\mu\delta} \quad \text{14.21}$$

Description

Equation 14.21 gives the volumetric flux through a membrane during ultrafiltration. The term *flux* simply means flow rate. Flow rate usually has such units as ft^3/min or m^3/s. In the case of Eq. 14.21 (the *Hagen-Poiseuille equation*), however, the flow rate is per unit area of membrane, such that $J_w = Q/A_m$. Therefore, flux has units of ft^3/ft^2-sec, the same as velocity (ft/sec).[10] ε is the porosity (that is, the fraction of the membrane area that is open), r is the average pore size, Δp is the net pressure drop across the filter, μ is the absolute viscosity, and δ is the filter thickness.

For other than pure solvents, the net pressure drop includes the effects of any foulants, but it also includes a reduction due to osmotic pressure, Π. That is, $\Delta p = p_{friction} - \Pi$. Osmotic pressure is zero when solids are being filtered from otherwise pure water. The pressure drop may also be referred to as the *transmembrane pressure*. Flux rate units of $L/m^2{\cdot}h$ are usually abbreviated as "Lmh," and gal/ft^2-day is abbreviated as "gfd."

[8]It is not clear if the subscript *f* in L_f in the *NCEES Handbook* refers to "filter" or "fluidized," but it is not "fouled," and L_f is not related to friction loss as is h_f.

[9]A μm is the same as a micron.

[10]The subscript *w* in J_w may refer to water or flux through a wall. The subscript *v* is more commonly used to indicate a volumetric flux.

Example

5

A fluid is being ultrafiltered through a membrane that has a porosity of 50%, a pore size of 8×10^{-5} in, and a thickness of 4×10^{-3} in. The fluid has a viscosity of 3×10^{-5} lbf-sec/ft^2 and experiences a 1.45 psi pressure drop during filtration. What is most nearly the volumetric flux during ultrafiltration?

(A) 0.058 ft/sec

(B) 0.091 ft/sec

(C) 0.35 ft/sec

(D) 0.58 ft/sec

Solution

The volumetric flux during ultrafiltration is

$$J_w = \frac{\varepsilon r^2 \int \Delta p}{8\mu\delta}$$

$$= \frac{(0.5)(8 \times 10^{-5}\ \text{in})^2\left(1.45\ \frac{\text{lbf}}{\text{in}^2}\right)\left(12\ \frac{\text{in}}{\text{ft}}\right)}{(8)\left(3 \times 10^{-5}\ \frac{\text{lbf-sec}}{\text{ft}^2}\right)(4 \times 10^{-3}\ \text{in})}$$

$$= 0.058\ \text{ft/sec}$$

The answer is (A).

23. ADSORPTION

Adsorption occurs when the target contaminant becomes attached to the surface of a medium, called the *adsorbent*. Typically, the adsorbent is *activated carbon*, which is available in both granular (GAC) or powdered (PAC) forms. The activated carbon is produced by starved air combustion of coal, coconut shells, wood, and other organic materials. Adsorption is effective in removing heavy metals such as lead and mercury, and in removing aromatic and aliphatic organic chemicals such as VOCs and dioxins. Adsorption processes may be employed to recover gases for reuse or recycling, or for destruction. Very high removal efficiencies, exceeding 99%, are possible.

Equation 14.22 and Eq. 14.23: Freundlich Isotherm

$$\frac{x}{m} = X = KC_e^{1/n} \qquad 14.22$$

$$\ln\frac{x}{m} = \frac{1}{n}\ln C_e + \ln K \qquad 14.23$$

Description

The activated carbon process provides essentially constant removal until *breakthrough* occurs. At breakthrough, the effluent concentration begins to increase and continues to do so until the carbon is saturated and no pollutant removal occurs. Because optimum operation uses carbon to saturation, adsorption processes are frequently operated in a *two-in-series lead-follow mode.* The carbon is regenerated or reactivated sometime during the interval between breakthrough and saturation. *Regeneration* is associated with contaminant recovery and is typically a chemical process. *Reactivation* is associated with contaminant destruction and is a thermal process.

Activated carbon adsorption process design is based on the $X = x/m$ ratio determined from an adsorption isotherm specific to the target contaminant. The x/m ratio represents the mass of chemical adsorbed per unit mass of carbon and is influenced by the gas temperature and moisture content. Many adsorption isotherm relationships have been developed, but one of the most common is the *Freundlich equation.*

Equation 14.22 is the Freundlich equation in its standard form, while Eq. 14.23 is the linearized form, obtained by taking the logarithm of both sides. C_e is the equilibrium concentration (mass per unit volume) of the solute to be adsorbed, essentially the amount remaining when the absorption process is functioning. X is the mass ratio, the mass of solute absorbed per unit mass of adsorbent. K and n are constants obtained by correlating data from *bottle point experiments.*

An *isotherm* is a plot of X versus C_e, so named because adsorption will be proportional to either T or $1/T$. For a linear Freundlich isotherm (i.e., one that can be described by a straight line), $n = 1$, and the plot is ln X versus ln C_e. Generally, the adsorption curve is considered to be favorable only for $n \geq 1$.

Equation 14.24 and Eq. 14.25: Langmuir Isotherm

$$\frac{x}{m} = X = \frac{aKC_e}{1 + KC_e} \qquad 14.24$$

$$\frac{m}{x} = \frac{1}{a} + \frac{1}{aK}\frac{1}{C_e} \qquad 14.25$$

Description

While the Freundlich equation is an accurate predictor of ion adsorption from solutions of intermediate concentrations, the Langmuir equation predicts higher and lower range concentration adsorption better. Equation 14.24 is the Langmuir equation in its standard form, while Eq. 14.25 is the linearized form, which plots

Environmental Engineering

$1/X$ versus $1/a$. a is the mass of solute required to completely saturate a unit mass of adsorbent. K is a correlation constant obtained from experimental data.

Example

Data for the adsorption of a polychlorinated biphenyl (PCB) by activated carbon are given in the table. The initial concentration of PCB is 1.0 MG/L.

equilibrium concentration of PCB (mg/L)	adsorption capacity (mass PCB adsorbed/ mass carbon)
0.01	0.01
0.1	0.05

Using the Langmuir equation, what is most nearly the adsorption capacity as a percentage of the carbon weight at saturation?

(A) 5.0%

(B) 9.0%

(C) 12%

(D) 15%

Solution

The adsorption capacity of the carbon is a, the mass of PCB required to completely saturate a unit mass of carbon. Use the experimental data to find a. Solve Eq. 14.25 for K. For the first observation,

$$K = \frac{X}{C_e(a-X)} = \frac{0.01}{(0.01)(a-0.01)} = \frac{1}{a-0.01}$$

From the second observation,

$$K = \frac{X}{C_e(a-X)} = \frac{0.05}{(0.1)(a-0.05)} = \frac{1}{2(a-0.05)}$$

Equate these two results and solve for a.

$$\frac{1}{a-0.01} = \frac{1}{2(a-0.05)}$$

$$a = 0.09 \quad (9.0\%)$$

The answer is (B).

Equation 14.26 and Eq. 14.27: Depth of Sorption Zone

$$Z_s = Z\left[\frac{V_Z}{V_T - 0.5V_Z}\right] \qquad 14.26$$

$$V_Z = V_T - V_B \qquad 14.27$$

Description

Consider a solute that is to be removed in an adsorbent (e.g., granular activated carbon) bed. When the adsorbent is fresh, all of the solute will be removed, and the exit stream will be pure solvent with zero solute concentration. Over time, as the adsorbent removes the solute and approaches saturation, the solute will begin appearing in the exit stream. A plot of the leaving solute concentration versus time is known as a *breakthrough curve*. If the solvent flow-through rate is constant, then the breakthrough curve can also be plotted as leaving concentration versus accumulated flow. Since the input solute concentration is constant, the plot can also be presented as the ratio of output to input concentrations versus time.

The *breakthrough point*, or *breakpoint*, occurs with adsorbent that is almost saturated, when the solute begins appearing in the exit stream. Any threshold concentration could be used, but 5% is chosen by convention. The breakpoint is defined as the time, volume, or condition at which the adsorbent is removing 95% of the solute. This is equivalent to 5% solute remaining in solution. The breakthrough concentration is represented as C_α, $C_{5\%}$ or $0.05C_0$, where C_0 represents the entering solute concentration.[11] The *exhaustion point* occurs when the adsorbent is essentially saturated, when the leaving concentration is 95% of the entering concentration. (See Fig. 14.4.)

Figure 14.4 *Adsorption Breakthrough Curve*

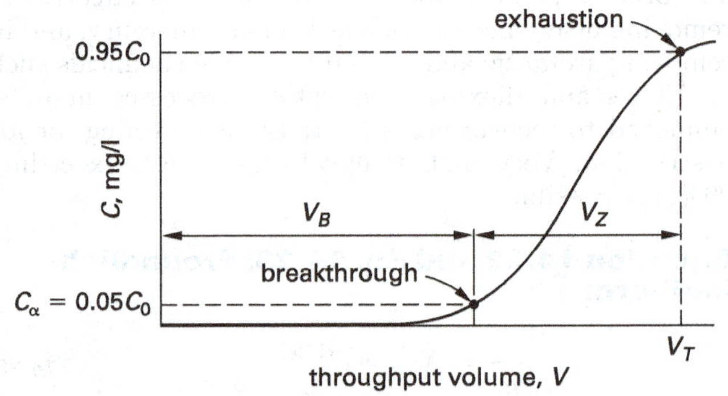

For a properly designed adsorption tower (filter or column) containing fresh adsorbent, all of the solute will have been removed by the time the solvent has passed

[11]The *NCEES Handbook* uses both 0 and O as subscripts to represent the entering condition.

through a thickness of Z_s. The remaining height, $Z - Z_s$, will initially see only pure solvent. As the top layer of adsorbent becomes saturated, the region of adsorption moves downward through the tower, but the thickness of active adsorption remains Z_s. This region is known as the *mass transfer zone*, and its thickness is the *depth of the sorption zone*.

Equation 14.26 calculates the depth of sorption, Z_s. Z is the total thickness (depth, height, etc.) of adsorbent. V_B is the volume of solvent that has passed through the process at breakthrough (i.e., leaving concentration is 5% of the incoming concentration, C_0). V_T is the volume of solvent that has passed through the process when the entire thickness, Z, has reached exhaustion (i.e., leaving concentration is 95% of the incoming concentration, C_0). The difference, $V_T - V_B$, is the amount of solvent that can be treated by the tower. Since the flow rate is constant, times could also be used instead of volumes.

24. PRECIPITATION SOFTENING

Precipitation softening using the *lime-soda ash process* adds lime (CaO), also known as *quicklime*, and soda ash (Na_2CO_3) to remove calcium and magnesium from hard water.[12] Granular quicklime is available with a minimum purity of 90%, and soda ash is available with a 98% purity.

Lime forms *slaked lime* (also known as *hydrated lime*), $Ca(OH)_2$, in an exothermic reaction when added to feed water. The slaked lime is delivered to the water supply as a *milk of lime* suspension.

Precipitation softening is relatively inexpensive for large quantities of water. Both alkalinity and total solids are reduced. The high pH and lime help disinfect the water.

However, the process produces large quantities of sludge that constitute a disposal problem. The intrinsic solubility of some of the compounds means that complete softening cannot be achieved. Flow rates and chemical feed rates must be closely monitored.

Equation 14.28 Through Eq. 14.35: Lime Soda Softening

Carbon dioxide removal:

$$CO_2 + Ca(OH)_2 \rightarrow CaCO_3(s) + H_2O \quad 14.28$$

Calcium carbonate hardness removal:

$$Ca(HCO_3)_2 + Ca(OH)_2 \rightarrow 2CaCO_3(s) + 2H_2O \quad 14.29$$

Magnesium carbonate hardness removal:

$$\begin{aligned} Mg(HCO_3)_2 + 2Ca(OH)_2 \rightarrow\ & 2CaCO_3(s) \\ & + Mg(OH)_2(s) + 2H_2O \end{aligned} \quad 14.30$$

Magnesium non-carbonate hardness removal:

$$\begin{aligned} MgSO_4 + Ca(OH)_2 + Na_2CO_3 \rightarrow\ & CaCO_3(s) \\ & + Mg(OH)_2(s) + 2Na^+ + SO_4^{2-} \end{aligned} \quad 14.31$$

Calcium non-carbonate hardness removal:

$$CaSO_4 + Na_2CO_3 \rightarrow CaCO_3(s) + 2Na^+ + SO_4^{-2} \quad 14.32$$

Destruction of excess alkalinity:

$$\begin{aligned} 2HCO_3^- + Ca(OH)_2 \rightarrow\ & CaCO_3(s) \\ & + CO_3^{2-} + 2H_2O \end{aligned} \quad 14.33$$

Recarbonation:

$$Ca^{2+} + 2OH^- + CO_2 \rightarrow CaCO_3(s) + H_2O \quad 14.34$$

Values

$$50 \text{ mg/L as } CaCO_3 \text{ equivalent} = 1 \text{ meq/L} \quad 14.35$$

Description

Equation 14.28 through Eq. 14.34 collectively constitute the process of lime soda softening.

Slaked lime reacts first with any carbon dioxide dissolved in the water (see Eq. 14.28). No softening occurs, but the carbon dioxide demand must be satisfied before any reactions involving calcium or magnesium can occur.

Lime next reacts with any carbonate hardness, precipitating calcium carbonate and magnesium hydroxide (see Eq. 14.29 and Eq. 14.30). The removal of carbonate hardness caused by magnesium requires two molecules of calcium hydroxide to precipitate calcium carbonate and magnesium hydroxide.

[12]Soda ash does not always need to be used. When lime alone is used, the process may be referred to as *lime softening*.

To remove noncarbonate hardness, it is necessary to add soda ash and more lime (see Eq. 14.31 and Eq. 14.32). The sodium sulfate that remains in solution does not contribute to hardness, for sodium is a single-valent ion (i.e., Na^+).

After softening, the water must be recarbonated to lower its pH and to reduce its scale-forming potential. This is accomplished by bubbling carbon dioxide gas through the water (see Eq. 14.34).

For molecular and equivalent weights, see Table 14.6.

Table 14.6 Molecular and Equivalent Weights

molecular formula	molecular weight	number of equivalents per mole, n	equivalent weight
CO_3^{2-}	60.0	2	30.0
CO_2	44.0	2	22.0
$Ca(OH)_2$	74.1	2	37.1
$CaCO_3$	100.1	2	50.0
$Ca(HCO_3)_2$	162.1	2	81.1
$CaSO_4$	136.1	2	68.1
Ca^{2+}	40.1	2	20.0
H^+	1.0	1	1.0
HCO_3^-	61.0	1	61.0
$Mg(HCO_3)_2$	146.3	2	73.2
$Mg(OH)_2$	58.3	2	29.2
$MgSO_4$	120.4	2	60.2
Mg^{2+}	24.3	2	12.2
Na^+	23.0	1	23.0
Na_2CO_3	106.0	2	53.0
OH^-	17.0	1	17.0
SO_4^{2-}	96.1	2	48.0

25. DISINFECTION

Chlorination is commonly used for disinfection. Chlorine can be added as a gas or as a liquid. If it is added to the water as a gas, it is stored as a liquid, which vaporizes around −31°F. Liquid chlorine is the primary form used since it is less expensive than calcium hypochlorite solid ($Ca(OCl)_2$) and sodium hypochlorite ($NaOCl$). When chlorine gas dissolves in water, it forms hydrochloric acid (HCl) and hypochlorous acid ($HOCl$).

Chlorine is corrosive and toxic. Special safety and handling procedures must be followed with its use.

Equation 14.36, Eq. 14.37, and Table 14.7: Disinfection Equations and Baffling Factors

$$T = \text{TDT} \times \text{BF} \qquad 14.36$$

$$\text{TDT} = C \times T \qquad 14.37$$

Table 14.7 Baffling Factors

baffling condition	baffling factor	baffling description
unbaffled (mixed flow)	0.1	none, agitated basin, very low length to width ratio, high inlet and outlet flow velocities
poor	0.3	single or multiple unbaffled inlets and outlets, no intra-basin baffles
average	0.5	baffled inlet or outlet with some intra-basin baffles
superior	0.7	perforated inlet baffle, serpentine or perforated intra-basin baffles, outlet weir or perforated launders
perfect (plug flow)	1.0	very high length to width ratio (pipeline flow), perforated inlet, outlet, and intra-basin baffles

Values

Chlorine contact chambers typically have length-to-width ratios between 20:1 and 50:1.

Description

Chlorine disinfection processes are often specified by their *CT values* (the product of the amount of disinfectant and the time the water is in contact with the disinfectant). The time that microorganisms and contaminants are in contact with the disinfectant is found from Eq. 14.36, where TDT is the theoretical detention time, and BF is the baffling factor. The *theoretical detention time* is the theoretical amount of time that it takes for wastewater to pass through a tank. The *baffling factor* is a factor describing the degree of mixing (i.e., absence of short-circuiting) within a basin. Baffling factors vary according to baffling conditions. Equation 14.37 is used to calculate the TDT, and the baffling factor is found from Table 14.7.

With constant-feed processes, the concentration of disinfectant will vary with dilution. Since water flow through wastewater treatment plants varies over time, the value of C is defined as a minimum value corresponding to the peak hourly flow. The concentration used is not the dose concentration, since some of the disinfectant will be used up neutralizing other compounds. The concentration used in calculating the CT value represents only the residual (free) disinfectant remaining after other reactions are complete.

Table 14.8 and Table 14.9: *CT* Values for Microorganism Inactivation

See Table 14.8 at the end of this chapter.

Table 14.9 *CT Values for 4-Log Inactivation of Viruses by Free Chlorine*

	CT value (min·mg/L)	
temperature (°C)	pH 6–9	pH 10
0.5	12	90
5	8	60
10	6	45
15	4	30
20	3	22
25	2	15

Source: *Guidance Manual LT1ESWTR Disinfection Profiling and Benchmarking*, U.S. Environmental Protection Agency, 2003.

Description

As described in Sec. 14.20, water regulations require a 3-log removal of *Giardia* and a 4-log removal of viruses. When chlorine is used for water disinfection, a *CT value* is used to calculate the amount of chlorine required. Table 14.8 gives CT values for 3-log inactivation of *Giardia* cysts, and Table 14.9 gives CT values for 4-log inactivation of viruses.

26. DEMINERALIZATION AND DESALINATION

Demineralization and desalination (*salt water conversion*) are required when only brackish water supplies are available.[13] These processes are carried out in distillation, electrodialysis, ion exchange, and membrane processes.

Distillation is a process whereby the raw water is vaporized, leaving the salt and minerals behind. The water vapor is reclaimed by condensation. Distillation cannot be used to economically provide large quantities of water.

Reverse osmosis is the least costly and most attractive membrane demineralization process. A thin membrane separates two solutions of different concentrations. Pore size is smaller (0.0001–0.001 μm) than with ultrafilter membranes, as salt ions are not permitted to pass through. Typical large-scale osmosis units operate at 150–500 psi.[14]

Nanofiltration is similar to ultrafiltration and reverse osmosis, with pore size (0.001 μm) and operating pressure (75–250 psig) intermediate between the two. Nanofilters are commonly referred to as *softening membranes*.

The *ion exchange* process is an excellent solution to demineralization and desalination.

In *electrodialysis*, positive and negative ions flow through selective membranes under the influence of an induced electrical current. Unlike pressure-driven filtration processes, however, the ions (not the water molecules) pass through the membrane. The ions removed from the water form a concentrate stream that is discarded.

The current, I, passing through an electrodialysis unit consisting of a stack of cell pairs[15] is induced by an applied direct-current (DC) electromotive force (voltage), E, which can be calculated from the current and the unit resistance, R. The resistance of the membrane stack is due to the friction of the ions with the membranes and the aqueous solution while being transferred from one solution to another. The current passing through the unit is given by the following equation, in which F is the *Faraday's constant*, 96 485 A·s/mol (same as C/mol and C/GEW, coulombs per gram equivalent weight), essentially the number of electrons in mole. Q is the flow rate, which would be in L/s if no conversion factors were used. N is the normality of the solution being processed. Normality is the number of moles (gram equivalent weights) of ions per liter of solution. n is the number of series-connected electrodialysis cell pairs in the stack. (Although the cells are connected in parallel from the viewpoint of the solution passing through the unit, the cell pairs are connected electrically in series.) Since the same electrons pass consecutively through each of the parallel cell pairs, the current is calculated from the volumetric flow rate, Q/n, through a single cell pair. E_1 is the *removal efficiency*, the fraction of generated charges that end up performing

Environmental Engineering

[13]In the United States, Florida, Arizona, and Texas are leaders in desalinization installations. Potable water is routinely made from sea water in the Middle East.

[14]Membrane processes operate in a *fixed flux condition*. In order to keep the yield constant over time, the pressure must be constantly increased in order to compensate for the effects of fouling and compaction. Fouling by inorganic substances and biofilms is the biggest problem with membranes.

[15]The *NCEES Handbook* refers to each cell pair as an electrodialysis "cell."

electrodialysis. E_2 is the electrical efficiency, or current efficiency, the fraction of electrons that perform useful work (such as pumping), as opposed to merely generating heat.[16]

$$I = \left(\frac{FQN}{n}\right)\left(\frac{E_1}{E_2}\right)$$

Electrical power is calculated from voltage and current as $P = IE$. Combining this with $E = IR$ results in the following equation.

$$P = IE = \frac{E^2}{R} = I^2R$$

Example

An electrodialysis unit with an electrical efficiency of 90% and a removal efficiency of 75% draws 100 A of current at 100 V. What is most nearly the power consumption of the unit?

(A) 6.0 kW

(B) 8.0 kW

(C) 10 kW

(D) 12 kW

Solution

The power consumption is

$$P = I^2R = I^2\left(\frac{E}{I}\right) = IE$$

$$= \frac{(100\ \text{A})(100\ \text{V})}{1000\ \frac{\text{W}}{\text{kW}}}$$

$$= 10\ \text{kW}$$

The answer is (C).

27. AIR STRIPPING

Air strippers are primarily used to remove volatile organic compounds (VOCs) or other target substances from water. In operation, contaminated water enters a stripping tower at the top, and fresh air enters at the bottom. (See Fig. 14.5.) The effectiveness of the process depends on the volatility of the compound, its temperature and concentration, and the liquid-air contact area. However, removal efficiencies of 80–90% (and above) are common for VOCs.

There are three types of stripping towers—*packed towers* filled with synthetic *packing media* (e.g., polypropylene balls, rings, and saddles as shown in Fig. 14.6), *spray towers*, and (less frequently for VOC removal) *tray towers* with horizontal trays arranged in vertical layers. *Redistribution rings* (*wall wipers*) prevent channeling down the inside of the tower. (*Channeling* is the flow of liquid through a few narrow paths rather than an over-the-bed packing.) The stripping air is generated by a small blower at the column base. A mist eliminator at the top eliminates entrained water from the air.

As the contaminated water passes over packing media in a packed tower, the target substance leaves the liquid and enters the air where the concentration is lower. The mole fraction of the target substance in the water decreases; the mole fraction of the target substance in the air increases.

The discharged air, known as *off-gas*, is discharged to a process that destroys or recovers the target substance. (Since the quantities are small, recovery is rarer.)

Figure 14.5 Schematic of Air Stripping Operation

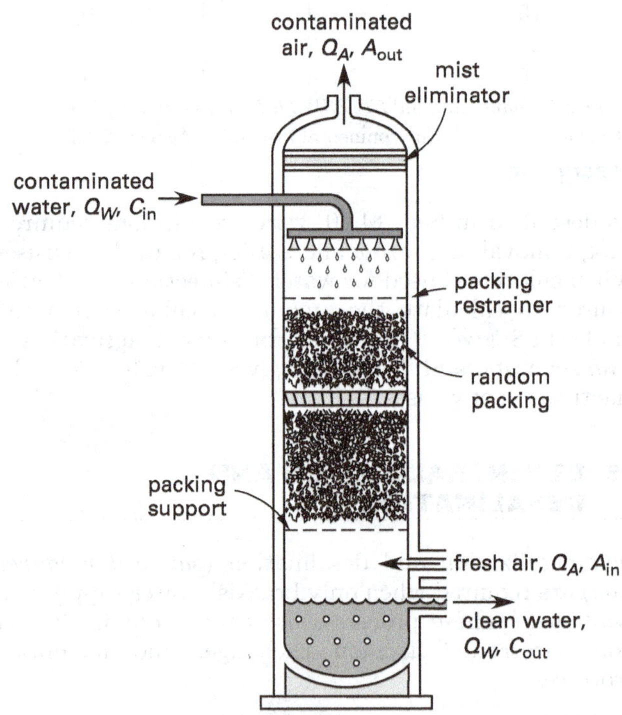

Destruction of the target substance can be accomplished by flaring, carbon absorption, and incineration. Since flaring is dangerous and carbon absorption creates a secondary waste if the activated carbon is not regenerated, incineration is often preferred.

[16]The *NCEES Handbook* uses the same symbol, E, for both voltage and efficiency in the same context, which can be confusing. Many texts use η for efficiency.

Figure 14.6 *Packing Media Types*

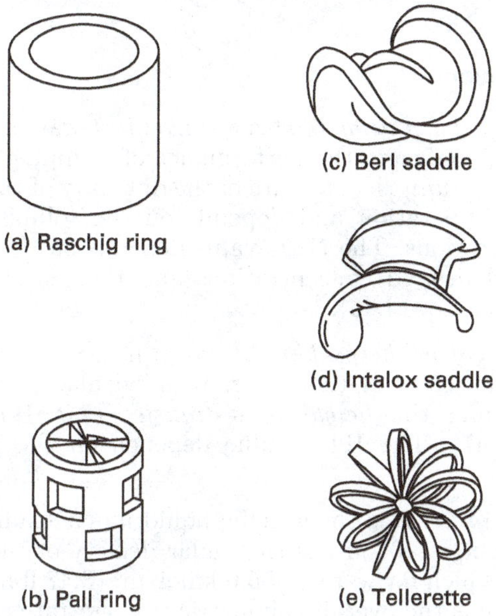

Equation 14.38 Through Eq. 14.44: Air Stripping

$$p_i = py_i = hx_i \quad \text{14.38}$$

$$p_i = HC_i \quad \text{[Henry's law]} \quad \text{14.39}$$

$$A_{\text{out}} = H'C_{\text{in}} \quad \text{14.40}$$

$$H' = H/RT \quad \text{14.41}$$

$$Q_W \cdot C_{\text{in}} = Q_A H' C_{\text{in}} \quad \text{14.42}$$

$$Q_W = Q_A H' \quad \text{14.43}$$

$$H'(Q_A/Q_W) = 1 \quad \text{14.44}$$

Description

Henry's law, as it applies to water treatment, states that at equilibrium, the vapor pressure of a target substance is directly proportional to the target substance's mole fraction, y. (The maximum mole fraction is the substance's *solubility*.) As Eq. 14.38 shows, this is written as $p_i = py_i$, where p is the total pressure and y is the mole fraction of the compound in the gas phase.[17] If pressure is in atmospheres, the value of p is almost always 1.0.

At a constant temperature, for a component i, the partial pressure of that component, p_i, is directly proportional to the concentration of that component, C_i, in the gas phase. (C_i has units of moles per unit volume, such as kmol/m^3.) Equation 14.39 shows this relationship, which is also known as *Henry's law*. H is *Henry's law constant*. This term is ambiguous because there are four different common formulations of Henry's law, and the constant is expressed in different units in each of them. For Eq. 14.39, the units of H are pressure over concentration, usually presented as atm·m^3/kmol or L·atm/mol.

Equation 14.38 relates the partial pressure of the compound in the gas phase to the mole fraction, x, in the liquid phase, but a different form of Henry's law constant, h, is needed. Since mole fraction is dimensionless, this Henry's law constant has units of atm. Since p is essentially always 1.0 atm, Eq. 14.38 is sometimes written $y_i = hx_i$, which hides the proper units of h.

In an air stripper, the compound removed from the liquid phase is transferred to the gas phase. In the case of wastewater treatment, the liquid is water and the gas is air. Equation 14.40 calculates the concentration of the compound in the exit air stream.[18] Henry's law constant is again involved, but this time, the units are different. To distinguish the dimensionless constant used in Eq. 14.40 from the constants used in Eq. 14.38 and Eq. 14.39, H and h, a different symbol, H', is used. Equation 14.41 can be used to calculate H' from the ideal gas law (because $pV = RT$). R is the *universal gas constant*, which, for the purpose of air stripping, in common SI units, has a value of 0.08206 m^3·atm/kmol·K, although any convenient set of units can be used.

Equation 14.42, Eq. 14.43, and Eq. 14.44 are used to relate the volumetric *loading* (i.e., flow) *rate* (in m^3/s, for example) of the liquid (subscript W for water) to the volumetric flow rate (also in m^3/s) of the gas (subscript A for air).[19] Concentration, C, has units of moles per unit volume (e.g., kmol/m^3) to match the units of flow rate. Any convenient set of units can be used with these equations, although the known values of Henry's law constant generally determine what units are used.

Example

Henry's law constant for dimethyl sulfide is 7.1 L·atm/mol. The mole fraction of dimethyl sulfide in water that is in equilibrium with a concentration

[17]Dalton's law of partial pressures is usually written $p_i = x_i p_{\text{total}}$. Stripping theory uses x and y for the liquid and gas mole fraction variables.

[18]A stands for air, not area. The *NCEES Handbook* uses C to represent the concentration in the liquid and A for the concentration in the gas, even though the gas is not always air.

[19]The *NCEES Handbook* uses subscripts to differentiate the flow rates of the liquid and gas phases. In traditional stripping theory, the variables used for liquid and gas flow rates are L and G, respectively.

of 10^{-6} parts per billion volume (ppbv) dimethyl sulfide in air is most nearly

(A) 1.4×10^{-13}

(B) 7.1×10^{-12}

(C) 1.0×10^{-11}

(D) 2.3×10^{-11}

Solution

For gases, the volumetric ratio is the same as the mole ratio, since all gases have the same molar volume (i.e., 22.4 L/mol). Therefore,

$$y = (10^{-6}\ \text{ppb})\left(10^{-6}\ \frac{1}{\text{ppb}}\right) = 10^{-12}$$

Equation 14.38 relates the mole fractions in the liquid and gas phases.

$$p_i = hx_i$$

$$x = \frac{py}{h} = \frac{(1\ \text{atm})(10^{-12})}{7.1\ \frac{\text{L·atm}}{\text{mol}}} = 1.408 \times 10^{-13} \quad (1.4 \times 10^{-13})$$

The answer is (A).

Equation 14.45: Stripping Factor

$$R_S = H'(Q_A/Q_W) \qquad 14.45$$

Description

The *stripping factor*, R_S, is the reciprocal of the *absorption factor*.[20] Units for the *gas and liquid loading rates*, Q_A and Q_W, respectively, depend on the correlations used to solve the stripping equations. The air flow rate is limited by the acceptable pressure drop through the tower. The ratio Q_A/Q_W is known as the *air-water ratio* and is dimensionless.

Equation 14.46 Through Eq. 14.48: Transfer Unit Method

$$\text{NTU} = \left(\frac{R_S}{R_S - 1}\right)\ln\left(\frac{(C_{\text{in}}/C_{\text{out}})(R_S - 1) + 1}{R_S}\right) \qquad 14.46$$

$$\text{HTU} = \frac{L}{M_W K_L a} \qquad 14.47$$

$$Z = \text{HTU} \times \text{NTU} \qquad 14.48$$

Description

The *transfer unit method* is a convenient way of designing and analyzing the performance of stripping towers. A *transfer unit* is a measure of the difficulty of the mass-transfer operation and depends on the solubility and concentrations. The NTU value depends on the incoming and desired concentrations and the material flow rates (see Eq. 14.46).[21]

The *height of the packing*, Z, given in Eq. 14.48, is the effective height of the tower. It is calculated from the NTUs and the *height of a transfer unit*, HTU (see Eq. 14.47). The HTU value depends on the packing media.

In Eq. 14.47, L is the specific liquid molar loading rate (e.g., mol/m²·s); M_W is the molar density of the liquid phase, which for water is 55.6 kmol/m³ (3.47 lbmol/ft³); and $K_L a$ is the overall volumetric transfer rate constant with units of m³/m³·s or 1/s. The quantity L/M_W is the liquid *surface hydraulic loading* of the packing material in m³/m²·s, or m/s.

Example

An air stripper is used to remove a solute from water. The outgoing concentration of solute in water is 0.5 ppm, and the incoming concentration of solute in water is 100 ppm. The dimensionless Henry's law constant, H', is 10. The average airflow rate is 6000 ft³/sec, while the average water flow rate is 600 ft³/sec. If the height of the transfer unit is 10 ft, the height of the column is most nearly

(A) 16 ft

(B) 53 ft

(C) 75 ft

(D) 250 ft

Solution

Find the stripping factor, R_S, using Eq. 14.45.

$$R_S = H'(Q_A/Q_W) = (10)\left(\frac{6000\ \frac{\text{ft}^3}{\text{sec}}}{600\ \frac{\text{ft}^3}{\text{sec}}}\right) = 100$$

[20] The symbol used more commonly by other authorities for the stripping factor is traditionally *S*. *R* is generally reserved for the gas constant.

[21] Stripping is one of many mass-transfer operations studied by chemical engineers. Determining the number of transfer units required and the height of a single transfer unit are typical chemical engineering calculations.

From Eq. 14.46, the number of transfer units is

$$\begin{aligned}\text{NTU} &= \left(\frac{R_S}{R_S-1}\right)\ln\left(\frac{(C_{\text{in}}/C_{\text{out}})(R_S-1)+1}{R_S}\right) \\ &= \left(\frac{100}{100-1}\right)\ln\left(\frac{\left(\frac{100\text{ ppm}}{0.5\text{ ppm}}\right)(100-1)+1}{100}\right) \\ &= 5.34\end{aligned}$$

Use Eq. 14.48 to calculate the stripper packing height.

$$\begin{aligned}Z &= \text{HTU} \times \text{NTU} \\ &= (10\text{ ft})(5.34) \\ &= 53.4\text{ ft} \quad (53\text{ ft})\end{aligned}$$

The answer is (B).

28. REVERSE OSMOSIS

Reverse osmosis is a type of diffusion in which molecules move under pressure from a solvent to a solution. Reverse osmosis is the least costly and most attractive membrane demineralization process. A thin membrane separates two solutions of different concentrations. Pore size is smaller (0.0001–0.001 μm) than with ultrafilter membranes, as salt ions are not permitted to pass through. Typical large-scale osmosis units operate at 150–500 psi.[22]

Equation 14.49: Osmotic Pressure of Solutions of Electrolytes

$$\Pi = \phi\nu\frac{n}{V}RT \qquad 14.49$$

Description

Equation 14.49 is used to calculate the osmotic pressure of solutions in electrolytes. In Eq. 14.49, Π is the osmotic pressure (in Pa or other pressure units consistent with R). ϕ is the *osmotic coefficient*. ν is the *van't Hoff factor*, also known as the *dissociation factor*, equal to the total number of ions (anions plus cations) formed when one molecule of electrolyte dissociates in the solvent.[23] n is the number of moles of electrolyte in the volume V, so that the ratio n/V is the molar density of electrolyte (in kmol/m^3 or similar). R is the universal gas constant (in Pa·m^3/kmol·K or any other appropriate set of units). T is the absolute temperature (in K).[24] Depending on units, the ratio of $n\nu/V$ is the total molar density (from all ions) of the solution, M, so Eq. 14.49 could also be written as

$$\Pi = \phi MRT$$

The osmotic coefficient, ϕ, is the ratio of the actual osmotic pressure to the ideal osmotic pressure of a pure solvent. It approaches 1.0 for dilute solutions. The osmotic coefficient is between 0.85 and 0.95 for most ionic compounds. For most organic compounds that do not dissociate, it is approximately 1.01–1.02.

Equation 14.50 Through Eq. 14.52: Salt Flux Through a Membrane

$$J_s = K_p(C_{\text{in}} - C_{\text{out}}) \qquad 14.50$$

$$J_s = (D_sK_s/\Delta Z)(C_{\text{in}} - C_{\text{out}}) \qquad 14.51$$

$$K_p = \frac{D_sK_s}{\Delta Z} \qquad 14.52$$

Description

In reverse osmosis, pure solvent (e.g., water) passes through the membrane, and ions of the solute (e.g., salt) are left behind. If the solute is allowed to build up on the membrane wall, the membrane will become fouled and processing will degrade. For that reason, the solute is continually removed from the membrane in the reject flow. Equation 14.50 calculates the *solute flux*, J_s (also called *salt flux*, *permeation rate*, *permeate rate*, *transport rate*, or *rejection rate*), which is the rate at which solute is removed in this flow. The solute flux is the molar flow rate per unit area (typical units are kmol/m^2·s). As indicated by Eq. 14.50, the solute flux is proportional to the difference between the salt concentrations on the two sides of the membrane, $C_{\text{in}} - C_{\text{out}}$, as well as to the membrane's *solute mass transfer coefficient*, K_p (also known as the salt *transport coefficient* and *permeability constant*, with typical units of m/s).[25] The solute flux is not greatly affected by the pressure difference.

Equation 14.52 calculates the mass transfer coefficient from three factors. The membrane thickness is ΔZ (typical units are m). D_s is the membrane's *diffusivity* (or *diffusion coefficient*) for that solute (typical units are

Environmental Engineering

[22]Membrane processes operate in a *fixed flux condition*. In order to keep the yield constant over time, the pressure must be constantly increased in order to compensate for the effects of fouling and compaction. Fouling by inorganic substances and biofilms is the biggest problem with membranes.
[23]The symbol i is more commonly used for the *van't Hoff factor*.
[24]The *NCEES Handbook* refers to V as the specific volume and gives it units of m^3/kmol, but this is incorrect. The ratio n/V has units of mol/m^3 or kmol/m^3.
[25]$C_{\text{in}} - C_{\text{out}}$ is the difference between concentrations on the two sides of the membrane. The "in" and "out" subscripts refer to "inside" and "outside", across the membrane. The subscripts would refer to the concentrations entering and leaving the desalination unit only if average conditions along the unit length are used.

m²/s). K_s is the dimensionless solute *distribution coefficient* (also known as the *partition constant*, *partition ratio*, and *distribution ratio*). The distribution coefficient is the ratio of solute concentration in the membrane to the feed or permeate (reject) concentrations, and is independent of concentrations.

$$K_s = \frac{C_{\text{solute,membrane,feed side}}}{C_{\text{solute,feed}}} = \frac{C_{\text{solute,membrane,permeate side}}}{C_{\text{solute,reject}}}$$

Equation 14.53 Through Eq. 14.55: Water Flux Across a Membrane

$$J_w = W_p(\Delta p - \Delta\pi) \qquad 14.53$$

$$\Delta p = p_{\text{in}} - p_{\text{out}} \qquad 14.54$$

$$\Delta\pi = \pi_{\text{in}} - \pi_{\text{out}} \qquad 14.55$$

Description

Equation 14.53 calculates the *solvent flux* (also called *water flux*), J_w, which is the molar solvent generation rate. (Its units depend on the units of pressure used; typical units are kmol/m²·s·Pa.) The solvent flux is proportional to the driving pressure across the membrane (known as the *transmembrane pressure difference* and *transmembrane hydraulic pressure*), Δp, less the osmotic pressure difference across the membrane, $\Delta\pi$.[26] The *coefficient of permeation*, W_p, is a function of the membrane used and can be calculated from its properties in a manner similar to Eq. 14.52. Depending on the units of pressure, typical units of the coefficient of permeation are kmol/m²·s.

Example

A reverse-osmosis unit is used to obtain pure water from saline water. The coefficient of water permeation is 0.1 mol/cm²·s·atm. The pressure differential across the membrane is 0.2 atm, and the osmotic pressure differential is 0.05 atm. The water flux across the membrane is most nearly

(A) 0.0050 mol/cm²·s

(B) 0.015 mol/cm²·s

(C) 0.050 mol/cm²·s

(D) 0.15 mol/cm²·s

Solution

From Eq. 14.53, the water flux is

$$\begin{aligned} J_w &= W_p(\Delta p - \Delta\pi) \\ &= \left(0.1\ \frac{\text{mol}}{\text{cm}^2\cdot\text{s}\cdot\text{atm}}\right)(0.2\ \text{atm} - 0.05\ \text{atm}) \\ &= 0.015\ \text{mol/cm}^2\cdot\text{s} \end{aligned}$$

The answer is (B).

[26]The *NCEES Handbook* uses both Π (see Eq. 14.49) and π (see Eq. 14.53 and Eq. 14.55) to represent osmotic pressure.

Table 14.8 CT Values For 3-Log Inactivation Of Giardia Cysts By Free Chlorine

CT value (min·mg/L)

chlorine concentration (mg/L)	temperature ≤ 0.5°C							temperature = 5°C							temperature = 10°C						
	pH							pH							pH						
	≤ 6.0	6.5	7.0	7.5	8.0	8.5	9.0	≤ 6.0	6.5	7.0	7.5	8.0	8.5	9.0	≤ 6.0	6.5	7.0	7.5	8.0	8.5	9.0
≤ 0.4	137	163	195	237	277	329	390	97	117	139	166	198	236	279	73	88	104	125	149	177	209
0.6	141	168	200	239	286	342	407	100	120	143	171	204	244	291	75	90	107	128	153	183	218
0.8	145	172	205	246	295	354	422	103	122	146	175	210	252	301	78	92	110	131	158	189	226
1.0	148	176	210	253	304	365	437	105	125	149	179	216	260	312	79	94	112	134	162	195	234
1.2	152	180	215	259	313	376	451	107	127	152	183	221	267	320	80	95	114	137	166	200	240
1.4	155	184	221	266	321	387	464	109	130	155	187	227	274	329	82	98	116	140	170	206	247
1.6	157	189	226	273	329	397	477	111	132	158	192	232	281	337	83	99	119	144	174	211	253
1.8	162	193	231	279	338	407	489	114	135	162	196	238	287	345	86	101	122	147	179	215	259
2.0	165	197	236	286	346	417	500	116	138	165	200	243	294	353	87	104	124	150	182	221	265
2.2	169	201	242	297	353	426	511	118	140	169	204	248	300	361	89	105	127	153	186	225	271
2.4	172	205	247	298	361	435	522	120	143	172	209	253	306	368	90	107	129	157	190	230	276
2.6	175	209	252	304	368	444	533	122	146	175	213	258	312	375	92	110	131	160	194	234	281
2.8	178	213	257	310	375	452	543	124	148	178	217	263	318	382	93	111	134	163	197	239	287
3.0	181	217	261	316	382	460	552	126	151	182	221	268	324	389	95	113	137	166	201	243	292

CT value (min·mg/L)

chlorine concentration (mg/L)	temperature = 15°C							temperature = 20°C							temperature = 25°C						
	pH							pH							pH						
	≤ 6.0	6.5	7.0	7.5	8.0	8.5	9.0	≤ 6.0	6.5	7.0	7.5	8.0	8.5	9.0	≤ 6.0	6.5	7.0	7.5	8.0	8.5	9.0
≤ 0.4	49	59	70	83	99	118	140	36	44	52	62	74	89	105	24	29	35	42	50	59	70
0.6	50	60	72	86	102	122	146	38	45	54	64	77	92	109	25	30	36	43	51	61	73
0.8	52	61	73	88	105	126	151	39	46	55	66	79	95	113	26	31	37	44	53	63	75
1.0	53	63	75	90	108	130	156	39	47	56	67	81	98	117	26	31	37	45	54	65	78
1.2	54	64	76	92	111	134	160	40	48	57	69	83	100	120	27	32	38	46	55	67	80
1.4	55	65	78	94	114	137	165	41	49	58	70	85	103	123	27	33	39	47	57	69	82
1.6	56	66	79	96	116	141	169	42	50	59	72	87	105	126	28	33	40	48	58	70	84
1.8	57	68	81	98	119	144	173	43	51	61	74	89	106	129	29	34	41	49	60	72	86
2.0	58	69	83	100	122	147	177	44	52	62	75	91	110	132	29	35	41	50	61	74	88
2.2	59	70	85	102	124	150	181	44	53	63	77	93	113	135	30	35	42	51	62	75	90
2.4	60	72	86	105	127	153	184	45	54	65	78	95	115	138	30	36	43	52	63	77	92
2.6	61	73	88	107	129	156	188	46	55	66	80	97	117	141	31	37	44	53	65	78	94
2.8	62	74	89	109	132	159	191	47	56	67	81	99	119	143	31	37	45	54	66	80	96
3.0	63	76	91	111	134	162	195	47	57	68	83	101	122	146	32	38	46	55	67	81	97

Source: *Guidance Manual LT1ESWTR Disinfection Profiling and Benchmarking*, U.S. Environmental Protection Agency, 2003.

15 Wastewater Collection and Treatment

Nomenclature

A	area	ft^2	m^2
Q	flow rate	ft^3/min	m^3/min
R	recycle ratio	–	–
S	BOD or COD concentration	mg/L	mg/L
Vol	volume	ft^3	m^3
X	solids concentration	mg/L	mg/L

Subscripts

0	influent
A	aeration tank
e	effluent
i	influent
M	medium
o	original
R	recycle

1. WASTEWATER TREATMENT PLANTS

For traditional *wastewater treatment plants* (WWTP), *preliminary treatment* of the wastewater stream is essentially a mechanical process intended to remove large objects, rags, and wood. Heavy solids and excessive oils and grease are also eliminated. Damage to pumps and other equipment would occur without preliminary treatment.

Odor control through chlorination or ozonation, freshening of septic waste by aeration, and flow equalization in holding basins can also be loosely categorized as preliminary processes.

After preliminary treatment, there are three "levels" of wastewater treatment: primary, secondary, and tertiary. *Primary treatment* is a mechanical (settling) process used to remove oil and most (i.e., approximately 50%) of the settleable solids. With domestic wastewater, a 25–35% reduction in BOD is also achieved, but BOD reduction is not the goal of primary treatment.

In the United States, secondary treatment is mandatory for all publicly owned wastewater treatment plants. *Secondary treatment* involves biological treatment in trickling filters, rotating contactors, biological beds, and activated sludge processes. Processing typically reduces the suspended solids and BOD content by more than 85%, volatile solids by 50%, total nitrogen by about 25%, and phosphorus by 20%.

Tertiary treatment (also known as *advanced wastewater treatment*, AWT) is targeted at specific pollutants or wastewater characteristics that have passed through previous processes in concentrations that are not allowed in the discharge. *Suspended solids* are removed by microstrainers or polishing filter beds. *Phosphorus* is removed by chemical precipitation. Aluminum and iron coagulants, as well as lime, are effective in removing phosphates. *Ammonia* can be removed by air stripping, biological denitrification, breakpoint chlorination, anion exchange, and algae ponds. Ions from *inorganic salts* can be removed by electrodialysis and reverse osmosis. The so-called *trace organics* or *refractory substances*, *dissolved organic solids* that are resistant to biological processes, can be removed by filtering through carbon or ozonation.

Equation 15.1 and Eq. 15.2: Organic Loading Rates (OLRs)

$$\text{organic loading rate (volumetric)} = Q_0S_0/\text{Vol} \quad (15.1)$$

$$\text{organic loading rate (surface area)} = Q_0S_0/A_M \quad (15.2)$$

Environmental Engineering

Description

Organic loading rate (OLR) is the rate at which organic materials are added to a system. The value is dependent on the hydraulic loading rate and the amount of organic material in the water source. Various units can be used.

Loading rates, including the organic loading rate, can be presented on either a volumetric or surface area basis. In Eq. 15.1 and Eq. 15.2, Q_0 and S_0 are the incoming volumetric flow rate and incoming BOD (or COD) concentration, respectively. Vol is the volume of the basin, tank, reactor, filter, or cell. In the case of a biological process, A_M is the surface area of the medium. In other processes, the plan area may be used. Any convenient set of units can be used, but concentrations are almost always reported in mg/L.[1] Conversions may be required to obtain the units of loading rate desired.

Example

A 50,000 ft^3 trickling filter has an influent BOD of 50 mg/L. The flow rate of the influent is 2×10^6 gal/day. What is most nearly the organic loading rate?

(A) 4.4×10^{-5} lbm/gal-day

(B) 2.2×10^{-3} lbm/gal-day

(C) 1.7×10^{-2} lbm/gal-day

(D) 2.7×10^{-1} lbm/gal-day

Solution

From Eq. 15.1,

$$\begin{aligned}\text{organic loading rate (volumetric)} &= Q_0S_0/\text{Vol}\\ &= \frac{\left(2\times10^6\ \frac{\text{gal}}{\text{day}}\right)\left(50\ \frac{\text{mg}}{\text{L}}\right)\left(3.785\ \frac{\text{L}}{\text{gal}}\right)}{(50{,}000\ \text{ft}^3)\left(7.48\ \frac{\text{gal}}{\text{ft}^3}\right)\left(454\ \frac{\text{g}}{\text{lbm}}\right)\times\left(10^3\ \frac{\text{mg}}{\text{g}}\right)}\\ &= 2.23\times10^{-3}\ \text{lbm/gal-day}\quad(2.2\times10^{-3}\ \text{lbm/gal-day})\end{aligned}$$

The answer is (B).

Equation 15.3: Solids Loading Rate[2]

$$\text{solids loading rate} = QX_A \qquad 15.3$$

Description

The solids loading rate is the surface loading rate of solids, usually for a sedimentation basin (clarifier). In Eq. 15.3, X is the solids concentration. In some cases, the solids concentration might be that of suspended inorganic solids. For biological processes, such as with activated sludge processes, X could be a biomass concentration (i.e., MLSS or MLVSS). Basically, the context determines which type of solids is of interest. The flow rate, Q, might be the influent flow rate, Q_0, but for an activated sludge process it would include the recycle flow, Q_R, if the solids loading rate of the secondary clarifier was being calculated.

2. FLOW EQUALIZATION

Equalization tanks or ponds are used to smooth out variations in flow that would otherwise overload wastewater processes. Graphical or tabular techniques similar to those used in reservoir sizing can be used to size equalization ponds. (See Fig. 15.1.) In practice, up to 25% excess capacity is added as a safety factor.

Figure 15.1 *Equalization Volume: Mass Diagram Method*

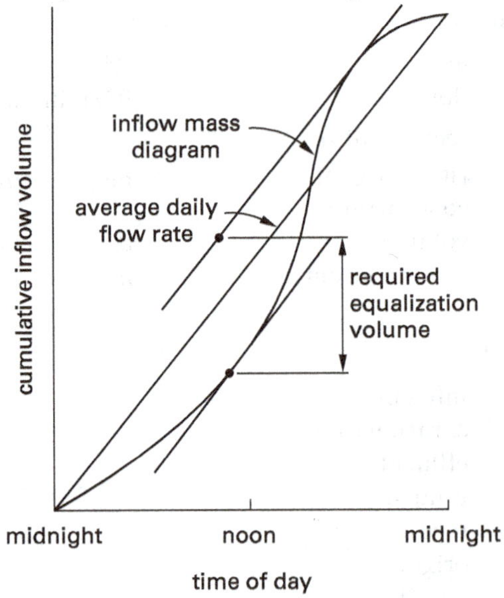

In a pure flow equalization process, there is no settling. Mechanical aerators provide the turbulence necessary to keep the solids in suspension while providing oxygen to prevent putrefaction.

[1]The NCEES *FE Reference Handbook* (*NCEES Handbook*) specifies the units of S_0 are kg/m^3.
[2]The *NCEES Handbook* is inconsistent in its definitions of loading rate. Organic loading rates are specified on an area or volumetric basis. (That is, the total amount is divided by the surface area or volume of the process.) BOD loading rate is similarly defined. However, solids loading rate is defined in Eq. 15.3 as a total amount over the entire process, which is inconsistent with values in the tables provided in the *NCEES Handbook*. Without a specification of units, it is not possible determine the meaning intended.

3. STABILIZATION PONDS

The term *stabilization pond* (*oxidation pond* or *stabilization lagoon*) refers to a pond used to treat organic waste by biological and physical processes. Aquatic plants, weeds, algae, and microorganisms stabilize the organic matter. The algae give off oxygen that is used by microorganisms to digest the organic matter. The microorganisms give off carbon dioxide, ammonia, and phosphates that the algae use. Even in modern times, such ponds may be necessary in remote areas (e.g., national parks and campgrounds). To keep toxic substances from leaching into the ground, ponds should not be used without strictly enforced industrial pretreatment requirements.

There are several types of stabilization ponds. *Aerobic ponds* are shallow ponds, less than 4 ft in depth, where dissolved oxygen is maintained throughout the entire depth, mainly by action of photosynthesis. *Facultative ponds* have an anaerobic lower zone, a facultative middle zone, and an aerobic upper zone. The upper zone is maintained in an aerobic condition by photosynthesis and, in some cases, mechanical aeration at the surface. *Anaerobic ponds* are so deep and receive such a high organic loading that anaerobic conditions prevail throughout the entire pond depth. *Maturation ponds* (*tertiary ponds* or *polishing ponds*) are used for polishing effluent from secondary biological processes. Dissolved oxygen is furnished through photosynthesis and mechanical aeration. *Aerated lagoons* are oxygenated through the action of surface or diffused air aeration. They are often used with activated sludge processes.

4. FACULTATIVE PONDS

Facultative ponds are the most common pond type selected for small communities. Approximately 25% of the municipal wastewater treatment plants in this country use ponds, and about 90% of these are located in communities of 5000 people or fewer. Long retention times and large volumes easily handle large fluctuations in wastewater flow and strength with no significant effect on effluent quality. Also, capital, operating, and maintenance costs are less than for other biological systems that provide equivalent treatment.

In a facultative pond, raw wastewater enters at the center of the pond. Suspended solids contained in the wastewater settle to the pond bottom, where an anaerobic layer develops. A facultative zone develops just above the anaerobic zone. Molecular oxygen is not available in the region at all times. Generally, the zone is aerobic during the daylight hours and anaerobic during the hours of darkness.

An aerobic zone with molecular oxygen present at all times exists above the facultative zone. Some oxygen is supplied from diffusion across the pond surface, but the majority is supplied through algal photosynthesis.

General guidelines are used to design facultative ponds. Ponds may be round, square, or rectangular. Usually, there are three cells, piped to permit operation in series or in parallel. Two of the three cells should be identical, each capable of handling half of the peak design flow. The third cell should have a minimum volume of one-third of the peak design flow.

Equation 15.4: Facultative Pond BOD Loading

$$M_{\text{lbm/day}} = C_{\text{mg/L}} \times Q_{\text{MGD}} \times 8.34 \text{ lbm-L/(mg-MG)} \qquad 15.4$$

Description

Equation 15.4 is used for the BOD loading of a facultative pond. To use Eq. 15.4, the total system must be less than or equal to 35 lbm BOD_5/acre-day, there must be a minimum of three ponds, the depth of the pond must be between 3 ft and 8 ft, and the minimum time must be between 90 days and 120 days.

5. AERATED LAGOONS

An *aerated lagoon* is a stabilization pond that is mechanically aerated. Such lagoons are typically deeper and have shorter detention times than nonaerated ponds. In warm climates and with floating aerators, one acre can support several hundred pounds of BOD per day.

The basis for the design of aerated lagoons is typically the organic loading and/or detention time. The detention time will depend on the desired BOD removal fraction and the reaction constant. Other factors that must be considered in the design process are solids removal requirements, oxygen requirements, temperature effects, and energy for mixing.

6. RACKS AND SCREENS

Trash racks or *coarse screens* with openings 2 in or larger should precede pumps to prevent clogging. *Medium screens* ($^1/_2$–$1^1/_2$ in openings) and *fine screens* ($^1/_{16}$–$^1/_8$ in) are also used to relieve the load on grit chambers and sedimentation basins. Fine screens are rare except when used with selected industrial waste processing plants. Screens in all but the smallest municipal plants are cleaned by automatic scraping arms. A minimum of two screen units is advisable.

Screen capacities and head losses are specified by the manufacturer. Although the flow velocity must be sufficient to maintain sediment in suspension, the approach velocity should be limited to 3 ft/sec to prevent debris from being forced through the screen.

7. GRIT CHAMBERS

Abrasive *grit* can erode pumps, clog pipes, and accumulate in excessive volumes. In a *grit chamber* (also known as a *grit clarifier* or *detritus tank*), the wastewater is slowed, allowing the grit to settle out but allowing the organic matter to continue through. Grit can be manually or mechanically removed with buckets or screw conveyors. A minimum of two units is needed.

Horizontal flow grit chambers are designed to keep the flow velocity as close to 1 ft/sec as possible. If an analytical design based on settling velocity is required, the *scouring velocity* should not be exceeded. Scouring is the dislodging of particles that have already settled. Scouring velocity is not the same as settling velocity.

8. AERATED GRIT CHAMBERS

An *aerated grit chamber* is a bottom-hoppered tank, as shown in Fig. 15.2, with a short detention time. Diffused aeration from one side of the tank rolls the water and keeps the organics in suspension while the grit drops into the hopper. The water spirals or rolls through the tank. Influent enters through the side, and degritted wastewater leaves over the outlet weir. A minimum of two units is needed.

Figure 15.2 *Aerated Grit Chamber*

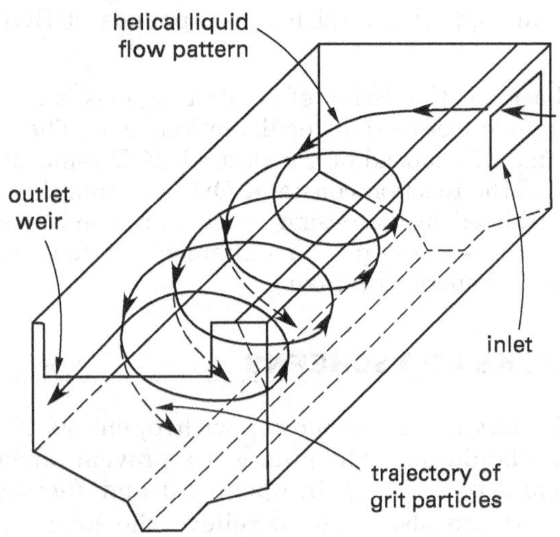

Solids are removed by pump, screw conveyer, bucket elevator, or gravity flow. However, the grit will have a significant organic content. A grit washer or cyclone separator can be used to clean the grit.

9. SHREDDERS

Shredders (also called *comminutors*) cut waste solids to approximately 1/4 in in size, reducing the amount of screenings that must be disposed of. Shreddings stay with the flow for later settling.

10. CHEMICAL SEDIMENTATION BASINS/ CLARIFIERS

Chemical flocculation (*clarification* or *coagulation*) operations in chemical sedimentation basins are similar to those encountered in the treatment of water supplies except that the coagulant doses are greater. Chemical precipitation may be used when plain sedimentation is insufficient, or occasionally when the stream into which the outfall discharges is running low, or when there is a large increase in sewage flow. As with water treatment, the five coagulants used most often are (a) aluminum sulfate, $Al_2(SO_4)_3$; (b) ferric chloride, $FeCl_3$; (c) ferric sulfate, $Fe_2(SO_4)_3$; (d) ferrous sulfate, $FeSO_4$; and (e) chlorinated copperas. Lime and sulfuric acid may be used to adjust the pH for proper coagulation.

11. TRICKLING FILTERS

Trickling filters (also known as *biological beds* and *fixed media filters*) consist of beds of rounded river rocks with approximate diameters of 2–5 in, wooden slats, or modern synthetic media. Wastewater from primary sedimentation processing is sprayed intermittently over the bed. The biological and microbial slime growth attached to the bed purifies the wastewater as it trickles down. The water is introduced into the filter by rotating arms that move by virtue of spray reaction (reaction-type) or motors (motor-type). The clarified water is collected by an underdrain system.

The distribution rate is sometimes given by an *SK rating*, where SK is the water depth in mm deposited per pass of the distributor. Though there is strong evidence that rotational speeds of 1–2 rev/hr (high SK) produce significant operational improvement, traditional distribution arms revolve at 1–5 rev/min (low SK).

On the average, one acre of *low-rate filter* (also referred to as *standard-rate filter*) is needed for each 20,000 people served. Trickling filters can remove 70–90% of the suspended solids, 65–85% of the BOD, and 70–95% of the bacteria. Although low-rate filters have rocks to a depth of 6 ft, most of the reduction occurs in the first few feet of bed, and organisms in the lower part of the bed may be in a near-starvation condition.

Due to the low concentration of carbonaceous material in the water near the bottom of the filter, nitrogenous bacteria produce a highly nitrified effluent from low-rate filters. With low-rate filters, the bed will periodically slough off (unload) parts of its slime coating. Therefore, sedimentation after filtering is necessary. *Filter flies* are a major problem with low-rate filters, since fly larvae are provided with an undisturbed environment in which to breed.

Since there are limits to the heights of trickling filters, longer contact times can be achieved by returning some of the collected filter water back to the filter. This is known as *recirculation* or *recycling*. Recirculation is also used to keep the filter medium from drying out and to smooth out fluctuations in the hydraulic loading.

High-rate filters are used in most facilities. The higher hydraulic loading flushes the bed and inhibits excess biological growth. High-rate stone filters may be only 3–6 ft deep. The high rate is possible because much of the filter discharge is recirculated. With the high flow rates, fly larvae are washed out, minimizing the filter fly problem. Since the biofilm is less thick and provided with carbon-based nutrients at a high rate, the effluent is nitrified only when the filter experiences low loading.

Super high-rate filters (*oxidation towers*) using synthetic media may be up to 40 ft tall. High-rate and super high-rate trickling filters may be used as *roughing filters*, receiving wastewater at high hydraulic or organic loading and providing intermediate treatment or the first step of a multistage biological treatment process.

12. TWO-STAGE TRICKLING FILTERS

If a higher BOD or solids removal fraction is needed, then two filters can be connected in series with an optional intermediate settling tank to form a *two-stage filter* system (see Fig. 15.3). The efficiency of the second-stage filter is considerably less than that of the first-stage filter because much of the biological food has been removed from the flow.

Figure 15.3 *Trickling Filter Process*

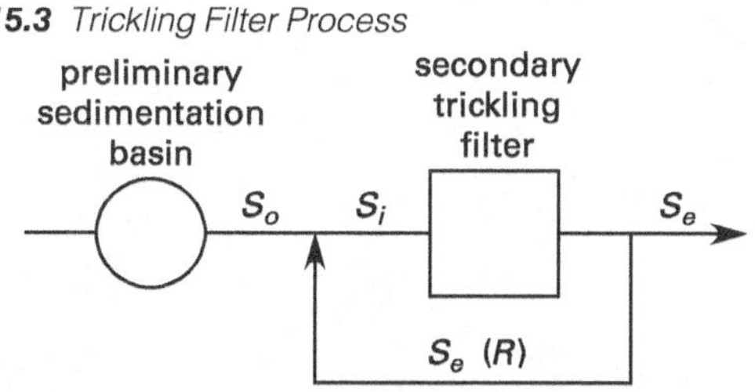

13. ROTATING BIOLOGICAL CONTACTORS

Rotating biological contactors, RBCs (also known as *rotating biological reactors*), consist of large-diameter plastic disks, partially immersed in wastewater, on which biofilm is allowed to grow. The disks are mounted on shafts that turn slowly. The rotation progressively wets the disks, alternately exposing the biofilm to organic material in the wastewater and to oxygen in the air. The biofilm population, since it is well oxygenated, efficiently removes organic solids from the wastewater. RBCs are primarily used for carbonaceous BOD removal, although they can also be used for nitrification or a combination of both.

The primary design criterion is hydraulic loading, not organic (BOD) loading. For a specific hydraulic loading, the BOD removal efficiency will be essentially constant, regardless of variations in BOD.

RBC operation is more efficient when several stages are used. Recirculation is not common with RBC processes. The process can be placed in series or in parallel with existing trickling filter or activated sludge processes.

14. SAND FILTERS

For small populations, a *slow sand filter* (*intermittent sand filter*) can be used. Because of the lower flow rate, the filter area per person is higher than in the case of a trickling filter. Roughly one acre is needed for each 1000 people. The filter is constructed as a sand bed 2–3 ft deep over a 6–12 in gravel bed. The filter is alternately exposed to water from a settling tank and to air (hence the term intermittent). Straining and aerobic decomposition clean the water. Application rates are usually 2–2.5 gal/day-ft^2. Up to 95% of the BOD can be satisfied in an intermittent sand filter. The filter is cleaned by removing the top layer of clogged sand.

If the water is applied continuously as a final process following secondary treatment, the filter is known as a *polishing filter* or *rapid sand filter*. The water rate of a polishing filter is typically 2–8 gal/min-ft^2 but may be as high as 10 gal/min-ft^2. Although the designs are similar to those used in water supply treatment, coarser media are used since the turbidity requirements are less stringent. Backwashing is required more frequently and is more aggressive than with water supply treatment.

15. PHOSPHORUS REMOVAL: PRECIPITATION

Phosphorus concentrations of 5–15 MG/L (as P) are experienced in untreated wastewater, most of which originates from synthetic detergents and human waste. Approximately 10% of the total phosphorus is insoluble and can be removed in primary settling. The amount that is removed by absorption in conventional biological processes is small. The remaining phosphorus is soluble and must be removed by converting it into an insoluble precipitate.

Soluble phosphorus is removed by precipitation and settling. Aluminum sulfate, ferric chloride ($FeCl_3$), and lime may be used depending on the nature of the phosphorus radical. Aluminum sulfate is more desirable since lime reacts with hardness and forms large quantities of additional precipitates. (Hardness removal is not as important as in water supply treatment.) However, the process requires about 10 lbm of aluminum sulfate for each kilogram of phosphorus removed. The process also produces a chemical sludge that is difficult to dewater, handle, and dispose of.

Due to the many other possible reactions the compounds can participate in, the dosage should be determined from testing. The stoichiometric chemical

Environmental Engineering

reactions describe how the phosphorus is removed, but they do not accurately predict the quantities of coagulants needed.

16. AMMONIA REMOVAL: AIR STRIPPING

Ammonia may be removed by either biological processing or air stripping. In the biological *nitrification and denitrification process*, ammonia is first aerobically converted to nitrite and then to nitrate (nitrification) by bacteria. Then, the nitrates are converted to nitrogen gas, which escapes (denitrification).

In the *air-stripping* (*ammonia-stripping*) method, lime is added to water to increase its pH to about 10. This causes the ammonium ions, NH_4^+, to change to dissolved ammonia gas, NH_3. The water is passed through a packed tower into which air is blown at high rates. The air strips the ammonia gas out of the water. Recarbonation follows to remove the excess lime.

17. CARBON ADSORPTION

Adsorption uses high surface-area activated carbon to remove organic contaminants. Adsorption can use *granular activated carbon* (GAC) in column or fluidized-bed reactors or *powdered activated carbon* (PAC) in complete-mix reactors. Activated carbon is relatively nonspecific, and it will remove a wide variety of refractory organics as well as some inorganic contaminants. It should generally be considered for organic contaminants that are nonpolar, have low solubility, or have high molecular weights.

The most common problems associated with columns are breakthrough, excessive headloss due to plugging, and premature exhaustion. *Breakthrough* occurs when the carbon becomes saturated with the target compound and can hold no more. *Plugging* occurs when biological growth blocks the spaces between carbon particles. *Premature exhaustion* occurs when large and high molecular-weight molecules block the internal pores in the carbon particles. These latter two problems can be prevented by locating the carbon columns downstream of filtration.

18. CHLORINATION

Chlorination to disinfect and deodorize is one of the final steps prior to discharge. Vacuum-type feeders are used predominantly with chlorine gas. Chlorine under vacuum is combined with wastewater to produce a chlorine solution. A *flow-pacing* chlorinator will reduce the chlorine solution feed rate when the wastewater flow decreases (e.g., at night).

The size of the *contact tank* varies, depending on economics and other factors. An average design detention time is 30 min at average flow, some of which can occur in the plant outfall after the contact basin. Contact tanks are baffled to prevent short-circuiting that would otherwise reduce chlorination time and effectiveness.

Alternatives to disinfection by chlorine include sodium hypochlorite, ozone, ultraviolet light, bromine (as bromine chloride), chlorine dioxide, and hydrogen peroxide. All alternatives have one or more disadvantages when compared to chlorine gas.

19. DECHLORINATION

Toxicity, by-products, and strict limits on *total residual oxidants* (TROs) now make dechlorination mandatory at many installations. Sulfur dioxide (SO_2) and sodium thiosulfite ($Na_2S_2SO_3$) are the primary compounds used as dechlorinators today. Other compounds seeing limited use are sodium metabisulfate ($Na_2S_2O_5$), sodium bisulfate ($NaHSO_3$), sodium sulfite (Na_2SO_3), hydrogen peroxide (H_2O_2), and granular activated carbon.

Though reaeration was at one time thought to replace oxygen in water depleted by sulfur dioxide, this is now considered to be unnecessary unless required to meet effluent discharge requirements.

16 Activated Sludge and Sludge Processing

Nomenclature

a	mass flow rate	lbm/day	kg/d
D	depth	ft	m
D	dissolved oxygen deficit	mg/L	mg/L
DO	dissolved oxygen	mg/L	mg/L
F	food arrival rate	mg/day	mg/d
F:M	food-to-microorganism ratio	day^{-1}	d^{-1}
G	fraction of solids that is biodegradable	–	–
k	first-order reaction constant	–	–
k	treatability constant	–	–
k_d	microbial death ratio	day^{-1}	d^{-1}
K_t	oxygen transfer coefficient	1/hr	1/h
M	mass	lbm	kg
MLSS	total mixed liquor suspended solids	mg/L	mg/L
p_1	absolute inlet pressure	lbf/in^2	Pa
p_2	absolute outlet pressure	lbf/in^2	Pa
P_v	volatile fraction of digester suspended solids	–	–
P_W	power requirement	hp	kW
q	hydraulic loading	ft^3/ft^2-day	$\text{m}^3/\text{m}^2{\cdot}\text{d}$
Q	flow rate	ft^3/day	m^3/d
R	recycle ratio	–	–
s	gravimetric fractional solids content	–	–
S	growth-limited substrate concentration	mg/L	mg/L
SVI	sludge volume index	mL/g	mL/g
t	time	day	d
TSS	total suspended solids	mg/L	mg/L
V	volume	ft^3	m^3
$\dot{V}$	volumetric flow rate	ft^3/day	m^3/d
Vol	volume	ft^3	m^3
X	biomass concentration	mg/L	mg/L
X	mixed liquor volatile suspended solids	mg/L	mg/L
Y	yield coefficient	lbm/lbm	mg/mg

Symbols

β	oxygen saturation coefficient	–	–
η	efficiency	–	–
θ	residence/detention time	day	d
μ_m	maximum specific growthrate	day^{-1}	d^{-1}
ρ	density	lbm/ft^3	kg/m^3

Subscripts

0	influent
5	5-day
a	applied
A	aeration tank
c	cell
e	effluent
r	residence
R	recycle
s	sludge, solids, or storage
t	thickening or transfer
u	ultimate
v	volatile
w	wasted

1. SLUDGE

Sludge is the mixture of water, organic and inorganic solids, and treatment chemicals that accumulates in settling tanks. The term is also used to refer to the dried residue (screenings, grit, filter cake, and drying bed scraping) from separation and drying processes, although the term *biosolids* is becoming more common in this regard. (The term *residuals* is also used, though this term more commonly refers to sludge from water treatment plants.)

2. ACTIVATED SLUDGE PROCESS

The *activated sludge process* is a secondary biological wastewater treatment process in which a mixture of wastewater and sludge solids is aerated. (See Fig. 16.1.) The sludge mixture produced during this oxidation process contains an extremely high concentration of aerobic bacteria, most of which are near starvation. This condition makes the sludge an ideal medium for the destruction of any organic material in the mixture. Since the bacteria are voraciously active, the sludge is called *activated sludge.*

The well-aerated mixture of wastewater and sludge, known as *mixed liquor*, flows from the aeration tank to a secondary clarifier, where the sludge solids settle out. Most of the settled sludge solids are returned to the aeration tank in order to maintain the high population of bacteria needed for rapid breakdown of the organic material. However, because more sludge is produced than is needed, some of the return sludge is diverted ("wasted") for subsequent treatment and disposal. This wasted sludge is referred to as *waste activated sludge*, WAS. The volume of sludge returned to the aeration basin is typically 20–30% of the wastewater flow. The liquid fraction removed from the secondary clarifier weir is chlorinated and discharged.

Though diffused aeration and mechanical aeration are the most common methods of oxygenating the mixed liquor, various methods of staging the aeration are used, each having its own characteristic ranges of operating parameters. (See Table 16.1 for additional information.)

In a traditional activated sludge plant using conventional aeration, the wastewater is typically aerated for 6–8 hours in long, rectangular aeration basins. Sufficient air, about eight volumes for each volume of wastewater treated, is provided to keep the sludge in suspension. The air is injected near the bottom of the aeration tank through a system of diffusers.

Equation 16.1: Blower Power Requirement

$$P_W = \frac{WRT_1}{Cne}\left[\left(\frac{p_2}{p_1}\right)^{0.283} - 1\right] \qquad 16.1$$

Values

e is the efficiency, and is usually between 0.70 and 0.90.

Description

Air for aeration in an activated sludge process is provided by compressors, commonly referred to as *blowers*. The *blower power* can be found from Eq. 16.1.[1] Equation 16.1 is derived from basic thermodynamic principles and can be used with SI or U.S. units. R is the specific gas constant for air, 53.5 ft-lbf/lbm-°R or 287 J/kg·K. The standard ratio of specific heats is 1.4 for air. n is defined by

$$n = \frac{k-1}{k}$$

n is 0.283 for air. When using U.S. units, C has a value of 550 ft-lbf/sec-hp, and the resulting power will be in horsepower.[2]

3. SECONDARY CLARIFIERS

Secondary clarifiers are used in activated sludge processes to separate and thicken sludge, enabling recycling of thickened, activated sludge. Secondary clarifiers are essentially sedimentation basins that follow activated sludge processes. Typically, a fraction of the settled sludge is returned to the aeration basin and the remainder is wasted. The operating principles of the secondary clarifier are the same as those of a sedimentation basin.

Equation 16.2: Activated Sludge Secondary Clarifiers

$$Q = Q_0 + Q_R \qquad 16.2$$

[1]In Eq. 16.1, the subscript W stands for wire (not water or work or watts), implying that this is the amount of electrical power drawn. (Other engineering sources use the subscript "purchased" for the concept of electrical power drawn from the electrical grid.) The quantity $(k-1)/k$ derived from ideal gas relationships for adiabatic compression has been replaced with n in the denominator, while the actual value of $(k-1)/k$ for air, 0.283, has been used as an exponent. (This value uses $k = 1.395$. In fact, if the traditional value of $k = 1.4$ is used to calculate n, the result will be 0.2857, not 0.283, so an inconsistency is inevitable unless the exact value of k is used. The true value of k depends on the compression temperature.) As it is likely that the only substance being used for aeration is air, either 0.283 or n should be used consistently. The variable R is defined as the "engineering" gas constant instead of the specific gas constant. The use of e as the symbol for efficiency is inconsistent with other equations in the NCEES *FE Reference Handbook* (*NCEES Handbook*), but that is because it represents the *standard aeration efficiency* (also known as the *wire-to-air efficiency*), not a mechanical or electrical efficiency. The variable W is defined as a weight flow, not a weight flow rate. In any case, the correct definition is mass flow rate, because a weight flow rate (obtained by multiplying a mass flow rate by g) doesn't work in the SI version of the equation. The implicit assumption that compression by the blower is adiabatic (isentropic) has not been stated. In practice, compression is never adiabatic.

[2]In Eq. 16.1, the *NCEES Handbook* specifies a value of 29.7 for C as a "constant for SI unit conversion." This number is close to the molecular weight of air (29.0), which would require the value of the specific gas constant to be given in kJ/kmol·K instead of kJ/kg·K.

Figure 16.1 *Typical Activated Sludge Plant*

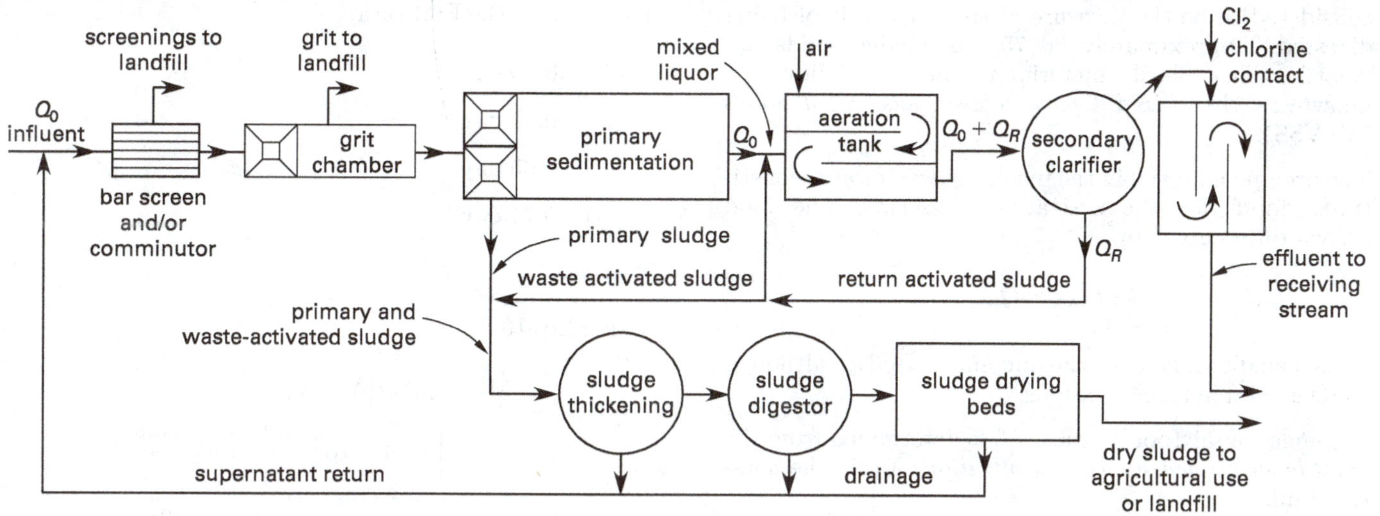

Table 16.1 *Design and Operational Parameters for Activated-Sludge Treatment of Municipal Wastewater*

type of process	mean cell residence time, θ_c (d)	food-to-mass ratio [(kg BOD_5/(d·kg MLSS)]	volumetric loading (kg BOD_5/m^3)	hydraulic residence time in aeration basin, θ (h)	mixed liquor suspended solids, MLSS (mg/L)	recycle ratio Q_R/Q	flow regime*	BOD_5 removal efficiency (%)	air supplied (m^3/kg BOD_5)
tapered aeration	5–15	0.2–0.4	0.3–0.6	4–8	1500–3000	0.25–0.5	PF	85–95	45–90
conventional	4–15	0.2–0.4	0.3–0.6	4–8	1500–3000	0.25–0.5	PF	85–95	45–90
step aeration	4–15	0.2–0.4	0.6–1.0	3–5	2000–3500	0.25–0.75	PF	85–95	45–90
completely mixed	4–15	0.2–0.4	0.8–2.0	3–5	3000–6000	0.25–1.0	CM	85–95	45–90
contact stabilization	4–15	0.2–0.6	1.0–1.2			0.25–1.0			45–90
contact basin				0.5–1.0	1000–3000		PF	80–90	
stabilization basin				4–6	4000–10 000		PF		
high-rate aeration	4–15	0.4–1.5	1.6–16	0.5–2.0	4000–10 000	1.0–5.0	CM	75–90	25–45
pure oxygen	8–20	0.2–1.0	1.6–4	1–3	6000–8000	0.25–0.5	CM	85–95	
extended aeration	20–30	0.05–0.15	0.16–0.40	18–24	3000–6000	0.75–1.50	CM	75–90	90–125

*PF = plug flow, CM = completely mixed.

Description

Equation 16.2 is used to calculate the hydraulic loading rate for activated sludge secondary clarifiers.

Example

An activated sludge secondary clarifier has a hydraulic loading rate of 290,000 ft^3/day and an observed recycle flow rate of 160,000 ft^3/day. What is most nearly the influent flow rate?

(A) 110,000 ft^3/day

(B) 130,000 ft^3/day

(C) 170,000 ft^3/day

(D) 190,000 ft^3/day

Solution

From Eq. 16.2, the influent flow rate is

$$Q_0 = Q - Q_R = 290{,}000\ \frac{\text{ft}^3}{\text{day}} - 160{,}000\ \frac{\text{ft}^3}{\text{day}}$$
$$= 130{,}000\ \text{ft}^3/\text{day}$$

The answer is (B).

4. SLUDGE PARAMETERS

The bacteria and other suspended material in the mixed liquor is known as *mixed liquor suspended solids* (MLSS) and is measured in mg/L. Suspended solids are further divided into fixed solids and volatile solids. *Fixed solids* (also referred to as *nonvolatile solids*) are those inert solids that are left behind after being fired in a furnace. *Volatile solids* are essentially those carbonaceous solids that are consumed in the furnace. The volatile solids are considered to be the measure of solids capable of being digested. Approximately 60–75% of sludge solids are volatile. The volatile material in the mixed liquor is known as the *mixed liquor volatile suspended solids* (MLVSS).

The organic material in the incoming wastewater constitutes "food" for the activated organisms. The food arrival rate is given by

$$F = S_0 Q_0$$

S_0 is usually taken as the incoming BOD_5, although COD is used in rare situations.

The mass of microorganisms, M, is determined from the *volatile suspended solids concentration*, X_A, in the aeration tank.

$$M = V_A X_A$$

Equation 16.3 and Eq. 16.4: Food-to-Microorganism Ratio and Hydraulic Detention Time

$$\text{F:M} = Q_0 S_0 / (\text{Vol } X_A) \qquad 16.3$$

$$\theta = V/Q \qquad 16.4$$

Variation

$$\text{F:M} = \frac{S_o}{\theta X_A}$$

Description

The *food-to-microorganism ratio*, F:M, is given by Eq. 16.3. θ in Eq. 16.4 is the *hydraulic detention time*, also called the *hydraulic residence time*. The hydraulic detention time is given in units of days and is the average time that a water molecule will spend in the reactor. Hydraulic detention time is one of the critical parameters for sizing the aeration basin.

In Eq. 16.3, Q_0 and S_0 are the incoming volumetric flow rate and incoming BOD (or COD) concentration, respectively. Vol is the volume of the basin, tank, reactor, filter, or cell. X_A is the biomass concentration (i.e., MLSS or MLVSS) in the aeration tank.

Example

A wastewater reactor has an influent flow rate of 1.36×10^5 ft^3, an influent BOD_5 concentration of 400 mg/L, a volume of 1.7×10^5 ft^3, and a mixed liquor suspended solids concentration of 4500 mg/L. What is most nearly the F:M ratio?

(A) 0.49 day^{-1}

(B) 0.57 day^{-1}

(C) 0.65 day^{-1}

(D) 0.71 day^{-1}

Solution

Using Eq. 16.3,

$$\begin{aligned}\text{F:M} &= Q_0 S_0 / (\text{Vol } X_A) \\ &= \frac{\left(1.36 \times 10^6 \ \frac{\text{ft}^3}{\text{day}}\right)\left(400 \ \frac{\text{mg}}{\text{L}}\right)}{(1.7 \times 10^5 \ \text{ft}^3)\left(4500 \ \frac{\text{mg}}{\text{L}}\right)} \\ &= 0.71 \ \text{day}^{-1}\end{aligned}$$

The answer is (D).

Equation 16.5: Mean Cell Residence Time

$$\theta_c = \frac{V(X_A)}{Q_w X_w + Q_e X_e} \qquad 16.5$$

Description

The liquid fraction of the wastewater passes through an activated sludge process in a matter of hours, the *detention time*, θ. However, the sludge solids are recycled continuously and have an average stay much longer in duration. There are two measures of *sludge age*: the *mean cell residence time* (also known as the *age of the suspended solids* and *solids residence time*), essentially the age of the microorganisms, θ_c, and *age of the BOD*, essentially the age of the food. For conventional aeration, typical values of the mean cell residence time, θ_c, are 6–15 days for high-quality effluent and sludge.

Equation 16.6 and Eq. 16.7: Sludge Volume

$$Q_s = \frac{M(100)}{\rho_s(\%\ \text{solids})} \qquad 16.6$$

$$\text{SVI} = \frac{\text{sludge volume after settling (mL/L)} * 1000}{\text{MLSS (mg/L)}} \qquad 16.7$$

Description

The liquid (wet) sludge volume generation rate, Q_s, is calculated from Eq. 16.6.[3] This is basically a mass over sludge density calculation modified by the gravimetric percentage of solids in the liquid sludge. The solids percentage is typically less than 3%, although it may be as high as 10% for primary sedimentation sludge. M is the dry mass of sludge solids produced per unit time. ρ_s is the wet sludge density, which will be slightly greater than the density of water due to the solids content.[4] The factor of 100 in the numerator is not needed if the solids fraction is reported in decimal form.

The *sludge volume index*, SVI, is a measure of the sludge's settleability. SVI can be used to determine the tendency toward *sludge bulking*. (See Sec. 16.9.) SVI is determined by taking 1 L of mixed liquor and measuring the volume of settled solids after 30 minutes. SVI is the volume in mL occupied by 1 g of settled volatile and nonvolatile suspended solids.

The concentration of total suspended solids, TSS, in the recirculated sludge can be found from

$$\text{TSS}_{\text{mL/g}} = \frac{\left(1000\ \frac{\text{mg}}{\text{g}}\right)\left(1000\ \frac{\text{mL}}{\text{L}}\right)}{\text{SVI}_{\text{mL/g}}}$$

This equation is useful in calculating the solids concentration for any sludge that is wasted from the return sludge line. (Suspended sludge solids wasted include both fixed and volatile portions.)

Example 14

A sample of influent wastewater has a total suspended solids concentration of 4000 mg/L, which settles to a volume of 325 mL in 30 min in a 1 L cylinder. What is most nearly the sludge volume index?

(A) 0.081 mL/g

(B) 0.81 mL/g

(C) 8.1 mL/g

(D) 81 mL/g

Solution

From Eq. 16.7,

$$\begin{aligned}\text{SVI} &= \frac{\text{sludge volume after settling (mL/L)} * 1000}{\text{MLSS (mg/L)}} \\ &= \frac{\left(325\ \frac{\text{mL}}{\text{L}}\right)\left(1000\ \frac{\text{mg}}{\text{g}}\right)}{4000\ \frac{\text{mg}}{\text{L}}} \\ &= 81\ \text{mL/g}\end{aligned}$$

The answer is (D).

Equation 16.8: Average Concentration of Microorganisms in Aeration Tank

$$X_A = \frac{\theta_c Y(S_0 - S_e)}{\theta(1 + k_d\theta_c)} \qquad 16.8$$

Description

The biomass concentration (concentration of microorganisms), X_A, in the aeration tank is given by Eq. 16.8.

The liquid in the aeration tank is referred to as *mixed liquor*. The mixed liquor carries the organic solids in a suspension, hence the term mixed liquor suspended solids, MLSS. Biological organisms are able to access (process) only a portion (70–80%) of the MLSS known as the *volatile solids*. The nonvolatile solids are inorganic material that cannot be metabolized. Therefore, the *mixed liquor volatile suspended solids*, MLVSS, concentration is more important in calculating the food-to-microorganism ratio.

In Eq. 16.8, θ_c is the cell residence time, and θ is the hydraulic residence time. S_0 and S_e are the incoming and exit BOD (or COD, depending on the context) concentrations. Y is the yield coefficient, a dimensionless ratio of mass of MLVSS (or MLSS) biomass produced per mass of BOD (or COD) consumed. Y is typically in the range of 0.4–1.2. k_d is a kinetic constant representing the *microbial death ratio*, also known as the *endogenous decay coefficient*, the fraction of microorganisms that die off in the designated time period, with typical values between 0.01 and 0.1 per day.

Example 15

Activated sludge is treated in an aeration tank. The hydraulic residence time is 4 hr, and the solids residence time is 5 days. The influent BOD concentration is 190 mg/L, and the effluent BOD concentration is 4.0 mg/L. With a yield coefficient of 1.2 and a microbial

Environmental Engineering

[3]The *NCEES Handbook* defines Q_s as the daily sludge production rate, but this will depend on the units used for the dry sludge production rate, M.
[4]The *NCEES Handbook* uses the same symbol, ρ_s, for both dry solids density and wet sludge density. In the case of Eq. 16.6, the wet sludge density is required.

death ratio of 0.2 day^{-1}, what is most nearly the biomass concentration in the tank?

(A) 1800 mg/L

(B) 2700 mg/L

(C) 3300 mg/L

(D) 4100 mg/L

Solution

From Eq. 16.8, the biomass concentration is

$$X_A = \frac{\theta_c Y(S_0 - S_e)}{\theta(1 + k_d\theta_c)}$$

$$= \frac{(5 \text{ days})(1.2) \times \left(190 \ \frac{\text{mg}}{\text{L}} - 4.0 \ \frac{\text{mg}}{\text{L}}\right)\left(24 \ \frac{\text{hr}}{\text{day}}\right)}{(4 \text{ hr})\left(1 + (0.2 \text{ day}^{-1})(5 \text{ days})\right)}$$

$$= 3348 \text{ mg/L} \quad (3300 \text{ mg/L})$$

The answer is (C).

5. ATMOSPHERIC AIR

Atmospheric air is a mixture of oxygen, nitrogen, and small amounts of carbon dioxide, water vapor, argon, and other inert gases. If all constituents except oxygen are grouped with the nitrogen, the air composition is as given in Table 16.2. It is necessary to supply by weight $1/0.2315 = 4.32$ masses of air to obtain one mass of oxygen. The average molecular weight of air is 28.9.

***Table 16.2** Composition of Air*[a]

component	fraction by mass	fraction by volume
oxygen	0.2315	0.209
nitrogen	0.7685	0.791
ratio of nitrogen to oxygen	3.320	3.773[b]
ratio of air to oxygen	4.320	4.773

[a]Inert gases and CO_2 are included in N_2.
[b]This value is also reported by various sources as 3.76, 3.78, and 3.784.

6. AERATION TANKS

The *aeration period* is the same as the *hydraulic detention time* and is calculated without considering recirculation.

$$\theta = \frac{V_A}{Q_0}$$

The organic (BOD) *volumetric loading* rate (with units of kg BOD_5/d·m^3) for the aeration tank is given by

$$\text{organic loading rate} = \frac{Q_o S_o}{\text{Vol}}$$

In the United States, volumetric loading is often specified in lbm/day-1000 ft^3.

7. RECYCLE RATIO AND RECIRCULATION RATE

For any given wastewater and treatment facility, the influent flow rate, BOD, and tank volume cannot be changed. Of all the variables appearing in Eq. 16.3, only the suspended solids can be controlled. Therefore, the primary process control variable is the amount of organic material in the wastewater. This is controlled by the sludge return rate.

Equation 16.9 and Eq. 16.10: Recycle Ratio and Recycle Flow Rate

$$R = Q_R/Q_0 \qquad 16.9$$

$$Q_R = Q_0 R \qquad 16.10$$

Values

The recycle ratio, R, is typically 0.20–0.30.

Description

The sludge *recycle ratio*, R, is given by Eq. 16.9. Equation 16.10 gives the sludge return (recycle flow) rate.

Example

A wastewater treatment process has a recycle flow rate of 0.263×10^6 ft^3/day and a recycle ratio of 0.30. What is most nearly the influent flow rate?

(A) 0.88×10^6 ft^3/day

(B) 1.1×10^6 ft^3/day

(C) 2.0×10^6 ft^3/day

(D) 2.2×10^6 ft^3/day

Solution

From Eq. 16.10, the influent flow rate is

$$Q_R = Q_0 R$$

$$Q_0 = \frac{Q_R}{R} = \frac{0.263\times 10^6\ \frac{\text{ft}^3}{\text{day}}}{0.30} = 876{,}667\ \text{ft}^3/\text{day} \quad (0.88\times 10^6\ \text{ft}^3/\text{day})$$

The answer is (A).

Equation 16.11: Suspended Solids Mass Balance

$$(Q_0 + Q_R)X_A = Q_eX_e + Q_RX_r + Q_wX_w \qquad 16.11$$

Description

The actual *recirculation rate* can be found by writing a suspended solids mass balance around the inlet to the reactor.

8. SLUDGE WASTING

The activated sludge process is a biological process. Influent brings in BOD ("biomass" or organic food), and the return activated sludge (RAS) brings in microorganisms (bacteria) to digest the BOD. The food-to-microorganism ratio (F:M) is balanced by adjusting the aeration and how much sludge is recirculated and how much is wasted. As the microorganisms grow and are mixed with air, they clump together (flocculate) and are readily settled as sludge in the secondary clarifier. More sludge is produced in the secondary clarifier than is needed, and without wasting, the sludge would continue to build up.

Sludge wasting is the main control of the effluent quality and microorganism population size. The wasting rate affects the mixed liquor suspended solids (MLSS) concentration and the mean cell residence time. Excess sludge may be wasted either from the sludge return line or directly from the aeration tank as mixed liquor. Wasting from the aeration tank is preferred, as the sludge concentration is fairly steady in that case. The wasted volume is normally 40% to 60% of the wastewater flow. When wasting is properly adjusted, the MLSS level will remain steady. With an excessive F:M ratio, sludge production will increase, and eventually, sludge will appear in the clarifier effluent.

9. OPERATIONAL DIFFICULTIES

Sludge bulking refers to a condition in which the sludge does not settle out. Since the solids do not settle, they leave the sedimentation tank and cause problems in subsequent processes. The sludge volume index can often (but not always) be used as a measure of settling characteristics. If the SVI is less than 100, the settling process is probably operating satisfactorily. If SVI is greater than 150, the sludge is bulking. Remedies include addition of lime, chlorination, additional aeration, and a reduction in MLSS.

Some sludge may float back to the surface after settling, a condition known as *rising sludge*. Rising sludge occurs when nitrogen gas is produced from denitrification of the nitrates and nitrites. Remedies include increasing the return sludge rate and decreasing the mean cell residence time. Increasing the speed of the sludge scraper mechanism to dislodge nitrogen bubbles may also help.

It is generally held that after 30 min of settling, the sludge should settle to between 20% and 70% of its original volume. If it occupies more than 70%, there are too many solids in the aeration basin, and more sludge should be wasted. If the sludge occupies less than 20%, less sludge should be wasted.

Sludge washout (*solids washout*) can occur during a period of peak flow or even with excess recirculation. Washout is the loss of solids from the sludge blanket in the settling tank. This often happens with insufficient wasting, such that solids build up (to more than 70% in a settling test), hinder settling in the clarifier, and cannot be contained.

10. BIOTOWER

Also called a *trickling filter*, a *biotower* operates by having the wastewater fall through a packed bed or tower filled with permeable packing. The packing has both aerobic and anaerobic microorganisms growing on it. Material balance equations can be formulated using the flux from both the connective and diffusive processes. The rate of reaction is formulated by using the Monod kinetic model. The biotowers can be operated either with or without recycling.

Equation 16.12: Fixed-Film Equation Without Recycle

$$\frac{S_e}{S_0} = e^{-kD/q^n} \qquad 16.12$$

Description

Equation 16.12 is used for fixed-film biotowers with no recycling. Equation 16.12 uses the same nomenclature as Eq. 16.13.

Equation 16.13 Through Eq. 16.16: Fixed-Film Equation with Recycle

$$\frac{S_e}{S_a} = \frac{e^{-kD/q^n}}{(1+R) - R(e^{-kD/q^n})} \qquad 16.13$$

Environmental Engineering

$$S_a = \frac{S_0 + RS_e}{1+R} \quad (16.14)$$

$$k_T = k_{20}(1.035)^{T-20} \quad (16.15)$$

$$q = (Q_0 + RQ_0)/A_{\text{plan}} \quad \text{[with recycle]} \quad (16.16)$$

Description

Q_0 is the incoming volumetric flow rate.[5] S_e is the effluent substrate concentration BOD$_5$; S_a is the BOD$_5$ of the mixture of raw and recycled applied material, found from a simple mass balance (see Eq. 16.14); D is the depth; q is the hydraulic loading; k is a treatability constant; n is an exponent related to tower media, with $n = 0.5$ being applicable to plastic media; and R is the recycle ratio.

Typical values of the *treatability constant*, k, are 0.01–0.1 min^{-1} at 20°C. For modular plastic media and municipal wastewater, $k \approx 0.06$ min^{-1} at 20°C. Equation 16.15 is used to convert standard 20°C treatability constant values (represented by k_{20}) to other temperatures.

11. AEROBIC DIGESTION

Aerobic digestion occurs in an open holding tank digester and is preferable for stabilized primary and combined primary-secondary sludges. Up to 70% (typically 40–50%) of the volatile solids can be removed in an aerobic digester. Mechanical aerators are used in a manner similar to aerated lagoons. Construction details of aerobic digesters are similar to those of aerated lagoons. Design criteria for aerobic digesters is given in Table 16.3.

Equation 16.17: Tank Volume

$$V = \frac{Q_i(X_i + FS_i)}{X_d(k_d P_v + 1/\theta_c)} \quad (16.17)$$

Description

The aerobic digestion process is similar to the activated sludge process. Aeration is accomplished by means of diffusing equipment. The process can be either batch or continuous.

For a continuous process with residence time θ_c, the tank volume is found from Eq. 16.17. In Eq. 16.17, Q_i is the volumetric sludge rate entering the digester. This quantity will be different from the wastewater treatment plan influent rate, Q_0. X_i is the incoming suspended (biological) solids concentration from secondary processes. If the digester also processes raw primary sludge, the decimal fraction of primary sludge, F, multiplied by the primary sludge's BOD, S_i, is included. X_d is the digester's suspended (biological) solids concentration. k_d is a digestion reaction rate constant. P_v is the decimal fraction of the suspended solids that is volatile.[6] θ_c is the sludge age (solids residence time). Any consistent set of units can be used in Eq. 16.17. The volume unit will come from the volume units of Q_i. The time unit for Q_i must be the same as the time unit for k_d and θ_c.

Table 16.3 Design Criteria for Aerobic Digesters

parameter	value
sludge retention time	
at 20°C	40 days
at 15°C	60 days
solids loading	0.1–0.3 lbm volatile solids/ft^3-day
oxygen requirements	
cell tissue	~2.3 lbm O_2/lbm solids destroyed
BOD$_5$ in primary sludge	1.6–1.9 lbm O_2/lbm solids destroyed
energy requirements for mixing	
mechanical aerators	0.7–1.50 hp/10^3 ft^3
diffused-air mixing	20–40 ft^3/10^3 ft^3-min
dissolved-oxygen residual in liquid	1–2 mg/L
reduction in volatile suspended solids	40–50%

Example 17

An aerobic digester has an inflow of 98 ft^3/day. The concentration of influent suspended solids is 5 mg/L. The digester suspended solids concentration is 10 mg/L. The reaction rate constant is 0.5 day^{-1}. The volatile fraction is 0.1. The solids retention time is 2 days. The fraction of influent 5 day BOD consisting of raw primary sludge is 0.5, and primary sludge BOD$_5$ is 200 mg/L. Under these conditions, the volume of the aerobic digester is most nearly

(A) 200 ft^3

(B) 990 ft^3

(C) 1200 ft^3

(D) 1900 ft^3

[5]The *NCEES Handbook* uses subscripts 0 and *o* interchangeably in this section.
[6]*P* is not a percentage.

Solution

Use Eq. 16.17 to find the aerobic digester tank volume.

$$
\begin{aligned}
V &= \frac{Q_i(X_i + FS_i)}{X_d(k_d P_v + 1/\theta_c)} \\
&= \frac{\left(98\ \frac{\text{ft}^3}{\text{day}}\right)\left(5\ \frac{\text{mg}}{\text{L}} + (0.5)\left(200\ \frac{\text{mg}}{\text{L}}\right)\right)}{\left(10\ \frac{\text{mg}}{\text{L}}\right)\left((0.5\ \text{day}^{-1})(0.1) + \frac{1}{2\ \text{days}}\right)} \\
&= 1871\ \text{ft}^3 \quad (1900\ \text{ft}^3)
\end{aligned}
$$

The answer is (D).

12. ANAEROBIC DIGESTION

Anaerobic digestion occurs in the absence of oxygen. Anaerobic digestion is more complex and more easily upset than aerobic digestion. However, it has a lower operating cost. Table 16.4 gives typical characteristics of anaerobic digesters.

***Table 16.4** Design Parameters for Anaerobic Digesters*

parameter	standard-rate	high-rate
solids residence time	30–90 d	10–20 d
volatile solids loading	0.5–1.6 kg/m^3/d	1.6–6.4 kg/m^3/d
digested solids concentration	4–6%	4–6%
volatile solids reduction	35–50%	45–55%
gas production	0.5–0.55 mg^3/kg VSS added	0.6–0.65 mg^3/kg VSS added
methane content	65%	65%

Three types of anaerobic bacteria are involved. The first type converts organic compounds into simple fatty or amino acids. The second group of *acid-formed bacteria* converts these compounds into simple organic acids, such as acetic acid. The third group of *acid-splitting bacteria* converts the organic acids to methane, carbon dioxide, and some hydrogen sulfide. The third phase takes the longest time and sets the rate and loadings. The pH should be 6.7–7.8, and temperature should be in the mesophilic range of 85–100°F for the methane-producing bacteria to be effective. Sufficient alkalinity must be added to buffer acid production.

A simple single-stage sludge digestion tank consists of an inlet pipe, outlet pipes for removing both the digested sludge and the clear supernatant liquid, a dome, an outlet pipe for collecting and removing the digester gas, and a series of heating coils for circulating hot water. (See Fig. 16.2.)

***Figure 16.2** Simple Anaerobic Digester*

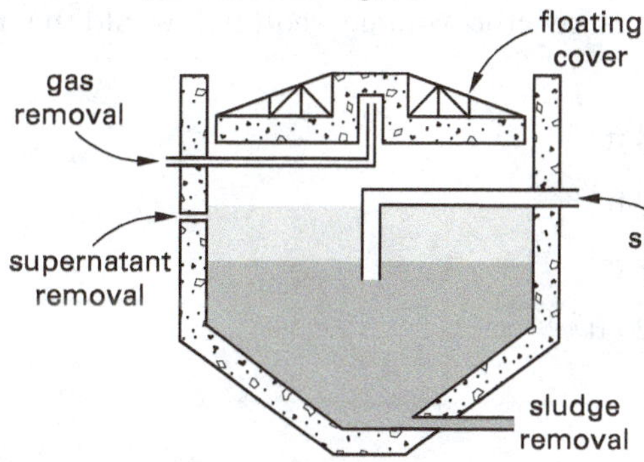

In a single-stage, floating-cover digester, sludge is brought into the tank at the top. The contents of the digester stratify into four layers: scum on top, clear supernatant, a layer of actively digesting sludge, and a bottom layer of concentrated sludge. Some of the contents may be withdrawn, heated, and returned in order to maintain a proper digestion temperature.

Supernatant is removed along the periphery of the digester and returned to the input of the processing plant. Digested sludge is removed from the bottom and dewatered prior to disposal. Digester gas is removed from the gas dome. Heat from burning the methane can be used to warm the sludge that is withdrawn or to warm raw sludge prior to entry.

A reasonable loading and well-mixed digester can produce a well-stabilized sludge in 15 days. With poorer mixing, 30–60 days may be required.

A single-stage digester performs the functions of digestion, gravity thickening, and storage in one tank. In a two-stage process, two digesters in series are used. Heating and mechanical mixing occur in the first digester. Since the sludge is continually mixed, it will not settle. Settling and further digestion occur in the unheated second tank.

Equation 16.18: Standard Rate Reactor Volume

$$\text{reactor volume} = \frac{V_1 + V_2}{2} t_r + V_2 t_s \qquad 16.18$$

Description

A standard rate digester must be sized to accommodate the raw sludge input, V_1, and the digested sludge accumulation, V_2, for the time for the sludge to digest and thicken (i.e., for the residence time, t_r) as well as to hold the accumulation for the period it is stored, t_s.

Example

The reactor volume of a standard-rate anaerobic digester needs to be estimated for a given treatment process. The raw sludge input rate is 350 ft^3/day. The digested sludge accumulation is 1 ft^3/day. If the storage time is 12 hr and the time to react in a high-rate digester is 8 hr, the reactor volume required would be most nearly

(A) 18 ft^3

(B) 59 ft^3

(C) 82 ft^3

(D) 120 ft^3

Solution

From Eq. 16.18,

$$\text{reactor volume} = \frac{V_1 + V_2}{2} t_r + V_2 t_s$$

$$= \frac{\left(\dfrac{350 \ \dfrac{\text{ft}^3}{\text{day}} + 1 \ \dfrac{\text{ft}^3}{\text{day}}}{2} \right)(8 \text{ hr}) + \left(1 \ \dfrac{\text{ft}^3}{\text{day}} \right)(12 \text{ hr})}{24 \ \dfrac{\text{hr}}{\text{day}}}$$

$$= 59 \text{ ft}^3$$

The answer is (B).

Equation 16.19 and Eq. 16.20: High Rate Reactor Volume

$$\text{reactor volume} = V_1 t_r \quad \text{[first stage]} \qquad 16.19$$

$$\text{reactor volume} = \frac{V_1 + V_2}{2} t_t + V_2 t_s \quad \text{[second stage]} \qquad 16.20$$

Description

The *first-stage reactor volume* is selected to hold the raw sludge for as long as it takes for digestion to occur (see Eq. 16.19).

The *second-stage reactor volume* is selected based on the time it takes for thickening to occur, t_t (see Eq. 16.20).

Environmental Engineering

17 Air Quality

1. CLEAN AIR ACT

Air quality issues have been addressed by federal legislation since 1955, and many updated statutes and amendments have followed. The statute providing regulatory authority for national air quality issues is the Clean Air Act (CAA) and amendments. Federal air pollution regulations address air quality issues through a variety of enforcement and incentive mechanisms.

The CAA was amended in 1990 to require the establishment of *National Ambient Air Quality Standards* (NAAQS) in the United States. The NAAQS implements a uniform national air pollution control program. For each pollutant included in the NAAQS, primary and secondary standards are defined. *Primary standards* are intended to protect human health, and *secondary standards* are intended to protect public welfare. The NAAQS are applied to a limited number of substances, known as the *criteria pollutants*. The NAAQS and the criteria pollutants to which they apply are presented in Table 17.1.

To ensure application of the NAAQS in each of the 50 states, the United States Environmental Protection Agency (EPA) has required the states to prepare and submit *State Implementation Plans* (SIP). Each SIP must address a control strategy for each of the criteria pollutants included in the NAAQS. If a state is unable or unwilling to submit an SIP acceptable to the EPA, the EPA has the authority to develop an SIP for the state and then require the state to enforce it.

Prevention of significant deterioration (PSD) is a program that is designed to protect air quality in clean-air areas that already meet the NAAQS. To apply the PSD principle, the EPA has developed a regional classification system. Class I regions, generally including national monuments and parks and national wilderness areas, allow very little air quality deterioration and, thereby, limit economic development. Class II regions allow moderate air quality deterioration, and class III regions allow deterioration to the secondary NAAQS.

Because of geographic conditions, population density, climate, and other factors, some areas in the United States are unable to meet the NAAQS. These areas are designated as *nonattainment areas* (NA). Any new or modified sources in nonattainment areas are required to control emissions at the *lowest achievable emission rate* (LAER). Recognizing the economic problems posed by the CAA in nonattainment areas, the EPA has allowed some flexibility, through *emissions trading*, in how industry goes about meeting the compliance requirements. Emissions trading can occur by the following methods.

- *Emission Reduction Credit:* By applying a higher level of treatment than required by regulation, businesses earn credits that can be redeemed through other CAA programs.
- *Offset:* The offset allows industrial expansion by letting businesses compensate for new emissions by offsetting them with credits acquired by other businesses already existing in the region.
- *Bubble:* All activities in a single plant or among a group of proximate industries can emit at various rates as long as the resulting total emission does not exceed the allowable emission for an individual source.
- *Netting:* Businesses can expand without acquiring a new permit as long as the net increase in emissions is not significant. Consequently, by applying improved control technology, businesses can earn credits that can be subsequently applied to plant expansion with no net increase in emissions.

New source performance standards (NSPS) are applied to specific source categories (e.g., asphalt concrete plants, incinerators) to prevent the emergence of other air pollution problems from new construction. The NSPS are intended to reflect the best emission control measures achievable at reasonable cost, and are to be applied during initial construction of new industrial facilities.

National Emission Standards for Hazardous Air Pollutants (NESHAP) focus on hazardous pollutants not included as criteria pollutants. *Hazardous pollutants* are compounds that pose particular, usually localized, hazards to the exposed population. On a localized level, their impact is more severe than that of the criteria pollutants. Compounds identified as hazardous pollutants are limited to asbestos, inorganic arsenic, benzene, mercury, beryllium, radionuclides, vinyl chloride, and coke oven emissions. The list is short because onerous demands are placed on the EPA by the CAA for

Table 17.1 National Ambient Air Quality Standards[a]

pollutant	time basis	primary standard[b]	secondary standard
carbon monoxide (CO)	8 h	9 ppm (10 MG/m^3)	none
	1 h	35 ppm (40 MG/m^3)	none
nitrogen dioxide (NO_2)	1 h	100 ppb	none
	mean annual	53 ppb (100 μg/m^3)	same
ground-level ozone (O_3)	8 h	0.075 ppm (147 μg/m^3)	same
sulfur dioxide (SO_2)	3 h	none	0.50 ppm (1300 μg/m^3)
	1 h	0.075 ppm	none
PM10	24 h	150 μg/m^3	same
PM2.5	annual	12 μg/m^3	15 μg/m^3
	24 h	35 μg/m^3	same
lead (Pb)	3 mo	0.15 μg/m^3	same
hydrocarbons[c]	–	–	–

[a]Typical values. The NAAQS are subject to ongoing legislation, and current values may be revised from those presented here.
[b]Equivalents in parentheses may or may not be part of the NAAQS.
[c]Hydrocarbon standards have been revoked by the EPA since hydrocarbons, as precursors of ozone, are adequately regulated by ozone standards.
Source: National Ambient Air Quality Standards (NAAQS), U.S. Environmental Protection Agency, epa.gov.

including pollutants and because a very high level of abatement, based on risk to the exposed population, is required once a pollutant is listed. The 1990 CAA amendments listed 187 other *hazardous air pollutants* (HAPs) whose abatement is based on the *maximum achievable control technology* (MACT), a less demanding criterion than risk. Many states have adopted their own regulations to designate and control other hazardous air pollutants.

The 1990 CAA amendments placed limits on allowable emissions of sulfur oxides (SO_x) and nitrogen oxides (NO_x)—two categories of compounds associated with *acid rain.* These limits require acid deposition emission controls, which are implemented through a system of emission allowances applied to fossil fuel-fired boilers in the 48 contiguous states. Control options may include such things as use of low-sulfur fuels, flue-gas desulfurization, and replacing older facilities.

The 1977 CAA amendments authorized the EPA to regulate any substance or practice with reasonable potential to damage the stratosphere, especially with regard to *ozone.* In response to this regulatory authority, the EPA has closely regulated the production and sale of chlorinated fluorocarbons (CFCs) through out-and-out bans on their production and use, taxes on emissions from their use, and a permitting program.

Criteria Pollutants

The EPA uses the following six criteria pollutants as indicators of air quality. It has established a maximum concentration for each pollutant that, when exceeded, may negatively impact human health.

- ozone
- carbon monoxide
- nitrogen dioxide
- sulfur dioxide
- particulate matter
- lead

The threshold concentrations for criteria pollutants are called *national ambient air quality standards* (NAAQS). (See Table 17.1.) If these standards are exceeded, the area may be designated as being in nonattainment. A *nonattainment area* is defined in the Clean Air Act and Amendments of 1990 as a locality where air pollution levels persistently exceed the national ambient air quality standards, or a locality that contributes to ambient air quality in a nearby area that fails to meet standards.

Nitrogen Oxides

Nitrogen oxides (NO_x) incorporate the sum total of the oxides of nitrogen, the most significant of which are nitric oxide (NO) and nitrogen dioxide (NO_2). NO_x are emitted from motor vehicles, power plants, and other combustion operations. NO_x gases play a major role in the formation of ozone and nitrogen-bearing particles, which are both associated with adverse health effects. NO_x also contribute to the formation of acid rain.

Adverse environmental effects of NO_x include visibility impairment, acidification of freshwater bodies, increases in levels of toxins harmful to fish and other aquatic life, and changes in the populations of some species of vegetation in wetland and terrestrial systems. Exposure to NO_2 may lead to changes in airway responsiveness and lung function in individuals with pre-existing respiratory illnesses and increases in respiratory illnesses in children (5–12 yr). Long-term exposures to NO_2 may lead to increased susceptibility to respiratory infection and may cause alterations in the lung.

Ozone

Ground level ozone (O_3) is a primary component in smog and may adversely affect human health. This is in contrast to *stratospheric ozone*, which occurs naturally and provides a protective layer against ultraviolet radiation. O_3 is not emitted directly into the air, but is formed by the reaction of volatile organic compounds (VOCs) and NO_x in the presence of heat and sunlight. Ground level ozone forms readily in the atmosphere, usually during hot summer weather. Because O_3 is formed in the atmosphere over time, the highest levels of ozone typically occur 50–100 km away from the location of the highest ozone precursor (NO_x and VOCs) emissions. VOCs are emitted from a variety of sources, including motor vehicles, chemical plants, refineries, factories, consumer and commercial products, and other industrial sources. O_3 concentrations are largely affected by climate and weather patterns.

Health effects from O_3 exposure include increased susceptibility to respiratory infection, lung inflammation, aggravation of pre-existing respiratory diseases such as asthma, significant decreases in lung function, and increased respiratory symptoms, such as chest pain and cough. Environmental impacts of ground level ozone include reductions in agricultural and commercial forest yields, reduced growth and survivability of tree seedlings, and increased plant susceptibility to disease, pests, and other environmental stresses. In 1997, the EPA revised the national ambient air quality standards for ozone by replacing the one hour ozone standard of 0.12 parts per million (ppm) with a new eight hour standard of 0.08 ppm.

Acid Rain

Acid rain describes a mixture of wet and dry atmospheric deposition containing higher than normal amounts of nitric and sulfuric acids. The precursors, or chemical forerunners, of acid rain result from both natural sources, such as volcanoes and decaying vegetation, and from manufactured sources, primarily sulfur dioxide (SO_2) and nitrogen oxide (NO_x) emissions from fossil fuel combustion. Acid rain is associated with the acidification of soils, lakes, and streams, accelerated corrosion of buildings and monuments, and reduced visibility. In the United States, about two-thirds of all SO_2 and one-quarter of all NO_x come principally from coal burning power plants. Acid rain occurs when SO_2 or NO_x gases react in the atmosphere with water, oxygen, and other chemicals to form mild solutions of sulfuric or nitric acid.

Wet deposition is acidic rain, fog, mist, and snow. It affects a variety of plants and animals, with the ultimate impact determined by factors including the acidity of the liquid solutions, the chemistry and buffering capacity of the soils involved, and the types of fish, trees, and other living organisms dependent on the water.

Dry deposition is the incorporation of acids into dust or smoke, which then accumulate on the ground, buildings, homes, cars, and trees. Dry deposited acids and particles can be washed from these surfaces by rainstorms, leading to increased acidity in the runoff. About half of the acidity in the atmosphere falls back to earth through dry deposition.

Carbon Monoxide

Carbon monoxide (CO) is a colorless, odorless, and, at high levels, poisonous gas, formed when the carbon in fuel is not burned completely. It is a component of motor vehicle exhaust, which contributes about 60% of all CO emissions nationwide and 95% of all emissions in cities. High concentrations of CO generally occur in areas with heavy traffic congestion. Other sources of CO emissions include refineries and industrial processes and non-transportation fuel combustion, including wood and refuse burning. Peak atmospheric CO concentrations typically occur in winter months, when CO automotive emissions are greater and nighttime atmospheric inversion conditions are more frequent. *Atmospheric inversions* occur when pollutants are trapped in a cold layer of air beneath a warmer one. High concentrations of CO enter the bloodstream through the lungs and reduce oxygen delivery to the body, which can cause visual impairment and reduce work capacity, manual dexterity, learning ability, and performance in complex tasks.

Smog

Smog is essentially a mixture of pollutants and water and is considered air pollution. *Classic smog* results from large amounts of coal burning in a geographical area and is caused by a mixture of smoke and sulfur dioxide (SO_2). *Photochemical smog* is a mixture of air pollutants, such as nitrogen oxides (NO_x), tropospheric ozone, volatile organic compounds (VOCs), peroxyacyl nitrates, and aldehydes. Photochemical smog components are generally highly reactive and oxidizing. Photochemical smog is caused by a reaction between sunlight and the various chemical components listed. Smog is of principal concern in urban areas.

Lead

Lead (Pb) is a metallic element that occurs naturally in soil, rocks, water, and food. Lead was largely used as an additive to paints and gasoline until the 1970s, when its use was significantly reduced. Current sources of Pb are lead-acid car batteries and industrial coatings. Approximately 70% of Pb pollution comes from smelters, power plants fueled by coal, and lead used in the processing of oil shale. Exposure to Pb can occur through inhalation of air containing Pb and ingestion of Pb in food, water, soil, or dust. Excessive Pb exposure can cause seizures, mental retardation, and/or behavioral disorders.

Radon

Radon is a radioactive noble gas formed by the natural decay of uranium, which is found in nearly all soils. Radon is considered a health hazard because it causes lung cancer. It is colorless, odorless, and tasteless. Radon gas can accumulate in buildings and drinking water. It causes an estimated 20,000 deaths per year in the United States. Radon is a significant contaminant that impacts indoor air quality worldwide, because it enters buildings through cracks and holes in the foundation, or through well water. Radon contamination inside of homes is of particular concern and cannot be remedied once detected. Radon concentrations in the air are typically expressed in units of picocuries per liter (pCi/L) of air. Concentrations can also be expressed in *working levels* (WL) rather than picocuries per liter (1 pCi/L = 0.004 WL).

PCBs

Polychlorinated biphenyls (PCBs) are manufactured mixtures of up to 209 individual chlorinated compounds (congeners). Because of their nonflammability, chemical stability, high boiling point, and electrical insulating properties, PCBs were used in hundreds of industrial and commercial applications, including electrical, heat transfer, and hydraulic equipment. They were also used as plasticizers in paints, plastics, and rubber products, as well as in pigments, dyes, carbonless copy paper, and many other applications. PCBs can be either oily liquids or solids, colorless to light yellow, and have no odor or taste. Some PCBs can exist as air vapor. However, the production of PCBs was stopped in the United States in 1977 because of PCB buildup in the environment, which gave rise to adverse health effects. Products made before 1977 that may still contain PCBs include old fluorescent lighting fixtures, transformers and capacitors, and cutting and hydraulic oils.

PCB waste is defined by the EPA as waste containing PCBs as a result of a spill, release, or other unauthorized disposal. The wastes are categorized according to the concentrations of PCBs that they contain and their date of disposal or spillage. PCB contaminated sites are most often cleaned up using *incineration* processes. Alternative technologies are allowed if the performance of these technologies is equivalent to incineration's *destruction and removal efficiency* (DRE) of 99.9999%. Both stationary and mobile incinerators can be used. The choice is often based on cost. Approximately 10 commercial incinerators have been approved for PCB disposal in the United States, and incineration has been used at approximately 65 hazardous waste sites with PCB contaminated soils. Alternative PCB treatment technologies include disposal in chemical waste landfills, high-efficiency boilers, thermal desorption, chemical dehalogenation or dechlorination, solvent extraction, soil washing, vitrification, and bioremediation.

Particulate Matter

Particulate matter (PM), or particle pollution, is a complex mixture of extremely small particles and liquid droplets and is one of the primary forms of air pollution. PM is emitted from numerous industrial, mobile (e.g., cars), residential, and even natural sources. PM is made up of a number of components, including acids (such as nitrates and sulfates), organic chemicals, metals, and soil or dust particles. Particle size is linked to the potential for causing health problems. The EPA focuses on particles that are 10 μm in diameter or smaller because particles of this size, which generally pass through the throat and nose and enter the lungs, can cause adverse health effects. The EPA groups particle pollution into two categories: *inhalable coarse particles* (between 2.5 μm and 10 μm in diameter) and *fine particles* (≤ 2.5 μm diameter). Inhalable particles are found near roadways and dusty industries, while fine particles are found in smoke and haze. PM is generally designated by putting the maximum size (in numbers of microns) in the subscript. For example, $PM_{2.5}$ designates fine particles.

Sulfur Dioxide

Sulfur dioxide (SO_2) is the main sulfur-containing compound produced in the combustion of coal or petroleum. Oxidation of SO_2 in the presence of nitrogen oxides (NO_x) in the atmosphere produces hydrogen sulfide (H_2S), an acid that ultimately produces acid rain. Over 65% of the SO_2 released into the air in the United States comes from electrical utilities, especially those that burn coal. Other sources of SO_2 are industrial facilities that derive their products from raw materials including metallic ore, coal, and crude oil, or that burn coal or oil to produce process heat (petroleum refineries, cement manufacturing, and metal processing facilities). Health and environmental impacts of SO_2 are respiratory illness, visibility impairment, acid rain, plant and water damage, and aesthetic damage to structures.

Dioxins

Dioxins are halogenated organic compounds, most commonly polychlorinated dibenzofurans (PCDFs) and polychlorinated dibenzodioxins (PCDDs). The most toxic dioxin is 2,3,7,8-tetrachlorodibenzo-*p*-dioxin (TCDD). Dioxins are formed as an unintentional by-product of

Environmental Engineering

many industrial processes involving chlorine, such as waste incineration, chemical and pesticide manufacturing, and pulp and paper bleaching. Two dioxins, polybrominated dibenzofurans and dibenzodioxins, have been discovered as impurities in brominated flame retardants. Dioxins are persistent in the environment and readily bioaccumulate up the food chain because they are fat soluble. Dioxins are of particular concern at sites where indirect exposure pathways (e.g., ingestion of fruits/vegetables, meat, dairy products, and breast milk) are of concern. The toxicity of other dioxins and chemicals, such as PCBs, which act like dioxins, are measured in relation to TCDD using *toxic equivalence factors* (TEFs). *2,3,7,8-TCDD toxic equivalents* (TEQs) are determined by multiplying the compound concentrations by their respective TEFs and summing them.

2. AIR QUALITY INDEX

The *Air Quality Index* (AQI)[1] was developed by the EPA to quantify and describe the daily air quality for the general public. The Air Quality Index is divided into six categories of health concern. Each category has a range of AQIs associated with it. A higher AQI indicates lower air quality. The categories are color-coded to simplify identification of hazardous air quality conditions by the public. For example, green corresponds to good air quality, while maroon signalizes hazardous air quality. The EPA requires the AQI to be calculated and reported daily for cities with a population exceeding 350,000 people.

The AQI is determined by five of the six criteria pollutants[2] regulated by the Clean Air Act: ground level ozone (O_3), total suspended particulates (TSP), carbon monoxide (CO), sulfur dioxide (SO_2), and nitrogen dioxide (NO_2). Each AQI category is a separate scale for each criteria pollutant that is based on the concentrations deemed permissible by the NAAQS. AQI values greater than 100 indicate unhealthy air quality, while values less than 100 indicate satisfactory air quality. The AQI of an air sample can be calculated based on the pollutants present.

Each AQI category has a range of pollutant concentrations associated with it. The upper and lower limits of an AQI category's pollutant concentrations are called *breakpoints*. The low and high breakpoint concentrations are the lowest and highest concentrations acceptable in each category, respectively. For example, an air sample is determined to have a pollutant concentration[3] of 8.2 ppm of CO, which falls in Table 17.2 between 4.5 ppm and 9.4 ppm. Therefore, the sample is in the moderate AQI category.

The AQI of an air sample is determined using linear interpolation. If an air sample has several pollutants present, the AQI is calculated for each pollutant, and the greatest AQI is selected (i.e., reported) as the daily AQI.

Table 17.2 *Air Quality Index Categories*

AQI category	color	AQI range	description
good	green	0 to 50	Air quality is considered satisfactory, and air pollution poses little or no risk.
moderate	yellow	51 to 100	Air quality is acceptable. However, for some pollutants, there may be a moderate health concern for a very small number of people who are unusually sensitive to air pollution.
unhealthy for sensitive groups	orange	101 to 150	Members of sensitive groups may experience health effects. The general public is not likely to be affected.
unhealthy	red	151 to 200	Everyone may begin to experience health effects. Members of sensitive groups may experience more serious health effects.
very unhealthy	purple	201 to 300	A health alert is issued that everyone may experience more serious health effects.
hazardous	maroon	301 to 500	A health warning of emergency conditions is issued. The entire population is more likely to be affected.

Environmental Engineering

[1]The EPA created the Air Quality Index as a replacement for the *Pollutant Standards Index* (PSI) in 1999.
[2]Lead is currently the only criteria pollutant not monitored using the AQI.
[3]It is common practice to round the pollutant concentration up to the nearest whole number.

3. AIR POLLUTION PREVENTION

Pollution prevention, also commonly referred to as *source reduction*, was promulgated in 1990 in the Pollution Prevention Act. *Pollution prevention* is the process of reducing or eliminating waste at the source; the modification of waste-producing processes that enter a waste stream; promoting the uses of nontoxic or less toxic substances; implementing conservation techniques; and reusing materials rather than wasting them. This act also detailed pollution prevention practices (such as recycling, green engineering, and sustainable agriculture) that improve efficiency in the use of energy, water, and other natural resources.

Air pollution generally results from the combustion of hydrocarbon-based fuels, but may also result from dust at construction sites and other disturbed areas. Particle removal is dependent on particle size, but the calculation of collection efficiency is based only on the mass percentage collected. A variety of processes and control devices have been developed to remove particulate matter from gas streams, including cyclones, wet scrubbers, electrostatic precipitators, and baghouses. (See Table 17.3 for removal efficiencies of control devices.) Electrostatic precipitators and baghouses are among the most efficient technologies for removing particulate matter and are thus generally used for air pollution prevention.

Table 17.3 *Typical Removal Efficiencies of Various Particulate Matter Control Devices*

	removal efficiency (%)			
control device	<1 μm	1–3 μm	3–10 μm	>10 μm
electrostatic precipitator	96.5	98.3	99.1	99.5
fabric filter	100	99.8	>99.9	>99.9
venturi scrubber	>70.0	99.5	>99.8	>99.8
cyclone	11	54	85	95

Environmental Engineering

Topic VI: Geotechnical Engineering

Chapter

18 Soil Properties and Testing

Nomenclature

a	area	ft^2	m^2
A	area	ft^2	m^2
c	cohesion	lbf/ft^2	Pa
c_v	coefficient of consolidation	$\text{ft}^2\text{/day}$	$\text{m}^2\text{/d}$
C	coefficient	–	–
C_C	compression index	–	–
C_R	recompression index	–	–
D	diameter	ft	m
D_r	relative density	–	–
e	void ratio	–	–
F	percentage fines	%	%
FS	factor of safety	–	–
g	gravitational acceleration, 32.2 (9.81)	ft/sec^2	m/s^2
g_c	gravitational constant, 32.2	ft-lbm/lbf-sec^2	–
G_S	specific gravity	–	–
GI	group index	–	–
h	head	ft	m
h	liquid level	ft	m
H	hydraulic head differential	ft	m
H	thickness	ft	m
H_{dr}	length of drainage path	ft	m
i	hydraulic gradient	–	–
I	stress influence value	–	–
k	coefficient of permeability	ft/sec	m/s
L	length	–	–
LI	liquidity index	–	–
LL	liquid limit	–	–
m	mass	lbm	kg
n	porosity	–	–
N	number	–	–
OCR	overconsolidation ratio	–	–
p	pressure	lbf/ft^2	Pa
p'	effective pressure	lbf/ft^2	Pa
P	load	lbf	N
PI	plasticity index	–	–
PL	plastic limit	–	–
q_U	unconfined compressive strength	–	–
Q	flow quantity	$\text{ft}^3\text{/sec}$	$\text{m}^3\text{/s}$
r	radius	ft	m
RC	relative compaction	%	%
RQD	rock quality designation	%	%
S	degree of saturation	%	%
S	settlement	ft	m
S	strength	lbf/ft^2	Pa
SI	shrinkage index	–	–
SL	shrinkage limit	–	–
t	time	sec	s
T	shearing force	lbf	N
T_v	time factor	–	–
u	pore pressure	lbf/ft^2	Pa
U	degree of consolidation	%	%
v	velocity	ft/sec	m/s
V	volume	ft^3	m^3
W	weight	lbf	N
z	depth	ft	m

Symbols

α	angle of failure plane	deg	deg
γ	specific (unit) weight	lbf/ft^3	N/m^3
γ'	submerged unit weight	lbf/ft^3	N/m^3

Geotechnical Engineering

δ	displacement	ft	m
ε	strain	–	–
ρ	density	lbm/ft^3	kg/m^3
σ	total stress	lbf/ft^2	Pa
σ'	effective stress	lbf/ft^2	Pa
τ	shear stress	lbf/ft^2	Pa
ϕ	angle of internal friction	deg	deg
ω	water content	%	%

Subscripts

0	initial
ave	average
A	air or axial
AV	average
b	buoyant
c	critical
C	curvature
d	dry
D	deviator or drop
e	exit
F	failure or flow
h	horizontal
i	initial
max	maximum
min	minimum
N	normal
o	initial or overburden
R	radial
s	seepage or surface
sat	saturated
S	sliding, soil, or solids
TOT	total
u	unconfined, undrained, or uplift
U	uniformity
ULT	ultimate
v	vertical
V	voids or volume
w	water
W	water

1. CLASSIFICATION OF ROCKS

Geologists classify rocks into three groups, according to the major Earth processes that formed them. The three rock groups are igneous, sedimentary, and metamorphic rocks.

Igneous rock is formed from melted rock that has cooled and solidified. When rock is buried deep within the Earth, it melts due to the high pressure and temperature. When magma cools slowly, usually at depths of thousands of feet, crystals grow from the molten liquid, forming a coarse-grained rock. When magma cools rapidly, usually at or near the Earth's surface, the crystals are extremely small, and a fine-grained rock results. Rocks formed from cooling magma are known as *intrusive igneous rocks*; rocks formed from cooling lava are known as *extrusive igneous rocks*. A wide variety of rocks are formed by different cooling rates and different chemical compositions of the original magma. Obsidian (volcanic glass), granite, basalt, and andesite porphyry are four of the many types of igneous rock. *Granite* is composed of three main minerals: quartz (grey), mica (black), and feldspar (white).

Sedimentary rocks such as chalk, sandstone, and clay are formed at the surface of the Earth. They are layered accumulations of sediments—fragments of rocks, minerals, or animal or plant material. Temperatures and pressures are low at the Earth's surface, and sedimentary rocks are easily identified by their appearance and by the diversity of minerals they contain. Most sedimentary rocks become cemented together by minerals and chemicals or are held together by electrical attraction. Some, however, remain loose and virtually unconsolidated. The layers are originally parallel to the Earth's surface. Sand and gravel on beaches or in river bars consolidate into *sandstone*. Compacted and dried mud flats harden into *shale*. Sediments of mud and shells settling on the floors of lagoons form sedimentary *chalk*.

There are several ways to classify sedimentary rocks, but a common one is according to the process which led to their deposition. Rocks formed from particles of older eroded rocks are known as *clastic sedimentary rocks*. These include sandstones and clays. Rocks formed from plant and animal remains are known as *organic sedimentary rocks*. Examples include limestone, chalk, and coal. Rocks formed from chemical action are known as *chemical sedimentary rocks*. These include sedimentary iron ores, evaporites such as rock salt (halite), and to some extent, flint, limestone, and chert.

Sometimes sedimentary and igneous rocks buried in the Earth's crust are subjected to pressures so intense or heat so high that the rocks are completely changed. They then become *metamorphic rocks*. Common metamorphic rocks include slate, schist, gneiss, and marble. The process of metamorphism does not melt the rocks, but it does transform them into denser, more compact rocks. New minerals are created, either by rearrangement of mineral components or by reactions with fluids that enter the rocks. Pressure or temperature can even change previously metamorphosed rocks into new types. In this way, limestone can become marble and shale can be converted into slate.

2. CHARACTERIZING ROCK MASS QUALITY

The suitability of rock for certain exposures (e.g., water scour, tunneling, and foundation compression) has been characterized by its *rock quality designation* (RQD) using a method proposed by D.U. Deere in 1967. RQD is a rough measure of the degree of fracturing (jointing) in a rock mass, measured as a percentage of drill core lengths that are unfragmented. Rock cores are taken using ASTM D6032 procedures. Then, the lengths of naturally occurring individual core segments longer than 4 in and that are hard and sound are summed. The RQD is the percentage of the core run that consists of pieces longer than 4 in.

$$\text{RQD} = \frac{\sum\limits_{L_i > 4\text{ in}} L_i}{L_{\text{core}}} \times 100\%$$

If pieces are broken by the recovery or handling process, the broken segments are assembled and counted as single lengths. Judgment is required in some cases. To minimize core recovery damage, cores should have diameters larger than 2⅛ in (corresponding to the NX core size) and be obtained from double-tube core barrels, although decades of experience has shown that use of smaller diameter NQ cores provides reliable samples.

Rock quality is characterized as very good (RQD > 90%), good (90–75%), fair (75–50%), poor (50–25%), or very poor (RQD < 25%). Soil is assumed to be soil-like with regard to scour potential when RQD < 50%. Table 18.1 contains recommendations for allowable bearing pressure for RQD categories.

RQD is not the only method of characterizing rock mass quality. Other characteristics include unconfined compressive strength (ASTM D2938), slake durability (useful on metamorphic and sedimentary rocks such as slate and shale), soundness (AASHTO T104), and abrasion (AASHTO T96, the "Los Angeles Abrasion" test). RQD has also been incorporated (as one of multiple quantified parameters) into two *rock mass rating systems*: the *Rock Mass Rating (RMR) System* (ca. 1972) and the *Q-system* (ca. 1974).

3. SOIL PARTICLE SIZE DISTRIBUTION

Soil is an aggregate of loose mineral and organic particles. *Rock*, on the other hand, exhibits strong and permanent cohesive forces between the mineral particles. From an engineering perspective, the distinction between soil and rock relates to the workability of materials. For example, one practical definition is that soil can be excavated with a backhoe, while rock needs to be blasted. The distinction between soil and rock can also be made based on strength, density, and other quantifiable parameters. A geologist or soil scientist might be more interested in how a material has been formed than its workability, and thus might distinguish between soil and rock differently.

Table 18.1 *Recommended Allowable Bearing Pressure for Footings on Rock*

material and RQD	allowable contact pressure (tsf)
igneous and sedimentary rock, including crystalline bedrock, including granite, diorite, gneiss, and traprock; hard limestone and dolomite*	
75–100%	120
50–75%	65
25–50%	30
< 25%	10
metamorphic rock, including foliated rocks (e.g., schist and slate); bedded limestone*	
> 50%	40
< 50%	10
sedimentary rocks, including hard shale and sandstone*	
> 50%	25
< 50%	10
soft or broken bedrock (excluding shale); soft limestone*	
> 50%	12
< 50%	8
soft shale	4

*in sound condition
(Multiply tsf by 95.8 to obtain kPa.)

Source: Federal Highway Administration memorandum, July 19 1991, from Stanley Gordon, FHA Bridge Division Chief, to Regional Federal Highway Administrators.

The primary mineral components of any soil are gravel, sand, silt, and clay. Organic material can also be present in surface samples. Gravel and sand are classified as coarse-grained soils, while inorganic silt and clay are classified as fine-grained soils.

The particle size distribution for a coarse soil is found from a sieve test. In a *sieve test*, the soil is passed through a set of sieves of consecutively smaller openings.

(See Table 18.2.) The particle size distribution for the finer particles of a soil is determined from a *hydrometer test.* This test is based on Stokes' law, which relates the speed of a particle falling out of suspension to its diameter and solid density. The results of both the sieve and hydrometer tests are graphed as a *particle size distribution,* as shown in Fig. 18.1.

Table 18.2 *Sieve Size Designations (nominal)**

alternative designation	standard designation (mm)
4 in	100
3 in	75
2 in	50
1½ in	37.5
1 in	25
¾ in	19
½ in	12.5
⅜ in	9.5
no. 4	4.75
no. 6	3.35
no. 8	2.36
no. 10	2.00
no. 16	1.18
no. 20	0.850
no. 30	0.600
no. 40	0.425
no. 50	0.300
no. 60	0.250
no. 70	0.212
no. 100	0.150
no. 140	0.106
no. 200	0.075

*Actual opening dimensions specified in ASTM E-11 differ slightly from the nominal designations.

Figure 18.1 *Typical Particle Size Distribution*

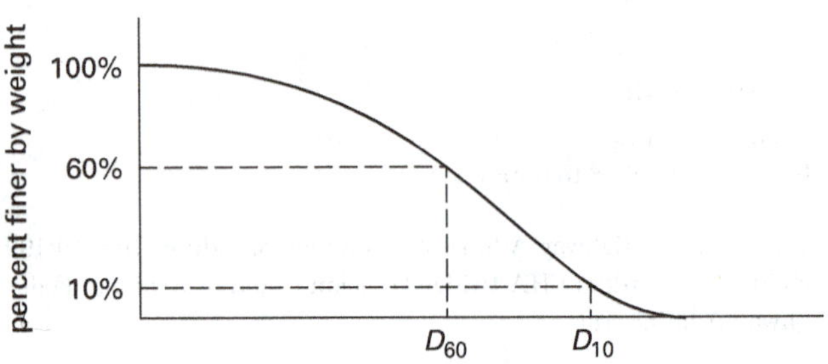

Equation 18.1: Hazen Uniformity Coefficient

$$C_U = D_{60}/D_{10} \qquad 18.1$$

Values

Table 18.3 *Typical Uniformity and Curvature Coefficients*

soil	C_U	C_C
gravel	>4	1–3
fine sand	5–10	1–3
coarse sand	4–6	–
mixture of silty sand and gravel	15–300	–
mixture of clay, sand, silt, and gravel	25–1000	–

Description

The *Hazen uniformity coefficient,* C_U, given by Eq. 18.1, indicates the general shape of the particle size distribution.[1] The diameter for which only 10% of the particles are finer is designated as D_{10} and is known as the *effective grain size.* The diameter for which 60% of the particles are finer is designated as D_{60}.

If the Hazen uniformity coefficient is less than 4 or 5, the soil is considered uniform in particle size, which means that all particle sizes fall within a narrow range. *Well-graded soils* have uniformity coefficients greater than 10 and have a continuous, wide range of particle sizes. *Gap-graded soils* are missing one or more ranges of particle sizes. Table 18.3 gives typical uniformity coefficients.

Example

A soil's grain-size distribution curve is shown.

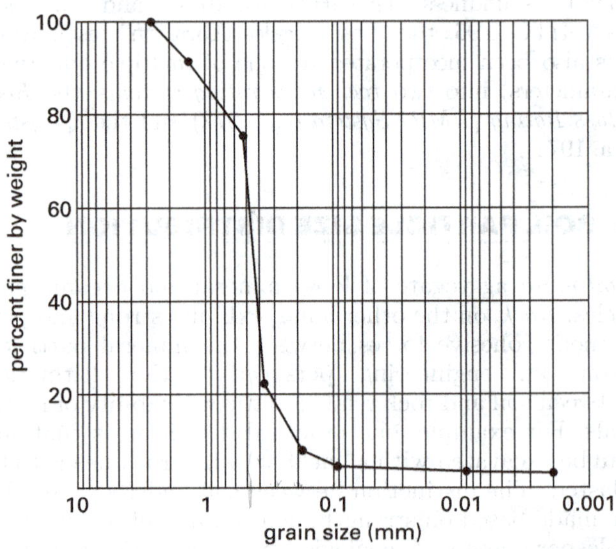

[1]While the NCEES *FE Reference Handbook* (*NCEES Handboook*) uses uppercase U as the subscript for the uniformity coefficient, it is more commonly denoted as lowercase u.

The uniformity coefficient is most nearly

(A) 1.6

(B) 2.1

(C) 2.6

(D) 3.2

Solution

As read from the distribution curve, $D_{60} = 0.49$ mm, and $D_{10} = 0.19$ mm. The uniformity coefficient is

$$C_U = D_{60}/D_{10} = \frac{0.49 \text{ mm}}{0.19 \text{ mm}} = 2.58 \quad (2.6)$$

The answer is (C).

Equation 18.2: Coefficient of Curvature

$$C_C = (D_{30})^2/(D_{10} \times D_{60}) \qquad 18.2$$

Description

The *coefficient of curvature* (also known as the *coefficient of gradation* or *coefficient of concavity*), C_C is defined in Eq. 18.2. Typical values are given in Table 18.3. The diameter for which 30% of the particles are finer is designated as D_{30}.

Equation 18.2, as well as various soil classification schemes, requires knowing percentages of soil passing through specific sieve sizes.[2] When sieve data are incomplete, the needed values can be interpolated by plotting known data on a *particle size distribution chart*, such as Fig. 18.2.

Example

A soil has a grain size distribution as shown in the sieve analysis chart.

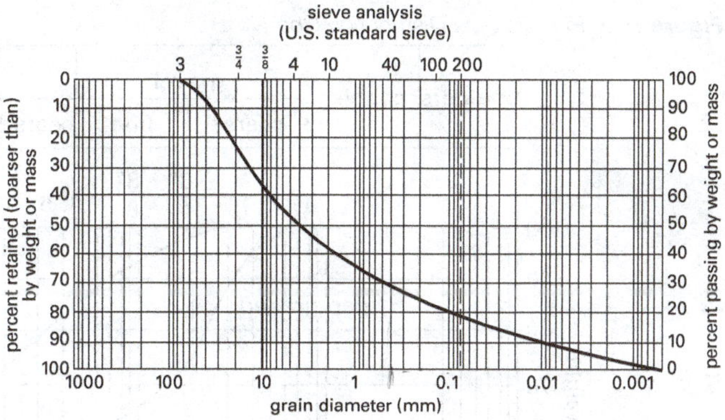

The coefficient of curvature for this soil is most nearly

(A) 2.6

(B) 3.5

(C) 4.3

(D) 4.5

Solution

From the grain size distribution, D_{10} is 0.009 mm, D_{30} is 0.45 mm, and D_{60} is 8.5 mm. The coefficient of curvature is

$$C_C = (D_{30})^2/(D_{10} \times D_{60}) = \frac{(0.45 \text{ mm})^2}{(0.009 \text{ mm})(8.5 \text{ mm})} = 2.6$$

The answer is (A).

4. SOIL CLASSIFICATION

Formal soil classification schemes have been established by a number of organizations. These schemes standardize the way that soils are described and group similar soils by the characteristics that are important in determining behavior. The classification of a soil depends mostly on the percentages of gravel, sand, silt, and clay. Usually, it will also depend on special characteristics of the silt and clay fractions.

A distinction is made between silt and clay. What are called *clays* for classification purposes depends on the existence of clay minerals in the *fines fraction* (silt and clay sizes) of the soil, as well as the size of the particles. These clay minerals have different compositions and behaviors than silts and coarse-grained soils.

Geotechnical Engineering

[2]While the *NCEES Handbook* uses uppercase C as the subscript for the coefficient of curvature, it is more commonly denoted as lowercase c.

Figure 18.2 Particle Size Distribution Chart

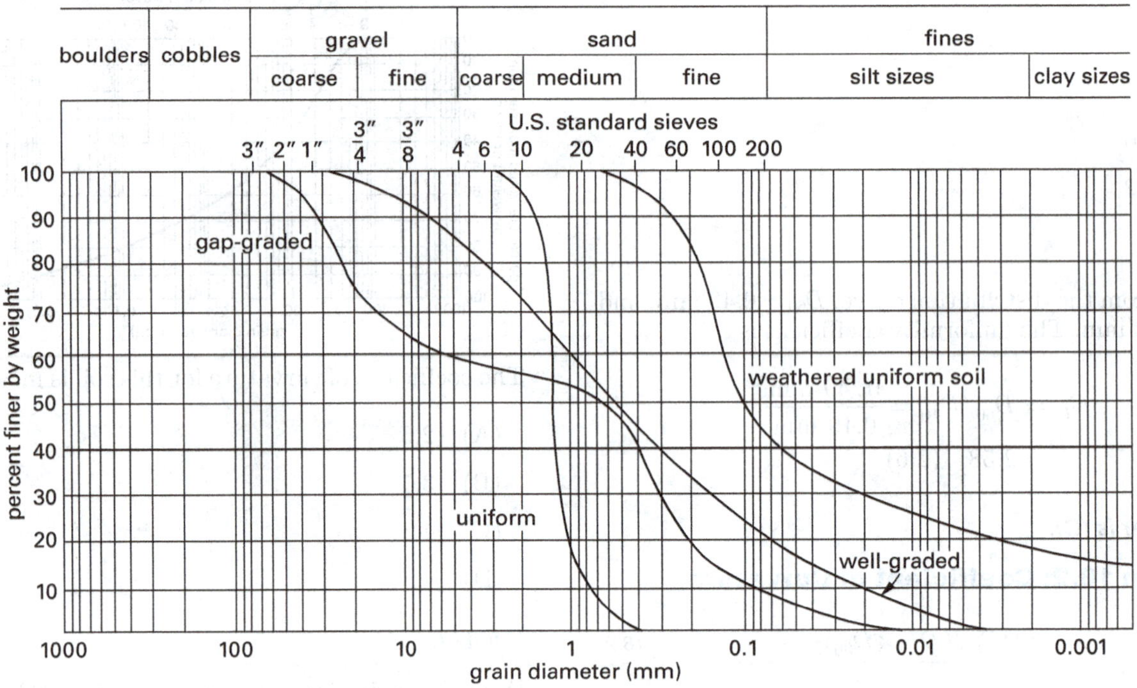

Clays display the characteristic of *plasticity.* For the most part, particles that are clay size are also clay minerals. The plasticity characteristics of the fines fraction of a soil is measured in the laboratory by the *Atterberg limit tests.*

5. ATTERBERG LIMIT TESTS

A clay soil can behave like a solid, semi-solid, plastic solid, or liquid, depending on the water content. The water contents corresponding to the transitions between these states are known as the *Atterberg limits* (*consistency limits*).[3] Each of the Atterberg limits varies with the clay content, type of clay mineral, and ions (cations) contained in the clay. Tests that determine two of these Atterberg limits—the plastic limit and the liquid limit—are frequently used to classify clay soils.

The *liquid limit* test is performed with a special apparatus. A soil sample is placed in a shallow container and the sample is parted in half with a grooving tool. The container is dropped a distance of 0.4 in repeatedly until the sample has rejoined for a length of $\frac{1}{2}$ in. The liquid limit is defined as the water content at which the soil rejoins at exactly 25 blows. The test is repeated at different water contents, and the water content corresponding to the liquid limit is found by interpolation using a *flow curve* (a plot of water content versus logarithm of number of blows).

When a soil has a liquid limit of 100, the weight of water equals the weight of the dry soil (i.e., $\omega = 100\%$). A liquid limit of 50 means that the soil at the liquid limit is two-thirds soil and one-third water.

Sandy soils have low liquid limits—on the order of 20. In such soils, the test is of little significance in judging load-carrying capacities. Silts and clays can have significant liquid limits—most clays have liquid limits less than 100, but they can be as high as 1000.

The *plastic limit* test consists of rolling a soil sample into a $\frac{1}{8}$ in thread. The sample will crumble at that diameter when it is at the plastic limit. The sample is remolded to remove moisture and rolled into a thread repeatedly until the plastic limit is reached. The water content is determined for three such samples, and the average value is the plastic limit.

Sands and most silts have no plastic limit at all. They are known as *nonplastic soils.* The test is of little significance in judging the relative load-carrying capacities of such soils. The plastic limit of clays and plastic silts can be from 0 to 100 or more, but it is usually less than 40.

Equation 18.3: Plasticity Index

$$\mathrm{PI} = \mathrm{LL} - \mathrm{PL} \qquad 18.3$$

[3]Although standard practice varies, it is suggested that Atterberg limits be reported as whole numbers rather than as percentages, to emphasize that they are an index of behavior, not a soil characteristic.

Description

The *plastic limit* (PL) is the water content corresponding to the transition between the semi-solid and plastic state. The *liquid limit* (LL) is the water content corresponding to the transition between the plastic and liquid state. Although Atterberg limits are moisture contents that could be presented as either decimal fractions or percentages, in practice they are presented as whole numbers without using the percentage symbol.

The difference between the liquid and plastic limits is known as the *plasticity index*, PI. The plasticity index indicates the range in moisture content over which the soil is in a plastic condition. In this condition it can be deformed and still hold together without crumbling. A large plasticity index (i.e., greater than 20) shows that considerable water can be added before the soil becomes liquid. The plasticity index correlates with strength, deformation properties, and insensitivity.

Example

34% of a soil sample passes through a no. 200 sieve. The soil's liquid limit is 39, and its plastic limit is 29. What is the AASHTO classification for this soil?

(A) A-2-4

(B) A-2-5

(C) A-4

(D) A-6

Solution

From the AASHTO soil classification system, when the percentage of fines is 34%, the soil is first classified as granular material. The plasticity index is

$$\begin{aligned}\text{PI} &= \text{LL} - \text{PL} = 39 - 29\\ &= 10\end{aligned}$$

For an LL of 39 and a PI of 10, the classification is A-2-4.

The answer is (A).

6. AASHTO SOIL CLASSIFICATION

The American Association of State Highway Transportation Officials' (AASHTO) classification system is based on the sieve analysis, liquid limit, and plasticity index. The best soils suitable for use as roadway subgrades are classified as A-1. Highly organic soils not suitable for roadway subgrades are classified as A-8. Soils can also be classified into subgroups. The AASHTO classification methodology is given in Table 18.4.

Equation 18.4: AASHTO Group Index

$$\begin{aligned}\text{GI} = (F-35)[0.2 + 0.005(\text{LL}-40)]\\ + 0.01(F-15)(\text{PI}-10)\end{aligned} \qquad 18.4$$

Description

The *group index*, GI, given by Eq. 18.4, may be added to the group classification. The group index is a means of comparing soils within a group, not between groups. A soil with a group index of zero is a good subgrade material within its particular group. Group indexes of 20 or higher represent poor subgrade materials. The group index is reported to the nearest whole number but is reported as zero if it is calculated to be negative.

F is the percentage of soil that passes through a no. 200 sieve (i.e., the percentage of *fines*). The liquid limit, LL, and plasticity index, PI, are discussed in Sec. 18.5. Both the quantities $\text{LL} - 40$ and $\text{PI} - 10$ may be negative. For the A-2-6 and A-2-7 subgroups, only the second term in Eq. 18.4 is used in calculating the group index.

Example

An inorganic soil has the given characteristics.

soil size (mm)	fraction retained on sieve	
< 0.002	0.19	$\text{LL} = 53$
0.002–0.005	0.12	$\text{PL} = 22$
0.005–0.05	0.36	$\text{PI} = 31$
0.05–0.075	0.04	$F_{200} = (0.04 + 0.36 + 0.12 + 0.19) \times 100 = 71$
0.075–2.0	0.29	
> 2.0	0	

Using the AASHTO classification system, what is the classification?

(A) A-7-5 (21)

(B) A-7-5 (43)

(C) A-7-6 (21)

(D) A-7-6 (43)

Solution

From Table 18.2, the no. 200 sieve has an opening of 0.075 mm. F_{200} is 0.71, so from Table 18.4, the soil is first classified as a silt-clay material (more than 35% passing a no. 200 sieve). For a liquid limit of 53 and a plasticity index of 31, the classification is A-7-5 or A-7-6. Since the

*Table 18.4 AASHTO Soil Classification System**

	granular materials (35% or less passing no. 200 sieve)							silt-clay materials (more than 35% passing no. 200 sieve)				
	A-1		A-3	A-2				A-4	A-5	A-6	A-7	A-8
	A-1-a	A-1-b		A-2-4	A-2-5	A-2-6	A-2-7				A-7-5 or A-7-6	
sieve analysis: % passing												
no. 10	50 max											
no. 40	30 max	50 max	51 min									
no. 200	15 max	25 max	10 max	35 max	35 max	35 max	35 max	36 min	36 min	36 min	36 min	
characteristics of fraction passing no. 40:												
LL: liquid limit				40 max	41 min	40 max	41 min	40 max	41 min	40 max	41 min	
PI: plasticity index	6 max		NP	10 max	10 max	11 min	11 min	10 max	10 max	11 min	11 min	
usual types of significant constituents	stone fragments gravel and sand		fine sand	silty or clayey gravel and sand				silty soils		clayey soils		peat, highly organic soils
general subgrade rating	excellent to good					fair to poor						unsatisfactory

**Classification procedure:* Using the test data, proceed from left to right in the chart. The correct group will be found by process of elimination. The first group from the left consistent with the test data is the correct classification. The A-7 group is subdivided into A-7-5 or A-7-6, depending on the plastic limit. For plastic limit PL = LL − PI less than 30, the classification is A-7-6. For plastic limit PL = LL − PI greater than or equal to 30, it is A-7-5. NP means non-plastic.

plastic limit is 22 (less than 30), the classification is A-7-6. From Eq. 18.4, the group index is

$$\begin{aligned}\text{GI} &= (F-35)[0.2+0.005(\text{LL}-40)] \\ &\quad +0.01(F-15)(\text{PI}-10) \\ &= (71-35)\big(0.2+(0.005)(53-40)\big) \\ &\quad +(0.01)(71-15)(31-10) \\ &= 21.3\end{aligned}$$

Round 21.3 to the nearest whole number. The soil classification is A-7-6 (21).

The answer is (C).

7. UNIFIED SOIL CLASSIFICATION

Figure 18.3 and Table 18.5: Classification Coefficients

Description

The *Unified Soil Classification System* (USCS) is described in Fig. 18.3 and Table 18.5. Like the AASHTO system, it is based on the grain size distribution, liquid limit, and plasticity index of the soil.

*Figure 18.3 Plasticity Chart**

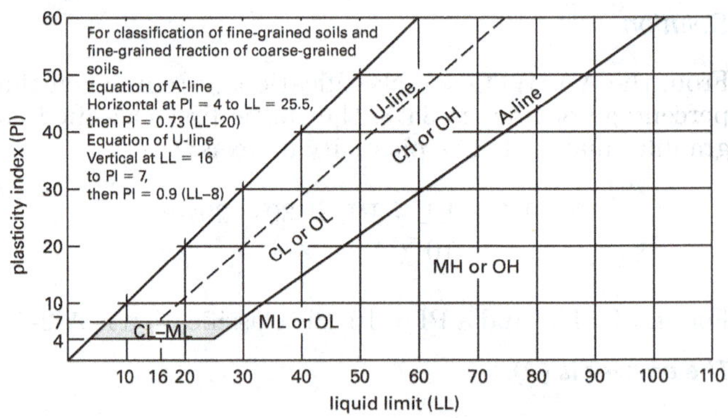

*ASTM D2487 (Unified Soil Classification System) must be used with this figure. (See Table 17.5.)

Soils are classified into USCS groups that are designated by a group symbol and a corresponding group name. The symbols each contain two letters: The first represents the most significant particle size fraction, and the second is a descriptive modifier. Some categories require dual symbols.

Table 18.5 *ASTM D2487-11 Standard Practice for Classification of Soils for Engineering Purposes (Unified Soil Classification System)*[a]

criteria for assigning group symbols and group names using laboratory tests[b]				soil classification: group symbol	soil classification: group name[c]
coarse-grained soils	gravels (more than 50% of coarse fraction retained on no. 4 sieve)	clean gravels (< 5% fines)[d]	$C_U \ge 4$ and $1 \le C_C \le 3$[e]	GW	well-graded gravel[f]
			$C_U < 4$ and/or ($C_C < 1$ or $C_C > 3$)[e]	GP	poorly graded gravel[f]
		gravel with fines (>12% fines)[d]	fines classify as ML or MH	GM	silty gravel[f,g,h]
			fines classify as CL or CH	GC	clayey gravel[f,g,h]
more than 50% retained on no. 200 sieve	sands (50% or more of coarse fraction passes no. 4 sieve)	clean sands (<5% fines)[i]	$C_U \ge 6$ and $1 \le C_C \le 3$[e]	SW	well-graded sand[j]
			$C_U < 6$ and/or ($C_C < 1$ or $C_C > 3$)[e]	SP	poorly graded sand[j]
		sands with fines (> 12% fines)[i]	fines classify as ML or MH	SM	silty sand[g,h,j]
			fines classify as CL or CH	SC	clayey sand[g,h,j]
fine-grained soils	silts and clays	inorganic	PI > 7 and plots on or above A-line[k]	CL	lean clay[l,m,n]
			PI < 4 or plots below A-line[k]	ML	silt[l,m,n]
	LL< 50	organic	LL (oven dried)/LL (not dried)	OL	organic clay[l,m,n,o]
			< 0.75	OL	organic silt[l,m,n,o,p]
50% or more passes no. 200 sieve	silts and clays	inorganic	PI plots on or above A-line	CH	fat clay[l,m,n]
			PI plots below A-line	MH	elastic silt[l,m,n]
	LL ≥ 50	organic	LL (oven dried)/LL (not dried)	OH	organic clay[l,m,n,q]
			< 0.75		organic silt[l,m,n,r]
highly organic soils	primarily organic matter, dark in color, and organic odor			PT	peat

[a]The plasticity chart must be used with this table. (See Fig. 17.3.)
[b]based on the material passing the 3 in (75 mm) sieve
[c]If the field sample contained cobbles or boulders, or both, add "with cobbles or boulders, or both" to the group name.
[d]Gravels with 5% to 12% fines require dual symbols: GW-GM = well-graded gravel with silt; GW-GC = well-graded gravel with clay; GP-GM = poorly graded gravel with silt; GP-GC = poorly graded gravel with clay.
[e]See Eq. 17.1 and Eq. 17.2.
[f]If the soil contains ≥ 15% sand, add "with sand" to the group name.
[g]If the fines classify as CL-ML, use the dual symbol GC-GM or SC-SM.
[h]If the fines are organic, add "with organic fines" to the group name.
[i]Sands with 5% to 12% fines require dual symbols: SW-SM = well-graded sand with silt; SW-SC = well-graded sand with clay; SP-SM = poorly graded sand with silt; SP-SC = poorly graded sand with clay.
[j]If the soil contains ≥ 15% gravel, add "with gravel" to the group name.
[k]If the Atterberg limits plot in the hatched area, the soil is a CL-ML, silty clay.
[l]If the soil contains 15% to 30% plus no. 200, add "with sand" or "with gravel," whichever is predominant.
[m]If the soil contains ≥ 30% plus no. 200, predominantly sand, add "sand" to the group name.
[n]If the soil contains ≥ 30% plus no. 200, predominantly gravel, add "gravelly" to the group name.
[o]PI ≥ 4 and plots on or above A-line
[p]PI < 4 or plots below A-line
[q]PI plots on or above A-line
[r]PI plots below A-line

Coarse-grained soils are divided into two categories: gravel soils (symbol G) and sand soils (symbol S). Sands and gravels are further subdivided into four subcategories as follows.

symbol W:	well-graded, fairly clean
symbol C:	significant amounts of clay
symbol P:	poorly graded, fairly clean
symbol M:	significant amounts of silt

Fine-grained soils are divided into three categories: inorganic silts (symbol M), inorganic clays (symbol C), and organic silts and clays (symbol O).[4] These three are subdivided into two subcategories as follows.

symbol L:	low compressibilities (LL less than 50)
symbol H:	high compressibilities (LL 50 or greater)

Example

An inorganic soil has the given characteristics.

soil size (mm)	fraction retained on sieve	
< 0.002	0.19	$LL = 53$
0.002–0.005	0.12	$PL = 22$
0.005–0.05	0.36	$PI = 31$
0.05–0.075	0.04	$F_{200} = (0.04 + 0.36 + 0.12 + 0.19) \times 100 = 71$
0.075–2.0	0.29	
> 2.0	0	

Using the USCS classification system, what is the classification of the soil?

(A) CH

(B) MH

(C) CL

(D) ML

Solution

From Fig. 18.3 and Table 18.5, the soil is first classified as a fine-grained soil (more than 50% passing the no. 200 sieve). The liquid limit is greater than 50 ($LL = 53$).

Since the soil plots above the A-line, it is classified as CH: a highly plastic clay.

The answer is (A).

8. MASS-VOLUME RELATIONSHIPS

Soil consists of solid soil particles separated by voids. The voids can be filled with either air or water. Thus, soil is a three-phase material (solids, water, and air), as shown in Fig. 18.4. The percentages by volume, mass, and weight of these three constituents are used to calculate the aggregate properties.

Figure 18.4 *Phase Diagram*

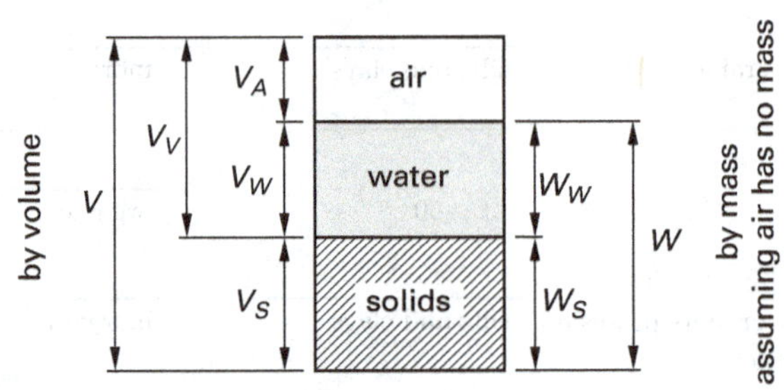

Equation 18.5: Volume of Voids

$$V_V = V_A + V_W \qquad 18.5$$

Description

Equation 18.5 shows that the volume of voids in a soil sample is equal to the sum of the volumes of air and water in the sample.[5]

Equation 18.6: Void Ratio

$$e = V_V / V_S \qquad 18.6$$

Variation

$$e = \frac{V_A + V_W}{V_S}$$

[4]The symbol M comes from the Swedish *mjala* (meaning silt) and *mo* (meaning very fine sand).

[5]While the *NCEES Handbook* uses uppercase *V*, *A*, and *S* as the subscripts for "voids," "air," and "solids," they are more commonly denoted using the lowercase letters *v*, *a*, and *s*, respectively.

Description

The *void ratio*, e, is the ratio of the volume of voids to the volume of solids.[6]

Example

A soil sample has a total mass of 23.3 g, a volume of 12 cm^3, and an oven-dry mass of 21.2 g. The volume of solids in the soil sample is 8.48 cm^3. The void ratio of this soil sample is most nearly

(A) 0.42

(B) 0.53

(C) 0.62

(D) 0.71

Solution

Find the volume of the voids. The volume of the voids is found by subtracting the volume of solids from the total volume.

$$\begin{aligned} V_V &= V - V_S \\ &= 12\ \text{cm}^3 - 8.48\ \text{cm}^3 \\ &= 3.52\ \text{cm}^3 \end{aligned}$$

The void ratio is

$$\begin{aligned} e = V_V/V_S &= \frac{3.52\ \text{cm}^3}{8.48\ \text{cm}^3} \\ &= 0.415 \quad (0.42) \end{aligned}$$

The answer is (A).

Equation 18.7: Porosity

$$n = V_V/V = e/(1+e) \qquad 18.7$$

Variation

$$n = \frac{V_A + V_W}{V_A + V_W + V_S}$$

Description

The *porosity*, n, is the ratio of the volume of voids to the total volume. In Eq. 18.7, V represents the total volume of the sample, incorporating air, water, and solid phases.[7]

Example

One cubic meter of a soil specimen has a volume of voids of 0.49 m^3. What is most nearly the porosity of the soil specimen?

(A) 0.23

(B) 0.36

(C) 0.49

(D) 0.57

Solution

The porosity is

$$\begin{aligned} n &= V_V/V \\ &= \frac{0.49\ \text{m}^3}{1.00\ \text{m}^3} \\ &= 0.49 \end{aligned}$$

The answer is (C).

Equation 18.8: Water Content

$$\omega = (W_W/W_S) \times 100\% \qquad 18.8$$

Variation

$$\omega = \frac{m_W}{m_S} \times 100\%$$

Description

The *water content* (*moisture content*), ω, is the ratio of the weight (mass) of water to the weight (mass) of solids.[8]

Example

A sample of saturated clay has a total mass of 1733 g and a dry mass of 1287 g. The specific gravity of the soil particles is 2.7. The water content of this soil is most nearly

(A) 13%

(B) 26%

(C) 35%

(D) 74%

[6] All of the soils parameters (known as *index properties*) can be stated as either percentages (e.g., 41%) or decimal fractions (e.g., 0.41). Some authorities present the void ratio, e, the degree of saturation, S, and the water content, ω, as decimal fractions, reserving percentage for porosity, n, since porosity is the only parameter with the total volume in the denominator. However, the *NCEES Handbook* presents e and n as decimal fractions and S and ω as percentages.

[7] Other sources may include a subscript t, tot, or total to designate "total" in soil phase diagrams.

[8] Although the *NCEES Handbook* uses the Greek letter ω for water content, other sources use w.

Solution

The water content is

$$\begin{aligned}\omega &= (W_W/W_S) \times 100\% \\ &= \frac{1733\ \text{g} - 1287\ \text{g}}{1287\ \text{g}} \times 100\% \\ &= 34.7\% \quad (35\%)\end{aligned}$$

The answer is (C).

Equation 18.9: Degree of Saturation

$$S = (V_W/V_V) \times 100\% \qquad \text{18.9}$$

Variation

$$S = \frac{V_W}{V_A + V_W} \times 100\%$$

Description

The *degree of saturation*, S, is the ratio of the volume of water to the volume of voids. This indicates how much of the void space is filled with water. If all the voids are filled with water, then the volume of air is zero and the sample's degree of saturation is 100%. The sample is said to be fully saturated.

Example

A soil specimen has a total water volume of 0.41 m^3 and a total voids volume of 0.49 m^3. The degree of saturation of the soil specimen is most nearly

(A) 57%

(B) 64%

(C) 71%

(D) 84%

Solution

The degree of saturation is

$$\begin{aligned}S &= (V_W/V_V) \times 100\% \\ &= \frac{0.41\ \text{m}^3}{0.49\ \text{m}^3} \times 100\% \\ &= 83.7\% \quad (84\%)\end{aligned}$$

The answer is (D).

Equation 18.10 Through Eq. 18.14: Unit Weight

$$\gamma = W/V \qquad \text{18.10}$$

$$\gamma' = \gamma_{\text{sat}} - \gamma_W \qquad \text{18.11}$$

$$\gamma_{\text{sat}} = (G_S + e)\gamma_W/(1+e) = \gamma(G_S + e)/(1+\omega) \qquad \text{18.12}$$

$$\gamma_D = W_S/V \qquad \text{18.13}$$

$$\gamma_S = W_S/V_S \qquad \text{18.14}$$

Values

$$\gamma_W = 9.81\ \text{kN/m}^3$$

Variations

$$\rho = \frac{m}{V} = \frac{m_W + m_S}{V_A + V_W + V_S}$$

$$\gamma = \rho g \quad [\text{SI only}]$$

$$\rho_D = \frac{m_S}{V} = \frac{m_S}{V_A + V_W + V_S}$$

$$\rho_D = \frac{m}{(1+\omega)V} = \frac{\rho}{1+\omega}$$

$$\gamma_D = G_S\gamma_W/(1+e) = \gamma/(1+\omega)$$

$$\rho_S = \frac{m_S}{V_S}$$

Description

Equation 18.10 through Eq. 18.14 calculate various (different) quantities sharing the common term "unit weight." Eq. 18.10, calculating γ, is the generic definition of *unit weight* (*specific weight*). Above the water table, Eq. 18.10 can be used to calculate the *total unit weight* (*bulk unit weight*, *moist unit weight*, or *wet unit weight*), so named because total weight includes the weight of any water in the unsaturated sample. Above the water table, with all of the voids filled with air, Eq. 18.13 calculates the *dry unit weight*, γ_D. For soils still above the water table but with all voids filled with water, the *saturated unit weight*, γ_{sat}, is calculated from the various forms of Eq. 18.12.[9] Below the water table, Eq. 18.11 is the *submerged unit weight*, also known as

[9]The *NCEES Handbook* uses G_S to designate the specific gravity of the soil solids. Other sources may use SG, sp.gr., or other symbols. However, the most traditional symbol used to represent specific gravity in soils-related calculations is G.

the *buoyant unit weight*. Equation 18.14, calculating γ_S, is the *solids specific weight* (i.e., the *unit weight of the solids*).

Strictly adhering to the use of proper units, the following equations are used in the United States to calculate weight and unit weight from mass and mass density. The *gravitational constant*, g_c (not the same as gravitational acceleration), always has a value of 32.2 ft-lbm/lbf-sec², regardless of the local gravitational acceleration, g, which depends on location.

$$W = \frac{mg}{g_c}$$

$$\gamma = \frac{\rho g}{g_c}$$

At the surface of the earth, the gravitational acceleration is essentially constant, and g and g_c have the same numerical values, although they have different units. So, although they have different units, specific weight and mass density also have the same numerical values. Because of this, in the United States, civil engineers use the term *unit weight* (also known as *specific weight* and *weight density*) interchangeably with the term *density*; they do not routinely differentiate between weight and mass.[10]

The unit weight of the solid constituents, γ_S, is the ratio of the mass of the solids to the volume of the solids. This would also be the unit weight of soil if there were no voids. For that reason, γ_S is also known as the *solid unit weight* and *zero-voids density*.

Example

A soil sample has a void ratio of 0.23. The specific gravity of the solids is 2.6. The saturated unit weight of the soil is most nearly

(A) 23 kN/m³

(B) 26 kN/m³

(C) 28 kN/m³

(D) 32 kN/m³

Solution

Use Eq. 18.12.

$$\begin{aligned}\gamma_{\text{sat}} &= (G_S + e)\gamma_W/(1+e)\\ &= \frac{(2.6+0.23)\left(9.81\ \frac{\text{kN}}{\text{m}^3}\right)}{1+0.23}\\ &= 22.6\ \text{kN/m}^3 \quad (23\ \text{kN/m}^3)\end{aligned}$$

The answer is (A).

Equation 18.15: Specific Gravity of Solids

$$G_S = (W_S/V_S)/\gamma_W \tag{18.15}$$

Variations

$$G_S = \frac{W_S}{V_S\gamma_W}$$

$$G_S = \frac{\rho_S}{\rho_W}$$

Description

The *specific gravity*, G_S, of the solid constituents is given by Eq. 18.15. For practical purposes, the specific weight of water, γ_W, is 62.4 lbf/ft³ (9.81 kN/m³). The specific gravity of most soils is within the range of 2.65–2.70, with clays as high as 2.9, and organic soils as low as 2.5.

Equation 18.16 and Eq. 18.17: Relative Density

$$D_r = [(e_{\text{max}} - e)/(e_{\text{max}} - e_{\text{min}})] \times 100\% \tag{18.16}$$

$$D_r = [(\gamma_{d\text{-field}} - \gamma_{d\text{-min}})/(\gamma_{d\text{-max}} - \gamma_{d\text{-min}})] \times [\gamma_{d\text{-max}}/\gamma_{d\text{-field}}] \times 100\% \tag{18.17}$$

Description

The *relative density*, D_r (also referred to as the *density index*), is a measure of the densification (compaction, settlement, etc.) achieved in a granular soil. Relative density is not a density at all. It is a fraction (percentage) representing how much a granular soil is compacted relative to the total possible compaction determined in the lab.[11] The *field unit weight* (*field density*), $\gamma_{d\text{-field}}$, is measured in the field, while the minimum and maximum densities, γ_{min} and γ_{max}, respectively, are measured in

[10]There is no concept of "weight" in the SI system. Use of the term "specific weight" for a quantity calculated from $W = mg$ and having units of N/m³ is a convention made necessary by using SI units in what, in the United States, has traditionally been a non-SI engineering subject.

[11]The *relative compaction* defined in Eq. 18.20 is a different concept, however.

the lab. Relative density is calculated from Eq. 18.16 with void ratios, while Eq. 18.17 uses unit weights.[12] This index is not applicable to clays because clays do not densify in this particular laboratory test. The relative density is equal to 1 for a very dense soil and 0 for a very loose soil.

Example

A sand has a minimum void ratio of 0.41 and a maximum void ratio of 0.78. The actual void ratio of the sand is 0.576. The relative density of the sand is most nearly

(A) 40%

(B) 55%

(C) 65%

(D) 80%

Solution

From Eq. 18.16, the relative density is

$$\begin{aligned} D_r &= [(e_{\text{max}} - e)/(e_{\text{max}} - e_{\text{min}})] \times 100\% \\ &= \frac{0.78 - 0.576}{0.78 - 0.41} \times 100\% \\ &= 55\% \end{aligned}$$

The answer is (B).

9. EFFECTIVE STRESS

Each of the three phases (solids, water, and air) in soil have different characteristics. Unlike other construction materials that essentially perform as homogeneous media, the three phases in soil interact with each other: Water and air can enter or leave the soil, and the soil grains can be pushed closer together. The water and air in the soil voids cannot support shear stress, so shear is supported entirely through grain-to-grain contact.

Equation 18.18 and Eq. 18.19: Effective Stress and Pore Water Pressure

$$\sigma' = \sigma - u \qquad 18.18$$

$$u = h_u \gamma_w \qquad 18.19$$

Description

The *effective stress*, σ', is the portion of the total stress that is supported through grain contact. It is the difference between the *total stress*, σ, and the *pore water pressure*, u. The effective stress is the average stress on a plane through the soil, not the actual contact stress between two soil particles (which can be much higher).

Pore water pressure can be interpreted as the hydrostatic pressure below the water table, and it is calculated in the same manner as hydrostatic pressure: as the product of the *submergence depth*, h_u, and water's specific weight, as shown in Eq. 18.19.[13] Essentially, the pore water pressure is the effect of buoyancy.

Example

The soil directly below a footing experiences a total compressive stress of 52 kPa. The surface of the water table is 2 m higher than the bottom of the footing. Most nearly, what is the effective stress under the footing?

(A) 13 kPa

(B) 19 kPa

(C) 32 kPa

(D) 38 kPa

Solution

From Eq. 18.18, the effective stress is

$$\begin{aligned} \sigma' &= \sigma - u = \sigma - h_u \gamma_w = 52 \text{ kPa} - (2 \text{ m})\left(9.81 \ \frac{\text{kN}}{\text{m}^3}\right) \\ &= 32.4 \text{ kPa} \quad (32 \text{ kPa}) \end{aligned}$$

The answer is (C).

10. PROCTOR TEST

Equation 18.20: Relative Compaction

$$\text{RC} = (\gamma_{d\text{-field}}/\gamma_{d\text{-max}}) \times 100\% \qquad 18.20$$

Description

Soils are compacted to increase stability and strength, enhance resistance to erosion, decrease permeability, and decrease compressibility. This is usually

[12]The variable e in Eq. 18.16 corresponds to the field condition, and to be consistent with Eq. 18.17, should be shown as e_{field}. Also, the hyphens in the subscripts of Eq. 18.17 are not minus signs; they should be interpreted as commas.

[13]The *NCEES Handbook* is inconsistent in its subscripting for water. The subscript w is used in Eq. 18.19, while W is the dominant subscript used elsewhere. Furthermore, in Eq. 18.19, the *NCEES Handbook* uses the subscript u and refers to the submergence depth as the "uplift" (pressure head). This is an uncommon convention, as uplift is only relevant below an impervious horizontal surface such as a footing, retaining wall base, or base of a dam. Uplift below impervious surfaces is not covered in the *NCEES Handbook*. The *NCEES Handbook* is inconsistent in its nomenclature regarding the submergence depth, using H_2 routinely for the same quantity as h_u.

accomplished by placing the soil in *lifts* (i.e., layers) of a few inches to a few feet thick, and then mechanically compacting the lifts. Compaction equipment can densify the soil by static loading, impact, vibration, and/or kneading actions.

The specification given to the grading contractor sets forth the minimum acceptable density as well as a range of acceptable water content values. The minimum density is specified as a *relative compaction* (RC), the percentage of the maximum value determined in the laboratory, $\gamma_{d\text{-max}}$.[14]

The basic laboratory test used to determine the maximum dry density of compacted soils is the *Proctor test.* In this procedure, a soil sample is compacted into a 1/30 ft^3 (944 cm^3) mold in 3 layers by 25 hammer blows on each layer. The hammer has a mass of 5.5 lbm (2.5 kg) and is dropped 12 in (305 mm). The dry density of the sample is calculated. This procedure is repeated for various water contents, and a graph similar to Fig. 18.5 is obtained.

$$\gamma_d = \frac{m}{V(1+\omega)}$$

Figure 18.5 *Proctor Test Curve*

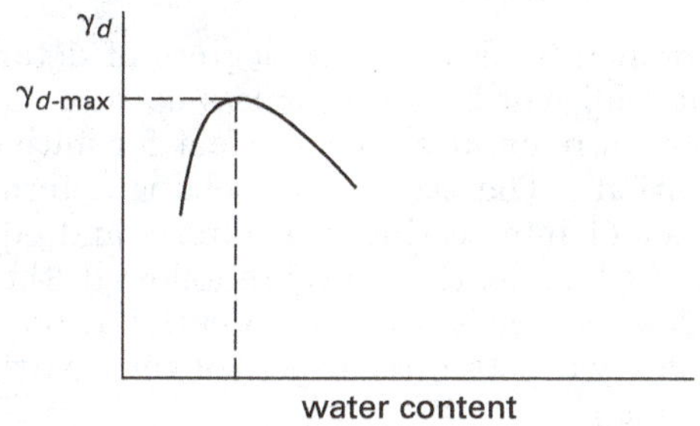

11. PERMEABILITY TESTS

Permeability of a soil is a measure of contiguous voids. It is not enough for a soil to have large voids. The voids must also be connected for water to flow through them. A permeable material supports a flow of water.

Clays are considered relatively impervious, while sands and gravels are pervious. For comparison, the permeability of concrete is approximately 10^{-10} cm/s.

Equation 18.21 and Eq. 18.22: Coefficient of Permeability

$$k = Q/(iAt_e) \quad \text{[constant head test]} \qquad 18.21$$

$$k = 2.303[(aL)/(At_e)]\log_{10}(h_1/h_2) \quad \text{[falling head test]} \qquad 18.22$$

Variations

$$k = \frac{QL}{hAt} \quad \text{[constant head]}$$

$$k = \frac{ah}{At_e}\ln\left(\frac{h_1}{h_2}\right) \quad \text{[falling head test]}$$

Description

Values for the coefficient of permeability can be calculated from controlled permeability tests using constant- or falling-head *permeameters.* (See Fig. 18.6.) In a *constant-head test*, the volume of water, V, percolating through the soil over time is measured while keeping the *hydraulic gradient*, dh/dL, constant.

Figure 18.6 *Permeameters*

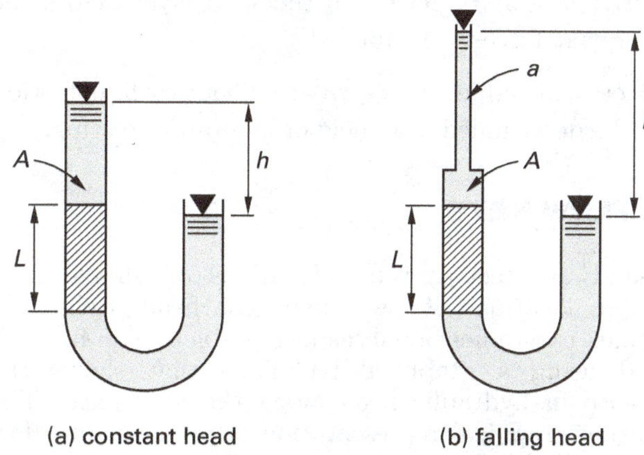

For a *constant-head permeator*, the permeability is calculated from Eq. 18.21 where A is the cross-sectional area of the permeator flow tube, and t_e is the (elapsed) time that a volume of water, Q, takes to flow through the sample. The *hydraulic gradient*, i, is the hydraulic head per unit length of the soil sample.[15]

For a *falling-head permeator*, the permeability is calculated from Eq. 18.22, where a is the cross-sectional area of the supply (reservoir) tube, A is the cross-sectional

[14]Relative compaction is not the same as relative density (see Eq. 18.16 and Eq. 18.17).

[15]The *NCEES Handbook* uses inconsistent nomenclature in Eq. 18.21 through Eq. 18.25. In Eq. 18.21, Q is a volume, while in Eq. 18.23, Q is a flow rate. In Eq. 18.23, H represents a difference in heads. In Eq. 18.25, the differential d operator indicates a difference in heads, h.

area of the permeator flow tube, and t_e is the (elapsed) time that it takes for the hydraulic head acting on the soil to decrease from h_1 to h_2.[16]

12. LIQUEFACTION

Liquefaction is a sudden drop in shear strength that can occur in soils of saturated cohesionless particles such as sand. The lower shear strength is manifested as a drop in bearing capacity. Repeated cycles of reversed shear in a saturated sand layer can cause pore pressures to increase, which in turn decreases the effective stress and shear strength. When the shear strength drops to zero, the sand liquefies. In effect, the soil turns into a liquid, allowing everything it previously supported to sink.

Conditions most likely to contribute to or indicate a potential for liquefaction include (a) a lightly loaded sand layer within 15–20 m of the surface, (b) uniform particles of medium size, (c) saturated conditions below the water table, and (d) a low-penetration test value (i.e., a low N-value).

The *cyclic stress ratio* is a numerical rating of the potential for liquefaction in sands with depths up to 12 m. This is the ratio of the average cyclic shear stress, τ_h, developed on the horizontal surfaces of the sand as a result of the earthquake loading to the initial vertical effective stress, σ_o', acting on the sand layer before the earthquake forces were applied.

The critical value of $\tau_{h,\text{ave}}/\sigma_o'$ that causes liquefaction must be determined from field or laboratory testing.

13. FLOW NETS

Groundwater seepage is from locations of high hydraulic head to locations of lower hydraulic head. Relatively complex two-dimensional seepage problems can be evaluated using a graphical technique that shows the decrease in hydraulic head along the flow path. The resulting graphical representation of pressure and flow path is called a *flow net.*

The graphical flow net concept is limited to cases where the flow is steady, two-dimensional, incompressible, and through a homogeneous medium, and where the liquid has a constant viscosity. This is the ideal case of groundwater seepage.

Flow nets are constructed from streamlines and equipotential lines. *Streamlines* (*flow lines*) show the path taken by the seepage. *Equipotential lines* (*equipotential pressure drop lines*) are contour lines of constant driving (differential) hydraulic head. (This head does not include static head, which varies with depth.)

A graphical flow net is a network of flow paths (outlined by the streamlines) and equal pressure drops (bordered by equipotential lines). No fluid flows across streamlines, and a constant amount of fluid flows between any two streamlines.

Flow nets are constructed graphically according to the following rules.

Rule 17.1: Streamlines enter and leave pervious surfaces perpendicular to those surfaces.

Rule 17.2: Streamlines approach the line of seepage (above which there is no hydrostatic pressure) asymptotically to (i.e., parallel but gradually approaching) that surface.

Rule 17.3: Streamlines are parallel to, but cannot touch, impervious surfaces that are streamlines.

Rule 17.4: Streamlines are parallel to the flow direction.

Rule 17.5: Equipotential lines are drawn perpendicular to streamlines such that the resulting cells are approximately square and the intersections are 90° angles. Theoretically, it should be possible to draw a perfect circle within each cell that touches all four boundaries, even though the cell is not actually square.

Rule 17.6: Equipotential lines enter and leave impervious surfaces perpendicular to those surfaces.

Many flow nets with differing degrees of detail can be drawn, and all will be more or less correct. Generally, three to five streamlines are sufficient for initial graphical evaluations. The size of the cells is determined by the number of intersecting streamlines and equipotential lines. As long as the rules are followed, the ratio of stream flow channels to equipotential drops will be approximately constant regardless of whether the grid is coarse or fine.

Figure 18.7 shows a flow net with three flow paths and 11 equipotential drops. A careful study of the flow nets will help to clarify the rules and conventions previously listed.

Figure 18.7 *Typical Flow Net*

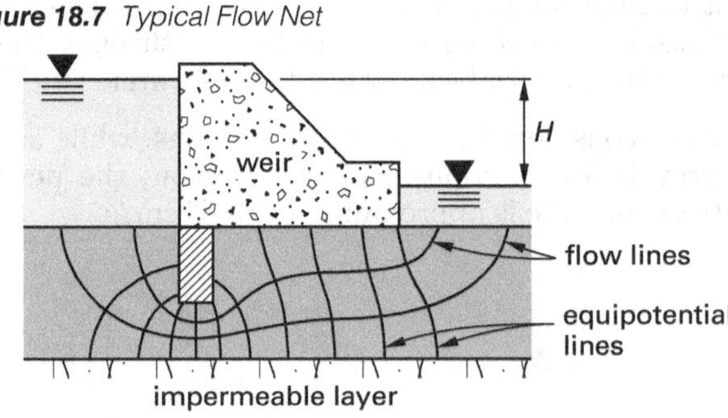

[16]The "2.303" constant term in Eq. 18.22 is a conversion from base-10 logarithms to natural logarithms. The *NCEES Handbook* could have presented Eq. 18.22 with a natural logarithm, as shown in the variation equation, without having to use an unexplained conversion factor.

14. SEEPAGE FROM FLOW NETS

Equation 18.23 Through Eq. 18.25: Seepage

$$Q = kH(N_f/N_d) \quad \text{[flow nets]} \tag{18.23}$$

$$\text{v} = ki \tag{18.24}$$

$$i = dh/dL \quad \text{[parallel to flow]} \tag{18.25}$$

Description

Once a flow net is drawn, it can be used to estimate the seepage. First, the number of flow channels, N_f, between the streamlines is counted. Then, the number of equipotential drops, N_d, between equipotential lines is counted. The total hydraulic head, H, is determined as the difference of the water surface levels.

$$H = H_1 - H_2$$

The velocity of water and contaminants moving through an aquifer, $\text{v} = Q/A$, is known as the *face velocity*, *Darcy velocity*, *Darcy flux velocity*, and *specific discharge*. It is improper and incorrect, however, to refer to the face velocity as the "seepage velocity." Eq. 18.24 is essentially a restatement of *Darcy's law* for slow, laminar flow through a porous medium.[17] The *hydraulic gradient*, i (also presented as Δh), in the direction of flow is given by Eq. 18.25.

Example

A soil sample with a coefficient of permeability of 5×10^{-6} cm/s will be subjected to a falling head test using the setup shown.

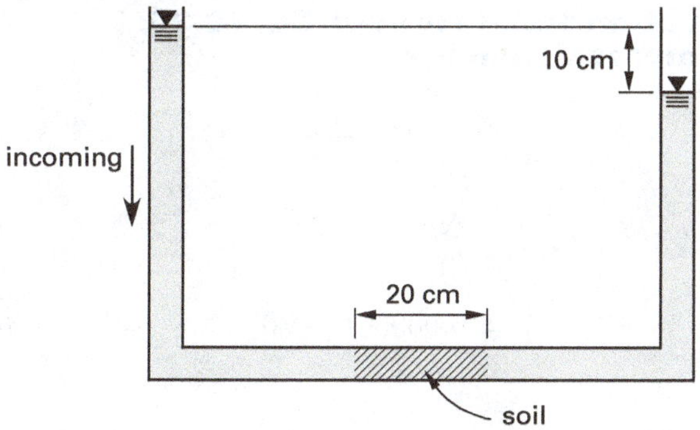

What is most nearly the face velocity of the water moving through the apparatus?

(A) 2.5×10^{-6} cm/s

(B) 5.0×10^{-6} cm/s

(C) 8.0×10^{-6} cm/s

(D) 3.0×10^{-5} cm/s

Solution

Use Eq. 18.25 to find the hydraulic gradient.

$$\begin{aligned} i = dh/dL &= \frac{10 \text{ cm}}{20 \text{ cm}} \\ &= 0.5 \text{ cm/cm} \end{aligned}$$

From Eq. 18.24, the face velocity is

$$\begin{aligned} \text{v} = ki &= \left(5 \times 10^{-6} \ \frac{\text{cm}}{\text{s}}\right)\left(0.5 \ \frac{\text{cm}}{\text{cm}}\right) \\ &= 2.5 \times 10^{-6} \text{ cm/s} \end{aligned}$$

The answer is (A).

Equation 18.26 and Eq. 18.27: Factor of Safety Against Liquefaction

$$\text{FS}_s = i_c/i_e \tag{18.26}$$

$$i_c = (\gamma_{\text{sat}} - \gamma_W)/\gamma_W \tag{18.27}$$

Description

In analyzing seepage under dams, levees, and embankments, the term *exit* (*hydraulic*) *gradient* refers to the hydraulic gradient causing flow through soil beneath the structure. When the exit gradient is high and the soil is cohesionless (sand), "quick" ground conditions and sand boiling may develop at the location of the seepage. Particularly in a cohesionless soil with fine sand and silt grains, with a high-enough exit gradient, the soil can become fluidized. The exit gradient corresponding to this fluidization is known as the *critical* (*hydraulic*) *gradient*. A catastrophic slope failure of the embankment can result. In a cohesionless foundation soil with a high percentage of larger particle sizes, although the structure stability may remain intact, the fine particles in the soil may be transported and deposited on the surface of the dry side as a *sand boil* with a corresponding increase in seepage rate through the more permeable large particles.

[17]Technically, *NCEES Handbook* Eq. 18.24 should be $\text{v} = -ki$ since flow is in the direction of decreasing hydraulic gradient.

Equation 18.26 gives the factor of safety, FS, against seepage liquefaction. i_c is the *critical hydraulic gradient*, and i_e is the *vertical seepage exit gradient*. The critical gradient can be found using Eq. 18.27.

Equation 18.27 shows that the critical exit gradient is the ratio of the buoyant unit weight (see Eq. 18.11) to the unit weight of water.

$$i_c = \frac{\gamma'}{\gamma_W}$$

15. CONSOLIDATION TESTS

Consolidation tests (also known as *confined compression tests* and *oedometer tests*) start with a disc of soil (usually clay) confined by a metal ring. (See Fig. 18.8.) The faces of the disc are covered with porous plates. The disc sandwich is loaded and submerged in water. Static loads are applied in increments, and the vertical displacement is measured with time for each load increment. The testing time is very long, usually 24 hours per load increment, since moisture content changes in clay soils are very slow. When the displacement rate levels off, the final void ratio is determined for that increment. The load versus the void ratio for all increments is plotted together as an *e-log p curve*.[18]

Figure 18.9 shows an *e*-log *p* curve for a soil sample from which the load has been removed at point *m*, allowing the clay to recover.

The line segment *m-r* is known as the *virgin compression line* or *virgin consolidation line*. In this region, the clay is considered to be *normally consolidated*, which means that the present load has never been exceeded.

Line *m-m'* is a *rebound curve*. Line *m'-r* is known as a *reloading curve*. Such curves result when a normally consolidated clay is unloaded and then reloaded. Since it has carried a higher load in the past, the soil is considered to be *overconsolidated* or *preloaded* in the rebound-reload region.

Point *m'* can only be reached by loading the soil to a pressure of p_1' and then removing the pressure. Although the pressure of the clay at p_2' is essentially the same as at the start of the test, the void ratio has been reduced.

The shape of the *e*-log *p* curve will depend on the degree of remolding or disturbance, as shown in Fig. 18.10. A highly disturbed soil will show a gradual transition between overconsolidated and normally consolidated behavior.

Figure 18.8 *Consolidation Test Apparatus*

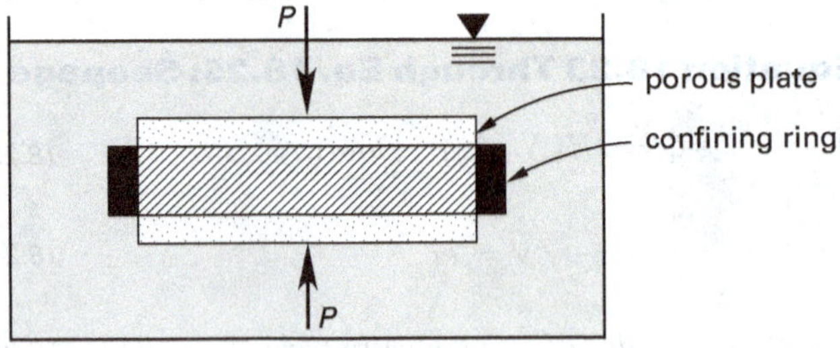

Figure 18.9 *e-log p Curve*

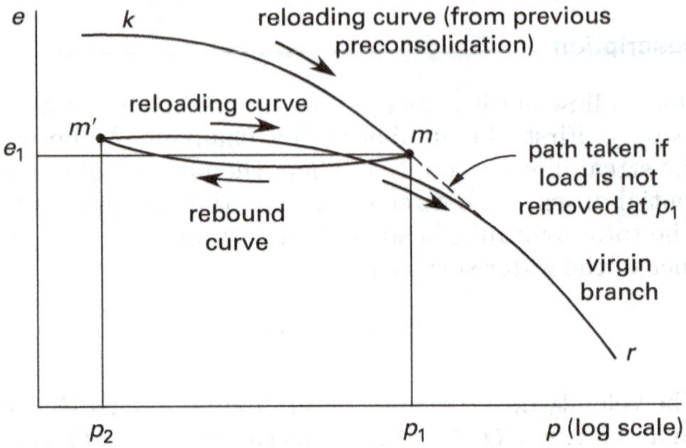

Figure 18.10 *Consolidation Curves*

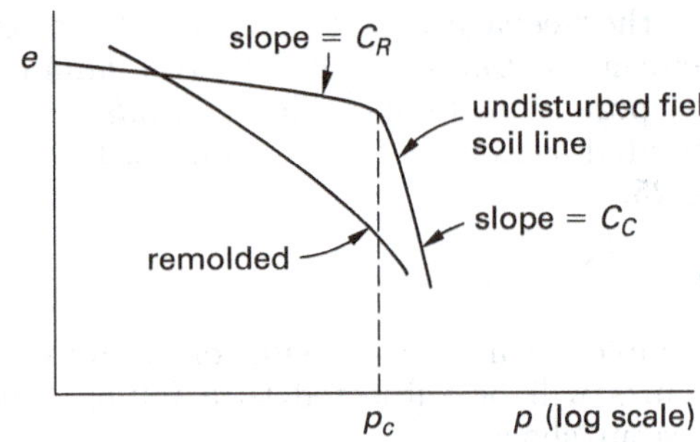

Equation 18.28 Through Eq. 18.30: Compression Index

$$C_C = \Delta e / \Delta \log p \qquad 18.28$$

$$\Delta p = I q_s \qquad 18.29$$

$$C_C = 0.009(\text{LL} - 10) \qquad 18.30$$

Variation

$$C_C = \frac{e_1 - e_2}{\log \frac{p_1}{p_2}}$$

[18]The use of pressure, p, instead of stress, σ, for the discussion on consolidation follows convention. Pressure refers to the external forces on the soil. Some soils texts substitute stress for pressure to emphasize that the soil response is dependent on internal forces in the soil, but otherwise the methodology is exactly the same.

Geotechnical Engineering

Description

The field virgin compression line, line k in Fig. 18.9, can be used to predict consolidation of the soil under any loading greater than p'_{max}. The *compression index*, C_C, or *compressibility index*, is the logarithmic slope of line m-r and is given by Eq. 18.28 and the variation equation, where points 1 and 2 correspond to any two points on line m-r.[19] Since the void ratio decreases as the pressure increases, the slope is negative. However, by long-standing convention, C_C is reported as a positive number.

Δp is the increase in pressure at the (midpoint of the) soil layer.[20] The increase in pressure can be calculated in any of the usual ways, such as from the Boussinesq equation, various influence diagrams, and reasonable assumptions about load spreading.[21]

If the clay is soft and near its liquid limit, the compression index can be approximated by Eq. 18.30.[22]

Example 14

A saturated, normally consolidated clay has a water content of 30 percent, a liquid limit of 40 percent, and a plastic limit of 20 percent. The specific gravity of solids is 2.70.

A saturated, normally consolidated clay has a water content of 30, a liquid limit of 40, and a plastic limit of 20. The specific gravity of solids in the clay is 2.70. The vertical stress increase caused by a newly constructed building causes the effective stress on the clay to double. Most nearly, how much will the void ratio of this clay decrease as the building settles?

(A) 0.013

(B) 0.081

(C) 0.19

(D) 0.50

Solution

Use Eq. 18.30 to find the approximate compression index.

$$\begin{aligned} C_C &\approx 0.009(\text{LL} - 10) = (0.009)(40 - 10) \\ &= 0.27 \end{aligned}$$

Rearrange Eq. 18.28 for the compression index as a function of the change in the void ratio.

$$\begin{aligned} C_C &= \Delta e / \Delta \log p \\ \Delta e &= C_C \Delta \log_{10} p = C_C \log_{10} \frac{p_1}{p_2} \\ &= 0.27 \log_{10} \frac{p_1}{2p_1} \\ &= 0.27 \log_{10} \frac{1}{2} \\ &= -0.081 \quad (0.081) \end{aligned}$$

The answer is (B).

Equation 18.31 and Eq. 18.32: Recompression Index

$$C_R = \Delta e / \Delta \log p \qquad 18.31$$

$$C_R = C_C / 6 \qquad 18.32$$

Variation

$$C_R = \frac{e_1 - e_2}{\log \frac{p_1}{p_2}}$$

[19]Although the compression index is, strictly speaking, the slope of the virgin compression line and, therefore, negative, it is typically reported and used as a positive number. This equation is a convenient form for calculating C_C as a positive number.

[20](1) Equation 18.28 is mathematically correct, but the *NCEES Handbook* uses an uncommon format ($\Delta\log p$) that makes one interpret p as Δp. The three traditional forms of Eq. 18.28 are (with the compression index represented by its more traditional notation, C_c, not C_C):

$$C_c = \frac{\Delta e}{\log \frac{p_1}{p_2}} = \frac{\Delta e}{\log(p_1 - p_2)} = \frac{\Delta e}{\log \Delta p}$$

(2) Although the *NCEES Handbook* does not include the "10" subscript in Eq. 18.28 (compared to Eq. 18.22), a base-10 logarithm is nevertheless required, since the compression index is intended to be the slope on e-log p graph paper. "log" is not a natural logarithm. (3) The pressures used in Eq. 18.28 are effective pressures as defined by Eq. 18.18. (The *NCEES Handbook* infers this by adding σ' after its definition of p and p_c.) (4) Equation 18.28 should use a symbol of p' or just σ' since they are the same.

[21]In Eq. 18.29, I is the stress influence value, usually read graphically from an influence diagram. q_s is the applied normal stress at the surface. The subscript s refers to "surface," not "solids" or "soil." Although q is commonly used for soil pressure under a footing, the *NCEES Handbook* does not use q for anything other than this equation. q is essentially the same as p and σ used throughout the *NCEES Handbook* for soil pressure (stress).

[22]Equation 18.30 cannot be derived from basic principles. Being one of the earliest correlative relationships, Eq. 18.30 is widely reported. However, it has distinct limitations on usage, which geotechnical engineers understand. In truth, the liquid limit has poor correlation with all consolidation properties. As originally intended, Eq. 18.30 is strictly for undisturbed clays with medium sensitivities, and even then, it is only a rough approximation.

Description

The *recompression index*, C_R, is the logarithmic slope of the recompression segment (line segment m'-m in Fig. 18.9). It is calculated from Eq. 18.31.[23] Like the compression index, C_R is traditionally reported as a positive number even though the slope is negative. Numerous correlations between the recompression index and other parameters have been proposed. Equation 18.32 can be used to obtain a very rough estimate.[24]

16. SETTLING

Settling is generally due to *consolidation* (i.e., a decrease in void fraction) of the supporting soil. There are three distinct periods of consolidation. (a) *Immediate settling*, also known as *elastic settling*, occurs immediately after the structure is constructed. Immediate settling is the major settling component in sandy soils. (b) In clayey soils, *primary consolidation* occurs gradually due to the extrusion of water from the void spaces. (c) *Secondary consolidation* occurs in clayey soils at a much slower rate after the primary consolidation has finished. Since plastic readjustment of the soil grains (including progressive fracture of the grains themselves) is the primary mechanism, the magnitude of secondary consolidation is considerably less than primary consolidation.

17. CLAY CONSOLIDATION

For clays, the method used to calculate consolidation depends on the clay condition. Virgin clay that has never experienced vertical stress higher than its current condition is known as *normally consolidated clay.* Clay that has previously experienced a stress that is no longer present is known as *overconsolidated clay.* Figure 18.11 shows a typical *consolidation curve* showing void ratio, e, versus applied effective pressure, p' (or effective stress, σ'), for a previously consolidated clayey soil, graphed on an e-log p curve. The curve shows a *recompression segment* (*preconsolidated* or *overconsolidated segment*) and the *virgin compression branch* (*normally consolidated segment*).

If the vertical pressure is increased from p_1' to p_2', the consolidation will be in the recompression zone, which has the lower slope, C_R. C_R is typically one-fifth to one-tenth of C_C. If the pressure is increased above p_2', consolidation is more dramatic, as the slope, C_C, is higher. If the pressure is increased from p_1' to p_3', the two consolidations are added together to obtain the total settlement.

Figure 18.11 *Consolidation Curve for Clay*

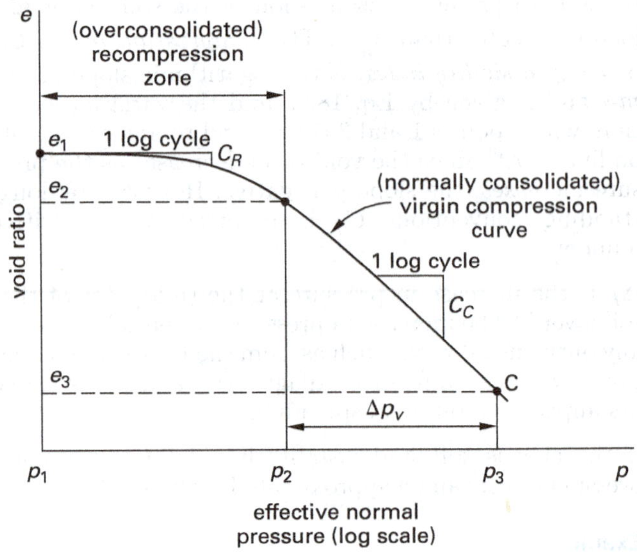

18. PRIMARY CONSOLIDATION

When a clay layer is loaded to a higher pressure, water is "squeezed" from the voids. Water is lost over a long period of time, though the rate of loss decreases with time. The loss of water results in a consolidation of the clay layer and a settlement of the soil surface. The long-term consolidation due to water loss is the *primary consolidation*, S_{primary}.

19. SETTLEMENT DUE TO CONSOLIDATION

If a structure is built on clayey soil layers, the structure loading will increase the stress in each of the layers below the foundation. This increase in stress, Δp, will gradually "squeeze" out water from the clay, resulting in a decrease in void ratio. The decrease in void ratio is referred to as consolidation, ΔH, and the decrease in structure elevation is referred to as settlement, S.

Equation 18.33 Through Eq. 18.35: Settlement

$$\Delta H = \frac{H_0}{1+e_0}\left[C_R \log\frac{p_0+\Delta p}{p_0}\right] \qquad 18.33$$

$$[(p_0 \text{ and } p_0+\Delta p) < p_c]$$

[23]See Ftn. 20. The three traditional forms of Eq. 18.31 are (with the recompression index represented by its more traditional notation, C_r, not C_R):

$$C_r = \frac{\Delta e}{\log\frac{p_1}{p_2}} = \frac{\Delta e}{\log(p_1 - p_2)} = \frac{\Delta e}{\log \Delta p}$$

[24]While it is generally true that the compression index, C_C, is generally somewhere between 5 and 10 times C_R, the *NCEES Handbook* should not state Eq. 18.32 as an equality.

$$\Delta H = \frac{H_0}{1+e_0}\left[C_C \log\frac{p_0+\Delta p}{p_0}\right] \qquad 18.34$$

$$[(p_0 \text{ and } p_0+\Delta p) > p_c]$$

$$\Delta H = \frac{H_0}{1+e_0}\left[C_R \log\frac{p_c}{p_0} + C_C \log\frac{p_0+\Delta p}{p_c}\right] \qquad 18.35$$

$$[p_0 < p_c < (p_0+\Delta p)]$$

Description

The amount of consolidation in a single clay layer depends on the thickness of the layer, H, and is directly related to the change in void ratio.

$$\Delta H = H\frac{\Delta e}{1+e_o}$$

Whether Eq. 18.28 or Eq. 18.31 (or both) is used to calculate the change in void ratio depends on the initial and final pressures, p_o and $p_o + \Delta p$, respectively.[25] As shown in Fig. 18.10, the pressure, p_c, corresponds to the knee of the consolidation curve and separates the two consolidation line segments. This pressure is known as the *preconsolidation pressure.* This pressure represents the maximum pressure that the soil saw prior to the new (structure) loading.[26] When $p < p_c$, the recompression index, C_R, is the slope of the curve, and Eq. 18.33 is used to calculate the consolidation. When $p > p_c$, the compression index, C_C, is the slope of the curve, and Eq. 18.34 is used to calculate the consolidation. If the initial pressure is less than p_c, but the final pressure is greater than p_c, then the consolidation is dependent on both curves, and Eq. 18.35 is used.[27,28]

For the purposes of calculation consolidation (settlement), the final consolidation pressure, $p = p_o + \Delta p$, used in Eq. 18.33 through Eq. 18.35, is the difference between the average pressure in the layer and the pressure at the midpoint of the clayey soil layer. Higher pressures will exist below the midpoint; lower pressure will exist above.

Equation 18.36: Vertical Strain

$$\varepsilon_v = \Delta e_{\text{TOT}}/(1+e_0) \qquad 18.36$$

Description

The total[29] *vertical soil strain,* ε_v, in a layer is the consolidation (on both the recompression and virgin segments) divided by the layer thickness. This is the basic concept expressed in Eq. 18.36.[30]

$$\varepsilon_v = \frac{\Delta H}{H} = \frac{\frac{H\Delta e}{1+e_o}}{H} = \frac{\Delta e}{1+e_o}$$

Equation 18.37: Settlement

$$S_{\text{ULT}} = \varepsilon_v H_S \qquad 18.37$$

Description

The settlement experienced by a structure due to consolidation on both the recompression and virgin segments can be calculated from the soil strain and the layer thickness using Eq. 18.37.[31] H_S is the layer thickness.[32]

Equation 18.38: Settlement vs. Time

$$S_T = U_{\text{AV}} S_{\text{ULT}} \qquad 18.38$$

[25]Although the *NCEES Handbook* uses p in Eq. 18.33 through Eq. 18.35, it uses σ in Eq. 18.18. Both represent the same quantity. The *NCEES Handbook* also uses both o and 0 as subscripts indicating "initial."

[26]In practice, p_c is not referred to as the "critical pressure," but the preconsolidation pressure. The *NCEES Handbook* labels the preconsolidation pressure, p_c, as "past maximum consolidation stress," which should be interpreted as "the previous maximum consolidation stress," not "past the maximum consolidation stress."

[27]Equation 18.35 must be used carefully. In the second term within the square brackets, used to calculate the consolidation of the virgin branch segment, Δp is the pressure change over the entire (recompression and virgin) compression range, not just the virgin compression. Both terms depend on the logarithm of the pressure ratio $p_{\text{final}}/p_{\text{initial}}$.

[28]The conditions specified by the *NCEES Handbook* for using Eq. 18.33 through Eq. 18.35 may be confusing. For example, in the *NCEES Handbook*, Eq. 18.33 is prefaced with "if (p_o and $p_o + \Delta p$) < p_c then..." which is intended to mean that the soil pressure always stays on the recompression segment. This is normally referred to as the *overconsolidated soil case.* Mathematically, if $p_o + \Delta p < p_c$, then $p_o < p_c$ is assured, so the condition is redundant and inadequate. Similarly, "if (p_o and $p_o + \Delta p$)>p_c then..." is intended to mean the "*normally consolidated soil*" case. Mathematically, if $p_o > p_c$, then $p_o + \Delta p > p_c$ is assured, so that condition is also redundant and inadequate.

[29]The term "total" refers to the fact that consolidation on both the recompression and virgin branches is included (even when only one is involved). "Total" does not refer to a summation over all soil layers, because each soil in a multi-layered foundation may have different characteristics.

[30]Because it is easy to confuse void ratio, e, with strain, ε, the concept of layer strain is rarely encountered in practical consolidation calculations. In Eq. 18.36, the subscript v denotes "vertical," not "void" or "average."

[31]Since consolidation occurs over time in different modes, the use of "ULT" (ultimate) in Eq. 18.37 could be misleading. Equation 18.37 is not a time-basis equation, and in fact, secondary consolidation will continue indefinitely, increasing the consolidation beyond what is calculated in Eq. 18.37. There is no "final" ultimate consolidation.

[32]In Eq. 18.37, the *NCEES Handbook* uses H_S as the thickness of the soil layer rather than the H that has been used in all preceding equations.

Description

Although sand will experience its consolidation (and settlement) almost instantly after a load is applied, clayey soils approach their eventual consolidations gradually. Equation 18.38 is used to calculate the settlement at a particular moment in time after the soil is loaded. U_{AV} is the average *degree of consolidation*.

S_{ULT} is the settlement calculated from Eq. 18.37. Usually, the degree of consolidation is calculated from the *time factor*, which in turn, depends on the time, layer thickness, and the *coefficient of consolidation*.[33]

Equation 18.39: Degree of Consolidation Time Factor

$$T_v = \frac{c_v t}{H_{dr}^2} \qquad 18.39$$

Description

The degree of consolidation, U, is the percentage of the total settlement of a clay layer achieved at a time, t. The degree of consolidation can be determined by dividing the change in excess pore pressure at time t by the initial excess pore pressure, u_0. The initial excess pore pressure is constant with the depth of the soil layer.

The time factor, T_v, given by Eq. 18.39, is a function of the average degree of consolidation, U (%), of a saturated clay layer at time t. Table 18.6 shows the variation of the time factor, T_v, with the degree of consolidation. The coefficient of consolidation, c_v, has units of square distance per time. The length of the average maximum drainage path, H_{dr}, is dependent on the soil type above and below the clay layer. Figure 18.12 shows the three types of drainage path conditions with a constant pore pressure u_0. Figure 18.12(a) depicts two-way drainage, with free-draining soil above and below the clay layer; in this case, the height of drainage, H_{dr}, is one-half the full height of the clay layer, with half the excess pore pressure both above and below the soil. Figure 18.12(b) and Fig. 18.12(c) depict one-way drainage, where either the soil layer above or below the clay layer are impervious; in these cases, the height of drainage is the full height of the clay layer.

Figure 18.12 Different Types of Drainage with u_0 Constant

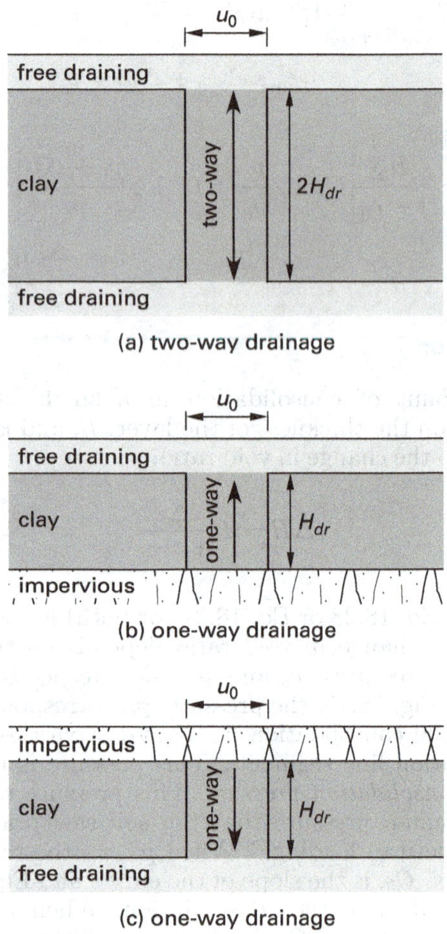

Example

A 10 ft thick clay layer has a permeable sand layer above and below it. The coefficient of consolidation is 0.25 ft²/day. Most nearly, what is the rate of consolidation?

(A) 55 days

(B) 68 days

(C) 81 days

(D) 85 days

Solution

Sand above and below the clay layer indicates two-way drainage, so the height of drainage, H_{dr}, is cut in half. From Table 18.6, for a degree of consolidation, U, of

[33] t_c is not related to the preconsolidation pressure, p_c, although they share the same subscript. t_c is the time since the load was applied. T is used to designate the time factor, although the traditional symbol is T_v.

*Table 18.6 Variation of Time Factor with Degree of Consolidation**

U (%)	T_v	U (%)	T_v	U (%)	T_v
0	0	34	0.0907	68	0.377
1	0.00008	35	0.0962	69	0.390
2	0.0003	36	0.102	70	0.403
3	0.00071	37	0.107	71	0.417
4	0.00126	38	0.113	72	0.431
5	0.00196	39	0.119	73	0.446
6	0.00283	40	0.126	74	0.461
7	0.00385	41	0.132	75	0.477
8	0.00502	42	0.138	76	0.493
9	0.00636	43	0.145	77	0.511
10	0.00785	44	0.152	78	0.529
11	0.0095	45	0.159	78	0.547
12	0.0113	46	0.166	80	0.567
13	0.0133	47	0.173	81	0.588
14	0.0154	48	0.181	82	0.610
15	0.0177	49	0.188	83	0.633
16	0.0201	50	0.197	84	0.658
17	0.0227	51	0.204	85	0.684
18	0.0254	52	0.212	86	0.712
19	0.0283	53	0.221	87	0.742
20	0.0314	54	0.230	88	0.774
21	0.0346	55	0.239	89	0.809
22	0.0380	56	0.248	90	0.848
23	0.0415	57	0.257	91	0.891
24	0.0452	58	0.267	92	0.938
25	0.0491	59	0.276	93	0.993
26	0.0531	60	0.286	94	1.055
27	0.0572	61	0.297	95	1.129
28	0.0615	62	0.307	96	1.219
29	0.0660	63	0.318	97	1.336
30	0.0707	64	0.329	98	1.500
31	0.0754	65	0.340	99	1.781
32	0.0803	66	0.352	100	∞
33	0.0855	67	0.364		

*u_0 constant with depth

90%, the time factor, T_v, is 0.848. The rate of consolidation is

$$T_v = \frac{c_v t}{H_{dr}^2}$$

$$\begin{aligned} t &= \frac{H_{dr}^2 T_v}{c_v} \\ &= \frac{(5\ \text{ft})^2(0.848)}{0.25\ \frac{\text{ft}^2}{\text{day}}} \\ &= 84.8\ \text{days} \quad (85\ \text{days}) \end{aligned}$$

The answer is (D).

20. DIRECT SHEAR TEST

The *direct shear test* is a relatively simple test used to determine the relationship of shear strength to consolidation stress. In this test, a disc of soil is inserted into the direct shear box. The box has a top half and a bottom half that can slide laterally with respect to each other. A normal stress, σ_N, is applied vertically, and then one half of the box is moved laterally relative to the other at a constant rate. Measurements of vertical and horizontal displacement, δ, and horizontal shear load, T, are taken. The test is usually repeated at three different vertical normal stresses. (See Fig. 18.13.)

Because of the box configuration, failure is forced to occur on a horizontal plane. Results from each test are plotted as horizontal displacement versus horizontal *shear stress*, τ (horizontal force divided by the nominal area). *Failure* is determined as the maximum value of horizontal stress achieved. The vertical normal stress and failure stress from each test are then plotted in Mohr's circle space of normal stress versus shear stress.

Figure 18.13 *Graphing Direct-Shear Test Results*

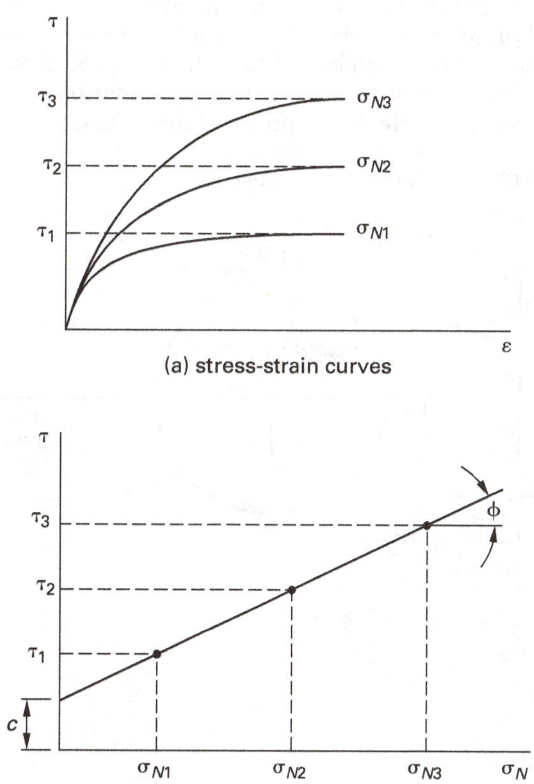

(b) Mohr's failure envelope

A line drawn through all of the test values is called the *failure envelope* (*failure line* or *rupture line*).

Equation 18.40 Through Eq. 18.42: Mohr-Coulomb Equation

$$\tau_F = c + \sigma_N \tan\phi \tag{18.40}$$

$$\sigma_N = P/A \tag{18.41}$$

$$\tau = T/A \tag{18.42}$$

Description

The equation for the failure envelope is given by the *Mohr-Coulomb equation*, Eq. 18.40, which relates the ultimate shear strength of the soil, τ_F, to the normal stress on the failure plane.[34,35,36] ϕ is the *angle of internal friction*.[37] The intercept c is the *cohesion*, a characteristic of cohesive soils.

Equation 18.40 is in the format of an equation of a straight line. Referring to an x-y plane, the "y-intercept" determines the cohesion, c, and the slope, $\tan\phi$, predicts the angle of internal friction, ϕ. These soil parameters and the failure line are shown in Fig. 18.14, superimposed on Mohr's circle.[38] Equation 18.40 can be used to determine the ultimate shear strength, S, from the unconfined compression test. Any combinations of stress that plot above the line represent failure modes.

Figure 18.14 Mohr-Couloumb Failure

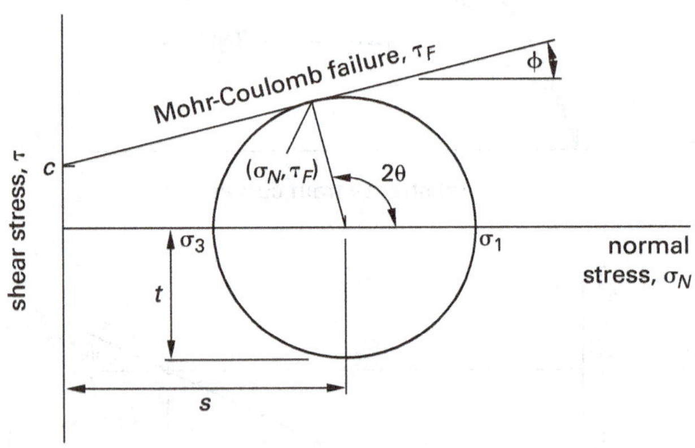

21. UNCONFINED COMPRESSIVE STRENGTH TEST

In an *unconfined compressive strength test*, a cylinder of cohesive soil is loaded axially to compressive failure. Since no horizontal confinement is used, this test can only be performed on soils that can stand without confinement, usually clays.

Considering the bearing capacity of foundations, dams, and footings, the most critical condition for a clayey soil usually occurs immediately after construction, when the soil is undrained and unconsolidated. When loads are first applied to saturated clay, the pore pressure increases. For a short period of time, the pore pressure does not dissipate, and the angle of internal friction is zero. This is known as the $\phi = 0°$ condition ($\phi = 0°$ case, *undrained case*, etc.), when the *undrained shear strength*, S_u, is equal to the cohesion, c. The undrained shear strength of clays is usually determined from the unconfined compression test as one-half the *unconfined compressive strength*, q_u.

$$S_{u,c} = c = \frac{q_u}{2}$$

[34]Equation 18.40 is also known as *Coulomb's equation*.

[35]The ultimate shear strength may be given the symbol S in some soils books.

[36]τ and σ in Coulomb's equation are the shear stress and normal stress, respectively, on the failure plane at failure.

[37]In a physical sense, the angle of internal friction for cohesionless soils is the angle from the horizontal naturally formed by a pile. For example, a uniform fine sand makes a pile with a slope of approximately 30°. For most soils, the natural angle of repose will not be the same as the angle of internal friction, due to the effects of cohesion.

[38]The *NCEES Handbook* defines t (as opposed to τ_{max}) in Fig. 18.14 as the *maximum shear stress*. The *mean normal stress*, the horizontal coordinate of the center of the circle, is designated s rather than σ_m or σ_{ave}. These maximum and average values are specifically for a single test, not any global value associated with past, current, or future field conditions. While it is likely that the parameters of the line (i.e., the intercept and slope, c and ϕ) would be unchanged, a shear test with a different normal stress would generate different maximum and average values.

19 Foundations

Nomenclature

A	area	ft^2	m^2
B	width	ft	m
c	cohesion	lbf/ft^2	Pa
D	depth	ft	m
FS	factor of safety	–	–
g	acceleration of gravity, 32.2 (9.81)	ft/sec^2	m/s^2
I	influence value	–	–
L	length	ft	m
N	bearing capacity factor	–	–
p	pressure	lbf/ft^2	Pa
P	load	lbf	N
q	bearing capacity	lbf/ft^2	Pa
r	radial distance	ft	m
R	radius	ft	m
z	depth	ft	m

Symbols

γ	specific (unit) weight	lbf/ft^3	N/m^3
ρ	density	lbm/ft^3	kg/m^3
σ	vertical stress	lbf/ft^2	Pa
σ'	effective stress	lbf/ft^2	Pa
ϕ	angle of internal friction	deg	deg

Subscripts

a	allowable
c	cohesion or critical
f	footing or foundation
q	depth or surcharge
s	surface
u	ultimate or undrained
ULT	ultimate
z	depth
γ	specific (unit) weight

1. INTRODUCTION[1]

A *foundation* is the part of an engineered structure that transmits the structure's forces into the soil or rock that supports it. The shape, depth, and materials of the foundation design depend on many factors, including the structural loads, the existing ground conditions, and local material availability.

The term *shallow foundation* refers to a foundation system in which the depth of the foundation is shallow relative to its width, usually $D_f/B \leq 1$. This category of foundation includes spread footings, continuous (or wall) footings, and mats.

The main considerations in designing shallow foundations are ensuring against bearing capacity failures and excessive settlements. *Bearing capacity* is the ability of the soil to support the foundation loads without shear failure. *Settlement* is the tendency of soils to deform (densify) under applied loads. Since structures can tolerate only a limited amount of settlement, foundation design will often be controlled by settlement criteria, because soil usually deforms significantly before it fails in shear. The most damaging settlements are *differential settlements*—those that are not uniform across the supported area. Excessive settlements may only lead to minor damage such as cracked floors and walls, windows and doors that do not operate correctly, and so on. However, bearing capacity failures have the potential to cause major damage or collapse.

With *general shear failures*, the soil resists an increased load until a sudden failure occurs. *Local shear failure* occurs in looser, more compressible soils and at high bearing pressures. The boundaries between these types of behavior are not distinct, and the methods of calculating general shear failure are commonly used for most soil conditions.

[1]Bearing capacity of foundations is not covered in the NCEES *FE Reference Handbook* (*NCEES Handbook*), but is an important geotechnical engineering subject.

2. SAND VERSUS CLAY

Ordinarily, sand makes a good foundation material. Sand is usually strong, and it drains quickly. It may have an initial, "immediate" settlement upon first loading, but this usually is small and is complete before most architectural elements (drywall, brick, glass), which are brittle and more susceptible to settlement damage, are constructed. However, it behaves poorly in excavations because it lacks cohesion. When sand is loose and saturated, it can become "quick" (i.e., liquefy), and a major loss in supporting strength occurs.

Clay is generally good in excavations but poor in foundations. Clay strength is usually lower than that of sand. Clays retain water, are relatively impermeable, and do not drain freely. Settlement in clays continues beyond the end of construction, and significant settlement can continue for years or even indefinitely. Large volume changes result from *consolidation*, which is the squeezing out of water from the pores as the soil comes to equilibrium with the applied loads.

3. GENERAL CONSIDERATIONS FOR FOOTINGS

A *footing* is an enlargement at the base of a load-supporting column that is designed to transmit forces to the soil. The area of the footing will depend on the load and the soil characteristics. Several types of footings are used, including spread and continuous footings, which are shown in Fig. 19.1. A *spread footing* is a footing used to support a single column. This is also known as an *individual column footing* or an *isolated footing*. A *continuous footing*, also known as a *wall footing* or *strip footing*, is a long footing supporting a continuous wall. A *combined footing* is a footing carrying more than one column. A *cantilever footing* is a combined footing that supports a column and an exterior wall or column.

4. PRESSURE FROM APPLIED POINT LOADS: BOUSSINESQ'S EQUATION

The vertical pressure (stress), σ_z, in a footing caused by an application of a point load, P, at the surface can be found from *Boussinesq's equation*. Boussinesq's equation assumes that radial distance, r, is small compared to the depth, z. (See Fig. 19.2.)

$$\sigma_z = \frac{3Pz^3}{2\pi(r^2+z^2)^{5/2}}$$

Example

A 1.5 kN point load is applied to the soil surface by the tire of a parked motorcycle. Using Boussinesq's equation, most nearly, what is the increase in vertical pressure at a point 1.0 m below the surface and 0.22 m radially from the tire?

(A) 470 Pa

(B) 540 Pa

(C) 640 Pa

(D) 710 Pa

Solution

The increase in vertical pressure is

$$\sigma_z = \frac{3Pz^3}{2\pi(r^2+z^2)^{5/2}} = \frac{(3)(1500\ \text{N})(1\ \text{m})^3}{2\pi\left((0.22\ \text{m})^2+(1\ \text{m})^2\right)^{5/2}} = 636\ \text{Pa} \quad (640\ \text{Pa})$$

The answer is (C).

Figure 19.1 *Types of Footings*

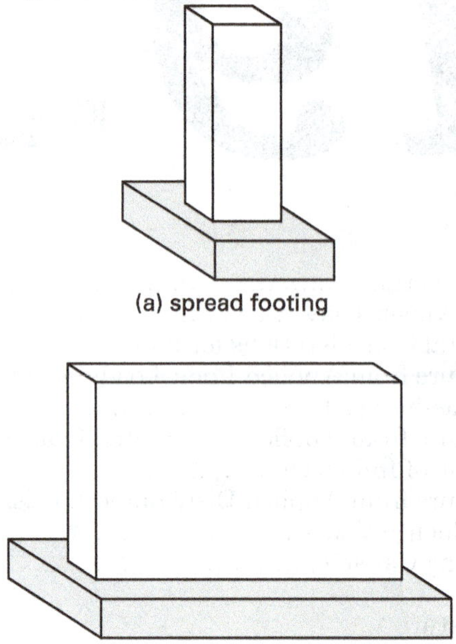

(b) continuous (wall, strip) footing

Figure 19.2 *Pressure at a Point*

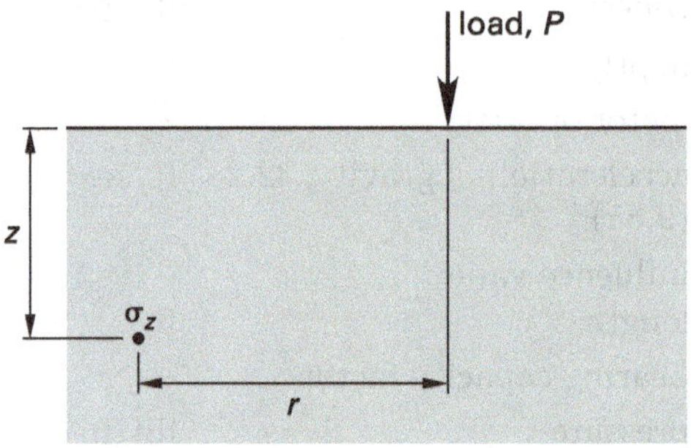

5. PRESSURE FROM APPLIED DISTRIBUTED LOADS: ZONE OF INFLUENCE

For a uniformly loaded rectangular surface area, the approximate stress at a depth can be determined by assuming the geometry of the affected area (i.e., the *zone of influence*). Approximate methods can be reasonably accurate in nonlayered homogeneous soils when $1.5 < z/x < 5$. (See Fig. 19.3.)

Figure 19.3 *Zone of Influence*

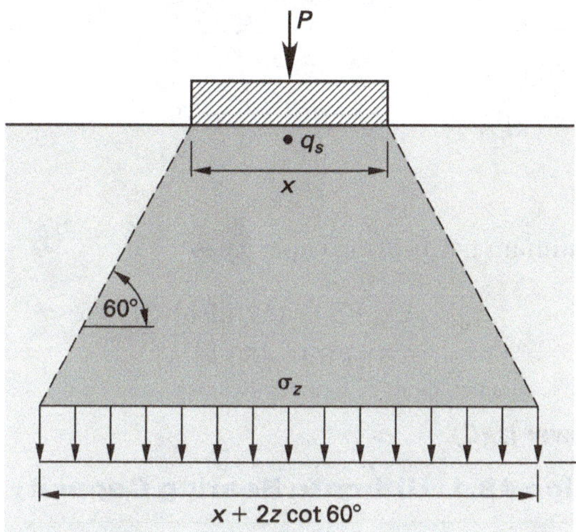

The zone of influence is defined by the angle of the *influence cone.* This angle is typically assumed to be 51° for point loads and 60° for uniformly loaded circular and rectangular footings. (Accordingly, this method is sometimes referred to as the *60° method.*) Alternatively, since 60° is an approximation anyway, for ease of computation, the angle is taken as 63.4°, corresponding to a 2:1 (vertical:horizontal) influence cone angle.

For any depth, the area of the cone of influence is the area of the horizontal plane enclosed by the influence boundaries, as determined geometrically. For a rectangular $x \times y$ footing, with the 60° method, the area of influence at depth z is

$$A_z = (x + 2z \cot 60°)(y + 2hz \cot 60°)$$

For a rectangular $x \times y$ footing, assuming a 2:1 influence, the area of influence at depth z is

$$A_z = (x + z)(y + z)$$

$$\sigma_z = \frac{P}{A_z} = \frac{xyP}{A_z xy} = \frac{xyq_s}{A_z}$$

6. PRESSURE FROM APPLIED DISTRIBUTED LOADS: INFLUENCE VALUE

The *influence value method* calculates the increase in pressure, $\Delta p = \sigma_z$, at a depth as a simple fraction of the applied surface pressure, q_s, calculated from the applied load and the area of the footing.

$$\sigma_z = I_z q_s = \frac{I_z P}{A}$$

The *influence value*, I_z, is invariably determined graphically using one of two methods. The first method is to use an *influence value chart* (*stress contour chart*) where curves are plotted below an accurately scaled (elevation view) representation of the footing. The second method, somewhat more laborious, is to draw a (plan view) scaled representation of the footing on an influence chart (i.e., a *Newmark influence chart*), and to count the number of squares covered. This is one of the few simple methods of determining the increase in soil pressure associated with irregularly shaped (noncircular, nonrectangular, etc.) footings.

The preparation of these two types of charts is dependent on the assumptions and soil theory (e.g., Boussinesq) used as well as on the plan view shape of the footing. Some charts accommodate virtually any combination of depth and radial distance from the footing; other charts are specifically for points below the center, edge, or corner.

Figure 19.4 is an influence value chart based on Boussinesq theory for use with circular spread footings. Figure 19.5 is an influence value chart based on Boussinesq theory for use with square and wall spread footings. (Figure 19.4 and Fig. 19.5 appear at the end of this chapter.)

Example

A large circular mat footing has a 5 m radius. The footing transmits a uniform load of 95 kPa to the surface of the soil. Using a Boussinesq influence chart, most nearly, what is the increase in soil pressure 15 m (radially) from the center and 20 m below the surface?

(A) 34 Pa

(B) 340 Pa

(C) 2700 Pa

(D) 4200 Pa

Solution

Use Fig. 19.4. Calculate each chart index.

$$\frac{z}{R} = \frac{20 \text{ m}}{5 \text{ m}} = 4$$
$$\frac{r}{R} = \frac{15 \text{ m}}{5 \text{ m}} = 3$$

From the Boussinesq stress contour chart for circular footings, the influence value, I_z, is approximately 0.028. The increase in soil pressure is

$$\begin{aligned}\Delta p = I_z q_s &= (0.028)(95 \text{ kPa})\left(1000 \ \frac{\text{Pa}}{\text{kPa}}\right) \\ &= 2660 \text{ kPa} \quad (2700 \text{ Pa})\end{aligned}$$

The answer is (C).

7. BEARING CAPACITY

Allowable Bearing Capacity

The *allowable bearing capacity* (also known as the *net allowable bearing pressure* or *safe bearing pressure*) is the net pressure in excess of the overburden stress that will not cause shear failure or excessive settlements. This is the soil pressure that is used to design the foundation. (The term "allowable" means that a factor of safety has already been applied.)

The allowable bearing capacity, q_a, is determined by dividing the net capacity by a factor of safety, FS. The safety factor accounts for the uncertainties in evaluating soil properties and anticipated loads, and also on the amount of risk involved in building the structure. A safety factor between 2 and 3 (based on net allowable pressure) is common for average conditions. Smaller safety factors are sometimes used for transient load conditions such as from wind and seismic forces.

$$q_a = \frac{q_{\text{net}}}{\text{FS}}$$

Example

A spread footing is to be built for a new construction project. Conditions for the spread footing require a factor of safety of 3. If the allowable bearing capacity of the soil under the footing is not at least 17 kPa, there will be a serious risk of shear failure. What is most nearly the minimum net bearing capacity the footing must have in order to meet the design requirements?

(A) 6.0 kPa

(B) 20 kPa

(C) 50 kPa

(D) 55 kPa

Solution

The minimum net bearing capacity is

$$\begin{aligned}q_{\text{net}} &= q_a\text{FS} = (17 \text{ kPa})(3) \\ &= 51 \text{ kPa} \quad (50 \text{ kPa})\end{aligned}$$

The answer is (C).

Equation 19.1: Ultimate Bearing Capacity

$$q_{\text{ULT}} = cN_c + \gamma' D_f N_q + \tfrac{1}{2}\gamma' B N_\gamma \qquad 19.1$$

Description

The *ultimate* (or *gross*) *bearing capacity* for a shallow continuous footing is given by Eq. 19.1. The equation is valid for both sandy and clayey soils.

Various researchers have made improvements on the theory supporting this equation, leading to somewhat different terms and sophistication in evaluating *bearing capacity factors* N_γ, N_c, and N_q. The approaches differ in the assumptions made of the shape of the failure zone beneath the footing. However, the general form of the equation is the same in most cases. The bearing capacity factors in Table 19.1 are based on Terzaghi's 1943 studies. Other values are also in use.

The Terzaghi-Meyerhof equation is appropriate for a foundation in a continuous wall footing. Corrections, called *shape factors*, for various footing geometries are presented in Table 19.2 and Table 19.3 using the parameters identified in Fig. 19.6. The bearing capacity factors N_c and N_γ are multiplied by the appropriate shape factors when they are used in the Terzaghi-Meyerhof equation.

*Table 19.1 Terzaghi Bearing Capacity Factors for General Shear**

ϕ	N_c	N_q	N_γ
0°	5.7	1.0	0.0
5°	7.3	1.6	0.5
10°	9.6	2.7	1.2
15°	12.9	4.4	2.5
20°	17.7	7.4	5.0
25°	25.1	12.7	9.7
30°	37.2	22.5	19.7
34°	52.6	36.5	35.0
35°	57.8	41.4	42.4
40°	95.7	81.3	100.4
45°	172.3	173.3	297.5
48°	258.3	287.9	780.1
50°	347.5	415.1	1153.2

*Curvilinear interpolation may be used. Do not use linear interpolation.

Table 19.2 N_c Bearing Capacity Factor Multipliers for Various Values of B/L

B/L	multiplier
1 (square)	1.25
0.5	1.12
0.2	1.05
0.0	1.00
1 (circular)	1.20

Table 19.3 N_γ Multipliers for Various Values of B/L

B/L	multiplier
1 (square)	0.85
0.5	0.90
0.2	0.95
0.0	1.00
1 (circular)	0.70

Figure 19.6 Spread Footing Dimensions

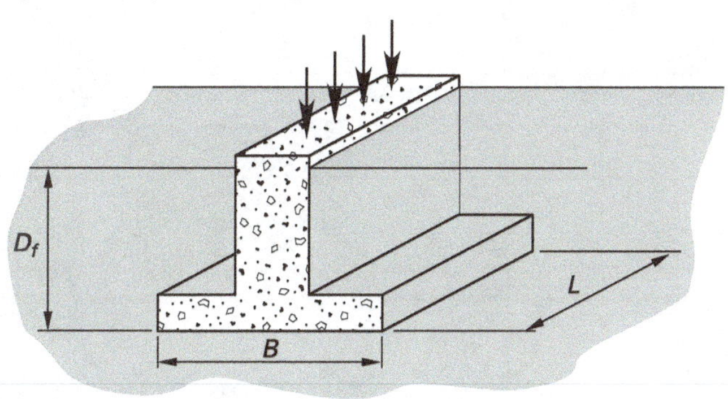

Example

A continuous footing is shown. Most nearly, what is the ultimate bearing capacity of the footing?

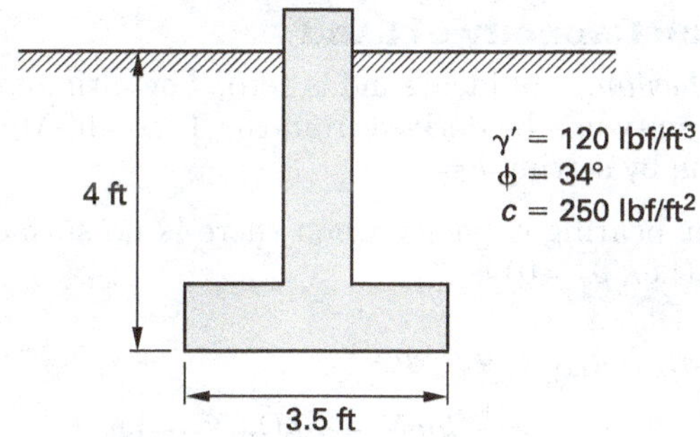

(A) 31,000 lbf/ft²

(B) 35,000 lbf/ft²

(C) 38,000 lbf/ft²

(D) 43,000 lbf/ft²

Solution

From Table 19.1, $N_c = 52.6$, $N_q = 36.5$, and $N_\gamma = 35.0$. From Eq. 19.1,

$$\begin{aligned} q_{\text{ULT}} &= cN_c + \gamma' D_f N_q + \tfrac{1}{2}\gamma' B N_\gamma \\ &= \left(250\ \frac{\text{lbf}}{\text{ft}^2}\right)(52.6) + \left(120\ \frac{\text{lbf}}{\text{ft}^3}\right)(4\ \text{ft})(36.5) \\ &\quad + \left(\frac{1}{2}\right)\left(120\ \frac{\text{lbf}}{\text{ft}^3}\right)(3.5\ \text{ft})(35.0) \\ &= 38{,}020\ \text{lbf/ft}^2 \quad (38{,}000\ \text{lbf/ft}^2) \end{aligned}$$

The answer is (C).

Bearing Capacity of Clay

Clay is often soft and fairly impermeable. The bearing capacity of clays should be designed for the undrained case, where $\phi = 0°$, and the cohesion is equal to half of the undrained compressive strength. If $\phi = 0°$, then $N_\gamma = 0$ and $N_q = 1$. If there is no surface surcharge (i.e., $p_q = 0$), the ultimate bearing capacity is given by the following equations.

$$q_u = cN_c + \rho g D_f$$

$$q_{\text{net}} = q_u - \rho g D_f = cN_c$$

The allowable clay loading is based on a factor of safety, which is typically taken as 3 for clay.

$$q_a = \frac{q_{\text{net}}}{\text{FS}}$$

From the equations for ultimate bearing capacity, it is evident that the cohesion term dominates the bearing capacity in cohesive soil.

Bearing Capacity of Sand

The *cohesion*, c, of ideal sand is zero. The ultimate bearing capacity can be derived from the Terzaghi-Meyerhof equation by setting $c = 0$.

The net bearing capacity when there is no surface surcharge (i.e., $p_q = 0$) is

$$\begin{aligned} q_{\text{net}} &= q_u - \rho g D_f \\ &= \tfrac{1}{2} B\rho g N_\gamma + \rho g D_f (N_q - 1) \end{aligned}$$

It is evident that the depth term, $\rho g D_f N_q$, dominates the bearing capacity in cohesionless soil. A small increase in depth increases the bearing capacity substantially.

The allowable bearing capacity is based on a factor of safety, which is typically taken as 2 for sand.

$$q_a = \frac{q_{\text{net}}}{\text{FS}} = \frac{B}{\text{FS}} \left(\tfrac{1}{2} \rho g N_\gamma + \rho g (N_q - 1) \left(\frac{D_f}{B} \right) \right)$$

Geotechnical Engineering

Figure 19.4 Boussinesq Stress Contour Chart for Uniformly Loaded Circular Footings

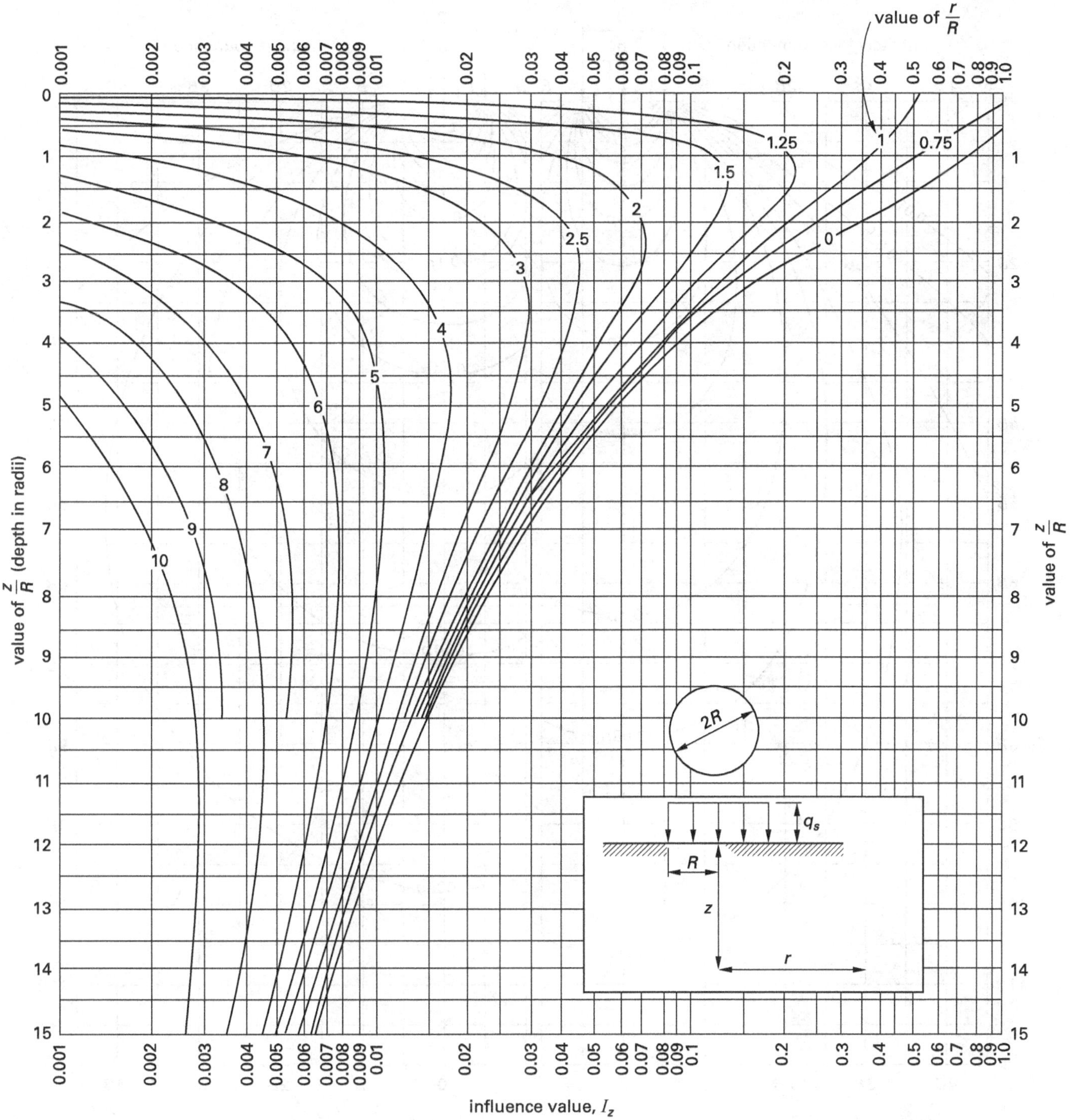

Geotechnical Engineering

Figure 19.5 Boussinesq Stress Contour Chart for Infinitely Long and Square Footings

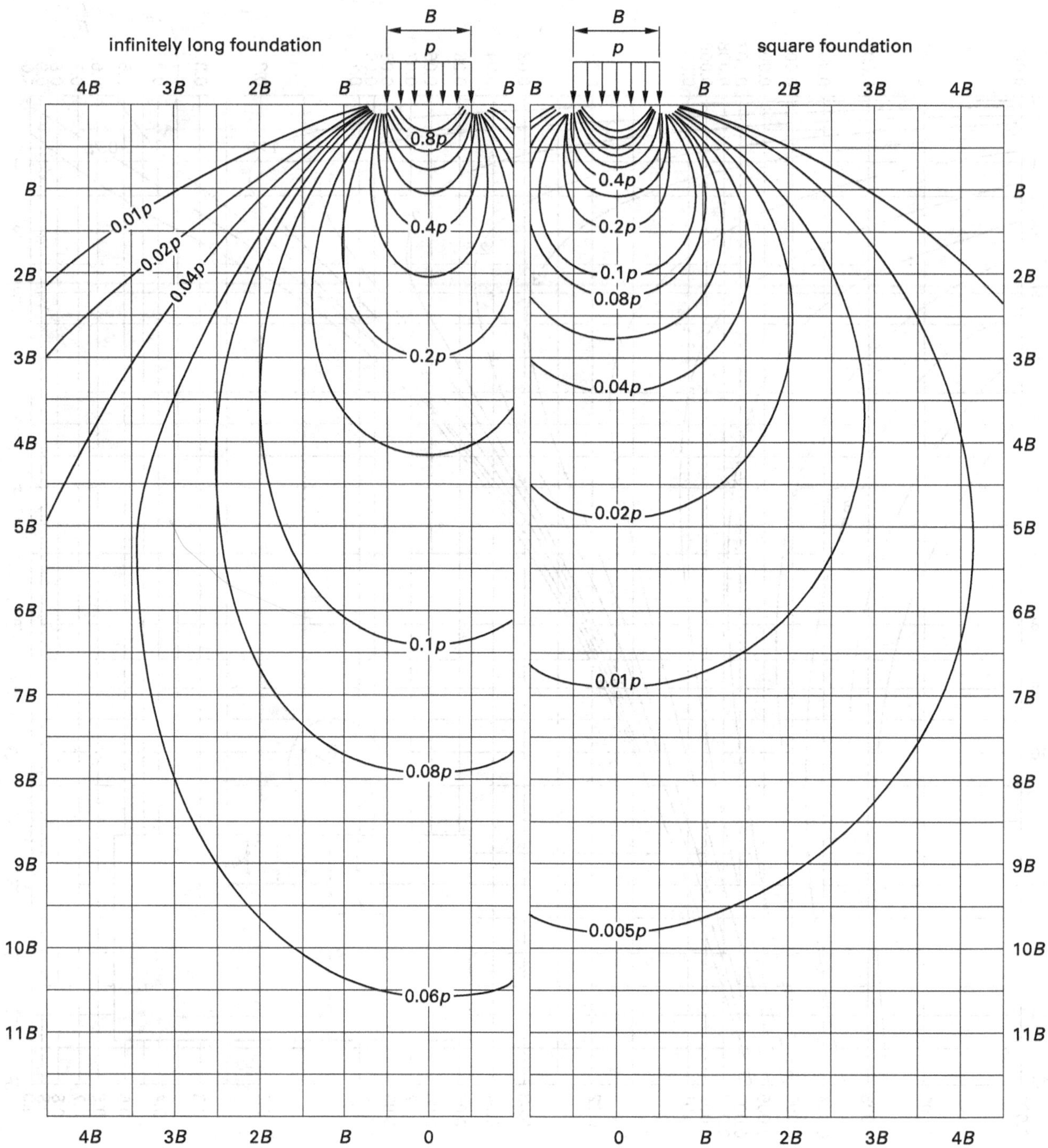

20 Rigid Retaining Walls

Nomenclature

A	area	ft^2	m^2
b	width	ft	m
B	width	ft	m
c	cohesion	lbf/ft^2	Pa
F	force	lbf	N
FS	factor of safety	–	–
g	gravitational acceleration, 32.2 (9.81)	ft/sec^2	m/s^2
H	height	ft	m
K	earth pressure coefficient	–	–
L	length	ft	m
M	moment	ft-lbf	N·m
p	pressure	lbf/ft^2	Pa
P	force	lbf	N
R	shear resistance	lbf	N
S	surcharge	lbf/ft^2	Pa
T	shear resistance	lbf	N
T_{FF}	sliding resistance	lbf	N
u	pore pressure	lbf/ft^2	Pa
W	weight	lbf	N
W_M	weight of wedge	lbf	N

Symbols

α	angle of failure plane	deg	deg
γ	specific (unit) weight	lbf/ft^3	N/m^3
ρ	density	lbm/ft^3	kg/m^3
σ	stress	lbf/ft^2	Pa
σ'	effective stress	lbf/ft^2	Pa
ϕ	angle of internal friction	deg	deg

Subscripts

ave	average
A	active
h	horizontal
M	above slip surface
MOB	mobilized
P	passive
R	resultant
sat	saturated
S	slip, slope, surcharge, or surface
u	uplift
V	vertical
w	water
W	water

1. TYPES OF RETAINING WALL STRUCTURES

A *gravity wall* is a high-bulk structure that relies on self-weight and the weight of the earth over the heel to resist overturning. *Semi-gravity walls* are similar though less massive. A *buttress wall* depends on compression ribs (*buttresses*) between the stem and the toe to resist flexure and overturning. *Counterfort walls* depend on tension ribs between the stem and the heel to resist flexure and overturning. A *cantilever wall* resists overturning through a combination of the soil weight over the heel and the resisting pressure under the base. Figure 20.1 illustrates types of retaining walls.

A *cantilever retaining wall* consists of a base, a stem, and an optional key.[1] The stem may have a constant thickness, or it may be tapered. The taper is known as *batter*. The *batter decrement* is the change in stem thickness per unit of vertical distance. Batter is used to "disguise" bending (deflection) that would otherwise make it appear as if the wall were failing. It also reduces the quantity of material needed at the top of the wall where less strength is needed.

Cantilever retaining walls are generally intended to be permanent and are made of cast-in-place poured concrete. However, retaining walls may also be constructed

[1]The use of a key may increase the cost of installing the retaining wall by more than just the cost of extending the thickness of the base by the depth of the key. To install a key, the contractor will have to hand-shovel the keyway or change buckets on the backhoe. Arranging any vertical key steel is also time consuming.

Geotechnical Engineering

from reinforced masonry block, stacked elements of various types, closely spaced driven members, railroad ties, and heavy lumber.

Figure 20.1 *Types of Retaining Walls*

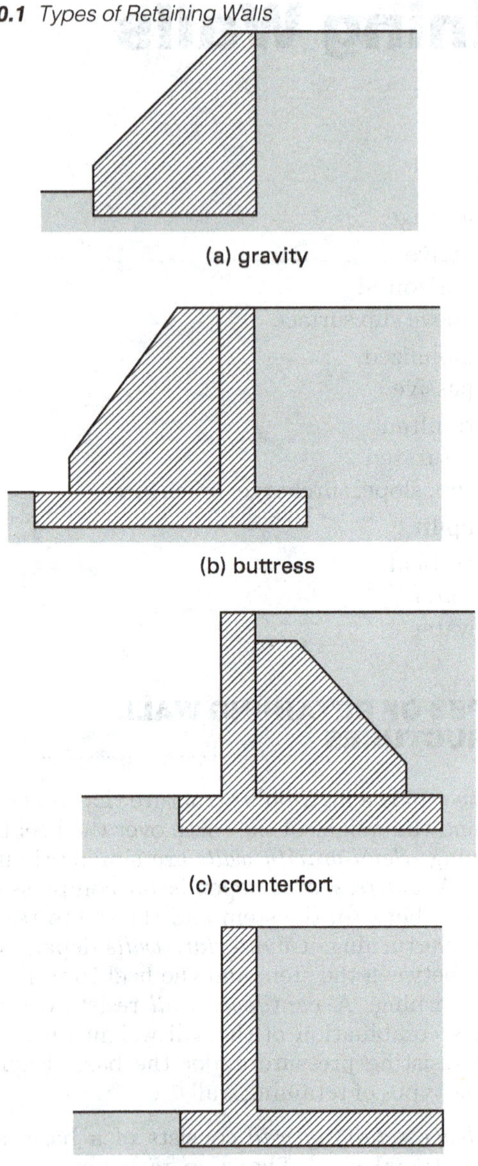

2. COHESIVE AND GRANULAR SOILS

The nature of the backfilled or retained soil greatly affects the design of retaining walls. The two main soil classifications are granular and cohesive soils. *Cohesive soils* are clay-type soils with angles of internal friction, ϕ, of close to zero. *Granular soils* (also referred to as *noncohesive soils* and *cohesionless soils*) are sand- and gravel-type soils with values of cohesion, c, of close to zero. Granular soils also encompass "moist sands" and "drained sands."

3. EFFECTIVE STRESS IN RETAINING WALLS

Rarely in a properly designed retaining wall is the *groundwater table* (GWT) above the base. However, with submerged construction or when drains become plugged, a significant water table can exist behind the wall.

The water provides a buoyant effect, since each sand particle is submerged. The *effective pressure* (generally referred to as the *effective stress*) is the difference between the *total pressure* and the pore pressure.

Equation 20.1: Pore Pressure

$$u = h_u \gamma_w \qquad 20.1$$

Description

The *pore pressure* (i.e., the *hydrostatic pressure*), u, behind the wall at a point h below the water table is given by Eq. 20.1.[2]

4. EARTH PRESSURE

Earth pressure is the force per unit area exerted by soil on the retaining wall. Generally, the term is understood to mean "horizontal earth pressure." *Active earth pressure* (also known as *tensioned soil pressure* and *forward soil pressure*) is present behind a retaining wall that moves away from and tensions the remaining soil. *Passive earth pressure* (also known as *backward soil pressure* and *compressed soil pressure*) is present in front of a retaining wall that moves toward and compresses the soil.

The active and passive earth pressure distributions shown in Fig. 20.2 are horizontal soil pressure distributions. Unlike a fluid, the horizontal earth pressure is not the same as the vertical earth pressure. However, the horizontal pressure can be derived from the vertical pressure.

[2]The NCEES *FE Reference Handbook* (*NCEES Handbook*) is inconsistent in its nomenclature for Eq. 20.1. The subscript w for water is used in Eq. 20.1, while W is the dominant subscript elsewhere. The *NCEES Handbook* is also inconsistent in its nomenclature regarding the submergence depth, using H_2 routinely for the same quantity as h_u. Also in Eq. 20.1, the *NCEES Handbook* uses the subscript u and refers to the submergence depth as the *uplift* (pressure head), an uncommon convention (as uplift is only relevant below an impervious horizontal surface such as a footing, retaining wall base, or base of a dam). Uplift below impervious surfaces is not included in the *NCEES Handbook*.

Figure 20.2 *Active and Passive Earth Pressure Distributions*

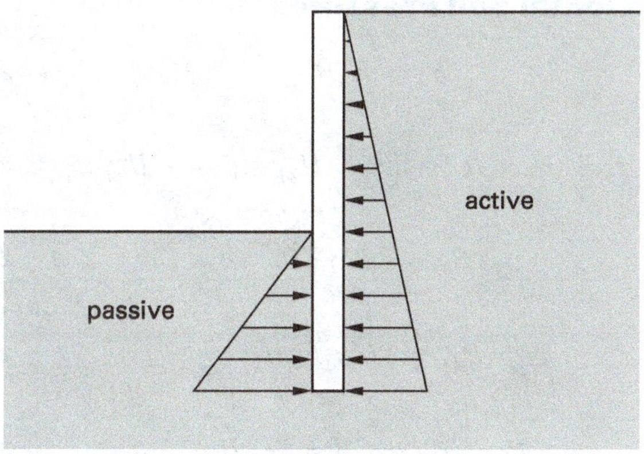

The vertical soil pressure is calculated from the soil specific (unit) weight and the depth similarly to how fluid pressure is calculated. Any reasonable symbol can be selected for pressure.[3] Above the water table, the vertical pressure at a depth H_1 below the free soil surface is

$$\sigma = \gamma H_1$$

Below the *water table*, the same concept is used with the buoyant specific weight. In effect, buoyancy supports some of the soil weight. This can be presented as a reduction in stress due to the pore pressure, but it is just a statement about buoyance, nonetheless. The effective vertical pressure at a depth H_2 below the water table is

$$\sigma' = \gamma_b H_2 = \sigma - u$$
$$\gamma_b = \gamma - \gamma_W$$

When a point is located below the water table, the contributions of the soil and the pore pressure distributions may be calculated separately. There are two ways to visualize this.

1. The vertical pressure distribution can be thought of as having three parts: (a) a pressure distribution due to the bulk specific weight, γ, extending from the soil free surface down the point, plus (b) a pressure distribution due to the saturated specific weight, γ_{sat}, extending from the water table down to the point, less (c) the pressure distribution due to the water specific weight, γ_W, extending from the water table down to the point.

$$\sigma' = \gamma_1 H_1 + \gamma_2 H_2 - \gamma_W H_2$$

2. Alternatively, the vertical pressure can be considered to be the result of (a) a pressure distribution due to the bulk specific weight, γ, extending from the soil free surface down to the point, plus (b) a pressure distribution due to (what is referred to as) the *effective specific (unit) weight* and *equivalent fluid specific (unit) weight of soil*, $\gamma_{\text{sat}} - \gamma_W$, extending from the water table down to the point.

$$\sigma' = \gamma_1 H_1 + (\gamma_2 - \gamma_W) H_2$$

Equation 20.2 Through Eq. 20.6: Vertical Soil Pressures

$$\sigma = \gamma_1 \times H_1 \qquad 20.2$$

$$\sigma' = \gamma_1 \times H_1 \qquad 20.3$$

$$u = \gamma_W \times H_2 \qquad 20.4$$

$$\sigma = (\gamma_1 \times H_1) + (\gamma_2 \times H_2) \qquad 20.5$$

$$\sigma' = (\gamma_1 \times H_1) + (\gamma_2 \times H_2) - (\gamma_W \times H_2) \qquad 20.6$$

Description

Equation 20.2 calculates the vertical stress in a soil above the water table. γ_1 is the bulk (moist, wet, etc.) specific (unit) weight of the soil above the water table. Equation 20.3 calculates the effective vertical stress in the soil above the water table. The following equation calculates the vertical stress component due to the soil below the water table. γ_2 is the saturated specific weight of the soil.

$$\sigma = \gamma_2 \times H_2$$

Equation 20.4 calculates the pore pressure component below the water table. γ_W is the specific weight of water, 9.81 N/m^3. Equation 20.6 combines Eq. 20.4 and Eq. 20.5 and calculates the total (effective) pressure behind the retaining wall.[4] (See Fig. 20.3.)

Figure 20.3 *Vertical Stress Profiles*

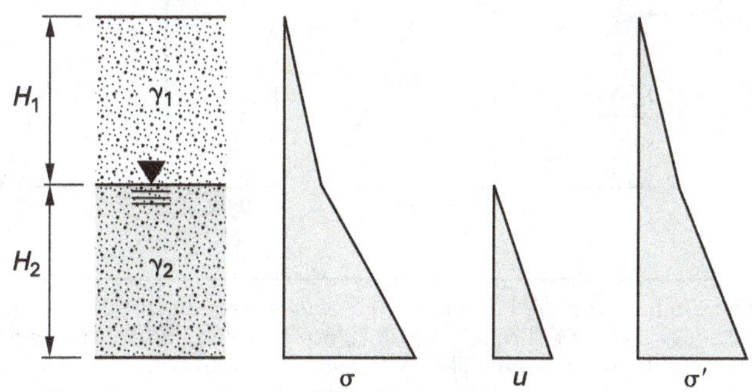

[3]The *NCEES Handbook* uses σ as the symbol for vertical pressure (stress). It is more commonly designated as *p*.

[4]By standard convention, γ_1 and γ_2 would be interpreted as the specific weights of two different soils, but that is not the case in Eq. 20.2 through Eq. 20.6. Instead, the symbols γ_1 and γ_2 are being used to designate bulk specific weight and saturated specific weight, respectively. The *NCEES Handbook* previously uses γ and γ_{sat} to designate the same quantities.

Equation 20.7 Through Eq. 20.10: Vertical Stress with Surcharge

$$\sigma = S + (\gamma_1 \times H_1) \qquad 20.7$$

$$\sigma' = S + (\gamma_1 \times H_1) \qquad 20.8$$

$$\sigma = S + [(\gamma_1 \times H_1) + (\gamma_2 \times H_2)] \qquad 20.9$$

$$\sigma' = S + [(\gamma_1 \times H_1) + (\gamma_2 \times H_2) - (\gamma_W \times H_2)] \qquad 20.10$$

Description

Any additional loading applied externally to the soil is known as a *surcharge*. There are several types of surcharges: point, line, and uniform (i.e., area). For example, the outrigger of a mobile crane might apply a point surcharge; a surface pipe would apply a line surcharge; and a layer of sand spread over the top of the soil would apply a uniform (area) surcharge.[5]

The method used to include a surcharge loading depends on the type of surcharge. In the simplest case of a uniform surcharge, the weight of the surcharge per unit area, S, increases the pressure everywhere below it by that amount. The vertical stress below a uniform surcharge is the unsurcharged pressure plus the surcharge. The surcharge simply increments the stress everywhere by S. Equation 20.7 through Eq. 20.10 describe the mathematics of adding S to the unsurcharged pressure distributions.[6] (See Fig. 20.4.) S does not need to be added to the pore pressure, so Eq. 20.4 may still be used with a surcharge.

Figure 20.4 Vertical Stress Profiles with Surcharge

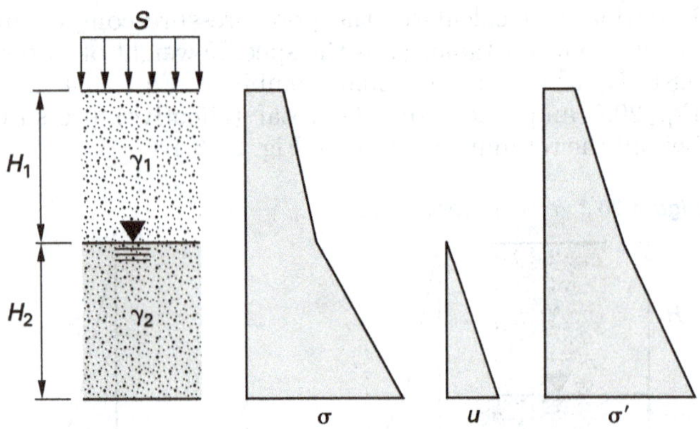

Equation 20.11 Through Eq. 20.15: Horizontal Soil Pressures

$$\sigma_h' = K_A H_2 \gamma' \qquad 20.11$$

$$\sigma_2' = (\gamma_1 \times H_1) + (\gamma_2 \times H_2) - (\gamma_W \times H_2) \qquad 20.12$$

$$P_W = 1/2\gamma_W H_2^2 \qquad 20.13$$

$$P_{A1} = 1/2\sigma_1' K_A H_1 \qquad 20.14$$

$$P_{A2} = 1/2(\sigma_1' + \sigma_2') K_A H_2 \qquad 20.15$$

Description

The horizontal pressure (stress, force, etc.) against the face of a vertical retaining wall is calculated from the vertical pressure. (See Fig. 20.5.) Since water exerts the same pressure in all directions, the horizontal pore pressure is the same as the vertical pore pressure. However, vertical pressure from moist, saturated, submerged, and similar soil must be converted to a horizontal pressure by multiplying the vertical pressure by an *earth pressure coefficient*, K. The *active earth pressure coefficient*, K_A, is used on the active side of the retaining wall; a different *passive earth pressure coefficient*, K_P, is used on the passive side. Without attempting to be rigorous, this is represented conceptually as

$$\sigma_h = u + K\sigma_{\text{soil},V}$$

Figure 20.5 Horizontal Stress Profiles and Forces

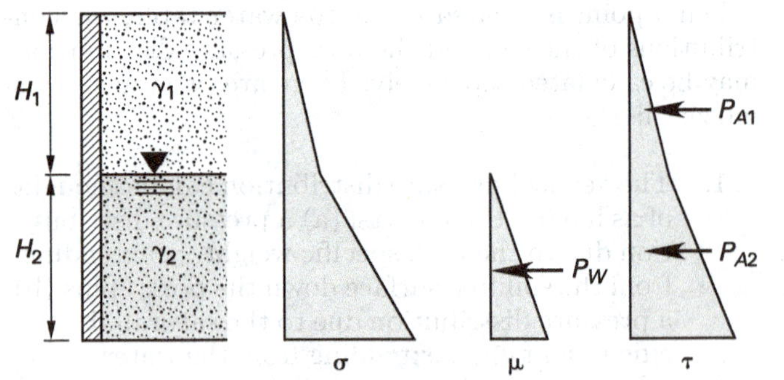

Equation 20.14 and Eq. 20.15 give the *active earth pressure resultant*, P_A, above and below the water table.[7] Equation 20.14 is derived from the triangular

[5]The surcharge covered in the *NCEES Handbook* is specifically limited to a uniform (area) surcharge, although this is not made explicit.

[6]Although Eq. 20.7 through Eq. 20.10 are written in a way that may be confusing, the concept that they portray is simple: The uniform surcharge simply adds to everything else. Although the *NCEES Handbook* previously uses q_s as the symbol for an applied surface stress, q_s is not used in these equations. Standard engineering practice is to designate the magnitude of a uniform surcharge as q.

[7](1) The active and passive force resultants are often represented by the variable R. (2) The *NCEES Handbook* forms of Eq. 20.14 and Eq. 20.15 are correct as shown, but they are nonstandard forms. The unsquared terms, H_1 and H_2, are correct since the height terms were incorporated into the calculations of the σ' terms. However, these equations are most commonly presented in terms of γ (not σ), where the H terms are squared. Since Eq. 20.13 does just that, Eq. 20.13, Eq. 20.14, and Eq. 20.15 are not parallel in their presentations, inviting misuse through inadvertent squaring of the H terms. Equation 20.13, Eq. 20.14, and Eq. 20.15 are misleading, as they are shown as "$1/2\gamma$..." and "$1/\sigma$..." They should not be interpreted as "one over 2γ...," and so on.

horizontal pressure distribution. For a triangular pressure distribution, the average vertical pressure occurs at midheight. The area of a retaining wall is the product of its height and width. However, earth pressure resultants are universally understood to be per unit width of wall, so the "area" contains only the height term.

$$P_{\text{horizontal}} = Kp_{V,\text{ave}}A = K\left(\gamma\frac{H}{2}\right)H$$

$$= \tfrac{1}{2}\gamma KH^2$$

The *passive earth pressure resultant*, P_P, is calculated similarly.[8] If the height of the passive soil is H_P, the passive earth pressure resultant is

$$P_P = \tfrac{1}{2}K_P\gamma H_P^2$$

The resultant active and passive horizontal forces, P, act through the centroids of their triangular pressure distributions. For triangular-shaped horizontal pressure distributions, this is two-thirds of the layer thickness down from the free soil surface.

For a *surcharge*, the resultant horizontal force, P_S, acts through the centroid of a rectangular pressure distribution, one-half of the layer thickness down. The surcharge pressure, S, does not vary with depth, so its maximum and average values are the same. The area of a retaining wall is the product of its height and width. However, earth pressure resultants are universally understood to be per unit width of wall, so the "area" contains only the height term.

$$P_S = Kp_VA = KSH$$

Equation 20.16 and Eq. 20.17: Rankine Earth Pressure Coefficients

$$K_A = \tan^2(45° - \phi/2) \quad \left[\begin{array}{r}\text{smooth wall, } c = 0,\\ \text{level backfill}\end{array}\right] \qquad 20.16$$

$$K_P = \tan^2(45° + \phi/2) \quad \left[\begin{array}{r}\text{smooth wall, } c = 0,\\ \text{level backfill}\end{array}\right] \qquad 20.17$$

Description

There are three common earth pressure theories. Of these, the simplest, the *Rankine earth pressure theory*, assumes that failure occurs along a flat plane behind the wall inclined at an angle α from the horizontal (counterclockwise being positive). The area above the failure plane is referred to as the *active zone*. The Rankine theory disregards friction between the wall and the soil.

Equation 20.16 is the equation for the Rankine active earth pressure coefficient. For saturated clays, the angle of internal friction, ϕ, is zero. As long as tension cracks do not develop near the top of the retaining wall, $K_A = 1$.

Equation 20.17 is the equation for the Rankine passive earth pressure coefficient.

Equation 20.16 and Eq. 20.17 assume that the retaining wall is smooth, the cohesion of the soil is 0, and the backfill is level.

Example

A smooth wall is acted upon by a horizontal force as shown.

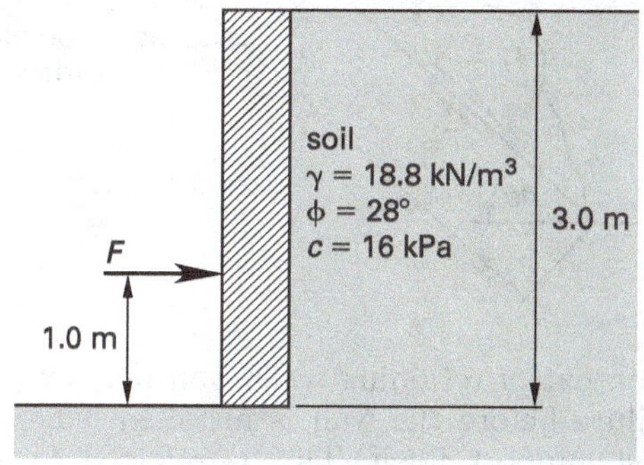

What is most nearly the Rankine earth coefficient for the soil behind the wall?

(A) 0.36

(B) 0.60

(C) 1.7

(D) 2.8

Solution

Since the soil is being compressed, the passive earth pressure coefficient is required. From Eq. 20.17, the passive earth coefficient is

$$K_P = \tan^2(45° + \phi/2)$$

$$= \tan^2\left(45° + \frac{28°}{2}\right)$$

$$= 2.77 \quad (2.8)$$

The answer is (D).

5. SLOPE FAILURE RESISTANCE

Soil is normally restrained by itself. For soil in level ground, there is nowhere for the soil to go. However, when excavations are opened in soil, it is possible for the

[8]Although the *NCEES Handbook* provides the formula for K_P, it does not provide the formula for calculating the passive earth pressure resultant. Past editions of the *NCEES Handbook* included the comment: "[other] forces computed in a similar manner."

soil to collapse into the excavation. *Slope failure* refers to the situation where a mass of soil slides off of itself. The mass of soil is often referred to as a *wedge*. Slope failure is a shear failure: the shear strength of the soil is exceeded. A *shear plane* is also referred to as a *failure plane*, *rupture plane*, and *slippage plane* (*plane of slippage*).[9] The maximum shearing force that a wedge can support is referred to as the available *shearing resistance*. (See Fig. 20.6.)

Figure 20.6 Slope Failure of Soil Wedge along Planar Surface

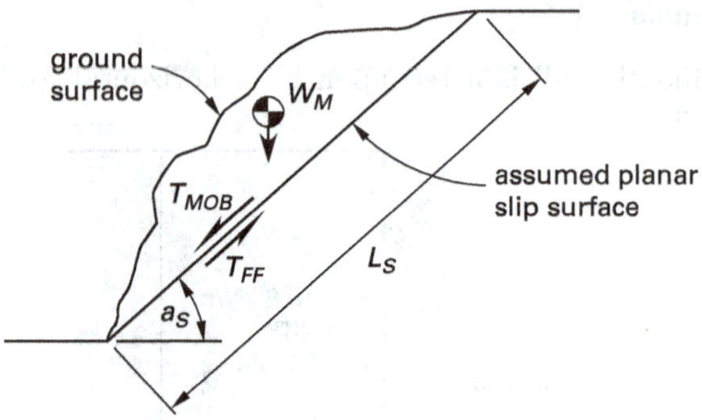

In the context of retaining walls, soil may experience a slope failure before the wall is installed, when the soil face must support itself. This is referred to as "local instability." Alternatively, the soil behind a retaining wall can fail with the failure plane extending before and after the retaining wall, a case referred to as "global instability."

For a soil wedge to slide, the force on the shear plane must overcome both the shear strength of the soil and the frictional force between the soil wedge and its base. The shear strength is referred to as the *mobilized shear strength* because the strength does not manifest itself until the soil begins to move. The undrained shear strength, S, has previously been shown to be equal to the cohesion, c. Shear failure occurs over an area of Lb, where L is the length of the shear plane, and b is the width of the wedge. If results are computed per unit width of wedge, $b=1$. The shear resistance, R, due to shear strength alone is

$$R_{\text{shear}} = SL = cL$$

The frictional force resisting sliding is calculated from the normal force and the coefficient of sliding friction. The normal force is the component of the wedge weight perpendicular to the shear plane. The coefficient of sliding friction is the tangent of the angle of internal friction.

$$R_{\text{friction}} = W\cos\alpha\tan\phi$$

The total resisting force is

$$R_{\text{total}} = R_{\text{shear}} + R_{\text{friction}} = cL + W\cos\alpha\tan\phi$$

Equation 20.18: Resistance Against Sliding

$$T_{\text{FF}} = cL_S + W_M\cos\alpha_S\tan\phi \qquad 20.18$$

Description

Equation 20.18 calculates the *resistance against sliding*, T_{FF}, for a soil wedge along a planar failure plane when the slope angle, α_S, and length of the failure plane, L_S, are known or assumed. W_M is the weight of the wedge.[10] The soil cohesion, c, is zero for granular soil.

Example

A cantilever retaining wall is installed in soil having a cohesion of 23 kPa. The slip surface of a trial soil wedge is 23 m long, and the weight of the soil above the slip surface is 18 kN/m. The angle of the failure plane is 15° from the horizontal, and the angle of internal friction is 30°. Most nearly, what is the available shear resistance per foot of soil?

(A) 540 kN/m

(B) 780 kN/m

(C) 930 kN/m

(D) 1100 kN/m

Solution

The available shear resistance is

$$\begin{aligned} T_{\text{FF}} &= cL_S + W_M\cos\alpha_S\tan\phi \\ &= (23\ \text{kPa})(23\ \text{m}) + \left(18\ \frac{\text{kN}}{\text{m}}\right)\cos 15°\ \tan 30° \\ &= 539\ \text{kN/m} \quad (540\ \text{kN/m}) \end{aligned}$$

The answer is (A).

---

[9]One of the difficulties with this kind of analysis is that the location of the failure plane is generally unknown in advance. Not only are the wedge angle and length of failure plane unknown, but also (1) the failure surface may not be planar, and (2), the fail plane may not be a toe-failure (that is, the failure plane may intercept the wedge above the toe, or it might include part of the toe). Therefore, this is a contrived situation that can only be solved if the failure plane is given as planar, and the angle and length of the failure plane are given, which may be the case when evaluating a slope failure after it has occurred.The assumption of a planar surface significantly overestimates the actual sliding resistance, so circular and log-spiral surfaces are preferred.

[10]Although the *NCEES Handbook* includes subscripts in the variables used in Eq. 20.18, it is standard engineering practice to use the symbols R, α, L, and W without subscripts in a soil stability analysis. Resistance is usually represented by R, or occasionally by F, P, Q, or S. The use of T_{FF} to designate sliding resistance seems to be unique to the *NCEES Handbook*. The subscript M in W_M is unknown (perhaps this is the "weight of the soil mass").

Equation 20.19: Mobilized Shear Force

$$T_{MOB} = W_M \sin\alpha_S \qquad 20.19$$

Description

The shear force *resisting* sliding is not present in a stationary wedge. It does not exist until the sliding begins (i.e., until the soil "mobilizes"). At failure, shear stress along the failure surface (*mobilized shear resistance*) reaches the shear strength. The force causing the soil wedge to slide, T_{MOB}, is known as the *mobilized shear force*. It is the component of weight parallel to the failure plane. The gravitational force on the shear wedge is present regardless of motion of the wedge.[11] Equation 20.19 gives the mobilized shear force along the slip surface.

Equation 20.20: Factor of Safety Against Sliding

$$FS = T_{FF}/T_{MOB} \qquad 20.20$$

Description

Equation 20.20 calculates the factor of safety against sliding.[12]

Example

An excavated slope in uniform soil is shown. The soil has a cohesion of 19.2 kPa and an internal friction angle of 15°. The weight of the soil above the trial failure plane is 3036 kN/m.

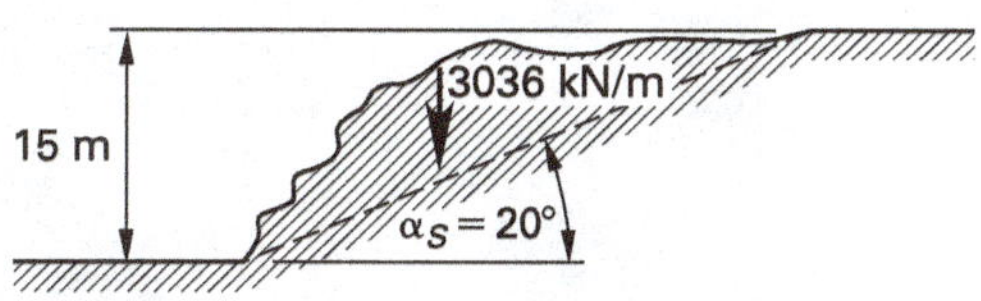

What is most nearly the factor of safety against sliding?

(A) 1.6

(B) 2.3

(C) 2.9

(D) 3.4

Solution

Find the length of the slip surface.

$$L_S = \frac{15\text{ m}}{\sin 20°} = 43.9\text{ m}$$

Find the resistance against sliding.

$$\begin{aligned} T_{FF} &= cL_S + W_M\cos\alpha_S\tan\phi \\ &= (19.2\text{ kPa})(43.9\text{ m}) + \left(3036\ \frac{\text{kN}}{\text{m}}\right)\cos 20°\tan 15° \\ &= 1606\text{ kN/m} \end{aligned}$$

Find the mobilized shear force.

$$\begin{aligned} T_{MOB} &= W_M\sin\alpha_S = \left(3036\ \frac{\text{kN}}{\text{m}}\right)\sin 20° \\ &= 1038\text{ kN/m} \end{aligned}$$

The factor of safety against sliding is

$$\begin{aligned} FS &= T_{FF}/T_{MOB} = \frac{1606\ \frac{\text{kN}}{\text{m}}}{1038\ \frac{\text{kN}}{\text{m}}} \\ &= 1.55 \quad (1.6) \end{aligned}$$

The answer is (A).

[11]Since the gravitational force (weight) on the wedge is always present, it is inappropriate to use the term "mobilized" in reference to it, as does the *NCEES Handbook*. No mobilization is required for the weight to exist.

[12]The term "mobilized" should be associated with the numerator term for Eq. 20.20, not the denominator term.

21 Excavations

Nomenclature

c	cohesion	lbf/ft^2	Pa
F_b	allowable bending stress	–	–
g	gravitational acceleration, 32.2 (9.81)	ft/sec^2	m/s^2
H	height	ft	m
K	earth pressure coefficient	–	–
M	moment	ft-lbf	N·m
p	pressure	lbf/ft^2	Pa
S	section modulus	ft^3	m^3

Symbols

γ	specific (unit) weight	lbf/ft^3	N/m^3
ρ	density	lbm/ft^3	kg/m^3
ϕ	angle of internal friction	deg	deg

Subscripts

A	active
b	bending
max	maximum
P	passive

1. EXCAVATION

An *excavation* is any area where soil has been removed from its original location. Generally, the soil surface elevation is significantly lowered in relation to the elevation of the surrounding soil. Excavations may be of any depth, size, or shape (i.e., plan area). However, if the excavation is long and narrow, the term *trench* is typically used.

To protect personnel in an excavated area, and to support structures and equipment located on adjacent unexcavated areas, the soil bordering the excavation must not be allowed to collapse into the excavation. Slope stability may be sufficiently provided in large, shallow excavations by shallow slopes and/or benched sides. For deeper excavations, support of the excavation slopes is needed. This support is provided by (the generic terms) *shoring*, *bulkheads*, *sheathing*, and *sheeting*, which commonly involves wood sheeting, metal sheet piles, and movable trench shields.

2. BRACED CUTS

Bracing is used when temporary trenches for water, sanitary, and other lines are opened in soil. A *braced cut* is an excavation in which the active earth pressure from one bulkhead is used to support the facing bulkhead. The bulkhead members in contact with the soil are known as sheeting, sheathing, sheet piling, lagging, and, rarely, *poling*. The *box-shoring* and *close-sheeting* methods of support are shown in Fig. 21.1. In box shoring, each of the upright-strut units is known as a *set*. Deeper braced cuts may be constructed with vertical *soldier piles* (typically steel H-sections) with horizontal timber sheeting members between the soldiers.

The load is transferred to the struts at various points, so the triangular active pressure distribution does not develop. Since struts are installed as the excavation goes down, the upper part of the wall deflects very little due to the strut restraint. The pressure on the upper part of the wall is considerably higher than is predicted by the active earth pressure equations. Failure in soils above the water table generally occurs by wale (*stringer*) crippling followed by strut buckling.

The soil removed from the excavation is known as the *spoils*. Spoils should be placed far enough from the edge of the cut so that they do not produce a surcharge lateral loading.

The bottom of the excavation is referred to as the *base of the cut*, *mud line* (or *mudline*), *dredge line*, and *toe of the excavation*.

Excavations below the water table should be dewatered prior to cutting.

Figure 21.1 Shoring of Braced Cuts

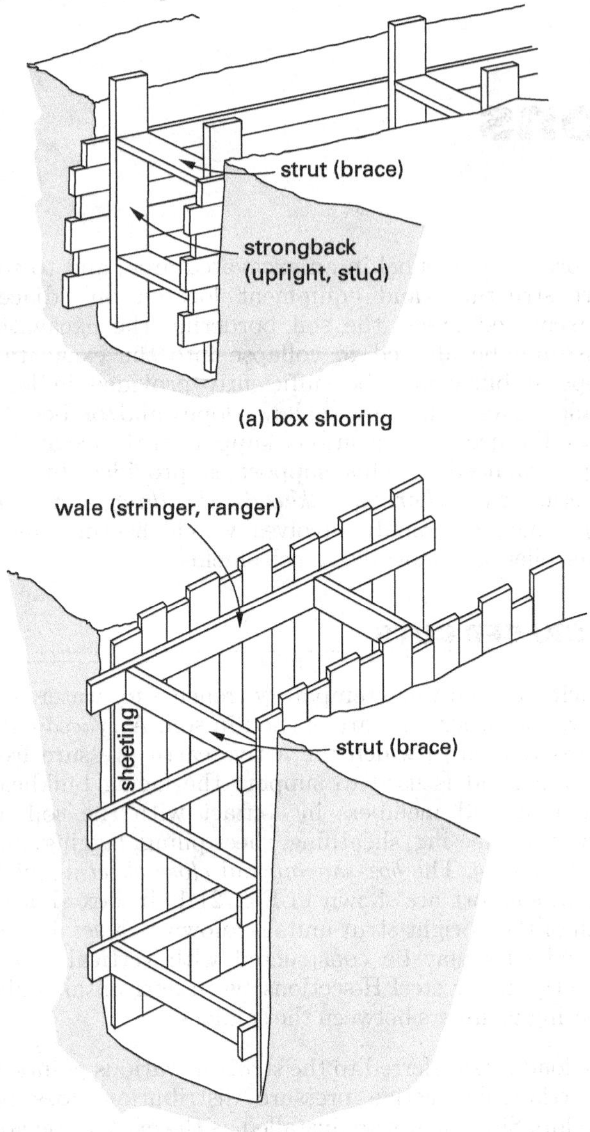

3. BRACED CUTS IN SAND

The analysis of braced cuts is approximate due to the extensive bending of the sheeting. For drained sand, the pressure distribution is approximately uniform with depth. (See Fig. 21.2.) For the simplest model of braced cut soil pressures, the maximum lateral pressure is given by

$$p_{\text{max}} = 0.65K_A\rho gH = 0.65K_A\gamma H$$

4. BRACED CUTS IN STIFF CLAY

For undrained clay (typical of cuts made rapidly in comparison to drainage times), $\phi = 0°$. In this case, the lateral pressure distribution depends on the average cohesion (undrained shear strength) of the clay. If

Figure 21.2 Cuts in Sand

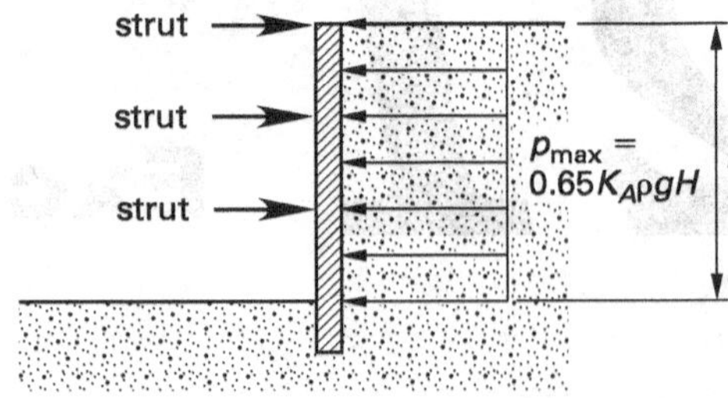

$\gamma H/c \leq 4$, the clay is stiff, and the pressure distribution is as given in Fig. 21.3. (The quantity $\gamma H/c$ is known as the *stability number* and is sometimes given by the variable N_o.) The value of p_{max} is affected by many variables. Use the lower range of values of p_{max} when movement is minimal or when the construction period is short.

$$0.2\rho gH \leq p_{\text{max}} \leq 0.4\rho gH$$

Figure 21.3 Cuts in Stiff Clay

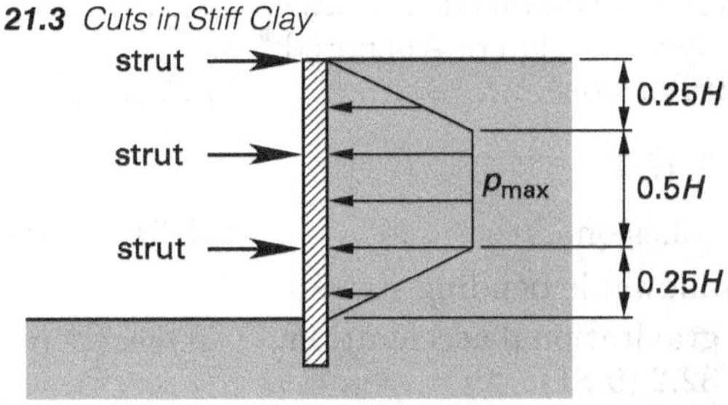

Except when the cut is underlain by deep, soft, normally consolidated clay, the maximum pressure can be estimated using the following equation.

$$p_{\text{max}} = K_A\rho gH = K_A\gamma H$$

$$K_A = 1 - \frac{4c}{\rho gH} = 1 - \frac{4c}{\gamma H}$$

5. BRACED CUTS IN SOFT CLAY

If $\gamma H/c \geq 6$, the clay is soft, and the lateral pressure distribution will be as shown in Fig. 21.4. Except for cuts underlain by deep, soft, normally consolidated clays, the maximum pressure is

$$p_{\text{max}} = \rho gH - 4c = \gamma H - 4c$$

Figure 21.4 Cuts in Soft Clay

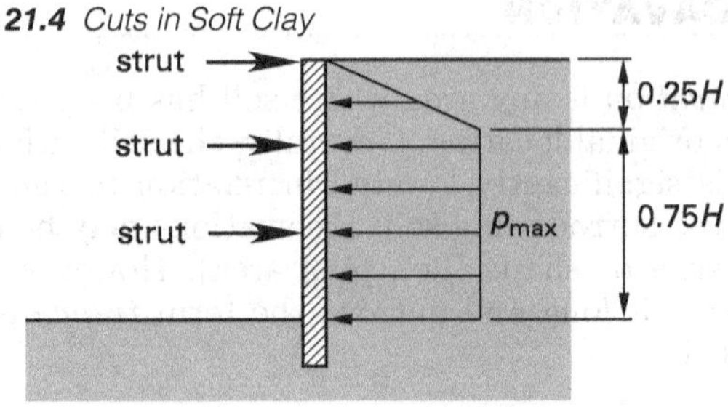

If $6 \leq \gamma H/c \leq 8$, the bearing capacity of the soil is probably sufficient to prevent shearing and upward heave. Simple braced cuts should not be attempted if $\gamma H/c > 8$.

6. BRACED CUTS IN MEDIUM CLAY

If $4 < \gamma H/c < 6$, the soft and stiff clay cases should both be evaluated. The case that results in the greater pressure should be used when designing the bracing.

7. SHEET PILING

Steel *sheet piles* (also known as *trench sheets*) are blade-like rolled steel sections that are driven into the ground to provide lateral support. Sheet piles are placed side by side and generally have interlocking edges ("clutches") that enable the individual sheet piles to act as a continuous wall. Sheet piles are driven into the unexcavated ground ahead of the trenching operation.

Sheet piling is selected according to its geometry (flat, waffle, Z-, etc.) and the required section modulus. ASTM A572 is a typical steel, with the grade indicating the minimum yield point in ksi. The allowable design (bending) stress is usually taken as 65% of the minimum yield stress, with some increases for temporary overstresses typically being allowed.

8. ANALYSIS/DESIGN OF BRACED EXCAVATIONS

Since braced excavations with more than one strut are statically indeterminate, strut forces and sheet piling moments may be evaluated by assuming continuous beam action or hinged beam action (i.e., with hinges at the strut points). In the absence of a true indeterminate analysis, the axial strut compression can be determined by making several strategic assumptions.

A hinge (point of zero moment) is assumed to exist at the second (from the top) strut support point. If moments above that point are taken, a force in the first strut will be derived. Next, a hinge is assumed at the third (from the top) strut support point, and moments are taken about that point (including moments from the top strut, now known), giving the force in the second strut. The process is repeated for all lower struts, finishing by assuming the existence of a fictitious support at the base of the cut in order to determine the force in the last real strut. The pressure distribution on any embedment below the base of the cut is disregarded.

The required section modulus of the sheet piling is given by

$$S = \frac{M_{\text{max,sheet piling}}}{F_b}$$

Struts are selected as column members in order to support the axial loads with acceptable slenderness ratios. Wales are selected as uniform-loaded beams, although cofferdam wales are simultaneously loaded in compression and are designed as beam-columns.

9. ANALYSIS/DESIGN OF FLEXIBLE BULKHEADS

There are two types of flexible bulkheads: untied and tied (i.e., anchored). If the bulkhead is untied, it is designed as a cantilever beam. This is known as a *cantilever wall* or *cantilever bulkhead.*

10. ANCHORED BULKHEADS

An *anchored bulkhead* (*tied bulkhead*), as shown in Fig. 21.5, is supported at its base by its embedment. The bulkhead is anchored near the top by *tie rods* (*tendons, ties, tiebacks,* etc.) projecting back into the soil. These rods can terminate at deadmen, vertical piles, walls, beams, or various other types of anchors. To be effective, the anchors must be located outside of the failure zone (i.e., behind the slip plane).

Anchored bulkheads can fail in several ways. (a) The soil's base layer can heave (i.e., fail due to inadequate bearing capacity). The soil will shear along a circular arc passing under the bulkhead. (b) The anchorage can fail. (c) The toe embedment at the base of the cut can fail. This is often referred to as *toe kick-out* and *toe wash-out.* (d) Rarely, the sheeting can fail.

Figure 21.5 *Anchored Bulkhead*

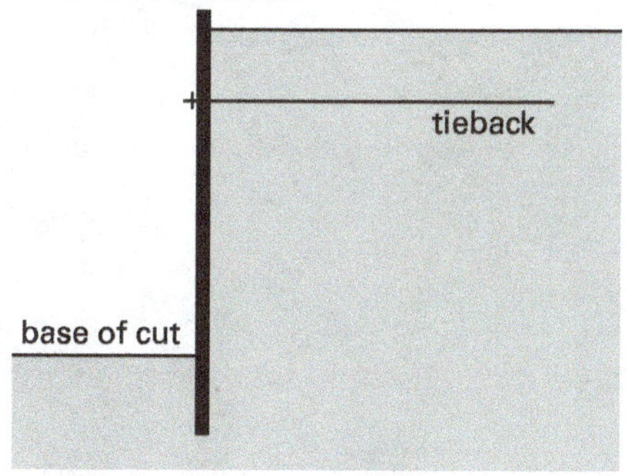

11. ANALYSIS/DESIGN OF ANCHORED BULKHEADS

The *anchor pull* (tension in the tie) is found by setting to zero the sum of all horizontal loads on the bulkhead, including the passive loads on the embedded portion.

If the dimensions are known, the factor of safety against toe failure is found by taking moments about the anchor point on the bulkhead. (Alternatively, if the dimensions are not known, the depth of embedment can be found by taking moments about the anchor point.)

Geotechnical Engineering

The maximum bending moment in the bulkhead sheeting is found by taking moments about the hinge point, where the shear is zero. If the dimensions are not known, the location of this point must be initially assumed. For firm and dense embedment soils, the hinge point is assumed to be at the base of the cut. For loose and weak soils, it is 1–2 ft (0.3–0.6 m) below the base of the cut. For a soft layer over a hard layer, it is at the depth of the hard layer. The stress in the bulkhead is

$$f = \frac{M}{S}$$

The required section modulus is

$$S = \frac{M}{F_b}$$

With multiple ties, the strut forces are statically indeterminate. Although exact methods (e.g., moment distribution) could be used, it is more expedient to assign portions of the active distribution to the struts based on the tributary areas. The area tributary to a strut is bounded by imaginary lines running midway between it and an adjacent strut, in all four directions.

Topic VII: Statics

Chapter

22 Systems of Forces and Moments

Nomenclature

d	distance	m
F	force	N
M	moment	N·m
r	distance	m
r	radius	m
R	resultant	N

Subscripts

θ	angle	deg

1. FORCES

Statics is the study of rigid bodies that are stationary. To be stationary, a rigid body must be in static equilibrium. In the language of statics, a stationary rigid body has no *unbalanced forces* acting on it.

Force is a push or a pull that one body exerts on another, including gravitational, electrostatic, magnetic, and contact influences. Force is a vector quantity, having a magnitude, direction, and point of application.

Strictly speaking, actions of other bodies on a rigid body are known as *external forces*. If unbalanced, an external force will cause motion of the body. *Internal forces* are the forces that hold together parts of a rigid body. Although internal forces can cause deformation of a body, motion is never caused by internal forces.

Forces are frequently represented in terms of unit vectors and force components. A *unit vector* is a vector of unit length directed along a coordinate axis. Unit vectors are used in vector equations to indicate direction without affecting magnitude. In the rectangular coordinate system, there are three unit vectors, **i**, **j**, and **k**.

Equation 22.1: Vector Form of a Two-Dimensional Force

$$\mathbf{F} = F_x\mathbf{i} + F_y\mathbf{j} \qquad 22.1$$

Description

The vector form of a two-dimensional force is described by Eq. 22.1.

Equation 22.2 and Eq. 22.3: Resultant of Two-Dimensional Forces

$$F = \left[\left(\sum_{i=1}^{n} F_{x,i}\right)^2 + \left(\sum_{i=1}^{n} F_{y,i}\right)^2\right]^{1/2} \qquad 22.2$$

$$\theta = \arctan\left(\sum_{i=1}^{n} F_{y,i} \Big/ \sum_{i=1}^{n} F_{x,i}\right) \qquad 22.3$$

Description

The *resultant*, or sum, F, of n two-dimensional forces is equal to the sum of the components. The direction of the resultant with respect to the x-axis is calculated from Eq. 22.3.

Equation 22.4 Through Eq. 22.9: Components of Force

$$F_x = F\cos\theta_x \qquad 22.4$$

$$F_y = F\cos\theta_y \qquad 22.5$$

$$F_z = F\cos\theta_z \qquad 22.6$$

$$\cos\theta_x = F_x/F \qquad 22.7$$

$$\cos\theta_y = F_y/F \qquad 22.8$$

$$\cos\theta_z = F_z/F \qquad 22.9$$

Description

The components of a two- or three-dimensional force can be found from its *direction cosines*, the cosines of the true angles made by the force vector with the x-, y-, and z-axes. (See Fig. 22.1.)

Figure 22.1 *Components and Direction Angles of a Force*

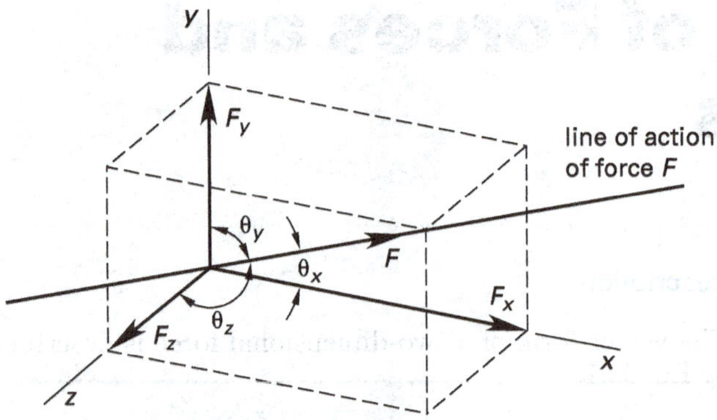

Example

What is most nearly the x-component of the 300 N force at point D on the member shown?

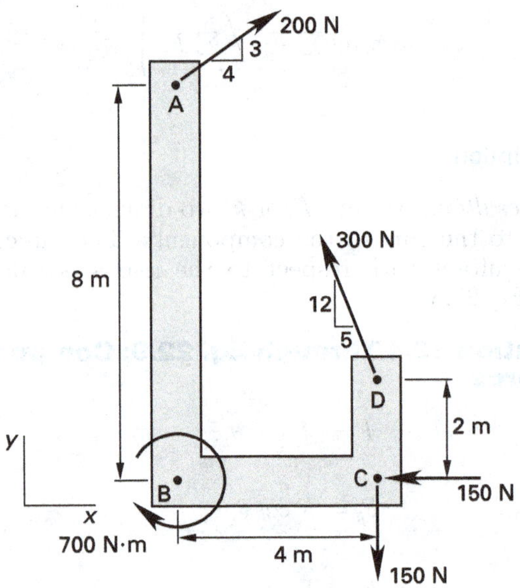

(A) 120 N

(B) 130 N

(C) 180 N

(D) 240 N

Solution

Use the Pythagorean theorem to calculate the hypotenuse of the inclined force triangle. (Alternatively, recognize that this is a 5-12-13 triangle.)

$$\sqrt{(12)^2 + (5)^2} = 13$$

The x-component of the force is

$$F_x = F\cos\theta_x = (300\ \text{N})\left(\frac{5}{13}\right)$$
$$= 115.4\ \text{N} \quad (120\ \text{N})$$

The answer is (A).

Equation 22.10 Through Eq. 22.13: Resultant Force

$$R = \sqrt{x^2 + y^2 + z^2} \qquad 22.10$$

$$F_x = (x/R)F \qquad 22.11$$

$$F_y = (y/R)F \qquad 22.12$$

$$F_z = (z/R)F \qquad 22.13$$

Description

When the x, y, and z components of a force are known, the resultant force is given by Eq. 22.10.

Example

Two forces of 20 N and 30 N act at right angles.

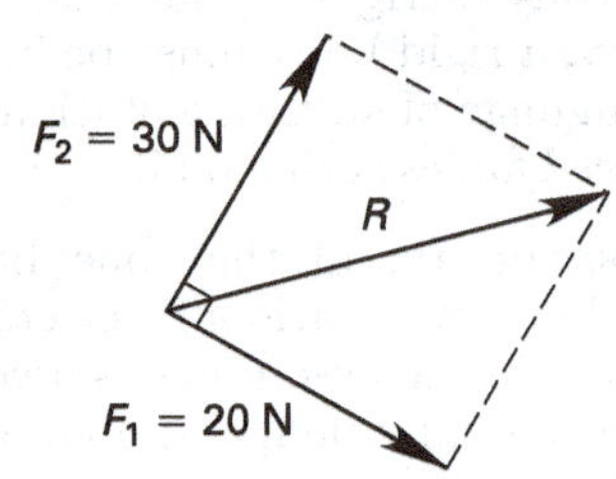

What is most nearly the magnitude of the resultant force?

(A) 7.0

(B) 36

(C) 50

(D) 75

Solution

Define the x-axis parallel to force F_1. From Eq. 22.10, the magnitude of the resultant force is

$$R = \sqrt{x^2 + y^2 + z^2} = \sqrt{(20\ \text{N})^2 + (30\ \text{N})^2 + (0\ \text{N})^2}$$
$$= 36$$

The answer is (B).

2. MOMENTS

Moment is the name given to the tendency of a force to rotate, turn, or twist a rigid body about an actual or assumed pivot point. (Another name for moment is *torque*, although torque is used mainly with shafts and other power-transmitting machines.) When acted upon by a moment, unrestrained bodies rotate. However, rotation is not required for the moment to exist. When a restrained body is acted upon by a moment, there is no rotation.

An object experiences a moment whenever a force is applied to it. Only when the line of action of the force passes through the center of rotation (i.e., the actual or assumed pivot point) will the moment be zero. (The moment may be zero, as when the moment arm length is zero, but there is a trivial moment nevertheless.)

Moments have primary dimensions of length × force. Typical units are foot-pounds, inch-pounds, and newton-meters. To avoid confusion with energy units, moments may be expressed as pound-feet, pound-inches, and newton-meters.

Equation 22.14: Moment Vector

$$\boldsymbol{M} = \boldsymbol{r} \times \boldsymbol{F} \qquad 22.14$$

Variation

$$M_{\text{O}} = |\boldsymbol{M}_{\text{O}}| = |\boldsymbol{r}||\boldsymbol{F}|\sin\theta = d|\boldsymbol{F}| \quad [\theta \leq 180°]$$

Description

Moments are vectors. The moment vector due to a vector force, $\boldsymbol{F}$, applied at a point P, about an axis passing through point O, is designated as $\boldsymbol{M}_{\text{O}}$. The moment also depends on the *position vector*, $\boldsymbol{r}$, from point O to point P. The moment is calculated as the *cross product*, $\boldsymbol{r} \times \boldsymbol{F}$. The axis of the moment will be perpendicular to the plane containing vectors $\boldsymbol{F}$ and $\boldsymbol{r}$. Any point could be chosen for point O, although it is usually convenient to select point O as the origin, and to put P in the horizontal x-y plane. In that case, $\boldsymbol{M}_{\text{O}}$ will be a moment about the vertical z-axis. The scalar product $|\boldsymbol{r}|\sin\theta$, shown in the variation equation, is known as the *moment arm*, *d*.

Example

What is most nearly the magnitude of the moment about the x-axis produced by a force of $\boldsymbol{F} = 10\mathbf{i} - 20\mathbf{j} + 40\mathbf{k}$ acting at the point $(2, 1, 1)$ with coordinates in meters?

(A) 30 N·m

(B) 40 N·m

(C) 50 N·m

(D) 60 N·m

Solution

The xyz coordinate axes are being used, so point O corresponds to the origin. Work in meters and newtons. Equation 22.14 calculates the moment about the vertical z-axis. The cross product can be calculated as a determinant.

$$\begin{aligned}
\boldsymbol{M}_{\text{O}} &= \boldsymbol{r} \times \boldsymbol{F} \\
&= (r_yF_z - r_zF_y)\mathbf{i} + (r_zF_x - r_xF_z)\mathbf{j} \\
&\quad + (r_xF_y - r_yF_x)\mathbf{k} \\
&= (2\mathbf{i} + \mathbf{j} + \mathbf{k}) \times (10\mathbf{i} - 20\mathbf{j} + 40\mathbf{k}) \\
&= ((1\text{ m})(40\text{ N}) - (1\text{ m})(-20\text{ N}))\mathbf{i} \\
&\quad + ((1\text{ m})(10\text{ N}) - (2\text{ m})(40\text{ N}))\mathbf{j} \\
&\quad + ((2\text{ m})(-20\text{ N}) - (1\text{ m})(10\text{ N}))\mathbf{k} \\
&= 60\mathbf{i} - 70\mathbf{j} - 50\mathbf{k}\ \text{N·m}
\end{aligned}$$

$\boldsymbol{M}_{\text{O}}$ is the moment about the origin, not the x-axis as requested. The moment about the x-axis is found as the projection of $\boldsymbol{M}_{\text{O}}$ onto the x-axis, which is the dot product operation.

$$\begin{aligned}
M_{\text{O},x} &= \mathbf{i} \cdot \boldsymbol{M}_{\text{O}} \\
&= (1\text{ m})(60\text{ N}) + (0\text{ m})(-70\text{ N}) + (0\text{ m})(-50\text{ N}) \\
&= 60\text{ N·m}
\end{aligned}$$

The answer is (D).

Right-Hand Rule

The line of action of the moment vector is normal to the plane containing the force vector and the position vector. The sense (i.e., the direction) of the moment is determined from the *right-hand rule*. (See Fig. 22.2.)

> *Right-hand rule:* Place the position and force vectors tail to tail. Close your right hand and position it over the pivot point. Rotate the position vector into the force vector and position your hand such that your fingers curl in the same direction as the position vector rotates. Your extended thumb will coincide with the direction of the moment.

Equation 22.15 Through Eq. 22.17: Components of a Moment

$$M_x = yF_z - zF_y \qquad 22.15$$

$$M_y = zF_x - xF_z \qquad 22.16$$

$$M_z = xF_y - yF_x \qquad 22.17$$

Figure 22.2 Right-Hand Rule

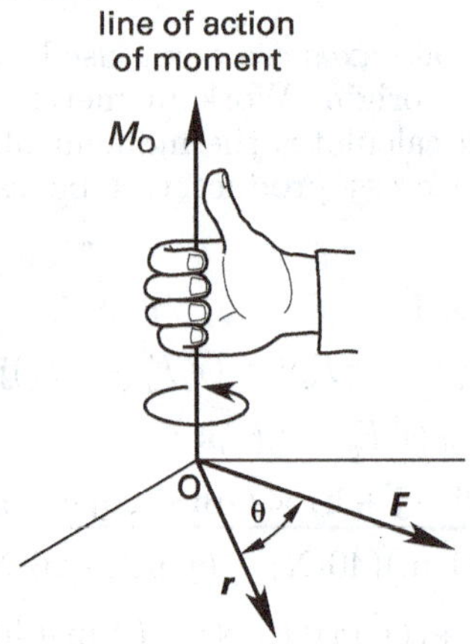

Variations

$$M_x = M\cos\theta_x$$

$$M_y = M\cos\theta_y$$

$$M_z = M\cos\theta_z$$

Description

Equation 22.15, Eq. 22.16, and Eq. 22.17 can be used to determine the components of the moment from the component of a force applied at point (x, y, z) referenced to an origin at $(0, 0, 0)$. The resultant moment magnitude can be reconstituted from its components.

$$M = \sqrt{M_x^2 + M_y^2 + M_z^2}$$

The direction cosines of a force can be used to determine the components of the moment about the coordinate axes, as shown in the variation equations.

Example

What is most nearly the magnitude of the moment about the x-axis produced by a force of $\mathbf{F} = 10\mathbf{i} - 20\mathbf{j} + 40\mathbf{k}$ N acting at the point $(2, 1, 1)$ with coordinates in meters?

(A) 30 N·m

(B) 40 N·m

(C) 50 N·m

(D) 60 N·m

Solution

This is the same as the previous example. Use Eq. 22.15.

$$\begin{aligned} M_x &= yF_z - zF_y = (1\text{ m})(40\text{ N}) - (1\text{ m})(-20\text{ N}) \\ &= 60\text{ N·m} \end{aligned}$$

The answer is (D).

Couples

Any pair of equal, opposite, and parallel forces constitutes a *couple*. A couple is equivalent to a single moment vector. Since the two forces are opposite in sign, the x-, y-, and z-components of the forces cancel out. Therefore, a body is induced to rotate without translation. A couple can be counteracted only by another couple. A couple can be moved to any location without affecting the equilibrium requirements. (Such a moment is known as a *free moment, moment of a couple*, or *coupling moment.*)

In Fig. 22.3, the equal but opposite forces produce a moment vector, $\mathbf{M}_O$, of magnitude Fd. The two forces can be replaced by this moment vector that can be moved to any location on a body.

$$M_O = 2rF\sin\theta = Fd$$

Figure 22.3 Couple

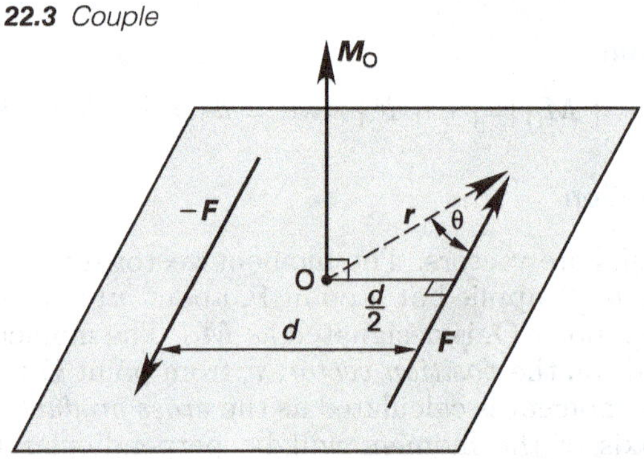

If a force, F, is moved a distance, d, from the original point of application, a couple, M, of magnitude Fd must be added to counteract the induced couple. The combination of the moved force and the couple is known as a *force-couple system*. Alternatively, a force-couple system can be replaced by a single force located a distance $d = M/F$ away.

3. SYSTEMS OF FORCES

Any collection of forces and moments in three-dimensional space is statically equivalent to a single resultant force vector plus a single resultant moment vector. (Either or both of these resultants can be zero.)

Equation 22.18 and Eq. 22.19[1]

$$F = \sum F_n \qquad 22.18$$

$$M = \sum (r_n \times F_n) \qquad 22.19$$

Description

The x-, y-, and z-components of the resultant force, given by Eq. 22.18, are the sums of the x-, y-, and z-components of the individual forces, respectively.

The resultant moment vector, given by Eq. 22.19, is more complex. It includes the moments of all system forces around the reference axes plus the components of all system moments.

Variations

$$R = \sum F_i$$

$$= i\sum_{i=1}^{n} F_{x,i} + j\sum_{i=1}^{n} F_{y,i} + k\sum_{i=1}^{n} F_{z,i} \quad \left[\text{three-dimensional}\right]$$

$$M = \sum M_i$$

$$M_x = \sum_i (yF_z - zF_y)_i + \sum_i (M\cos\theta_x)_i$$

$$M_y = \sum_i (zF_x - xF_z)_i + \sum_i (M\cos\theta_y)_i$$

$$M_z = \sum_i (xF_y - yF_x)_i + \sum_i (M\cos\theta_z)_i$$

Equation 22.20 and Eq. 22.21: Equilibrium Requirements

$$\sum F_n = 0 \qquad 22.20$$

$$\sum M_n = 0 \qquad 22.21$$

Description

An object is static when it is stationary. To be stationary, all of the forces and moments on the object must be in *equilibrium*. For an object to be in equilibrium, the resultant force and moment vectors must both be zero.

The following equations follow directly from Eq. 22.20 and Eq. 22.21.

$$\sum F_x = 0$$

$$\sum F_y = 0$$

$$\sum F_z = 0$$

$$\sum M_x = 0$$

$$\sum M_y = 0$$

$$\sum M_z = 0$$

These equations seem to imply that six simultaneous equations must be solved in order to determine whether a system is in equilibrium. While this is true for general three-dimensional systems, fewer equations are necessary for most problems.

Concurrent Forces

A *concurrent force system* is a category of force systems where all of the forces act at the same point.

If the forces on a body are all concurrent forces, then only force equilibrium is necessary to ensure complete equilibrium.

In two dimensions,

$$\sum F_x = 0$$

$$\sum F_y = 0$$

In three dimensions,

$$\sum F_x = 0$$

$$\sum F_y = 0$$

$$\sum F_z = 0$$

Two- and Three-Force Members

Members limited to loading by two or three forces in the same plane are special cases of equilibrium. A *two-force member* can be in equilibrium only if the two forces have the same line of action (i.e., are collinear) and are equal but opposite.

In most cases, two-force members are loaded axially, and the line of action coincides with the member's longitudinal axis. By choosing the coordinate system so that one axis coincides with the line of action, only one equilibrium equation is needed.

A *three-force member* can be in equilibrium only if the three forces are concurrent or parallel. Stated another way, the force polygon of a three-force member in

[1]The NCEES *FE Reference Handbook* (*NCEES Handbook*) uses both n and i for summation variables. Though i is traditionally used to indicate summation, n appears to be used as the summation variable in order to indicate that the summation is over all n of the forces and all n of the position vectors that make up the system (i.e., $i = 1$ to n).

equilibrium must close on itself. If the member is in equilibrium and two of the three forces are known, the third can be determined.

4. PROBLEM-SOLVING APPROACHES

Determinacy

When the equations of equilibrium are independent, a rigid body force system is said to be *statically determinate.* A statically determinate system can be solved for all unknowns, which are usually reactions supporting the body. Examples of determinate beam types are illustrated in Fig. 22.4.

Figure 22.4 *Types of Determinate Systems*

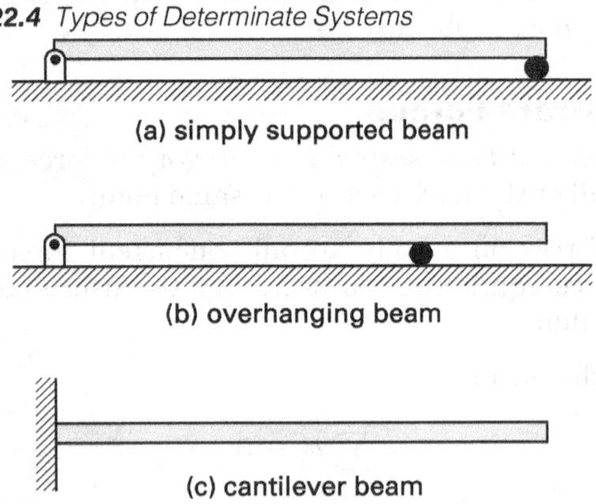

When the body has more supports than are necessary for equilibrium, the force system is said to be *statically indeterminate.* In a statically indeterminate system, one or more of the supports or members can be removed or reduced in restraint without affecting the equilibrium position. Those supports and members are known as *redundant supports* and *redundant members.* The number of redundant members is known as the *degree of indeterminacy.* Figure 22.5 illustrates several common indeterminate structures.

A body that is statically indeterminate requires additional equations to supplement the equilibrium equations. The additional equations typically involve deflections and depend on mechanical properties of the body or supports.

Free-Body Diagrams

A *free-body diagram* is a representation of a body in equilibrium, showing all applied forces, moments, and reactions. Free-body diagrams do not consider the internal structure or construction of the body, as Fig. 22.6 illustrates.

Since the body is in equilibrium, the resultants of all forces and moments on the free body are zero. In order to maintain equilibrium, any portions of the body that are conceptually removed must be replaced by the forces and moments those portions impart to the body. Typically, the body is isolated from its physical supports in

Figure 22.5 *Examples of Indeterminate Systems*

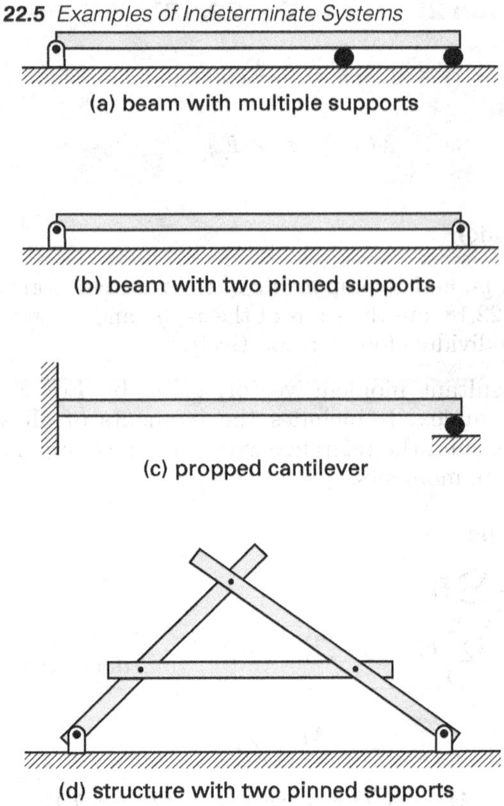

Figure 22.6 *Bodies and Free Bodies*

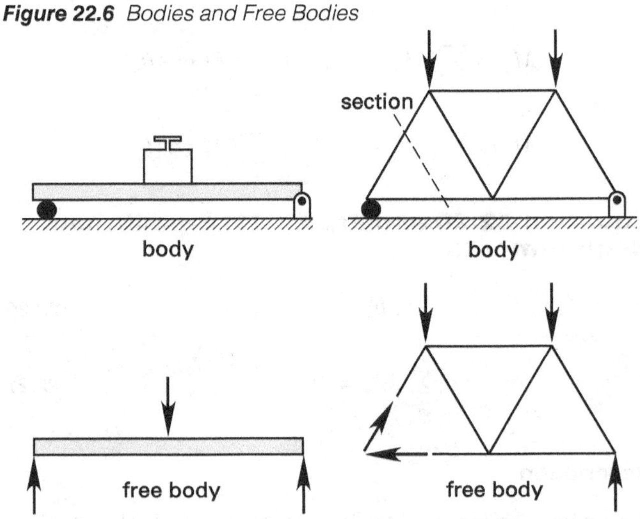

order to help evaluate the reaction forces. In other cases, the body may be sectioned (i.e., cut) in order to determine the forces at the section.

Reactions

The first step in solving most statics problems, after drawing the free-body diagram, is to determine the reaction forces (i.e., the *reactions*) supporting the body. The manner in which a body is supported determines the type, location, and direction of the reactions. Common support types are shown in Table 22.1.

Table 22.1 *Types of Two-Dimensional Supports*

type of support	reactions and moments	number of unknowns*
simple, roller, rocker, ball, or frictionless surface 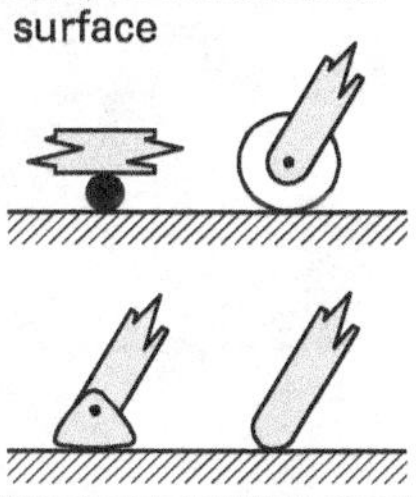	reaction normal to surface, no moment	1
cable in tension, or link 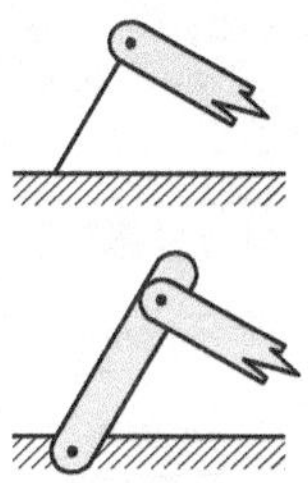	reaction in line with cable or link, no moment	1
frictionless guide or collar 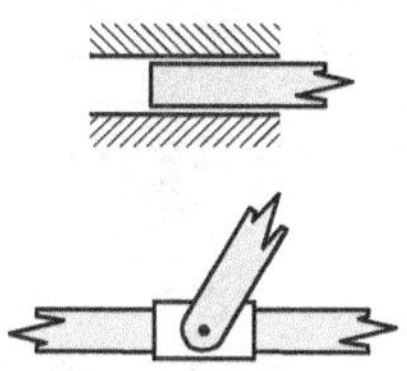	reaction normal to rail, no moment	1
built-in, fixed support 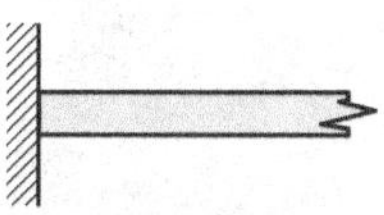	two reaction components, one moment	3
frictionless hinge, pin connection, or rough surface 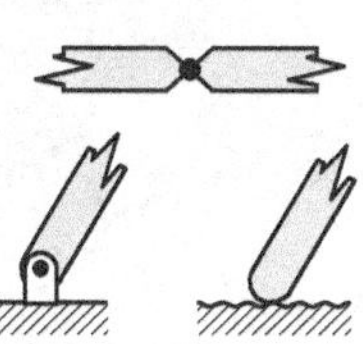	reaction in any direction, no moment	2

*The number of unknowns is valid for two-dimensional problems only.

For beams, the two most common types of supports are the roller support and the pinned support. The *roller support*, shown as a cylinder supporting the beam, supports vertical forces only. Rather than support a horizontal force, a roller support simply rolls into a new equilibrium position. Only one equilibrium equation (i.e., the sum of vertical forces) is needed at a roller support. Generally, the terms *simple support* and *simply supported* refer to a roller support.

The *pinned support*, shown as a pin and clevis, supports both vertical and horizontal forces. Two equilibrium equations are needed.

Generally, there will be vertical and horizontal components of a reaction when one body touches another. However, when a body is in contact with a *frictionless surface*, there is no frictional force component parallel to the surface, so the reaction is normal to the contact surfaces. The assumption of frictionless contact is particularly useful when dealing with systems of spheres and cylinders in contact with rigid supports. Frictionless contact is also assumed for roller and rocker supports.

The procedure for finding determinate reactions in two-dimensional problems is straightforward. Determinate structures will have either a roller support and pinned support or two roller supports.

step 1: Establish a convenient set of coordinate axes. (To simplify the analysis, one of the coordinate directions should coincide with the direction of the forces and reactions.)

step 2: Draw the free-body diagram.

step 3: Resolve the reaction at the pinned support (if any) into components normal and parallel to the coordinate axes.

step 4: Establish a positive direction of rotation (e.g., clockwise) for purposes of taking moments.

step 5: Write the equilibrium equation for moments about the pinned connection. (By choosing the pinned connection as the point about which to take moments, the pinned connection reactions do not enter into the equation.) This will usually determine the vertical reaction at the roller support.

step 6: Write the equilibrium equation for the forces in the vertical direction. Usually, this equation will have two unknown vertical reactions.

step 7: Substitute the known vertical reaction from step 5 into the equilibrium equation from step 6. This will determine the second vertical reaction.

step 8: Write the equilibrium equation for the forces in the horizontal direction. Since there is a minimum of one unknown reaction component in the horizontal direction, this step will determine that component.

step 9: If necessary, combine the vertical and horizontal force components at the pinned connection into a resultant reaction.

23 Trusses

Nomenclature

F	force	N
M	moment	N·m

1. STATICALLY DETERMINATE TRUSSES

A *truss* or *frame* is a set of pin-connected *axial members* (i.e., *two-force members*). The connection points are known as *joints*. Member weights are disregarded, and truss loads are applied only at joints. A *structural cell* consists of all members in a closed loop of members. For the truss to be a table (i.e., to be a *rigid truss*), all of the structural cells must be triangles. Figure 23.1 identifies *chords*, *end posts*, *panels*, and other elements of a typical *bridge truss*.

Figure 23.1 *Parts of a Bridge Truss*

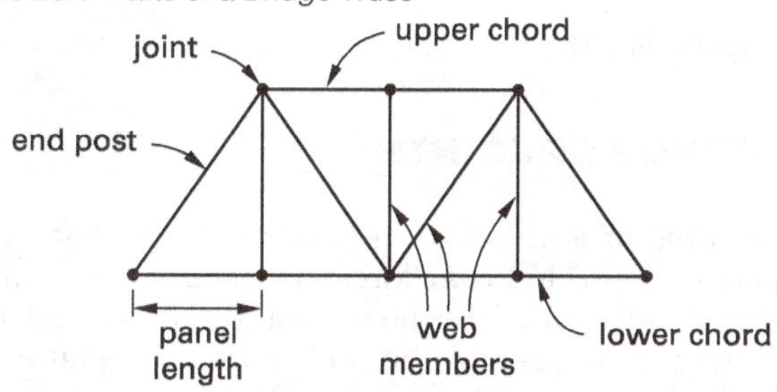

Several types of trusses have been given specific names. Some of the more common named trusses are shown in Fig. 23.2.

Truss loads are considered to act only in the plane of a truss, so trusses are analyzed as two-dimensional structures. Forces in truss members hold the various truss parts together and are known as *internal forces*. The internal forces are found by drawing free-body diagrams.

Although free-body diagrams of truss members can be drawn, this is not usually done. Instead, free-body diagrams of the pins (i.e., the joints) are drawn. A pin in compression will be shown with force arrows pointing toward the pin, away from the member. Similarly, a pin in tension will be shown with force arrows pointing away from the pin, toward the member.

Figure 23.2 *Special Types of Trusses*

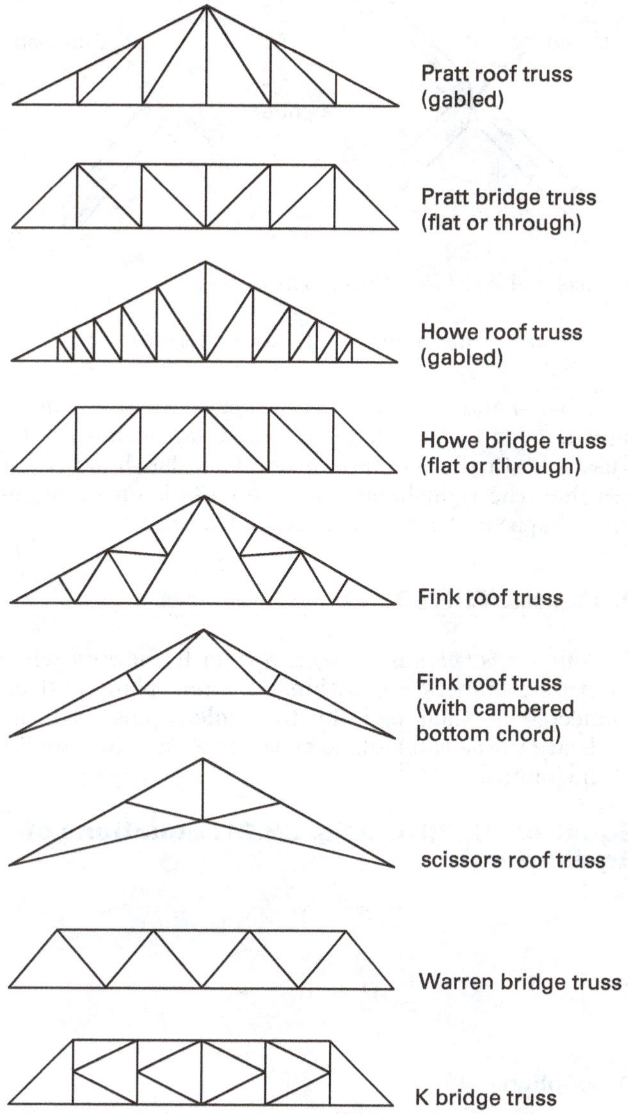

With typical bridge trusses supported at the ends and loaded downward at the joints, the upper chords are almost always in compression, and the end panels and lower chords are almost always in tension.

Since truss members are axial members, the forces on the truss joints are concurrent forces. Only force equilibrium needs to be enforced at each pin; the sum of the forces in each of the coordinate directions equals zero.

Forces in truss members can sometimes be determined by inspection. One of these cases is *zero-force members.* A third member framing into a joint already connecting two collinear members carries no internal force unless there is a load applied at that joint. Similarly, both members forming an apex of the truss are zero-force members unless there is a load applied at the apex. (See Fig. 23.3.)

Figure 23.3 *Zero-Force Members*

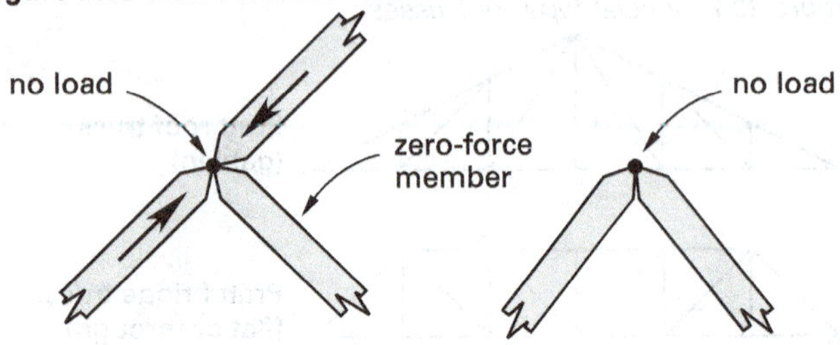

A truss will be *statically determinate* if

$$\text{no. of members} = (2)(\text{no. of joints}) - 3$$

If the left-hand side of the equation is greater than the right-hand side (i.e., there are *redundant members*), the truss is statically indeterminate. If the left-hand side is less than the right-hand side, the truss is unstable and will collapse under certain types of loading.

2. PLANE TRUSS

A *plane truss* (*planar truss*) is a rigid framework where all truss members are within the same plane and are connected at their ends by frictionless pins. External loads are in the same plane as the truss and are applied at the joints only.

Equation 22.20 and Eq. 22.21: Equations of Equilibrium[1]

$$\sum \mathbf{F}_n = 0 \qquad 23.1$$

$$\sum \mathbf{M}_n = 0 \qquad 23.2$$

Description

A plane truss is statically determinate if the truss reactions and member forces can be determined using the equations of equilibrium. If not, the truss is considered statically indeterminate.

Example

Most nearly, what are reactions F_1 and F_2 for the truss shown?

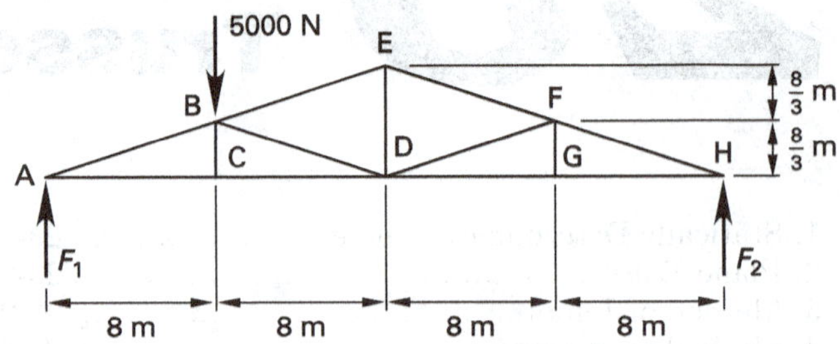

(A) $F_1 = 1000$ N; $F_2 = 4000$ N

(B) $F_1 = 1300$ N; $F_2 = 3800$ N

(C) $F_1 = 2500$ N; $F_2 = 2500$ N

(D) $F_1 = 3800$ N; $F_2 = 1300$ N

Solution

Calculate the reactions from Eq. 22.20 and Eq. 22.21. Let clockwise moments be positive.

$$\begin{aligned} \sum M_{\text{A}} &= 0 = (5000 \text{ N})(8 \text{ m}) - F_2(32 \text{ m}) \\ F_2 &= 1250 \text{ N} \quad (1300 \text{ N}) \quad [\text{upward}] \\ \sum F_y &= 0 = F_1 + 1250 \text{ N} - 5000 \text{ N} \\ F_1 &= 3750 \text{ N} \quad (3800 \text{ N}) \quad [\text{upward}] \end{aligned}$$

The answer is (D).

3. METHOD OF JOINTS

The *method of joints* is one of the methods that can be used to find the internal forces in each truss member. This method is useful when most or all of the truss member forces are to be calculated. Because this method advances from joint to adjacent joint, it is inconvenient when a single isolated member force is to be calculated.

The method of joints is a direct application of the equations of equilibrium in the x- and y-directions. Traditionally, the method begins by finding the reactions supporting the truss. Next the joint at one of the reactions is evaluated, which determines all the member forces framing into the joint. Then, knowing one or more of the member forces from the previous step, an adjacent joint is analyzed. The process is repeated until all the unknown quantities are determined.

[1]The NCEES *FE Reference Handbook* (*NCEES Handbook*) uses both n and i for summation variables. Though i is traditionally used to indicate summation, n appears to be used as the summation variable in order to indicate that the summation is over all n of the forces and all n of the position vectors that make up the system (i.e., $i = 1$ to n).

At a joint, there may be up to two unknown member forces, each of which can have dependent x- and y-components. Since there are two equilibrium equations, the two unknown forces can be determined. Even though determinate, however, the sense of a force will often be unknown. If the sense cannot be determined by logic, an arbitrary decision can be made. If the incorrect direction is chosen, the force will be negative.

Occasionally, there will be three unknown member forces. In that case, an additional equation must be derived from an adjacent joint.

4. METHOD OF SECTIONS

The *method of sections* is a direct approach to finding forces in any truss member. This method is convenient when only a few truss member forces are unknown.

As with the previous method, the first step is to find the support reactions. Then, a cut is made through the truss, passing through the unknown member. (Knowing where to cut the truss is the key part of this method. Such knowledge is developed only by repeated practice.) Finally, all three conditions of equilibrium are applied as needed to the remaining truss portion. Since there are three equilibrium equations, the cut cannot pass through more than three members in which the forces are unknown.

Example

A truss is loaded as shown. The support reactions have already been determined.

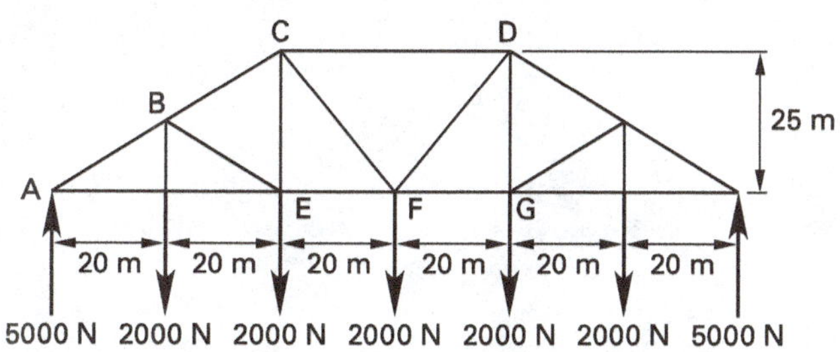

Most nearly, what is the force in member CE?

(A) 1000 N

(B) 2000 N

(C) 3000 N

(D) 4000 N

Solution

Cut the truss as shown.

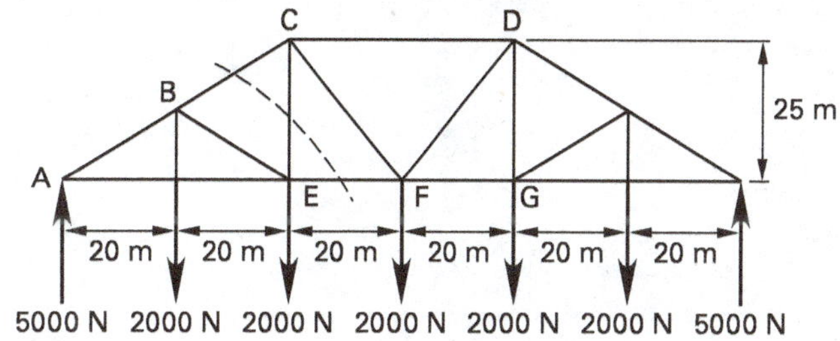

Draw the free body.

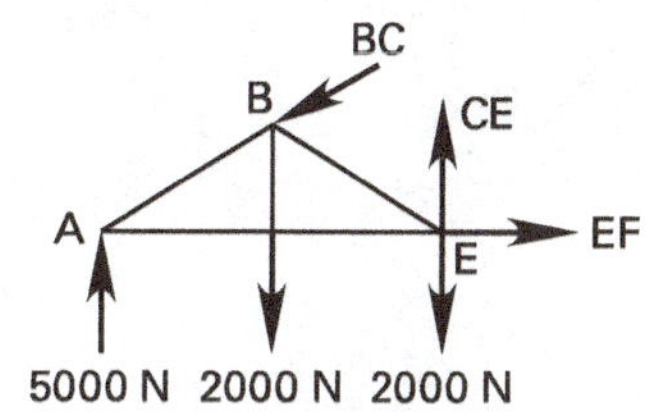

Taking moments about point A will eliminate all of the unknown forces except CE. Let clockwise moments be positive.

$$\sum M_A = 0$$

$$(2000\ \text{N})(20\ \text{m}) + (2000\ \text{N})(40\ \text{m}) - \text{CE}(40\ \text{m}) = 0$$

$$\text{CE} = 3000\ \text{N}$$

The answer is (C).

24 Pulleys, Cables, and Friction

Nomenclature

d	inside diameter	m
D	outside diameter	m
F	force	N
g	gravitational acceleration, 9.81	m/s^2
m	mass	kg
M	moment	N·m
n	number	–
N	normal force	N
p	pitch	m
P	power	W
r	radius	m
T	torque	N·m
v	velocity	m/s
W	weight	N

Symbols

α	pitch angle	deg
η	efficiency	–
θ	angle of wrap	radians
μ	coefficient of friction	–
ϕ	angle	deg

Subscripts

f	friction
k	kinetic
m	mechanical
s	static
t	tangential

1. PULLEYS

A *pulley* (also known as a *sheave*) is used to change the direction of an applied tensile force. A series of pulleys working together (known as a *block and tackle*) can also provide *pulley advantage* (i.e., *mechanical advantage*). (See Fig. 24.1.)

Figure 24.1 *Mechanical Advantage of Rope-Operated Machines*

	fixed sheave	free sheave	ordinary pulley block (n sheaves)	differential pulley block
F_{ideal}	W	$\frac{W}{2}$	$\frac{W}{n}$	$\frac{W}{2}\left(1-\frac{d}{D}\right)$

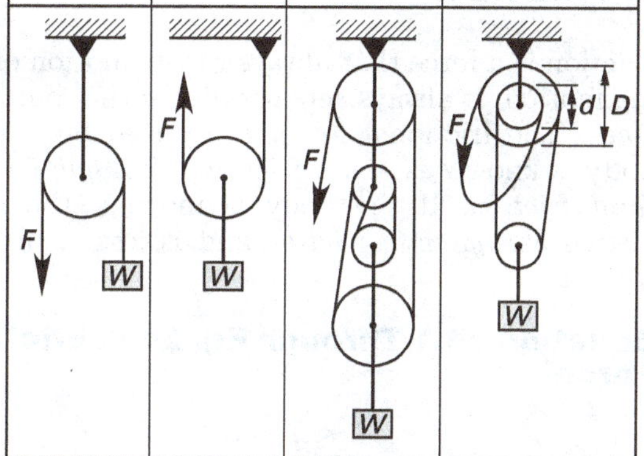

If the pulley is attached by a bracket to a fixed location, it is said to be a *fixed pulley*. If the pulley is attached to a load, or if the pulley is free to move, it is known as a *free pulley*.

Most simple problems disregard friction and assume that all ropes (fiber ropes, wire ropes, chains, belts, etc.) are parallel. In such cases, the pulley advantage is equal to the number of ropes coming to and going from the load-carrying pulley. The diameters of the pulleys are not factors in calculating the pulley advantage.

2. CABLES

An *ideal cable* is assumed to be completely flexible, massless, and incapable of elongation; it acts as an axial tension member between points of concentrated loading. The term *tension* or *tensile force* is commonly used in place of member force when dealing with cables.

The methods of joints and sections used in truss analysis can be used to determine the tensions in cables carrying concentrated loads. (See Fig. 24.2.) After separating the reactions into x- and y-components, it is particularly useful to sum moments about one of the reaction points. All cables will be found to be in tension, and (with vertical loads only) the horizontal tension component will be the same in all cable segments. Unlike the case of a rope passing over a series of pulleys, however, the total tension in the cable will not be the same in every cable segment.

Figure 24.2 Cable with Concentrated Load

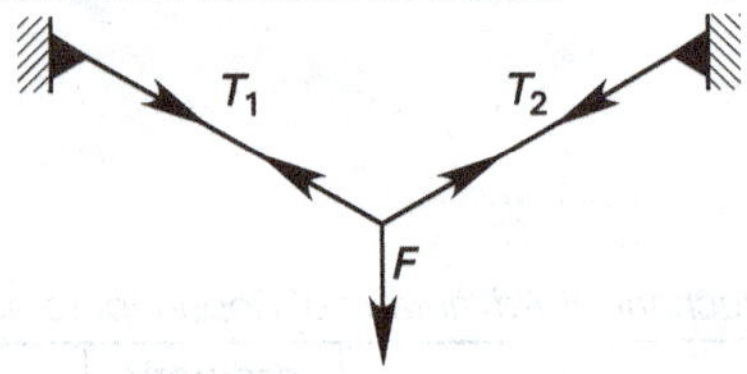

3. FRICTION

Friction is a force that always resists motion or impending motion. It always acts parallel to the contacting surfaces. The frictional force, F, exerted on a stationary body is known as *static friction*, *Coulomb friction*, and *fluid friction*. If the body is moving, the friction is known as *dynamic friction* and is less than the static friction.

Equation 24.1 Through Eq. 24.4: Frictional Force[1]

$$F \le \mu_s N \qquad 24.1$$

$$\mathbf{F} < \mu_s \mathbf{N} \quad \text{[no slip occurring]} \qquad 24.2$$

$$\mathbf{F} = \mu_s \mathbf{N} \quad \text{[point of impending slip]} \qquad 24.3$$

$$\mathbf{F} = \mu_k \mathbf{N} \quad \text{[slip occurring]} \qquad 24.4$$

Values

$$\mu_k \approx 0.75\mu_s$$

Description

The actual magnitude of the frictional force depends on the *normal force*, N, and the *coefficient of friction*, μ, between the body and the surface. The coefficient of kinetic friction, μ_k, is approximately 75% of the coefficient of static friction, μ_s.

Equation 24.1 is a general expression of the laws of friction. Several specific cases exist depending on whether slip is occurring or impending. Use Eq. 24.2 when no slip is occurring. Equation 24.3 is valid at the point of impending slip (or slippage), and Eq. 24.4 is valid when slip is occurring.

For a body resting on a horizontal surface, the normal force is the weight of the body.

$$N = mg$$

If a body rests on a plane inclined at an angle ϕ from the horizontal, the normal force is

$$N = mg \cos\phi$$

Example

A 35 kg block resting on the 30° incline is shown.

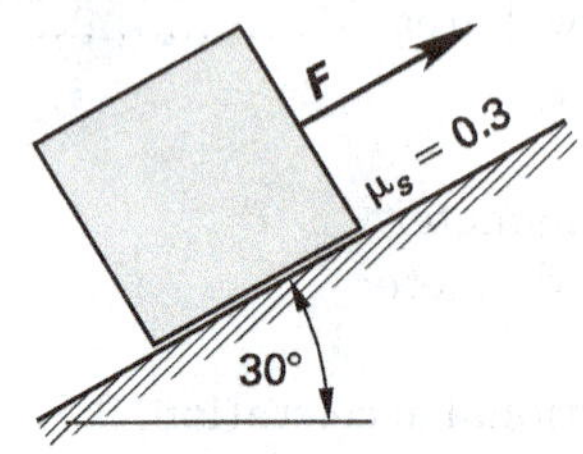

What is most nearly the frictional force at the point of impending slippage?

(A) 37 N

(B) 52 N

(C) 89 N

(D) 100 N

Solution

From Eq. 24.3, the frictional force is

$$\begin{aligned}\mathbf{F} &= \mu_s \mathbf{N} = \mu_s(mg\cos\phi) \\ &= (0.3)\left((35\ \text{kg})\left(9.81\ \frac{\text{m}}{\text{s}^2}\right)\cos 30°\right) \\ &= 89.2\ \text{N} \quad (89\ \text{N})\end{aligned}$$

The answer is (C).

4. BELT FRICTION

Friction between a belt, rope, or band wrapped around a pulley or sheave is responsible for the transfer of torque. Except when stationary, one side of the belt (the tight side) will have a higher tension than the other (the slack side). (See Fig. 24.3.)

[1]Although the NCEES *FE Reference Handbook* (*NCEES Handbook*) uses bold, Eq. 24.2 through Eq. 24.4 are not vector equations.

Figure 24.3 Belt Friction

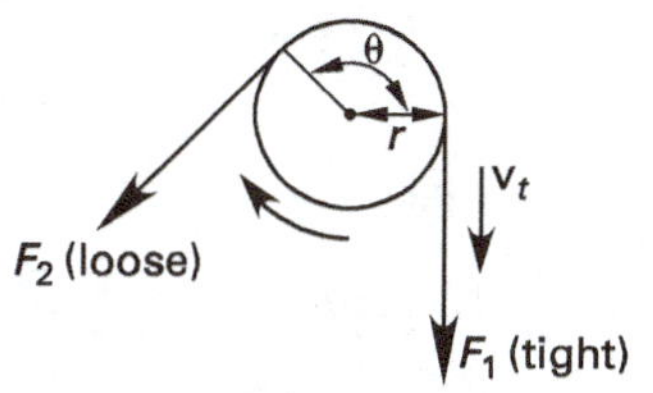

Equation 24.5: Belt Tension Relationship

$$F_1 = F_2 e^{\mu\theta} \tag{24.5}$$

Description

The basic relationship between the belt tensions and the coefficient of friction neglects centrifugal effects and is given by Eq. 24.5. F_1 is the tension on the tight side (direction of movement); F_2 is the tension on the other side. The *angle of wrap*, θ, must be expressed in radians.

The net transmitted torque is

$$T = (F_1 - F_2)r$$

The power transmitted to a belt running at tangential velocity v_t is

$$P = (F_1 - F_2)v_t$$

Example

A rope passes over a fixed sheave, as shown. The two rope ends are parallel. A fixed load on one end of the rope is supported by a constant force on the other end. The coefficient of friction between the rope and the sheave is 0.30.

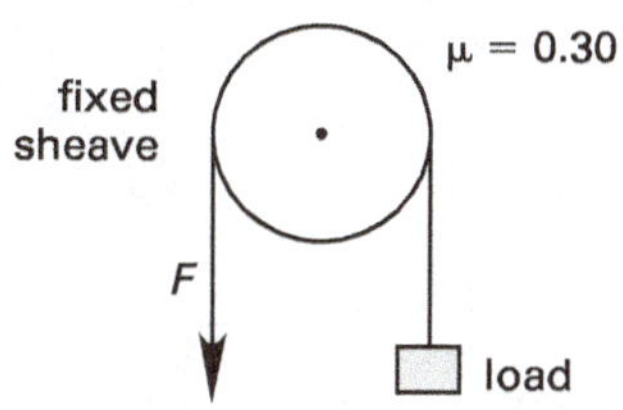

What is most nearly the maximum ratio of tensile forces in the two rope ends?

(A) 1.1

(B) 1.2

(C) 1.6

(D) 2.6

Solution

The angle of wrap, θ, is 180°, but it must be expressed in radians.

$$\theta = (180°)\left(\frac{2\pi \text{ rad}}{360°}\right) = \pi \text{ rad}$$

$$F_1 = F_2 e^{\mu\theta}$$

$$\frac{F_1}{F_2} = e^{(0.30)(\pi \text{ rad})} = 2.57 \quad (2.6)$$

Either side could be the tight side. Therefore, the restraining force could be 2.6 times smaller or larger than the load tension.

The answer is (D).

5. SQUARE SCREW THREADS

A *power screw* changes angular position into linear position (i.e., changes rotary motion into traversing motion). The linear positioning can be horizontal (as in vices and lathes) or vertical (as in a jack). Square, Acme, and 10-degree modified screw threads are commonly used in power screws. A square screw thread is shown in Fig. 24.4.

Figure 24.4 Square Screw Thread

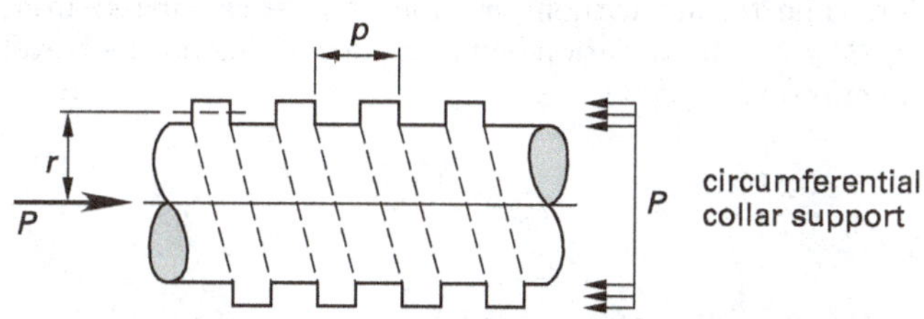

A square screw thread is designated by a mean radius, r, pitch, p, and *pitch angle*, α. The *pitch*, p, is the distance between corresponding points on a thread. The *lead* is the distance the screw advances each revolution. Often, double- and triple-threaded screws are used. The lead is one, two, or three times the pitch for single-, double-, and triple-threaded screws, respectively.

$$P = 2\pi r \tan\alpha$$

Equation 24.6 and Eq. 24.7: Coefficient of Friction and External Moment

$$\mu = \tan\phi \tag{24.6}$$

$$M = Pr\tan(\alpha \pm \phi) \tag{24.7}$$

Statics

Description

The coefficient of friction, μ, between the threads can be designated directly or by way of a *thread friction angle*, ϕ.

The torque or external moment, M, required to turn a square screw in motion against an axial force, P (i.e., "raise" the load), is found from Eq. 24.7.

r is the mean thread radius, M is the torque on the screw, and P is the tensile or compressive force in the screw (i.e., is the load being raised or lowered). The angles are added for tightening operations; they are subtracted for loosening. This equation assumes that all of the torque is used to raise or lower the load.

In Eq. 24.7, the "+" is used for screw tightening (i.e., when the load force is opposite in direction of the screw movement). The "−" is used for screw loosening (i.e., when the load force is in the same direction as the screw movement). If the torque is zero or negative (as it would be if the lead is large or friction is low), then the screw is not self-locking and the load will lower by itself, causing the screw to spin (i.e., it will "overhaul"). The screw will be self-locking when $\tan\alpha \leq \mu$.

The torque calculated in Eq. 24.7 is required to overcome thread friction and to raise the load (i.e., axially compress the screw). Typically, only 10–15% of the torque goes into axial compression of the screw. The remainder is used to overcome friction. The mechanical efficiency of the screw is the ratio of torque without friction to the torque with friction. The torque without friction can be calculated from Eq. 24.7 (or the variation equation, depending on the travel direction) using $\phi = 0$.

$$\eta_m = \frac{M_{f=0}}{M}$$

In the absence of an antifriction ring, an additional torque will be required to overcome friction in the collar. Since the collar is generally flat, the normal force is the jack load for the purpose of calculating the frictional force.

$$M_{\text{collar}} = N\mu_{\text{collar}} r_{\text{collar}}$$

Example

The nuts on a collar are each tightened to 18 N·m torque. 17% of this torque is used to overcome screw thread friction. The bolts have a nominal diameter of 10 mm. The threads are a simple square cut with a pitch angle of 15°. The coefficient of friction in the threads is 0.10.

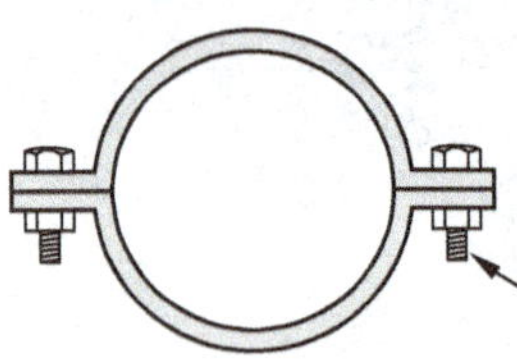

Most nearly, what is the approximate tensile force in each bolt?

(A) 130 N

(B) 200 N

(C) 410 N

(D) 1600 N

Solution

The friction angle, ϕ, is

$$\begin{aligned}\phi &= \arctan\mu = \arctan 0.10 \\ &= 5.71^\circ\end{aligned}$$

Use Eq. 24.7. Only the screw thread friction (17% of the total torque in this application) contributes to the tensile force in the bolt. The force in the bolt is

$$\begin{aligned}P &= \frac{M}{r\tan(\alpha+\phi)} \\ &= \frac{(0.17)(18\ \text{N·m})}{\left(\frac{0.01\ \text{m}}{2}\right)\tan(15^\circ + 5.71^\circ)} \\ &= 1619\ \text{N} \quad (1600\ \text{N})\end{aligned}$$

The answer is (D).

Statics

25 Centroids and Moments of Inertia

Nomenclature

a	subarea	m^2
a	length or radius	m
A	area	m^2
b	base	m
d	distance	m
h	height	m
I	moment of inertia	m^4
I_{xy}	product of inertia	m^4
J	polar moment of inertia	m^4
l	length	m
L	total length	m
m	mass	kg
M	statical moment	m^3
r	radius or radius of gyration	m
v	volume	m^3
V	volume	m^3

Symbols

θ	angle	deg

Subscripts

a	area
c	centroidal
deg	degrees
l	line
o	origin
p	polar
rad	radians
v	volume

1. CENTROIDS

The *centroid* of an area is often described as the point at which a thin, homogeneous plate would balance. This definition, however, combines the definitions of centroid and center of gravity, and implies gravity is required to identify the centroid, which is not true. Nonetheless, this definition provides some intuitive understanding of the centroid.

Centroids of continuous functions can be found by the methods of integral calculus. For most engineering applications, though, the functions to be integrated are regular shapes such as the rectangular, circular, or composite rectangular shapes of beams. For these shapes, simple formulas are readily available and should be used.

Equation 25.1 and Eq. 25.2: First Moment of an Area in the *x-y* Plane[1]

$$M_{ay} = \sum x_n a_n \qquad 25.1$$

$$M_{ax} = \sum y_n a_n \qquad 25.2$$

Variations

$$M_y = \int x\,dA = \sum x_i A_i$$

$$M_x = \int y\,dA = \sum y_i A_i$$

Description

The quantity $\sum x_n a_n$ is known as the *first moment of the area* or *first area moment* with respect to the y-axis. Similarly, $\sum y_n a_n$ is known as the first moment of the area with respect to the x-axis. Equation 25.1 and Eq. 25.2 apply to regular shapes with subareas a_n.

The two primary applications of the first moment are determining centroidal locations and shear stress distributions. In the latter application, the first moment of the area is known as the *statical moment.*

Statics

[1]The NCEES *FE Reference Handbook* (*NCEES Handbook*) deviates from conventional notation in several ways. Q is the most common symbol for the first area moment (then referred to as the *statical moment*), although symbols S and M are also encountered. To avoid confusion with the moment of a force, the subscript a is used to designate the moment of an area. The *NCEES Handbook* uses a lowercase a to designate the area of a subarea (instead of A_i). The *NCEES Handbook* uses n as a summation variable (instead of i), probably to indicate that the moment has to be calculated from all n of the subareas that make up the total area.

Centroid of Line Segments in the *x-y* Plane

For a composite line of total length L, the location of the centroid of a line is defined by the following equations.

$$x_c = \frac{\int x\,dL}{L} = \frac{\sum x_i L_i}{\sum L_i}$$

$$y_c = \frac{\int y\,dL}{L} = \frac{\sum y_i L_i}{\sum L_i}$$

Using the *NCEES Handbook* notation, the equations would be written as

$$L = \sum l_n$$

$$x_{lc} = \frac{\sum x_n l_n}{L}$$

$$y_{lc} = \frac{\sum y_n l_n}{L}$$

Example

Find the approximate x- and y-coordinates of the centroid of wire ABC.

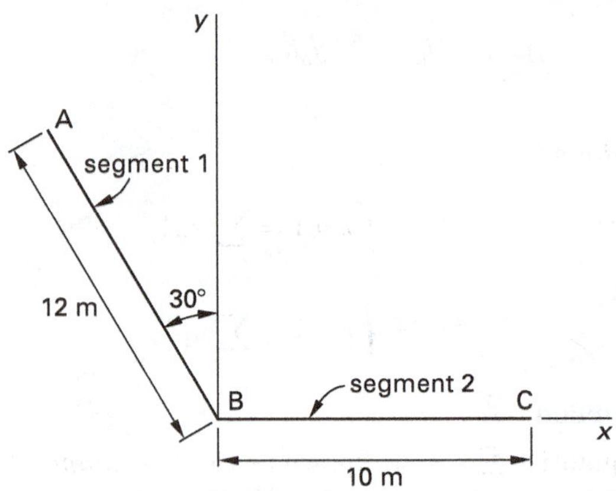

(A) 0.43 m; 1.3 m

(B) 0.64 m; 2.8 m

(C) 2.7 m; 1.5 m

(D) 3.3 m; 2.7 m

Solution

The total length of the line is

$$\sum L_i = 12 \text{ m} + 10 \text{ m} = 22 \text{ m}$$

The coordinates of the centroid of the line are

$$\begin{aligned} x_c &= \frac{\sum x_i L_i}{\sum L_i} \\ &= \frac{\left(\frac{(-12 \text{ m})\sin 30^\circ}{2}\right)(12 \text{ m}) + \left(\frac{10 \text{ m}}{2}\right)(10 \text{ m})}{22 \text{ m}} \\ &= 0.64 \text{ m} \end{aligned}$$

$$\begin{aligned} y_c &= \frac{\sum y_i L_i}{\sum L_i} \\ &= \frac{\left(\frac{(12 \text{ m})\cos 30^\circ}{2}\right)(12 \text{ m}) + (0 \text{ m})(10 \text{ m})}{22 \text{ m}} \\ &= 2.83 \text{ m} \quad (2.8 \text{ m}) \end{aligned}$$

The answer is (B).

Equation 25.3 Through Eq. 25.5: Centroid of an Area in the *x-y* Plane[2]

$$A = \sum a_n \qquad 25.3$$

$$x_{ac} = M_{ay}/A = \sum x_n a_n / A \qquad 25.4$$

$$y_{ac} = M_{ax}/A = \sum y_n a_n / A \qquad 25.5$$

Variations

$$x_c = \frac{\int x\,dA}{A} = \frac{\sum x_{c,i} A_i}{A}$$

$$y_c = \frac{\int y\,dA}{A} = \frac{\sum y_{c,i} A_i}{A}$$

Description

The centroid of an area A composed of subareas a_n (see Eq. 25.3) is located using Eq. 25.4 and Eq. 25.5. The location of the centroid of an area depends only on the geometry of the area, and it is identified by the coordinates (x_{ac}, y_{ac}), or, more commonly, (x_c, y_c).

[2]In Eq. 25.4 and Eq. 25.5, the subscript a is used to designate the centroid of an area, but this convention is largely omitted throughout the rest of the *NCEES Handbook*.

Example

What are the approximate x- and y-coordinates of the centroid of the area shown?

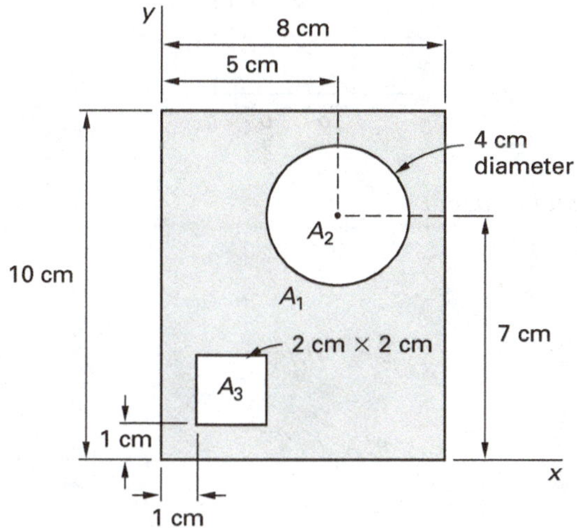

(A) 3.4 cm; 5.6 cm

(B) 3.5 cm; 5.5 cm

(C) 3.9 cm; 4.4 cm

(D) 3.9 cm; 4.8 cm

Solution

Calculate the total area.

$$
\begin{aligned}
A &= \sum a_n \\
&= (8\ \text{cm})(10\ \text{cm}) - \frac{\pi(4\ \text{cm})^2}{4} \\
&\quad -(2\ \text{cm})(2\ \text{cm}) \\
&= 63.43\ \text{cm}^2
\end{aligned}
$$

Find the first moments about the x-axis and y-axis.

$$
\begin{aligned}
M_{ax} &= \sum y_n a_n \\
&= (5\ \text{cm})(80\ \text{cm}^2) + \left(-\pi\left(\frac{4\ \text{cm}}{2}\right)^2(7\ \text{cm})\right) \\
&\quad +(-(2\ \text{cm})(4\ \text{cm}^2)) \\
&= 304.03\ \text{cm}^3
\end{aligned}
$$

$$
\begin{aligned}
M_{ay} &= \sum x_n a_n \\
&= (4\ \text{cm})(80\ \text{cm}^2) + \left(-(5\ \text{cm})\left(\frac{\pi}{4}\right)(4\ \text{cm})^2\right) \\
&\quad +(-(2\ \text{cm})(4\ \text{cm}^2)) \\
&= 249.17\ \text{cm}^3
\end{aligned}
$$

The x-coordinate of the centroid is

$$
\begin{aligned}
x_c &= M_{ay}/A \\
&= \frac{249.17\ \text{cm}^3}{63.43\ \text{cm}^2} \\
&= 3.93\ \text{cm} \quad (3.9\ \text{cm})
\end{aligned}
$$

The y-coordinate of the centroid is

$$
\begin{aligned}
y_c &= M_{ax}/A \\
&= \frac{304.03\ \text{cm}^3}{63.43\ \text{cm}^2} \\
&= 4.79\ \text{cm} \quad (4.8\ \text{cm})
\end{aligned}
$$

The answer is (D).

Equation 25.6 Through Eq. 25.9: Centroid of a Volume[3]

$$V = \sum v_n \qquad 25.6$$

$$x_{vc} = \left(\sum x_n v_n\right)/V \qquad 25.7$$

$$y_{vc} = \left(\sum y_n v_n\right)/V \qquad 25.8$$

$$z_{vc} = \left(\sum z_n v_n\right)/V \qquad 25.9$$

Description

The centroid of a volume V composed of subvolumes v_n (see Eq. 25.6) is located using Eq. 25.7 through Eq. 25.9, which are analogous to the equations used for centroids of areas and lines.

A solid body will have both a center of gravity and a centroid, but the locations of these two points will not necessarily coincide. The earth's attractive force, which is called *weight*, can be assumed to act through the *center of gravity* (also known as the *center of mass*). Only when the body is homogeneous will the *centroid of a volume* coincide with the center of gravity.

[3]The *NCEES Handbook* uses lowercase v to designate the area of a subvolume (instead of V_i). In Eq. 25.7 through Eq. 25.9, the subscript v is used to designate the centroid of a volume, but this convention is largely omitted throughout the rest of the *NCEES Handbook*.

Example

The structure shown is formed of three separate solid aluminum cylindrical rods, each with a 1 cm diameter.

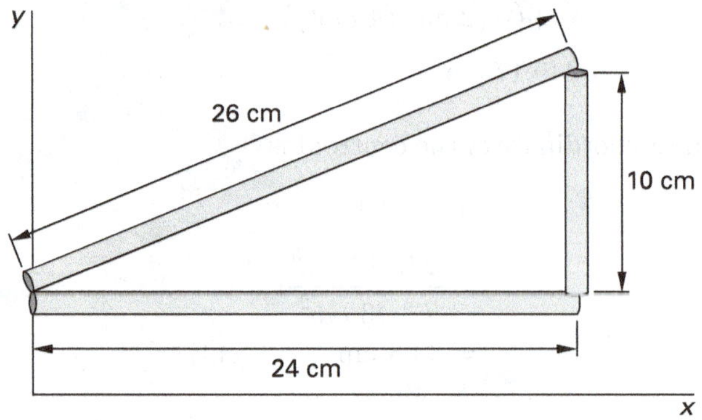

What is the approximate x-coordinate of the centroid of the structure?

(A) 14.0 cm

(B) 15.2 cm

(C) 15.9 cm

(D) 16.0 cm

Solution

Use Eq. 25.7.

$$V_1 = \left(\frac{\pi}{4}\right)(1\text{ cm})^2(24\text{ cm}) = 18.85\text{ cm}^3$$

$$V_2 = \left(\frac{\pi}{4}\right)(1\text{ cm})^2(10\text{ cm}) = 7.85\text{ cm}^3$$

$$V_3 = \left(\frac{\pi}{4}\right)(1\text{ cm})^2(26\text{ cm}) = 20.42\text{ cm}^3$$

$$V = 18.85\text{ cm}^3 + 7.85\text{ cm}^3 + 20.42\text{ cm}^3 = 47.12\text{ cm}^3$$

$$x_n = \left(\sum x_{vc}v_n\right)/V = \frac{\left(\frac{24\text{ cm}}{2}\right)(18.85\text{ cm}^3) + (24\text{ cm})(7.85\text{ cm}^3) + \left(\frac{24\text{ cm}}{2}\right)(20.42\text{ cm}^3)}{47.12\text{ cm}^3} = 14.0\text{ cm}$$

(The $\pi/4$ and area terms all cancel and could have been omitted.)

The answer is (A).

Equation 25.10 Through Eq. 25.50: Centroid and Area Moments of Inertia for Right Triangles

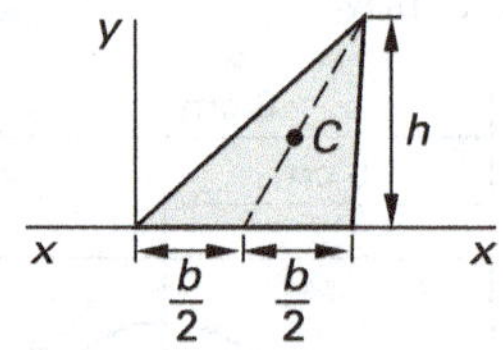

area and centroid

$$A = bh/2 \quad 25.10$$

$$x_c = 2b/3 \quad 25.11$$

$$y_c = h/3 \quad 25.12$$

area moment of inertia

$$I_{x_c} = bh^3/36 \quad 25.13$$

$$I_{y_c} = b^3h/36 \quad 25.14$$

$$I_x = bh^3/12 \quad 25.15$$

$$I_y = b^3h/4 \quad 25.16$$

(radius of gyration)^2

$$r_{x_c}^2 = h^2/18 \quad 25.17$$

$$r_{y_c}^2 = b^2/18 \quad 25.18$$

$$r_x^2 = h^2/6 \quad 25.19$$

$$r_y^2 = b^2/2 \quad 25.20$$

product of inertia

$$I_{x_cy_c} = Abh/36 = b^2h^2/72 \quad 25.21$$

Statics

$$I_{xy} = Abh/4 = b^2h^2/8 \quad 25.22$$

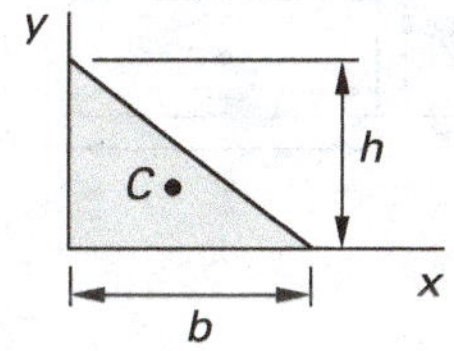

area and centroid

$$A = bh/2 \quad 25.23$$

$$x_c = b/3 \quad 25.24$$

$$y_c = h/3 \quad 25.25$$

area moment of inertia

$$I_{x_c} = bh^3/36 \quad 25.26$$

$$I_{y_c} = b^3h/36 \quad 25.27$$

$$I_x = bh^3/12 \quad 25.28$$

$$I_y = b^3h/12 \quad 25.29$$

(radius of gyration)^2

$$r_{x_c}^2 = h^2/18 \quad 25.30$$

$$r_{y_c}^2 = b^2/18 \quad 25.31$$

$$r_x^2 = h^2/6 \quad 25.32$$

$$r_y^2 = b^2/6 \quad 25.33$$

product of inertia

$$I_{x_cy_c} = -Abh/36 = -b^2h^2/72 \quad 25.34$$

$$I_{xy} = Abh/12 = b^2h^2/24 \quad 25.35$$

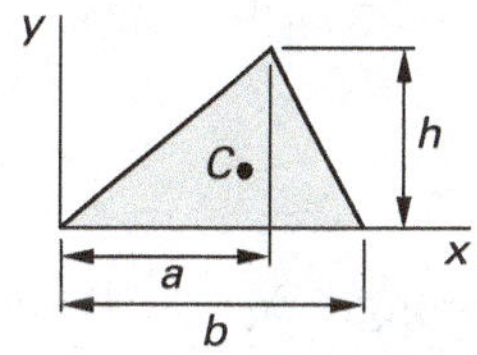

area and centroid

$$A = bh/2 \quad 25.36$$

$$x_c = (a+b)/3 \quad 25.37$$

$$y_c = h/3 \quad 25.38$$

area moment of inertia

$$I_{x_c} = bh^3/36 \quad 25.39$$

$$I_{y_c} = [bh(b^2 - ab + a^2)]/36 \quad 25.40$$

$$I_x = bh^3/12 \quad 25.41$$

$$I_y = [bh(b^2 + ab + a^2)]/12 \quad 25.42$$

(radius of gyration)^2

$$r_{x_c}^2 = h^2/18 \quad 25.43$$

$$r_{y_c}^2 = (b^2 - ab + a^2)/18 \quad 25.44$$

$$r_x^2 = h^2/6 \quad 25.45$$

$$r_y^2 = (b^2 + ab + a^2)/6 \quad 25.46$$

product of inertia

$$I_{x_cy_c} = [Ah(2a-b)]/36 \quad 25.47$$

$$I_{x_cy_c} = [bh^2(2a-b)]/72 \quad 25.48$$

Statics

$$I_{xy} = [Ah(2a+b)]/12 \quad 25.49$$

$$I_{xy} = [bh^2(2a+b)]/24 \quad 25.50$$

Description

Equation 25.10 through Eq. 25.50 give the areas, centroids, and moments of inertia for triangles.

The traditional moments of inertia, I_x and I_y (i.e., the second moments of the area), are always positive. However, the *product of inertia*, $I_{x_c y_c}$, listed in Eq. 25.34, is negative. Since the product of inertia is calculated as $I_{xy} = \sum x_i y_i A_i$, where the x_i and y_i are distances from the composite centroid to the subarea A_i, and since these distances can be either positive or negative depending on where the centroid is located, the product of inertia can be either positive or negative.

Example 13

If a triangle has a base of 13 cm and a height of 8 cm, what is most nearly the vertical distance between the centroid and the radius of gyration about the x-axis?

(A) 0.5 cm

(B) 0.6 cm

(C) 0.7 cm

(D) 0.8 cm

Solution

From Eq. 25.38, the y-component of the centroidal location is

$$y_c = h/3 = \frac{8 \text{ cm}}{3} = 2.667 \text{ cm}$$

From Eq. 25.45, the radius of gyration about the x-axis is

$$r_x = \sqrt{\frac{h^2}{6}} = \sqrt{\frac{(8 \text{ cm})^2}{6}} = 3.266 \text{ cm}$$

The vertical separation between these two points is

$$\begin{aligned} r_x - y_c &= 3.266 \text{ cm} - 2.667 \text{ cm} \\ &= 0.599 \text{ cm} \quad (0.6 \text{ cm}) \end{aligned}$$

The answer is (B).

Equation 25.51 Through Eq. 25.62: Centroid and Area Moments of Inertia for Rectangles

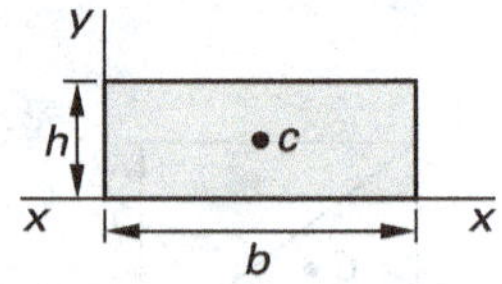

area and centroid

$$A = bh \quad 25.51$$

$$x_c = b/2 \quad 25.52$$

$$y_c = h/2 \quad 25.53$$

area moment of inertia

$$I_x = bh^3/3 \quad 25.54$$

$$I_{x_c} = bh^3/12 \quad 25.55$$

$$J = [bh(b^2+h^2)]/12 \quad 25.56$$

(radius of gyration)^2

$$r_x^2 = h^2/3 \quad 25.57$$

$$r_{x_c}^2 = h^2/12 \quad 25.58$$

$$r_y^2 = b^2/3 \quad 25.59$$

$$r_p^2 = (b^2+h^2)/12 \quad 25.60$$

product of inertia

$$I_{x_c y_c} = 0 \quad 25.61$$

$$I_{xy} = Abh/4 = b^2h^2/4 \quad 25.62$$

Description

Equation 25.51 through Eq. 25.62 give the area, centroids, and moments of inertia for rectangles.

Statics

Example

A 12 cm wide × 8 cm high rectangle is placed such that its centroid is located at the origin, (0, 0). What is the percentage change in the product of inertia if the rectangle is rotated 90° counterclockwise about the origin?

(A) −32% (decrease)

(B) 0%

(C) 32% (increase)

(D) 64% (increase)

Solution

The product of inertia is zero whenever one or more of the reference axes are lines of symmetry. In this case, both axes are lines of symmetry before and after the rotation. From Eq. 25.61, $I_{x_c y_c} = 0$.

The answer is (B).

Equation 25.63 Through Eq. 25.68: Centroid and Area Moments of Inertia for Trapezoids

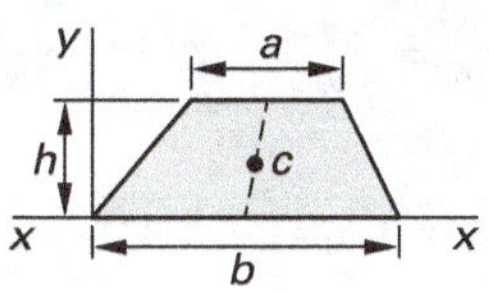

area and centroid

$$A = h(a+b)/2 \quad 25.63$$

$$y_c = \frac{h(2a+b)}{3(a+b)} \quad 25.64$$

area moment of inertia

$$I_{x_c} = \frac{h^3(a^2+4ab+b^2)}{36(a+b)} \quad 25.65$$

$$I_x = \frac{h^3(3a+b)}{12} \quad 25.66$$

(radius of gyration)²

$$r_{x_c}^2 = \frac{h^2(a^2+4ab+b^2)}{18(a+b)} \quad 25.67$$

$$r_x^2 = \frac{h^2(3a+b)}{6(a+b)} \quad 25.68$$

Description

Equation 25.63 through Eq. 25.68 give the area, centroids, and moments of inertia for trapezoids.

Example

What are most nearly the area and the y-coordinate, respectively, of the centroid of the trapezoid shown?

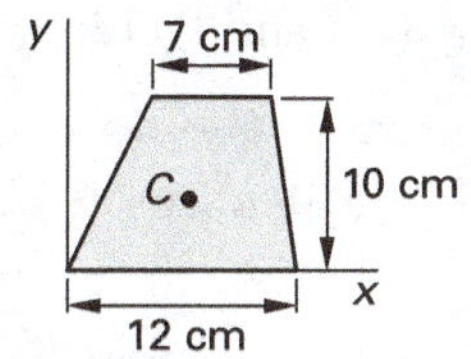

(A) 95 cm²; 4.6 cm

(B) 110 cm²; 5.4 cm

(C) 120 cm²; 6.1 cm

(D) 140 cm²; 7.2 cm

Solution

The area of the trapezoid is

$$A = h(a+b)/2 = \frac{(10\text{ cm})(7\text{ cm} + 12\text{ cm})}{2} = 95\text{ cm}^2$$

From Eq. 25.64, the y-coordinate of the centroid of the trapezoid is

$$\begin{aligned} y_c &= \frac{h(2a+b)}{3(a+b)} \\ &= \frac{(10\text{ cm})((2)(7\text{ cm}) + 12\text{ cm})}{(3)(7\text{ cm} + 12\text{ cm})} \\ &= 4.56\text{ cm} \quad (4.6\text{ cm}) \end{aligned}$$

The answer is (A).

Equation 25.69 Through Eq. 25.80: Centroid and Area Moments of Inertia for Rhomboids

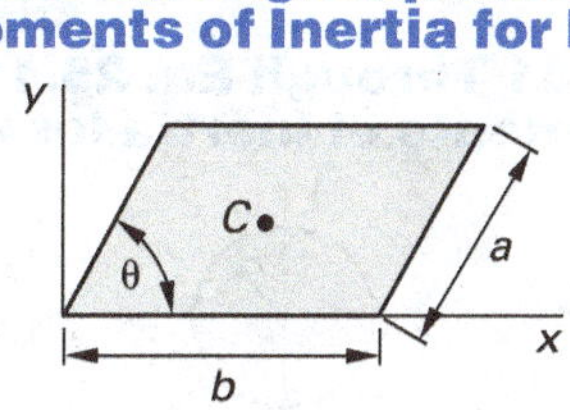

area and centroid

$$A = ab\sin\theta \quad 25.69$$

$$x_c = (b + a\cos\theta)/2 \quad 25.70$$

$$y_c = (a\sin\theta)/2 \quad 25.71$$

area moment of inertia

$$I_{x_c} = (a^3 b \sin^3\theta)/12 \quad 25.72$$

$$I_{y_c} = [ab\sin\theta(b^2 + a^2\cos^2\theta)]/12 \quad 25.73$$

$$I_x = (a^3 b \sin^3\theta)/3 \quad 25.74$$

$$I_y = [ab\sin\theta(b + a\cos\theta)^2]/3 - (a^2b^2\sin\theta\cos\theta)/6 \quad 25.75$$

(radius of gyration)^2

$$r_{x_c}^2 = (a\sin\theta)^2/12 \quad 25.76$$

$$r_{y_c}^2 = (b^2 + a^2\cos^2\theta)/12 \quad 25.77$$

$$r_x^2 = (a\sin\theta)^2/3 \quad 25.78$$

$$r_y^2 = (b + a\cos\theta)^2/3 - (ab\cos\theta)/6 \quad 25.79$$

product of inertia

$$I_{x_c y_c} = (a^3 b \sin^2\theta\cos\theta)/12 \quad 25.80$$

Description

Equation 25.69 through Eq. 25.80 give the area, centroids, and moments of inertia for rhomboids.

Equation 25.81 Through Eq. 25.116: Centroid and Area Moments of Inertia for Circles[4]

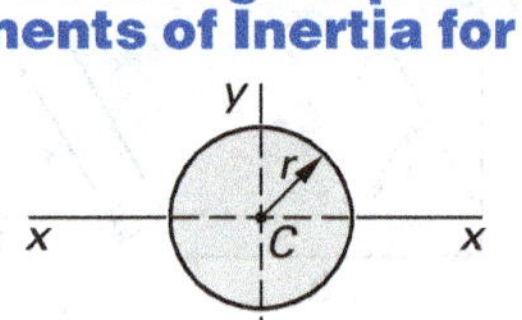

area and centroid

$$A = \pi a^2 \quad 25.81$$

$$x_c = a \quad 25.82$$

$$y_c = a \quad 25.83$$

area moment of inertia

$$I_{x_c} = I_{y_c} = \pi a^4/4 \quad 25.84$$

$$I_x = I_y = 5\pi a^4/4 \quad 25.85$$

$$J = \pi r^4/2 \quad 25.86$$

(radius of gyration)^2

$$r_{x_c}^2 = r_{y_c}^2 = a^2/4 \quad 25.87$$

$$r_x^2 = r_y^2 = 5a^2/4 \quad 25.88$$

$$r_p^2 = a^2/2 \quad 25.89$$

product of inertia

$$I_{x_c y_c} = 0 \quad 25.90$$

$$I_{xy} = Aa^2 \quad 25.91$$

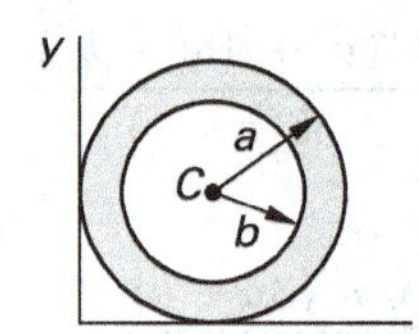

area and centroid

$$A = \pi(a^2 - b^2) \quad 25.92$$

$$x_c = a \quad 25.93$$

$$y_c = a \quad 25.94$$

[4]In Eq. 25.81 through Eq. 25.116, the *NCEES Handbook* designates the radius of a circle or circular segment as *a*, rather than as the conventional *r* or *R*, which are used almost everywhere else in the *NCEES Handbook*.

Statics

area moment of inertia

$$I_{x_c} = I_{y_c} = \pi(a^4 - b^4)/4 \qquad 25.95$$

$$I_x = I_y = \frac{5\pi a^4}{4} - \pi a^2 b^2 - \frac{\pi b^4}{4} \qquad 25.96$$

$$J = \pi(r_a^4 - r_b^4)/2 \qquad 25.97$$

(radius of gyration)^2

$$r_{x_c}^2 = r_{y_c}^2 = (a^2 + b^2)/4 \qquad 25.98$$

$$r_x^2 = r_y^2 = (5a^2 + b^2)/4 \qquad 25.99$$

$$r_p^2 = (a^2 + b^2)/2 \qquad 25.100$$

product of inertia

$$I_{x_c y_c} = 0 \qquad 25.101$$

$$I_{xy} = Aa^2 \qquad 25.102$$

$$I_{xy} = \pi a^2(a^2 - b^2) \qquad 25.103$$

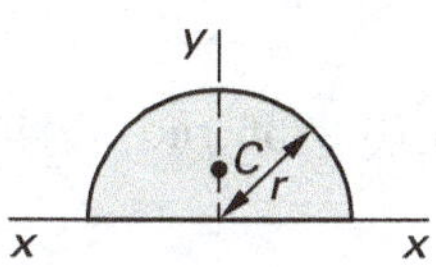

area and centroid

$$A = \pi a^2/2 \qquad 25.104$$

$$x_c = a \qquad 25.105$$

$$y_c = 4a/3\pi \qquad 25.106$$

area moment of inertia

$$I_{x_c} = \frac{a^4(9\pi^2 - 64)}{72\pi} \qquad 25.107$$

$$I_{y_c} = \pi a^4/8 \qquad 25.108$$

$$I_x = \pi a^4/8 \qquad 25.109$$

$$I_y = 5\pi a^4/8 \qquad 25.110$$

(radius of gyration)^2

$$r_{x_c}^2 = \frac{a^2(9\pi^2 - 64)}{36\pi^2} \qquad 25.111$$

$$r_{y_c}^2 = a^2/4 \qquad 25.112$$

$$r_x^2 = a^2/4 \qquad 25.113$$

$$r_y^2 = 5a^2/4 \qquad 25.114$$

product of inertia

$$I_{x_c y_c} = 0 \qquad 25.115$$

$$I_{xy} = 2a^4/3 \qquad 25.116$$

Description

Equation 25.81 through Eq. 25.116 give the area, centroids, and moments of inertia for circles.

Example

The center of a circle with a radius of 7 cm is located at $(x, y) = (3 \text{ cm}, 4 \text{ cm})$. Most nearly, what is the minimum distance that the origin of the x- and y-axes would have to be moved in order to reduce the product of inertia to its smallest absolute value?

(A) 3 cm

(B) 4 cm

(C) 5 cm

(D) 7 cm

Solution

The product of inertia of a circle can be a positive value, a negative value, or zero, depending on the location of the axes. The absolute value is zero (i.e., is minimized) when at least one of the axes coincides with a line of symmetry. Although this can be accomplished by moving the origin to the center of the circle (a distance of 5 cm recognizing that this is a 3-4-5 triangle), a shorter move results when the y-axis is moved 3 cm to the right.

Then, the y-axis passes through the centroid, which is sufficient to reduce the product of inertia to zero.

The answer is (A).

Equation 25.117 Through Eq. 25.125: Centroid and Area Moments of Inertia for Circular Sectors

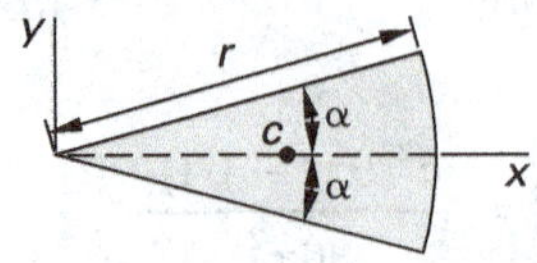

area and centroid

$$A = a^2\theta \qquad 25.117$$

$$x_c = \frac{2a}{3}\frac{\sin\theta}{\theta} \qquad 25.118$$

$$y_c = 0 \qquad 25.119$$

area moment of inertia

$$I_x = a^4(\theta - \sin\theta\cos\theta)/4 \qquad 25.120$$

$$I_y = a^4(\theta + \sin\theta\cos\theta)/4 \qquad 25.121$$

(radius of gyration)^2

$$r_x^2 = \frac{a^2}{4}\frac{(\theta - \sin\theta\cos\theta)}{\theta} \qquad 25.122$$

$$r_y^2 = \frac{a^2}{4}\frac{(\theta + \sin\theta\cos\theta)}{\theta} \qquad 25.123$$

product of inertia

$$I_{x_c y_c} = 0 \qquad 25.124$$

$$I_{xy} = 0 \qquad 25.125$$

Description

Equation 25.117 through Eq. 25.125 give the area, centroids, and moments of inertia for circular sectors. In order to incorporate θ into the calculations, as is done for some of the circular sector equations, the angle must be expressed in radians.

$$\theta_{\text{rad}} = \theta_{\text{deg}}\left(\frac{2\pi}{360°}\right)$$

Example

A grassy parcel of land shaped like a rhombus has sides measuring 120 m with a 65° included angle. A small, straight creek runs between the opposing acute corners. A goat is humanely tied to the bank of the creek at one of the acute corners by a 40 m long rope. Without crossing the creek, most nearly, on what area of grass can the goat graze?

(A) 450 m^2

(B) 710 m^2

(C) 910 m^2

(D) 26 000 m^2

Solution

The creek bisects the 65° angle. The goat sweeps out a circular sector with a 40 m radius.

$$\theta = \frac{\text{swept angle}}{2} = \left(\frac{\frac{65°}{2}}{2}\right)\left(\frac{2\pi}{360°}\right) = 0.2836 \text{ rad}$$

Use Eq. 25.117. The swept area is

$$A = a^2\theta = (40 \text{ m})^2(0.2836 \text{ rad}) = 453.8 \text{ m}^2 \quad (450 \text{ m}^2)$$

The answer is (A).

Equation 25.126 Through Eq. 25.134: Centroid and Area Moments of Inertia for Circular Segments

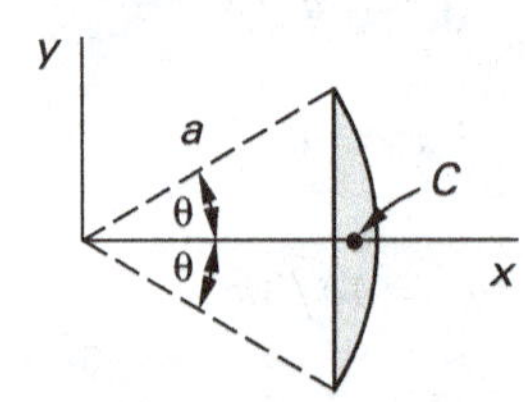

area and centroid

$$A = a^2\left[\theta - \frac{\sin 2\theta}{2}\right] \qquad 25.126$$

$$x_c = \frac{2a}{3}\frac{\sin^3\theta}{\theta - \sin\theta\cos\theta} \qquad 25.127$$

$$y_c = 0 \tag{25.128}$$

area moment of inertia

$$I_x = \frac{Aa^2}{4}\left[1 - \frac{2\sin^3\theta\cos\theta}{3\theta - 3\sin\theta\cos\theta}\right] \tag{25.129}$$

$$I_y = \frac{Aa^2}{4}\left[1 + \frac{2\sin^3\theta\cos\theta}{\theta - \sin\theta\cos\theta}\right] \tag{25.130}$$

(radius of gyration)^2

$$r_x^2 = \frac{a^2}{4}\left[1 - \frac{2\sin^3\theta\cos\theta}{3\theta - 3\sin\theta\cos\theta}\right] \tag{25.131}$$

$$r_y^2 = \frac{a^2}{4}\left[1 + \frac{2\sin^3\theta\cos\theta}{\theta - \sin\theta\cos\theta}\right] \tag{25.132}$$

product of inertia

$$I_{x_c y_c} = 0 \tag{25.133}$$

$$I_{xy} = 0 \tag{25.134}$$

Description

Equation 25.126 through Eq. 25.134 give the area, centroids, and moments of inertia for circular segments.

Equation 25.135 Through Eq. 25.145: Centroid and Area Moments of Inertia for Parabolas

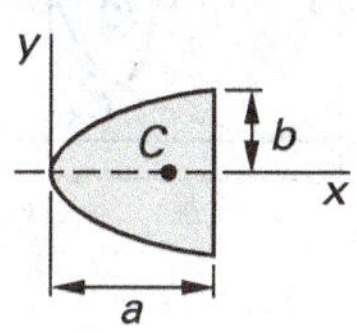

area and centroid

$$A = 4ab/3 \tag{25.135}$$

$$x_c = 3a/5 \tag{25.136}$$

$$y_c = 0 \tag{25.137}$$

area moment of inertia

$$I_{x_c} = I_x = 4ab^3/15 \tag{25.138}$$

$$I_{y_c} = 16a^3b/175 \tag{25.139}$$

$$I_y = 4a^3b/7 \tag{25.140}$$

(radius of gyration)^2

$$r_{x_c}^2 = r_x^2 = b^2/5 \tag{25.141}$$

$$r_{y_c}^2 = 12a^2/175 \tag{25.142}$$

$$r_y^2 = 3a^2/7 \tag{25.143}$$

product of inertia

$$I_{x_c y_c} = 0 \tag{25.144}$$

$$I_{xy} = 0 \tag{25.145}$$

Description

Equation 25.135 through Eq. 25.145 give the area, centroids, and moments of inertia for parabolas.

Example

The entrance freeway to a city passes under a decorative parabolic arch with a 28 m base and a 200 m height. A famous illusionist contacts the city with a plan to make the city disappear from behind a curtain draped down from the arch. If the drape spans the entire width and height of the opening, and if seams and reinforcement increase the material requirements by 15%, most nearly, how much drapery fabric will be needed?

(A) 3700 m^2

(B) 4300 m^2

(C) 7500 m^2

(D) 8600 m^2

Solution

b is half of the width of the arch.

$$b = \frac{28 \text{ m}}{2} = 14 \text{ m}$$

Use Eq. 25.135. Including the allowance for seams and reinforcement, the required area is

$$\begin{aligned} A &= (1+\text{allowance})\frac{4ab}{3} \\ &= (1+0.15)\left(\frac{(4)(200\text{ m})(14\text{ m})}{3}\right) \\ &= 4293\text{ m}^2 \quad (4300\text{ m}^2) \end{aligned}$$

The answer is (B).

Equation 25.146 Through Eq. 25.153: Centroid and Area Moments of Inertia for Semiparabolas

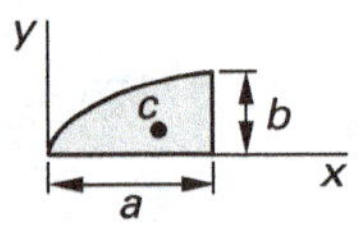

area and centroid

$$A = 2ab/3 \qquad 25.146$$

$$x_c = 3a/5 \qquad 25.147$$

$$y_c = 3b/8 \qquad 25.148$$

area moment of inertia

$$I_x = 2ab^3/15 \qquad 25.149$$

$$I_y = 2ba^3/7 \qquad 25.150$$

(radius of gyration)²

$$r_x^2 = b^2/5 \qquad 25.151$$

$$r_y^2 = 3a^2/7 \qquad 25.152$$

product of inertia

$$I_{xy} = Aab/4 = a^2b^2 \qquad 25.153$$

Description

Equation 25.146 through Eq. 25.153 give the area, centroids, and moments of inertia for semiparabolas.

Example

What is most nearly the area of the shaded section above the parabolic curve shown?

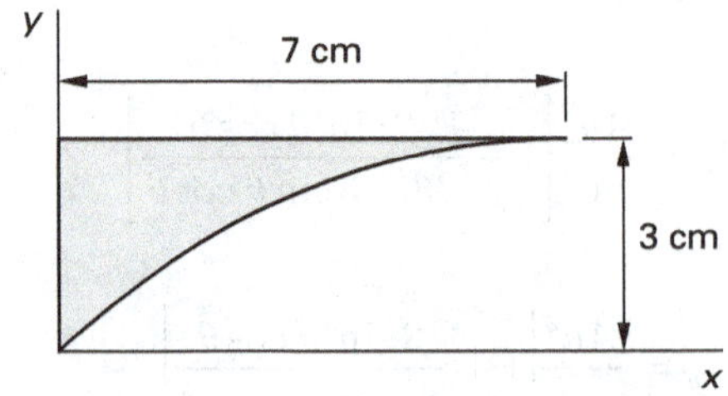

(A) 7 cm²

(B) 9 cm²

(C) 11 cm²

(D) 14 cm²

Solution

From Eq. 25.146, the semiparabolic area below the curve is

$$\begin{aligned} A_{\text{below}} &= 2ab/3 = \frac{(2)(3\text{ cm})(7\text{ cm})}{3} \\ &= 14\text{ cm}^2 \end{aligned}$$

The shaded area above the parabolic curve is

$$\begin{aligned} A_{\text{above}} &= A - A_{\text{below}} = (7\text{ cm})(3\text{ cm}) - 14\text{ cm}^2 \\ &= 7\text{ cm}^2 \end{aligned}$$

The answer is (A).

Equation 25.154 Through Eq. 25.160: Centroid and Area Moments of Inertia for General Spandrels (*n*th Degree Parabolas)

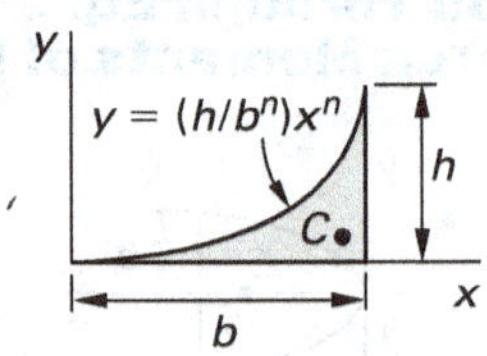

area and centroid

$$A = bh/(n+1) \qquad 25.154$$

$$x_c = \frac{n+1}{n+2}b \qquad 25.155$$

$$y_c = \frac{h}{2}\,\frac{n+1}{2n+1} \qquad 25.156$$

Statics

area moment of inertia

$$I_x = \frac{bh^3}{3(3n+1)} \quad 25.157$$

$$I_y = \frac{hb^3}{n+3} \quad 25.158$$

(radius of gyration)^2

$$r_x^2 = \frac{h^2(n+1)}{3(3n+1)} \quad 25.159$$

$$r_y^2 = \frac{n+1}{n+3}b^2 \quad 25.160$$

Description

Equation 25.154 through Eq. 25.160 give the area, centroids, and moments of inertia for general spandrels.

Example

For the curve $y = x^3$, what are the approximate coordinates of the centroid of the shaded area between $x = 0$ and $x = 3$ cm?

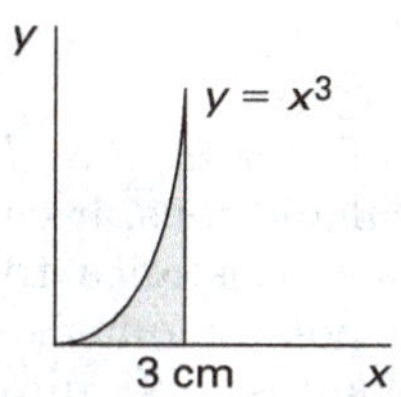

(A) 1.6 cm; 7.8 cm

(B) 1.8 cm; 5.8 cm

(C) 2.0 cm; 18 cm

(D) 2.4 cm; 7.7 cm

Solution

Treat x as the base and y as the height. n is 3 for this spandrel. The height is $h = x^3 = (3)^3 = 27$ cm. The x- and y-coordinates, respectively, are

$$x_c = \left(\frac{n+1}{n+2}\right)b = \left(\frac{3+1}{3+2}\right)(3\text{ cm}) = 2.4\text{ cm}$$

$$y_c = \left(\frac{h}{2}\right)\left(\frac{n+1}{2n+1}\right) = \left(\frac{27\text{ cm}}{2}\right)\left(\frac{3+1}{(2)(3)+1}\right) = 7.714\text{ cm} \quad (7.7\text{ cm})$$

The answer is (D).

Equation 25.161 Through Eq. 25.167: Centroids and Area Moments of Inertia for *n*th Degree Parabolas

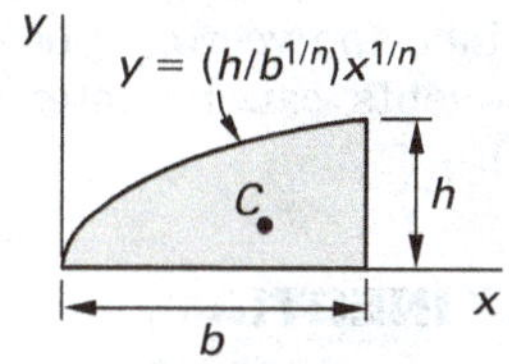

area and centroid

$$A = \frac{n}{n+1}bh \quad 25.161$$

$$x_c = \frac{n+1}{2n+1}b \quad 25.162$$

$$y_c = \frac{n+1}{2(n+2)}h \quad 25.163$$

area moment of inertia

$$I_x = \frac{n}{3(n+3)}bh^3 \quad 25.164$$

$$I_y = \frac{n}{3n+1}b^3h \quad 25.165$$

(radius of gyration)^2

$$r_x^2 = \frac{n+1}{3(n+1)}h^2 \quad 25.166$$

$$r_y^2 = \frac{n+1}{3n+1}b^2 \quad 25.167$$

Description

Equation 25.161 through Eq. 25.167 give the area, centroids, and moments of inertia for nth degree parabolas.

Equation 25.168: Centroid of a Volume

$$\mathbf{r}_c = \sum m_n \mathbf{r}_n / \sum m_n \qquad 25.168$$

Description

Equation 25.168 provides a convenient method of locating the centroid of an object that consists of several isolated component masses. The masses do not have to be contiguous and can be distributed throughout space. It is implicit that the vectors that terminate at the submasses' centroids are based at the origin, $(0, 0, 0)$. These vectors have the form of $r_x\mathbf{i} + r_y\mathbf{j} + r_z\mathbf{k}$. The end result is a vector, but since the vector is based at the origin, the vector components can be interpreted as coordinates, (r_{cx}, r_{cy}, r_{cz}).

2. MOMENT OF INERTIA

The *moment of inertia*, I, of an area is needed in mechanics of materials problems. It is convenient to think of the moment of inertia of a beam's cross-sectional area as a measure of the beam's ability to resist bending. Given equal loads, a beam with a small moment of inertia will bend more than a beam with a large moment of inertia.

Since the moment of inertia represents a resistance to bending, it is always positive. Since a beam can be asymmetric in cross section (e.g., a rectangular beam) and be stronger in one direction than another, the moment of inertia depends on orientation. A reference axis or direction must be specified.

The symbol I_x is used to represent a moment of inertia with respect to the x-axis. Similarly, I_y is the moment of inertia with respect to the y-axis. I_x and I_y do not combine and are not components of some resultant moment of inertia.

Any axis can be chosen as the reference axis, and the value of the moment of inertia will depend on the reference selected. The moment of inertia taken with respect to an axis passing through the area's centroid is known as the *centroidal moment of inertia*, I_{x_c} or I_{y_c}. The centroidal moment of inertia is the smallest possible moment of inertia for the area.

Equation 25.169 and Eq. 25.170: Second Moment of the Area

$$I_y = \int x^2 dA \qquad 25.169$$

$$I_x = \int y^2 dA \qquad 25.170$$

Description

Integration can be used to calculate the moment of inertia of a function that is bounded by the x- and y-axes and a curve $y = f(x)$. From Eq. 25.169 and Eq. 25.170, it is apparent why the moment of inertia is also known as the *second moment of the area* or *second area moment.*

Equation 25.171 and Eq. 25.172: Perpendicular Axis Theorem

$$I_z = J = I_y + I_x = \int (x^2 + y^2) dA \qquad 25.171$$

$$I_z = r_p^2 A \qquad 25.172$$

Variation

$$J_c = I_{x_c} + I_{y_c}$$

Description

The *polar moment of inertia*, J or I_z, is required in torsional shear stress calculations. It can be thought of as a measure of an area's resistance to torsion (twisting). The definition of a polar moment of inertia of a two-dimensional area requires three dimensions because the reference axis for a polar moment of inertia of a plane area is perpendicular to the plane area.

The polar moment of inertia can be derived from Eq. 25.171.

It is often easier to use the perpendicular axis theorem to quickly calculate the polar moment of inertia.

Perpendicular axis theorem: The moment of inertia of a plane area about an axis normal to the plane is equal to the sum of the moments of inertia about any two mutually perpendicular axes lying in the plane and passing through the given axis.

Since the two perpendicular axes can be chosen arbitrarily, it is most convenient to use the centroidal moments of inertia, as shown in the variation equation.

Example

For the composite plane area made up of two circles as shown, the moment of inertia about the y-axis is 4.7 cm^4, and the moment of inertia about the x-axis is 23.5 cm^4.

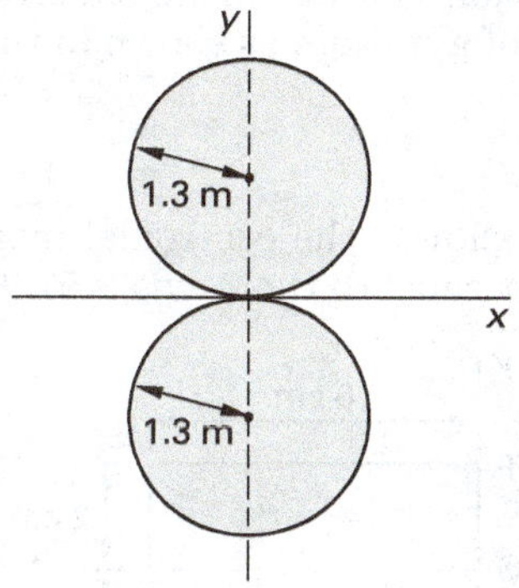

What is the approximate polar moment of inertia of the area taken about the intersection of the x- and y-axes?

(A) 0 cm^4

(B) 14 cm^4

(C) 28 cm^4

(D) 34 cm^4

Solution

Use the perpendicular axis theorem, as given by Eq. 25.171.

$$\begin{aligned} J &= I_y + I_x \\ &= 4.7\ \text{cm}^4 + 23.5\ \text{cm}^4 \\ &= 28.2\ \text{cm}^4 \quad (28\ \text{cm}^4) \end{aligned}$$

The answer is (C).

Equation 25.173 and Eq. 25.174: Parallel Axis Theorem

$$I_x' = I_{x_c} + d_y^2 A \qquad 25.173$$

$$I_y' = I_{y_c} + d_x^2 A \qquad 25.174$$

Description

If the moment of inertia is known with respect to one axis, the moment of inertia with respect to another, parallel axis can be calculated from the *parallel axis theorem*, also known as the *transfer axis theorem*. This theorem is used to evaluate the moment of inertia of areas that are composed of two or more basic shapes. d is the distance between the centroidal axis and the second, parallel axis.

The second term in Eq. 25.173 and Eq. 25.174 is often much larger than the first term in each equation, since areas close to the centroidal axis do not affect the moment of inertia considerably. This principle is exploited in the design of structural steel shapes that derive bending resistance from *flanges* located far from the centroidal axis. The *web* does not contribute significantly to the moment of inertia. (See Fig. 25.1.)

Figure 25.1 *Structural Steel Shape*

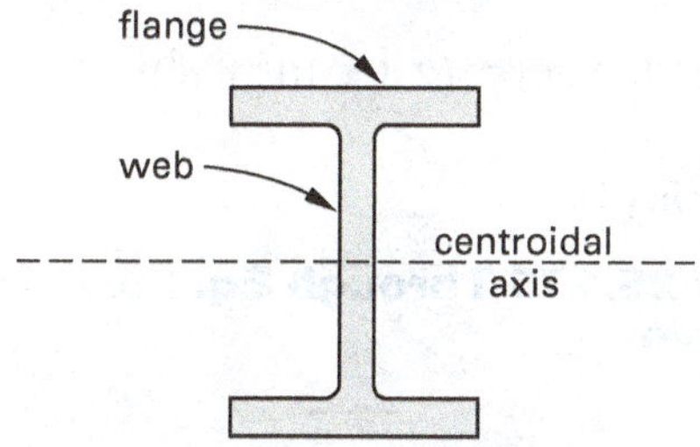

Example

The moment of inertia about the x'-axis of the cross section shown is 334 000 cm^4. The cross-sectional area is 86 cm^2, and the thicknesses of the web and the flanges are the same.

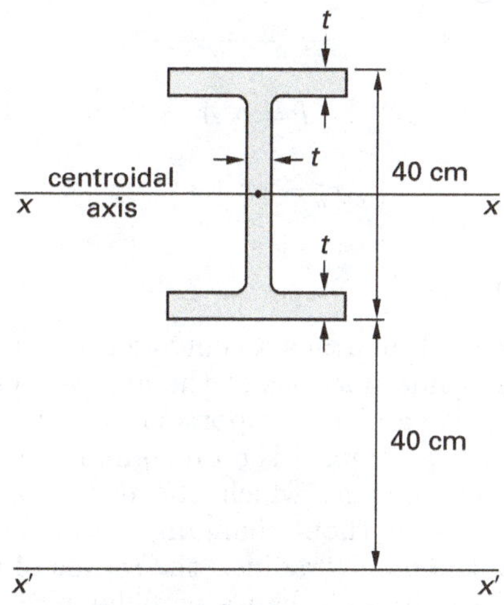

What is most nearly the moment of inertia about the centroidal axis?

(A) $2.4 \times 10^4\ \text{cm}^4$

(B) $7.4 \times 10^4\ \text{cm}^4$

(C) $2.0 \times 10^5\ \text{cm}^4$

(D) $6.4 \times 10^5\ \text{cm}^4$

Solution

Use Eq. 25.173. The moment of inertia around the centroidal axis is

$$I_x' = I_{x_c} + d_x^2 A$$

$$\begin{aligned} I_{x_c} &= I_x' - d_x^2 A \\ &= 334\,000\ \text{cm}^4 - (86\ \text{cm}^2)\left(40\ \text{cm} + \frac{40\ \text{cm}}{2}\right)^2 \\ &= 24\,400\ \text{cm}^4 \quad (2.4 \times 10^4\ \text{cm}^4) \end{aligned}$$

The answer is (A).

Equation 25.175 Through Eq. 25.177: Radius of Gyration

$$r_x = \sqrt{I_x/A} \qquad 25.175$$

$$r_y = \sqrt{I_y/A} \qquad 25.176$$

$$r_p = \sqrt{J/A} \qquad 25.177$$

Variations

$$I = r^2 A$$

$$r_p^2 = r_x^2 + r_y^2$$

Description

Every nontrivial area has a centroidal moment of inertia. Usually, some portions of the area are close to the centroidal axis, and other portions are farther away. The *radius of gyration*, r, is an imaginary distance from the centroidal axis at which the entire area can be assumed to exist without changing the moment of inertia. Despite the name "radius," the radius of gyration is not limited to circular shapes or polar axes. This concept is illustrated in Fig. 25.2.

Figure 25.2 *Radius of Gyration of Two Equivalent Areas*

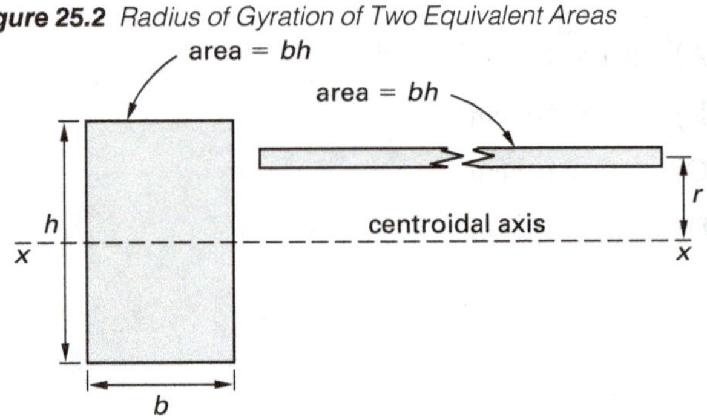

The radius of gyration, r, is given by Eq. 25.175 and Eq. 25.176. The analogous quantity in the polar system is calculated using Eq. 25.177.

Just as the polar moment of inertia, J, can be calculated from the two rectangular moments of inertia, the polar radius of gyration can be calculated from the two rectangular radii of gyration, as shown in the second variation equation.

Example

For the shape shown, the centroidal moment of inertia about an axis parallel to the x-axis is 57.9 cm^4.

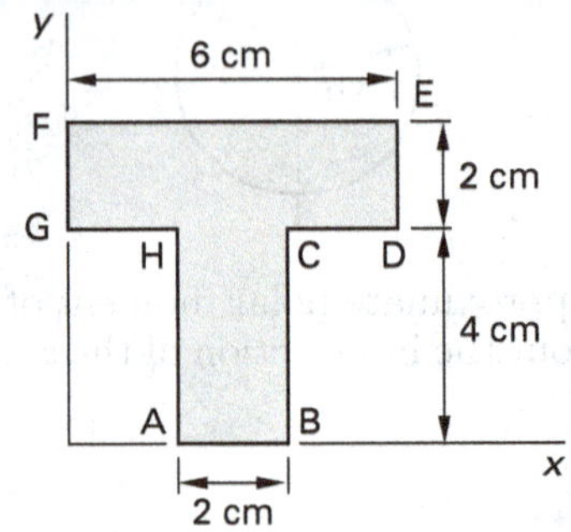

What is the approximate radius of gyration about a horizontal axis passing through the centroid?

(A) 0.86 cm

(B) 1.7 cm

(C) 2.3 cm

(D) 3.7 cm

Solution

The area is

$$\begin{aligned} A &= (2\ \text{cm})(4\ \text{cm}) + (2\ \text{cm})(6\ \text{cm}) \\ &= 20\ \text{cm}^2 \end{aligned}$$

By definition, the radius of gyration is calculated with respect to the centroidal axis. From Eq. 25.175,

$$\begin{aligned} r_x &= \sqrt{I_{x_c}/A} \\ &= \sqrt{\frac{57.9\ \text{cm}^4}{20\ \text{cm}^2}} \\ &= 1.70\ \text{cm} \quad (1.7\ \text{cm}) \end{aligned}$$

The answer is (B).

Equation 25.178 and Eq. 25.179: Product of Inertia

$$I_{xy} = \int xy\,dA \qquad 25.178$$

$$I_{xy}' = I_{x_c y_c} + d_x d_y A \qquad 25.179$$

Description

The *product of inertia*, I_{xy}, of a two-dimensional area is found by multiplying each differential element of area by its x- and y-coordinate and then summing over the entire area.

The product of inertia is zero when either axis is an axis of symmetry. Since the axes can be chosen arbitrarily, the area may be in one of the negative quadrants, and the product of inertia may be negative.

The transfer theorem for products of inertia is given by Eq. 25.179. (Both axes are allowed to move to new positions.) d_x and d_y are the distances to the centroid in the new coordinate system, and $I_{x_cy_c}$ is the centroidal product of inertia in the old system.

26 Indeterminate Statics

Nomenclature

A	area	m^2
b	width	m
c	distance from neutral axis to extreme fiber	m
COF	carryover factor	–
DF	distribution factor	–
E	modulus of elasticity	Pa
f	load	N
FEM	fixed end moment	N·m
F	force	N
I	moment of inertia	m^4
j	number of joints	–
k	stiffness	N/m
L	length	m
m	number of members (bars)	–
M	moment	N·m
P	load	N
Q	statical moment	m^3
r	number of reactions	–
s	number of special conditions	–
S	force	N
T	temperature	°C
u	force	N
U	strain energy	J
V	shear	N
w	distributed load	N/m
W	work	N·m

Symbols

α	coefficient of thermal expansion	1/°C
δ	deformation	m
Δ	displacement	m
θ	deflection angle	rad
θ	unit rotation	rad

1. INTRODUCTION TO INDETERMINATE STATICS

A structure that is *statically indeterminate* is one for which the equations of statics are not sufficient to determine all reactions, moments, and internal forces. Additional formulas involving deflection are required to completely determine these variables.

Although there are many configurations of statically indeterminate structures, this chapter is primarily concerned with beams on more than two supports and trusses with more members than are required for rigidity.

2. DEGREE OF INDETERMINACY

The *degree of indeterminacy* (*degree of redundancy*) is equal to the number of reactions or members that would have to be removed in order to make the structure statically determinate. For example, a two-span beam on three simple supports is indeterminate (redundant) to the first degree. The degree of indeterminacy of a pin-connected, two-dimensional truss is

$$I = r + m - 2j$$

The degree of indeterminacy of a pin-connected, three-dimensional truss is

$$I = r + m - 3j$$

Rigid frames have joints that transmit moments. The degree of indeterminacy of two-dimensional rigid plane frames is more complex. In the following equation, s is the number of *special conditions* (also known as the number of *equations of conditions*). s is 1 for each internal hinge or a shear release, 2 for each internal roller, and 0 if neither hinges nor rollers are present.

$$I = r + 3m - 3j - s$$

3. INDETERMINATE BEAMS

Three common configurations of beams can easily be recognized as being statically indeterminate. These are the *continuous beam*, *propped cantilever beam*, and *fixed-end beam*, as illustrated in Fig. 26.1.

Figure 26.1 *Types of Indeterminate Beams*

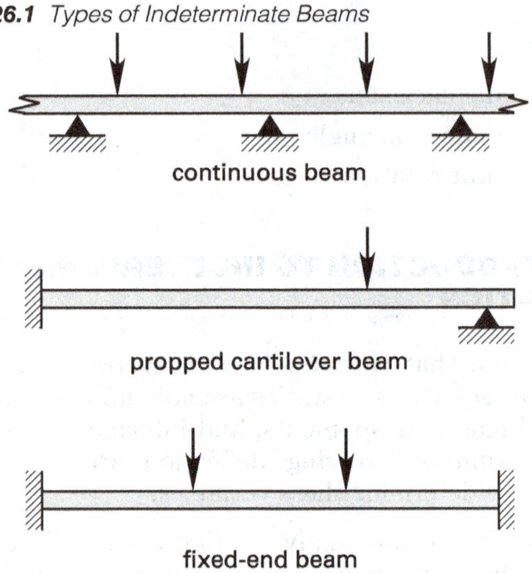

4. MEMBER FIXED-END MOMENTS (MAGNITUDES)

When the end of a beam is constrained against rotation, it is said to be a *fixed end* (also known as a *built-in end*). The ends of fixed-end beams are constrained to remain horizontal. Cantilever beams have a single fixed end. Some beams, as illustrated in Fig. 26.1, have two fixed ends and are known as *fixed-end beams*.[1]

Fixed-end beams are inherently indeterminate. To reduce the work required to find end moments and reactions, tables of fixed-end moments are often used.

Equation 26.1 Through Eq. 26.3

$$\text{FEM}_{\text{AB}} = \text{FEM}_{\text{BA}} = \frac{wL^2}{12} \qquad \textit{26.1}$$

$$\text{FEM}_{\text{AB}} = \frac{Pab^2}{L^2} \qquad \textit{26.2}$$

$$\text{FEM}_{\text{BA}} = \frac{Pa^2b}{L^2} \qquad \textit{26.3}$$

Description

Equation 26.1 gives the resisting *fixed-end moments* for a uniformly distributed load (see Fig. 26.2(a)). Equation 26.2 and Eq. 26.3 give the fixed-end moments for a concentrated load of magnitude P (see Fig. 26.2(b)).

Figure 26.2 *Fixed-End Beam Loading*

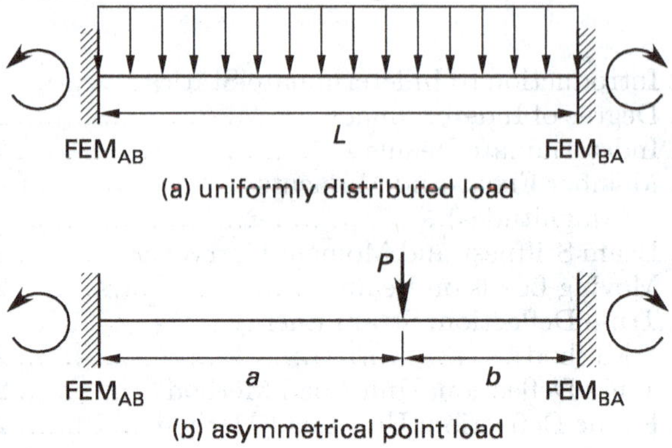

Example

What is most nearly the resisting fixed-end moment at support A for the beam shown?

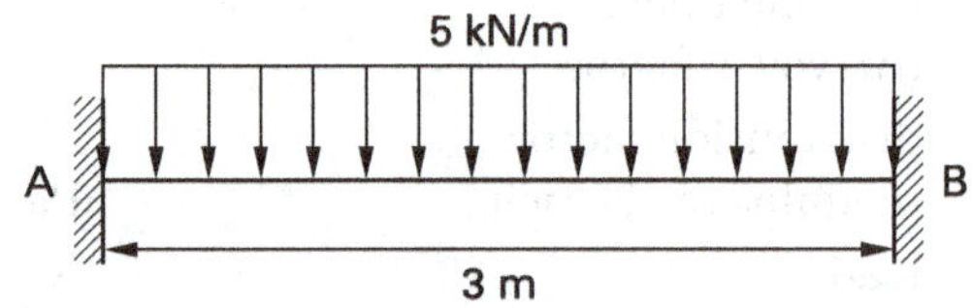

(A) 1.2 kN·m

(B) 3.8 kN·m

(C) 5.5 kN·m

(D) 45 kN·m

Solution

From Eq. 26.1, the fixed-end moment at A is

$$\text{FEM}_{\text{AB}} = \frac{wL^2}{12} = \frac{\left(5\ \frac{\text{kN}}{\text{m}}\right)(3\ \text{m})^2}{12}$$
$$= 3.75\ \text{kN·m} \quad (3.8\ \text{kN·m})$$

The answer is (B).

[1]The definition is loose. The term *fixed-end beam* can also be used to mean any indeterminate beam with at least one built-in end (e.g., a propped cantilever).

5. BEAM STIFFNESS AND MOMENT CARRYOVER

Equation 26.4 Through Eq. 26.6: Moment Distribution Method

$$M_B = M_A/2 \qquad 26.4$$

$$\theta = \frac{ML}{4EI} \qquad 26.5$$

$$M = \left(\frac{4EI}{L}\right)\theta = k_{AB}\theta \qquad 26.6$$

Description

Equation 26.4 through Eq. 26.6 are used in the *moment distribution method*, also known as the *Cross method*, which is used to calculate fixed-end moments for continuous beams and other indeterminate structures.

step 1: Calculate the moment at the end of each member, assuming all joints in the structure are fixed.

step 2: Calculate the stiffness, k, of each beam. Each beam may have a different modulus of elasticity and cross-sectional area. The stiffness is the moment that will induce a *unit rotation*, θ, at the far end of the beam when the beam end is fixed. (See Eq. 26.5 and Eq. 26.6.)

$$k = \frac{4EI}{L} \quad \text{[fixed end]}$$

$$k = \frac{3EI}{L} \quad \text{[pinned/roller end]}$$

step 3: All of the unbalanced moment (calculated in step 1) is given to a beam segment at the actual beam end. For each interior joint, distribute the negative of the total of any unbalanced fixed-end moments at a joint calculated in step 1 among the connecting members (two in the case of interior joints in continuous beams) in proportion to their relative stiffnesses. The relative stiffness is known as a *distribution factor*, DF. Special rules may be invoked when the actual beam end is fixed or simply supported.

$$\text{DF}_{\text{AB}} = \frac{k_{\text{AB}}}{\sum k_i}$$

step 4: Determine the *carryover factor*, COF, the ratio of the moment that appears at the far end of the beam in response, to the moment for each beam end. This is usually taken as ½ regardless of actual support type (see Fig. 26.3). Special rules may be invoked for pinned and cantilevered ends.

step 5: Carry over the distributed moments to the far beam ends by multiplying the distributed moments by the carryover factors. (See Eq. 26.4.)

This method is an iterative process, and should be repeated until the moment carryovers are small enough to be neglected. Then, add all moments at each end of each member. This yields the true moment at each end.

Figure 26.3 Moment Carryover

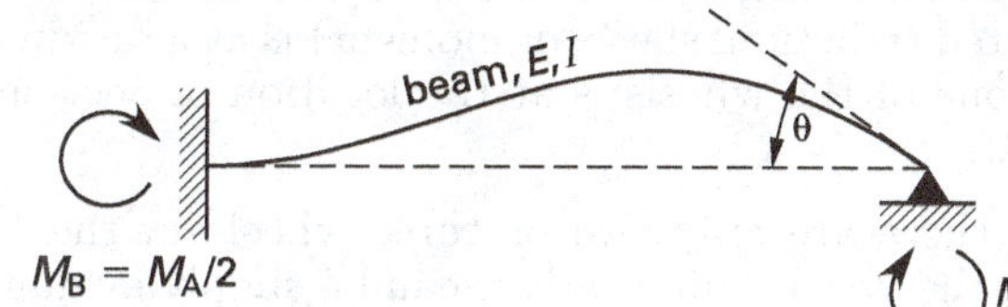

6. MOVING LOADS ON BEAMS

Global Maximum Moment and Shear Stresses Anywhere on Beam

If a beam supports a single moving load, the maximum bending and shearing stresses at any point can be found by drawing the moment and shear influence diagrams for that point. Once the positions of maximum moment and maximum shear are known, the stresses at the point in question can be found from Mc/I and QV/Ib.

If a simply supported beam carries a set of moving loads (which remain equidistant as they travel across the beam), the following procedure can be used to find the *dominant load.* (The dominant load is the one that occurs directly over the point of maximum moment.)

step 1: Calculate and locate the resultant of the load group.

step 2: Assume that one of the loads is dominant. Place the group on the beam such that the distance from one support to the assumed dominant load is equal to the distance from the other support to the resultant of the load group.

step 3: Check to see that all loads are on the span and that the shear changes sign under the assumed dominant load. If the shear does not change sign under the assumed dominant load, the maximum moment may occur when only some of the load group is on the beam. If it does change sign, calculate the bending moment under the assumed dominant load.

step 4: Repeat steps 2 and 3, assuming that the other loads are dominant.

Statics

step 5: Find the maximum shear by placing the load group such that the resultant is a minimum distance from a support.

Placement of Load Group to Maximize Local Moment

In the design of specific members or connections, it is necessary to place the load group in a position that will maximize the load on those members or connections. The procedure for finding these positions of local maximum loadings is different from the global maximum procedures.

The solution to the problem of local maximization is somewhat trial-and-error oriented. It is aided by use of the influence diagram. In general, the variable being evaluated (reaction, shear, or moment) is at a maximum when one of the wheels is at the location or section of interest.

When there are only two or three wheels in the load group, the various alternatives can be simply evaluated by using the influence diagram for the variable being evaluated. When there are many loads in the load group (e.g., a train loading), it may be advantageous to use heuristic rules for predicting the dominant wheel.

7. TRUSS DEFLECTION: STRAIN ENERGY METHOD

The deflection of a truss at the point of a single load application can be found by the *strain energy method* if all member forces are known.

When a point displaces Δ, in response to an applied force, F, the work done in the direction of the force is

$$W = F\Delta$$

The strain energy, U, stored by a member that displaces Δ in response to its internal load, F, is

$$U = \tfrac{1}{2}F\Delta = F\left(\frac{FL}{AE}\right) = \frac{F^2L}{AE}$$

From the work-energy principle,

$$W = U$$

8. TRUSS DEFLECTION: UNIT LOAD METHOD

The *unit load method* (also known as the *virtual work method*) is an extension of the strain energy method. It can be used to determine the deflection of any point on a truss. Truss joint displacement can be calculated by applying a unit load at the point corresponding to the desired displacement.

step 1: Draw the truss twice.

step 2: On the first truss, place all the actual loads.

step 3: Find the internal forces, F_i, due to the actual applied loads in all the members.

step 4: On the second truss, place a dummy one-unit load in the direction of the desired displacement.

step 5: Find the forces, f_i, due to the one-unit dummy load in all members.

step 6: Find the desired displacement from Eq. 26.7. The summation is over all truss members that have nonzero forces in *both* trusses.

Equation 26.7 Through Eq. 26.9: Displacement Equations[2]

$$\Delta_{\text{joint}} = \sum_{i=1}^{\text{members}} f_i(\Delta L)_i \qquad 26.7$$

$$(\Delta L)_i = \left(\frac{FL}{AE}\right)_i \qquad 26.8$$

$$(\Delta L)_i = \alpha L_i(\Delta T)_i \quad \left[\begin{array}{c}\text{for temperature}\\ \text{change in member}\end{array}\right] \qquad 26.9$$

Variation

$$\delta = \sum \frac{FfL}{AE}$$

Description

Equation 26.7 gives the *joint displacement* at the point where the unit load is applied. Equation 26.8 gives the displacement when the truss is put under a bar force caused by an external load. Equation 26.9 gives the displacement caused by temperature change in a member.

To avoid confusion with coordinate directions, the following sign conventions should be used: Δ_{joint} is positive in the direction of the applied load. Internal tensile forces, F and f, are positive, while internal compressive forces are negative. ΔL is positive when the member increases in length.

Example

A 10 cm long member in a roof truss experiences a 10°C temperature rise. The coefficient of thermal expansion

[2]The NCEES *FE Reference Handbook* (*NCEES Handbook*) uses Δ both to designate the spatial deformation (see Eq. 26.7) and as a mathematical operator indicating "change in" (see Eq. 26.9).

for the member is 1×10^{-5} 1/°C. Most nearly, what is the elongation of the member?

(A) 0.001 cm

(B) 0.010 cm

(C) 0.020 cm

(D) 0.100 cm

Solution

From Eq. 26.9, the elongation in the member is

$$\begin{aligned}(\Delta L)_i &= \alpha L_i (\Delta T)_i \\ &= \left(1 \times 10^{-5} \ \frac{1}{^\circ\text{C}}\right)(10 \text{ cm})(10^\circ\text{C}) \\ &= 0.001 \text{ cm}\end{aligned}$$

The answer is (A).

9. FRAME DEFLECTION: UNIT LOAD METHOD

Equation 26.10: Frame Displacement[3]

$$\Delta = \sum_{i=1}^{\text{members}} \int_{x=0}^{x=L_i} \frac{m_i M_i}{EI_i} dx \qquad 26.10$$

Description

Truss members have pinned ends, and they are considered to be axial members carrying tensile or compressive loads. Members of a *frame* have one or two built-in ends. They act like beams, responding by bending to applied loads. For any point on a frame, the displacement caused by external loads can be calculated by applying a unit load at that point corresponding to the desired displacement.

In Eq. 26.10, the summation is taken over all members that respond (i.e., have nonzero moments) to both the actual load and the unit load. M_i is the moment equation for member i due to the actual applied loads. m_i is the moment equation for member i due to the unit load.

Statics

[3]Equation 26.10 assumes that the modulus of elasticity, E, of all members is the same. However, this is not always the case.

Topic VIII: Dynamics

Chapter

27 Kinematics

Nomenclature

a	acceleration	m/s^2
f	coefficient of friction	–
f	frequency	Hz
g	gravitational acceleration, 9.81	m/s^2
r	position	m
r	radius	m
s	displacement	m
s	distance	m
s	position	m
t	time	s
v	velocity	m/s
x	horizontal distance	m
y	elevation	m

Symbols

α	angular acceleration	rad/s^2
θ	angular position	rad
ρ	radius of curvature	m
ω	angular velocity	rad/s

Subscripts

0	initial
c	constant
f	final
n	normal
r	radial
t	tangential
x	horizontal
y	vertical
θ	transverse

1. INTRODUCTION TO KINEMATICS

Dynamics is the study of moving objects. The subject is divided into kinematics and kinetics. *Kinematics* is the study of a body's motion independent of the forces on the body. It is a study of the geometry of motion without consideration of the causes of motion. Kinematics deals only with relationships among position, velocity, acceleration, and time.

2. PARTICLES AND RIGID BODIES

A body in motion can be considered a *particle* if rotation of the body is absent or insignificant. A particle does not possess rotational kinetic energy. All parts of a particle have the same instantaneous displacement, velocity, and acceleration.

A *rigid body* does not deform when loaded and can be considered a combination of two or more particles that remain at a fixed, finite distance from each other. At any given instant, the parts (particles) of a rigid body can have different displacements, velocities, and accelerations if the body has rotational as well as translational motion.

Equation 27.1 and Eq. 27.2: Instantaneous Velocity and Acceleration

$$\mathbf{v} = d\boldsymbol{r}/dt \tag{27.1}$$

$$\boldsymbol{a} = d\mathbf{v}/dt \tag{27.2}$$

Variation

$$\boldsymbol{a} = \frac{d^2\boldsymbol{r}}{dt^2}$$

Dynamics

Description

For the position vector of a particle, $\boldsymbol{r}$, the instantaneous velocity, $\mathbf{v}$, and acceleration, $\boldsymbol{a}$, are given by Eq. 27.1 and Eq. 27.2, respectively.

Example

The position of a particle moving along the x-axis is given by $r(t) = t^2 - t + 8$, where r is in units of meters and t is in seconds. What is most nearly the velocity of the particle when $t = 5$ s?

(A) 9.0 m/s

(B) 10 m/s

(C) 11 m/s

(D) 12 m/s

Solution

The velocity equation is the first derivative of the position equation with respect to time.

$$\begin{aligned}\mathbf{v}(t) &= d\boldsymbol{r}(t)/dt \\ &= \frac{d}{dt}(t^2 - t + 8) \\ &= 2t - 1 \\ \mathbf{v}(5) &= (2)(5) - 1 \\ &= 9.0 \text{ m/s}\end{aligned}$$

The answer is (A).

3. DISTANCE AND SPEED

The terms "displacement" and "distance" have different meanings in kinematics. *Displacement* (or *linear displacement*) is the net change in a particle's position as determined from the position function, $r(t)$. *Distance traveled* is the accumulated length of the path traveled during all direction reversals, and it can be found by adding the path lengths covered during periods in which the velocity sign does not change. Therefore, distance is always greater than or equal to displacement.

$$s = r(t_2) - r(t_1)$$

Similarly, "velocity" and "speed" have different meanings: *velocity* is a vector, having both magnitude and direction; *speed* is a scalar quantity, equal to the magnitude of velocity. When specifying speed, direction is not considered.

4. RECTANGULAR COORDINATES

The position of a particle is specified with reference to a coordinate system. Three coordinates are necessary to identify the position in three-dimensional space; in two dimensions, two coordinates are necessary. A coordinate can represent a linear position, as in the rectangular coordinate system, or it can represent an angular position, as in the polar system.

Consider the particle shown in Fig. 27.1. Its position, as well as its velocity and acceleration, can be specified in three primary forms: vector form, rectangular coordinate form, and unit vector form.

Figure 27.1 *Rectangular Coordinates*

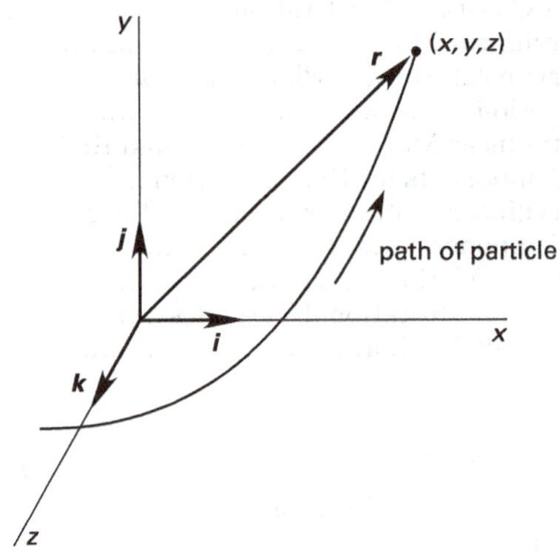

The *vector form* of the particle's position is $\boldsymbol{r}$, where the vector $\boldsymbol{r}$ has both magnitude and direction. The *Cartesian coordinate system form* (*rectangular coordinate form*) is (x, y, z).

Equation 27.3: Cartesian Unit Vector Form

$$\boldsymbol{r} = x\boldsymbol{i} + y\boldsymbol{j} + z\boldsymbol{k} \qquad 27.3$$

Description

The *unit vector form* of a position vector is given by Eq. 27.3.

Example

The position of a particle in Cartesian coordinates over time is $x = 5t$ in the x-direction, $y = 6t$ in the y-direction,

Dynamics

and $z = 5t$ in the z-direction. What is the vector form of the particle's position, $\mathbf{r}$?

(A) $\mathbf{r} = 5t\mathbf{i} + 6t\mathbf{j} + 5t\mathbf{k}$

(B) $\mathbf{r} = 5t\mathbf{i} + 6t\mathbf{j} + 6t\mathbf{k}$

(C) $\mathbf{r} = 6t\mathbf{i} + 5t\mathbf{j} + 5t\mathbf{k}$

(D) $\mathbf{r} = 6t\mathbf{i} + 5t\mathbf{j} + 6t\mathbf{k}$

Solution

Using Eq. 27.3, the vector form of the particle's position is

$$\begin{aligned} \mathbf{r} &= x\mathbf{i} + y\mathbf{j} + z\mathbf{k} \\ &= 5t\mathbf{i} + 6t\mathbf{j} + 5t\mathbf{k} \end{aligned}$$

The answer is (A).

5. RECTILINEAR MOTION

Equation 27.4 Through Eq. 27.9: Particle Rectilinear Motion

$$a = \frac{d\mathrm{v}}{dt} \quad \text{[general]} \qquad 27.4$$

$$\mathrm{v} = \frac{ds}{dt} \quad \text{[general]} \qquad 27.5$$

$$a\,ds = \mathrm{v}\,d\mathrm{v} \quad \text{[general]} \qquad 27.6$$

$$\mathrm{v} = \mathrm{v}_0 + a_ct \qquad 27.7$$

$$s = s_0 + \mathrm{v}_0t + \tfrac{1}{2}a_ct^2 \qquad 27.8$$

$$\mathrm{v}^2 = \mathrm{v}_0^2 + 2a_c(s - s_0) \qquad 27.9$$

Description

A *rectilinear system* is one in which particles move only in straight lines. (Another name is *linear system*.) The relationships among position, velocity, and acceleration for a linear system are given by Eq. 27.4 through Eq. 27.9. Equation 27.4 through Eq. 27.6 show relationships for general (including variable) acceleration of particles.[1] Equation 27.7 through Eq. 27.9 show relationships given constant acceleration, a_c.[2]

When values of time are substituted into these equations, the position, velocity, and acceleration are known as *instantaneous values*.

Equation 27.10 Through Eq. 27.13: Cartesian Velocity and Acceleration

$$\mathbf{v} = \dot{x}\mathbf{i} + \dot{y}\mathbf{j} + \dot{z}\mathbf{k} \qquad 27.10$$

$$\mathbf{a} = \ddot{x}\mathbf{i} + \ddot{y}\mathbf{j} + \ddot{z}\mathbf{k} \qquad 27.11$$

$$\dot{x} = dx/dt = \mathrm{v}_x \qquad 27.12$$

$$\ddot{x} = d^2x/dt^2 = a_x \qquad 27.13$$

Description

The velocity and acceleration are the first two derivatives of the position vector, as shown in Eq. 27.10 and Eq. 27.11.

6. CONSTANT ACCELERATION

Equation 27.14 Through Eq. 27.17: Velocity and Displacement with Constant Linear Acceleration

$$a(t) = a_0 \qquad 27.14$$

$$\mathrm{v}(t) = a_0(t - t_0) + \mathrm{v}_0 \qquad 27.15$$

$$s(t) = a_0(t - t_0)^2/2 + \mathrm{v}_0(t - t_0) + s_0 \qquad 27.16$$

$$\mathrm{v}^2 = \mathrm{v}_0^2 + 2a_0(s - s_0) \qquad 27.17$$

Variations

$$\mathrm{v}(t) = a_0\int dt$$

$$s(t) = a_0\iint dt^2$$

[1]Equation 27.6 can be derived from $a\,dt = d\mathrm{v}$ and $\mathrm{v}\,dt = ds$ by eliminating dt. One scenario where the acceleration depends on position is a particle being accelerated (or decelerated) by a compression spring. The spring force depends on the spring extension, so the acceleration does also.

[2]The NCEES *FE Reference Handbook* (*NCEES Handbook*) is inconsistent in what it uses subscripts to designate. For example, in its Dynamics section, it uses subscripts to designate the location of the accelerating point (e.g., c in a_c for acceleration of the centroid), the direction or related axis (e.g., x in a_x for acceleration in the x-direction), the type of acceleration (e.g., n in a_n for normal acceleration), and the moment in time (e.g., 0 in a_0 for initial acceleration). In Eq. 27.7 through Eq. 27.9, the *NCEES Handbook* uses subscripts to designate the nature of the acceleration (i.e., the subscript c indicates constant acceleration). Elsewhere in the *NCEES Handbook*, the subscript c is used to designate centroid and mass center.

Description

Acceleration is a constant in many cases, such as a free-falling body with constant acceleration g. If the acceleration is constant, the acceleration term can be taken out of the integrals shown in Sec. 27.5. The initial distance from the origin is s_0; the initial velocity is a constant, v_0; and a constant acceleration is denoted a_0.

Example

In standard gravity, block A exerts a force of 10 000 N, and block B exerts a force of 7500 N. Both blocks are initially held stationary. There is no friction, and the pulleys have no mass. Block A has an acceleration of 1.4 m/s² once the blocks are released.

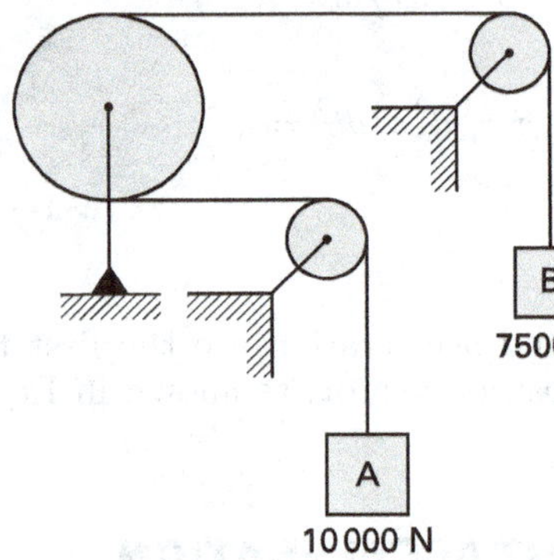

What is most nearly the velocity of block A 2.5 s after the blocks are released?

(A) 0 m/s

(B) 3.5 m/s

(C) 4.4 m/s

(D) 4.9 m/s

Solution

Use Eq. 27.15 to solve for the velocity of block A.

$$v_A = a_A(t - t_0) + v_0 = \left(1.4\ \frac{\text{m}}{\text{s}^2}\right)(2.5\ \text{s} - 0\ \text{s}) + 0\ \frac{\text{m}}{\text{s}}$$
$$= 3.5\ \text{m/s}$$

The answer is (B).

Equation 27.18 Through Eq. 27.21: Velocity and Displacement with Constant Angular Acceleration

$$\alpha(t) = \alpha_0 \qquad 27.18$$

$$\omega(t) = \alpha_0(t - t_0) + \omega_0 \qquad 27.19$$

$$\theta(t) = \alpha_0(t - t_0)^2/2 + \omega_0(t - t_0) + \theta_0 \qquad 27.20$$

$$\omega^2 = \omega_0^2 + 2\alpha_0(\theta - \theta_0) \qquad 27.21$$

Description

Equation 27.18 through Eq. 27.21 give the equations for constant angular acceleration.

Example

A flywheel rotates at 7200 rpm when the power is suddenly cut off. The flywheel decelerates at a constant rate of 2.1 rad/s² and comes to rest 6 min later. What is most nearly the angular displacement of the flywheel?

(A) 43×10^3 rad

(B) 93×10^3 rad

(C) 140×10^3 rad

(D) 270×10^3 rad

Solution

From Eq. 27.20, the angular displacement is

$$\theta(t) = \alpha_0(t - t_0)^2/2 + \omega_0(t - t_0) + \theta_0$$
$$= \frac{\left(-2.1\ \frac{\text{rad}}{\text{s}^2}\right)\left(60\ \frac{\text{s}}{\text{min}}\right)^2(6\ \text{min} - 0\ \text{min})^2}{2}$$
$$+ \left(7200\ \frac{\text{rev}}{\text{min}}\right)\left(2\pi\ \frac{\text{rad}}{\text{rev}}\right)(6\ \text{min} - 0\ \text{min})$$
$$+ 0\ \text{rad}$$
$$= 135.4 \times 10^3\ \text{rad} \quad (140 \times 10^3\ \text{rad})$$

The answer is (C).

7. NON-CONSTANT ACCELERATION

Equation 27.22 and Eq. 27.23: Velocity and Displacement for Non-Constant Acceleration

$$v(t) = \int_{t_0}^{t} a(t)\,dt + v_{t_0} \qquad 27.22$$

$$s(t) = \int_{t_0}^{t} v(t)\,dt + s_{t_0} \qquad 27.23$$

Description

The velocity and displacement, respectively, for non-constant acceleration, $a(t)$, are calculated using Eq. 27.22 and Eq. 27.23.

Example

A particle initially traveling at 10 m/s experiences a linear increase in acceleration in the direction of motion as shown. The particle reaches an acceleration of 20 m/s^2 in 6 seconds.

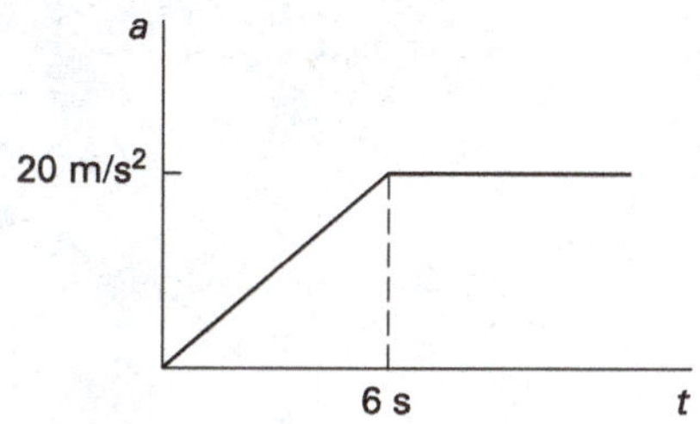

Most nearly, what is the distance traveled by the particle during those 6 seconds?

(A) 60 m

(B) 70 m

(C) 120 m

(D) 180 m

Solution

The expression for the acceleration as a function of time is

$$a(t) = \left(\frac{20\ \frac{\text{m}}{\text{s}^2}}{6\ \text{s}}\right)t = \frac{20\ \frac{\text{m}}{\text{s}^3}}{6}t$$

From Eq. 27.22, the velocity function is

$$\text{v}(t) = \int a(t)\,dt = \int \tfrac{20}{6}t\,dt = \tfrac{20}{12}t^2 + C_1$$

Since $\text{v}(0) = 10$, $C_1 = 10$.

From Eq. 27.23, the position function is

$$\begin{aligned} s(t) &= \int \text{v}(t)\,dt = \int (\tfrac{20}{12}t^2 + 10)\,dt \\ &= \tfrac{20}{36}t^3 + 10t + C_2 \end{aligned}$$

In a calculation of distance traveled, the initial distance (position) is $s(0) = 0$, so $C_2 = 0$. The distance traveled during the first 6 seconds is

$$\begin{aligned} s(6) &= \int_0^6 \text{v}(t)\,dt = \tfrac{20}{36}t^3 + 10t\Big|_0^6 \\ &= 180\ \text{m} - 0\ \text{m} \\ &= 180\ \text{m} \end{aligned}$$

The answer is (D).

Equation 27.24 and Eq. 27.25: Variable Angular Acceleration

$$\omega(t) = \int_{t_0}^{t} \alpha(t)\,dt + \omega_{t_0} \qquad 27.24$$

$$\theta(t) = \int_{t_0}^{t} \omega(t)\,dt + \theta_{t_0} \qquad 27.25$$

Description

For non-constant angular acceleration, $\alpha(t)$, the angular velocity, ω, and angular displacement, θ, can be calculated from Eq. 27.24 and Eq. 27.25.

8. CURVILINEAR MOTION

Curvilinear motion describes the motion of a particle along a path that is not a straight line. Special examples of curvilinear motion include plane circular motion and projectile motion. For particles traveling along curvilinear paths, the position, velocity, and acceleration may be specified in rectangular coordinates as they were for rectilinear motion, or it may be more convenient to express the kinematic variables in terms of other coordinate systems (e.g., polar coordinates).

9. CURVILINEAR MOTION: PLANE CIRCULAR MOTION

Plane circular motion (also known as *rotational particle motion, angular motion,* or *circular motion*) is the motion of a particle around a fixed circular path. (See Fig. 27.2.)

Figure 27.2 Plane Circular Motion

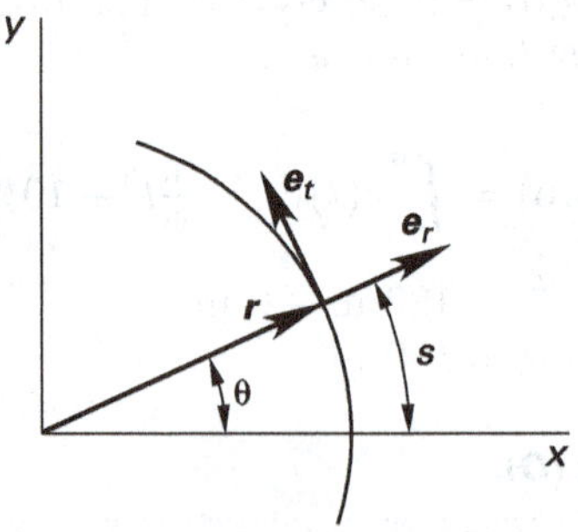

Equation 27.26 Through Eq. 27.31: *x*, *y*, *z* Coordinates

$$v_x = \dot{x} \quad 27.26$$

$$v_y = \dot{y} \quad 27.27$$

$$v_z = \dot{z} \quad 27.28$$

$$a_x = \ddot{x} \quad 27.29$$

$$a_y = \ddot{y} \quad 27.30$$

$$a_z = \ddot{z} \quad 27.31$$

Description

Equation 27.26 through Eq. 27.31 give the relationships between acceleration, velocity, and the Cartesian coordinates of a particle in plane circular motion.

Equation 27.32 Through Eq. 27.37: Polar Coordinates

$$v_r = \dot{r} \quad 27.32$$

$$v_\theta = r\dot{\theta} \quad 27.33$$

$$v_z = \dot{z} \quad 27.34$$

$$a_r = \ddot{r} - r\dot{\theta}^2 \quad 27.35$$

$$a_\theta = r\ddot{\theta} + 2\dot{r}\dot{\theta} \quad 27.36$$

$$a_z = \ddot{z} \quad 27.37$$

Description

In *polar coordinates*, the position of a particle is described by a radius, r, and an angle, θ. Equation 27.32 through Eq. 27.37 give the relationships between velocity and acceleration for particles in plane circular motion in a polar coordinate system.

Equation 27.38 Through Eq. 27.41: Rectilinear Forms of Curvilinear Motion

$$v = \dot{s} \quad 27.38$$

$$a_t = \dot{v} = \frac{dv}{ds} \quad 27.39$$

$$a_n = \frac{v^2}{\rho} \quad 27.40$$

$$\rho = \frac{\left[1 + (dy/dx)^2\right]^{3/2}}{\left|\frac{d^2y}{dx^2}\right|} \quad 27.41$$

Description

The relationship between acceleration, velocity, and position in an *ntb coordinate system* is given by Eq. 27.38 through Eq. 27.41.

Equation 27.42 Through Eq. 27.44: Particle Angular Motion

$$\omega = d\theta/dt \quad 27.42$$

$$\alpha = d\omega/dt \quad 27.43$$

$$\alpha d\theta = \omega d\omega \quad 27.44$$

Variation

$$\alpha = \frac{d^2\theta}{dt^2}$$

Description

The behavior of a rotating particle is defined by its angular position, θ, angular velocity, ω, and angular acceleration, α. These variables are analogous to the s, v, and a variables for linear systems. Angular variables can be substituted one-for-one for linear variables in most equations.

Dynamics

Example 6

The position of a car traveling around a curve is described by the following function of time (in seconds).

$$\theta(t) = t^3 - 2t^2 - 4t + 10$$

What is most nearly the angular velocity after 3 s of travel?

(A) −16 rad/s

(B) −4.0 rad/s

(C) 11 rad/s

(D) 15 rad/s

Solution

The angular velocity is

$$\omega(t) = \frac{d\theta}{dt} = 3t^2 - 4t - 4$$

$$\omega(3) = (3)(3)^2 - (4)(3) - 4 = 11 \text{ rad/s}$$

The answer is (C).

10. CURVILINEAR MOTION: TRANSVERSE AND RADIAL COMPONENTS FOR PLANAR MOTION

Equation 27.45 Through Eq. 27.49: Polar Coordinate Forms of Curvilinear Motion

$$\mathbf{r} = r\mathbf{e}_r \quad 27.45$$

$$\mathrm{v} = \dot{r}\mathbf{e}_r + r\dot{\theta}\mathbf{e}_\theta \quad 27.46$$

$$\mathbf{a} = (\ddot{r} - r\dot{\theta}^2)\mathbf{e}_r + (r\ddot{\theta} + 2\dot{r}\dot{\theta})\mathbf{e}_\theta \quad 27.47$$

$$\dot{r} = dr/dt \quad 27.48$$

$$\ddot{r} = d^2r/dt^2 \quad 27.49$$

Variations

$$\mathrm{v} = \mathrm{v}_r\mathbf{e}_r + \mathrm{v}_\theta\mathbf{e}_\theta$$

$$\mathbf{a} = a_r\mathbf{e}_r + a_\theta\mathbf{e}_\theta$$

Description

The position of a particle in a polar coordinate system may also be expressed as a vector of magnitude r and direction specified by unit vector $\mathbf{e}_r$. Since the velocity of a particle is not usually directed radially out from the center of the coordinate system, it can be divided into two components, called *radial* and *transverse*, which are parallel and perpendicular, respectively, to the unit radial vector. Figure 27.3 illustrates the radial and transverse components of velocity in a polar coordinate system, and the unit radial and unit transverse vectors, $\mathbf{e}_r$ and $\mathbf{e}_\theta$, used in the vector forms of the motion equations.

***Figure 27.3** Radial and Transverse Coordinates*

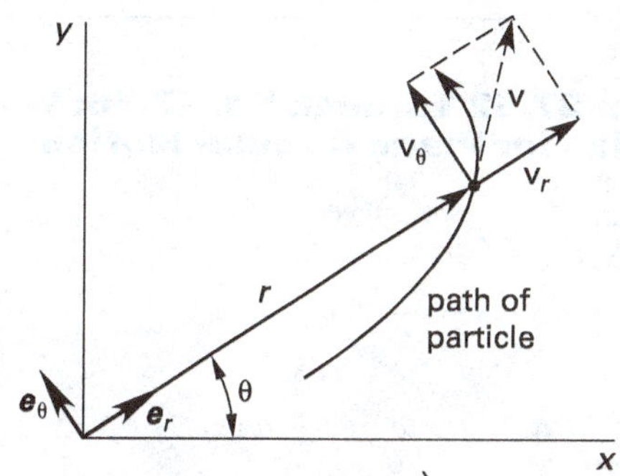

11. CURVILINEAR MOTION: NORMAL AND TANGENTIAL COMPONENTS

Equation 27.50 and Eq. 27.51: Velocity and Resultant Acceleration

$$\mathrm{v} = \mathrm{v}(t)\mathbf{e}_t \quad 27.50$$

$$\mathbf{a} = a(t)\mathbf{e}_t + (\mathrm{v}_t^2/\rho)\mathbf{e}_n \quad 27.51$$

Variation

$$\mathbf{a} = \frac{d\mathrm{v}_t}{dt}\mathbf{e}_t + \frac{\mathrm{v}_t^2}{\rho}\mathbf{e}_n$$

Description

A particle moving in a curvilinear path will have instantaneous linear velocity and linear acceleration. These linear variables will be directed tangentially to the path, and are known as *tangential velocity*, v_t, and *tangential acceleration*, a_t, respectively. The force that constrains the particle to the curved path will generally be directed toward the center of rotation, and the particle will experience an inward acceleration perpendicular to the tangential velocity and acceleration, known as the *normal acceleration*, a_n. The resultant acceleration, $\mathbf{a}$, is the vector sum of the tangential and normal accelerations. Normal and tangential components of acceleration are

Dynamics

illustrated in Fig. 27.4. The unit vectors $\boldsymbol{e}_n$ and $\boldsymbol{e}_t$ are normal and tangential to the path, respectively. ρ is the instantaneous *radius of curvature*.

Figure 27.4 Normal and Tangential Components

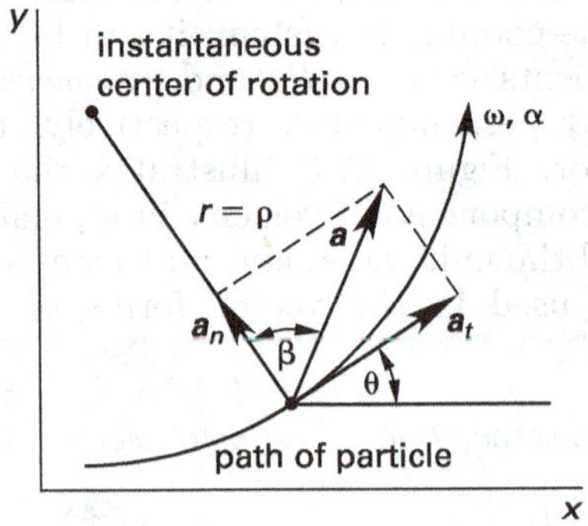

Equation 27.52 Through Eq. 27.56: Vector Quantities for Plane Circular Motion

$$\boldsymbol{r} = r\boldsymbol{e}_r \quad 27.52$$

$$\mathbf{v} = r\omega\boldsymbol{e}_t \quad 27.53$$

$$\boldsymbol{a} = (-r\omega^2)\boldsymbol{e}_r + r\alpha\boldsymbol{e}_t \quad 27.54$$

$$\omega = \dot{\theta} \quad 27.55$$

$$\alpha = \dot{\omega} = \ddot{\theta} \quad 27.56$$

Description

For plane circular motion, the vector forms of position, velocity, and acceleration are given by Eq. 27.52, Eq. 27.53, and Eq. 27.54. The magnitudes of the angular velocity and angular acceleration are defined by Eq. 27.55 and Eq. 27.56.

12. RELATIVE MOTION

The term *relative motion* is used when motion of a particle is described with respect to something else in motion. The particle's position, velocity, and acceleration may be specified with respect to another moving particle or with respect to a moving frame of reference, known as a *Newtonian* or *inertial frame of reference*.

Equation 27.57 Through Eq. 27.59: Relative Motion with Translating Axis

$$\boldsymbol{r}_A = \boldsymbol{r}_B + \boldsymbol{r}_{A/B} \quad 27.57$$

$$\mathbf{v}_A = \mathbf{v}_B + \omega \times \boldsymbol{r}_{A/B} = \mathbf{v}_B + \mathbf{v}_{A/B} \quad 27.58$$

$$\begin{aligned} \boldsymbol{a}_A &= \boldsymbol{a}_B + \alpha \times \boldsymbol{r}_{A/B} + \omega \times (\omega \times \boldsymbol{r}_{A/B}) \\ &= \boldsymbol{a}_B + \boldsymbol{a}_{A/B} \end{aligned} \quad 27.59$$

Description

The relative position, $\boldsymbol{r}_A$, velocity, $\boldsymbol{v}_A$, and acceleration, $\boldsymbol{a}_A$, with respect to a translating axis can be calculated from Eq. 27.57, Eq. 27.58, and Eq. 27.59, respectively. The angular velocity, ω, and angular acceleration, α, are the magnitudes of the relative position vector, $\boldsymbol{r}_{A/B}$. (See Fig. 27.5.)

Figure 27.5 Translating Axis

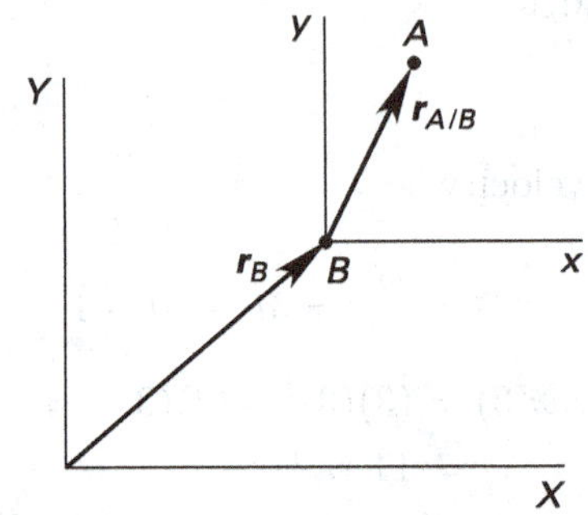

Equation 27.60 Through Eq. 27.62: Relative Motion with Rotating Axis

$$\boldsymbol{r}_A = \boldsymbol{r}_B + \boldsymbol{r}_{A/B} \quad 27.60$$

$$\mathbf{v}_A = \mathbf{v}_B + \omega \times \boldsymbol{r}_{A/B} + \mathbf{v}_{A/B} \quad 27.61$$

$$\begin{aligned} \boldsymbol{a}_A = \boldsymbol{a}_B &+ \alpha \times \boldsymbol{r}_{A/B} + \omega \times (\omega \times \boldsymbol{r}_{A/B}) \\ &+ 2\omega \times \mathbf{v}_{A/B} + \boldsymbol{a}_{A/B} \end{aligned} \quad 27.62$$

Description

Equation 27.60, Eq. 27.61, and Eq. 27.62 give the relative position, $\boldsymbol{r}_A$, velocity, $\boldsymbol{v}_A$, and acceleration, $\boldsymbol{a}_A$, with respect to a rotating axis, respectively. (See Fig. 27.6.)

Figure 27.6 Rotating Axis

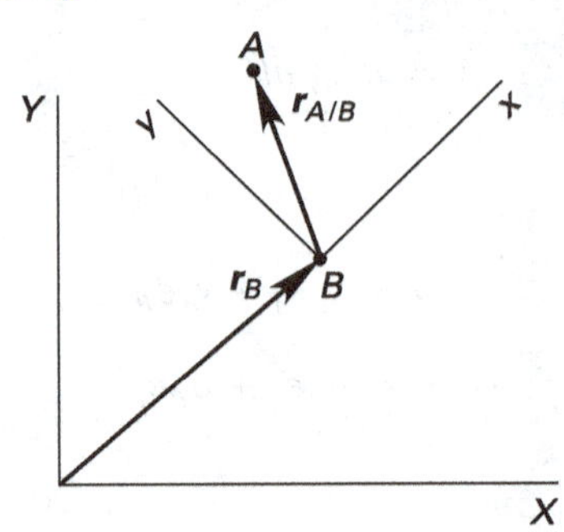

Dynamics

13. LINEAR AND ROTATIONAL VARIABLES

A particle moving in a curvilinear path will also have instantaneous linear velocity and linear acceleration. These linear variables will be directed tangentially to the path and, therefore, are known as *tangential velocity* and *tangential acceleration*, respectively. (See Fig. 27.7.) In general, the linear variables can be obtained by multiplying the rotational variables by the path radius.

Figure 27.7 *Tangential Variables*

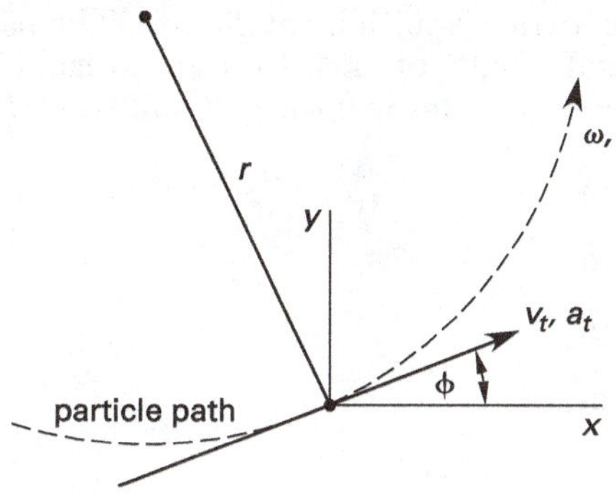

Equation 27.63 Through Eq. 27.66: Relationships Between Linear and Angular Variables

$$v_t = r\omega \tag{27.63}$$

$$a_t = r\alpha \tag{27.64}$$

$$a_n = -r\omega^2 \quad \left[\begin{array}{c}\text{toward the center}\\ \text{of the circle}\end{array}\right] \tag{27.65}$$

$$s = r\theta \tag{27.66}$$

Variations

$$v_t = r(2\pi f)$$

$$a_t = \frac{dv_t}{dt}$$

$$a_n = \frac{v_t^2}{r}$$

Description

Equation 27.63 through Eq. 27.65 are used to calculate tangential velocity, v_t, tangential acceleration, a_t, and normal acceleration, a_n, respectively, from their corresponding angular variables. If the path radius, r, is constant, as it would be in rotational motion, the linear distance (i.e., the *arc length*) traveled, s, is calculated from Eq. 27.66.

Example

For the reciprocating pump shown, the radius of the crank is 0.3 m, and the rotational speed is 350 rpm. Two seconds after the pump is activated, the angular position of point A is 35 rad.

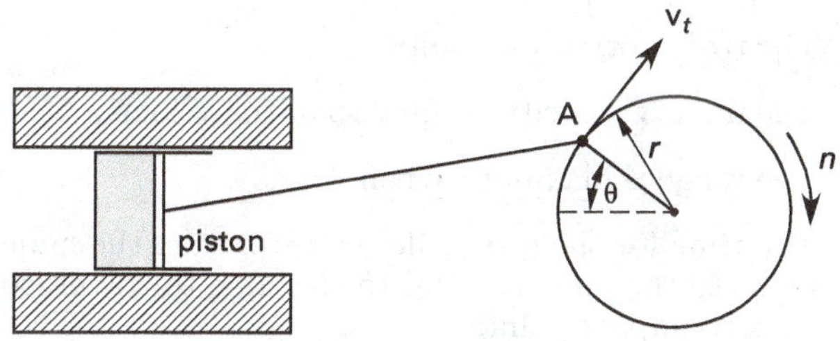

What is most nearly the tangential velocity of point A two seconds after the reciprocating pump is activated?

(A) 0 m/s

(B) 1.1 m/s

(C) 10 m/s

(D) 11 m/s

Solution

Use the relationship between the tangential and angular variables.

$$\begin{aligned}\omega &= \text{angular velocity of the crank in rad/s}\\ &= \frac{\left(350\ \frac{\text{rev}}{\text{min}}\right)\left(2\pi\ \frac{\text{rad}}{\text{rev}}\right)}{60\ \frac{\text{s}}{\text{min}}}\\ &= 36.65\ \text{rad/s}\end{aligned}$$

Use Eq. 27.63.

$$v_t = r\omega = (0.3\ \text{m})\left(36.65\ \frac{\text{rad}}{\text{s}}\right) = 11\ \text{m/s}$$

The tangential velocity is the same for any point on the crank at $r = 0.3$ m.

The answer is (D).

14. PROJECTILE MOTION

A projectile is placed into motion by an initial impulse. (Kinematics deals only with dynamics during the flight. The force acting on the projectile during the launch phase is covered in kinetics.) Neglecting air drag, once the projectile is in motion, it is acted upon only by the downward gravitational acceleration (i.e., its own weight). Projectile motion is a special case of motion under constant acceleration.

Consider a general projectile set into motion at an angle θ from the horizontal plane and initial velocity, v_0, as shown in Fig. 27.8. The *apex* is the point where the projectile is at its maximum elevation. In the absence of air drag, the following rules apply to the case of travel over a horizontal plane.

- The trajectory is parabolic.
- The impact velocity is equal to initial velocity, v_0.
- The range is maximum when $\theta = 45°$.
- The time for the projectile to travel from the launch point to the apex is equal to the time to travel from apex to impact point.
- The time for the projectile to travel from the apex of its flight path to impact is the same time an initially stationary object would take to fall straight down from that height.

Figure 27.8 Projectile Motion

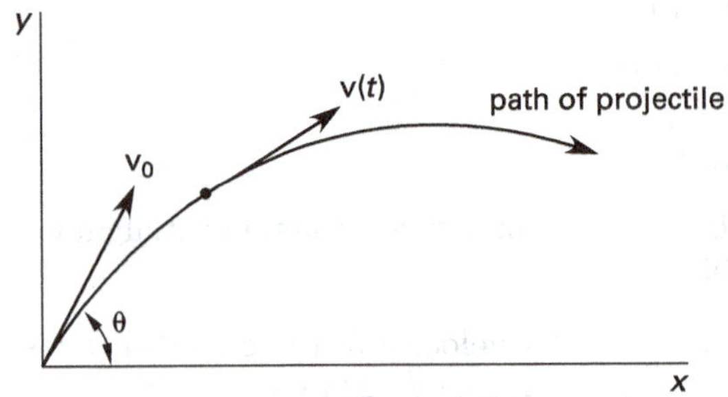

Equation 27.67 Through Eq. 27.72: Equations of Projectile Motion[3]

$$a_x = 0 \qquad 27.67$$

$$a_y = -g \qquad 27.68$$

$$v_x = v_0 \cos(\theta) \qquad 27.69$$

$$v_y = -gt + v_0 \sin(\theta) \qquad 27.70$$

$$x = v_0 \cos(\theta)t + x_0 \qquad 27.71$$

$$y = -gt^2/2 + v_0 \sin(\theta)t + y_0 \qquad 27.72$$

Variations

$$v_y(t) = v_{y,0} - gt$$

$$y(t) = v_{y,0}t - \tfrac{1}{2}gt^2$$

Description

The equations of projectile motion are derived from the laws of uniform acceleration and conservation of energy.

Example

A golfer on level ground at the edge of a 50 m wide pond attempts to drive a golf ball across the pond, hitting the ball so that it travels initially at 25 m/s. The ball travels at an initial angle of 45° to the horizontal plane. Approximately how far will the golf ball travel?

(A) 32 m

(B) 45 m

(C) 58 m

(D) 64 m

Solution

To determine the distance traveled by the golf ball, the time of impact must be found. At a time of 0 s and the time of impact, the elevation of the ball is known to be 0 m. Rearrange Eq. 27.72 to solve for time, substituting a value of 0 m for the elevation at a time of 0 s and the time of impact.

$$\begin{aligned} y &= -gt^2/2 + v_0 \sin(\theta)t + y_0 \\ 0\ \text{m} &= \frac{-gt^2}{2} + v_0 \sin(\theta)t + 0\ \text{m} \\ t &= \frac{2v_0 \sin\theta}{g} \end{aligned}$$

Substitute the expression for the time of impact into Eq. 27.71 and solve. The starting position is 0 m.

$$\begin{aligned} x &= v_0 \cos(\theta)t + x_0 \\ &= v_0 \cos\theta\left(\frac{2v_0 \sin\theta}{g}\right) + x_0 \\ &= \left(25\ \frac{\text{m}}{\text{s}}\right)\cos 45°\left(\frac{(2)\left(25\ \frac{\text{m}}{\text{s}}\right)\sin 45°}{9.81\ \frac{\text{m}}{\text{s}^2}}\right) + 0\ \text{m} \\ &= 63.7\ \text{m} \quad (64\ \text{m}) \end{aligned}$$

The answer is (D).

[3]As formulated in the *NCEES Handbook*, it is easy to misinterpret the second term in Eq. 27.72. $v_0 \sin(\theta)t$ is the product of three terms: v_0, $\sin\theta$, and t. $\sin(\theta)t$ is not the sine of a function of t. A clearer formulation would have been $v_0 t \sin\theta$.

28 Kinetics

Nomenclature

a	acceleration	m/s^2
F	force	N
g	gravitational acceleration, 9.81	m/s^2
I	mass moment of inertia	$kg \cdot m^2$
m	mass	kg
M	moment	N·m
N	normal force	N
p	momentum	N·s
R	resultant	N
t	time	s
v	velocity	m/s
W	weight	N
x	displacement or position	m

Symbols

α	angular acceleration	rad/s^2
μ	coefficient of friction	–
ρ	radius of curvature	m
ϕ	angle	deg

Subscripts

0	initial
c	centroidal
f	final or frictional
i	initial
k	dynamic
n	normal
pc	from point p to point c
r	radial
R	resultant
s	static
t	tangential
θ	transverse

1. INTRODUCTION

Kinetics is the study of motion and the forces that cause motion. Kinetics includes an analysis of the relationship between force and mass for translational motion and between torque and moment of inertia for rotational motion. Newton's laws form the basis of the governing theory in the subject of kinetics.

2. MOMENTUM

The vector *linear momentum* (*momentum*), **p**, is defined by the following equation. It has the same direction as the velocity vector from which it is calculated. Momentum has units of force × time (e.g., N·s).

$$\mathbf{p} = m\mathbf{v}$$

Momentum is conserved when no external forces act on a particle. If no forces act on a particle, the velocity and direction of the particle are unchanged. The *law of conservation of momentum* states that the linear momentum is unchanged if no unbalanced forces act on the particle. This does not prohibit the mass and velocity from changing, however. Only the product of mass and velocity is constant.

3. NEWTON'S FIRST AND SECOND LAWS OF MOTION

Newton's first law of motion states that a particle will remain in a state of rest or will continue to move with constant velocity unless an unbalanced external force acts on it.

This law can also be stated in terms of conservation of momentum: If the resultant external force acting on a particle is zero, then the linear momentum of the particle is constant.

Newton's second law of motion (conservation of momentum) states that the acceleration of a particle is directly proportional to the force acting on it and is inversely proportional to the particle mass. The direction of acceleration is the same as the direction of force.

Equation 28.1 and Eq. 28.2: Newton's Second Law for a Particle

$$\sum \mathbf{F} = d(m\mathbf{v})/dt \quad 28.1$$

$$\sum \mathbf{F} = m\,d\mathbf{v}/dt = m\mathbf{a} \quad \text{[constant mass]} \quad 28.2$$

Variation

$$F = \frac{d\mathbf{p}}{dt}$$

Description

Newton's second law can be stated in terms of the force vector required to cause a change in momentum. The resultant force is equal to the rate of change of linear momentum. For a constant mass, Eq. 28.2 applies.

Example

A 3 kg block is moving at a speed of 5 m/s. The force required to bring the block to a stop in 8×10^{-4} s is most nearly

(A) 10 kN

(B) 13 kN

(C) 15 kN

(D) 19 kN

Solution

From Newton's second law, Eq. 28.2, the force required to stop a constant mass of 3 kg moving at a speed of 5 m/s is

$$\begin{aligned}\sum \mathbf{F} &= m\,d\mathbf{v}/dt = m(\Delta\mathrm{v}/\Delta t)\\ &= (3\ \text{kg})\left(\frac{5\ \frac{\text{m}}{\text{s}} - 0\ \frac{\text{m}}{\text{s}}}{(8 \times 10^{-4}\ \text{s})\left(1000\ \frac{\text{N}}{\text{kN}}\right)}\right)\\ &= 18.75\ \text{kN} \quad (19\ \text{kN})\end{aligned}$$

The answer is (D).

Equation 28.3 Through Eq. 28.5: Newton's Second Law for a Rigid Body[1]

$$\sum \mathbf{F} = m\mathbf{a}_c \quad 28.3$$

$$\sum \mathbf{M}_c = I_c\boldsymbol{\alpha} \quad 28.4$$

$$\sum \mathbf{M}_p = I_c\boldsymbol{\alpha} + \boldsymbol{\rho}_{pc} \times m\mathbf{a}_c \quad 28.5$$

Description

A *rigid body* is a complex shape that cannot be described as a particle. Generally, a rigid body is nonhomogeneous (i.e., the center of mass does not coincide with the volumetric center) or is constructed of subcomponents. In those cases, applying an unbalanced force will cause rotation as well as translation. Newton's second law of motion (conservation of momentum) can be applied to a rigid body, but the law must be applied twice: once for linear momentum and once for angular momentum. Equation 28.3 pertains to linear momentum and relates the net (resultant) force, $\mathbf{F}$, on an object in any direction to the acceleration, $\mathbf{a}_c$, of the object's centroid in that direction.[2] The acceleration is "resisted" by the object's inertial mass, m. Equation 28.4 pertains to angular momentum and relates the net (resultant) moment or torque, $\mathbf{M}_c$, on an object about a centroidal axis to the angular rotational acceleration, $\boldsymbol{\alpha}$, around the centroidal axis.[3] The angular acceleration is resisted by the object's centroidal mass moment of inertia, I_c.

In pure rotation, the object rotates about a centroidal axis. The centroid remains stationary as elements of the rigid body. Equation 28.5 pertains to rotation about any particular axis, p, where $\boldsymbol{\rho}_{pc}$ is the perpendicular vector from axis p to the object's centroidal axis.

Example

A net unbalanced torque acts on a 50 kg cylinder that is allowed to rotate around its longitudinal centroidal axis on frictionless bearings. The cylinder has a radius of 40 cm and a mass moment of inertia of 4 kg·m². The

[1]In Eq. 28.3 through Eq. 28.5, the NCEES *FE Reference Handbook* (*NCEES Handbook*) uses bold characters to designate vector quantities (i.e., $\mathbf{F}$, $\mathbf{M}$, $\mathbf{a}$, $\boldsymbol{\alpha}$, and $\boldsymbol{\rho}$). Rectilinear components of vectors may be added; and cross-products are used for multiplication. In most calculations, however, the vector nature of these quantities is disregarded, and only the magnitudes of the quantities are used.

[2]The *NCEES Handbook* is inconsistent in its meaning of a_c. In Eq. 28.3, a_c refers to the acceleration of the centroid, which the *NCEES Handbook* calls "mass center." a_c does not mean constant acceleration as it did earlier in the *NCEES Handbook* Dynamics section.

[3]The *NCEES Handbook* is inconsistent in designating the centroidal parameters. Whereas a_c represents the acceleration of the centroid in Eq. 28.3, and I_c represents the centroidal moment of inertia in Eq. 28.4, the subscript c has been omitted on $\boldsymbol{\alpha}$, the angular acceleration about the centroidal axis, in Eq. 28.5.

cylinder accelerates from a standstill with an angular acceleration of 5 rad/s^2.

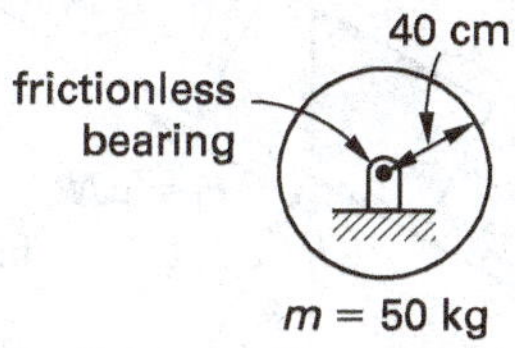

What is most nearly the unbalanced torque on the cylinder?

(A) 20 N·m

(B) 40 N·m

(C) 200 N·m

(D) 2000 N·m

Solution

Using Eq. 28.4, the magnitude of the moment acting on the cylinder is

$$\begin{aligned}\sum M_c &= I_c\alpha \\ &= (4\ \text{kg·m}^2)\left(5\ \frac{\text{rad}}{\text{s}^2}\right) \\ &= 20\ \text{N·m}\end{aligned}$$

The answer is (A).

Equation 28.6 Through Eq. 28.12: Rectilinear Equations for Rigid Bodies

$$\sum F_x = ma_{xc} \quad 28.6$$

$$\sum F_y = ma_{yc} \quad 28.7$$

$$\sum M_{zc} = I_{zc}\alpha \quad 28.8$$

$$\sum F_x = m(a_G)_x \quad 28.9$$

$$\sum F_y = m(a_G)_y \quad 28.10$$

$$\sum M_G = I_G\alpha \quad 28.11$$

$$\sum M_P = \sum (M_k)_P \quad 28.12$$

Description

These equations are the scalar forms of Newton's second law equations, assuming the rigid body is constrained to move in an x-y plane. The subscript zc describes the z-axis passing through the body's centroid. Placing the origin at the body's centroid, the acceleration of the body in the x- and y-directions is a_{xc} and a_{yc}, respectively. α is the angular acceleration of the body about the z-axis. Equation 28.6 through Eq. 28.12 are limited to motion in the x-y plane (i.e., two dimensions). Equation 28.11 calculates the sum of moments about a rigid body's center of gravity (mass center, etc.), G. Equation 28.12 calculates the sum of moments about any point, P.[4]

4. WEIGHT

Equation 28.13: Weight of an Object

$$W = mg \quad 28.13$$

Description

The *weight*, W, of an object is the force the object exerts due to its position in a gravitational field.[5]

[4](1) Equation 28.6 through Eq. 28.8 are prefaced in the *NCEES Handbook* with, "Without loss of generality, the body may be assumed to be in the x-y plane." This statement sounds as though all bodies can be simplified to planar motion, which is not true. The more general three-dimensional case is not specifically presented, so there is no generality to lose. In fact, Eq. 28.8 represents the sum of moments about any point, so this equation *is* the more general case, not the less general case. (2) Equation 28.9, Eq. 28.10, and Eq. 28.11 are functionally the same as Eq. 28.6, Eq. 28.7, and Eq. 28.8 and are redundant. Both sets of equations are limited to the x-y plane. (3) The subscripts c (centroidal or center of mass) and G (center of gravity) refer to the same thing. The change in notation is unnecessary. (4) The subscripts G and P are not defined. (5) The subscript k is not defined, but probably represents an uncommon choice for the first summation variable, normally i. Since k does not appear in the summation symbol, the meaning of M_k must be inferred. (6) Equation 28.11 specifies the point through which the rotational axis passes, but it does not specify an axis, as does Eq. 28.8. Since the equations are limited to the x-y plane, the rotational axis can only be parallel to the z-axis, as in Eq. 28.8. (7) The subscript c has been omitted on α, the angular acceleration about the center of mass, in Eq. 28.8 and Eq. 28.11.

[5](1) The *NCEES Handbook* introduces Eq. 28.13 with the section heading, "Concept of Weight." Units of weight are specified as newtons. In fact, the concept of weight is entirely absent in the SI system. Only the concepts of mass and force are used. The SI system does not support the concept of "body weight" in newtons. It only supports the concept of the force needed to accelerate a body. In presenting Eq. 28.13, the *NCEES Handbook* perpetuates the incorrect ideas that mass and weight are synonyms, and that weight is a fixed property of a body. (2) The *NCEES Handbook* includes a parenthetical "(lbf)" as the unit of weight for U.S. equations. However, Eq. 28.13 cannot be used with customary and normal U.S. units (i.e., mass in pounds) without including the gravitational constant, g_c. In order to make Eq. 28.13 consistent, the *NCEES Handbook* is forced to specify the unit of mass for U.S. equations as lbf-sec^2/ft. This (essentially now obsolete) unit of mass is known as a *slug*, something that is not called out in the *NCEES Handbook*. Since a slug is 32.2 times larger than a pound, an examinee using Eq. 28.13 with customary and normal U.S. units could easily be misdirected by the lbf label.

Example

A man weighs himself twice in an elevator. When the elevator is at rest, he weighs 713 N; when the elevator starts moving upward, he weighs 816 N. What is most nearly the man's actual mass?

(A) 70 kg

(B) 73 kg

(C) 78 kg

(D) 83 kg

Solution

The mass of the man can be determined from his weight at rest.

$$W = mg$$

$$m = \frac{W}{g} = \frac{713\ \text{N}}{9.81\ \frac{\text{m}}{\text{s}^2}}$$

$$= 72.7\ \text{kg} \quad (73\ \text{kg})$$

The answer is (B).

5. FRICTION

Friction is a force that always resists motion or impending motion. It always acts parallel to the contacting surfaces. If the body is moving, the friction is known as *dynamic friction.* If the body is stationary, friction is known as *static friction.*

The magnitude of the frictional force depends on the normal force, N, and the *coefficient of friction*, μ, between the body and the contacting surface.

$$F_f = \mu N$$

The static coefficient of friction is usually denoted with the subscript s, while the dynamic (i.e., kinetic) coefficient of friction is denoted with the subscript k. μ_k is often assumed to be 75% of the value of μ_s. These coefficients are complex functions of surface properties. Experimentally determined values for various contacting conditions can be found in handbooks.

For a body resting on a horizontal surface, the *normal force*, N, is the weight, W, of the body. If the body rests on an inclined surface, the normal force is calculated as the component of weight normal to that surface, as illustrated in Fig. 28.1. Axes in Fig. 28.1 are defined as parallel and perpendicular to the inclined plane.

$$N = mg\cos\phi = W\cos\phi$$

***Figure 28.1** Frictional and Normal Forces*

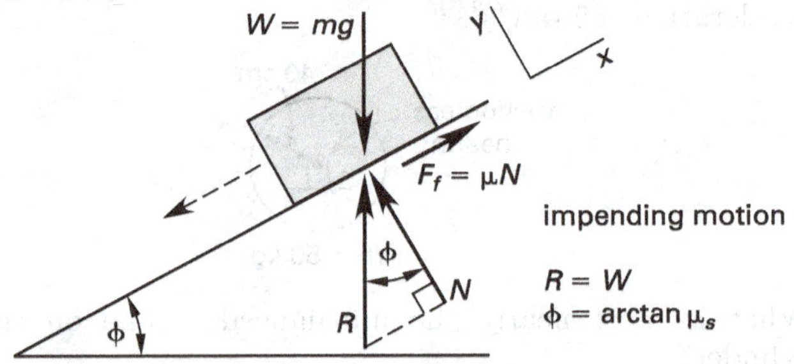

The frictional force acts only in response to a disturbing force, and it increases as the disturbing force increases. The motion of a stationary body is impending when the disturbing force reaches the maximum frictional force, $\mu_s N$. Figure 28.1 shows the condition of impending motion for a block on a plane. Just before motion starts, the resultant, R, of the frictional force and normal force equals the weight of the block. The angle at which motion is just impending can be calculated from the coefficient of static friction.

$$\phi = \arctan \mu_s$$

Once motion begins, the coefficient of friction drops slightly, and a lower frictional force opposes movement. This is illustrated in Fig. 28.2.

***Figure 28.2** Frictional Force Versus Disturbing Force*

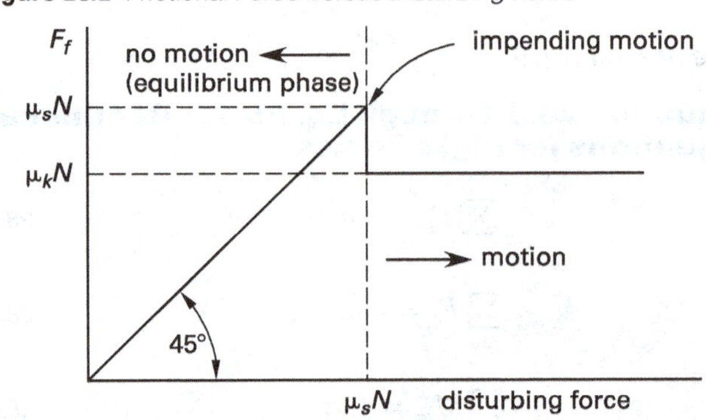

Equation 28.14 Through Eq. 28.17: Laws of Friction

$$F \le \mu_s N \qquad 28.14$$

$$\boldsymbol{F} < \mu_s \boldsymbol{N} \quad \text{[no slip occurring]} \qquad 28.15$$

$$\boldsymbol{F} = \mu_s \boldsymbol{N} \quad \text{[point of impending slip]} \qquad 28.16$$

$$\boldsymbol{F} = \mu_k \boldsymbol{N} \quad \text{[slip occurring]} \qquad 28.17$$

Values

$$\mu_k \approx 0.75\mu_s$$

Description

The laws of friction state that the maximum value of the total friction force, $\boldsymbol{F}$, is independent of the magnitude of the area of contact. The maximum total friction force is proportional to the normal force, $\boldsymbol{N}$. For low velocities of sliding, the maximum total frictional force is nearly independent of the velocity. However, experiments show that the force necessary to initiate slip is greater than that necessary to maintain the motion.

Example 12

A boy pulls a sled with a mass of 35 kg horizontally over a surface with a dynamic coefficient of friction of 0.15. What is most nearly the force required for the boy to pull the sled?

(A) 49 N

(B) 52 N

(C) 55 N

(D) 58 N

Solution

N is the normal force, and μ_k is the dynamic coefficient of friction. The force that the boy must pull with, F_b, must be large enough to overcome the frictional force. From Eq. 28.17,

$$\begin{aligned} F_b &= F_f = \mu_k N = \mu_k mg \\ &= (0.15)(35\ \text{kg})\left(9.81\ \frac{\text{m}}{\text{s}^2}\right) \\ &= 51.5\ \text{kg·m/s}^2 \quad (52\ \text{N}) \end{aligned}$$

The answer is (B).

6. KINETICS OF A PARTICLE

Newton's second law can be applied separately to any direction in which forces are resolved into components. The law can be expressed in rectangular coordinate form (i.e., in terms of x- and y-component forces), in polar coordinate form (i.e., in tangential and normal components), or in radial and transverse component form.

Equation 28.18: Newton's Second Law

$$a_x = F_x/m \qquad 28.18$$

Variation

$$F_x = ma_x$$

Description

Equation 28.18 is Newton's second law in rectangular coordinate form and refers to motion in the x-direction. Similar equations can be written for the y-direction or any other coordinate direction. In general, F_x may be a function of time, displacement, and/or velocity.

Example 13

A car moving at 70 km/h has a mass of 1700 kg. The force necessary to decelerate it at a rate of 40 cm/s² is most nearly

(A) 0.68 N

(B) 42 N

(C) 680 N

(D) 4200 N

Solution

Use Newton's second law.

$$\begin{aligned} a_x &= F_x/m \\ F_x &= ma_x \\ &= \frac{(1700\ \text{kg})\left(40\ \frac{\text{cm}}{\text{s}^2}\right)}{100\ \frac{\text{cm}}{\text{m}}} \\ &= 680\ \text{kg·m/s}^2 \quad (680\ \text{N}) \end{aligned}$$

The answer is (C).

Equation 28.19 Through Eq. 28.21: Equations of Motion with Constant Mass and Force as a Function of Time

$$a_x(t) = F_x(t)/m \qquad 28.19$$

$$\text{v}_x(t) = \int_{t_0}^{t} a_x(t)\,dt + \text{v}_{xt_0} \qquad 28.20$$

$$x(t) = \int_{t_0}^{t} \text{v}_x(t)\,dt + x_{t_0} \qquad 28.21$$

Variation

$$\text{v}_x(t) = \int_{t_i}^{t_f} \frac{F_x(t)}{m}\,dt + \text{v}_{x,0}$$

Description

If F_x is a function of time only, then the equations of motion are given by Eq. 28.19, Eq. 28.20, and Eq. 28.21.

Equation 28.22 Through Eq. 28.24: Equations of Motion with Constant Mass and Force

$$a_x = F_x/m \qquad 28.22$$

$$v_x = a_x(t-t_0) + v_{xt_0} \qquad 28.23$$

$$x = a_x(t-t_0)^2/2 + v_{xt_0}(t-t_0) + x_{t_0} \qquad 28.24$$

Variations

$$F_x = ma_x$$

$$v_x(t) = v_{x,0} + \left(\frac{F_x}{m}\right)(t-t_0)$$

$$x(t) = x_0 + v_{x,0}(t-t_0) + \frac{F_x(t-t_0)^2}{2m}$$

Description

If F_x is constant (i.e., is independent of time, displacement, or velocity) and mass is constant, then the equations of motion are given by Eq. 28.22, Eq. 28.23, and Eq. 28.24.

Example

A force of 15 N acts on a 16 kg body for 2 s. If the body is initially at rest, approximately how far is it displaced by the force?

(A) 1.1 m

(B) 1.5 m

(C) 1.9 m

(D) 2.1 m

Solution

The acceleration is found using Newton's second law, Eq. 28.22.

$$a_x = F_x/m = \frac{15\ \text{N}}{16\ \text{kg}} = 0.94\ \text{m/s}^2$$

For a body undergoing constant acceleration, with an initial velocity of 0 m/s, an initial time of 0 s, and a total elapsed time of 2 s, the horizontal displacement is found from Eq. 28.24.

$$\begin{aligned} x &= a_x(t-t_0)^2/2 + v_{xt_0}(t-t_0) + x_{t_0} \\ &= \frac{\left(0.94\ \frac{\text{m}}{\text{s}^2}\right)(2\ \text{s} - 0\ \text{s})^2}{2} + \left(0\ \frac{\text{m}}{\text{s}}\right)(2\ \text{s} - 0\ \text{s}) + 0\ \text{m} \\ &= 1.88\ \text{m} \quad (1.9\ \text{m}) \end{aligned}$$

The answer is (C).

Equation 28.25 and Eq. 28.26: Tangential and Normal Components

$$\sum F_t = ma_t = m\,dv_t/dt \qquad 28.25$$

$$\sum F_n = ma_n = m(v_t^2/\rho) \qquad 28.26$$

Description

For a particle moving along a circular path, the tangential and normal components of force, acceleration, and velocity are related.

Radial and Transverse Components

For a particle moving along a circular path, the radial and transverse components of force are

$$\sum F_r = ma_r$$

$$\sum F_\theta = ma_\theta$$

29 Kinetics of Rotational Motion

Nomenclature

a	acceleration	m/s^2
A	area	m^2
c	number of instantaneous centers	–
d	length	m
F	force	N
g	gravitational acceleration, 9.81	m/s^2
h	height	m
H	angular momentum	N·m·s
I	mass moment of inertia	kg·m^2
l	length	m
L	length	m
m	mass	kg
M	mass	kg
M	moment	N·m
n	quantity	–
r	radius of gyration	m
R	mean radius	m
t	time	s
v	velocity	m/s
W	weight	N

Symbols

α	angular acceleration	rad/s^2
θ	angular position	rad
μ	coefficient of friction	–
ρ	density	kg/m^3
ω	angular velocity	rad/s

Subscripts

0	initial
c	centrifugal or centroidal
f	frictional
G	center of gravity
m	mass
n	normal
O	origin or center
s	static
t	tangential

1. MASS[1] MOMENT OF INERTIA

Equation 29.1 Through Eq. 29.4: Mass Moment of Inertia

$$I = \int r^2 dm \qquad 29.1$$

$$I_x = \int (y^2 + z^2)\,dm \qquad 29.2$$

$$I_y = \int (x^2 + z^2)\,dm \qquad 29.3$$

$$I_z = \int (x^2 + y^2)\,dm \qquad 29.4$$

Description

The *mass moment of inertia* measures a solid object's resistance to changes in rotational speed about a specific axis. Equation 29.1 shows that the mass moment of inertia is calculated as the second moment of the mass.[2] When the origin of a coordinate system is located at the object's center of mass, the radius, r, to the differential element can be calculated from the components of position as

$$r = \sqrt{x^2 + y^2 + z^2}$$

[1]The NCEES *FE Reference Handbook* (*NCEES Handbook*) is inconsistent in its nomenclature usage. It uses both m and M to designate the mass of a object. It generally uses uppercase M to designate the total mass of non-particles (i.e., cylinders). Care must be taken when solving problems involving both mass and moment, as equations for both quantities use the same symbol.

[2](1) There are two closely adjacent sections in the *NCEES Handbook* labeled "Mass Moment of Inertia," each covering the same topic. (2) The integral shown in Eq. 29.1 is implicitly a triple integral (volume integral), more properly shown as $\int_V$ or $\iiint$.

For a homogeneous body with density ρ, Eq. 29.1 can be written as

$$I = \rho \int_V r^2 dV$$

I_x, I_y, and I_z are the mass moments of inertia with respect to the x-, y-, and z-axes, respectively. They are not components of a resultant value.

Equation 29.5 and Eq. 29.6: Parallel Axis Theorem

$$I_{\text{new}} = I_c + md^2 \qquad 29.5$$

$$I = I_G + md^2 \qquad 29.6$$

Variation

$$I = I_{c,1} + m_1 d_1^2 + I_{c,2} + m_2 d_2^2 + \cdots$$

Description

The *centroidal mass moment of inertia*, I_c, is obtained when the origin of the axes coincides with the object's center of gravity.[3] The *parallel axis theorem*, also known as the *transfer axis theorem*, is used to find the mass moment of inertia about any axis. In Eq. 29.5, d is the distance from the center of mass to the new axis.

For a composite object, the parallel axis theorem must be applied for each of the constituent objects, as shown in the variation equation.

Example

The 5 cm long uniform slender rod shown has a mass of 20 g. The origin of the y-axis corresponds with the rod's center of gravity. The centroidal mass moment of inertia is 42 g·cm^2.

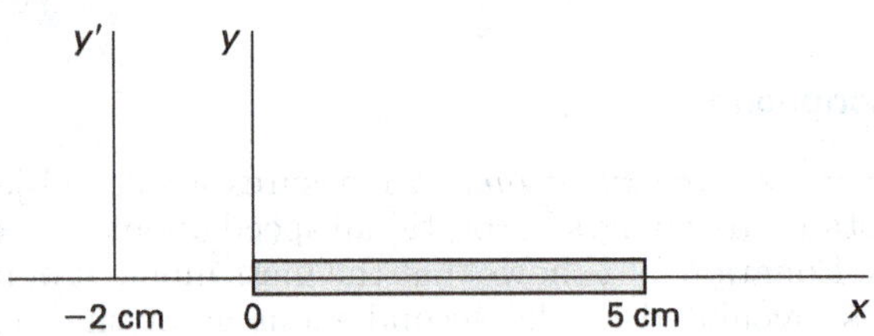

What is most nearly the mass moment of inertia of the rod about the y' axis 2 cm to the left?

(A) 0.12 kg·cm^2

(B) 0.33 kg·cm^2

(C) 0.45 kg·cm^2

(D) 0.91 kg·cm^2

Solution

The y' axis is 2 cm from the y-axis. The center of gravity of the rod is located halfway along its length. Use Eq. 29.5.

$$I_{y'} = I_c + md^2 = \frac{42 \text{ g·cm}^2 + (20 \text{ g})(2.5 \text{ cm} + 2 \text{ cm})^2}{1000 \,\frac{\text{g}}{\text{kg}}}$$

$$= 0.45 \text{ kg·cm}^2$$

The answer is (C).

Equation 29.7: Mass Radius of Gyration

$$r_m = \sqrt{I/m} \qquad 29.7$$

Variation

$$I = r^2 m$$

Description

The *mass radius of gyration*, r_m, of a solid object represents the distance from the rotational axis at which the object's entire mass could be located without changing the mass moment of inertia.

Equation 29.8 Through Eq. 29.19: Properties of Uniform Slender Rods

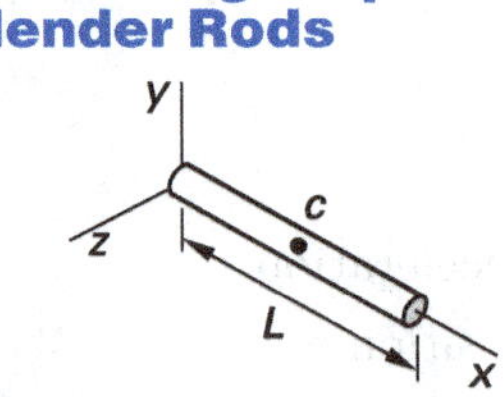

mass and centroid

$$M = \rho LA \qquad 29.8$$

$$x_c = L/2 \qquad 29.9$$

$$y_c = 0 \qquad 29.10$$

[3]Equation 29.5 and Eq. 29.6 both appear on the same page in the *NCEES Handbook* using different notation. The inconsistent subscripts c and G both refer to the same concept: centroidal (center of gravity, center of mass, etc.).

$$z_c = 0 \quad (29.11)$$

mass moment of inertia

$$I_x = I_{x_c} = 0 \quad (29.12)$$

$$I_{y_c} = I_{z_c} = ML^2/12 \quad (29.13)$$

$$I_y = I_z = ML^2/3 \quad (29.14)$$

(radius of gyration)^2

$$r_x^2 = r_{x_c}^2 = 0 \quad (29.15)$$

$$r_{y_c}^2 = r_{z_c}^2 = L^2/12 \quad (29.16)$$

$$r_y^2 = r_z^2 = L^2/3 \quad (29.17)$$

product of inertia

$$I_{x_c y_c} = 0 \quad (29.18)$$

$$I_{xy} = 0 \quad (29.19)$$

Description

Equation 29.8 through Eq. 29.19 give the properties of slender rods. The center of mass (center of gravity) is located at (x_c, y_c, z_c), designated point c. M is the total mass; A is the cross-sectional area perpendicular to the longitudinal axis; ρ is the mass density, equal to the mass divided by the volume; I is the mass moment of inertia about the subscripted axis, used in calculating rotational acceleration and moments about that axis; and r is the radius of gyration, a distance from the designated axis from the centroid where all of the mass can be assumed to be concentrated. I_{xy} is the *product of inertia*, a measure of symmetry, with respect to a plane containing the subscripted axes. The product of inertia is zero if the object is symmetrical about an axis perpendicular to the plane defined by the subscripted axes.

Example

A uniform rod is 2.0 m long and has a mass of 15 kg. What is most nearly the rod's mass moment of inertia?

(A) 5.0 kg·m^2

(B) 20 kg·m^2

(C) 27 kg·m^2

(D) 31 kg·m^2

Solution

From Eq. 29.14, the mass moment of inertia of the rod is

$$I_{\text{rod}} = ML^2/3 = \frac{(15\ \text{kg})(2.0\ \text{m})^2}{3} = 20\ \text{kg}\cdot\text{m}^2$$

The answer is (B).

Equation 29.20 Through Eq. 29.34: Properties of Slender Rings

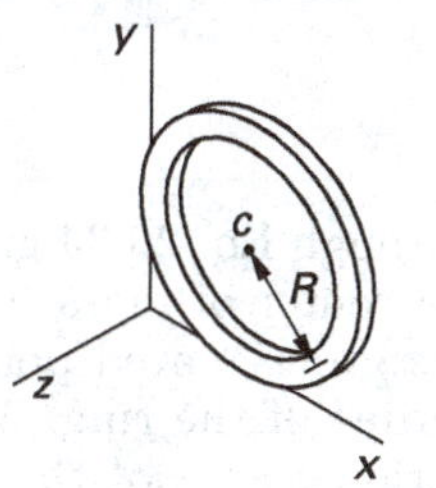

mass and centroid

$$M = 2\pi R\rho A \quad (29.20)$$

$$x_c = R \quad (29.21)$$

$$y_c = R \quad (29.22)$$

$$z_c = 0 \quad (29.23)$$

mass moment of inertia

$$I_{x_c} = I_{y_c} = MR^2/2 \quad (29.24)$$

$$I_{z_c} = MR^2 \quad (29.25)$$

$$I_x = I_y = 3MR^2/2 \quad (29.26)$$

$$I_z = 3MR^2 \quad (29.27)$$

(radius of gyration)^2

$$r_{x_c}^2 = r_{y_c}^2 = R^2/2 \quad \text{29.28}$$

$$r_{z_c}^2 = R^2 \quad \text{29.29}$$

$$r_x^2 = r_y^2 = 3R^2/2 \quad \text{29.30}$$

$$r_z^2 = 3R^2 \quad \text{29.31}$$

product of inertia

$$I_{x_c y_c} = 0 \quad \text{29.32}$$

$$I_{y_c z_c} = MR^2 \quad \text{29.33}$$

$$I_{xz} = I_{yz} = 0 \quad \text{29.34}$$

Description

Equation 29.20 through Eq. 29.34 give the properties of slender rings. The center of mass (center of gravity) is located at (x_c, y_c, z_c), designated point c, and measured from the mean radius of the ring. M is the total mass; A is the cross-sectional area of the ring; ρ is the mass density, equal to the mass divided by the volume; I is the mass moment of inertia about the subscripted axis, used in calculating rotational acceleration and moments about that axis; and r is the radius of gyration, a distance from the designated axis from the centroid where all of the mass can be assumed to be concentrated. r_{zc} is the radius of gyration of the ring about an axis parallel to the z-axis and passing through the centroid. I_{xy} is the product of inertia, a measure of symmetry, with respect to a plane containing the subscripted axes. The product of inertia is zero if the object is symmetrical about an axis perpendicular to the plane defined by the subscripted axes.

Example

The period of oscillation of a clock balance wheel is 0.3 s. The wheel is constructed as a slender ring with its 30 g mass uniformly distributed around a 0.6 cm radius. What is most nearly the wheel's moment of inertia?

(A) 1.1×10^{-6} kg·m^2

(B) 1.6×10^{-6} kg·m^2

(C) 2.1×10^{-6} kg·m^2

(D) 2.6×10^{-6} kg·m^2

Solution

From Eq. 29.25, the wheel's moment of inertia is

$$\begin{aligned} I &= MR^2 \\ &= \left(\frac{30 \text{ g}}{10^3 \frac{\text{g}}{\text{kg}}}\right)\left(\frac{0.6 \text{ cm}}{100 \frac{\text{cm}}{\text{m}}}\right)^2 \\ &= 1.08 \times 10^{-6} \text{ kg·m}^2 \quad (1.1 \times 10^{-6} \text{ kg·m}^2) \end{aligned}$$

The answer is (A).

Equation 29.35 Through Eq. 29.46: Properties of Cylinders

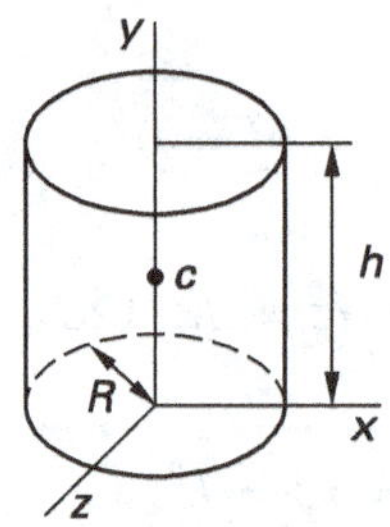

mass and centroid

$$M = \pi R^2 \rho h \quad \text{29.35}$$

$$x_c = 0 \quad \text{29.36}$$

$$y_c = h/2 \quad \text{29.37}$$

$$z_c = 0 \quad \text{29.38}$$

mass moment of inertia

$$I_{x_c} = I_{z_c} = M(3R^2 + h^2)/12 \quad \text{29.39}$$

$$I_{y_c} = I_y = MR^2/2 \quad \text{29.40}$$

$$I_x = I_z = M(3R^2 + 4h^2)/12 \quad \text{29.41}$$

(radius of gyration)^2

$$r_{x_c}^2 = r_{z_c}^2 = (3R^2 + h^2)/12 \quad \text{29.42}$$

$$r_{y_c}^2 = r_y^2 = R^2/2 \quad \text{29.43}$$

$$r_x^2 = r_z^2 = (3R^2 + 4h^2)/12 \qquad 29.44$$

product of inertia

$$I_{x_c y_c} = 0 \qquad 29.45$$

$$I_{xy} = 0 \qquad 29.46$$

Description

Equation 29.35 through Eq. 29.46 give the properties of solid (right) cylinders. The center of mass (center of gravity) is located at (x_c, y_c, z_c), designated point c. M is the total mass; ρ is the mass density, equal to the mass divided by the volume; I is the mass moment of inertia about the subscripted axis, used in calculating rotational acceleration and moments about that axis; and r is the radius of gyration, a distance from the designated axis from the centroid where all of the mass can be assumed to be concentrated. r_{y_c} is the radius of gyration of the cylinder about an axis parallel to the y-axis and passing through the centroid. I_{xy} is the product of inertia, a measure of symmetry, with respect to a plane containing the subscripted axes. The product of inertia is zero if the object is symmetrical about an axis perpendicular to the plane defined by the subscripted axes.

Example

A 50 kg solid cylinder has a height of 3 m and a radius of 0.5 m. The cylinder sits on the x-axis and is oriented with its longitudinal axis parallel to the y-axis. What is most nearly the mass moment of inertia about the x-axis?

(A) 4.1 kg·m^2

(B) 16 kg·m^2

(C) 41 kg·m^2

(D) 150 kg·m^2

Solution

Find the mass moment of inertia using Eq. 29.41.

$$\begin{aligned} I_x &= M(3R^2 + 4h^2)/12 \\ &= \frac{(50\text{ kg})\left((3)(0.5\text{ m})^2 + (4)(3\text{ m})^2\right)}{12} \\ &= 153.1\text{ kg·m}^2 \quad (150\text{ kg·m}^2) \end{aligned}$$

The answer is (D).

Equation 29.47 Through Eq. 29.58: Properties of Hollow Cylinders

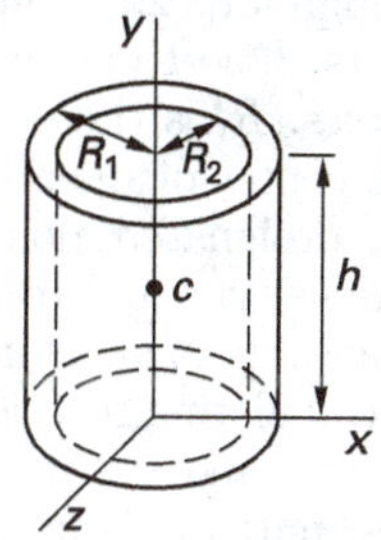

mass and centroid

$$M = \pi(R_1^2 - R_2^2)\rho h \qquad 29.47$$

$$x_c = 0 \qquad 29.48$$

$$y_c = h/2 \qquad 29.49$$

$$z_c = 0 \qquad 29.50$$

mass moment of inertia

$$I_{x_c} = I_{z_c} = M(3R_1^2 + 3R_2^2 + h^2)/12 \qquad 29.51$$

$$I_{y_c} = I_y = M(R_1^2 + R_2^2)/2 \qquad 29.52$$

$$I_x = I_z = M(3R_1^2 + 3R_2^2 + 4h^2)/12 \qquad 29.53$$

(radius of gyration)^2

$$r_{x_c}^2 = r_{z_c}^2 = (3R_1^2 + 3R_2^2 + h^2)/12 \qquad 29.54$$

$$r_{y_c}^2 = r_y^2 = (R_1^2 + R_2^2)/2 \qquad 29.55$$

$$r_x^2 = r_z^2 = (3R_1^2 + 3R_2^2 + 4h^2)/12 \qquad 29.56$$

product of inertia

$$I_{x_c y_c} = 0 \qquad 29.57$$

$$I_{xy} = 0 \qquad 29.58$$

Description

Equation 29.47 through Eq. 29.58 give the properties of hollow (right) cylinders. Due to symmetry, the properties are the same for all axes. R_1 is the outer radius, and R_2 is the inner radius. The center of mass (center of gravity) is located at (x_c, y_c, z_c), designated point c. M is the total mass; ρ is the mass density, equal to the mass divided by the volume; I is the mass moment of inertia about the subscripted axis, used in calculating rotational acceleration and moments about that axis; and r is the radius of gyration, a distance from the designated axis from the centroid where all of the mass can be assumed to be concentrated. r_{y_c} is the radius of gyration of the hollow cylinder about an axis parallel to the y-axis and passing through the centroid. I_{xy} is the product of inertia, a measure of symmetry, with respect to a plane containing the subscripted axes. The product of inertia is zero if the object is symmetrical about an axis perpendicular to the plane defined by the subscripted axes.

Example 19

A hollow cylinder has a mass of 2 kg, a height of 1 m, an outer diameter of 1 m, and an inner diameter of 0.8 m.

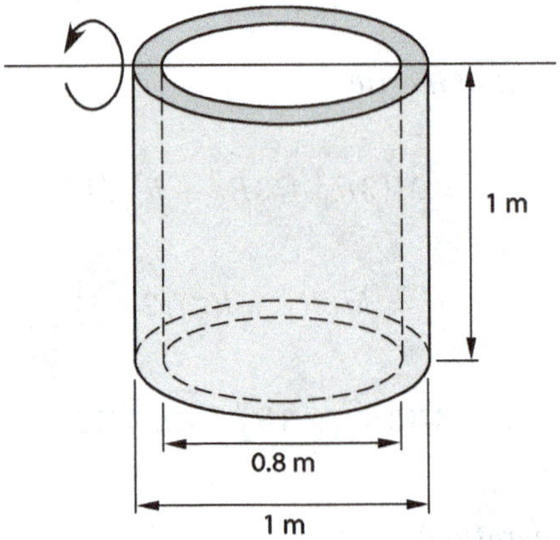

What is most nearly the cylinder's mass moment of inertia about an axis perpendicular to the cylinder's longitudinal axis and located at the cylinder's end?

(A) 0.41 kg·m²

(B) 0.79 kg·m²

(C) 0.87 kg·m²

(D) 1.5 kg·m²

Solution

The outer radius, R_1, and inner radius, R_2, are

$$R_1 = \frac{1 \text{ m}}{2} = 0.5 \text{ m}$$

$$R_2 = \frac{0.8 \text{ m}}{2} = 0.4 \text{ m}$$

Use Eq. 29.53.

$$\begin{aligned} I &= M(3R_1^2 + 3R_2^2 + 4h^2)/12 \\ &= \frac{(2 \text{ kg})\big((3)(0.5 \text{ m})^2 + (3)(0.4 \text{ m})^2 + (4)(1 \text{ m})^2\big)}{12} \\ &= 0.87 \text{ kg·m}^2 \end{aligned}$$

The answer is (C).

Equation 29.59 Through Eq. 29.69: Properties of Spheres

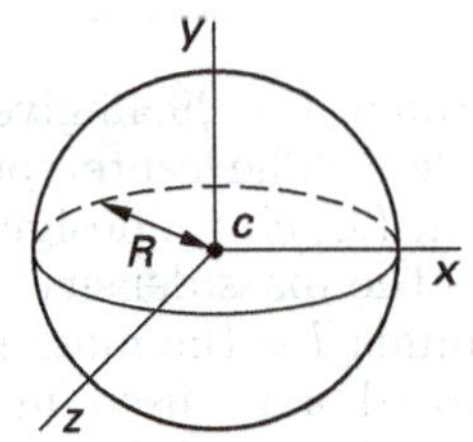

mass and centroid

$$M = \tfrac{4}{3}\pi R^3 \rho \qquad 29.59$$

$$x_c = 0 \qquad 29.60$$

$$y_c = 0 \qquad 29.61$$

$$z_c = 0 \qquad 29.62$$

mass moment of inertia

$$I_{x_c} = I_x = 2MR^2/5 \qquad 29.63$$

$$I_{y_c} = I_y = 2MR^2/5 \qquad 29.64$$

$$I_{z_c} = I_z = 2MR^2/5 \qquad 29.65$$

(radius of gyration)²

$$r_{x_c}^2 = r_x^2 = 2R^2/5 \qquad 29.66$$

$$r_{y_c}^2 = r_y^2 = 2R^2/5 \qquad 29.67$$

$$r_{z_c}^2 = r_z^2 = 2R^2/5 \qquad 29.68$$

Dynamics

product of inertia

$$I_{x_c y_c} = 0 \qquad 29.69$$

Description

Equation 29.59 through Eq. 29.69 give the properties of spheres. The center of mass (center of gravity) is located at (x_c, y_c, z_c), designated point c. M is the total mass; ρ is the mass density, equal to the mass divided by the volume; I is the mass moment of inertia about the subscripted axis, used in calculating rotational acceleration and moments about that axis; and r is the radius of gyration, a distance from the designated axis from the centroid where all of the mass can be assumed to be concentrated. The product of inertia for any plane passing through the centroid is zero because the object is symmetrical about an axis perpendicular to that plane.

2. PLANE MOTION OF A RIGID BODY

General rigid body plane motion, such as rolling wheels, gear sets, and linkages, can be represented in two dimensions (i.e., the plane of motion). Plane motion can be considered as the sum of a translational component and a rotation about a fixed axis, as illustrated in Fig. 29.1.

Figure 29.1 *Components of Plane Motion*

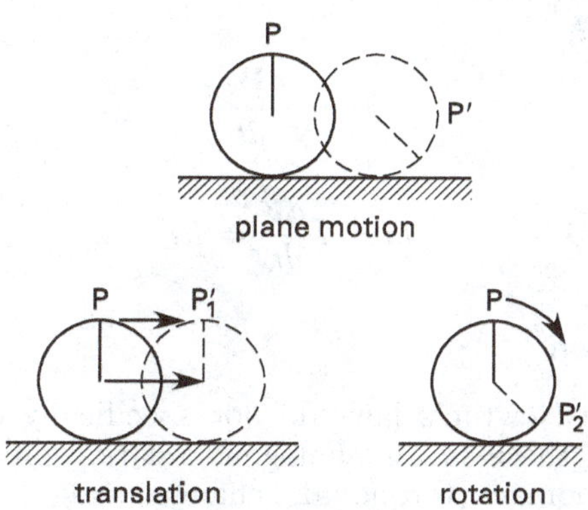

3. ROTATION ABOUT A FIXED AXIS

Instantaneous Center of Rotation

Analysis of the rotational component of a rigid body's plane motion can sometimes be simplified if the location of the body's *instantaneous center* is known. Using the instantaneous center reduces many relative motion problems to simple geometry. The instantaneous center (also known as the *instant center* and IC) is a point at which the body could be fixed (pinned) without changing the instantaneous angular velocities of any point on the body. For angular velocities, the body seems to rotate about a fixed, instantaneous center.

The instantaneous center is located by finding two points for which the absolute velocity directions are known. Lines drawn perpendicular to these two velocities will intersect at the instantaneous center. (This graphic procedure is slightly different if the two velocities are parallel, as Fig. 29.2 shows.) For a rolling wheel, the instantaneous center is the point of contact with the supporting surface.

Figure 29.2 *Graphic Method of Finding the Instantaneous Center*

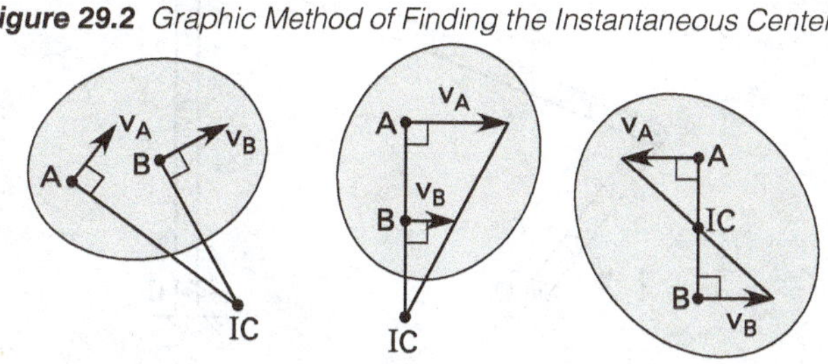

The absolute velocity of any point, P, on a wheel rolling (see Fig. 29.3) with translational velocity, v_O, can be found by geometry. Assume that the wheel is pinned at point C and rotates with its actual angular velocity, $\dot{\theta} = \omega = v_O/r$. The direction of the point's velocity will be perpendicular to the line of length, l, between the instantaneous center and the point.

$$v = l\omega = \frac{l v_O}{r}$$

Figure 29.3 *Instantaneous Center of a Rolling Wheel*

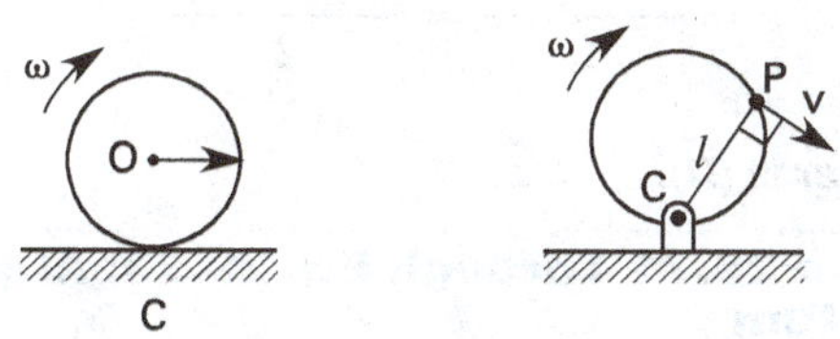

Equation 29.70: Kennedy's Rule

$$c = \frac{n(n-1)}{2} \qquad 29.70$$

Description

The location of the instantaneous center can be found by inspection for many mechanisms, such as simple pinned pulleys and rolling/rotating objects. *Kennedy's rule* (law, theorem, etc.) can be used to help find the instantaneous centers when they are not obvious, such as with slider-crank and bar linkage mechanisms. Kennedy's rule states that any three links (bodies), designated as 1, 2, and 3, of a mechanism (that may have more than three links), and undergoing motion relative to one another, will have exactly three associated instantaneous centers, IC_{12}, IC_{13}, and IC_{23}, and those three instant centers will lie on a straight line. Equation 29.70 calculates the number of instantaneous centers for any number of links. c is the number of instantaneous centers, and n is the number of links.

Example

How many instantaneous centers does the linkage shown have?

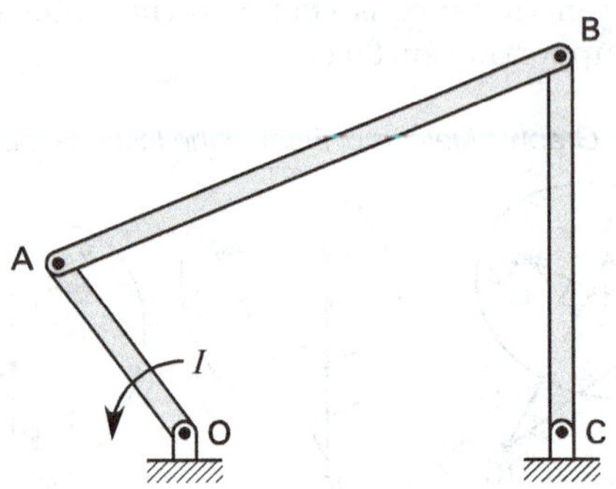

(A) 3

(B) 4

(C) 5

(D) 6

Solution

This is a four-bar linkage. The fourth bar consists of the fixed link between points O and C. Use Eq. 29.70. The number of instantaneous centers is

$$c = \frac{n(n-1)}{2} = \frac{(4)(4-1)}{2} = 6$$

The answer is (D).

Equation 29.71 Through Eq. 29.73: Angular Momentum

$$\mathbf{H}_0 = \mathbf{r} \times m\mathbf{v} \qquad 29.71$$

$$\mathrm{H}_0 = I_0 \omega \qquad 29.72$$

$$\sum(\text{syst. } \mathbf{H})_1 = \sum(\text{syst. } \mathbf{H})_2 \qquad 29.73$$

Description

The *angular momentum* taken about a point 0 is the moment of the linear momentum vector. Angular momentum has units of distance × force × time (e.g., N·m·s). It has the same direction as the rotation vector and can be determined from the vectors by use of the right-hand rule (cross product). (See Eq. 29.71.)

For a rigid body rotating about an axis passing through its center of gravity located at point 0, the scalar value of angular momentum is given by Eq. 29.72.[4]

The *law of conservation of angular momentum* states that if no external torque acts upon an object, the angular momentum cannot change. The angular momentum before and after an internal torque is applied is the same. Equation 29.73 expresses the angular momentum conservation law for a system consisting of multiple masses.[5]

Equation 29.74 and Eq. 29.75: Change in Angular Momentum

$$\dot{\mathbf{H}}_0 = d(I_0 \omega)/dt = \mathbf{M} \qquad 29.74$$

$$\sum(\mathbf{H}_{0i})_{t_2} = \sum(\mathbf{H}_{0i})_{t_1} + \sum \int_{t_1}^{t_2} \mathbf{M}_{0i} dt \qquad 29.75$$

Variations

$$\mathbf{M} = \frac{d\mathbf{H}_{\mathrm{O}}}{dt}$$

$$M = I\frac{d\omega}{dt} = I\alpha$$

Description

Although Newton's laws do not specifically deal with rotation, there is an analogous relationship between applied moment (torque) and change in angular momentum. For a rotating body, the moment (torque), **M**, required to change the angular momentum is given by Eq. 29.74.

The rotation of a rigid body will be about the center of gravity unless the body is constrained otherwise. The scalar form of Eq. 29.74 for a constant moment of inertia is shown in the second variation.

[4]The *NCEES Handbook* is inconsistent in its representation of the centroidal mass moment of inertia. I_0 and I_c are both used in the Dynamics section for the same concept.

[5]In Eq. 29.73, the nonstandard notation "syst" should be interpreted as the limits of summation (i.e., summation over all masses in the system). This would normally be written as $\sum_{\text{system}}$ or something similar.

For a collection of particles, Eq. 29.74 may be expanded as shown in Eq. 29.75. Equation 29.75 determines the angular momentum at time t_2 from the angular momentum at time t_1, $\sum(\mathbf{H}_{0i})_{t_1}$, and the angular impulse of the moment between t_1 and t_2, $\sum\int_{t_1}^{t_2}\mathbf{M}_{0i}dt$.

Equation 29.76 Through Eq. 29.86: Rotation About an Arbitrary Fixed Axis

$$\sum M_q = I_q\alpha \tag{29.76}$$

$$\alpha = \frac{d\omega}{dt} \quad \text{[general]} \tag{29.77}$$

$$\omega = \frac{d\theta}{dt} \quad \text{[general]} \tag{29.78}$$

$$\omega\, d\omega = \alpha\, d\theta \quad \text{[general]} \tag{29.79}$$

$$\omega = \omega_0 + \alpha_c t \tag{29.80}$$

$$\theta = \theta_0 + \omega_0 t + \tfrac{1}{2}\alpha_c t^2 \tag{29.81}$$

$$\omega^2 = \omega_0^2 + 2\alpha_c(\theta - \theta_0) \tag{29.82}$$

$$\alpha = M_q/I_q \tag{29.83}$$

$$\omega = \omega_0 + \alpha t \tag{29.84}$$

$$\theta = \theta_0 + \omega_0 t + \alpha t^2/2 \tag{29.85}$$

$$I_q\omega^2/2 = I_q\omega_0^2/2 + \int_{\theta_0}^{\theta} M_q d\theta \tag{29.86}$$

Variations

$$\omega = \int \alpha dt = \omega_0 + \left(\frac{M}{I}\right)t$$

$$\theta = \iint \alpha dt^2 = \theta_0 + \omega_0 t + \left(\frac{M}{2I}\right)t^2$$

Description

The rotation about an arbitrary fixed axis q is found from Eq. 29.76. Equation 29.77 through Eq. 29.79 apply when the angular acceleration of the rotating body is variable. Equation 29.80 through Eq. 29.82 apply when the angular acceleration of the rotating body is constant.[6] Equation 29.83 through Eq. 29.85 apply when the moment applied to the fixed axis is constant. The change in kinetic energy (i.e., the work done to accelerate from ω_0 to ω) is calculated using Eq. 29.86.

Example

A 50 N wheel has a mass moment of inertia of 2 kg·m². The wheel is subjected to a constant 1 N·m torque. What is most nearly the angular velocity of the wheel 5 s after the torque is applied?

(A) 0.5 rad/s

(B) 3 rad/s

(C) 5 rad/s

(D) 10 rad/s

Solution

Use Eq. 29.83 to find the angular acceleration of the wheel when subjected to a 1 N·m moment.

$$\begin{aligned}\alpha = M_q/I_q &= \frac{1\ \text{N·m}}{2\ \text{kg·m}^2}\\ &= 0.5\ \text{rad/s}^2\end{aligned}$$

From Eq. 29.84, the angular velocity after 5 s is

$$\begin{aligned}\omega = \omega_0 + \alpha t &= 0\ \frac{\text{rad}}{\text{s}} + \left(0.5\ \frac{\text{rad}}{\text{s}^2}\right)(5\ \text{s})\\ &= 2.5\ \text{rad/s}\quad (3\ \text{rad/s})\end{aligned}$$

The answer is (B).

4. CENTRIPETAL AND CENTRIFUGAL FORCES

Newton's second law states that there is a force for every acceleration that a body experiences. For a body moving around a curved path, the total acceleration can be separated into tangential and normal components. By Newton's second law, there are corresponding forces in the tangential and normal directions. The force associated with the normal acceleration is known as the *centripetal force*. The centripetal force is a real force on the body toward the center of rotation. The so-called

[6]The use of subscript c in the *NCEES Handbook* Dynamics section to designate a constant angular acceleration is not a normal and customary engineering usage. Since subscript c is routinely used in dynamics to designate centroidal (mass center), the subscript is easily misinterpreted.

centrifugal force is an apparent force on the body directed away from the center of rotation. The centripetal and centrifugal forces are equal in magnitude but opposite in sign.

The centrifugal force on a body of mass m with distance r from the center of rotation to the center of mass is

$$F_c = ma_n = \frac{m\text{v}_t^2}{r} = mr\omega^2$$

5. BANKING OF CURVES

If a vehicle travels in a circular path on a flat plane with instantaneous radius r and tangential velocity v_t, it will experience an apparent centrifugal force. The centrifugal force is resisted by a combination of roadway banking (superelevation) and sideways friction. The vehicle weight, W, corresponds to the normal force. For small banking angles, the maximum frictional force is

$$F_f = \mu_s N = \mu_s W$$

For large banking angles, the centrifugal force contributes to the normal force. If the roadway is banked so that friction is not required to resist the centrifugal force, the superelevation angle, θ, can be calculated from

$$\tan\theta = \frac{\text{v}_t^2}{gr}$$

Dynamics

30 Energy and Work

Nomenclature

e	coefficient of restitution	–
E	energy	J
F	force	N
g	gravitational acceleration, 9.81	m/s^2
h	height	m
I	moment of inertia	in^4
k	spring constant	N/m
m	mass	kg
M	moment	N·m
p	linear momentum	kg·m/s
P	power	W
r	distance	m
s	position	m
t	time	s
T	kinetic energy	J
U	potential energy	J
v	velocity	m/s
W	work	J
x	displacement	m
y	horizontal displacement	m

Symbols

ε	efficiency	–
θ	angle	deg
ω	angular velocity	rad/s

Subscripts

$1\rightarrow2$	moving from state 1 to state 2
c	centroidal or constant
e	elastic
f	final or frictional
F	force
g	gravity
IC	instantaneous center
M	moment
n	normal
s	spring
W	weight

1. INTRODUCTION

The *energy* of a mass represents the capacity of the mass to do work. Such energy can be stored and released. There are many forms that the stored energy can take, including mechanical, thermal, electrical, and magnetic energies. Energy is a positive, scalar quantity, although the change in energy can be either positive or negative. *Work*, W, is the act of changing the energy of a mass. Work is a signed, scalar quantity. Work is positive when a force acts in the direction of motion and moves a mass from one location to another. Work is negative when a force acts to oppose motion. (Friction, for example, always opposes the direction of motion and can only do negative work.) The net work done on a mass by more than one force can be found by superposition.

Equation 30.1 Through Eq. 30.6: Work[1]

$$W = \int \mathbf{F} \cdot d\mathbf{r} \qquad 30.1$$

$$U_F = \int F \cos\theta ds \quad \text{[variable force]} \qquad 30.2$$

$$U_F = (F_c \cos\theta)\Delta s \quad \text{[constant force]} \qquad 30.3$$

$$U_W = -W\Delta y \quad \text{[weight]} \qquad 30.4$$

$$U_s = -\tfrac{1}{2}k(s_2^2 - s_1^2) \quad \text{[spring]} \qquad 30.5$$

$$U_M = M\Delta\theta \quad \text{[couple moment]} \qquad 30.6$$

[1]In Eq. 30.2 through Eq. 30.6, the variable for work is given as U for consistency with the NCEES *FE Reference Handbook* (*NCEES Handbook*). Normally, it is given as W.

Description

The work performed by a force is calculated as a dot product of the force vector acting through a displacement vector, as shown in Eq. 30.1. Since the dot product of two vectors is a scalar, work is a scalar quantity. The integral in Eq. 30.1 is essentially a summation over all forces acting at all distances.[2] Only the component of force in the direction of motion does work. In Eq. 30.2 and Eq. 30.3, the component of force in the direction of motion is $F\cos\theta$, where θ represents the acute angle between the force and the direction vectors. For a single constant force (or a force resultant), the integral can be dropped, and the differential ds replaced with Δs, as in Eq. 30.3.[3] Equation 30.4 represents the work done in a moving weight, W, a vertical distance, Δy, against earth's gravitational field.[4] Equation 30.5 represents the work associated with an extension or compression of a spring with a spring constant k.[5] Equation 30.6 is the work performed by a couple (i.e., a moment), M, rotating through an angle θ.[6]

2. KINETIC ENERGY

Kinetic energy is a form of mechanical energy associated with a moving or rotating body.

Equation 30.7: Linear Kinetic Energy[7]

$$T = m\text{v}^2/2 \qquad 30.7$$

Description

The *linear kinetic energy* of a body moving with instantaneous linear velocity v is calculated from Eq. 30.7.

Example 22

A 3500 kg car traveling at 65 km/h skids. The car hits a wall 3 s later. The coefficient of friction between the tires and the road is 0.60, and the speed of the car when it hits the wall is 0.20 m/s. What is most nearly the energy that the bumper must absorb in order to prevent damage to the rest of the car?

(A) 70 J

(B) 140 J

(C) 220 J

(D) 360 kJ

Solution

Using Eq. 30.7, the kinetic energy of the car is

$$T = m\text{v}^2/2 = \frac{(3500\ \text{kg})\left(0.20\ \frac{\text{m}}{\text{s}}\right)^2}{2} = 70\ \text{J}$$

The answer is (A).

[2](1) The *NCEES Handbook* attempts to distinguish between work and stored energy. For example, the work-energy principle (called the principle of work and energy in the *NCEES Handbook*) is essentially presented as $U_2 - U_1 = W$. However, there is no energy storage associated with a force, say, moving a box across a frictionless surface, which is one of the possible applications of Eq. 30.3. (2) When denoting work associated with a translating body, the *NCEES Handbook* uses both W and U. (3) The *NCEES Handbook* is inconsistent in its use of the variable U, which has three meanings: work, stored energy, and change in stored energy. (4) The *NCEES Handbook* is inconsistent in the variable used to indicate position or distance. Both r and s are used in this section.

[3]The *NCEES Handbook* is inconsistent in its use of the subscript c. In Eq. 30.3, F_c means a constant force. F_c is not a force directed through the centroid.

[4](1) The subscript W is not associated with work, but rather is associated with the object (i.e., weight) that is moved. (2) The meaning of the negative sign is ambiguous. If Δy is assumed to mean $y_2 - y_1$, then the negative sign would support a thermodynamic first law interpretation (i.e., work is negative when the surroundings do work on the system). However, Eq. 30.1, Eq. 30.2, Eq. 30.3, and Eq. 30.6 do not have negative signs, so these equations do not seem to be written to be consistent with a thermodynamic sign convention. Δy could mean $y_1 - y_2$, and the negative sign may represent mere algebraic convenience.

[5](1) Equation 30.5 is incorrectly presented in the *NCEES Handbook*. The 1/2 multiplier is incorrectly shown inside the parentheses. (2) There is no mathematical reason why the spring constant, k, cannot be brought outside of the parentheses. (3) Whereas the subscripts F, W, and M in Eq. 30.3, Eq. 30.4, and Eq. 30.6 are uppercase, the subscript s in Eq. 30.5 is lowercase. (4) Whereas the subscripts F, W, and M in Eq. 30.3, Eq. 30.4, and Eq. 30.6 are derived from the source of the energy change (i.e., from the item that moves), the subscript s in Eq. 30.5 is derived from the independent variable that changes. If a similar convention had been followed with Eq. 30.3, Eq. 30.4, and Eq. 30.6, the variables in those equations would have been U_r, U_s, and U_θ. (5) For a compression spring acted upon by an increasing force, $s_2 < s_1$, so the negative sign is incorrect for this application from a thermodynamic system standpoint. (6) This equation is shown in a subsequent column of this section of the *NCEES Handbook* as $U_2 - U_1 = k(x_2^2 - x_1^2)/2$, which is not only a different format, but uses x instead of s, and changes the meaning of U from change in energy to stored energy.

[6]The *NCEES Handbook* associates Eq. 30.6 with a "couple moment," an uncommon term. A property of a couple is the moment it imparts, so it is appropriate to speak of the moment of a couple. Similarly, a property of a hurricane is its wind speed, but referring to the hurricane itself as a hurricane speed would be improper. If Eq. 30.6 is meant to describe a pure moment causing rotation without translation, the terms *couple*, *pure moment*, or *torque* would all be appropriate.

[7](1) The *NCEES Handbook* uses different variables to represent kinetic energy. In its section on Units, KE is used. In its Dynamics section, T is used. (2) In its description of Eq. 30.7, the *NCEES Handbook* uses bold **v**, indicating a vector quantity, but subsequently does not indicate a vector quantity in the equation. Kinetic energy is not a vector quantity, and a vector velocity is not required to calculate kinetic energy.

Equation 30.8: Rotational Kinetic Energy

$$T = I_{IC}\omega^2/2 \qquad 30.8$$

Description

The *rotational kinetic energy* of a body moving with instantaneous angular velocity ω about an instantaneous center is described by Eq. 30.8. I_{IC} is the mass moment of inertia about the *instantaneous center* of rotation.[8]

Example

A 10 kg homogeneous disk of 5 cm radius rotates on a massless axle AB of length 0.5 m and rotates about a fixed point A. The disk is constrained to roll on a horizontal floor.

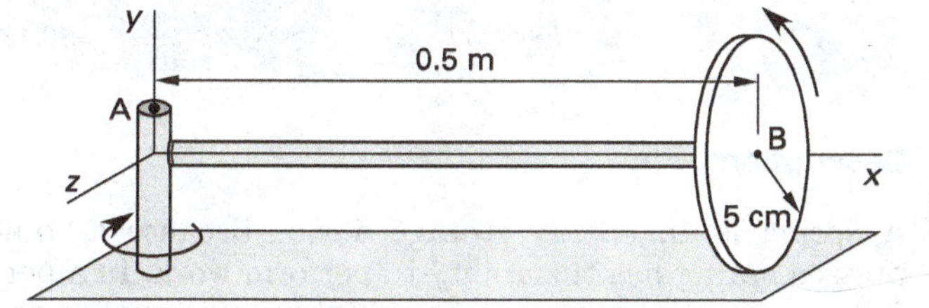

Given an angular velocity of 30 rad/s about the x-axis and -3 rad/s about the y-axis, the kinetic energy of the disk is most nearly

(A) 0.62 J

(B) 17 J

(C) 18 J

(D) 34 J

Solution

Assuming the axle is part of the disk, the disk has a fixed point at A. Since the x-, y-, and z-axes are principal axes of inertia for the disk, the kinetic energy is most nearly

$$T = I_{IC}\omega^2/2 = \frac{I_x\omega_x^2}{2} + \frac{I_y\omega_y^2}{2} + \frac{I_z\omega_z^2}{2}$$

$$= \frac{\frac{1}{2}mr^2\omega_x^2}{2} + \frac{\left(mL^2 + \frac{1}{4}mr^2\right)\omega_y^2}{2} + 0$$

$$= \frac{\left(\frac{1}{2}\right)(10\text{ kg})\left(\frac{5\text{ cm}}{100\,\frac{\text{cm}}{\text{m}}}\right)^2\left(30\,\frac{\text{rad}}{\text{s}}\right)^2}{2}$$

$$+ \frac{\left((10\text{ kg})(0.5\text{ m})^2 + \left(\frac{1}{4}\right)(10\text{ kg})\left(\frac{5\text{ cm}}{100\,\frac{\text{cm}}{\text{m}}}\right)^2\right)\left(-3\,\frac{\text{rad}}{\text{s}}\right)^2}{2}$$

$$+0$$

$$= 16.9\text{ J} \quad (17\text{ J})$$

The answer is (B).

Equation 30.9 and Eq. 30.10: Kinetic Energy of Rigid Bodies

$$T = m\text{v}^2/2 + I_c\omega^2/2 \qquad 30.9$$

$$T = m(\text{v}_{cx}^2 + \text{v}_{cy}^2)/2 + I_c\omega_z^2/2 \qquad 30.10$$

Description

Equation 30.9 gives the kinetic energy of a rigid body. For general plane motion in which there are translational and rotational components, the kinetic energy is the sum of the translational and rotational forms. Equation 30.10 gives the kinetic energy for motion in the x-y plane.

Example

A uniform disk with a mass of 10 kg and a diameter of 0.5 m rolls without slipping on a flat horizontal surface, as shown.

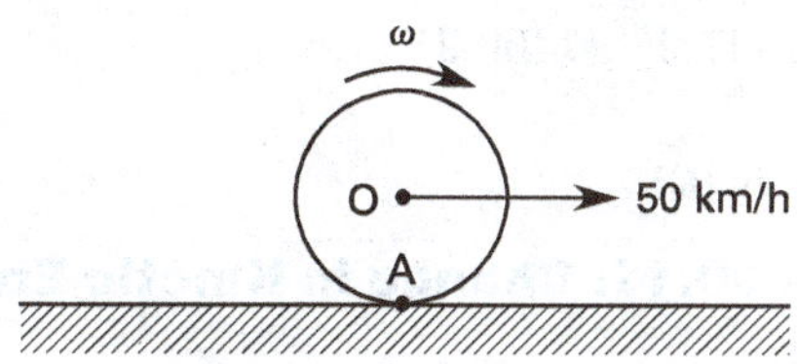

[8]Depending on the configuration, in addition to I, the *NCEES Handbook* uses different variables for the mass moment of inertia of bodies with rotational components. For rotation about an arbitrary axis, q, the *NCEES Handbook* uses I_q. For bodies rotating around axes other than their own centroidal axes, it uses I_{IC}. For unconstrained bodies rotating around axes passing through their own centroidal axes, I_{IC} in Eq. 30.8 is the same as I_c in Eq. 30.9 and Eq. 30.10.

Dynamics

When its horizontal velocity is 50 km/h, the total kinetic energy of the disk is most nearly

(A) 1000 J

(B) 1200 J

(C) 1400 J

(D) 1600 J

Solution

The linear velocity is

$$\mathrm{v}_0 = \frac{\left(50\ \frac{\text{km}}{\text{h}}\right)\left(1000\ \frac{\text{m}}{\text{km}}\right)}{\left(60\ \frac{\text{s}}{\text{min}}\right)\left(60\ \frac{\text{min}}{\text{h}}\right)} = 13.89\ \text{m/s}$$

The angular velocity is

$$\omega = \frac{\mathrm{v}_0}{r} = \frac{13.89\ \frac{\text{m}}{\text{s}}}{\frac{0.5\ \text{m}}{2}} = 55.56\ \text{rad/s}$$

Using Eq. 30.9, the total kinetic energy is

$$\begin{aligned} T &= m\mathrm{v}_0^2/2 + I_c\omega^2/2 \\ &= \frac{m\mathrm{v}_0^2}{2} + \frac{\left(\frac{1}{2}mR^2\right)\omega^2}{2} \\ &= \frac{(10\ \text{kg})\left(13.89\ \frac{\text{m}}{\text{s}}\right)^2}{2} \\ &\quad + \frac{\left(\left(\frac{1}{2}\right)(10\ \text{kg})\left(\frac{0.5\ \text{m}}{2}\right)^2\right)\left(55.56\ \frac{\text{rad}}{\text{s}}\right)^2}{2} \\ &= 1447\ \text{J} \quad (1400\ \text{J}) \end{aligned}$$

The answer is (C).

Equation 30.11: Change in Kinetic Energy

$$T_2 - T_1 = m(\mathrm{v}_2^2 - \mathrm{v}_1^2)/2 \qquad \textit{30.11}$$

Description

The change in kinetic energy is calculated from the difference of squares of velocity, not from the square of the velocity difference (i.e., $m(\mathrm{v}_2^2 - \mathrm{v}_1^2)/2 \neq m(\mathrm{v}_2 - \mathrm{v}_1)^2/2$).

3. POTENTIAL ENERGY

Equation 30.12: Potential Energy in Gravity Field

$$U = mgh \qquad \textit{30.12}$$

Description

Potential energy (also known as *gravitational potential energy*), U, is a form of mechanical energy possessed by a mass due to its relative position in a gravitational field. Potential energy is lost when the elevation of a mass decreases. The lost potential energy usually is converted to kinetic energy or heat.

Equation 30.13: Force in a Spring (Hooke's Law)

$$F_s = kx \qquad \textit{30.13}$$

Description

A spring is an energy storage device because a compressed spring has the ability to perform work. In a perfect spring, the amount of energy stored is equal to the work required to compress the spring initially. The stored spring energy does not depend on the mass of the spring.

Equation 30.13 gives the force in a spring, which is the product of the *spring constant* (*stiffness*), k, and the displacement of the spring from its original position, x.

Example

A spring has a constant of 50 N/m. The spring is hung vertically, and a mass is attached to its end. The spring end displaces 30 cm from its equilibrium position. The same mass is removed from the first spring and attached to the end of a second (different) spring, and the displacement is 25 cm. What is most nearly the spring constant of the second spring?

(A) 46 N/m

(B) 56 N/m

(C) 60 N/m

(D) 63 N/m

Solution

The gravitational force on the mass is the same for both springs. From Hooke's law,

$$F_s = k_1x_1 = k_2x_2$$

$$k_2 = \frac{k_1x_1}{x_2} = \frac{\left(50\ \frac{\text{N}}{\text{m}}\right)(30\ \text{cm})}{25\ \text{cm}} = 60\ \text{N/m}$$

The answer is (C).

Equation 30.14: Elastic Potential Energy

$$U = kx^2/2 \qquad 30.14$$

Description

Given a linear spring with spring constant (stiffness), k, the spring's *elastic potential energy* is calculated from Eq. 30.14.

Example

The 40 kg mass, m, shown is acted upon by a spring and guided by a frictionless rail. When the compressed spring is released, the mass stops slightly short of point A. The spring constant, k, is 3000 N/m, and the spring is compressed 0.5 m.

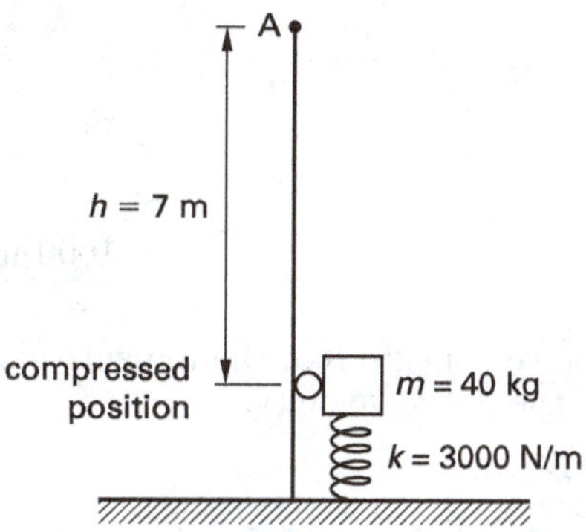

What is most nearly the energy stored in the spring?

(A) 380 J

(B) 750 J

(C) 1500 J

(D) 2100 J

Solution

Using Eq. 30.14, the potential energy is

$$U = kx^2/2 = \frac{\left(3000\ \frac{\text{N}}{\text{m}}\right)(0.5\ \text{m})^2}{2}$$
$$= 375\ \text{J} \quad (380\ \text{J})$$

The answer is (A).

Equation 30.15: Change in Potential Energy

$$U_2 - U_1 = k(x_2^2 - x_1^2)/2 \qquad 30.15$$

Description

The change in potential energy stored in the spring when the deformation in the spring changes from position x_1 to position x_2 is found from Eq. 30.15.[9]

Equivalent Spring Constant

The entire applied load is felt by each spring in a series of springs linked end-to-end. The *equivalent* (*composite*) *spring constant* for springs in series is

$$\frac{1}{k_{\text{eq}}} = \frac{1}{k_1} + \frac{1}{k_2} + \frac{1}{k_3} + \cdots \quad \left[\begin{array}{c}\text{series}\\ \text{springs}\end{array}\right]$$

Springs in parallel (e.g., concentric springs) share the applied load. The equivalent spring constant for springs in parallel is

$$k_{\text{eq}} = k_1 + k_2 + k_3 + \cdots \quad \left[\begin{array}{c}\text{parallel}\\ \text{springs}\end{array}\right]$$

Equation 30.16 Through Eq. 30.18: Combined Potential Energy

$$V = V_g + V_e \qquad 30.16$$

$$V_g = \pm Wy \qquad 30.17$$

$$V_e = +1/2ks^2 \qquad 30.18$$

Description

In mechanical systems, there are two common components of what is normally referred to as potential energy: gravitational potential energy and strain energy.[10] For a

[9]The *NCEES Handbook* uses the notation x to denote position, as in Eq. 30.14, and to denote change in length (i.e., a change in position Δx), as in Eq. 30.15. In Eq. 30.5 and Eq. 30.18, the *NCEES Handbook* uses s instead of x as in Eq. 30.15, but the meaning is the same.

[10]Equation 30.16 is not limited to mechanical systems. Potential energy storage exists in electrical, magnetic, fluid, pneumatic, and thermal systems also.

system containing a linear, elastic spring that is located at some elevation in a gravitational field, Eq. 30.16 gives the total of these two components.[11] Equation 30.17 gives the potential energy of a weight in a gravitational field.[12] Equation 30.18 gives the strain energy in a linear, elastic spring.[13]

4. ENERGY CONSERVATION PRINCIPLE

According to the *energy conservation principle*, energy cannot be created or destroyed. However, energy can be transformed into different forms. Therefore, the sum of all energy forms of a system is constant.

$$\sum E = \text{constant}$$

Because energy can be neither created nor destroyed, external work performed on a conservative system must go into changing the systems total energy. This is known as the *work-energy principle*.

$$W = E_2 - E_1$$

Generally, the principle of conservation of energy is applied to mechanical energy problems (i.e., conversion of work into kinetic or potential energy).

Conversion of one form of energy into another does not violate the conservation of energy law. Most problems involving conversion of energy are really special cases. For example, consider a falling body that is acted upon by a gravitational force. The conversion of potential energy into kinetic energy can be interpreted as equating the work done by the constant gravitational force to the change in kinetic energy.

Equation 30.19: Law of Conservation of Energy (Conservative Systems)

$$T_2 + U_2 = T_1 + U_1 \qquad 30.19$$

Description

For *conservative systems* where there is no energy dissipation or gain, the total energy of the mass is equal to the sum of the kinetic and potential (gravitational and elastic) energies.

Example

A projectile with a mass of 10 kg is fired directly upward from ground level with an initial velocity of 1000 m/s. Neglecting the effects of air resistance, what will be the speed of the projectile when it impacts the ground?

(A) 710 m/s

(B) 980 m/s

(C) 1000 m/s

(D) 1400 m/s

Solution

Use the law of conservation of energy.

$$T_2 + U_2 = T_1 + U_1$$

$$\frac{m\mathrm{v}_2^2}{2} + mgh_2 = \frac{m\mathrm{v}_1^2}{2} + mgh_1$$

$$(mgh_1 - mgh_2) + \frac{m\mathrm{v}_1^2 - m\mathrm{v}_2^2}{2} = 0$$

$$0 + \frac{m(\mathrm{v}_1^2 - \mathrm{v}_2^2)}{2} = 0$$

$$\mathrm{v}_2^2 = \mathrm{v}_1^2$$

$$\mathrm{v}_2 = \mathrm{v}_1 = 1000 \text{ m/s}$$

If air resistance is neglected, the impact velocity will be the same as the initial velocity.

The answer is (C).

Equation 30.20: Law of Conservation of Energy (Nonconservative Systems)

$$T_2 + U_2 = T_1 + U_1 + W_{1\to2} \qquad 30.20$$

---

[11](1) Although PE is used in the Units section of the *NCEES Handbook* to identify potential energy, and U is defined as energy in the Dynamics section, here the *NCEES Handbook* introduces a new variable, V, for potential energy. Outside of the conservation of energy equation, this new variable does not seem to be used anywhere else in the *NCEES Handbook*. (2) V_g and V_e have previously (in the *NCEES Handbook*) been represented by U_W and U_s, among others. (3) The subscripts g and e are undefined, but gravitational acceleration is implied for g. The meaning of e is unclear but almost certainly refers to an elastic strain energy.

[12](1) The *NCEES Handbook* is inconsistent in the variable used to represent deflection. Equation 30.17 uses y while other equations in the Dynamics section of the *NCEES Handbook* use Δy and h. y is implicitly the distance from some arbitrary elevation for which $y = 0$ is assigned. (2) This use of $\pm$ is inconsistent with a thermodynamic interpretation of energy, as was apparently used in Eq. 30.4. The sign of the energy would normally be derived from the position, which can be positive or negative. By using $\pm$, the implication is that y is always a positive quantity, regardless of whether the mass is above or below the reference datum.

[13](1) Equation 30.18 has previously been presented with different variables in this section as Eq. 30.14. (2) Equation 30.18 uses s while other equations in the Dynamics section of the *NCEES Handbook* use x. (3) The + symbol is ambiguous, but probably should be interpreted as meaning kinetic energy is always positive. The + is redundant, because the s^2 term is always positive. (4) Equation 30.18 should not be interpreted as one over two times ks^2.

Description

Nonconservative forces (e.g., friction) are accounted for by the work done by the nonconservative forces in moving between state 1 and state 2, $W_{1\rightarrow 2}$. If the nonconservative forces increase the energy of the system, $W_{1\rightarrow 2}$ is positive. If the nonconservative forces decrease the energy of the system, $W_{1\rightarrow 2}$ is negative.

5. LINEAR IMPULSE

Impulse is a vector quantity equal to the change in vector momentum. Units of linear impulse are the same as those for linear momentum: N·s. Figure 30.1 illustrates that impulse is represented by the area under the force-time curve.

$$\mathbf{Imp} = \int_{t_1}^{t_2} \mathbf{F}dt$$

Figure 30.1 *Impulse*

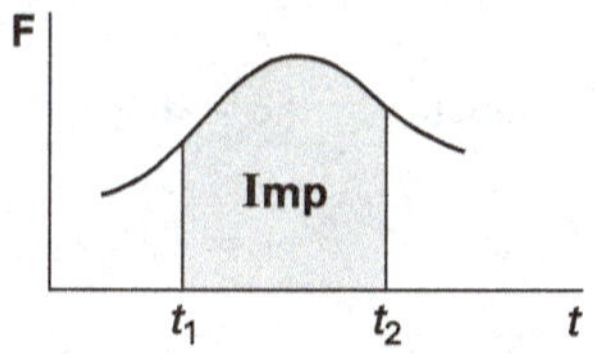

If the applied force is constant, impulse is easily calculated.

$$\mathbf{Imp} = \mathbf{F}(t_2 - t_1)$$

The change in momentum is equal to the impulse. This is known as the *impulse-momentum principle*. For a linear system with constant force and mass,

$$\mathbf{Imp} = \Delta \mathbf{p}$$

Equation 30.21 and Eq. 30.22: Impulse-Momentum Principle for a Particle

$$m\,d\mathbf{v}/dt = \mathbf{F} \qquad 30.21$$

$$m\,d\mathbf{v} = \mathbf{F}dt \qquad 30.22$$

Variation

$$\mathbf{F}(t_2 - t_1) = \Delta(m\mathbf{v})$$

Description

The impulse-momentum principle for a constant force and mass demonstrates that the impulse-momentum principle follows directly from Newton's second law.

Example

A 60 000 kg railcar moving at 1 km/h is coupled to a second, stationary railcar. If the velocity of the two cars after coupling is 0.2 m/s (in the original direction of motion) and the coupling is completed in 0.5 s, what is most nearly the average impulsive force on the railcar?

(A) 520 N

(B) 990 N

(C) 3100 N

(D) 9300 N

Solution

The original velocity of the 60 000 kg railcar is

$$\mathrm{v} = \frac{\left(1\ \frac{\text{km}}{\text{h}}\right)\left(1000\ \frac{\text{m}}{\text{km}}\right)}{\left(60\ \frac{\text{s}}{\text{min}}\right)\left(60\ \frac{\text{min}}{\text{h}}\right)} = 0.2777\ \text{m/s}$$

Use the impulse-momentum principle.

$$F\Delta t = m\Delta \mathrm{v}$$

$$F = \frac{m(\mathrm{v}_1 - \mathrm{v}_2)}{t_1 - t_2} = \frac{(60\,000\ \text{kg})\left(0.2777\ \frac{\text{m}}{\text{s}} - 0.2\ \frac{\text{m}}{\text{s}}\right)}{0\ \text{s} - 0.5\ \text{s}}$$
$$= -9324\ \text{N} \quad (9300\ \text{N}) \quad [\text{opposite original direction}]$$

The answer is (D).

Equation 30.23: Impulse-Momentum Principle for a System of Particles

$$\sum m_i(\mathbf{v}_i)_{t_2} = \sum m_i(\mathbf{v}_i)_{t_1} + \sum \int_{t_1}^{t_2} \mathbf{F}_i dt \qquad 30.23$$

Description

$\sum m_i(\mathbf{v}_i)_{t_1}$ and $\sum m_i(\mathbf{v}_i)_{t_2}$ are the linear momentum at time t_1 and time t_2, respectively, for a system (i.e., collection) of particles. The impulse of the forces $\mathbf{F}$ from time t_1 to time t_2 is

$$\sum \int_{t_1}^{t_2} \mathbf{F}_i dt$$

6. IMPACTS

According to Newton's second law, momentum is conserved unless a body is acted upon by an external force such as gravity or friction. In an impact or collision, contact is very brief, and the effect of external forces is insignificant. Therefore, momentum is conserved, even though energy may be lost through heat generation and deforming the bodies.

Consider two particles, initially moving with velocities v_1 and v_2 on a collision path, as shown in Fig. 30.2. The conservation of momentum equation can be used to find the velocities after impact, v_1' and v_2'.

Figure 30.2 *Direct Central Impact*

The impact is said to be an *inelastic impact* if kinetic energy is lost. The impact is said to be *perfectly inelastic* or *perfectly plastic* if the two particles stick together and move on with the same final velocity. The impact is said to be an *elastic impact* only if kinetic energy is conserved.

$$m_1 v_1^2 + m_2 v_2^2 = m_1 v_1'^2 + m_2 v_2'^2 \quad \text{[elastic only]}$$

Equation 30.24: Conservation of Momentum

$$m_1\mathbf{v}_1 + m_2\mathbf{v}_2 = m_1\mathbf{v}_1' + m_2\mathbf{v}_2' \qquad 30.24$$

Description

The *conservation of momentum* equation is used to find the velocity of two particles after collision. $\mathbf{v}_1$ and $\mathbf{v}_2$ are the initial velocities of the particles, and $\mathbf{v}_1'$ and $\mathbf{v}_2'$ are the velocities after impact.

Example

A 60 000 kg railcar moving at 1 km/h is instantaneously coupled to a stationary 40 000 kg railcar. What is most nearly the speed of the coupled cars?

(A) 0.40 km/h

(B) 0.60 km/h

(C) 0.88 km/h

(D) 1.0 km/h

Solution

Use the conservation of momentum principle.

$$m_1\mathbf{v}_1 + m_2\mathbf{v}_2 = (m_1 + m_2)\mathbf{v}'$$

$$(60\,000\ \text{kg})\left(1\ \frac{\text{km}}{\text{h}}\right) + (40\,000\ \text{kg})(0) = (60\,000\ \text{kg} + 40\,000\ \text{kg})\mathbf{v}'$$

$$\mathbf{v}' = 0.60\ \text{km/h}$$

The answer is (B).

Equation 30.25: Coefficient of Restitution

$$e = \frac{(v_2')_n - (v_1')_n}{(v_1)_n - (v_2)_n} \qquad 30.25$$

Values

inelastic	$e < 1.0$
perfectly inelastic (plastic)	$e = 0$
perfectly elastic	$e = 1.0$

Description

The *coefficient of restitution*, *e*, is the ratio of relative velocity differences along a mutual straight line. When both impact velocities are not directed along the same straight line, the coefficient of restitution should be calculated separately for each velocity component.

In Eq. 30.25, the subscript *n* indicates that the velocity to be used in calculating the coefficient of restitution should be the velocity component normal to the plane of impact.

When an object rebounds from a stationary object (an infinitely massive plane), the stationary object's initial and final velocities are zero. In that case, the *rebound velocity* can be calculated from only the object's velocities.

$$e = \left|\frac{v_1'}{v_1}\right|$$

The value of the coefficient of restitution can be used to categorize the collision as elastic or inelastic. For a perfectly inelastic collision (i.e., a plastic collision), as when two particles stick together, the coefficient of restitution is zero. For a perfectly elastic collision, the coefficient of restitution is 1.0. For most collisions, the coefficient of restitution will be between zero and 1.0, indicating a (partially) inelastic collision.

Example 30

A 2 kg clay ball moving at a rate of 40 m/s collides with a 5 kg ball of clay moving in the same direction at a rate of 10 m/s. What is most nearly the final velocity of both balls if they stick together after colliding?

(A) 10 m/s

(B) 12 m/s

(C) 15 m/s

(D) 19 m/s

Solution

From the coefficient of restitution definition, Eq. 30.25,

$$e = \frac{(v_2')_n - (v_1')_n}{(v_1)_n - (v_2)_n} = 0$$

$$v_2' = v_1' = v'$$

From the conservation of momentum, Eq. 30.24,

$$\begin{aligned} m_1 v_1 + m_2 v_2 &= (m_1 + m_2)v' \\ v' &= \frac{m_1 v_1 + m_2 v_2}{m_1 + m_2} \\ &= \frac{(2\text{ kg})\left(40\ \frac{\text{m}}{\text{s}}\right) + (5\text{ kg})\left(10\ \frac{\text{m}}{\text{s}}\right)}{2\text{ kg} + 5\text{ kg}} \\ &= 18.6\text{ m/s} \quad (19\text{ m/s}) \end{aligned}$$

The answer is (D).

Equation 30.26 and Eq. 30.27: Velocity After Impact

$$(v_1')_n = \frac{m_2(v_2)_n(1+e) + (m_1 - em_2)(v_1)_n}{m_1 + m_2} \qquad 30.26$$

$$(v_2')_n = \frac{m_1(v_1)_n(1+e) - (em_1 - m_2)(v_2)_n}{m_1 + m_2} \qquad 30.27$$

Description

If the coefficient of restitution is known, Eq. 30.26 and Eq. 30.27 may be used to calculate the velocities after impact.

Topic IX: Mechanics of Materials

31 Stresses and Strains

Nomenclature

A	area	m^2
C	stress at center of Mohr's circle	MPa
d	diameter	m
E	modulus of elasticity	MPa
F	force	N
FS	factor of safety	–
g	gravitational acceleration, 9.81	m/s^2
G	shear modulus	MPa
k	stress concentration factor	–
L	length	m
P	force	N
R	radius	m
S	strength	MPa
u	strain energy per unit volume	MPa
U	energy	N·m
W	work	N·m

Symbols

γ	shear strain	–
δ	deformation	m
δ	elongation	m
ε	linear strain	–
θ	angle	rad
ν	Poisson's ratio	–
σ	normal stress	MPa
τ	shear stress	MPa

Subscripts

a	allowable or alternating
b	bending
e	endurance
eff	effective
f	final
m	mean
r	range
s	shear
t	tension
u	ultimate
y	yield

1. DEFINITIONS

Mechanics of materials deals with the elastic behavior of materials and the stability of members. Mechanics of materials concepts are used to determine the stress and deformation of axially loaded members, connections, torsional members, thin-walled pressure vessels, beams, eccentrically loaded members, and columns.

Equation 31.1: Stress

$$\sigma = \frac{P}{A} \qquad 31.1$$

Variation

$$\tau = \frac{P_{\text{parallel to area}}}{A}$$

Description

Stress is force per unit area. Typical units of stress are lbf/in^2, ksi, and MPa. There are two primary types of stress: *normal stress* and *shear stress*. With normal stress, σ, the force is normal to the surface area. With shear stress, τ, the force is parallel to the surface area.

Equation 31.1 describes the normal stress. Shear stress is given by the variation equation.

In mechanics of materials, stresses have a specific *sign convention*. Tensile stresses make a part elongate in the direction of application; tensile stresses are given a positive sign. Compressive stresses make a part shrink in the direction of application; compressive stresses are given a negative sign.

Example

A steel bar with a cross-sectional area of 6 cm^2 is subjected to axial tensile forces of 50 kN applied at each end of the bar. What is most nearly the stress in the bar?

(A) 67 MPa

(B) 78 MPa

(C) 83 MPa

(D) 94 MPa

Solution

From Eq. 31.1,

$$\sigma_{\text{axial}} = \frac{P}{A} = \frac{(50\ \text{kN})\left(1000\ \frac{\text{N}}{\text{kN}}\right)\left(100\ \frac{\text{cm}}{\text{m}}\right)^2}{6\ \text{cm}^2}$$
$$= 8.33 \times 10^7\ \text{Pa} \quad (83\ \text{MPa})$$

The answer is (C).

Equation 31.2: Linear Strain

$$\varepsilon = \delta/L \qquad 31.2$$

Description

Linear strain (*normal strain*, *longitudinal strain*, *axial strain*, *engineering strain*), ε, is a change of length per unit of length. Linear strain may be listed as having units of in/in, mm/mm, percent, or no units at all. *Shear strain*, γ, is an angular deformation resulting from shear stress. Shear strain may be presented in units of radians, percent, or no units at all.

Equation 31.2 shows the relationship between engineering strain, ε, and elongation, δ.

Example

A 200 m cable is suspended vertically. At any point along the cable, the strain is proportional to the length of the cable below that point. If the strain at the top of the cable is 0.001, what is most nearly the total elongation of the cable?

(A) 0.05 m

(B) 0.10 m

(C) 0.15 m

(D) 0.20 m

Solution

Since the strain is proportional to the cable length, it varies from 0 at the end to the maximum value of 0.001 at the supports. The average engineering strain is

$$\varepsilon_{\text{ave}} = \frac{\varepsilon_{\text{max}}}{2} = \frac{0.001}{2}$$
$$= 0.0005$$

From Eq. 31.2, the total elongation is

$$\varepsilon = \delta/L$$
$$\delta = \varepsilon_{\text{ave}}L = (0.0005)(200\ \text{m})$$
$$= 0.10\ \text{m}$$

The answer is (B).

Equation 31.3: Hooke's Law

$$E = \sigma/\varepsilon = \frac{P/A}{\delta/L} \qquad 31.3$$

Variation

$$\sigma = E\varepsilon$$

Values

material	units*	steel	aluminum	cast iron	wood (fir)
modulus of	Mpsi	29	10	14.5	1.6
elasticity, E	GPa	200	69	100	11

*Mpsi = millions of pounds per square inch

Description

Hooke's law is a simple mathematical statement of the relationship between elastic stress and strain: stress is proportional to strain. For normal stress, the constant of proportionality is the *modulus of elasticity* (*Young's modulus*), *E*.

An *isotropic material* has the same properties in all directions. For example, steel is generally considered to be isotropic, and its modulus of elasticity is invariant with respect to the direction of loading. The properties of an *anisotropic material* vary with the direction of loading.

3

Example

A 2 m long aluminum bar (modulus of elasticity = 69 GPa) is subjected to a tensile stress of 175 MPa. What is most nearly the elongation?

(A) 4 mm

(B) 5 mm

(C) 8 mm

(D) 9 mm

Solution

From Hooke's law,

$$E = \sigma/\varepsilon = \frac{P/A}{\delta/L}$$

$$\delta = \frac{\sigma L}{E}$$

$$= \frac{(175\ \text{MPa})\left(10^6\ \frac{\text{Pa}}{\text{MPa}}\right)(2\ \text{m})}{(69\ \text{GPa})\left(10^9\ \frac{\text{Pa}}{\text{GPa}}\right)}$$

$$= 0.00507\ \text{m} \quad (5\ \text{mm})$$

The answer is (B).

Equation 31.4: Poisson's Ratio

$$\nu = -(\text{lateral strain})/(\text{longitudinal strain}) \qquad 31.4$$

Variation

$$\nu = -\frac{\varepsilon_{\text{lateral}}}{\varepsilon_{\text{axial}}}$$

Values

Theoretically, Poisson's ratio could vary from 0 to 0.5; *typical values* are shown.

material	steel	aluminum	cast iron	wood (fir)
Poisson's ratio, ν	0.30	0.33	0.21	0.33

Description

Poisson's ratio, ν, is a constant that relates the lateral strain to the axial (longitudinal) strain for axially loaded members.

4

Example

A 1 m × 1 m × 0.01 m square steel plate (modulus of elasticity = 200 GPa) is loaded in tension parallel to one of its long edges. The resulting 30 MPa stress is uniform across the 1 m × 0.01 m cross section. Poisson's ratio of steel is 0.29. Neglecting the change in plate thickness, the new dimensions of the plate are most nearly

(A) 1000.15 mm × 999.852 mm

(B) 1000.15 mm × 999.957 mm

(C) 1000.15 mm × 1000.05 mm

(D) 1000.20 mm × 1000.50 mm

Solution

Use Eq. 31.3. The elongation in the direction of the tensile stress is

$$E = \sigma/\varepsilon = \frac{P/A}{\delta/L}$$

$$\delta_{\text{axial}} = \frac{\sigma L_{\text{axial}}}{E}$$

$$= \frac{(30\ \text{MPa})\left(10^6\ \frac{\text{Pa}}{\text{MPa}}\right)(1\ \text{m})\left(1000\ \frac{\text{mm}}{\text{m}}\right)}{(200\ \text{GPa})\left(10^9\ \frac{\text{Pa}}{\text{GPa}}\right)}$$

$$= 0.15\ \text{mm}$$

Use Eq. 31.2 and the variation of Eq. 31.4. The elongation perpendicular to the tensile stress is

$$\delta_{\text{lateral}} = L_{\text{lateral}}\varepsilon_{\text{lateral}} = -L_{\text{lateral}}\nu\varepsilon_{\text{axial}}$$

$$= -L_{\text{lateral}}\nu\left(\frac{\delta_{\text{axial}}}{L_{\text{axial}}}\right)$$

Since the plate is square, $L_{\text{lateral}} = L_{\text{axial}}$.

$$\delta_{\text{lateral}} = -\nu\delta_{\text{axial}} = -(0.29)(0.15\ \text{mm})$$
$$= -0.0435\ \text{mm}$$

The dimensions of the plate under stress are

$$\left((1\ \text{m})\left(1000\ \frac{\text{mm}}{\text{m}}\right) + 0.15\ \text{mm}\right)$$
$$\times\left((1\ \text{m})\left(1000\ \frac{\text{mm}}{\text{m}}\right) - 0.0435\ \text{mm}\right)$$
$$= 1000.15\ \text{mm} \times 999.957\ \text{mm}$$

The answer is (B).

Equation 31.5: Shear Strain

$$\gamma = \tau / G \qquad 31.5$$

Values

material	units*	steel	aluminum	cast iron	wood (fir)
modulus of	Mpsi	11.5	3.8	6.0	0.6
rigidity, G	GPa	80.0	26.0	41.4	4.1

*Mpsi = millions of pounds per square inch

Description

Hooke's law applies also to a plane element in pure shear. For such an element, the shear stress is linearly related to the shear strain, γ, by the *shear modulus* (also known as the *modulus of rigidity*), G. Shear strain is a deflection angle measured from the vertical. In effect, a cubical element is deformed into a rhombohedron. In Eq. 31.5, shear strain is measured in radians.

Example 5

Given a shear stress, τ, of 12 000 kPa and a shear modulus, G, of 87 GPa, the shear strain is most nearly

(A) 0.73×10^{-5} rad

(B) 1.4×10^{-4} rad

(C) 2.5×10^{-4} rad

(D) 5.5×10^{-4} rad

Solution

The shear strain is

$$\gamma = \tau / G = \frac{12\,000 \text{ kPa}}{(87 \text{ GPa})\left(10^6 \frac{\text{kPa}}{\text{GPa}}\right)} = 1.38 \times 10^{-4} \text{ rad} \quad (1.4 \times 10^{-4} \text{ rad})$$

The answer is (B).

Equation 31.6: Shear Modulus

$$G = \frac{E}{2(1+\nu)} \qquad 31.6$$

Description

For an elastic, isotropic material, the modulus of elasticity, shear modulus, and Poisson's ratio are related by Eq. 31.6.

Example 6

What is most nearly the shear modulus for a material with a modulus of elasticity of 25.55 GPa and a Poisson's ratio of 0.25?

(A) 10.07 GPa

(B) 10.09 GPa

(C) 10.11 GPa

(D) 10.22 GPa

Solution

From Eq. 31.6,

$$G = \frac{E}{2(1+\nu)} = \frac{25.55 \text{ GPa}}{(2)(1+0.25)} = 10.22 \text{ GPa}$$

The answer is (D).

2. UNIAXIAL LOADING AND DEFORMATION

Equation 31.7: Deformation

$$\delta = \frac{PL}{AE} \qquad 31.7$$

Variations

$$\delta = L\varepsilon = L\left(\frac{\sigma}{E}\right)$$

$$\delta = \frac{mgL}{AE}$$

$$\delta = \sum \frac{PL}{AE} = P\sum \frac{L}{AE} \quad \left[\begin{array}{c}\text{segments differing in}\\ \text{cross-sectional}\\ \text{area or composition}\end{array}\right]$$

$$\delta = \int \frac{PdL}{AE} = P\int \frac{dL}{AE} \quad \left[\begin{array}{c}\text{one variable varies}\\ \text{continuously}\\ \text{along the length}\end{array}\right]$$

Description

The deformation, δ, of an axially loaded member of original length L can be derived from Hooke's law. Tension loading is considered to be positive; compressive loading is negative. The sign of the deformation will be the same as the sign of the loading.

When an axial member has distinct sections differing in cross-sectional area or composition, superposition is used to calculate the total deformation as the sum of individual deformations.

Example

A 10 kg mass is supported axially by an aluminum alloy pipe with an outside diameter of 10 cm and an inside diameter of 9.6 cm. The pipe is 1.2 m long. The modulus of elasticity for the aluminum alloy is 7.5×10^4 MPa. Approximately how much does the pipe compress?

(A) 0.00025 mm

(B) 0.0025 mm

(C) 0.11 mm

(D) 25 mm

Solution

The modulus of elasticity is

$$E = (7.5 \times 10^4 \text{ MPa})\left(10^6 \frac{\text{Pa}}{\text{MPa}}\right) = 7.5 \times 10^{10} \text{ Pa}$$

Calculate the pipe's deformation using Eq. 31.7.

$$\begin{aligned} \delta &= \frac{PL}{AE} = \frac{mgL}{AE} \\ &= \frac{(10 \text{ kg})\left(9.81 \dfrac{\text{m}}{\text{s}^2}\right)(1.2 \text{ m})}{\left(\dfrac{\pi}{4}\right)\left(\left(\dfrac{10 \text{ cm}}{100 \dfrac{\text{cm}}{\text{m}}}\right)^2 - \left(\dfrac{9.6 \text{ cm}}{100 \dfrac{\text{cm}}{\text{m}}}\right)^2\right)(7.5 \times 10^{10} \text{ Pa})} \\ &= 2.5 \times 10^{-6} \text{ m} \quad (0.0025 \text{ mm}) \end{aligned}$$

The answer is (B).

Equation 31.8 and Eq. 31.9: Strain Energy

$$U = W = P\delta/2 \qquad 31.8$$

$$u = U/AL = \sigma^2/2E \qquad 31.9$$

Variations

$$\text{work per volume} = \int \frac{PdL}{AL}$$

$$W = \int \sigma d\varepsilon$$

$$U = \frac{P^2 L}{2AE}$$

Description

Strain energy, also known as *internal work*, is the energy per unit volume stored in a deformed material. The strain energy is equivalent to the work done by the applied force. Simple work is calculated as the product of a force moving through a distance. For an axially loaded member below the proportionality limit, the total strain energy depends on the average force ($P/2$) and is given by Eq. 31.8.

Work per unit volume corresponds to the area under the stress-strain curve. Units are N·m/m^3, usually written as Pa (N/m^2).

The strain energy per unit volume is given by Eq. 31.9.

Example

A 25 cm long piece of elastic material is placed under a 25 000 N tensile force and elongated by 0.01 m. The material is stressed within its proportional limit. What is most nearly the elastic strain energy stored in the steel?

(A) 130 J

(B) 150 J

(C) 180 J

(D) 250 J

Solution

From Eq. 31.8, the strain energy is

$$\begin{aligned} U = W = P\delta/2 &= \frac{(25\,000 \text{ N})(0.01 \text{ m})}{2} \\ &= 125 \text{ J} \quad (130 \text{ J}) \end{aligned}$$

The answer is (A).

3. TRIAXIAL AND BIAXIAL LOADING

Triaxial loading on an infinitesimal solid element is illustrated in Fig. 31.1.

Loading is rarely confined to a single direction. All real members are three dimensional, and most experience *triaxial loading* (see Fig. 31.1). Most problems can be analyzed with two dimensions because the normal stresses in one direction are either zero or negligible. This two-dimensional loading of the member is called *plane stress* or *biaxial loading*. Positive stresses are defined as shown (see Fig. 31.2).

Biaxial loading on an infinitesimal element is illustrated in Fig. 31.2.

Figure 31.1 Sign Conventions for Positive Stresses in Three Dimensions

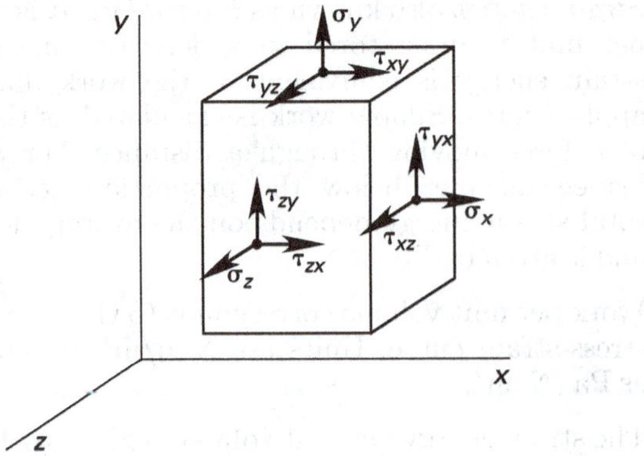

Figure 31.2 Sign Conventions for Positive Stresses in Two Dimensions

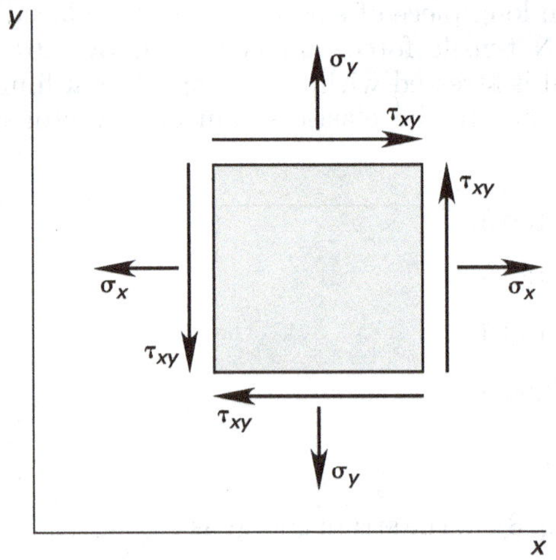

4. PRINCIPAL STRESSES

For any point in a loaded specimen, a plane can be found where the shear stress is zero. The normal stresses associated with this plane are known as the *principal stresses*, σ_a and σ_b, which are the maximum and minimum normal stresses acting at that point.

Equation 31.10 and Eq. 31.11: Maximum and Minimum Normal Stresses (Biaxial Loading)[1]

$$\sigma_a,\sigma_b = \frac{\sigma_x+\sigma_y}{2} \pm \sqrt{\left(\frac{\sigma_x-\sigma_y}{2}\right)^2 + \tau_{xy}^2} \qquad 31.10$$

$$\sigma_c = 0 \qquad 31.11$$

Variations

$$\sigma_a,\sigma_b = \frac{\sigma_x+\sigma_y}{2} \pm \tfrac{1}{2}\sqrt{(\sigma_x-\sigma_y)^2 + (2\tau_{xy})^2}$$

$$\sigma_a,\sigma_b = \tfrac{1}{2}(\sigma_x+\sigma_y) \pm \tau_{\text{max}}$$

Description

The maximum and minimum normal stresses may be found from Eq. 31.10. The maximum and minimum shear stresses (on a different plane) may be found from

$$\tau_1,\tau_2 = \pm\sqrt{\left(\frac{\sigma_x-\sigma_y}{2}\right)^2 + \tau_{xy}^2} = \pm\frac{\sigma_1-\sigma_2}{2}$$

The angles of the planes on which the principal stresses act are given by

$$\theta_{\sigma_1,\sigma_2} = \tfrac{1}{2}\arctan\frac{2\tau_{xy}}{\sigma_x-\sigma_y}$$

θ is measured from the x-axis, clockwise if positive. This equation will yield two angles, 90° apart.

Example

For the element of plane stress shown, what are most nearly the principal stresses?

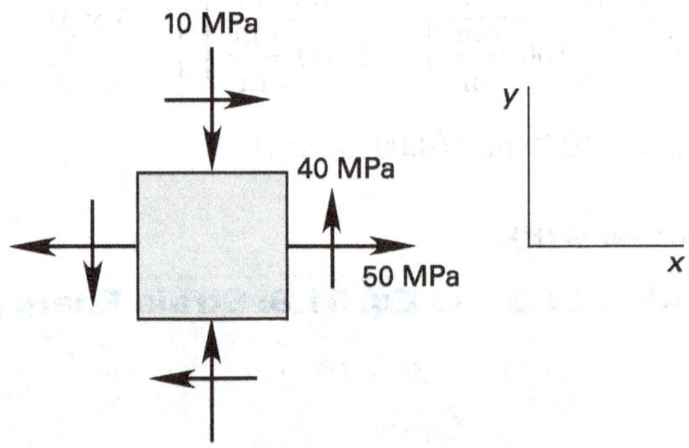

(A) $\sigma_{\text{max}} = 35$ MPa, $\sigma_{\text{min}} = -25$ MPa

(B) $\sigma_{\text{max}} = 45$ MPa, $\sigma_{\text{min}} = 55$ MPa

(C) $\sigma_{\text{max}} = 70$ MPa, $\sigma_{\text{min}} = -30$ MPa

(D) $\sigma_{\text{max}} = 85$ MPa, $\sigma_{\text{min}} = 15$ MPa

Solution

The stresses on the element are $\sigma_x = 50$ MPa (tensile), $\sigma_y = -10$ MPa (compressive), and $\tau_{xy} = 40$ MPa (positive because the direction of stress is consistent with Fig. 31.2).

[1]The NCEES *FE Reference Handbook* (*NCEES Handbook*) is inconsistent in its subscripting for principal stresses. In Eq. 31.10, Eq. 31.11, Eq. 31.14, and Eq. 31.15, lowercase letters are used (e.g., σ_a, σ_b, and σ_c). In Eq. 31.16, numbers are used (e.g., σ_1, σ_2, and σ_3). In Eq. 31.28, uppercase letters are used (e.g., σ_A, σ_B, and σ_C). In Eq. 31.31 and Eq. 31.33, the subscript a refers to an alternating stress, not to a principal stress.

From Eq. 31.10,

$$\sigma_a,\sigma_b = \frac{\sigma_x+\sigma_y}{2} \pm \sqrt{\left(\frac{\sigma_x-\sigma_y}{2}\right)^2+\tau_{xy}^2}$$

$$= \frac{50\text{ MPa}+(-10\text{ MPa})}{2} \pm \sqrt{\left(\frac{50\text{ MPa}-(-10\text{ MPa})}{2}\right)^2+(40\text{ MPa})^2}$$

$$= 20\text{ MPa} \pm 50\text{ MPa}$$

$$= 70\text{ MPa or } -30\text{ MPa}$$

The answer is (C).

5. MOHR'S CIRCLE

Mohr's circle can be constructed to graphically determine the principal normal and shear stresses. (See Fig. 31.3.) In some cases, this procedure may be faster than using the preceding equations, but a solely graphical procedure is less accurate. By convention, tensile stresses are positive; compressive stresses are negative. Clockwise shear stresses are positive; counterclockwise shear stresses are negative.

step 1: Determine the applied stresses: σ_x, σ_y, and τ_{xy}. Observe the correct sign conventions.

step 2: Draw a set of σ-τ axes.

step 3: Locate the center of the circle, point C, using Eq. 31.12.

step 4: Locate the point $p_1 = (\sigma_x, -\tau_{xy})$. (Alternatively, locate p_1' at $(\sigma_y, +\tau_{xy})$.)

step 5: Draw a line from point p_1 through the center, C, and extend it an equal distance above the σ axis to p_1'. This is the diameter of the circle.

step 6: Using the center, C, and point p_1, draw the circle. An alternative method is to draw a circle of radius R about point C (see Eq. 31.13).

step 7: Point p_2 defines the smaller principal stress, σ_b. Point p_3 defines the larger principal stress, σ_a.

step 8: Determine the angle θ as half of the angle 2θ on the circle. This angle corresponds to the larger principal stress, σ_a. On Mohr's circle, angle 2θ is measured from the $p_1 - p_1'$ line to the horizontal axis. θ (measured from the x-axis to the plane of principal stress) is in the same direction as 2θ (measured from line $p_1 - p_1'$ to the σ-axis).

step 9: The top and bottom of the circle define the largest and smallest shear stresses.

Figure 31.3 *Mohr's Circle for Stress*

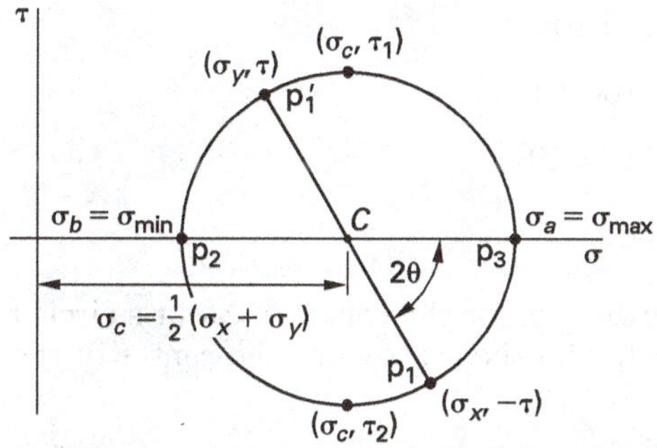

Equation 31.12 Through Eq. 31.16: Mohr's Circle Equations[2]

$$C = \frac{\sigma_x+\sigma_y}{2} \qquad 31.12$$

$$R = \sqrt{\left(\frac{\sigma_x-\sigma_y}{2}\right)^2+\tau_{xy}^2} \qquad 31.13$$

$$\sigma_a = C + R \qquad 31.14$$

$$\sigma_b = C - R \qquad 31.15$$

$$\tau_{\max} = \frac{\sigma_1-\sigma_3}{2} \qquad 31.16$$

Variation

$$R = \sqrt{\tfrac{1}{4}(\sigma_x-\sigma_y)^2+\tau_{xy}^2}$$

Description

C is the stress at the center of the circle and is found from Eq. 31.12. The radius, R, about the center C is calculated from Eq. 31.13.

Equation 31.14 and Eq. 31.15 determine the principal stresses, and Eq. 31.16 determines the maximum shear stress.

Example

A plane element has an axial stress of $\sigma_x = \sigma_y = 100$ MPa and shear stress of $\tau_{xy} = 50$ MPa. What are

[2]In defining the normal stress at point C, the *NCEES Handbook* confuses the identifier of the point, C, with the stress at the point. After referring to the component stresses σ_x and σ_y, in Eq. 31.12, the *NCEES Handbook* refers to the stress at point C as C, rather than as σ_C.

most nearly the (x, y) coordinates for the center of the Mohr's circle in MPa?

(A) $(0, 100)$

(B) $(50, 0)$

(C) $(100, 0)$

(D) $(100, 50)$

Solution

The coordinates for the center of Mohr's circle for stress are (C, τ). The shear stress at the center of the circle is $\tau = 0$.

From Eq. 31.12, the stress at the center of the circle, C, is

$$C = \frac{\sigma_x + \sigma_y}{2} = \frac{100\text{ MPa} + 100\text{ MPa}}{2} = 100\text{ MPa}$$

The answer is (C).

6. GENERAL STRAIN (THREE-DIMENSIONAL STRAIN)

Equation 31.17 Through Eq. 31.27

$$\varepsilon_x = (1/E)[\sigma_x - \nu(\sigma_y + \sigma_z)] \quad 31.17$$

$$\varepsilon_x = (1/E)(\sigma_x - \nu\sigma_y) \quad 31.18$$

$$\varepsilon_y = (1/E)[\sigma_y - \nu(\sigma_z + \sigma_x)] \quad 31.19$$

$$\varepsilon_y = (1/E)(\sigma_y - \nu\sigma_x) \quad 31.20$$

$$\varepsilon_z = (1/E)[\sigma_z - \nu(\sigma_x + \sigma_y)] \quad 31.21$$

$$\varepsilon_z = -(1/E)(\nu\sigma_x + \nu\sigma_y) \quad 31.22$$

$$\gamma_{xy} = \tau_{xy}/G \quad 31.23$$

$$\gamma_{yz} = \tau_{yz}/G \quad 31.24$$

$$\gamma_{zx} = \tau_{zx}/G \quad 31.25$$

$$\begin{Bmatrix} \sigma_x \\ \sigma_y \\ \tau_{xy} \end{Bmatrix} = \frac{E}{1-\nu^2} \begin{bmatrix} 1 & \nu & 0 \\ \nu & 1 & 0 \\ 0 & 0 & \dfrac{1-\nu}{2} \end{bmatrix} \begin{Bmatrix} \varepsilon_x \\ \varepsilon_y \\ \gamma_{xy} \end{Bmatrix} \quad 31.26$$

$$\text{uniaxial case } (\sigma_y = \sigma_z = 0): \sigma_x = E\varepsilon_x \text{ or } \sigma = E\varepsilon \quad 31.27$$

Description

Hooke's law, previously defined for axial loads and for pure shear, can be extended to three-dimensional stress-strain relationships and written in terms of the three elastic constants, E, G, and ν. Equation 31.17 through Eq. 31.27 can be used to find the stresses and strains on the differential element in Fig. 31.1.

Example

The plane element shown is acted upon by combined stresses. The material has a modulus of elasticity of 200 GPa and a Poisson's ratio of 0.27.

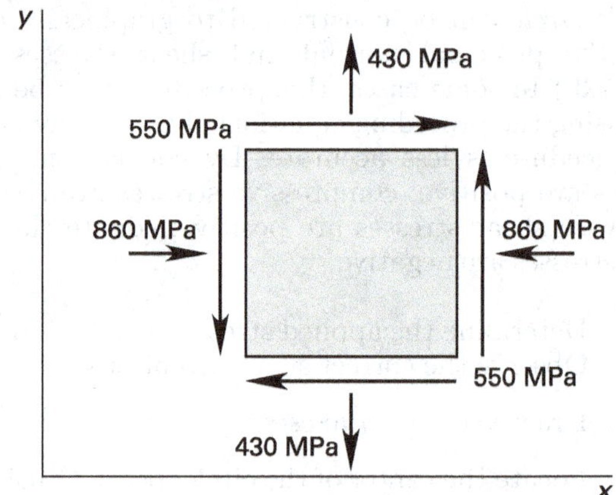

What is the approximate strain in the y-direction?

(A) -4.9×10^{-3}

(B) 0.58×10^{-3}

(C) 0.99×10^{-3}

(D) 3.3×10^{-3}

Solution

The normal stress in the y-direction is tensile, so σ_y is positive. The normal stress in the x-direction is compressive, so σ_x is negative.

The modulus of elasticity is

$$E = (200\text{ GPa})\left(1000\ \frac{\text{MPa}}{\text{GPa}}\right) = 2 \times 10^5\text{ MPa}$$

Use Eq. 31.19. The axial strain is

$$\begin{aligned}\varepsilon_y &= (1/E)[\sigma_y - \nu(\sigma_z + \sigma_x)] \\ &= \left(\frac{1}{2\times 10^5\ \text{MPa}}\right)\left(\begin{array}{l}430\ \text{MPa} \\ \quad -(0.27)\left(\begin{array}{l}0\ \text{MPa} \\ \quad -860\ \text{MPa}\end{array}\right)\end{array}\right) \\ &= 3.31\times 10^{-3} \quad (3.3\times 10^{-3})\end{aligned}$$

The answer is (D).

7. FAILURE THEORIES

Maximum Normal Stress Theory

The *maximum normal stress theory* predicts the failure stress reasonably well for brittle materials under static biaxial loading. Failure is assumed to occur if the largest tensile principal stress, σ_1, is greater than the ultimate tensile strength, or if the largest compressive principal stress, σ_2, is greater than the ultimate compressive strength. Brittle materials generally have much higher compressive than tensile strengths, so both tensile and compressive stresses must be checked.

Stress concentration factors are applicable to brittle materials under static loading. The *factor of safety*, FS, is the ultimate strength, S_u, divided by the actual stress, σ. Where a factor of safety is known in advance, the *allowable stress*, S_a, can be calculated by dividing the ultimate strength by FS.

$$\text{FS} = \frac{S_u}{\sigma}$$

$$S_a = \frac{S_u}{\text{FS}}$$

The failure criterion is

$$\sigma > \frac{S_u}{\text{FS}}$$

Maximum Shear Stress Theory

For ductile materials (e.g., steel) under static loading (the conservative *maximum shear stress theory*), shear stress can be used to predict yielding (i.e., failure). Despite the theory's name, however, loading is not limited to shear and torsion. Loading can include normal stresses as well as shear stresses. According to the maximum shear stress theory, yielding occurs when the maximum shear stress exceeds the yield strength in shear. It is implicit in this theory that the yield strength in shear is half of the tensile yield strength.

$$S_{ys} = \frac{S_{yt}}{2}$$

The maximum shear stress, τ_{max}, is the maximum of the three combined shear stresses. (For biaxial loading, only the equation for τ_{12} is used.)

$$\tau_{12} = \frac{\sigma_1 - \sigma_2}{2}$$

$$\tau_{23} = \frac{\sigma_2 - \sigma_3}{2}$$

$$\tau_{31} = \frac{\sigma_3 - \sigma_1}{2}$$

$$\tau_{\text{max}} = \max(\tau_{12}, \tau_{23}, \tau_{31})$$

The failure criterion is

$$\tau_{\text{max}} > S_{ys}$$

The factor of safety with the maximum shear stress theory is

$$\text{FS} = \frac{S_{yt}}{2\tau_{\text{max}}} = \frac{S_{ys}}{\tau_{\text{max}}}$$

Equation 31.28 Through Eq. 31.30: von Mises Stress Equations

$$\sigma' = \left(\sigma_A^2 - \sigma_A\sigma_B + \sigma_B^2\right)^{1/2} \qquad 31.28$$

$$\sigma' = \left(\sigma_x^2 - \sigma_x\sigma_y + \sigma_y^2 + 3\tau_{xy}^2\right)^{1/2} \qquad 31.29$$

$$\left[\frac{(\sigma_1-\sigma_2)^2 + (\sigma_2-\sigma_3)^2 + (\sigma_1-\sigma_3)^2}{2}\right]^{1/2} \geq S_y \qquad 31.30$$

Description

Whereas the maximum shear stress theory is conservative, the *distortion energy theory* is commonly used to accurately predict tensile and shear failure in steel and other ductile parts subjected to static loading.

The *von Mises stress* (also known as the *effective stress*), σ', is calculated for biaxial loading from the principal stresses using Eq. 31.28 or Eq. 31.29. The von Mises stress for triaxial loading is calculated from Eq. 31.30. The failure criterion is given by Eq. 31.30.

The factor of safety is

$$\text{FS} = \frac{S_{yt}}{\sigma'}$$

If the loading is pure torsion at failure (as with a shaft loaded purely in torsion), then $\sigma_1 = \sigma_2 = \pm\tau_{\text{max}}$, and $\sigma_3 = 0$. If τ_{max} is substituted for σ in Eq. 31.29 (with $\sigma_3 = 0$), an expression for the yield strength in shear is derived. The following equation predicts a larger yield strength in shear than did the maximum shear stress theory ($0.5S_{yt}$).

$$S_{ys} = \tau_{\text{max,failure}} = \frac{S_{yt}}{\sqrt{3}} = 0.577S_{yt}$$

8. VARIABLE LOADING FAILURE THEORIES

Equation 31.31 and Eq. 31.32: Modified Goodman Theory

$$\frac{\sigma_a}{S_e} + \frac{\sigma_m}{S_{ut}} \geq 1 \quad [\sigma_m \geq 0] \qquad 31.31$$

$$\frac{\sigma_{\text{max}}}{S_y} \geq 1 \qquad 31.32$$

Description

Many parts are subjected to a combination of static and reversed loadings, as illustrated in Fig. 31.4 for sinusoidal loadings. For these parts, failure cannot be determined solely by comparing stresses with the yield strength or endurance limit. The combined effects of the average stress and the amplitude of the reversal must be considered. This is done graphically on a diagram that plots the mean stress versus the alternating stresses.

Figure 31.4 *Sinusoidal Fluctuating Stress*

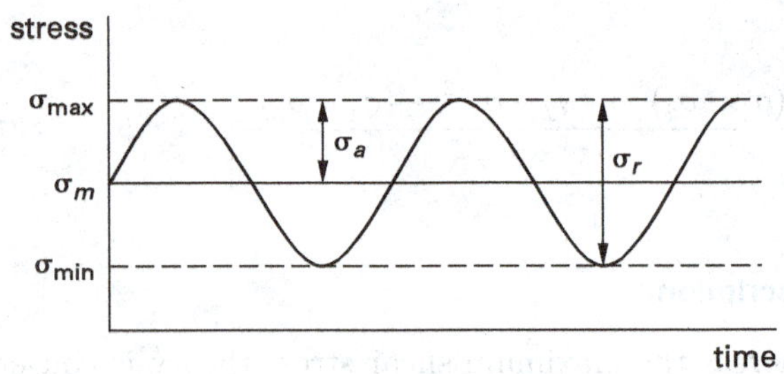

The *mean stress* is

$$\sigma_m = \frac{\sigma_{\text{max}} + \sigma_{\text{min}}}{2}$$

The *alternating stress* is half of the *range stress.*

$$\sigma_r = \sigma_{\text{max}} - \sigma_{\text{min}}$$

$$\sigma_a = \tfrac{1}{2}\sigma_r = \tfrac{1}{2}(\sigma_{\text{max}} - \sigma_{\text{min}})$$

Under the modified Goodman theory, fatigue failure will occur whenever the conditions given in Eq. 31.31 and Eq. 31.32 occur.

Equation 31.33: Soderberg Theory

$$\frac{\sigma_a}{S_e} + \frac{\sigma_m}{S_y} \geq 1 \quad [\sigma_m \geq 0] \qquad 31.33$$

Description

Under the Soderberg theory, fatigue failure will occur whenever the conditions given in Eq. 31.33 are true.

32 Thermal, Hoop, and Torsional Stress

Nomenclature

A	area	m^2
d	diameter	m
E	modulus of elasticity	MPa
F	force	N
g	gravitational acceleration, 9.81	m/s^2
G	shear modulus	MPa
J	polar moment of inertia	m^4
k	spring constant	N/m
L	length	m
p	pressure	MPa
q	shear flow	N/m
r	radius	m
t	thickness	m
T	temperature	°C
T	torque	N·m

Symbols

α	coefficient of linear thermal expansion	1/°C
δ	deformation	m
ε	strain	–
ν	Poisson's ratio	–
ρ	density	kg/m^3
σ	normal stress	MPa
τ	shear stress	MPa
ϕ	angle of twist	rad

Subscripts

0	initial
a	axial
i	inner
m	mean
o	initial or outer
t	tangential, thermal, or total
th	thermal

1. THERMAL STRESS

Equation 32.1: Coefficient of Linear Thermal Expansion[1]

$$\delta_t = \alpha L(T - T_o) \qquad 32.1$$

Values

Table 32.1 Average Coefficients of Linear Thermal Expansion

substance	1/°F	1/°C
aluminum	13.1	23.6
cast iron	6.7	12.1
steel	6.5	11.7
wood (fir)	1.7	3.0

(Multiply all values by 10^{-6}.)

Description

If the temperature of an object is changed, the object will experience length, area, and volume changes. The magnitude of these changes will depend on the *coefficient of linear thermal expansion*, α. (See Table 32.1.) The change in length in any direction is given by Eq. 32.1.

Changes in temperature affect all dimensions the same way. An increase in temperature will cause an increase in the dimensions, and likewise, a decrease in temperature will cause a decrease in the dimensions. It is a common misconception that a hole in a plate will decrease in size when the plate is heated (because the surrounding material squeezes in on the hole). In this case, the circumference of the hole is a linear dimension that follows Eq. 32.1. As the circumference increases, the hole area also increases. (See Fig. 32.1.)

If Eq. 32.1 is rearranged, an expression for the *thermal strain* is obtained.

$$\varepsilon_t = \frac{\delta_t}{L} = \alpha(T - T_0)$$

[1]The NCEES *FE Reference Handbook* (*NCEES Handbook*) uses two variables in Eq. 32.1 that are easily confused with related topics. In Eq. 32.1, T_o indicates the temperature at time zero (i.e., the initial temperature), not the outer temperature. Equation 32.1 is used to calculate the thermally induced elongation, δ_t, which is the same variable used to represent the total or tangential elongation in this subject.

Figure 32.1 Thermal Expansion of an Area

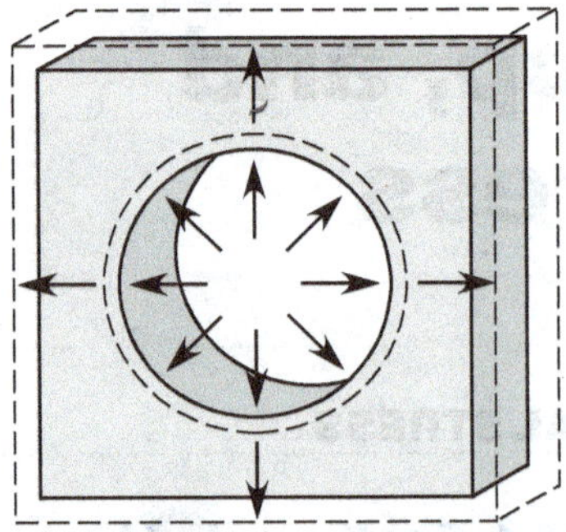

Thermal strain is handled in the same manner as strain due to an applied load. For example, if a bar is heated but is not allowed to expand, the *thermal stress* can be calculated from the thermal strain and Hooke's law.

$$\sigma_t = E\varepsilon_t$$

Low values of the coefficient of expansion, such as with Pyrex™ glassware, result in low thermally induced stresses and insensitivity to temperature extremes. Intentional differences in the coefficients of expansion of two materials are used in *bimetallic elements*, such as thermostatic springs and strips.

Example

A 30 cm long rod ($E = 3 \times 10^7$ N/cm², $\alpha = 6 \times 10^{-6}$ cm/cm·°C) with a 2 cm² cross section is fixed at both ends. If the rod is heated to 60°C above the neutral temperature, what is most nearly the stress?

(A) 110 N/cm²

(B) 11 000 N/cm²

(C) 36 000 N/cm²

(D) 57 000 N/cm²

Solution

From Eq. 32.1,

$$\delta_t = \alpha L(T - T_o)$$

$$\varepsilon_t = \frac{\delta_t}{L} = \alpha(T - T_o) = \left(6 \times 10^{-6}\ \frac{\text{cm}}{\text{cm}\cdot{}^\circ\text{C}}\right)(60^\circ\text{C} - 0^\circ\text{C})$$

$$= 0.00036$$

$$\sigma_t = E\varepsilon_t = \left(3 \times 10^7\ \frac{\text{N}}{\text{cm}^2}\right)(0.00036)$$

$$= 10\,800\ \text{N/cm}^2 \quad (11\,000\ \text{N/cm}^2)$$

The answer is (B).

Equation 32.2: Thermally Induced Normal Strain

$$\varepsilon_x = \frac{1}{E}\left[\sigma_x - \nu(\sigma_y + \sigma_z)\right] + \alpha(T_f - T_i) \qquad 32.2$$

Description

Materials that are linearly elastic and isotropic produce only normal strain, ε, when a temperature change is applied. Equation 32.2 is the generalized equation of Hooke's law in three dimensions for normal stresses when a temperature change is applied. Equation 32.2 can be rearranged for strain in the y- and z- directions. The modulus of elasticity, E, Poisson's ratio, ν, and coefficient of linear thermal expansion, α, are elastic constants that represent physical properties of the material. These relationships are valid within the linear region of the material's stress-strain response. For triaxial deformation, tension is considered positive and compression is considered negative.

Example

A steel block is subjected to triaxial loading with uniformly distributed stress equal to 20 MPa tension in the x-direction, 5 MPa compression in the y-direction, and 10 MPa compression in the z-direction. The steel has a modulus of elasticity of 200,000 MPa, a Poisson's ratio of 0.30, and a coefficient of linear thermal expansion of 11.7×10^{-6} 1/°C. Most nearly, what is the strain in the x-direction if the steel block experiences a temperature drop of 30°C?

(A) 0.000038

(B) 0.000423

(C) 0.000582

(D) 0.000651

Solution

For triaxial deformation, tension is considered positive and compression is considered negative. Using the values given and Eq. 32.2, the strain is

$$\begin{aligned}\varepsilon_x &= \frac{1}{E}\left[\sigma_x - \nu(\sigma_y + \sigma_z)\right] + \alpha(T_f - T_i)\\ &= \left(\frac{1}{200{,}000\ \text{MPa}}\right)\big(20\ \text{MPa} - (0.30)(-5\ \text{MPa} + (-10\ \text{MPa}))\big)\\ &\quad + \left(0.00001170\ \frac{\frac{\text{in}}{\text{in}}}{{}^\circ\text{C}}\right)(-30^\circ\text{C})\\ &= 0.000651\end{aligned}$$

The answer is (D).

2. CYLINDRICAL THIN-WALLED TANKS

Cylindrical tanks under internal pressure experience circumferential, longitudinal, and radial stresses. (See Fig. 32.2.) If the wall thickness is small, the radial stress component is negligible and can be disregarded. A cylindrical tank can be assumed to be a *thin-walled tank* if the ratio of thickness-to-internal radius is less than approximately 0.1.

$$\frac{t}{r_i} < 0.1 \quad \text{[thin-walled]}$$

A cylindrical tank with a wall thickness-to-radius ratio greater than 0.1 should be considered a *thick-walled pressure vessel.* In thick-walled tanks, radial stress is significant and cannot be disregarded, and for this reason, the radial and circumferential stresses vary with location through the tank wall.

Tanks under external pressure usually fail by buckling, not by yielding. For this reason, thin-wall equations cannot be used for tanks under external pressure.

Figure 32.2 *Stresses in a Thin-Walled Tank*

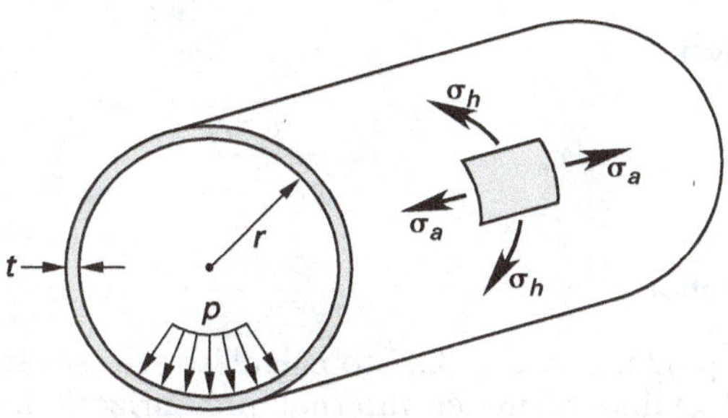

Equation 32.3 and Eq. 32.4: Hoop Stress

$$\sigma_t = \frac{p_i r}{t} \qquad 32.3$$

$$r = \frac{r_i + r_o}{2} \qquad 32.4$$

Variation

$$\sigma_t = \frac{pd}{2t}$$

Description

The *hoop stress*, σ_t, also known as *circumferential stress* and *tangential stress*, for a cylindrical thin-walled tank under internal pressure, p, is derived from the free-body diagram of a cylinder. If the cylinder tank is truly thin walled, it is not important which radius, r (e.g., inner, mean, or outer), is used in Eq. 32.3. Although the inner diameter is used by common convention, the mean diameter will provide more accurate values as the wall thickness increases. The hoop stress is given by Eq. 32.3, where r is the radius as given by Eq. 32.4.

Example

The pressure gauge in an air cylinder reads 850 kPa. The cylinder is constructed of 6 mm rolled plate steel with an internal radius of 0.175 m. What is most nearly the tangential stress in the tank?

(A) 2.1 MPa

(B) 12 MPa

(C) 17 MPa

(D) 25 MPa

Solution

Tangential stress is the same as hoop stress. Use Eq. 32.3.

$$\sigma_t = \frac{p_i r}{t} = \frac{(850\ \text{kPa})\left(10^3\ \frac{\text{Pa}}{\text{kPa}}\right)\left(10^3\ \frac{\text{mm}}{\text{m}}\right)(0.175\ \text{m})}{6\ \text{mm}}$$
$$= 2.479 \times 10^7\ \text{Pa} \quad (25\ \text{MPa})$$

The answer is (D).

Equation 32.5: Axial Stress

$$\sigma_a = \frac{p_i r}{2t} \qquad 32.5$$

Variation

$$\sigma_a = \frac{F}{A} = \frac{pd}{4t} = \frac{\sigma_t}{2}$$

Description

When the cylindrical tank is closed at the ends like a soft drink can, the axial force on the ends produces a stress directed along the longitudinal axis known as the *longitudinal*, *long*, or *axial stress*, σ_a.

Example

A small cylindrical pressure tank has an internal gauge pressure of 1600 Pa. The inside diameter is 69 mm, and the wall thickness is 3 mm. What is most nearly the axial stress of the tank?

(A) 7700 Pa

(B) 9200 Pa

(C) 11 000 Pa

(D) 18 000 Pa

Solution

Check the ratio of wall thickness to radius.

$$\frac{t}{r} \approx \frac{3\ \text{mm}}{\frac{69\ \text{mm}}{2}} = 0.087$$

Since $t/r < 0.1$, this can be evaluated as a thin-walled tank.

The axial tensile stress is

$$\sigma_a = \frac{p_i r}{2t} = \frac{(1600\ \text{Pa})\left(\frac{69\ \text{mm}}{2}\right)}{(2)(3\ \text{mm})}$$
$$= 9200\ \text{Pa}$$

The answer is (B).

Principal Stresses in Tanks

The hoop and axial stresses are the principal stresses for pressure vessels when internal pressure is the only loading. It is not necessary to use the combined stress equations. If a three-dimensional portion of the shell is considered, the stress on the outside surface is zero. For this reason, the largest shear stress in three dimensions is $\sigma_h/2$ and is oriented at 45° to the surface.

Thin-Walled Spherical Tanks

Because of symmetry, the surface (tangential) stress of a spherical tank is the same in all directions.

$$\sigma = \frac{pd}{4t} = \frac{pr}{2t}$$

3. THICK-WALLED PRESSURE VESSELS

A thick-walled cylinder has a wall thickness-to-radius ratio greater than 0.1. In thick-walled tanks, radial stress is significant and cannot be disregarded. In *Lame's solution*, a thick-walled cylinder is assumed to be made up of thin laminar rings. This method shows that the radial and tangential (circumferential or hoop) stresses vary with location within the tank wall.

At every point in the cylinder, the tangential, radial, and long stresses are the principal stresses. Unless an external torsional shear stress is added, it is not necessary to use the combined stress equations.

The maximum radial, tangential, and shear stresses occur at the inner surface for both internal and external pressurization. (The terms *tangential stress* and *circumferential stress* are preferred over *hoop stress* when dealing with thick-walled cylinders.) Compressive stresses are negative.

Equation 32.6 Through Eq. 32.8: Thick-Walled Cylinder with Internal Pressurization

$$\sigma_t = p_i \frac{r_o^2 + r_i^2}{r_o^2 - r_i^2} \qquad 32.6$$

$$\sigma_r = -p_i \qquad 32.7$$

$$\sigma_a = p_i \frac{r_i^2}{r_o^2 - r_i^2} \qquad 32.8$$

Variation

$$\sigma_{\text{axial}} = \frac{F}{A} = \frac{p_i \pi r_i^2}{\pi(r_o^2 - r_i^2)}$$

Description

Use Eq. 32.6 and Eq. 32.7 to calculate stresses in thick-walled cylinders under internal pressurization. Cylinders under internal pressurization will also experience an axial stress in the direction of the end caps (see Eq. 32.8 and the variation equation). This axial stress is calculated as the axial force divided by the annular area of the wall material.

Equation 32.9 and Eq. 32.10: Thick-Walled Cylinder with External Pressurization

$$\sigma_t = -p_o \frac{r_o^2 + r_i^2}{r_o^2 - r_i^2} \qquad 32.9$$

$$\sigma_r = -p_o \qquad 32.10$$

Description

Equation 32.9 and Eq. 32.10 are used for cylinders with external pressurization.

4. TORSIONAL STRESS

Equation 32.11 Through Eq. 32.13: Shafts

$$\tau = \frac{Tr}{J} \quad [t > 0.1r] \qquad \text{32.11}$$

$$J = \pi r^4/2 \qquad \text{32.12}$$

$$J = \pi(r_a^4 - r_b^4)/2 \qquad \text{32.13}$$

Variations

$$J = \frac{\pi d^4}{32} \quad \text{[solid]}$$

$$J = \frac{\pi}{32}(d_o^4 - d_i^4) \quad \text{[hollow]}$$

Description

Shear stress occurs when a shaft is placed in *torsion.* The shear stress at the outer surface of a bar of radius r, which is torsionally loaded by a torque, T, is calculated from Eq. 32.11.

The *polar moment of inertia*, J, of a solid round shaft is found from Eq. 32.12. For a hollow shaft, use Eq. 32.13.

Example

A solid steel shaft of 200 mm diameter experiences a torque of 135.6 kN·m. What is most nearly the maximum shear stress in the shaft?

(A) 86 MPa

(B) 110 MPa

(C) 160 MPa

(D) 190 MPa

Solution

Calculate the maximum shear stress using Eq. 32.11.

$$\tau = \frac{Tr}{J} = \frac{Tr}{\pi r^4/2} = \frac{2T}{\pi r^3} = \frac{(2)(135.6 \text{ kN·m})}{\pi\left(\dfrac{200 \text{ mm}}{(2)\left(1000 \ \dfrac{\text{mm}}{\text{m}}\right)}\right)^3}$$

$$= 86\,326 \text{ kPa} \quad (86 \text{ MPa})$$

The answer is (A).

Equation 32.14 and Eq. 32.15: Angle of Twist

$$\phi = \int_0^L \frac{T}{GJ}\,dz = \frac{TL}{GJ} \qquad \text{32.14}$$

$$T = G(d\phi/dz)\int_A r^2 dA = GJ(d\phi/dz) \qquad \text{32.15}$$

Description

If a shaft of length L carries a torque T, the angle of twist (in radians) can be found from Eq. 32.14.

Example

An aluminum bar 17 m long and 0.6 m in diameter is acted upon by an 11 kN·m torque. The shear modulus of elasticity, G, is 26 GPa. Neglect bending.

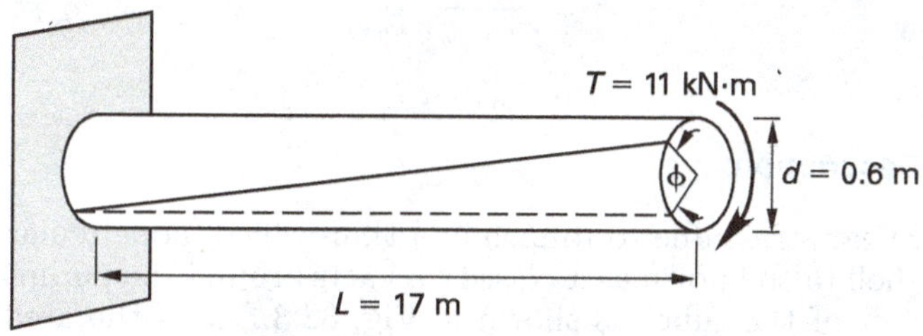

What is most nearly the angle of twist, ϕ, for the aluminum bar?

(A) 0.000 57°

(B) 0.0057°

(C) 0.032°

(D) 0.082°

Solution

From Eq. 32.14, and using the relationship $J = \pi d^4/32$ for a solid shaft, the angle of twist is

$$\phi = \frac{TL}{GJ} = \frac{TL}{G\left(\dfrac{\pi d^4}{32}\right)}$$

$$= \frac{(11 \text{ kN·m})\left(10^3 \ \dfrac{\text{N}}{\text{kN}}\right)(17 \text{ m})\left(\dfrac{180°}{\pi \text{ rad}}\right)}{(26 \text{ GPa})\left(10^9 \ \dfrac{\text{Pa}}{\text{GPa}}\right)\left(\dfrac{\pi(0.6 \text{ m})^4}{32}\right)}$$

$$= 0.032°$$

The answer is (C).

Equation 32.16: Torsional Stiffness

$$k_t = GJ/L \qquad 32.16$$

Variation

$$k_t = \frac{T}{\phi}$$

Description

The *torsional stiffness* (*torsional spring constant* or *twisting moment per radian of twist*) is given by Eq. 32.16.

Equation 32.17: Torsion in Hollow, Thin-Walled Shells

$$\tau = \frac{T}{2A_m t} \qquad 32.17$$

Description

Shear stress due to torsion in a thin-walled, noncircular shell (also known as a *closed box*) acts around the perimeter of the tube, as shown in Fig. 32.3. A_m is the area enclosed by the centerline of the shell. The shear stress, τ, is given by Eq. 32.17.

Figure 32.3 *Torsion in Thin-Walled Shells*

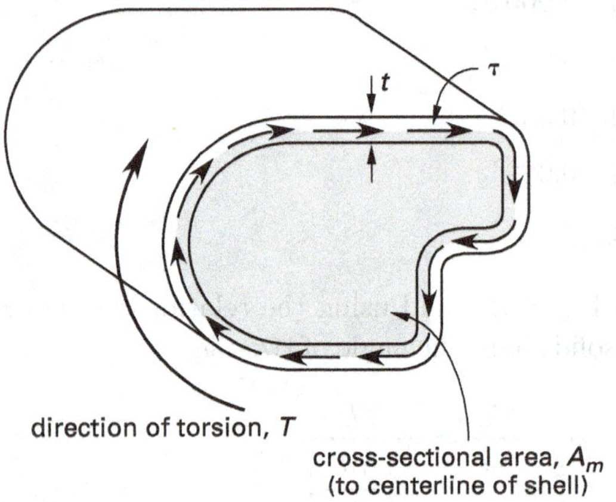

The shear stress at any point is not proportional to the distance from the centroid of the cross section. Rather, the *shear flow*, q, around the shell is constant, regardless of whether the wall thickness is constant or variable. The shear flow is the shear per unit length of the centerline path. At any point where the shell thickness is t,

$$q = \tau t = \frac{T}{2A_m} \quad \text{[constant]}$$

Example

A hollow, thin-walled shell has a wall thickness of 12.5 mm. The shell is acted upon by a 280 N·m torque.

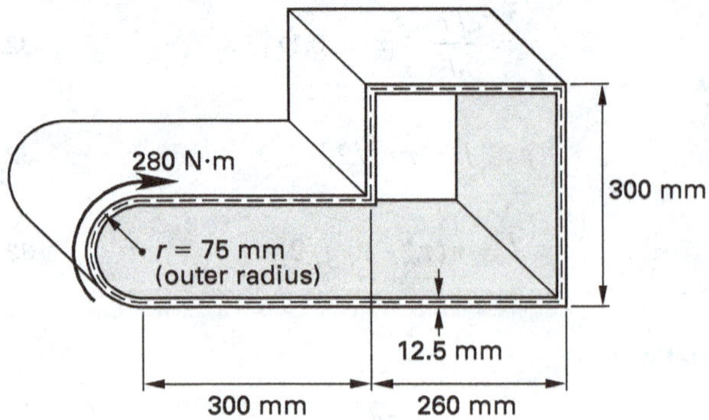

Find the approximate torsional shear stress in the shell's wall.

(A) 14 kPa

(B) 44 kPa

(C) 59 kPa

(D) 92 kPa

Solution

The enclosed area to the centerline of the shell is

$$A = \frac{\begin{gathered}(300 \text{ mm} - 12.5 \text{ mm})\left(260 \text{ mm} - \frac{12.5 \text{ mm}}{2}\right) \\ +((2)(75 \text{ mm}) - 12.5 \text{ mm})(300 \text{ mm}) \\ +\left(\frac{\pi}{2}\right)\left(75 \text{ mm} - \frac{12.5 \text{ mm}}{2}\right)^2\end{gathered}}{\left(1000 \ \frac{\text{mm}}{\text{m}}\right)^2}$$

$$= 0.1216 \text{ m}^2$$

The torsional shear stress is

$$\tau = \frac{T}{2A_m t} = \frac{280 \text{ N·m}}{(2)(0.1216 \text{ m}^2)\left(\frac{12.5 \text{ mm}}{1000 \ \frac{\text{mm}}{\text{m}}}\right)}$$

$$= 92\,084 \text{ Pa} \quad (92 \text{ kPa})$$

The answer is (D).

33 Beams

Nomenclature

A	area	m^2
b	width	m
c	distance from neutral axis to extreme fiber	m
C	couple	N·m
d	depth	m
d	distance	m
e	eccentricity	m
E	modulus of elasticity	MPa
F	force	N
g	gravitational acceleration, 9.81	m/s^2
h	height	m
I	moment of inertia	m^4
L	length	m
m	mass per unit length	kg/m
M	moment	N·m
n	modular ratio	–
P	load	N
q	shear flow	N/m
Q	statical moment	m^3
r	radius	m
R	reaction	N
s	elastic section modulus	m^3
t	thickness	m
v	deflection	m
V	shear	N
w	load	N
w	load per unit length	N/m
x	distance	m
y	deflection	m
y	depth	m
y'	slope	–

Symbols

ε	axial strain	–
θ	angle	deg
ρ	radius of curvature	m
σ	normal stress	MPa
τ	shear stress	MPa
ϕ	angle	deg

Subscripts

b	bending
c	centroidal
o	original
t	transformed
T	transformed
x	in x-direction
y	in y-direction

1. SHEARING FORCE AND BENDING MOMENT

Sign Conventions

The internal *shear* at a section is the sum of all shearing (e.g., vertical) forces acting on an object up to that section. It has units of pounds, kips, newtons, and so on. Shear is not the same as shear stress, since the area of the object is not considered.

The most typical application is shear, V, at a section on a horizontal beam defined as the sum of all vertical forces between the section and one of the ends. The direction (i.e., to the left or right of the section) in which the summation proceeds is not important. Since the values of shear will differ only in sign for summation to the left and right ends, the direction that results in the fewest calculations should be selected.

$$V = \sum_{\substack{\text{section to}\\ \text{one end}}} F_i$$

For beams, shear is positive when there is a net upward force to the left of a section, and it is negative when there is a net downward force to the left of the section. (See Fig. 33.1.)

Figure 33.1 Shear Sign Conventions

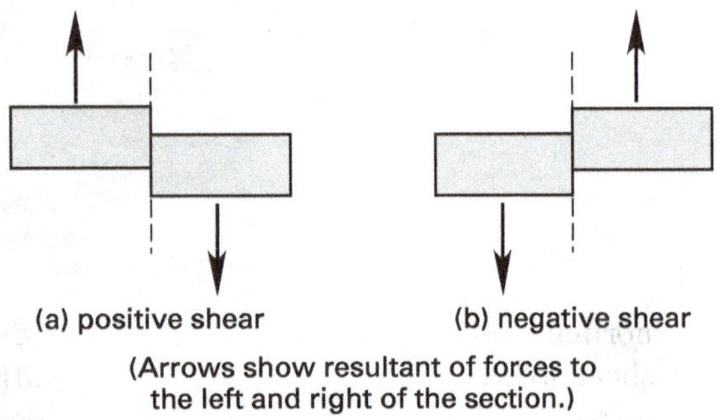

The *moment*, M, will be the algebraic sum of all moments and couples located between the section and one of the ends.

$$M = \underset{\text{section to one end}}{\sum F_i x_i} + \underset{\text{section to one end}}{\sum C_i}$$

Moments in a beam are positive when the upper surface of the beam is in compression and the lower surface is in tension. Positive moments cause lengthening of the lower surface and shortening of the upper surface. A useful image with which to remember this convention is to imagine the beam smiling when the moment is positive. (See Fig. 33.2.)

Figure 33.2 Bending Moment Sign Conventions

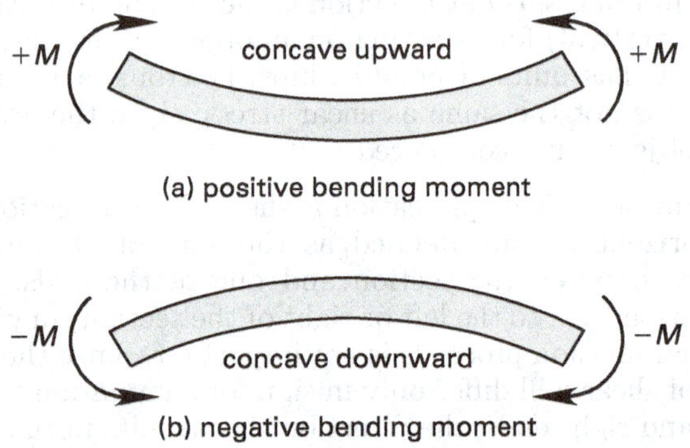

Equation 33.1 and Eq. 33.2: Change in Shear Magnitude

$$V_2 - V_1 = \int_{x_1}^{x_2} [-w(x)]\,dx \qquad 33.1$$

$$w(x) = -\frac{dV(x)}{dx} \qquad 33.2$$

Description

The change in magnitude of the shear between two points is the integral of the *load function*, $w(x)$, or the area under the load diagram between those points.

Equation 33.3 and Eq. 33.4: Change in Moment Magnitude

$$M_2 - M_1 = \int_{x_1}^{x_2} V(x)\,dx \qquad 33.3$$

$$V = \frac{dM(x)}{dx} \qquad 33.4$$

Description

The change in magnitude of the moment between two points is equal to the integral of the *shear function*, $V(x)$, or the area under the shear diagram between those points.

Shear and Moment Diagrams

Both shear and moment can be described mathematically for simple loadings, but the formulas become discontinuous as the loadings become more complex. It is more convenient to describe complex shear and moment functions graphically. Graphs of shear and moment as functions of position along the beam are known as *shear and moment diagrams*. The following guidelines and conventions should be observed when constructing a *shear diagram*.

- The shear at any section is equal to the sum of the loads and reactions from the section to the left end.
- The magnitude of the shear at any section is equal to the slope of the moment function at that section.
- Loads and reactions acting upward are positive.
- The shear diagram is straight and sloping for uniformly distributed loads.
- The shear diagram is straight and horizontal between concentrated loads.
- The shear is undefined at points of concentrated loads.

The following guidelines and conventions should be observed when constructing a *moment diagram*. By convention, the moment diagram is drawn on the compression side of the beam.

- The moment at any section is equal to the sum of the moments and couples from the section to the left end.
- The change in magnitude of the moment at any section is the integral of the shear diagram, or the area under the shear diagram. A concentrated moment will produce a jump or discontinuity in the moment diagram.
- The maximum or minimum moment occurs where the shear is either zero or passes through zero.

- The moment diagram is parabolic and is curved downward (i.e., is convex) for downward uniformly distributed loads.

Example

The shear diagram for a simply supported beam is as shown.

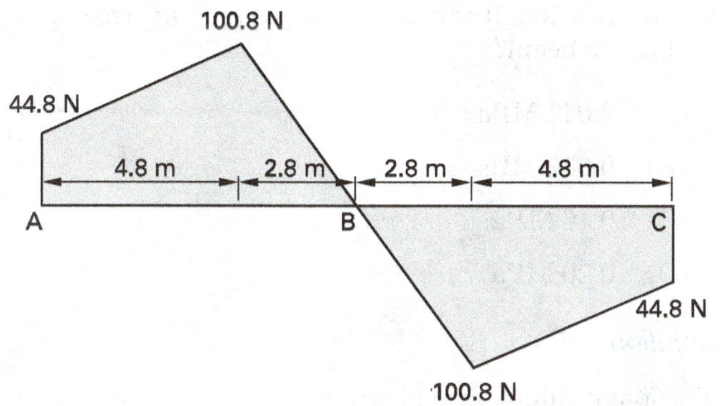

What is most nearly the maximum moment in the beam?

(A) 100 N·m

(B) 430 N·m

(C) 490 N·m

(D) 740 N·m

Solution

The maximum moment occurs at point B where the shear is zero.

$$\begin{aligned} M_B &= (44.8\ \text{N})(4.8\ \text{m}) \\ &\quad + \left(\frac{1}{2}\right)(100.8\ \text{N} - 44.8\ \text{N})(4.8\ \text{m}) \\ &\quad + \left(\frac{1}{2}\right)(100.8\ \text{N})(2.8\ \text{m}) \\ &= 491\ \text{N·m} \quad (490\ \text{N·m}) \end{aligned}$$

The answer is (C).

2. STRESSES IN BEAMS

Equation 33.5 Through Eq. 33.8: Bending Stress[1]

$$\sigma_x = -My/I \qquad 33.5$$

$$\sigma_{x,\text{max}} = \pm Mc/I \qquad 33.6$$

$$s = I/c \qquad 33.7$$

$$\sigma_{x,\text{max}} = -M/s \qquad 33.8$$

Description

Normal stress is distributed triangularly in a bending beam as shown in Fig. 33.3. Although it is a normal stress, the term *bending stress* or *flexural stress* is used to indicate the source of the stress. For a positive bending moment, the lower surface of the beam experiences tensile stress while the upper surface of the beam experiences compressive stress. The bending stress distribution passes through zero at the centroid, or *neutral axis*, of the cross section. The distance from the neutral axis is y, and the distance from the neutral axis to the *extreme fiber* (i.e., the top or bottom surface most distant from the neutral axis) is c.

Figure 33.3 *Bending Stress Distribution at a Section in a Beam*

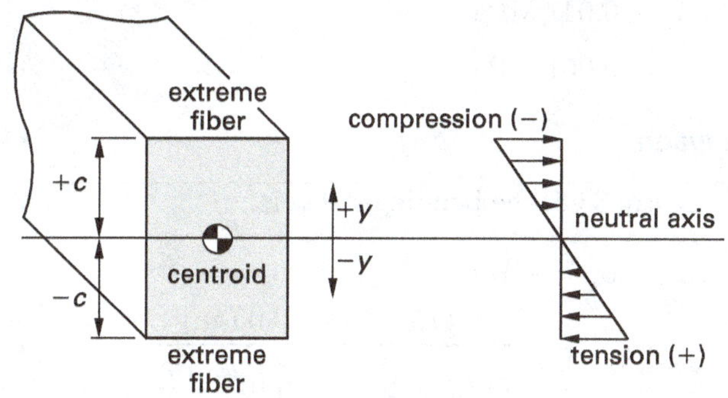

Bending stress varies with location (depth) within the beam. It is zero at the neutral axis, and increases linearly with distance from the neutral axis, as predicted by Eq. 33.5. In Eq. 33.5, I is the centroidal area moment of inertia of the beam. The negative sign in Eq. 33.5, required by the convention that compression is negative, is commonly omitted.

Since the maximum stress will govern the design, y can be set equal to c to obtain the extreme fiber stress. Equation 33.6 shows that the maximum bending stress will occur at the section where the moment is maximum.

For standard structural shapes, I and c are fixed. For design, the *elastic section modulus*, s, given by Eq. 33.7, is often used.

[1]The NCEES *FE Reference Handbook* (*NCEES Handbook*) presents Eq. 33.6 as "σ_x," without "max." While the moment on the beam can vary with the location, x, the maximum stress for any particular location will be derived when y is maximum (i.e., when $y = c$).

Mechanics of Materials

Example 20

For the beam shown, the moment, M, at the cross section is 15.7 N·m, and the moment of inertia, I, is 1.91×10^{-4} m^4.

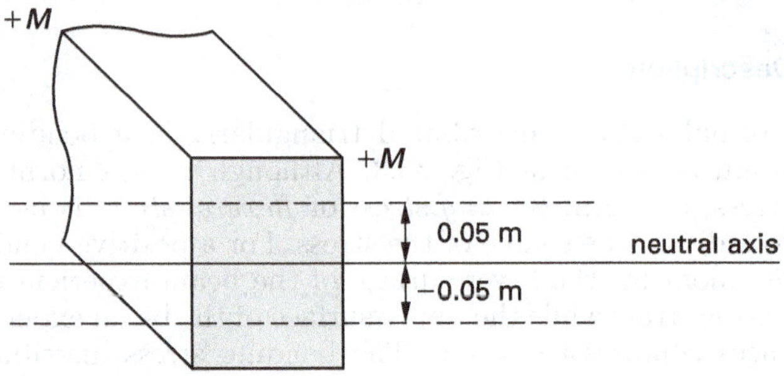

The bending stress that the beam experiences 0.05 m above the neutral axis is most nearly

(A) −0.0051 MPa

(B) −0.0041 MPa

(C) −0.041 MPa

(D) −0.051 MPa

Solution

Using Eq. 33.5, the bending stress is

$$\begin{aligned}\sigma_x &= -My/I\\ &= -\frac{(15.7\ \text{N·m})(0.05\ \text{m})}{(1.91\times 10^{-4}\ \text{m}^4)\left(10^6\ \frac{\text{Pa}}{\text{MPa}}\right)}\\ &= -0.0041\ \text{MPa}\quad [\text{compression}]\end{aligned}$$

The answer is (B).

Stresses in Beams

Hooke's law is valid for any point within a beam, and any distance y from the neutral axis.

$$\sigma = E\varepsilon$$

At any point, x, a loaded beam that is oriented with its longitudinal axis parallel to the x-direction will have an instantaneous radius of curvature of ρ and an instantaneous strain in the x-direction of ε_x.

$$\varepsilon_x = -y/\rho$$

$$\sigma_x = -Ey/\rho$$

$$\frac{1}{\rho} = \frac{\varepsilon_{\max}}{c} = \frac{d^2y}{dx^2} = \frac{d\theta}{dx} = \frac{M}{EI}$$

$$\varepsilon_{\max} = \frac{c}{\rho}$$

Example 21

At a particular point within a beam, the longitudinal strain is 0.000284 and Poisson's ratio is 0.29. The modulus of elasticity of the beam is 200 MPa. What is most nearly the longitudinal normal stress at that point within the beam?

(A) 0.042 MPa

(B) 0.057 MPa

(C) 0.16 MPa

(D) 0.20 MPa

Solution

The longitudinal normal stress is

$$\begin{aligned}\sigma &= E\varepsilon = (200\ \text{MPa})(0.000284)\\ &= 0.0568\ \text{MPa}\quad (0.057\ \text{MPa})\end{aligned}$$

The answer is (B).

Equation 33.9: Centroidal Moment of Inertia

$$I_{x_c} = bh^3/12 \qquad 33.9$$

Description

Equation 33.9 is the centroidal moment of inertia for a rectangular $b \times h$ section. The section modulus for a rectangular $b \times h$ section is

$$s_{\text{rectangular}} = \frac{bh^2}{6}$$

Example 22

What is most nearly the moment of inertia of a rectangular 46 cm × 61 cm beam installed in its strongest vertical orientation?

(A) 4.9×10^{-3} m^4

(B) 6.1×10^{-3} m^4

(C) 8.7×10^{-3} m^4

(D) 3.5×10^{-2} m^4

Solution

Using Eq. 33.9,

$$I_{x_c} = bh^3/12 = \frac{(46 \text{ cm})(61 \text{ cm})^3}{(12)\left(100 \ \frac{\text{cm}}{\text{m}}\right)^4} = 8.7 \times 10^{-3} \text{ m}^4$$

The answer is (C).

Equation 33.10: Shear Stress

$$\tau_{xy} = VQ/Ib \qquad 33.10$$

Variations

$$\tau_{\text{max,rectangular}} = \frac{3V}{2A} = \frac{3V}{2bh} = 1.5\tau_{\text{ave}}$$

$$\tau_{\text{max,circular}} = \frac{4V}{3A} = \frac{4V}{3\pi r^2}$$

$$\tau_{\text{ave}} = \frac{V}{A_{\text{web}}} = \frac{V}{dt_{\text{web}}} \quad \left[\begin{array}{c}\text{web shear stress; steel beam} \\ \text{with web thickness} \\ t_{\text{web}} \text{ and depth } d\end{array}\right]$$

Description

The shear stresses in a vertical section of a beam consist of both horizontal and transverse (vertical) shear stresses.

The exact value of shear stress is dependent on the location, y, within the depth of the beam. The shear stress distribution is given by Eq. 33.10. The shear stress is zero at the top and bottom surfaces of the beam. For a regular shaped beam, the shear stress is maximum at the neutral axis. (See Fig. 33.4.)

***Figure 33.4** Dimensions for Shear Stress Calculations*

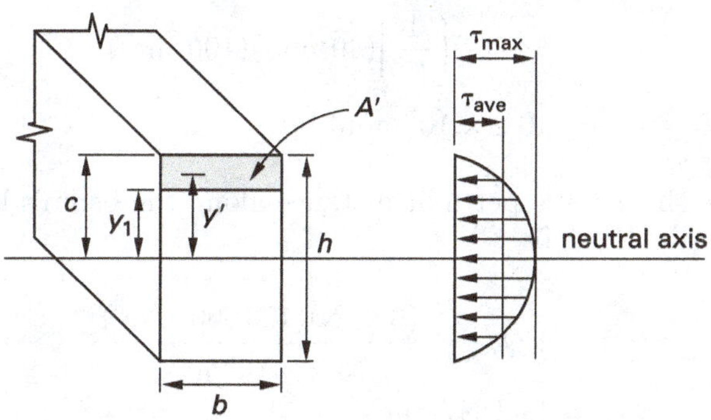

In Eq. 33.10, I is the area moment of inertia, and b is the width or thickness of the beam at the depth, y, within the beam where the shear stress is to be found.

The variation equations give simplifications of Eq. 33.10 for rectangular beams, beams with circular cross sections, and steel beams with web thicknesses and depths shown in Fig. 33.5, respectively.

***Figure 33.5** Dimensions of a Steel Beam*

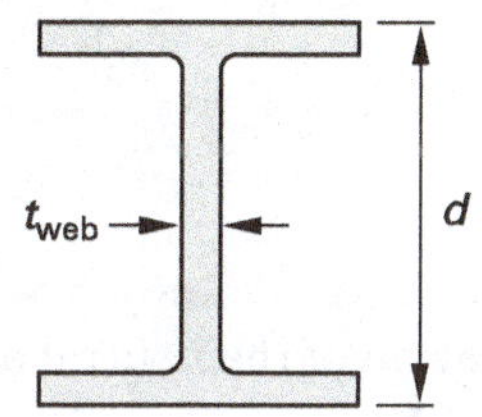

Equation 33.11: Moment of the Area

$$Q = A'\bar{y}' \qquad 33.11$$

Variation

$$Q = \int_{y_1}^{c} y\,dA \quad \left[\begin{array}{c}\text{first moment of the area with} \\ \text{respect to neutral axis}\end{array}\right]$$

Description

The *first* (or *statical*) *moment of the area* of the beam with respect to the neutral axis, Q, is defined by the variation equation.

For rectangular beams, $dA = bdy$. Then, the moment of the area, A', above layer y is equal to the product of the area and the distance from the centroidal axis to the centroid of the area.

Example

A composite cross section is made up of two identical members—horizontally oriented member A and vertically oriented member B—as shown. The neutral axis passes through member B.

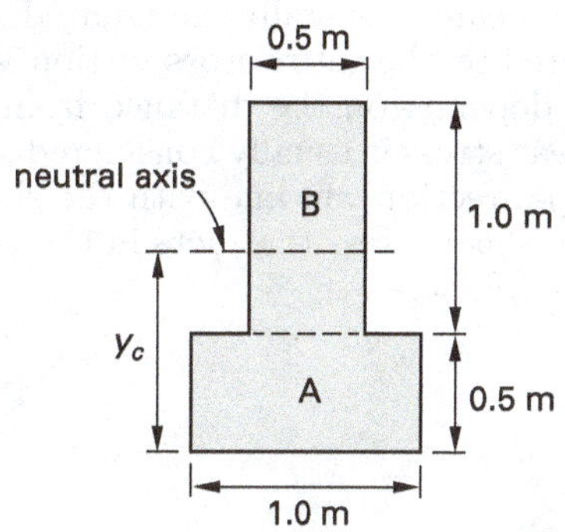

What is most nearly the first moment of the area for member A?

(A) 0.12 m^3

(B) 0.19 m^3

(C) 0.24 m^3

(D) 0.31 m^3

Solution

Determine the location of the neutral axis.

$$y_c = \frac{\sum y_i A_i}{\sum A_i} = \frac{\left(\frac{0.5 \text{ m}}{2}\right)(1.0 \text{ m})(0.5 \text{ m}) + \left(0.5 \text{ m} + \frac{1.0 \text{ m}}{2}\right)(0.5 \text{ m})(1.0 \text{ m})}{(2)(0.5 \text{ m})(1.0 \text{ m})}$$
$$= 0.625 \text{ m}$$

Use Eq. 33.11. The moment of the area is

$$Q = A'\bar{y}' = (1.0 \text{ m})(0.5 \text{ m})\left(0.625 \text{ m} - \frac{0.5 \text{ m}}{2}\right)$$
$$= 0.1875 \text{ m}^3 \quad (0.19 \text{ m}^3)$$

The answer is (B).

Equation 33.12: Shear Flow

$$q = VQ/I \qquad 33.12$$

Description

The *shear flow*, q, is the shear per unit length. In Eq. 33.12, the vertical shear, V, is a function of location, x, along the beam, generally designated as V_x. This shear is resisted by the entire cross section, although the shear stress depends on the distance from the neutral axis. The shear stress is usually considered to be vertical (i.e., in the y-direction), in line with the shearing force, but the same shear stress that acts in the y-z plane also acts on the x-z plane.

Example

An I-beam is made of three planks, each 20 mm × 100 mm in cross section, nailed together with a single row of nails on top and bottom as shown.

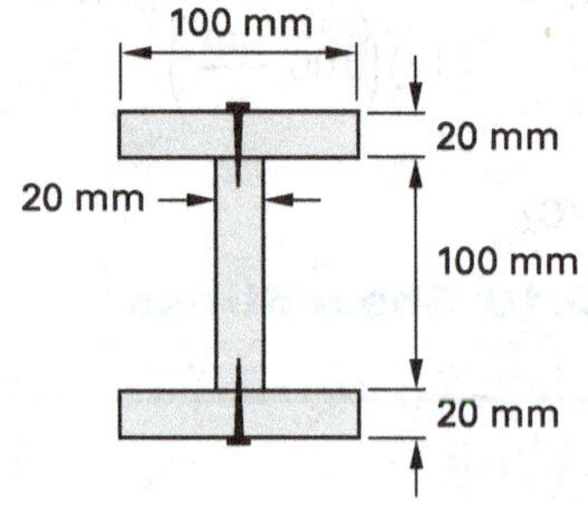

If the longitudinal spacing between the nails is 25 mm, and the vertical shear force acting on the cross section is 600 N, what is most nearly the shear per nail?

(A) 56 N

(B) 76 N

(C) 110 N

(D) 160 N

Solution

The statical moment of each flange is

$$Q = A'\bar{y}' = (100 \text{ mm})(20 \text{ mm})\left(\frac{100 \text{ mm}}{2} + \frac{20 \text{ mm}}{2}\right)$$
$$= 120\,000 \text{ mm}^3$$

The moment of inertia is

$$I = \frac{bh^3}{12}$$
$$= \left(\frac{1}{12}\right)(100 \text{ mm})(140 \text{ mm})^3 - (2)\left(\frac{1}{12}\right)(40 \text{ mm})(100 \text{ mm})^3$$
$$= 16.2 \times 10^6 \text{ mm}^4$$

The shear force per unit distance along the beam's longitudinal axis is

$$q = VQ/I = \frac{(600 \text{ N})(120\,000 \text{ mm}^3)}{16.2 \times 10^6 \text{ mm}^4}$$
$$= 4.44 \text{ N/mm}$$

The shear per nail spaced at d is

$$\begin{aligned} F = qd &= \left(4.44\ \frac{\text{N}}{\text{mm}}\right)(25\ \text{mm}) \\ &= 111\ \text{N} \quad (110\ \text{N}) \end{aligned}$$

The answer is (C).

3. DEFLECTION OF BEAMS

Equation 33.13 Through Eq. 33.15: Deflection and Slope Relationships

$$EI\frac{d^2y}{dx^2} = M \tag{33.13}$$

$$EI\frac{d^3y}{dx^3} = dM(x)/dx = V \tag{33.14}$$

$$EI\frac{d^4y}{dx^4} = dV(x)/dx = -w \tag{33.15}$$

Variations

$$y' = \frac{dy}{dx} = \text{slope}$$

$$y'' = \frac{d^2y}{dx^2} = \frac{M(x)}{EI}$$

$$y''' = \frac{d^3y}{dx^3} = \frac{V(x)}{EI}$$

$$y'''' = \frac{d^4y}{dx^4} = \frac{w(x)}{EI}$$

Description

The deflection, y, and slope, y', of a loaded beam are related to the moment $M(x)$, shear $V(x)$, and load $w(x)$ by Eq. 33.13 through Eq. 33.15 and the variation equations.

Equation 33.16 and Eq. 33.17: Deflection on a Beam Section

$$EI(dy/dx) = \int M(x)\,dx \tag{33.16}$$

$$EIy = \int\left[\int M(x)\,dx\right]dx \tag{33.17}$$

Description

If the moment function, $M(x)$, is known for a section of the beam, the deflection at any point on that section can be found from Eq. 33.16. The constants of integration are determined from the beam boundary conditions shown in Table 33.1.

Table 33.1 *Beam Boundary Conditions*

end condition	y	y'	y''	V	M
simple support	0				0
built-in support	0	0			
free end			0	0	0
hinge					0

Example

A beam of length L carries a concentrated load, P, at point C.

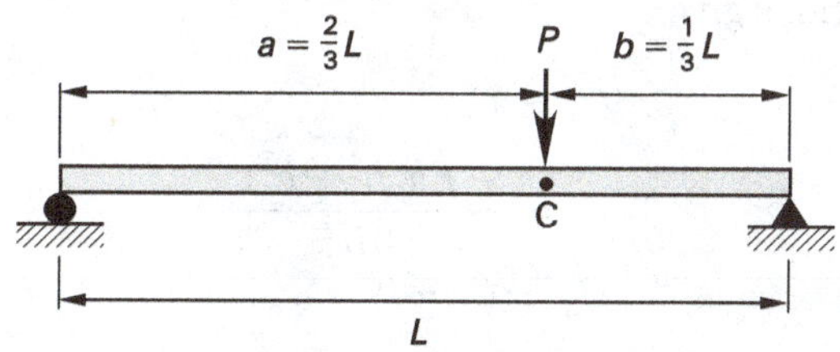

Determine the deflection at point C in terms of P, L, E, and I, where E is the modulus of elasticity, and I is the moment of inertia.

(A) $\dfrac{2PL^3}{243EI}$

(B) $\dfrac{4PL^3}{243EI}$

(C) $\dfrac{PL^3}{27EI}$

(D) $\dfrac{PL^3}{9EI}$

Solution

The equation for bending moment in the beam is

$$EI\frac{d^2y}{dx^2} = M$$

Computing M for the different beam sections,

$$EI\frac{d^2y}{dx^2} = \frac{Pbx}{L} \quad [0 \le x \le a]$$

$$EI\frac{d^2y}{dx^2} = \frac{Pbx}{L} - P(x-a) \quad [a \le x \le L]$$

Integrating each equation twice gives

$$EIy = \frac{Pbx^3}{6L} + C_1x + C_3 \quad [0 \le x \le a]$$

$$EIy = \frac{Pbx^3}{6L} - \frac{P(x-a)^3}{6} + C_2x + C_4 \quad [a \le x \le L]$$

The constants are determined by the following conditions: (1) at $x=a$, the slopes dy/dx and deflections y are equal; (2) at $x=0$ and $x=L$, the deflection $y=0$. These conditions give

$$C_1 = C_2 = \frac{-Pb(L^2+b^2)}{6L}$$

$$C_3 = C_4 = 0$$

Evaluating the equation for $(0 \le x \le a)$ at $x=a={}^2\!/_3L$ and $b={}^1\!/_3L$,

$$EIy = \left(\frac{P\left(\frac{1}{3}L\right)\left(\frac{2}{3}L\right)}{6L}\right)\left(L^2 - \left(\frac{L}{3}\right)^2 - \left(\frac{2L}{3}\right)^2\right)$$

$$y = \frac{4PL^3}{243EI}$$

The answer is (B).

Equation 33.18 Through Eq. 33.41: Simply Supported Beam Slopes and Deflections[2]

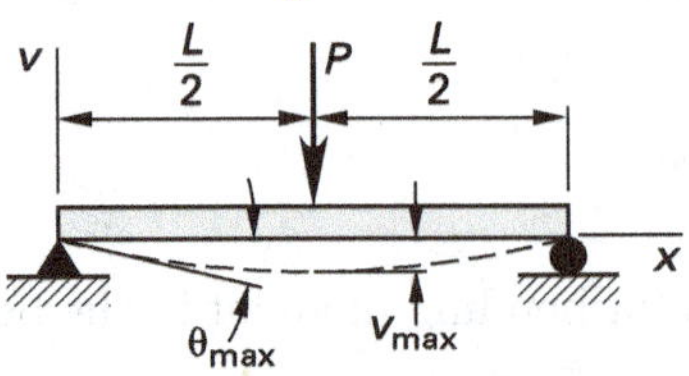

slope

$$\theta_{max} = \frac{-PL^2}{16EI} \qquad 33.18$$

deflection

$$v_{max} = \frac{-PL^3}{48EI} \qquad 33.19$$

elastic curve

$$v = \frac{-Px}{48EI}(3L^2 - 4x^2) \quad [0 \le x \le L/2] \qquad 33.20$$

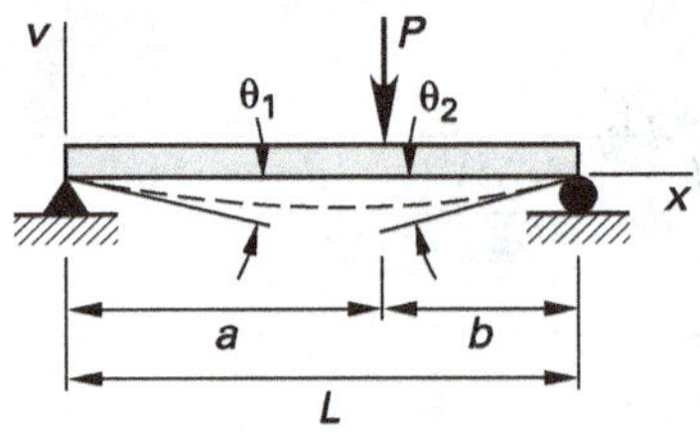

slope

$$\theta_1 = \frac{-Pab(L+b)}{6EIL} \qquad 33.21$$

$$\theta_2 = \frac{Pab(L+a)}{6EIL} \qquad 33.22$$

deflection

$$v\big|_{x=a} = \frac{-Pba}{6EIL}(L^2 - b^2 - a^2) \qquad 33.23$$

[2]Deflection in the vertical direction in the *NCEES Handbook* beam deflection table (see Eq. 33.18 Through Eq. 33.41) is given a new symbol, v. This is the same quantity that the *NCEES Handbook* refers to as y in other places. The deflection angles in Eq. 33.18 through Eq. 33.41, referred to as θ, have units of radians, and so, are more conveniently thought of as slopes, corresponding to y' or $\tan\theta$ if variables consistent with the rest of the chapter are used.

elastic curve

$$v = \frac{-Pbx}{6EIL}(L^2 - b^2 - x^2) \quad [0 \le x \le a] \qquad 33.24$$

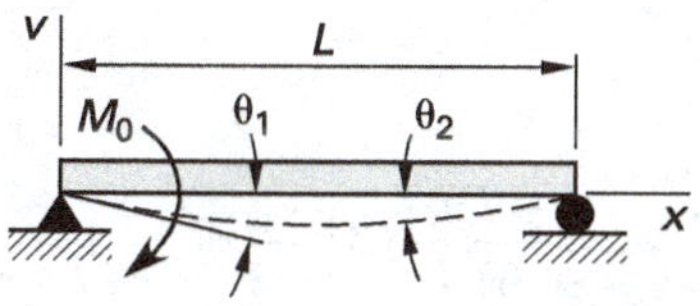

slope

$$\theta_1 = \frac{-M_0 L}{3EI} \qquad 33.25$$

$$\theta_2 = \frac{M_0 L}{6EI} \qquad 33.26$$

deflection

$$v_{\text{max}} = \frac{-M_0 L^2}{\sqrt{243}\,EI} \qquad 33.27$$

elastic curve

$$v = \frac{-M_0 x}{6EIL}(x^2 - 3Lx + 2L^2) \qquad 33.28$$

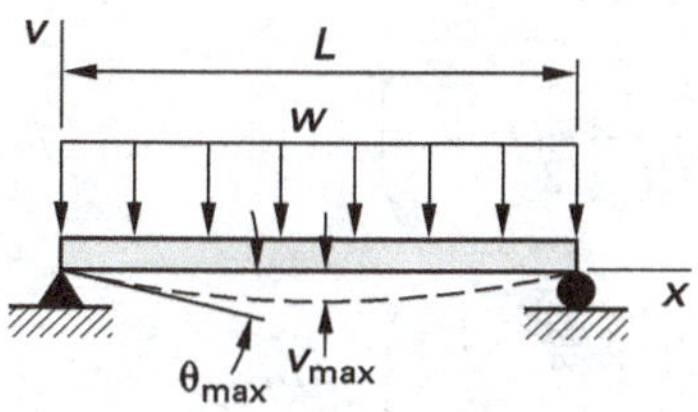

slope

$$\theta_{\text{max}} = \frac{-wL^3}{24EI} \qquad 33.29$$

deflection

$$v_{\text{max}} = \frac{-5wL^4}{384EI} \qquad 33.30$$

elastic curve

$$v = \frac{-wx}{24EI}(x^3 - 3Lx^2 + L^3) \qquad 33.31$$

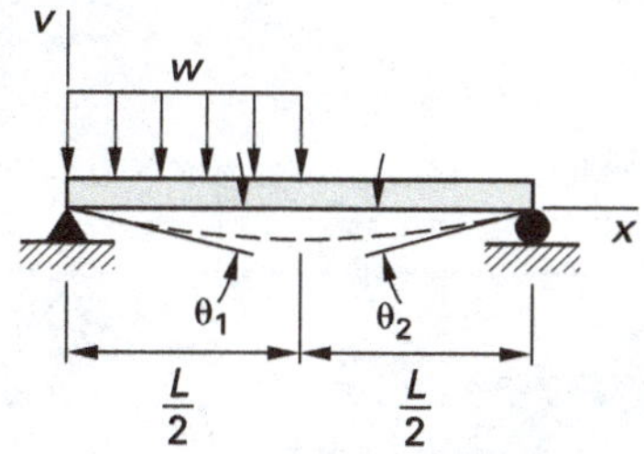

slope

$$\theta_1 = \frac{-3wL^3}{128EI} \qquad 33.32$$

$$\theta_2 = \frac{7wL^3}{384EI} \qquad 33.33$$

deflection

$$v\big|_{x=L/2} = \frac{-5wL^4}{768EI} \qquad 33.34$$

$$v_{\text{max}} = -0.006563\frac{wL^4}{EI} \quad [\text{at } x = 0.4598L] \qquad 33.35$$

elastic curve

$$v = \frac{-wx}{384EI}(16x^3 - 24Lx^2 + 9L^3) \quad [0 \le x \le L/2] \qquad 33.36$$

$$v = \frac{-wL}{384EI}(8x^3 - 24Lx^2 + 17L^2x - L^3) \quad [L/2 \le x < L] \qquad 33.37$$

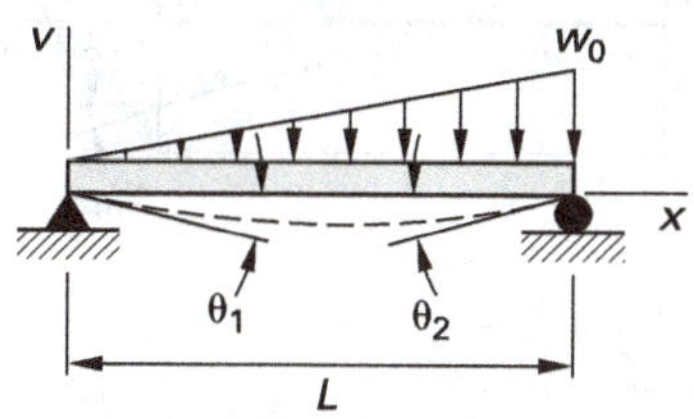

slope

$$\theta_1 = \frac{-7w_0L^3}{360EI} \qquad 33.38$$

$$\theta_2 = \frac{w_0L^3}{45EI} \qquad 33.39$$

deflection

$$v_{\text{max}} = -0.00652\frac{w_0L^4}{EI} \qquad 33.40$$

$$[\text{at } x = 0.5193]$$

elastic curve

$$v = \frac{-w_0x}{360EIL}(3x^4 - 10L^2x^2 + 7L^4) \qquad 33.41$$

Description

Commonly used beam deflection formulas are given in Eq. 33.18 through Eq. 33.41. These formulas never need to be derived and should be used whenever possible.

Superposition

When multiple loads act simultaneously on a beam, all of the loads contribute to deflection. The principle of *superposition* permits the deflections at a point to be calculated as the sum of the deflections from each individual load acting singly. Superposition can also be used to calculate the shear and moment at a point and to draw the shear and moment diagrams. This principle is valid as long as the normal stress and strain are related by the modulus of elasticity, E (i.e., as long as Hooke's law is valid). Generally, this is true when the deflections are not excessive and all stresses are kept less than the yield point of the beam material.

Equation 33.42 Through Eq. 33.61: Cantilevered Beam Slopes and Deflections

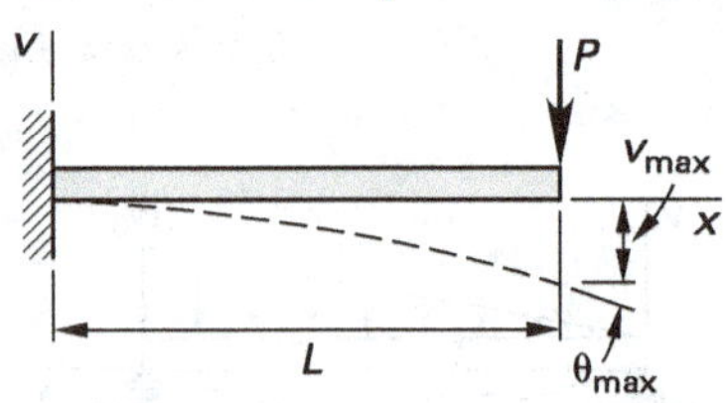

slope

$$\theta_{\text{max}} = \frac{-PL^2}{2EI} \qquad 33.42$$

deflection

$$v_{\text{max}} = \frac{-PL^3}{3EI} \qquad 33.43$$

elastic curve

$$v = \frac{-Px^2}{6EI}(3L - x) \qquad 33.44$$

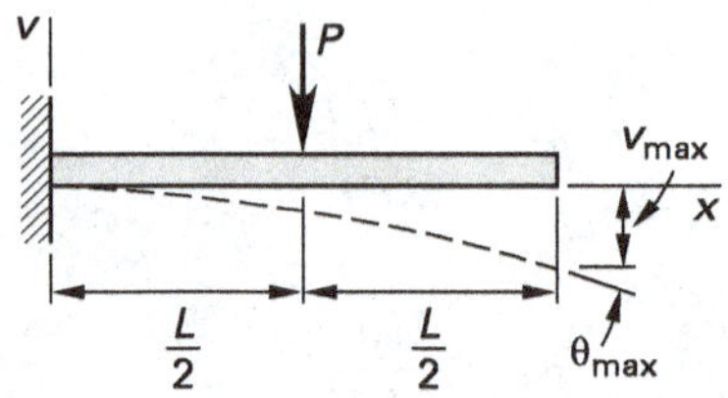

slope

$$\theta_{\text{max}} = \frac{-PL^2}{8EI} \qquad 33.45$$

deflection

$$v_{\text{max}} = \frac{-5PL^3}{48EI} \qquad 33.46$$

elastic curve

$$v = \frac{-Px^2}{6EI}\left(\frac{3}{2}L - x\right) \quad [0 \le x \le L/2] \qquad 33.47$$

$$v = \frac{-PL^2}{24EI}\left(3x - \frac{1}{2}L\right) \quad [L/2 \le x \le L] \qquad 33.48$$

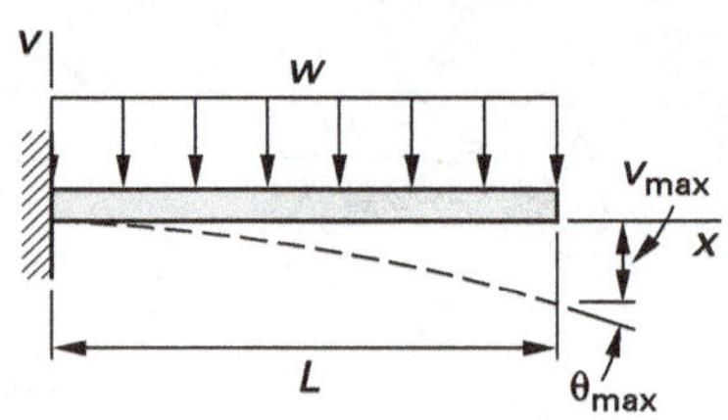

slope

$$\theta_{\text{max}} = \frac{-wL^3}{6EI} \qquad 33.49$$

deflection

$$v_{\text{max}} = \frac{-wL^4}{8EI} \qquad 33.50$$

elastic curve

$$v = \frac{-wx^2}{24EI}(x^2 - 4Lx + 6L^2) \qquad 33.51$$

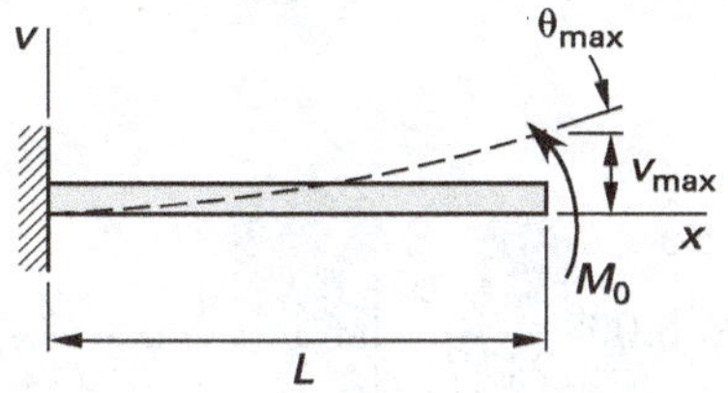

slope

$$\theta_{\text{max}} = \frac{M_0L}{EI} \qquad 33.52$$

deflection

$$v_{\text{max}} = \frac{M_0L^2}{2EI} \qquad 33.53$$

elastic curve

$$v = \frac{M_0x^2}{2EI} \qquad 33.54$$

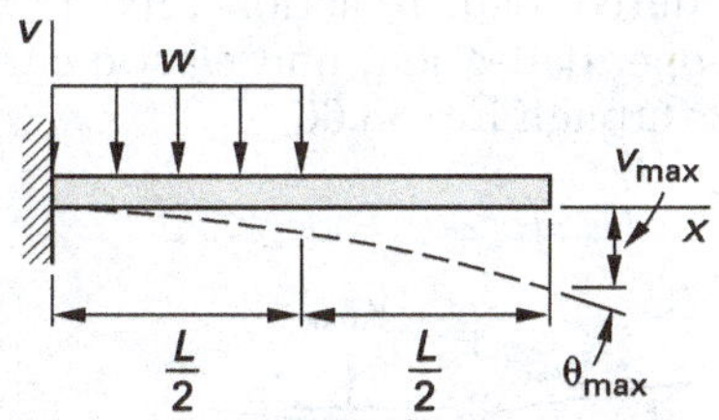

slope

$$\theta_{\text{max}} = \frac{-wL^3}{48EI} \qquad 33.55$$

deflection

$$v_{\text{max}} = \frac{-7wL^4}{384EI} \qquad 33.56$$

elastic curve

$$v_{\text{max}} = \frac{-wx^2}{24EI}\left(x^2 - 2Lx + \frac{3}{2}L^2\right) \quad [0 \le x \le L/2] \qquad 33.57$$

$$v = \frac{-wL^3}{192EI}(4x - L/2) \quad [L/2 \le x \le L] \qquad 33.58$$

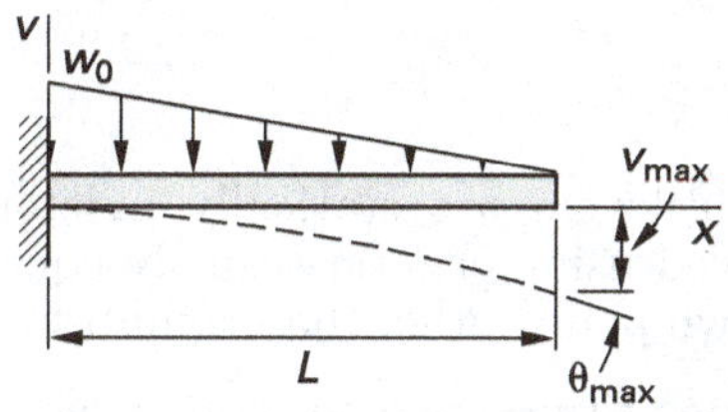

slope

$$\theta_{\text{max}} = \frac{-w_0L^3}{24EI} \qquad 33.59$$

deflection

$$v_{\text{max}} = \frac{-w_0L^4}{30EI} \qquad 33.60$$

elastic curve

$$v = \frac{-w_0x^2}{120EIL}(10L^3 - 10L^2x + 5Lx^2 - x^3) \qquad 33.61$$

Description

Commonly used cantilevered beam slopes and deflections are compiled into Eq. 33.42 through Eq. 33.61.

Example

The unloaded propped cantilever shown is fixed at one end and simply supported at the other end. The beam has a mass of 30.6 kg/m. The modulus of elasticity of the beam is 210 GPa; the moment of inertia is 2880 cm^4.

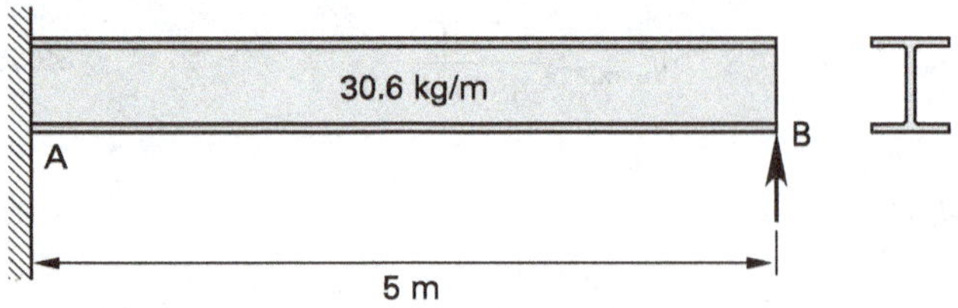

What is most nearly the reaction at the simply supported end?

(A) 72 N

(B) 510 N

(C) 560 N

(D) 770 N

Solution

Propped cantilevers are statically indeterminate and must be solved using criteria (usually equal deflections at some known point) other than equilibrium.

The deflection at the supported end is known to be zero. Therefore, the deflection due to the distributed self-weight load combined with the deflection due to the concentrated reaction load must sum to zero.

The deflection for a distributed load is found from Eq. 33.50.

$$v_1 = \frac{-wL^4}{8EI} \quad \text{[downward]}$$

The deflection due to a concentrated load (i.e., the reaction at point B) is found from Eq. 33.43, with $x = L$.

$$v_2 = \frac{-PL^3}{3EI} \quad \text{[upward]}$$

Since $v_1 + v_2 = 0$,

$$\frac{wL^4}{8EI} = \frac{PL^3}{3EI}$$

$$P = \frac{3wL}{8} = \frac{3mgL}{8} = \frac{(3)\left(30.6\ \frac{\text{kg}}{\text{m}}\right)\left(9.81\ \frac{\text{m}}{\text{s}^2}\right)(5\ \text{m})}{8}$$
$$= 562.8\ \text{N} \quad (560\ \text{N})$$

The answer is (C).

Equation 33.62 Through Eq. 33.66: Fixed-Supported Beam Slopes and Deflections[3]

reactions

$$R_1 = R_2 = \frac{wL}{2} \qquad 33.62$$

moments

$$M_1 = M_2 = \frac{wL^2}{12} \qquad 33.63$$

slope

$$|\theta_{\text{max}}| = 0.008\frac{wL^3}{24EI} \quad \left[\text{at } x = \frac{1}{2} \pm \frac{L}{\sqrt{12}}\right] \qquad 33.64$$

deflection

$$|v_{\text{max}}| = \frac{wL^4}{384EI} \quad \left[\text{at } x = \frac{L}{2}\right] \qquad 33.65$$

elastic curve

$$v_{\text{max}}(x) = \frac{wx^2}{24EI}(L^2 - Lx + x^2) \qquad 33.66$$

Description

A beam is rigidly fixed when the slope at both ends is zero and moments develop at the supports. Figure 33.6 shows a beam fixed at both end supports with a uniformly distributed load. Reactions, fixed end moments, maximum slope, deflection, and elastic curve are given by Eq. 33.62 through Eq. 33.66.

***Figure 33.6** Beam Fixed at Both Ends*

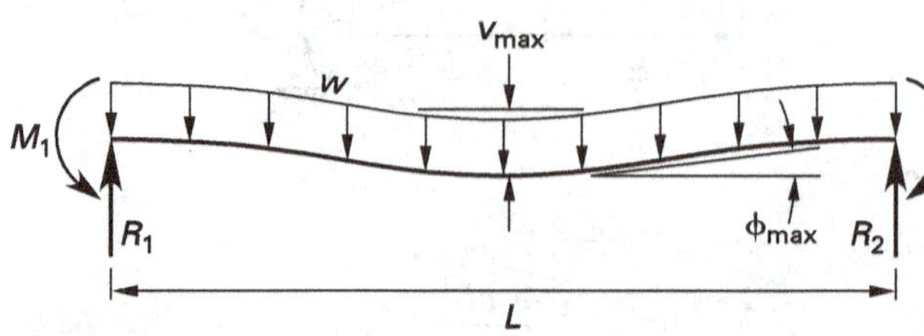

[3]The *NCEES Handbook* uses the term "piping segment" instead of "fixed-supported beam." However, given the equations and diagram used, the term "fixed supported beam" is more appropriate, so that term is used throughout this section.

Example

A steel beam is shown.

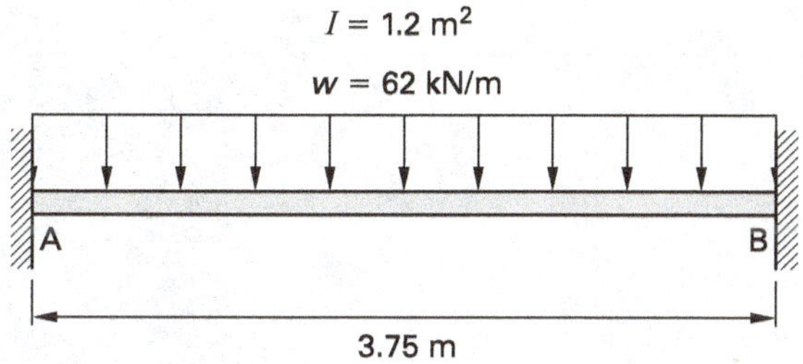

Most nearly, what is the moment at support B?

(A) 65 kN·m

(B) 73 kN·m

(C) 80 kN·m

(D) 87 kN·m

Solution

From Eq. 33.63,

$$M_A = M_B = \frac{wL^2}{12} = \frac{\left(62\ \frac{\text{kN}}{\text{m}}\right)(3.75\ \text{m})^2}{12} = 73\ \text{kN·m}$$

The answer is (B).

4. COMPOSITE BEAMS

A *composite structure* is one in which two or more different materials are used. Each material carries part of the applied load. Examples of composite structures include steel-reinforced concrete and timber beams with bolted-on steel plates.

Most simple composite structures can be analyzed using the *method of consistent deformations*, also known as the *transformation method*. This method assumes that the strains are the same in both materials at the interface between them. Although the strains are the same, the stresses in the two adjacent materials are not equal, since stresses are proportional to the moduli of elasticity.

The transformation method starts by determining the modulus of elasticity for each (usually two in number) of the materials in the composite beam and then calculating the *modular ratio*, *n*. E_{weaker} is the smaller modulus of elasticity.

$$n = \frac{E}{E_{\text{weaker}}}$$

The area of the stronger material is increased by a factor of n. The transformed area is used to calculate the transformed composite area, A_{ct}, or transformed moment of inertia, I_{ct}. For compression and tension members, the stresses in the weaker and stronger materials are

$$\sigma_{\text{weaker}} = \frac{F}{A_{ct}}$$

$$\sigma_{\text{stronger}} = \frac{nF}{A_{ct}}$$

Equation 33.67 and Eq. 33.68: Transformation Method for Beams in Bending

$$\sigma_1 = -nMy/I_T \qquad 33.67$$

$$\sigma_2 = -My/I_T \qquad 33.68$$

Variations

$$\sigma_{\text{weaker}} = \frac{Mc_{\text{weaker}}}{I_{ct}}$$

$$\sigma_{\text{stronger}} = \frac{nMc_{\text{stronger}}}{I_{ct}}$$

Description

For beams in bending, the bending stresses in the stronger and weaker materials, respectively, are given by Eq. 33.67 and Eq. 33.68. The transformed section is composed of a single material. (See Fig. 33.7.)

***Figure 33.7** Composite Section Transformation*

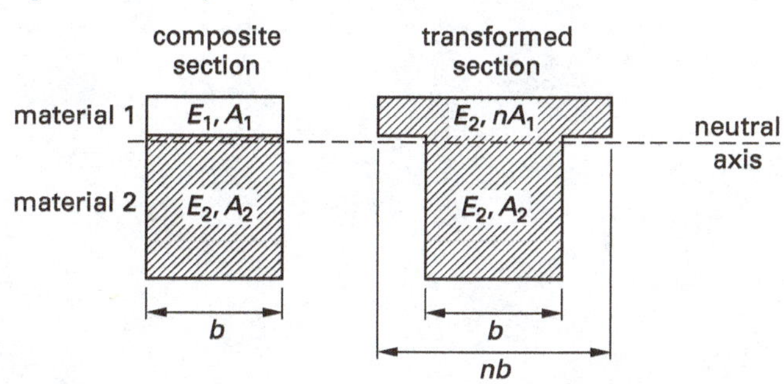

34 Columns

Nomenclature

A	area	m^2
b	width	m
c	distance to extreme fiber	m
e	eccentricity	m
E	modulus of elasticity	MPa
F	force	N
h	thickness	m
I	moment of inertia	m^4
K	effective length factor	–
ℓ	unbraced length	m
L	column length	m
M	moment	N·m
P	force	N
r	radius of gyration	m
S	strength	MPa

Symbols

δ	deformation	m
σ	normal stress	MPa

Subscripts

c	compressive
cr	critical
col	column
t	tensile
y	yield

1. BEAM-COLUMNS

If a load is applied through the centroid of a tension or compression member's cross section, the loading is said to be *axial loading* or *concentric loading*. *Eccentric loading* occurs when the load is not applied through the centroid. In Fig. 34.1, distance e is known as the *eccentricity*.

If an axial member is loaded eccentrically, it will bend and experience bending stress in the same manner as a beam. Since the member experiences both axial stress and bending stress, it is known as a *beam-column*. Beam-columns may be horizontal or vertical members.

Figure 34.1 *Eccentric Loading of a Beam-Column*

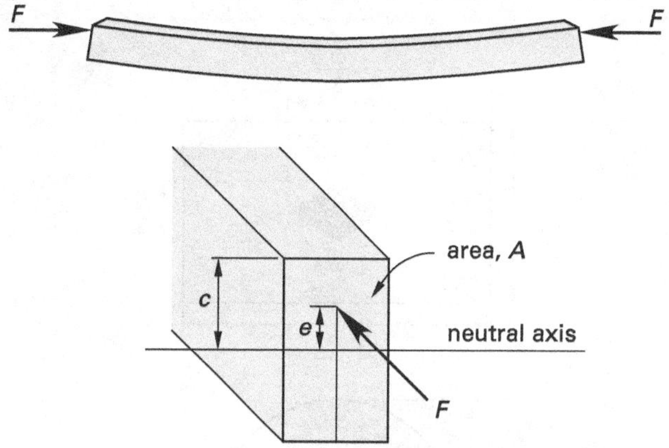

Both the axial stress and bending stress are normal stresses oriented in the same direction; therefore, simple addition can be used to combine them.

$$\sigma_{\text{max,min}} = \frac{F}{A} \pm \frac{Mc}{I} = \frac{F}{A} \pm \frac{Fec}{I}$$

$$M = Fe$$

If a vertical pier or column (primarily designed as a compression member) is loaded with an eccentric compressive load, part of the section can still be in tension. Tension will exist when the Mc/I term is larger than the F/A term. It is particularly important to eliminate or severely limit tensile stresses in concrete and masonry piers, since these materials cannot support much tension.

Regardless of the size of the load, there will be no tension as long as the eccentricity is low enough. In a rectangular member, the load must be kept within a rhombus-shaped area formed from the middle thirds of the centroidal axes. This area is known as the *core*, *kern*, or *kernel*. Figure 34.2 illustrates the kernel for various cross sections.

2. LONG COLUMNS

Short columns, called *piers* or *pedestals*, will fail by compression of the material. *Long columns* will *buckle* in the transverse direction that has the smallest radius of gyration. Buckling failure is sudden, often without

Figure 34.2 Kernel for Various Column Shapes

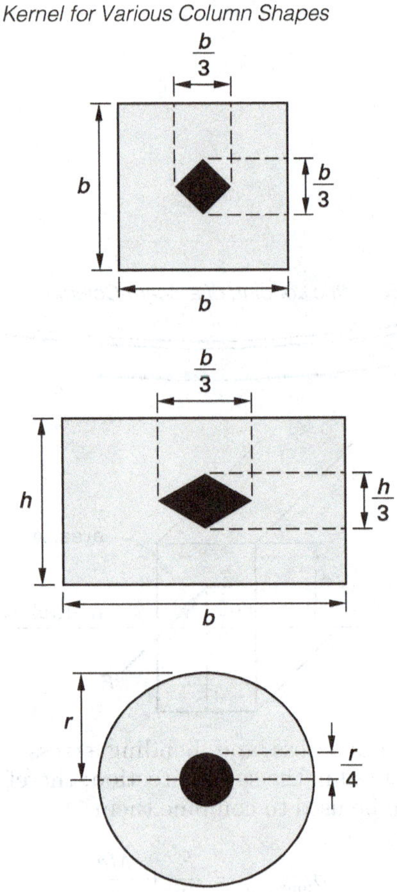

significant warning. If the material is wood or concrete, the material will usually fracture (because the yield stress is low); however, if the column is made of steel, the column will usually fail by local buckling, followed later by twisting and general yielding failure. Intermediate length columns will usually fail by a combination of crushing and buckling.

Equation 34.1: Euler's Formula for Pinned or Frictionless Ends[1]

$$P_{cr} = \frac{\pi^2 EI}{(K\ell)^2} \qquad 34.1$$

Variations

$$P_{cr} = \frac{\pi^2 EI}{\ell^2}$$

$$P_{cr} = \frac{\pi^2 EI}{L_{\text{col}}^2}$$

Description

The load at which a long column fails is known as the *critical load* or *Euler load.* The Euler load is the theoretical maximum load that an initially straight column can support without transverse buckling. For columns with unrestrained or pinned ends, this load is given by the first variation equation, known as *Euler's formula.*

When a column is not braced along its entire length, the unbraced length is equal to the length (height) of the column: $\ell = L$. As shown in Table 34.1, if the column has pinned or frictionless ends, the effective length factor, K, is 1.00. In that case, the effective length of the column is simply the column length, as presented in the second variation equation.

ℓ is the longest unbraced column length. If a column is braced against buckling at some point between its two ends, the column is known as a *braced column*, and ℓ will be less than the full column height.

Columns do not usually have unrestrained or pinned ends. Often, a column will be fixed at its top and base. In such cases, the *effective length*, $K\ell$, the distance between inflection points on the column, must be used in place of ℓ.

K is the *effective length factor* (*end-restraint coefficient*), which theoretically varies from 0.5 to 2.0 according to Table 34.1. For design, values of K should be modified using engineering judgment based on realistic assumptions regarding end fixity.

[1]The NCEES *FE Reference Handbook* (*NCEES Handbook*) uses ℓ to represent unbraced length, but the symbol L is also used in the structural design parts of the *NCEES Handbook* to represent the same quantity for beams and columns. The effective length may be represented by ℓ_{eff}.

Table 34.1 Effective Length Factors

illus.	end conditions	K theoretical	K design
(a)	both ends pinned	1	1.00
(b)	both ends built in	0.5	0.65
(c)	one end pinned, one end built in	0.7	0.8
(d)	one end built in, one end free	2	2.10
(e)	one end built in, one end fixed against rotation but free	1	1.20
(f)	one end pinned, one end fixed against rotation but free	2	2.0

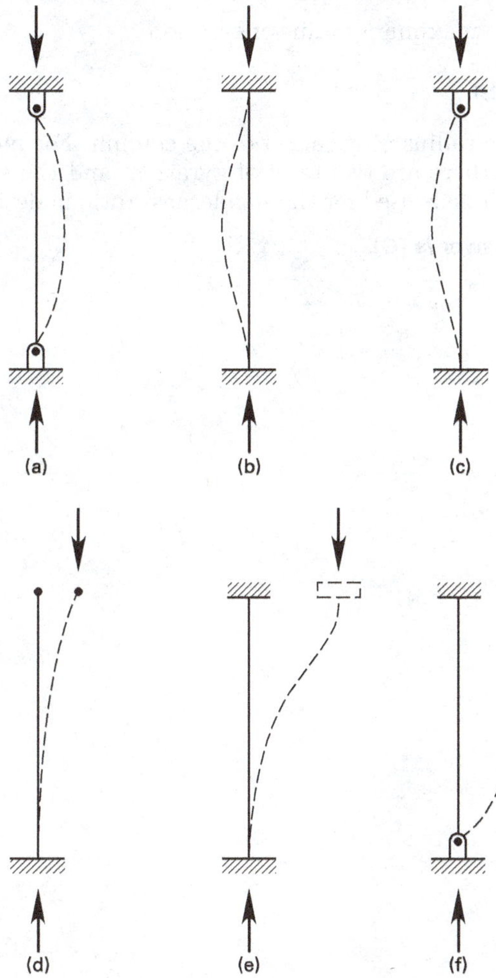

Example

A real (i.e., nonideal) rectangular steel bar supports a concentric load of 58.5 kN. Both ends are fixed (i.e., built in).

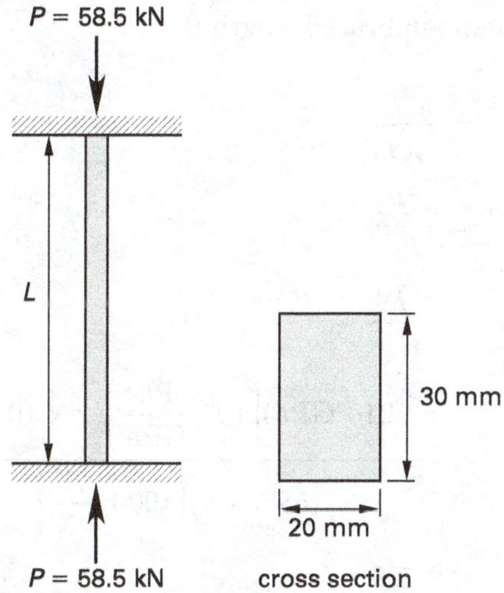

If the modulus of elasticity is 210 GPa, what is most nearly the maximum unbraced length the rod can be without experiencing buckling failure?

(A) 1.3 m

(B) 1.7 m

(C) 4.9 m

(D) 12 m

Solution

Since the column is fixed at the top and base, use the effective length.

$$P_{cr} = \frac{\pi^2 EI}{(K\ell)^2}$$

For a nonideal column fixed at both ends, use the design value of $K = 0.65$. For the cross-sectional area of the bar, the smaller dimension is the height. The moment of inertia is

$$\begin{aligned} I &= \frac{bh^3}{12} \\ &= \frac{(30 \text{ mm})(20 \text{ mm})^3}{(12)\left(1000 \ \frac{\text{mm}}{\text{m}}\right)^4} \\ &= 2 \times 10^{-8} \text{ m}^4 \end{aligned}$$

The maximum unbraced length is

$$P_{cr} = \frac{\pi^2 EI}{(K\ell)^2}$$

$$(K\ell)^2 = \frac{\pi^2 EI}{P_{cr}}$$

$$(0.65\ell)^2 = \frac{\pi^2 EI}{P_{cr}}$$

$$= \frac{\pi^2(210\text{ GPa})\left(10^9\ \frac{\text{Pa}}{\text{GPa}}\right)(2\times 10^{-8}\text{ m}^4)}{(58.5\text{ kN})\left(1000\ \frac{\text{N}}{\text{kN}}\right)}$$

$$= 0.709\text{ m}^2$$

$$\ell = 1.3\text{ m}$$

The answer is (A).

Equation 34.2 and Eq. 34.3: Critical Column Stress

$$\sigma_{cr} = \frac{P_{cr}}{A} = \frac{\pi^2 E}{(K\ell/r)^2} \qquad \textit{34.2}$$

$$r = \sqrt{I/A} \qquad \textit{34.3}$$

Description

The column stress corresponding to the Euler load is given by Eq. 34.2. This stress cannot exceed the yield strength of the column material.

The quantity $K\ell/r$ is known as the *effective slenderness ratio*. Long columns have high effective slenderness ratios. The smallest effective slenderness ratio for which Eq. 34.2 is valid is the *critical slenderness ratio*, which can be calculated from the material's yield strength and modulus of elasticity. Typical critical slenderness ratios range from 80 to 120. The critical slenderness ratio becomes smaller as the compressive yield strength increases.

Noncircular columns have two radii of gyration, r_x and r_y, and therefore, have two effective slenderness ratios. The effective slenderness ratio (i.e., the smallest radius of gyration) will govern the design.

Example

The slenderness ratio, $K\ell$, of a column divided by r is one of the terms in the equation for the buckling of a column subjected to compression loads. What does r stand for in the $K\ell/r$ ratio?

(A) radius of the column

(B) radius of gyration

(C) least radius of gyration

(D) maximum radius of gyration

Solution

r is the radius of gyration of the column. For most columns, there are two radii of gyration, and the smallest (least) one is used for the slenderness ratio in design.

The answer is (C).

Topic X: Materials

Chapter

35 Material Properties and Testing

Nomenclature

a	crack length	m
A	area	m^2
A	constant	–
BHN	Brinell hardness number	–
c	specific heat	J/kg·°C
C	capacitance	F
C	molar specific heat	J/mol·°C
C_V	impact energy	J
d	diameter	m
d	distance	m
D	diameter	m
E	energy	eV
E	modulus of elasticity	MPa
F	force	N
F	load	N
g	gravitational acceleration, 9.81	m/s^2
G	modulus of rigidity	GPa
J	flux	$1/m^2·s$
k	reduction factor	–
K_{IC}	fracture toughness	$MPa\cdot\sqrt{m}$
L	length	m
m	mass	kg
n	stress sensitivity exponent	–
N	number of cycles	–
P	load	N
q	charge	C
q	reduction in area	–
Q	activation energy	J/mol
Q	heat	J
R	resistance	Ω
R	universal gas constant, 8.314	J/mol·K
S	strength	MPa
S_e'	endurance limit	MPa
t	thickness	m
t	time	s
T	temperature	K
TS	tensile strength	MPa
V	voltage	V
V	volume	m^3
w	width	m
Y	geometrical factor	–

Symbols

α	thermal expansion coefficient	1/°C
γ	conductivity	W/m·K
δ	deformation	m
ε	engineering strain	–
ε	permittivity	F/m or $C^2/N·m^2$
ε_0	permittivity of a vacuum, 8.85×10^{-12}	F/m or $C^2/N·m^2$
κ	dielectric constant	–
λ	conductivity	W/m·K
μ	ductility	–
ν	Poisson's ratio	–
ρ	density	kg/m^3
ρ	resistivity	Ω·m
σ	engineering stress	MPa
ϕ	work function	eV

Subscripts

0	initial
a	activation or surface
b	size

c	conduction, critical, or load
d	diffusion or temperature
e	effects or endurance
eff	effective
f	failure, final, or fracture
g	gap or glass transition
i	intrinsic
I	intensity
o	original
p	constant pressure or particular
t	tensile or total
T	total or true
u	ultimate
ut	ultimate tensile
v	valence
v	constant volume
y	yield

1. MATERIALS SELECTION

Materials selection is the process of selecting materials used to design and manufacture a part or product. Materials selection is an important component in the design process, as materials must be carefully selected with product performance and manufacturing processes in mind. The goal of materials selection is to meet product performance goals (e.g., strength, ductility, safety) while minimizing costs and waste.

Materials selection typically begins by considering the ideal properties the material would exhibit based on the product's specifications. Then, materials that best exemplify those needs are selected, and a comparison of the selected materials, including costs, is performed. Because many kinds of materials are available, the process often starts by considering broad categories of materials before zeroing in on a specific choice. These general types of materials include

- *ceramics:* glass ceramics, glasses, graphite, and diamond
- *composites:* reinforced plastics, metal-matrix composites, and honeycomb structures
- *ferrous metals:* carbon, alloy, stainless steel, and tool and die steels
- *nonferrous metals and alloys:* aluminum, magnesium, copper, nickel, titanium, superalloys, refractory metals, beryllium, zirconium, low-melting alloys, and precious metals
- *plastics:* thermoplastics, thermosets, and elastomers

The materials selection process is often iterative; selections and comparisons may be done multiple times before finding the optimal material for a given use.

2. MATERIALS SCIENCE

Materials science is the study of materials to understand their properties, limits, and uses. *Material properties* are key characteristics of a material commonly classified into five main categories: chemical, electrical, mechanical, physical, and thermal.

Mechanical properties describe the relationship between properties and mechanical (i.e., physical) forces, such as stresses, strains, and applied force. Examples of mechanical properties include strength, toughness, ductility, hardness, fatigue, and creep.

Thermal properties are properties that are observed when heat energy is applied to a material. Examples of thermal properties include thermal conductivity, thermal diffusivity, the heat of fusion, and the glass transition temperature.

Electrical properties define the reaction of a material to an electric field. Typical electrical properties are dielectric strength, conductivity, permeability, permittivity, and electrical resistance.

Chemical properties are properties that are evident only when a substance is changed chemically. Common chemical properties include corrosivity, toxicity, and flammability.

Unlike chemical properties, *physical properties* can be observed without altering the material or its structure. Common physical properties include density, melting point, and specific heat.

Some common properties of various materials are given in Table 35.1 and Table 35.2. Additional mechanical properties are given in Table 35.5.

3. ELECTRICAL PROPERTIES

Equation 35.1 Through Eq. 35.3: Capacitance

$$q = CV \tag{35.1}$$

$$C = \frac{\varepsilon A}{d} \tag{35.2}$$

$$\varepsilon = \kappa\varepsilon_0 \tag{35.3}$$

Value

$\varepsilon_0 = 8.85 \times 10^{-12}$ F/m (same as $C^2/N{\cdot}m^2$)

Description

A *capacitor* is a device that stores electric charge. A capacitor is constructed as two conducting surfaces separated by an insulator, such as oiled paper, mica, or air.

Materials

*Table 35.1 Typical Material Properties**

material	modulus of elasticity, E (Mpsi (GPa))	modulus of rigidity, G (Mpsi (GPa))	Poisson's ratio, ν	coefficient of thermal expansion, α (10^{-6}/°F (10^{-6}/°C))	density, ρ (lbm/in^3 (Mg/m^3))
steel	29.0 (200.0)	11.5 (80.0)	0.30	6.5 (11.7)	0.282 (7.8)
aluminum	10.0 (69.0)	3.8 (26.0)	0.33	13.1 (23.6)	0.098 (2.7)
cast iron	14.5 (100.0)	6.0 (41.4)	0.21	6.7 (12.1)	0.246–0.282 (6.8–7.8)
wood (fir)	1.6 (11.0)	0.6 (4.1)	0.33	1.7 (3.0)	–
brass	14.8–18.1 (102–125)	5.8 (40)	0.33	10.4 (18.7)	0.303–0.313 (8.4–8.7)
copper	17 (117)	6.5 (45)	0.36	9.3 (16.6)	0.322 (8.9)
bronze	13.9–17.4 (96–120)	6.5 (45)	0.34	10.0 (18.0)	0.278–0.314 (7.7–8.7)
magnesium	6.5 (45)	2.4 (16.5)	0.35	14 (25)	0.061 (1.7)
glass	10.2 (70)	–	0.22	5.0 (9.0)	0.090 (2.5)
polystyrene	0.3 (2)	–	0.34	38.9 (70.0)	0.038 (1.05)
polyvinyl chloride (PVC)	<0.6 (<4)	–	–	28.0 (50.4)	0.047 (1.3)
alumina fiber	58 (400)	–	–	–	0.141 (3.9)
aramide fiber	18.1 (125)	–	–	–	0.047 (1.3)
boron fiber	58 (400)	–	–	–	0.083 (2.3)
beryllium fiber	43.5 (300)	–	–	–	0.069 (1.9)
BeO fiber	58 (400)	–	–	–	0.108 (3.0)
carbon fiber	101.5 (700)	–	–	–	0.083 (2.3)
silicon carbide fiber	58 (400)	–	–	–	0.116 (3.2)

*Use these values if the specific alloy and temper are not listed in Table 35.5.

A *parallel plate capacitor* is a simple type of capacitor constructed as two parallel plates. If the plates are connected across a voltage potential, charges of opposite polarity will build up on the plates and create an electric field between the plates. The amount of charge, q, built up is proportional to the applied voltage, V, as shown in Eq. 35.1. The constant of proportionality, C, is the *capacitance* in farads (F) and depends on the capacitor construction. Capacitance represents the ability to store charge; the greater the capacitance, the greater the charge stored.

Equation 35.2 gives the capacitance of two parallel plates of equal area A separated by distance d. ε is the permittivity of the medium separating the plates. The permittivity may also be expressed as the product of the *dielectric constant* (*relative permittivity*), κ, and the *permittivity of a vacuum* (also known as the *permittivity of free space*), ε_0, as shown in Eq. 35.3.

Materials

Table 35.2 Properties of Metals

metal	symbol	atomic weight	density, ρ (kg/m^3) water = 1000	melting point (°C)	melting point (°F)	specific heat (J/kg·K)	electrical resistivity (10^{-8} Ω·m) at 0°C (273.2K)[a]	heat conductivity,[b] λ (W/m·K) at 0°C (273.2K)
aluminum	Al	26.98	2698	660	1220	895.9	2.5	236
antimony	Sb	121.75	6692	630	1166	209.3	39	25.5
arsenic	As	74.92	5776	subl. 613	subl. 1135	347.5	26	–
barium	Ba	137.33	3594	710	1310	284.7	36	–
beryllium	Be	9.012	1846	1285	2345	2051.5	2.8	218
bismuth	Bi	208.98	9803	271	519	125.6	107	8.2
cadmium	Cd	112.41	8647	321	609	234.5	6.8	97
caesium	Cs	132.91	1900	29	84	217.7	18.8	36
calcium	Ca	40.08	1530	840	1544	636.4	3.2	–
cerium	Ce	140.12	6711	800	1472	188.4	7.3	11
chromium	Cr	52	7194	1860	3380	406.5	12.7	96.5
cobalt	Co	58.93	8800	1494	2721	431.2	5.6	105
copper	Cu	63.54	8933	1084	1983	389.4	1.55	403
gallium	Ga	69.72	5905	30	86	330.7	13.6	41
gold	Au	196.97	19 281	1064	1947	129.8	2.05	319
indium	In	114.82	7290	156	312	238.6	8	84
iridium	Ir	192.22	22 550	2447	4436	138.2	4.7	147
iron	Fe	55.85	7873	1540	2804	456.4	8.9	83.5
lead	Pb	207.2	11 343	327	620	129.8	19.2	36
lithium	Li	6.94	533	180	356	4576.2	8.55	86
magnesium	Mg	24.31	1738	650	1202	1046.7	3.94	157
manganese	Mn	54.94	7473	1250	2282	502.4	138	8
mercury	Hg	200.59	13 547	−39	−38	142.3	94.1	7.8
molybdenum	Mo	95.94	10 222	2620	4748	272.1	5	139
nickel	Ni	58.69	8907	1455	2651	439.6	6.2	94
niobium	Nb	92.91	8578	2425	4397	267.9	15.2	53
osmium	Os	190.2	22 580	3030	5486	129.8	8.1	88
palladium	Pd	106.4	11 995	1554	2829	230.3	10	72
platinum	Pt	195.08	21 450	1772	3221	134	9.81	72
potassium	K	39.09	862	63	145	753.6	6.1	104
rhodium	Rh	102.91	12 420	1963	3565	242.8	4.3	151
rubidium	Rb	85.47	1533	38.8	102	330.7	11	58
ruthenium	Ru	101.07	12 360	2310	4190	255.4	7.1	117
silver	Ag	107.87	10 500	961	1760	234.5	1.47	428
sodium	Na	22.989	966	97.8	208	1235.1	4.2	142
strontium	Sr	87.62	2583	770	1418	–	20	–
tantalum	Ta	180.95	16 670	3000	5432	150.7	12.3	57
thallium	Tl	204.38	11 871	304	579	138.2	10	10
thorium	Th	232.04	11 725	1700	3092	117.2	14.7	54
tin	Sn	118.69	7285	232	449	230.3	11.5	68
titanium	Ti	47.88	4508	1670	3038	527.5	39	22
tungsten	W	183.85	19 254	3387	6128	142.8	4.9	177
uranium	U	238.03	19 050	1135	2075	117.2	28	27
vanadium	V	50.94	6090	1920	3488	481.5	18.2	31
zinc	Zn	65.38	7135	419	786	393.5	5.5	117
zirconium	Zr	91.22	6507	1850	3362	284.7	40	23

[a]This is a rounded value. In its Units and Environmental Engineering sections, the NCEES *FE Reference Handbook* (*NCEES Handbook*) correctly lists the offset between degrees Celsius and kelvins as 273.15°.

[b]In this table, the *NCEES Handbook* uses lambda, λ, as the symbol for thermal (heat) conductivity. While this usage is not unheard of, it is less common than the use of k, and it is inconsistent with the symbol k used elsewhere in the *NCEES Handbook*.

Example

Two square parallel plates (0.04 m × 0.04 m) are separated by a 0.1 cm thick insulator with a dielectric constant of 3.4. What is most nearly the capacitance?

(A) 1.2×10^{-12} F

(B) 1.4×10^{-11} F

(C) 4.8×10^{-11} F

(D) 1.1×10^{-10} F

Solution

From Eq. 35.2 and Eq. 35.3, the capacitance is

$$C = \frac{\varepsilon A}{d} = \frac{\kappa\varepsilon_0 A}{d}$$
$$= \frac{(3.4)\left(8.85\times10^{-12}\ \frac{\text{F}}{\text{m}}\right)(0.04\ \text{m})^2\left(100\ \frac{\text{cm}}{\text{m}}\right)}{0.1\ \text{cm}}$$
$$= 4.814\times10^{-11}\ \text{F} \quad (4.8\times10^{-11}\ \text{F})$$

The answer is (C).

Equation 35.4: Resistivity and Resistance

$$R = \frac{\rho L}{A} \qquad 35.4$$

Description

Resistance, R (measured in ohms, Ω), is the property of a circuit or circuit element to oppose current flow. A circuit with zero resistance is a *short circuit*, whereas an *open circuit* has infinite resistance.

Resistors are usually constructed from carbon compounds, ceramics, oxides, or coiled wire. Resistance depends on the *resistivity*, ρ (in Ω·m), which is a material property, and the length and cross-sectional area of the resistor. (The resistivities of metals at 0°C are given in Table 35.2.) Resistors with larger cross-sectional areas have more free electrons available to carry charge and have less resistance. Each of the free electrons has a limited ability to move, so the electromotive force must overcome the limited mobility for the entire length of the resistor. The resistance increases with the length of the resistor.

Resistivity depends on temperature. For most conductors, it increases with temperature. For most semiconductors, resistivity decreases with temperature.

Example

A standard copper wire has a diameter of 1.6 mm. What is most nearly the resistance of 150 m of wire at 0°C?

(A) 0.91 Ω

(B) 1.2 Ω

(C) 1.5 Ω

(D) 1.7 Ω

Solution

From Table 35.2, the resistivity of copper at 0°C is 1.55×10^{-8} Ω·m. From Eq. 35.4, the resistance is

$$R = \frac{\rho L}{A} = \frac{(1.55\times10^{-8}\ \Omega\cdot\text{m})(150\ \text{m})\left(1000\ \frac{\text{mm}}{\text{m}}\right)^2}{\left(\frac{\pi}{4}\right)(1.6\ \text{mm})^2}$$
$$= 1.156\ \Omega \quad (1.2\ \Omega)$$

The answer is (B).

Semiconductors

Conductors or *semiconductors* are materials through which charges flow more or less easily. When a semiconductor is pure, it is called an *intrinsic semiconductor*. When minor amounts of impurities called *dopants* are added, the materials are termed *extrinsic semiconductors*. The solubility of a dopant determines how well the dopant can *diffuse* (move into areas with low dopant concentration) within the material.

The electrical conductivity of semiconductor materials is affected by temperature, light, electromagnetic field, and the concentration of dopants (impurities). The *solubility of dopant atoms* (i.e., the concentration, typically given in atoms/cm^3) increases very slightly with increasing temperature, reaching a relatively constant maximum in the 1000°C to 1200°C range. Higher concentrations result in precipitation of the doping element into a solid phase. Table 35.3 lists maximum values of dopant solubility. However, there may be limited value in achieving the maximum values, since some of the dopant atoms may not be electrically active. For example, arsenic (As) has a maximum solubility in *p*-type silicon of approximately 5×10^{-20} atoms/cm^3, but the maximum useful electrical solubility is approximately 2×10^{-20} atoms/cm^3. As calculated from *Fick's first law of diffusion*, the concentration gradient, dC/dx, is a major factor in determining the *electrical flux* (i.e., current), J.

$$J = -D\frac{dC}{dx}$$

Table 35.3 Some Extrinsic, Elemental Semiconductors

element	dopant	periodic table group of dopant	maximum solid solubility of dopant (atoms/m^3)
Si	B	III A	600×10^{24}
	Al	III A	20×10^{24}
	Ga	III A	40×10^{24}
	P	V A	1000×10^{24}
	As	V A	2000×10^{24}
	Sb	V A	70×10^{24}
Ge	Al	III A	400×10^{24}
	Ga	III A	500×10^{24}
	In	III A	4×10^{24}
	As	V A	80×10^{24}
	Sb	V A	10×10^{24}

Electrons in semiconductors may be bonded or free. Bonding electrons occupy states in the atoms' *valence bands.* Free electrons occupy states in the *conduction bands. Holes* are empty states in the valence band. Both holes and electrons can move around, so both are known as *carriers* or *charge carriers.*

Often, a small amount of energy (usually available thermally or provided electrically) is required to fill an *energy gap*, E_g, in order to initiate carrier movement through a semiconductor. The energy gap, often referred to as an *ionization energy*, is the difference in energy between the highest point in the valence band, E_v, and the lowest point in the conduction band, E_c. (The valence band energy may be referred to as the *intrinsic band* energy, and be given the symbol E_i.) The energy gap in insulators is relatively large compared to conductor or semiconductor materials.

$$E_g = E_v - E_c$$

Intrinsic semiconductors are those that occur naturally. When an electron in an intrinsic semiconductor receives enough energy, it can jump to the conduction band and leave behind a hole, a process known as *electron-hole pair production.* For an intrinsic material, electrons and holes are always created in pairs. Therefore, the *activation energy* is half of the energy gap.[1]

$$E_a = \tfrac{1}{2}E_g \quad \text{[intrinsic]}$$

An *extrinsic semiconductor* is created by artificially introducing dopants into otherwise "perfect" crystals. The analysis of energy levels is similar, except that the dopant energies are within the energy band gap, effectively reducing the energy required to overcome the gap. The valence band energy of the dopants may be referred to as a *donor level* (for *n*-type semiconductors) or an *acceptor level* (for *p*-type semiconductors).

At high temperatures, the carrier density approaches the intrinsic carrier concentration. Therefore, for extrinsic semiconductors at high temperatures, the activation energy (ionization energy) is the same as for intrinsic semiconductors, half of the difference in ionization energies. At low temperatures, including normal room temperatures, the carrier density is dominated by the ionization of the donors. At lower temperatures, the activation energy is equal to the difference in ionization energies.

$$\begin{aligned} E_a &= \tfrac{1}{2}(E_g - E_d) \\ &\approx \tfrac{1}{2}(E_v - E_c) \quad \text{[extrinsic, high temperatures]} \end{aligned}$$

$$E_a = E_g - E_d \quad \text{[extrinsic, low temperatures]}$$

Table 35.4 lists ionization energy differences, $E_g - E_d$, and activation energies, E_a, for various extrinsic semiconductors.

Table 35.4 Impurity Energy Levels for Extrinsic Semiconductors

semiconductor	dopant	$E_g - E_d$ (eV)	E_a (eV)
Si	P	0.044	–
	As	0.049	–
	Sb	0.039	–
	Bi	0.069	–
	B	–	0.045
	Al	–	0.057
	Ga	–	0.065
	In	–	0.160
	Tl	–	0.260
Ge	P	0.012	–
	As	0.013	–
	Sb	0.096	–
	B	–	0.010
	Al	–	0.010
	Ga	–	0.010
	In	–	0.011
	Tl	–	0.010
GaAs	Se	0.005	–
	Te	0.003	–
	Zn	–	0.024
	Cd	–	0.021

[1]The *NCEES Handbook* is inconsistent in the symbols used for activation energy. E_a in Table 35.4 is the same as Q in Eq. 35.19. A common symbol used in practice for diffusion activation energy is Q_d, where the subscript clarifies that the activation energy is for diffusion.

Photoelectric Effect

The *work function*, ϕ, is a measure of the energy required to remove an electron from the surface of a metal. It is usually given in terms of electron volts, eV. It is specifically the minimum energy necessary to move an electron from the *Fermi level* of a metal (an energy level below which all available energy levels are filled, and above which all are empty, at 0K) to infinity, that is, the vacuum level. This energy level must be reached in order to move electrons from semiconductor devices into the metal conductors that constitute the remainder of an electrical circuit. It is also the energy level of importance in the design of optical electronic devices.

In photosensitive electronic devices, an incoming photon provides the energy to release an electron and make it available to the circuit, that is, free it so that it may move under the influence of an electric field. This phenomenon whereby a short wavelength photon interacts with an atom and releases an electron is called the *photoelectric effect.*

Transduction Principles

The *transduction* principle of a given *transducer* (a device that converts a signal to a different energy form) determines nearly all its other characteristics. There are three *self-generating* transduction types: photovoltaic, piezoelectric, and electromagnetic. All other types require the use of an external excitation power source.

In *photovoltaic transduction* (*photoelectric transduction*), light is directed onto the junction of two dissimilar metals, generating a voltage. This type of transduction is used primarily in optical sensors. It can also be used with the measured quantity (known as the *measurand*) controlling a mechanical-displacement shutter that varies the intensity of the built-in light source. *Piezoelectric transduction* occurs because certain crystals generate an electrostatic charge or potential when mechanical forces are applied to the material (i.e., the material is placed in compression or tension or bending forces are applied to it). In *electromagnetic transduction*, the measured quantity is converted into a voltage by a change in magnetic flux that occurs when magnetic material moves relative to a coil with a ferrous core. These self-generating types of transduction are illustrated in Fig. 35.1.

Figure 35.1 *Self-Generating Transducers*

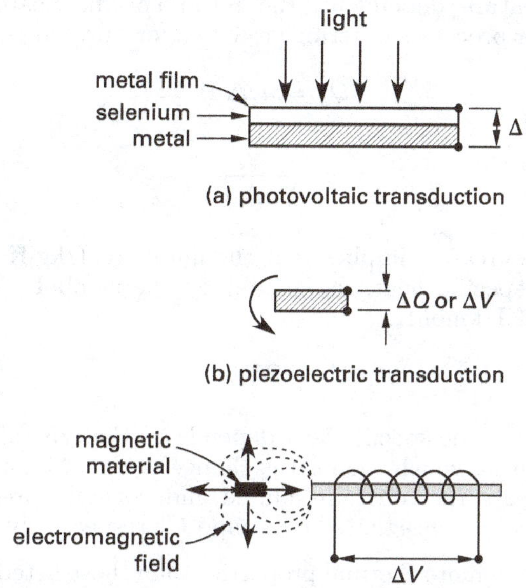

4. THERMAL PROPERTIES

Equation 35.5 and Eq. 35.6: Specific Heat

$$Q = C_p \Delta T \quad \text{[constant pressure]} \qquad 35.5$$

$$Q = C_v \Delta T \quad \text{[constant volume]} \qquad 35.6$$

Description

An increase in internal energy is needed to cause a rise in temperature. Different substances differ in the quantity of heat needed to produce a given temperature increase.

The *specific heat* (known as the *specific heat capacity*), c, of a substance is the heat energy, q, required to change the temperature of one unit mass of the substance by one degree. The *molar specific heat*, conventionally designated by C, is the heat energy, Q, required to change the temperature of one mole of the substance by one degree. Specific heat capacity can be presented on a volume basis (e.g., J/m^3·°C), but a *volumetric heat capacity* is rarely encountered in practice outside of composite materials.[2] Even then, values of the volumetric heat capacity must usually be calculated from specific heats (by mass) and densities. The total heat energy required, Q_t, depends on the total mass or total number of moles.[3]

[2]The *NCEES Handbook* introduces C_v in reference to "constant volume," then follows it closely with a statement that "the heat capacity of a material can be reported as energy per degree per unit mass or per unit volume." The volumetric heat capacity is not related to C_v and is so rarely encountered that it doesn't have a common differentiating symbol other than "VHC."

[3]The *NCEES Handbook* describes Eq. 35.5 and Eq. 35.6 as the "...amount of heat required to raise the temperature of something..." "Something" here means "an entire object" as opposed to "some material." Without a definition of "something," the equations are ambiguous and misleading, as they are implicitly valid otherwise only for one unit mass or one mole.

Because specific heats of solids and liquids are slightly temperature dependent, the mean specific heats are used for processes covering large temperature ranges.

$$Q_t = mc\Delta T$$

$$c = \frac{Q_t}{m\Delta T}$$

The lowercase c implies that the units are J/kg·K. The molar specific heat, designated by the symbol C, has units of J/kmol·K.

$$C = \text{MW} \times c$$

For gases, the specific heat depends on the type of process during which the heat exchange occurs. Molar specific heats for constant-volume and constant-pressure processes are designated by C_v and C_p, respectively.

There are more thermal properties than those listed.

Equation 35.7: Coefficient of Thermal Expansion

$$\alpha = \frac{\varepsilon}{\Delta T} \quad 35.7$$

Variation

$$\alpha = \frac{\Delta L}{L_0 \Delta T}$$

Description

If the temperature of an object is changed, the object will experience length, area, and volume changes. The magnitude of these changes will depend on the *thermal expansion coefficient* (*coefficient of linear thermal expansion*), α, calculated from the engineering strain, ε, and the change in temperature, ΔT.

5. MECHANICAL PROPERTIES

Mechanical properties are those that describe how a material will react to external forces. Materials are commonly classified by their mechanical properties, including strength, hardness, and roughness. Typical design values of various mechanical properties are given in Table 35.5. Various mechanical properties are covered in the following sections.

6. CLASSIFICATION OF MATERIALS

When used to describe engineering materials, the terms "strong" and "tough" are not synonymous. Similarly, "weak," "soft," and "brittle" have different engineering meanings. A *strong material* has a high ultimate strength, whereas a *weak material* has a low ultimate strength. A *tough material* will yield greatly before breaking, whereas a *brittle material* will not. (A brittle material is one whose strain at fracture is less than approximately 0.5%.) A *hard material* has a high modulus of elasticity, whereas a *soft material* does not. Figure 35.2 illustrates some of the possible combinations of these classifications, comparing the material's stress, σ, and strain, ε.

Figure 35.2 *Types of Engineering Materials*

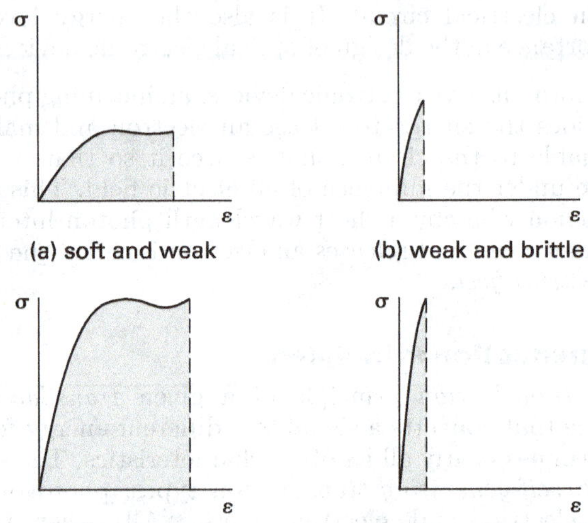

7. ENGINEERING STRESS AND STRAIN

Figure 35.3 shows a *load-elongation curve* of *tensile test* data for a ductile ferrous material (e.g., low-carbon steel or other body-centered cubic (BCC) transition metal). In this test, a prepared material sample (i.e., a *specimen*) is axially loaded in tension, and the resulting elongation, ΔL, is measured as the load, F, increases.

Figure 35.3 *Typical Tensile Test of a Ductile Material*

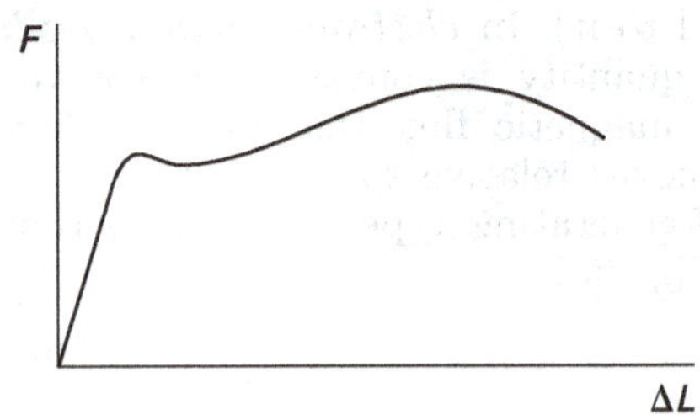

When elongation is plotted against the applied load, the graph is applicable only to an object with the same length and area as the test specimen. To generalize the test results, the data are converted to stresses and strains.

Table 35.5 Average Mechanical Properties of Typical Engineering Materials (customary U.S. units)[a,b]

materials		specific weight, γ (lbf/in^3)	modulus of elasticity, E (10^3 ksi)	modulus of rigidity, G (10^3 ksi)	yield strength, σ_y (ksi)[c]			ultimate strength, σ_u (ksi)[c]			% elongation in 2 in specimen	Poisson's ratio, ν	coefficient of thermal expansion, α $(10^{-6})/°F$
					tens.	comp.	shear	tens.	comp.	shear			
metallic													
aluminum wrought alloys	2014-T6	0.101	10.6	3.9	60	60	25	68	68	42	10	0.35	12.8
	6061-T6	0.098	10.0	3.7	37	37	19	42	42	27	12	0.35	13.1
cast iron alloys	gray ASTM 20	0.260	10.0	3.9	–	–	–	26	97	–	0.6	0.28	6.70
	malleable ASTM A197	0.263	25.0	9.8	–	–	–	40	83	–	5	0.28	6.60
copper alloys	red brass C83400	0.316	14.6	5.4	11.4	11.4	–	35	35	–	35	0.35	9.80
	bronze C86100	0.319	15.0	5.6	50	50	–	95	95	–	20	0.34	9.60
magnesium alloy	Am 1004-T61	0.066	6.48	2.5	22	22	–	40	40	22	1	0.30	14.3
steel alloys	structural A36	0.284	29.0	11.0	36	36	–	58	58	–	30	0.32	6.60
	stainless 304	0.284	28.0	11.0	30	30	–	75	75	–	40	0.27	9.60
	tool L2	0.295	29.0	11.0	102	102	–	116	116	–	22	0.32	6.50
titanium alloy	Ti-6Al-4V	0.160	17.4	6.4	134	134	–	145	145	–	16	0.36	5.20
nonmetallic													
concrete	low strength	0.086	3.20	–	–	–	1.8	–	–	–	–	0.15	6.0
	high strength	0.086	4.20	–	–	–	5.5	–	–	–	–	0.15	6.0
plastic reinforced	Kevlar 49	0.0524	19.0	–	–	–	–	104	70	10.2	2.8	0.34	–
	30% glass	0.0524	10.5	–	–	–	–	13	19	–	–	0.34	–
wood select structural grade	Douglas Fir	0.017	1.90	–	–	–	–	0.30[d]	3.78[e]	0.90[e]	–	0.29[f]	–
	White Spruce	0.130	1.40	–	–	–	–	0.36[d]	5.18[e]	0.97[e]	–	0.31[f]	–

[a]Use these values for the specific alloys and temper listed. For all other materials, refer to Table 35.1.

[b]Specific values may vary for a particular material due to alloy or mineral composition, mechanical working of the specimen, or heat treatment. For a more exact value, reference books for the material should be consulted.

[c]The yield and ultimate strengths for ductile materials can be assumed to be equal for both tension and compression.

[d]Measured perpendicular to the grain.

[e]Measured parallel to the grain.

[f]Deformation measured perpendicular to the grain when the load is applied along the grain.

Source: Hibbeler, R. C., *Mechanics of Materials*, 4th ed., Prentice Hall, 2000.

Equation 35.8 and Eq. 35.9: Engineering Stress and Strain

$$\sigma = \frac{F}{A_0} \tag{35.8}$$

$$\varepsilon = \frac{\Delta L}{L_0} \tag{35.9}$$

Description

Equation 35.8 describes *engineering stress*, σ (usually called *stress*), which is the load per unit original area. Typical units of engineering stress are MPa.

Equation 35.9 describes *engineering strain*, ε (usually called *strain*), which is the elongation of the test specimen expressed as a percentage or decimal fraction of the original length. The units m/m are also sometimes used for strain.[4]

If the stress-strain data are plotted, the shape of the resulting line will be essentially the same as the force-elongation curve, although the scales will differ.

[4]In the NCEES *FE Reference Handbook* (*NCEES Handbook*), strain, ε, is the same as creep but is unrelated to permittivity. All three share the same symbol in this section.

Example 3

A 100 mm gage length is marked on an aluminum rod. The rod is strained so that the gage marks are 109 mm apart. The strain is most nearly

(A) 0.001

(B) 0.01

(C) 0.1

(D) 1.0

Solution

From Eq. 35.9, the strain is

$$\varepsilon = \frac{\Delta L}{L_0} = \frac{109\ \text{mm} - 100\ \text{mm}}{100\ \text{mm}} = 0.09 \quad (0.1)$$

The answer is (C).

Equation 35.10 Through Eq. 35.12: True Stress and Strain

$$\sigma_T = \frac{F}{A} \qquad 35.10$$

$$\varepsilon_T = \frac{dL}{L} \qquad 35.11$$

$$\varepsilon_T = \ln(1+\varepsilon) \qquad 35.12$$

Description

As the stress increases during a tensile test, the length of a specimen increases, and the area decreases. The engineering stress and strain are not *true stress and strain parameters*, σ_T and ε_T, which must be calculated from instantaneous values of length, L, and area, A.[5] Figure 35.4 illustrates engineering and true stresses and strains for a ferrous alloy. Although true stress and strain are more accurate, most engineering work has traditionally been based on engineering stress and strain, which is justifiable for two reasons: (1) design using ductile materials is limited to the elastic region where engineering and true values differ little, and (2) the reduction in area of most parts at their service stresses is not known; only the original area is known.

Figure 35.4 *True and Engineering Stresses and Strains for a Ferrous Alloy*

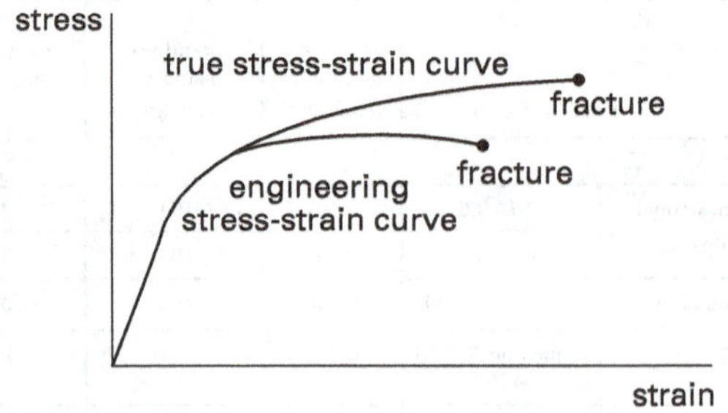

8. STRESS-STRAIN CURVES

Equation 35.13: Hooke's Law

$$\sigma = E\varepsilon \qquad 35.13$$

Variation

$$E = \frac{F/A_0}{\Delta L/L_0} = \frac{FL_0}{A_0\Delta L}$$

Description

Segment OA in Fig. 35.5 is a straight line. The relationship between the stress and the strain in this linear region is given by *Hooke's law*, Eq. 35.13.

Figure 35.5 *Typical Stress-Strain Curve for Steel*

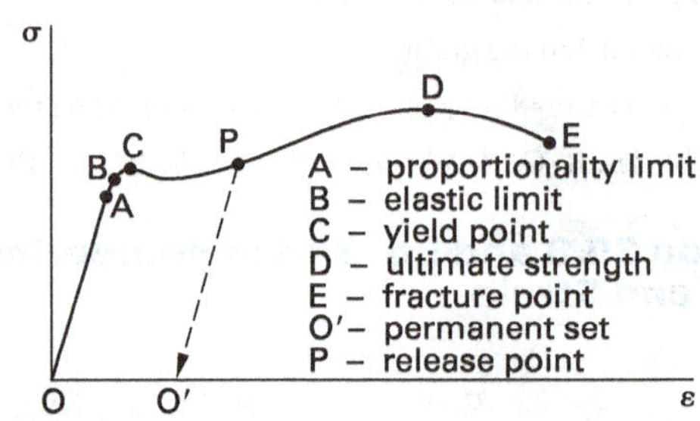

The slope of the line segment OA is the *modulus of elasticity*, E, also known as *Young's modulus* or the *elastic modulus*. Table 35.5 lists approximate values of the modulus of elasticity for materials at room temperature. The modulus of elasticity will be lower at higher temperatures.

Example 4

A test specimen with a circular cross section has an initial gage length of 500 mm and an initial diameter of 60 mm. The specimen is placed in a tensile test apparatus. When the instantaneous tensile force in the specimen is 50 kN, the specimen has a longitudinal

[5]The *NCEES Handbook* is inconsistent in representing change in length. ΔL in Eq. 35.9 is the same as dL in Eq. 35.11.

elongation of 0.16 mm and a lateral decrease in diameter of 0.01505 mm. What is most nearly the modulus of elasticity?

(A) 30×10^9 Pa

(B) 46×10^9 Pa

(C) 55×10^9 Pa

(D) 70×10^9 Pa

Solution

The area of the 60 mm bar is

$$A_0 = \frac{\pi d_0^2}{4} = \frac{\pi\left(\frac{60\ \text{mm}}{1000\ \frac{\text{mm}}{\text{m}}}\right)^2}{4} = 2.827 \times 10^{-3}\ \text{m}^2$$

Using Eq. 35.13 and its variation, the modulus of elasticity is

$$\sigma = E\varepsilon$$

$$\begin{aligned} E &= \frac{\sigma}{\varepsilon} = \frac{FL_0}{A_0\Delta L} = \frac{(50\ \text{kN})\left(1000\ \frac{\text{N}}{\text{kN}}\right)(500\ \text{mm})}{(2.827 \times 10^{-3}\ \text{m}^2)(0.16\ \text{mm})} \\ &= 55.26 \times 10^9\ \text{N/m}^2 \quad (55 \times 10^9\ \text{Pa}) \end{aligned}$$

The answer is (C).

9. POINTS ALONG THE STRESS-STRAIN CURVE

The stress at point A in Fig. 35.5 is known as the *proportionality limit* (i.e., the maximum stress for which the linear relationship is valid). Strain in the *proportional region* is called *proportional* (or *linear*) *strain.*

The *elastic limit*, point B in Fig. 35.5, is slightly higher than the proportionality limit. As long as the stress is kept below the elastic limit, there will be no *permanent set* (*permanent deformation*) when the stress is removed. Strain that disappears when the stress is removed is known as *elastic strain*, and the stress is said to be in the *elastic region.* When the applied stress is removed, the *recovery* is 100%, and the material follows the original curve back to the origin.

If the applied stress exceeds the elastic limit, the recovery will be along a line parallel to the straight line portion of the curve, as shown in the line segment PO′. The strain that results (line OO′) is permanent set (i.e., a permanent deformation). The terms *plastic strain* and *inelastic strain* are used to distinguish this behavior from the elastic strain.

For steel, the *yield point*, point C, is very close to the elastic limit. For all practical purposes, the *yield strength* or *yield stress*, S_y, can be taken as the stress that accompanies the beginning of plastic strain. Yield strengths are reported in MPa.

Most nonferrous materials, such as aluminum, magnesium, copper, and other face-centered cubic (FCC) and hexagonal close-packed (HCP) metals, do not have well-defined yield points. In such cases, the yield point is usually taken as the stress that will cause a 0.2% *parallel offset* (i.e., a plastic strain of 0.002), shown in Fig. 35.6. However, the yield strength can also be defined by other offset values or by total strain characteristics.

Figure 35.6 *Yield Strength of a Nonferrous Metal*

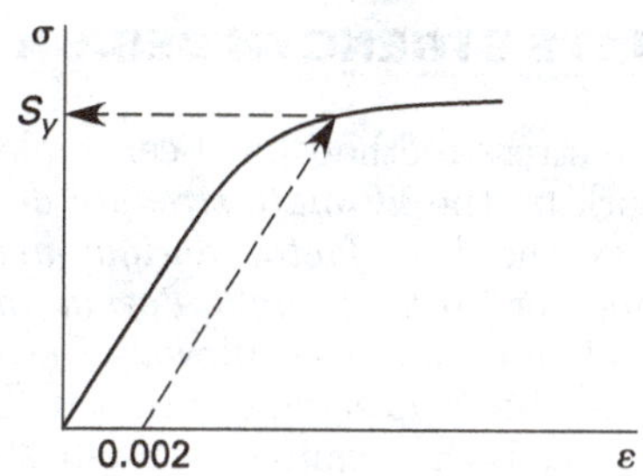

The *ultimate strength* or *tensile strength*, S_u, point D in Fig. 35.5, is the maximum stress the material can support without failure. This property is seldom used in the design of ductile material, since stresses near the ultimate strength are accompanied by large plastic strains.

The *breaking strength* or *fracture strength*, S_f, is the stress at which the material actually fails (point E in Fig. 35.5). For ductile materials, the breaking strength is less than the ultimate strength, due to the necking down in cross-sectional area that accompanies high plastic strains.

10. ALLOWABLE STRESS DESIGN

Once an actual stress has been determined, it can be compared to the *allowable stress.* In engineering design, the term "allowable" always means that a factor of safety has been applied to the governing material strength.

$$\text{allowable stress} = \frac{\text{material strength}}{\text{factor of safety}}$$

For ductile materials, the material strength used is the yield strength. For steel, the factor of safety, FS, ranges from 1.5 to 2.5, depending on the type of steel and the application. Higher factors of safety are seldom necessary in normal, noncritical applications, due to steel's predictable and reliable performance.

$$\sigma_a = \frac{S_y}{\text{FS}} \quad \text{[ductile]}$$

For brittle materials, the material strength used is the ultimate strength. Since brittle failure is sudden and unpredictable, the factor of safety is high (e.g., in the 6 to 10 range).

$$\sigma_a = \frac{S_u}{\text{FS}} \quad \text{[brittle]}$$

If an actual stress is less than the allowable stress, the design is considered acceptable. This is the principle of the *allowable stress design method*, also known as the *working stress design method.*

$$\sigma_{\text{actual}} \leq \sigma_a$$

11. ULTIMATE STRENGTH DESIGN

The allowable stress method has been replaced in most structural work by the *ultimate strength design method*, also known as the *load factor design method*, *plastic design method*, or just *strength design method.* This design method does not use allowable stresses at all. Rather, the member is designed so that its actual *nominal strength* exceeds the required ultimate strength.[6]

The *ultimate strength* (i.e., the required strength) of a member is calculated from the actual *service loads* and multiplicative factors known as *overload factors* or *load factors.* Usually, a distinction is made between dead loads and live loads.[7] For example, the required ultimate moment-carrying capacity in a concrete beam designed according to the American Concrete Institute's *Building Code Requirements for Structural Concrete* (ACI 318) would be[8]

$$M_u = 1.2M_{\text{dead load}} + 1.6M_{\text{live load}}$$

The *nominal strength* (i.e., the actual ultimate strength) of a member is calculated from the dimensions and materials. A *capacity reduction factor*, ϕ, of 0.70 to 0.90 is included in the calculation to account for typical workmanship and increase required strength. The moment criteria for an acceptable design is

$$M_n \geq \frac{M_u}{\phi}$$

12. DUCTILE AND BRITTLE BEHAVIOR

Ductility is the ability of a material to yield and deform prior to failure.[9] Not all materials are ductile. *Brittle materials*, such as glass, cast iron, and ceramics, can support only small strains before they fail catastrophically, without warning. As the stress is increased, the elongation is linear, and Hooke's law can be used to predict the strain. Failure occurs within the linear region, and there is very little, if any, *necking down* (i.e., a localized decrease in cross-sectional area). Since the failure occurs at a low strain, brittle materials are not ductile.

Figure 35.7 illustrates typical stress-strain curves for ductile and brittle materials.

Figure 35.7 *Stress-Strain Curves for Ductile and Brittle Materials*

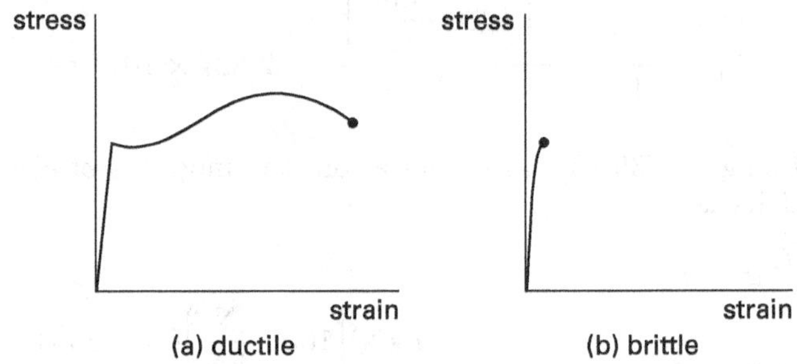

Ductility, μ, is a ratio of two quantities—one of which is related to catastrophic failure (i.e., collapse), and the other related to the loss of serviceability (i.e., yielding). For example, a building's ductility might be the ratio of earthquake energy it takes to collapse the building to the earthquake energy that just causes the beams and columns to buckle or doorframes to warp.

$$\mu = \frac{\text{energy at collapse}}{\text{energy at loss of serviceability}}$$

In contrast to the ductility of an entire building, various measures of ductility are calculated from test specimens for engineering materials. Definitions based on length, area, and the volume of the test specimen are in use. If ductility is to be based on test specimen length, the following definition might be used.

$$\mu = \frac{L_u}{L_y} = \frac{\varepsilon_u}{\varepsilon_y}$$

[6]It is a characteristic of the ultimate strength design method that the term "strength" actually means load, shear, or moment. Strength seldom, if ever, refers to stress. Therefore, the nominal strength of a member might be the load (in newtons) or moment (in N·m) that the member supports at plastic failure.

[7]*Dead load* is an inert, inactive load, primarily due to the structure's own weight. *Live load* is the weight of all nonpermanent objects, including people and furniture, in the structure.

[8]ACI 318 has been adopted as the source of concrete design rules in the United States.

[9]The *NCEES Handbook* gives an incorrect and misleading statement when it says "Ductility (also called percent elongation) [is the] permanent engineering strain after failure." Ductility is a ratio of two quantities, not a percentage. Although there are many measures of ductility, none of them involve the permanent (snapped-back) set of a failed member. Percent elongation at failure might be used to categorize a ductile material, but it is not the same as ductility.

Equation 35.14: Percent Elongation

$$\%\ \text{elongation} = \left(\frac{\Delta L}{L_o}\right) \times 100\% \qquad 35.14$$

Variation

$$\%\ \text{elongation} = \frac{L_f - L_0}{L_0} \times 100\% = \varepsilon_f \times 100\%$$

Description

In contrast to ductility, the *percent elongation at failure* is based on the fracture length, as in Eq. 35.14, or fracture strain, ε_f, shown in Fig. 35.8 and the variation of Eq. 35.14. Percent elongation at failure might be used to categorize a ductile material, but it is not ductility. Since ductility is a ratio of energy absorbed at two points on the loading curve (as represented by the area under the curve), the ultimate length (area, volume, etc.) is measured just prior to fracture, not after. The ultimate length is not the same as what is commonly referred to as "fracture length." The *fracture length* is the length obtained after failure by measuring the two pieces of the failed specimen placed together end-to-end. This length includes the permanent, plastic strain but does not include the recovered elastic strain. The work (energy) required to elastically strain the failed member is not considered with this measure of fracture length.

If area is the measured parameter, the term *reduction in area*, *q*, is used. At failure, the reduction in area due to necking down will be 50% or greater for ductile materials and less than 10% for brittle materials. Reduction in area can be used to categorize a ductile material, but it is not ductility.

$$q_f = \frac{A_0 - A_f}{A_0}$$

Figure 35.8 *Fracture and Ultimate Strain*

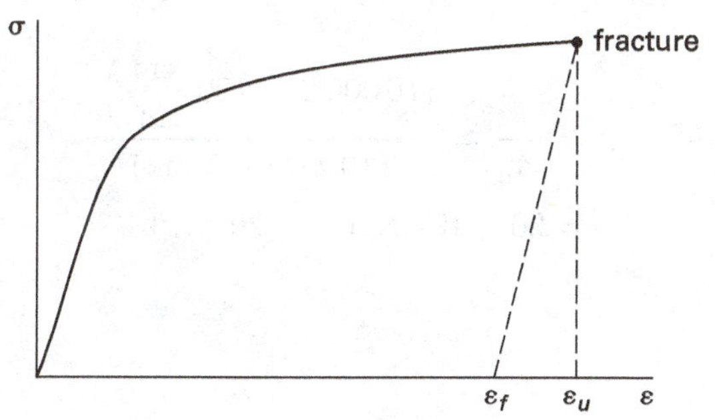

13. CRACK PROPAGATION IN BRITTLE MATERIALS

If a material contains a crack, stress is concentrated at the tip or tips of the crack. A crack in the surface of the material will have one tip (i.e., a stress concentration point); an internal crack in the material will have two tips. This increase in stress can cause the crack to propagate (grow) and can significantly reduce the material's ability to bear loads. Other things being equal, a crack in the surface of a material has a more damaging effect.

There are three modes of *crack propagation*, as illustrated by Fig. 35.9.

- *opening or tensile:* forces act perpendicular to the crack, which pulls the crack open, as shown in Fig. 35.9(a). This is known as mode I.
- *in-plane shear or sliding:* forces act parallel to the crack, which causes the crack to slide along itself, as shown in Fig. 35.9(b). This is known as mode II.
- *out-of-plane shear or pushing (pulling):* forces act perpendicular to the crack, tearing the crack apart, as shown in Fig. 35.9(c). This is known as mode III.

Figure 35.9 *Crack Propagation Modes*

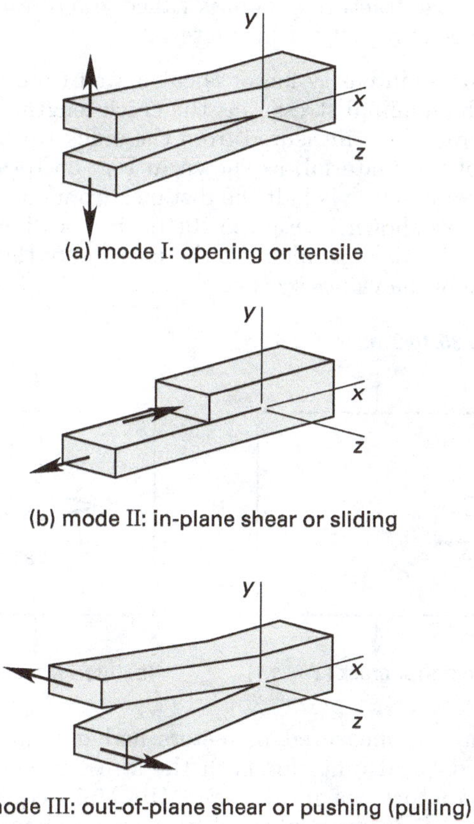

Equation 35.15: Fracture Toughness

$$K_{IC} = Y\sigma\sqrt{\pi a} \qquad \textit{35.15}$$

Values

crack location	geometrical factor, Y
internal	1.0
surface (exterior)	1.1

Description

Fracture toughness is the amount of energy required to propagate a preexisting flaw. Fracture toughness is quantified by a *stress intensity factor*, K, a measure of energy required to grow a thin crack. For a mode I crack (see Fig. 35.9(a)), the stress intensity factor for a crack is designated K_{IC} or K_{Ic}. The stress intensity factor can be used to predict whether an existing crack will propagate through the material. When K_{IC} reaches a critical value, *fast fracture* occurs. The crack suddenly begins to propagate through the material at the speed of sound, leading to catastrophic failure. This critical value at which fast fracture occurs is called *fracture toughness*, and it is a property of the material.

The stress intensity factor is calculated from Eq. 35.15. σ is the nominal stress. a is the crack length. For a surface crack, a is measured from the crack tip to the surface of the material, as shown in Fig. 35.10(a). For an internal crack, a is half the distance from one tip to the other, as shown in Fig. 35.10(b). Y is a dimensionless factor that is dependent on the location of the crack, as shown in the values section.

Figure 35.10 Crack Length

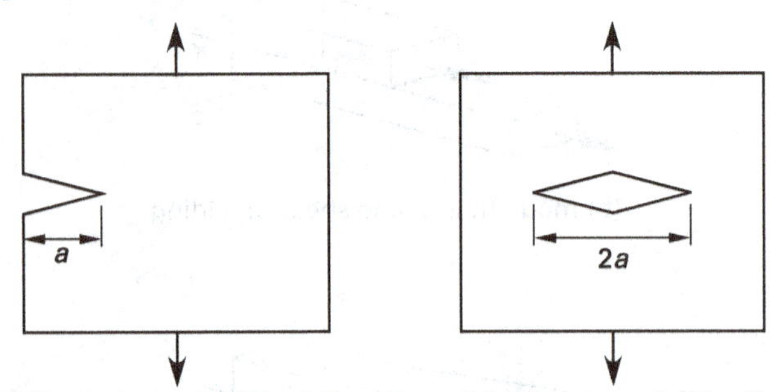

(a) exterior crack (Y = 1.1) (b) interior crack (Y = 1)

When a is measured in meters and σ is measured in MPa, typical units for both the stress intensity factor and fracture toughness are $\text{MPa}\cdot\sqrt{\text{m}}$, equivalent to $\text{MN/m}^{3/2}$. Typical values of fracture toughness for various materials are given in Table 35.6.

Table 35.6 Representative Values of Fracture Toughness

material	K_{IC} ($\text{MPa}\cdot\sqrt{\text{m}}$)	K_{IC} ($\text{ksi}\cdot\sqrt{\text{in}}$)
Al 2014-T651	24.2	22
Al 2024-T3	44	40
52100 steel	14.3	13
4340 steel	46	42
alumina	4.5	4.1
silicon carbide	3.5	3.2

Example

An aluminum alloy plate containing a 2 cm long crack is 10 cm wide and 0.5 cm thick. The plate is pulled with a uniform tensile force of 10 000 N.

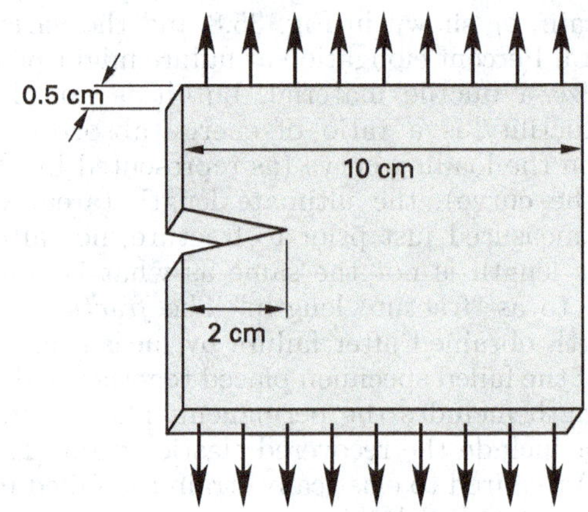

What is most nearly the stress intensity factor at the end of the crack?

(A) $2.1\ \text{MPa}\cdot\sqrt{\text{m}}$

(B) $5.5\ \text{MPa}\cdot\sqrt{\text{m}}$

(C) $12\ \text{MPa}\cdot\sqrt{\text{m}}$

(D) $21\ \text{MPa}\cdot\sqrt{\text{m}}$

Solution

From Eq. 35.8, the nominal stress is

$$\sigma = \frac{F}{A_0} = \frac{(10\,000\ \text{N})\left(100\ \frac{\text{cm}}{\text{m}}\right)^2}{(10\ \text{cm})(0.5\ \text{cm})} = 20\times10^6\ \text{N/m}^2 \quad (20\ \text{MPa})$$

Since this is an exterior crack, $Y = 1.1$. Using Eq. 35.15, the stress intensity factor is

$$\begin{aligned} K_{IC} &= Y\sigma\sqrt{\pi a} \\ &= (1.1)(20\ \text{MPa})\sqrt{\pi\left(\frac{2\ \text{cm}}{100\ \frac{\text{cm}}{\text{m}}}\right)} \\ &= 5.51\ \text{MPa}\cdot\sqrt{\text{m}}\quad (5.5\ \text{MPa}\cdot\sqrt{\text{m}}) \end{aligned}$$

The answer is (B).

14. FATIGUE

A material can fail after repeated stress loadings even if the stress level never exceeds the ultimate strength, a condition known as *fatigue failure.*

The behavior of a material under repeated loadings is evaluated by an *endurance test* (or *fatigue test*). A specimen is loaded repeatedly to a specific stress amplitude, S, and the number of applications of that stress required to cause failure, N, is counted. *Rotating beam tests* that load the specimen in bending, as shown in Fig. 35.11, are more common than alternating deflection and push-pull tests, but are limited to round specimens. The *mean stress* is zero in rotating beam tests.

Figure 35.11 *Rotating Beam Test*

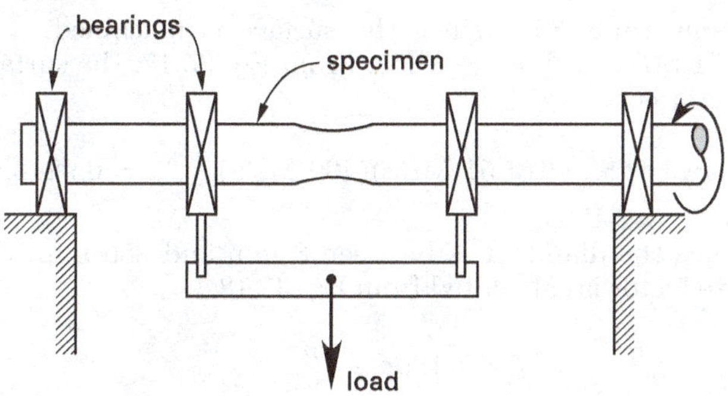

This procedure is repeated for different stresses using different specimens. The results of these tests are graphed on a semi-log plot, resulting in the *S-N curve* shown in Fig. 35.12.

Figure 35.12 *Typical S-N Curve for Steel*

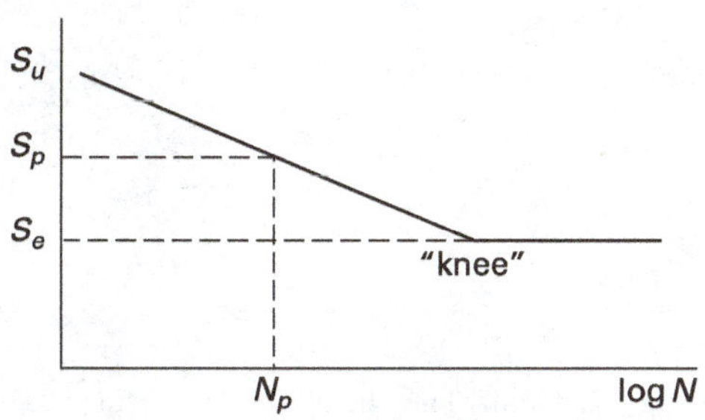

For a particular stress level, such as S_p in Fig. 35.12, the number of cycles required to cause failure, N_p, is the *fatigue life.* S_p is the *fatigue strength* corresponding to N_p.

For steel that is subjected to fewer than approximately 10^3 loadings, the fatigue strength approximately equals the ultimate strength. (Although *low-cycle fatigue theory* has its own peculiarities, a part experiencing a small number of cycles can usually be designed or analyzed as for static loading.) The curve is linear between 10^3 and approximately 10^6 cycles if a logarithmic N-scale is used. Above 10^6 cycles, there is no further decrease in strength.

Below a certain stress level, called the *endurance limit*, *endurance stress*, or *fatigue limit*, S_e', the material will withstand an almost infinite number of loadings without experiencing failure. This is characteristic of steel and titanium. If a dynamically loaded part is to have an infinite life, the stress must be kept below the endurance limit.

The yield strength is an irrelevant factor in cyclic loading. Fatigue failures are fracture failures, not yielding failures. They start with microscopic cracks at the material surface. Some of the cracks are present initially; others form when repeated cold working reduces the ductility in strain-hardened areas. These cracks grow minutely with each loading. Since cracks start at the location of surface defects, the endurance limit is increased by proper treatment of the surface. Such treatments include polishing, surface hardening, shot peening, and filleting joints.

Equation 35.16 Through Eq. 35.18: Endurance Limit Modifying Factors

$$S_e = k_a k_b k_c k_d k_e S_e' \qquad 35.16$$

$$k_a = aS_{\text{ut}}^b \qquad 35.17$$

$$k_b = 1.189 d_{\text{eff}}^{-0.097} \quad [8\ \text{mm} \le d \le 250\ \text{mm}] \qquad 35.18$$

Description

The endurance limit is not a true property of the material since the other significant influences, particularly surface finish, are never eliminated. However, representative values of S_e' obtained from ground and polished specimens provide a baseline to which other factors can be applied to account for the effects of surface finish, temperature, stress concentration, notch sensitivity, size, environment, and desired reliability. These other influences are accounted for by *endurance limit modifying factors* that are used to calculate a working endurance strength, S_e, for the material.

The *surface factor*, k_a, is calculated from Eq. 35.17 using values of the factors a and b found from Table 35.7.

Table 35.7 Factors for Calculating k_a

	a		
surface finish	(kpsi)	(MPa)	b
ground	1.34	1.58	−0.085
machined or cold-drawn (CD)	2.70	4.51	−0.265
hot rolled	14.4	57.7	−0.718
as forged	39.9	272.0	−0.995

The *size factor*, k_b, and *load factor*, k_c, are determined for axial loadings from Table 35.8 and for bending and torsion from Table 35.9. For bending and torsion where the diameter, d, is between 8 mm and 250 mm, k_b is calculated from Eq. 35.18.

As the size gets larger, the endurance limit decreases due to the increased number of defects in a larger volume. Since the endurance strength, S_e', is derived from a circular specimen with a diameter of 7.6 mm, the size modification factor is 1.0 for bars of that size.[10] d_{eff} is the effective dimension.[11] Simplistically, for noncircular cross-sections, the smallest cross-sectional dimension should be used, and for a solid circular specimen in rotating bending, $d_{\text{eff}} = d$. For a nonrotating or noncircular cross section, d_{eff} is obtained by equating the area of material stressed above 95% of the maximum stress to the same area in the rotating-beam specimen of the same length. That area is designated $A_{0.95\sigma}$. For a nonrotating solid rectangular section with width w and thickness t, the effective dimension is

$$d_{\text{eff}} = 0.808\sqrt{wt}$$

Table 35.8 Endurance Limit Modifying Factors for Axial Loading

size factor, k_b	1
load factor, k_c	
$S_{\text{ut}} \le 1520$ MPa	0.923
$S_{\text{ut}} < 1520$ MPa	1

Table 35.9 Endurance Limit Modifying Factors for Bending and Torsion

size factor, k_b	
$d \le 8$ mm	1
8 mm $\le d \le$ 250 mm	use Eq. 35.18
$d > 250$ mm	between 0.6 and 0.75
load factor, k_c	
bending	1
torsion	0.577

Values of the *temperature factor*, k_d, and the *miscellaneous effects factor*, k_e, are found from Table 35.10. The miscellaneous effects factor is used to account for various factors that reduce strength, such as corrosion, plating, and residual stress.

Table 35.10 Additional Endurance Limit Modifying Factors

temperature factor, k_d	1 [$T \le 450$°C]
miscellaneous effects factor, k_e	1, unless otherwise specified

Example 6

A 25 mm diameter machined bar is exposed to a fluctuating bending load in a 200°C environment. The bar is made from ASTM A36 steel, which has a yield strength of 250 MPa, an ultimate tensile strength of 400 MPa, and a density of 7.8 g/cm^3. The endurance limit is determined to be 200 MPa. What is most nearly the fatigue strength of the steel?

(A) 95 MPa

(B) 130 MPa

(C) 160 MPa

(D) 200 MPa

Solution

Determine the endurance limit modifying factors.

From Table 35.7, since the surface is machined, a = 4.51 MPa, and $b = -0.265$. From Eq. 35.17, the surface factor is

$$k_a = aS_{\text{ut}}^b = (4.51 \text{ MPa})(400 \text{ MPa})^{-0.265} = 0.9218$$

Since the diameter is between 8 mm and 250 mm, the size factor is calculated from Eq. 35.18.

$$\begin{aligned} k_b &= 1.189d_{\text{eff}}^{-0.097} \\ &= (1.189)(25 \text{ mm})^{-0.097} \\ &= 0.8701 \end{aligned}$$

From Table 35.9, $k_c = 1$ for bending stress. From Table 35.10, the temperature is less than 450°C, so $k_d = 1$. $k_e = 1$ since it was not specified otherwise.

[10]In Table 35.9, the *NCEES Handbook* gives the limits for the use of Eq. 35.18 as "8 mm < $d \le$ 250 mm." This should be "8 mm $\le d \le$ 250 mm" so as to be unambiguous at $d = 8$ mm.

[11]The *NCEES Handbook* does not give any explanation or guidance in determining the effective dimension.

Using Eq. 35.16, the approximate fatigue strength is

$$\begin{aligned} S_e &= k_a k_b k_c k_d k_e S_e' \\ &= (0.9218)(0.8701)(1)(1)(1)(200 \text{ MPa}) \\ &= 160.4 \text{ MPa} \quad (160 \text{ MPa}) \end{aligned}$$

The answer is (C).

15. TOUGHNESS

Toughness is a measure of a material's ability to yield and absorb highly localized and rapidly applied stress. A tough material will be able to withstand occasional high stresses without fracturing. Products subjected to sudden loading, such as chains, crane hooks, railroad couplings, and so on, should be tough. One measure of a material's toughness is the *modulus of toughness*, which is the *strain energy* or work per unit volume required to cause fracture. This is the total area under the stress-strain curve. Another measure is the *notch toughness*, which is evaluated by measuring the *impact energy* that causes a notched sample to fail. At 21°C, the energy required to cause failure ranges from 60 J for carbon steels to approximately 150 J for chromium-manganese steels.

16. CHARPY TEST

In the *Charpy test* (*Charpy V-notch test*), which is popular in the United States, a standardized beam specimen is given a 45° notch. The specimen is then centered on simple supports with the notch down. (See Fig. 35.13.) A falling pendulum striker hits the center of the specimen. This test is performed several times with different heights and different specimens until a sample fractures.

Figure 35.13 *Charpy Test*

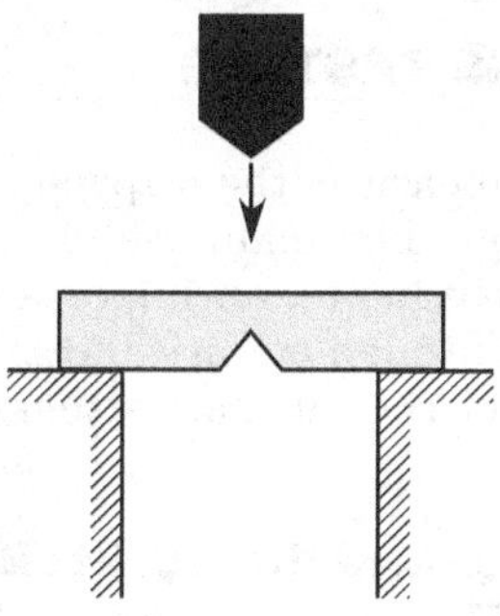

The kinetic energy expended at impact, equal to the initial potential energy, is calculated from the height. It is designated C_V and is expressed in joules (J). The energy required to cause failure is a measure of toughness. Without a notch, the specimen would experience uniaxial stress (tension and compression) at impact. The notch allows triaxial stresses to develop. Most materials become more brittle under triaxial stresses than under uniaxial stresses.

17. DUCTILE-BRITTLE TRANSITION

As temperature is reduced, the toughness of a material decreases. In BCC metals, such as steel, at a low enough temperature the toughness will decrease sharply. The transition from high-energy ductile failures to low-energy brittle failures begins at the *fracture transition plastic* (FTP) *temperature.*

Since the transition occurs over a wide temperature range, the *transition temperature* (also known as the *ductility transition temperature*) is taken as the temperature at which an impact of 20 J will cause failure. (See Table 35.11.) This occurs at approximately −1°C for low-carbon steel.

Table 35.11 *Approximate Ductile Transition Temperatures*

type of steel	ductile transition temperature (°C)
carbon steel	−1
high-strength, low-alloy steel	−18 to −1
heat-treated, high-strength, carbon steel	−32
heat-treated, construction alloy steel	−40 to −62

The appearance of the fractured surface is also used to evaluate the transition temperature. The fracture can be fibrous (from shear fracture) or granular (from cleavage fracture), or a mixture of both. The fracture planes are studied, and the percentages of ductile failure are plotted against temperature. The temperature at which the failure is 50% fibrous and 50% granular is known as *fracture appearance transition temperature* (FATT).

Not all materials have a ductile-brittle transition. Aluminum, copper, other face-centered cubic (FCC) metals, and most hexagonal close-packed (HCP) metals do not lose their toughness abruptly. Figure 35.14 illustrates the failure energy curves for several materials.

Figure 35.14 *Failure Energy versus Temperature*

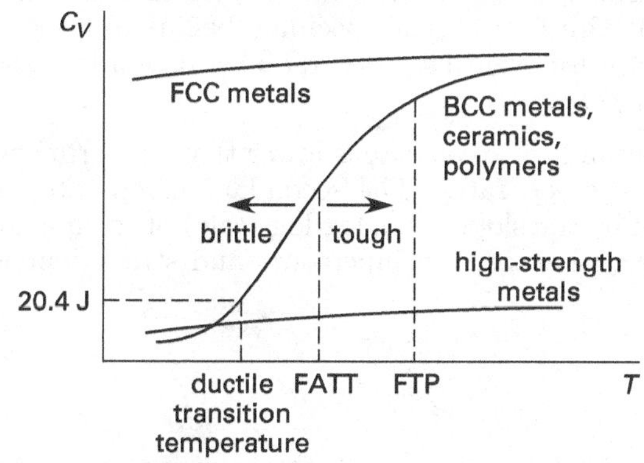

18. CREEP TEST

Creep or *creep strain* is the continuous yielding of a material under constant stress. For metals, creep is negligible at low temperatures (i.e., less than half of the absolute melting temperature), although the usefulness of nonreinforced plastics as structural materials is seriously limited by creep at room temperature.

During a *creep test*, a low tensile load of constant magnitude is applied to a specimen, and the strain is measured as a function of time. The *creep strength* is the stress that results in a specific creep rate, usually 0.001% or 0.0001% per hour. The *rupture strength*, determined from a *stress-rupture test*, is the stress that results in a failure after a given amount of time, usually 100, 1000, or 10,000 hours.

If strain is plotted as a function of time, three different curvatures will be apparent following the initial elastic extension.[12] (See Fig. 35.15.) During the first stage, the *creep rate* ($d\varepsilon/dt$) decreases since strain hardening (dislocation generation and interaction with grain boundaries and other barriers) is occurring at a greater rate than annealing (annihilation of dislocations, climb, cross-slip, and some recrystallization). This is known as *primary creep*.

Figure 35.15 *Stages of Creep*

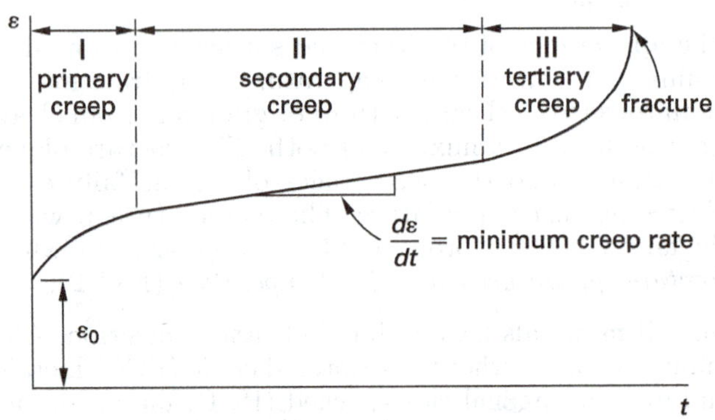

During the second stage, the creep rate is constant, with strain hardening and annealing occurring at the same rate. This is known as *secondary creep* or *cold flow*. During the third stage, the specimen begins to neck down, and rupture eventually occurs. This region is known as *tertiary creep*.

The secondary creep rate is lower than the primary and tertiary creep rates. The secondary creep rate, represented by the slope (on a log-log scale) of the line during the second stage, is temperature and stress dependent. This slope increases at higher temperatures and stresses. The creep rate curve can be represented by the following empirical equation, known as *Andrade's equation*.

$$\varepsilon = \varepsilon_0(1+\beta t^{1/3})e^{kt}$$

Dislocation climb (glide and creep) is the primary creep mechanism, although diffusion creep and grain boundary sliding also contribute to creep on a microscopic level. On a larger scale, the mechanisms of creep involve slip, subgrain formation, and grain-boundary sliding.

Equation 35.19: Creep

$$\frac{d\varepsilon}{dt} = A\sigma^n e^{-Q/RT} \qquad 35.19$$

Value

$\overline{R} = 8314\ \text{J/kmol·K}$

Description

Equation 35.19 calculates creep from the strain, ε, time, t, a constant, A, universal gas constant, $\overline{R}$, absolute temperature, T, applied stress, σ, activation energy, Q, and stress sensitivity, n.[13] The activation energy and stress sensitivity are dependent on the material type and glass transition temperature, T_g, as shown in Table 35.12. The exponent $-Q/\overline{R}T$ is unitless.

Table 35.12 *Creep Parameters*

material	n	Q
polymer		
$<T_g$	2–4	$\geq$ 100 kJ/mol
$>T_g$	6–10	approx. 30 kJ/mol
metals and ceramics	3–10	80–200 kJ/mol

19. HARDNESS TESTING

Hardness tests measure the capacity of a surface to resist deformation. The main use of hardness testing is to verify heat treatments, an important factor in product service life. Through empirical correlations, it is also possible to predict the ultimate strength and toughness of some materials.

Equation 35.20 and Eq. 35.21: Brinell Hardness Test

$$\text{TS}_{\text{MPa}} \approx 3.5(\text{BHN}) \qquad 35.20$$

$$\text{TS}_{\text{psi}} \approx 500(\text{BHN}) \qquad 35.21$$

[12]In Great Britain, the initial elastic elongation, ε_0, is considered the first stage. Therefore, creep has four stages in British nomenclature.

[13]The *NCEES Handbook* is inconsistent in the variable it uses for the universal gas constant. R in Eq. 35.19 is the same as $\overline{R}$ in its Units section and used almost everywhere in the *NCEES Handbook*.

Description

The *Brinell hardness test* is used primarily with iron and steel castings, although it can be used with softer materials. The *Brinell hardness number*, BHN (or HB or H_B), is determined by pressing a hardened steel ball into the surface of a specimen. The diameter of the resulting depression is correlated to the hardness. The standard ball is 10 mm in diameter and loads are 500 kg and 3000 kg for soft and hard materials, respectively.

The Brinell hardness number is the load per unit contact area. If a load, P (in kilograms), is applied through a steel ball of diameter, D (in millimeters), and produces a depression of diameter, d (in millimeters), and depth, t (in millimeters), the Brinell hardness number can be calculated from

$$\begin{aligned} \text{BHN} &= \frac{P}{A_{\text{contact}}} = \frac{P}{\pi D t} \\ &= \frac{2P}{\pi D(D - \sqrt{D^2 - d^2})} \end{aligned}$$

For heat-treated plain-carbon and medium-alloy steels, the ultimate tensile strength, TS, can be approximately calculated from the steel's Brinell hardness number, as shown in Eq. 35.20 and Eq. 35.21.

Other Hardness Tests

The *scratch hardness test*, also known as the *Mohs test*, compares the hardness of the material to that of minerals. Minerals of increasing hardness are used to scratch the sample. The resulting *Mohs scale* hardness can be used or correlated to other hardness scales.

The *file hardness test* is a combination of the cutting and scratch tests. Files of known hardness are drawn across the sample. The file ceases to cut the material when the material and file hardnesses are the same.

The *Rockwell hardness test* is similar to the Brinell test. A steel ball or diamond spheroconical penetrator (known as a *brale indenter*) is pressed into the material. The machine applies an initial load (60 kgf, 100 kgf, or 150 kgf) that sets the penetrator below surface imperfections.[14] Then, a significant load is applied. The Rockwell hardness, R (or HR or H_R), is determined from the depth of penetration and is read directly from a dial.

Although a number of Rockwell scales (A through G) exist, the B and C scales are commonly used for steel. The *Rockwell B scale* is used with a steel ball for mild steel and high-strength aluminum. The *Rockwell C scale* is used with the brale indentor for hard steels having ultimate tensile strengths up to 2 GPa. The *Rockwell A scale* has a wide range and can be used with both soft materials (such as annealed brass) and hard materials (such as cemented carbides).

Other penetration hardness tests include the *Meyer*, *Vickers*, *Meyer-Vickers*, and *Knoop* tests.

Materials

[14]Other Rockwell tests use 15 kgf, 30 kgf, and 45 kgf. The use of kgf units is traditional, and even modern test equipment is calibrated in kgf. Multiply kgf by 9.80665 to get newtons.

36 Engineering Materials

Nomenclature

c	specific heat	kJ/kg·K
C	number of components	–
D	diffusion coefficient	m^2/s
D	distance	m
D_o	proportionality constant	m^2/s
DP	degree of polymerization	–
E	energy	kJ
E	modulus of elasticity	GPa
E_o	oxidation potential	V
f	volumetric fraction	–
f_r	modulus of rupture	MPa
f'_c	compressive strength	MPa
F	degrees of freedom	–
L	length	m
m	mass	kg
M	Martensite transformation temperature	°C
MC	moisture content	–
MW	molecular weight	kg/kmol
n	grain size	–
n	number	–
N	number of grains per unit area	$1/m^2$
P	number of phases	–
P_L	points per unit length	1/m
Q	activation energy	kJ/kmol
R	universal gas constant, 8314	J/kmol·K
R_C	Rockwell hardness (C-scale)	–
S_V	surface area per unit volume	1/m
T	absolute temperature	K
W	water content	%
x	gravimetric fraction	–

Symbols

ρ	density	kg/m^3
σ	stress	MPa

Subscripts

a	activation
ave	average
c	composite
d	diffusion
f	finish
g	glass
i	individual
m	melting
o	original or oxidation
qe	quenched end
s	start

1. CHARACTERISTICS OF METALS

Metals are the most frequently used materials in engineering design. Steel is the most prevalent engineering metal because of the abundance of iron ore, simplicity of production, low cost, and predictable performance. However, other metals play equally important parts in specific products.

Metallurgy is the subject that encompasses the procurement and production of metals. *Extractive metallurgy* is the subject that covers the refinement of pure metals from their ores. Most metals are characterized by the properties in Table 36.1.

Table 36.1 Properties of Most Metals and Alloys

high thermal conductivity (low thermal resistance)
high electrical conductivity (low electrical resistance)
high chemical reactivity[a]
high strength
high ductility[b]
high density
high radiation resistance
highly magnetic (ferrous alloys)
optically opaque
electromagnetically opaque

[a]Some alloys, such as stainless steel, are more resistant to chemical attack than pure metals.
[b]Brittle metals, such as some cast irons, are not ductile.

2. UNIFIED NUMBERING SYSTEM

The *Unified Numbering System* (UNS) was introduced in the mid-1970s to provide a consistent identification of metals and alloys for use throughout the world. The UNS designation consists of one of seventeen single uppercase letter prefixes followed by five digits. Many of the letters are suggestive of the family of metals, as Table 36.2 indicates.

Table 36.2 UNS Alloy Prefixes

A	aluminum
C	copper
E	rare-earth metals
F	cast irons
G	AISI and SAE carbon and alloy steels
H	AISI and SAE H-steels
J	cast steels (except tool steels)
K	miscellaneous steels and ferrous alloys
L	low-melting metals
M	miscellaneous nonferrous metals
N	nickel
P	precious metals
R	reactive and refractory metals
S	heat- and corrosion-resistant steels (stainless and valve steels and superalloys)
T	tool steels (wrought and cast)
W	welding filler metals
Z	zinc

3. FERROUS METALS

Steel and Alloy Steel Grades

The properties of steel can be adjusted by the addition of *alloying ingredients.* Some steels are basically mixtures of iron and carbon. Other steels are produced with a variety of ingredients.

The simplest and most common grades of steel belong to the group of *carbon steels.* Carbon is the primary noniron element, although sulfur, phosphorus, and manganese can also be present. Carbon steel can be subcategorized into *plain carbon steel* (*nonsulfurized carbon steel*), *free-machining steel* (*resulfurized carbon steel*), and *resulfurized and rephosphorized carbon steel.* Plain carbon steel is subcategorized into *low-carbon steel* (less than 0.30% carbon), *medium-carbon steel* (0.30% to 0.70% carbon), and *high-carbon steel* (0.70% to 1.40% carbon).

Low-carbon steels are used for wire, structural shapes, and screw machine parts. Medium-carbon steels are used for axles, gears, and similar parts requiring medium to high hardness and high strength. High-carbon steels are used for drills, cutting tools, and knives.

Low-alloy steels (containing less than 8.0% total alloying ingredients) include the majority of steel alloys but exclude the high-chromium content *corrosion-resistant* (*stainless*) *steels.* Generally, low-alloy steels will have higher strength (e.g., double the yield strength) of plain carbon steel. *Structural steel, high-strength steel,* and *ultrahigh-strength steel* are general types of low-alloy steel.[1]

High-alloy steels contain more than 8.0% total alloying ingredients.

Table 36.3 lists typical alloying ingredients and their effects on steel properties. The percentages represent typical values, not maximum solubilities.

Tool Steel

Each grade of *tool steel* is designed for a specific purpose. As such, there are few generalizations that can be made about tool steel. Each tool steel exhibits its own blend of the three main performance criteria: toughness, wear resistance, and hot hardness.[2]

Some of the generalizations possible are listed as follows.

- An increase in carbon content increases wear resistance and reduces toughness.
- An increase in wear resistance reduces toughness.
- Hot hardness is independent of toughness.
- Hot hardness is independent of carbon content.

[1]The ultrahigh-strength steels, also known as *maraging steels*, are very low-carbon (less than 0.03%) steels, with 15–25% nickel and small amounts of cobalt, molybdenum, titanium, and aluminum. With precipitation hardening, ultimate tensile strengths up to 2.8 GPa, yield strengths up to 1.7 GPa, and elongations in excess of 10% are achieved. Maraging steels are used for rocket motor cases, aircraft and missile turbine housings, aircraft landing gear, and other applications requiring high strength, low weight, and toughness.
[2]The ability of a steel to resist softening at high temperatures is known as *hot hardness* and *red hardness*.

Table 36.3 Steel Alloying Ingredients

ingredient	range (%)	purpose
aluminum	–	deoxidation
boron	0.001–0.003	increase hardness
carbon	0.1–4.0	increase hardness and strength
chromium	0.5–2	increase hardness and strength
	4–18	increase corrosion resistance
copper	0.1–0.4	increase atmospheric corrosion resistance
iron sulfide	–	increase brittleness
manganese	0.23–0.4	reduce brittleness, combine with sulfur
	> 1.0	increase hardness
manganese sulfide	0.8–0.15	increase machinability
molybdenum	0.2–5	increase dynamic and high-temperature strength and hardness
nickel	2–5	increase toughness, increase hardness
	12–20	increase corrosion resistance
	> 30	reduce thermal expansion
phosphorus	0.04–0.15	increase hardness and corrosion resistance
silicon	0.2–0.7	increase strength
	2	increase spring steel strength
	1–5	improve magnetic properties
sulfur	–	(see *iron sulfide* and *manganese sulfide*)
titanium	–	fix carbon in inert particles; reduce martensitic hardness
tungsten	–	increase high-temperature hardness
vanadium	0.15	increase strength

Group A steels are air-hardened, medium-alloy cold-work tool steels. Air-hardening allows the tool to develop a homogeneous hardness throughout, without distortion. This hardness is achieved by large amounts of alloying elements and comes at the expense of wear resistance.

Group D steels are high-carbon, high-chromium tool steels suitable for cold-working applications. These steels are high in abrasion resistance but low in machinability and ductility. Some steels in this group are air hardened, while others are oil quenched. Typical uses are blanking and cold-forming punches.

Group H steels are hot-work tool steels, capable of being used in the 600–1100°C range. They possess good wear resistance, hot hardness, shock resistance, and resistance to surface cracking. Carbon content is low, between 0.35% and 0.65%. This group is subdivided according to the three primary alloying ingredients: chromium, tungsten, or molybdenum. For example, a particular steel might be designated as a "chromium hot-work tool steel."

Group M steels are molybdenum high-speed steels. Properties are very similar to the group T steels, but group M steels are less expensive since one part molybdenum can replace two parts tungsten. For that reason, most high-speed steel in common use is produced from group M. Cobalt is added in large percentages (5–12%) to increase high-temperature cutting efficiency in heavy-cutting (high-pressure cutting) applications.

Group O steels are oil-hardened, cold-work tool steels. These high-carbon steels use alloying elements to permit oil quenching of large tools and are sometimes referred to as *nondeforming steels*. Chromium, tungsten, and silicon are typical alloying elements.

Group S steels are shock-resistant tool steels. Toughness (not hardness) is the main characteristic, and either water or oil may be used for quenching. Group S steels contain chromium and tungsten as alloying ingredients. Typical uses are hot header dies, shear blades, and chipping chisels.

Group T steels are tungsten high-speed tool steels that maintain a sharp hard cutting edge at temperatures in excess of 550°C. The ubiquitous 18-4-1 grade T1 (named after the percentages of tungsten, chromium, and vanadium, respectively) is part of this group. Increases in hot hardness are achieved by simultaneous increases in carbon and vanadium (the key ingredient in these tool steels) and special, multiple-step heat treatments.[3]

Group W steels are water-hardened tool steels. These are plain high-carbon steels (modified with small amounts of vanadium or chromium, resulting in high surface hardness but low hardenability). The combination of high surface hardness and ductile core makes group W steels ideal for rock drills, pneumatic tools, and cold header dies. The limitation on this tool steel group is the loss of hardness that begins at temperatures above 150°C and is complete at 300°C.

[3]For example, the 18-4-1 grade is heated to approximately 550°C for two hours, air cooled, and then heated again to the same temperature. The term *double-tempered steel* is used in reference to this process. Most heat treatments are more complex.

Stainless Steel

Adding chromium improves steel's corrosion resistance. Moderate corrosion resistance is obtained by adding 4–6% chromium to low-carbon steel. (Other elements, specifically less than 1% each of silicon and molybdenum, are also usually added.)

For superior corrosion resistance, larger amounts of chromium are needed. At a minimum level of 12% chromium, steel is *passivated* (i.e., an inert film of chromic oxide forms over the metal and inhibits further oxidation). The formation of this protective coating is the basis of the corrosion resistance of *stainless steel*.[4]

Passivity is enhanced by oxidizers and aeration but is reduced by abrasion that wears off the protective oxide coating. An increase in temperature may increase or decrease the passivity, depending on the abundance of oxygen.

Stainless steels are generally categorized into ferritic, martensitic (heat-treatable), austenitic, duplex, and high-alloy stainless steels.[5]

Ferritic stainless steels contain more than 10% to 27% chromium. The body-centered cubic (BCC) ferrite structure is stable (i.e., does not transform to austenite, a face-centered cubic (FCC) structure) at all temperatures. For this reason, ferritic steels cannot be hardened significantly. Since ferritic stainless steels contain no nickel, they are less expensive than austenitic steels. Turbine blades are typical of the heat-resisting products manufactured from ferritic stainless steels.

The so-called *superferritics* are highly resistant to chloride pitting and crevice corrosion. Superferritics have been incorporated into marine tubing and heat exchangers for power plant condensers. Like all ferritics, however, superferritics experience embrittlement above 475°C.

The *martensitic* (*heat-treatable*) *stainless steels* contain no nickel and differ from ferritic stainless steels primarily in higher carbon contents. Cutlery and surgical instruments are typical applications requiring both corrosion resistance and hardness.

The *austenitic stainless steels* are commonly used for general corrosive applications. The stability of the austenite (a face-centered cubic structure) depends primarily on 4–22% nickel as an alloying ingredient. The basic composition is approximately 18% chromium and 8% nickel.

The so-called *superaustenitics* achieve superior corrosion resistance by adding molybdenum (typically up to about 7%) or nitrogen (typically up to about 14%).

Cast Iron and Wrought Iron

Cast iron is a general name given to a wide range of alloys containing iron, carbon, and silicon, and to a lesser extent, manganese, phosphorus, and sulfur. Generally, the carbon content will exceed 2%. The properties of cast iron depend on the amount of carbon present, as well as the form (i.e., graphite or carbide) of the carbon.

The most common type of cast iron is *gray cast iron*. The carbon in gray cast iron is in the form of graphite flakes. Graphite flakes are very soft and constitute points of weakness in the metal, which simultaneously improve machinability and decrease ductility. Compressive strength of gray cast iron is three to five times the tensile strength.

Magnesium and cerium can be added to improve the ductility of gray cast iron. The resulting *nodular cast iron* (also known as *ductile cast iron*) has the best tensile and yield strengths of all the cast irons. It also has good ductility (typically 5%) and machinability. Because of these properties, it is often used for automobile crankshafts.

White cast iron has been cooled quickly from a molten state. No graphite is produced from the cementite, and the carbon remains in the form of a carbide, Fe_3C.[6] The carbide is hard and is the reason that white cast iron is difficult to machine. White cast iron is used primarily in the production of malleable cast iron.

Malleable cast iron is produced by reheating white cast iron to between 800°C and 1000°C for several days, followed by slow cooling. During this treatment, the carbide is partially converted to nodules of graphitic carbon known as *temper carbon*. The tensile strength is increased to approximately 380 MPa, and the elongation at fracture increases to approximately 18%.

Mottled cast iron contains both cementite and graphite and is between white and gray cast irons in composition and performance.

Compacted graphitic iron (CGI) is a unique form of cast iron with worm-shaped graphite particles. The shape of the graphite particles gives CGI the best properties of both gray and ductile cast iron: twice the strength of gray cast iron and half the cost of aluminum. The higher strength permits thinner sections. (Some engine blocks are 25% lighter than gray iron castings.) Using computer-controlled refining, volume production of CGI with the consistency needed for commercial applications is possible.

---

[4]Stainless steels are corrosion resistant in oxidizing environments. In reducing environments (such as with exposure to hydrochloric and other halide acids and salts), the steel will corrode.

[5](1) There is a fifth category, that of *precipitation-hardened stainless steels*, widely used in the aircraft industry. (Precipitation hardening is also known as *age hardening*.) (2) The *sigma phase* structure that appears at very high chromium levels (e.g., 24–50%) is usually undesirable in stainless steels because it reduces corrosion resistance and impact strength. A notable exception is in the manufacture of automobile engine valves.

[6]White and gray cast irons get their names from the coloration at a fracture.

Wrought iron is low-carbon (less than 0.1%) iron with small amounts (approximately 3%) of slag and gangue in the form of fibrous inclusions. It has good ductility and corrosion resistance. Prior to the use of steel, wrought iron was the most important structural metal.

4. NONFERROUS METALS

Aluminum and Its Alloys

Aluminum satisfies applications requiring low weight, corrosion resistance, and good electrical and thermal conductivities. Its corrosion resistance derives from the oxide film that forms over the raw metal, inhibiting further oxidation. The primary disadvantages of aluminum are its cost and low strength.

In pure form, aluminum is soft, ductile, and not very strong. Except for use in electrical work, most aluminum is alloyed with other elements. Copper, manganese, magnesium, and silicon can be added to increase its strength, at the expense of other properties, primarily corrosion resistance.[7] Aluminum is hardened by the *precipitation hardening* (*age hardening*) process.

Silicon occurs as a normal impurity in aluminum, and in natural amounts (less than 0.4%), it has little effect on properties. If moderate quantities (above 3%) of silicon are added, the molten aluminum will have high fluidity, making it ideal for castings. Above 12%, silicon improves the hardness and wear resistance of the alloy. When combined with copper and magnesium (as Mg_2Si and AlCuMgSi) in the alloy, silicon improves age hardenability. Silicon has negligible effect on the corrosion resistance of aluminum.

Copper improves the age hardenability of aluminum, particularly in conjunction with silicon and magnesium. Therefore, copper is a primary element in achieving high mechanical strength in aluminum alloys at elevated temperatures. Copper also increases the conductivity of aluminum, but decreases its corrosion resistance.

Magnesium is highly soluble in aluminum and is used to increase strength by improving age hardenability. Magnesium improves corrosion resistance and may be added when exposure to saltwater is anticipated.

Copper and Its Alloys

Zinc is the most common alloying ingredient in copper. It constitutes a significant part (up to 40% zinc) in brass.[8] (Brazing rod contains even more, approximately 45% to 50%, zinc.) Zinc increases copper's hardness and tensile strength. Up to approximately 30%, it increases the percent elongation at fracture. It decreases electrical conductivity considerably. *Dezincification*, a loss of zinc in the presence of certain corrosive media or at high temperatures, is a special problem that occurs in brasses containing more than 15% zinc.

Tin constitutes a major (up to 20%) component in most bronzes. Tin increases fluidity, which improves casting performance. In moderate amounts, corrosion resistance in saltwater is improved. (*Admiralty metal* has approximately 1%; *government bronze* and *phosphorus bronze* have approximately 10% tin.) In moderate amounts (less than 10%), tin increases the alloy's strength without sacrificing ductility. Above 15%, however, the alloy becomes brittle. For this reason, most bronzes contain less than 12% tin. Tin is more expensive than zinc as an alloying ingredient.

Lead is practically insoluble in solid copper. When present in small to moderate amounts, it forms minute soft particles that greatly improve machinability (2–3% lead) and wearing (bearing) properties (10% lead).

Silicon increases the mechanical properties of copper by a considerable amount. On a per unit basis, silicon is the most effective alloying ingredient in increasing hardness. *Silicon bronze* (96% copper, 3% silicon, 1% zinc) is used where high strength combined with corrosion resistance is needed (e.g., in boilers).

If aluminum is added in amounts of 9–10%, copper becomes extremely hard. Therefore, *aluminum bronze* (as an example) trades an increase in brittleness for increased wearing qualities. Aluminum in solution with the copper makes it possible to precipitation harden the alloy.

Beryllium in small amounts (less than 2%) improves the strength and fatigue properties of copper. These properties make precipitation-hardened *copper-beryllium* (*beryllium-copper*, *beryllium bronze*, etc.) ideal for small springs. These alloys are also used for producing nonsparking tools.

Nickel and Its Alloys

Like aluminum, nickel is largely hardened by precipitation hardening. Nickel is similar to iron in many of its properties, except that it has higher corrosion resistance and a higher cost. Also, nickel alloys have special electrical and magnetic properties.

Copper and iron are completely miscible with nickel. Copper increases formability. Iron improves electrical and magnetic properties markedly.

[7]One ingenious method of having both corrosion resistance and strength is to produce a composite material. *Alclad* is the name given to aluminum alloy that has a layer of pure aluminum bonded to the surface. The alloy provides the strength, and the pure aluminum provides the corrosion resistance.

[8]*Brass* is an alloy of copper and zinc. *Bronze* is an alloy of copper and tin. Unfortunately, brasses are often named for the color of the alloys, leading to some very misleading names. For example, *nickel silver*, *commercial bronze*, and *manganese bronze* are all brasses.

Some of the better-known nickel alloys are *monel metal* (30% copper, used hot-rolled where saltwater corrosion resistance is needed), *K-monel metal*[9] (29% copper, 3% aluminum, precipitation-hardened for use in valve stems), *inconel* (14% chromium, 6% iron, used hot-rolled in gas turbine parts), and *inconel-X* (15% chromium, 7% iron, 2.5% titanium, aged after hot rolling for springs and bolts subjected to corrosion). *Hastelloy* (22% chromium) is another well-known nickel alloy.

Nichrome (15–20% chromium) has high electrical resistance, high corrosion resistance, and high strength at red heat temperatures, making it useful in resistance heating. *Constantan* (40% to 60% copper, the rest nickel) also has high electrical resistance and is used in thermocouples.

Alnico (14% nickel, 8% aluminum, 24% cobalt, 3% copper, the rest iron) and *cunife* (20% nickel, 60% copper, the rest iron) are two well-known nickel alloys with magnetic properties ideal for permanent magnets. Other magnetic nickel alloys are *permalloy* and *permivar*.

Invar, *Nilvar*, and *Elinvar* are nickel alloys with low or zero thermal expansion and are used in thermostats, instruments, and surveyors' measuring tapes.

Refractory Metals

Reactive and *refractory metals* include alloys based on titanium, tantalum, zirconium, molybdenum, niobium (also known as columbium), and tungsten. These metals are used when superior properties (i.e., corrosion resistance) are needed. They are most often used where high-strength acids are used or manufactured.

5. AMORPHOUS MATERIALS

Amorphous materials are materials that lack a crystalline structure like liquids, but are rigid and maintain their shape like solids. Glass is a distinct kind of amorphous material that exhibits a glass transition (see Sec. 36.8). Figure 36.1 illustrates the difference between amorphous and crystalline solids over the glass transition temperature, T_g, and melting temperature, T_m.

Figure 36.1 *Volume-Temperature Curve for Amorphous Materials*

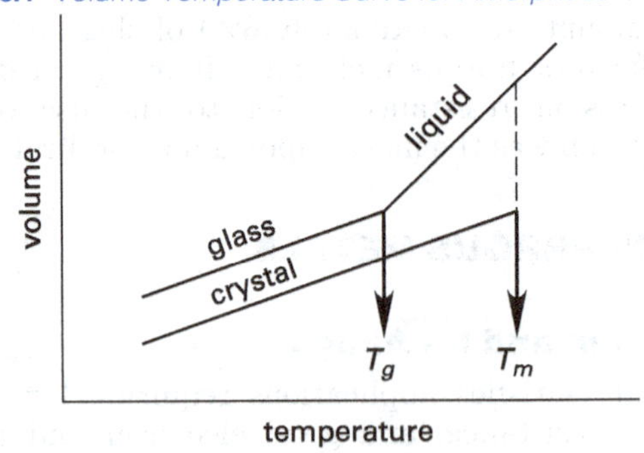

6. POLYMERS

Natural Polymers

A *polymer* is a large molecule in the form of a long chain of repeating units. The basic repeating unit is called a *monomer* or just *mer*. (A large molecule with two alternating mers is known as a *copolymer* or *interpolymer*. Vinyl chloride and vinyl acetate form one important family of copolymer plastics.)

Many of the natural organic materials (e.g., rubber and asphalt) are polymers. (Polymers with elastic properties similar to rubber are known as *elastomers*.) Natural rubber is a polymer of the *isoprene latex* mer (formula $[C_5H_8]_n$, repeating unit of $CH_2{=}CCH_3{-}CH{=}CH_2$, systematic name of 2-methyl-1,3-butadiene). The strength of natural polymers can be increased by causing the polymer chains to cross-link, restricting the motion of the polymers within the solid.

Cross-linking of natural rubber is known as *vulcanization*. Vulcanization is accomplished by heating raw rubber with small amounts of sulfur. The process raises the tensile strength of the material from approximately 2.1 MPa to approximately 21 MPa. The addition of carbon black as a reinforcing *filler* raises this value to approximately 31 MPa and provides tear resistance and toughness.

The amount of cross-linking between the mers determines the properties of the solid. Figure 36.2 shows how sulfur joins two adjacent isoprene (natural rubber) mers in *complete cross-linking*.[10] If sulfur does not replace both of the double carbon bonds, *partial cross-linking* is said to have occurred.

Degree of Polymerization

The *degree of polymerization*, DP, is the average number of mers in the molecule, typically several hundred to several thousand.[11] (In general, compounds with degrees of less than ten are called *telenomers* or *oligomers*.) The degree of polymerization can be calculated from the mer and polymer molecular weights, MW.

$$\mathrm{DP} = \frac{\mathrm{MW}_{\mathrm{polymer}}}{\mathrm{MW}_{\mathrm{mer}}}$$

A polymer batch usually will contain molecules with different length chains. Therefore, the degree of polymerization will vary from molecule to molecule, and an average degree of polymerization is reported.

[9] K-monel is one of four special forms of monel metal. There are also H-monel, S-monel, and R-monel forms.
[10] A tire tread may contain 3–4% sulfur. Hard rubber products, which do not require flexibility, may contain as much as 40–50% sulfur.
[11] Degrees of polymerization for commercial plastics are usually less than 1000.

Figure 36.2 *Vulcanization of Natural Rubber*

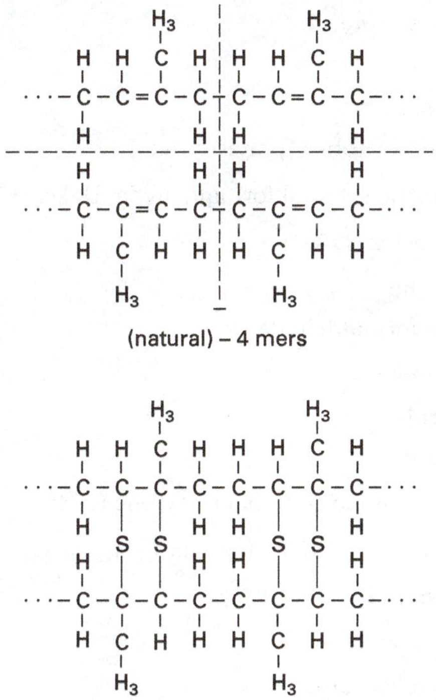

The stiffness and hardness of polymers vary with their degrees. Polymers with low degrees are liquids or oils. With increasing degree, they go through waxy to hard resin stages. High-degree polymers have hardness and strength qualities that make them useful for engineering applications. Tensile strength and melting (softening) point also increase with increasing degree of polymerization.

Synthetic Polymers

Table 36.4 lists some of the common mers. Polymers are named by adding the prefix "poly" to the name of the basic mer. For example, C_2H_4 is the chemical formula for ethylene. Chains of C_2H_4 are called polyethylene.

Polymers are able to form when double (covalent) bonds break and produce reaction sites. The number of bonds in the mer that can be broken open for attachment to other mers is known as the *functionality* of the mer.

Vinyl polymers are a class of polymers where a different element or molecule, R, has been substituted for one of the hydrogen atoms in the repeating units (i.e., $-[CH_2–CHR]-$). For example, the repeating unit of polyethylene is $-[CH_2]-$ or $-[CH_2–CH_2]-$, but the repeating unit of polyvinyl chloride is $-[CH_2–CHCl]-$. *Copolymers* contain two repeating units (i.e., *comonomers*). One of the units may alternate with the other, or it may appear on a regular basis (i.e., periodically) according to some specific statistic or structure. In an *addition polymer*, all of the atoms in the combining molecules appear in the resulting polymer molecule. In a *condensation*

Table 36.4 *Names of Common Mers*

name	repeating unit	combined formula
ethylene	CH_2CH_2	C_2H_4
propylene	$CH_2(HCCH_3)$	C_3H_6
styrene	$CH_2CH(C_6H_5)$	C_8H_8
vinyl acetate	$CH_2CH(C_2H_3O_2)$	$C_4H_6O_2$
vinyl chloride	CH_2CHCl	C_2H_3Cl
isobutylene	$CH_2C(CH_3)_2$	C_4H_8
methyl methacrylate	$CH_2C(CH_3)(COOCH_3)$	$C_5H_8O_2$
acrylonitrile	CH_2CHCN	C_3H_3N
epoxide (ethoxylene)	CH_2CH_2O	C_2H_4O
amide (nylon)	$CONH_2$ or CONH	$CONH_2$ or CONH

polymer, the resulting structure has fewer atoms than the monomers from which it was created. (The missing atoms or fragments "condense out.") Usually, some energy source (e.g., heat) is required to form condensation polymers.

Example

Polyester repeating unit $-[CO(CH_2)_4CO–OCH_2CH_2O]-$ is formed by combining the two repeating units $-[HO_2C–(CH_2)_4–CO_2H]-$ and $-[HO–CH_2CH_2–OH]-$. Most accurately, what type of polymer is the polyester?

(A) addition polymer

(B) comonomer polymer

(C) condensation polymer

(D) vinyl polymer

Solution

Count the number of each atom in the combining and polyester molecules.

element	atoms in combining units	atoms in polyester repeating unit	discrepancy
carbon (C)	$6+2=8$	8	0
hydrogen (H)	$10+6=16$	12	4
oxygen (O)	$4+2=6$	4	2

Four hydrogen atoms and two oxygen atoms (the equivalent of two water molecules) are missing from the resulting polymer unit. Polyester is a condensation polymer.

The answer is (C).

Thermosetting and Thermoplastic Polymers

Most polymers can be softened and formed by applying heat and pressure. These are known by various terms, including *thermoplastics*, *thermoplastic resins*, and *thermoplastic polymers*. Thermoplastics may be either semicrystalline or amorphous. Polymers that are resistant to heat (and that actually harden or "kick over" through the formation of permanent cross-linking upon heating) are known as *thermosetting plastics*. Table 36.5 lists the common polymers in each category. Thermoplastic polymers retain their chain structures and do not experience any chemical change (i.e., bonding) upon repeated heating and subsequent cooling. Thermoplastics can be formed in a cavity mold, but the mold must be cooled before the product is removed. Thermoplastics are particularly suitable for injection molding. The mold is kept relatively cool, and the polymer solidifies almost instantly.

Figure 36.3 illustrates the relationship between temperature and the strength, σ, or modulus of elasticity, E, of thermoplastics.

Thermosetting polymers form complex, three-dimensional networks. Thus, the complexity of the polymer increases dramatically, and a product manufactured from a thermosetting polymer may be thought of as one big molecule. Thermosetting plastics are rarely used with injection molding processes.

Thermosetting compounds are purchased in liquid form, which makes them easy to combine with additives. Thermoplastic materials are commonly purchased in granular form. They are mixed with additives in a *muller* (i.e., a bulk mixer) before transfer to the feed hoppers. Thermoplastic materials can also be molded into small pellets called *preforms* for easier handling in subsequent melting operations. Common additives for plastics include

- *plasticizers:* vegetable oils, low molecular weight polymers or monomers
- *fillers:* talc, chopped glass fibers
- *flame retardants:* halogenated paraffins, zinc borate, chlorinated phosphates
- *ultraviolet or visible light resistance:* carbon black
- *oxidation resistance:* phenols, aldehydes

Table 36.5 *Thermosetting and Thermoplastic Polymers*

thermosetting

- epoxy
- melamine
- natural rubber (polyisoprene)
- phenolic (phenol formaldehyde, Bakelite®)
- polyester (DAP)
- silicone
- urea formaldehyde

thermoplastic

- acetal
- acrylic
- acrylonitrile-butadiene-styrene (ABS)
- cellulosics (e.g., cellophane)
- polyamide (nylon)
- polyarylate
- polycarbonate
- polyester (PBT and PET)
- polyethylene
- polymethyl-methacrylate (Plexiglas®, Lucite®)
- polypropylene
- polystyrene
- polytetrafluoroethylene (Teflon®)
- polyurethane
- polyvinyl chloride (PVC)
- synthetic rubber (Neoprene®)
- vinyl

Bakelite® is a trademark of Momentive Specialty Chemicals Inc. Plexiglas® is a trademark of Altuglas International. Lucite® is a trademark of Lucite International. Teflon® and Neoprene® are trademarks of DuPont.

Fluoropolymers

Fluoropolymers (*fluoroplastics*) are a class of paraffinic, thermoplastic polymers in which some or all of the hydrogens have been replaced by fluorine.[12] There are seven major types of fluoropolymers, with overlapping characteristics and applications. They include the fully fluorinated fluorocarbon polymers of Teflon® polytetrafluoroethylene (PTFE), fluorinated ethylene propylene (FEP), and perfluoroalkoxy (PFA), as well as the partially fluorinated

[12] *Fluoroelastomers* are uniquely different from fluoropolymers. They have their own areas of application.

Figure 36.3 Temperature Dependent Strength or Modulus for Thermoplastic Polymers

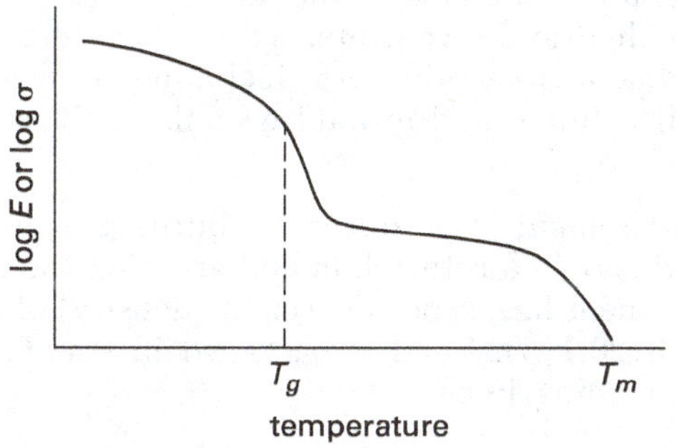

polymers of polychloro-trifluoroethylene (PCTFE), ethylene tetrafluoroethylene (ETFE), ethylene chlorotrifluoroethylene (ECTFE), and polyvinylidene fluoride (PVDF).

Fluoropolymers compete with metals, glass, and other polymers in providing corrosion resistance. Choosing the right fluoropolymer depends on the operating environment, including temperature, chemical exposure, and mechanical stress.

PTFE, the first available fluoropolymer, is probably the most inert compound known. It has been used extensively for pipe and tank linings, fittings, gaskets, valves, and pump parts. It has the highest operating temperature—approximately 260°C. Unlike the other fluoropolymers, however, it is not a melt-processed polymer. Like a powdered metallurgy product, PTFE is processed by compression and isostatic molding, followed by sintering. PTFE is also the weakest of all the fluoropolymers.

7. WOOD

Woods are classified broadly as softwoods or hardwoods, although it is difficult to define these terms exactly. *Softwoods* contain tube-like fibers (*tracheids*) oriented with the longitudinal axis (grain) and cemented together with *lignin*. *Hardwoods* contain more complex structures (e.g., storage cells) in addition to longitudinal fibers. Fibers in hardwoods are also much smaller and shorter than those in softwoods.

The mechanical properties of woods are influenced by moisture content and grain orientation. (Strengths of dry woods are approximately twice those of wet or green woods. Longitudinal strengths may be as much as 40 times higher than cross-grain strengths.) *Moisture content*, MC, is defined by

$$\text{MC} = \frac{m_{\text{wet}} - m_{\text{oven-dry}}}{m_{\text{oven-dry}}}$$

Wood is considered to be green if its moisture content is above 19%. Wood is considered to be dry when it has reached its *equilibrium moisture content*, generally between 12% and 15% moisture. Therefore, moisture is not totally absent in dry wood.[13]

8. GLASS

Glass is a term used to designate any material that has a volumetric expansion characteristic similar to Fig. 36.4. Glasses are sometimes considered to be *supercooled liquids* because their crystalline structures solidify in random orientation when cooled below their melting points. It is a direct result of the high liquid viscosities of oxides, silicates, borates, and phosphates that the molecules cannot move sufficiently to form large crystals with cleavage planes. Glass is considered an amorphous material.

As a liquid glass is cooled, its atoms develop more efficient packing arrangements. This leads to a rapid decrease in volume (i.e., a steep slope on the temperature-volume curve). Since no crystallization occurs, the liquid glass simply solidifies without molecular change when cooled below the melting point. (This is known as *vitrification*.) The more efficient packing continues past the point of solidification.

Figure 36.4 Behavior of a Glass

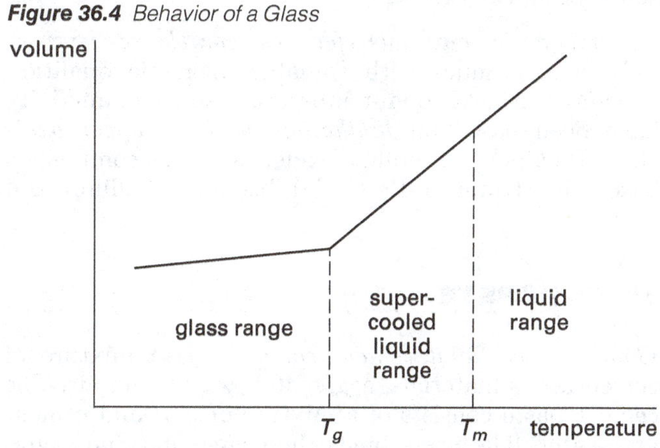

At the *glass transition temperature* (*fictive temperature*), T_g, the glass viscosity increases suddenly by several orders of magnitude. Since the molecules are more restrained in movement, efficient atomic rearrangement is curtailed, and the volume-temperature curve changes slope. This temperature also divides the region into flexible and brittle regions. At the glass transition temperature, there is a 100-fold to 1000-fold increase in stiffness (modulus of elasticity).

Both organic and inorganic compounds may behave as glasses. *Common glasses* are mixtures of SiO_2, B_2O_3, and various other compounds to modify performance.[14]

[13]Oven-dry lumber is not used in construction.
[14]This excludes lead-alkali glasses that contain 30–60% PbO.

9. CERAMICS

Ceramics are compounds of metallic and nonmetallic elements. Ceramics form crystalline structures but have no free valence electrons. All electrons are shared ionically or in covalent bonds. Common examples include brick, portland cement, refractories, and abrasives. (Glass is also considered a ceramic even though it does not crystallize.)

Although perfect ceramic crystals have extremely high tensile strengths (e.g., some glass fibers have ultimate strengths of 700 MPa), the multiplicity of cracks and other defects in natural crystals reduces their tensile strengths to near-zero levels.

Due to the absence of free electrons, ceramics are typically poor conductors of electrical current, although some (e.g., magnetite, Fe_3O_4) possess semiconductor properties. Other ceramics, such as $BaTiO_3$, SiO_2, and $PbZrO_3$, have *piezoelectric* (*ferroelectric*) *qualities* (i.e., generate a voltage when compressed).

Polymorphs are compounds that have the same chemical formula but have different physical structures. Some ceramics, of which *silica* (SiO_2) is a common example, exhibit *polymorphism*. At room temperature, silica is in the form of *quartz*. At 875°C, the structure changes to *tridymite*. A change to a third structure, that of *cristobalite*, occurs at 1470°C.

Ferrimagnetic materials (*ferrites*, *spinels*, or *ferrispinels*) are ceramics with valuable magnetic qualities. Advances in near-room-temperature superconductivity have been based on *lanthanum barium copper oxide* ($La_{2-x}Ba_xCuO_4$), a ceramic oxide, as well as compounds based on yttrium (Y-Ba-Cu-O), bismuth, thallium, and others.

10. CONCRETE

Concrete (*portland cement concrete*) is a mixture of cementitious materials, aggregates, water, and air. The cement paste consists of a mixture of portland cement and water. The paste binds the coarse and fine aggregates into a rock-like mass as the paste hardens during the chemical reaction (*hydration*). Table 36.6 lists the approximate volumetric percentage of each ingredient.

Table 36.6 *Typical Volumetric Proportions of Concrete Ingredients*

component	air-entrained	non-air-entrained
coarse aggregate	31%	31%
fine aggregate	28%	30%
water	18%	21%
portland cement	15%	15%
air	8%	3%

Cementitious Materials

Cementitious materials include portland cement, blended hydraulic cements, expansive cement, and other cementitious additives, including fly ash, pozzolans, silica fume, and ground granulated blast-furnace slag.

Portland cement is produced by burning a mixture of lime and clay in a rotary kiln and grinding the resulting mass. Cement has a specific weight (density) of approximately 3120 kg/m^3 and is packaged in standard sacks (bags) weighing 40 kg.

Aggregate

Because aggregate makes up 60–75% of the total concrete volume, its properties influence the behavior of freshly mixed concrete and the properties of hardened concrete. Aggregates should consist of particles with sufficient strength and resistance to exposure conditions such as freezing and thawing cycles. Also, they should not contain materials that will cause the concrete to deteriorate.

Most sand and rock aggregate has a specific weight of approximately 2640 kg/m^3 corresponding to a specific gravity of 2.64.

Water

Water in concrete has three functions: (1) Water reacts chemically with the cement. This chemical reaction is known as *hydration*. (2) Water wets the aggregate. (3) The water and cement mixture, which is known as *cement paste*, lubricates the concrete mixture and allows it to flow.

Water has a standard specific weight of 1000 kg/m^3. 1000 L of water occupy 1 m^3.

Potable water that conforms to ASTM C1602 and that has no pronounced odor or taste can be used for producing concrete. (With some quality restrictions, the American Concrete Institute (ACI) code also allows nonpotable water to be used in concrete mixing.) Impurities in water may affect the setting time, strength, and corrosion resistance. Water used in mixing concrete should be clean and free from injurious amounts of oils, acids, alkalis, salt, organic materials, and other substances that could damage the concrete or reinforcing steel.

Admixtures

Admixtures are routinely used to modify the performance of concrete. Advantages include higher strength, durability, chemical resistance, and workability; controlled rate of hydration; and reduced shrinkage and cracking. Accelerating and retarding admixtures fall into several different categories, as classified by ASTM C494.

Type A: water-reducing
Type B: set-retarding
Type C: set-accelerating
Type D: water-reducing and set-retarding
Type E: water-reducing and set-accelerating
Type F: high-range water-reducing
Type G: high-range water-reducing and set-retarding

Slump

The four basic concrete components (cement, sand, coarse aggregate, and water) are mixed together to produce a homogeneous concrete mixture. The *consistency* and *workability* of the mixture affect the concrete's ability to be placed, consolidated, and finished without segregation or bleeding. The slump test is commonly used to determine consistency and workability.

The *slump test* consists of completely filling a slump cone mold in three layers of about one-third of the mold volume. Each layer is rodded 25 times with a round, spherical-nosed steel rod of 16 mm diameter. When rodding the subsequent layers, the previous layers beneath are not penetrated by the rod. After rodding, the mold is removed by raising it carefully in the vertical direction. The slump is the difference in the mold height and the resulting concrete pile height. Typical values are 25–100 mm.

Concrete mixtures that do not slump appreciably are known as *stiff mixtures*. Stiff mixtures are inexpensive because of the large amounts of coarse aggregate. However, placing time and workability are impaired. Mixtures with large slumps are known as *wet mixtures* (*watery mixtures*) and are needed for thin castings and structures with extensive reinforcing. Slumps for concrete that is machine-vibrated during placement can be approximately one-third less than for concrete that is consolidated manually.

Density

The density, also known as *weight density*, *unit weight*, and *specific weight*, of normal-weight concrete varies from about 2240 kg/m^3 to 2560 kg/m^3, depending on the specific gravities of the constituents. For most calculations involving normal-weight concrete, the density may be taken as 2320 kg/m^3 to 2400 kg/m^3. Lightweight concrete can have a density as low as 1450 kg/m^3. Although steel has a density of more than three times that of concrete, due to the variability in concrete density values and the relatively small volume of steel, the density of steel-reinforced concrete is typically taken as 2400 kg/m^3 without any refinement for exact component contributions.

Compressive Strength

The concrete's *compressive strength*, f'_c, is the maximum stress a concrete specimen can sustain in compressive axial loading. It is also the primary parameter used in ordering concrete. When one speaks of "6000 psi concrete," the compressive strength is being referred to. Compressive strength is expressed in psi or MPa. SI compressive strength may be written as "Cxx" (e.g., "C20"), where xx is the compressive strength in MPa. (MPa is equivalent to N/mm^2, which is also commonly quoted.)

Typical compressive strengths range from 4000 psi to 6000 psi for traditional structural concrete, though concrete for residential slabs-on-grade and foundations will be lower in strength (e.g., 3000 psi). 6000 psi concrete is used in the manufacture of some concrete pipes, particularly those that are jacked in.

Cost is approximately proportional to concrete's compressive strength—a rule that applies to high-performance concrete as well as traditional concrete. For example, if 5000 psi concrete costs $100 per cubic yard, then 14,000 psi concrete will cost approximately $280 per cubic yard.

Compressive strength is controlled by selective proportioning of the cement, coarse and fine aggregates, water, and various admixtures. However, the compressive strength of traditional concrete is primarily dependent on the mixture's water-cement ratio. (See Fig. 36.5.) Provided that the mix is of a workable consistency, strength varies directly with the water-cement ratio. (This is *Abrams' strength law*, named after Dr. Duff Abrams, who formulated the law in 1918.)

Figure 36.5 *Water-Cement (W/C) Ratio**

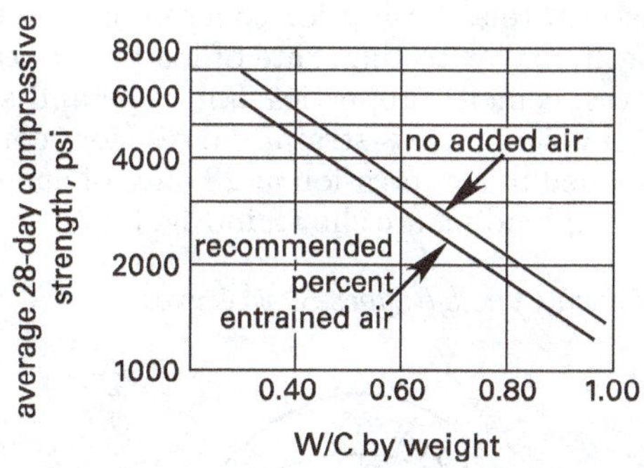

*Concrete strength decreases with increases in water-cement ratio for concrete with and without entrained air.

Concrete Manual, 8th ed., U.S. Bureau of Reclamation, 1975.

Compressive strength is normally measured on the 28th day after the specimens are cast. Since the strength of concrete increases with time, all values of f'_c must be stated with respect to a known age. If no age is given, a strength at a standard 28-day age is assumed.

The effect of the water-cement ratio on compressive strength (i.e., the more water the mix contains, the lower the compressive strength will be) is a different issue than the use of large amounts of surface water to cool the concrete during curing (i.e., *moist-curing*). (See Fig. 36.6.) The strength of newly poured concrete can be increased significantly (e.g., doubled) if the concrete is kept cool during part or all of curing. This is often accomplished by covering new concrete with wet burlap or by spraying with water. Although best results occur when the concrete is moist-cured for 28 days, it is seldom economical to do so. A substantial strength increase can be achieved if the concrete is kept moist for as little as three days. Externally applied curing retardants can also be used.

Figure 36.6 *Concrete Compressive Strength**

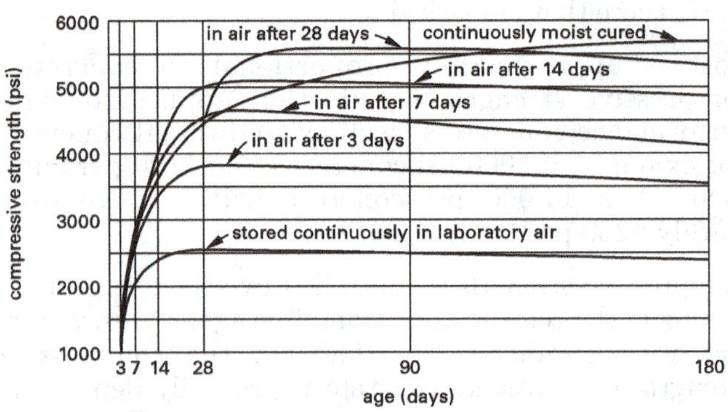

*Concrete compressive strength varies with moist-curing conditions. Mixes tested had a water-cement ratio of 0.50, a slump of 3.5 in, cement content of 556 lbf/yd^3, sand content of 36%, and air content of 4%.

Stress-Strain Relationship

The stress-strain relationship for concrete is dependent on its strength, age at testing, rate of loading, nature of the aggregates, cement properties, and type and size of specimens. Typical stress-strain curves for concrete specimens loaded in compression at 28 days of age under a normal rate of loading are shown in Fig. 36.7.

Figure 36.7 *Typical Concrete Stress-Strain Curves*

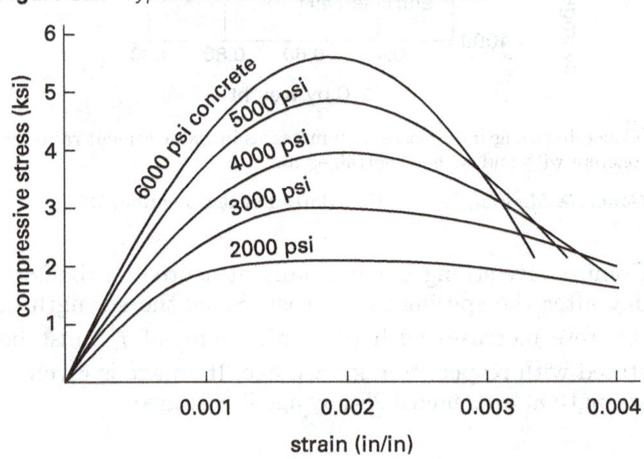

Modulus of Elasticity

The *modulus of elasticity* (also known as *Young's modulus*) is defined as the ratio of stress to strain in the elastic region. Unlike steel, the modulus of elasticity of concrete varies with compressive strength. Since the slope of the stress-strain curve varies with the applied stress, there are several ways of calculating the modulus of elasticity. Figure 36.8 shows a typical stress-strain curve for concrete with the *initial modulus*, the *tangent modulus*, and the *secant modulus* indicated.

Figure 36.8 *Concrete Moduli of Elasticity*

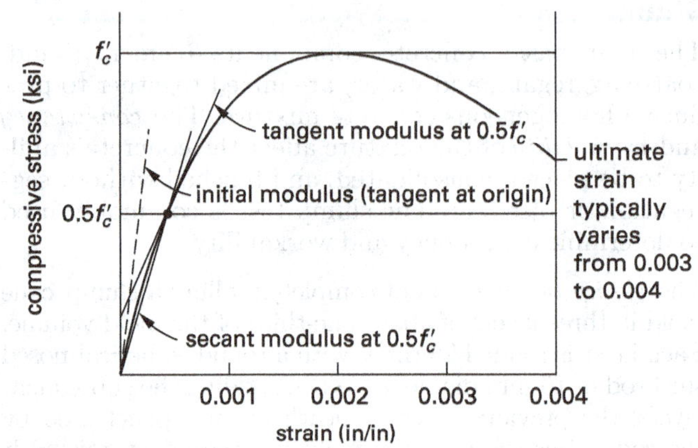

Splitting Tensile Strength

The extent and size of cracking in concrete structures are affected to a great extent by the tensile strength of the concrete. Lightweight concrete has a lower tensile strength than normal weight concrete, even if both have the same compressive strength.

Modulus of Rupture

The tensile strength of concrete in flexure is known as the *modulus of rupture*, f_r, and is an important parameter for evaluating cracking and deflection in beams. The tensile strength of concrete is relatively low, about 10–15% (and occasionally up to 20%) of the compressive strength.

Shear Strength

Concrete's true *shear strength* is difficult to determine in the laboratory because shear failure is seldom pure and is typically affected by other stresses in addition to the shear stress. Reported values of shear strength vary greatly with the test method used, but they are a small percentage (e.g., 25% or less) of the ultimate compressive strength.

Poisson's Ratio

Poisson's ratio is the ratio of the lateral strain to the axial strain. It varies in concrete from 0.11 to 0.23, with typical values being from 0.17 to 0.21.

11. COMPOSITE MATERIALS

There are many types of modern composite material systems, including dispersion-strengthened, particle-strengthened, and fiber-strengthened materials. High-performance composites are generally produced by dispersing large numbers of particles or whiskers of a strengthening component in a lightweight binder. (Steel-reinforced concrete and steel-plate-on-wood systems are also composite systems. However, these are designed and analyzed according to various building codes rather than to the theoretical methods presented in this section.)

Assuming a well-dispersed, well-bonded, and homogeneous mixture of components, the mechanical and thermal properties of a composite material can be predicted as volumetrically weighted fractions ($0 < f_i < 1.0$) of the properties of the individual components. This is known as the *rule of mixtures.*

Table 36.7 lists properties of common components used in producing composite materials.

Table 36.7 *Properties of Components of Composite Materials*

	density, ρ (Mg/m^3)	modulus of elasticity, E (GPa)	E/ρ (N·m/g)
binders/matrix			
polystyrene	1.05	2	2700
polyvinyl chloride	1.3	<4	3500
strengtheners			
alumina fiber	3.9	400	100 000
aluminum	2.7	70	26 000
aramide fiber	1.3	125	100 000
BeO fiber	3.0	400	130 000
beryllium fiber	1.9	300	160 000
boron fiber	2.3	400	170 000
carbon fiber	2.3	700	300 000
glass	2.5	70	28 000
magnesium	1.7	45	26 000
silicon carbide fiber	3.2	400	120 000
steel	7.8	205	26 000

Equation 36.1 Through Eq. 36.5: Properties of a Composite Material

$$\rho_c = \sum f_i \rho_i \qquad 36.1$$

$$C_c = \sum f_i c_i \qquad 36.2$$

$$\left[\sum \frac{f_i}{E_i}\right]^{-1} \le E_c \le \sum f_i E_i \qquad 36.3$$

$$\sigma_c = \sum f_i \sigma_i \qquad 36.4$$

$$(\Delta L/L)_1 = (\Delta L/L)_2 \qquad 36.5$$

Description

Equation 36.1 gives the density of a composite material, ρ_c, typically expressed in units of kg/m^3. f_i is the volumetric fraction of each individual material, and ρ_i is the density of each material.

The heat capacity of a composite material per unit volume, C_c, is calculated from Eq. 36.2.[15]

Equation 36.3 is used to calculate the *modulus of elasticity* (*Young's modulus*) of a composite material, E_c, calculated from the volumetric fraction of each material, f_i, and modulus of elasticity of each material, E_i.

The ultimate tensile strength of a composite material parallel to the fiber direction, σ_c, is calculated from Eq. 36.4.[16]

Assuming perfect bonding, the strain in two adjacent components (e.g., strengthening whiskers and the supporting matrix) will be the same. This is expressed in Eq. 36.5.

Example

A composite material is 57% resin (density of 2.3 g/cm^3) and 43% unidirectionally placed carbon fibers (1.05 g/cm^3) by volume. The material has a composite modulus of

[15]Normally, lowercase c is used to represent specific heat capacity on a per unit mass basis, and uppercase C is used for molar specific heat. In Eq. 36.2, the NCEES *FE Reference Handbook* (*NCEES Handbook*) uses both uppercase and lowercase to represent specific heat capacity. However, it is more common to use c_c for the composite specific heat for clarity. Equation 36.2 is valid only for the composite *volumetric heat capacity* (VHC) based on component VHCs. This equation cannot be used to calculate the specific heat on a per unit mass basis because a gravimetric fraction (not volumetric fraction) would be needed.

[16]Although the term "strength" is correctly used in reference to the ultimate tensile strength, the common symbol for stress, σ, is unfortunately used. The ultimate tensile strength can indeed be predicted by the rule of mixtures, although the composite strength will be greatly optimistic. Stress is generally weighted by area (or a combination of area and modulus of elasticity).

elasticity of 400 GPa parallel to the carbon fibers. What is most nearly the density of the composite material?

(A) 1.3 g/cm^3

(B) 1.8 g/cm^3

(C) 1.9 g/cm^3

(D) 2.2 g/cm^3

Solution

Using Eq. 36.1, the density of the composite material is

$$\begin{aligned}\rho_c &= \sum f_i \rho_i \\ &= (0.57)\left(2.3\ \frac{\text{g}}{\text{cm}^3}\right) + (0.43)\left(1.05\ \frac{\text{g}}{\text{cm}^3}\right) \\ &= 1.763\ \text{g/cm}^3 \quad (1.8\ \text{g/cm}^3)\end{aligned}$$

The answer is (B).

12. CORROSION

Corrosion is an undesirable degradation of a material resulting from a chemical or physical reaction with the environment. *Galvanic action* results from a difference in oxidation potentials of metallic ions. The greater the difference in oxidation potentials, the greater the galvanic corrosion will be. If two metals with different oxidation potentials are placed in an *electrolytic medium* (e.g., seawater), a *galvanic cell* (*voltaic cell*) will be created. The more electropositive metal will act as an anode and will corrode. The metal with the lower potential, being the cathode, will be unchanged.

A galvanic cell is a device that produces electrical current by way of an oxidation-reduction reaction—that is, chemical energy is converted into electrical energy. Galvanic cells typically have the following characteristics.

- The oxidizing agent is separate from the reducing agent.
- Each agent has its own electrolyte and metallic electrode, and the combination is known as a *half-cell.*
- Each agent can be in solid, liquid, or gaseous form, or can consist simply of the electrode.
- The ions can pass between the electrolytes of the two half-cells. The connection can be through a porous substance, salt bridge, another electrolyte, or other method.

The amount of current generated by a half-cell depends on the electrode material and the oxidation-reduction reaction taking place in the cell. The current-producing ability is known as the *oxidation potential*, *reduction potential*, or *half-cell potential.* *Standard oxidation potentials* have a zero reference voltage corresponding to the potential of a *standard hydrogen electrode.*

To specify their tendency to corrode, metals are often classified according to their position in the galvanic series listed in Table 36.8. As expected, the metals in this series are in approximately the same order as their half-cell potentials. However, alloys and proprietary metals are also included in the series.

Table 36.8 *Galvanic Series in Seawater (top to bottom anodic (sacrificial, active) to cathodic (noble, passive))*

magnesium
zinc
Alclad 3S
cadmium
2024 aluminum alloy
low-carbon steel
cast iron
stainless steels (active)
  no. 410
  no. 430
  no. 404
  no. 316
Hastelloy A
lead
lead-tin alloys
tin
nickel
brass (copper-zinc)
copper
bronze (copper-tin)
90/10 copper-nickel
70/30 copper-nickel
Inconel
silver solder
silver
stainless steels (passive)
Monel metal
Hastelloy C
titanium
graphite
gold

It is not necessary that two dissimilar metals be in contact for corrosion by galvanic action to occur. Different regions within a metal may have different half-cell potentials. The difference in potential can be due to different phases within the metal (creating very small galvanic cells), heat treatment, cold working, and so on.

In addition to corrosion caused by galvanic action, there is also *stress corrosion*, *fretting corrosion*, and *cavitation*. Conditions within the crystalline structure can accentuate or retard corrosion. In one extreme type of intergranular corrosion, *exfoliation*, open endgrains separate into layers.

Table 36.9 gives the oxidation potentials for common corrosion reactions.

Equation 36.6 Through Eq. 36.9: Oxidation-Reduction Corrosion Reactions

$$M^{o} \rightarrow M^{n+} + ne^{-} \quad \text{[at the anode]} \qquad 36.6$$

$$\tfrac{1}{2}O_2 + 2e^{-} + H_2O \rightarrow 2OH^{-} \quad \text{[at the cathode]} \qquad 36.7$$

$$\tfrac{1}{2}O_2 + 2e^{-} + 2H_3O^{+} \rightarrow 3H_2O \quad \text{[at the cathode]} \qquad 36.8$$

$$2e^{-} + 2H_3O^{+} \rightarrow 2H_2O + H_2 \quad \text{[at the cathode]} \qquad 36.9$$

Description

In an oxidation-reduction reaction, such as corrosion, one substance is oxidized and the other is reduced. The oxidized substance loses electrons and becomes less negative; the reduced substance gains electrons and becomes more negative.

Oxidation occurs at the *anode* (positive terminal) in an electrolytic reaction. Equation 36.6 shows the oxidation reaction (or *anode reaction*) of a typical metal, M. The superscript "o" is used to designate the standard, natural state of the atom.

Reduction occurs at the *cathode* (negative terminal) in an electrolytic reaction. Equation 36.7 through Eq. 36.9 list some reduction reactions (or *cathode reactions*) involving hydrogen and oxygen.

13. DIFFUSION OF DEFECTS

Real crystals possess a variety of imperfections and defects that affect *structure-sensitive properties*. Such properties include electrical conductivity, yield and ultimate strengths, creep strength, and semiconductor properties. Most imperfections can be categorized into *point*, *line*, and *planar* (*grain boundary*) imperfections. As shown in Fig. 36.9, *point defects* include vacant lattice sites, ion vacancies, substitutions of foreign atoms into lattice points or interstitial points, and occupation of interstitial points by atoms. *Line defects* consist of imperfections that are repeated consistently in many adjacent cells and have extension in a particular direction. *Grain boundary defects* are the interfaces between two or more crystals. This interface is almost always a mismatch in crystalline structures.

All point defects can move individually and independently from one position to another through *diffusion*. The *activation energy* for such diffusion generally comes from heat and/or strain (i.e., bending or forming). In the absence of the activation energy, the defect will move very slowly, if at all.

Diffusion of defects is governed by *Fick's laws*.

Table 36.9 *Standard Oxidation Potentials for Corrosion Reactions**

corrosion reaction	potential, E_o (volts), versus normal hydrogen electrode
$Au \rightarrow Au^{3+} + 3e^{-}$	–1.498
$2H_2O \rightarrow O_2 + 4H^{+} + 4e^{-}$	–1.229
$Pt \rightarrow Pt^{2+} + 2e^{-}$	–1.200
$Pd \rightarrow Pd^{2+} + 2e^{-}$	–0.987
$Ag \rightarrow Ag^{+} + e^{-}$	–0.799
$2Hg \rightarrow Hg_2^{2+} + 2e^{-}$	–0.788
$Fe^{2+} \rightarrow Fe^{3+} + e^{-}$	–0.771
$4(OH)^{-} \rightarrow O_2 + 2H_2O + 4e^{-}$	–0.401
$Cu \rightarrow Cu^{2+} + 2e^{-}$	–0.337
$Sn^{2+} \rightarrow Sn^{4+} + 2e^{-}$	–0.150
$H_2 \rightarrow 2H^{+} + 2e^{-}$	0.000
$Pb \rightarrow Pb^{2+} + 2e^{-}$	+0.126
$Sn \rightarrow Sn^{2+} + 2e^{-}$	+0.136
$Ni \rightarrow Ni^{2+} + 2e^{-}$	+0.250
$Co \rightarrow Co^{2+} + 2e^{-}$	+0.277
$Cd \rightarrow Cd^{2+} + 2e^{-}$	+0.403
$Fe \rightarrow Fe^{2+} + 2e^{-}$	+0.440
$Cr \rightarrow Cr^{3+} + 3e^{-}$	+0.744
$Zn \rightarrow Zn^{2+} + 2e^{-}$	+0.763
$Al \rightarrow Al^{3+} + 3e^{-}$	+1.662
$Mg \rightarrow Mg^{2+} + 2e^{-}$	+2.363
$Na \rightarrow Na^{+} + e^{-}$	+2.714
$K \rightarrow K^{+} + e^{-}$	+2.925

*Measured at 25°C. Reactions are written as anode half-cells. Arrows are reversed for cathode half-cells.

NOTE: In some chemistry texts, the reactions and the signs of the values (in this table) are reversed; for example, the half-cell potential of zinc is given as –0.763 V for the reaction $Zn^{2+} + 2e^{-} \rightarrow Zn$. When the potential E_o is positive, the reaction proceeds spontaneously as written.

Figure 36.9 Point Defects

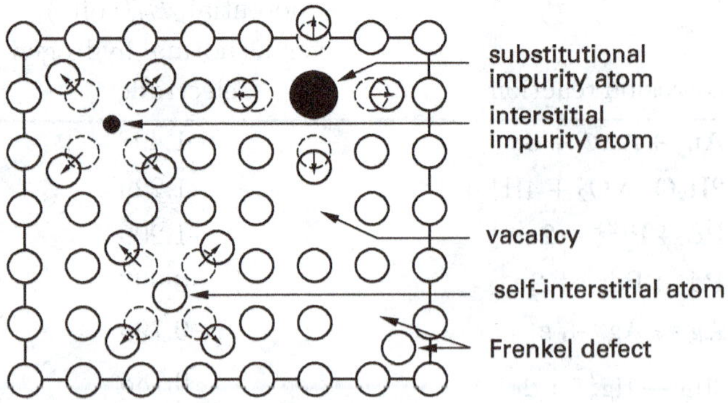

Equation 36.10: Diffusion Coefficient

$$D = D_o e^{-Q/(RT)} \qquad 36.10$$

Values

$\overline{R} = 8314$ J/kmol·K

Description

Equation 36.10 is used to determine the *diffusion coefficient*, D (also known as the *diffusivity*), expressed in units of square meters per second. The diffusion coefficient is dependent on the material, activation energy, and temperature. It is calculated from the *proportionality constant*, D_o, the activation energy, Q, the universal gas constant, $\overline{R}$, and the absolute temperature, T.[17] Since $e^{-Q/(RT)}$ is a number less than 1.0, the proportionality constant is actually the maximum possible value of the diffusion coefficient, which would occur at an infinite temperature.[18] The exponent $-Q/(\overline{R}\,T)$ in Eq. 36.10 must be unitless.

Example

The activation energy, Q, for aluminum in a copper solvent at 575°C is 1.6×10^5 kJ/kmol. What is most nearly the diffusion coefficient, D, if the proportionality constant, D_o, is 7×10^{-6} m²/s?

(A) 4.0×10^{-47} m²/s

(B) 2.0×10^{-20} m²/s

(C) 9.8×10^{-16} m²/s

(D) 2.3×10^{-5} m²/s

Solution

The absolute temperature is

$$T = 575°\text{C} + 273° = 848\text{K}$$

To use Eq. 36.10, the units in the exponent must cancel. Since the universal gas constant, $\overline{R}$, is given in units of joules per kmol, the activation energy, Q, must also have those units.

$$Q = \left(1.6 \times 10^5 \ \frac{\text{kJ}}{\text{kmol}}\right)\left(1000 \ \frac{\text{J}}{\text{kJ}}\right) = 1.6 \times 10^8 \ \text{J/kmol}$$

The diffusion coefficient is

$$\begin{aligned} D &= D_o e^{-Q/(RT)} \\ &= \left(7 \times 10^{-6} \ \frac{\text{m}^2}{\text{s}}\right) \times e^{-(1.6\times10^8 \ \text{J/kmol}/(8314 \ \text{J/kmol·K})(848\text{K}))} \\ &= 9.753 \times 10^{-16} \ \text{m}^2/\text{s} \quad (9.8 \times 10^{-16} \ \text{m}^2/\text{s}) \end{aligned}$$

The answer is (C).

14. BINARY PHASE DIAGRAMS

Most engineering materials are not pure elements but are alloys of two or more elements. Alloys of two elements are known as *binary alloys*. Steel, for example, is an alloy of primarily iron and carbon. Usually one of the elements is present in a much smaller amount, and this element is known as the *alloying ingredient*. The primary ingredient is known as the *host ingredient*, *base metal*, or *parent ingredient*.

Sometimes, such as with alloys of copper and nickel, the alloying ingredient is 100% soluble in the parent ingredient. Nickel-copper alloy is said to be a *completely miscible alloy* or a *solid-solution alloy*.

The presence of the alloying ingredient changes the thermodynamic properties, notably the freezing (or melting) temperatures of both elements. Usually the freezing temperatures decrease as the percentage of alloying ingredient is increased. Because the freezing points of the two elements are not the same, one of them will start to solidify at a higher temperature than the other.

[17](1) The *NCEES Handbook* is inconsistent in the variable it uses for the universal gas constant. R in Eq. 36.10 is the same as $\overline{R}$ defined in its *Units* section and used almost everywhere in the *NCEES Handbook*. (2) There is no mathematical significance to the parentheses around the denominator of the exponent in Eq. 36.10.

[18](1) The *NCEES Handbook* uses a subscript letter o for the proportionality constant, which should be interpreted as the subscript zero, 0, normally used in references. (2) The *NCEES Handbook* is inconsistent in the symbols used for activation energy. Q in Eq. 36.10 is the same as $E_g - E_d$ and E_a in Table 35.4. A common symbol used in practice for diffusion activation energy is Q_d, where the subscript clarifies that the activation energy is for diffusion.

For any given composition, the alloy might consist of all liquid, all solid, or a combination of solid and liquid, depending on the temperature.

A *phase* of a material at a specific temperature will have a specific composition and crystalline structure and distinct physical, electrical, and thermodynamic properties. (In metallurgy, the word "phase" refers to more than just solid, liquid, and gas phases.)

The regions of an *equilibrium diagram*, also known as a *phase diagram*, illustrate the various alloy phases. The phases are plotted against temperature and composition. The composition is usually a gravimetric fraction of the alloying ingredient. Only one ingredient's gravimetric fraction needs to be plotted for a binary alloy.

It is important to recognize that the equilibrium conditions do not occur instantaneously and that an equilibrium diagram is applicable only to the case of slow cooling.

Figure 36.10 is an equilibrium diagram for a copper-nickel alloy. (Most equilibrium diagrams are much more complex.) The *liquidus line* is the boundary above which no solid can exist. The *solidus line* is the boundary below which no liquid can exist. The area between these two lines represents a mixture of solid and liquid phase materials.

Figure 36.10 *Copper-Nickel Phase Diagram*

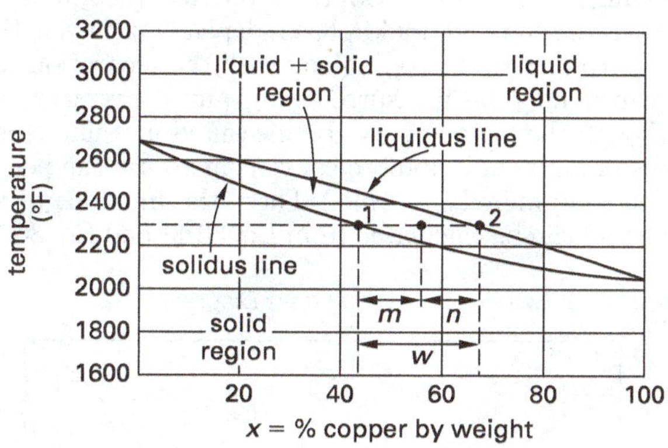

Just as only a limited amount of salt can be dissolved in water, there are many instances where a limited amount of the alloying ingredient can be absorbed by the solid mixture. The elements of a binary alloy may be completely soluble in the liquid state but only partially soluble in the solid state.

When the alloying ingredient is present in an amount above the maximum solubility percentage, the alloying ingredient precipitates out. In aqueous solutions, the precipitate falls to the bottom of the container. In metallic alloys, the precipitate remains suspended as pure crystals dispersed throughout the primary metal.

In chemistry, a *mixture* is different from a *solution*. Salt in water forms a solution. Sugar crystals mixed with salt crystals form a mixture.

Figure 36.11 is typical of an equilibrium diagram for ingredients displaying a limited solubility.

Figure 36.11 *Equilibrium Diagram of a Limited Solubility Alloy*

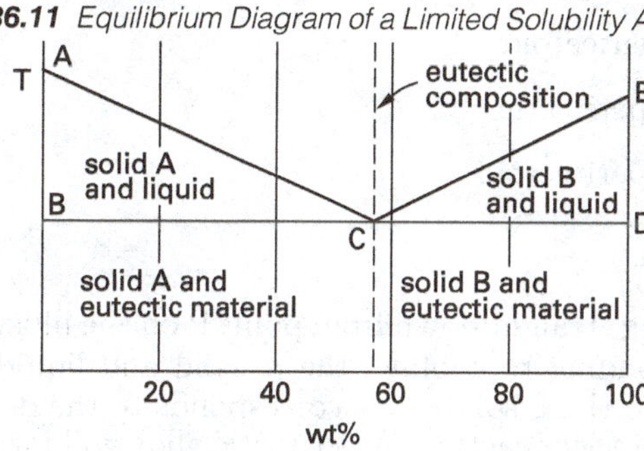

In Fig. 36.11, the components are perfectly miscible at point C only. This point is known as the *eutectic composition*. A *eutectic alloy* is an alloy having the composition of its eutectic point. The material in the region ABC consists of a mixture of solid component A crystals in a liquid of components A and B. This liquid is known as the *eutectic material*, and it will not solidify until the line BD (the *eutectic line*, *eutectic point*, or *eutectic temperature*)—the lowest point at which the eutectic material can exist in liquid form—is reached.

Since the two ingredients do not mix, reducing the temperature below the eutectic line results in crystals (layers or plates) of both pure ingredients forming. This is the microstructure of a solid eutectic alloy: alternating pure crystals of the two ingredients. Since two solid substances are produced from a single liquid substance, the process could be written in chemical reaction format as liquid $\rightarrow$ solid α + solid β. (Alternatively, upon heating, the reaction would be solid α + solid $\beta \rightarrow$ liquid.) For this reason, the phase change is called a *eutectic reaction*.

There are similar reactions involving other phases and states. Table 36.10 and Fig. 36.12 illustrate these.

Example

A binary solution with the characteristics and composition of point P is cooled.

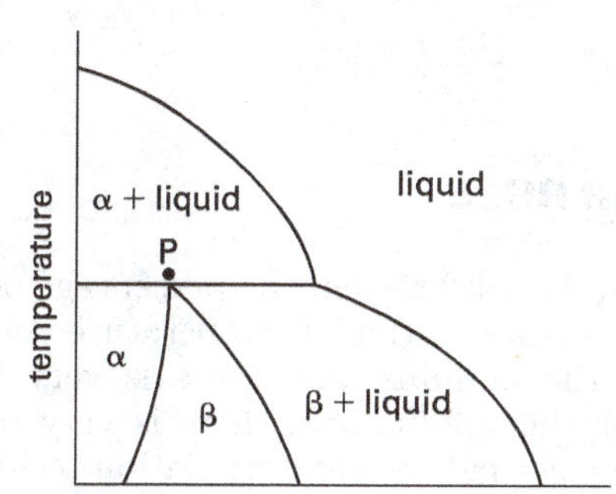

What type of alloy will result?

(A) eutectic

(B) eutectoid

(C) peritectic

(D) peritectoid

Solution

Dropping straight down from point P on the illustration, corresponding to cooling, the α solid and liquid transform into the β solid. This corresponds to the definition of a peritectic reaction. A peritectic alloy will form.

The answer is (C).

Table 36.10 *Types of Equilibrium Reactions*

reaction name	type of reaction upon cooling
eutectic	liquid $\rightarrow$ solid α + solid β
peritectic	liquid + solid $\alpha \rightarrow$ solid β
eutectoid	solid $\gamma \rightarrow$ solid α + solid β
peritectoid	solid α + solid $\gamma \rightarrow$ solid β

Figure 36.12 *Typical Appearance of Equilibrium Diagram at Reaction Points*

reaction name	phase reaction	phase diagram
eutectic	$L \rightarrow \alpha(s) + \beta(s)$ cooling	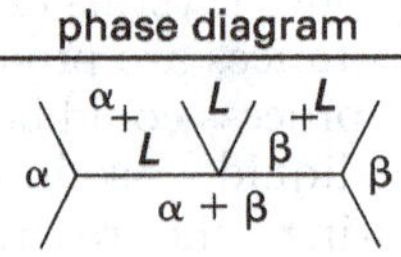
peritectic	$L + \alpha(s) \rightarrow \beta(s)$ cooling	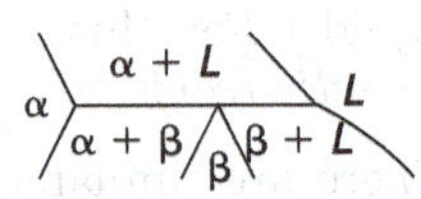
eutectoid	$\gamma(s) \rightarrow \alpha(s) + \beta(s)$ cooling	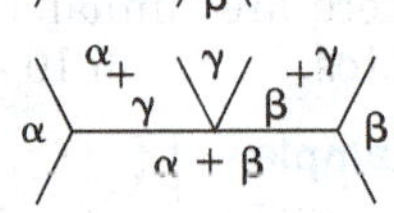
peritectoid	$\alpha(s) + \gamma(s) \rightarrow \beta(s)$ cooling	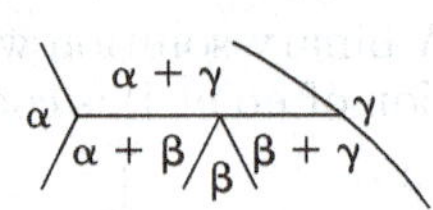

15. LEVER RULE

Within a liquid-solid region, the percentage of solid and liquid phases is a function of temperature and composition. Near the liquidus line, there is very little solid phase. Near the solidus line, there is very little liquid phase. The *lever rule* is an interpolation technique used to find the relative amounts of solid and liquid phase at any composition. These percentages are given in fraction (or percent) by weight.

Figure 36.10 shows an alloy with an average composition of 55% copper at 2300°F. (A horizontal line representing different conditions at a single temperature is known as a *tie line*.) The liquid composition is defined by point 2, and the solid composition is defined by point 1. Referring to Fig. 36.10, the gravimetric fraction of solid and liquid can be determined from the lever rule using the line segment lengths m, n, and $w = m + n$ in a method that is analogous to determining steam quality.

$$\text{fraction solid} = 1 - \text{fraction liquid} = \frac{n}{m+n} = \frac{n}{w}$$

$$\text{fraction liquid} = 1 - \text{fraction solid} = \frac{m}{m+n} = \frac{m}{w}$$

Equation 36.11 and Eq. 36.12: Gravimetric Component Fraction

$$\text{wt\% } \alpha = \frac{x_\beta - x}{x_\beta - x_\alpha} \times 100\% \quad \text{36.11}$$

$$\text{wt\% } \beta = \frac{x - x_\alpha}{x_\beta - x_\alpha} \times 100\% \quad \text{36.12}$$

Description

Referring to Fig. 36.13, from the lever rule, the gravimetric fractions of solid and liquid phases depend on the lengths of the lines $x - x_\alpha$ and $x_\beta - x$, along with the separation, $x_\beta - x_\alpha$, which may be measured using any convenient scale. (Although the distances can be measured in millimeters or tenths of an inch, it is more convenient to use the percentage alloying ingredient scale.) Then, the fractions of solid and liquid can be calculated from Eq. 36.11 and Eq. 36.12.

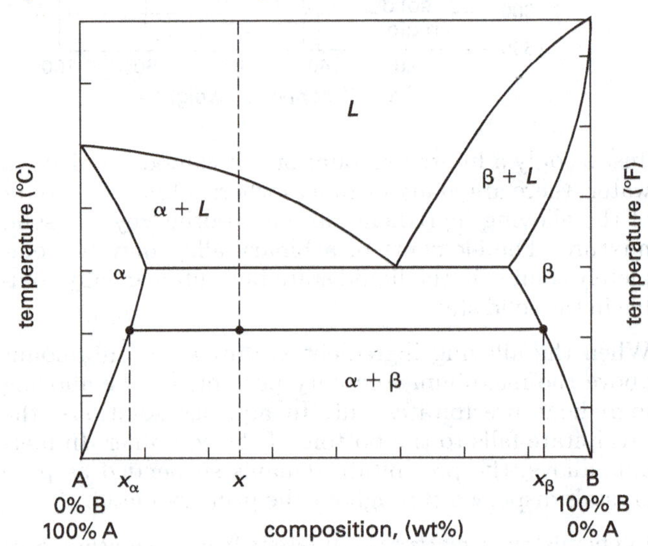

Figure 36.13 *Two-Phase System Phase Diagram*

The lever rule and method of determining the composition of the two components are applicable to any solution or mixture, liquid or solid, in which two phases are present.

Example

Consider the Ag-Cu phase diagram given.

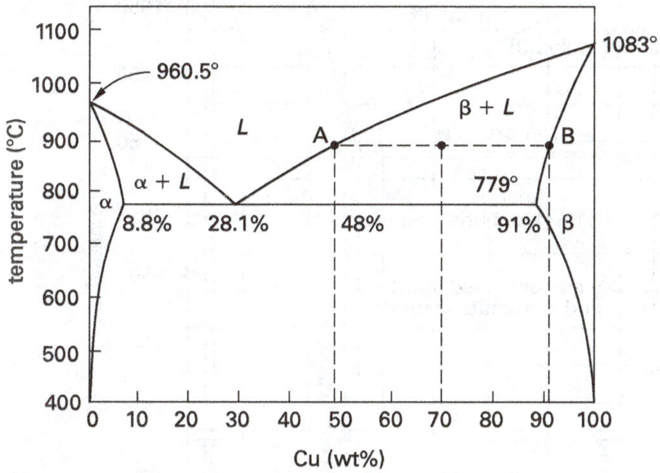

What is most nearly the equilibrium percentage of β in an alloy of 30% Ag, 70% Cu at 900°C?

(A) 0.0%

(B) 22%

(C) 51%

(D) 59%

Solution

Draw the horizontal tie line at 900°C between the liquid, L, phase at point A and the solid, β, at point B. Draw the vertical line that identifies the alloy at $x = 70\%$ Cu. Rather than measure the lengths of the lines, use the horizontal Cu percentage scale. Point A is located at $x_L = 48\%$ Cu, and point B is located at $x_\beta = 91\%$ Cu. Use Eq. 36.12.

$$\begin{aligned}\text{wt\% } \beta &= \frac{x - x_L}{x_\beta - x_L} \times 100\% \\ &= \frac{70\% - 48\%}{91\% - 48\%} \times 100\% \\ &= 51.2\% \quad (51\%)\end{aligned}$$

The answer is (C).

16. IRON-CARBON PHASE DIAGRAM

The iron-carbon phase diagram (see Fig. 36.14) is much more complex than idealized equilibrium diagrams due to the existence of many different phases. Each of these phases has a different microstructure and different mechanical properties. By treating the steel in such a manner as to force the occurrence of particular phases, steel with desired wear and endurance properties can be produced.

Allotropes have the same composition but different atomic structures (microstructures), volumes, electrical resistances, and magnetic properties. *Allotropic changes* are reversible changes that occur at the *critical points* (i.e., *critical temperatures*).

Iron exists in three primary allotropic forms: alpha-iron, delta-iron, and gamma-iron. The changes are brought about by varying the temperature of the iron. Heating pure iron from room temperature changes its structure from body-centered cubic (BCC) *alpha-iron* (−273–970°C), also known as *ferrite*, to face-centered cubic (FCC) *gamma-iron* (910–1400°C), to BCC *delta-iron* (above 1400°C).

Iron-carbon mixtures are categorized into *steel* (less than 2% carbon) and *cast iron* (more than 2% carbon) according to the amounts of carbon in the mixtures.

The most important eutectic reaction in the iron-carbon system is the formation of a solid mixture of austenite and cementite at approximately 1129°C. *Austenite* is a solid solution of carbon in gamma-iron. It is nonmagnetic, decomposes on slow cooling, and does not normally exist below 723°C, although it can be partially preserved by extremely rapid cooling.

Cementite (Fe_3C), also known as *carbide* or *iron carbide*, has approximately 6.67% carbon. Cementite is the hardest of all forms of iron, has low tensile strength, and is quite brittle.

The most important eutectoid reaction in the iron-carbon system is the formation of *pearlite* from the decomposition of austentite at approximately 723°C. Pearlite is actually a mixture of two solid components, ferrite and cementite, with the common *lamellar* (*layered*) *appearance.*

Ferrite is essentially pure iron (less than 0.025% carbon) in BCC alpha-iron structure. It is magnetic and has properties complementary to cementite, since it has low hardness, high tensile strength, and high ductility.

17. EQUILIBRIUM MIXTURES

Equation 36.13: Gibbs' Phase Rule

$$P + F = C + 2 \qquad 36.13$$

Variation

$$P + F = C + 1 \quad \Big| \text{ constant pressure, constant temperature, or constant composition}$$

Figure 36.14 Iron-Carbon Diagram

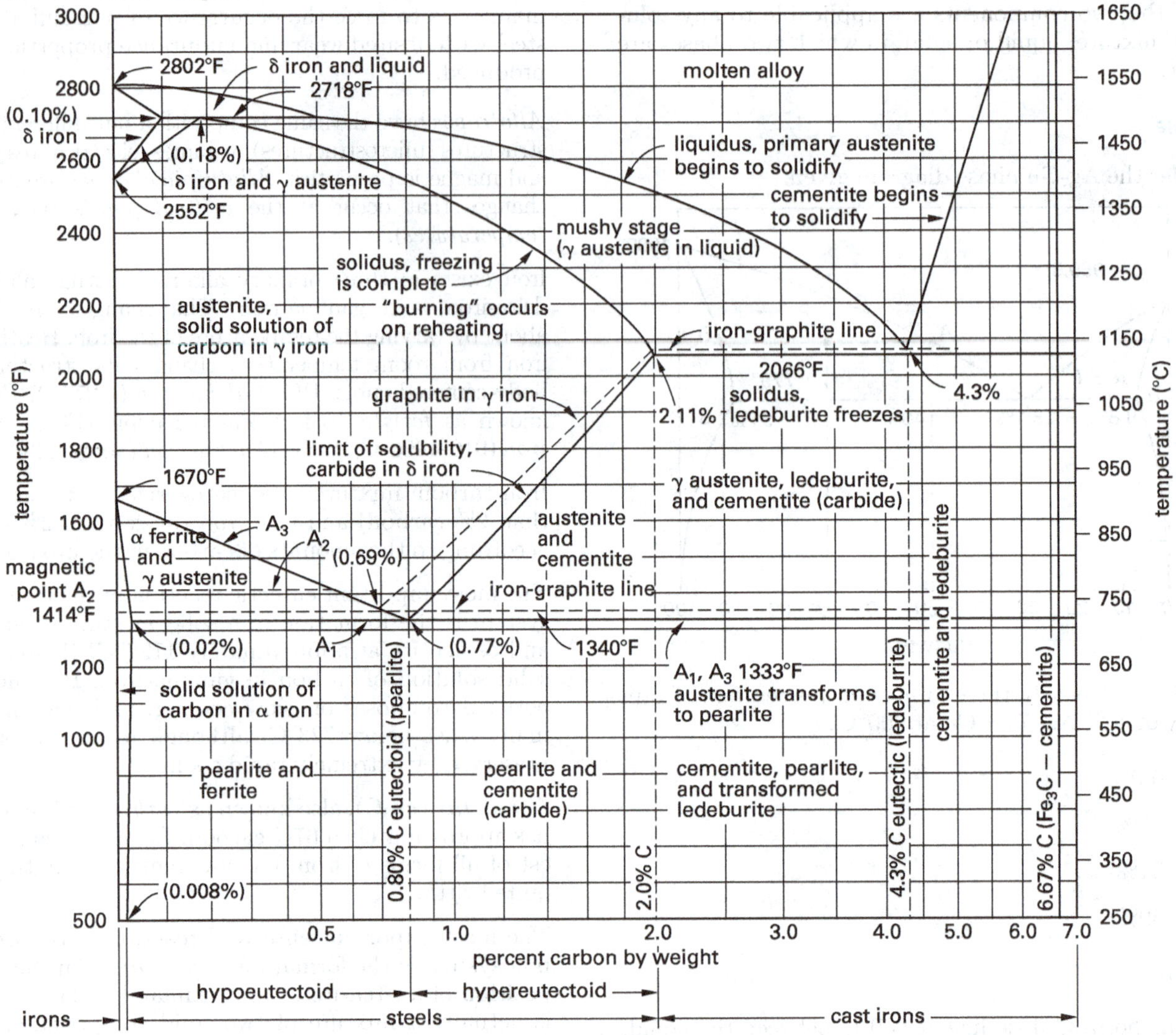

Description

Gibbs' phase rule defines the relationship between the number of phases and elements in an equilibrium mixture. For such an equilibrium mixture to exist, the alloy must have been slowly cooled and thermodynamic equilibrium must have been achieved along the way.

At equilibrium, and considering both temperature and pressure to be independent variables, Gibbs' phase rule is Eq. 36.13.

P is the number of phases existing simultaneously; F is the number of independent variables, known as *degrees of freedom*; and C is the number of elements in the alloy. Composition, temperature, and pressure are examples of degrees of freedom that can be varied.

For example, if water is to be stored in a condition where three phases (solid, liquid, gas) are present simultaneously, then $P = 3$, $C = 1$, and $F = 0$. That is, neither pressure nor temperature can be varied. This state corresponds to the *triple point* of water.

If pressure is constant, then the number of degrees of freedom is reduced by one, and Gibbs' phase rule can be rewritten as shown in the variation.

If Gibbs' rule predicts $F = 0$, then an alloy can exist with only one composition.

Example

A system consisting of an open bucket containing a mixture of ice and water is to be warmed from 0°C to 20°C. How many degrees of freedom does the system have?

(A) 0

(B) 1

(C) 2

(D) 3

Solution

Degrees of freedom and the system properties are instantaneous values. What is happening and what is going to happen to the system are not relevant. Only the current, instantaneous equilibrium properties are relevant. In this case, the system consists of an open bucket containing ice and water, so the number of phases is $P = 2$. The only substance in the system is water, so the number of components is $C = 1$.

$$\begin{aligned} P + F &= C + 2 \\ F &= C + 2 - P = 1 + 2 - 2 \\ &= 1 \end{aligned}$$

The answer is (B).

18. THERMAL PROCESSING

Thermal processing, including hot working, heat treating, and quenching, is used to obtain a part with desirable mechanical properties.

Cold and hot working are both forming processes (rolling, bending, forging, extruding, etc.). The term *hot working* implies that the forming process occurs above the *recrystallization temperature.* (The actual temperature depends on the rate of strain and the cooling period, if any.) *Cold working* (also known as *work hardening* and *strain hardening*) occurs below the recrystallization temperature.

Above the recrystallization temperature, almost all of the internal defects and imperfections caused by hot working are eliminated. In effect, hot working is a "self-healing" operation. A hot-worked part remains softer and has a greater ductility than a cold-worked part. Large strains are possible without strain hardening. Hot working is preferred when the part must go through a series of forming operations (passes or steps), or when large changes in size and shape are needed.

The hardness and toughness of a cold-worked part will be higher than that of a hot-worked part. Because the part's temperature during the cold working is uniform, the final microstructure will also be uniform. There are many times when these characteristics are desirable, and hot working is not always the preferred forming method. In many cases, cold working will be the final operation after several steps of hot working.

Once a part has been worked, its temperature can be raised to slightly above the recrystallization temperature. This *heat treatment* operation is known as *annealing* and is used to relieve stresses, increase grain size, and recrystallize the grains. Stress relief is also known as *recovery.*

Quenching is used to control the microstructure of steel by preventing the formation of equilibrium phases with undesirable characteristics. The usual desired result is hard steel, which resists plastic deformation. The quenching can be performed with gases (usually air), oil, water, or brine. Agitation or spraying of these fluids during the quenching process increases the severity of the quenching.

Time-temperature-transformation (TTT) *curves* are used to determine how fast an alloy should be cooled to obtain a desired microstructure. Although these curves show different phases, they are not equilibrium diagrams. On the contrary, they show the microstructures that are produced with controlled temperatures or when quenching interrupts the equilibrium process.

TTT curves are determined under ideal, isothermal conditions. However, the curves are more readily available than experimentally determined *controlled-cooling-transformation* (CCT) *curves.* Both curves are similar in shape, although the CCT curves are displaced downward and to the right from TTT curves.

Figure 36.15 shows a TTT diagram for a high-carbon (0.80% carbon) steel. Curve 1 represents extremely rapid quenching. The transformation begins at 216°C, and continues for 8–30 seconds, changing all of the austenite to martensite. The martensitic transformation does not depend on diffusion. Since martensite has almost no ductility, martensitic microstructures are used in applications such as springs and hardened tools where a high elastic modulus and low ductility are needed.

Figure 36.15 *TTT Diagram for High-Carbon Steel*

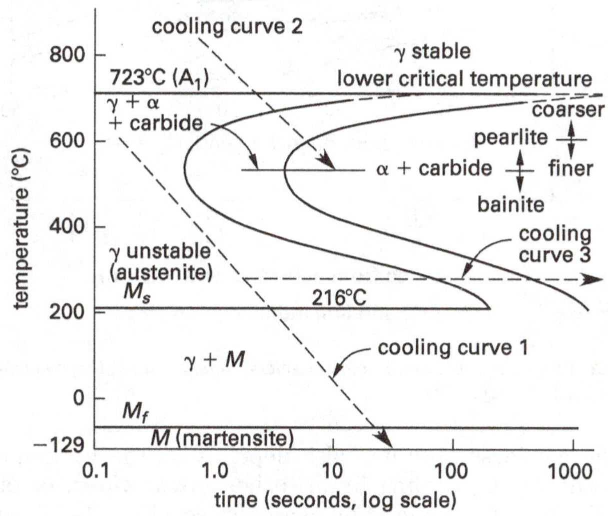

Curve 2 represents a slower quench that converts all of the austenite to fine pearlite.

A horizontal line below the critical temperature is a *tempering* process. If the temperature is decreased rapidly along curve 1 to 270°C and is then held constant along cooling curve 3, bainite is produced. This is the principle of *austempering.* Bainite is not as hard as martensite, but it does have good impact strength and fairly high hardness. Performing the same procedure at 180–200°C is *martempering*, which produces *tempered martensite*, a soft and tough steel.

19. HARDNESS AND HARDENABILITY

Hardness is the measure of resistance a material has to plastic deformation. Various *hardness tests* (e.g., Brinell, Rockwell, Meyer, Vickers, and Knoop) are used to determine hardness. These tests generally measure the depth of an impression made by a hardened penetrator set into the surface by a standard force.

Hardenability is a relative measure of the ease by which a material can be hardened. Some materials are easier to harden than others. See Fig. 36.16 for sample hardenability curves for steel.

Figure 36.16 *Jominy Hardenability Curves for Six Steel Alloys*

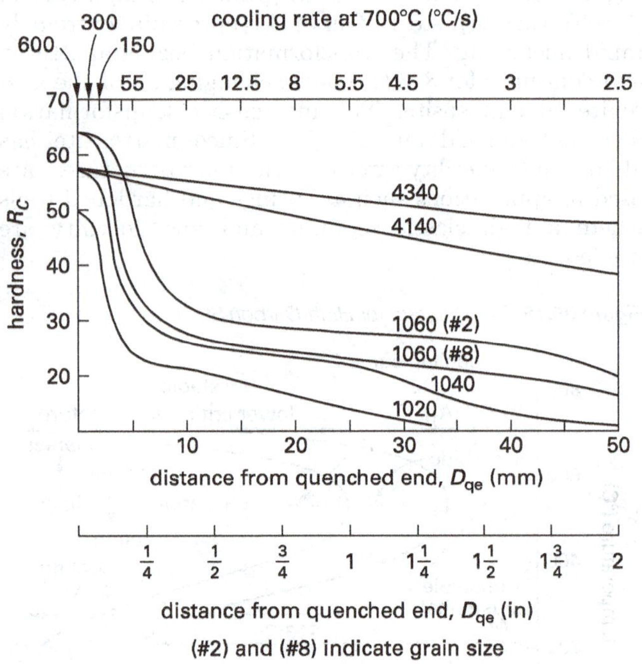

Van Vlack, L., *Elements of Materials Science and Engineering*, Addison-Wesley, 1989.

The hardness obtained also depends on the hardening method (e.g., cooling in air, other gases, water, or oil) and rate of cooling. For example, see Fig. 36.17 and Fig. 36.18. Since the hardness obtained depends on the material, hardening data is often presented graphically. There are a variety of curve types used for this purpose.

Figure 36.17 *Cooling Rates for Bars Quenched in Agitated Water*

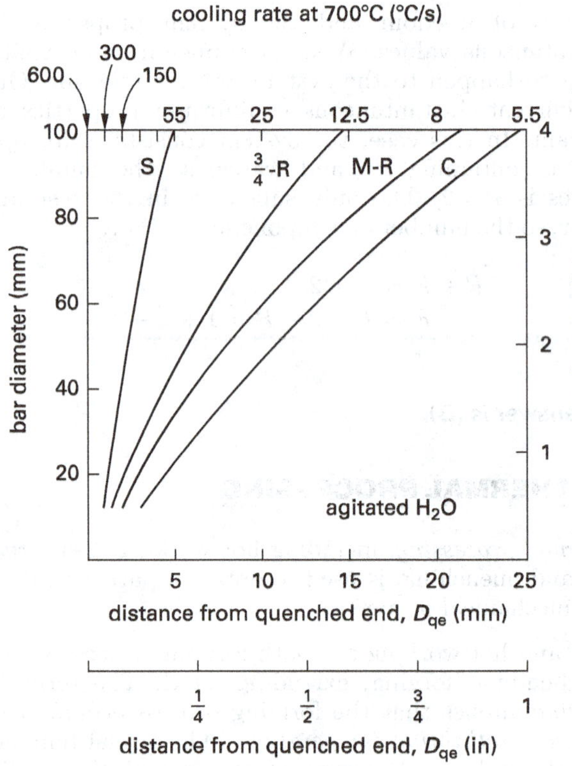

Van Vlack, L., *Elements of Materials Science and Engineering*, Addison-Wesley, 1989.

Hardenability is not the same as hardness. *Hardness* refers to the ability to resist deformation when a load is applied, whereas *hardenability* refers to the ability to be hardened to a particular depth.

In the *Jominy end quench test*, a cylindrical steel specimen is heated long enough to obtain a uniform grain structure. Then, one end of the specimen is cooled ("quenched") in a water spray, while the remainder of the specimen is allowed to cool by conduction. The cooling rate decreases with increasing distance from the quenched end. When the specimen has entirely cooled, the hardness is determined at various distances from the quenched end. The hardness at the quenched end corresponds to water cooling, while the hardness at the opposite end corresponds to air cooling. The same test can be performed with different alloys. *Rockwell hardness* (C-scale) (also known as *Rockwell C hardness*), R_C, is plotted on the vertical axis, while distance from the quenched end, D_{qe}, is plotted on the horizontal scale.[19] The horizontal scale can also be correlated to and calibrated as *cooling rate*. The *Jominy hardenability curves* are used to select an alloy and heat treatment that

[19]In ASTM A255, "Standard Test Methods for Determining Hardenability of Steel," the distance from the quenched end is given the symbol J and is referred to as the *Jominy distance*. Rockwell C hardness is given the standard symbol "HRC." The *initial hardness* at the $J = 1/16$ in position is given the symbol "IH."

Figure 36.18 Cooling Rates for Bars Quenched in Agitated Oil

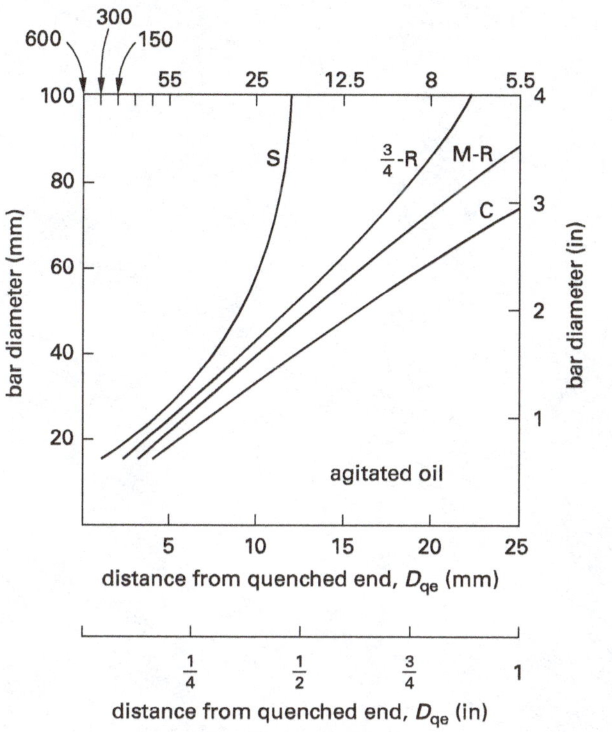

Van Vlack, L., *Elements of Materials Science and Engineering*, Addison-Wesley, 1989.

minimizes distortion during manufacturing. Figure 36.16 illustrates the results of Jominy hardenability tests of six steel alloys.

The position within the bar (see Fig. 36.19) is indicated by the following nomenclature.

- C = center of the bar
- M-R = halfway between the center of the bar and the surface of the bar
- $\frac{3}{4}$-R = three-quarters (75%) of the distance between C and S
- S = surface of the bar

20. METAL GRAIN SIZE

One of the factors affecting hardness and hardenability is the average metal *grain size*. Grain size refers to the diameter of a three-dimensional spherical grain as determined from a two-dimensional micrograph of the metal.

The size of the grains formed depends on the number of nuclei formed during the solidification process. If there are many nuclei, as would occur when cooling is rapid, there will be many small grains. However, if cooling is

Figure 36.19 Bar Positions

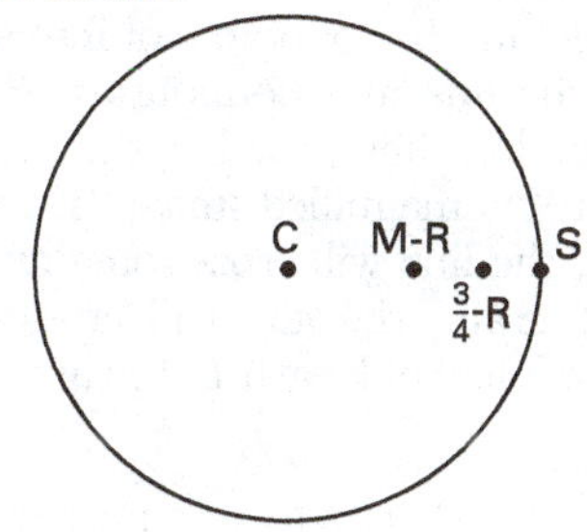

slow, a smaller number of larger grains will be produced. Fast cooling produces fine-grained material, and slow cooling produces coarse-grained material.

At moderate temperatures and strain rates, fine-grained materials deform less (i.e., are harder and tougher) while coarse-grained materials deform more. For ease of cold-formed manufacturing, coarse-grained materials may be preferred. However, appearance and strength may suffer.

It is difficult to measure grain size because the grains are varied in size and shape and because a two-dimensional image does not reveal volume. Semi-empirical methods have been developed to automatically or semi-automatically correlate grain size with the number of intersections observed in samples. ASTM E112, "Determining Average Grain Size" (and the related ASTM E1382), describes a planimetric procedure for metallic and some nonmetallic materials that exist primarily in a single phase.

Equation 36.14 Through Eq. 36.16: ASTM Grain Size, *n*

$$\frac{N_{\text{actual}}}{\text{actual area}} = \frac{N}{0.0645\ \text{mm}^2} \qquad 36.14$$

$$N_{(0.0645\ \text{mm}^2)} = 2^{(n-1)} \qquad 36.15$$

$$S_V = 2P_L \qquad 36.16$$

Description

Data on grain size is obtained by counting the number of grains in any small two-dimensional area. The number of grains, N, in a standard area (0.0645 mm^2) can be extrapolated from the observations using Eq. 36.14.

The standard number of grains, N, is the number of grains per square inch in an image of a polished specimen magnified 100 times. Equation 36.15 calculates the ASTM *grain size number* (also known as the ASTM *grain size*), n, from the standard number of grains, N, as the nearest integer greater than 1.

$$n = \frac{\log N + \log 2}{\log 2}$$

The grain-boundary surface area per unit volume, S_V, is taken as twice the number of points of intersection per unit length between the line and boundaries, P_L, as shown by Eq. 36.16. If a random line (any length, any orientation) is drawn across a 100× magnified image (i.e., a *photomicrograph*) of grains, the line will cross some number of grains, say N. For each grain, the line will cross over two grain boundaries, so for a line of length L, the average grain diameter will be

$$D_{ave} = \frac{L}{2N} = \frac{1}{S_V}$$

Example 10

Eight grains are observed in a 0.0645 mm² area of a polished metal surface. What is most nearly the number of grains in an area of 710 mm²?

(A) 16 grains

(B) 120 grains

(C) 1100 grains

(D) 88 000 grains

Solution

From Eq. 36.14, and rearranging to solve for N_{actual}, the number of grains in an area of 710 mm² is

$$\frac{N_{actual}}{\text{actual area}} = \frac{N}{0.0645\ \text{mm}^2}$$

$$\begin{aligned} N_{actual} &= (\text{actual area})\left(\frac{N}{0.0645\ \text{mm}^2}\right) \\ &= (710\ \text{mm}^2)\left(\frac{8\ \text{grains}}{0.0645\ \text{mm}^2}\right) \\ &= 88\,062\ \text{grains} \quad (88\,000\ \text{grains}) \end{aligned}$$

The answer is (D).

Topic XI: Structural Design

37 Structural Design: Materials and Basic Concepts

Nomenclature

A	area	in^2
b	width	in
C	coefficient	–
D	dead load	lbf
E	earthquake load	lbf
E	modulus of elasticity	lbf/in^2
f	computed stress	$kips/in^2$
f'_c	compressive strength	lbf/in^2
F	fluid load	lbf
F	allowable stress or strength	$kips/in^2$
G	shear modulus	$kips/in^2$
h	overall height (thickness)	in
H	earth pressure	lbf
K	effective length factor	–
K_{LL}	ratio of area of influence to tributary area	–
L	length	in
L_b	length between laterally braced points	in
L_p	limiting laterally unbraced length for yielding	in
L_r	limiting laterally unbraced length for buckling	in
L	live load	lbf
M	moment	in-kips
r	radius of gyration	in
R	rain load	lbf
S	elastic section modulus	in^3
S	snow load	lbf
S'_e	endurance strength	lbf/in^2
T	temperature, shrinkage, creep, and settlement loads	lbf
U	required (ultimate) strength	lbf
V	shear	kips
w	unit weight (specific weight)	lbf/ft^3
W	wind load	lbf
Z	plastic section modulus	in^3

Symbols

α	coefficient of thermal expansion	1/°F
ε	strain	–
ν	Poisson's ratio	–
ρ	density	lbm/ft^3
ϕ	resistance factor	–
ϕ	strength reduction factor	–
Ω	safety factor	–

Subscripts

b	bending or braced
c	compressive or concrete
cr	critical
f	flange
n	nominal
p	bearing or plastic
r	roof
t	tensile
T	tributary
u	ultimate
x	strong axis
y	weak axis or yield

1. INTRODUCTION

The design of reinforced concrete beams is governed by the provisions of the American Concrete Institute code ACI 318. Many of the code equations are not homogeneous—the units of the result cannot be derived from those of the input. In particular, the quantity $\sqrt{f'_c}$ is treated as though it has units of lbf/in^2, even though that does not follow from the expression.

2. SPECIFICATIONS AND BUILDING CODES

Although the terms "specification" and "building code" are often used interchangeably in the context of design, there is a difference between the two. A *specification* is a set of guidelines or recommendations put forth by a group of experts in research and design with the intent of ensuring safety. A specification is not legally enforceable unless it is a part of a building code.

The design of steel buildings in the United States is principally based on the specifications of the American Institute of Steel Construction (AISC), a nonprofit trade association representing and serving the fabricated structural steel industry in the United States. AISC's *Steel Construction Manual* (referred to as the *AISC Manual* in this book) describes and defines the details of the two permitted methods of design: *load and resistance factor design* (LRFD) and *allowable strength design* (ASD) (see Sec. 37.15).[1] The *AISC Manual* includes the AISC *Specification for Structural Steel Buildings* (referred to as *AISC Specifications* in this book). In addition to the ASD and LRFD specifications, the *AISC Manual* contains a wealth of information on products available, design aids, examples, erection guidelines, and other applicable specifications.

A *building code* is a broad-based document covering all facets of safety, such as design loadings, occupancy limits, plumbing, electrical requirements, and fire protection. Building codes are adopted by states, cities, or other government bodies as a legally enforceable means of protecting public safety and welfare. Although other codes have seen widespread use, since 2000, the predominant building code in the United States has been the *International Building Code* (IBC), maintained and published by the International Code Council (ICC).[2] The ICC adopts AISC methodologies, both by reference and by duplication, with and without modifications.

3. ALLOWABLE STRESS VS. STRENGTH DESIGN

Structural members can be sized using two alternative design procedures. With the *allowable stress method*, stresses induced in the concrete and the steel are estimated assuming that the behavior is linearly elastic. The member is sized so that the computed stresses do not exceed certain predetermined values. With the *strength design method*, the actual stresses are not the major concern. Rather, a strength is provided that will support factored (i.e., amplified) loads. Although both methods produce adequate designs, strength design is considered to be more rational. The structural chapters in this book only cover strength design provisions. ACI 318 no longer contains provisions for *allowable stress design*.

It is important to recognize that even though beam cross sections are sized on the basis of strength, the effects (i.e., moments, shears, deflections, etc.) of the factored loads are computed using elastic analysis.

ACI 318 contains two options for using strength design in the design of reinforced and prestressed concrete members: The *"unified" methods* in ACI 318 Chap. 9 ("Strength and Serviceability Requirements") and the older ACI 318 App. C ("Alternative Load and Strength Reduction Factors"). These methods differ in how the factored loads are calculated, as well as specify different strength reduction factors, ϕ.

ACI 318 Chap. 9 uses the following abbreviations for load types: D, dead load; L, live load; W, wind load; E, earthquake load; H, earth pressure; T, temperature, shrinkage, creep, and settlement loads; F, fluid load; L_r, live roof load; S, snow load; and R, rain load.

ACI 318 Chap. 9 methods determine strength reduction factors on the basis of the strain conditions in the reinforcement farthest from the extreme compression face. The methods of ACI 318 Chap. 9 are consistent with ASCE7 and the IBC, which have diverged from the previous ACI 318 methods. The unified methods bring uniformity and eliminate many of the inconsistencies in the previous design requirements.

ACI 318 App. C uses only the following abbreviations for load types: D, dead load; L, live load; W, wind load; E, earthquake load; H, earth pressure; and T, temperature, shrinkage, creep, and settlement loads.

ACI 318 App. C methods determine strength reduction factors solely on the basis of the type of loading (axial, flexural, or both) on the section. From ACI 318 sec. RC.9.1.1, the methods in ACI 318 App. C have "... evolved since the early 1960s and are considered to be reliable for concrete construction." ACI 318 App. C allows the use of the old (ACI 318-89 through ACI 318-99) load factors and ϕ values, but not a combination of

[1]For over 100 years, steel structures in the United States were designed to keep induced stresses less than allowable stresses. This method was universally referred to as the *allowable stress design* (ASD) method. Since the introduction of design methods (e.g., LRFD) based on ultimate strength, allowable stress design has become disfavored, even though it is explicitly permitted in the *AISC Manual*. Accordingly, in a move to make all design methods "strength" related processes, starting with the 13th edition of the *AISC Manual*, ASD was renamed as the *allowable strength design* method, even though ASD does not utilize ultimate strength concepts at all. The name *allowable stress design* is still used by other authorities, including ASCE7.

[2]Prior to the widespread adoption of the IBC, three model building codes were in use in the United States: the *Uniform Building Code* (UBC) from the International Conference of Building Officials (ICBO), the *National Building Code* (NBC) from the Building Officials and Codes Administrators International, Inc. (BOCA), and the *Standard Building Code* (SBC) from the Southern Building Code Congress International (SBCCI).

new load factors and old ϕ values. However, the 0.75 reduction times the $D+L+W$ load combination, allowed in ACI 318-99, cannot be taken.

4. DESIGN STRENGTH AND CRITERIA

ACI 318 uses the subscript n to indicate a "nominal" quantity. A *nominal value* can be interpreted as being in accordance with theory for the specified dimensions and material properties. The nominal moment and shear strengths are designated M_n and V_n, respectively.

The *design strength* is the result of multiplying the *nominal strength* by a *strength reduction factor* (also known as a *capacity reduction factor* or *resistance factor*), ϕ.

$$\text{design strength} = \phi(\text{nominal strength})$$

The design criteria for all sections in a beam are

$$\phi M_n \geq M_u$$

$$\phi V_n \geq V_u$$

Some values of ϕ are given in Table 37.1.

Table 37.1 Strength Reduction Factors

	ϕ
tension-controlled sections ($\varepsilon_t \geq 0.005$)	0.9
compression-controlled sections ($\varepsilon_t \leq 0.002$)	
members with spiral reinforcement	0.70
members with tied reinforcement	0.65
transition sections ($0.002 < \varepsilon_t < 0.005$)	
members with spiral reinforcement	$0.57 + 67\varepsilon_t$
members with tied reinforcement	$0.48 + 83\varepsilon_t$
shear and torsion	0.75
bearing on concrete	0.65

5. CONCRETE

Concrete (*portland cement concrete*) is a mixture of cementitious materials, aggregates, water, and air. The cement paste consists of a mixture of portland cement and water. The paste binds the coarse and fine aggregates into a rock-like mass as the paste hardens during the chemical reaction (*hydration*). Table 37.2 lists the typical volumetric percentage of each ingredient.

Table 37.2 Typical Volumetric Proportions of Concrete Ingredients

component	air-entrained	non-air-entrained
coarse aggregate	31%	31%
fine aggregate	28%	30%
water	18%	21%
portland cement	15%	15%
air	8%	3%

6. CONCRETE MIX DESIGN CONSIDERATIONS

Concrete can be designed for compressive strength or durability in its hardened state. In its wet state, concrete should have good workability. *Workability* relates to the effort required to transport, place, and finish wet concrete without segregation or bleeding. Workability is often closely correlated with *slump*. Table 37.3 gives typical slumps by application. All other design requirements being met, the most economical mix should be selected.

Durability is defined as the ability of concrete to resist environmental exposure or service loadings. One of the most destructive environmental factors is the freeze/thaw cycle. ASTM C666, "Standard Test Method for Resistance of Concrete to Rapid Freezing and Thawing," is the standard laboratory procedure for determining the freeze-thaw durability of hardened concrete. This test determines a *durability factor*, the number of freeze-thaw cycles required to produce a certain amount of deterioration.

ACI 318 Sec. 4.3.1 places maximum limits on the *water-cement ratio* and minimum limits on the strength for concrete with special exposures, including concrete exposed to freeze-thaw cycles, deicing chemicals, and chloride, and installations requiring low permeability.

Specifying an *air-entrained concrete* will improve the durability of concrete subject to freeze-thaw cycles or deicing chemicals. The amount of entrained air needed will depend on the exposure conditions and the size of coarse aggregate, as prescribed in ACI 318 Sec. 4.4.1.

Table 37.3 *Typical Slumps by Application*

application	slump (in) maximum	slump (in) minimum
reinforced footings and foundations	3	1
plain footings and substructure walls	3	1
slabs, beams, and reinforced walls	4	1
reinforced columns	4	1
pavements and slabs	3	1
heavy mass construction	2	1
roller-compacted concrete	0	0

(Multiply in by 25.4 to obtain mm.)

7. CEMENTITIOUS MATERIALS

Cementitious materials include portland cement, blended hydraulic cements, expansive cement, and other cementitious additives, including fly ash, pozzolans, silica fume, and ground granulated blast-furnace slag.

Portland cement is produced by burning a mixture of lime and clay in a rotary kiln and grinding the resulting mass. Cement has a specific weight (density) of approximately 195 lbf/ft^3 and is packaged in standard sacks (bags) weighing 94 lbf.

ASTM C150 describes the five classifications of portland cement.

Type I—Normal portland cement: This is a general-purpose cement used whenever sulfate hazards are absent and when the heat of hydration will not produce a significant rise in the temperature of the cement. Typical uses are sidewalks, pavement, beams, columns, and culverts.

Type II—Modified portland cement: This cement has a moderate sulfate resistance, but is generally used in hot weather for the construction of large structures. Its heat rate and total heat generation are lower than those of normal portland cement.

Type III—High-early strength portland cement: This type develops its strength quickly. It is suitable for use when a structure must be put into early use or when long-term protection against cold temperatures is not feasible. Its shrinkage rate, however, is higher than those of types I and II, and extensive cracking may result.

Type IV—Low-heat portland cement: For massive concrete structures such as gravity dams, low-heat cement is required to maintain a low temperature during curing. The ultimate strength also develops more slowly than for the other types.

Type V—Sulfate-resistant portland cement: This type of cement is appropriate when exposure to sulfate concentration is expected. This typically occurs in regions having highly alkaline soils.

Types I, II, and III are available in two varieties: normal and air-entraining (designated by an "A" suffix). The compositions of the three types of air-entraining portland cement (types IA, IIA, and IIIA) are similar to types I, II, and III, respectively, with the exception that an air-entraining admixture is added.

Many state departments of transportation use modified concrete mixes in critical locations in order to reduce *concrete-disintegration cracking* (*D-cracking*) caused by the freeze-thaw cycle. Coarse aggregates are the primary cause of D-cracking, so the maximum coarse aggregate size is reduced. However, a higher cement paste content causes *shrinkage cracking* during setting, leading to increased water penetration and corrosion of reinforcing steel. The cracking can be reduced or eliminated by using *shrinkage-compensating cement*, known as *type-K cement* (named after ASTM C845 type E-1(K)).

Type-K cement (often used in bridge decks) contains an aluminate that expands during setting, offsetting the shrinkage. The net volume change is near zero. The resulting concrete is referred to as *shrinkage-compensating concrete.*

Special cement formulations are needed to reduce *alkali-aggregate reactivity* (AAR)—the reaction of the alkalis in cement with compounds in the sand and gravel aggregate. AAR produces long-term distress in the forms of network cracking and spalling (popouts) in otherwise well-designed structures. AAR takes on two forms: the more common *alkali-silica reaction* (ASR) and the less-common *alkali-carbonate reaction* (ACR). ASR is countered by using low-alkali cement (ASTM C150) with an equivalent alkali content of less than 0.60% (as sodium oxide), using lithium-based admixtures, or "sweetening" the mixture by replacing approximately 30% of the aggregate with crushed limestone. ACR is not effectively controlled by using low-alkali cements. Careful selection, blending, and sizing of the aggregate are needed to minimize ACR.

A waste product of coal-burning power-generation stations, *fly ash*, is the most common *pozzolanic additive.* As cement sets, calcium silicate hydrate and calcium hydroxide are formed. While the former is a binder that holds concrete together, calcium hydroxide does not contribute to binding. However, fly ash reacts with some of the calcium hydroxide to increase binding. Also, since fly ash acts as a microfiller between cement particles, strength and durability are increased while permeability is reduced. When used as a replacement for less than 45% of the portland cement, fly ash meeting ASTM C618 enhances resistance to scaling from road-deicing chemicals.

Microsilica (*silica fume*) is an extremely fine particulate material, approximately 1/100th the size of cement particles. It is a waste product of electric arc furnaces. It acts as a "super pozzolan." Adding 5–15% microsilica will increase the pozzolanic reaction as well as provide a microfiller to reduce permeability.

Microsilica reacts with calcium hydroxide in the same manner as fly ash. It is customarily used to achieve concrete compressive strengths in the 8000–9000 psi range.

8. STRENGTH ACCEPTANCE TESTING FOR CONCRETE

When extensive statistical data are not available, acceptance testing of laboratory-cured specimens can be evaluated per ACI 318 sec. 5.6.3. Cylinder sizes of either 4 in × 8 in or 6 in × 12 in are permitted. A *single strength test* is defined as the arithmetic average of the compressive strengths of at least two cylinders measuring 6 in × 12 in or at least three cylinders measuring 4 in × 8 in. Single strength tests should be conducted at 28 days unless otherwise specified. The compressive strength of concrete is considered satisfactory if both criteria are met: (a) No single strength test falls below the specified compressive strength, f'_c, by more than 500 psi when f'_c is 5000 psi or less, or by more than $0.10f'_c$ when f'_c is more than 5000 psi; (b) the average of every three consecutive strength tests equals or exceeds the specified compressive strength.

9. MODULUS OF ELASTICITY

The *modulus of elasticity* (also known as *Young's modulus*) is defined as the ratio of stress to strain in the elastic region. Unlike steel, the modulus of elasticity of concrete varies with compressive strength. Since the slope of the stress-strain curve varies with the applied stress, there are several ways of calculating the modulus of elasticity. Figure 37.1 shows a typical stress-strain curve for concrete with the *initial modulus*, the *tangent modulus*, and the *secant modulus* indicated.

Figure 37.1 *Concrete Moduli of Elasticity*

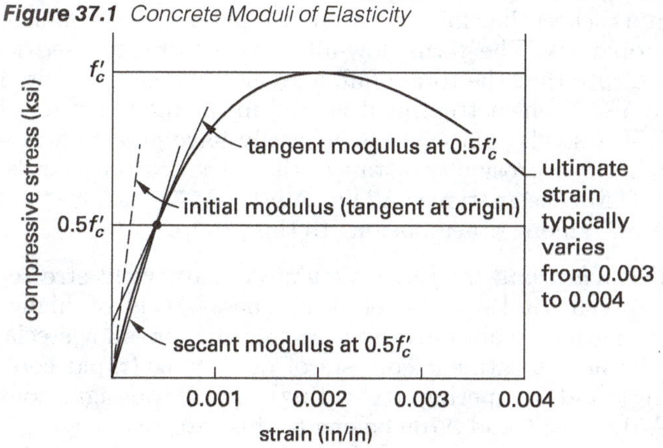

Equation 37.1: Secant Modulus of Elasticity

$$E_c = 33w_c^{1.5}\sqrt{f'_c} \quad \text{[U.S. only]} \tag{37.1}$$

Values

For normal weight unreinforced concrete, ACI 318 prescribes a standard specific weight, w, of approximately 145 lbf/ft^3 [ACI 318 Sec. 8.5.1]. The unit weight of reinforced normal weight concrete is usually assumed to be 150 lbf/ft^3.

Description

The *secant modulus of elasticity* is specified by ACI 318 for use with specific weights that are between 90 lbf/ft^3 and 160 lbf/ft^3. Equation 37.1 is used for both instantaneous and long-term deflection calculations. w_c is in lbf/ft^3, and E_c and f'_c are in lbf/in^2 [ACI 318 sec. 8.5.1]. Equation 37.1 is not dimensionally homogeneous.

Example

An unreinforced concrete beam of normal weight concrete and spanning 12 ft has a compressive strength of 3000 lbf/in^2 . What is most nearly the modulus of elasticity of the beam?

(A) 2600 kips/in^2

(B) 3200 kips/in^2

(C) 3300 kips/in^2

(D) 3600 kips/in^2

Solution

The modulus of elasticity is

$$E_c = 33w_c^{1.5}\sqrt{f'_c} = \frac{(33)\left(145\ \frac{\text{lbf}}{\text{ft}^3}\right)^{1.5}\sqrt{3000\ \frac{\text{lbf}}{\text{in}^2}}}{1000\ \frac{\text{lbf}}{\text{kip}}}$$

$$= 3156\ \text{kips/in}^2 \quad (3200\ \text{kips/in}^2)$$

The answer is (B).

10. REINFORCING BARS

Reinforcing steel for use in steel-reinforced concrete may be formed from billet steel, axle steel, or rail steel. Most modern reinforcing bars are made from new billet steel. (Special bars of titanium, stainless steel, corrosion-resistant alloys, and glass fiber composites may see extremely limited use in corrosion-sensitive applications.) The following ASTM designations are used for steel reinforcing bars. (See Table 37.4 for ASTM standards.)

ASTM A615: carbon steel, grades 40, 60, and 75 (symbol "S")

ASTM A996: rail steel, grades 50 and 60 (symbols "R" and "⊥"; only "R" is permitted to be used by ACI 318), and axle steel, grades 40 and 60 (symbol "A")

ASTM A706: low-alloy steel, grade 60 (symbol "W")

ASTM A955: stainless steel (mechanical property requirements are the same as carbon-steel bars under ASTM A615)

ASTM A1035: low-carbon, chromium, steel bars (permitted only as transverse or as spiral reinforcement)

Reinforcing steel used for concrete structures comes in the form of bars (known as *rebar*), welded wire reinforcement, and wires. Reinforcing bars can be plain or deformed; however, most bars are manufactured deformed to increase the bond between concrete and steel.

Table 37.4 ASTM Standards for Reinforcing Bars

customary U.S. bar no.	nominal diameter (in)	nominal area (in^2)	nominal weight (lbf/ft)
3	0.375	0.11	0.376
4	0.500	0.20	0.668
5	0.625	0.31	1.043
6	0.750	0.44	1.502
7	0.875	0.60	2.044
8	1.000	0.79	2.670
9	1.128	1.00	3.400
10	1.270	1.27	4.303
11	1.410	1.56	5.313
14	1.693	2.25	7.650
18	2.257	4.00	13.600

(Multiply in by 25.4 to obtain mm.)
(Multiply in^2 by 6.452 to obtain cm^2.)
(Multiply lbf/ft by 1.488 to obtain kg/m.)

11. STEEL NOMENCLATURE AND UNITS

It is traditional in steel design to use the uppercase letter F to indicate strength or allowable stress. Furthermore, such strengths or maximum stresses are specified in ksi (MPa in metricated countries). For example, $F_y =$ 36 ksi is a steel that has a yield stress of 36 ksi. Similarly, V_n is the nominal shear strength, and ϕM_n is the flexural design strength, both in ksi. Actual or computed stresses are given the symbol of lowercase f. Computed stresses are also specified in ksi. For example, f_t is a computed tensile stress in ksi.

In the United States, steel design is carried out exclusively in customary U.S. (inch-pound) units.

12. TYPES OF STRUCTURAL STEEL

The term *structural steel* refers to a number of steels that, because of their economy and desirable mechanical properties, are suitable for load-carrying members in structures. In the United States, the customary way to specify a structural steel is to use an ASTM (American Society for Testing and Materials) designation. For ferrous metals, the designation has the prefix letter "A" followed by two or three numerical digits (e.g., ASTM A36, ASTM A992). The general requirements for such steels are covered under ASTM A6 specifications. Basically, three groups of hot-rolled structural steels are available for use in buildings: carbon steels, high-strength low-alloy steels, and quenched and tempered alloy steels.

Carbon steels use carbon as the chief strengthening element. These are divided into four categories based on the percentages of carbon: *low-carbon* (less than 0.15%), *mild-carbon* (0.15–0.29%), *medium-carbon* (0.30–0.59%), and *high-carbon* (0.60–1.70%). ASTM A36 belongs to the mild-carbon category, and it has a maximum carbon content varying from 0.25–0.29%, depending on thickness. Carbon steels used in structures have minimum yield stresses ranging from 36 ksi to 55 ksi. An increase in carbon content raises the yield stress but reduces ductility, making welding more difficult. The maximum percentages of other elements of carbon steels are: 1.65% manganese, 0.60% silicon, and 0.60% copper.

High-strength low-alloy steels (HSLA) having yield stresses from 40 ksi to 70 ksi are available under several ASTM designations. In addition to carbon and manganese, these steels contain one or more alloying elements (e.g., columbium, vanadium, chromium, silicon, copper, and nickel) that improve strength and other mechanical properties. The term "low-alloy" is arbitrarily used to indicate that the total of all alloying elements is limited to 5%. No heat treatment is used in the manufacture of HSLA steels. These steels generally have greater atmospheric corrosion resistance than the carbon steels. ASTM designations A242, A441, A572, A588, and A992, among others, belong to this group.

Quenched and tempered alloy steels have yield stresses between 70 ksi and 100 ksi. These steels of higher strengths are obtained by heat-treating low-alloy steels. The heat treatment consists of quenching (rapid cooling) and tempering (reheating). ASTM designations A514, A852, and A709 belong to this category.

13. STEEL PROPERTIES

Each structural steel is produced to specified minimum mechanical properties as required by the specific ASTM designation by which it is identified. Some properties of steel (such as the modulus of elasticity and density) are essentially independent of the type of steel. Other properties (such as the tensile strength and the yield stress) depend not only on the type of steel, but also on the size or thickness of the piece. The mechanical properties of structural steel are generally determined from tension tests on small specimens in accordance with standard ASTM procedures. Table 37.5 gives typical properties of structural steels, and Fig. 37.2 shows typical diagrams of three different types of steels.

Table 37.5 *Typical Properties of Structural Steels*

	A992[a]/A572, grade 50	A36
modulus of elasticity, E	29,000 ksi[b]	29,000 ksi[b]
tensile yield strength, F_y	50 ksi	36 ksi (up to 8 in thickness)
tensile strength, F_u	65 ksi (min.)	58 ksi (min.)
endurance strength, S_e'	30 ksi (approx.)	30 ksi (approx.)
density, ρ	490 lbm/ft^3	490 lbm/ft^3
Poisson's ratio, ν	0.30 (ave.)	0.30 (ave.)
shear modulus, G	11,200 ksi[b]	11,200 ksi[b]
coefficient of thermal expansion, α	6.5×10^{-6} 1/°F (ave.)	6.5×10^{-6} 1/°F (ave.)
specific heat (32–212°F)	0.107 Btu/lbm-°F	0.107 Btu/lbm-°F

(Multiply ksi by 6.9 to obtain MPa.)
(Multiply in by 25.4 to obtain mm.)
(Multiply lbm/ft^3 by 16 to obtain kg/m^3.)
(Multiply °F^{-1} by 9/5 to obtain °C^{-1}.)

[a]A992 steel is the *de facto* material for rolled W-shapes, having replaced A36 and A572 in most designs for new structures.
[b]as designated by AISC

Figure 37.2 *Typical Stress-Strain Curves for Structural Steels*

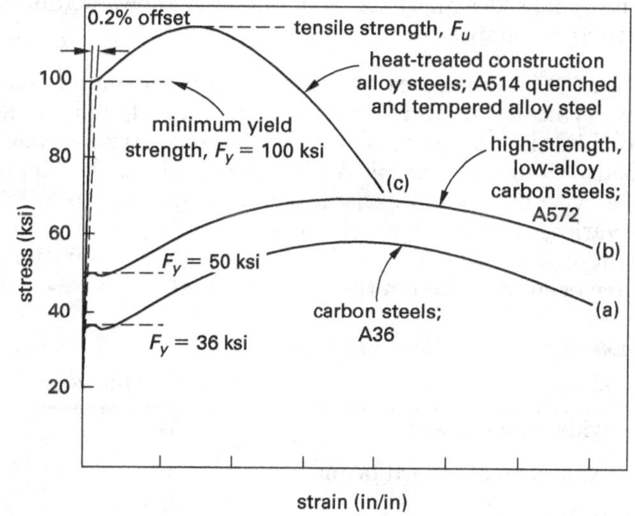

Yield stress, F_y, is that unit tensile stress at which the stress-strain curve exhibits a well-defined increase in strain without an increase in stress. The yield stress property is extremely important in structural design because it serves as a limiting value of a member's usefulness.

Ultimate tensile strength, or just *tensile strength*, F_u, is the largest unit stress that the material achieves in a tension test.

The *modulus of elasticity*, E, is the slope of the initial straight-line portion of the stress-strain diagram. For all structural steels, it is usually taken as 29,000 ksi for design calculations.

Ductility is the ability of the material to undergo large inelastic deformations without fracture. In a tension test, it is generally measured by percent elongation at fracture for a specified gage length (usually 2 in or 8 in).

Toughness is the ability of a specimen to absorb energy and is characterized by the area under a stress-strain curve.

Poisson's ratio is the ratio of transverse strain to longitudinal strain. Poisson's ratio is essentially the same for all structural steels and has a value of 0.3 in the elastic range.

Shear modulus is the ratio of shearing stress to shearing strain during the initial elastic behavior.

14. STRUCTURAL STEEL SHAPES

Many different structural shapes are available. The dimension and weight is added to the designation to uniquely identify the shape. For example, W30 × 132 refers to a W-shape with an overall depth of approximately 30 in that weighs 132 lbf/ft. The term *hollow structural sections* (HSS) is used to describe round and rectangular tubular members, which are often used as struts in trusses and space frames. Table 37.6 lists structural shape designations, and Fig. 37.3 shows common structural shapes.

Figure 37.4 illustrates two combinations of shapes that are typically used in construction. The double-angle combination is particularly useful for carrying axial loads. Combinations of W-shapes and channels, channels with channels, or channels with angles are used for a variety of special applications, including struts and light crane rails. Properties for certain combinations have been tabulated in the *AISC Manual.*

***Table 37.6** Structural Shape Designations*

shape	designation
wide flange beam	W
American standard beam	S
bearing piles	HP
miscellaneous (those that cannot be classified as W, S, or HP)	M
American standard channel	C
miscellaneous channel	MC
angle	L
structural tee (cut from W, S, or M)	WT or ST
structural tubing	TS
round and rectangular tubing	HSS
pipe	pipe or P
plate	PL
bar	bar

***Figure 37.3** Structural Shapes*

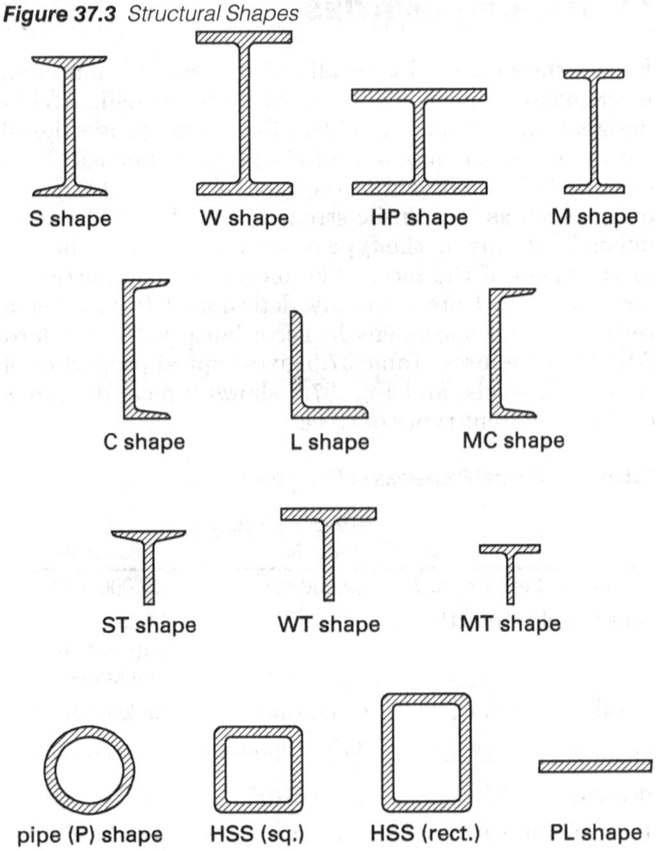

***Figure 37.4** Typical Combined Sections*

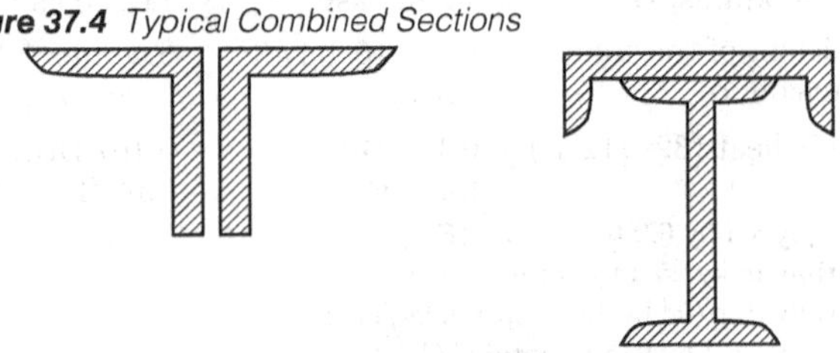

15. PHILOSOPHIES OF STEEL DESIGN

The ASD design philosophy is based on the premise that structural members remain elastic when subject to applied loads. According to this philosophy, a structural member is designed so that its computed strength under service or working loads does not exceed available strength. The available strengths are prescribed by the building codes or specifications to provide a factor of safety against attaining some limiting strength such as that defined by yielding or buckling.

The LRFD design philosophy, also referred to as *limit states design*, is the predominant design philosophy for concrete structures. The philosophy was first introduced for steel structures in 1986 with the publication of the first edition of AISC's *LRFD Manual of Steel Construction.* "Limit state" is a general term meaning a condition at which a structure or some part of it ceases to fulfill its intended function. There are two categories of limit states: strength and serviceability. The *strength limit states* that are of primary concern to designers include plastic strength, fracture, buckling, fatigue, and so on. These affect the safety or load-carrying capacity of a structure. The *serviceability limit states* refer to the performance under normal service loads and pertain to uses and/or occupancy of structures, including excessive deflection, drift, vibration, and cracking. This book uses the LRFD method in examples unless otherwise noted.

16. SERVICE LOADS, FACTORED LOADS, AND LOAD COMBINATIONS

The objective in all concrete and steel designs is to ensure that the safety margin against any possible collapse under the service loads is adequate. This margin is achieved by designing member strength to be equal to or greater than the effect of the service loads amplified by appropriate load factors.

The term *service loads* designates the loads (forces or moments) that are expected to be actually imposed on a structure during its service life, and for design purposes are taken from building codes. The term *factored loads* designates the service loads increased by various amplifying *load factors*. The load factors depend both on the uncertainty of the various loads as well as on the load combination being considered.

Required (ultimate) strength, U, must be at least equal to the factored loads in Eq. 37.2 through Eq. 37.9.

Equation 37.2 Through Eq. 37.9: Required Strength

$$U = 1.4D \qquad \text{37.2}$$

$$U = 1.2D + 1.6L \qquad \text{37.3}$$

$$U = 1.2D + 1.6L + 0.5(L_r/S/R) \qquad \text{37.4}$$

$$U = 1.2D + 1.6(L_r/S/R) + (L \text{ or } 0.8W) \qquad \text{37.5}$$

$$U = 1.2D + 1.6W + L + 0.5(L_r/S/R) \qquad \text{37.6}$$

$$U = 0.9D + 1.6W \qquad \text{37.7}$$

$$U = 1.2D + 1.0E + L + 0.2S \qquad \text{37.8}$$

$$U = 0.9D + 1.0E \qquad \text{37.9}$$

Description

Equation 37.2 and Eq. 37.3 give the required strength when only dead and live loads are considered. Both must be satisfied.

Equation 37.4 through Eq. 37.7 give the required strength when dead, live (floor and roof), wind, rain, and snow loads are considered. L_r represents the roof load, S represents the snow load, and R represents the rain load. The notation $L_r/S/R$ indicates that only the largest of the three load values should be used. Similarly, the notation "L or $0.8W$" indicates that the larger of L and $0.8W$ should be used.

When earthquake forces are considered, the earthquake loading has a factor of 1.0, as it is derived from factored loads in the seismic code. Members must also satisfy the requirements of Eq. 37.8 and Eq. 37.9.

Example 2

A steel W-section is uniformly loaded to produce bending about its strong axis. The beam supports a uniformly distributed service dead load of 2.5 kips/ft and a service live load of 1.8 kips/ft on a 36 ft simple span. What is most nearly the design load?

(A) 2.5 kips/ft

(B) 3.5 kips/ft

(C) 5.9 kips/ft

(D) 6.2 kips/ft

Solution

Use Eq. 37.2.

$$U = 1.4D = (1.4)\left(2.5\ \frac{\text{kips}}{\text{ft}}\right) = 3.5\ \text{kips/ft}$$

Use Eq. 37.3.

$$\begin{aligned} U &= 1.2D + 1.6L \\ &= (1.2)\left(2.5\ \frac{\text{kips}}{\text{ft}}\right) + (1.6)\left(1.8\ \frac{\text{kips}}{\text{ft}}\right) \\ &= 5.88\ \text{kips/ft} \quad (5.9\ \text{kips/ft}) \end{aligned}$$

The larger value, 5.9 kips/ft, controls.

The answer is (C).

Equation 37.10: Live Load Reduction

$$L_{\text{reduced}} = L_{\text{nominal}}\left(0.25 + \frac{15}{\sqrt{K_{LL}A_T}}\right) \geq 0.4L_{\text{nominal}} \qquad \text{37.10}$$

Description

Under certain conditions, the live load can be reduced. The load factor on L can be reduced to 0.5 for all structures except garages, areas occupied as places of public assembly, and all areas where the live load is greater than 100 lbf/ft^2. Equation 37.10 is a common reduction model prescribed by ASCE7 and many building codes. It reduces the live load based on the loaded floor area supported by the member. L_{nominal} is the nominal live load as required ASCE7 or a building code; A_T is the tributary floor area supported by the member; K_{LL} is the ratio of the area of influence to the tributary area, typically 2 for beams and girders, and 4 for columns. $K_{LL}A_T$ is the area of influence supported by the member. For a more detailed account of live load reduction, see ASCE7 Sec. 4.8.

Structural Design

38 Reinforced Concrete: Beams

Nomenclature

a	depth of rectangular stress block	in
A	area	in^2
A_s'	compression steel area	in^2
b	width	in
c	distance from neutral axis to extreme compression fiber	in
C	coefficient	–
C	compressive force	lbf
d	beam effective depth (to tension steel centroid)	in
d	diameter	in
E	modulus of elasticity	lbf/in^2
f_c'	compressive strength	lbf/in^2
f_s	stress in tension steel	lbf/in^2
f_y	tension steel yield strength	lbf/in^2
h	overall height (thickness)	in
L	live load	lbf
L	total span length	in
P	live load	lbf
M	moment	in-lbf
s	stirrup spacing	in
T	tensile force	lbf
V	shear strength	lbf
w	unit weight (specific weight)	lbf/ft^3

Symbols

β_1	ratio of depth of equivalent stress block to depth of actual neutral axis	–
ε	strain	–
θ	angle	deg
λ	distance from centroid of compressed area to extreme compression fiber	in
λ	lightweight aggregate factor	–
λ	long-term amplification deflection factor	–
ρ	reinforcement (steel) ratio	–
ϕ	strength reduction factor	–

Subscripts

b	balanced condition or bar
c	concrete, cover, or critical
D	dead
e	effective
f	flange
g	gross
L	live
max	maximum
min	minimum
n	nominal (strength)
r	rectangle
s	reinforcement or steel
t	tensile
u	ultimate (factored)
v	shear
y	yield
w	web

1. STEEL REINFORCING

Figure 38.1 illustrates typical reinforcing used in a simply supported reinforced concrete beam. The straight longitudinal bars resist the tension induced by bending, and the *stirrups* resist the diagonal tension resulting from shear stresses. As shown in Fig. 38.1(b), it is also possible to bend up the longitudinal bars to resist diagonal tension near the supports. This alternative, however, is rarely used because the saving in stirrups tends to be offset by the added cost associated with bending the longitudinal bars. In either case, the stirrups are usually passed underneath the bottom steel for anchoring. Prior to pouring the concrete, the horizontal steel is supported on *bolsters* (*chairs*), of which there are a variety of designs.

Figure 38.1 *Typical Reinforcement in Beams*

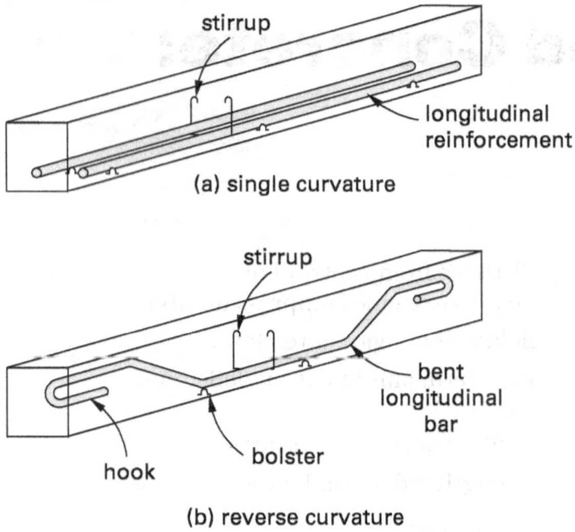

2. TYPES OF BEAM CROSS SECTIONS

The design of reinforced concrete beams is based on the assumption that concrete does not resist any tensile stress. A consequence of this assumption is that the effective shape of a cross section is determined by the part of the cross section that is in compression. Consider the design of monolithic slab-and-girder systems. As Fig. 38.2 shows, moments are negative (i.e., there is tension on the top fibers) near the columns. The effective section in the region of negative moments is rectangular, as in Fig. 38.2(b). Elsewhere, moments are positive, and the effective section may be either rectangular or T-shaped, depending on the depth of the compressed region, as in Fig. 38.2(c).

Reinforced concrete sections can be singly or doubly reinforced. Doubly reinforced sections are used when concrete in the compression zone cannot develop the required moment strength. Doubly reinforced sections are also used to reduce long-term deformations resulting from creep and shrinkage.

Depending on the ratio of beam clear span to the depth of the cross section, beams may be considered "regular" or deep.

3. MOMENTS IN CONTINUOUS BEAMS

With certain restrictions, ACI 318 permits continuous beams and slabs constructed of reinforced concrete to be designed using tabulated moment coefficients. This section deals specifically with beams and one-way slabs (slabs that support moments in one direction only) [ACI 318 Sec. 8.3]. Approximate design of two-way slabs, known as the *direct design method*, follows different rules and is not covered in this section [ACI 318 Sec. 13.6].

Figure 38.2 *Slab-Beam Floor System*

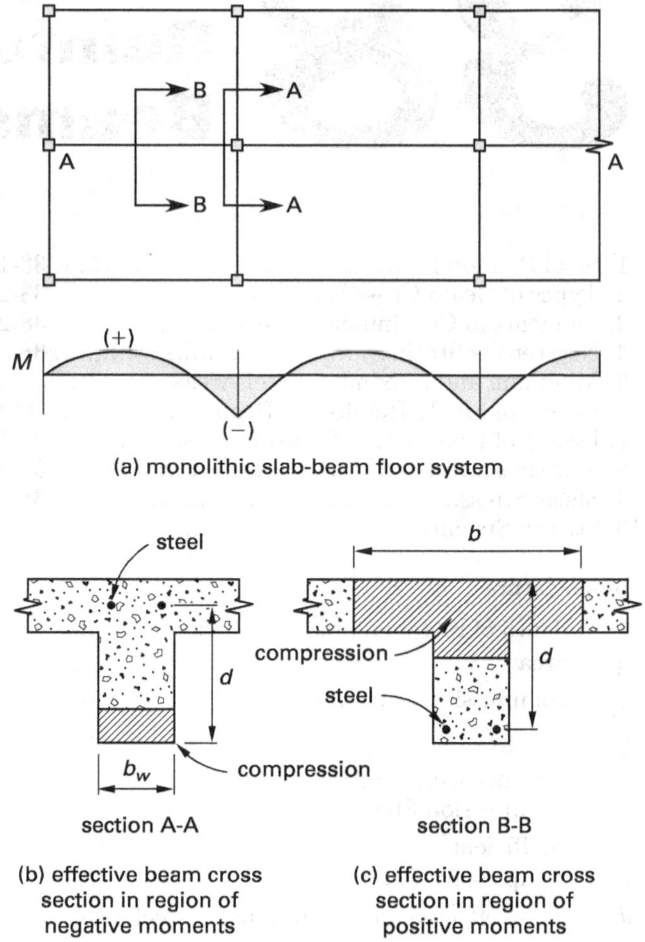

Moment Coefficient

The maximum moments at the ends and midpoints of continuously loaded spans are taken as fractions of the distributed load function, $w_uL_n^2$, where w_u is the factored load per unit beam length. L_n is the nominal beam length, which is taken as the clear span (between column faces) when the beam is acted upon by positive moments (which make the beam deflect downward). L_n is taken as the average adjacent clear span length for negative moments.

Although exact methods can always be used to determine the moments on a beam, an approximate method can be used provided that certain conditions are met. The simplified method can be used for uniform loading over three or more spans, when the beam sections have the same area, moment of inertia, and modulus of elasticity, and when the span lengths between points of support are approximately the same (specifically, when the span lengths of adjacent spans do not differ by more than 20% of the shorter), and the live load is or more than three times the dead load. The moment at some point along the beam is obtained from the *moment coefficient*, *C*. (See Fig. 38.3.)

$$M_u = Cw_uL_n^2$$

***Figure 38.3** Moment Coefficients*

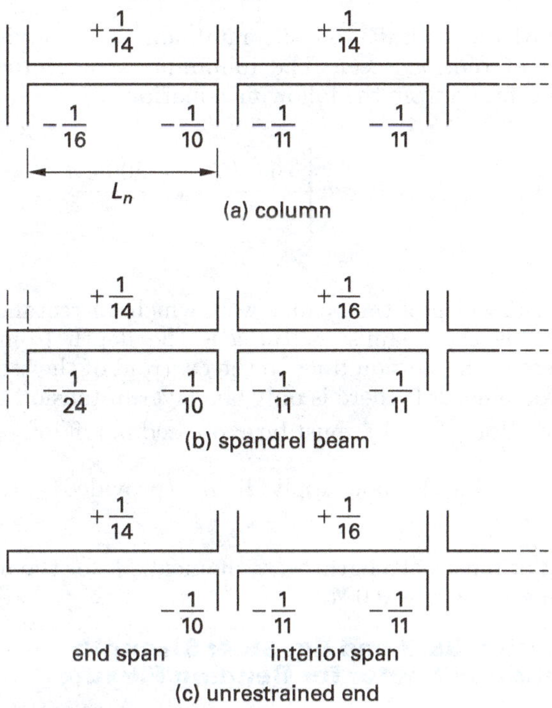

4. NET TENSILE STRAIN

Figure 38.4 depicts the location of the *extreme tension steel* for two sections with different reinforcement arrangements, both of which assume the top fiber of the section is the extreme compression fiber. The distance from the extreme compression fiber to the centroid of the extreme tension steel is denoted in the illustration as d_t. The *net tensile strain* in the extreme tension steel due to the external loads can be determined from a strain compatibility analysis for sections with multiple layers of reinforcement. For sections with one layer of reinforcement, the net tensile strain can be determined from the strain diagram by using similar triangles.

***Figure 38.4** Location of Extreme Tension Steel and Net Tensile Strain at Nominal Strength*

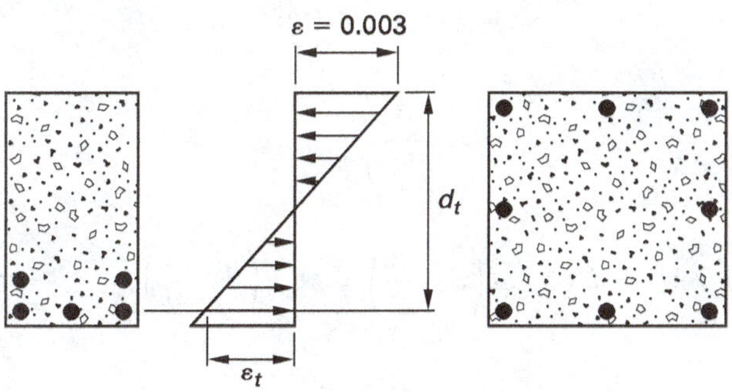

Equation 38.1 and Eq. 38.2: Net Tensile Strain

$$a = \beta_1 c \qquad 38.1$$

$$\varepsilon_t = \frac{0.003(d_t - c)}{c} = \frac{0.003(\beta_1 d_t - a)}{a} \qquad 38.2$$

Description

Net tensile strain, ε_t, is the tensile strain in the extreme tension steel at nominal strength, exclusive of strains due to effective prestress, creep, shrinkage, and temperature. The net tensile strain is caused by external axial loads and/or bending moments at a section when the concrete strain at the extreme compression fiber reaches its assumed limit of 0.003. Generally speaking, the net tensile strain can be used as a measure of cracking or deflection. (See Fig. 38.5.)

Extreme tension steel is the reinforcement (prestressed or nonprestressed) that is the farthest from the extreme compression fiber. In Eq. 38.2, variable a is the depth of the equivalent rectangular stress block, as shown in Fig. 38.5. It can be calculated from the distance from the neutral axis to the extreme compression fiber, c, if β_1 is known. (See Eq. 38.5.) The net tensile strain, ε_t, in the extreme tension steel at nominal strength can be calculated from the distance from the extreme compression fiber to the centroid of the highest stress tension steel layer, d_t, and c.[1]

***Figure 38.5** Internal Forces and Strain*

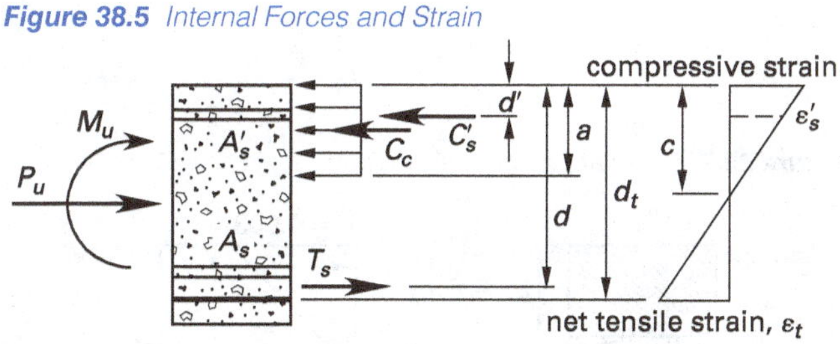

The unified design provisions limit the maximum reinforcement in a flexural member (with factored axial load less than $0.1f'_cA_g$) to that which would result in a net tensile strain, ε_t, not less than 0.004 [ACI 318 Sec. 10.3.5]. For a singly reinforced section with grade 60 reinforcement, this is equivalent to a maximum reinforcement ratio of $0.714\rho_b$. The ACI App. B limit of $\rho_{max} = 0.75\rho_b$ results in a net tensile strain of 0.00376. At the net tensile strain limit of 0.004, the strength reduction factor of ACI Sec. 9.3.2, ϕ, is reduced to 0.817 (with $\varepsilon_t = 0.004$).

It is almost always advantageous to limit the net tensile strain in flexural members to a minimum of 0.005, as shown in Fig. 38.6, which is equivalent to a maximum

[1]The NCEES *FE Reference Handbook* (*NCEES Handbook*) refers to d_t as the distance to the extreme tension steel. However, this distance is to the centroid of the layer, not to the lowest extent of the layer.

reinforcement ratio of $0.63\rho_b$ [ACI 318 Sec. R.9.3.2.2], even though the code permits higher amounts of reinforcement that produce lower net tensile strains. Where member size is limited and extra strength is needed, it is best to use compression reinforcement to limit the net tensile strain so that the section is tension controlled.

Figure 38.6 Strain Conditions

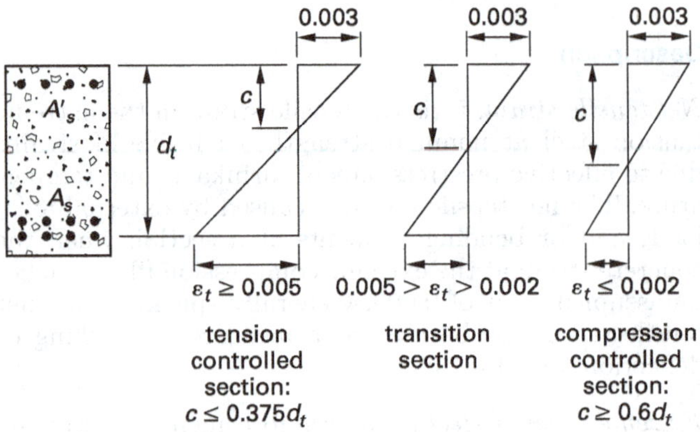

The *compression-controlled strain limit* is the net tensile strain corresponding to a balanced condition. The definition of a balanced strain condition, given in ACI 318 Sec. 10.3.2, is unchanged from previous editions of the code. A balanced strain condition exists in a cross section when tension reinforcement reaches the strain corresponding to its specified yield strength just as the concrete strain in the extreme compression fiber reaches its assumed limit of 0.003. (See Fig. 38.7.)

$$\varepsilon_t = \varepsilon_y = \frac{f_y}{E_s}$$

Figure 38.7 Beam at Balanced Condition

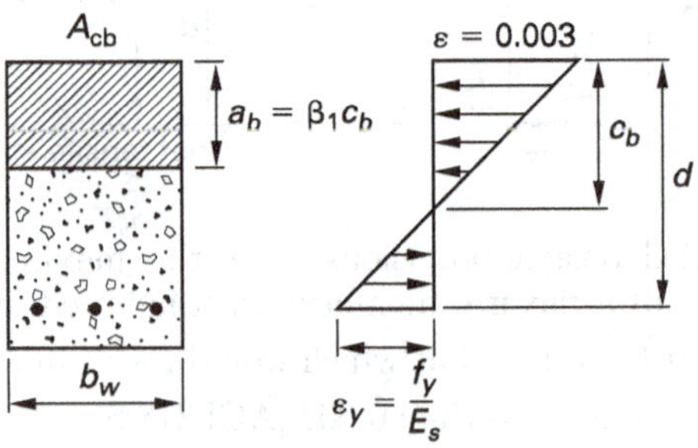

5. MINIMUM AND MAXIMUM STEEL AREAS

ACI 318 sets limits on the minimum and maximum areas of tension steel. The minimum area of tension steel is found from the following equation.

$$A_{s,\min} = \text{larger}\begin{cases} \dfrac{3\sqrt{f'_c}\,b_w d}{f_y} \quad \text{or} \quad \dfrac{200 b_w d}{f_y} \end{cases}$$

b_w is the width of the beam's web, which for rectangular beams is the beam's width. d is the depth from the extreme compression fiber to the centroid of the tension reinforcement. If there is only one layer of tension steel, $d = d_t$. Both f'_c and f_y must be expressed in lbf/in^2.

$A_{s,\min}$ limits do not apply if A_s (provided) $\geq 1.33A_s$ (required).

The net tensile strength, ε_t, at moment M_n for the maximum area of A_s is 0.004.

Equation 38.3 and Eq. 38.4: Strength Reduction Factor for Bending/Flexure

$$\phi = 0.9 \quad [\varepsilon_t \geq 0.005] \qquad 38.3$$

$$\phi = 0.48 + 83\varepsilon_t \quad [0.004 \leq \varepsilon_t < 0.005] \qquad 38.4$$

Description

The *strength reduction factor*, ϕ, for design moment strength, ϕM_n, is dependent on the tensile strain.

6. DESIGN OF SINGLY REINFORCED BEAMS

Equation 38.5 Through Eq. 38.7: Singly Reinforced Beams

$$\beta_1 = 0.85 \geq 0.85 - 0.05\left(\frac{f'_c - 4000}{1000}\right) \geq 0.65 \qquad 38.5$$

$$a = \frac{A_s f_y}{0.85 f'_c b} \qquad 38.6$$

$$M_n = 0.85 f'_c ab\left(d - \frac{a}{2}\right) = A_s f_y\left(d - \frac{a}{2}\right) \qquad 38.7$$

Description

In singly reinforced beams, steel is used to withstand only tensile forces. The nominal moment capacity, M_n, can be computed for any singly reinforced section with the following procedure.

step 1: Compute the tension force.

$$T = f_y A_s$$

step 2: Calculate the area of concrete, $A_c = ab$, that, at a stress of $0.85f_c'$, gives a force equal to T. Use Eq. 38.6.

step 3: Locate the centroid of the area A_c. Designating the distance from the centroid of the compression region to the most compressed fiber as $\lambda = d - \frac{1}{2}a$, calculate the nominal moment capacity. Use Eq. 38.7.

The conditions and maximum moment are shown in Fig. 38.8.

Figure 38.8 *Conditions at Maximum Moment*

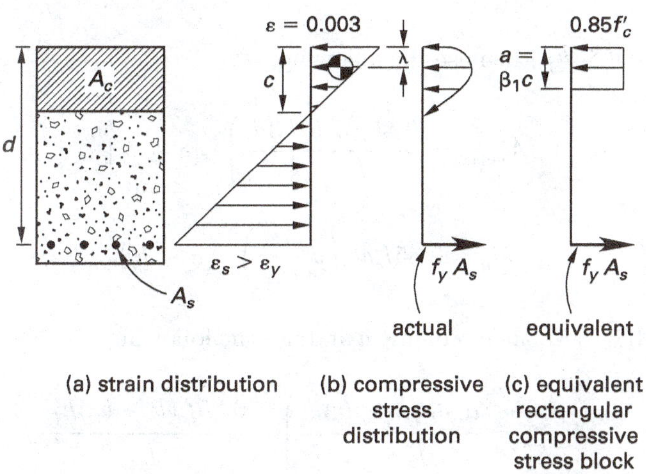

Example

The cross section of a beam is reinforced with 3 in^2 of steel. $f_c' = 4000$ lbf/in^2, and $f_y = 60{,}000$ lbf/in^2. Assume that the tension steel yields at maximum moment. Most nearly, what is the area of concrete required to balance the steel force when the steel yields?

(A) 15 in^2

(B) 18 in^2

(C) 45 in^2

(D) 53 in^2

Solution

Use Eq. 38.6. The area of concrete required to balance the steel force at yield is $A_c = ab$.

$$a = \frac{A_s f_y}{0.85 f_c' b}$$

$$A_c = ab = \frac{A_s f_y}{0.85 f_c'} = \frac{(3\ \text{in}^2)\left(60{,}000\ \frac{\text{lbf}}{\text{in}^2}\right)}{(0.85)\left(4000\ \frac{\text{lbf}}{\text{in}^2}\right)} = 52.94\ \text{in}^2 \quad (53\ \text{in}^2)$$

The answer is (D).

7. DESIGN OF DOUBLY REINFORCED BEAMS

Sections that have steel in the tension and the compression regions are referred to as *doubly reinforced beams*. (See Fig. 38.9.) The most common situation where a doubly reinforced beam is needed is found when the moment capacity of the singly reinforced section, with the maximum area of steel permitted, is not sufficient to carry the applied moment. In such cases, steel can be provided in excess of the maximum allowed for a singly reinforced section if the added tensile force is balanced by steel in compression.

Figure 38.9 *Parameters for Doubly Reinforced Beam*

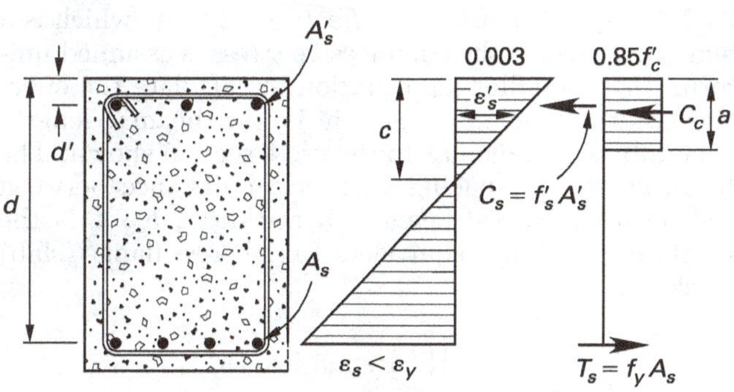

$$A_s - A_s' \geq \frac{0.85\beta_1 f_c' d' b}{f_y}\left(\frac{87{,}000}{87{,}000 - f_y}\right) \quad \left[\begin{array}{c}\text{if compression}\\ \text{steel yields}\end{array}\right]$$

$$A_{s,\text{max}} = \frac{0.85 f_c' \beta_1 b}{f_y}\left(\frac{3d_t}{7}\right) + A_s' \quad \left[\begin{array}{c}\text{area of tension}\\ \text{reinforcement if}\\ \text{compression yields}\end{array}\right]$$

$$a = \frac{(A_s - A_s')f_y}{0.85 f_c' b} \quad \left[\begin{array}{c}\text{depth of equivalent rectangular}\\ \text{stress block if compression yields}\end{array}\right]$$

$$M_n = f_y\left((A_s - A_s')\left(d - \frac{a}{2}\right) + A_s'(d - d')\right)$$

$$\left[\begin{array}{c}\text{nominal moment strength}\\ \text{if compression yields}\end{array}\right]$$

$$c^2 + \left(\frac{(87{,}000 - 0.85f_c')A_s' - A_sf_y}{0.85f_c'\beta_1 b}\right)c - \frac{87{,}000A_s'd'}{0.85f_c'\beta_1 b} = 0$$

$$\left[\begin{array}{c}\text{If compression steel does}\\ \text{not yield, solve for } c\end{array}\right]$$

$$M_n = 0.85bc\beta_1 f_c'\left(d - \frac{\beta_1 c}{2}\right) + A_s'\left(\frac{c - d'}{c}\right)(d - d')87{,}000$$

$$\left[\begin{array}{c}\text{If compression steel does not yield, use } c\\ \text{to find nominal moment strength}\end{array}\right]$$

8. T-BEAMS

Although a beam can be specifically cast with a flange, that is rarely done. Flanged-beam behavior usually occurs in monolithic beam-slab (one-way) systems. Determining the width of the flange (i.e., the slab) that is effective in resisting compressive loads is code-sensitive. ACI 318 also specifies requirements pertaining to the distribution of the reinforcement.

The compressive stresses in the flange of a T-beam decrease with distance from the centerline of the web. ACI 318 specifies using an *effective width*, b_e, which is a width over which the compressive stress is assumed uniform. Use the following equation to calculate the effective width. The span length is the beam's length, generally perpendicular to the cross section shown. The beam centerline spacing is the clear distance between the beams plus one beam web thickness, b_w. h_f is the height of the flange, equivalent to the gross flange (slab) thickness.

$$b_e = \text{smallest}\begin{cases}\left(\frac{1}{4}\right)(\text{span length})\\ b_w + 16h_f\\ \text{beam centerline spacing}\end{cases}$$

It is assumed that a stress of $0.85f_c'$ acting over the effective width provides approximately the same compressive force as that realized from the actual variable stress distribution over the total flange width. The effective width of the flange for T-shaped beams is illustrated in Fig. 38.10. The effective width is the smallest value from the options listed [ACI 318 Sec. 8.12]. Effective width (including the compression area of the stem) for a T-beam is the minimum of one-fourth of the beam's span length, the stem width plus 16 times the thickness of the slab, or the beam spacing.

Figure 38.10 *Effective Flange Width (one-way reinforced slab systems) (ACI 318 Sec. 8.12)*

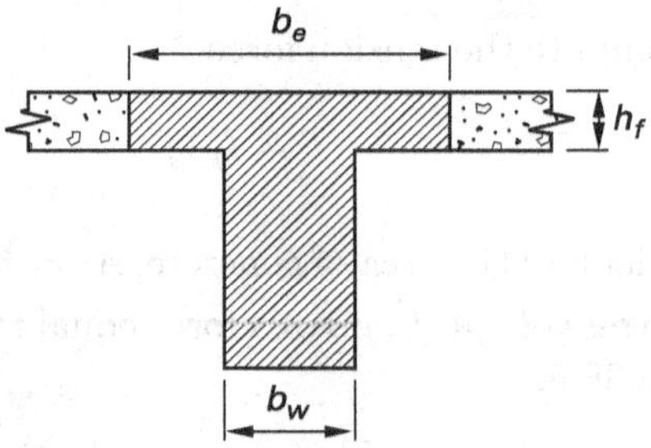

The design moment strength of a T-beam is contingent on the value of a, as shown in the following equations.

$$a = \frac{A_sf_y}{0.85f_c'b_e}$$

$$a = \frac{A_sf_y}{0.85f_c'b_w} - \frac{h_f(b_e - b_w)}{b_w} \quad [\text{redefine } a]$$

If $a \le h_f$, then use the following.

$$A_{s,\max} = \frac{0.85f_c'\beta_1 b_e}{f_y}\left(\frac{3d_t}{7}\right) \quad [a \le h_f]$$

$$M_n = 0.85f_c'ab_e\left(d - \frac{a}{2}\right) \quad [a \le h_f]$$

If $a > h_f$, then redefine a and use the following.

$$A_{s,\max} = \frac{0.85f_c'\beta_1 b_w}{f_y}\left(\frac{3d_t}{7}\right) + \frac{0.85f_c'(b_e - b_w)h_f}{f_y} \quad [a > h_f]$$

$$M_n = 0.85f_c'\left(h_f(b_e - b_w)\left(d - \frac{h_f}{2}\right) + ab_w\left(d - \frac{a}{2}\right)\right)$$

9. SHEAR STRESS

In addition to producing bending moments, beam loads also produce shear forces. Shear forces induce diagonal tension stresses that lead to diagonal cracking. The typical pattern of cracks induced by shear forces is depicted in Fig. 38.11. If a diagonal crack forms and the beam does not contain shear reinforcement, failure will occur abruptly. In order to have ductile beams (where the controlling mechanism is the ductile yielding of the tension steel), shear reinforcement is designed conservatively.

Figure 38.11 *Typical Pattern of Shear Cracks*

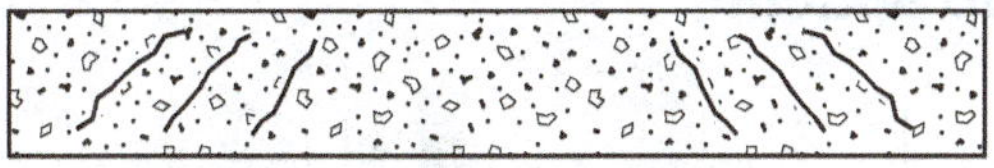

The maximum shear at a given location does not necessarily result from placing the live load over the complete beam.

In the design of beams, it is customary to call out the span length as the distance from the centerline of supports. Although the shear computed is largest immediately adjacent to a reaction, shears corresponding to locations that are within the support width are not physically meaningful. Furthermore, for the typical case where the beam is supported from underneath, the compression induced by the reaction increases the shear strength in the vicinity of the support.

ACI 318 Sec. 11.1.3 takes this enhanced strength into account by specifying that the region of the beam from a distance d to the face of the support can be designed for the same shear force that exists at a distance d from the support face.

The shear capacity provided by the concrete, V_c, results from the strength in shear of the uncracked compression zone, a contribution from aggregate interlock across the cracks, and dowel action from the tension steel that crosses the diagonal crack.

Equation 38.8 Through Eq. 38.11: Nominal Shear Strength

$$V_n = V_c + V_s \tag{38.8}$$

$$V_c = 2b_w d\sqrt{f'_c} \tag{38.9}$$

$$V_u \le \frac{\phi V_c}{2} \quad \text{[no stirrups required]} \tag{38.10}$$

$$V_u > \frac{\phi V_c}{2} \quad \text{[stirrups required]} \tag{38.11}$$

Description

The beam width used in shear equations is specified by

$$b_w = \begin{cases} b & \text{[rectangular beams]} \\ b_w & \text{[T-beams]} \end{cases}$$

Shear is carried only by a vertical strip with width equal to the width of the web of a T-beam or flanged beam. The strip extends upward through the flange. The entire effective width of the flange, b_e, is considered to be effective in resisting moment, but not in resisting shear.

The beam's shear capacity, V_n, is determined semi-empirically as the sum of the concrete shear capacity, V_c, and the shear capacity that derives from the presence of shear reinforcement, V_s.

The term *nominal shear strength* is ambiguous in concrete design. The concrete shear strength, V_c, is often referred to as the *nominal concrete shear strength.* The sum of $V_c + V_s$ is referred to as the *nominal beam shear strength.* Common usage attributes the term "nominal" to both.

Equation 38.9 is specifically for normal weight concrete, not for lightweight concrete.

If the factored shear loading, V_u, is low enough, the concrete will have sufficient shear capacity, V_c, by itself, and steel in the form of stirrups will not be required. As Eq. 38.10 indicates, less than half of the concrete's shear capacity is considered reliable. Stirrups must be provided to increase the shear capacity when Eq. 38.11 is valid.

Example

A reinforced concrete beam of normal weight concrete with an effective depth of 16 in and a width of 12 in is reinforced with grade 60 bars and has a concrete compressive strength of 3000 lbf/in^2. The factored shear force is $V_u = 7$ kips, the factored moment is $M_u = 20$ ft-kips, and the reinforcement ratio is $\rho_w = 0.015$. What is most nearly the nominal shear reinforcement capacity required?

(A) 0 kips

(B) 0.9 kips

(C) 1.1 kips

(D) 1.6 kips

Solution

The nominal shear strength provided by the concrete is given by Eq. 38.9 as

$$V_c = 2b_w d\sqrt{f'_c} = \frac{(2)(12\text{ in})(16\text{ in})\sqrt{3000\ \frac{\text{lbf}}{\text{in}^2}}}{1000\ \frac{\text{lbf}}{\text{kip}}} = 21\text{ kips}$$

Use Eq. 38.10 to calculate the shear strength required. The capacity reduction factor, ϕ, is 0.75 for shear.

$$\frac{\phi V_c}{2} = \frac{(0.75)(21\text{ kips})}{2} = 7.9\text{ kips}$$

Since 7 kips < 7.9 kips, no shear reinforcement is required.

The answer is (A).

Structural Design

Equation 38.12: Shear Strength Provided by Shear Reinforcement

$$V_s = \frac{A_v f_y d}{s} \quad \left[\text{may not exceed } 8 b_w d\sqrt{f'_c}\right] \qquad 38.12$$

Description

The shear capacity provided by the steel reinforcement, V_s, is derived by considering a free-body diagram of the beam with an idealized diagonal crack at 45° and computing the vertical component of the force developed by the reinforcement intersecting the crack. It is assumed that all the steel is yielding when the strength is attained. For the typical case of vertical stirrups, as shown in Fig. 38.12, the shear capacity is found from Eq. 38.12.

Figure 38.12 Contribution of Vertical Stirrups to Shear Capacity

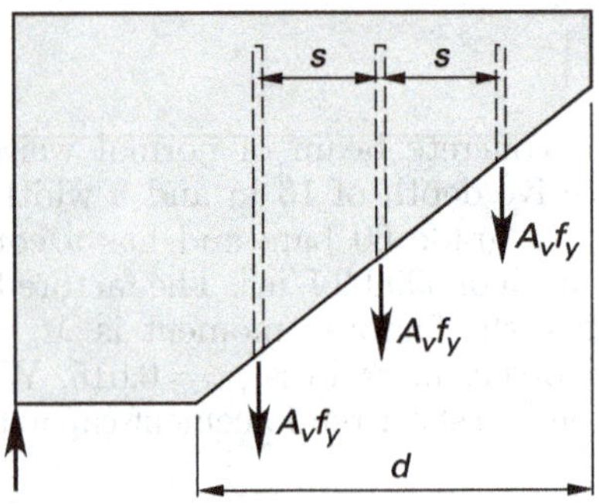

10. STIRRUP SPACING

Whenever stirrups are required, ACI 318 Sec. 11.4.4 and Sec. 11.4.5 specify that the spacing, s, shall not be larger than calculated in Eq. 38.19 through Eq. 38.22. This maximum spacing ensures that at least one stirrup crosses each potential diagonal crack. The limit in Eq. 38.21, which applies when the shear is high, ensures that at least two stirrups cross each potential shear crack.

To ensure that the stirrups have sufficient strength (i.e., a minimum strength of at least $0.75\sqrt{f'_c}$, but no less than 50 lbf/in^2), ACI 318 Sec. 11.4.6.3 requires all stirrups to have a minimum area of

$$[\text{U.S. only}]\; A_{v,\min} = 0.75\sqrt{f'_c}\left(\frac{b_w s}{f_{yt}}\right) \ge \frac{50 b_w s}{f_y}$$

Equation 38.13 Through Eq. 38.22: Required Stirrup Spacing

	$\frac{\phi V_c}{2} < V_u \le \phi V_c$	$V_u > \phi V_c$
required spacing	smaller of: $s = \frac{A_v f_y}{50 b_w}$ 38.13 $s = \frac{A_v f_y}{0.75 b_w \sqrt{f'_c}}$ 38.14	$V_s \le 4 b_w d\sqrt{f'_c}$ $V_s = \frac{V_u}{\phi} - V_c$ 38.15 $s = \frac{A_v f_y d}{V_s}$ 38.16
maximum permitted spacing	smaller of: $s = \frac{d}{2}$ 38.17 $s = 24''$ 38.18	$V_s \le 4 b_w d\sqrt{f'_c}$ smaller of: $s = \frac{d}{2}$ 38.19 $s = 24''$ 38.20 $V_s > 4 b_w d\sqrt{f'_c}$ smaller of: $s = \frac{d}{4}$ 38.21 $s = 12''$ 38.22

Description

To determine the stirrup spacing, s, use Eq. 38.13 through Eq. 38.22.

Example

The simply supported beam shown spans 20 ft and, in addition to its own weight, carries a uniformly distributed service dead load of 1.75 kips/ft and a uniformly distributed live load of 3.0 kips/ft. Five no. 9 bars running the full length of the beam are used as flexural reinforcement. $f_c' = 4000$ lbf/in^2, and $f_y = 60{,}000$ lbf/in^2. Number 3 stirrups are used. The concrete cover on longitudinal steel is 2.5 in.

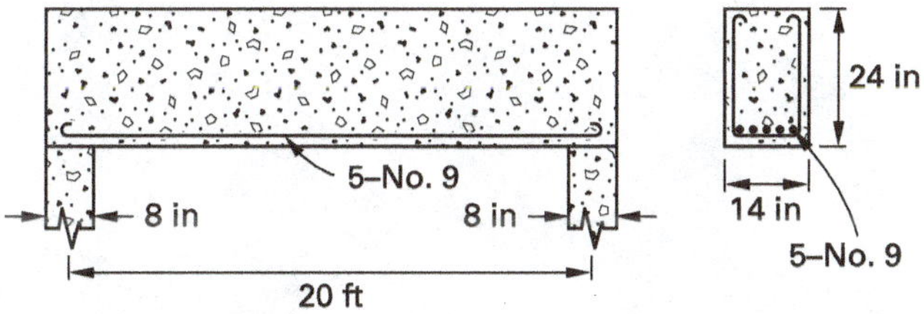

What is most nearly the theoretical spacing of the stirrups at a point where the required shear strength is 60.20 kips?

(A) 2.4 in

(B) 2.9 in

(C) 6.7 in

(D) 13 in

Solution

Calculate the effective reinforcement depth.

$$d = 24 \text{ in} - 2.5 \text{ in} = 21.5 \text{ in}$$

From Eq. 38.9,

$$\begin{aligned} V_c = 2b_w d\sqrt{f_c'} &= \frac{(2)(14 \text{ in})(21.5 \text{ in})\sqrt{4000 \dfrac{\text{lbf}}{\text{in}^2}}}{1000 \dfrac{\text{lbf}}{\text{kip}}} \\ &= 38.07 \text{ kips} \end{aligned}$$

From Eq. 38.15,

$$V_s = \frac{V_u}{\phi} - V_c = \frac{60.20 \text{ kips}}{0.75} - 38.07 \text{ kips} = 42.20 \text{ kips}$$

The stirrups are no. 3 bars which have a cross-sectional area of 0.11 in^2. Each U-shaped stirrup consists of two vertical legs. The area of reinforcement per stirrup is

$$A_v = 2A_{\text{bar}} = (2)(0.11 \text{ in}^2) = 0.22 \text{ in}^2$$

From Eq. 38.16, the stirrup spacing is

$$\begin{aligned} s = \frac{A_v f_y d}{V_s} &= \frac{(0.22 \text{ in}^2)\left(60 \dfrac{\text{kips}}{\text{in}^2}\right)(21.5 \text{ in})}{42.20 \text{ kips}} \\ &= 6.73 \text{ in} \quad (6.7 \text{ in}) \end{aligned}$$

The answer is (C).

Structural Design

39 Reinforced Concrete: Columns

Nomenclature

a	depth of rectangular stress block	in
A	area	in^2
c	neutral axis depth	in
C	compressive force	kips
C_m	moment correction parameter	–
d	diameter	in
d'	distance from extreme compression fiber to the centroid of the compression reinforcement	in
d_s	distance from extreme tension fiber to the centroid of the tension reinforcement	in
D	circular column diameter	in
D	dead load	kips
e	eccentricity	in
E	modulus of elasticity	lbf/in^2
f'_c	compressive strength of concrete	lbf/in^2
f_y	yield strength	lbf/in^2
F	lateral load	kips
h	dimension perpendicular to the axis of bending	in
I	moment of inertia	in^4
K	effective length factor	–
K_n	interaction diagram parameter	–
L	clear height of column	in
M	end moment	in-kips
M_b	moment that together with P_b leads to a balanced strain diagram	in-kips
M_c	amplified moment for bracedconditions	in-kips
M_1	smaller (in absolute value) of the two end moments acting on a column	in-kips
M_2	larger (in absolute value) of the two end moments acting on a column	in-kips
N	number	–
P	axial load	kips
P_b	axial force that together with M_b leads to a balanced strain diagram	kips
P_o	maximum nominal axial load capacity	kips
P_n	factored axial load	kips
r	radius of gyration	in
R_n	interaction diagram parameter	–
T	tensile force	kips

Symbols

β	column strength factor	–
γ	ratio of confined to gross lateral dimensions	–
Δ	deflection	in
δ	amplification factor	–
ε	strain	in/in
ρ_g	longitudinal reinforcement ratio	–
ϕ	strength reduction factor	–

Subscripts

b	balanced or braced
c	compressive, core, or critical
col	column
D	dead
g	gross
L	live
n	nominal
s	spiral or steel
st	longitudinal steel
t	tensile or tension
tc	tension-controlled
u	ultimate or unbraced
y	yield

1. COLUMNS

Columns are vertical members whose primary purpose is to transfer axial compression to lower members. In many practical situations, columns are subjected not only to axial compression but also to significant bending moments. Another factor sometimes increasing column loads is the *P-Δ effect.* The *P*-Δ effect is the increase in moment that results when the column sways, adding eccentricity. Columns where the P-Δ effect is significant are known as *long columns*. Columns where the P-Δ effect can be neglected are known as *short columns*.

2. SHORT COLUMNS

Equation 39.1: Column Longitudinal Reinforcement Limits

$$\rho_g = \frac{A_{st}}{A_g} \quad [0.01 \le \rho_g \le 0.08] \qquad 39.1$$

Description

Equation 39.1 gives the limits of the reinforcement ratio for short columns.

Example

A short, concentrically loaded, reinforced concrete column has an 18 in diameter and an area of reinforcing steel of 2.54 in^2. What is most nearly the reinforcement ratio?

(A) 0.005

(B) 0.01

(C) 0.02

(D) 0.04

Solution

The reinforcement ratio is

$$\rho_g = \frac{A_{st}}{A_g} = \frac{2.54\ \text{in}^2}{\frac{\pi(18\ \text{in})^2}{4}} = 0.00998 \quad (0.01)$$

The answer is (B).

Slenderness Ratio

Columns that are part of braced structures are considered to be short columns (also called *piers* or *pedestals*) if the following equation holds.

$$\frac{KL}{r} \le 34 - \frac{12M_1}{M_2}$$

$$KL = L_{col} \quad [\text{clear height of column, assume } K = 1.0]$$

The quantity KL/r is the column's *slenderness ratio*. For rectangular columns, use $r = 0.288h$, where h is the side length in the plane of buckling. For circular columns, use $r = 0.25h$, where h is the diameter.

M_1 is the smaller (in absolute value) of the two end moments acting on a column. M_2 is the larger (in absolute value) of the two end moments acting on a column. The ratio M_1/M_2 is positive if the member is bent in single curvature, and it is negative if the member is bent in reverse curvature. The ratio M_1/M_2 cannot be less than -0.5. It is always conservative to take $K=1$ [ACI 318 Sec. 10.10.1]. For short columns that are part of an unbraced structure, $M_1 = M_2 = 0$ and $KL/r \le 22$.

Equation 39.2: Tied Columns

$$\phi P_n = 0.80\phi\left[0.85f_c'(A_g - A_{st}) + A_{st}f_y\right] \qquad 39.2$$

Values

$\phi = 0.65$ for tied columns.

Description

Reinforced concrete columns with transverse reinforcement in the form of closed ties or hoops are known as *tied columns*. Figure 39.1 illustrates a number of typical configurations. Equation 39.2 calculates the axial compressive load capacity for design (i.e., the *design axial load capacity*), ϕP_n, for a tied column from the *nominal axial load capacity*, P_n.

Figure 39.1 *Types of Tied Columns*

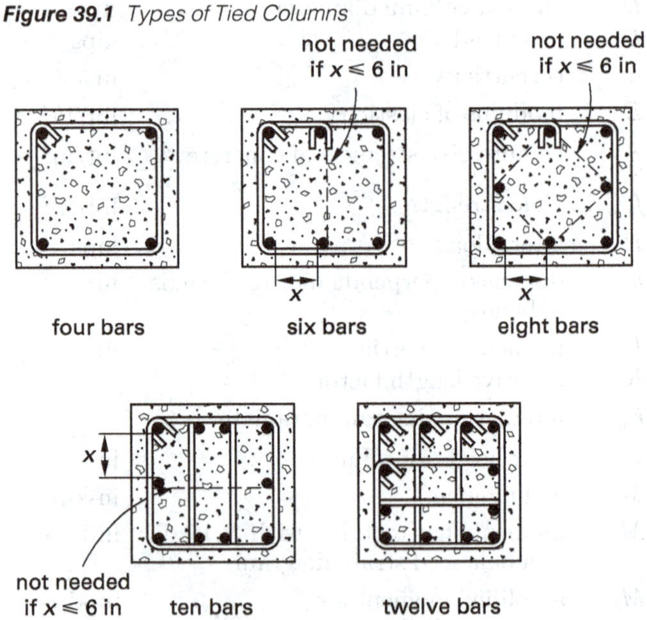

The following construction details are specified by ACI 318 for tied columns.

- Longitudinal bars must have a clear distance between bars of at least 1.5 times the bar diameter, but not less than 1.5 in [ACI 318 Sec. 7.6.3].
- Ties must be at least no. 3 if the longitudinal reinforcement consists of bars no. 10 in size or smaller. For bars larger than no. 10, or when bundles are used in the longitudinal reinforcement, the minimum tie size is no. 4 [ACI 318 Sec. 7.10.5.1]. The maximum practical tie size is normally no. 5.
- The specified concrete cover must be at least 1.5 in over the outermost surface of the tie steel [ACI 318 Sec. 7.7.1].
- At least four longitudinal bars are needed for columns with square or circular ties [ACI 318 Sec. 10.9.2].
- The ratio of longitudinal steel area to the gross column area (given in Eq. 39.1) must be between 0.01 and 0.08 [ACI 318 Sec. 10.9.1]. The lower limit keeps the column from behaving like a plain concrete member. The upper limit keeps the column from being too congested.

$$0.01 \leq \rho_g \leq 0.08$$

- Center-to-center spacing of ties must not exceed the smallest of 16 longitudinal bar diameters, 48 diameters of the tie, or the least column dimension [ACI 318 Sec. 7.10.5.2].
- Every corner and alternating longitudinal bar must be supported by a tie corner. The included angle of the tie cannot be more than 135° [ACI 318 Sec. 7.10.5.3]. (Tie corners are not relevant to tied columns with longitudinal bars placed in a circular pattern.)
- No longitudinal bar can be more than 6 in away from a bar that is properly restrained by a corner [ACI 318 Sec. 7.10.5.3].

Example

A short, square tied column has a minimum area of reinforcing steel of 3.24 in^2 and a gross area of 18 in^2. The compressive strength of the concrete is 4000 lbf/in^2, and the yield strength of the concrete is 60,000 lbf/in^2. What is most nearly the design axial compressive load capacity?

(A) 130 kips

(B) 140 kips

(C) 400 kips

(D) 570 kips

Solution

Use Eq. 39.2. For axial compression with tied reinforcement, ϕ . The design axial load capacity is

$$\phi P_n = 0.80\phi\left[0.85f'_c(A_g - A_{st}) + A_{st}f_y\right]$$

$$= \frac{(0.80)(0.65)\left(\begin{array}{l}(0.85)\left(4000\ \frac{\text{lbf}}{\text{in}^2}\right) \\ \quad\times(18\ \text{in}^2 - 3.24\ \text{in}^2) \\ \quad+(3.24\ \text{in}^2)\left(60{,}000\ \frac{\text{lbf}}{\text{in}^2}\right)\end{array}\right)}{1000\ \frac{\text{lbf}}{\text{kip}}}$$

$$= 127\ \text{kips} \quad (130\ \text{kips})$$

The answer is (A).

Spiral Columns

Reinforced concrete columns with transverse reinforcement in the form of a continuous spiral are known as *spiral columns*. (See Fig. 39.2.)

The design axial load capacity for spiral columns is

$$\phi P_n = 0.85\phi\left(0.85f'_c(A_g - A_{st}) + A_{st}f_y\right)$$

$\phi = 0.70$ for spiral columns.

The following construction details are specified by ACI 318 for spiral columns.

- Longitudinal bars must have a clear distance between bars of at least 1.5 times the bar diameter but not less than 1.5 in [ACI 318 Sec. 7.6.3].
- The minimum spiral wire diameter is 3/8 in [ACI 318 Sec. 7.10.4.2]. The maximum practical spiral diameter is 5/8 in.
- The clear distance between spirals cannot exceed 3 in or be less than 1 in [ACI 318 Sec. 7.10.4.3].
- The specified concrete cover should be at least 1.5 in from the outermost surface of the spiral steel [ACI 318 Sec. 7.7.1].
- At least six longitudinal bars are to be used for spiral columns [ACI 318 Sec. 10.9.2].
- The ratio of the longitudinal steel area to the gross column area must be between 0.01 and 0.08 [ACI 318 Sec. 10.9.1].
- Splicing of spiral reinforcement is covered by ACI 318 Sec. 7.10.4.5.

Figure 39.2 Tied and Spiral Columns

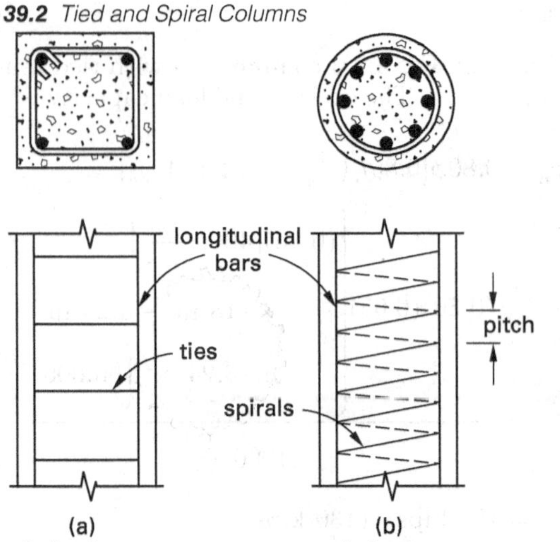

Short Columns with End Moments

Use the following equation to find the end moment due to fixed ends or eccentric loading.

$$M_u = M_2 \text{ or } M_u = P_u e$$

Interaction Diagrams

Based on approximate eccentricity limits, when the maximum moment acting on a column is larger than approximately $0.1P_u h$ for tied columns or $0.05P_u h$ for spiral columns, the design must account explicitly for the effect of bending. In practice, design of columns subjected to significant bending moments is based on interaction diagrams. An *interaction diagram* is a plot of the values of bending moments and axial forces corresponding to the strength (nominal or design) of a reinforced concrete section.

Figure 39.3 illustrates that, for small compressive loads, the moment capacity initially increases, then at a certain axial load it reaches a maximum and subsequently decreases, becoming zero when the axial load equals P_o, where P_o is given by the following equation.

$$P_o = 0.85f'_c(A_g - A_{st}) + f_y A_{st}$$

The point in the interaction diagram where the moment is maximum is referred to as the *balanced point* because at that point, the yielding of the extreme tension reinforcement coincides with the attainment of a strain of 0.003 at the extreme compression fiber of concrete.

Figure 39.4 illustrates the conversion of a nominal interaction diagram to the design interaction diagram per ACI 318 Chap. 9. Note that the balanced point and the point corresponding to the beginning of compression-controlled sections (those with ε_t equal to 0.002) are the same. The subscript "tc" in the figure refers to the beginning of tension-controlled sections (those with ε_t equal to 0.005).

Figure 39.3 Nominal Interaction Diagram

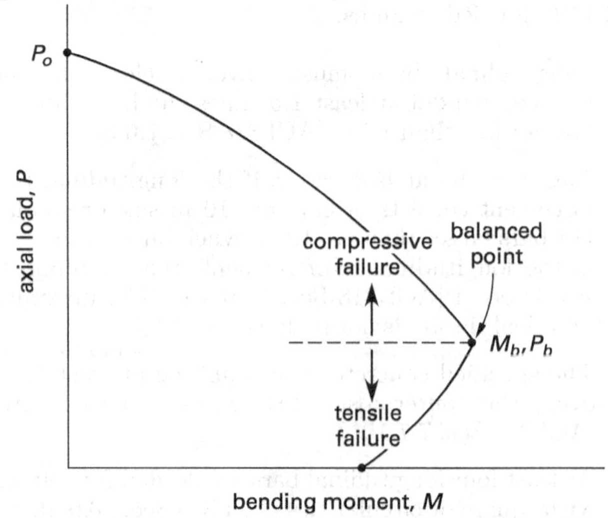

Figure 39.4 Relationship Between Nominal and Design Interaction Diagrams (per ACI 318 Chap. 9)

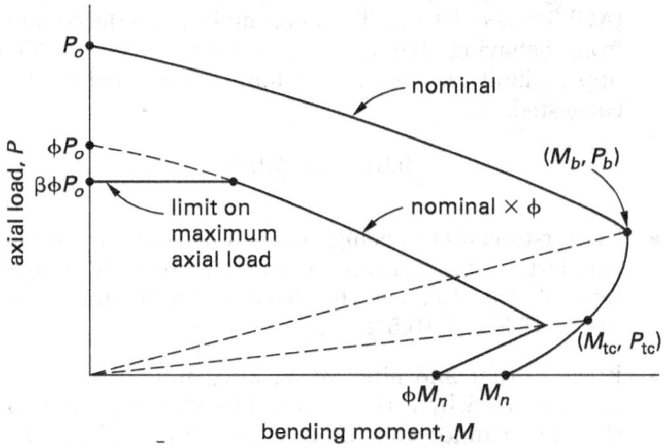

Design for Large Eccentricity

Although interaction diagrams for columns with unusual shapes need to be drawn on a case-by-case basis, routine design of rectangular or circular columns is carried out using dimensionless interaction diagrams available from several sources. (See Fig. 39.5 and Fig. 39.6 at the end of this chapter.)

The use of dimensionless interaction diagrams is straightforward. Each diagram includes envelope lines (most of the diagrams have eight lines), and each line corresponds to a percentage of steel from the minimum of 1% to the maximum of 8%. Each diagram corresponds to a particular set of material properties and to a particular value of γ that is the ratio of the parallel distance between the outer steel layers to the gross section dimension perpendicular to the neutral axis. For practical purposes, γ can be estimated as $(h-5)/h$, where h is in inches. For rectangular cross sections, there is also a need to determine if the steel is to be placed on the two faces parallel to the neutral axis or if a uniform distribution around the column is to be used. In this regard, it is worth noting that the circles represent only the reinforcing pattern, not the exact positioning.

The following procedure can be used to design short columns for large eccentricity using standard interaction diagrams.

step 1: If the column size is not known, an initial estimate can be obtained from the reinforcement ratio. For spiral columns,

$$A_g = \frac{\phi P_n}{0.85\phi\left(0.85f'_c(1-\rho_g)+\rho_g f_y\right)}$$

For tied columns,

$$A_g = \frac{\phi P_n}{0.80\phi\left(0.85f'_c(1-\rho_g)+\rho_g f_y\right)}$$

Use an arbitrarily selected value of ρ_g between the limits allowed. Since the moment will increase the steel demand above that for no moment, a value of ρ_g somewhat smaller than the desired target should be selected.

step 2: Select the type of reinforcement pattern desired and estimate the value of γ. Choose the appropriate interaction diagram. Using a diagram for a γ value smaller than the actual leads to conservative results.

step 3: Evaluate the coordinates of $P_u/\phi f'_c A_g$ and of $P_u e/\phi f'_c A_g h$ (where $P_u e = M_u$) and plot this point on the diagram, assuming a value of ϕ. The dashed line corresponding to $\varepsilon_t = 0.002$ (compression-controlled section) and to $\varepsilon_t = 0.005$ (tension-controlled section) on the diagrams can be used as guides when assuming a value for the strength reduction factor. If the point is outside the last curve for $\rho_g = 0.08$, the section size is too small. In that case, the section size should be increased and this step repeated. If the point is inside the curve for $\rho_g = 0.01$, the section is oversized. In that case, the dimensions should be reduced and this step should be repeated.

step 4: Verify the assumed ϕ value. In most cases, the section will turn out to be clearly compression controlled or tension controlled, and verification will involve no effort. In rare situations where this is not the case, calculate the strain in the extreme tension steel from a strain compatibility analysis for the selected size and arrangement of reinforcement. Determine the strength reduction factor on the basis of the magnitude of this strain and compare it to the value assumed in step 3. If the calculated and assumed values vary by more than a few percentage points, repeat the preceding steps based on the calculated strength reduction factor from this step.

Once the size and steel reinforcement are selected, the design is completed by selecting the transverse reinforcement. This step is carried out exactly the same way as in the case of columns with small eccentricity. Although seldom an issue in the design of columns, the adequacy of the cross section in shear should also be checked when significant bending moments exist.

Example

A reinforced concrete tied column is subjected to a design axial compression force of 875 kips and a design bending moment about its strong axis of 740 ft-kips. The column's cross section measures 30 in × 18 in, and the distance from edge of column to center of steel in each face is 3 in. The concrete's specified compressive strength is 4000 psi, and the reinforcing steel is grade 60. The required area of longitudinal steel is most nearly

(A) 4.0 in²

(B) 8.0 in²

(C) 12 in²

(D) 14 in²

Solution

For a column that measures 30 in in the direction that resists bending, the steel placement constant, γ, is

$$\gamma = \frac{h-2c'}{h} = \frac{30\text{ in} - (2)(3\text{ in})}{30\text{ in}} = 0.8$$

Use Fig. 39.5. The reference interaction diagram requires two parameters.

$$\begin{aligned} R_n &= \frac{P_u e}{\phi f'_c A_g h} = \frac{M_u}{\phi f'_c A_g h} \\ &= \frac{(740\text{ ft-kips})\left(12\ \frac{\text{in}}{\text{ft}}\right)}{(0.65)\left(4\ \frac{\text{kips}}{\text{in}^2}\right)(30\text{ in})(18\text{ in})(30\text{ in})} \\ &= 0.21 \end{aligned}$$

$$K_n = \frac{P_u}{\phi f'_c A_g} = \frac{875\text{ kips}}{(0.65)\left(4\ \frac{\text{kips}}{\text{in}^2}\right)(30\text{ in})(18\text{ in})} = 0.62$$

From the interaction curves, interpolation at the point (0.21, 0.62) gives a longitudinal steel ratio, ρ_g, of 0.026.

Structural Design

The required steel area is

$$A_{st} = \rho_g A_g = (0.026)(30 \text{ in})(18 \text{ in})$$
$$= 14.04 \text{ in}^2 \quad (14 \text{ in}^2)$$

The answer is (D).

3. LONG COLUMNS

A typical assumption used in the calculation of internal forces in structures is that the change in geometry resulting from deformations is sufficiently small to be neglected. This type of analysis, known as a *first-order analysis*, is an approximation that is often sufficiently accurate. There are other instances, however, where the changes in geometry resulting from the structural deformations have a notable effect and must be considered.

ACI 318 requires that the design of slender columns, restraining beams, and other supporting members be based on the factored loading from a second-order analysis, which must satisfy one of three potential analysis approaches: *nonlinear second-order analysis*, *elastic second-order analysis*, or *moment magnification*.

If the following equation holds, the column is a long column.

$$\frac{KL}{r} > 34 - \frac{12M_1}{M_2}$$

Braced and Unbraced Columns

A fundamental determination that has to be made in the process of computing the amplification factors for columns is whether the column is part of a braced (nonsway) or an unbraced (sway) structure. The magnitude of the vertical load required to induce buckling when sway is possible is typically much smaller than when sway is restrained.

Conceptually, a column is considered braced if its buckling mode shape does not involve translation of the end points. A structure may be braced even though *diagonal bracing* is not provided. For example, if the frame in Fig. 39.7(a) is assumed elastic and the vertical loading is progressively increased, columns AB and DE will eventually buckle without joint translation. Wall CF provides the necessary bracing in this case.

Because floor systems are essentially rigid against distortions in their own planes, it is not even necessary for a wall to be located in the same plane of a column to provide adequate bracing. The only requirement is that the arrangement of walls be such that significant torsional stiffness is provided so that the slab cannot rotate about a wall while a certain frame buckles.

Figure 39.7 *Braced and Unbraced Frames*

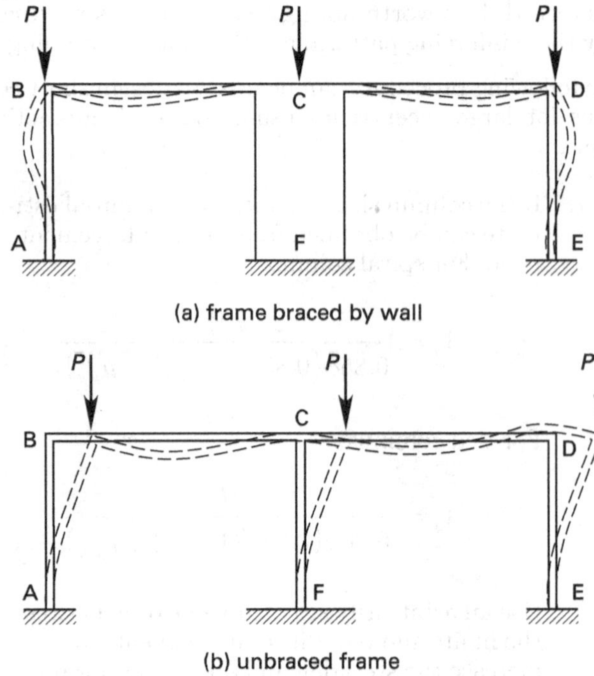

When the wall is replaced by a column, however, buckling takes place with lateral translation of the joints, so the structure must be considered unbraced, as in Fig. 39.7(b).

Critical Load

Short columns will fail by compression of the material. *Long columns* will *buckle* in the transverse direction that has the smallest radius of gyration. Buckling failure is sudden, often without significant warning. If the material is wood or concrete, the material will usually fracture (because the yield stress is low); however, if the column is made of steel, the column will usually fail by local buckling, followed later by twisting and general yielding failure. Intermediate length columns will usually fail by a combination of crushing and buckling.

The load at which a long column fails is known as the *critical load* or *Euler load*. The Euler load is the theoretical maximum load that an initially straight column can support without transverse buckling. For columns with frictionless or pinned ends, this load is given by the following equation, known as *Euler's formula*.

$$P_c = \frac{\pi^2 EI}{(KL)^2} = \frac{\pi^2 EI}{L_{\text{col}}^2}$$

$$EI = 0.25 E_c I_g$$

E_c is the modulus of elasticity of the concrete, as calculated from Eq. 37.1 using the specific weight for normal weight concrete. I_g is the gross moment of inertia based on the exterior dimensions of the column. Only 25% of the product is considered reliable or effective in resisting column loads.

Example 9

A 5 m long rectangular steel column has a moment of inertia of 2.5×10^{-5} m^4. The column is fixed at one end and free at the other. The modulus of elasticity of the column is 3.2×10^5 MPa. What is most nearly the maximum theoretical vertical load the column can support without buckling?

(A) 0.7 MN

(B) 0.8 MN

(C) 2 MN

(D) 3 MN

Solution

Since the column is fixed at one end and free at the other, the theoretical end-restraint coefficient, K, is 2. Find the critical buckling force.

$$\begin{aligned} P_c &= \frac{\pi^2 EI}{(KL)^2} \\ &= \frac{\pi^2 (3.2 \times 10^5 \text{ MPa})\left(10^6 \frac{\text{Pa}}{\text{MPa}}\right)(2.5 \times 10^{-5} \text{ m}^4)}{\left((2)(5 \text{ m})\right)^2} \\ &= 7.90 \times 10^5 \text{ N} \quad (0.8 \text{ MN}) \end{aligned}$$

The answer is (B).

Concentrically Loaded Long Columns

For positive curvature, the minimum value for the moment is given by

$$\begin{aligned} M_1 &= M_2 \\ &= P_u e_{\text{min}} \quad \text{[positive curvature]} \end{aligned}$$

$$e_{\text{min}} = 0.6 + 0.03h$$

Columns that are part of a braced structure are considered to be long columns if $KL / r > 22$.

When long columns in a story are braced, the design is carried out using a factored axial load obtained from a first-order analysis and for a moment, M_c, given by

$$M_c = \frac{M_2}{1 - \frac{P_u}{0.75 P_c}}$$

Long Columns with End Moments

When the columns in a story are braced, the design is carried out using a factored axial load obtained from a first-order analysis and for a moment, M_c, given by ACI 318 Eq. 10-11. M_2 is the larger factored end moment from a first-order solution.

$$M_c = \frac{C_m M_2}{1 - \frac{P_u}{0.75 P_c}} \geq M_2$$

The parameter C_m is used to account for the shape of the first-order moment diagram. For columns with transverse loads between the endpoints, $C_m = 1$. For columns without transverse loads between the end-points, the following equation can be used.

$$C_m = 0.6 + \frac{0.4 M_1}{M_2} \geq 0.4$$

The quantity M_1/M_2 is negative when the column is bent in double curvature. (When $M_1 = M_2 = 0$, use $M_1/M_2 = 1$ [ACI 318 Sec. 10.10.6.4]).

Structural Design

Figure 39.5 Reinforced Concrete Interaction Diagram, $\gamma = 0.80$ (uniplane, 4 ksi concrete, 60 ksi steel)

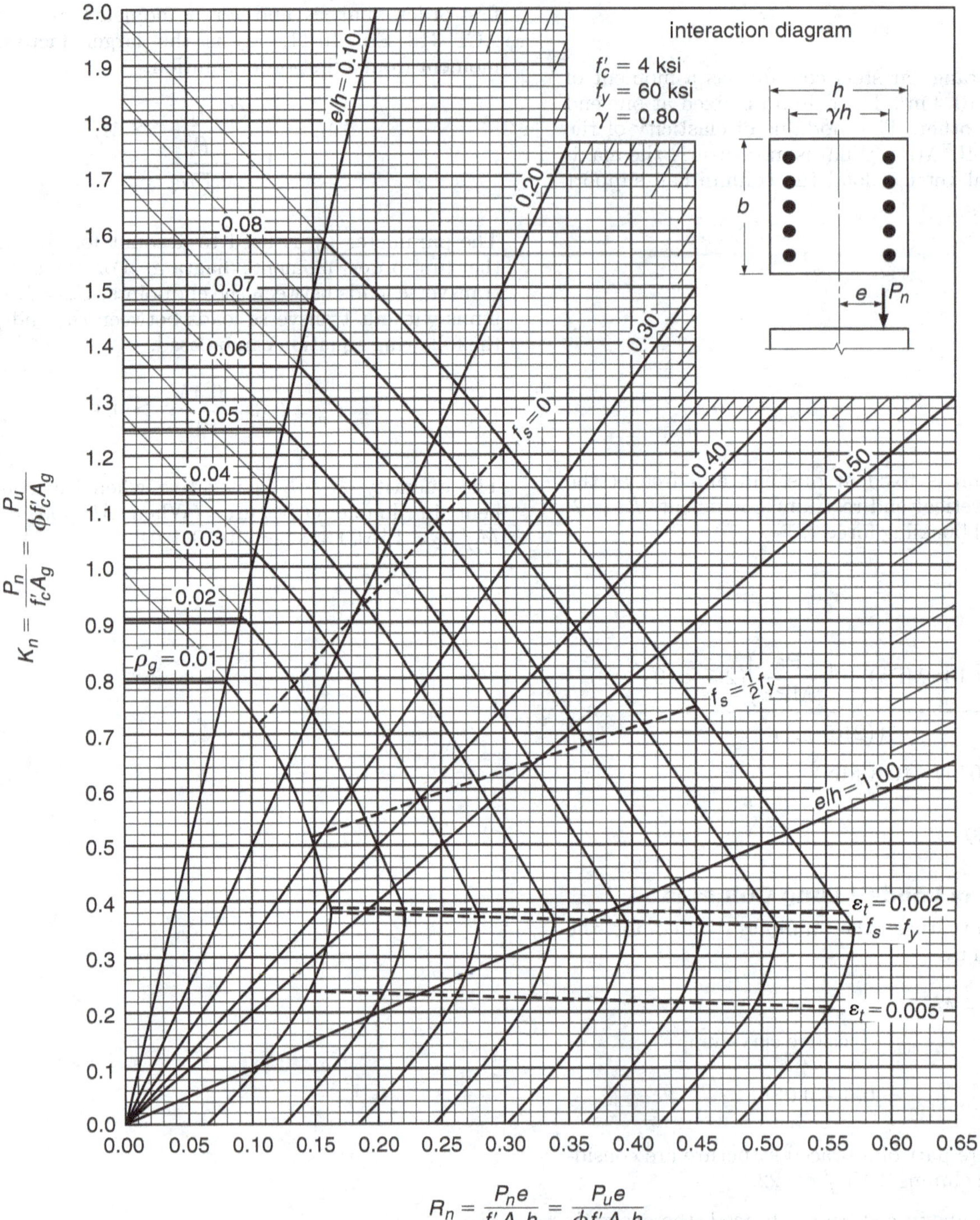

Figure 39.6 *Reinforced Concrete Interaction Diagram, γ = 0.80 (round, 4 ksi concrete, 60 ksi steel)*

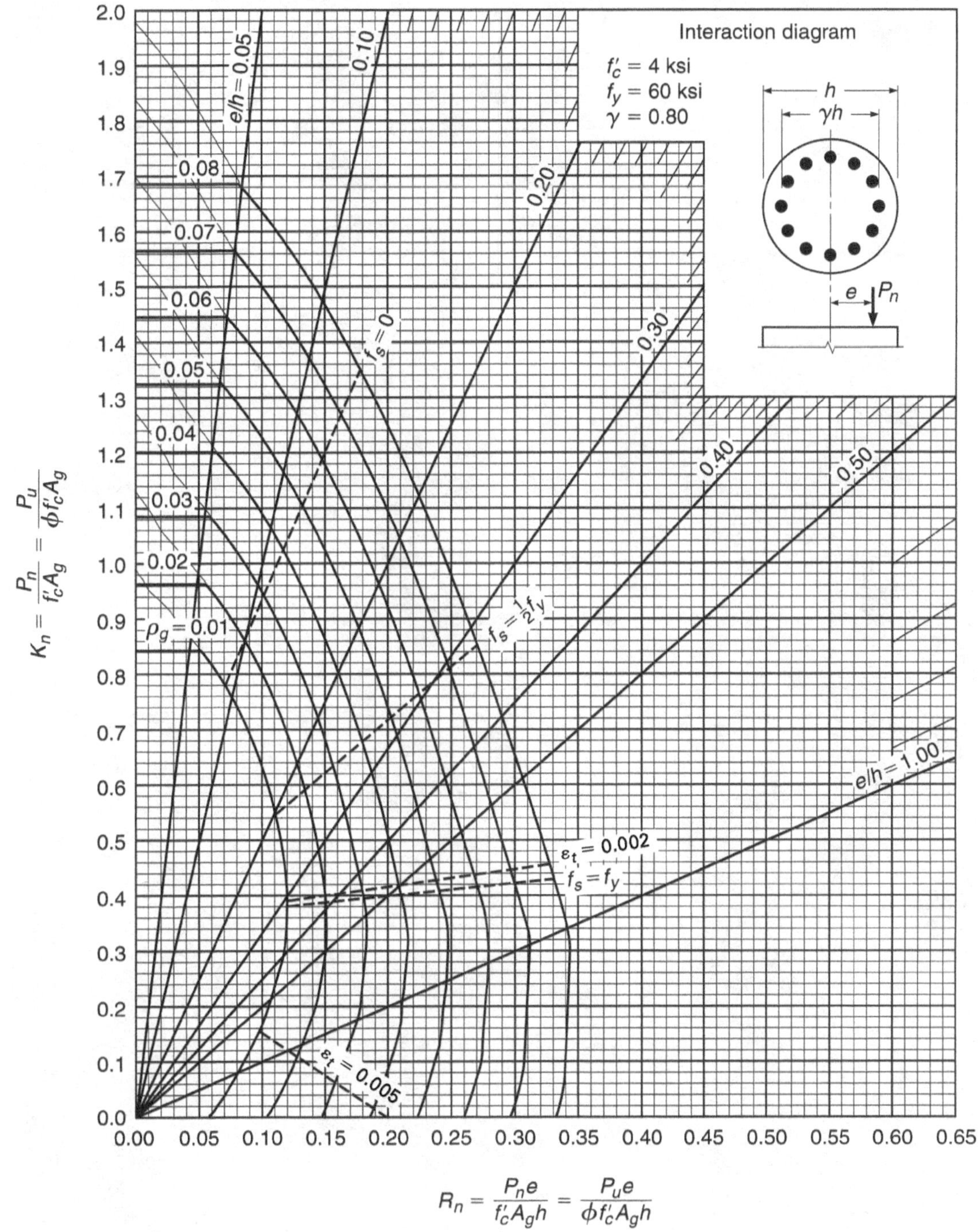

Structural Design

40 Reinforced Concrete: Slabs

Nomenclature

b	width of compressive member	in
d	effective depth	in
E	modulus of elasticity	lbf/in^2
f_y	yield strength	lbf/in^2
h	overall slab depth (thickness)	in
I	gross moment of inertia	in^4
l	span length for one-way slab (centerline-to-centerline of supports)	in
l_1	span length in the direction moments are being computed	in
l_2	length of panel in the direction normal to that for which moments are being computed	in
t	thickness of slab	in
V	shear strength	lbf/in^2
w_u	factored uniform load per unit area of beam or one-way slab	lbf/in^2

Symbols

α	ratio of flexural stiffness of beams in comparison to slab	–
ϕ	strength reduction factor	–

Subscripts

b	bar or beam
c	concrete
s	slab or steel
u	ultimate or uniform
w	web

1. INTRODUCTION

Slabs are structural elements whose lengths and widths are large in comparison to their thicknesses. Shear is generally carried by the concrete without the aid of shear reinforcement (which is difficult to place and anchor in shallow slabs). Longitudinal reinforcement is used to resist bending moments. Slab thickness is typically governed by deflection criteria or fire rating requirements.

A primary issue associated with the design of slabs is the computation of the moments induced by the applied loads. Slabs are highly indeterminate two-dimensional structures, so an exact analysis to obtain the distribution of moments is impractical. Fortunately, moments can be obtained using simplified techniques. In fact, in many instances, slabs can be designed assuming that all the load is carried by moments in one direction only. These slabs are known as one-way slabs and are discussed in detail starting in Sec. 40.2. A practical approach for obtaining the distribution of moments in the more general two-dimensional case is presented starting in Sec. 40.8.

2. ONE-WAY SLABS

Floor slabs are typically supported on all four sides. If the slab length is more than twice the slab width, a uniform load will produce a deformed surface that has little curvature in the direction parallel to the long dimension. Given that moments are proportional to curvature, bending moments will be significant only in the short direction. Slabs designed under the assumption that bending takes place in only one direction are known as *one-way slabs.*

In one-way slabs, the internal forces are computed by taking a strip of unit width and treating it like a beam. The torsional restraint introduced by the supporting beams is typically neglected. For example, the moments per unit width in the short direction for the floor system shown in Fig. 40.1(a) would be obtained by analyzing the four-span continuous beam depicted in Fig. 40.1(b).

3. TEMPERATURE STEEL

Although one-way slabs are designed to be capable of carrying the full applied load by spanning in a single direction, ACI 318 Sec. 7.12 requires that reinforcement

Figure 40.1 One-Way Slab

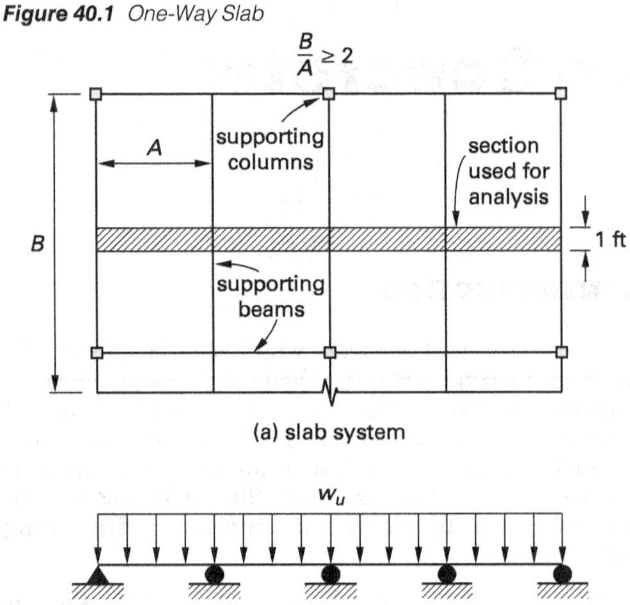

(a) slab system

(b) model used to obtain moments and shears

for shrinkage and temperature stresses be provided normal to the main flexural steel. This is known as *temperature steel*. Temperature steel is required only in structural slabs, not slabs cast against the earth and retaining wall footings. The minimum reinforcement ratio for shrinkage and temperature control (based on the gross area) is 0.0020 (grade 40 and grade 50 steel) and 0.0018 (grade 60 steel).

The minimum reinforcement ratio cannot be less than 0.0014.

Although slabs are typically designed for uniform loads, slabs also experience significant concentrated loads. Temperature steel makes one-way slabs less vulnerable to excessive cracking from moments parallel to the long dimension that are induced by such concentrated loads. In cases where restraint is significant, shrinkage and temperature effects can cause substantial internal forces and displacements. The amount of reinforcing steel required for shrinkage and temperature could exceed that required for flexure. Temperature steel must be spaced no farther apart than five times the slab thickness nor 18 in [ACI 318 Sec. 7.12.2.2].

Although no. 3 bars may be used, it is common practice to use no. 4 bars, which are stiffer and, therefore, bend less during construction handling and installation.

4. MINIMUM THICKNESS FOR DEFLECTION CONTROL

Deflections of one-way slabs can be computed using the procedures for beams. If the slab is not supporting or attached to partitions that are likely to be damaged by large deflections, the computation of deflections can be avoided if the thickness, t, equals or exceeds the values in Table 40.1.

Table 40.1 Minimum One-Way Slab Thickness[a,b] (unless deflections are computed)

construction	minimum thickness, t (fraction of span length, l[c])
simply supported	$\frac{l}{20}$
one end continuous	$\frac{l}{24}$
both ends continuous	$\frac{l}{28}$
cantilever	$\frac{l}{10}$

[a]ACI 318 Table 9.5(a)

[b]For normal weight concrete that is reinforced with $f_y = 60{,}000$ lbf/in^2 steel.

[c]For slabs built integrally with supports, the minimum depth can be based on the clear span [ACI 318 Sec. 8.9.3]. For slabs that are not built integrally with supports, the span length (for the purposes of this table) equals the clear span plus the thickness of the slab but need not exceed the centerline-to-centerline distance [ACI 318 Sec. 8.9.1].

5. ANALYSIS USING ACI COEFFICIENTS

Although the shear and moments in one-way slabs can be computed using standard indeterminate structural analysis, ACI 318 Sec. 8.3.3 specifies a simplified method that can be used when the following conditions are satisfied: (a) There are two or more spans. (b) Spans are approximately equal, with the longer of two adjacent spans not longer than the shorter by more than 20%. (c) The loads are uniformly distributed. (d) The ratio of live to dead loads is no more than 3. (e) The slab has a uniform thickness.

6. SLAB DESIGN FOR FLEXURE

The procedure for selecting steel reinforcement is identical to that for beams, except that moments per unit width are used. The maximum spacing for the main steel is three times the slab thickness but no more than 18 in [ACI 318 Sec. 7.6.5]. The smallest bar size typically used for flexural resistance is no. 4.

Although the design procedures for beams and slabs are similar, there are some differences in detailing. (a) From ACI 318 sec. 7.7.1, the specified cover for the steel is ¾ in when size no. 11 or smaller bars are used. For no. 14 and no. 18 bars, the specified cover is 1.5 in. (b) The minimum steel ratio is equal to that used for shrinkage and temperature. The equations for minimum steel used in beams do not apply.

7. SLAB DESIGN FOR SHEAR

Due to the small thickness, shear reinforcement is difficult to anchor and is seldom used in one-way slabs. Because one-way slabs are wide members, the requirement for no shear reinforcement is $\phi V_c \geq V_u$. As in beams, the critical section for shear is at a distance, d, from the face of the support if the support induces compression in the vertical direction.

8. TWO-WAY SLABS

Slabs are classified as *two-way slabs* when the ratio of long-to-short sides is less than two. A two-way slab supported on a column grid without the use of beams is known as a *flat plate*. (See Fig. 40.2(a).) A modified version of a flat plate, where the shear capacity around the columns is increased by thickening the slab in those regions, is known as a *flat slab*. (See Fig. 40.2(b).) The thickened part of the flat plate is known as a *drop panel*. Another two-way slab often used without beams between column lines is a *waffle slab*. (See Fig. 40.2(c).) In a waffle slab, the forms used to create the voids are omitted around the columns to increase resistance to punching shear. A two-way slab system that is supported on beams is illustrated in Fig. 40.2(d).

Figure 40.2 *Two-Way Slabs*

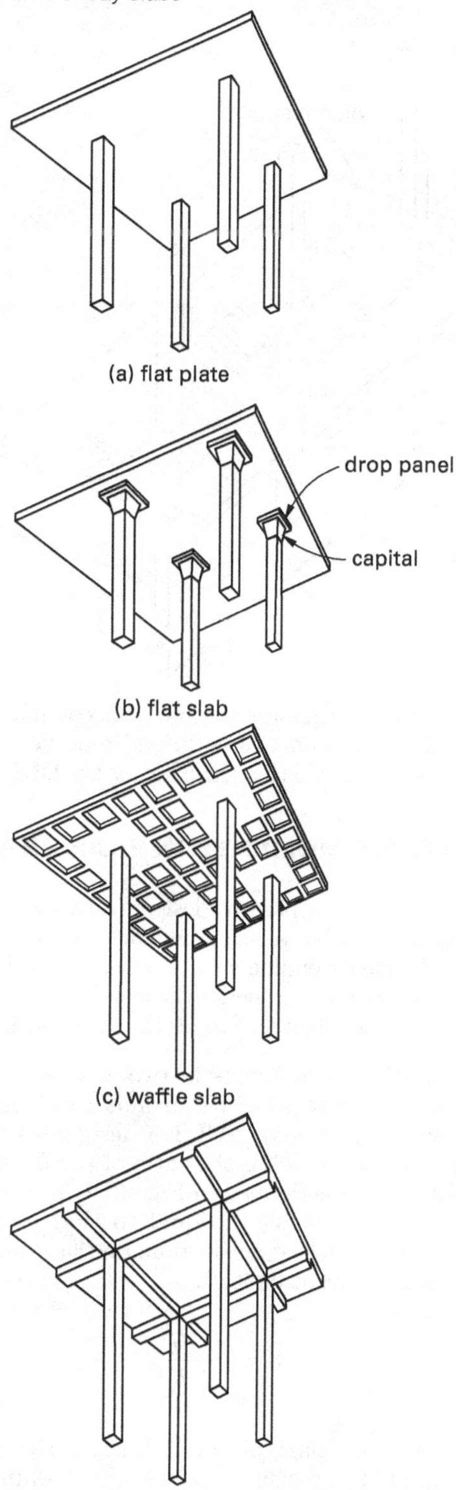

The moments used to design the reinforcement in two-way slabs, whether with or without beams, are obtained in the same manner as for beams and one-way slabs: The slab system and supporting columns are reduced to a series of one-dimensional frames running in both directions. As Fig. 40.3 illustrates, the beams in the frames are wide elements whose edges are defined by cuts midway between the columns. In ACI 318, the dimensions associated with the direction in which moments are being computed are identified by the subscript 1 and those in the transverse direction by the subscript 2. Figure 40.3 also shows that the width of the wide beam is the average of the transverse panel dimensions on either side of the column line. For the edge frames, the width is half the width (centerline-to-centerline) of the first panel.

If the small effect of edge panels is neglected, and if the slab system is assumed to be uniformly loaded (from symmetry), the vertical shears and torques in the edges of the wide beams will be zero. There are two steps in the analysis of two-way slabs using the procedure in ACI 318: (a) calculating the longitudinal distribution of the moments (variation in the direction l_1), and (b) distributing the moment at any cross section across the width of the wide beams.

ACI 318 provides two alternative methods for computing the longitudinal distribution of moments. In the *equivalent frame method* (EFM), the moments are obtained from a structural analysis of the wide beam frame. In this analysis, it is customary to treat the columns as fixed one level above and below the level being considered. (Special provisions that apply to the definition of the stiffnesses of the elements in the equivalent frame, and to other related aspects, are presented in ACI 318 Sec. 13.7.) In the second alternative, known as the *direct design method* (DDM), the moments are obtained using a simplified procedure conceptually similar to that introduced for one-way slabs [ACI 318 Sec. 13.6].

Figure 40.3 Wide Beam Frame (used in the analysis of two-way slabs)

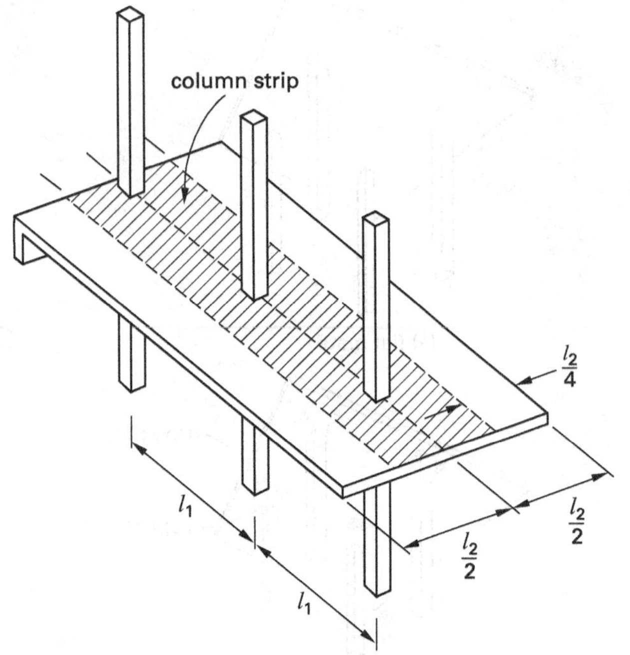

The procedure to distribute the moment computed at any section of the wide beam across the width of the wide beam is not dependent on whether the EFM or the DDM is used.

9. FACTORED MOMENTS IN SLAB BEAMS

A "beam," for the purpose of designing two-way slabs, is illustrated in Fig. 40.4. A beam includes the portion of the slab on each side extending a distance equal to the projection of the beam above or below the slab, whichever is largest, but not greater than four times the slab thickness.

The distribution of moments in two-way slabs depends on the relative stiffness of the beams (with respect to the slab without beams). ACI 318 designates this relative stiffness as α, which is the ratio of the flexural stiffness of the beam (as defined in Fig. 40.4) to the flexural stiffness of a slab of width equal to that of the wide beam (i.e., the width of a slab bounded laterally by the centerlines of adjacent panels).

$$\alpha = \frac{E_{cb}I_b}{E_{cs}I_s}$$

When beams are part of the column strip, they are proportioned to resist 85% of the column strip moment if $\alpha_1 l_2/l_1 \geq 1$. For values of $\alpha_1 l_2/l_1$ between 0 and 1, linear interpolation is used to select the moment to be assigned to the beam, with zero percent assigned when $\alpha_1 l_2/l_1 = 0$. The value of l_2 in the previous expressions refers to the average of the values for the panels on either side of the column line (i.e., to the width of the wide beam).

Figure 40.4 Two-Way Slab Beams (monolithic or fully composite construction) (ACI 318 Sec. 13.2.4)

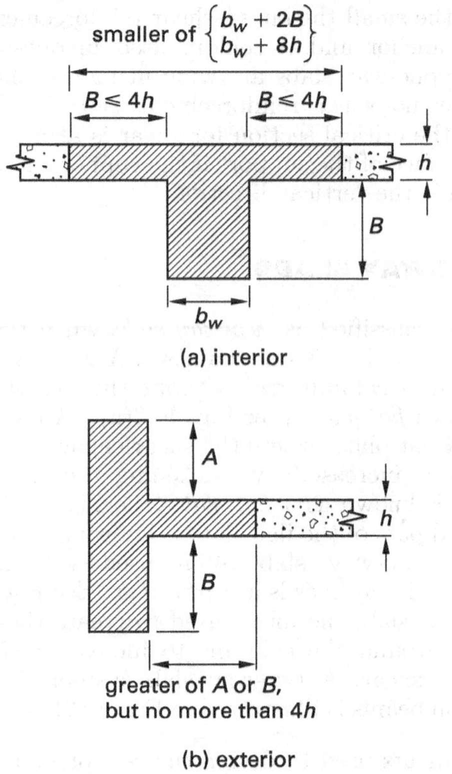

41 Reinforced Concrete: Walls

Nomenclature

f_y	yield strength	lbf/ft²
h	overall thickness of wall	ft

1. NONBEARING WALLS

Nonbearing walls support their own weight and, occasionally, lateral wind and seismic loads. The following minimum details are specified by ACI 318 for designing nonbearing walls.

Minimum vertical reinforcement [ACI 318 Sec. 14.3.2]

1. 0.0012 times the gross concrete area for deformed no. 5 bars or smaller and $f_y \geq 60{,}000$ psi
2. 0.0015 times the gross concrete area for other deformed bars
3. 0.0012 times gross concrete area for smooth or deformed welded wire reinforcement not larger than W31 or D31

Minimum horizontal reinforcement [ACI 318 Sec. 14.3.3]

1. 0.0020 times the gross concrete area for deformed no. 5 bars or smaller and $f_y \geq 60{,}000$ psi
2. 0.0025 times the gross concrete area for other deformed bars
3. 0.0020 times the gross concrete area for smooth or deformed welded wire reinforcement not larger than W31 or D31

Number of reinforcing layers [ACI 318 Sec. 14.3.4]

1. Wall thickness > 10 in: two layers; one layer containing from $\frac{1}{2}$ to $\frac{2}{3}$ of the total steel placed not less than 2 in and not more than $h/3$ from the exterior surface; the other layer placed at a distance not less than $\frac{3}{4}$ in and no more than $h/3$ from the interior surface.
2. Wall thickness ≤ 10 in: ACI 318 does not specify two layers of reinforcement.

Spacing of vertical and horizontal reinforcement [ACI 318 Sec. 14.3.5]

1. The spacing of vertical and horizontal reinforcement may not exceed three times the wall thickness or 18 in.

Need for ties [ACI 318 Sec. 14.3.6]

1. The vertical reinforcement does not have to be enclosed by ties unless (a) the vertical reinforcement is greater than 0.01 times the gross concrete area, or (b) the vertical reinforcement is not required as compression reinforcing.

Thickness [ACI 318 Sec. 14.6.1]

1. Thickness cannot be less than 4 in or $\frac{1}{30}$ times the least distance between members that provide lateral support.

2. BEARING WALLS: EMPIRICAL METHOD

The majority of concrete walls in buildings are *bearing walls* (*load-carrying walls*). In addition to their own weights, bearing walls support vertical and lateral loads. There are two methods for designing bearing walls. They may be designed (a) by the ACI empirical design method, (b) by the ACI alternate design method, or (c) as compression members using the strength design provisions for flexure and axial loads. The minimum design requirements for nonbearing walls must also be satisfied for bearing walls.

The empirical method may be used only if the resultant of all factored loads falls within the middle third of the wall thickness. This is the case with short vertical walls with approximately concentric loads. Use of the ACI empirical equation is further limited to the following conditions. The wall thickness, h, must not be less than $\frac{1}{25}$ times the supported length or height, whichever is shorter, nor may it be less than 4 in. Exterior basement walls and foundation walls must be at least 7.5 in thick. The following additional provisions apply to all walls designed by ACI 318 Chap. 14. (a) To be considered effective for beam reaction or other concentrated load, the length of the wall must not exceed the minimum of the center-to-center distance between reactions and the width of bearing plus four times the wall thickness

Structural Design

[ACI 318 Sec. 14.2.4]. (b) The wall must be anchored to the floors or to columns and other structural elements of the building [ACI 318 Sec. 14.2.6].

3. BEARING WALLS: STRENGTH DESIGN METHOD

If the wall has a nonrectangular cross section (as in ribbed wall panels) or if the eccentricity of the force resultant is greater than one sixth of the wall thickness, the wall must be designed as a column subject to axial loading and bending [ACI 318 Sec. 14.5.1].

4. SHEAR WALLS

Shear walls are designed to resist lateral wind and seismic loads. Reinforced concrete walls have high in-plane stiffness, making them suitable for resisting lateral forces. The ductility provided by the reinforcing steel ensures a ductile-type failure. Ductile failure is the desirable failure mode because it gives warning and enough time for the occupants of the building to escape prior to total collapse.

42 Reinforced Concrete: Footings

Nomenclature

A	area	in^2
b_1	length of the critical area parallel to the axis of the applied column moment	in
b_2	length of the critical area normal to the axis of the applied column moment	in
B	one of the plan dimensions of a rectangular footing	in
d	effective footing depth (to steel layer)	in
d_b	diameter of a bar	in
e	distance from the critical section to the edge	in
f_c'	compressive strength of concrete	lbf/in^2
h	overall height (thickness) of footing	in
H	distance from soil surface to footing base	in
l	length of critical section	in
L	one of the plan dimensions of a rectangular footing	in
M_s	service moment	ft-kips
P	axial load	kips
q	footing load per unit length	lbf/ft
t	wall thickness	in
v_c	nominal concrete shear stress	lbf/in^2
v_u	shear stress due to factored loads	lbf/in^2

Symbols

λ	lightweight aggregate factor	–
ϕ	capacity reduction factor	–

Subscripts

D	dead or development
L	live
p	bearing or punching shear
s	service, soil, or steel
u	factored or ultimate

1. INTRODUCTION

Footings are designed in two steps: (a) selecting the footing area, and (b) selecting the footing thickness and reinforcement. The footing area is chosen so that the soil contact pressure is within limits. The footing thickness and the reinforcement are chosen to keep the shear and bending stresses in the footing within permissible limits.

Although the footing area is obtained from the unfactored service loads, the footing thickness and reinforcement are calculated from factored loads. This is because the weight of the footing and the overburden contribute to the contact stress, but these forces do not induce shears or bending moments in the footing.

Figure 42.1 illustrates the general case of a column footing with dimensions L and B carrying a vertical (downward) axial service load, P_s, and a service moment, M_s.

Figure 42.1 *General Footing Loading*

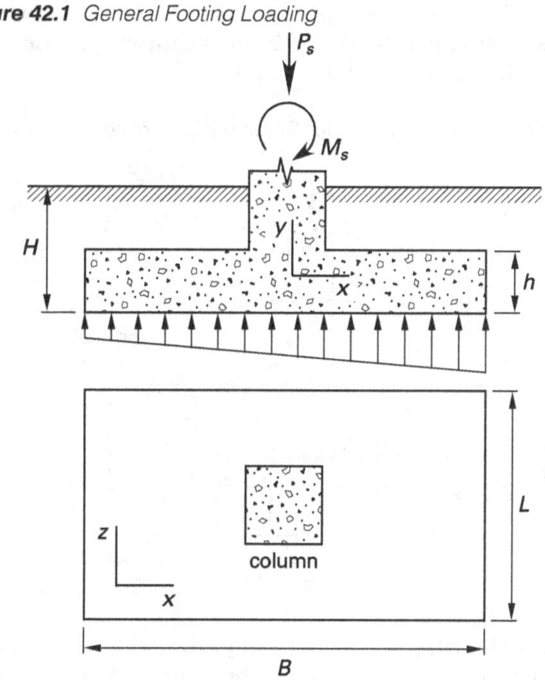

Structural Design

2. WALL FOOTINGS

For a wall footing, the factored load per unit length at failure is

$$q_u = \frac{P_u}{B}$$

The factored load is

$$P_u = 1.2P_D + 1.6P_L$$

The critical plane for shear is assumed to be located a distance, d (the effective footing depth), from the wall face, where d is measured to the center of the flexural steel layer. (See Fig. 42.2.) v_u is the shear stress at a distance, d, from the face of the wall; v_c is the shear stress of the concrete. For shear, $\phi = 0.75$. The thickness of wall footings is selected to satisfy the shear stress requirement of the following inequality. From ACI 318 Sec. 11.1.1,

$$\phi v_c \geq v_u$$

From ACI 318 Sec. 11.2.1.1,

$$v_c = 2\lambda\sqrt{f_c'}$$

The *lightweight aggregate factor*, λ, is 0.75 for all-lightweight concrete, 0.85 for sand-lightweight concrete, and 1.0 for normal weight concrete.

Figure 42.2 *Critical Section for Shear in Wall Footing*

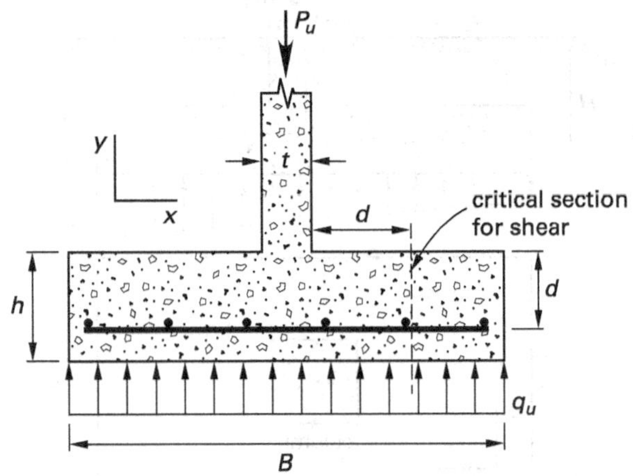

In a wall footing, the main steel (x-direction) resists flexure and is perpendicular to the wall face. Orthogonal steel (z-direction) runs parallel to the wall and is placed in contact with the main steel.

The specified cover (below the bars) for cast-in-place concrete cast against and permanently exposed to earth is 3 in [ACI 318 Sec. 7.7.1]. The minimum depth above the bottom reinforcement is 6 in [ACI 318 Sec. 15.7]. Therefore, the minimum total thickness of a footing is

$$\begin{aligned} h \geq\ & 6 \text{ in} + \text{diameter of } x \text{ bars} \\ & + \text{diameter of } z \text{ bars} + 3 \text{ in} \end{aligned}$$

3. COLUMN FOOTINGS

The effective depth, d (and thickness, h), of column footings is controlled by shear strength. Two shear failure mechanisms are considered: one-way shear and two-way shear. The footing depth is taken as the larger of the two values calculated. In the majority of cases, the required depth is controlled by two-way shear.

For *one-way shear* (also known as *single-action shear* and *wide-beam shear*), the critical sections are a distance, d, from the face of the column. (See Fig. 42.3.) The following equation applies for the case of uniform pressure distribution (i.e., where there is no moment). The section with the largest values of e controls.

$$v_u = \frac{q_u e}{d}$$

Figure 42.3 *Critical Section for One-Way Shear*

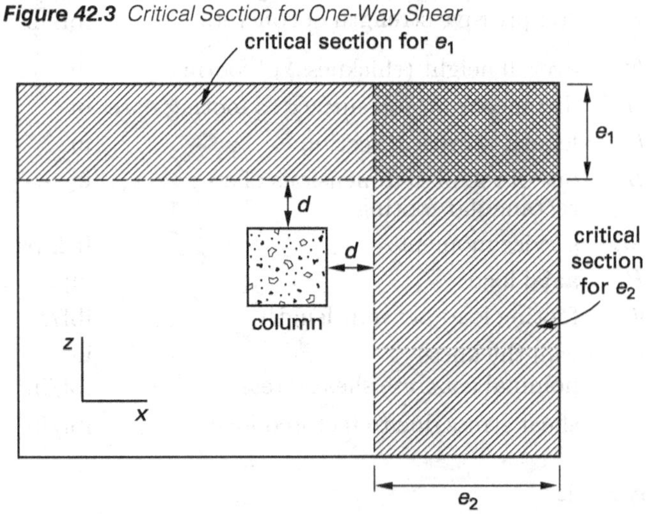

In addition to the footing failing in shear as a wide beam, failure can also occur in *two-way shear* (also known as *double-action shear* and *punching shear*). In this failure mode, the column and an attached concrete piece punch through the footing. Although the failure plane is actually inclined outward, the failure surface is assumed for simplicity to consist of vertical planes located a distance $d/2$ from the column sides. (See Fig. 42.4.)

The area in punching is given by the following equation. b_1 is the length of the critical area parallel to the axis of the applied column moment. Similarly, b_2 is the length of the critical area normal to the axis of the applied column moment.

$$A_p = 2(b_1 + b_2)d$$

Figure 42.4 Critical Section for Two-Way Shear

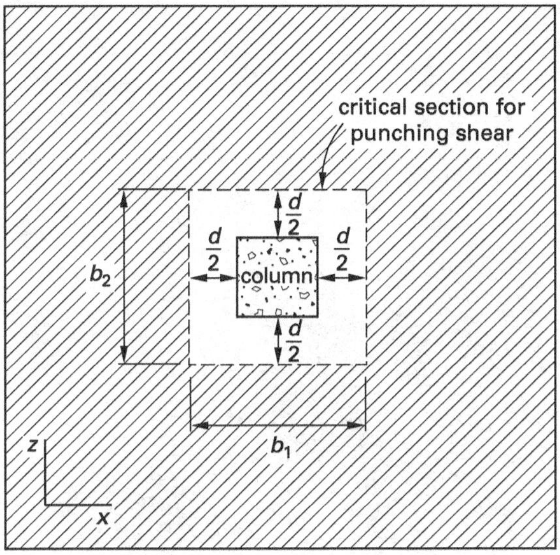

Determining the footing thickness is accomplished through trial and error.

The effective depth for shear is typically assumed to be the distance from the top of the footing to the average depth of the two perpendicular steel reinforcement layers. (The two steel layers may use different bar diameters, so the average depth may not correspond to the contact point.) The total thickness of a column footing is

$$h = d + \tfrac{1}{2}(\text{diameter of } x \text{ bars} + \text{diameter of } z \text{ bars}) + \text{cover}$$

As with wall footings, the specified cover (below the bars) for cast-in-place concrete cast against and permanently exposed to earth is 3 in [ACI 318 Sec. 7.7.1]. The minimum depth above the bottom reinforcement is 6 in [ACI 318 Sec. 15.7]. Therefore, the minimum total thickness of a footing is

$$h \geq 6 \text{ in} + \text{diameter of } x \text{ bars} + \text{diameter of } z \text{ bars} + 3 \text{ in}$$

4. SELECTION OF FLEXURAL REINFORCEMENT

For the purpose of designing flexural reinforcement, spread footings are treated as cantilever beams carrying an upward-acting, nonuniform distributed load. The critical section for flexure is located at the face of the column or wall. (See Fig. 42.5.)

The effective depth of the flexural reinforcement will differ in each perpendicular direction. However, it is customary to design both layers using a single effective depth. While the conservative approach is to use the smaller effective depth for both layers, it is not unusual in practice to design the steel in the x and z directions based on the average effective depth.

Figure 42.5 Critical Section for Moment

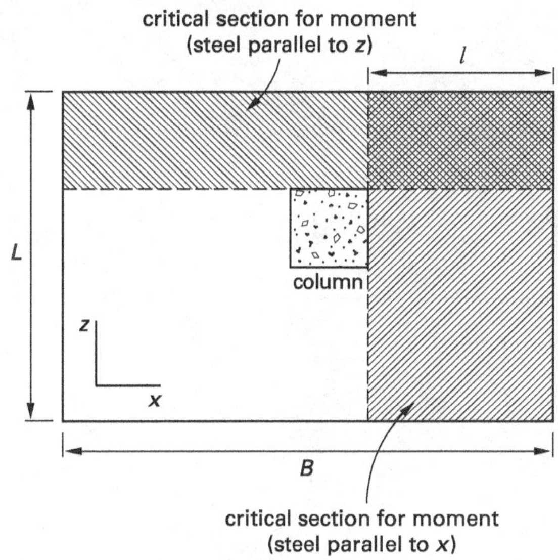

5. TRANSFER OF FORCE AT COLUMN BASE

Column and wall forces are transferred to footings by direct bearing and by forces in steel bars. The steel bars used to transfer forces from one member to the other are known as *dowels* or *dowel bars*. Dowel bars are left to extend above the footing after the footing concrete has hardened, and they are spliced into the column or wall reinforcement prior to pouring the column or wall concrete.

Structural Design

43 Structural Steel: Beams

Nomenclature

A	area	in^2
b	width	in
B	width of base plate	in
C	coefficient	–
C_b	lateral torsion buckling factor	–
C_w	warping constant	in^6
d	depth	in
E	modulus of elasticity	kips/in^2
F	allowable stress or strength	kips/in^2
h	clear distance between flanges	in
h_o	distance between flange centroids	in
I	moment of inertia	in^4
J	torsional constant	in^4
L	span length	in
L_b	length between laterally braced points	in
L_p	limiting laterally unbraced length for yielding	in
L_r	limiting laterally unbraced length for buckling	in
M	moment	in-kips
M_s	bending moment produced by factored loads acting on the cut-back structure	in-kips
P	load	kips
r_{ts}	effective radius of gyration of the compression flange	in
R_n	nominal strength	kips
S	elastic section modulus	in^3
t	thickness	in
V	shear	kips
w	load per unit length	kips/in
x	distance	in
Z	plastic section modulus	in^3

Symbols

λ	slenderness parameter	–
ϕ	resistance factor	–

Subscripts

A	quarter point
b	bending or braced
B	midpoint
c	centroidal, compressive, or concrete
cr	critical
C	three-quarter point
D	dead
f	flange
L	live
max	maximum
n	net or nominal
p	bearing, compact, or plastic
u	ultimate
v	shear
w	warping or web
x	strong axis
y	weak axis or yield

1. TYPES OF BEAMS

Beams primarily support transverse loads (i.e., loads that are applied at right angles to the longitudinal axis of the member). They are subjected primarily to flexure (bending). Although some axial loading is unavoidable in any structural member, the effect of axial loads is generally negligible, and the member can be treated strictly as a beam. If an axial compressive load of substantial magnitude is also present with transverse loads, the member is called a *beam-column.*

Beams are often designated by names that are representative of some specialized functions: A *girder* is a major beam that often provides supports for other beams. A *stringer* is a main longitudinal beam, usually supporting bridge decks. A *floor beam* is a transverse beam in bridge decks. A *joist* is a light beam that supports a floor. A *lintel* is a beam spanning an opening (a door or window), usually in masonry construction. A *spandrel* is a beam on the outside perimeter of a building that supports, among other loads, the exterior wall. A *purlin* is a beam that supports a roof and frames between or over supports, such as roof trusses or rigid frames. A *girt* is a light beam that supports only the lightweight exterior sides of a building, as is typical in pre-engineered metal buildings.

Commonly used beam cross sections are standard hot-rolled shapes including the W, S, M, C, T, and L shapes. Doubly symmetrical shapes, such as W, S, and M sections, are the most efficient. They have excellent flexural strength and relatively good lateral strength for their weight. Channels have reasonably good flexural stre ngth but poor lateral strength, and they require horizontal bracing or some other lateral support. Tees and angles are suitable only for light load. Nomenclature and terminology for W shapes are shown in Fig. 43.1. Dimensions and properties of W shapes are given in Table 43.1 and Table 43.2 at the end of this chapter.

Figure 43.1 *Nomenclature and Terminology for Steel W-Shape Beams*

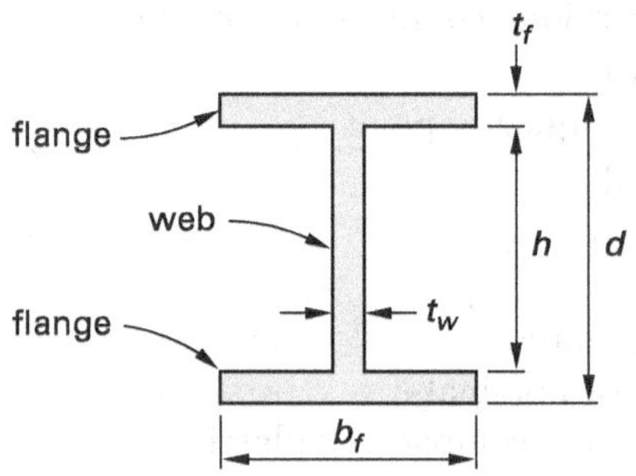

The flexural strength of a rolled section can be improved by adding flange plates. But if the loadings are too great or the spans are too long for a standard rolled section, a plate girder may be necessary. *Plate girders* are built up from plates in I (most common), H, or box shapes of any depth. A *box shape* is used if depth is restricted or if intrinsic lateral stability is needed.

2. DESIGN OF STEEL BEAMS

Steel beams can be designed and analyzed with either the *allowable strength design* (ASD) method (also known as *allowable stress design* and, occasionally, *elastic design*) or *load and resistance factor design* (LRFD) (also known as *strength design, ultimate strength design, plastic design, inelastic design, load factor design, limit state design,* and *mechanism method*). LRFD may also be specified as being "in accordance with Appendix 1 of the *AISC Specification.*" Simple, single-span beams are usually designed by ASD since there is little advantage to using LRFD, which is well-suited for continuous spans and frames. LRFD should not be used with noncompact shapes, crane runway rails, members constructed from A514 steel, or steels with yield strengths in excess of 65 ksi [*AISC Specification* Part 16.1, App. 1, Sec. 1.2].

The *AISC Manual* also provides three design aids for choosing beams: the Design Selection Tables (AISC Tables 3-2 through 3-5), the Maximum Total Uniform Load Tables (AISC Tables 3-6 through 3-9), and the Plots of Available Flexural Strength vs. Unbraced Length Tables (AISC Tables 3-10 through 3-11).

3. FLEXURAL STRENGTH IN STEEL BEAMS

Design and analysis of simple beams are based on comparing the moment due to the loads to a nominal flexural strength modified by a safety factor. For compact sections with adequate lateral support, the nominal flexural strength is given by

$$M_n = F_y Z_x \quad \text{[AISC Eq. F2-1]}$$

The nominal strength is modified by the resistance factor, ϕ_b, for bending. The design equation is

$$M_u = 1.6M_L + 1.2M_D \leq \phi_b M_n$$

M_u is the factored ultimate moment due to the loads, and $\phi_b = 0.90$ (*AISC Specification* Sec. F1).

4. STEEL BEAMS: COMPACT SECTIONS

All of the compression elements of *compact sections* satisfy the limits of *AISC Specification* Table B4.1. Beams whose compression elements are compact are known as compact beams. Compact beams achieve higher strengths and may be more highly loaded than noncompact beams. Compactness requires more than compression element geometry. For a beam to be compact, its flanges must be attached to the web continuously along the beam length. Therefore, a built-up section or plate girder constructed with intermittent welds does not qualify.

Compact Section Criteria

The following equations must be satisfied by standard rolled shapes without flange stiffeners.

$$\frac{b_f}{2t_f} \leq \lambda_{pf} = \frac{64.7}{\sqrt{F_y}} \quad \begin{bmatrix}\text{flanges in flexural}\\ \text{compression only}\end{bmatrix}$$

$$\frac{h}{t_w} \leq \lambda_{pw} = \frac{640.3}{\sqrt{F_y}} \quad \begin{bmatrix}\text{webs in flexural}\\ \text{compression only}\end{bmatrix}$$

The first equation applies only to flanges in flexural compression. The second equation applies only to webs in flexural compression [*AISC Specification* Table B4.1, with $b = b_f/2$].

Compactness depends on the steel strength. Most rolled W shapes are compact at lower values of F_y. A 36 ksi beam may be compact, while the same beam in 50 ksi steel may not be.

Example

Establish whether a W21 × 55 beam of A992 steel is compact.

(A) $\frac{b_f}{2t_f} < \lambda_{pf}$ and $\frac{h}{t_w} < \lambda_{pw}$; beam is compact

(B) $\frac{b_f}{2t_f} < \lambda_{pf}$ and $\frac{h}{t_w} > \lambda_{pw}$; beam is not compact

(C) $\frac{b_f}{2t_f} > \lambda_{pf}$ and $\frac{h}{t_w} < \lambda_{pw}$; beam is compact

(D) $\frac{b_f}{2t_f} > \lambda_{pf}$ and $\frac{h}{t_w} > \lambda_{pw}$; beam is not compact

Solution

For A992 steel, $F_y = 50$ ksi. Obtain b_f, t_f, d, and t_w values for a W21 × 55 beam from Table 43.1.

Check for compactness. For the flange,

$$\frac{b_f}{2t_f} = \frac{8.22 \text{ in}}{(2)(0.522 \text{ in})} = 7.87$$

$$\lambda_{pf} = \frac{64.7}{\sqrt{F_y}} = \frac{64.7}{\sqrt{50 \frac{\text{kips}}{\text{in}^2}}} = 9.15$$

$$\frac{b_f}{2t_f} < \lambda_{pf}$$

For the web,

$$\frac{h}{t_w} = \frac{d - 2t_f}{t_w} = \frac{20.8 \text{ in} - (2)(0.522 \text{ in})}{0.375 \text{ in}} = 52.7$$

$$\lambda_{pw} = \frac{640.3}{\sqrt{F_y}} = \frac{640.3}{\sqrt{50 \frac{\text{kips}}{\text{in}^2}}} = 90.55$$

$$\frac{h}{t_w} < \lambda_{pw}$$

The shape is compact.

The answer is (A).

5. LATERAL BRACING

To prevent *lateral torsional buckling* (illustrated in Fig. 43.2), a beam's compression flange must be supported at frequent intervals. Complete support is achieved when a beam is fully encased in concrete or has its flange welded or bolted along its full length. (See Fig. 43.3.) In many designs, however, lateral support is provided only at regularly spaced intervals. The actual spacing between points of lateral bracing is designated as L_b.

Figure 43.2 *Lateral Buckling in a Beam*

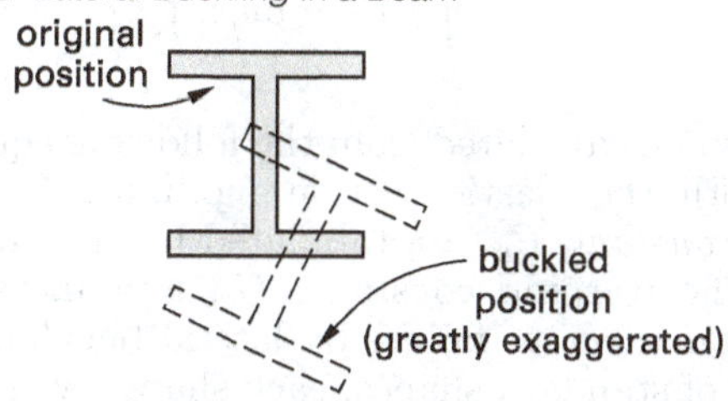

Figure 43.3 *Compression Flange Bracing Using Headed Studs*

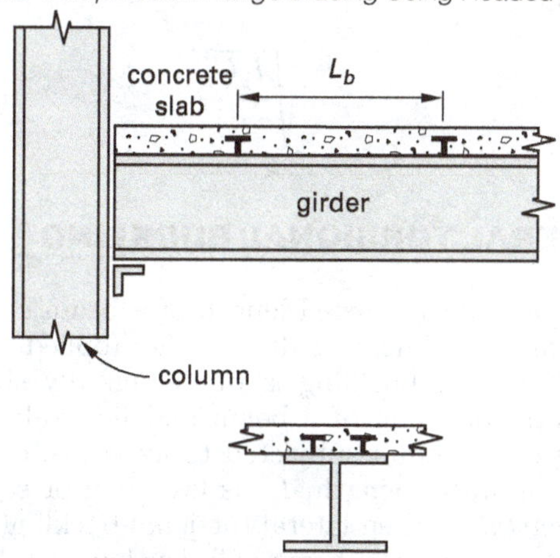

For the purpose of determining flexural design strength ($\phi_b M_n$), two spacing categories are distinguished: $L_p < L_b \le L_r$ and $L_b > L_r$. For most shapes, L_p and L_r are calculated from

$$L_p = 1.76 r_y \sqrt{\frac{E}{F_y}} \quad \text{[AISC Eq. F2-5]}$$

$$L_r = 1.95 r_{ts} \left(\frac{E}{0.7F_y}\right) \sqrt{\frac{Jc}{S_x h_o}} \times \sqrt{1 + \sqrt{1 + 6.76 \left(\left(\frac{0.7F_y}{E}\right)\left(\frac{S_x h_o}{Jc}\right)\right)^2}} \quad \text{[AISC Eq. F2-6]}$$

Effective Radius of Gyration

r_{ts} is the *effective radius of gyration* used to determine L_r for the lateral torsional buckling state. For doubly symmetrical, compact shapes bending about the major axis, it is accurately (and conservatively) estimated as the radius of gyration of the compression flange plus one-sixth of the radius of gyration of the web. It is tabulated in the shape tables in *AISC Manual* Part 1.

$$r_{ts} = \frac{b_f}{\sqrt{12\left(1 + \dfrac{ht_w}{6b_ft_f}\right)}}$$

r_{ts} can also be calculated from the following equation, in which both the elastic section modulus, S_x, and the *warping constant*, C_w, are tabulated in the AISC shape tables. The warping constant, C_w, is a measure of a shape's resistance to failure by lateral buckling. For an assembly of standard shapes, each shape's warping constant can be summed to give the assembly's warping coefficient.

$$r_{ts} = \sqrt[4]{\frac{I_yC_w}{S_x^2}}$$

6. LATERAL TORSIONAL BUCKLING

The laterally unsupported length of a beam can fail in lateral torsional buckling due to the applied moment. Lateral torsional buckling is fundamentally similar to the flexural buckling of a beam and flexural torsional buckling of a column subjected to axial loading. If the laterally unbraced length, L_b, is less than or equal to a *plastic length*, L_p, then lateral torsional buckling will not be a problem, and the beam will develop its full plastic strength, M_p. However, if L_b is greater than L_p, then lateral torsional buckling will occur and the moment capacity of the beam will be reduced below the *plastic strength*, M_p.

Equation 43.1: Available Moment Capacity

$$M_n = M_p = F_yZ_x \qquad 43.1$$

Values

$\phi = 0.90$ for bending.

Description

The Z_x beam selection tables (AISC Table 3-2 through Table 3-5) are easy to use, and they provide a method for quickly selecting economical beams. Their use assumes that either the required plastic section modulus, Z, or the required design plastic moment, $M_u = \phi M_p$, is known, and one of these two criteria is used to select the beam.

Example

A W21 × 48 beam of A992 steel has a compression flange braced at 6 ft intervals. Assume the beam is compact. What is most nearly the available plastic moment capacity of the beam?

(A) 360 ft-kips

(B) 400 ft-kips

(C) 410 ft-kips

(D) 420 ft-kips

Solution

From Table 43.1, for a W21 × 48 beam, $Z_x = 107 \text{ in}^3$. From Eq. 43.1, the available plastic moment capacity is

$$\phi M_p = \phi F_yZ_x$$

$$= \frac{(0.90)\left(50 \ \dfrac{\text{kips}}{\text{in}^2}\right)(107 \text{ in}^3)}{12 \ \dfrac{\text{in}}{\text{ft}}}$$

$$= 401.3 \text{ ft-kips} \quad (400 \text{ ft-kips})$$

The answer is (B).

Equation 43.2: Lateral Torsional Buckling Modification Factor

$$C_b = \frac{12.5M_{\text{max}}}{2.5M_{\text{max}} + 3M_A + 4M_B + 3M_C} \le 3.0 \qquad 43.2$$

Values

$1.0 \le C_b \le 3.0$

Description

Usually, the bending moment varies along the unbraced length of a beam. The worst case scenario is for beams subjected to uniform bending moments along their unbraced lengths. For this situation, the *AISC Manual* specifies a value of $C_b = 1.0$ for the *lateral torsional buckling modification factor*, also known as the *beam bending coefficient*, the *moment modification factor*, and, in the past, the *moment gradient multiplier*. For nonuniform moment loadings, C_b is greater than 1.0, effectively increasing the strength of the beam. C_b is equal to 1.0 for uniform bending moments, and is always greater than 1.0 for nonuniform bending moments. C_b may not exceed a value of 3.0. Of course, the increased moment capacity for the nonuniform moment case cannot be more than M_p. If the calculated strength value, M_n, is greater than M_p, it must be reduced to M_p.

C_b can always be conservatively assumed as 1.0. In fact, for cantilever beams where the free ends are unbraced, $C_b = 1.0$. However, with C_b values being as high as 3, substantial material savings are possible by not being conservative.

Various theoretical and experimental methods exist for calculating C_b. The most convenient method is to read C_b directly from a table. (See Fig. 43.4.) *AISC Specification* Eq. F1-1 provides a method for calculating C_b when both ends of the unsupported segment are braced and the beam is bent in single or double curvature. In Eq. 43.2, M_{max} is the magnitude of maximum bending moment in L_b. M_A, M_B, and M_C are the magnitudes of the bending moments at the quarter point, midpoint, and three-quarter point, respectively, along the unbraced segment of length L_b.

R_m is the *cross-section monosymmetric parameter* having a value of 1.0 for doubly symmetric members, as well as singly symmetric members bent in single curvature. Refer to *AISC Specification* Sec. F1 for other cases.

Figure 43.4 Values of C_b for Simply Supported Beams

load	lateral bracing along span	C_b
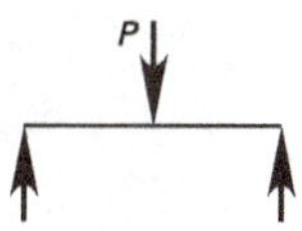 P	none load at midpoint	1.32
	at load point	1.67 1.67
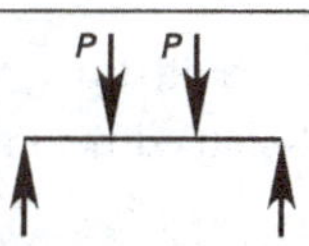 P P	none load at third point	1.14
	at load point loads symmetrically placed	1.67 1.00 1.67
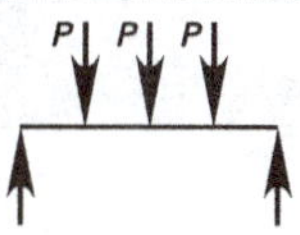 P P P	none load at quarter point	1.14
	at load point loads at quarter points	1.67 1.11 1.11 1.67
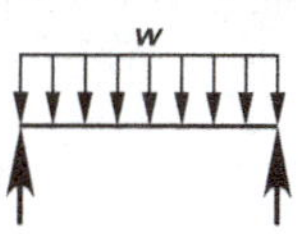 w	none	1.14
	at midpoints	1.30 1.30
	at third points	1.45 1.01 1.45
	at quarter points	1.52 1.06 1.06 1.52
	at fifth points	1.56 1.12 1.00 1.12 1.56

Note: Lateral bracing must always be provided at points of support per *AISC Specification* Chap. F.

Adapted from *Steel Construction Manual*, 14th ed., 2011, AISC.

7. PROCESSES OF FLEXURAL FAILURE

Strength failure due to excessive stress (as opposed to serviceability failure due to excessive deflection) of steel beams in flexure can occur through three different processes (known as *failure modes* corresponding to their respective limit states) depending on the unbraced length, L_b: (1) failure by yielding (material failure), when $L_b < L_p$; (2) failure by inelastic buckling, when $L_p < L_b < L_r$; and (3) failure by elastic buckling, when $L_b > L_r$. These three regions are shown in Fig. 43.5. Three different AISC equations are used to calculate values of M_n corresponding to the actual unbraced length. However, in most cases, values of M_n are read directly from tables in the *AISC Manual*.

Figure 43.5 *Available Moment vs. Unbraced Length*

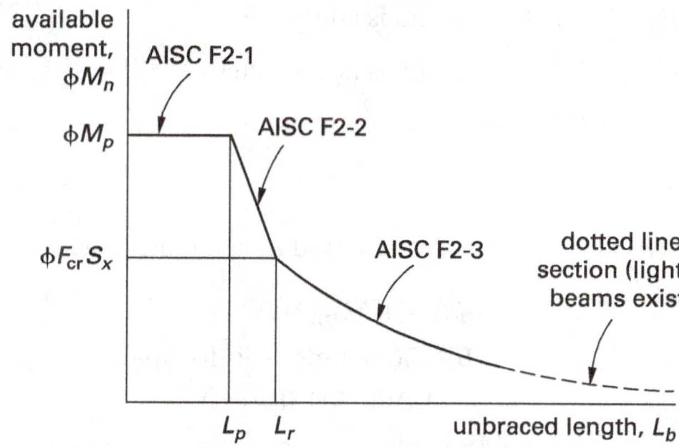

Equation 43.3: Flexural Design Strength

$$M_n = C_b\left[M_p - (M_p - 0.7F_yS_x)\left(\frac{L_b - L_p}{L_r - L_p}\right)\right] \le M_p \quad [L_p < L_b \le L_r] \qquad 43.3$$

Description

If $L_b \le L_p$, or if the bracing is continuous and the beam is compact, the nominal flexural strength, M_n, is M_p. If $L_p < L_b \le L_r$, then M_n is given by Eq. 43.3. If $L_b > L_r$, or if there is no bracing at all between support points, then M_n is given by

$$M_n = C_b\left(\frac{\pi^2 E}{\left(\frac{L_b}{r_{ts}}\right)^2}\sqrt{1 + 0.078\left(\frac{J}{S_z h_o}\right)\left(\frac{L_b}{r_{ts}}\right)^2}\right) \quad [L_b > L_r]$$

$$S_x \le M_p$$

$$h_o = d - t_f$$

The symbol J represents the *torsional constant,* which is a measure of the shape's resistance to failure by twisting. Values of J are tabulated in the shape tables in *AISC Manual* Part 1 for rolled shapes and angles.

Example

A W21 × 68 beam is 32 ft long. The beam is made of A992 steel, which has a yield strength of 50 ksi. The compression flange is braced at the ends and at intervals of 8 ft. C_b is 1.0, and ϕM_p is 600 ft-kips. L_p is 6.36 ft, and L_r is 18.7 ft. The service loading results in a 256 ft-kips moment due to uniform loading (live), an 80 ft-kips moment due to point loads (live), and an 8.704 ft-kips moment due to the beam's self-weight. What is most nearly the available moment strength, and is the beam adequate?

(A) 550 ft-kips; beam is not adequate

(B) 550 ft-kips; beam is adequate

(C) 570 ft-kips; beam is not adequate

(D) 570 ft-kips; beam is adequate

Solution

Design for moment. The required moment strength is

$$\begin{aligned} M_u &= 1.6M_L + 1.2M_D \\ &= (1.6)(256\text{ ft-kips} + 80\text{ ft-kips}) \\ &\quad + (1.2)(8.704\text{ ft-kips}) \\ &= 548\text{ ft-kips} \end{aligned}$$

From Table 43.1, for a W21 × 68 beam, $S_x = 140\text{ in}^3$.

Multiply the nominal moment strength from Eq. 43.3 by the capacity reduction factor, $\phi = 0.90$, to determine the available strength.

$$\begin{aligned} \phi M_n &= C_b\left(\phi M_p - \left(\phi M_p - \phi(0.7F_yS_x)\right)\left(\frac{L_b - L_p}{L_r - L_p}\right)\right) \\ &= (1.0) \\ &\quad \times \left(\begin{array}{l} 600\text{ ft-kips} \\ - \left(\begin{array}{l} (600\text{ ft-kips}) \\ -(0.90)(0.7)\left(50\ \dfrac{\text{kips}}{\text{in}^2}\right) \\ \times\left(\dfrac{140\text{ in}^3}{12\ \dfrac{\text{in}}{\text{ft}}}\right) \end{array}\right) \\ \times\left(\dfrac{8\text{ ft} - 6.36\text{ ft}}{18.7\text{ ft} - 6.36\text{ ft}}\right) \end{array}\right) \\ &= 569\text{ ft-kips} \le 600\text{ ft-kips}\quad\text{[OK]} \end{aligned}$$

The available moment strength, ϕM_n, is approximately 570. Since $\phi M_n > M_u$, the beam is adequate.

The answer is (D).

8. SHEAR IN UNSTIFFENED BEAMS

Equation 43.4: Shear Design Strength

$$V_n = 0.6F_y(dt_w)\quad\text{[rolled I shape]} \qquad 43.4$$

Values

For rolled I shapes with $h/t_w \le 2.24\sqrt{\varepsilon/F_y}$, $\phi = 1.0$.

Description

Although there are different requirements for plate girders, bolts, and rivets, the *nominal shear strength* of unstiffened or stiffened webs of singly- and doubly-symmetrical beams is given by Eq. 43.4 for rolled I shapes. The following equations give the nominal shear stress for built-up I shapes.

$$\frac{h}{t_w} \le \frac{418}{\sqrt{F_y}}:\ \phi V_n = \phi(0.6F_y)A_w\quad\left[\begin{array}{c}\text{built-up}\\ \text{I-shaped beam}\end{array}\right]$$

$$\frac{418}{\sqrt{F_y}} < \frac{h}{t_w} \le \frac{522}{\sqrt{F_y}}:$$

$$V_n = (0.6F_y)(dt_w)\left(\frac{418}{\left(\dfrac{h}{t_w}\right)\sqrt{F_y}}\right)\quad\left[\begin{array}{c}\text{built-up}\\ \text{I-shaped beam}\end{array}\right]$$

$$\frac{h}{t_w} > \frac{522}{\sqrt{F_y}}:$$

$$V_n = (0.6F_y)(dt_w)\left(\frac{220{,}000}{\left(\dfrac{h}{t_w}\right)^2 F_y}\right)\quad\left[\begin{array}{c}\text{built-up}\\ \text{I-shaped beam}\end{array}\right]$$

Table 43.1 W Shapes (Dimensions and Properties)

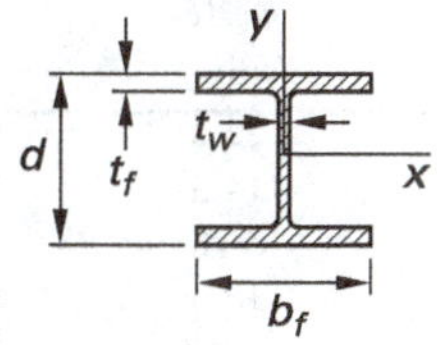

	web	depth	web	flange		axis x-x				axis y-y	
shape	A (in^2)	d (in)	t_w (in)	b_f (in)	t_f (in)	I (in^4)	S (in^3)	r (in)	Z (in^3)	I (in^4)	r (in)
W24×68	20.1	23.7	0.415	8.97	0.585	1830	154	9.55	177	70.4	1.87
W24×62	18.2	23.7	0.430	7.04	0.590	1550	131	9.23	153	34.5	1.38
W24×55	16.3	23.6	0.395	7.01	0.505	1350	114	9.11	134	29.1	1.34
W21×73	21.5	21.2	0.455	8.30	0.740	1600	151	8.64	172	70.6	1.81
W21×68	20.0	21.1	0.430	8.27	0.685	1480	140	8.60	160	64.7	1.80
W21×62	18.3	21.0	0.400	8.24	0.615	1330	127	8.54	144	57.5	1.77
W21×55	16.2	20.8	0.375	8.22	0.522	1140	110	8.40	126	48.4	1.73
W21×57	16.7	21.1	0.405	6.56	0.650	1170	111	8.36	129	30.6	1.35
W21×50	14.7	20.8	0.380	6.53	0.535	984	94.5	8.18	110	24.9	1.30
W21×48	14.1	20.6	0.350	8.14	0.430	959	93.0	8.24	107	38.7	1.66
W21×44	13.0	20.7	0.350	6.50	0.450	843	81.6	8.06	95.4	20.7	1.26
W18×71	20.8	18.5	0.495	7.64	0.810	1170	127	7.50	146	60.3	1.70
W18×65	19.1	18.4	0.450	7.59	0.750	1070	117	7.49	133	54.8	1.69
W18×60	17.6	18.2	0.415	7.56	0.695	984	108	7.47	123	50.1	1.68
W18×55	16.2	18.1	0.390	7.53	0.630	890	98.3	7.41	112	44.9	1.67
W18×50	14.7	18.0	0.355	7.50	0.570	800	88.9	7.38	101	40.1	1.65
W18×46	13.5	18.1	0.360	6.06	0.605	712	78.8	7.25	90.7	22.5	1.29
W18×40	11.8	17.9	0.315	6.02	0.525	612	68.4	7.21	78.4	19.1	1.27
W16×67	19.7	16.3	0.395	10.2	0.67	954	117	6.96	130	119	2.46
W16×57	16.8	16.4	0.430	7.12	0.715	758	92.2	6.72	105	43.1	1.60
W16×50	14.7	16.3	0.380	7.07	0.630	659	81.0	6.68	92.0	37.2	1.59
W16×45	13.3	16.1	0.345	7.04	0.565	586	72.7	6.65	82.3	32.8	1.57
W16×40	11.8	16.0	0.305	7.00	0.505	518	64.7	6.63	73.0	28.9	1.57
W16×36	10.6	15.9	0.295	6.99	0.430	448	56.5	6.51	64.0	24.5	1.52
W14×74	21.8	14.2	0.450	10.1	0.785	795	112	6.04	126	134	2.48
W14×68	20.0	14.0	0.415	10.0	0.720	722	103	6.01	115	121	2.46
W14×61	17.9	13.9	0.375	9.99	0.645	640	92.1	5.98	102	107	2.45
W14×53	15.6	13.9	0.370	8.06	0.660	541	77.8	5.89	87.1	57.7	1.92
W14×48	14.1	13.8	0.340	8.03	0.595	484	70.2	5.85	78.4	51.4	1.91
W12×79	23.2	12.4	0.470	12.1	0.735	662	107	5.34	119	216	3.05
W12×72	21.1	12.3	0.430	12.0	0.670	597	97.4	5.31	108	195	3.04
W12×65	19.1	12.1	0.390	12.0	0.605	533	87.9	5.28	96.8	174	3.02
W12×58	17.0	12.2	0.360	10.0	0.640	475	78.0	5.28	86.4	107	2.51
W12×53	15.6	12.1	0.345	9.99	0.575	425	70.6	5.23	77.9	95.8	2.48
W12×50	14.6	12.2	0.370	8.08	0.640	391	64.2	5.18	71.9	56.3	1.96
W12×45	13.1	12.1	0.335	8.05	0.575	348	57.7	5.15	64.2	50.0	1.95
W12×40	11.7	11.9	0.295	8.01	0.515	307	51.5	5.13	57.0	44.1	1.94
W10×60	17.6	10.2	0.420	10.1	0.680	341	66.7	4.39	74.6	116	2.57
W10×54	15.8	10.1	0.370	10.0	0.615	303	60.0	4.37	66.6	103	2.56
W10×49	14.4	10.0	0.340	10.0	0.560	272	54.6	4.35	60.4	93.4	2.54
W10×45	13.3	10.1	0.350	8.02	0.620	248	49.1	4.32	54.9	53.4	2.01
W10×39	11.5	9.92	0.315	7.99	0.530	209	42.1	4.27	46.8	45.0	1.98

Adapted from *Steel Construction Manual*, 14th ed., 2011, AISC, Table 1.1.

Table 43.2 *W Shapes (Selection by Z_x)* F_y = 50 ksi ϕ_b = 0.90 ϕ_v = 1.00*

shape	Z_x (in^3)	$\phi_b M_{px}$ (ft-kips)	$\phi_b M_{rx}$ (ft-kips)	$\phi_b BF$ (kips)	L_p (ft)	L_r (ft)	I_x (in^4)	$\phi_v V_{nx}$ (kips)
W24×55	**134**	**503**	**299**	**22.2**	**4.73**	**13.9**	**1350**	**251**
W18×65	133	499	307	14.9	5.97	18.8	1070	248
W12×87	132	495	310	5.76	10.8	43.0	740	194
W16×67	130	488	307	10.4	8.69	26.1	954	194
W10×100	130	488	294	4.01	9.36	57.7	623	226
W21×57	129	484	291	20.1	4.77	14.3	1170	256
W21×55	**126**	**473**	**289**	**16.3**	**6.11**	**17.4**	**1140**	**234**
W14×74	126	473	294	8.03	8.76	31.0	795	191
W18×60	123	461	284	14.5	5.93	18.2	984	227
W12×79	119	446	281	5.67	10.8	39.9	662	175
W14×68	115	431	270	7.81	8.69	29.3	722	175
W10×88	113	424	259	3.95	9.29	51.1	534	197
W18×55	**112**	**420**	**258**	**13.9**	**5.90**	**17.5**	**890**	**212**
W21×50	**110**	**413**	**248**	**18.3**	**4.59**	**13.6**	**984**	**237**
W12×72	108	405	256	5.59	10.7	37.4	597	158
W21×48	**107**	**398**	**244**	**14.7**	**6.09**	**16.6**	**959**	**217**
W16×57	105	394	242	12.0	5.56	18.3	758	212
W14×61	102	383	242	7.46	8.65	27.5	640	156
W18×50	101	379	233	13.1	5.83	17.0	800	192
W10×77	97.6	366	225	3.90	9.18	45.2	455	169
W12×65	96.8	356	231	5.41	11.9	35.1	533	142
W21×44	**95.4**	**358**	**214**	**16.8**	**4.45**	**13.0**	**843**	**217**
W16×50	92.0	345	213	11.4	5.62	17.2	659	185
W18×46	90.7	340	207	14.6	4.56	13.7	712	195
W14×53	87.1	327	204	7.93	6.78	22.2	541	155
W12×58	86.4	324	205	5.66	8.87	29.9	475	132
W10×68	85.3	320	199	3.86	9.15	40.6	394	147
W16×45	82.3	309	191	10.8	5.55	16.5	586	167
W18×40	**78.4**	**294**	**180**	**13.3**	**4.49**	**13.1**	**612**	**169**
W14×48	78.4	294	184	7.66	6.75	21.1	484	141
W12×53	77.9	292	185	5.48	8.76	28.2	425	125
W10×60	74.6	280	175	3.80	9.08	36.6	341	129
W16×40	**73.0**	**274**	**170**	**10.1**	**5.55**	**15.9**	**518**	**146**
W12×50	71.9	270	169	5.97	6.92	23.9	391	135
W8×67	70.1	263	159	2.60	7.49	47.7	272	154
W14×43	69.6	261	164	7.24	6.68	20.0	428	125
W10×54	66.6	250	158	3.74	9.04	33.7	303	112
W18×35	**66.5**	**249**	**151**	**12.1**	**4.31**	**12.4**	**510**	**159**
W12×45	64.2	241	151	5.75	6.89	22.4	348	121
W16×36	64.0	240	148	9.31	5.37	15.2	448	140
W14×38	61.5	231	143	8.10	5.47	16.2	385	131
W10×49	60.4	227	143	3.67	8.97	31.6	272	102
W8×58	59.8	224	137	2.56	7.42	41.7	228	134
W12×40	57.0	214	135	5.50	6.85	21.1	307	106
W10×45	54.9	206	129	3.89	7.10	26.9	248	106

*M_r in AISC Table 3-2, the *NCEES Handbook*, and this table is the required nominal moment resistance, which would normally be designated M_n. The subscript r is used throughout the *AISC Manual* to designate a condition just prior to the onset of plastic yielding. Since a condition of yielding is usually equivalent to failure, the required moment strength must be greater than M_r.

Adapted from *Steel Construction Manual*, 14th ed., 2011, AISC, Table 3-2.

44 Structural Steel: Columns

Nomenclature

A	cross-sectional area	in^2
E	modulus of elasticity	kips/in^2
F	allowable stress or strength	kips/in^2
G	end condition coefficient	–
I	moment of inertia	in^4
K	effective length factor	–
l	critical base cantilever dimension	in
L	column length or dimension in base plate calculations	in
P	axial load	kips
r	radius of gyration	in

Symbols

ϕ	resistance factor	–

Subscripts

a	area, available, or axial
c	column or compression
cr	critical
e	effective, elastic, or Euler
g	gross
n	nominal
u	ultimate
x	strong axis
y	weak axis or yield

1. INTRODUCTION

Structural members subjected to axial compressive loads are often called by names identifying their functions. Of these, the best-known are *columns*, the main vertical compression members in a building frame.

For building columns, *W shapes* having nominal depths of 14 in or less are commonly used. These sections, being rather square in shape, are more efficient than others for carrying compressive loads. (Deeper sections are more efficient as beams.) *Pipe sections* are satisfactory for small or medium loads. Pipes are often used as columns in long series of windows, in warehouses, and in basements and garages. In the past, square and *rectangular tubing* saw limited use, primarily due to the difficulty in making bolted or riveted connections at the ends. Modern welding techniques have essentially eliminated this problem.

2. EULER'S COLUMN BUCKLING THEORY

Column design and analysis are based on the Euler buckling load theory. However, specific factors of safety and slenderness ratio limitations distinguish design and analysis procedures from purely theoretical concepts.

An *ideal column* is initially perfectly straight, isotropic, and free of residual stresses. When loaded to the *buckling* (or *Euler*) *load*, a column will fail by sudden buckling (bending). The Euler column buckling load for an ideal, pin-ended, concentrically loaded column is given by the following equation. The modulus of elasticity term, E, implies that the following equation is valid as long as the loading remains in the elastic region.

$$P_e = \frac{\pi^2 EI}{L^2}$$

The *buckling stress*, F_e, is derived by dividing both sides of the previous equation by the area, A, and using the relationship $I = Ar^2$, where r is the radius of gyration.

$$F_e = \frac{P_e}{A} = \frac{\pi^2 E}{\left(\frac{L}{r}\right)^2}$$

The buckling stress is not a function of the material strength, but rather, it is a function of the ratio L/r, known as the *slenderness ratio*. As the slenderness ratio increases, the buckling stress decreases, meaning that as a column becomes longer and more slender, the load that causes buckling becomes smaller.

Structural Design

3. EFFECTIVE LENGTH

Real columns usually do not have pin-connected ends. The restraints placed on a column's ends greatly affect its stability. Therefore, an *effective length factor*, K (also known as an *end-restraint factor*), is used to modify the unbraced length. The product, KL, of the effective length factor and the unbraced length is known as the *effective length* of the column. The effective length approximates the length over which a column actually buckles. The effective length can be longer or shorter than the actual unbraced length.

$$F_e = \frac{\pi^2 E}{\left(\frac{KL}{r}\right)^2} \quad \text{[AISC Specification Eq. E3-4]}$$

Values of the effective length factor depend upon the rotational restraint at the column ends as well as on the resistance provided against any lateral movement along the column length (i.e., whether or not the column is braced against sidesway). Theoretically, restraint at each end can range from complete *fixity* (though this is impossible to achieve in practice) to zero fixity (e.g., as with a freestanding signpost or flagpole end). Table 44.1 gives recommended values of K for use with steel columns. For braced columns (sidesway inhibited), $K \le 1$, whereas for unbraced columns (sidesway uninhibited), $K > 1$.

The values of K do not require prior knowledge of the column size or shape. However, if the columns and beams of an existing design are known, the alignment charts in Fig. 44.1 can be used to obtain a more accurate effective length factor.

To use Fig. 44.1 [*AISC Commentary* Fig. C-C2.3 and Fig. C-C2.4], the *end condition coefficients*, G_A and G_B, are first calculated for the two column ends. (The alignment charts are symmetrical. Either end can be designated A or B.)

Table 44.1 *Effective Length Factors*

illus.	end conditions	K theoretical	K design
(a)	both ends pinned	1	1.00
(b)	both ends built in	0.5	0.65
(c)	one end pinned, one end built in	0.7	0.8
(d)	one end built in, one end free	2	2.10
(e)	one end built in, one end fixed against rotation but free	1	1.20
(f)	one end pinned, one end fixed against rotation but free	2	2.0

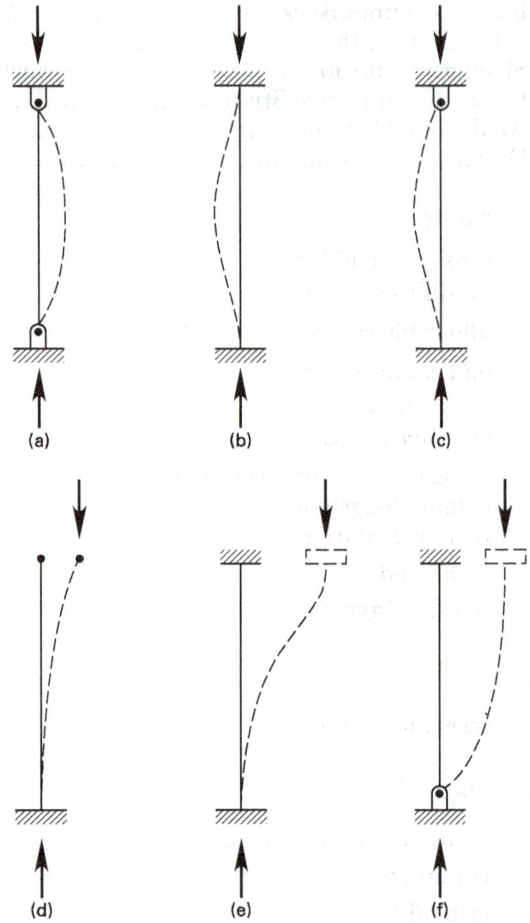

4. SLENDERNESS RATIO

Equation 44.1: Slenderness Ratio

$$\text{slenderness ratio} = \frac{KL}{r} \qquad 44.1$$

Values

The slenderness ratio of compressive members preferably should not exceed 200 [*AISC Specification* Sec. E2].

Description

Steel columns are categorized as long columns and intermediate columns, depending on their slenderness ratios. Since there are two values of the radius of gyration, r (and two values of K and L), corresponding to the x- and y-directions, there will be two slenderness ratios.

Example

A 3400 lbf sign is supported by three equally spaced cables from a W10 × 39 beam. The beam, in turn, is supported by two side wires and a threaded tie rod. Assume all beam, wire, and rod connections are pinned.

Figure 44.1 Effective Length Factor Alignment Charts

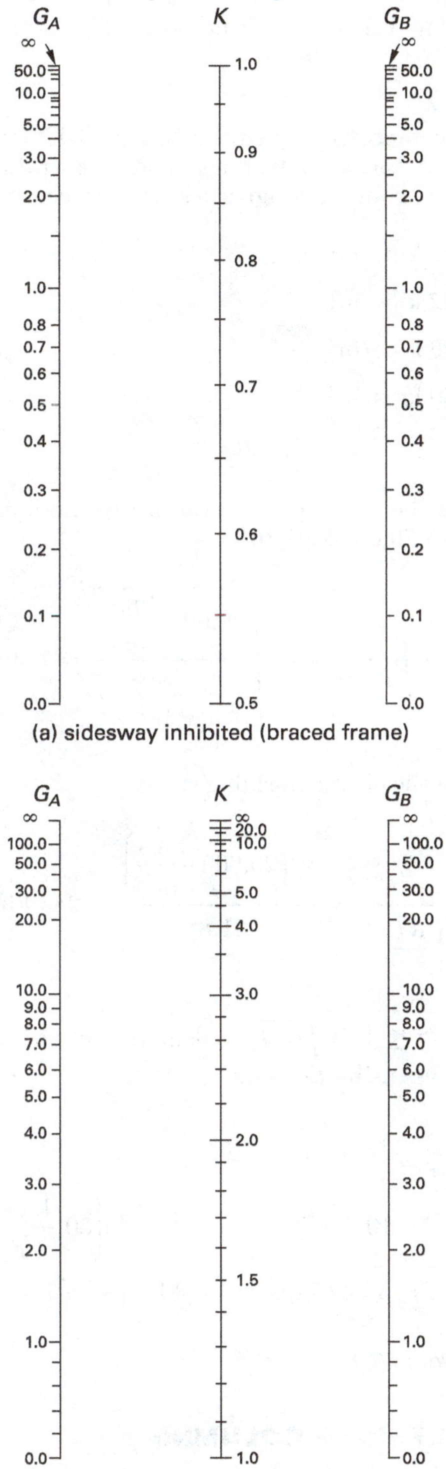

(a) sidesway inhibited (braced frame)

(b) sidesway uninhibited (moment frame)

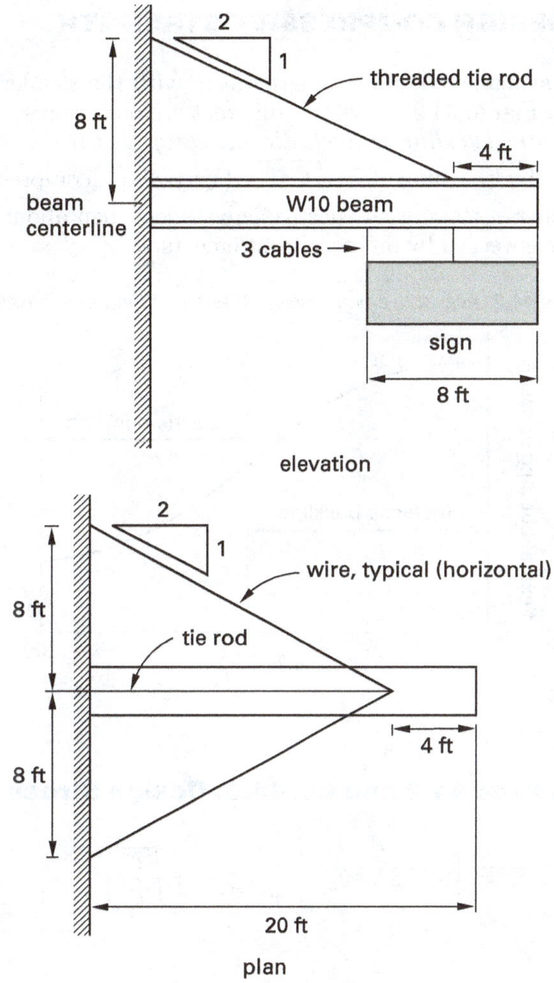

What is most nearly the slenderness ratio of the W10 beam?

(A) 8.0

(B) 60

(C) 97

(D) 190

Solution

Compression strength is controlled by the y-axis.

For a W10 × 39, from Table 43.1, $r_y = 1.98$ in. From Table 44.1, $K = 1.00$.

$$\text{slenderness ratio} = \frac{KL}{r} = \frac{(1.00)(20\ \text{ft} - 4\ \text{ft})\left(12\ \frac{\text{in}}{\text{ft}}\right)}{1.98\ \text{in}} = 96.97 \quad (97)$$

The answer is (C).

5. DESIGN COMPRESSIVE STRENGTH

The available column strength varies with the slenderness ratio. Figure 44.2 shows two different regions, representing *inelastic buckling* and *elastic buckling*, which are separated by the limit $4.71\sqrt{E/F_y}$. (Very short compression members—those with effective lengths less than about 2 ft —are governed by different requirements.)

Figure 44.2 *Available Compressive Stress vs. Slenderness Ratio*

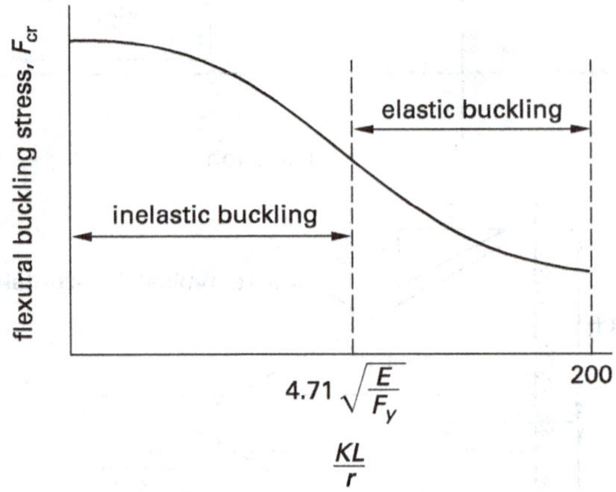

Equation 44.2 and Eq. 44.3: Design Stress

$$F_{cr} = \left[0.658^{\frac{F_y}{F_e}}\right]F_y \quad \left[\frac{KL}{r} \leq 4.71\sqrt{\frac{E}{F_y}}\right] \tag{44.2}$$

$$F_{cr} = 0.877F_e \quad \left[\frac{KL}{r} > 4.71\sqrt{\frac{E}{F_y}}\right] \tag{44.3}$$

Values

$$E = 29{,}000 \text{ kips/in}^2$$

Description

The Euler buckling load, F_e, is often referred to as the critical load. AISC reserves the term "critical stress" (i.e., the critical load divided by the cross-sectional area) for a stress that is calculated from one of its equations. For example, the AISC critical stresses can be calculated from Eq. 44.2 or Eq. 44.3, depending on the slenderness ratio. It is not actually necessary to use these two equations, however, as the values can be read directly for 50 ksi steel members from AISC Table 4-22, presented at the end of this chapter as Table 44.2.

For intermediate columns ($KL/r \leq 4.71\sqrt{E/F_y}$) that fail by inelastic buckling, Eq. 44.2 gives the design stress.

The parabolic curve of Eq. 44.2 accounts for the inelastic behavior due to residual stresses and initial crookedness in a real column.

For long columns ($KL/r > 4.71\sqrt{E/F_y}$) that fail by elastic buckling, Eq. 44.3 must be used.

Example

A steel compression member has a yield strength of 50 kips/in^2 and a controlling slenderness ratio of 31.8. The critical column design stress is most nearly

(A) 40 kips/in^2

(B) 42 kips/in^2

(C) 46 kips/in^2

(D) 50 kips/in^2

Solution

Calculate the limiting stress ratio. The modulus of elasticity, E, is 29,000 kips/in^2.

$$4.71\sqrt{\frac{E}{F_y}} = 4.71\sqrt{\frac{29{,}000\ \frac{\text{kips}}{\text{in}^2}}{50\ \frac{\text{kips}}{\text{in}^2}}} = 113.43$$

Calculate the elastic buckling stress.

$$F_e = \frac{\pi^2 E}{\left(\frac{KL}{r}\right)^2} = \frac{\pi^2\left(29{,}000\ \frac{\text{kips}}{\text{in}^2}\right)}{31.8^2} = 283 \text{ kips/in}^2$$

Since $KL/r \leq 4.71\sqrt{E/F_y}$, calculate the critical column design stress using Eq. 44.2.

$$\begin{aligned} F_{cr} &= \left[0.658^{\frac{F_y}{F_e}}\right]F_y \\ &= \left(0.658^{50\ \text{kips/in}^2/283\ \text{kips/in}^2}\right)\left(50\ \frac{\text{kips}}{\text{in}^2}\right) \\ &= 46.44 \text{ kips/in}^2 \quad (46 \text{ kips/in}^2) \end{aligned}$$

The answer is (C).

6. ANALYSIS OF COLUMNS

Equation 44.4: Design Compressive Strength

$$P_n = F_{cr}A_g \tag{44.4}$$

Description

Columns are analyzed by verifying that the required strength, P_u, does not exceed the design strength, $\phi_c P_n$. Equation 44.4 is the nominal axial compressive strength, P_n. In Eq. 43.4, A_g is the gross column area.

Column Analysis Procedure

step 1: Determine the shape properties A, r_x, and r_y from the *AISC Manual.* Determine the unbraced lengths, L_x and L_y.

step 2: Obtain K_x and K_y from Table 44.1 or Fig. 44.1.

step 3: Calculate the maximum slenderness ratio as

$$\text{slenderness ratio} = \text{larger of} \begin{Bmatrix} \dfrac{K_x L_x}{r_x} \\ \dfrac{K_y L_y}{r_y} \end{Bmatrix}$$

step 4: Calculate the limit, $4.71\sqrt{\dfrac{E}{F_y}}$.

step 5: Using the slenderness ratio, determine the design stress, F_{cr}, from either Table 44.2 or Table 44.3. Alternatively, calculate the design stress, F_{cr}, from either Eq. 44.2 or Eq. 44.3, depending on the value of the slenderness ratio.

step 6: Compute the allowable strength or the design compressive strength and compare against the required strength.

7. DESIGN OF STEEL COLUMNS

Trial-and-error column selection is difficult. Accordingly, Part 4 of the *AISC Manual* contains column selection tables [AISC Table 4-2] that make it fairly easy to select a column based on the required column capacity. These tables assume that buckling will occur first about the minor axis. Steps 4 through 7 in the following procedure accommodate the case where buckling occurs first about the major axis.

step 1: Determine the load to be carried. Include an allowance for the column weight.

step 2: Determine the effective length factors, K_y and K_x, for the column. Calculate the effective length assuming that buckling will be about the minor axis.

$$\text{effective length} = K_y L_y$$

step 3: Enter the table and locate a column that will support the required load with an effective length of $K_y L_y$.

step 4: Check for buckling in the strong direction. Calculate $K_x L_x'$ from the following equation. (The ratio r_x/r_y is tabulated in the column tables.)

$$K_x L_x' = \frac{K_x L_x}{\dfrac{r_x}{r_y}}$$

step 5: If $K_x L_x' < K_y L_y$, the column is adequate, and the procedure is complete. Go to step 8.

step 6: If $K_x L_x' > K_y L_y$ but the column chosen can support the load at a length of $K_x L_x'$, the column is adequate, and the procedure is complete. Go to step 8.

step 7: If $K_x L_x' > K_y L_y$ but the column chosen cannot support the load at a length of $K_x L_x'$, choose a larger member that will support the load at a length of $K_x L_x'$. (The ratio r_x/r_y is essentially constant.)

step 8: If sufficient information on other members framing into the column is available, use the alignment charts to check the values of K.

Table 44.2 Available Critical Stress (ϕF_{cr}) for Compression Members (F_y = 50 ksi, ϕ_c = 0.90)

KL/r	ϕF_{cr} (ksi)	KL/r	ϕF_{cr} (ksi)	KL/r	ϕF_{cr} (ksi)	KL/r	ϕF_{cr} (ksi)	KL/r	ϕF_{cr} (ksi)
1	45.0	41	39.8	81	27.9	121	15.4	161	8.72
2	45.0	42	39.5	82	27.5	122	15.2	162	8.61
3	45.0	43	39.3	83	27.2	123	14.9	163	8.50
4	44.9	44	39.1	84	26.9	124	14.7	164	8.40
5	44.9	45	38.8	85	26.5	125	14.5	165	8.30
6	44.9	46	38.5	86	26.2	126	14.2	166	8.20
7	44.8	47	38.3	87	25.9	127	14.0	167	8.10
8	44.8	48	38.0	88	25.5	128	13.8	168	8.00
9	44.7	49	37.7	89	25.2	129	13.6	169	7.89
10	44.7	50	37.5	90	24.9	130	13.4	170	7.82
11	44.6	51	37.2	91	24.6	131	13.2	171	7.73
12	44.5	52	36.9	92	24.2	132	13.0	172	7.64
13	44.4	53	36.7	93	23.9	133	12.8	173	7.55
14	44.4	54	36.4	94	23.6	134	12.6	174	7.46
15	44.3	55	36.1	95	23.3	135	12.4	175	7.38
16	44.2	56	35.8	96	22.9	136	12.2	176	7.29
17	44.1	57	35.5	97	22.6	137	12.0	177	7.21
18	43.9	58	35.2	98	22.3	138	11.9	178	7.13
19	43.8	59	34.9	99	22.0	139	11.7	179	7.05
20	43.7	60	34.6	100	21.7	140	11.5	180	6.97
21	43.6	61	34.3	101	21.3	141	11.4	181	6.90
22	43.4	62	34.0	102	21.0	142	11.2	182	6.82
23	43.3	63	33.7	103	20.7	143	11.0	183	6.75
24	43.1	64	33.4	104	20.4	144	10.9	184	6.67
25	43.0	65	33.0	105	20.1	145	10.7	185	6.60
26	42.8	66	32.7	106	19.8	146	10.6	186	6.53
27	42.7	67	32.4	107	19.5	147	10.5	187	6.46
28	42.5	68	32.1	108	19.2	148	10.3	188	6.39
29	42.3	69	31.8	109	18.9	149	10.2	189	6.32
30	42.1	70	31.4	110	18.6	150	10.0	190	6.26
31	41.9	71	31.1	111	18.3	151	9.91	191	6.19
32	41.8	72	30.8	112	18.0	152	9.78	192	6.13
33	41.6	73	30.5	113	17.7	153	9.65	193	6.06
34	41.4	74	30.2	114	17.4	154	9.53	194	6.00
35	41.2	75	29.8	115	17.1	155	9.40	195	5.94
36	40.9	76	29.5	116	16.8	156	9.28	196	5.88
37	40.7	77	29.2	117	16.5	157	9.17	197	5.82
38	40.5	78	28.8	118	16.2	158	9.05	198	5.76
39	40.3	79	28.5	119	16.0	159	8.94	199	5.70
40	40.0	80	28.2	120	15.7	160	8.82	200	5.65*

*Adapted from *Steel Construction Manual*, 14th ed., 2011, AISC, Table 4-22.

Table 44.3 Available Strength in Axial Compression, W Shapes

selected W14, W12, W10	available strength in axial compression (kips), W shapes* LRFD: ϕP_n														$F_y = 50$ ksi $\phi_c = 0.90$
shape	W14					W12					W10				
(wt/ft)	74	68	61	53	48	58	53	50	45	40	60	54	49	45	39
effective length KL (ft) with respect to least radius of gyration, r_y: 0	980	899	806	702	636	767	701	657	590	526	794	712	649	597	516
6	922	844	757	633	573	722	659	595	534	475	750	672	612	543	469
7	901	826	740	610	552	707	644	574	516	458	734	658	599	525	452
8	878	804	721	585	529	689	628	551	495	439	717	643	585	505	435
9	853	781	700	557	504	670	610	526	472	419	698	625	569	483	415
10	826	755	677	528	477	649	590	499	448	397	677	607	551	460	395
11	797	728	652	497	449	627	569	471	422	375	655	586	533	435	373
12	766	700	626	465	420	603	547	443	396	351	631	565	513	410	351
13	734	670	599	433	391	578	525	413	370	328	606	543	493	384	328
14	701	639	572	401	361	553	501	384	343	304	581	520	471	358	305
15	667	608	543	369	332	527	477	354	317	280	555	496	450	332	282
16	632	576	515	338	304	500	452	326	291	257	528	472	428	306	260
17	598	544	486	308	276	473	427	297	265	234	501	448	405	281	238
18	563	512	457	278	250	446	402	270	241	212	474	423	383	256	216
19	528	480	428	250	224	420	378	244	217	191	447	399	360	233	195
20	494	448	400	226	202	393	353	220	196	172	420	375	338	210	176
22	428	387	345	186	167	342	306	182	162	142	367	327	295	174	146
24	365	329	293	157	140	293	261	153	136	120	317	282	254	146	122
26	311	281	250	133	120	249	222	130	116	102	270	241	216	124	104
28	268	242	215	115	103	215	192	112	99.8	88.0	233	208	186	107	90.0
30	234	211	187	100	89.9	187	167	97.7	87.0	76.6	203	181	162	93.4	78.4
32	205	185	165	88.1		165	147	82.9	76.4	67.3	179	159	143	82.1	68.9
34	182	164	146			146	130				158	141	126		
36	162	146	130			130	116				141	126	113		
38	146	131	117			117	104				127	113	101		
40	131	119	105			105	93.9				114	102	91.3		

*Heavy line indicates KL/r is greater than or equal to 200.

Adapted from *Steel Construction Manual*, 14th ed., 2011, AISC, Table 4-1.

Structural Design

45 Structural Steel: Tension Members

Nomenclature

A	cross-sectional area	in^2
b	width	in
d	depth or diameter	in
F	allowable stress or strength	kips/in^2
g	lateral gage spacing of adjacent holes	in
L	member length	in
P	applied load or load capacity	kips
R	block shear rupture strength	kips
s	pitch spacing of adjacent holes	in
t	thickness	in
T	block or tensile shear strength	kips
U	shear lag reduction coefficient	–
w	plate width	in
$\bar{x}$	connection eccentricity	in

Symbols

ϕ	resistance factor	–

Subscripts

a	axial
b	bolt
bs	block shear
e	effective
f	fracture
g	gross
h	hole
n	net or nominal
t	tensile
u	ultimate
v	shear
y	yield

1. INTRODUCTION

Tension members occur in a variety of structures, including bridge and roof trusses, transmission towers, wind bracing systems in multistoried buildings, and suspension and cable-stayed bridges. Wire cables, rods, eyebars, structural shapes, and built-up members are typically used as tension members. (See Fig. 45.1.)

Figure 45.1 Common Tension Members

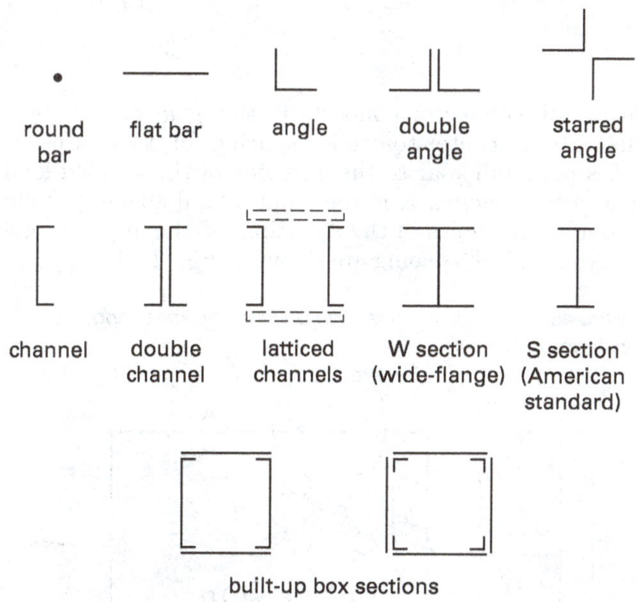

2. GROSS AREA

Equation 45.1: Gross Area

$$A_g = b_g t \quad \text{[use tabulated areas for angles]} \qquad 45.1$$

Description

The gross area, A_g, is the original unaltered cross-sectional area of the member. The gross area for flat bars and angles, bolted or welded, is given by Eq. 45.1. b_g is the gross width of the member without reductions for holes, and t is the member thickness.

3. NET AREA

Plates and shapes are connected to other plates and shapes by means of welds, rivets, and bolts. If rivets or bolts are used, holes must be punched or drilled in the member. As a result, the member's cross-sectional area at the connection is reduced. The load-carrying capacity of the member may also be reduced, depending on the size and location of the holes. The *net area* available to resist tension is illustrated in Fig. 45.2(b).

Figure 45.2 *Tension Member with Unstaggered Rows of Holes*

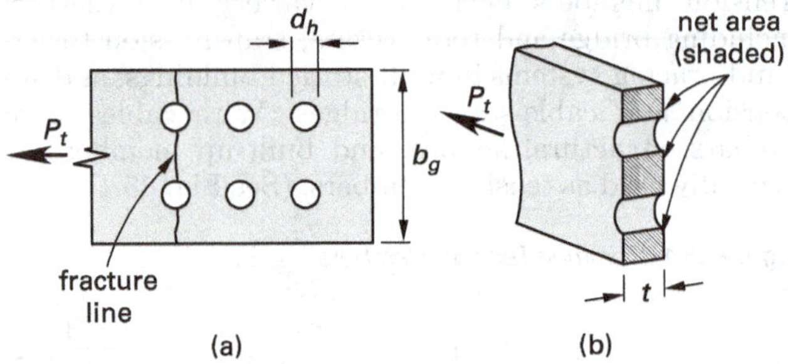

In a multiconnector connection, the *gage spacing*, g, is the lateral center-to-center spacing of two adjacent holes perpendicular to the direction of the applied load. The *pitch spacing*, s, is the longitudinal spacing of any two adjacent holes in the direction for the applied load. Gage and pitch spacing are shown in Fig. 45.3.

Figure 45.3 *Tension Member with Uniform Thickness and Unstaggered Holes*

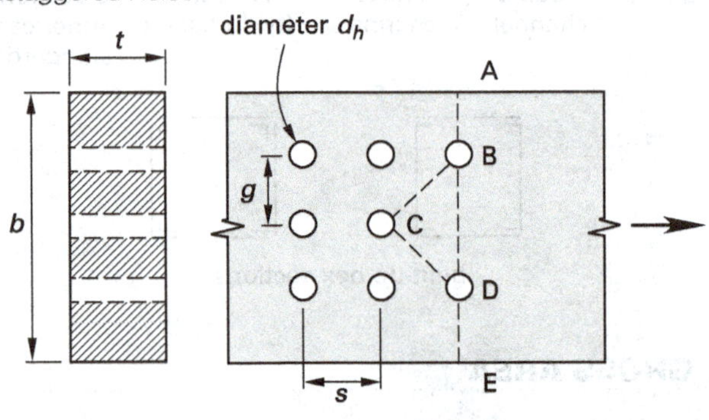

When space limitations (such as a limit on dimension x in Fig. 45.4(a), or odd connection geometry as in Fig. 45.4(b)) make it necessary to use more than one row of fasteners, the reduction in net cross-sectional area can be minimized by using a staggered arrangement of fasteners.

Figure 45.4 *Tension Members with Staggered Holes*

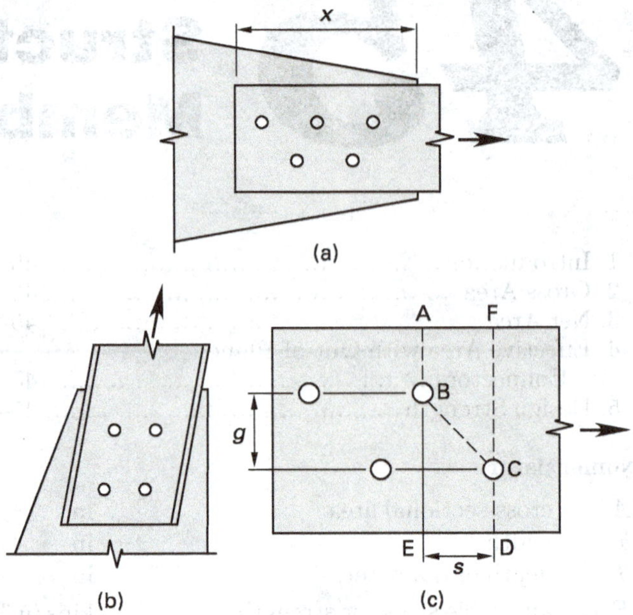

Equation 45.2 Through Eq. 45.4: Net Area

$$d_h = d_b + 1/16'' \qquad 45.2$$

$$A_n = \left[b_g - \sum\left(d_h + \frac{1}{16}''\right)\right]t \quad \text{[parallel holes]} \qquad 45.3$$

$$A_n = \left[b_g - \sum\left(d_h + \frac{1}{16}''\right) + \sum\frac{s^2}{4g}\right]t \quad \text{[staggered holes]} \qquad 45.4$$

Description

The net area for tension members is given by Eq. 45.3, for parallel holes, and by Eq. 45.4, for staggered holes.

Punched bolt holes should be $1/16$ in larger than nominal fastener dimensions. (This is sometimes referred to as a *standard hole* [*AISC Specification* Sec. B2].) The *effective hole diameter*, d_h, is given by Eq. 45.2.

When holes are drilled or punched, some of the material immediately adjacent to the hole may be damaged. AISC *Specification* B4.3b takes this into consideration by specifying that, when computing net area for tension and shear, the nominal diameter of the hole should be increased by $1/16$ in. (The term "nominal" is used by AISC for both the bolt and hole diameters.) This does not change the definition of a standard hole, but it does affect the calculation of net area.

$$d = d_h + \frac{1}{16}\text{ in} = d_b + \frac{1}{8}\text{ in}$$

The net area is the member thickness, t, times the net width. For multiple rows of fasteners (staggered or not) at a connection, more than one potential fracture line may exist. For each possible fracture line, the *net width*, b_n, is calculated by subtracting the hole diameters, d_h, in the potential failure path from the gross width, b_g, and then adding a correction factor, $s^2/4g$, for each diagonal leg in the failure path. (There is one diagonal leg, BC, in the failure path ABCD shown in Fig. 45.4(c), so the term $s^2/4g$ would be added once.)

Example

A steel flat bar tension member has a tensile yield strength of 50 ksi and an ultimate strength of 65 ksi. Holes for a bolted connection are punched at the end of the member. The effective hole diameter is 1.0 in.

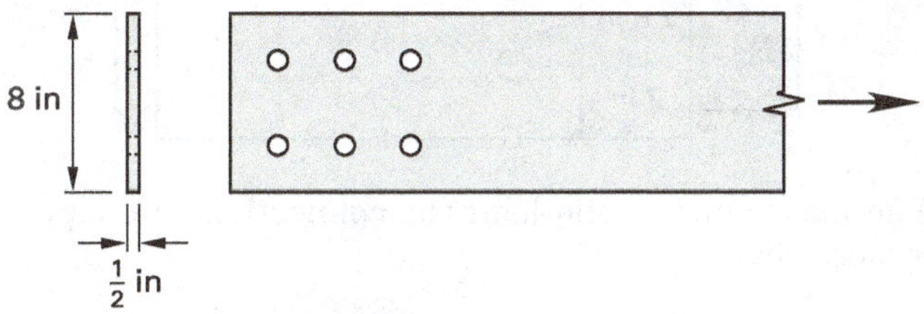

The net tensile area is most nearly

(A) 1.0 in²

(B) 2.5 in²

(C) 3.0 in²

(D) 4.0 in²

Solution

Tensile failure in a tension member with multiple parallel rows of (not staggered) adjacent holes will usually be along a path through adjacent holes across the member. From Eq. 45.3, the net area is

$$\begin{aligned} A_n &= \left[b_g - \sum\left(d_h + \frac{1}{16}''\right)\right]t \\ &= \left(8.0 \text{ in} - (2 \text{ holes})\left(1.0 \ \frac{\text{in}}{\text{hole}} + \frac{1}{16} \text{ in}\right)\right)(0.5 \text{ in}) \\ &= 2.94 \text{ in}^2 \quad (3.0 \text{ in}^2) \end{aligned}$$

The answer is (C).

4. EFFECTIVE AREA WITH OUT-OF-PLANE CONNECTORS

When a tension member frames into a supporting member, some of the load-carrying ability will be lost unless all connectors are in the same plane and all elements of the tension member are connected to the support. (An angle connected to its support by only one of its legs is a case in which load-carrying ability would be lost.) A *shear lag reduction coefficient*, U, is used to calculate the *effective net area.* U can be taken as 1.0 if all cross-sectional elements are connected to the support to transmit the tensile force.

Equation 45.5: Effective Area (Bolted and Welded Members)

$$A_e = UA_n \qquad 45.5$$

Values

Bolted members

- flat bars: $U = 1.0$
- angles: $U = 1 - \bar{x}/L$

Welded members

- flat bars or angles with transverse welds: $U = 1.0$
- flat bars of width w with longitudinal welds of length L:

$$\begin{aligned} U &= 1.0 \quad [L \geq 2w] \\ U &= 0.87 \quad [2w > L \geq 1.5w] \\ U &= 0.75 \quad [1.5w > L > w] \end{aligned}$$

- angles with longitudinal welds only: $U = 1 - \bar{x}/L$

Description

When the load is transmitted by bolts through only some of the cross-sectional elements of the tension member, the *effective net area*, A_e, is computed from Eq. 45.5 [*AISC Specification* Eq. D3-1]. Values of U for various configurations are given. $\bar{x}$ is the connection eccentricity, equal to the distance from the plane of the connection to the centroid of the connected member, often tabulated in shape tables. (See *AISC Commentary* Fig. C-D3.1 for guidance on establishing $\bar{x}$.) For bolted members, L is the length of the connection in the direction of loading, measured for bolted connections as the center-to-center distance between the two end connectors in a line.

Bolted tension elements rarely have efficiencies greater than 85%, even when holes represent a small percentage of the gross area. Therefore, the net area used in Eq. 45.5 may not exceed 85% of A_g.

Example

A bolted steel tension member with flat bars has a net area of 2 in². What is most nearly the effective net area in tension for this plate?

(A) 1.3 in²

(B) 1.5 in²

(C) 1.7 in²

(D) 2.0 in²

Solution

Since the steel member is bolted with flat bars, the shear lag reduction coefficient is 1.0. The effective net area is

$$A_e = UA_n = (1.0)(2\ \text{in}^2) = 2.0\ \text{in}^2$$

The answer is (D).

5. DESIGN STRENGTH

The design axial strength of a tensile member is calculated from the nominal strength, T_n. The design strength must be greater than the required strength.

$$[\text{LRFD; design strength}]\quad \phi_t T_n \geq T_u$$

The value of ϕ_t depends on the nature of the tensile area under consideration.

For calculations of nominal strength based on gross area, A_g, $\phi_t = \phi_y = 0.90$, for yielding. For calculations involving nominal strength based on net area, A_n, or effective net area, A_e, $\phi_t = \phi_f = 0.75$ for fracture.

Equation 45.6 and Eq. 45.7: Design Tensile Strength

$$\phi T_n = \phi_y F_y A_g\quad [\text{yielding}] \qquad 45.6$$

$$\phi T_n = \phi_f F_u A_e\quad [\text{fracture}] \qquad 45.7$$

Variations

$$P_n = A_g F_y \quad \left[\begin{array}{l}\text{yielding criterion;}\\ \textit{AISC Specification}\ \text{Eq. D2-1}\end{array}\right]$$

$$P_n = A_e F_u = UA_n F_u \quad \left[\begin{array}{l}\text{fracture criterion;}\\ \textit{AISC Specification}\ \text{Eq. D2-1}\end{array}\right]$$

Description

Equation 45.6 gives the available tensile strength considering failure from yielding.

Equation 45.7 gives the available tensile strength considering failure from fracture.

Example 17

A long tensile member is constructed by connecting two plates as shown. Each plate is ½ in × 9 in, A572 grade-50 steel with an ultimate strength of 65 ksi. The fasteners are ½ in diameter bolts. Disregard the shear strength of the bolts, and disregard eccentricity.

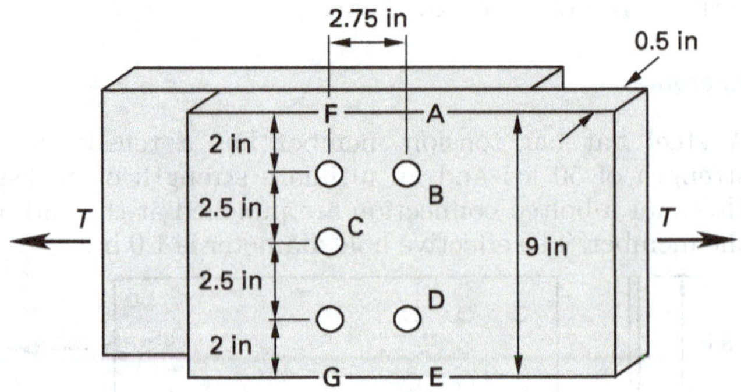

The maximum tensile load the connection can support is most nearly

(A) 140 kips

(B) 150 kips

(C) 170 kips

(D) 200 kips

Solution

The gross area of the plate is

$$A_g = b_g t = (9\ \text{in})(0.5\ \text{in}) = 4.5\ \text{in}^2$$

The effective hole diameter includes 1/16 in allowance for clearance and manufacturing tolerances.

$$d_h = 0.5\ \text{in} + \frac{1}{16}\ \text{in} = 0.5625\ \text{in}$$

The net area of the connection must be evaluated in three ways: paths ABDE, ABCDE, and FCG. Path ABDE does not have any diagonal runs. The net area is

$$\begin{aligned}A_{n,\text{ABDE}} &= \left(b_g - \sum\left(d_h + \frac{1}{16}''\right) + \sum\frac{s^2}{4g}\right)t\\ &= \left(\begin{array}{l}9\ \text{in} - (2)\\ \quad\times\left(0.5625\ \text{in} + \dfrac{1}{16}\ \text{in}\right)\end{array}\right)(0.5\ \text{in})\\ &= 3.875\ \text{in}^2\end{aligned}$$

To determine the net area of path ABCDE, the quantity $s^2/4g$ must be calculated. The *pitch*, s, is 2.75 in, and the *gage*, g, is 2.5 in.

$$\frac{s^2}{4g} = \frac{(2.75\ \text{in})^2}{(4)(2.5\ \text{in})} = 0.756\ \text{in}$$

There are two diagonals in path ABCDE. Using Eq. 45.4, the net area is

$$A_{n,\text{ABCDE}} = \left(b_g - \sum\left(d_h + \frac{1}{16}''\right) + \sum\frac{s^2}{4g}\right)t$$

$$= \left(\begin{array}{l} 9\text{ in} - (3)\left(0.5625\text{ in} + \dfrac{1}{16}\text{ in}\right) \\ \quad + (2\text{ diagonals})\left(0.756\,\dfrac{\text{in}}{\text{diagonal}}\right)\end{array}\right) \times (0.5\text{ in})$$

$$= 4.381\text{ in}^2$$

The net area of path FCG is

$$A_{n,\text{FCG}} = \left(b_g - \sum\left(d_h + \frac{1}{16}''\right) + \sum\frac{s^2}{4g}\right)t$$

$$= \left(\begin{array}{l} 9\text{ in} - (3) \\ \quad \times \left(0.5625\text{ in} + \dfrac{1}{16}\text{ in}\right)\end{array}\right)(0.5\text{ in})$$

$$= 3.5625\text{ in}^2$$

The smallest area is $A_{n,\text{FCG}}$, which is less than 85% of A_g. The design tensile strength, ϕT_n, of the connection based on the gross section is found with Eq. 45.6. ϕ is 0.90.

$$\phi T_n = \phi F_y A_g$$

$$= (0.90)\left(50\,\frac{\text{kips}}{\text{in}^2}\right)(4.5\text{ in}^2)$$

$$= 202.5\text{ kips}$$

Since all of the connections are in the same plane, $U = 1.0$. Based on the net section, the design tensile strength is determined from Eq. 45.5. ϕ is 0.75.

$$A_e = UA_n = (1.0)(3.5625\text{ in}^2)$$

$$= 3.5625\text{ in}^2$$

$$\phi T_n = \phi F_u A_e$$

$$= (0.75)\left(65\,\frac{\text{kips}}{\text{in}^2}\right)(3.5625\text{ in}^2)$$

$$= 173.6\text{ kips} \quad (170\text{ kips})$$

The design tensile strength is the smaller value of 170 kips.

The answer is (C).

Equation 45.8: Block Shear Strength

$$\phi T_n = \text{smaller}\begin{cases} 0.75F_u[0.6A_{nv} + U_{bs}A_{nt}] \\ 0.75[0.6F_yA_{gv} + U_{bs}F_uA_{nt}] \end{cases} \qquad 45.8$$

Values

Where the tensile stress is uniform, $U_{bs} = 1.0$, and where the tensile stress is nonuniform, $U_{bs} = 0.5$. $\phi = 0.75$.

Description

For tension members, another possible failure mode exists in addition to yielding and fracture. In this mode, a segment or "block" of material at the end of the member tears out, as shown in Fig. 45.5. Depending on the connection at the end of a tension member, such *block shear failure* can occur in either the tension member itself or in the member to which it is attached (e.g., a gusset plate). Block shear failure represents a combination of two failures—a shear failure along a plane through the bolt holes or welds and a tension failure along a perpendicular plane—occurring simultaneously.

Block shear strength, T_n, is computed from Eq. 45.8 [*AISC Specification* Eq. J4-5].[1] A_{gv} is the gross area subject to shear, A_{nt} is the net area subject to tension, and A_{nv} is the net area subject to shear.

Figure 45.5 *Block Shear Failures*

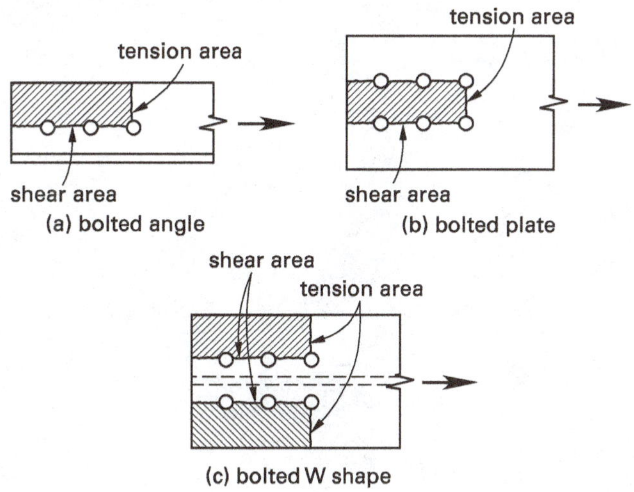

Structural Design

[1]Block shear failure is a combination of shear and tension failure. The NCEES *FE Reference Handbook* uses its own symbol, T_n, for the nominal block shear rupture strength, which might be interpreted as indicating failure is by tension. *AISC Specification* J4-5 uses the standard nomenclature of R_n for this quantity.

46 Structural Steel: Beam-Columns

Nomenclature

A	cross-sectional area	in^2
b	shape factor	(in-kips)$^{-1}$
B_1	second order effect flexural amplifier	–
C_m	lateral load coefficient	–
E	modulus of elasticity	kips/in^2
I	moment of inertia	in^4
K	effective length factor	–
L	live load	kips
M	moment	in-kips
M_c	available flexural strength	in-kips
M_r	required flexural strength	in-kips
p	shape factor	kips^{-1}
P	axial load	kips
P_c	available axial strength	kips
P_r	required axial strength	kips
S	elastic section modulus	in^3
Z	plastic section modulus	in^3

Symbols

α	factor used in calculating B_1	–
Δ	deflection	in
ϕ	resistance factor	–

Subscripts

b	bending or unbraced
c	available
e	Euler
n	nominal
nt	no translation
p	bearing or plastic
r	required
u	ultimate
x	strong axis
y	weak axis or yield

1. INTRODUCTION

A member that is acted upon by a compressive force and a bending moment is known as a *beam-column*. The bending moment can be due to an eccentric load or a true lateral load.

For pure beams and columns, second-order effects are neglected. Only first-order effects based on undeformed geometry need to be considered (i.e., for a member subjected to a moment only, or when the same member carries an axial compressive load alone).

When a member is acted upon by both bending moment and an axial load, the two stresses cannot be added directly to obtain the so-called combined stresses. Additional stresses resulting from a secondary moment must be taken into account, especially when the axial compressive load is large. The secondary moment is caused by the *P-Δ effect* and has a magnitude of $P\Delta$. The deflection, Δ, results from the initial lateral deflection due to the bending moment, but causes further bending of its own and produces secondary stresses.

The secondary moment and the associated stress are known as *second-order effects*, since they are dependent on the deformed geometry of the member. *Moment amplification factors* are used to account for the second-order effects.

Design and analysis of beam-columns generally attempt to transform moments into equivalent axial loads (i.e., the *equivalent axial compression method*) or make use of *interaction equations*. Interaction-type equations are best suited for beam-column analysis and validation, since so much (e.g., area, moment of inertia) needs to be known about a shape. Equivalent axial compression methods are better suited for design.

2. FLEXURAL/AXIAL COMPRESSION

Strength Limit States

The following equations give the flexural/compressive force relationships (also known as flexural/force *interaction equations*) for beam-columns where there is no sidesway, bending occurs only around the x-axis, there are no end moments, and ends are pinned, allowing rotation in the bending plane. ϕP_n is the compression strength with respect to the weak axis (y-axis). ϕM_n is the bending strength with respect to the bending axis (x-axis).

For large compression, where $P_r/\phi P_n \geq 0.2$, the first equation applies. For small compression, where $P_r/\phi P_n < 0.2$, the second equation applies.

$$\frac{P_r}{\phi P_n} \geq 0.2\text{:}\quad \frac{P_r}{\phi P_n} + \frac{8}{9}\frac{M_r}{\phi M_{nx}} \leq 1.0$$

$$\frac{P_r}{\phi P_n} < 0.2\text{:}\quad \frac{P_r}{2(\phi P_n)} + \frac{M_r}{\phi M_{nx}} \leq 1.0$$

The following variations are more general flexural/compressive force relationships, as presented in *AISC Specification* Sec. H1.1. For $P_r/P_c \geq 0.2$, AISC Eq. H1-1a applies and for $P_r/P_c < 0.2$, AISC Eq. H1-1b applies.

$$P_c = \phi P_n$$

$$\frac{P_r}{P_c} + \left(\frac{8}{9}\right)\left(\frac{M_{rx}}{M_{cx}} + \frac{M_{ry}}{M_{cy}}\right) \leq 1.0 \quad \text{[AISC Eq. H1-1a]}$$

$$\frac{P_r}{2P_c} + \frac{M_{rx}}{M_{cx}} + \frac{M_{ry}}{M_{cy}} \leq 1.0 \quad \text{[AISC Eq. H1-1b]}$$

Example 18

An A992 W14 × 48 member has a compressive strength requirement of 170 kips and a moment requirement of 32.5 ft-kips. The effective length with respect to the weak axis is 9 ft. The beam-column is subjected to

(A) small compression and is adequate

(B) large compression and is adequate

(C) small compression and is inadequate

(D) large compression and is inadequate

Solution

From Table 44.3, ϕP_n is 504 kips. From Table 43.2, ϕM_{nx} is 184 ft-kips.

$$\frac{P_r}{\phi P_n} = \frac{170 \text{ kips}}{504 \text{ kips}} = 0.337$$

Since $P_r/\phi P_n = 0.337 > 0.2$, the member is subjected to large compression.

$$\frac{P_r}{\phi P_n} + \frac{8}{9}\frac{M_r}{\phi M_{nx}} = \frac{170 \text{ kips}}{504 \text{ kips}} + \left(\frac{8}{9}\right)\left(\frac{32.5 \text{ ft-kips}}{184 \text{ ft-kips}}\right) = 0.494 \leq 1.0$$

The beam-column is adequate.

The answer is (B).

3. SECOND-ORDER EFFECTS (DEFLECTION)

In frames and structures subject to sway, the *effect of deflection* (i.e., *second-order effects*) on the required strengths of members, M_r, subjected to both axial and bending loads is given by

$$M_r = B_1 M_{nt}$$

$$B_1 = \frac{C_m}{1 - \frac{P_r}{P_{e1}}}$$

B_1 is the *flexural amplification factor* (*flexural magnifier*) and is applied to the required elastic nominal moment strength. P_r and M_r are the required axial and flexural strengths, respectively, calculated from the applied loads and the appropriate load factors. P_{nt} is the required first-order axial design strength assuming no translation (no sidesway, the restrained case). M_{nt} is the required first-order moment design strength assuming no translation. B_1 is not applied to the axial strength, the required strength of which is given by

$$P_r = P_{nt}$$

P_{e1} is the *Euler buckling load* calculated with $K = 1$ for the end restraint factor in the plane of buckling and assuming zero sidesway.[1] (See *AISC Specification* Sec. C2.)

$$P_{e1} = \frac{\pi^2 EI}{(KL)_x^2}$$

$C_m = 1.0$ for beam-columns where there is no sidesway, bending occurs only around the x-axis, there are no end moments, applied moments come from transverse loading (such as wind), and ends are pinned, allowing rotation in the bending plane. For steel, the modulus of elasticity, E, is 29,000 kips/in^2.

Example 19

A steel W14 × 132, 50 ksi member carries an axial live load of 200 kips and a live moment of 250 ft-kips about the strong axis, caused by lateral loading. The unsupported

[1]The numeral 1 in the subscript refers to the use of $K = 1$ when calculating the Euler buckling load.

length is 20 ft. $K = 1$, and $I_x = 1530$ in^4. The x-x axis flexural magnifier for the member is most nearly

(A) 1.0

(B) 1.2

(C) 1.5

(D) 1.7

Solution

Find the Euler buckling load.

$$P_{e1} = \frac{\pi^2 EI}{(KL)_x^2} = \frac{\pi^2\left(29{,}000\ \frac{\text{kips}}{\text{in}^2}\right)(1530\ \text{in}^4)}{\left((1)(20\ \text{ft})\left(12\ \frac{\text{in}}{\text{ft}}\right)\right)^2} = 7603\ \text{kips}$$

The required strength is

$$P_r = (1.6)(200\ \text{kips}) = 320\ \text{kips}$$

$C_m = 1.0$ for this condition. The x-x axis flexural magnifier is

$$B_1 = \frac{C_m}{1 - \frac{P_r}{P_{e1}}} = \frac{1.0}{1 - \frac{320\ \text{kips}}{7603\ \text{kips}}} = 1.044 \quad (1.0)$$

The answer is (A).

4. ANALYSIS OF BEAM-COLUMNS

The following procedure can be used for beam-column analysis.

step 1: Use appropriate load combinations to calculate the required axial compression strength, P_r.

step 2: Calculate the slenderness ratios for both bending modes.

step 3: Based on the larger slenderness ratio, determine the design axial compressive strength, ϕP_n.

step 4: Calculate the ratio $P_r/\phi P_n$.

step 5: Calculate the moment magnifier, B_1.

step 6: Calculate the required flexural strength, M_{rx}.

step 7: Calculate the design flexural strength, M_{cx}.

step 8: Determine the adequacy of the design using the applicable interaction equation. (See Sec. 46.2.)

47 Structural Steel: Connectors

Nomenclature

A	nominal area	in^2
d	diameter	in
F	allowable stress or strength	kips/in^2
L_c	clear distance between edge of hole and edge of adjacent hole	in
L_e	distance from center of end hole to end edge of member	in
r_n	slip resistance per bolt	kips/in
s	center-to-center spacing of interior holes	in

Symbols

μ	slip (friction) coefficient	–
ϕ	resistance factor	–

Subscripts

b	bolt
h	hole
n	nominal
u	ultimate

1. INTRODUCTION

There are two types of structural *connectors* (also referred to as *fasteners*): rivets and bolts. *Rivets*, however, have become virtually obsolete because of their low strength, high cost of installation, installed variability, and other disadvantages. The two types of *bolts* generally used in the connections of steel structures are common bolts and high-strength bolts. *Common bolts* are classified by ASTM as A307 bolts. The two basic types of *high-strength bolts* have ASTM designations of A325 and A490. There are also twist-off bolt types that are equivalent to A325 and A490: F1852 is equivalent to A325, and F2280 is equivalent to A490. Figure 47.1 shows typical bolted connections.

A *concentric connection* is one for which the applied load passes through the centroid of the fastener group. If the load is not directed through the fastener group centroid, the connection is said to be an *eccentric connection.*

At low loading, the distribution of forces among the fasteners is very nonuniform, since friction carries some of the load. However, at higher stresses (i.e., those near yielding), the load is carried equally by all fasteners in the group. Stress concentration factors are not normally applied to connections with multiple redundancy.

In many connection designs, materials with different strengths will be used. The item manufactured from the material with the minimum strength is known as the *critical part*, and the critical part controls the design.

Connections using bolts and rivets are analyzed and designed similarly. Such connections can place fasteners in direct shear, torsional shear, tension, or any combination of shear and tension. In accordance with *AISC Specification* Sec. B3.1, theoretical design by elastic, inelastic, or plastic analysis is permitted.

2. HOLE SPACING AND EDGE DISTANCES

The minimum distance between centers of standard, oversized, or slotted holes is $2\frac{2}{3}$ times the nominal fastener diameter; a distance of $3d$ is preferred [*AISC Specification* Sec. J3.3]. The longitudinal spacing between bolts along the line of action of the force is specified in Sec. J3.5 of the *AISC Specification* and is a function of painted surfaces and material.

The minimum distance from the hole center to the edge of a member is approximately 1.75 times the nominal diameter for sheared edges, and approximately 1.25 times the nominal diameter for rolled or gas-cut edges, both rounded to the nearest $\frac{1}{8}$ in [*AISC Specification* Sec. J3.4]. (For exact values, refer to Table J3.4 in the *AISC Specification.*)

For parts in contact, the maximum edge distance from the center of a fastener to the edge in contact is 12 times the plate thickness or 6 in, whichever is less [*AISC Specification* Table J3.5].

Figure 47.1 *Typical Bolted Connections*

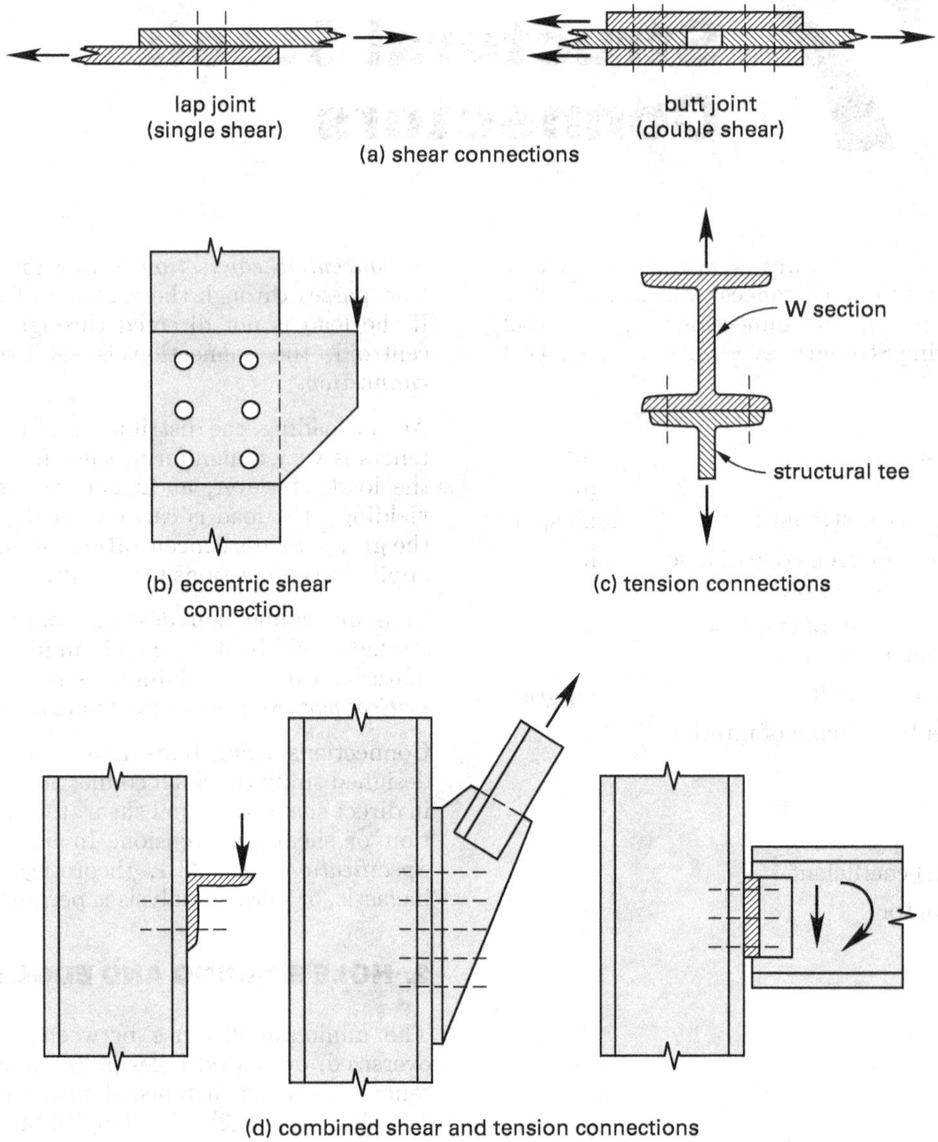

3. BEARING AND SLIP-CRITICAL CONNECTIONS

A distinction is made between bearing and slip-critical connections. A *bearing connection* (which includes the categories of snug-tightened and pretensioned connections) relies on the shearing resistance of the fasteners to resist loading. In effect, it is assumed that the fasteners are loose enough to allow the plates to slide slightly, bringing the fastener shanks into contact with the holes. The area surrounding the hole goes into bearing, hence the name. Connections using rivets, welded studs, and A307 bolts are always considered to be bearing connections. However, high-strength bolts can also be used in bearing connections.

If the fasteners are constructed from high-strength steel, a high preload can be placed on the bolts. This preload will clamp the plates together, and friction alone will keep the plates from sliding. Such connections are known as *slip-critical connections*. The fastener shanks never come into contact with the plate holes. Bolts constructed from A325 and A490 steels are suitable for slip-critical connections.

Bolt type and connection type designations may be combined (e.g., A325SC, A325N, and A325X). A325 refers to the bolt type, while the SC, N, and X refer to the connection type and geometry. SC is used for a slip-critical connection. N is used to indicate that bolt threads are located in the shear plane, while X indicates no bolt threads are in the shear plane. Shear strength is

Table 47.1 Allowable Bolt Strengths

	bolt designation	bolt size (in)		
		$^3/_4$	$^7/_8$	1
tension (kips)	A325	29.8	40.6	53.0
	A307	14.9	20.3	26.5
shear (bearing-type connection, (kips))	A325N	15.9	21.6	28.3
	A325X	19.9	27.1	35.3
	A307	7.95	10.8	14.1
shear (slip-critical connection, serviceability limit state, (kips))	A325SC	11.1	15.4	20.2
shear (slip-critical connection, strength limit state, (kips))	A325SC	9.41	13.1	17.1

reduced if threads are located within the shear plane. See Table 47.1 for the allowable bolt strengths of various bolt designations and sizes.

In a slip-critical connection, the two clamped surface areas in contact are known as *faying surfaces.* (The verb *fay* means to join tightly.) *Coated faying surfaces* include surfaces that have been primed, primed and painted, and protected against corrosion, but not galvanized. *Galvanized faying surfaces* are specifically referred to by name. The condition of the faying surfaces affects the frictional shear force that can develop. The friction coefficient, μ, is referred to as the *slip coefficient* and is the ratio of the total frictional shear load to the normal clamping force.

Faying surfaces are categorized as Class A and Class B. (A previous designation, Class C for roughened galvanized surfaces, is obsolete, and such surfaces have been merged into Class A.) *Class A surfaces* include unpainted clean mill scale surfaces, blast-cleaned (wire-brushed or sand-blasted) surfaces with Class A coatings (including phosphate conversion and zinc silicate paint), and galvanized surfaces that have been roughed by hand wire brushing. Power wire brushing is not permitted. *Class B surfaces* include unpainted blast-cleaned steel surfaces and blast-cleaned surfaces with Class B coatings. If qualified by testing, many commercial primers meet the requirements for Class B coatings. However, most epoxy top coats do not, and faying surfaces must be taped off when fabrication shops apply epoxy top coats. Corrosion on uncoated blast-cleaned steel surfaces due to normal atmospheric exposure up to a year or more between the time of fabrication and subsequent erection is not detrimental and may actually increase the slip resistance of the joint.

4. AVAILABLE BEARING STRENGTH

Bearing strength should be evaluated in bearing connections. Theoretically, bearing should not be a problem in friction-type connections because the bolts never bear on the pieces assembled. However, in the event there is slippage due to insufficient tension in the connectors, bearing should be checked anyway.

Bearing Strength of Connected Member at Bolt Hole

The available bearing strength of a connected member at the bolt hole is given by

$$\phi r_n = \phi 1.2 L_c F_u \leq \phi 2.4 d_b F_u$$

The resistance factor, ϕ, is 0.75 for bearing. The value calculated is per bolt per inch of connection thickness. Values can be read directly from Table 47.1 for various bolt sizes, materials, and connection types. L_c is the clear distance between the edge of hole and the edge of the adjacent hole, or edge of member, in direction of force.

$$L_c = s - d_h \quad \text{[interior hole]}$$

$$L_c = L_e - \frac{d_h}{2} \quad \text{[end hole]}$$

L_e is the end distance (center of end hole to end edge of member). s is the longitudinal hole spacing, in the direction of loading. d_h is found from the following equation.

$$d_h = d_b + \frac{1}{16} \text{ in}$$

Table 47.2 gives the available bearing strength for various bolt sizes and spacing.

Table 47.2 Available Bearing Strength, ϕr_n, at Bolt Holes

	F_u, connected member (ksi)	bolt size (in)		
		$^3/_4$	$^7/_8$	1
$s = 2^2/_3 d_b$ (min.)	58	62.0	72.9	83.7
	65	69.5	81.7	93.8
$s = 3$ in	58	78.3	91.3	101
	65	87.7	102	113
$L_e = 1^1/_4$ in	58	44.0	40.8	37.5
	65	49.4	45.7	42.0
$L_e = 2$ in	58	78.3	79.9	76.7
	65	87.7	89.6	85.9

Structural Design

Example

Flanges of a W8 × 24 grade A36 steel section are to be connected to a steel bracket by $^3/_4$ in diameter bolts. What is the appropriate diameter for the bolt holes?

(A) $^3/_4$ in

(B) $^{13}/_{16}$ in

(C) $^7/_8$ in

(D) 1 in

Solution

The appropriate diameter is

$$d_h = d_b + \frac{1}{16}\text{ in} = \frac{3}{4}\text{ in} + \frac{1}{16}\text{ in} = \frac{13}{16}\text{ in}$$

The answer is (B).

Topic XII: Transportation and Surveying

Chapter

48 Transportation Capacity and Planning

Nomenclature

a	acceleration	ft/sec^2
a	coefficient	–
A_j	number of trips attracted to zone j	–
D	density	pc/mi/ln, veh/mi, veh/mi-ln
E	passenger car equivalent	–
F_{HV}	heavy vehicle factor	–
F_{ij}	adjustment factor	–
F_{LC}	adjustment factor for right-side lateral clearance	mi/hr
F_{LW}	adjustment factor for lane width	mi/hr
F_p	passenger familiarity factor	–
FFS	free flow speed	mi/hr
G	grade	decimal
k	traffic density	veh/mi
K_{ij}	socioeconomic adjustment factor	–
l	vehicle length	ft
L	expected customers in system	–
L_q	expected customers in queue	–
n	number	–
N	number of lanes	–
P	percentage	%
P	probability	–
PHF	peak hour factor	–
P_I	total number of trips generated in zone i	–
P_n	probability of n customers in system	–
q	traffic volume	veh/hr
r	red clearance interval	sec
s	number of parallel servers	–
S	speed	mi/hr
t	time	sec
T_{ij}	number of trips generated in zone i and attracted to zone j	–
TRD	total ramp density	–
U	utility	–
v	speed	mi/hr
v_p	demand flow rate	pc/hr/ln
V	volume	veh/hr
W	expected time in system (includes service)	sec
W	intersection width	ft
W_q	expected time in queue	sec
X	attribute	–
y	yellow interval length	sec

Symbols

λ	mean arrival rate	1/sec
μ	mean service rate per server	1/sec
ρ	utilization factor (λ/μ)	–

Subscripts

0	initial
A	auto
f	final, first, or free flow
j	jam
m	maximum
o	optimum or original
O	optimum
R	recreational vehicles
T	transit
T	trucks/buses

1. TRANSPORTATION PLANNING

Transportation planning involves travel demand analysis and forecasting and utilizes mathematical models to predict the volume of trips between activity centers. The Federal Highway Administration (FHWA) has a standardized process for all regions with a population of greater than 50,000.[1] The FHWA works with a region's existing plans for land use, environmental, or economic development to establish a transportation system's goals (irreducible values) and objectives (attainable targets). From these goals and objectives, the movements of people and products are mathematically modeled and calibrated to historical patterns prevalent in the region in order to project future traffic patterns. Alternative strategies for managing travel demand are proposed, tested, and evaluated using the modeled projections and the goals and objectives. Changes are made as needed to produce an optimal strategy to implement the proposed transportation system. The costs of alternative strategies are compiled, and a draft report is prepared for public comment. Simultaneously, a *draft environmental impact statement* (DEIS), commonly known as an *environmental impact report* (EIR), is prepared using the possible strategies to determine the projected environmental effects of each alternative. After a period of public input and assessment by the appropriate decision makers, a strategy is selected, the funding stream is secured, and the system's final design begins. Traffic volumes along planned travel paths, which were developed from the travel demand modeling, are used to design the facility to meet the required capacity.

Transportation planning is a process that persists over the life of a project, rather than a single task completed at the start of a project. The process allows transportation planners to make necessary improvements as the project evolves. There are certain considerations, however, that transportation planners must always keep in mind when planning a new project. For example, federal legislation requires that transportation planners follow certain environmental guidelines in every new project. Several important acts pertaining to transportation planning include the 1991 *Intermodal Surface Transportation Efficiency Act* (ISTEA), the 1998–2003 *Transportation Equity Act for the 21st Century* (TEA-21), and the 2005–2009 *Safe, Accountable, Flexible, Efficient Transportation Equity Act: A Legacy for Users* (SAFETEA-LU). Many of the environmental considerations in these acts are based on the *Clean Air Act*,[2] which places responsibility on the states to set pollution limits. The challenge transportation planning often faces is decreasing pollution while devising strategies that will meet the capacity needs of future transport.

2. TRAFFIC STUDIES

A *travel time and delay study* assesses the time it takes a vehicle to travel between two points along a given route (referred to as the *study segment*). The data collected from these studies are used by traffic engineers and analysts to identify a route's problem locations and to determine design or operational improvements that will better facilitate traffic flow.

Traffic studies are most frequently conducted on major arterials leading to and from a region's *central business district* (CBD), but a study segment can include any type of roadway.

Traffic volume studies determine the numbers, movements, and classifications of vehicles (and/or bicycles and pedestrians) at specific locations and times. They are essential for many types of traffic analysis, such as determining the influence of pedestrians on traffic flow or calculating annual average daily traffic. Volumes are usually counted over a set time period, which can range from a few minutes to more than one year. The count period should represent the typical conditions (such as time of day, day of month, and month) applicable to the desired analysis. For example, studies at schools should not usually be conducted on weekends.

The tools and equipment used to determine the traffic volume can be as simple as tally sheets or manual count boards or as complex as electronic detectors and computer data generation programs. Regardless of the tools used, the purpose of a traffic volume study remains the same: to accurately establish the actual amount of traffic passing a given location during a unit of time.

3. TRIP GENERATION

Trip generation requires estimating the number of trips that will result from a particular population or occupancy. Estimates can be obtained from general or area-specific tables or correlations. There are many general trip generation correlations that can be used if specific, targeted data are not available. Since predictive formulas are highly dependent on the characteristics of the local area and population, as well as on time of day and year, numerous assumptions must be made and verified before such correlations are relied upon.

[1]Local and state governments set standards for areas with populations of fewer than 50,000.

[2]The *Clean Air Act* was enacted by Congress in 1990 and defines the Environmental Protection Agency's (EPA's) responsibilities for protecting and improving the nation's air quality. The EPA sets air pollution limits for the United States, and states are prohibited from having less stringent pollution controls than those set for the nation. However, while the EPA sets the national standards, states are responsible for carrying out the Act, as combating pollution requires regional knowledge.

4. TRIP DISTRIBUTION MODELS

Trip distribution estimates the number of trips between study regions (*zones*). Trip distribution models are made for various trip purposes using trip generation data. Two common trip distribution models are the gravity model and the logit model.

Equation 48.1: Gravity Model

$$T_{ij} = P_i\left[\frac{A_{ij}F_{ij}K_{ij}}{\sum_j A_j F_{ij} K_{ij}}\right] \qquad 48.1$$

Description

The *gravity model*, named after Newton's theory of gravity, makes two main assumptions: (1) trips generated from one zone will be proportional to trips ending in another zone, and (2) the number of trips between two zones is inversely proportional to the distance between zones.

Equation 48.1 identifies the zone that generates trips as zone i and the zone that attracts trips as zone j. T_{ij} is the number of trips from zone i to zone j and is calculated from the trips generated in zone i, P_i, and the trips attracted to zone j, A_{ij}. Factors accounting for the inverse proportionality of trips and distance, F_{ij}, and socioeconomic factors, K_{ij}, are applied to the trip generation-attraction data.

Equation 48.2 Through Eq. 48.4: Logit Models

$$U_x = \sum_{i=1}^{n} a_i X_i \qquad 48.2$$

$$P(x) = \frac{e^{U_x}}{\sum_{e=1}^{n} e^{U_{xi}}} \qquad 48.3$$

$$P(A) = \frac{e^{U_A}}{e^{U_A} + e^{U_T}} \qquad 48.4$$

Description

Logit models determine the utility of a mode of transportation. *Utility* is the satisfaction or value a user receives from using a mode of transportation and is calculated from Eq. 48.2. X_i represents various attribute values i, such as time or cost. Coefficient values, a_i, for those attributes are used to weight the attribute values. *Disutilities* have negative coefficient values.

The probability of choosing one mode, x, over another can be calculated from Eq. 48.3. Equation 48.4 is a more specific version of Eq. 48.3 and calculates the probability of choosing to drive (*auto mode*), A, over taking public transit, T.

5. TRANSPORTATION FACILITIES TERMINOLOGY

Although common usage does not always distinguish between highways, freeways, and other types of roadways, the major traffic/transportation references are more specific. For example, "vehicle" and "car" have different meanings. "Vehicle" encompasses trucks, buses, and recreational vehicles, as well as passenger cars.

A *freeway* is a divided corridor with at least two lanes in each direction that operates in an *uninterrupted flow* mode (i.e., without *fixed elements* such as signals, stop signs, and at-grade intersections). *Access points* are limited to ramp locations. Since grades, curves, and other features can change along a freeway, performance measures (e.g., capacity) are evaluated along shorter *freeway segments*.

Multilane and two-lane *highways*, on the other hand, contain some fixed elements and access points from at-grade intersections, though relatively uninterrupted flow can occur if signal spacing is greater than 2 mi. Where signal spacing is less than 2 mi, the roadway is classified as an *urban street* (or *arterial*), and flow is considered to be interrupted. An urban street has a significant amount of driveway access, while an arterial does not. *Divided highways* have separate roadbeds for the opposing directions, whereas *undivided highways* do not.

Smaller roadways are classified as *local roads* and *streets*. All roadways can be classified as *urban*, *suburban*, or *rural*, depending on the surrounding population density. *Urban areas* have populations greater than 5000. *Rural areas* are outside the boundaries of urban areas.

6. TRAFFIC CAPACITY ANALYSIS

Capacity planning analysis is used to estimate the traffic carrying ability of transportation network segments at prescribed levels of operation. Criteria for operational levels are characterized by a level of service. A *level of service* (LOS) is a user's quality of service through or over a specific facility (e.g., over a highway, through an intersection, across a crosswalk). Levels of service are designated A through F. Level A represents unimpeded flow, which is ideal but only possible when the volume of traffic is small. Level F represents a highly impeded, packed condition. Generally, level E will have the maximum flow rate (i.e., capacity).

Equation 48.5: Free Flow Speed

$$\text{FFS} = 75.4\ \frac{\text{mi}}{\text{hr}} - f_{\text{LW}} - f_{\text{LC}} - 3.22\text{TRD}^{0.84} \qquad 48.5$$

Description

A basic freeway segment is a length of freeway between merging, diverging, or weaving areas. Equation 48.5 gives the free flow speed, FFS. The *free flow speed* is the theoretical speed (mi/hr) of traffic flow in a given freeway segment during low-volume conditions when traffic density is zero. Free flow speed is determined by taking the base free flow speed, given as 75.4 mi/h for basic freeway segments, and subtracting adjustments for lane width, f_{LW}; *right-side lateral clearance*, f_{LC}; and *total ramp density*, TRD. The adjustments for lane width and right-side lateral clearance are both found from tables in the "Basic Freeway Segment Highway Capacity" section of the *NCEES Handbook*; adjustments for lane width are required only for lane widths less than 12 ft, and adjustments for right-side lateral clearance are required only for shoulder widths less than 6 ft. The adjustment for total ramp density is the ramp density along the basic freeway segment in ramps/mi. Free flow speed is always rounded to the nearest 5 mi/hr.

Example

A freeway segment has three 11 ft wide lanes in each direction. The segment has a 5 ft right side lateral clearance, a ramp density of 2 ramps per mile, and a peak hour factor of 0.90. Most nearly, what is the free flow speed of the freeway segment?

(A) 65 mi/h

(B) 69 mi/h

(C) 70 mi/h

(D) 75 mi/h

Solution

From the *NCEES Handbook*, a freeway segment with 11 ft lane widths requires an adjustment for lane width of 1.9 mi/hr, and a 5 ft right side lateral clearance with 3 lanes in each direction requires an adjustment of 0.4 mi/hr. From Eq. 48.5,

$$\begin{aligned}\text{FFS} &= 75.4\ \frac{\text{mi}}{\text{hr}} - f_{\text{LW}} - f_{\text{LC}} - 3.22\text{TRD}^{0.84} \\ &= 75.4 - 1.9\ \frac{\text{mi}}{\text{hr}} - 0.4\ \frac{\text{mi}}{\text{hr}} - (3.22)(2)^{0.84} \\ &= 69.02\ \text{mi/hr} \quad (69\ \text{mi/hr})\end{aligned}$$

The answer is (B).

Equation 48.6: Demand Flow Rate

$$v_p = \frac{V}{\text{PHF} \times N \times f_{\text{HV}} \times f_p} \qquad 48.6$$

Description

The *demand flow rate*, v_p, is also referred to as the *15 minute passenger car equivalent flow rate*, expressed in passenger cars per hour per lane (pc/hr/ln). The volume, V, is expressed in vehicles/hr for each travel direction. The peak hour factor, PHF, is calculated using Eq. 48.10. N is the number of lanes in each travel direction. The heavy vehicle factor, f_{HV}, is calculated using Eq. 48.7. The *driver familiarity factor*, f_p, is taken as 1.0 for familiar drivers unless a lower value is provided with sufficient evidence. In practical situations, a value of 1.0 should be used, reflecting daily commuters or drivers accustomed to the area. In situations where greater accuracy is required, field studies comparing commuter and recreational traffic flow characteristics should be used.

Example

A freeway segment has three 12 ft lanes in each direction. The segment has a traffic volume of 1300 veh/hr, a peak hour factor of 0.75, and a heavy vehicle factor of 0.90. Studies suggest a driver familiarity factor of 0.75 be used. Most nearly, what is the demand flow rate of the freeway segment?

(A) 400 pc/hr/ln

(B) 600 pc/hr/ln

(C) 750 pc/hr/ln

(D) 860 pc/hr/ln

Solution

From Eq. 48.6,

$$\begin{aligned}v_p &= \frac{V}{\text{PHF} \times N \times f_{\text{HV}} \times f_p} \\ &= \frac{1300\ \frac{\text{veh}}{\text{hr}}}{(0.75)(3)(0.90)(0.75)} \\ &= 856\ \text{pc/hr/ln} \quad (860\ \text{pc/hr/ln})\end{aligned}$$

The answer is (D).

Equation 48.7: Heavy Vehicle Adjustment Factor

$$f_{\text{HV}} = \frac{1}{1 + P_T(E_T - 1) + P_R(E_R - 1)} \qquad 48.7$$

Description

The *heavy vehicle factor*, f_{HV}, given by Eq. 48.7, is a way to equate trucks, buses, and recreational vehicles to passenger cars, and is a function of the percentage of trucks, buses, and recreational vehicles on the freeway segment. The percentage of trucks/buses, P_T, and percentage of recreational vehicles, P_R, are functions of the prevailing traffic conditions. The passenger car equivalents of trucks and buses, E_T, and recreational vehicles, E_R, are functions of roadway grades and type of terrain, and are determined using the table of passenger car equivalents in the *NCEES Handbook.* For the purposes of calculating the heavy vehicle factor, light delivery vans do not count as trucks.

Example

A freeway segment built through mountainous terrain has two 12 ft wide lanes in each direction. 10% of the vehicles on the segment are trucks or buses, and 3% of the vehicles on the freeway are recreational vehicles. Most nearly, what is the heavy vehicle factor?

(A) 0.54

(B) 0.69

(C) 0.73

(D) 0.81

Solution

From the *NCEES Handbook*, in mountainous terrain, the passenger car equivalent for trucks/buses is 4.5, and the passenger car equivalent for recreational vehicles is 4.0. From Eq. 48.7,

$$\begin{aligned} f_{HV} &= \frac{1}{1+P_T(E_T-1)+P_R(E_R-1)} \\ &= \frac{1}{1+(0.10)(4.5-1)+(0.03)(4.0-1)} \\ &= 0.694 \quad (0.69) \end{aligned}$$

The answer is (B).

Equation 48.8: Vehicle Density

$$D = \frac{v_p}{S} \qquad 48.8$$

Description

The *vehicle density*, *D*, given by Eq. 48.8, is the quantity of vehicles within a given freeway segment at a given moment in time, expressed in pc/mi/ln. v_p is the demand flow rate in pc/hr/ln, and *S* is the mean speed of traffic under base conditions in mi/h. The mean speed of traffic is taken as the free flow speed for flow rates under the breakpoint, and is determined by Eq. 48.5.

Example

A freeway segment has 2 lanes of traffic in each direction, with a demand flow rate of 625 pc/hr/ln. The mean speed of travel in the segment is 65 mi/hr. Most nearly, what is the density of the freeway segment?

(A) 9.6 pc/mi/ln

(B) 10.0 pc/mi/ln

(C) 12.0 pc/mi/ln

(D) 13.0 pc/mi/ln

Solution

From Eq. 48.8,

$$\begin{aligned} D &= \frac{v_p}{S} = \frac{625\ \frac{\frac{\text{pc}}{\text{hr}}}{\text{ln}}}{65\ \frac{\text{mi}}{\text{hr}}} \\ &= 9.62\ \text{pc/mi/ln} \quad (9.6\ \text{pc/mi/ln}) \end{aligned}$$

The answer is (A).

7. SPEED, DENSITY, AND VOLUME

Several measures of vehicle speed are used in highway design and capacity calculations. Most measures will not be needed in every capacity calculation. The *design speed* is the maximum safe speed that can be maintained over a specified section of roadway when conditions are so favorable that the design features of the roadway govern. Most elements of roadway design depend on the design speed. The *legal speed* on a roadway section is often set at approximately the *85th percentile speed*, determined by observation of a sizable sample of vehicles. In suburban and urban areas, the legal speed limit is often influenced by additional considerations such as visibility at intersections, the presence of driveways, parking and pedestrian activity, population density, and other local factors.

With uninterrupted flow, the speed of travel decreases as the number of cars occupying a freeway or multilane highway increases (i.e., the volume and density of cars increases). Since speed, density, and volume are closely related, knowing any two values can determine the third. The *volume*, *q*, is the number of vehicles passing a given point in an hour. *Density*, *k*, is the number of vehicles per unit length (miles or kilometers) per lane.

Equation 48.9: Volume-Density-Speed

$$q = kv \qquad 48.9$$

Description

The basic relationship between speed, v, density, k, and volume, q, is given in Eq. 48.9 and illustrated graphically in Fig. 48.1.

Figure 48.1 Flow-Density-Speed Relationships

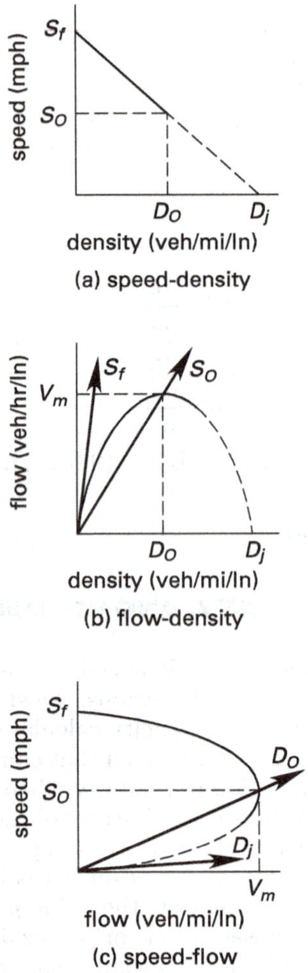

Example

Given the volume-density graph shown, what is the approximate vehicle mean speed at a point where the density is 175 veh/mi and the volume is 10,000 veh/hr?

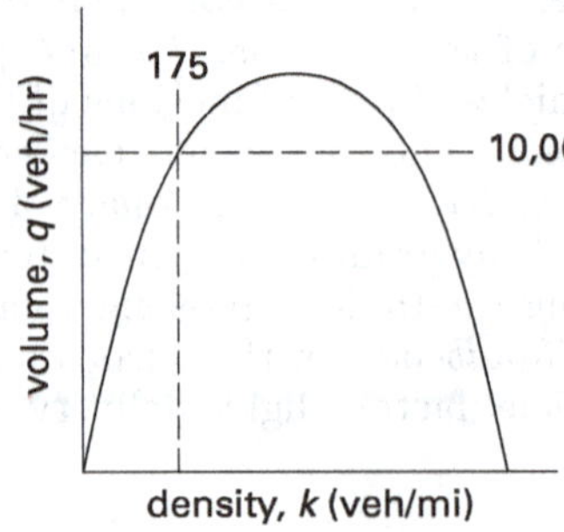

(A) 49 mph

(B) 52 mph

(C) 57 mph

(D) 60 mph

Solution

Rearrange Eq. 48.9 to solve for the speed, v.

$$q = k\text{v}$$

$$\text{v} = \frac{q}{k} = \frac{10{,}000 \ \frac{\text{veh}}{\text{hr}}}{175 \ \frac{\text{veh}}{\text{mi}}} = 57.14 \text{ mi/hr} \quad (57 \text{ mph})$$

The answer is (C).

Equation 48.10: Peak Hour Factor

$$\text{PHF} = \frac{\text{hourly volume}}{\text{hourly flow rate}} = \frac{\text{hourly volume}}{4 \times \text{peak 15 minute volume}} \qquad 48.10$$

Description

Transportation engineers use peak-hour traffic volumes when evaluating traffic parameters, including capacity and level of service. The *peak hour factor*, PHF, given by Eq. 48.10, is a measure of traffic fluctuations within the peak hour of flow. The peak hour factor is the peak hour volume divided by the hourly flow rate. Industry practice uses 15 minute peak intervals; therefore, the hourly flow rate can be taken as four times the peak 15 minute volume within the peak hour. Lower values of PHF are typical of low-volume rural conditions; urban and suburban roads typically have higher values of PHF.

6

Example

15 minute interval traffic counts for a road are shown. Most nearly, what is the peak hour factor?

time	vehicles
4:00–4:15 pm	130
4:15–4:30 pm	120
4:30–4:45 pm	140
4:45–5:00 pm	150
5:00–5:15 pm	160
5:15–5:30 pm	140
5:30–5:45 pm	170
5:45–6:00 pm	180

(A) 0.50

(B) 0.65

(C) 0.80

(D) 0.90

Solution

Tabulate the hourly volumes.

time	hourly volume (vehicles)
4:00–5:00 pm	540
4:15–5:15 pm	570
4:30–5:30 pm	590
4:45–5:45 pm	620
5:00–6:00 pm	650

The peak hour occurs between 5:00 pm and 6:00 pm, with a peak hour volume of 650 vehicles. The peak 15 minute volume within the peak hour is 180 vehicles. From Eq. 48.10, the peak hour factor is

$$\begin{aligned}\text{PHF} &= \frac{\text{hourly volume}}{4 \times \text{peak 15 minute volume}} \\ &= \frac{650 \text{ vehicles}}{(4)(180 \text{ vehicles})} \\ &= 0.903 \quad (0.90)\end{aligned}$$

The answer is (D).

Equation 48.11 Through Eq. 48.14: Traffic Speed and Density Relationships

$$S = S_f - \frac{S_f}{D_j}D \qquad 48.11$$

$$V = S_f D - \frac{S_f}{D_j}D^2 \qquad 48.12$$

$$V_m = \frac{D_j S_f}{4} \qquad 48.13$$

$$D_o = \frac{D_j}{2} \qquad 48.14$$

Description

Traffic models are used to investigate the relationship when one parameter is changed with respect to another. The most important traffic model is the one developed between speed and density. Greenshields proposed a linear speed-density relationship, representing the first and most simple of the speed and density models. This relationship, often referred to as the *Greenshields' model,* is given by Eq. 48.11, where S is the mean speed at density D, S_f is the free flow speed, and D_j is the *jam density.* The Greenshields' model demonstrates that as density approaches zero, speed approaches free flow.

Knowing that flow, V, is equal to the product of speed and density, the relationship between flow and density can be derived. Substituting this relationship into Eq. 48.11 creates a parabolic model, given by Eq. 48.12. The maximum flow, V_m, given by Eq. 48.13 is one-fourth the product of free flow speed and jam density. The optimum density at maximum flow, D_o, given by Eq. 48.14, is sometimes called the *critical density,* and is expressed as half of the jam density.

Example 7

A data set for a freeway segment shows a free flow speed of 55 mi/hr and a jam density of 150 veh/mi. The observed density of the segment is 75 veh/mi. Most nearly, what is the travel speed?

(A) 28 mi/hr

(B) 35 mi/hr

(C) 50 mi/hr

(D) 55 mi/hr

Solution

From Eq. 48.11,

$$S = S_f - \frac{S_f}{D_j} D$$

$$= 55\ \frac{\text{mi}}{\text{hr}} - \left(\frac{55\ \frac{\text{mi}}{\text{hr}}}{150\ \frac{\text{veh}}{\text{mi}}}\right)\left(75\ \frac{\text{veh}}{\text{mi}}\right)$$

$$= 27.5\ \text{mi/hr} \quad (28\ \text{mi/hr})$$

The answer is (A).

8. SIGNALIZED INTERSECTIONS

Signalized intersections are controlled by signals operating in two or more phases. Each phase consists of three intervals: green, amber (i.e., "yellow"), and red. For a typical intersection with two streets crossing at 90° to each other, a *two-phase signal* is one that has one phase for each axis of travel (e.g., one phase of north-south movements and one phase of east-west movements). A *three-phase signal* provides one of the roads with a left-turn phase. In a *four-phase signal*, both roads have left-turn phases.

Level of service for signalized intersections is defined in terms of control delay in the intersection. *Control delay* consists of only the portion of delay attributable to the control facility (e.g., initial deceleration delay, queue move-up time, stopped delay, and final acceleration delay), but not geometric delay or incident delay.

Geometric delay is caused by the need for vehicles to slow down in order to negotiate an intersection. For example, a vehicle would need to slow down in order to enter a separate left-turn lane or negotiate a roundabout, regardless of the traffic already in the intersection. *Queueing delay* is the delay caused by the need to slow down in order to avoid other vehicles already present. All or parts of queueing delay may be included in geometric delay. For example, the time a driver takes to determine if there is traffic that needs to be avoided overlaps both geometric and queueing delay. *Incident delay* occurs when there is an incident (i.e., an accident). Delay can be measured in the field, or it can be estimated.

9. TRAFFIC SIGNAL TIMING AND INTERVALS

Equation 48.15: Yellow Interval

$$y = t + \frac{\text{v}}{2a \pm 64.4G} \qquad 48.15$$

Variations

$$y = t + \frac{\text{v}}{2(g + a)}$$

$$y = t + \frac{1.47\text{v}_{\text{mi/hr}}}{2a \pm 64.4G}$$

Description

The *yellow interval* is the length of time a design-driver, once the signal turns yellow, needs to stop a vehicle before entering the intersection. The driver is assumed to do nothing except think and react during the first t seconds, then to decelerate uniformly to a stop. A downhill grade, G, will increase the time needed, while an uphill grade will decrease the time needed.

Equation 48.15 gives an initial estimate of the yellow interval length, y, calculated from the vehicle approach speed, v, in ft/sec, rate of deceleration, a, in ft/sec^2, driver reaction time, t, and grade, G, expressed as a decimal. Uphill grade is expressed as a positive value (i.e., $2a + 64.4G$), while downhill grade is negative. 64.4 has units of ft/sec^2. The yellow interval length is rounded to the nearest 0.1 sec.

Example

One approach to an intersection has a downhill grade of 2%. Using an average deceleration rate of 10 ft/sec^2 and a driver reaction time of 1 sec, what is most nearly the yellow interval length required to accommodate a vehicle approaching the intersection at 35 mi/hr?

(A) 1.1 sec

(B) 2.9 sec

(C) 3.4 sec

(D) 3.7 sec

Solution

The yellow interval is

$$y = t + \frac{\text{v}}{2a \pm 64.4G}$$

$$= 1\ \text{sec} + \frac{\left(\frac{35\ \frac{\text{mi}}{\text{hr}}}{\left(60\ \frac{\text{sec}}{\text{min}}\right)\left(60\ \frac{\text{min}}{\text{hr}}\right)}\right)\left(5280\ \frac{\text{ft}}{\text{mi}}\right)}{(2)\left(10\ \frac{\text{ft}}{\text{sec}^2}\right) - \left(64.4\ \frac{\text{ft}}{\text{sec}^2}\right)(0.02)}$$

$$= 3.74\ \text{sec} \quad (3.7\ \text{sec})$$

The answer is (D).

Equation 48.16: Red Clearance Interval

$$r = \frac{W + l}{\mathrm{v}} \qquad 48.16$$

Description

The *red clearance interval*, r, also known as the *all-red interval*, is the length of time a design-driver needs to completely clear the intersection. During this period, the signals for all directions will be red.

The red clearance interval is calculated from the intersection width, W, measured curb-to-curb, the vehicle length, l, and the vehicle's approach speed, v. The red clearance interval length is rounded to the nearest 0.1 sec.

Example

A driver approaches the intersection shown at the posted speed limit of 40 mi/hr.

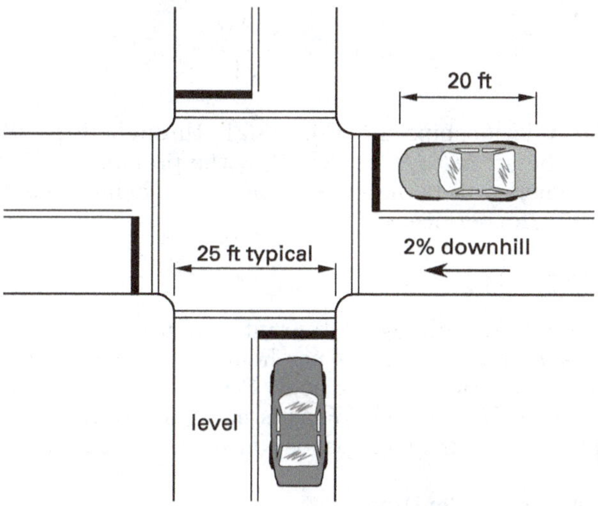

Using a driver reaction time of 1 sec and a deceleration rate of 10 ft/sec^2, what is most nearly the red clearance interval required for the driver to clear the intersection?

(A) 0.5 sec

(B) 0.6 sec

(C) 0.8 sec

(D) 1.0 sec

Solution

The red clearance interval is

$$r = \frac{W + l}{\mathrm{v}} = \frac{(25\ \mathrm{ft} + 20\ \mathrm{ft})\left(60\ \frac{\mathrm{sec}}{\mathrm{min}}\right)\left(60\ \frac{\mathrm{min}}{\mathrm{hr}}\right)}{\left(40\ \frac{\mathrm{mi}}{\mathrm{hr}}\right)\left(5280\ \frac{\mathrm{ft}}{\mathrm{mi}}\right)}$$

$$= 0.767\ \mathrm{sec} \quad (0.8\ \mathrm{sec})$$

The answer is (C).

10. QUEUEING

A *queue* is a waiting line. Time spent in a system (i.e., the *system time*) includes the *waiting time* (time spent waiting to be served) and the *service time*. *Queuing theory* can be used to predict the system time, W, the time spent in the queue, W_q, the average queue length, L_q, and the probability that a given number of users will be in the queue, P_n.

Different queuing models accommodate different service policies and populations. For example, some multi-server processes (e.g., tellers drawing from a line of banking customers) draw from a single queue. Other processes (e.g., bridge tolltakers) have their own queues. Most models predict performance only for steady-state operation, which means that the service facility has been open and in operation for some time. Start-up performance often must be evaluated by simulation.

Most queuing models are mathematically complex and fairly specialized. However, two models are important because they adequately predict the performance of simple queuing processes drawing from typical populations. These are the $M/M/1$ single-server model and the $M/M/s$ multi-server model.

Equation 48.17 Through Eq. 48.20: Fundamental Queuing Relationships

$$L = \lambda W \qquad 48.17$$

$$L_q = \lambda W_q \qquad 48.18$$

$$W = W_q + 1/\mu \qquad 48.19$$

$$\rho = \lambda/(s\mu) \qquad 48.20$$

Description

Equation 48.17 through Eq. 48.20 are valid equations for all queuing models predicting steady-state performance. Equation 48.17 is known as *Little's law* and calculates the expected number of customers in the system, L. Equation 48.18 calculates the expected number of customers in the queue, L_q.

λ is the mean arrival rate of customers per hour, and W is the expected waiting time for each customer. $1/\lambda$ is the average time between arrivals, and $1/\mu$ is the average service time per server.

For a system to be stable and viable (i.e., to reach a steady-state equilibrium queue length), λ must be less than μs, where s is the number of servers. This is known as the *undersaturated condition*. The *utilization factor*, ρ, is defined as the ratio λ/μ.

Example 10

In a queueing system that has an arrival rate of 5 customers per hour, the expected waiting time for any customer in the system, including service time, is 40 min. What is the expected number of customers in the system under steady-state conditions?

(A) 5/40 customers

(B) 2/15 customers

(C) 10/3 customers

(D) 8 customers

Solution

Using Little's law, Eq. 48.17, the expected number of customers in the system is

$$L = \lambda W = \left(5\ \frac{\text{customers}}{\text{hr}}\right)\left(\frac{40\ \text{min}}{60\ \frac{\text{min}}{\text{hr}}}\right) = 10/3\ \text{customers}$$

The answer is (C).

Equation 48.21 Through Eq. 48.26: *M/M/1* Single-Server Model

$$P_0 = 1 - \lambda/\mu = 1 - \rho \qquad 48.21$$

$$P_n = (1-\rho)\rho^n = P_0\rho^n \qquad 48.22$$

$$L = \rho/(1-\rho) = \lambda/(\mu-\lambda) \qquad 48.23$$

$$L_q = \lambda^2/[\mu(\mu-\lambda)] \qquad 48.24$$

$$W = 1/[\mu(1-\rho)] = 1/(\mu-\lambda) \qquad 48.25$$

$$W_q = W - 1/\mu = \lambda/[\mu(\mu-\lambda)] \qquad 48.26$$

Description

In the *M/M/1 single-server model*, a single server ($s=1$) draws from an infinite calling population. The service times are exponentially distributed with mean μ. The specific service time distribution is defined by

$$f(t) = \mu e^{-\mu t}$$

As a consequence of using the exponential distribution, the probability of a customer's remaining service time exceeding h (after already spending time with the server) is given by the following equation. The probability is not affected by the time a customer has already spent with the server.

$$P\{t > h\} = e^{-\mu h}$$

Arrival rates are described by a Poisson distribution with mean λ. The probability of x customers arriving in any period is

$$P\{x\} = \frac{e^{-\lambda}\lambda^x}{x!}$$

The relationships in Eq. 48.21 through Eq. 48.26 describe an $M/M/1$ model. P_n is the probability that n customers will be in the system, either being served or waiting for service.

Example 11

A small fine dining restaurant can accommodate an average of 12 customers per hour. Customers arrive at the average rate of 7 per hour. Performance is described by an $M/M/1$ model. What is most nearly the steady-state time a customer spends waiting to be seated?

(A) 0.2 hr/customer

(B) 0.5 hr/customer

(C) 2 hr/customer

(D) 5 hr/customer

Solution

Using Eq. 48.25, the average time spent waiting in the queue and being served is

$$W = 1/(\mu-\lambda) = \frac{1}{12\ \frac{\text{customers}}{\text{hr}} - 7\ \frac{\text{customers}}{\text{hr}}} = 0.2\ \text{hr/customer}$$

The answer is (A).

Transportation/ Surveying

Equation 48.27 Through Eq. 48.41: *M/M/s* Multi-Server Model

$$P_0 = \left[\sum_{n=0}^{s-1} \frac{\left(\frac{\lambda}{\mu}\right)^n}{n!} + \frac{\left(\frac{\lambda}{\mu}\right)^s}{s!} \left(\frac{1}{1 - \frac{\lambda}{s\mu}} \right) \right]^{-1} \quad (48.27)$$

$$P_0 = 1 \Bigg/ \left[\sum_{n=0}^{s-1} \frac{(s\rho)^n}{n!} + \frac{(s\rho)^s}{s!\,(1-\rho)} \right] \quad (48.28)$$

$$L_q = \frac{P_0 \left(\frac{\lambda}{\mu}\right)^s \rho}{s!\,(1-\rho)^2} \quad (48.29)$$

$$L_q = \frac{P_0 s^s \rho^{s+1}}{s!\,(1-\rho)^2} \quad (48.30)$$

$$P_n = P_0 (\lambda/\mu)^n / n! \quad [0 \leq n \leq s] \quad (48.31)$$

$$P_n = P_0 (\lambda/\mu)^n / (s! s^{n-s}) \quad [n \geq s] \quad (48.32)$$

$$W_q = L_q/\lambda \quad (48.33)$$

$$W = W_q + 1/\mu \quad (48.34)$$

$$L = L_q + \lambda/\mu \quad (48.35)$$

$$P_0 = 1 - \rho \quad [s = 1] \quad (48.36)$$

$$L_q = \rho^2/(1-\rho) \quad [s = 1] \quad (48.37)$$

$$P_0 = (1-\rho)/(1+\rho) \quad [s = 2] \quad (48.38)$$

$$L_q = 2\rho^3/(1-\rho^2) \quad [s = 2] \quad (48.39)$$

$$P_0 = \frac{2(1-\rho)}{2 + 4\rho + 3\rho^2} \quad [s = 3] \quad (48.40)$$

$$L_q = \frac{9\rho^4}{2 + 2\rho - \rho^2 - 3\rho^3} \quad [s = 3] \quad (48.41)$$

Description

Similar assumptions are made for the *M/M/s multi-server model* as for the *M/M/1* model (i.e., exponentially distributed service times and Poisson arrival rates), except that there are s servers instead of 1. Each server has an average service rate of μ and draws from a single common calling line. Therefore, the first person in line goes to the first server that is available. Each server does not have its own line. (This model can be used to predict the performance of a multiple-server system where each server has its own line if customers are allowed to change lines so that they go to any available server.) Equation 48.27 through Eq. 48.41 describe the steady-state performance of an *M/M/s* system with $\rho = \lambda/\mu s$. P_0 and L_q can be time-consuming to calculate, so the simplified equations, Eq. 48.36 through Eq. 48.41, for 1, 2, and 3 servers may be used.

49 Plane Surveying

Nomenclature

A	area	ft^2
A	azimuth	deg
B	bearing angle	deg
BS	backsight	ft
d	distance	ft
D	distance	ft
DMD	double meridian distance	ft
FS	foresight	ft
h	correction	ft
h	height	ft
HI	height of instrument	ft
k	number of observations	–
L	length	ft
LCL	lower confidence limit	various
n	number	–
N	number	–
R	rod reading	ft
w	width	ft
x	distance	ft
x	location	ft
X	location	ft
y	distance	ft
y	elevation	ft
Y	elevation	ft

Symbols

α	angle	deg
β	angle	deg
θ	angle	deg

Subscripts

a	actual
c	curvature
k	number of observations
N	north
r	refraction
S	south

1. TYPES OF SURVEYS

Plane surveys disregard the curvature of the earth. A plane survey is appropriate if the area is small. This is true when the area is not more than approximately 12 mi in any one direction. *Geodetic surveys* consider the curvature of the earth. *Zoned surveys*, as used in various *State Plane Coordinate Systems* and in the *Universal Transverse Mercator* (UTM) system, allow computations to be performed as if on a plane while accommodating larger areas.

2. SURVEYING METHODS

A *stadia survey* requires the use of a transit, theodolite, or engineer's level, as well as a rod for reading elevation differences and a tape for measuring horizontal distances. Stadia surveys are limited by the sighting capabilities of the instrument as well as by the terrain ruggedness.

In a *plane table survey*, a plane table is used in conjunction with a *telescopic instrument*. The plane table is a drawing board mounted on a tripod in such a way that the board can be leveled and rotated without disturbing the azimuth. The primary use of the combination of the plane table and telescope is in field compilation of maps, for which the plane table is much more versatile than the transit.

Total station surveys integrate theodolites, electronic distance measurement (EDM), and data recorders, collecting vertical and horizontal data in a single operation. *Manual total stations* use conventional optical-reading theodolites. *Automatic total stations* use electronic theodolites. *Data collectors* work in conjunction with total stations to store data electronically. Previously determined data can be downloaded to the data recorders for use in the field to stake out or field-locate construction control points and boundaries.

Triangulation is a method of surveying in which the positions of survey points are determined by measuring the angles of triangles defined by the points. Each survey point or monument is at a corner of one or more triangles. The survey lines form a network of triangles. The three angles of each triangle are measured. Lengths of triangle sides are calculated from trigonometry. The positions of the points are established from the measured angles and the computed sides.

Triangulation is used primarily for geodetic surveys, such as those performed by the National Geodetic Survey. Most first- and second-order control points in the national control network have been established by triangulation procedures. The use of triangulation for transportation surveys is minimal. Generally, triangulation is limited to strengthening traverses for control surveys.

Trilateration is similar to triangulation in that the survey lines form triangles. In trilateration, however, the lengths of the triangles' sides are measured. The angles are calculated from the side lengths. Orientation of the survey is established by selected sides whose directions are known or measured. The positions of trilaterated points are determined from the measured distances and the computed angles.

Photogrammetric surveys are conducted using aerial photographs. The advantages of using photogrammetry are speed of compilation, reduction in the amount of surveying required to control the mapping, high accuracy, faithful reproduction of the configuration of the ground by continuously traced contour lines, and freedom from interference by adverse weather and inaccessible terrain. Disadvantages include difficulty in areas containing heavy ground cover, high cost of mapping small areas, difficulty of locating contour lines in flat terrain, and the necessity for field editing and field completion. Photogrammetry works with ground control panels that are three to four times farther apart than with conventional surveys. But a conventional survey is still required in order to establish control initially.

Airborne LIDAR (*Light Detection and Ranging*) units are aircraft-mounted laser systems designed to measure the 3D coordinates of a passive target. This is achieved by combining a laser with positioning and orientation measurements. The laser measures the distance to the ground surface or target, and when combined with the position and orientation of the sensors, yields the 3D position of the target. Unlike EDM, which uses phase shifts of a continuous laser beam, LIDAR measures the time of flight for a single laser pulse to make the round trip from the source to receiver. LIDAR systems typically use a single wavelength, either 1064 nm or 532 nm, corresponding to the infrared and green areas of the electronic spectrum, respectively. Time can be measured to approximately $\frac{1}{3}$ ns, corresponding to a distance resolution of approximately 5 cm.

3. GLOBAL POSITIONING SYSTEM

The NAVSTAR (Navigation Satellite Timing and Ranging) *Global Positioning System* (GPS) is a one-way (satellite to receiver) ranging system. The advantages of GPS are speed and access, while the main disadvantages are cost of equipment and dependence on satellite availability/visibility. The GPS system is functional with 24 satellites arranged in six planes of four, although approximately 32 satellites are in the system performing various functions. At any moment, a minimum of six satellites are visible from any point on the earth with an unobstructed view of the sky, and up to 12 satellites may occasionally be visible at one time in particular locations. Four satellites must be visible for 3D work, and an unobstructed view of the sky is needed.

GPS determines positions without reference to any other point. It has the advantage of allowing work to proceed day or night, rain or shine, in fair weather or foul. It is not necessary to have clear lines of sight between stations, or to provide mountaintop stations for *inversability*.

GPS uses precisely synchronized clocks on each end—atomic clocks in the satellites and quartz clocks in the receivers. Frequency shift (as predicted by the Doppler effect) data is used to determine positions. Accurate positions anywhere on the earth's surface can easily be determined.

4. INERTIAL SURVEY SYSTEMS

Inertial survey systems (ISSs) determine a position on the earth by analyzing the movement of a transport vehicle, usually a light-duty truck or helicopter in which the equipment is installed. An ISS consists of a precise clock, a computer, a recording device, sensitive accelerometers, and gyroscopes. The equipment measures the acceleration of the vehicle in all three axes and converts the acceleration into distance. The production rate is 8–10 mph when mounted in tired vehicles. Helicopter ISSs can proceed at 50–70 mph.

5. GEOGRAPHIC INFORMATION SYSTEMS

A *Geographic Information System* (GIS) is a computerized database management system used to capture, store, retrieve, analyze, and display spatial data in map and overlay form. GIS replaces rolls and piles of time-bleached paper prints stored in drawers. The integrated database contains spatial information, literal information (e.g., descriptions), and characteristics (e.g., land cover and soil type). Users can pose questions involving both position and characteristics.

6. UNITS

In most of the United States, survey work is conducted using the foot and decimal parts for distance. The SI system, using meters to measure distance, is widely used throughout the world. In the United States, the SI system has been adopted by only some state highway departments.

In the United States, angles are measured and reported in the *sexagesimal system*, using degrees, minutes, and seconds, abbreviated as °, ′, and ″. 60 seconds make up a minute, and 60 minutes make up a degree. Depending on convention and convenience, angles may be measured and reported in decimal values, with tenths, hundredths, thousandths, etc., of degrees.

7. POSITIONS

Modern equipment permits surveyors to obtain positioning information directly from the theodolite. Two methods are used to specify positions: (a) by latitude and longitude, and (b) by rectangular (Cartesian) coordinates measured from a reference point. The *state plane coordinate systems* are rectangular systems that use a partial latitude/longitude system for baseline references.

8. BENCHMARKS

Benchmark is the common name given to permanent monuments of known vertical positions. Monuments with known horizontal positions are referred to as *control stations* or *triangulation stations.* A nail or hub in the pavement, the flange bolts of a fire hydrant, or the top of a concrete feature (e.g., curb) can be used as a *temporary benchmark.* The elevations of temporary benchmarks are generally found in field notes and local official filings. *Official benchmarks* (monuments) installed by surveyors and engineers generally consist of a bronze (or other inert material) disk set in the top of a concrete post/pillar.

Vertical positions (i.e., elevations) are measured above a reference surface or *datum*, often taken as *mean sea level.* For small projects, a local datum on a temporary permanent benchmark can be used.

9. DISTANCE MEASUREMENT

Gunter's chain (developed around 1620), also known as a *surveyor's chain*, consisted of 100 links totaling 66 ft. It was superseded by steel tape in the 1900s. However, steel tapes (i.e., "engineers' chains") are still sometimes referred to as "chains," and measuring distance with a steel tape is still called "chaining." The earliest tapes were marked at each whole foot, with only the last foot being subdivided into hundredths, similar to a Gunter's chain. Tapes exactly 100 ft long were called *cut chains.* Other tapes were 101 ft long and were called *add chains.* Tapes are now almost universally subdivided into hundreds along their entire lengths. In practice, this enables the rear chainman to hold the zero mark on the point being measured from, while the head chainman reads the distance directly off the face of the tape.

Tapes can be marked with customary U.S. units, SI units, or both (on either side). They come in a variety of lengths, though 100 ft tapes are the most common.

Tachyometric distance measurement involves sighting through a small angle at a distant scale. The angle may be fixed and the length measured (*stadia method*), or the length may be fixed and the angle measured (*European method*). Stadia measurement consists of observing the apparent locations of the horizontal crosshairs on a distant stadia rod. The interval between the two rod readings is called the *stadia interval* or the *stadia reading.*

Long-range laser (infrared and microwave in older units) *electronic distance measurement* (EDM) equipment is capable of measuring lines up to 10 mi long in less than 1 sec with an accuracy of better than ⅓ ft. Shorter ranges are even more precise. EDM can measure lines 1 mi long in less than 5 sec with an accuracy of about 0.01 ft. Light beams are reflected back by glass *retroreflectors* (precisely ground trihedral prisms). Common EDM accuracy ranges up to 2 mi are ±3–7 mm and ±2–7 ppm. Shorter distances enjoy even better accuracies.

An *optical plummet* allows the instrument or retroprism to be placed precisely over a point on the ground, in spite of winds that would make a string *plumb bob* swing.

10. STATIONING

In route surveying, lengths are divided into 100 ft sections called *stations.* The word "station" can mean both a location and a distance. A station is a length of 100 ft, and the unit of measure is frequently abbreviated "sta." When a length is the intended meaning, the unit of measure will come after the numerical value; for example, "the length of curve is 4 sta." When a location is the intended meaning, the unit of measure will come before the number. For example, "the point of intersection is at

Transportation/ Surveying

sta 4." The location is actually a distance from the starting point. Therefore, the two meanings are similar and related.

Interval *stakes* along an established route are ordinarily laid down at *full station* intervals. If a marker stake is placed anywhere else along the line, it is called a *plus station* and labeled accordingly. A stake placed 825 ft from station 0+00 is labeled "8+25." Similarly, a stake placed 2896 ft from a reference point at station 10+00 is labeled "28+96."

11. LEVELS

Automatic and self-leveling laser *levels* have all but replaced the optical level for leveling and horizontal alignment. A tripod-mounted laser can rotate and project a laser plane 300 ft to 2000 ft in diameter. A receiver mounted on a rod with visual and audible signals will alert the rodman to below-, at-, and above-grade conditions.

12. ELEVATION MEASUREMENT

Leveling is the act of using an engineer's level (or other leveling instrument) and rod to measure a vertical distance (*elevation*) from an arbitrary level surface.

If a level sighting is taken on an object with actual height h_a, the *curvature of the earth* will cause the object to appear taller by an amount h_c. (See Fig. 49.1.) In the following equation, x is measured in feet along the curved surface of the earth.

Figure 49.1 *Curvature and Refraction Effects*

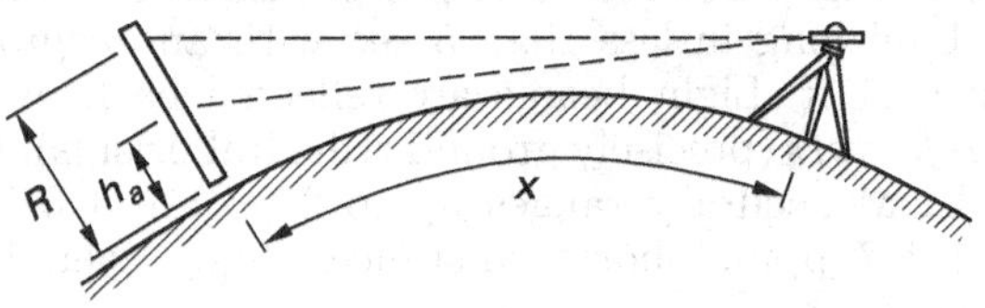

$$h_c = \left(2.39 \times 10^{-8} \ \frac{1}{\text{ft}}\right) x_{\text{ft}}^2$$

Atmospheric refraction will make the object appear shorter by an amount h_r.

$$h_r = \left(3.3 \times 10^{-9} \ \frac{1}{\text{ft}}\right) x_{\text{ft}}^2$$

The effects of curvature and refraction are combined into a single correction, h_{rc}. As shown in the following equation, the correction is positive by convention. Whether the correction increases or reduces a measured quantity depends on the nature of the quantity. To determine if rod readings, elevations at the rod base, and elevations at the instrument should be increased or decreased by h_{rc}, it is helpful to draw the geometry.

$$h_{rc} = |h_r - h_c| = \left(2.1 \times 10^{-8} \ \frac{1}{\text{ft}}\right) x_{\text{ft}}^2$$

The corrected rod reading (actual height) is

$$h_a = R_{\text{observed}} - h_{rc}$$

13. ELEVATION MEASUREMENT: DIRECT LEVELING

A *telescope* is part of the sighting instrument. A *transit* or *theodolite* is often referred to as a "telescope" even though the telescope is only a part of the instrument. (Early telescoping theodolites were so long that it was not possible to invert, or "transit," them. Hence the name "transit" is used for inverting telescopes.) An *engineer's level* (*dumpy level*) is a sighting device with a telescope rigidly attached to the level bar. A sensitive level vial is used to ensure level operation. The engineer's level can be rotated, but in its basic form it cannot be elevated. An *alidade* consists of the base and telescopic part of the transit, but not the tripod or leveling equipment.

In a *semi-precise level*, also known as a *prism level*, the level vial is visible from the eyepiece end. In other respects, it is similar to the engineer's level. The *precise* or *geodetic level* has even better control of horizontal angles. The bubble vial is magnified for greater accuracy.

Transits and levels are used with rods. *Leveling rods* are used to measure the vertical distance between the line of sight and the point being observed. The *standard rod* is typically made of fiberglass and is extendable. It is sometimes referred to as a *Philadelphia rod*. *Precise rods*, made of wood-mounted invar, are typically constructed in one piece and are spring-loaded in tension to avoid sagging.

With *direct leveling*, a level is set up at a point approximately midway between the two points whose difference in elevation is desired. The vertical *backsight* (*plus sight*) and *foresight* (*minus sight*) are read directly from the rod. HI is the height of the instrument above the ground, and h_{rc} is the correction for refraction and curvature.

Referring to Fig. 49.2, and recalling that h_{rc} is positive,

$$y_{\text{A-L}} = R_{\text{A}} - h_{rc,\text{A-L}} - \text{HI}$$

$$y_{\text{L-B}} = \text{HI} + h_{rc,\text{L-B}} - R_{\text{B}}$$

Figure 49.2 Direct Leveling

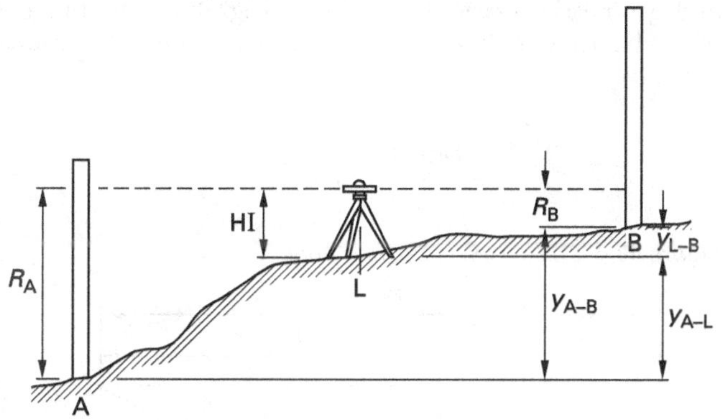

The difference in elevations between points A and B is

$$\begin{aligned} y_{A\text{-}B} &= y_{A\text{-}L} + y_{L\text{-}B} \\ &= R_A - R_B + h_{rc,L\text{-}B} - h_{rc,A\text{-}L} \end{aligned}$$

If the backsight and foresight distances are equal (or approximately so), then the effects of refraction and curvature cancel.

$$y_{A\text{-}B} = R_A - R_B$$

14. ELEVATION MEASUREMENT: DIFFERENTIAL LEVELING

Differential leveling is the consecutive application of direct leveling to the measurement of large differences in elevation. There is usually no attempt to exactly balance the foresights and backsights. Therefore, there is no record made of the exact locations of the level positions. Furthermore, the path taken between points need not be along a straight line connecting them, as only the elevation differences are relevant. If greater accuracy is desired without having to accurately balance the foresight and backsight distance pairs, it is possible to eliminate most of the curvature and refraction error by balancing the sum of the foresights against the sum of the backsights.

The following abbreviations are used with differential leveling.

BM	benchmark or monument
TP	turning point
FS	foresight (also known as a minus sight)
BS	backsight (also known as a plus sight)
HI	height of the instrument
L	level position

HI (or H.I.) is the distance between the instrument axis and the datum. For differential leveling, the datum established for all elevations is used to define the HI. However, in stadia measurements, HI may be used (and recorded in the notes) to represent the height of the instrument axis above the ground.

15. ELEVATION MEASUREMENT: INDIRECT LEVELING

Indirect leveling does not require a backsight. (A backsight reading can still be taken to eliminate the effects of curvature and refraction.) In Fig. 49.3, distance AC has been determined. Within the limits of ordinary practice, angle ACB is 90°. Including the effects of curvature and refraction, the elevation difference between points A and B is

$$[\text{U.S.}]\; y_{A\text{-}B} = \text{AC}\tan\alpha + 2.1\times10^{-8}(\text{AC})^2$$

If a backsight is taken from B to A and angle β is measured, then

$$[\text{U.S. only}]\; y_{A\text{-}B} = \text{AC}\tan\beta - 2.1\times10^{-8}(\text{AC})^2$$

Adding these results and dividing by 2,

$$y_{A\text{-}B} = \tfrac{1}{2}\text{AC}(\tan\alpha + \tan\beta)$$

Figure 49.3 Indirect Leveling

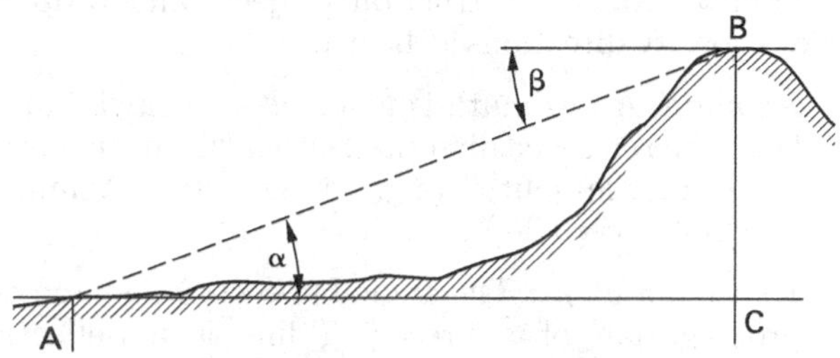

16. EQUIPMENT AND METHODS USED TO MEASURE ANGLES

A *transit* is a telescope that measures vertical angles as well as horizontal angles. The conventional *engineer's transit* (*surveyor's transit*) used in the past had a 6¼ in diameter metal circle and vernier. It could be clamped vertically and used as a level. Few instruments of this nature are still available.

With a *theodolite*, horizontal and vertical angles are measured by looking into viewpieces. There may be up to four viewpieces, one each for the sighting-in telescope, the compass, the horizontal and vertical angles, and the optical plummet. Theodolites may be scale-reading optical, micrometer-reading optical, or electronic. Angles read electronically interface with EDM equipment and data recorders.

Transportation/ Surveying

17. DIRECTION SPECIFICATION

The direction of any line can be specified by an angle between it and some other reference line, known as a *meridian*. If the meridian is arbitrarily chosen, it is called an *assumed meridian*. If the meridian is a true north-to-south line passing through the true north pole, it is called a *true meridian*. If the meridian is parallel to the earth's magnetic field, it is known as a *magnetic meridian*. A rectangular grid may be drawn over a map with any arbitrary orientation. If so, the vertical lines are referred to as *grid meridians*.

A true meridian differs from a magnetic meridian by the *declination* (*magnetic declination* or *variation*). An *isogonic line* connects locations in a geographical region that have the same magnetic declination. If the north end of a compass points to the west of true meridian, the declination is referred to as a *west declination* or *minus declination*. Otherwise, it is referred to as an *east declination* or *plus declination*. Plus declinations are added to the magnetic compass azimuth to obtain the true azimuth. Minus declinations are subtracted from the magnetic compass azimuth.

A direction (i.e., the variation of a line from its meridian) may be specified in several ways. Directions are normally specified as either bearings or azimuths referenced to either north or south. In the United States, most control work is performed using directions stated as azimuths. Most construction projects and property surveys specify directions as bearings.

Azimuth: An azimuth is given as a clockwise angle from the reference direction, either "from the north" or "from the south" (e.g., "NAz 320"). Azimuths may not exceed 360°.

Deflection angle: The angle between a line and the prolongation of a preceding line is a deflection angle. Such measurements must be labeled as "right" for clockwise angles and "left" for counterclockwise angles.

Angle to the right: An angle to the right is a clockwise deflection angle measured from the preceding to the following line.

Azimuths from the back line: Same as angle to the right.

Bearing: The bearing of a line is referenced to the quadrant in which the line falls and the angle that the line makes with the meridian in that quadrant. It is necessary to specify the two cardinal directions that define the quadrant in which the line is found. The north and south directions are always specified first. A bearing contains an angle and a direction from a reference line, either north or south (e.g., "N45°E"). A bearing may not have an angular component exceeding 90°.

Bearings and azimuths are calculated from trigonometric and geometric relationships. From Fig. 49.4, the tangent of the bearing angle B of the line between the two points is given by

$$\tan B = \frac{x_2 - x_1}{y_2 - y_1}$$

***Figure 49.4** Calculation of Bearing Angle*

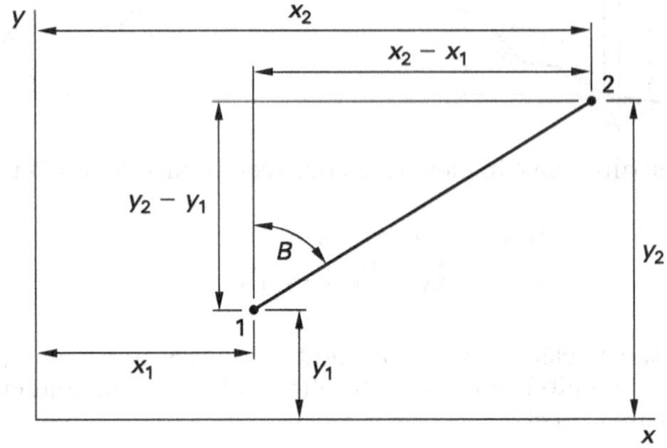

If the numerator is positive, the line from point 1 to 2 bears east. If the denominator is positive, the line bears north. Determining the bearing and length of a line from the coordinates of the two points is called *inversing the line*. (See Fig. 49.5.)

***Figure 49.5** Equivalent Angle Measurements*

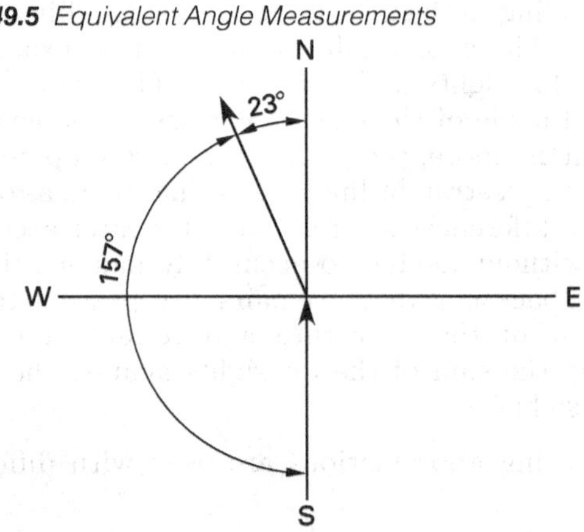

azimuth from the south: 157°
azimuth from the north: 337°
deflection angle: 23° L
angle to the right: 157°
bearing: N23° W

The distance, D, between the two points is given by

$$D = \sqrt{(y_2 - y_1)^2 + (x_2 - x_1)^2}$$

The *azimuth*, A_N, of a line from point 1 to 2 measured from the north meridian is given by

$$\tan A_N = \frac{x_2 - x_1}{y_2 - y_1}$$

The azimuth, A_S, of the same line from point 1 to 2 measured from the south meridian is given by

$$\tan A_S = \frac{x_1 - x_2}{y_1 - y_2}$$

18. LATITUDES AND DEPARTURES

The *latitude* of a line is the distance that the line extends in a north or south direction. A line that runs toward the north has a positive latitude; a line that runs toward the south has a negative latitude. The *departure* of a line is the distance that the line extends in an east or west direction. A line that runs toward the east has a positive departure; a line that runs toward the west has a negative departure. (See Fig. 49.6.)

Figure 49.6 *Latitudes and Departures*

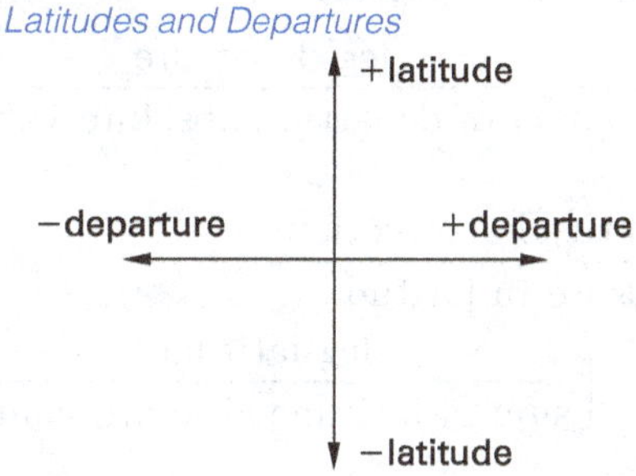

If the x- and y-coordinates of two points are known, the latitude and departure of the line between the two points are determined by

$$\text{latitude of line from point A to B: } y_B - y_A$$

$$\text{departure of line from point A to B: } x_B - x_A$$

19. TRAVERSES

A *traverse* is a series of straight lines whose lengths and directions are known. A traverse that does not come back to its starting point is an *open traverse.* A traverse that comes back to its starting point is a *closed traverse.* The polygon that results from closing a traverse is governed by the following two geometric requirements.

1. The sum of the deflection angles is 360°.
2. The sum of the interior angles of a polygon with n sides is $(n - 2)(180°)$.

There are three ways to specify angles in traverses: *deflection angles, interior angles* (also known as *explement angles*), and *station angles.* These are illustrated in Fig. 49.7.

Figure 49.7 *Angles Used in Defining Traverses*

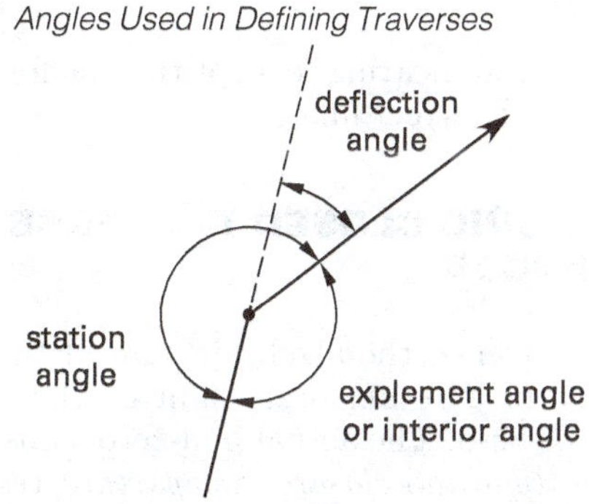

20. BALANCING

Refined field measurements are not the goal of measurement *balancing* (also known as *adjusting*). Balancing is used simply to accommodate the mathematical necessity of having a balanced column of figures. Balancing is considered inappropriate by some surveyors, since errors are not eliminated, only distributed to all of the measurements. Balancing does not make inaccurate measurements more accurate—it only makes inaccurate measurements more inaccurate. It is particularly inappropriate when a *blunder* (a measurement grossly in error that should be remeasured) is fixed by balancing.

21. BALANCING CLOSED TRAVERSE ANGLES

Due to measurement errors, variations in magnetic declination, and local magnetic attractions, it is likely that the sum of angles making up the interior angles will not exactly equal $(n - 2)(180°)$. The following adjusting procedure can be used to distribute the angle error of closure among the angles.

step 1: Calculate the interior angle at each vertex from the observed bearings.

step 2: Subtract $(n - 2)(180°)$ from the sum of the interior angles.

step 3: Unless additional information in the form of numbers of observations or probable errors is available, assume the angle error of closure can be divided equally among all angles. Divide the error by the number of angles.

step 4: Find a line whose bearing is assumed correct; that is, find a line whose bearing appears unaffected by errors, variations in magnetic declination, and local attractions. Such a line may be chosen as one for which the forward and back bearings are the same. If there is no such line, take the line whose difference in forward and back bearings is the smallest.

step 5: Start with the assumed correct line and add (or subtract) the prorated error to each interior angle.

step 6: Correct all bearings except the one for the assumed correct line.

22. BALANCING CLOSED TRAVERSE DISTANCES

In a closed traverse, the algebraic sum of the latitudes should be zero. The algebraic sum of the departures should also be zero. The actual non-zero sums are called *closure in latitude* and *closure in departure*, respectively.

The *traverse closure* is the line that will exactly close the traverse. Since latitudes and departures are orthogonal, the closure in latitude and closure in departure can be considered as the rectangular coordinates and the traverse closure length calculated from the Pythagorean theorem. The coordinates will have signs opposite the closures in departure and latitude. That is, if the closure in departure is positive, point A will lie to the left of point A′, as shown in Fig. 49.8.

Figure 49.8 *Traverse Closure*

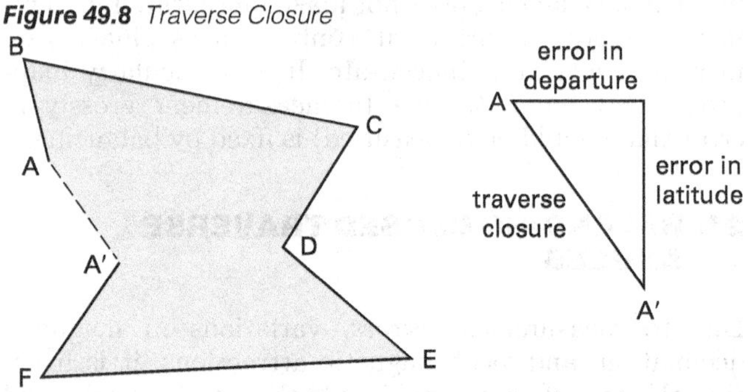

The length of a traverse closure is calculated from the Pythagorean theorem.

$$L = \sqrt{(\text{closure in departure})^2 + (\text{closure in latitude})^2}$$

To balance a closed traverse, the traverse closure must be divided among the various legs of the traverse. (Of course, if it is known that one leg was poorly measured due to difficult terrain, all of the error may be given to that leg.) This correction requires that the latitudes and departures be known for each leg of the traverse.

Computer-assisted traverse balancing systems offer at least four balancing methods: compass rule, transit rule, Crandall method, and least squares. The use of one rule over another is often arbitrary or a matter of convention.

The most common balancing method is the *compass rule*, also known as the *Bowditch method:* the ratio of a leg's correction to the total traverse correction is equal to the ratio of leg length to the total traverse length, with the signs reversed. The compass rule is used when the angles and distances in the traverse are considered equally precise.

$$\frac{\text{leg departure correction}}{\text{closure in departure}} = \frac{-\text{leg length}}{\text{total traverse length}}$$

$$\frac{\text{leg latitude correction}}{\text{closure in latitude}} = \frac{-\text{leg length}}{\text{total traverse length}}$$

If the angles are precise but the distances are less precise (such as when the distances have been determined by taping through rugged terrain), the *transit rule* is a preferred method of distributing the correction to the traverse legs. This rule distributes the closure error in proportion to the absolute values of the latitudes and departures.

$$\frac{\text{leg departure correction}}{\text{closure in departure}} = -\left(\frac{\text{leg departure}}{\text{sum of departure absolute values}}\right)$$

$$\frac{\text{leg latitude correction}}{\text{closure in latitude}} = -\left(\frac{\text{leg latitude}}{\text{sum of latitude absolute values}}\right)$$

With the *Crandall rule*, adjustments assume that the closing errors are random and normally distributed, and that angular error is negligible or has already been eliminated (i.e., adjusted out). The closure error is distributed throughout the traverse by adjusting only the traverse distances, not the angles. The adjustment made to each leg is such that the sum of the squares is a minimum. However, the Crandall rule is not the same as the method of least squares.

The *method of least squares* balancing has significant advantages over the other methods, although the computational burden makes it difficult to implement with manual calculations. This method reduces the sum of the squares of the differences between the unadjusted and adjusted measurements (angles and distances) to a minimum. Particularly when calculations are implemented in surveying equipment or analysis software, the least squares method accommodates weighting of measurements and the inclusion of redundant data to strengthen some measurements. With this method, adjusted coordinate positions are computed using estimated precisions of observations' coordinates to reconcile differences between observations and the inverses to their adjusted coordinates. The least squares method also produces adjustment statistics that indicate the strengths of computed locations. The strength of the computed location translates directly into measurement confidence and can also be helpful in detecting input or measurement errors.

23. TRAVERSE AREA

Equation 49.1: Area by Coordinates

$$\text{area} = \left| \begin{array}{l} X_A(Y_B - Y_N) + X_B(Y_C - Y_A) \\ \quad +X_C(Y_D - Y_B) + \cdots \\ \quad +X_N(Y_A - Y_{N-1}) \end{array} \right| /2 \qquad 49.1$$

Variation

$$A = \tfrac{1}{2} \left| \begin{array}{l} \Sigma \text{ of full line products} \\ \quad -\Sigma \text{ of broken line products} \end{array} \right|$$

Description

The area of a simple traverse can be found by dividing the traverse into a number of geometric shapes and summing their areas. If the coordinates of the traverse leg end points are known, the *area by coordinates method* (*method of coordinates*) can be used. The coordinates can be x-y coordinates referenced to some arbitrary set of axes, or they can be sets of departure and latitude.

The area calculation is simplified if the coordinates are written in the following form.

$$\frac{X_A}{Y_A} \times \frac{X_B}{Y_B} \times \frac{X_C}{Y_C} \times \frac{X_D}{Y_D} \times \frac{X_A}{Y_A} \cdots$$

The simplified area calculation is given in the variation equation.

Example

Using the area by coordinates method, what is most nearly the area enclosed by the lines shown?

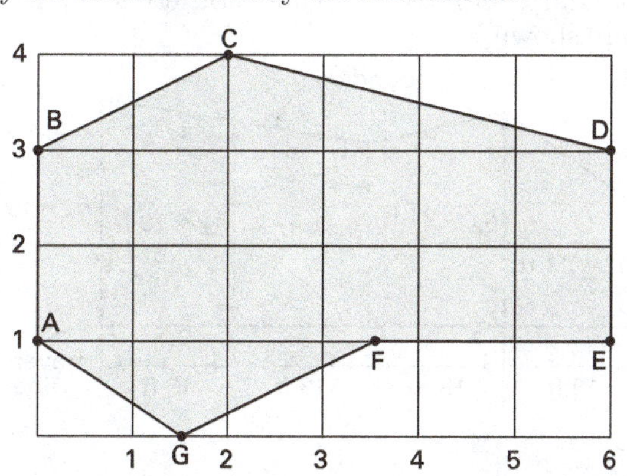

(A) 16.6

(B) 16.8

(C) 17.1

(D) 17.3

Solution

The coordinates of the enclosed area are

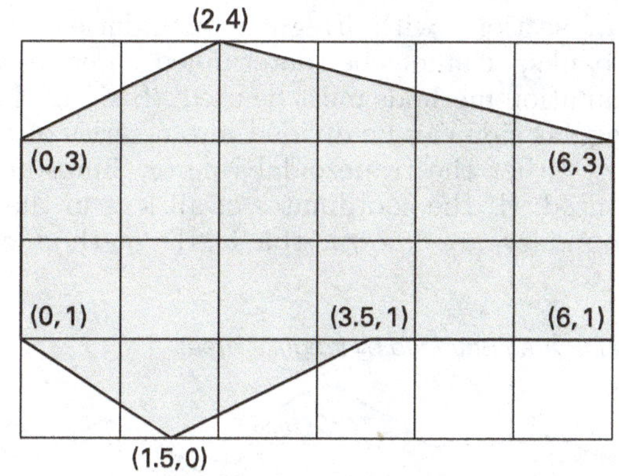

The enclosed area is

$$\begin{aligned} \text{area} &= \frac{\begin{array}{l} X_A(Y_B - Y_G) + X_B(Y_C - Y_A) \\ \quad +X_C(Y_D - Y_B) + \cdots \\ \quad +X_G(Y_A - Y_F) \end{array}}{2} \\ &= \frac{\begin{array}{l} (0)(3-0) + (0)(4-1) \\ \quad +(2)(3-3) + (6)(1-4) \\ \quad +(6)(1-3) + (3.5)(0-1) \\ \quad +(1.5)(1-1) \end{array}}{2} \\ &= 16.75 \quad (16.8) \end{aligned}$$

The answer is (B).

Double Meridian Distance

If the latitudes and departures are known, the *double meridian distance* (DMD) method can be used to calculate the area of a traverse. A double meridian distance for a leg is defined by

$$\text{DMD}_{\text{leg } i} = \text{DMD}_{\text{leg } i-1} + \text{departure}_{\text{leg } i-1} + \text{departure}_{\text{leg } i}$$

Special rules are required to handle the first and last legs. The DMD of the first course is defined as the departure of that course. The DMD of the last course is the negative of its own departure. A tabular approach is the preferred manual method of using the DMD method. The traverse area is calculated from

$$A = \tfrac{1}{2}\left|\sum(\text{latitude}_{\text{leg}\,i} \times \text{DMD}_{\text{leg}\,i})\right|$$

24. AREAS BOUNDED BY IRREGULAR BOUNDARIES

Areas of sections with irregular boundaries, such as creek banks, cannot be determined precisely, and approximation methods must be used. (See Fig. 49.9.) If the irregular side can be divided into a series of cells of width w, either the trapezoidal rule or Simpson's rule can be used. (If the coordinates of all legs in the irregular boundary are known, the DMD method can be used.)

Figure 49.9 *Area Bounded by Irregular Areas*

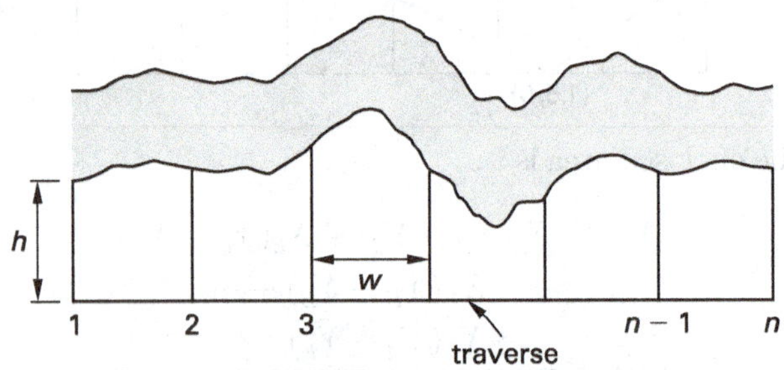

Equation 49.2: Trapezoidal Rule

$$\text{area} = w\left(\frac{h_1 + h_n}{2} + h_2 + h_3 + h_4 + \cdots + h_{n-1}\right) \qquad 49.2$$

Description

If the irregular side of each cell is fairly straight, the *trapezoidal rule*, as given in Eq. 49.2, can be used.

Example

The boundary and traverse line of an irregular area are shown.

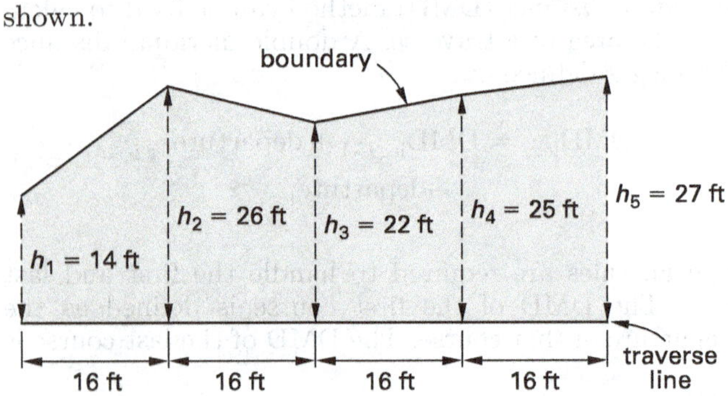

Using the trapezoidal rule, the total area between the irregular boundary and the traverse line is most nearly

(A) 1500 ft^2

(B) 1520 ft^2

(C) 1590 ft^2

(D) 1630 ft^2

Solution

By the trapezoidal rule, the area is

$$\begin{aligned}\text{area} &= w\left(\frac{h_1 + h_5}{2} + h_2 + h_3 + h_4\right)\\ &= (16\text{ ft})\left(\frac{14\text{ ft} + 27\text{ ft}}{2} + 26\text{ ft} + 22\text{ ft} + 25\text{ ft}\right)\\ &= 1496\text{ ft}^2 \quad (1500\text{ ft}^2)\end{aligned}$$

The answer is (A).

Equation 49.3: Simpson's Rule

$$\text{area} = w\left[h_1 + 2\left(\sum_{k=3,5,\ldots}^{n-2} h_k\right) + 4\left(\sum_{k=2,4,\ldots}^{n-1} h_k\right) + h_n\right]\Big/3 \qquad 49.3$$

Variation

$$A = \frac{d}{3}(h_1 + 4h_2 + 2h_3 + 4h_4 + \cdots + h_n)$$

Description

If the irregular side of each cell is curved or parabolic, then *Simpson's rule* (sometimes referred to as *Simpson's ⅓ rule*), as given in Eq. 49.3, can be used. n must be an odd number of measurements.

Example

A boundary and traverse line bordering an irregular area are shown.

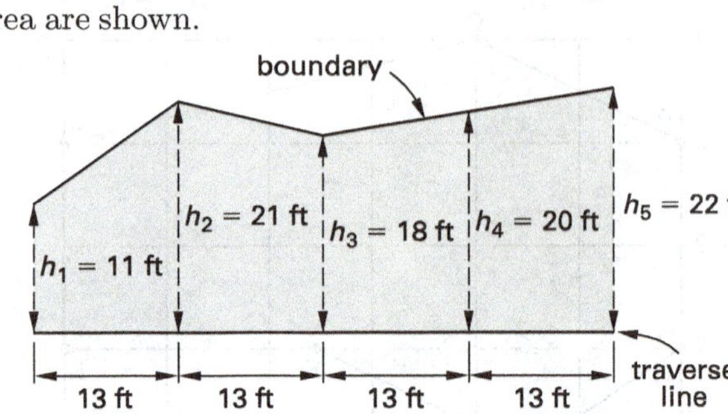

Using Simpson's ⅓ rule, the total area between the boundary and the traverse line is most nearly

(A) 900 ft^2

(B) 1000 ft^2

(C) 3000 ft^2

(D) 4000 ft^2

Solution

By Simpson's ⅓ rule,

$$\text{area} = w\left[h_1 + 2\left(\sum_{k=3,5,\ldots}^{n-2} h_k\right) + 4\left(\sum_{k=2,4,\ldots}^{n-1} h_k\right) + h_n\right]/3$$

$$= \frac{(13\text{ ft})\left(\begin{array}{l}11\text{ ft} + (2)(18\text{ ft}) \\ \quad +(4)(21\text{ ft} + 20\text{ ft}) + 22\text{ ft}\end{array}\right)}{3}$$

$$= 1010\text{ ft}^2 \quad (1000\text{ ft}^2)$$

The answer is (B).

Transportation/Surveying

50 Geometric Design

Nomenclature

a	acceleration	ft/sec^2
a	parabola constant	–
A	absolute value of the algebraic grade difference	%
c	length of subchord	ft
C	clearance	ft
C	increase in rate of lateral acceleration	ft/sec^3
d	angle of subchord	deg
D	degree of curve	deg
e	superelevation (rate)	%
E	external distance	ft
E	tangent offset	ft
f	side friction factor	–
g	acceleration due to gravity, 32.2	ft/sec^2
g	grade	decimal
G	grade	decimal
h	height	ft
HSO	horizontal sightline offset	ft
I	intersection angle	deg
K	rate of vertical curvature	ft/%
l	length of curve (subchord)	ft
L	length of curve from PC to PT	ft
LC	long chord	ft
M	middle ordinate	ft
r	rate of change in grade	%/sta
R	curve radius	ft
S	sight distance	ft
SSD	stopping sight distance	ft
t	time	sec
T	tangent distance	ft
v	velocity	ft/sec
v	vehicle speed	mi/hr
V	vertex	ft
x	distance from PVC	sta
x	tangent distance	ft
y	tangent offset	ft
Y	elevation	ft

Symbols

θ	angle	deg

Subscripts

eff	effective
m	maximum or minimum
s	spiral

1. HORIZONTAL CURVES

A *horizontal circular curve* is a circular arc between two straight lines known as *tangents*. When traveling in a particular direction, the first tangent encountered is the *back tangent* (*approach tangent*), and the second tangent encountered is the *forward tangent* (*departure tangent* or *ahead tangent*).

The geometric elements of a horizontal curve are shown in Fig. 50.1. Table 50.1 lists the standard terms and abbreviations used to describe the elements.

The *intersection angle* (*deflection angle*, curve *central angle*, or *interior angle*), I, has units of degrees unless indicated otherwise.

Curves are specified by either the radius, R, or degree of curve, D. The equations in this chapter that use the degree of curve are valid for customary U.S. units only.

Table 50.1 *Horizontal Curve Abbreviations*

PC	point of curvature (the point where the back tangent ends and the curve begins); same as beginning of curve (BC)
PI	point of intersection of back and forward tangents
PT	point of tangency (the point where the curve ends and the forward tangent begins); same as end of curve (EC)

Figure 50.1 Horizontal Curve Elements

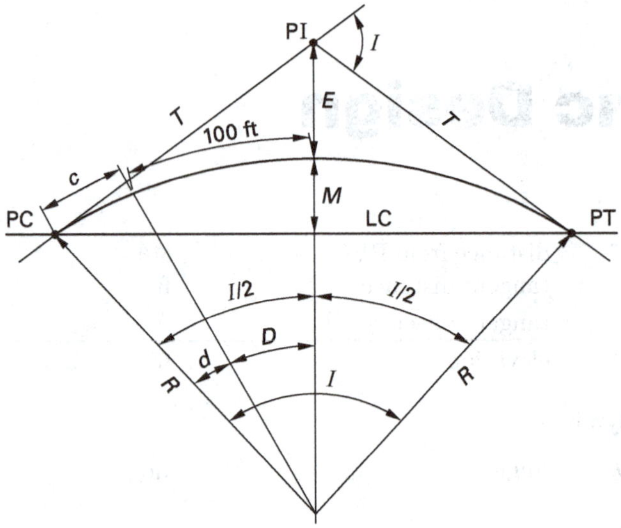

Equation 50.1 and Eq. 50.2: Radius of a Curve

$$R = \frac{5729.58}{D} \tag{50.1}$$

$$R = \frac{\text{LC}}{2\sin(I/2)} \tag{50.2}$$

Description

Equation 50.1 is used to calculate the *radius of a curve* when the *degree of curve*, D, is known. 5729.58, sometimes also given as 5729.578, has units of feet per degree of curve.

The radius of a curve can also be calculated from the length of the *long chord*, LC, and *intersection angle*, I, as shown in Eq. 50.2.

Example

A circular curve has a long chord length of 200 ft and an intersection angle of 20°. What is most nearly the radius of the curve?

(A) 290 ft

(B) 410 ft

(C) 580 ft

(D) 590 ft

Solution

From Eq. 50.2, the curve radius is

$$R = \frac{\text{LC}}{2\sin(I/2)} = \frac{200 \text{ ft}}{2\sin\left(\frac{20°}{2}\right)} = 575.88 \text{ ft} \quad (580 \text{ ft})$$

The answer is (C).

Equation 50.3: Length of a Curve

$$L = RI\frac{\pi}{180} = \frac{I}{D}100 \tag{50.3}$$

Variations

$$L = RI_{\text{radians}}$$

$$L = \frac{2\pi RI}{360°}$$

Description

The *length of curve* is the distance from the *point of curve*, PC, to the *point of tangent*, PT, along the arc. The length of curve can be calculated using the curve radius, R, or degree of curve, D, and the intersection angle, I. In Eq. 50.3, 180 has units of degrees, and 100 has units of feet.

Example

What is most nearly the curve length of a horizontal curve with a radius of 2080 ft and an intersection angle of 60°?

(A) 690 ft

(B) 1800 ft

(C) 2200 ft

(D) 3100 ft

Solution

The length of the curve is

$$\begin{aligned} L &= RI\frac{\pi}{180} \\ &= (2080 \text{ ft})(60°)\left(\frac{\pi}{180°}\right) \\ &= 2178.17 \text{ ft} \quad (2200 \text{ ft}) \end{aligned}$$

The answer is (C).

Equation 50.4: Tangent Distance

$$T = R\tan(I/2) = \frac{\text{LC}}{2\cos(I/2)} \qquad 50.4$$

Description

The *tangent distance*, T, is the distance between the PC or PT and the *point of intersection*, PI.

Example

A horizontal curve has an intersection angle of 12° and a curve radius of 1700 ft. What is most nearly the tangent distance?

(A) 92 ft

(B) 160 ft

(C) 180 ft

(D) 360 ft

Solution

The tangent of the curve is

$$\begin{aligned} T &= R\tan(I/2) \\ &= (1700\text{ ft})\tan\left(\frac{12^\circ}{2}\right) \\ &= 178.68\text{ ft} \quad (180\text{ ft}) \end{aligned}$$

The answer is (C).

Equation 50.5 and Eq. 50.6: External Distance

$$E = R\left[\frac{1}{\cos(I/2)} - 1\right] \qquad 50.5$$

$$\frac{R-M}{R} = \cos(I/2) \qquad 50.6$$

Variation

$$E = R\left(\sec\frac{I}{2} - 1\right) = R\tan\frac{I}{2}\tan\frac{I}{4}$$

Description

The *external distance*, E, is the distance between the midpoint of the curve and the point of intersection, PI. The relationship $\cos(I/2)$ is described by the curve radius and middle ordinate as shown in Eq. 50.6.

Example

A horizontal curve has an intersection angle of 26°12′ and a curve radius of 1150 ft. What is most nearly the external distance?

(A) 28 ft

(B) 31 ft

(C) 44 ft

(D) 53 ft

Solution

The intersection angle, I, in decimal degrees is

$$\begin{aligned} I &= 26^\circ + \frac{12\text{ min}}{60\,\frac{\text{min}}{\text{deg}}} \\ &= 26.2^\circ \end{aligned}$$

From Eq. 50.5, the external distance is

$$\begin{aligned} E &= R\left[\frac{1}{\cos(I/2)} - 1\right] = (1150\text{ ft})\left(\frac{1}{\cos\left(\frac{26.2^\circ}{2}\right)} - 1\right) \\ &= 30.73\text{ ft} \quad (31\text{ ft}) \end{aligned}$$

The answer is (B).

Equation 50.7 and Eq. 50.8: Middle Ordinate Length

$$M = R[1 - \cos(I/2)] \qquad 50.7$$

$$\frac{R}{E+R} = \cos(I/2) \qquad 50.8$$

Variation

$$M = \frac{\text{LC}}{2}\tan\frac{I}{4}$$

Description

The length of the *middle ordinate*, M, is the distance between the midpoint of the arc and the midpoint of the long chord. The relationship $\cos(I/2)$ is described by the curve radius and external length as shown in Eq. 50.8.

Transportation/ Surveying

Example

A horizontal curve has a curve radius of 2500 ft and an external distance of 460 ft. What is most nearly the middle ordinate?

(A) 390 ft

(B) 440 ft

(C) 520 ft

(D) 610 ft

Solution

Using Eq. 50.7 and the relationship in Eq. 50.8, the middle ordinate is

$$\begin{aligned} M &= R[1-\cos(I/2)] \\ &= R\left[1-\frac{R}{E+R}\right] \\ &= (2500\text{ ft})\left(1-\frac{2500\text{ ft}}{460\text{ ft}+2500\text{ ft}}\right) \\ &= 388.51\text{ ft} \quad (390\text{ ft}) \end{aligned}$$

The answer is (A).

Equation 50.9 Through Eq. 50.11: Subchords

$$c = 2R\sin(d/2) \qquad 50.9$$

$$d = D/2 \qquad 50.10$$

$$l = Rd\left(\frac{\pi}{180}\right) \qquad 50.11$$

Description

A *subchord* is any straight line that can be drawn from one point on a curve to another. The subchord length is calculated from the curve radius and the subtended angle of the subchord, d. The *deflection angle* from the tangent is $d/2$. The angle of the subchord per 100 ft of length, d, is equal to one-half of the degree of curve, D, as shown in Eq. 50.10. (d is not always equal to $D/2$ as Eq. 50.10 implies. Equation 50.10 is valid only when the chord length is 100 ft.) The curve length of the subchord, l, is found from Eq. 50.11.

Example

A subchord deflects 9.9° from the tangent. If the curve radius is 800 ft, what is most nearly the length of the subchord?

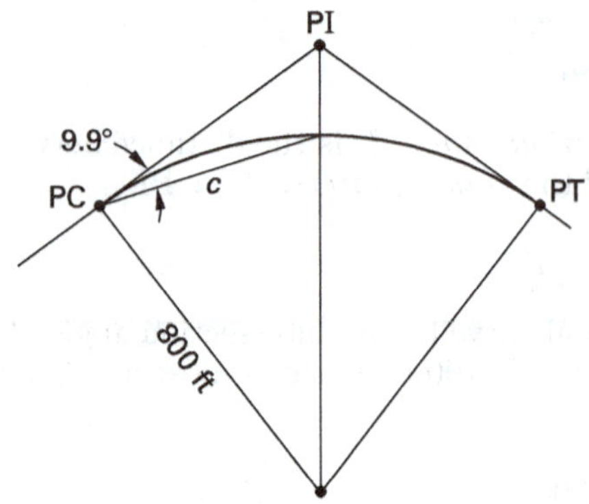

(not to scale)

(A) 220 ft

(B) 240 ft

(C) 260 ft

(D) 280 ft

Solution

From Eq. 50.10, the deflection angle is $d/2 = 9.9°$. Using Eq. 50.9, the subchord length is

$$\begin{aligned} c &= 2R\sin(d/2) = (2)(800\text{ ft})\sin(9.9°) \\ &= 275\text{ ft} \quad (280\text{ ft}) \end{aligned}$$

The answer is (D).

2. STATIONING ON A HORIZONTAL CURVE

When the route is initially laid out between PIs, the curve is undefined. The "route" distance is measured from PI to PI. The route distance changes, though, when the curve is laid out. Stationing along the curve is continuous, as a vehicle's odometer would record the distance. The PT station is equal to the PC station plus the curve length.

$$\text{sta PT} = \text{sta PC} + L$$

$$\text{sta PC} = \text{sta PI} - T$$

In route surveying, stationing is carried ahead continuously from a starting point or hub designated as station 0+00, and called *station zero plus zero zero.* The term *station* is applied to each subsequent 100 ft length, where a stake is normally set. Also, the term *station* is applied to any point whose position is given by its total distance from the beginning hub. Thus, station 8+33.2 is a unique point 833.2 ft from the starting point, measuring along the survey line. The partial length beyond the full station 8 is 33.2 ft and is termed a *plus.* Moving or looking toward increasing stations is called *ahead stationing.* Moving or looking toward decreasing stations is called *back stationing.* Offsets from the centerline are either left or right looking ahead on stationing.

Normal curve layout follows the convention of showing increased stationing (ahead stationing) from left to right on the plan sheet, or from the bottom to the top of the sheet. Moving ahead on stationing, the first point of the curve is the *point of curvature* (PC). This point is alternatively called the point of *beginning of curve* (BC). Stationing is carried ahead along the arc of the curve to the end point of the curve, called the *point of tangency* (PT). This end point is alternatively called the point of *end of curve* (EC). The PI is stationed ahead from the PC. Therefore, the difference in stationing along the back curve tangent is the curve tangent length. The ahead tangent is rarely stationed to avoid the confusion of creating two stations for the PC.

3. SPIRAL CURVES

Spiral curves are introduced at the ends of a circular curve (i.e., at the PC and PT) in order to provide a transition, or *easement*, between the straight tangents and the circular parts of the curve. Without an easement, the sudden change from no lateral acceleration along the tangent to the lateral acceleration necessary to travel around a curve causes an increase in lateral force on the vehicle. Spiral curves were first used on railroads to reduce the jerk caused at the ends of curves. The principle of easement is a gradual increase in lateral acceleration by decreasing the curve radius until the central curve radius is reached. *Superelevation*, which is the difference between the inner and outer elevation of a track or roadway, is also transitioned along the easement, providing a smooth and comfortable ride for passengers by changing the cross slope in proportion to the change in curve radius.

The elements of a spiral curve are shown in Fig. 50.2. Table 50.2 gives a list of spiral curve abbreviations.

Equation 50.12: Spiral Transition Length

$$L_s = \frac{3.15\text{v}^3}{RC} \qquad 50.12$$

Value

$C = 1\ \text{ft/sec}^3$, unless otherwise stated.

Figure 50.2 *Fully Spiraled Circular Curve Layout*

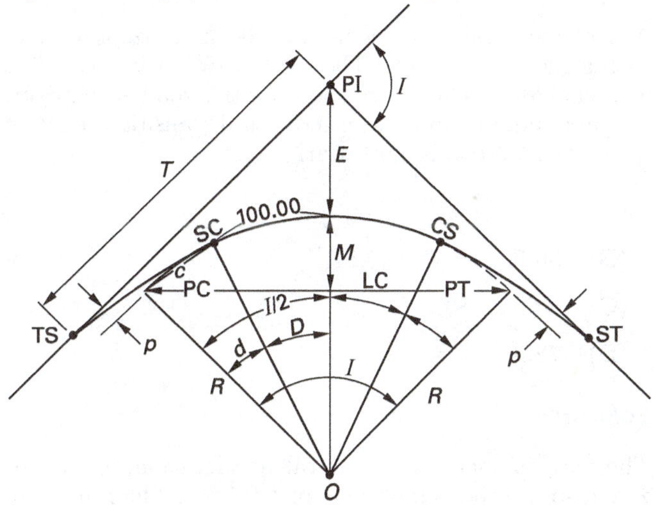

Table 50.2 *Spiral Curve Abbreviations*

CS	curve to spiral point
I_s	spiral angle
L_c	curve length
L_P	distance to any point, P
L_s	length of spiral
Q	offset of ghost PC to new tangent (tangent shift)
R_c	radius of circular curve
SC	spiral to curve point
SPI	spiral to point of intersection
ST	spiral to tangent point
TS	tangent to spiral point

Description

The minimum spiral transition length is based on driver comfort and lateral shifts in the position of a vehicle within the curve. Spirals should be long enough that the increase in lateral acceleration as a vehicle enters the curve is comfortable to the driver. Spirals should also be long enough that the shift they cause in a vehicle's lateral position is similar to the shift produced by a vehicle's natural path.

The minimum spiral curve length, L_s, is calculated from the speed, v (in mi/hr), radius of the curve, R (in ft), and the increase in the rate of lateral acceleration, C (in ft/sec^3). Equation 50.12 is also known as the 1909 *Shortt equation.*[1]

Transportation/ Surveying

[1]Although not mentioned in the NCEES *FE Reference Hanbook* (*NCEES Handbook*), Eq. 50.12 is only valid for U.S. units, since the constant term consolidates the conversion factors between miles and hours to feet and seconds, respectively. The units of the 3.15 term are $\text{ft}^3\text{-hr}^3/\text{mi}^3\text{-sec}^3$.

Example 20

A one-lane rural road includes a 700 ft horizontal curve with a radius of 4010 ft. The road is 15 ft wide with 9 ft wide shoulders. The design speed for this road is 45 mi/hr. The minimum required length of spiral transition between the curve and road is most nearly

(A) 28 ft

(B) 36 ft

(C) 44 ft

(D) 72 ft

Solution

The rate of increase in lateral acceleration, C, is not given, so use the default rate of 1 ft/sec^3. The minimum required length of spiral transition onto this road is

$$L_s = \frac{3.15\text{v}^3}{RC} = \frac{\left(3.15\ \frac{\text{ft}^3\text{-hr}^3}{\text{mi}^3\text{-sec}^3}\right)\left(45\ \frac{\text{mi}}{\text{hr}}\right)^3}{(4010\ \text{ft})\left(1\ \frac{\text{ft}}{\text{sec}^3}\right)}$$
$$= 71.58\ \text{ft} \quad (72\ \text{ft})$$

The answer is (D).

Equation 50.13: Superelevation and Side Friction Factor

$$0.01e + f = \frac{\text{v}^2}{15R} \qquad 50.13$$

Description

Equation 50.13 illustrates the relationship between the superelevation rate, e; side friction factor, f; speed, v; and curve radius, R. The side friction factor varies, but is commonly taken as 0.16 for speeds less than 30 mi/hr. Equation 50.13 is a dimensionally inconsistent equation and is valid for customary U.S. units only. The coefficient 0.01 in Eq. 50.13 converts superelevation from percent to ft/ft. The coefficient is not required if superelevation is given in decimal form.

Example 21

A one-lane rural road has a spiral curve with a curve radius of 4320 ft. The curve extends for 820 ft along its centerline. The road is 16 ft wide with 10 ft wide shoulders. The design speed for this road is 45 mi/hr. The superelevation required so that side friction is NOT needed is most nearly

(A) 0.00050

(B) 0.031

(C) 0.072

(D) 0.10

Solution

For a side friction factor of $f = 0$, the superelevation is

$$e + f = e = \frac{\text{v}^2}{15R} = \frac{\left(45\ \frac{\text{mi}}{\text{hr}}\right)^2}{(15)(4320\ \text{ft})}$$
$$= 0.0313 \quad (0.031)$$

The answer is (B).

4. VERTICAL CURVES

Vertical curves are used to connect two vertical tangents in order to change the grade of a highway. Most vertical curve layouts use *parabolic curves*, which can be symmetrical or unsymmetrical. Symmetrical parabolic curves are most often used because they have a constant rate of grade change, which simplifies calculation. Calculations may also be simplified by basing control points on offsets from the vertical tangents. All calculations are based on a horizontal plane, projecting horizontal measurements up or down to the road surface, regardless of the slope of the grade. Slope distances are generally only used for quantity estimation purposes. Vertical alignment is shown as a vertical surface cut along the *horizontal control line*, which is normally the centerline of the roadway. Vertical curves are usually identified by the station of the point of vertical intersection, PVI, and the PVI elevation, such as "vertical curve 67+84 at elevation 100." The location and elevation of the PVI are the most important geometric points of a vertical curve, as the PVI location controls the grade lines. The elements of a vertical curve are shown in Fig. 50.3. Table 50.3 lists the standard abbreviations used to describe geometric elements of vertical curves.

A *sag vertical curve* is concave upward, and a *crest vertical curve* is concave downward. Grades are shown as positive (+) if the profile rises with advancing stations and negative (−) if the profile falls with advancing stations. Figure 50.4(a) illustrates a symmetrical parabolic crest vertical curve, and Fig. 50.4(b) illustrates a sag vertical curve. Vertical curves that connect positive slopes with negative slopes, such as those shown in Fig. 50.4, have *turning points*, locations where the slope is zero (i.e., horizontal). The station of a turning point

Figure 50.3 *Vertical Curve Elements*

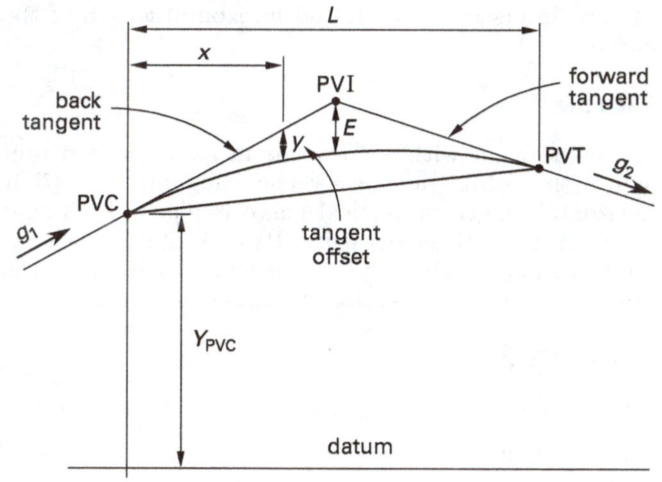

(not to scale)

Table 50.3 *Vertical Curve Abbreviations*

PVC	(beginning) point on vertical curve
PVI	point of vertical curve intersection
PVT	(ending) point on vertical tangent
V	vertex (the intersection of the two tangents)

Figure 50.4 *Symmetrical Parabolic Vertical Curve*

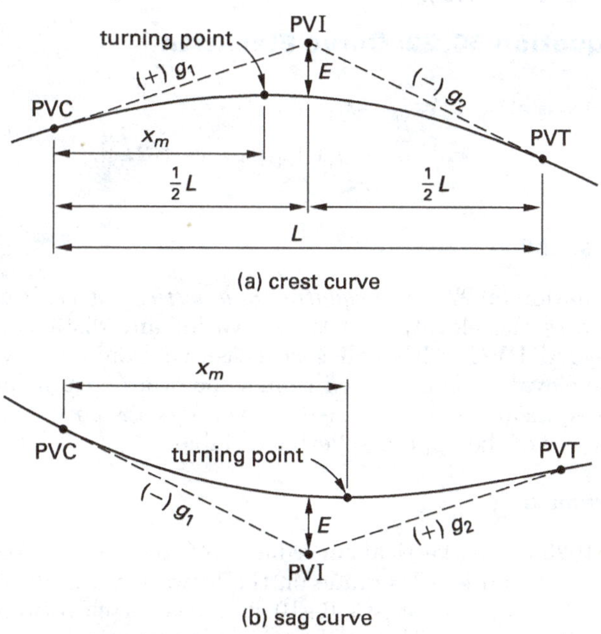

generally does not coincide with the station of the vertex. Catch basins and other drains in sag vertical curves should be located at the turning point.

The principle of a parabolic curve is applied directly to curve calculations. A parabolic curve consists of three basic points: the *point of vertical curve* (PVC), the *point of vertical intersection* (PVI), and the *point of vertical tangency* (PVT). The PVC is the *beginning of the curve* (also known as BVC), and the PVT is the *end of the curve* (also called EVC). The PVI is the intersection of the two tangents and is also referred to as the vertex, V. Offsets to a tangent, y, of a parabola are proportional to the squares of the distances from the point of tangency (i.e., the point where the curve and tangent meet), as shown in Fig. 50.5.

Figure 50.5 *Tangent Offset Relationship to Distance in a Parabolic Curve*

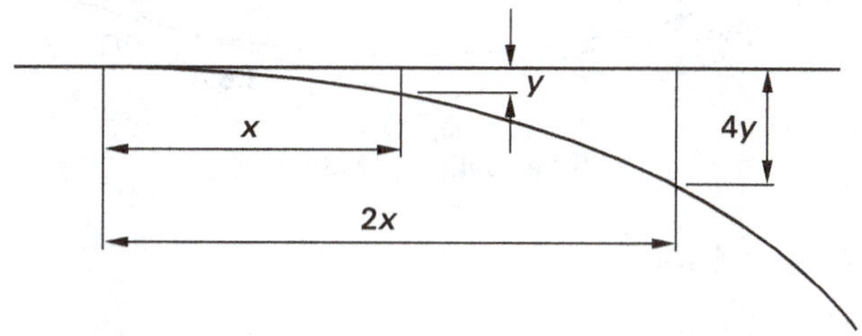

Equation 50.14 Through Eq. 50.17: Tangent Offsets

$$y = ax^2 \qquad 50.14$$

$$E = a\left(\frac{L}{2}\right)^2 \qquad 50.15$$

$$E = \frac{AL}{800} \quad [L \text{ in feet}] \qquad 50.16$$

$$a = \frac{g_2 - g_1}{2L} \qquad 50.17$$

Description

The tangent offset, y, is found from the *parabola constant*, a, calculated from Eq. 50.17, and the horizontal distance from the PVC to any given point on the curve, x. Offsets to a tangent are proportional to the squares of the distances from the point of tangency (i.e., the point where the curve and tangent meet), as shown in Eq. 50.14. The tangent offset at PVI, E, is calculated from Eq. 50.15. E may also be calculated from Eq. 50.16 from the absolute value of the algebraic difference in grades, A, and the curve length, L, in feet.

Equation 50.18: Horizontal Distance

$$x_m = -\frac{g_1}{2a} = \frac{g_1 L}{g_1 - g_2} \qquad 50.18$$

Description

Equation 50.18 calculates the horizontal distance to the minimum or maximum elevation on a curve, x_m, known as the *turning point*. The turning point is also called the *high point* for crest curves and the *low point* for sag curves. g_2 is the grade out of the curve, and g_1 is the grade into the curve.

Example

What is most nearly the horizontal distance to the low point for the sag curve shown?

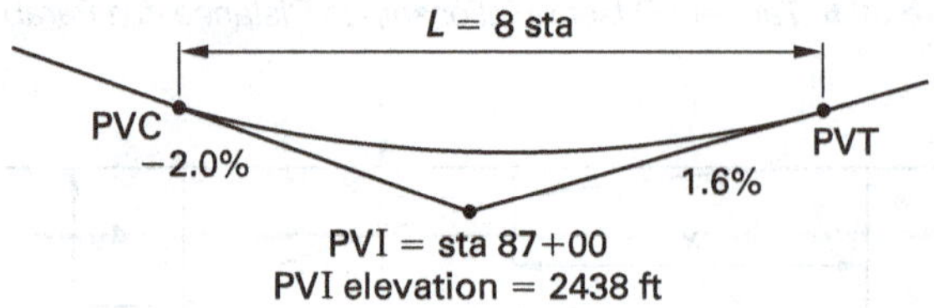

(A) 4.4 sta

(B) 5.6 sta

(C) 6.5 sta

(D) 7.1 sta

Solution

The horizontal distance to the low point is

$$x_m = \frac{g_1 L}{g_1 - g_2} = \frac{(-0.02)(8\text{ sta})}{-0.02 - 0.016} = 4.44\text{ sta} \quad (4.4\text{ sta})$$

The answer is (A).

Equation 50.19: Rate of Change in Grade

$$r = \frac{g_2 - g_1}{L} \qquad 50.19$$

Description

The *rate of change in grade*, r, is calculated per station. The rate of change will be negative for crest vertical curves and positive for sag vertical curves.

Equation 50.20 and Eq. 50.21: Tangent Elevation

$$\text{tangent elev} = Y_{\text{PVC}} + g_1 x \quad \text{[back tangent]} \qquad 50.20$$

$$\text{tangent elev} = Y_{\text{PVI}} + g_2(x - L/2) \quad \begin{bmatrix}\text{forward}\\ \text{tangent}\end{bmatrix} \qquad 50.21$$

Description

The *tangent elevation* is found from Eq. 50.20 by adding the product of the grade and distance to the elevation of the point of vertical curvature, PVC, if calculating along the back tangent. For the forward tangent, the elevation is calculated using Eq. 50.21 from the point of vertical intersection, PVI, and horizontal length of the curve, L.

Example

A back tangent with a 7% grade meets a forward tangent with a −5% grade on a vertical alignment. A 975 ft horizontal length of vertical curve is placed such that the point of vertical curvature (PVC) is at sta 10+35 at an elevation of 210 ft. The tangent elevation at the point of vertical intersection (PVI) is most nearly

(A) 170 ft

(B) 190 ft

(C) 220 ft

(D) 240 ft

Solution

The PVI is located at $x = L/2 = (975\text{ ft})/2 = 487.5\text{ ft}$ from the PVC. From Eq. 50.20, the tangent elevation is

$$\begin{aligned}\text{tangent elev} &= Y_{\text{PVC}} + g_1 x = 210\text{ ft} + (0.07)(487.5\text{ ft})\\ &= 244\text{ ft} \quad (240\text{ ft})\end{aligned}$$

The answer is (D).

Equation 50.22: Curve Elevation

$$\begin{aligned}\text{curve elev} &= Y_{\text{PVC}} + g_1 x + a x^2\\ &= Y_{\text{PVC}} + g_1 x + [(g_2 - g_1)/(2L)]x^2\end{aligned} \qquad 50.22$$

Description

Equation 50.22 is the *equation of a vertical curve.* It calculates the elevation of the curve for any distance, x, beyond PVC. Although x is measured from the PVC, the elevation is measured from some other datum (perhaps, mean sea level). x is relative; Y is absolute, hence the use of the uppercase letter variable.

Example

A 1025 ft long vertical curve has a 5% grade on the back tangent and a −2% grade on the forward tangent. The PVC is located at sta 6+10 and has an elevation of 195 ft. At sta 7+31, the elevation of the vertical curve is most nearly

(A) 170 ft

(B) 200 ft

(C) 280 ft

(D) 310 ft

Solution

From the PVC at sta 6+10 to the location of desired elevation at sta 7+31, $x = 731\text{ ft} - 610\text{ ft} = 121\text{ ft}$. The curve elevation at sta 7+31 is

$$\begin{aligned}\text{curve elev}_{\text{sta}\,7+31} &= Y_{\text{PVC}} + g_1 x \\ &\quad + [(g_2 - g_1)/(2L)]x^2 \\ &= 195\text{ ft} + (0.05)(121\text{ ft}) \\ &\quad + \left(\frac{-0.02 - 0.05}{(2)(1025\text{ ft})}\right)(121\text{ ft})^2 \\ &= 200.6\text{ ft} \quad (200\text{ ft})\end{aligned}$$

The answer is (B).

5. SIGHT DISTANCE

Sight distance is the length of roadway a driver can see ahead of the vehicle. Within this sight distance, a driver must analyze upcoming road conditions and traffic situations, select appropriate actions or maneuvers, and then complete these actions or maneuvers. The design speed of horizontal curves and vertical curves and a driver's ability to see around obstructions are based on the maximum sight distance available to a driver.

American Association of State Highway and Transportation Officials (AASHTO) design criteria use three types of sight distance: stopping sight distance, decision sight distance, and passing sight distance.

Safe *stopping sight distance* is the total distance required for a driver traveling at design speed to stop a vehicle before reaching an object in its path. Stopping sight distance comprises two distances: the *perception-reaction distance* and the braking distance. The *perception-reaction distance* (also called the *PIEV distance*) is the distance traveled at a constant approach speed during the *perception-reaction time* (known as the *PIEV time*), which is measured from the moment the driver sees an object requiring a stop to the time the brakes are applied. AASHTO refers to the PIEV distance as the *brake reaction distance* and uses an average of 2.5 sec as the *brake reaction time*, which is the time it takes a driver to apply the brakes after seeing an object. Stopping sight distance calculations are determined using the driver's eye height set at 3.5 ft and the object height at 2.0 ft, which is equivalent to the height of a passenger car's taillight.

Decision sight distance is appropriate where hazards exist that require drivers to make decisions to perform maneuvers other than a stop, such as lane changes or exit ramp selections, in order for traffic to proceed in an orderly and smooth fashion. These decisions often include multiple actions to be taken simultaneously and may involve selection from several choices of action to be performed. Examples include approaches to complex intersections, multiple interchange ramps, toll booth plazas, restrictive sight distance locations, and instances in which the driver may need to be prepared for further alternative maneuvers in quick succession. Only providing sufficient sight distance for a hurried stop may increase the danger to other motorists, while not providing enough time to make an appropriate selection of an alternate path or course of action for an evasive maneuver. More decision time may also be needed where visual clutter exists, such as advertising signs and busy commercial activity found along a commercial corridor. Since decision sight distances include a margin of error in addition to the time necessary to make evasive maneuvers, decision sight distance values are typically much larger than stopping sight distances.

Passing sight distance is applicable on two-lane highways when there are sufficient gaps in opposing flows to allow passing maneuvers to occur and when there are few access points, with only occasional entering traffic and sufficient sight distance to observe traffic in both lanes and both directions. Occasional access includes residential and rural driveways, minor side roads, and driveways to adjoining land uses. Passing sight distance is not applicable to multilane highways.

Transportation/ Surveying

Equation 50.23: *K*-Value Method

$$K = \frac{L}{A} \qquad 50.23$$

Variation

$$K = \frac{L}{|g_2 - g_1|}$$

Description

The *K-value method* of analysis used in the *AASHTO Geometric Design of Highways and Streets* (*Green Book*) is a simplified and more conservative method of choosing a stopping sight distance for a crest vertical curve. The length of the vertical curve per percent grade difference, K, is the ratio of the curve length, L, to grade difference, A. K is always a positive value.

Equation 50.24: Straight-Ahead Stopping Sight Distance[2]

$$\text{SSD} = 1.47\text{v}t + \frac{\text{v}^2}{30\left(\left(\frac{a}{32.2}\right) \pm G\right)} \qquad 50.24$$

[2]The *NCEES Handbook* uses SSD to represent stopping sight distance in Eq. 50.24. SSD in Eq. 50.24 is the same as S used in Eq. 50.25 through Eq. 50.30.

Description

Equation 50.24 calculates the stopping sight distance, SSD, for straight-line travel on a constant grade. The grade, G, is in decimal form and is negative if the roadway is downhill. The perception-reaction time, t, often is taken as 2.5 sec. v is the vehicle's approach speed, a is the deceleration rate, and 32.2 is the gravitational acceleration in ft/sec^2. Equation 50.24 applies for customary U.S. units only.

Example

A vehicle travels at 50 mi/hr on a straight, level highway. Using a driver reaction time of 2.5 sec and a deceleration rate of 20 ft/sec^2, what is most nearly the stopping sight distance?

(A) 180 ft

(B) 220 ft

(C) 290 ft

(D) 320 ft

Solution

The stopping sight distance is

$$\begin{aligned}\text{SSD} &= 1.47\text{v}t + \frac{\text{v}^2}{30\left(\left(\frac{a}{32.2}\right) \pm G\right)} \\ &= (1.47)\left(50\ \frac{\text{mi}}{\text{hr}}\right)(2.5\ \text{sec}) + \frac{\left(50\ \frac{\text{mi}}{\text{hr}}\right)^2}{(30)\left(\frac{20\ \frac{\text{ft}}{\text{sec}^2}}{32.2\ \frac{\text{ft}}{\text{sec}^2}} + 0\right)} \\ &= 317.9\ \text{ft} \quad (320\ \text{ft})\end{aligned}$$

The answer is (D).

Equation 50.25 and Eq. 50.26: Crest Vertical Curve Length Based on Stopping Sight Distance

$$L = \frac{AS^2}{100\%\left(\sqrt{2h_1} + \sqrt{2h_2}\right)^2} \quad [S \le L] \qquad 50.25$$

$$L = 2S - \frac{200\%\left(\sqrt{h_1} + \sqrt{h_2}\right)^2}{A} \quad [S > L] \qquad 50.26$$

Description

Equation 50.25 and Eq. 50.26 are used to determine the curve length based on acceptable sight distance on crest vertical curves. The appropriate equation is chosen based on whether the sight distance is shorter than the length of the vertical curve ($S \le L$) or is longer than the length ($S > L$). A is the absolute value of the algebraic difference in grades, g, expressed as a percentage.

$$A = |g_2 - g_1| \times 100\%$$

S is the stopping sight distance, in feet, and h_1 and h_2 are the driver's eye height and the object height, respectively, measured from the roadway surface.

Example

A crest on a section of highway consists of a vertical curve with a 1600 ft radius and a 1% grade followed by a −3% grade. The design requirements are as follows.

design speed = 50 mi/hr
driver eye height = 3.9 ft
object (to be avoided) height = 0.6 ft
stopping sight distance = 980 ft

The minimum required length of vertical curve needed to satisfy the design stopping sight distance is most nearly

(A) 2200 ft

(B) 2300 ft

(C) 2500 ft

(D) 2800 ft

Solution

The curve length must be calculated using both Eq. 50.25 and Eq. 50.26. From Eq. 50.25, with the assumption of $S \le L$,

$$\begin{aligned}L &= \frac{AS^2}{100\%\left(\sqrt{2h_1} + \sqrt{2h_2}\right)^2} \\ &= \frac{|g_2 - g_1|\, S^2}{100\%\left(\sqrt{2h_1} + \sqrt{2h_2}\right)^2} \\ &= \frac{|-3\% - 1\%|\,(980\ \text{ft})^2}{(100\%)\left(\sqrt{(2)(3.9\ \text{ft})} + \sqrt{(2)(0.6\ \text{ft})}\right)^2} \\ &= 2541\ \text{ft}\end{aligned}$$

From Eq. 50.26, with the assumption of $S>L$,

$$\begin{aligned} L &= 2S - \frac{200\%\left(\sqrt{h_1}+\sqrt{h_2}\right)^2}{A} \\ &= 2S - \frac{200\%\left(\sqrt{h_1}+\sqrt{h_2}\right)^2}{|g_2 - g_1|} \\ &= (2)(980 \text{ ft}) - \frac{(200\%)\left(\sqrt{3.9 \text{ ft}}+\sqrt{0.6 \text{ ft}}\right)^2}{|-3\% - 1\%|} \\ &= 1582 \text{ ft} \end{aligned}$$

The assumption of $S>L$ is invalid, so the minimum required vertical curve length is

$$L = 2541 \text{ ft} \quad (2500 \text{ ft})$$

The answer is (C).

Equation 50.27 and Eq. 50.28: Crest Vertical Curve Length Based on Stopping Sight Distance

$$L = \frac{AS^2}{2158} \quad [S \leq L] \qquad 50.27$$

$$L = 2S - \frac{2158}{A} \quad [S > L] \qquad 50.28$$

Description

Equation 50.27 and Eq. 50.28 are used to calculate the minimum crest vertical curve length based on stopping sight distance. These equations are based on a driver's eye height, h_1, of 3.5 ft and an object height, h_2, of 2.0 ft.

Equation 50.29 and Eq. 50.30: Sag Vertical Curve Length Based on Stopping Sight Distance

$$L = \frac{AS^2}{400 + 3.5S} \quad [S \leq L] \qquad 50.29$$

$$L = 2S - \left(\frac{400 + 3.5S}{A}\right) \quad [S > L] \qquad 50.30$$

Description

Equation 50.29 and Eq. 50.30 are used to calculate the minimum sag vertical curve length based on stopping sight distance. These equations are based on a driver's eye height, h_1, of 3.5 ft and an object height, h_2, of 2.0 ft. However, 2 and 3.5 are generally used without units in calculations, since the equations are dimensionally inconsistent.

Example 27

A section of highway has grades of −1% followed by 3%. The stopping sight distance is 430 ft for a design speed of 50 mi/hr. The required length of vertical curve needed to satisfy the stopping sight distance for this design speed is most nearly

(A) 270 ft

(B) 380 ft

(C) 410 ft

(D) 450 ft

Solution

Since there is a negative grade preceding a positive grade, this is a sag vertical curve. The algebraic difference between grades is

$$\begin{aligned} A &= |g_2 - g_1| \\ &= |3\% - (-1\%)| \\ &= 4\% \end{aligned}$$

From Eq. 50.29, where the stopping sight distance, S, is less than or equal to the vertical curve length, L,

$$\begin{aligned} L &= \frac{AS^2}{400 + 3.5S} \\ &= \frac{(4\%)(430 \text{ ft})^2}{400 \text{ ft} + (3.5)(430 \text{ ft})} \\ &= 388.2 \text{ ft} \end{aligned}$$

From Eq. 50.30, where $S>L$,

$$\begin{aligned} L &= 2S - \left(\frac{400 + 3.5S}{A}\right) \\ &= (2)(430 \text{ ft}) - \left(\frac{400 \text{ ft} + (3.5)(430 \text{ ft})}{4\%}\right) \\ &= 383.8 \text{ ft} \end{aligned}$$

From the two values for length of curve, it can be seen that the stopping sight distance is greater than the curve length. Therefore, the required vertical length of curve is

$$L = 383.8 \text{ ft} \quad (380 \text{ ft})$$

The answer is (B).

Transportation/ Surveying

Equation 50.31: Sag Vertical Curve Length Based on Riding Comfort

$$L = \frac{Av^2}{46.5} \qquad 50.31$$

Description

With sag curves, both gravitational and centrifugal forces act on the driver and passengers, making comfort the controlling factor in the design. Equation 50.31 can be used to calculate the length of the curve so that the added acceleration is kept below 1 ft/sec^2.

Equation 50.32 and Eq. 50.33: Sight Distance Under an Overhead Structure

$$L = \frac{AS^2}{800\left(C - \frac{h_1 + h_2}{2}\right)} \quad [S \le L] \qquad 50.32$$

$$L = 2S - \frac{800}{A}\left(C - \frac{h_1 + h_2}{2}\right) \quad [S > L] \qquad 50.33$$

Description

The sight distance under an overhead structure, such as an overpass, to see an object beyond a sag vertical curve is given by Eq. 50.32 and Eq. 50.33. C is the vertical clearance for an overhead structure located within 200 ft of the midpoint of the curve.

Equation 50.34: Sight Distance Around an Obstruction

$$\text{HSO} = R\left[1 - \cos\left(\frac{28.65S}{R}\right)\right] \qquad 50.34$$

Variation

$$\text{HSO} = R(1 - \cos\theta) = R\left(1 - \cos\frac{DS}{200}\right)$$

Description

A horizontal circular curve is shown in Fig. 50.6. Obstructions along the inside of curves, such as retaining walls, cut slopes, trees, buildings, and bridge piers, can limit the available (chord) sight distance. Often, a curve must be designed that will simultaneously provide the required stopping sight distance while maintaining a clearance from a roadside obstruction.

Figure 50.6 *Horizontal Curve with Obstructions*

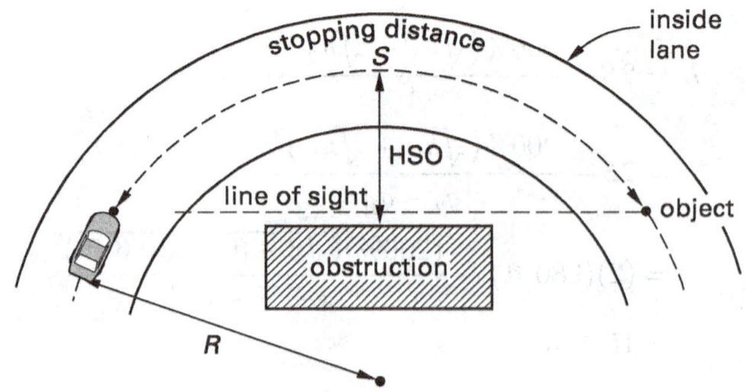

The equations for calculating the geometry of a horizontal curve to see around an obstruction assume that the stopping sight distance is less than the curve length (i.e., $S \le L$). The stopping sight distance and length along the centerline of the inside lane of the curve are the same. Given a stopping sight distance and a curve radius, the clear area *horizontal sightline offset*, HSO, can be calculated. The angles are given in degrees, not radians.

51 Earthwork

Nomenclature

a	dimension	ft
A	area	ft^2
b	dimension	ft
h	height	ft
L	length	ft
L	load factor	–
N	number	–
r	radius	ft
V	volume	ft^3

Symbols

ϕ	angle of repose	deg

Subscripts

b	bank measure
c	compacted
l	loose measure
m	mean or midsection

1. EARTHWORK

Earthwork is the excavation, hauling, and placing of soil, rock, gravel, or other material found below the surface of the earth. This definition also includes the measurement of such material in the field, the computation in the office of the volume of such material, and the determination of the most economical method of performing such work.

2. UNIT OF MEASURE

In the United States, the *cubic yard* (i.e., the "yard") is the unit of measure for earthwork. The cubic meter is used in metricated countries.

3. SWELL AND SHRINKAGE

The volume and density of earth changes under natural conditions and during the operations of excavation, hauling, and placing. The volume of a loose pile of excavated earth will be greater than the original, in-place (i.e., *in situ*) natural volume. If the earth is compacted after it is placed, the volume may be less than its original volume.

The volume of the earth in its natural state is known as *bank-measure*. The volume during transport is known as *loose-measure*. The volume after compaction is known as *compacted-measure*. In the United States, the terms *bank cubic yards* (BCY), *loose cubic yards* (LCY), and *compacted cubic yards* (CCY) may be encountered.

When earth is excavated, it increases in volume because of an increase in voids. The change in volume of earth from its natural to loose state is known as *swell*. Swell is expressed as a percentage of the natural volume. A soil's *load factor*, L, in a particular excavation environment is the inverse of the *swell factor* (also known as the *bulking factor*), the sum of 1 and the decimal swell.

$$V_l = \left(\frac{100\% + \%\,\text{swell}}{100\%}\right) V_b = \frac{V_b}{L}$$

The decrease in volume of earth from its natural state to its compacted state is known as *shrinkage*. Shrinkage also is expressed as a percent decrease from the natural state.

$$V_c = \left(\frac{100\% - \%\,\text{shrinkage}}{100\%}\right) V_b$$

As an example, 1 yd^3 in the ground may become 1.2 yd^3 loose-measure and 0.85 yd^3 after compaction. The swell would be 20%, and the shrinkage would be 15%. Swell and shrinkage vary with soil types.

4. CLASSIFICATION OF MATERIALS

Excavated material is usually classified as *common excavation* or *rock excavation*. Common excavation is soil.

In highway construction, common road excavation is soil found in the roadway. *Common borrow* is soil found outside the roadway and brought in to the roadway. Borrow is necessary where there is not enough material in the roadway excavation to provide for the embankment.

5. CUT AND FILL

Earthwork that is to be excavated is known as *cut*. Excavation that is placed in embankment is known as *fill*.

Payment for earthwork is normally either for cut and not for fill, or for fill and not for cut. In highway work, payment is usually for cut; in dam work, payment is usually for fill. To pay for both would require measuring two different volumes and paying for moving the same earth twice.

6. FIELD MEASUREMENT

Cut and fill volumes can be computed from slope-stake notes, from plan cross sections, or by photogrammetric methods.

7. PLANIMETERS

A *planimeter* is a table-mounted device used to measure irregular areas on a map, photograph, or illustration through mechanical integration. (See Fig. 51.1.) It typically has two arms. The pole arm rotates freely around the pole, which is fixed on the table. The tracer arm rotates around the pivot, which is where it joins the polar arm. The pointer on the other end of the tracer arm is used to trace the perimeter of the region. Near the pivot is a wheel which simply rolls and slides along the table. In operation, the area to be measured is traced clockwise with the tracer. As the area is traced, the measuring wheel rolls and accumulates the total distance on a dial. An optional support wheel may maintain balance. The number of turns recorded on the dial is proportional to the area of the region traced out.

Green's theorem (*Green's law*) relates a double integral over a closed region to a line integral over its boundary. Although the mathematics necessary to prove how planimeters measure area functions can be complex, the actual usage is not. The following equation is the governing equation of planimeter area. r is the radius of the wheel, N is the number of wheel revolutions, and L is the length of the tracer arm. The traced area, A, has the

Figure 51.1 *Planimeter*

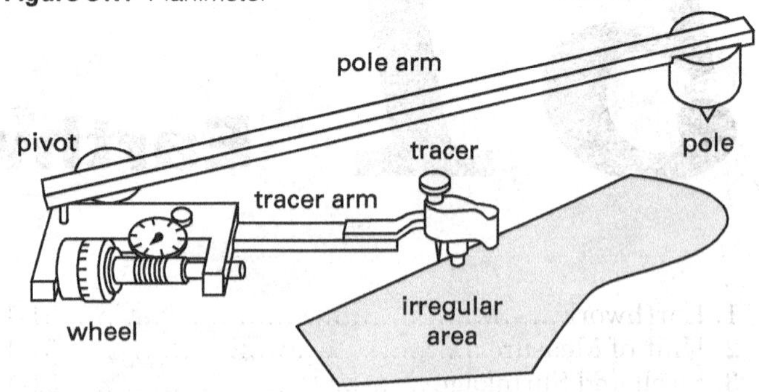

same units as r and L. This area is multiplied by the square of the illustration scale to determine the feature's true area.

$$A = 2\pi rNL$$

8. CROSS SECTIONS

Cross sections are profiles of the earth taken at right angles to the centerline of an engineering project (such as a highway, canal, dam, or railroad). A cross section for a highway is shown in Fig. 51.2.

Figure 51.2 *Typical Highway Cross Section*

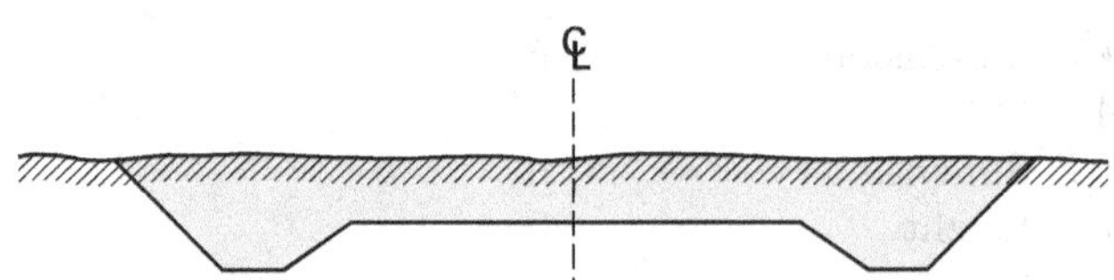

9. ORIGINAL AND FINAL CROSS SECTIONS

To obtain volume measurement, cross sections are taken before construction begins and after it is completed. By plotting the cross section at a particular station both before and after construction, a sectional view of the change in the profile of the earth along a certain line is obtained. The change along this line appears on the plan as an area. By using these areas at various intervals along the centerline, and by using distance between the areas, volume can be calculated.

10. TYPICAL SECTIONS

Typical sections show the cross section view of the project as it will look on completion, including all dimensions. Highway projects usually show several typical sections including cut sections, fill sections, and sections showing both cut and fill. Interstate highway plans also show access-road sections and sections at ramps.

11. DISTANCE BETWEEN CROSS SECTIONS

Cross sections are usually taken at each full station and at breaks in the ground along the centerline. In taking cross sections, the change in the earth's surface from one cross section to the next is assumed to be uniform, and a section halfway between the cross sections is taken as an average of the two. If the ground breaks appreciably between any two full-stations, one or more cross sections between full-stations must be taken. This is referred to as *taking sections at pluses.* Figure 51.3 shows eleven stations at which cross sections should be taken.

Figure 51.3 *Cross Section Locations*

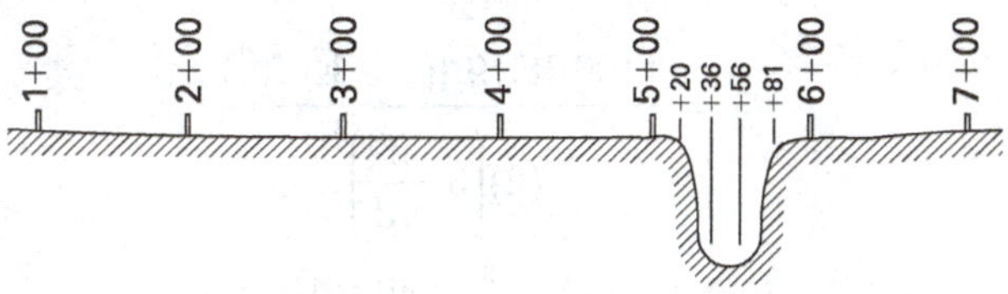

In rock excavation, or any other expensive operation, cross sections should be taken at intervals of 50 ft or less. Cross sections should always be taken at the point of curvature (PC) and point of tangency (PT) of a curve. Plans should also show a section on each end of a project (where no construction is to take place) so that changes caused by construction will not be abrupt.

Where a cut section of a highway is to change to a fill section, several additional cross sections are needed. Such sections are shown in Fig. 51.4.

12. GRADE POINT

A point where a cut or fill section meets the natural ground (where a cut section begins) is known as the *grade point.* Similarly, a locus of grade points is known as a *grade line.*

13. VOLUMES OF PILES

The volumes of piles of soil, recycled pavement, and paving materials can be calculated if the pile shapes are assumed. The *angle of repose*, ϕ, depends on the material, but it is approximately 30° for smooth gravel, 40° for sharp gravel, 25–35° for dry sand, 30–45° for moist sand, 20–40° for wet sand, and 37° for cement.

Equation 51.1: Pyramid or Cone Volume

$$V = h(\text{area of base})/3 \qquad 51.1$$

Description

The volume of a pyramid or cone (see Fig. 51.5) can be calculated from Eq. 51.1.

Figure 51.4 *Cut Changing to Fill*

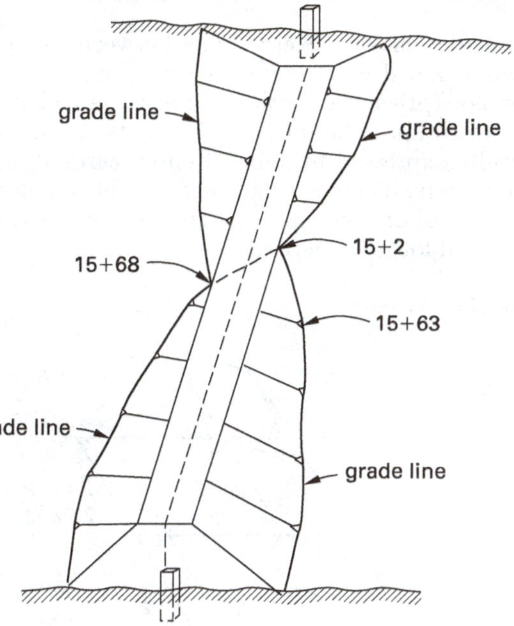

Figure 51.5 *Cone Pile Shape*

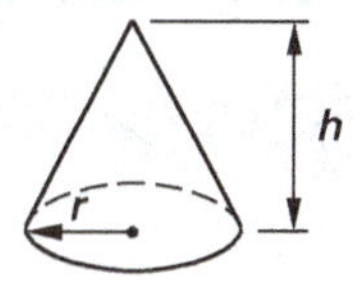

Example

An 80 ft tall conical pile of soil has a base with an area of 1100 ft^2. What is most nearly the volume of the pile?

(A) 1100 yd^3

(B) 1300 yd^3

(C) 1600 yd^3

(D) 2900 yd^3

Solution

The pile volume is

$$V = h(\text{area of base})/3 = \frac{(80 \text{ ft})(1100 \text{ ft}^2)}{(3)\left(3 \dfrac{\text{ft}}{\text{yd}}\right)^3}$$

$$= 1086 \text{ yd}^3 \quad (1100 \text{ yd}^3)$$

The answer is (A).

14. EARTHWORK VOLUMES

A three-dimensional soil volume between two points is known as a soil *prismoid* or *prism*. (See Fig. 51.6.) The prismoid (prismatic) volume must be calculated in order to estimate hauling requirements. Such volume is generally expressed in units of cubic yards ("yards") or cubic meters. There are two methods of calculating the prismoid volume: the average end area method and the prismoidal formula method.

Figure 51.6 *Prismoid*

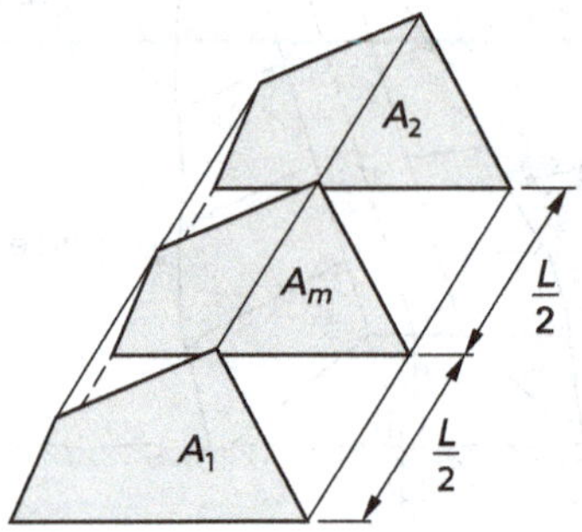

Equation 51.2: Average End Area Method

$$V = L(A_1 + A_2)/2 \qquad 51.2$$

Description

With the *average end area method*, the volume is calculated by averaging the two end areas and multiplying by the prism length. This disregards the slopes and orientations of the ends and sides, but is sufficiently accurate for most earthwork calculations. When the end area is complex, it may be necessary to use a planimeter or to plot the area on fine grid paper and simply count the squares. The average end area method usually overestimates the actual soils volume, favoring the contractor in earthwork cost estimates.

The precision obtained from the average end area method is generally sufficient unless one of the end areas is very small or zero. In that case, the volume should be calculated as a pyramid or truncated pyramid using Eq. 51.1.

Example

The total cut area and total fill area for two 100 ft stations (1 and 2) along a roadway are as follows.

station	total cut area (ft^2)	total fill area (ft^2)
1	9	24
2	15	32

What are most nearly the volumes of the cut, V_{cut}, and fill, V_{fill}, between the two stations?

(A) $V_{cut} = 0.4\ yd^3$; $V_{fill} = 1\ yd^3$

(B) $V_{cut} = 10\ yd^3$; $V_{fill} = 30\ yd^3$

(C) $V_{cut} = 40\ yd^3$; $V_{fill} = 100\ yd^3$

(D) $V_{cut} = 60\ yd^3$; $V_{fill} = 90\ yd^3$

Solution

The volume of cut is

$$\begin{aligned} V_{cut} &= L(A_1 + A_2)/2 \\ &= \frac{(100\ ft)(9\ ft^2 + 15\ ft^2)}{(2)\left(3\ \frac{ft}{yd}\right)^3} \\ &= 44.44\ yd^3 \quad (40\ yd^3) \end{aligned}$$

The volume of fill is

$$\begin{aligned} V_{fill} &= L(A_1 + A_2)/2 = \frac{(100\ ft)(24\ ft^2 + 32\ ft^2)}{(2)\left(3\ \frac{ft}{yd}\right)^3} \\ &= 103.70\ yd^3 \quad (100\ yd^3) \end{aligned}$$

The answer is (C).

Equation 51.3: Prismoidal Formula Method

$$V = L(A_1 + 4A_m + A_2)/6 \qquad 51.3$$

Description

The *prismoidal formula* is preferred when the two end areas differ greatly or when the ground surface is irregular. It generally produces a smaller volume than the average end area method and, thus, favors the owner-developer in earthwork cost estimating.

The prismoidal formula uses the mean area, A_m, midway between the two end sections. In the absence of actual observed measurements, the dimensions of the middle area can be found by averaging the similar dimensions of the two end areas. The middle area is not found by averaging the two end areas.

When using the prismoidal formula, the volume is not found as LA_m, although that quantity is usually sufficiently accurate for estimating purposes.

Example

Using the prismoidal formula and 100 ft stations, what is most nearly the prismoidal volume if the two end areas are 21 ft^2 and 76 ft^2, respectively, and the mean area is 52 ft^2?

(A) 50 yd^3

(B) 170 yd^3

(C) 190 yd^3

(D) 380 yd^3

Solution

The prismoid volume is

$$V = L(A_1 + 4A_m + A_2)/6$$
$$= \frac{(100\text{ ft})(21\text{ ft}^2 + (4)(52\text{ ft}^2) + 76\text{ ft}^2)}{(6)\left(3\ \frac{\text{ft}}{\text{yd}}\right)^3}$$
$$= 188.3\text{ yd}^3 \quad (190\text{ yd}^3)$$

The answer is (C).

15. BORROW PIT GEOMETRY

It is often necessary to borrow earth from an adjacent area to construct embankments. Normally, the *borrow pit* area is laid out in a rectangular grid with 10 ft, 50 ft, or even 100 ft squares. Elevations are determined at the corners of each square by leveling before and after excavation so that the cut at each corner can be calculated. (See Fig. 51.7.)

Points outside the cut area are established on the grid lines so that the lines can be reestablished after excavation is completed.

Figure 51.7 *Depth of Excavation (Cut) in a Borrow Pit Area*

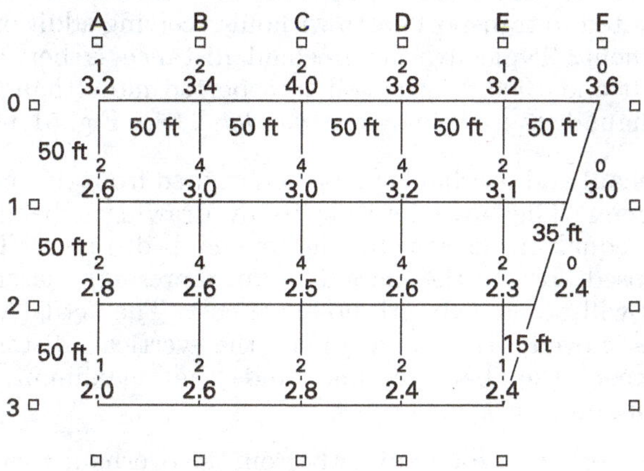

16. MASS DIAGRAMS

A *profile diagram* is a cross section of the existing ground elevation along a route alignment. The elevation of the route is superimposed to identify the *grade points*, the points where the final grade coincides with the natural elevation.

A *mass diagram* is a record of the cumulative earthwork volume moved along an alignment, usually plotted below profile sections of the original ground and finished grade. The mass diagram can be used to establish a finished grade that balances cut-and-fill volumes and minimizes long hauls.

After volumes of cut and fill between stations have been calculated, they are tabulated as shown in Table 51.1. The cuts and fills are then added, and the cumulative yardage at each station is recorded in the table. It is this cumulative yardage that is plotted as an ordinate. In Table 51.1, the baseline serves as the x-axis, and cumulative yardage that has a plus sign is plotted above the baseline. Cumulative yardage that has a minus sign is plotted below the baseline. The scale is chosen for convenience. 1 in = 5 ft is typical, although 1 in = 10 ft or 20 ft can be used in steep terrain.

Table 51.1 *Typical Cut and Fill Calculations (all volumes are in cubic yards)*

sta	cut +	fill −	cum sum
0			0
	184		
1			+184
	622		
2			+806
	1035		
3			+1841
	1268		
4			+3109
	1231		
5			+4340
	919		
6			+5259
	503		
7			+5762
	164	21	
8			+5905
	12	190	
9			+5727
		616	
10			+5111

In Fig. 51.8, the mass diagram is plotted on the lower half of the sheet, and the centerline profile of the project is plotted on the upper half.

Figure 51.8 Baseline and Centerline Profile Mass Diagram

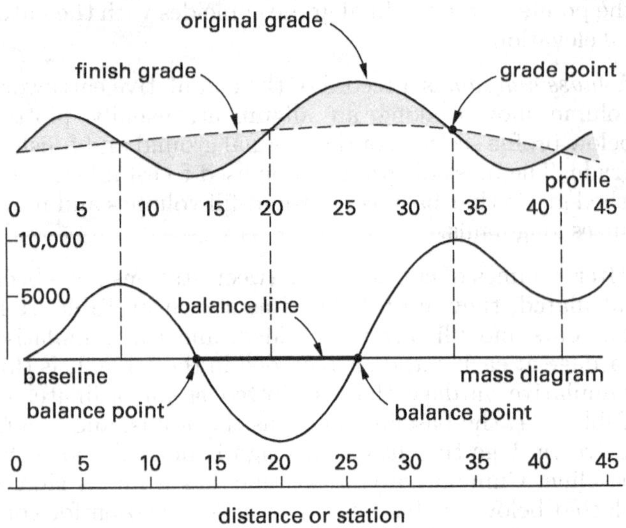

A rising line on the mass diagram represents areas of excavation; a falling line represents areas of embankment. The local minima and maxima on the mass diagram identify the grade points. Vertical distances on a mass diagram represent volumes of material (areas on the profile diagram). Areas in a mass diagram represent the product of volume and distance.

A *balance line* is a horizontal line drawn between two adjacent *balance points* in a crest or sag area. The volumes of excavation and embankment between the points are equal. Therefore, a contractor can plan on using the earth volume from the excavation on one side of the grade point for the embankment on the other side of the grade point. In Fig. 51.9, a balance line has been drawn that intersects the mass diagram at two points. These points represent the inclusive stations for which the cut-and-fill volumes are equal. That is, the cut soil (area B) can be used for the fill soil (area A).

It is important in planning and construction to know the points along the centerline of a particular section of cut that will balance a particular section of fill. For example, assume that a cut section extends from sta 12+25 to sta 18+65, and a fill section extends from sta 18+65 to sta 26+80. Also, assume that the excavated material will exactly provide the material needed to make the embankment. Then, cut balances fill, and sta 12+25 and sta 26+80 are balance points.

A balance point occurs where the mass curve crosses the baseline. In Fig. 51.10, a balance point falls between sta 13 and 14. The ordinate of 13 is +1519; the ordinate of 14 is −254. Therefore, the curve fell 1519 yd^3 + 254 yd^3 = 1773 yd^3 in 100 ft or 17.73 yd^3/ft. The curve crosses the baseline at a distance of 1519 $\text{yd}^3/(17.73\ \text{yd}^3/\text{ft})$ = 86 ft from sta 13 (13+86).

Figure 51.9 Balance Line Between Two Points

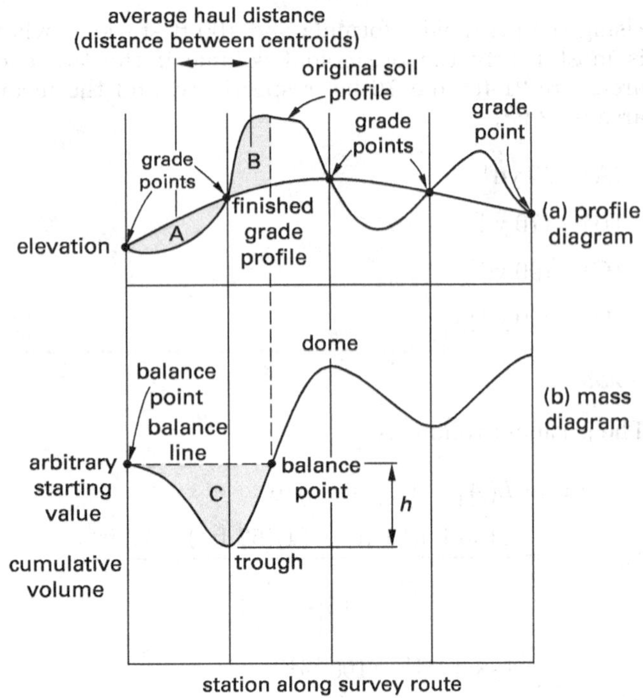

The average *haul distance* is the separation of the centroids of the excavation and embankment areas on the profile diagram. This distance is usually determined for each section of earthwork rather than for the entire mass of earthwork in a project.

The average haul can also be calculated from the mass diagram. For customary U.S. usage, the curve area under a balance line represents the total haul in *yard-stations*, which is the number of cubic yards moved in 100 ft. This figure can be used by the contractor to determine hauling costs. The height of the curve, or the distance between the balance line and the minimum or maximum, represents the *solidity* or the volume of the total haul. The average haul distance is found by dividing the curve area (area C) by the curve height, h.

The *freehaul distance* is the maximum distance, as specified in the construction contract, that the contractor is expected to transport earth without receiving additional payment. Typically, the freehaul distance is between 500 ft and 1000 ft. Any soil transported more than the freehaul distance is known as *overhaul.* (See Fig. 51.11.)

Freehaul and overhaul can be determined from the mass diagram. The procedure begins by drawing a balance line equal in length to the freehaul distance. The enclosed area on the mass diagram represents material that will be hauled with no extra cost. The actual volume moved (the solidity) is the vertical distance between the balance line and the maximum or minimum.

The overhaul is found directly from the overhaul area on the mass diagram. It can also be calculated indirectly from the overhaul volume and overhaul distance. The

Figure 51.10 Mass Diagram Showing Subbases

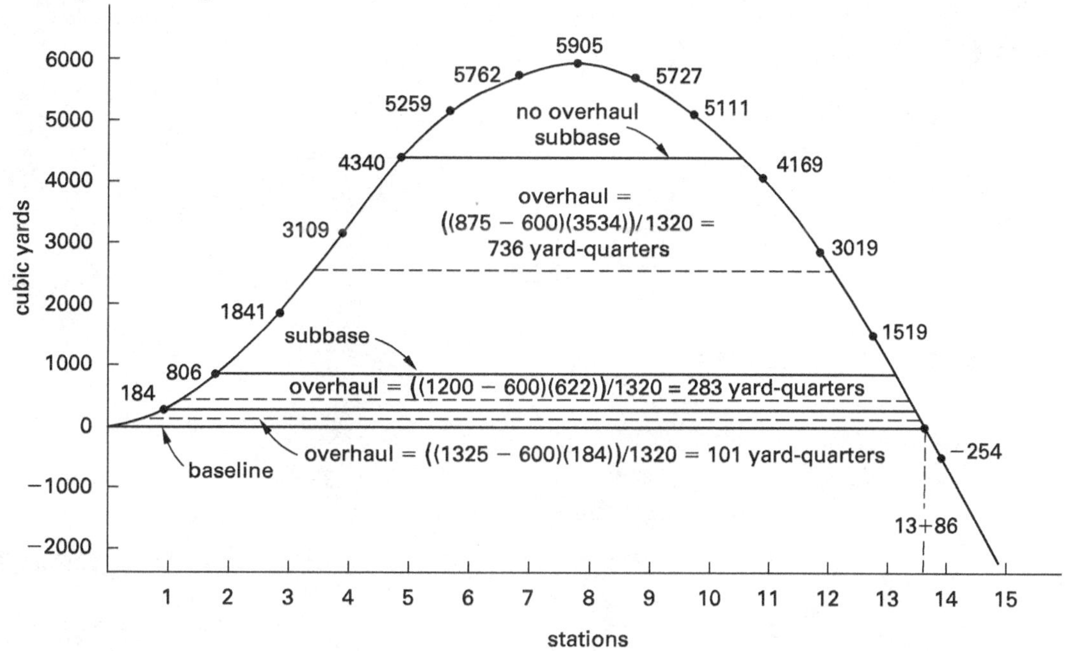

Figure 51.11 Freehaul and Overhaul

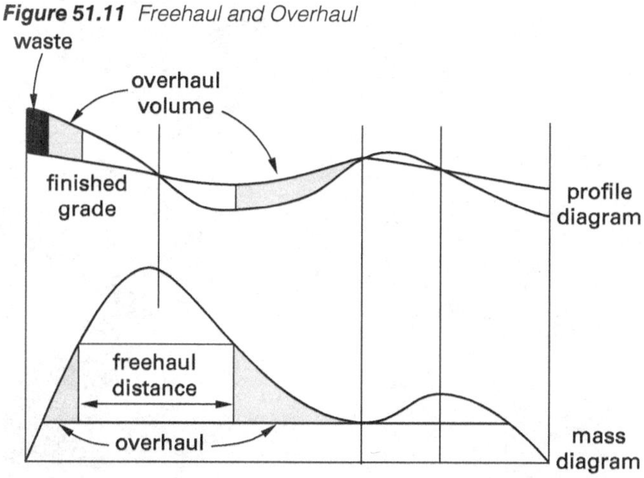

overhaul volume is determined from the maximum height of the overhaul area on the mass diagram, or from the overhaul area on the profile diagram. The overhaul distance is found as the separation in overhaul centroids on the profile diagram.

Subbases are horizontal balance lines that divide an area of the mass diagram between two balance points into trapezoids for the purpose of more accurately calculating overhaul. In Fig. 51.10, the top subbase is less than 600 ft in length, the freehaul distance for this project. Therefore, the volume of earth represented by the area above this line will be hauled a distance less than the free-haul distance, and no payment will be made for overhaul. All the volume represented by the area below this top subbase will receive payment for overhaul.

The area between two subbases is nearly trapezoidal. The average length of the bases of a trapezoid can be measured in feet, and the altitude of the trapezoid can be measured in cubic yards of earth. The product of these two quantities can be expressed in *yard-quarters*. A "yard-quarter" is a cubic yard of earthwork transported one quarter mile. If free haul is subtracted from the length, the quantity can be expressed as overhaul.

Subbases are drawn at distinct breaks in the mass curve. Distinct breaks in Fig. 51.10 can be seen at sta 1+00 (184 yd^3), sta 2+00 (806 yd^3), and sta 5+00 (4340 yd^3). After subbases are drawn, a horizontal line is drawn midway between the subbases. This line represents the average haul for the volume of earth between the two subbases. If a horizontal scale is used, the length of the average haul can be determined by scaling. This line, shown as a dashed line in Fig. 51.10, scales 875 ft for the area between the top subbases.

The volume of earth between the two subbases is found by subtracting the ordinate of the lower subbase from the ordinate of the upper subbase.

Multiplying average haul minus free haul in feet by volume of earth in cubic yards gives overhaul in yard-feet. Dividing by the number of feet in one-quarter mile, 1320 ft, gives yard-quarters.

When factors such as *shrinkage*, *swell*, *loss during transport*, and *subsidence* can be estimated, they are included in determining embankment volumes. A 5–15% excess is usually included in the calculations. This is achieved by increasing all fill volumes by the necessary percentage.

52 Pavement Design

Nomenclature

a	layer coefficient	–	–
A	oven-dried weight	lbf	kg
B	weight at SSD conditions	lbf	kg
C	submerged weight of aggregate in water	lbf	kg
D	distribution factor	–	–
D	thickness	in	mm
g	growth rate	%	%
G	specific gravity	–	–
G^*	complex shear modulus	lbf/ft²	Pa
GF	growth factor	–	–
m	mass	lbm	kg
m	m-value (creep curve slope)	–	–
N	number	–	–
p	pavement serviceability	–	–
P	percentage	%	%
S	soil sample weight	lbf	kg
S	standard deviation	–	–
S	stiffness	lbf/ft²	Pa
SN	structural number	–	–
V	volume	ft³	m³
V	volumetric ratio	–	–
VFA	voids filled with asphalt	%	%
VMA	voids in mineral aggregate	%	%
VTM	volume of air voids	%	%
w_{18}	design lane traffic	–	–
$\widehat{w}_{18}$	all-lane, two-directional traffic	–	–

Symbols

γ	unit weight	lbf/ft³	N/m³
δ	shear phase angle	deg	deg
ρ	density	lbm/ft³	kg/m³

Subscripts

a	air voids
b	asphalt or binder
ba	absorbed asphalt
be	effective asphalt
des	design
D	directional
fa	fine aggregate
int	intermediate
L	lane
max	maximum
mb	bulk
mm	maximum (zero air voids)
o	actual initial or overall
s	aggregate or supervision
sa	apparent
sb	bulk
se	effective
t	terminal

1. INTRODUCTION

Pavement design involves matching a project's strength and durability requirements to a given material. The *base soil*, or *foundation*, is analyzed, and the depths of various layers of pavement are designed based on the loads applied, the subgrade strength, and the frost penetration depth.

Pavement design's primary concern is vertical loads plus impacts. Designs are based on a combination of maximum expected wheel or axle loadings and the expected frequency of groups of axle loadings. *Horizontal loads*, such as those resulting from braking, acceleration, or turning movements, are not usually considered for concrete, but are included for Superpave asphalt. *Pavement loadings* in areas where vehicles will stand for extended periods of time, such as in parking lots, do not need to account for the frequency of wheel-load repetitions, but are instead governed by the maximum load applied for concrete and the distribution of load for asphalt.

2. PAVEMENT THICKNESS DESIGN

Concrete slab thickness is determined primarily by the maximum single load applied or by the repetition of loads over a long period of time, depending on which is greater. Generally, the repetition of loads over a period of time requires greater strength than a maximum single load because a highway slab with a depth of no more than 152 mm can occasionally sustain axle loadings at low speeds considerably in excess of legal limits. A slab thickness of 127 mm is the nominal minimum thickness for lightly traveled streets and driveways. Additionally, the relative strength of the subbase using Hveem's resistance values (R-values) or California bearing ratio (CBR) values influences slab thickness.

Highway pavement thickness is commonly specified in 25 mm increments from 130 mm to 250 mm or greater. For heavily traveled interstates, pavement thicknesses of 300 mm or more are used to help prevent costly traffic congestion caused by pavement rehabilitation. Where frost is a factor, highway subbase thickness should be at least 152 mm and can range up to a foot or more depending on the subsoil and load applied.

In order to simplify the variety of vehicle combinations riding on highways, vehicle traffic for pavement thickness design is usually grouped according to axle load weights, spacing of axle combinations, and frequency of application. Because of the exponential relationship of axle load to pavement life, the heaviest and most frequent load combinations are used to determine the design strength. Traffic is commonly broken down into classifications of automobiles, light trucks, heavy single-unit trucks, and heavy multiple-unit trucks. Truck traffic can be further classified as commercial or industrial to indicate expected loading needs along with appropriate axle load weight classifications.

Pavement thickness design is a complex and involved process that is thoroughly covered in publications written by the American Association of State Highway and Transportation Officials (AASHTO), the American Concrete Pavement Association, the National Pavement Association, the Portland Cement Association (PCA), and the Asphalt Institute.

3. SOILS AND SUBGRADE

Because the soil layer immediately below a pavement base has a direct effect on the performance of the pavement, the characteristics of this soil must be known in order to design a pavement that is economical and able to carry the expected loads. The best soil types have good load strengths, drain well, are not affected by frost or other types of heaving, are free of organic material, and can be shaped and compacted into a stable mass. In the AASHTO classification system, soils in the A-1 and A-2 classifications are considered proper subgrade materials. Soils in the A-1 and A-2 classifications have less than 35% *fines*, or material passing the no. 200 sieve.

Soil classification information may be obtained from a number of different systems. Generally, both pavement bearing pressure and design charts are based on Hveem's resistance value (*R-value*) or the *California bearing ratio* (CBR) index. When a classification is given in one system, its name in another system may be approximated using charts such as the USDA triangle or graphs relating one system to another. PCA publishes *A Soil Primer*, which also contains conversions between classification systems.

Usually, subgrade soil is specified to be carefully excavated, compacted, and graded as a separate operation, often with its own unit price for payment. If suitable soil is not available for subgrade, cement treating or other modifications to available soil may be necessary. Once a subgrade design strength is identified, the design of the granular drainage base, subbase, and surface pavement may proceed.

4. PAVEMENT MATERIALS

Highway pavement materials are relatively few in number, yet their makeup and composition are directly responsible for the successful performance of a paved surface. All high-type pavements have coarse aggregate, fine aggregate, and a cementitious material to bond the aggregates.

The most common choice for high strength and durability is *concrete pavement*. Concrete is highly resistant to abrasion, chemical attack, and spilled fuel solvents. When properly placed, concrete has been known to last well over 100 years with little or no maintenance. However, concrete requires specialized knowledge to handle its unforgiving qualities (i.e., once placed, setting and curing occur within a few hours, and any mistakes made are permanently captured), and it is susceptible to the abuse of excessive deicing salts, which soften the surface and weaken the joint edges. It also can have a higher cost compared with bituminous paving mixtures with equal strength or extended life cycles. Figure 52.1 illustrates the typical concrete surfacing layers. Table 52.1 gives commonly used maximum aggregate sizes.

Figure 52.1 *Typical Concrete Surfacing Layers*

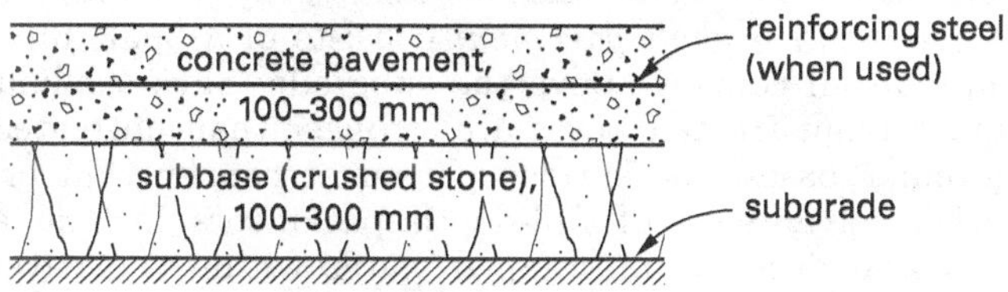

Table 52.1 *Typical Aggregate Maximum Sizes for Concrete Pavements*

aggregate type	size (mm)
concrete pavement	37.5–50
subbase	64
subgrade	100

(Multiply mm by 0.03937 to obtain in.)

Bituminous pavement (commonly referred to as *asphalt*, *asphalt pavement*, and *flexible pavement*) is composed of upper layers of aggregate mixed with a bituminous binder, such as asphalt. It is often preferred to concrete because it can be rapidly placed and requires less skill. It may also be installed on weaker subgrade sections, which decreases the amount of overlaying needed and defers some capital cost. However, if asphalt is not placed well, its edges can take on a ragged appearance. It also raises environmental concerns due to its hydrocarbon solvent emissions and safety concerns for workers who can become injured due to the high temperatures and irritating solvents used during the installation process. Additionally, asphalt is highly susceptible to sun damage in low traffic applications and to solvent action resulting from fuel spills. Construction costs vary regionally, making it difficult to select either asphalt or concrete pavement based on empirical data and cost alone. Table 52.2 gives commonly used maximum aggregate sizes. Figure 52.2 illustrates the layers of asphalt paving surfaces, though not all layers are present in every design.

Table 52.2 *Typical Aggregate Maximum Sizes for Asphalt Pavements*

layer	size (mm)
wearing course	12.5
binder course	19
base course	25–50
stone base	75
subgrade	100

(Multiply mm by 0.03937 to obtain in.)

Asphalt pavement is called *flexible pavement* because the surface material adjusts to the load condition imposed, then elastically rebounds. Asphalt (including bitumen) is used to bond aggregate together to avoid permanent displacement under loads. Asphalt also

Figure 52.2 *Typical Asphalt Surfacing Layers*

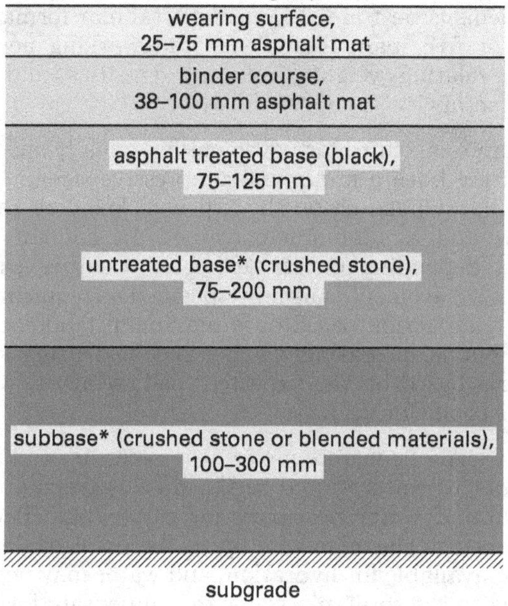

*These layers may be combined according to local practice.

creates an impermeable layer, protecting the base and subbase from water. The most common causes of asphalt pavement failure include

- water penetration through surface cracks
- subgrade/subbase failure due to overload
- surface cracking due to aging caused by oxidation of the asphalt cement and the effects of the sun

5. MATERIALS PROCEDURES

Pavement design usually requires calculating weights for each ingredient based on a predicted range of outcomes that are selected from known data. Bituminous mixes have a relatively narrow range of strengths and applications. Therefore, design becomes more a matter of total in-place (in situ) volume. Concrete, on the other hand, can be used in a variety of ways and can have a wide range of strengths based on how its ingredients are proportioned. A commonly used reference is the AASHTO *Guide for Design of Pavement Structures.* Additional reference material is published by trade organizations that concentrate on providing technical data for their members. For concrete, PCA's *Design and Control of Concrete Mixtures* is often used as a manual of concrete mix design and should be a part of any review of concrete mix design procedures. For asphalt pavements, the Asphalt Institute's *Mix Design Methods for Asphalt Concrete and Other Hot-Mix Types* (*MS-2*) is often referenced.

The procedure for listing data and calculating material problems is best aided by using a tabular format of the four or five main ingredients and working across the page, relating weights, volumes, densities, and correction factors.

Concrete mixtures for pavement are usually specified as requiring both a minimum compressive strength of 17–34 MPa, depending on the expected loads on the pavement, and a minimum entrained air content of 1.0–7.5%, depending on the expected weather exposure. Asphalt pavement is usually specified as requiring a percent compaction based on a maximum laboratory density, but is increasingly being specified using strength criteria based on the expected load, weather, and subgrade conditions.

For design of concrete mixes, it is necessary to know the amount of water stored in the bulk aggregate and the amount of water necessary for saturation. The water absorbed in the aggregate up to the point of saturation is not available for hydration, and water may need to be added to the total mix water to compensate for absorption. Water contained in a bulk pile that exceeds the saturation requirements is available for hydration and will need to be subtracted from the amount of water needed before water is added to the mix.

The amount of free water available for hydration is called the *water-cement ratio.* The water-cement ratio is used to determine strength qualities, workability, and dissolvability of a mix. Higher water-cement ratios improve workability, but reduce strength. Lower water-cement ratios increase strength, but require more power for equipment to be used in the placing and finishing of concrete slabs. Water-cement ratios of 0.4–0.6 by weight are common in concrete pavement design.

Water must be removed from bituminous mixtures and so is not a factor in design. Bituminous pavement design is instead performed based on the quality of the bituminous material and the quality of the aggregate available for the mix.

6. ASPHALT CONCRETE PAVEMENT

Hot mix asphalt (HMA), commonly referred to as *asphalt concrete* (AC), *bituminous mix* (BM), and sometimes *hot mix asphaltic concrete* (HMAC), is a mixture of *asphalt cement* (*asphalt binder*) and well-graded, high-quality aggregate. The mixture is heated and compacted by a paving machine into a uniform dense mass.

HMA pavement is used in the construction of traffic lanes, auxiliary lanes, ramps, parking areas, frontage roads, and shoulders. HMA pavement adjusts to limited amounts of differential settlement. It is easily repaired, and additional thicknesses can be placed at any time to withstand increased usage and loading. Over time, its nonskid properties do not significantly deteriorate. However, as the asphalt oxidizes, it loses some of its flexibility and cohesion, often requiring resurfacing sooner than would be needed with portland cement concrete. HMA is not normally chosen where water is expected to permeate the surface layer.

Most HMA pavement is applied over several other structural layers, not all of which are necessarily HMA layers. A *full-depth asphalt pavement* consists of asphalt mixtures in all courses above the subgrade. Since asphalt and asphalt-treated bases are stronger than untreated granular bases, the surface pavement can be thinner. Other advantages of full-depth pavements include a potential decrease in trapped water within the pavement, a decrease in the moisture content of the subgrade, and little or no reduction in subgrade strength.

If the asphalt layer is thicker than approximately 102 mm and is placed all in one lift, or if lift layers are thicker than 102 mm, the construction is said to be *deep strength asphalt pavement.* Using deep lifts to place layers of HMA is advantageous for several reasons. (a) Thicker layers hold heat longer, making it easier to roll the layer to the required density. (b) Lifts can be placed in cooler weather. (c) One lift of a given thickness is more economical to place than multiple lifts equaling the same thickness. (d) Placing one lift is faster than placing several lifts. (e) Less distortion of the asphalt course will result than if thin lifts are rolled.

Modern advances in asphalt concrete paving include *Superpave*™ and *stone matrix asphalt.* Both of these are outgrowths of the United States Strategic Highway Research Program (SHRP), although stone matrix asphalt has its roots in the German Autobahn.

7. ASPHALT GRADES

Asphalt varies significantly in its properties. In general, softer grades are used in colder climates to resist the expansion and contraction of the asphalt concrete caused by thermal changes. Harder grades of asphalt are specified in warmer climates to protect against rutting.

In the past, asphalt cement has been graded by viscosity grading and penetration grading methods. *Penetration grading* is based upon the penetration of a standard-sized needle loaded with a mass of 100 g in 5 sec at 25°C. A penetration of 40–50 (in units of mm) is graded as hard, and a penetration of 200–300 is soft. Intermediate ranges of 60–70, 85–100, and 120–150 have also been established. A penetration-graded asphalt is identified by its range and the word "pen" (e.g., "120–150 pen").

Viscosity grading is based on a measure of the absolute viscosity (tendency to flow) at 60°C. A viscosity measurement of AC-40 is graded as hard, and a measurement of AC-2.5 is soft. Intermediate ranges of AC-5, AC-10, and AC-20 have been established.

To meet the minimum specified viscosity requirements, oil suppliers have typically added more "light ends" (liquid petroleum distillates) to soften asphalt, regardless of the effects that such additions might have on structural characteristics. Concerns about the effects of such additions include oily buildup in mixing plant baghouses, light-end evaporation during pavement life, and possible reduced roadway performance.

Most state transportation and highway departments have stopped using penetration grading and have switched to performance grading. The *performance grading* (PG) system, which is used in Superpave and other designs, eliminates the structural concerns by specifying other characteristics. PG grading is specified by two numbers, such as PG 64-22. These two numbers represent the high and low (seven-day average) pavement temperatures in degrees Celsius that the project location will likely experience in its lifetime. Both high and low temperature ratings have been established in 6° increments. (See Table 52.3.)

PG ratings can be achieved with or without polymers, at the supplier's preference. However, the "rule of 90" says that if the sum of the absolute values of the two PG numbers is 90 or greater, then a polymer-modified asphalt is needed. Asphalts without polymers can be manufactured "straight-run."

The use of *recycled asphalt pavement* (RAP) in the mixture has an effect on the asphalt grade specified. However, specifying less than 15–25% RAP generally does not require a different asphalt cement grade to be used.

Table 52.3 *Superpave Binder Grades (°C)**

high-temperature grades	low-temperature grades
PG 46	−34, −40, −46
PG 52	−10, −16, −22, −28, −34, −40, −46
PG 58	−16, −22, −28, −34, −40
PG 64	−10, −16, −22, −28, −34, −40
PG 70	−10, −16, −22, −28, −34, −40
PG 76	−10, −16, −22, −28, −34
PG 82	−10, −16, −22, −28, −34

*For example, PG 52-16 would be applicable in the range of 52°C down to −16°C.

Example

An asphaltic road mixture component has been graded as PG 58-22. What is the lowest seven-day average temperature that pavement using this asphalt is expected to experience?

(A) −58°C

(B) −22°C

(C) 22°C

(D) 58°C

Solution

A performance-graded asphalt is graded by its high and low seven-day temperature ratings. An asphalt graded PG 58-22 has ratings of 58°C and −22°C.

The answer is (B).

8. AGGREGATE

The *mineral aggregate* component of the asphalt mixture comprises 90–95% of the weight and 75–85% of the mix volume. Mineral aggregate consists of sand, gravel, or crushed stone. The size and grading of the aggregate is important, as the minimum lift thickness depends on the maximum aggregate size. Generally, the minimum lift thickness should be at least three times the nominal maximum aggregate size. However, compaction is not an issue with open-graded mixes, since it is intended that the final result be very open. Therefore, the maximum-size aggregate can be as much as 80% of the lift thickness.

Generally, *coarse aggregate* is material retained on a no. 8 sieve (2.36 mm openings), *fine aggregate* is material passing through a no. 8 sieve (2.36 mm openings), and *mineral filler* is fine aggregate for which at least 70% passes through a no. 200 sieve (0.075 mm openings). The fine aggregate should not contain organic materials.

Aggregate size grading is done by sieving. Results may be expressed as either the percent passing through the sieve or the percent retained on the sieve, where the percent passing and the percent retained add up to 100%.

The maximum size of the aggregate is determined from the smallest sieve through which 100% of the aggregate passes. The *nominal maximum size* is designated as the largest sieve that retains some (but not more than 10%) of the aggregate.

An *asphalt mix* is specified by its nominal size and a range of acceptable passing percentages for each relevant sieve size, such as is presented in Table 52.4. The properties of the final mixture are greatly affected by the grading of aggregate. Aggregate grading can be affected by stockpile handling, cold-feed proportioning, degradation at impact points (more of an issue with batch plants), dust collection, and the addition of baghouse fines to the mix. Aggregates not only can become segregated; they also can be contaminated with other materials in adjacent stockpiles.

Dense-graded mixtures contain enough fine, small, and medium particles to fill the majority of void space between the largest particles without preventing direct contact between all of the largest particles. Aggregate for most HMA pavement is *open graded*, which means that insufficient fines and sand are available to fill all of the volume between the aggregate. Asphalt cement is expected to fill

Table 52.4 Percent Passing Criteria (Control Points)

standard sieve size (mm)	nominal maximum sieve size (mm)				
	9.5	12.5	19.0	25.0	37.5
50					100
37.5				100	90–100
25.0			100	90–100	
19.0		100	90–100		
12.5	100	90–100			
9.5	90–100				
2.36	32–67	28–58	23–49	19–45	15–41
0.075	2–10	2–10	2–8	1–7	0–6

Source: Federal Highway Administration Report, "Background of Superpave Asphalt Mixture Design and Analysis," FHWA-SA-95-003, Feb. 1995.

such voids. *Gap-graded mixtures* (as used in stone matrix asphalt and Superpave) are mixtures where two sizes (very large and very small) predominate.

Open-graded friction courses (OGFCs) have experienced problems with asphalt stripping, drain-down, and raveling. Properly formulated, mixes perform reasonably well, though they are susceptible to problems when the voids become plugged, retain moisture, or become oxidized.

9. PAVEMENT PROPERTIES

Several properties are desirable in an asphalt mixture, including stability, durability, flexibility, fatigue resistance, skid resistance, impermeability, and workability. *Stability* refers to the ability to resist permanent deformation, usually at high temperatures over a long time period. It depends on the amount of internal friction present in the mixture, which is in turn dependent on aggregate shape and surface texture, mix density, and asphalt viscosity. Lack of stability results in rutting in wheel paths, shoving at intersections, bleeding or flushing, and difficulty in compaction.

Durability refers to the ability of a mixture to resist disintegration by weathering. A mixture's durability is enhanced by a high asphalt content, dense aggregate gradation, and high density. Lack of durability results in raveling, early aging of the asphalt cement, and stripping.

A mixture's *flexibility* refers to its ability to conform to the gradual movement due to temperature changes or settlement of the underlying pavement layers. Flexibility is enhanced by a high asphalt content, open gradation, and low asphalt viscosity. A lack of flexibility can result in transverse or block cracking and shear failure cracking.

Fatigue resistance refers to the ability to withstand repeated wheel loads. A mixture's fatigue resistance can be enhanced with the presence of dense aggregate gradation and high asphalt content. Lack of fatigue resistance will result in fatigue or alligator cracking.

Skid resistance refers to the ability to resist tire slipping or skidding. An asphalt mixture with the optimum asphalt content, angular surface aggregates, and hard, durable aggregates will provide adequate skid resistance. The *polishing characteristics* of the aggregate need to be determined before use, either by lab polishing, insoluble residue tests, or previous service records. ASTM procedure C131, "Resistance to Degradation of Small-Size Coarse Aggregate by Abrasion and Impact in Los Angeles Machine," known as the *Los Angeles Abrasion Test* and the *Los Angeles Rattler Test*, is used to determine the durability of the aggregate and quality of long-term skid resistance. Prepared aggregate is placed in a drum with steel balls. The drum is rotated a specific number of times (e.g., 500), and the mass loss is measured.

A mixture's *impermeability* refers to its resistance to the passage of water and air. Impermeability is enhanced with a dense aggregate gradation, high asphalt content, and increased compaction. Permeable mixtures may result in stripping, raveling, and early hardening.

Workability refers to the ease of placement and compaction. A mixture's workability is enhanced by a proper asphalt content, smaller-size coarse aggregates, and proper mixing and compacting temperatures.

10. PROBLEMS AND DEFECTS

Various terms are used to describe the problems and defects associated with the mixing, placing, and functionality of HMA pavements.

- *alligator cracks:* interconnected cracks forming a series of small blocks resembling the marking on alligator skin or chicken wire
- *bleeding:* forming a thin layer of asphalt that has migrated upward to the surface (same as "flushing")
- *blowing:* see "pumping"
- *blow-up:* a disintegration of the pavement in a limited area
- *channeling:* see "rutting"
- *cold-cracking:* separation of pavement caused by temperature extremes, usually observed as cracks running perpendicular to the road's centerline
- *corrugations:* plastic deformation characterized by ripples across the pavement
- *cracking:* separation of pavement caused by loading, temperature extremes, and fatigue

- *disintegration:* breakup of pavement into small, loose pieces
- *drain-down:* liquid asphalt draining through the aggregate in a molten stage
- *faulting:* a difference in the elevations of the edges of two adjacent slabs
- *flushing:* see "bleeding"
- *lateral spreading cracking:* longitudinal cracking that occurs when the two edges of a road embankment supporting a pavement move away from the center
- *placement problems:* tender mixes
- *plastic instability:* excessive displacement under traffic
- *polishing:* aggregate surfaces becoming smooth and rounded (and subsequently, slippery) under the action of traffic
- *pothole:* a bowl-shaped hole in the pavement
- *pumping:* a bellows-like movement of the pavement, causing trapped water to be forced or ejected through cracks and joints
- *raveling:* a gradual, progressive loss of surface material caused by the loss of fine and increasingly larger aggregate from the surface, leaving a pock-marked surface
- *reflective cracking (crack reflection):* cracking in asphalt overlays that follow the crack or joint pattern of layers underneath
- *rutting:* channelized depressions that occur in the normal paths of wheel travel
- *scaling:* the peeling away of an upper layer
- *shoving:* pushing the pavement around during heavy loading, resulting in bulging where the pavement abuts an immobile edge
- *slippage cracking:* cracks in the pavement surface, sometimes crescent-shaped, that point in the direction of the wheel forces
- *spalling:* breaking or chipping of the pavement at joints, cracks, and edges
- *streaking:* alternating areas (longitudinal or transverse) of asphalt caused by uneven spraying
- *stripping:* asphalt not sticking to aggregate, water susceptibility
- *tender mix:* a slow-setting pavement that is difficult to roll or compact, and that shoves (slips or scuffs) under normal loading
- *upheaving:* a local upward displacement of the pavement due to swelling of layers below
- *washboarding:* see "corrugations"

Example

During wet weather, deep cracks in pavement lead to potholes due to a process known as

(A) flushing

(B) bleeding

(C) drain-down

(D) pumping

Solution

Pumping occurs when water that has penetrated to the base and subbase is compressed by the weight of a passing vehicle. The compressed water is forced through voids in the pavement and soil.

The answer is (D).

11. WEIGHT-VOLUME RELATIONSHIPS

The objectives of an asphalt mixture design are to provide enough asphalt cement to adequately coat, waterproof, and bind the aggregate; provide adequate stability for traffic demands; provide enough air voids to avoid bleeding and loss of stability; and produce a mixture with adequate workability to permit efficient placement and compaction.

Figure 52.3 illustrates the weight-volume relationships for asphalt concrete. The total weight of an asphalt mixture is the sum of the weight of the asphalt and the aggregate. The total volume is the sum of the volume of aggregate and the asphalt not absorbed by the aggregate, plus the air voids.

Although the specific gravity of the asphalt cement can vary widely, the percentage of the total mixture weight contributed by asphalt cement varies from 3–8% (by weight) for large nominal maximum sizes (e.g., 37.5 mm) and from 5–12% for small nominal maximum sizes (e.g., 9.5 mm).

The *surface area method* can be used as a starting point in mix design for determining the percent of asphalt needed. The percentage of asphalt binder, P_b, is

$$\begin{aligned} P_b = &\text{(aggregate surface area)} \\ &\times\text{(asphalt thickness)} \\ &\times\text{(specific weight asphalt)} \times 100\% \end{aligned}$$

(The aggregate surface area is measured in units of m^2/kg, the asphalt thickness is in in or mm, and the specific weight of asphalt is in kg/m^3.)

The aggregate surface area is obtained by multiplying the weight (mass) of the aggregate by a *surface area factor* (m^2/kg). The factor must be known, as it is different for each type of aggregate and for each sieve size. In

Transportation/ Surveying

Figure 52.3 Weight-Volume Relationships for Asphalt Mixtures

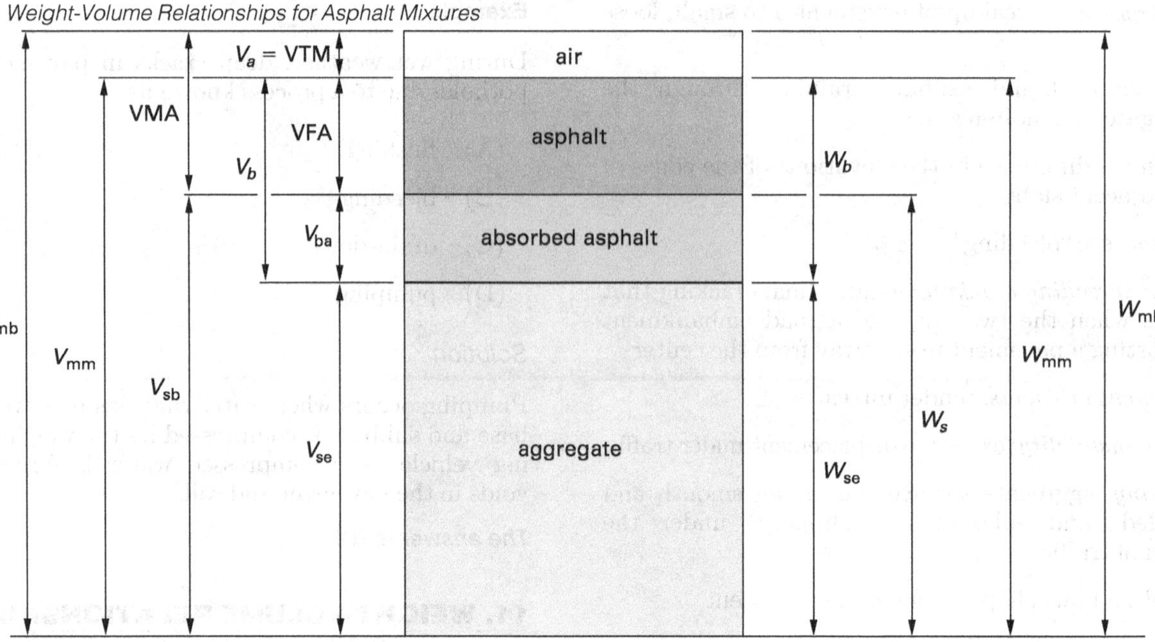

VMA = volume of voids in mineral aggregate
V_{mb} = bulk volume of compacted mix
V_{mm} = voidless volume of paving mix
V_a = volume of air voids
V_b = volume of asphalt
V_{ba} = volume of absorbed asphalt
V_{sb} = volume of mineral aggregate (by bulk specific gravity)
V_{se} = volume of mineral aggregate (by effective specific gravity)

practice, each surface area factor is multiplied by the percent (converted to a decimal) passing each associated sieve, not by the actual aggregate weight, and the products are summed. The total is the surface area in m^2/kg for the aggregate mixture.

12. CHARACTERISTICS OF ASPHALT CONCRETE

There are three specific gravity terms used to describe aggregate. The *apparent specific gravity*, G_{sa}, of an aggregate (coarse or fine) is

$$G_{sa} = \frac{m}{V_{aggregate}\rho_{water}}$$

$$G_{sa} = \frac{W}{V_{aggregate}\gamma_{water}}$$

Each of the components, including the coarse aggregate, fine aggregate, and mineral filler, has its own specific gravity, G, and proportion, P, by weight in the total mixture. The *bulk specific gravity*, G_{sb}, of the combined aggregate is

$$G_{sb} = \frac{P_1 + P_2 + \cdots + P_n}{\frac{P_1}{G_1} + \frac{P_2}{G_2} + \cdots + \frac{P_n}{G_n}}$$

ASTM procedure C127, "Specific Gravity and Absorption of Coarse Aggregate," is used to measure the specific gravity of coarse aggregate without having to measure its volume. In the equations given, A represents the oven-dried weight, B is the weight at saturated-surface dry conditions, and C is the submerged weight of the aggregate in water.

$$G_{sa} = \frac{A}{A - C}$$

$$G_{sb} = \frac{A}{B - C}$$

$$\text{absorption} = \frac{B - A}{A} \times 100\%$$

ASTM procedure C128, "Specific Gravity and Absorption of Fine Aggregate," is used for the fine aggregate. In the equations given, A is the oven-dried weight, B represents the weight of a *pycnometer* filled with water, S is the weight of the soil sample (standardized as 500 g), and C is the weight of a pycnometer filled with the soil sample and water added to the calibration mark. The *bulk specific gravity*, G_{sb}, of the aggregate mixture is found from

$$G_{sb} = \frac{A}{B+S-C}$$

The *apparent specific gravity* of the aggregate mixture, G_{sa}, is calculated from the following equation, except that the apparent specific gravities are used in place of the bulk specific gravities.

$$G_{sa} = \frac{A}{B+A-C}$$

$$\text{absorption} = \frac{S-A}{A} \times 100\%$$

The *effective specific gravity* of the aggregate, G_{se}, is calculated from the maximum specific gravity, the specific gravity of the asphalt, G_b, and the proportion of the asphalt in the total mixture, P_b, expressed as a percentage. The effective specific gravity is always between the bulk and apparent specific gravities.

$$G_{se} = \frac{100\% - P_b}{\dfrac{100\%}{G_{mm}} - \dfrac{P_b}{G_b}}$$

ASTM procedure D2041, "Theoretical Maximum Density (Rice Method)," is used to measure the *maximum specific gravity*, G_{mm}, of the paving mixture, which is the specific gravity if the mix had no air voids. However, in practice, some air voids always remain and this maximum specific gravity cannot be achieved.

The maximum specific gravity is calculated from the proportions of aggregate and asphalt, P_s and P_b, respectively, in percent, and the specific gravity of the asphalt, G_b.

$$G_{mm} = \frac{100\%}{\dfrac{P_s}{G_{se}} + \dfrac{P_b}{G_b}}$$

If ASTM D2041 is used to determine the maximum specific gravity, G_{mm}, the following equation is used. A is the mass of an oven-dried sample, D is the mass of a container filled with water at 25°C, and E is the mass of the container filled with the sample and water at 25°C.

$$G_{mm} = \frac{A}{A+D-E}$$

By convention, the *absorbed asphalt*, P_{ba}, is expressed as a percentage by weight of the aggregate, not as a percentage of the total mixture.

$$P_{ba} = \frac{G_b(G_{se} - G_{sb})}{G_{sb}G_{se}} \times 100\%$$

The *effective asphalt content* of the paving mixture, P_{be}, is the total percentage of asphalt in the mixture less the percentage of asphalt lost by absorption into the aggregate.

$$P_{be} = P_b - \frac{P_{ba}P_s}{100\%}$$

The *bulk specific gravity of the compacted mixture*, G_{mb}, is determined through testing according to ASTM procedure D2726, "Bulk Density of Cores (SSD)."

The *percent VMA* (*voids in mineral aggregate*) in the compacted paving mixture is the total of the air voids volume and the volume of aggregate available for binding.

$$\text{VMA} = 100\% - \frac{G_{mb}P_s}{G_{sb}}$$

The percent *total air voids*, P_a, in the compacted mixture represents the air spaces between coated aggregate particles. The total air voids in the compacted mixture should be 3–5%. With higher void percentages, voids can interconnect and allow air and moisture to permeate the pavement. If air voids are less than 3%, there will be inadequate room for expansion of the asphalt binder in hot weather. Below 2%, the asphalt becomes plastic and unstable.

$$P_a = \text{VTM} = \frac{G_{mm} - G_{mb}}{G_{mm}} \times 100\%$$

The *voids filled with asphalt*, VFA, indicates how much of the VMA contains asphalt and is found using

$$\text{VFA} = \frac{\text{VMA} - \text{VTM}}{\text{VMA}} \times 100\%$$

Nondestructive *nuclear gauge testing* (with the gauge in the surface or backscatter position) can be used to determine actual relative density or percent air voids of new pavement. *Extraction methods* (which use dangerous and expensive chemicals) of determining density and air voids are now essentially outdated.

Example

Which of the following relationships relating to asphaltic pavement is valid?

(A) VTM = VFA + VMA

(B) VFA = VTM + VMA

(C) VMA = VTM + VFA

(D) VTA = VFA/VFM

Solution

When aggregate is compressed into a pavement, there will be voids between the pieces. The voids in the mineral aggregate, VMA, can be filled with air or with asphalt. The voids filled with asphalt are designated VFA; the remaining air voids in the total mix designated VTM.

The answer is (C).

13. MARSHALL MIX DESIGN

The *Marshall method* was developed in the late 1930s and was subsequently modified by the U.S. Army Corps of Engineers for higher airfield tire pressures and loads. It was subsequently adopted by the entire U.S. military. Although the method disregards shear strength, it considers strength, durability, and voids. It also has the advantages of being a fairly simple procedure, not requiring complex equipment, and being portable.

The primary goal of the Marshall method is to determine the optimum asphalt content. The method starts by determining routine properties of the chosen asphalt and aggregate. The specific gravities of the components are determined by standard methods, as are the mixing and blending temperatures from the asphalt viscosities. (See Fig. 52.4.) Trial blends with varying asphalt contents are formulated, and samples are heated and compacted. Density and voids are determined.

The stability portion of the Marshall procedure measures the maximum load supported by the test specimen at a loading rate of 50.8 mm/min. The load is increased until it reaches a maximum, then when the load just begins to decrease, the loading is stopped and the maximum load is recorded. During the loading, an attached dial gauge measures the specimen's plastic flow. The flow value is recorded in 0.25 mm increments at the same time the maximum load is recorded.

Figure 52.4 *Determining Asphalt Mixing and Compaction Temperatures**

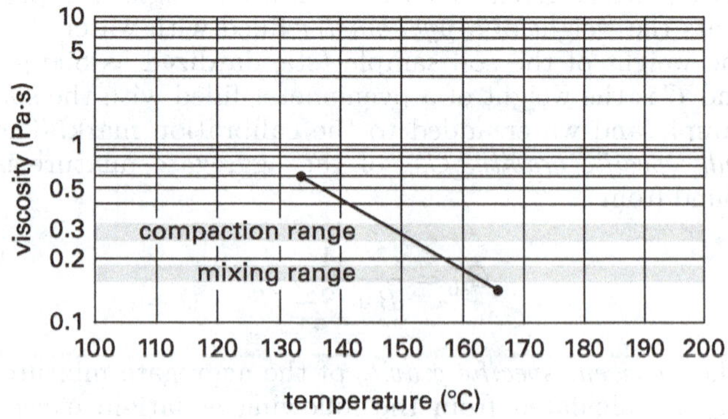

*Compaction and mixing temperature are found by connecting two known temperature-viscosity points with a straight line.

Various parameters are used to determine the best mix. Three asphalt contents are averaged to determine the *target optimum asphalt content*: the 4% air voids content, the maximum stability content, and the maximum density (unit weight) content. The target optimum asphalt content is subsequently validated by checking against flow and VMA minimums.

14. HVEEM MIX DESIGN

The *Hveem mix design method* was developed in the 1920s and has been extensively used in the western states. Like the Marshall method, the goal is to determine the optimum asphalt content. Unlike the Marshall method, it has the additional sophistications of measuring resistance to shear and considering asphalt absorption by aggregates. It has the disadvantage of requiring more specialized and nonportable equipment for mixing, compaction, and testing.

The method makes three assumptions. (1) The optimum asphalt binder content depends on the aggregate surface area and absorption. (2) Stability is a function of aggregate particle friction and mix cohesion. (3) HMA durability increases with asphalt binder content.

After selecting the materials, the *centrifuge kerosene equivalent* (CKE) of the fine aggregate and the retained surface oil content of the coarse aggregate are determined. These measures of surface absorption are used to estimate the optimum asphalt content.[1] Trial blends are formulated over the asphalt range of [CKE−1%, CKE +2%] in 0.5% increments. Then, specimens of the trial blends are prepared by heating and densifying in a California kneading compactor.[2]

[1]It is important that the trial blends include asphalt contents greater than and less than the optimum content. The purpose of determining the CKE and surface oil content is to estimate the optimum content prior to formulating trial blends. However, it is also practical to simply formulate trial blends within a range of 4-7% of asphalt, varying each blend by an increment such as 0.5%. Because of this alternative methodology of bracketing the optimum asphalt content, the CKE process has become essentially obsolete.

[2]The kneading compactor produces compactions that are more similar to pavements that have been roller-compacted with steel- and rubber-tired rollers.

A *Hveem stabilometer*, a closed-system triaxial test device, is used to determine the stability (i.e., the horizontal deformation under axial load). The stabilometer applies an increasing load to the top of the sample at a predetermined rate. As the load increases, the lateral pressure is read at specified intervals.

Visual observation, volumetrics (air voids, voids filled with asphalt, and voids in the mineral aggregate), and stability are used to determine the optimum asphalt content. The design asphalt content is selected as the content that produces the highest durability without dropping below a minimum allowable stability. Essentially, as much asphalt binder as possible is used while still meeting minimum stability requirements. The *pyramid method* can be used to select the optimum asphalt content.

15. SUPERPAVE

Superpave (Superior Performing Asphalt Pavement) mixture design is one of the products of the FHWA's 1988–1993 *Strategic Highway Research Program* (SHRP). Superpave has rapidly become the standard design methodology for HMA highway pavements in the United States and Canada. It greatly increases the stone-on-stone contact in the pavement, thus improving its loadbearing capacity. More than 80% of new highway construction in the United States uses Superpave mixture design.

There are three elements that make up the Superpave concept. Level 1 mix design includes specifying asphalts by a set of performance-based *binder specifications*. Specifications include stiffness, dynamic shear (stiffness at high and medium temperatures), bending (stiffness at low temperatures), and in some cases, direct tension. The binder specifications relate laboratory tests to actual field performance. (See Table 52.5.)

Level 1 mix design is used by most highway agencies and in most new highway construction and has replaced older methods of asphalt call-outs. Performance-graded (PG) asphalts are selected based on the climate and traffic expected. The physical property requirements are the same for all performance grades. The distinction between the various asphalt grades is the minimum and maximum temperatures at which the requirements must be met. For example, an asphalt classified as a PG 64-22 will meet the high temperature physical property requirements of the pavement up to a temperature of 64°C and the low temperature physical property requirements of the pavement down to −22°C. Selection of the asphalt grade depends on the expected temperatures, as well as the nature of the loading. (See Table 52.6.) Although the highest pavement temperature in the United States is about 70°C, a PG 70-22 specification may be insufficient. In general, one grade increase (e.g., to PG 76-22) should be considered for 20-yr ESALs greater than 20 million and slow moving traffic. Two grade increases (e.g., to PG 82-22) should be considered for standing traffic and stationary loads.

Level 2 mix design includes performance-based tests that measure primary mixture performance factors, such as fatigue, cracking, permanent deformation, low-temperature (thermal) cracking, aging, and water sensitivity. Much of Superpave's success is related to its strict requirements regarding aggregate blending and quality. Mixes are designed at 9.5 mm, 12.5 mm, 19.0 mm, and 25.0 mm according to their nominal maximum aggregate size. (In general, surface courses should use 9.5 mm or 12.5 mm mixes.) Evaluations include the aggregate's *dust-to-binder ratio* (also known as the *fineness-to-effective asphalt ratio*),[3] $P_{0.075}/P_{be}$, coarse and fine aggregate angularities, percentage of flat or elongated particles with length/thickness ratios greater than 5, and sand-equivalent clay content. Dust-to-binder ratios are normally 0.6–1.2, but a ratio of up to 1.6 may be used in some cases. The percentage of flat or elongated particles is limited to approximately 10%. Angularity and clay content specifications depend on traffic level (ESALs) and distance from the wearing surface.

Level 3 mix design is a computer-aided volumetric mix design and analysis that incorporates test results, geographical location, and climatological data, as well as mix-testing technology such as the *Superpave gyratory compactor*, SGC. The level 3 mix design proceeds similarly to the Marshall mix design method, with the exception that the Superpave gyratory compactor is used instead of the Marshall hammer for compacting mixture specimens. The Superpave gyratory compactor is a laboratory compaction device that orients the aggregate particles in a manner similar to standard rollers, resulting in laboratory densities equivalent to field densities. Specimens with a range of asphalt contents less than and greater than the anticipated optimum value are compacted in the gyratory compactor based on a specified number of gyrations determined by design traffic levels and design high air temperatures.

N_{des}, the design number of gyrations, is a function of climate and traffic level. Climate condition is a function of the average high air temperature determined using the average seven-day maximum air temperature for the project conditions. This is the equivalent temperature at 50% reliability. The traffic level is the design ESALs, taken from actual traffic data or from other nearby data and adapted to the project location. $N_{initial}$ is an estimation of the mixture's ability for compaction. N_{max} is a laboratory density determined by using additional SGC specimens of the selected design as a check to help guard against plastic failure caused by traffic levels that exceed the design level.

[3]What Superpave refers to as "dust" is referred to as "silt-clay" by AASHTO. It is the portion of the aggregate that passes through a no. 200 (0.075 mm) sieve.

Transportation/ Surveying

Table 52.5 Performance-Graded (PG) Binder Grading System

	PG 52							PG 58					PG 64				
performance grade	−10	−16	−22	−28	−34	−40	−46	−16	−22	−28	−34	−40	−16	−22	−28	−34	−40
average 7-day maximum pavement design temperature, °C[a]	<52							<58					<64				
minimum pavement design temperature, °C[a]	>−10	>−16	>−22	>−28	>−34	>−40	>−46	>−16	>−22	>−28	>−34	>−40	>−16	>−22	>−28	>−34	>−40
original binder																	
flash point temp, T48: minimum °C	230																
viscosity, ASTM D 4402:[b] maximum, 3 Pa·s (3000 cP), test temp, °C	135																
dynamic shear, TP5:[c] $G^*/\sin\delta$, minimum, 1.00 kPa test temp @ 10 rad/sec., °C	52							58					64				
rolling thin film oven (T240) or thin film oven (T179) residue																	
mass loss, maximum, %	1.00																
dynamic shear, TP5: $G^*/\sin\delta$, minimum, 2.20 kPa test temp @ 10 rad/sec., °C	52							58					64				
pressure aging vessel residue (PP1)																	
PAV aging temperature, °C[d]	90							100					100				
dynamic shear, TP5: $G^*/\sin\delta$, maximum, 5000 kPa test temp @ 10 rad/sec., °C	25	22	19	16	13	10	7	25	22	19	16	13	28	25	22	19	16
physical hardening[e]	report																
creep stiffness, TP1:[f] S, maximum, 300 MPa m-value, minimum, 0.300 test temp, @ 60 sec., °C	0	−6	−12	−18	−24	−30	−36	−6	−12	−18	−24	−30	−6	−12	−18	−24	−30
direct tension, TP3:[f] failure strain, minimum, 1.0% test temp @ 1.0 mm/min, °C	0	−6	−12	−18	−24	−30	−36	−6	−12	−18	−24	−30	−6	−12	−18	−24	−30

[a]Pavement temperatures can be estimated from air temperatures using an algorithm contained in the SUPERPAVE software program or may be provided by the specifying agency, or by following the procedures as outlined in PPX.

[b]This requirement may be waived at the discretion of the specifying agency if the supplier warrants that the asphalt binder can be adequately pumped and mixed at temperatures that meet all applicable safety standards.

[c]For quality control of unmodified asphalt cement production, measurement of the viscosity of the original asphalt cement may be substituted for dynamic shear measurements of $G^*/\sin\delta$ at test temperatures where the asphalt is a Newtonian fluid. Any suitable standard means of viscosity measurement may be used, including capillary or rotational viscometry (AASHTO T201 or T202).

[d]The PAV aging temperature is based on simulated climatic conditions and is one of three temperatures 90°C, 100°C or 110°C. The PAV aging temperature is 100°C for PG 58- and above, except in desert climates, where it is 110°C.

[e]Physical Hardening—TP1 is performed on a set of asphalt beams according to Section 13.1, except the conditioning time is extended to 24 hrs ± 10 minutes at 10°C above the minimum performance temperature. The 24-hour stiffness and m-value are reported for information purposes only.

[f]If the creep stiffness is below 300 MPa, the direct tension test is not required. If the creep stiffness is between 300 MPa and 600 MPa, the direct tension failure strain requirement can be used in lieu of the creep stiffness requirement. The m-value requirement must be satisfied in both cases.

Source: Federal Highway Administration Report FHWA-SA-95-003, Background of Superpave Asphalt Mixture Design and Analysis, Feb. 1995.

Because Superpave greatly increases stone-on-stone contact in the pavement, Superpave mixtures pay close attention to voids in the mineral aggregate (VMA), the percentage of air voids in the total mix (VTM), and voids filled with asphalt (VFA). Like the Marshall mix design method, the Superpave mix design method uses volumetrics (e.g., G_{mm}; G_{mb}; V_a or VTM; VMA; and VFA) to select the optimum asphalt content corresponding to 4% air voids that satisfies all other DOT requirements (e.g., minimum VMA and a range of VFA) when compacted N_{des} gyrations. A minimum density of 96% of maximum is required for the specimens compacted to N_{des}. Unlike the Marshall and Hveem methods, there are no standard performance (e.g., stability, flow, or cohesion) tests. (See Table 52.7 and Table 52.8.) Additional Superpave requirements are given in Table 52.9 through Table 52.12.

Table 52.6 Superpave Mixture Designations

Superpave designation (mm)	nominal max. size (mm)	max. size (mm)
9.5	9.5	12.5
12.5	12.5	19.0
19.0	19.0	25.0
25.0	25.0	37.5
37.5	37.5	50.0

Table 52.7 VFA Range Requirements at 4% Air Voids

Source: Federal Highway Administration Report FHWA-SA-95-003, "Background of Superpave Asphalt Mixture Design and Analysis," Feb. 1995.

traffic loading (millions of ESALs)	design VFA range (%)
< 0.3	70–80
< 1	65–78
< 3	65–78
< 10	65–75
< 30	65–75
< 100	65–75
≥ 100	65–75

Table 52.8 Minimum VMA Requirements at 4% Air Voids

nominal max. aggregate (mm)	9.5	12.5	19.0	25.0	37.5
minimum VMA (%)	15	14	13	12	11

Source: Federal Highway Administration Report FHWA-SA-95-003, "Background of Superpave Asphalt Mixture Design and Analysis," Feb. 1995.

Example

An asphalt concrete mixture having a compacted density of 144 lbm/ft^3 is being considered as a possible Superpave mixture. The mixture contains coarse and fine aggregate, filler (dust), and asphalt in the gravimetric percentages presented in the table.

component	specific gravity	gravimetric percentage of aggregate	gravimetric percentage of mixture
coarse aggregate	2.70	70%	
fine aggregate	2.65	25%	95%
filler	2.60	5%	
asphalt	1.00	–	5%

What are most nearly the percentage of voids in the mineral aggregate, VMA, and the percentage of air voids in the mixture, VTM?

(A) VTM = 3.7%; VMA = 14%

(B) VTM = 4.3%; VMA = 16%

(C) VTM = 5.7%; VMA = 17%

(D) VTM = 6.8%; VMA = 18%

Solution

VMA is a volumetric measure, so the gravimetric percentages must be converted. The three forms of mineral matter constitute 95% of the finished weight. Calculate the component volumes from the relationship $V = m/\rho$.

The total volume represented by the aggregates, filler, and asphalt in 1 ft^3 of mixture is

$$0.568\ \text{ft}^3 + 0.207\ \text{ft}^3 + 0.042\ \text{ft}^3 + 0.115\ \text{ft}^3 = 0.932\ \text{ft}^3$$

Since the volume occupied by 144 lbm of mixture is actually 1.000 ft^3, the difference between 1.000 ft^3 and the total volume is the percentage of air voids.

$$\text{VTM} = \left(1000\ \frac{\text{ft}^3}{\text{ft}^3} - 0.932\ \frac{\text{ft}^3}{\text{ft}^3}\right) \times 100\% = 6.8\%$$

component	m/ρ	V (per ft^3)
coarse aggregate	$\dfrac{(0.70)(0.95\ \text{ft}^3)\left(144\ \dfrac{\text{lbm}}{\text{ft}^3}\right)}{(2.70)\left(62.4\ \dfrac{\text{lbm}}{\text{ft}^3}\right)}$	0.568 ft^3
fine aggregate	$\dfrac{(0.25)(0.95\ \text{ft}^3)\left(144\ \dfrac{\text{lbm}}{\text{ft}^3}\right)}{(2.65)\left(62.4\ \dfrac{\text{lbm}}{\text{ft}^3}\right)}$	0.207 ft^3
filler	$\dfrac{(0.05)(0.95\ \text{ft}^3)\left(144\ \dfrac{\text{lbm}}{\text{ft}^3}\right)}{(2.60)\left(62.4\ \dfrac{\text{lbm}}{\text{ft}^3}\right)}$	0.042 ft^3
asphalt	$\dfrac{(0.05)(1.00\ \text{ft}^3)\left(144\ \dfrac{\text{lbm}}{\text{ft}^3}\right)}{(1.00)\left(62.4\ \dfrac{\text{lbm}}{\text{ft}^3}\right)}$	0.115 ft^3

Transportation/Surveying

The percentage of voids in the mixture is

$$\text{VMA} = \left(1000 \text{ ft}^3 - \left(\begin{array}{c} 0.568 \text{ ft}^3 + 0.207 \text{ ft}^3 \\ + 0.042 \text{ ft}^3 \end{array} \right) \right) \times 100\%$$
$$= 18.3\% \quad (18\%)$$

Since VTM > 4%, this probably would not qualify as a Superpave mixture.

The answer is (D).

Table 52.9 *Additional Superpave Requirements*

fineness-to-effective asphalt ratio (wt/wt)	0.6–1.2
short-term oven aging at 135°C (hr)	4
tensile strength ratio T283 (min)	0.80
traffic (for designation) based on (yr)	15

Source: Federal Highway Administration Report FHWA-SA-95-003, "Background of Superpave Asphalt Mixture Design and Analysis," Feb. 1995.

16. FLEXIBLE PAVEMENT STRUCTURAL DESIGN METHODS

The goal of flexible pavement structural design methods is to specify the thicknesses of all structural layers in a pavement. Different methodologies are used for full-depth asphalt pavements, asphalt courses over an aggregate base, asphalt courses over emulsified base, and overlays. Aside from generic catalog methods, design methods can be divided into two types: state-of-the-practice empirical methods and state-of-the-art mechanistic and mechanistic-empirical (ME) methods.[4] Like the choice of HMA mix design methods, the methodology used at the DOT level varies from state to state. All of the methods can be implemented manually or by computer.

Empirical methods, which include the 1993 AASHTO design procedure contained in the AASHTO *Guide for Design of Pavement Structures*, the 1998 *Supplement* containing a modified AASHTO design procedure, Asphalt Institute methods, and the Modified Texas Triaxial Design Method (not covered in this book), are based on the extrapolated performance of experimental test tracks.

ME methods build on multilayer elastic theory, finite element analysis, and simulation. They typically take into consideration reliability, climate, and life-cycle costs. Traffic is characterized by its spectrum, rather than in the form of a single (ESALs) number. Although ME methods have been implemented in Washington and Minnesota by WSDOT and MnDOT, respectively, for all practical purposes, ME design now refers only to the National Cooperative Highway Research Program (NCHRP) 1-37A method, as presented in the 2008 AASHTO *Mechanistic-Empirical Pavement Design Guide*. ME methods require more training and situational awareness (environmental knowledge) than empirical methods and, unless implemented on a computer, tend to be dauntingly complex. Even then, computer "runs" can take hours.

Design of overlays of existing pavements can employ both empirical and mechanistic-empirical methods, as well as surface deflection methods using the results of *falling weight deflectometer* tests. These methods tend to use *back-calculation* of parameters obtained from in situ testing.

The design of an HMA pavement requires knowledge of climate, traffic, subgrade soils support, and drainage. Stiffness of the asphalt layer varies with temperature; and, unbound layers (aggregate and subgrade) are affected by freeze-thaw cycles. Climate is characterized by the *mean annual air temperature*, MAAT, of the design area. MAATs of 7°C or less require attention to frost effect; at 16°C, frost effects are considered possible; and at 24°C, frost effects can be neglected. In empirical methods, the traffic is characterized by the numbers and weights of truck and bus axle loads expected during a given period of time, specified in ESALs. ESALs are calculated by multiplying the number of vehicles in each weight class by the appropriate truck factor and summing the products. Consideration must be given to subsurface drainage (e.g., installation of underdrains and/or interceptor drains) where high water tables occur or where water may accumulate in low areas. Good surface drainage, obtained through proper crown design, is also essential.

Example

To what do the compactions of a Superpave mixture at N_{int}, N_{des}, and N_{max} gyratory cycles relate?

(A) workability during three stages of asphalt-aggregate mixing

(B) pavement performance at low, design, and maximum ambient temperatures

(C) strength at low, design, and maximum vehicle loading

(D) density after a single compaction pass, after several years, and at maximum pavement life

[4]Mechanistic methods are based on familiar mechanics (strengths) of materials concepts that relate wheel loading to pavement stresses, strains, and deflections according to the material properties. The term "mechanistic-empirical" acknowledges that the theory is calibrated and corrected according to observed performance.

Table 52.10 Superpave Aggregate Requirements

traffic, million equivalent single-axle loads (ESALs)	coarse aggregate angularity		fine aggregate angularity		flat and elongated particles	clay content
	depth from surface		depth from surface		maximum percentage	minimum sand equivalent
	≤100 mm	100 mm	≤100 mm	100 mm		
<0.3	55/–	–/–	–	–	–	40
<1	65/–	–/–	40	–	–	40
<3	75/–	50/–	40	40	10	40
<10	85/80	60/–	45	40	10	45
<30	95/90	80/75	45	40	10	45
<100	100/100	95/90	45	45	10	50
≥ 100	100/100	100/100	45	45	10	50

Source: Federal Highway Administration Report FHWA-SA-95-003, "Background of Superpave Asphalt Mixture Design and Analysis," Feb. 1995.

Table 52.11 Superpave Gyratory Compaction Effort

traffic (millions of ESALs)	average design high air temperature											
	< 39°C			39–40°C			41–42°C			42–43°C		
	N_{int}	N_{des}	N_{max}	N_{int}	N_{des}	N_{max}	N_{int}	N_{des}	N_{max}	N_{int}	N_{des}	N_{max}
< 0.3	7	68	104	7	74	114	7	78	121	7	82	127
< 1	7	76	117	7	83	129	7	88	138	8	93	146
< 3	7	86	134	8	95	150	8	100	158	8	105	167
< 10	8	96	152	8	106	169	8	113	181	9	119	192
< 30	8	109	174	9	121	195	9	128	208	9	135	220
< 100	9	126	204	9	139	228	9	146	240	10	153	253
≥ 100	9	142	233	10	158	262	10	165	275	10	177	288

Source: Federal Highway Administration Report FHWA-SA-95-003, "Background of Superpave Asphalt Mixture Design and Analysis," Feb. 1995.

Table 52.12 Superpave Gyratory Compaction Key

superpave gyratory compaction	N_{int}	N_{des}	N_{max}
percentage of G_{mm}	≤89%	96%	≤98%

Source: Federal Highway Administration Report FHWA-SA-95-003, "Background of Superpave Asphalt Mixture Design and Analysis," Feb. 1995.

Solution

The intermediate, design, and maximum percentages of maximum compaction are used to predict the density of pavement after a single compaction pass, when "mature" after several years, and at maximum pavement life.

The answer is (D).

17. TRAFFIC

(The information included in this section is consistent with the *AASHTO Guide for Design of Pavement Structures* (*AASHTO Guide*), 1993.)

The AASHTO pavement design method requires that all traffic be converted into *equivalent single-axle loads* (ESALs). This is the number of 18,000 lbf single axles (with dual tires) on pavements of specified strength that would produce the same amount of traffic damage over the design life of the pavement.

Appendix D of the *AASHTO Guide* gives *load equivalency factors* (LEFs) for flexible pavements for various *terminal serviceability indices*, p_t. A p_t value of 2.5 is assumed unless other information is available. Typical load equivalency factors for various gross axle loads and axle formations are given in Table 52.13.

$$\text{ESALs} = (\text{no. of axles})(\text{LEF})$$

Since the structural number, SN, is not known until the design is complete, initial values of 3 and 5 are assumed for the structural number of low-volume roads and high-volume roads, respectively. Once the design is complete, the ESALs can be recalculated and the design verified.

18. TRUCK FACTORS

Truck factors, TFs, are the average LEF for a given class of vehicle and are calculated using *loadometer data*. The truck factor is calculated as the total ESALs for all axles divided by the number of trucks.

$$\text{TF} = \frac{\text{ESALs}}{\text{no. of trucks}}$$

19. DESIGN TRAFFIC

Once the truck factors are calculated, the design ESALs can be calculated from the distribution of vehicle classes in the average annual daily traffic (AADT) and the expected growth rate. If the 20 yr ESAL (i.e., the total number of vehicles over 20 years) is to be predicted from the current (first) year ESAL and if a constant growth rate of $g\%$ per year is assumed, traffic *growth factors*, GFs (also known as *projection factors*), for growth rates, g, can be read as the $(F/A,\ g\%, 20)$ factors from economic analysis tables. ESAL_{20} is the cumulative number of vehicles over 20 years, not the 20th year traffic.

$$\text{ESAL}_{20} = (\text{ESAL}_{\text{first year}})(\text{GF})$$

The *design traffic* is calculated as the product of the AADT, the fraction of AADT that represents truck traffic, the days in one year, and the growth factor over the design life. The ESAL is obtained by multiplying by the truck factor.

The *directional distribution factor*, D_D, is used to account for the differences in loading according to travel direction. It is usually assumed to be 50%. Table 52.14 gives recommended values for *lane distribution factors*, D_L, on multilane facilities. The ESALs for the design lane are calculated from

$$w_{18} = D_D D_L \widehat{w}_{18}$$

Example

A highway is being designed to last 40 years before rehabilitation. Traffic currently consists of 200,000 ESALs per year but is expected to grow by 4% each year. Most nearly, how many ESALs will be experienced over 40 years?

(A) 3.9×10^6

(B) 19×10^6

(C) 22×10^6

(D) 31×10^6

Solution

From economic analysis tables, $(F/A, 4\%, 40) = 95.0255$. The 40 year ESALs are (200,000 ESALs)(95.0255) = 19×10^6.

The answer is (B).

20. AASHTO METHOD OF FLEXIBLE PAVEMENT DESIGN

The *AASHTO Guide* is the basis of the conventional flexible pavement design method presented in this chapter. It is a conservative methodology, so average values can be used for all design variables. The design is based on four main design variables: time, traffic, reliability, and environment. Performance criteria, material properties, and pavement structural characteristics are also considered.

Time: The *analysis period* is the length of time that a given design strategy covers (the *design life* or *design period*). The *performance period* is the time that the initial pavement structure is expected to perform adequately before needing rehabilitation. For example, on a high-volume urban roadway, the analysis period should be 30 to 50 years, which may include the initial performance period and several *rehabilitation periods* following overlays or maintenance operations.

Traffic: The traffic counts are converted into standard 18 kip ESALs.

Reliability: Reliability considerations ensure that the structure will last for the designated design period. They take into consideration variations in traffic and performance predictions. Facilities that are considered to be more critical are designed using higher reliability factors. Table 52.15 gives suggested levels of reliability for various functional classifications. It is necessary to select an *overall standard deviation*, S_o, for reliability to account for traffic and pavement performance that is representative of local conditions. Based on historical information obtained during the AASHTO Road Test, appropriate standard deviations for flexible pavements are 0.4–0.5.

Table 52.13 Load Equivalency Factors (LEFs)

gross axle load		load equivalency factors		gross axle load		load equivalency factors	
kN	lbf	single axles	tandem axles	kN	lbf	single axles	tandem axles
4.45	1000	0.00002		187	42,000	25.64	2.51
8.9	2000	0.00018		195.7	44,000	31	3
17.8	4000	0.00209		200	45,000	34	3.27
22.25	5000	0.005		204.5	46,000	37.24	3.55
26.7	6000	0.01043		213.5	48,000	44.5	4.17
35.6	8000	0.0343		222.4	50,000	52.88	4.86
44.5	10,000	0.0877	0.00688	231.3	52,000		5.63
53.4	12,000	0.189	0.0144	240.2	54,000		6.47
62.3	14,000	0.36	0.027	244.6	55,000		6.93
66.7	15,000	0.478	0.036	249	56,000		7.41
71.2	16,000	0.623	0.0472	258	58,000		8.45
80	18,000	1	0.0773	267	60,000		9.59
89	20,000	1.51	0.1206	275.8	62,000		10.84
97.8	22,000	2.18	0.18	284.5	64,000		12.22
106.8	24,000	3.03	0.26	289	65,000		12.96
111.2	25,000	3.53	0.308	293.5	66,000		13.73
115.6	26,000	4.09	0.364	302.5	68,000		15.38
124.5	28,000	5.39	0.495	311.5	70,000		17.19
133.5	30,000	6.97	0.658	320	72,000		19.16
142.3	32,000	8.88	0.857	329	74,000		21.32
151.2	34,000	11.18	1.095	333.5	75,000		22.47
155.7	35,000	12.5	1.23	338	76,000		23.66
160	36,000	13.93	1.38	347	78,000		26.22
169	38,000	17.2	1.7	356	80,000		28.99
178	40,000	21.08	2.08				

Note: kN converted to lbf are within 0.1% of lbf shown.

Table 52.14 Lane Distribution Factors, D_L

no. of lanes in each direction	fraction of ESALs in design lane
1	1.00
2	0.80–1.00
3	0.60–0.80
4	0.50–0.75

From *AASHTO Guide for Design of Pavement Structures*, p. II-9, copyright © 1993, by the American Association of State Highway and Transportation Officials, Washington, D.C. Used by permission.

21. PERFORMANCE CRITERIA

The *terminal pavement serviceability index*, p_t, represents the lowest pavement serviceability index that can be experienced before rehabilitation, resurfacing, or reconstruction is required. Suggested levels are between 2 and 3, with 2.5 recommended for major highways and 2.0 recommended for less important roads. If costs are to be kept low, the design traffic volume or design period should be reduced. Terminal serviceability should not be reduced, as small changes in it will result in large differences in pavement design.

The *actual initial pavement serviceability*, p_o, represents the actual ride quality of the new roadway immediately after it is installed. This value is not usually known during the initial design, but it may be assumed to be 4.2 for flexible pavement and 4.5 for rigid pavement.

The *change in pavement serviceability index*, ΔPSI, is calculated as the difference between the initial pavement serviceability index and the terminal pavement serviceability index.

$$\Delta\text{PSI} = p_o - p_t$$

Transportation/ Surveying

Table 52.15 Suggested Levels of Reliability for Various Functional Classifications

functional classification	recommended level of reliability (%) urban	rural
interstates and freeways	85–99.9	80–99.9
principal arterials	80–99	75–95
collectors	80–95	75–95
local	50–80	50–80

From *AASHTO Guide for Design of Pavement Structures*, Sec. II, p. II-9, Table 2.2, copyright © 1993 by the American Association of State Highway and Transportation Officials, Washington, D.C. Used by permission.

22. PAVEMENT STRUCTURAL NUMBER

Equation 52.1: Structural Number

$$SN = a_1D_1 + a_2D_2 + \cdots + a_nD_n \qquad 52.1$$

Description

AASHTO combines pavement layer properties and thicknesses into one variable called the design *structural number*, SN. Once the structural number for a pavement is determined, a set of pavement layer thicknesses are chosen. When combined, these layer thicknesses must provide the load-carrying capacity corresponding to the design structural number.

Equation 52.1 is the AASHTO *layer-thickness equation.* D_1, D_2, and D_3 represent actual thicknesses (in inches) of surface, base, and subbase courses, respectively. (If a subbase layer is not used, the third term is omitted.) a_i are the *layer coefficients*, also known as *strength coefficients.* The layer coefficients vary from material to material, but the values given in Table 52.16 can be used for general calculations.

Theoretically, any combinations of thicknesses that satisfy Eq. 52.1 will work. However, minimum layer thicknesses result from construction techniques and strength requirements. Therefore, the thickness of the flexible pavement layers should be rounded to the nearest ½ in. When selecting layer thicknesses, cost effectiveness as well as placement and compaction issues must be considered to avoid impractical designs. (See Table 52.17.)

Flexible pavements are layered systems and are designed accordingly. First, the required structural number over the native soil is determined, followed by the structural number over the subbase and base layers, using applicable strength values for each. The maximum allowable thickness of any layer can be calculated from the differences between the calculated structural numbers. (See Fig. 52.5.)

Table 52.16 Typical Layer Strength Coefficients[a] (1/in)

	value	range
subbase coefficient, a_3		
sandy gravel[b]	0.11	
sand, sandy clay		0.05–0.10
lime-treated soil	0.11	
lime-treated clay, gravel		0.14–0.18
base coefficient, a_2		
sandy gravel[b]	0.07	
crushed stone	0.14	0.08–0.14
cement treated base (CTB)	0.27	0.15–0.29
seven-day $f'_c > 650$ psi	0.23	
400–650 psi	0.20	
< 400 psi	0.15	
bituminous treated base (BTB)		
coarse	0.34	
sand	0.30	
lime-treated base		0.15–0.30
soil cement	0.20	
lime/fly ash base		0.25–0.30
surface course coefficient, a_1		
plant mix[b]	0.44	
recycled AC,		
3 in or less	0.40	0.40–0.44
4 in or more	0.42	0.40–0.44
road mix	0.20	
sand-asphalt	0.40	

(Multiply 1/in by 0.0394 to obtain 1/mm.)

(Multiply psi by 6.89 to obtain kPa.)

[a]The AASHTO method correlates layer coefficients with resilient modulus.

[b]The average value for materials used in the original AASHTO road tests were:

asphaltic concrete surface course: 0.44
crushed stone base course: 0.14
sandy gravel subbase: 0.11

Compiled from a variety of sources.

This method should not be used to determine the structural number required above subbase or base materials having moduli of resilience greater than 40,000 psi. Layer thicknesses for materials above high-strength subbases and bases should be based on cost effectiveness and minimum practical thickness considerations.

Example

The American Association of State and Highway and Transportation Officials' (AASHTO) design structural number for a road is 4. The material specifications are as follows.

material	layer thickness (in)	layer coefficient
sandy gravel subbase	10	0.11
crushed stone base course	6	0.14

If a high-stability plant mix asphalt concrete surface course with a layer coefficient of 0.44 is to be placed on top of the specified subbase and base course materials, the required surface course thickness is most nearly

(A) 3 in

(B) 4 in

(C) 5 in

(D) 6 in

Solution

The AASHTO structural number equation can be rearranged to solve for the surface thickness, D_1.

$$\begin{aligned} \text{SN} &= a_1D_1 + a_2D_2 + a_3D_3 \\ D_1 &= \frac{\text{SN} - a_2D_2 - a_3D_3}{a_1} \\ &= \frac{4 - (0.14)(6 \text{ in}) - (0.11)(10 \text{ in})}{0.44} \\ &= 4.68 \text{ in} \quad (5 \text{ in}) \end{aligned}$$

The answer is (C).

23. PAVEMENT REHABILITATION AND RECYCLING

There are a variety of processes that collectively compose *recycled asphalt pavement* (RAP) technology. The primary benefit of recycling is cost effectiveness, although strength and reliability do not appear to be affected. *Full-depth reclamation* turns a roadway into the base material for a new surface. The original roadway pavement and some of the underlying material are removed, pulverized, and reused.

Profiling, also known as *surface recycling*, is a modification of the visible surface of the pavement. *Cold planing* is the removal of asphalt pavement by special pavement milling and planing machines (i.e., *profiling equipment*). The resulting pavement with long striations provides a good skid-resistant surface. Cold planing is used to restore a road to an even surface. The material removed is 95% aggregate and 5% asphalt cement. Conventional milling machines run at 1.5–3 m/min with a 25–50 mm cut. High-capacity machines can run at up to 24 m/min with a 100–150 mm cut (though half this speed is more typical).

Cold in-place recycling employs a *cold train* of equipment to mill 25–150 mm of pavement, crush, size, add asphalt rejuvenating agents and/or light asphalt oil in a

Table 52.17 *Minimum Thickness*[a]

traffic (ESALs)	asphalt concrete (in)	aggregate base[b] (in)
< 50,000	1.0 (or surface treatment)	4
50,001–150,000	2	4
150,001–500,000	2.5	4
500,001–2,000,000	3	6
2,000,001–7,000,000	3.5	6
> 7,000,000	4	6

(Multiply in by 25.4 to obtain mm.)

[a]Minimum thicknesses may also be specified by local agencies and by contract.

[b]Includes cement-, lime-, and asphalt-treated bases and subbases.

From *AASHTO Guide for Design of Pavement and Structures*, Sec. II, p. II-35, copyright © 1993 by the American Association of State Highway and Transportation Officials, Washington, D.C. Used by permission.From *AASHTO Guide for Design of Pavement and Structures*, Sec. II, p. II-35, copyright © 1993 by the American Association of State Highway and Transportation Officials, Washington, D.C. Used by permission.

Figure 52.5 *Procedure for Determining Thicknesses of Layers Using a Layered Analysis Approach**

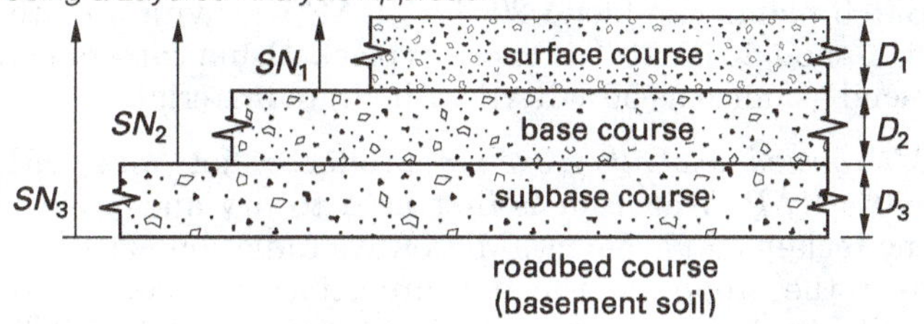

$$D_1^* > \frac{SN_1}{a_1}$$

$$SN_1^* = a_1D_1^* > SN_1$$

$$D_2^* > \frac{SN_2 - SN_1^*}{a_2m_2}$$

$$SN_1^* + SN_2^* > SN_2$$

$$D_3^* > \frac{SN_3 - \left(SN_1^* + SN_2^*\right)}{a_3m_3}$$

*Asterisks indicate values actually used, which must be equal to or greater than the required values.

pug mill, and finally redeposit and recompact the mixture in a conventional manner. A cold-recycled mix will have at least 6% voids, requiring some type of seal or overlay to keep air and water from entering. The overlay can be a sand, slurry, or chip seal. To achieve the required density, roller compaction is required.

Hot in-place recycling, also known as *hot in-place remixing* and *surface-recycling*, begins by heating the pavement to above 120°C with high-intensity indirect propane heaters. Then, a scarifier loosens and removes a layer of up to 50 mm of softened asphalt concrete. The removed material is mixed with a rejuvenator in a traveling pugmill, augured out laterally, redistributed, and compressed by a screed and rollers. A petroleum-based agent may be subsequently applied to restore the asphalt cement's adhesive qualities, or up to 50 mm of additional new hot-mix asphalt cement can be applied over the replaced surface. Alternatively, a heavier petroleum product can be applied to help the recycled surface withstand direct traffic. The entire paving train runs at approximately 0.6 m/min and the pavement can be used almost immediately.

Hot mix recycling is used in the majority of pavement rehabilitation projects. This process removes the existing asphalt pavement, hauls it to an offsite plant, and blends it with asphalt cement and virgin rock aggregate. Batch plants can blend 20–40% RAP with virgin material, though 10–20% is more typical. Drum mixers can blend as much as 50% RAP with virgin material.

Microwave asphalt recycling produces hot mix with 100% RAP. Warm air is first used to dry and preheat the reclaimed pavement. Microwave radiation is used to heat the stones in the mixture (since asphalt is not easily heated by microwaves) to approximately 150°C without coking any hydrocarbons. Rejuvenating agents restore the pavement's original characteristics, or new aggregate can be used to improve performance.

24. SUBGRADE DRAINAGE

Subgrade drains are applicable whenever the following conditions exist: (a) high groundwater levels that reduce subgrade stability and provide a source of water for frost action, (b) subgrade soils of silts and very fine sands that become quick or spongy when saturated, (c) water seepage from underlying water-bearing strata, (d) drainage path of higher elevations intercepts sag curves with low-permeability subgrade soils below.

Figure 52.6 illustrates typical subgrade drain placement. In general, drains should not be located too close to the pavement (to prevent damage to one while the other is being worked on), and some provision should be made to prevent the infiltration of silt and fines into the drain. Roofing felt or geotextile sleeves can be placed around the drains for this purpose.

Figure 52.6 *Typical Subgrade Drain Details*

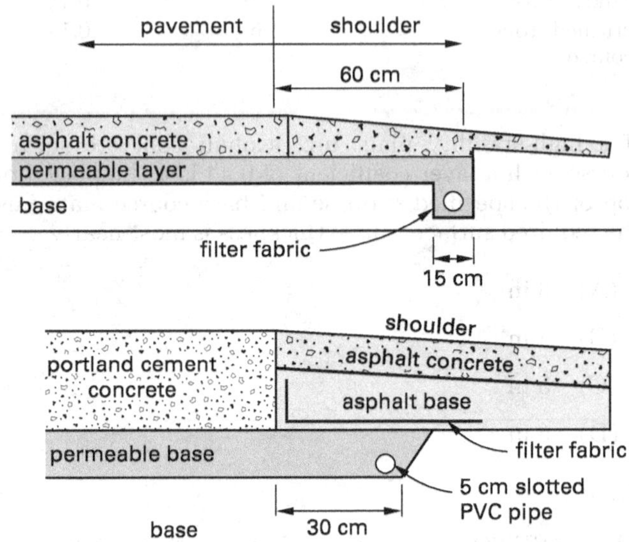

Transportation/ Surveying

53 Traffic Safety

Nomenclature

a	acceleration	mi/hr
A	number of crashes	–
ADT	average daily traffic	veh
CR	crash reduction factor	–
F	force	lbf
g_c	gravitational constant, 32.2	lbm-ft/lbf-sec^2
m	mass	lbm
N	number	–
R	crash rate	various
RMEV	crash rate per million entering vehicles	various
RMVM	crash rate per million vehicle-miles	various
S	stopping distance	ft
t	time	sec
V	volume	veh/hr
VMT	vehicle-miles of travel	veh-mi

Subscripts

i	specific countermeasure
m	number of countermeasures

1. TRAFFIC SAFETY AND ROADWAY DESIGN ANALYSIS

Traffic safety is a fundamental part of transportation design and affects multiple aspects of planning, including capacity analysis, design standards, construction, operations, and maintenance. Safe design practices include providing adequate roadside clearances, analyzing and planning for traffic management in conflict zones, and providing work zone safety during construction and maintenance activity. Once a roadway network is in place, the emphasis on safety continues in order to mitigate crashes, which often result in property damage, injury, and/or fatalities. Local governments implement *countermeasures* to reduce the severity of crashes or prevent them altogether. When crashes do occur, crash causation analysis should be undertaken to identify factors and mechanisms that increase crash risk and to prioritize changes and improvements that will prevent crashes in the future.

Traffic safety begins with applying highway safety principles during design. The comfort, convenience, and safety of a highway can be greatly improved by designs that work to eliminate surprises, such as sudden sharp curves or rapid lane shifts. One of the greatest past improvements to highway safety was the elimination of at-grade intersection crossings so that all traffic enters and leaves the main roadways via high-speed ramps, and all traffic crossing the roadway is carried on grade-separated bridges. This allows the mainline flow to continue unimpeded (i.e., in "free flow") in all but high-density traffic conditions (hence the term "freeway").

Observations of consistent driver behavior patterns are quantified by the American Association of State Highway and Transportation Officials (AASHTO) and in design guides. In particular, AASHTO's *A Policy on Geometric Design of Highways and Streets* (*Green Book*) provides charts, tables, and graphs for acceleration rates, braking rates, comfortable travel speeds, perception-reaction times, signage visibility, and blending of horizontal and vertical curve geometry. This information is beneficial to designing roadways with safety in mind. AASHTO has adopted the design philosophy that accommodating 85% of the usual driving behavior is a cost-effective goal. Although accommodating the upper or lower 5–10% of behavior beyond the normal range is not necessarily ignored, unusual or erratic behavior and extreme driving patterns are generally not accounted for due to cost or other considerations.

2. ANALYSIS OF ACCIDENT DATA

Accident data are compiled and evaluated to identify hazardous features and locations, set priorities for safety improvements, support economic analyses, and identify patterns, causes (i.e., driver, highway, or vehicle), and possible countermeasures.

Accidents are classified into three *severity categories*, depending on whether there is (a) property damage only (referred to as *PDO accidents*), (b) personal injury, or (c) fatalities. The *severity ratio* is defined as the ratio of the number of injury and fatal accidents divided by the total number of all accidents (including PDO accidents).

To identify the most dangerous features, it is necessary to determine what the accidents at an intersection or along a route had in common. This can be done in a number of ways, one of which is by drawing a collision diagram. A *collision diagram* is essentially a true-scale cumulative drawing of the intersection showing details of all of the accidents in the study. Dates, times, weather conditions, types of maneuvers, directions, severity, types of accidents, and approximate locations are recorded on the diagram.

Common *accident-type* categories include: rear-end, right-angle, left-turn, fixed-object, sideswipe, parked-vehicle, run off road, head-on, bicycle-related, and pedestrian-related.

Hypothesis testing can be used to determine if installed countermeasures are being effective. The *null hypothesis* is that the highway improvements have brought no significant decrease in accident rate. The counter-hypothesis is that the highway improvements have brought significant decreases. A *chi-squared test* with a 5% confidence level is used to evaluate the two hypotheses.

Equation 53.1 and Eq. 53.2: Crash Rate per Million Entering Vehicles (RMEV)

$$\text{RMEV} = \frac{A \times 1{,}000{,}000}{V} \quad 53.1$$

$$V = \text{ADT} \times 365 \quad 53.2$$

Variation

$$R_{\text{RMEV}} = \frac{(\text{no. of accidents})(10^6)}{(\text{ADT})(\text{no. of years})\left(365\ \frac{\text{days}}{\text{yr}}\right)}$$

Description

It is common to prioritize intersections according to the *accident rate*. The accident rate may be determined for PDO, personal injury, and fatal accidents, or the total thereof. The accident ratio is the ratio of the number of accidents per year to the average daily traffic, ADT. The rate is reported as RMEV (rate per million entering vehicles) taking into consideration vehicles entering an intersection from all directions.

The number of crashes, A, can be either the total number of crashes in an intersection or the number of crashes of a particular type (e.g., PDO) during a given year. The volume, V, is calculated from the average daily traffic entering the intersection, ADT, as shown in Eq. 53.2.

Example 31

A particularly treacherous stretch of highway is 18 miles long. Over a period of 12 years, the average daily traffic has been 1500 vehicles. During that same period, there have been 7 fatal accidents, 16 nonfatal personal injury accidents, and 23 property damage only accidents. What is most nearly the accident rate per million entering vehicles (MEV)?

(A) 0.39 accidents/MEV

(B) 3.5 accidents/MEV

(C) 7.0 accidents/MEV

(D) 84 accidents/MEV

Solution

All types of accidents are included in the accident rate, so $A = 7 + 16 + 23 = 46$ accidents. From Eq. 53.1 and Eq. 53.2,

$$\begin{aligned}\text{RMEV} &= \frac{A \times 1{,}000{,}000}{V} = \frac{A \times 1{,}000{,}000}{\text{ADT} \times 365} \\ &= \frac{(46\ \text{accidents})\left(10^6\ \frac{\text{veh}}{\text{MEV}}\right)}{\left(1500\ \frac{\text{veh}}{\text{day}}\right)(12\ \text{yr})\left(365\ \frac{\text{days}}{\text{yr}}\right)} \\ &= 7.0\ \text{accidents/MEV}\end{aligned}$$

The answer is (C).

Equation 53.3 and Eq. 53.4: Crash Rate per Million Vehicle-Miles (RMVM)

$$\text{RMVM} = \frac{A \times 1{,}000{,}000}{\text{VMT}} \quad 53.3$$

$$\begin{aligned}\text{VMT} = \text{ADT} &\times (\text{number of days in study period}) \\ &\times (\text{length of road})\end{aligned} \quad 53.4$$

Variation

$$R_{\text{MVM}} = \frac{(\text{no. of accidents})(10^6)}{(\text{ADT})(\text{no. of years})\left(365\ \frac{\text{days}}{\text{yr}}\right)L_{\text{mi}}}$$

Description

Routes (segments or links) between points are prioritized according to the accident rate per mile (per kilometer), calculated as the ratio of the number of accidents per year to the ADT per mile of length, counting traffic from all directions in the intersection. For convenience, the rate may be calculated per million vehicle-miles, RMVM, as shown in Eq. 53.3. RMVM is calculated from the vehicle-miles of travel during the study period, VMT, calculated from Eq. 53.4.

Equation 53.5 and Eq. 53.6: Crashes Prevented

$$\text{crashes prevented} = N \times \text{CR}\frac{\text{ADT after improvement}}{\text{ADT before improvement}} \qquad 53.5$$

$$\begin{aligned}\text{CR} = \text{CR}_1 &+ (1 - \text{CR}_1)\text{CR}_2 \\ &+ (1 - \text{CR}_1)(1 - \text{CR}_2)\text{CR}_3 + \cdots \\ &+ (1 - \text{CR}_1)\cdots(1 - \text{CR}_{m-1})\text{CR}_m\end{aligned} \qquad 53.6$$

Description

The effectiveness of countermeasures for a given site can be calculated from the estimated number of crashes prevented, Eq. 53.5. N is the expected number of crashes if countermeasures are not implemented using current traffic volumes. The crash reduction factor, CR, is determined for specific countermeasures, and the overall crash reduction factor, based on m improvements at a site, can be found from Eq. 53.6.

3. ROAD SAFETY FEATURES

Road safety features are installed to protect public life and property and to reduce traffic-related lawsuits against highway and transportation departments. The most common actions include the installation of illumination, guardrails, and impact attenuators, as well as the relocation of dangerous facilities.

Guardrails are used on roadways where there is a severe slope or vertical dropoff to the side of the road, ditches, permanent bodies of water, embankments, and roadside obstacles (e.g., boulders, retaining walls, and sign and signal supports). Guardrails with turned-down ends are now prohibited in new installations and should be upgraded. Such guardrails, rather than protecting motorists from a fixed impact or spearing, often cause *vaulting* (also known as *launching* or *ramping*) and subsequent rollover.

Impact attenuators are used to provide crash protection at bridge pillars, center piers, gore areas, butterfly signs, light posts, and ends of concrete median barriers. These are mainly of the *crash cushion* variety (sand barrel, water barrel, and water-tube arrays), sandwich-type units, and crushable cartridges.

Slip- and breakaway- (frangible coupling) bases are used on poles of roadside luminaires and signs to prevent vehicles from "wrapping around" the poles. Hinged and weakened bases for high-voltage power poles, however, may violate other codes (e.g., the *National Electric Code*) designed to prevent high-voltage wires from dropping on the public below. Other solutions, such as placing utility lines underground, increasing lateral offset or longitudinal pole spacing, or dividing the load over multiple poles, should be considered.

4. ROADSIDE SAFETY RAILINGS

Properly designed and placed highway safety barriers and guardrails reduce the number of injuries and deaths associated with vehicles leaving the traveled surface. The most common type of highway guardrail used in the United States is the solid guardrail, made of hot dipped galvanized steel and aluminum. The steel variety is found in areas that present a higher level of danger, such as the edges of ravines and bodies of water. The W-beam rail is the most widely used safety system. W-beams are available with strong-post, weak-post, and wood-block cushions.

In locations where snow drifts may accumulate, tension cable guardrails can be used. Unlike solid guard railing, the cable allows snow to pass through, reducing accumulations. Tension cable railing also "gives" elastically to impacts received from vehicles.

Concrete barriers are often used as permanent median barriers as well as temporary construction barriers on both sides of a lane. They have initial costs that are three times that of the galvanized guardrail, but they are cost effective because they last longer and are more easily installed and replaced. The earliest concrete barrier was the *GM barrier* (*General Motors barrier*). It is similar in appearance to modern designs, but it is clearly fatter. The barrier was larger than modern design, primarily because vehicles were larger. The GM barrier is obsolete, and there are four main types of modern concrete barriers. The *Jersey barrier* (also known as a *K-rail*) is 32 in tall. It rises from a 55° angle to an 85° tapered top. The *Ontario barrier* is the same as the Jersey barrier, except it is 42 in tall. The *constant slope barrier* is also 42 in tall and has a slope of 79°. The *F-shape barrier* resembles the Jersey barrier except for the slope. A reinforced version of the 42 in Jersey barrier is used as a median divider and is known as a *heavy vehicle median barrier*. (See Fig. 53.1.)

Example 32

Which of the following end-treatments would generally be used at the end of a bridge median traffic barrier?

(A) no end treatment

(B) sloped or curved, rising from ground to full barrier height

(C) cantilevered top railing without substantial vertical support

(D) cantilevered blunt-nosed head with sliding rail

Transportation/Surveying

Figure 53.1 Ontario/Heavy Vehicle Median Barrier

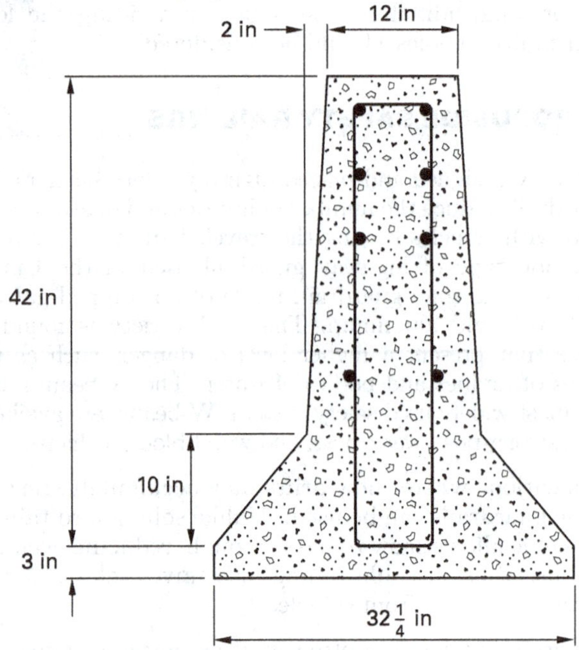

Solution

Cantilever barriers and barriers without end treatment can penetrate a vehicle upon impact. Barriers that slope up from the ground can cause a vehicle to roll over. For head-on impacts, it is best for the car's bumper to make contact with a blunt plate that is supported by a sliding rail. Friction between the sliding rail and its track, in addition to a few break-away bolts, is used to dissipate the vehicle's kinetic energy.

The answer is (D).

5. MODELING VEHICLE ACCIDENTS

Traffic accident analysis, modeling, and simulation draw on many concepts. What can be done by hand (i.e., with a pencil and calculator) for a single body at a time differs dramatically with what can be done in computer analyses that use simulation techniques and draw upon finite element analysis and multi-body system models.

Modeling the vehicle dynamics (kinematics) of a body is the first step in modeling an accident. For manual calculations, common assumptions include rigid bodies (lumped masses); one or two dimensional movement; one or two degrees of freedom; constant masses; constant friction coefficients; constant (including zero), linear, and instantaneous forces; linear stiffness; linear damping; absence of atmospheric (wind, rain, and weather) and other secondary effects; and either constant speeds, constant accelerations, or decelerations. Sophisticated computer simulations relax or eliminate all of these assumptions, and they are able to "step" through the time domain in small discrete time increments. As an added bonus, comparisons of the actual vehicle motions (obtained from the scene of the accident) with the simulation results can be used to adjust input parameters.

As a comparison, a simple model intended for manual calculations might use uniform acceleration formulas to determine acceleration, speed, and position of a rigid body in a single direction at one or two times or locations. It might use basic concepts such as conservation of momentum and work-energy principles. A simulation model, on the other hand, would determine acceleration, speed, position, and orientation for each moving body, in each of three dimensions, and for each degree of freedom (e.g., straight ahead movement, flat spinning, lateral rolling, and tumbling end for end) at each discrete time period.

Example

The front end of a car traveling at 30 mi/hr crashes into a massive tree. The tree trunk compresses the car length by 1 ft. The vehicle's driver weighs 160 lbm and is wearing a seatbelt that stretches during the accident. The stretching seatbelt allows the driver to move forward 0.5 ft from the snug condition. Most nearly, what force does the driver experience?

(A) 3200 lbf

(B) 5700 lbf

(C) 33,000 lbf

(D) 100,000 lbf

Solution

The stretching seatbelt effectively increases the stopping distance to

$$S = 1\text{ ft} + 0.5\text{ ft} = 1.5\text{ ft}$$

The vehicle is initially traveling at 30 mi/hr, but its final velocity is zero. The average velocity is 15 mi/hr. The time taken for the driver to travel 1.5 ft during the accident is

$$t = \frac{S}{\text{v}} = \frac{1.5\text{ ft}}{\left(15\ \frac{\text{mi}}{\text{hr}}\right)\left(\frac{5280\ \frac{\text{ft}}{\text{mi}}}{3600\ \frac{\text{sec}}{\text{hr}}}\right)}$$
$$= 0.068\text{ sec}$$

Use Newton's second law.

$$F = \frac{ma}{g_c} = \frac{m\Delta \text{v}}{g_c \Delta t}$$

$$= \frac{(160 \text{ lbm})\left(30 \ \frac{\text{mi}}{\text{hr}} - 0 \ \frac{\text{mi}}{\text{hr}}\right)\left(5280 \ \frac{\text{ft}}{\text{mi}}\right)}{\left(32.2 \ \frac{\text{lbm-ft}}{\text{lbf-sec}^2}\right)(0.068 \text{ sec})\left(3600 \ \frac{\text{sec}}{\text{hr}}\right)}$$

$$= 3215 \text{ lbf} \quad (3200 \text{ lbf})$$

The answer is (A).

6. DRIVER BEHAVIOR AND PERFORMANCE

Driving is an acquired skill that relies on auditory and visual awareness, motor skill coordination, and the ability to judge situations and react accordingly. Therefore, *driver education* plays an important role in expected on-road driver behavior. Skills are taught and demonstrated, then studied and practiced. Although the quality of instruction is important in driver training, the driver's attitude toward the rules of the road has a greater effect on how well the driver's actions will conform to general traffic behavior. Attempts to change nonconforming behavior can range from passive measures, such as road designs and signs, to active measures, such as law enforcement.

Transportation systems should be designed to accommodate the *design driver*, a term that encompasses a range of drivers whose differences and limitations are accounted for during road, vehicle, and traffic control designs. Historically, designers have used the 85th percentile of driver behavior as the cutoff to determine transportation system designs and controls. The 85th percentile is used to represent the average driver, or the *reasonable worst case driver*, though this percentile may need to be adjusted depending on regional population characteristics. For instance, regions with large elderly populations may require more careful consideration of sign placement and sign lettering, as well as an increase in standard parking stall size and signal timing design.

Driver behavior is also influenced by vehicle design. Drivers of large commercial vehicles tend to drive conservatively due to the size of the vehicle, limited maneuverability, and the nature and value of the carried cargo. Similarly, bus drivers generally make driving decisions based on the comfort and safety of the passengers. Of standard passenger cars, the faster acceleration and maneuverability of performance cars allow and even encourage their drivers to engage in higher-risk driving behaviors than drivers of sedan-like cars.

Of the three major elements making up a highway system—the driver, the vehicle, and the road—the driver is perhaps the most variable and least predictable. Therefore, it is important to understand how the following internal and external factors influence driver performance so that they can be accounted for in a transportation system's design.

Perception-Reaction Responses

Perception-reaction refers to how quickly a driver can react to a situation. Most studies divide the perception-reaction response into four subprocesses—perception, identification, emotion, and volition—also known as PIEV. *Perception* first occurs when the driver sees a control device, sign, object, or another vehicle on the road. *Identification* follows perception when the driver classifies the nature of the perceived object. *Emotion* encompasses a driver's decision-making process to determine what action needs to be taken. Finally, *volition*, or *reaction*, occurs at the point at which the driver executes the action determined to be necessary.

These subprocesses take small but measurable amounts of time, together known as the *perception-reaction time* (PRT). The average necessary time to react and execute a response to a roadway condition or object message must be accounted for in the design process, such as placement of warning signs at the proper distance ahead of the condition being warned of.

PRT is important in determining the necessary sight distance on curves and for setting the amber phase of traffic signals. It is also a necessary factor to include when determining the safe following distance between vehicles. PRT increases with age, but is not uniform across the driver population at any age group. Experienced and alert drivers can react in 1.5 sec or less, while more passive drivers may take as long as 3 sec to respond to the same situation. AASHTO design guides generally use 2.5 sec as an average reaction time for determining stopping distances. However, PRT can vary significantly depending on roadway types and conditions. 2.5 sec may not be adequate under adverse or very complex conditions. For instance, an intersection approach may be complicated enough that the amber phase must be increased by 1–2 sec to eliminate running the red phase. Other instances where a longer PRT may exist include confusing or incorrect information on warning or directional signs. Drivers have generally learned to expect that adequate warning time will be given in order for them to react to a situation and make adjustments in vehicle operation. When drivers come across a situation that deviates from this norm (i.e., detour signs or unanticipated construction), they will likely need more time to react than usual, as the situation will come as a surprise. A *surprise* is when the time allowed for adjustment to a change in road condition, traffic condition, or visibility is less than the required normal perception-reaction time. The goal of good highway design is to eliminate surprises whenever possible.

Example

What is the difference between perception-identification-emotion-volition (PIEV) times and perception-reaction times (PRT)?

(A) PIEV includes time for the startle reflex, while PRT does not.

(B) PIEV includes time for surprise, while PRT does not.

(C) PIEV includes time for volition, while PRT does not.

(D) There is no difference.

Solution

Perception-reaction time includes time to perceive, identify, think, and consciously initiate an action. "Volition" is synonymous with acting consciously.

The answer is (D).

Visual and Recognition Acuity

Visual acuity refers to the sharpness with which a person can see an object, while *recognition acuity* requires the viewer to not only see an object, but recognize the meaning of the object as well. For a normal eye, visual acuity is greatest near the visual center of the retina, which is along the visual axis of the eye. The cone of greatest visual acuity is about 3°, or about 1.5° in each direction from the center axis of vision. *Peripheral vision* can extend as far as 180° or more in some people, but 160° is the accepted normal limit of peripheral vision. In the *peripheral zone*, objects are visible and important for orientation and awareness, but may be blurred compared to the central cone of vision. Traffic signs and critical road markings should be placed so that they are within the normal central cone of vision in order to be clearly read by drivers.

The standard *Snellen chart* is used universally for checking visual acuity during an eye examination. A person reads letters of different heights on the chart from a specified distance. Charts used to determine visual acuity are designed so that a person with normal vision can read lettering $\frac{1}{3}$ in in height from a distance of 20 ft. This ratio is called *20/20 vision.* When a person needs an object twice as large to be visible at 20 ft, the ratio is given as 20/40. Visual and recognition acuity can deteriorate because of age, fatigue, eye disease, medication side effects, and/or alcohol and drug use.

The criterion for letter height on signs is often cited as 1 in for each 50 ft of viewing distance. However, this criterion is incomplete without including letter spacing and letter shape. The letter shapes used on interstate highway signs were developed after extensive visibility research during the development of the early interstate highway system. The Federal Highway Administration (FHWA) has developed the *federal alphabet* with respect to within letter spaces and between letter spaces. *Within letter spaces* are the openings in loop letters such as a, p, and so on. *Between letter spaces* are the spaces that exist between letters when they are put together to form words. Each letter in the federal alphabet has been created with an optimum amount of white space around and, if necessary, within it. By this design, when letters are put together, the white space between them optically balances with the white space within them, producing optimum readability and legibility. These guidelines allow traffic signs to be viewed from a distance of 100 ft for each 1 in of letter height, or twice the viewing distance when using other lettering styles. This standard has not been replicated in most current electronic digital lettering signs, meaning the viewing distance of many of these changeable lettering signs is much less than the fixed lettering version. Standards for a sign's height, color, letter spacing, message content, panels, layout, and location are given in further detail in the *Manual on Uniform Traffic Control Devices* (*MUTCD*).

Color Vision

The colors chosen for traffic control devices and informational signs were selected to minimize the effects of color blindness on their readability. *Color blindness* refers to a difficulty in perceiving differences between colors. Driver color blindness exists in various types and degrees that range from a difficulty in distinguishing between minor hue differences to a complete lack of color definition, in which the driver sees everything as varying intensities of gray. Background and letter coloring have been researched to provide the greatest amount of contrast so that lettering does not disappear when viewed by someone with color-impaired vision. For example, the green background with white lettering on roadway direction signs has become a standard on U.S. federal highways, eliminating the effect of color versus contrast sensitivities. Similarly, traffic signals place the red, amber, and green lenses in standardized positions so that a driver with total or red/green color blindness will still be able to determine signal phases.

Glare Vision

Glare can be caused by a direct shaft of light in the field of vision or by reflected light. When too much light enters the eye, the pupil closes to reduce the amount of incoming light. *Glare recovery* can take 6 sec or more for a normal eye, though factors such as age, congenital disabilities, or drug and alcohol use can increase the eye's recovery time.

Glare also affects the eye's visual contrast ratio between light and dark objects. The human eye cannot discern a light-intensity-contrast ratio greater than 3 to 1. *Glare blindness* occurs when the presence of an intense light source at one part of the visual plane causes images in the rest of the visual plane to disappear. For example, glare blindness can occur when driving at night along a dark road. Objects are illuminated by the headlight's radius, but objects appear invisible outside that radius.

When designing transportation systems, designers must account for changes between lit and unlit sections of the roadway and allow sufficient time for drivers' eyes to acclimate to new conditions before introducing critical movement, such as exit ramps or abrupt geometry conditions. Sources of light off of, but near to, roadways, such as advertising lighting and lighting in shopping center parking lots, can also create glare and affect the safety of traffic.

Example

At night, the intensity of lighted, flashing work-zone signs is normally

(A) reduced to increase depth perception

(B) reduced to eliminate glare

(C) increased to increase visibility from a distance

(D) increased to increase contrast

Solution

During the day, the intensity of lighted, flashing work-zone signs must be high to overcome ambient light levels. At night, when there is little or no ambient light, the intensity can be reduced in order to eliminate glare.

The answer is (B).

Depth Perception

Depth perception allows a person to estimate the distance to an object by using either stereo vision for objects closer to the eye or by estimating the relative size, position, and movement of objects farther away. The human eye has difficulty measuring the speed and the relative distances of objects that are more than a few feet away. This is especially true of approaching objects that are moving in a straight line, such as two oncoming vehicles occupying the same lane. Traffic signs and marking devices have standardized locations to account for the driver's viewing distance, which helps maintain uniformity and predictability of approaching roadway conditions. One example is the placement of a stop sign on the corner of a minor street intersecting a major street.

Hearing

Sounds received by the human ear can help a driver adjust speed, vehicle location, and other factors in relation to certain circumstances, such as an oncoming vehicle or the sound of a car horn. They can also warn drivers of unusual occurrences, such as an approaching emergency vehicle. Drivers with limited or no hearing are generally able to drive quite well, as long as they take extra precautions by using more visual feedback to compensate for the lack of hearing.

Driver Error and Unavoidable Situations

When a driver encounters traffic or road conditions that are outside of the range of normal conditions, collisions, off-road excursions, or other out-of-the-ordinary travel may occur. A pattern of similar crashes, particularly if involving drivers unfamiliar with the local conditions, may indicate a specific need for countermeasure development.

For example, if drivers frequently miss a turn at an intersection and drive off the side of the road, there may be a sight-limiting condition that is blocking the view at the driver's eye level. If the condition is primarily weather-related, such as icing, there may be a need to redirect drainage or recontour the intersection.

When a specific countermeasure cannot be developed to prevent a crash, or when crashes are random, countermeasures may include providing a safer environment for the most damaging or injury-prone type of occurrence.

7. WORK ZONE SAFETY

Work zones increase crashes and congestion, and account for approximately 2% of all roadway fatalities each year. With the majority of the fatalities being motorists, the FHWA created the National Highway Work Zone Safety Program, which focuses on standardization and evaluation to improve work zone safety. Work zone standardization is set by the FHWA in both traffic control and work zone safety devices. All safety standards are contained in the *MUTCD*.

The National Work Zone Safety Information Clearinghouse is a public outreach organization that provides information and resources about work zone safety. Managed by the FHWA Office of Safety, it contains the largest online database of accident and crash rate data and statistics, training programs and materials, research services, latest technologies and equipment, best practices, safety engineer contact information, laws and regulations, and public education.

The FHWA is also responsible for National Cooperative Highway Research Program (NCHRP) *Report 350* (NCHRP 350), which contains the federal standards and guidelines for all work zone safety devices, including traffic barriers, end treatments, crash cushions, and breakaway devices. NCHRP 350 provides a wide range of test procedures to evaluate these devices, with different test levels defined for various classes of roadside safety features. It provides enhanced measurement techniques related to occupant risk, as well as guidelines for device installation and test instrumentation. Evaluation criteria to assess the performance of work zone safety devices include occupant risk, occupant compartment integrity, test article debris, and vehicle stability.

Topic XIII: Construction

Chapter

54 Construction Management, Scheduling, and Estimating

Nomenclature

a	optimistic duration	days
b	most likely duration	days
c	pessimistic duration	days
CP	critical path	days
d	duration	days
T	project duration	days

Symbols

μ	mean
σ	standard deviation

1. PROJECT MANAGEMENT

Project management is the coordination of the entire process of completing a job, from its inception to final move-in and post-occupancy follow-up. In many cases, project management is the responsibility of one person. Large projects can be managed with *partnering.* With this method, the various stakeholders of a project, such as the architect, owner, contractor, engineer, vendors, and others, are brought into the decision-making process. Partnering can produce much closer communication on a project and shared responsibilities. However, the day-to-day management of a project may be difficult with so many people involved. A clear line of communications and delegation of responsibility should be established and agreed to before the project begins.

Many project managers follow the procedures outlined in *A Guide to the Project Management Body of Knowledge* (PMBOK Guide), published by the Project Management Institute. The PMBOK Guide is an internationally recognized standard (IEEE Std 1490) that defines the fundamentals of project management as they apply to a wide range of projects, including construction, engineering, software, and many other industries. The PMBOK Guide is process-based, meaning it describes projects as being the outcome of multiple processes. Processes overlap and interact throughout the various phases of a project. Each process occurs within one of five *process groups*, which are related as shown in Fig. 54.1. The five PMBOK process groups are (1) initiating, (2) planning, (3) controlling and monitoring, (4) executing, and (5) closing.

Figure 54.1 *PMBOK Process Groups*

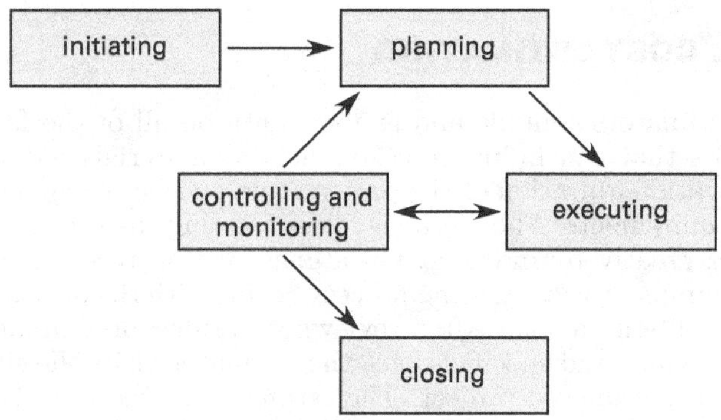

The PMBOK Guide identifies nine project management knowledge areas that are typical of nearly all projects. The nine knowledge areas and their respective processes are summarized as follows.

1. *Integration:* develop project charter, project scope statement, and management plan; direct and manage execution; monitor and control work; integrate change control; close project
2. *Scope:* plan, define, create work breakdown structure (WBS), verify, and control scope
3. *Time:* define, sequence, estimate resources, and estimate duration of activities; develop and control schedule
4. *Cost:* estimate, budget, and control costs
5. *Quality:* plan; perform quality assurance and quality control
6. *Human Resources:* plan; acquire, develop, and manage project team
7. *Communications:* plan; distribute information; report on performance; manage stakeholders

8. *Risk:* identify risks; plan risk management and response; perform qualitative and quantitative risk analysis; monitor and control
9. *Procurement:* plan purchases, acquisitions, and contracts; request seller responses; select sellers; administer and close contracts

Each of the processes also falls into one of the five basic process groups, creating a matrix so that every process is related to one knowledge area and one process group. Additionally, processes are described in terms of inputs (documents, plans, designs, etc.), tools and techniques (mechanisms applied to inputs), and outputs (documents, products, etc.).

Establishing a budget (knowledge area 4) and scheduling design and construction (knowledge area 3) are two of the most important parts of project management because they influence many of the design decisions to follow and can determine whether a project is even feasible.

2. COST ESTIMATING

Estimators compile and analyze data on all of the factors that can influence costs, such as materials, labor, location, duration of the project, and special machinery requirements. The methods for estimating costs can differ greatly by industry. On a construction project, for example, the estimating process begins with the decision to submit a bid. After reviewing various preliminary drawings and specifications, the estimator visits the site of the proposed project. The estimator needs to gather information on access to the site; the availability of electricity, water, and other services; and surface topography and drainage. The estimator usually records this information in a signed report that is included in the final project estimate.

After the site visit, the estimator determines the quantity of materials and labor the firm will need to furnish. This process, called the quantity survey or "takeoff," involves completing standard estimating forms, filling in dimensions, numbers of units, and other information. Table 54.1 is a small part of a larger takeoff report illustrating the degree of detail needed to estimate the project cost. A cost estimator working for a general contractor, for example, uses a construction project's plans and specifications to estimate the materials dimensions and count the quantities of all items associated with the project.

Though the quantity takeoff process can be done manually using a printout, a red pen, and a clicker, it can also be done with a digitizer that enables the user to take measurements from paper bid documents, or with an integrated *takeoff viewer* program that interprets electronic bid documents. In any case, the objective is to generate a set of takeoff elements (counts, measurements, and other conditions that affect cost) that is used to establish cost estimates. Table 54.1 is an example of a typical quantity takeoff report for lumber needed for a construction project.

Although subcontractors estimate their costs as part of their own bidding process, the general contractor's cost estimator often analyzes bids made by subcontractors. Also during the takeoff process, the estimator must make decisions concerning equipment needs, the sequence of operations, the size of the crew required, and physical constraints at the site. Allowances for wasted materials, inclement weather, shipping delays, and other factors that may increase costs also must be incorporated in the estimate. After completing the quantity surveys, the estimator prepares a cost summary for the entire project, including the costs of labor, equipment, materials, subcontracts, overhead, taxes, insurance, markup, and any other costs that may affect the project. The chief estimator then prepares the bid proposal for submission to the owner. Construction cost estimators also may be employed by the project's architect or owner to estimate costs or to track actual costs relative to bid specifications as the project develops.

Estimators often specialize in large construction companies employing more than one estimator. For example, one may estimate only electrical work and another may concentrate on excavation, concrete, and forms.

Computers play an integral role in cost estimation because estimating often involves numerous mathematical calculations requiring access to various historical databases. For example, to undertake a parametric analysis (a process used to estimate costs per unit based on square footage or other specific requirements of a project), cost estimators use a computer database containing information on the costs and conditions of many other similar projects. Although computers cannot be used for the entire estimating process, they can relieve estimators of much of the drudgery associated with routine, repetitive, and time-consuming calculations.

3. SCHEDULING

There are two primary elements that affect a project's schedule: design sequencing and construction sequencing. The architect has control over design and the production of contract documents, while the contractor has control over construction. The entire project should be scheduled for the best course of action to meet the client's goals. For example, if the client must move by a certain date and normal design and construction sequences make this impossible, the engineer, architect, or contractor may recommend a fast-track schedule or some other approach to meet the deadline.

Table 54.1 Partial Lumber and Hardware Take-Off Report

size	description	usage	pieces	total length
foundation framing				
2 × 4	DF STD/BTR	stud	38	8
2 × 6	DF #2/BTR	stud	107	8
2 × 6	PTDF	mudsill	–	RL
2 × 6	DF #2/BTR	bracing	–	RL
2 × 4	DF STD/BTR	blocking	–	RL
$11\frac{7}{8}''$	TJI/250	floor joist	11	18
$11\frac{7}{8}''$	TJI/250	floor joist	12	10
2 × 4	DF STD/BTR	plate	–	RL
2 × 6	DF #2/BTR	plate	–	RL
4 × 4	PTDF	posts	7	4
4 × 6	PTDF	posts	23	4
6 × 6	PTDF	posts	1	4
2 × 12	DF #2/BTR	rim	–	RL
$\frac{15}{32}''$	CDX plywood	subfloor	12	4 × 8
pre-cut doors				
$3\frac{1}{2}'' \times 7\frac{1}{4}''$	LVL	header	1	100″
$3\frac{1}{2}'' \times 7\frac{1}{4}''$	LVL	header	1	130″
$3\frac{1}{2}'' \times 7\frac{1}{4}''$	LVL	header	10	24″
exterior sheathing and shear wall				
$\frac{1}{2}''$	CDX plywood	shear wall	100	4 × 8
$\frac{1}{2}''$	CDX plywood	exterior sheathing	89	4 × 8
roof sheathing				
$\frac{1}{2}''$	CDX plywood	roof sheathing	67	4 × 8
building A & B hardware				
Simpson HD64	6 pieces each			
Simpson HD22	23 pieces each			

Design Sequencing

Prior to creating a design, the architect needs to gather information about a client's specific goals and objectives, as well as analyze any additional factors that may influence a project's design. (This gathering of information is known as *programming.*)

Once this preliminary information has been gathered, the design process may begin. The design process normally consists of several clearly defined phases, each of which must be substantially finished and approved by the client before the next phase may begin. These phases are outlined in the "Owner-Architect Agreement," which is published by the American Institute of Architects (AIA), as well as in other AIA documents. The architectural profession commonly refers to the phases as follows.

1. *Schematic Design Phase:* develops the general layout of the project through schematic design drawings, along with any preliminary alternate studies for materials and building systems

2. *Design Development Phase:* refines and further develops any decisions made during the schematic design phase; preliminary or outline specifications

are written and a detailed project budget is created

3. *Construction Documents Phase:* final working drawings, as well as the project manual and any bidding or contract documents, are solidified
4. *Bidding or Negotiation Phase:* bids from several contractors are obtained and analyzed; negotiations with a contractor begin and a contractor is selected
5. *Construction Phase:* see the following section, "Construction Sequencing"

The time required for each of these five phases is highly variable and depends on the following factors.

- *Size and complexity of the project:* A 500,000 ft^2 hospital will take much longer to design than a 30,000 ft^2 office building.
- *Number of people working on the project:* Although adding more people to the job can shorten the schedule, there is a point of diminishing returns. Having too many people only creates a management and coordination problem, and for some phases only a few people are needed, even for very large jobs.
- *Abilities and design methodology of the project team:* Younger, less-experienced designers will usually need more time to do the same amount of work than would more senior staff members.
- *Type of client, client decision-making, and approval processes of the client:* Large corporations or public agencies are likely to have a multilayered decision-making and approval process. Getting necessary information or approval for one phase from a large client may take weeks or even months, while a small, single-authority client might make the same decision in a matter of days.

The construction schedule may be established by the contractor or construction manager, or it may be estimated by the architect during the programming phase so that the client has some idea of the total time required from project conception to move-in.

Many variables can affect construction time. Most can be controlled in one way or another, but others, like weather, are independent of anyone's control. Beyond the obvious variables of size and complexity, the following is a partial list of some of the more common variables.

- management ability of the contractor to coordinate the work of direct employees with that of any subcontractors
- material delivery times
- quality and completeness of the architect's drawings and specifications
- weather
- labor availability and labor disputes
- new construction or remodeling (remodeling generally takes more time and coordination than for new buildings of equal areas)
- site conditions (construction sites or those with subsurface problems usually take more time to build on)
- characteristics of the architect (some professionals are more diligent than others in performing their duties during construction)
- lender approvals
- agency and governmental approvals

Construction Sequencing

Construction sequencing involves creating and following a work schedule that balances the timing and sequencing of land disturbance activities (e.g., earthwork) and the installation of *erosion and sedimentation control* (ESC) measures. The objective of construction sequencing is to reduce on-site erosion and off-site sedimentation that might affect the water quality of nearby water bodies.

The project manager should confirm that the general construction schedule and the construction sequencing schedule are compatible. Key construction activities and associated ESC measures are listed in Table 54.2.

Time-Cost Trade-Off

A project's completion time and its cost are intricately related. Though some costs are not directly related to the time a project takes, many costs are. This is the essence of the time-cost trade-off: The cost increases as the project time is decreased and vice versa. A project manager's roles include understanding the time-cost relationship, optimizing a project's pace for minimal cost, and predicting the impact of a schedule change on project cost.

The costs associated with a project can be classified as direct costs or indirect costs. The project cost is the sum of the direct and indirect costs.

Direct costs, also known as *variable costs*, *operating costs*, *prime costs*, and *on costs*, are costs that vary directly with the level of output (e.g., labor, fuel, power, and the cost of raw material). Generally, direct costs increase as a project's completion time is decreased, since more resources need to be allocated to increase the pace.

Indirect costs, also known as *fixed costs*, are costs that are not directly related to a particular function or product. Indirect costs include taxes, administration, personnel, and security costs. Such costs tend to be relatively steady over the life of the project and decrease as the project duration decreases.

Table 54.2 Construction Activities and ESC Measures

construction activity	ESC measures
designate site access	Stabilize exposed areas with gravel and/or temporary vegetation. Immediately apply stabilization to areas exposed throughout site development.
protect runoff outlets and conveyance systems	Install principal sediment traps, fences, and basins prior to grading; stabilize stream banks and install storm drains, channels, etc.
land clearing	Mark trees and buffer areas for preservation.
site grading	Install additional ESC measures as needed during grading.
site stabilization	Install temporary and permanent seeding, mulching, sodding, riprap, etc.
building construction and utilities installation	Install additional ESC measures as needed during construction.
landscaping and final site stabilization*	Remove all temporary control measures; install topsoil, trees and shrubs, permanent seeding, mulching, sodding, riprap, etc.; stabilize all open areas, including borrow and spoil areas.

*This is the last construction phase.

The time required to complete a project is determined by the critical path, so to compress (or "crash") a project schedule (accelerate the project activities in order to complete the project sooner), a project manager must focus on critical path activities.

A procedure for determining the optimal project time, or time-cost trade-off, is to determine the normal completion time and direct cost for each critical path activity and compare it to its respective "crash time" and direct cost. The *crash time* is the shortest time in which an activity can be completed. If a new critical path emerges, consider this in subsequent time reductions. In this way, one can step through the critical path activities and calculate the total direct project cost versus the project time. (To minimize the cost, those activities that are not on the critical path can be extended without increasing the project completion time.) The indirect, direct, and total project costs can then be calculated for different project durations. The optimal duration is the one with the lowest cost. This model assumes that the normal cost for an activity is lower than the crash cost, the time and cost are linearly related, and the resources needed to shorten an activity are available. If these assumptions are not true, then the model would need to be adapted. Other cost considerations include incentive payments, marketing initiatives, and the like.

Fast Tracking

Besides efficient scheduling, construction time can be compressed with *fast-track scheduling.* This method overlaps the design and construction phases of a project. Ordering of long-lead materials and equipment can occur, and work on the site and foundations can begin before all the details of the building are completely worked out. With fast-track scheduling, separate contracts are established so that each major system can be bid and awarded by itself to avoid delaying other construction.

Although the fast-track method requires close coordination between the architect, contractor, subcontractors, owner, and others, it makes it possible to construct a high-quality building in 10% to 30% less time than with a conventional construction contract.

Schedule Management

Several methods are used to schedule and monitor projects. The most common and easiest is the *bar chart* or *Gantt chart*, such as Fig. 54.2. The various activities of the schedule are listed along the vertical axis. Each activity is given a starting and finishing date, and overlaps are indicated by drawing the bars for each activity so that they overlap. Bar charts are simple to make and understand and are suitable for small to midsize projects. However, they cannot show all the sequences and dependencies of one activity on another.

Critical path techniques are used to graphically represent the multiple relationships between stages in a complicated project. The graphical network shows the *precedence relationships* between the various activities. The graphical network can be used to control and monitor the progress, cost, and resources of a project. A critical path technique will also identify the most critical activities in the project.

Critical path techniques use *directed graphs* to represent a project. These graphs are made up of *arcs* (arrows) and *nodes* (junctions). The placement of the arcs and nodes completely specifies the precedences of the project. Durations and precedences are usually given in a *precedence table* (matrix).

Resource Leveling

Resource leveling is used to address *overallocation* (i.e., situations that demand more resources than are available). Usually, people and equipment are the limited resources, although project funding may also be limited if it becomes available in stages. Two common ways are

Figure 54.2 Gantt Chart

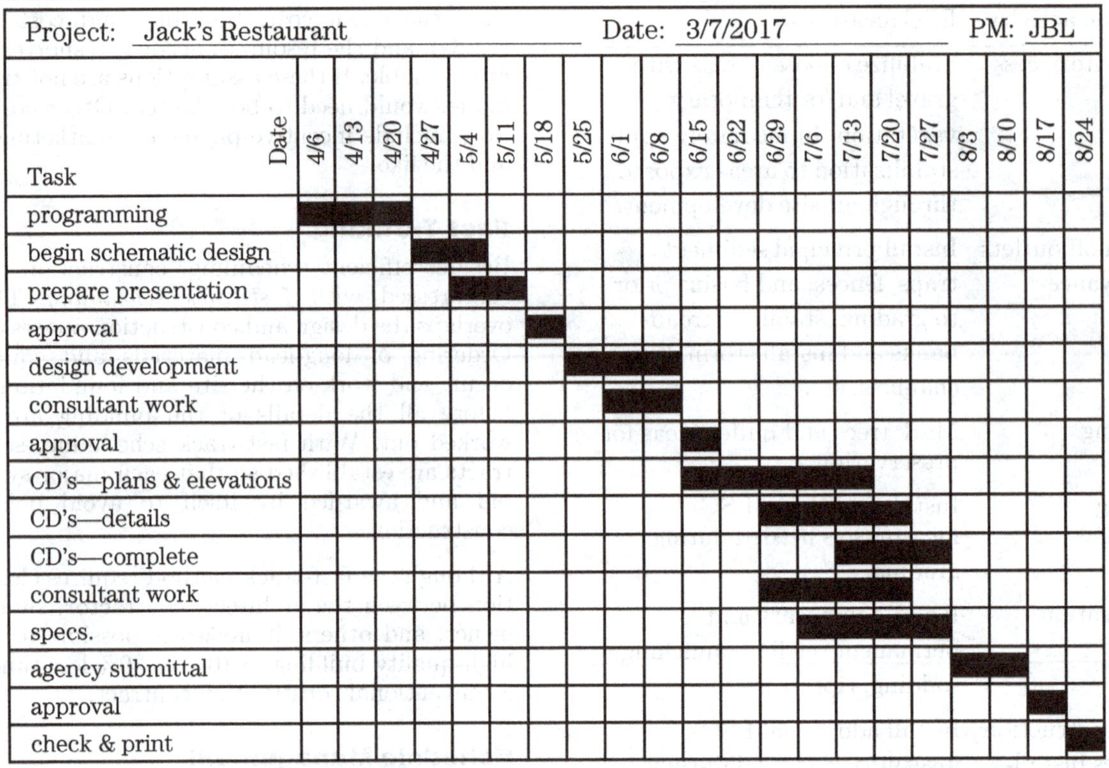

used to level resources: (1) Tasks can be delayed (by either postponing their start dates or extending their completion dates) until resources become available. (2) Tasks can be split so that the parts are completed when planned and the remainders are completed when resources become available. The methods used depend on the limitations of the project, including budget, resource availability, finish date, and the amount of flexibility available for scheduling tasks. If resource leveling is used with tasks on a project's critical path, the project's completion date will inevitably be extended.

Example

Because of a shortage of reusable forms, a geotechnical contractor decides to build a 250 ft long concrete wall in 25 ft segments. Each segment requires the following crews and times.

- set forms: 3 carpenters; 2 days
- pour concrete: 3 carpenters; 1 day
- strip forms: 4 carpenters; 1 day

The contractor has budgeted for only three carpenters. Compared to the ideal schedule, most nearly how many additional days will it take to construct the wall if the number of carpenters is leveled to three?

(A) 4 days

(B) 5 days

(C) 6 days

(D) 7 days

Solution

Having only three carpenters affects the form stripping operation, but it does not affect the form setting and concrete pouring operations. If the contractor had four carpenters, stripping the forms from each 25 ft segment would take 1 day; stripping forms for the entire 250 ft wall would take

$$(10\text{ segments})\left(1\ \frac{\text{day}}{\text{segment}}\right) = 10\text{ days}$$

With only three carpenters, form stripping productivity is reduced to $3/4$ of the four-carpenter rate. Each form stripping operation takes $4/3$ days, and the entire wall takes

$$(10\text{ segments})\left(\frac{4}{3}\ \frac{\text{days}}{\text{segment}}\right) = 13.3\text{ days}$$

The increase in time is

$$13.3 \text{ days} - 10 \text{ days} = 3.3 \text{ days} \quad (4 \text{ days})$$

The answer is (A).

4. ACTIVITY-ON-NODE NETWORKS

The *critical path method* (CPM) is one of several critical path techniques that uses a directed graph to describe the precedence of project activities. CPM requires that all activity durations be specified by single values. That is, CPM is a *deterministic method* that does not intrinsically support activity durations that are distributed as random variables.

Another characteristic of CPM is that each activity (task) is traditionally represented by a node (junction), hence the name *activity-on-node network*. Each node is typically drawn on the graph as a square box and labeled with a capital letter, although these are not absolute or universally observed conventions. Each activity can be thought of as a continuum of work, each with its own implicit "start" and "finish" *events*. For example, the activity "grub building site" starts with the event of a bulldozer arriving at the native site, followed by several days of bulldozing, and ending with the bulldozer leaving the cleaned site. An activity, including its start and finish events, occurs completely within its box (node).

Each activity in a CPM diagram is connected by arcs (connecting arrows, lines, etc.). The arcs merely show precedence and dependencies. Events are not represented on the graph, other than, perhaps, as the heads of tails of the arcs. Nothing happens along the arcs, and the arcs have zero durations. Because of this, arcs are not labeled. (See Fig. 54.3.)

For convenience, when a project starts or ends with multiple simultaneous activities, *dummy nodes* with zero durations may be used to specify the start and/or finish of the entire project. Dummy nodes do not add time to the project. They are included for convenience. Since all CPM arcs have zero durations, the arcs connecting dummy nodes are not different from arcs connecting other activities.

Figure 54.3 *Activity-on-Node Network*

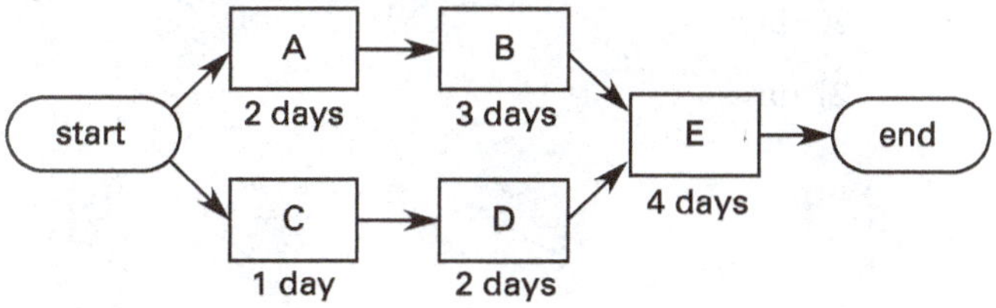

A CPM graph depicts the activities required to complete a project and the sequence in which the activities must be completed. No activity can begin until all of the activities with arcs leading into it have been completed.

The duration and various dates are associated with each node, and these dates can be written on the CPM graph for convenience. These dates include the earliest possible start date, ES; the latest possible start date, LS; the earliest possible finish date, EF; and, the latest possible finish date, LF.

After the ES, EF, LS, and LF dates have been determined for each activity, they can be used to identify the critical path. The *critical path* is the sequence of activities (the *critical activities*) that must all be started and finished exactly on time in order to not delay the project. Delaying the starting time of any activity on the critical path, or increasing its duration, will delay the entire project. The critical path is the longest path through the network. The project duration is the sum of all activity times along the critical path (see Eq. 54.4). If it is desired that the project be completed sooner than expected, then one or more activities in the critical path must be shortened. The critical path is generally identified on the network with heavier (thicker) arcs.

Equation 54.1 Through Eq. 54.3: Nomenclature

$$\text{EF} = \text{ES} + \text{duration} \tag{54.1}$$

$$\text{LS} = \text{LF} - \text{duration} \tag{54.2}$$

$$\text{float} = \text{LS} - \text{ES} = \text{LF} - \text{EF} \tag{54.3}$$

Description

The ES, EF, LS, and LF dates are calculated from the durations and the activity interdependencies, as indicated by Eq. 54.1 through Eq. 54.3. ES and EF dates are calculated working from the start of the project; LS and LF dates are calculated working backward from the end of the project. (See Sec. 54.5.)

Activities not on the critical path are known as *noncritical activities*. Other paths through the network will require less time than the critical path, and hence, will have inherent delays. The noncritical activities can begin or finish earlier or later (within limits) without affecting the overall schedule. The amount of time that an activity can be delayed without affecting the overall schedule is known as the *float* (*float time* or *slack time*). Float is calculated as indicated by Eq. 54.3.

Float is zero along the critical path. This fact can be used to identify the activities along the critical path from their ES, EF, LS, and LF dates.

Construction

5. SOLVING A CPM PROBLEM WITH AN ACTIVITY-ON-NODE NETWORK

As previously described, the solution to a critical path method problem reveals the earliest and latest times that an activity can be started and finished, and it also identifies the critical path and generates the float for each activity.

The following procedure may be used to solve a CPM problem. To facilitate the solution, each node should be replaced by a square that has been quartered. The compartments have the meanings indicated by the key.

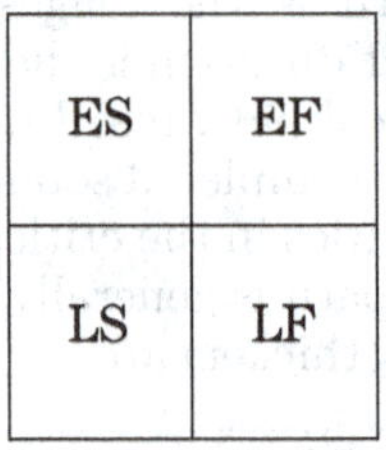

key

ES: Earliest Start
EF: Earliest Finish
LS: Latest Start
LF: Latest Finish

step 1: Place the project start time or date in the **ES** and **EF** positions of the start activity. The start time is zero for relative calculations.

step 2: Consider any unmarked activity, all of whose predecessors have been marked in the **EF** and **ES** positions. (Go to step 4 if there are none.) Mark in its **ES** position the largest number marked in the **EF** position of those predecessors.

step 3: Add the activity time to the **ES** time and write this in the **EF** box. Go to step 2.

step 4: Place the value of the latest finish date in the **LS** and **LF** boxes of the finish mode.

step 5: Consider unmarked predecessors whose successors have all been marked. Their **LF** is the smallest **LS** of the successors. Go to step 7 if there are no unmarked predecessors.

step 6: The **LS** for the new node is **LF** minus its activity time. Go to step 5.

step 7: The float for each node is **LS** – **ES** and **LF** – **EF**.

step 8: The critical path encompasses nodes for which the float equals **LS** – **ES** from the start node. There may be more than one critical path.

6. ACTIVITY-ON-ARROW NETWORKS

Equation 54.4 calculates the *duration* of the project as the sum of all durations of activities on the *critical path.* The notation used to designate activity durations is for an *activity-on-arrow* (or *activity-on-arc*) *network.* The activity is identified by its starting and ending nodes. For example, $d_{3,4}$ is the duration of the activity having starting node 3 and ending node 4. The nodes themselves do not have durations. Nodes are used to indicate the starts and ends of activities. Industry labeling of arcs (activities) and nodes is inconsistent. Letters and numbers are both used for arcs; numbers and letters are both used for nodes. Care is required when numbers are used for arcs, nodes, and durations.

Equation 54.4: Project Duration

$$T = \sum_{(i,j) \in CP} d_{ij} \qquad \text{54.4}$$

Description

It is essential to maintain a distinction between time, date, length, and duration. Like the timeline for engineering economics problems, all projects start at time = 0, but this corresponds to starting on day = 1. That is, work starts on the first day. Work ends on the last day of the project, but this end rarely corresponds to midnight. So, if a project has a critical path length (duration-to-completion) of 15 days and starts on (at the beginning of) May 1, it will finish on (at the end of) May 15, not May 16. The solution to a critical path method problem reveals the earliest and latest times that an activity can be started and finished, and it also identifies the critical path and generates the float for each activity.

Example

An activity-on-arrow diagram for a project is

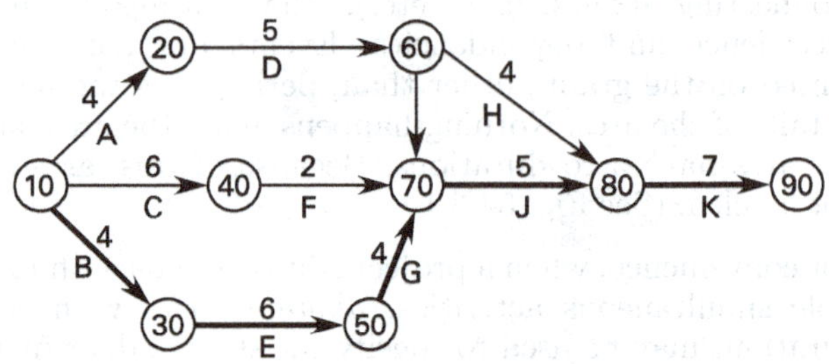

If the critical path is B-E-G-J-K, what is most nearly the duration of the project?

(A) 20 days

(B) 21 days

(C) 26 days

(D) 27 days

Solution

From Eq. 54.4, the duration of the project is

$$\begin{aligned} T &= \sum_{(i,j)\in CP} d_{ij} = d_{\text{B}} + d_{\text{E}} + d_{\text{G}} + d_{\text{J}} + d_{\text{K}} \\ &= 4 \text{ days} + 6 \text{ days} + 4 \text{ days} \\ &\quad + 5 \text{ days} + 7 \text{ days} \\ &= 26 \text{ days} \end{aligned}$$

The answer is (C).

7. STOCHASTIC CRITICAL PATH MODELS

Stochastic models differ from deterministic models only in the way in which the activity durations are found. Whereas durations are known explicitly for the deterministic model, the time for a stochastic activity is distributed as a random variable.

This stochastic nature complicates the problem greatly since the actual distribution is often unknown. Such problems are solved as a deterministic model using the mean of an assumed duration distribution as the activity duration. The most common stochastic critical path model is PERT, which stands for *program evaluation and review technique*. PERT is commonly used for large projects.

Equation 54.5 and Eq. 54.6: Mean and Variance of a Project

$$\mu = \sum_{(i,j)\in CP} \mu_{ij} \qquad 54.5$$

$$\sigma^2 = \sum_{(i,j)\in CP} \sigma_{ij}^2 \qquad 54.6$$

Description

In the PERT model, all duration variables are assumed to come from a *beta distribution*, with the mean and variance of the project duration given by Eq. 54.5 and Eq. 54.6, respectively.

Equation 54.7 and Eq. 54.8: Mean and Standard Deviation of an Activity

$$\mu_{ij} = \frac{a_{ij} + 4b_{ij} + c_{ij}}{6} \qquad 54.7$$

$$\sigma_{ij} = \frac{c_{ij} - a_{ij}}{6} \qquad 54.8$$

Description

Equation 54.7 and Eq. 54.8 are the mean duration and standard deviation, respectively, of activity (i, j). a_{ij} is the *most optimistic duration* of activity (i, j), b_{ij} is the *most likely duration* of activity (i, j), and c_{ij} is the *most pessimistic duration* of activity (i, j).

Example

Durations for a project's four key activities are shown.

	duration (hr)		
activity	optimistic	most likely	pessimistic
1-2	5	5	12
1-3	5	6	9
2-4	4	4	4
3-4	1	4	4

What is most nearly the mean duration of activity 1-2?

(A) 3.5 hr

(B) 4.0 hr

(C) 6.2 hr

(D) 7.1 hr

Solution

Use the project evaluation and review technique (PERT) mean equation. For activity 1-2,

$$\begin{aligned} \mu_{1,2} &= \frac{a_{1,2} + 4b_{1,2} + c_{1,2}}{6} \\ &= \frac{5 \text{ hr} + (4)(5 \text{ hr}) + 12 \text{ hr}}{6 \text{ hr}} \\ &= 6.17 \text{ hr} \quad (6.2 \text{ hr}) \end{aligned}$$

The answer is (C).

8. MONITORING

Monitoring is keeping track of the progress of the job to see if the planned aspects of time, fee, and quality are being accomplished. The original fee projections can be monitored by comparing weekly timesheets with the original estimate. This can be done manually or with project management software. A manual method is shown in Fig. 54.4.

Figure 54.4 Project Monitoring Chart

Project: Mini-mall Phase/People/Departments	fee dollars	time 1	2	3	4	5	6	7	8	9	10	11	12	total
schematic design	budgeted	1320	2640											
	actual	2000	2900											
design development	budgeted			2510	4090									
	actual			3200										
construction docs.	budgeted					2156	1716	4136	5456	2376				
	actual													
	budgeted													
	actual													
	budgeted													
	actual													
	budgeted													
	actual													
	budgeted													
	actual													
total (cumulative)	budgeted	1320	3960	6470	10,560	12,716	14,432	18,568	24,024	26,400				
	actual	2000	4900	8100										

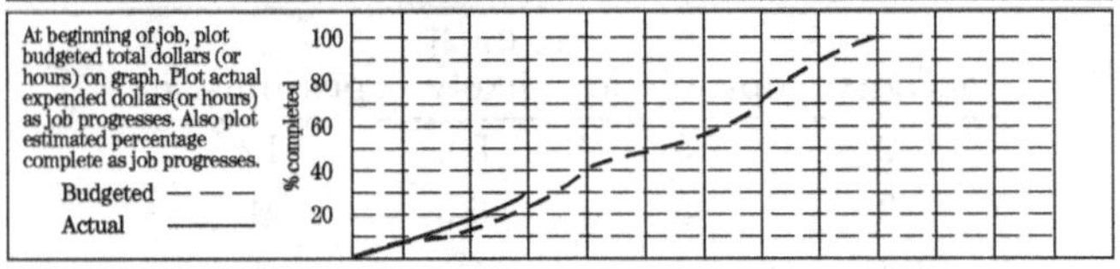

In Fig. 54.4, the budgeted weekly costs are placed in the table under the appropriate time-period column and phase-of-work row. The actual costs expended are written next to them. At the bottom of the chart, a cumulative graph is plotted that shows the actual money expended against the budgeted fees. The cumulative ratio of percentage completion to cost can also be plotted.

Monitoring quality is more difficult. At regular times during a project, the project manager, designers, and office principals should review the progress of the job to determine if the original project goals are being met and if the job is being produced according to the client's and design firm's expectations. The work in progress can also be reviewed to see whether it is technically correct and if all the contractual obligations are being met.

9. COORDINATING

During the project, the project manager must constantly coordinate the various people involved: the architect's staff, the consultants, the client, the building code officials, firm management, and, of course, the construction contractors. This may be done on a weekly, or even daily, basis to make sure the schedule is being maintained and the necessary work is getting done.

The coordination can be done by using checklists, holding weekly project meetings to discuss issues and assign work, and exchanging drawings or project files among the consultants.

10. EARNED VALUE METHOD

The *earned value method* (EVM), also known as *earned value management*, is a project management technique that correlates actual *project value* (PV) with *earned value* (EV).[1] *Value* is generally defined in dollars, but it may also be defined in hours. Close monitoring of earned value makes it possible to forecast cost and schedule overruns early in a project. In its simplest form, the method monitors the project plan, actual work performed, expenses, and cumulative value to see if the project is on track. Earned value shows how much of the budget and time should have been spent, with regard to the amount of work actually completed. The earned value method differs from typical budget versus expenses models by requiring the cost of work in progress to be quantified. Because of its complexity, the method is best implemented in its entirety in very large projects.

A *work breakdown structure* (WBS) is at the core of the method. A WBS is a hierarchical structure used to organize tasks for reporting schedules and tracking costs. For monitoring, the WBS is subsequently broken down into manageable *work packages*—small sets of activities at lower and lowest levels of the WBS that collectively constitute the overall project scope. Each work package has a relatively short completion time and can be divided into a series of milestones whose status can be objectively measured. Each work package has start and finish dates and a budget value.

The earned value method uses specific terminology for otherwise common project management and accounting principles. There are three primary measures of project performance: BCWS, ACWP, and BCWP. The *budgeted cost of work scheduled* (BCWS) is a spending plan for the project as a function of schedule and performance. For any specified time period, the *cumulative planned expenditures* is the total amount budgeted for the project up to that point. With EVM, the spending plan serves as a performance baseline for making predictions about cost and schedule variance and estimates of completion.

The *actual cost of work performed* (ACWP) is the actual spending as a function of time or performance. It is the cumulative actual expenditures on the project viewed at regular intervals within the project duration. The *budgeted cost of work performed* (BCWP) is the actual earned value based on the technical accomplishment. BCWP is the cumulative budgeted value of the work actually completed. It may be calculated as the sum of the values budgeted for the work packages actually completed, or it may be calculated by multiplying the fraction of work completed by the planned cost of the project.

These three primary measures are used to derive secondary measures that give a different view of the project's current and future health. Figure 54.5 illustrates how some of these measures can be presented graphically.

[1]The earned value method may also be referred to as *performance measurement*, *management by objectives*, *budgeted cost of work performed*, and *cost/schedule control systems*.

Figure 54.5 *Earned Value Management Measures*

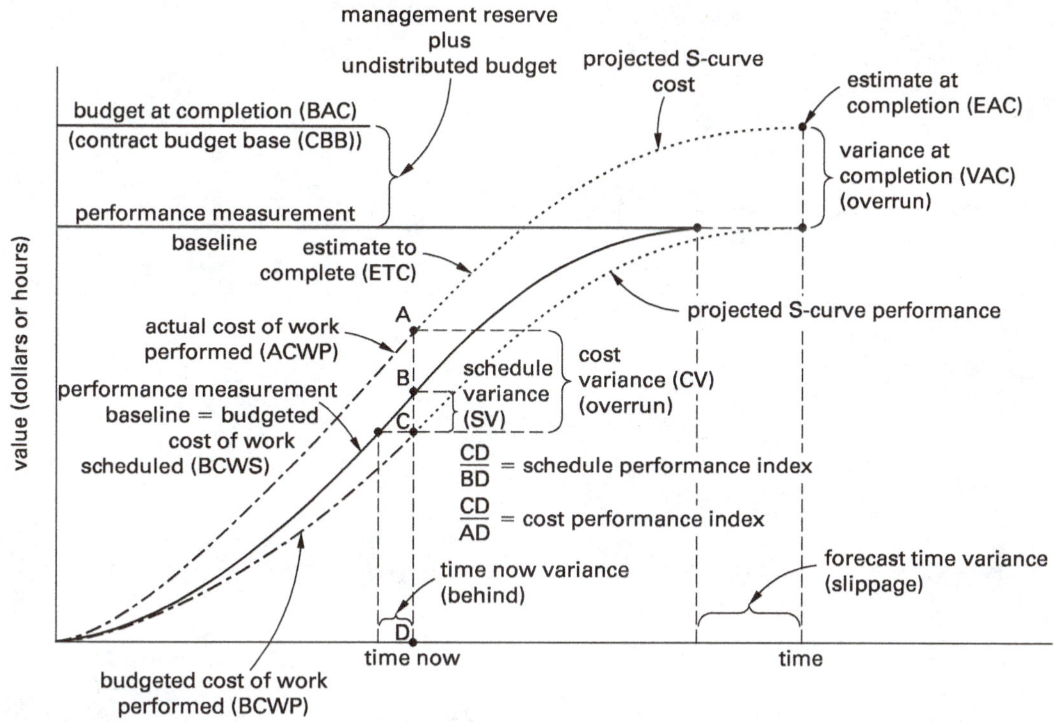

Equation 54.9 and Eq. 54.10: Variances

$$CV = BCWP - ACWP \quad (54.9)$$

$$SV = BCWP - BCWS \quad (54.10)$$

Description

Cost variance (CV) is the difference between the planned and actual costs of the work completed, as indicated by Eq. 54.9.

Schedule variance (SV) is the difference between the value of work accomplished for a given period and the value of the work planned, as indicated by Eq. 54.10. It measures how much a project is ahead of or behind schedule. Schedule variance is measured in value (e.g., dollars), not time.

Equation 54.11 and Eq. 54.12: Indices

$$CPI = \frac{BCWP}{ACWP} \quad (54.11)$$

$$SPI = \frac{BCWP}{BCWS} \quad (54.12)$$

Description

Cost performance index (CPI) is a cost efficiency factor representing the relationship between the actual cost expended and the earned value, as indicated by Eq. 54.11. A CPI ≥ 1 suggests efficiency.

Schedule performance index (SPI) is a measure of schedule effectiveness, determined from the earned value and the initial planned schedule, as indicated by Eq. 54.12. SPI ≥ 1 suggests work is ahead of schedule.

Equation 54.13 and Eq. 54.14: Forecasting

$$ETC = \frac{BAC - BCWP}{CPI} \quad (54.13)$$

$$EAC = ACWP + ETC \quad (54.14)$$

Description

Budget at completion (BAC), also known as *performance measurement baseline*, is the sum total of the time-phased budget.

Estimate to complete (ETC) is a calculated value representing the cost of work required to complete remaining project tasks, as indicated by Eq. 54.13.

Estimate at completion (EAC) is a calculated value that represents the projected total final cost of work when completed, as indicated by Eq. 54.14.

55 Procurement and Project Delivery Methods

1. INTRODUCTION

The *procurement* stage of a project is when a project manager plans purchases, acquisitions, and contracts; requests supplier and contractor responses; selects suppliers and contractors; and awards, administers, and closes contracts. There are three types of contract structure that may be used to deliver the project: design-bid-build, design-build, and management contracting. Each contract type involves the same parties: the design professional, the owner, and the contractor (who is also responsible for coordinating the work of any subcontractors, fabricators, and suppliers).

2. DESIGN-BID-BUILD

The *design-bid-build method* is the traditional approach to project delivery, where the engineer designs the project and prepares the construction documents, which are then sent out for bidding by interested contractors. The lowest bid is selected, and either the owner or a contracted construction manager enters into a single contract with the successful bidder. This method of project delivery generally produces the lowest cost, with the responsibility for meeting the construction cost and schedule falling on the contractor. However, this process takes longer than others because the construction documentation must be completely finished before bidding.

Multiple-prime project delivery is a specific form of the design-bid-build method. Many states require using this method for public projects. Once a project's design documents are complete, the owner or a contracted construction manager creates a bid package for each subdivision of work (e.g., heating, ventilating, and air conditioning (HVAC); framing; plumbing; electrical). Bidding is commonly done on a unit-pricing basis. The lowest bidder for each work subdivision is awarded the respective contract. Each subcontractor is responsible for coordinating and delivering the work included within the scope of his or her particular contract, but is not responsible for the entire project. The owner or construction manager coordinates the efforts of all subcontractors.

When done well, multiple-prime project delivery is a "fast-track" approach that can reduce a project's overall duration and cost. That said, multiple contracts complicate project administration, so multiple-prime project delivery is less appropriate for extremely complex projects or for inexperienced project managers.

3. DESIGN-BUILD

The *design-build method* places responsibility for both design and construction on one entity. The lead professional may be an engineer that also owns a construction company or a construction company that employs engineers and architects. With this method, the design and construction personnel work together without the usual adversarial roles found under the traditional design-bid-build method. Because the same entity is pricing the construction and design work, it can offer the owner a fixed, guaranteed price and assume all responsibility for design and construction.

4. MANAGEMENT CONTRACTING

A *construction manager* (CM) is a third party who is an expert in the construction, costing, and management of the construction process. A CM may be hired by the owner simply as a consultant to advise on costs and construction methods, but not to participate in the construction of the project. A CM may also act as an agent of the owner and be responsible for hiring contractors, advising on costs, and managing the construction process. Finally, a CM can act as the contractor, guaranteeing the price of the project.

When a CM is involved in the delivery process, the traditional design-bid-build method can be used, with the CM consulting early in the project on cost and constructability details. More commonly, however, the CM will act as an agent of the owner, advising the engineer on material selection, costs, and constructability; selecting contractors and subcontractors; negotiating their contracts and construction pricing; and coordinating construction. This approach is often used with fast-track construction, where construction may begin before all design and contract document production is completed. For example, structural, mechanical, and plumbing design may be completed and work started on those elements before architectural woodwork drawings and finish selection are complete.

5. AWARDING CONTRACTS

Contracts can be awarded through either negotiation or competitive bidding.

Negotiation

With a *negotiated* contract, the owner, with the assistance of the construction manager or engineer, works out the final contract price and conditions with each contractor. The contractor with which the owner negotiates may be selected in one of two ways. In the first, the owner may know precisely which contractor he or she wants to complete the project. This knowledge may come from having worked with the contractor before, through a referral, or by reputation. In the second method, the owner may select several possible contractors to be interviewed. Each of the contractors is interviewed, and one is selected based on qualifications and possibly a fee proposal. During the negotiation process, the contractor may point out problems, make suggestions, or propose changes in the design or specifications to reduce the cost of the project. If the agreement is negotiated with a general contractor, the subcontracts may be open to competitive bidding.

Competitive Bidding

The bid is usually awarded to the *lowest responsible bidder*, the lowest bidder whose offer best responds in quality, fitness, and capacity to fulfill the particular requirements of the proposed project, and who can fulfill these requirements with the qualifications needed to complete the job in accordance with the terms of the contract.

The ethical guidelines for dealing with other engineers presented here and in more detailed codes of ethics no longer include a prohibition on *competitive bidding* for engineering services. Until 1978, most codes of ethics for engineers considered competitive bidding detrimental to public welfare, since cost-cutting normally results in a lower quality design.

However, in a 1978 case against the National Society of Professional Engineers (NSPE) that went all the way to the U.S. Supreme Court, the prohibition against competitive bidding was determined to be a violation of the Sherman Antitrust Act (i.e., it was an unreasonable restraint of trade).

The opinion of the Supreme Court does not *require* competitive bidding—it merely forbids a prohibition against competitive bidding in NSPE's code of ethics. The following points must be considered.

- Engineers and design firms may individually continue to refuse to bid competitively on engineering services.
- Clients are not required to seek competitive bids for design services.
- Federal, state, and local statutes governing the procedures for procuring engineering design services, even those statutes that prohibit competitive bidding, are not affected.
- Any prohibitions against competitive bidding in individual state engineering registration laws remain unaffected.
- Engineers and their societies may actively and aggressively lobby for legislation that would prohibit competitive bidding for design services by public agencies.

Example

Ethical codes and state legislation forbidding competitive bidding by design engineers are

(A) enforceable in some states

(B) not enforceable on public (nonfederal) projects

(C) enforceable for projects costing less than \$5 million

(D) not enforceable

Solution

Ethical bans on competitive bidding are not enforceable. The National Society of Professional Engineers' (NSPE) ethical ban on competitive bidding was struck down by the U.S. Supreme Court in 1978 as a violation of the Sherman Antitrust Act of 1890.

The answer is (D).

Construction

56 Construction Documents

1. INTRODUCTION

Construction plans, specifications, and any necessary contracts or supporting documents must be produced at the start of a project. Collectively, these documents are known as the *construction documents.* In general, construction documents must be of sufficient clarity to indicate the location, nature, and extent of the proposed work and how it will conform to the code. This normally includes the standard types of floor plan drawings, elevations, sections, and details for architectural, structural, mechanical, electrical, and other specialty construction, as well as applicable schedules and specifications.

2. DOCUMENTATION

Everything that is done on a project must be documented in writing. This documentation provides a record in case legal problems develop and serves as a project history to use for future jobs. Documentation is also a vital part of communication. An email or written memo is more accurate, communicates more clearly, and is more difficult to forget than a simple phone call, for example.

Most design firms have standard forms or project management software for documents such as transmittals, job observation reports, timesheets, and the like. Such software makes it easy to record the necessary information. In addition, all meetings should be documented with meeting notes. Phone call logs (listing date, time, participants, and discussion topics), emails, personal daily logs, and formal communications like letters and memos should also be generated and preserved to serve as documentation.

Two types of documents, *change orders* (also known as *record drawings*) due to unexpected conditions or changes to the plans after bidding, and *as-built construction documents* to record what was actually installed (as opposed to what was shown in the original construction documents) are particularly important. As-built construction documents must be certified by the engineer involved with the project. This is commonly referred to as "stamping" or "sealing" and includes dating and signing the design documents.

Example

Record drawings ("as-builts") of a buried sewer line installation that will be submitted to the client should be certified by the

(A) contractor

(B) building official

(C) architect

(D) engineer

Solution

The engineer of record certifies engineering design. Architects do not certify engineering work.

The answer is (D).

3. CONSTRUCTION PLANS

Construction plans, also called *plan views*, *floor plans*, and *partition plans*, are the most common type of floor plan and are required for every project regardless of size or complexity. Plan views are views seen from the top down. A plan view shows the building configuration, including all walls, dimensions, existing construction to remain, references to elevations and details drawn elsewhere, room names (and numbers, if used), floor material indications, millwork, plumbing fixtures, built-in fixtures, stairs, special equipment, and notes as required to explain items on the plan. (See Fig. 56.1.) Construction plans are usually drawn at the scale of 1/8 in = 1 ft 0 in (1:100) or 1/4 in = 1 ft 0 in (1:50). If large-scale plans are required for very complex areas, they are typically drawn at a 1/2 in scale (1:25). If there are other plans in the set of drawings, they should be drawn at the same scale as the primary construction plan.

4. SPECIFICATIONS

Construction specifications dictate which materials and methods must be used for a given construction project.

Figure 56.1 *Partial Floor Plan*

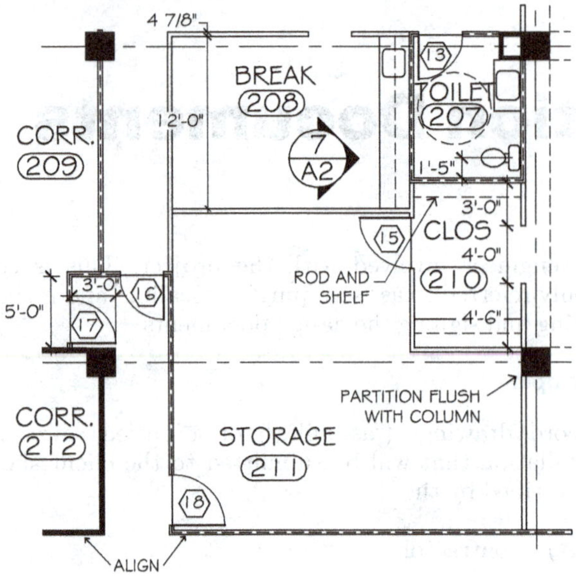

A *specifications document*, separate from the plans and drawings, is usually prepared by the design professional. The specifications document becomes part of the construction contract. The document assembles and consolidates all of the specific and technical details (e.g., materials specifications, standard processes and methods, finishes, and specific part descriptions) that apply to a project. It is often necessary to use both plans and specifications in order for a contractor to satisfy the contractual obligations.

Specifications may be created by a consulting engineer as part of the design process, but it is also possible to adopt commercial documents that use standard information, language, and content. For example, the Construction Specifications Institute (CSI) MasterFormat™ document is a consensus document that is jointly sponsored by the Construction Specifications Institute (in the United States) and the Construction Specifications Canada organizations.

CSI MasterFormat

Construction projects involve many different kinds of products, activities, and installation methods. MasterFormat, a publication of CSI, facilitates communication among engineers, architects, specifiers, contractors, and suppliers in the United States and Canada by standardizing information and terminology.[1] MasterFormat is not a building code or a regulation, but rather, it is a standardized list of titles and numbers used to organize construction documents (specifications) and other project information for most commercial building design and construction projects in North America. It lists titles and section numbers for organizing data about construction requirements, products, and activities.

MasterFormat 2012 is organized into the 50 divisions given in Table 56.1, although many are reserved for future use. Each division is divided into up to four layers of subdivisions, which are identified numerically. For example, the material specification 03 52 16.13 subdivision hierarchy is

03 Concrete
    (50 Cast Decks and Underlayment)—not designated
        52 Lightweight Concrete Roof Insulation
            16 Lightweight Insulating Concrete
                13 Lightweight Cellular Insulating Concrete

Example

Which division(s) of the Construction Specifications Institute's MasterFormat deal(s) with construction concrete?

(A) Division 02

(B) Division 03

(C) Divisions 04–07

(D) Divisions 13–14

Solution

Only Division 03 of the Construction Specifications Institute's MasterFormat deals with concrete. Division 01 deals with general requirements, Division 02 deals with site construction, and Divisions 04–14 deal with architectural (not engineering) features.

The answer is (B).

[1]The *SpecsIntact* (*Specifications Kept Intact*), also known as *Master Text* and *Masters* standardization schemes used in projects involving the U.S. government or U.S. military, integrates MasterFormat.

Table 56.1 MasterFormat Divisions

Procurement and Contracting Requirements Group:
Division 00 – Procurement and Contracting Requirements

Specifications Group:
General Requirements Subgroup:
Division 01 – General Requirements

Facility Construction Subgroup:
Division 02 – Existing Conditions
Division 03 – Concrete
Division 04 – Masonry
Division 05 – Metals
Division 06 – Wood, Plastics, and Composites
Division 07 – Thermal and Moisture Protection
Division 08 – Openings
Division 09 – Finishes
Division 10 – Specialties
Division 11 – Equipment
Division 12 – Furnishings
Division 13 – Special Construction
Division 14 – Conveying Equipment
Division 15 – Reserved for future expansion
Division 16 – Reserved for future expansion
Division 17 – Reserved for future expansion
Division 18 – Reserved for future expansion
Division 19 – Reserved for future expansion

Facility Services Subgroup:
Division 20 – Reserved for future expansion
Division 21 – Fire Suppression
Division 22 – Plumbing
Division 23 – Heating, Ventilating, and Air Conditioning
Division 24 – Reserved for future expansion
Division 25 – Integrated Automation
Division 26 – Electrical
Division 27 – Communications
Division 28 – Electronic Safety and Security
Division 29 – Reserved for future expansion

Site and Infrastructure Subgroup:
Division 30 – Reserved for future expansion
Division 31 – Earthwork
Division 32 – Exterior Improvements
Division 33 – Utilities
Division 34 – Transportation
Division 35 – Waterway and Marine Construction
Division 36 – Reserved for future expansion
Division 37 – Reserved for future expansion
Division 38 – Reserved for future expansion
Division 39 – Reserved for future expansion

Process Equipment Subgroup:
Division 40 – Process Integration
Division 41 – Material Processing and Handling Equipment
Division 42 – Process Heating, Cooling, and Drying Equipment
Division 43 – Process Gas and Liquid Handling, Purification, and Storage Equipment
Division 44 – Pollution and Waste Control Equipment
Division 45 – Industry-Specific Manufacturing Equipment
Division 46 – Water and Wastewater Equipment
Division 47 – Reserved for future expansion
Division 48 – Electrical Power Generation
Division 49 – Reserved for future expansion

57 Construction Operations and Management

Nomenclature

C	compressive force	lbf
D	distance	ft
h	height	ft
l	length	ft
P	load or tipping load	lbf
r	radius	ft
T	tensile force	lbf
V	volume	yd^3
w	width	ft
W	weight	lbf

Symbols

ϕ	angle	deg

Subscripts

b	boom
c	center, compacted, or crane
CL	centerline
e	end
f	fulcrum
g	center of gravity
h	hook
l	lift or load
r	roller pass
s	superstructure
sb	spreader bar
sl	sling distance
sw	swing

1. EQUIPMENT PRODUCTIVITY AND SELECTION

Equipment productivity and selection has a major influence on the efficiency and profitability of a construction operation. Equipment is a factor in determining a construction project's cost. Selecting the appropriate type and size of equipment can reduce the time, money, and labor required to complete a project. The most important considerations in equipment selection, whether renting or buying, are the equipment's ability to perform the work, its efficiency, and its cost.

In order to maximize the return on an investment, equipment is evaluated by the *cost per unit of production* (CPP); that is, the amount of time it takes to complete the work relative to the cost of renting, operating, and maintaining the equipment for that period of time.

An equipment's production capability is based on how quickly the work is completed. The basic method for calculating the production capability of earthmoving equipment is based on the equipment's capacity and cycle time. The exact method for calculating the production capability of equipment varies depending on the equipment type.

An equipment's *cycle time* is the amount of time that passes between the beginning of the action the equipment performs (hauling, excavating, etc.) and the beginning of the next repetition of the action. The length of a cycle must take into account all steps of the process (e.g., a backhoe's cycle time includes the time to move the full bucket to the borrow, empty it, and return to the digging site) and will also include time spent on actions related to the work that do not contribute directly, such as breaks, refueling, repositioning equipment, and so on.

There are two approaches to estimating the cycle time. The first method is to calculate the number of effective working minutes per hour, taking into account time for known delays. The second method is to multiply the theoretical number of cycles per hour by a numerical efficiency, representing the percentage of work time that is actually spent performing work (rather than time spent waiting in traffic, replacing tires, negotiating obstacles on the work site, etc.). A numerical efficiency for the equipment is usually recommended by the equipment manufacturer, but in practice, a construction engineer's prior experience or conditions at and around the work site will often alter this value. For example, a manufacturer may suggest that a given excavator generally works at 90% efficiency, but if intermittent storms are causing work delays or there is heavy foot traffic on the work site, a lower efficiency should be assumed.

Excavation and Loading Equipment

Excavation and loading equipment is used to excavate material for large earthwork cuts, to install underground utilities, to construct foundations, and so on. The production capability of excavation and loading equipment is evaluated based on how much material it can move in an hour.

With the exception of dozers, finding the production capability of excavation and loading equipment requires knowing the equipment's load capacity. (A dozer's production capability is based on its blade capacity, which is similar to load capacity but requires a different calculation.) The load capacity for excavating and loading equipment is measured in *loose cubic yards* (LCY). All excavation and loading equipment has both a *struck capacity*, the capacity when the bucket is filled only to the rim, and a *heaped capacity*, the capacity when material is mounded above the rim at a slope of 2:1. Table 57.1 gives descriptions and load capacities for the most common equipment in excavating and loading operations. Load capacities are assumed to be heaped.

The production capability of a backhoe is measured in loose cubic yards of material moved per hour. Production capability of a backhoe is always based on the backhoe's heaped capacity. Calculating a given backhoe's production capability requires the cycle time for the backhoe and the *fill factor* for the bucket. A backhoe's cycle time includes time for loading the bucket, swinging the load over the haul equipment, dumping the load, and returning the bucket to the excavation position. The fill factor is the ratio of the actual loose volume of the material being moved and the heaped capacity of the bucket. Fill factors range from 80% to 100%, depending on the material being excavated—solid, heavy materials like rock will typically have a much lower fill factor than a soft material like clay or sand.

The production capability of an excavator is calculated similarly to that of a backhoe, and is also based on the backhoe's heaped capacity and measured in loose cubic yards moved per hour. The cycle time for an excavator includes all the same activities as that of a backhoe. An excavator's production capability must also take into account the ratio of the excavator's angle of swing to its depth cut correction factor. A chart showing this ratio is supplied by the equipment manufacturer.

For dozers, the cycle time includes the time it takes to maneuver the dozer into position, the time it takes to move the load, and the time spent returning the dozer to the location of the loads to be moved. Instead of struck or heaped capacity, a dozer's capacity is measured in terms of blade capacity. The dozer's specifications often include the blade capacity, but in cases where the manufacturer's information is unavailable or greater accuracy is required, the blade capacity of a dozer can be calculated in the field. To determine the blade capacity, soil is first loaded onto the dozer's blade, and the dimensions (length, height, and width) of the soil pile are measured. Because the soil pushed by the dozer is a rounded pile, not a regular cubic mass, a dimensionless shape factor of 0.375 should also be applied to the blade capacity calculation.

Table 57.1 *Major Pieces of Equipment Used in Excavating and Loading Operations*

equipment	description	typical load capacity
backhoe	A backhoe consists of a digging bucket on the end of a two-part articulated arm. It is typically mounted on the back of a tractor or front loader.	1 yd^3 to 3 yd^3
excavator	Excavators were developed to excavate below the level of the tracks (e.g., in trenches or below grade foundations). They are useful in confined locations and can be used as small cranes for lifting and laying pipe and installing trench shores.	1 yd^3 to 5 yd^3
dragline	A dragline bucket system consists of a large bucket suspended from a boom with wire ropes. The bucket is maneuvered with ropes and chains. A hoist rope supports the bucket and hoist-coupler assembly from the boom. A dragrope is used to draw the bucket assembly horizontally.	40 yd^3 to 80 yd^3
dozer	A dozer is a versatile machine used for clearing and grubbing, shallow excavation, pushing scrapers, maintaining haul roads, spreading and grading, and ripping.	not applicable
loader	A loader can serve as both a fixed-position excavator and a transporter of material over short distances. Loaders have either wheels or tracks.	1 yd^3 to 9 yd^3

Loaders are used to handle and transport bulk carriers, to load trucks, and to excavate earth. They vary in the capacity of their buckets and/or the weight the buckets can lift. A loader is only able to operate above the wheel/track level. The production capability of a loader is calculated based on the heaped capacity. A fill factor dependent on the material is applied to the bucket capacity, as with backhoes and excavators. In addition to the fill factor, the capacity of the bucket must be determined before the production capability of the loader can be calculated. Cycle time of a loader includes time to load the bucket, position the loader, travel to the fill location, and return to the borrow.

Example

The pile of soil in front of a dozer blade is measured. The longest length of the pile is 3.8 m. The height of the pile varies between 1.5 m and 1.6 m. The width of the pile varies from 2.0 m to 2.1 m. What is most nearly the blade capacity in loose cubic yards (LCY)?

(A) 5.1 LCY

(B) 6.0 LCY

(C) 13 LCY

(D) 16 LCY

Solution

Since the height and width of the pile vary, use the average dimensions. To account for the irregular shape, multiply the pile dimensions by a shape factor of 0.375.

$$\begin{aligned} V &= 0.375hwl \\ &= \frac{(0.375)\left(\frac{1.5\text{ m}+1.6\text{ m}}{2}\right)\left(\frac{2.0\text{ m}+2.1\text{ m}}{2}\right)(3.8\text{ m})}{\left(0.3048\ \frac{\text{m}}{\text{ft}}\right)^3\left(3\ \frac{\text{ft}}{\text{yd}}\right)^3} \\ &= 5.92\text{ yd}^3 \quad (6.0\text{ LCY}) \end{aligned}$$

The answer is (B).

Hauling and Placing Equipment

Hauling and placing equipment is used to transport soil from one location on a site to another. Equipment selection depends on a variety of factors, including site conditions, the volume and type of material to be moved, and the time to complete the work. The most common types of hauling and placing equipment are listed in Table 57.2.

The method for finding the production capability of a scraper depends on whether the scraper is the only equipment being used for the work, or is being *push-loaded*—that is, having its load moved into its bowl by a tractor, referred to in scraper operations as a *pusher*. Pushers are used in situations where the scraper cannot gain enough traction at the hauling site to easily or efficiently lift a load without assistance. The decision of whether or not to use a pusher depends on the needs of the project and the type of scraper being used. Table 57.3 describes the types of scrapers in more detail, as well as the reasons to use or not use a pusher.

Table 57.2 *Common Hauling and Placing Equipment*

equipment	description	reasonable transport distances (ft)
loader	Loaders can serve as fixed-position excavators and transporters of material over short distances. Tracked loaders are genuine excavators; wheeled loaders are more suited to stockpiling and digging in loose soils. Tracked loaders can apply more traction than wheeled loaders and are better for conditions in which it is hard to get traction with wheels, such as in clay or other soft materials.	tracked: 10–250 wheeled: 10–650
scraper	Scrapers were developed for hauling over medium distances. Scrapers are able to load, haul, and discharge material by themselves. The material is cut and loaded directly into the scraper box, transported to the discharge area, and spread in layers. The whole process takes place in a continuous cycle.	towed: 300–1000 elevator: 300–3000 motorized: 600–10,000
truck	There is a wide range of trucks available to suit any type of work. The most popular trucks are supported on two axles to improve maneuverability and reduce turning radii.	2500–30,000

A scraper's load capacity is measured in both heaped capacity and struck capacity. For a scraper, the heaped capacity is measured when the material to be hauled is heaped at a 1:1 slope, rather than the 2:1 slope used for other equipment. The cycle time for a scraper includes time for loading the scraper, hauling the load, dumping and spreading the load, and returning to the loading

Construction

Table 57.3 Types of Scrapers

equipment	description
push-loaded, single-powered axle	Loading power is limited by weight on drive wheels (approximately 50% of gross vehicle weight) and coefficient of traction for rubber tire, so this type of scraper is push-loaded by a tractor (pusher).
push-loaded, tandem-powered axle	Engines power each axle (front and rear), but a pusher may still be required. These machines are more expensive than single-powered scrapers, and so are only used for very steep uphill grades or very soft ground.
push-pull, tandem-powered axle	Similar to a push-loaded scraper with tandem-powered axles, but with a hook and bail on the front above the standard push block. The hook and bail allows a pair of scrapers to help each other load, thus eliminating the need for a pusher. The front scraper loads with the rear scraper pushing, then the rear scraper loads with the front scraper pulling.
elevating	Self-loads with a paddle-wheel elevator mechanism. Elevator mechanism is dead weight while hauling, so elevating scrapers are only economical for short haul distances. Cannot be used for rock or rocky material.

site. On work items where multiple scrapers with different load capacities are used, each scraper's production capability must be calculated separately.

Finding the production capability of a given truck or group of trucks requires finding the load capacity and the cycle time of all trucks used to complete the work. In addition to a struck capacity and a heaped capacity, a truck has a *gravimetric capacity* equal to the weight of the hauled load. A truck's cycle time takes into account time spent loading, hauling the load, dumping the load, spreading it as needed for the project, and returning the truck to the loading site. If multiple trucks with different cycle times or load capacities are employed, the production capability must be calculated separately for each set of trucks.

Table 57.4 Compaction Equipment

equipment	description and use
smooth wheel roller	Consists of a frame and smooth steel cylinder filled with sand or water to increase self-weight. Effective on most types of granular soils.
rubber tired roller	Usually self-propelled. Most widely used for rolling base courses on roads and fill for large earthworks. Effective on loamy soil.
sheep's foot roller	Has projecting feet mounted on the surface, resembling the shape of a sheep's foot. The small surface area of each foot transmits high pressure to knead the soil particles together. Most effective on highly cohesive soil materials.
vibratory plate	Walk-behind compactor, very maneuverable, ideal for small-scale work in narrow areas, in trenches, and for placement of pipes or footings. Most effective on gravel and sand.
tamper	Walk-behind compactor, used on small scale works, compacting foundations, trenches, and so on. Most effective on gravel and sand.

Compaction Equipment

Compaction equipment uses a variety of forces to mechanically increase the density of soil. Selecting the equipment to use for a compaction job depends largely on the properties of the soil to be compacted and the thickness of the soil layer to be created (known as the *lift*), as well as the scale of the compaction job and the type of construction work being done. The most common compaction equipment is described in further detail in Table 57.4.

The compaction equipment selected for a project should have a production capability that matches the excavating and hauling equipment. Production capability for compaction equipment is given in compacted cubic yards per hour.

Finding the production capability of a roller compactor requires knowing the compacted thickness of the lift, the average velocity of the roller, the width of soil compacted per roller pass, and the quantity of roller passes required to compact the soil to the density needed for the project.

Vibratory plates are similar to roller compactors, but use a large, vibrating plate, rather than a static roller, to compact the soil. The production capability of a vibratory plate must take into account the dimensions of the vibratory plate in the same way the production capability of a roller compactor takes into account the width of soil compacted per pass.

Grading and Finishing Equipment

Grading and finishing equipment is used to bring earthwork to the shape and elevation specified by the project. There are two major types of grading equipment, described in Table 57.5, which are used for different types of work. Selection of grading equipment is a matter of production capability.

Table 57.5 *Major Equipment Used in Grading and Finishing Operations*

equipment	description and use
grader	Also referred to as a road grader or a motor grader. A construction machine with a long blade used to create a flat surface. Developed for trimming subgrade and subbase on roads, road cuttings and banks, smoothing off the walls on earth-fill dams, and maintaining haul roads.
asphalt paver	Used to lay asphalt on roadways. Fed by a dump truck or a conveyor discharge.

A grader cuts and moves material with a blade known as a *moldboard.* A grader can move small amounts of material, but the strength limitations of the moldboard and its placement on the grader make it unable to perform dozer-type work. The production capability of a grader is measured in square yards per hour. Finding the production capability of a grader requires knowing the effective width graded per pass and the average velocity of the grader.

Unlike other equipment, an asphalt paver's production is measured in tons of asphalt per foot rather than units per hour. This is because an asphalt paver's production cannot exceed the production capability of the plant producing material for the paver. Instead of finding production capability for a paver, a construction engineer must find the maximum velocity at which the paver can be operated. Once the rate of production for the paver is known, the maximum velocity of the paver can be found.

After an asphalt paver has laid down the asphalt required for the project, it is followed by an asphalt roller that flattens the asphalt before it hardens. For maximum efficiency, the velocity of the asphalt roller should be matched with the paver velocity.

2. CRANES

A crane is one of the most common pieces of equipment used on a construction site. Cranes can increase the speed of transporting materials for small structures and are necessary to lift larger or heavier materials into place. Construction engineers need to understand how to select the right crane for the right activity, and how to properly erect a crane and keep it stable on the construction site.

There are many considerations and calculations to be made when selecting a crane in order to ensure the crane is large enough, suited for the work site and its surrounding environment, and stable while lifting loads. An evaluation of a crane includes determining the type and size of crane needed, the load capacity required to perform all necessary lifting, and other specifications as required for the project. The crane must be properly positioned to ensure that it is safe and stable, with minimal risk to existing structures and the rest of the work site.

The first and most basic consideration is how large the crane must be to safely complete the project. *Crane size* is measured in tons and requires consideration of a number of factors. The larger the crane, the larger the load it can support, but larger cranes also occupy more space than smaller cranes and can require more time to assemble, disassemble, and move around the work site. A construction engineer must evaluate the size of the loads to be moved, the distance the loads must be moved, and the amount of space available for assembly, disassembly, and maneuvering, and determine the most efficient size of crane to use on the project. If a crane is to be rented or purchased (rather than already being owned by the construction company), the crane manufacturer will often be able to help with the evaluation of crane size.

The type of crane needed for the project must also be determined. Each crane type has different components and alignments, and requires different calculations to ensure that the crane is stable and able to support the loads required for the project. As shown in Table 57.6, each type of crane is suited to a different range of jobs. Cranes are subcategorized into *mobile cranes,* which can be driven to and around the work site without needing to be hauled or disassembled, and stationary *tower cranes.* These two basic classifications tell the construction engineer a great deal about the general capabilities of a crane, and determine how to calculate the specifications and capacities.

The major components for each crane type are described as follows.

Table 57.6 Crane Types

crane type	description and use
rough terrain	Designed for unimproved work sites. Able to perform pick and carry operations and usually includes three position outriggers.
truck mounted	Can be driven at highway speeds but has limited off-road capability. Able to work on several job sites in one day. Typically includes hydraulic booms that allow fast setup, but the weight of the boom reduces lifting capability.
all terrain	Combines features of rough terrain and truck mounted cranes: off-road capable, all-wheel steering, and able to drive at highway speeds.
crawler lattice	High capacity, long-reach lifts. Able to perform pick and carry operations. Multiple attachments provide great flexibility in boom configuration.
tower	Used when space is at a premium. Performs an up and over reach. A moving counterweight balances load. Usually resides on a fixed foundation or crawler w/ attachment.

A *mobile crane* is typically powered by a diesel engine and comprises a base frame, outriggers, a superstructure, a boom, and a hoisting tackle. The base frame is constructed from a welded steel channel to which the two machine axles are attached. The base frame supports the weight of the engine, gearing, winches, controls, cab, boom, and counterweight. *Outriggers* are built into the base frame to increase the operating range of the crane; they can be extended and retracted from the base. The superstructure consists of a revolving frame sitting on a large turntable, mounted on the base frame to allow rotation. The *boom* is assembled in two sections, each constructed from hollow rectangular sections of high-tensile steel. The sections are pin-connected, and the top section incorporates a *head sheave.* The hoist rope between the head sheave and hook block may be arranged to accommodate the load being raised and lowered. To overcome lifting problems, such as obstructions, a *fly-jib* may be attached to the boom. The fly-jib is of similar construction to the main boom and is usually available in various lengths depending on the duty and capacity of the crane. A fully assembled and extended mobile crane is shown in Fig. 57.1.

Tower cranes comprise a vertical standing lattice-frame central mast, which supports a horizontal boom in two parts. The large section of the boom is used for lifting and carries a trolley that travels on guides along the length of the boom. The major advantage of a tower

Figure 57.1 Mobile Crane

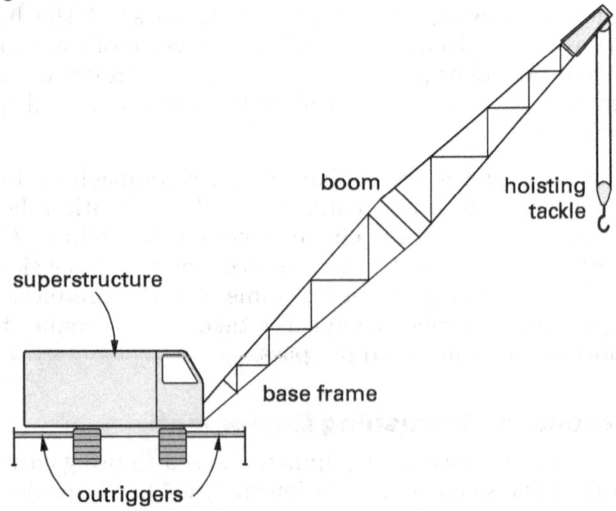

crane is that the boom can be set high enough off the ground to clear any obstructions, allowing the crane to be set very close to, or even inside, the structure being constructed. Tower cranes are most suitable for low-rise construction in areas that mobile cranes cannot easily access.

Once the type of crane needed has been determined, the construction engineer must determine the specifications required for a crane to perform the necessary work. Most crane specifications, including dimensions, weights, and working range, are given on the *crane data sheet.* Other specifications are dependent upon the configuration and placement of the crane, or vary depending on the crane manufacturer. The most common of these derived specifications are the crane's *load capacity,* the amount of weight it can safely lift, and its *tipping load,* the weight at which the crane will tip or otherwise become unsafe. For reasons of safety, the load capacity for a crane is never more than 75% of its tipping load.

Load capacities for a crane are provided on the crane's *load chart,* like that shown in Table 57.7. The load radius is the distance the load must be moved from the centerline. The weight of the block, rigging equipment, and cable below the boom must be subtracted from the load capacity to determine the actual load the crane can lift. The maximum capacity is always the maximum capacity over the shortest lift distance, usually over the rear of the crane, and with the outriggers fully

Along with a load chart and data sheet, the manufacturer will usually provide a *working range diagram.* Figure 57.2 shows a working range diagram for a crane with a 26 ft to 61 ft boom. The left axis of the diagram shows the boom lengths available for the crane, and the right axis shows the distance from the ground to the elevation where the load will be set (and, therefore, cease to be a load on the crane). Using the working range diagram and the load chart for a crane, construction engineers can determine the length of boom required to lift a load.

Table 57.7 Sample Load Chart

	boom length 26 ft			boom length 34 ft			boom length 43 ft			boom length 52 ft			boom length 61 ft			
load radius (ft)	loaded boom angle (deg)	over front (lbf)	360° (lbf)	loaded boom angle (deg)	over front (lbf)	360° (lbf)	loaded boom angle (deg)	over front (lbf)	360° (lbf)	loaded boom angle (deg)	over front (lbf)	360° (lbf)	loaded boom angle (deg)	over front (lbf)	360° (lbf)	load radius (ft)
10	58.7	50,000*	50,000*	66.3	45,300*	45,300*	71.5	42,900*	42,900*							10
12	53.3	44,000*	44,000*	62.5	40,900*	40,900*	68.7	38,600*	38,600*	72.5	36,900*	36,900*				12
15	44.4	36,100*	34,800*	56.6	35,900*	35,300*	64.3	33,600*	33,600*	69.0	32,100*	32,100*	72.2	29,600*	29,600*	15
20	23.9	25,600*	21,600	45.6	26,200*	22,100	56.5	26,500*	22,300	62.9	26,500*	22,400	67.2	24,200*	22,500	20
25	**			31.6	19,900*	14,900	48.0	20,300*	15,100	56.5	20,500*	15,200	62.0	20,500*	15,300	25
30				**			38.0	16,100*	11,100	49.5	16,400*	11,200	56.5	16,500*	11,300	30
35							24.7	13,100*	8400	41.7	13,400*	8600	50.6	13,500*	8700	35
40							**			32.3	11,100*	6800	44.1	11,300*	6900	40
45										19.1	9400*	5400	36.7	9600*	5600	45
50										**			27.7	8200*	4500	50
55													13.9	7000*	3700	55

Figure 57.2 *Sample Range Diagram (26 ft to 61 ft boom)*

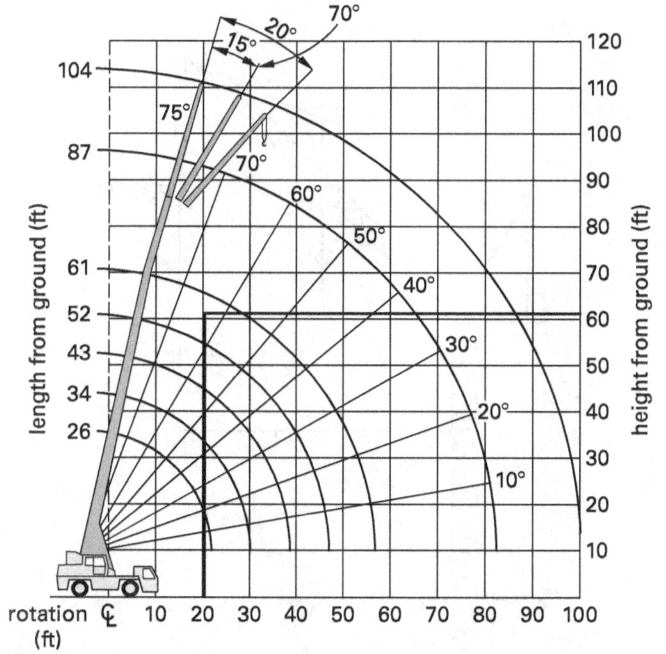

Figure 57.3 *Measurements for Boom Length*

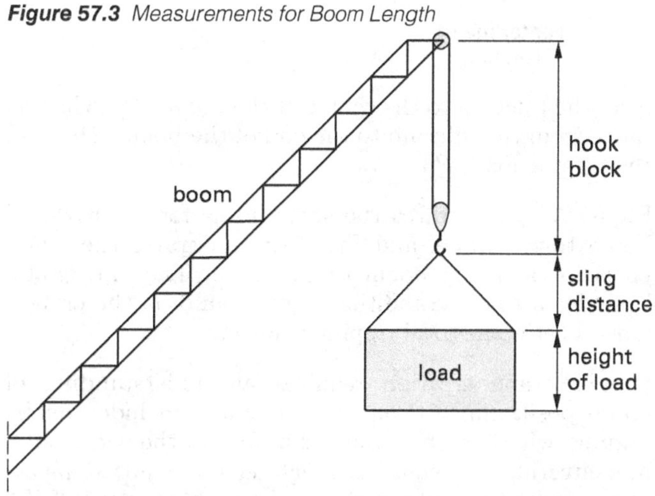

A working range diagram is a useful tool for calculating the required length of boom for a given load, but it is not always accurate. A more precise measurement of the necessary length requires knowing the load, the distance the load is to be lifted, the vertical height of the load, the distance between the top of the load and the crane hook (the *sling distance*), and the length of the hook block. (See Fig. 57.3 for an illustration of these measurements.)

For example, to determine the boom length required to lift a load to an elevation 50 ft above the ground, consider the following: The load is located 20 ft from the crane centerline of rotation, and the vertical height of the load is 5 ft. The sling distance between the top of the load and the hook is 4 ft, and the vertical distance of the hook block is 4.5 ft. The total height from the ground is 50 ft + 5 ft + 4 ft + 4.5 ft = 63.5 ft. From Fig. 57.2, the required boom length is 61 ft.

There are some situations in which a crane's specifications and diagrams do not tell a construction engineer all he or she needs to know about the crane in order to make a decision. The positioning of a crane can have a major effect on its load capacity and tipping load, and there are projects for which the construction engineer must know the exact tipping load of the crane in order to be certain the project is completed safely. In these instances, the construction engineer must determine the *static basis* for the crane.

Figure 57.4 illustrates the variables used to calculate the *tipping moment* of a mobile crane. The values shown in Fig. 57.4 can be found on the crane's data sheet, with four exceptions: the distance from the end of the superstructure to the center of the boom, $D_{s,c}$, the distance

Figure 57.4 Static Basis for a Mobile Crane

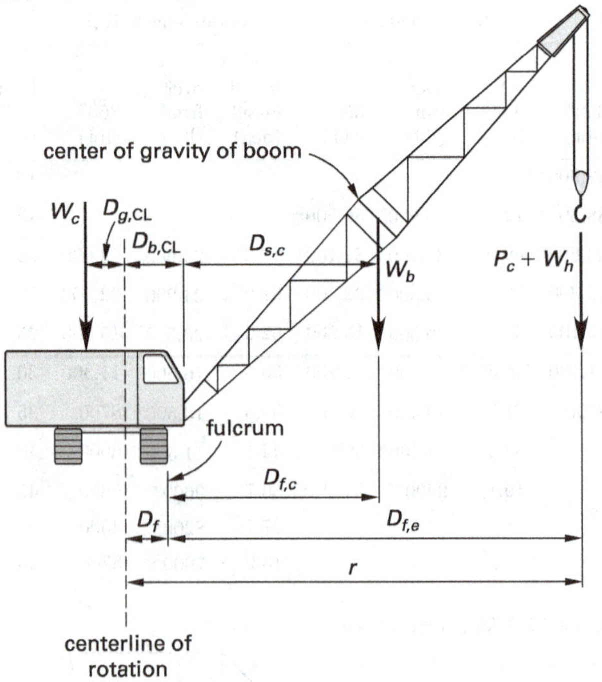

Figure 57.5 Horizontal Boom Tower Crane

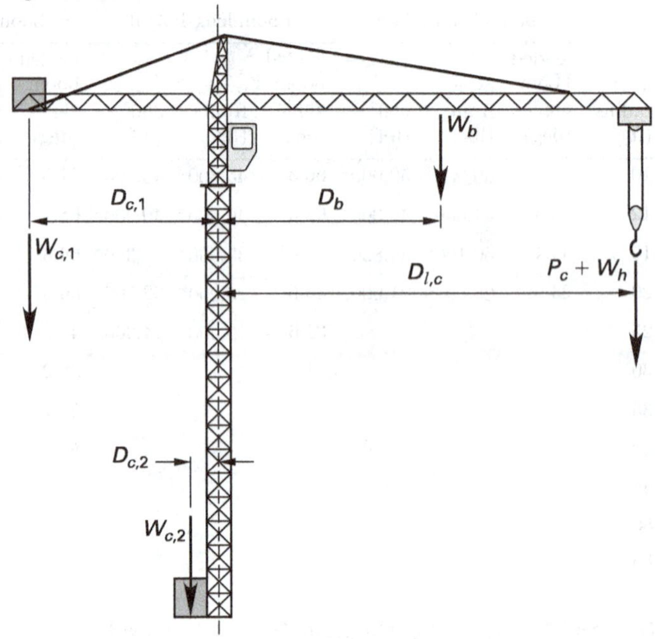

Figure 57.6 Lifting Boom Tower Crane

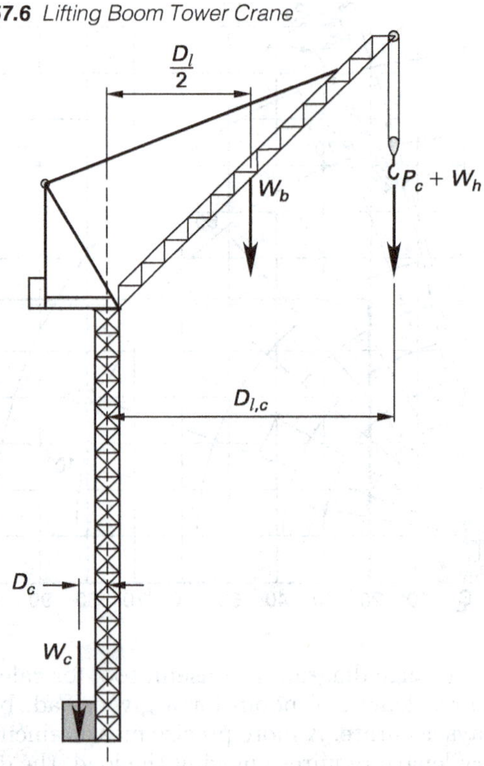

from the fulcrum to the center of the boom, $D_{f,c}$, the distance from the fulcrum to the end of the boom, $D_{f,e}$, and the tipping load, P_c.

Figure 57.5 illustrates the static basis for a horizontal boom tower crane, and Fig. 57.6 illustrates the static basis for a lifting boom tower crane. The only static basis for a tower crane that is not found on the crane's data sheet is the total tipping load, P_c.

Proper crane erection requires an understanding of crane positioning. Positioning a crane includes determining whether the crane can fit on the work site, maneuvering as required, reaching all required placement positions, and avoiding existing obstacles. OSHA requirements state the controlling contractor for the project must ensure the crane is placed on level ground, with a solid foundation that can support the crane's weight, and with adequate setback from any slopes on the work site. Construction engineers must also consider the addition of surcharge load to earth pressure on walls and the protection of underground utilities and structures, and ensure that the crane will cause only minimal interference with other construction operations.

Example

A construction project requires a crane that can lift a load of 16,000 lbf. The load must be moved 30 ft from the crane's centerline of rotation. All cranes available to the construction firm carry a load of 450 lbf from the block, rigging equipment, and cable. Using Table 57.7, how long must the boom be in order to safely perform the work necessary?

(A) 34 ft

(B) 43 ft

(C) 52 ft

(D) 61 ft

Solution

The total load the crane needs to carry is

$$W = 16{,}000 \text{ lbf} + 450 \text{ lbf} = 16{,}450 \text{ lbf}$$

Use Table 57.7 to find the extended boom length needed to carry the load. If the boom is extended 61 ft, the crane can move a 16,500 lbf load 30 ft, so a boom length of 61 ft will allow the crane to move a 16,450 lbf load without tipping.

The answer is (D).

3. RIGGING

There are a number of methods employed by construction engineers to attach the load to the hook block, including using slings, spreader beams, lifting brackets, and pincers. To move bundles (of reinforcement, timber, steel and concrete beams, etc.), slings are most commonly used.

Columns are usually rigged using a device that allows ironworkers to release the rigging from the ground by pulling on a rope. Beams are usually lifted using a choker wrapped around the beam at the center of gravity; however, a beam may also be hoisted with two cables.

Lifting a beam or column using a single cable is a simple matter of selecting a cable that can support the necessary weight and attaching it at the center of the beam. The cables between the head sheave and hook block may be arranged to accommodate the load, and the hoist rope may be arranged with multiple *falls* to accommodate load increases. A single tackle permits faster hoisting but limits the load compared to multiple tackles. (See Fig. 57.7.)

Figure 57.8 demonstrates the lifting of a single beam. The tensile force in the two lifting cables and the compressive force induced on the beam will vary depending on the length of the cables used.

Figure 57.7 *Rigging Statics*

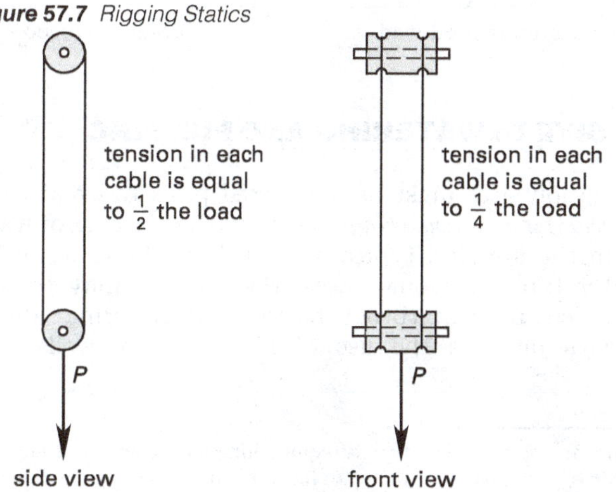

Figure 57.8 *Lifting a Single Beam*

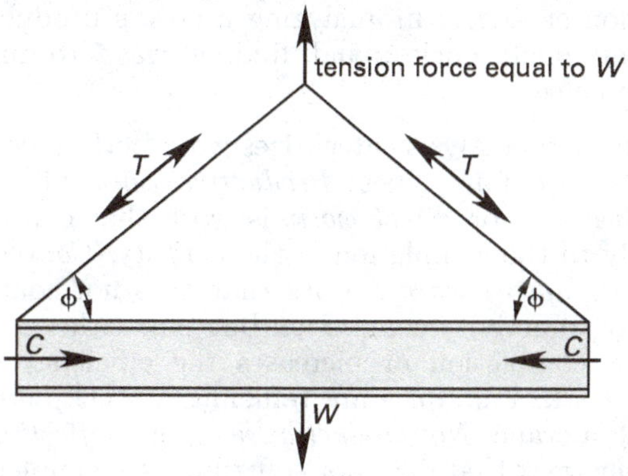

When multiple loads must be lifted at one time, or when there is concern about increasing productivity or reducing fatigue for the crane operator, the multiple lift rigging technique is used. A *spreader beam* is used to attach the multiple loads to the rigging. The multiple lift rigging technique increases the load on the crane, but reduces the number of lifts required to complete the activity. OSHA Subpart R has specific requirements for multiple lift rigging that must be followed at all times.

Multiple lift rigging requires the construction engineer to calculate the center of gravity of the load alone and the load in skewed cables of variable sling lengths, and to evaluate the bending moment capacity of the spreader beam. Figure 57.9 illustrates the variables used to perform the calculations. In all cases, the first step is to determine the center of gravity of the combined load. The lifting point is always directly above the center of gravity.

Figure 57.9 *Multiple Lift Rigging Technique*

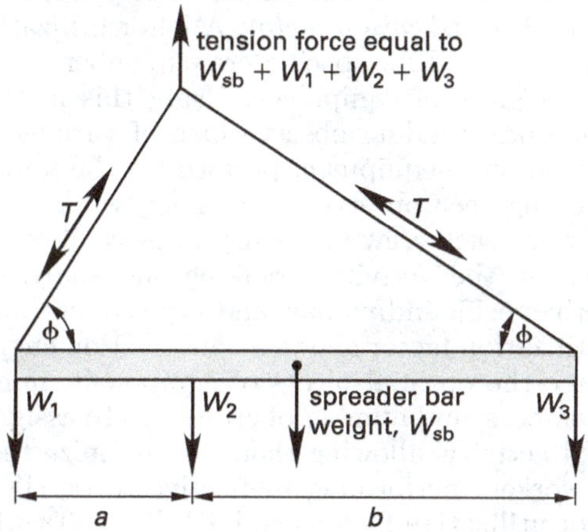

4. PRODUCTIVITY ANALYSIS AND IMPROVEMENT

A full understanding of a project's productivity requires a construction engineer to optimize both the production capability of the equipment used on the project and the production capability of the crew working on the project. A full analysis of a project crew's productivity requires careful observation and sound judgment, but there are some established methods that can aid construction engineers in analyzing a crew's productivity, spotting inefficiencies, and finding ways to improve performance.

Productivity analysis categorizes work performed on an activity into four types. *Productive work*, also called *effective work* or *direct work*, is work that contributes directly to the completion of the activity. *Contributory work* or *support work* is work that does not contribute directly to activity completion but that either is essential for completion or increases the efficiency of the direct work. Examples are refueling a truck or repositioning a crane. *Nonproductive work* or *ineffective work* is either work that does not contribute to completion or efficiency, or work that has already been done and must be redone due to errors or delays. (Engineering convention assumes work performed more than 35 ft from the work area is nonproductive.) *Obstructive work* is work that prevents the activity from being completed, either by causing delays or by preventing direct and contributory work from being done, such as a crane operator positioning a crane in such a way that a pipe fitter cannot continue working until the crane is moved.

The most basic method of analyzing the productivity of a crew is to observe the percentage of time a crew performs effective work on a given activity. This is a simplistic method of analyzing productivity that does not take into account contributory or support work. Furthermore, this percentage can only be ascertained once a portion of the work is done, meaning that inefficiencies may not be noticed until they have already caused a significant increase in costs.

A better method of assessing productivity is the calculation of a *labor utilization factor*. A labor utilization factor can be created for specific crew members, an entire crew, or a piece of equipment. Using this method, an engineer takes random observations of various project crews or pieces of equipment performing the same activity. The engineer observes how much work and what type of work each crew member and piece of equipment performs on the activity, breaking the analysis down into both specific individuals and types of craftsmen. A labor utilization factor allows a construction engineer to determine the optimal pieces of equipment, number of crew members, and number of craftsmen to assign to the analyzed activity, allowing them to minimize the number of workers performing ineffective work. Typically, the labor utilization factor will be 25% to 50%, depending on the activity.

Learning Curves

As a contractor works on more projects and gains more experience, he or she will improve at the job, resulting in a higher rate of productivity. This improvement over

Figure 57.10 *Learning Curve*

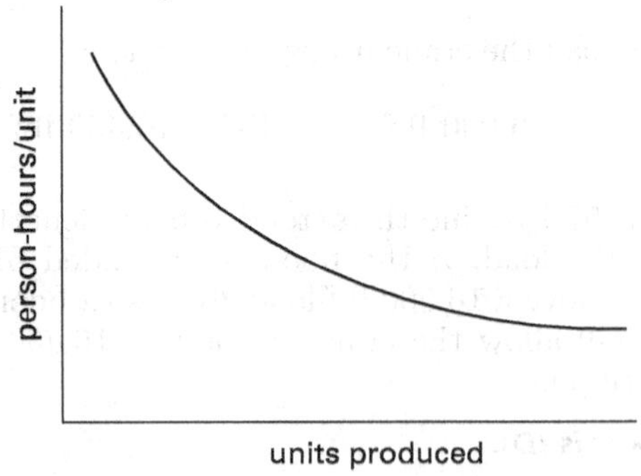

time is represented by a *learning curve*, a mathematical representation of time spent to produce a unit of work.[1] A typical learning curve is shown in Fig. 57.10.

Learning curve gains can be offset by boredom and carelessness on the part of a worker, so a construction engineer or superintendent should be certain to watch for these issues, rather than relying entirely on learning curves for guidance. Care should also be taken to restrict the use of learning curves when analyzing activities where the capacity for improvement is constrained by equipment limitations, supply of materials or equipment, or regulations or specifications (e.g., concrete must be allowed to cure before work can be continued).

Learning curves can be used to help estimate the cost of labor on a project. By integrating the area under the curve, an estimator can calculate the total hours of labor needed to complete an activity. (See Fig. 57.11.) Most budget estimates use an average of working hours to calculate the labor cost.

Figure 57.11 *Estimation Using Learning Curves*

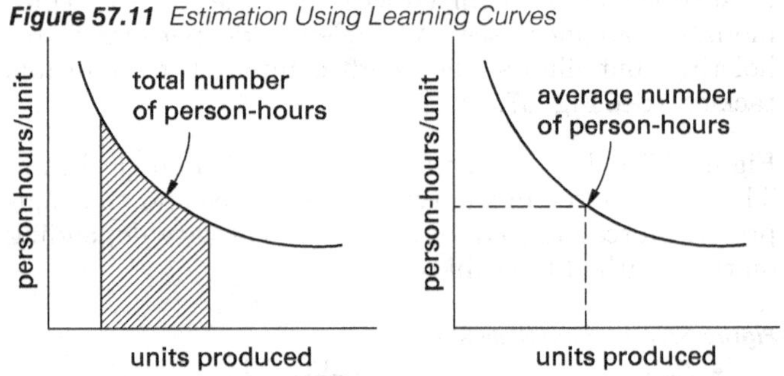

5. SITE DEWATERING AND PUMPING

Some job sites must be dewatered prior to excavation. *Dewatering* (*unwatering*) is simply the removal of water from the job site, typically requiring a lowering of the water table. In some cases, the job site may be surrounded by waterproof barriers. Dewatering almost always involves the use of high-capacity wells with

[1]A "unit of work" is a variable term that must be defined for each individual learning curve, as different tasks require different amounts of time to be completed in a reasonably efficient manner. For example, it would be unreasonable to expect a truck driver to finish hauling a load of earthmoving material 10 miles in the same amount of time it takes for a crane operator to lift a reinforced concrete column into place.

submersible pumps, or vacuum wellpoint methods. In addition to the cost of installation, operational costs are primarily a function of the volume of water removed.

Deep *wells* use submersible electric pumps to lower the water table, bringing the job site into the well's *cone of depression*. The soil's permeability and the pump's power determine the volume of water that can be removed. Typically, several deep wells surrounding or ringing the job site will be used. The *Dupuit equation* can be used to predict the steady-state drawdown and volumetric flow from a deep *penetrating well* (i.e., a well that penetrates all the way down to an impermeable layer). More advanced methods are required to predict the transient performance and the performance of a non-penetrating well.

Vacuum dewatering uses wellpoints to draw water up small-diameter (e.g., 2 in) tubes with side perforations (integral strainers) in the liquid water zone. Wellpoints are typically jetting into position. This method is typically used when the dewatering depth is less than about 15–20 ft, since the vertical lift is limited by the practical vacuum achievable. Wellpoints can be installed throughout a job site; however, they are typically installed around its periphery. Individual wellpoints are connected with flexible hoses to a manifold (header pipe) that is connected to the wellpoint vacuum pump. The wellpoint pump is selected to handle the liquid water, as well as entrained gasses found in the water. Sites can be dewatered to depths greater than 15–20 ft by using multiple wellpoint stages.

6. DEEP FOUNDATIONS

Chapter 19 has information on deep foundations.

7. EARTH RETAINING STRUCTURES

Chapter 20 has information on earth retaining structures.

8. SLOPE AND EROSION CONTROL FEATURES

Chapter 21 has information on slope and erosion control.

58 Construction Safety

1. INTRODUCTION

Common sense can go a long way in preventing many jobsite accidents. However, scheduling, economics, lack of concern, carelessness, and laxity are often more likely to guide actions than common sense. For that reason, the "science" of accident prevention is highly regulated. In the United States, workers' safety is regulated by the federal Occupational Safety and Health Act (OSHA). State divisions (e.g., Cal-OSHA) are charged with enforcing federal and state safety regulations. Surface and underground mines, which share many hazards with the construction industry and which have others specific to them, are regulated by the federal Mine Safety and Health Act (MSHA). Other countries may have more restrictive standards.

All of the federal OSHA regulations are published in the Congressional Federal Register (CFR). The two main categories of standards are the "1910 standards," which apply to general industry, and the "1926 standards," which apply to the construction industry. Table 58.1 contains some of the major subjects covered by the general industry OSHA standards. Table 58.2 lists OSHA standards for the construction industry.

In addition to 1910 and 1926 regulations, other pertinent regulations include 1903 regulations about inspections, citations, and proposed penalties, and 1904 regulations about recording and reporting occupational injuries and illnesses.

Table 58.1 *OSHA 29 CFR 1910 Subjects for General Industry*

Safety and Health Programs
Recordkeeping
Hazard Communication
Exit Routes
Emergency Action Plans
Fire Prevention
Fire Detection and Protection
Electrical
Flammable and Combustible Liquids
Lockout/Tagout
Machine Guarding
Walking and Working Surfaces
Welding, Cutting, and Brazing
Material Handling
Ergonomics
Permit-Required Confined Spaces
Personal Protective Equipment (PPE)
Industrial Hygiene
Blood-Borne Pathogens
Hand and Portable Power Tools and Equipment

2. SOIL CLASSIFICATION

Soils are classified into stable rock and types A, B, or C, with type C being the most unstable.

Type A soils are cohesive soils with an unconfined compressive strength of 1.5 tons per square foot or greater. Examples of type A cohesive soils are clay, silty clay, sandy clay, clay loam, and in some cases, silty clay loam and sandy clay loam. No soil is type A if it is fissured, is subject to vibration of any type, has previously been disturbed, is part of a sloped, layered system where the layers dip into the excavation on a slope of four horizontal to one vertical or greater, or has seeping water.

Type B soils are cohesive soils with an unconfined compressive strength greater than 0.5 tons per square foot but less than 1.5. Examples of type B soils are angular gravel; silt; silt loam; previously disturbed soils unless

Table 58.2 *OSHA 29 CFR 1926 Subjects for the Construction Industry*

Health Hazards in Construction
Hazard Communication
Fall Protection
Signs, Signals, and Barricades
Cranes
Rigging
Excavations and Trenching
Tools
Material Handling
Scaffolds
Walking and Working Surfaces
Stairways and Ladders
Hand and Power Tools
Welding and Cutting
Electrical
Fire Prevention
Concrete and Masonry
Confined Space Entry
Personal Protective Equipment (PPE)
Motor Vehicles

otherwise classified as type C; soils that meet the unconfined compressive strength or cementation requirements of type A soils but are fissured or subject to vibration; dry unstable rock; and layered systems sloping into the trench at a slope less than four horizontal to one vertical (but only if the material would be classified as a type B soil).

Type C soils are cohesive soils with an unconfined compressive strength of 0.5 tons per square foot or less. Type C soils include granular soils such as gravel, sand and loamy sand, submerged soil, soil from which water is freely seeping, and submerged rock that is not stable. Also included in this classification is material in a sloped, layered system where the layers dip into the excavation or have a slope of four horizontal to one vertical or greater.

Layered geological strata include soils that are configured in layers. Where a layered geologic structure exists, the soil must be classified on the basis of the soil classification of the weakest soil layer. Each layer may be classified individually if a more stable layer lies below a less stable layer (e.g., where a type C soil rests on top of stable rock).

Soils classified as types A and B may need to be reclassified as type C following rain and flooding [OSHA 1926.652, App. A to Subpart P].

OSHA (App. A to Subpart P) requires that at least one visual and one manual method be used to classify the soil type as A, B, and C. (Tests are not necessary if type C soil is assumed and protection is provided for type C soil conditions.) Manual methods are the *thumb penetration*, *pocket penetrometer*, and *torvane sheer vane tests.* The thumb penetration test is subjective and the least accurate of the three methods. Only one method is required, and although penetrometer and sheer vane tests are preferred because numerical readings can be recorded, the thumb penetration test is equally permissible.

The thumb penetration test should be conducted on an undisturbed soil sample, such as a large clump of soil, within five minutes after excavation to minimize moisture loss from drying. Using the thumb penetration test, soils that can be readily indented by the thumb (i.e., that have unconfined compressive strengths of at least 1.5 tsf), but can only be penetrated with great effort, are classified as type A soils. If the thumb penetrates no further than the length of the thumbnail, the soil is classified as type B. Type C soils (with unconfined strengths of 0.5 tsf or less) can be easily penetrated by the full length of the thumb and can be molded by light finger pressure. Figure 58.1 illustrates how the soil types correlate to other geotechnical qualities.

Figure 58.1 *Soil Type by Geotechnical Qualities*

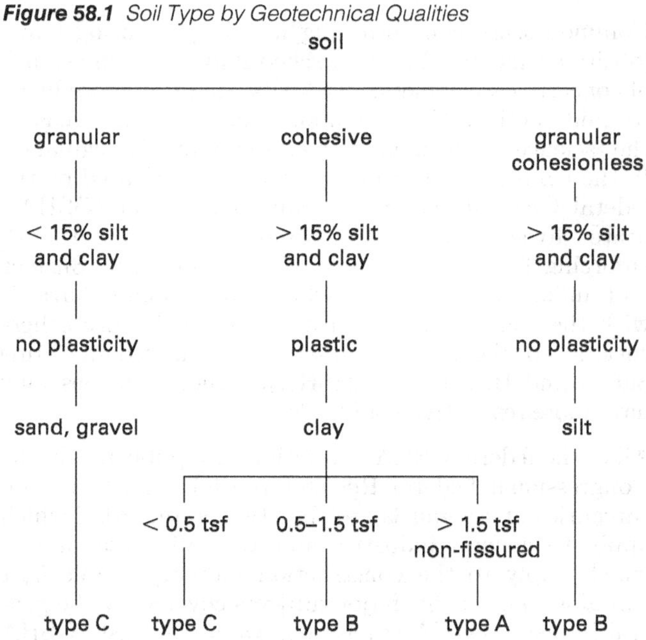

3. TRENCHING AND EXCAVATION

Except for excavations entirely in stable rock, excavations deeper than 5 ft in all types of earth must be protected from cave-in and collapse [OSHA 1926.652]. Excavations less than 5 ft deep are usually exempt but may also need to be protected when inspection indicates that hazardous ground movement is possible.

Timber and aluminum shoring (hydraulic, pneumatic, and screw-jacked) and trench shields that meet the requirements of OSHA 1926.652 App. C to Subpart P may be used in excavations up to 20 ft deep. (See Fig. 58.1.)

Sloping and benching the trench walls may be substituted for shoring. Sloped walls in excavations deeper than 20 ft must be designed by a professional engineer. Table 58.3 provides maximum slopes for excavations less than 20 ft deep. Greater slopes are permitted for short-term usage in excavations less than 12 ft deep.

Table 58.3 *Maximum Allowable Slopes*

soil or rock type	maximum allowable slopes (H:V)[a] for excavations less than 20 ft deep[b]	
stable rock	vertical	(90°)
type A[c]	$\frac{3}{4}$:1	(53°)
type B	1:1	(45°)
type C[d]	$1\frac{1}{2}$:1	(34°)[e]

[a]Numbers shown in parentheses next to maximum allowable slopes are angles expressed in degrees from the horizontal. Angles have been rounded off.

[b]Sloping or benching for excavations greater than 20 ft deep must be designed by a registered professional engineer.

[c]A short-term maximum allowable slope of $\frac{1}{2}$H:1V (63°) is allowed in excavations in type A soil that are 12 ft or less in depth. Short-term maximum allowable slopes for excavations greater than 12 ft in depth must be $\frac{3}{4}$H:1V (53°).

[d]These slopes must be reduced 50% if the soil shows signs of distress.

[e]Source: OSHA 1926.652 App. B

In trenches 4 ft deep or more, ladders, stairways, or ramps are required so that no more than 25 ft of lateral travel is required to reach an exit (i.e., with a maximum lateral spacing of 50 ft) [OSHA 1926.651(c)(2)].

Spoils and other equipment that could fall into a trench or an excavation must be kept at least 2 ft from the edge of a trench unless secured in some other fashion [OSHA 1926.651(l)].

Examples of excavations by soil types A, B, and C are illustrated in Fig. 58.2. Figure 58.3 illustrates excavations in layered soils.

4. COMPETENT PERSON: SOIL EXCAVATIONS

OSHA requirements are effective when they are interpreted and implemented by competent individuals. For the purpose of determining the safety of soil excavations, OSHA defines a *competent person* as someone with (1) training, experience, and knowledge of soil analysis, use of protective systems, and the OSHA regulations related to excavations (i.e., 29 CFR Part 1926 Subpart P); (2) the ability to detect conditions that could result in cave-ins, failures in protective systems, hazardous atmospheres, and other hazards including those associated with confined spaces; and (3) the authority to take prompt corrective measures to eliminate existing and predictable hazards and to stop work when required. A competent person is different from a *trained person*, who has been trained in specific tasks.

Example

The first person to enter an excavation should normally be a

(A) competent person

(B) trained person

(C) newly hired person

(D) supervisor

Solution

The first person to enter an excavation must be a trained person. The responsibilities of a competent person (as defined by OSHA) include soil analysis, assigning employee duties, and designing protective systems. However, the first person to enter an excavation does not necessarily have to be a competent person.

The answer is (B).

5. CHEMICAL HAZARDS

OSHA's Hazard Communication Standard [OSHA 1910.1200] requires that the dangers of all chemicals purchased, used, or manufactured be known to employees. The hazards are communicated in a variety of ways, including labeling containers, training employees, and providing ready access to *material safety data sheets* (MSDSs).

OSHA has suggested a nonmandatory standard form for the MSDS, but other forms are acceptable as long as they contain the same (or more) information. The information contained on an MSDS consists of the following categories.

Chemical identity: the identity of the substance as it appears on the label

Section I: Manufacturer's name and contact information: manufacturer's name, address, telephone number, and emergency phone number; date the MSDS was prepared; and, an optional signature of the preparer

Section II: Hazardous ingredients/identity information: list of the hazardous components by chemical identity and other common names; OSHA *permissible exposure limit* (PEL), ACGIH *threshold level value* (TLV), and other recommended exposure limits; percentage listings of the hazardous components is optional

Figure 58.2 Excavations by Soil Type

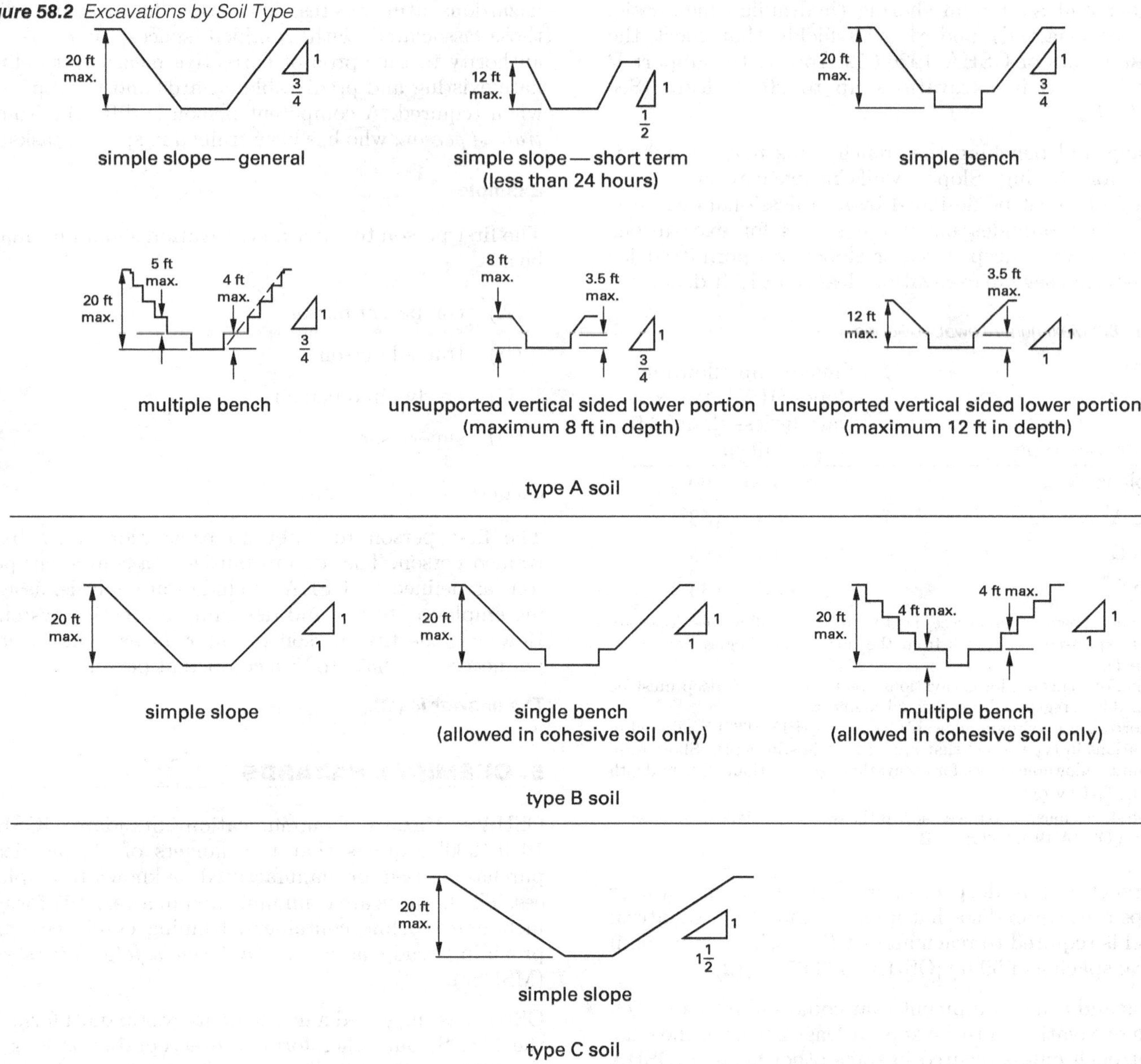

Section III: Physical/chemical characteristics: boiling point, vapor pressure, vapor density, specific gravity, melting point, evaporation rate, solubility in water, physical appearance, and odor

Section IV: Fire and explosion hazard data: flash point (and method used to determine it), flammability limits, extinguishing media, special fire-fighting procedures, and unusual fire and explosion hazards

Section V: Reactivity data: stability, conditions to avoid, incompatibility (materials to avoid), hazardous decomposition or by-products, and hazardous polymerization (and conditions to avoid)

Section VI: Health hazard data: routes of entry into the body (inhalation, skin, ingestion), health hazards (*acute* = immediate, or *chronic* = builds up over time), carcinogenicity, signs and symptoms of exposure, medical conditions generally aggravated by exposure, and emergency and first-aid procedures

Section VII: Precautions for safe handling and use: steps to be taken in case the material is released or spilled, waste disposal method, precautions to be taken in handling or storage, and other precautions

Section VIII: Control measures: respiratory protection (type to be specified), ventilation (local, mechanical exhaust, special, or other), protective gloves, eye protection, other protective clothing or equipment, and work/hygienic practices

Section IX: Special precautions and comments: safe storage and handling, types of labels or markings for containers, and Department of Transportation (DOT) policies for handling the material

Figure 58.3 Excavations in Layered Soils

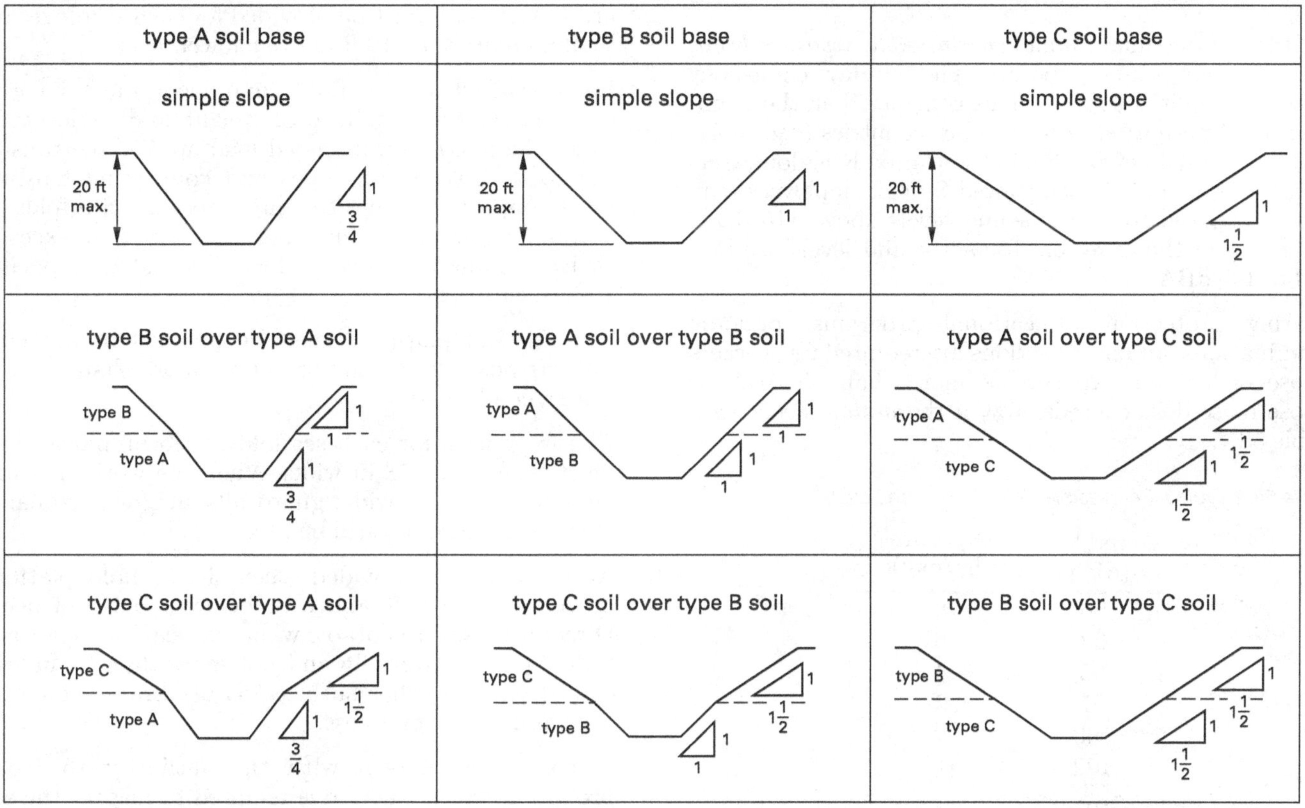

6. CONFINED SPACES AND HAZARDOUS ATMOSPHERES

Employees entering confined spaces (e.g., excavations, sewers, tanks) must be properly trained, supervised, and equipped. Atmospheres in confined spaces must be monitored for oxygen content and other harmful contaminants. Oxygen content in confined spaces must be maintained at 19.5% or higher unless a breathing apparatus is provided [OSHA 1910.146]. Employees entering deep confined excavations must wear harnesses with lifelines [OSHA 1926.651].

7. POWER LINE HAZARDS

Employees operating cranes or other overhead material-handling equipment (e.g., concrete boom trucks, backhoe arms, and raised dump truck boxes) must be aware of the possibility of inadvertent power line contact. Prior to operation, the site must be thoroughly inspected for the danger of power line contact. OSHA provides specific minimum requirements for safe operating distances. For example, for lines of 50 kV or less, all parts of the equipment must be kept at least 10 ft from the power line [OSHA 1926.550(a)(15)]. A good rule of thumb for voltages greater than 50 kV is a clearance of 35 ft.

8. FALL AND IMPACT PROTECTION

Fall protection can take the form of barricades, walkways, bridges (with guardrails), nets, and fall arrest systems. Personal fall arrest systems include lifelines, lanyards, and deceleration devices. Such equipment is attached to an anchorage at one end and to the body-belt or body hardness at the other. All equipment is to be properly used, certified, and maintained. Employees are to be properly trained in the equipment's use and operation.

Employees must be protected from impalement hazards from exposed rebar [OSHA 1926.701(b)]. A widely used method for covering rebar ends has been plastic *mushroom caps*, often orange or yellow in color. However, OSHA no longer considers plastic caps adequate for anything more than scratch protection. Commercially available steel-reinforced caps and wooden troughs capable of withstanding a 250 lbm/10 ft drop test without breakthrough can still be used.

Head protection, usually in the form of a helmet (*hard hat*), is part of a worker's *personal protective equipment* (PPE). Head protection is required where there is a danger of head injuries from impact, flying or falling objects, electrical shock, or burns [OSHA 1910.132(a) and (c)]. Head protection should be nonconductive when there is electrical or thermal danger [OSHA 1910.335(a)(1)(iv)].

9. NOISE

OSHA sets maximum limits (*permissible exposure limit*, PEL) on daily sound exposure. The "all-day" eight-hour noise level limit is 90 dBA. This is higher than the maximum level permitted in most other countries (e.g., noise control threshold of 85 dBA in the United Kingdom, Germany, and Japan). In the United States, employees may not be exposed to steady sound levels above 115 dBA, regardless of the duration. Impact sound levels are limited to 140 dBA.

Hearing protection, educational programs, periodic examinations, and other actions are required for workers whose eight-hour exposure is more than 85 dBA or whose noise dose exceeds 50% of the *action levels.* (See Table 58.4.)

Table 58.4 *Typical Permissible Noise Exposure Levels**

sound level (dBA)	exposure (hr/day)
90	8
92	6
95	4
97	3
100	2
102	$1\frac{1}{2}$
105	1
110	$\frac{1}{2}$
115	$\frac{1}{4}$ or less

*without hearing protection

Source: OSHA Sec. 1910.95, Table G-16

10. SCAFFOLDS

Scaffolds are any temporary elevated platform (supported or suspended) and its supporting structure (including points of anchorage) used for supporting employees, materials, or both. Construction and use of scaffolds are regulated in detail by OSHA Std. 1926.451. A few of the regulations are summarized in the following paragraphs.

Each employee who performs work on a scaffold must be trained by a person qualified to recognize the hazards associated with the type of scaffold used and to understand the procedures to control or minimize those hazards. The training must include such topics as the nature of any electrical hazards, fall hazards, falling object hazards, the maintenance and disassembly of the fall protection systems, the use of the scaffold, handling of materials, the capacity, and the maximum intended load.

Fall protection (guardrail systems and personal fall arrest systems) must be provided for each employee on a scaffold more than 10 ft above a lower level.[1]

Each scaffold and scaffold component must have the capacity to support its own weight and at least four times the maximum intended load applied or transmitted to it. Suspension ropes and connecting hardware must support six times the intended load. Scaffolds and scaffold components may not be loaded in excess of their maximum intended loads or rated capacities, whichever is less.

The scaffold platform must be planked or decked as fully as possible. It cannot deflect more than $\frac{1}{60}$ of the span when loaded.

The work area for each scaffold platform and walkway must be at least 18 in wide. When the work area must be less than 18 in wide, guardrails and/or personal fall arrest systems must still be used.

Access must be provided when the scaffold platforms are more than 2 ft above or below a point of access. Direct access is acceptable when the scaffold is not more than 14 in horizontally and not more than 24 in vertically from the other surfaces. Cross braces cannot be used as a means of access.

A competent person with the authority to require prompt corrective action is required to inspect the scaffold, scaffold components, and ropes on suspended scaffolds before each work shift and after any occurrence which could affect the structural integrity.

11. TEMPORARY STRUCTURES

Temporary structures are those that exist for only a period of time during construction and are removed prior to project completion. They include temporary buildings for occupancy and storage, shoring and underpinning, concrete formwork and slipforms, scaffolding (see Sec. 58.10), diaphragm and slurry walls, wharves and docks, temporary fuel and water tanks, cofferdams, earth-retaining structures, bridge falsework, tunneling supports, roadway decking, and ramps and inclines.

While a persistent school of thought permits temporary structures to be installed and operated with lower factors of safety than permanent structures, just the opposite is required unless the expected loading is entirely predictable and manageable. Temporary structures may or may not require separate permits and/or drawings. Most temporary structures need ongoing inspection during their existence, since connections loosen and members get moved, weaken, deform, and/or fail.

[1]OSHA prescribes in Subpart L (1926.451(g)(1)) a 10 ft fall protection for scaffolds. This differs from the 6 ft threshold for fall protection in Subpart M (1926.502(d)(16)(iii)) for other walking/working surfaces in construction, because scaffolds, unlike these other surfaces, are temporary structures erected to provide a work platform for employees who are constructing or demolishing other structures. The features that make scaffolds appropriate for short-term use in construction make them less amendable to the use of fall protection as the first level is being erected.

Temporary structures are often erected by a separate subcontractor, complicating the inspection tasks and responsibilities.

12. TRUCK AND TOWER CRANES

The main parts of a *hydraulic truck crane* are the boom, jib, rotex gear, outriggers, counterweights, wire rope, and hook. The *boom* is the arm that holds the load. Most hydraulic truck cranes have booms consisting of several telescoping sections. Some booms are equipped with a separate *jib*, which is a lattice structure that extends the boom's reach. Jibs may be fixed or luffing.[2] The free end of a *fixed jib* (*fixed-angle jib*) remains at a constant elevation above the ground. A *luffing jib* pivots where it attaches to the boom, allowing the free end to vary in elevation. ("Luffing" means "moving.") The *slewing platform* containing the operator's cab is mounted on a turntable bearing driven through the *rotex gear* by a bidirectional, hydraulic motor.

Tires don't offer the stability needed for the crane to achieve full capacity, so *outriggers* are used to lift the entire truck off the ground. Each outrigger consists of the beam (leg) and pad (foot). Wooden *floats* (blocks) may be placed under the pad to dissipate the outrigger down force over an area greater than the pad. To be effective, such blocks should have an area of at least three times the area of the outrigger pad. Detachable *counterweights* are placed on the back of the crane to prevent it from tipping during operation. The amount of counterweight needed for a particular lift is determined by the weight of the load, the radius of the boom, and the boom's angle during operation. Counterweights are only used during lifts and are removed when the crane is driven.

Wire rope typically runs from a winch on pulleys up and over the boom and jib. The lines attach to a *load block* or metal ball that keeps the lines pulled taut when no load is attached to the hook. With *standard hoisting wire hoist rope*, which has an approximate mass of 2 lbm/ft, each line can support an allowable *line pull* of 14,000 lbf. However, crane capacities also depend on the *reeving*, the path of the wire rope as it comes off the hoist drum and wraps around the various upper and lower sheaves. When the hoist line is not centered over the boom tip, *eccentric reeving* occurs, and the boom twists. The number of lines reeved on the main hoist block determines the capacity of the crane. As with simple pulleys, the lift capacity is proportional to the number of wire rope lines going to and coming from the hook. This number is known as *parts of the line* (*parts of line*, *n-part line*, etc.).

Tower cranes consist of a *base*, often bolted to a concrete pad, that supports the crane, a *mast* (*tower*), and a *slewing unit* (*turntable*) consisting of the gear and motor that allows the crane to rotate. (See Fig. 58.4.) Four parts are mounted on top of the slewing unit: (1) the *top tower* holds pulleys for the wire rope and bracing; (2) the horizontal *jib* (*boom* or *working arm*) is the portion of the crane that carries the load on a *trolley*, which runs along the jib and moves the load in and out from the crane's center; (3) a shorter horizontal *machinery arm* (*counterjib*), which contains the crane's motors and electronics as well as *counter weights*; and (4) the *operator's cab*.

Figure 58.4 *Tower Crane*

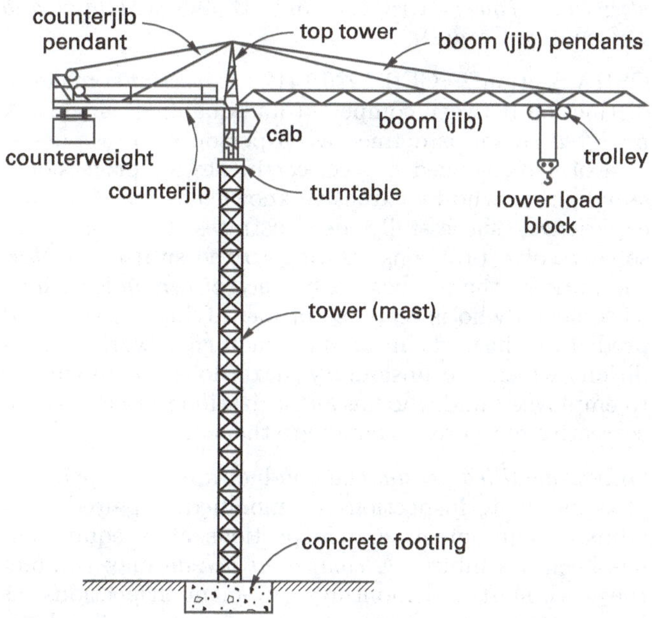

Example

What is the danger of eccentric boom reeving?

(A) boom twisting

(B) decreased jib capacity

(C) increased wire rope wear

(D) increased sheave wear

Solution

Eccentric reeving occurs when the hoist line is not centered over the boom tip. This causes torque (twisting) in the boom.

The answer is (A).

13. CRANE USE AND SAFETY

OSHA federal crane safety regulations for general industry are covered in 29 CFR Sections 1910.179 (*Overhead and Gantry Cranes*), 1910.180 (*Crawler Locomotive and*

[2]To reach all the corners of a site, the jib has to *oversail* adjacent property. Following a legal ruling in the 1970s, permission for oversailing must be given by the adjacent owner. If oversailing is not permitted, the jib can be fixed in a non-offending configuration.

Truck Cranes), and 1917.45 (*Cranes and Derricks*). Cranes used in construction are covered in 29 CFR 1926 Subparts CC (*Cranes and Derricks in Construction*) and DD (*Cranes and Derricks Used in Demolition and Underground Construction*). Some states have their own crane safety standards. In general, these regulations reference American National Standards Institute (ANSI)/American Society of Mechanical Engineers (ASME) Standards B-30.1 through B-30.6. Two ANSI/ASME standards delineate safe operation and maintenance practices for construction cranes: B-30.3 (*Construction Tower Cranes*) and B-30.5 (*Mobile and Locomotive Cranes*).

OSHA Section 29 CFR 1926.1412 (*Inspections*) makes a distinction between competent and qualified persons. A *qualified person* is defined as "a person who by possession of a recognized degree, certificate, or professional standing, or who by extensive knowledge, training, and experience, successfully demonstrates the ability to solve/resolve problems relating to the subject matter, the work, or the project." A *competent person* is defined as someone who is "capable of identifying existing and predictable hazards in the surroundings or working conditions which are unsanitary, hazardous, or dangerous to employees, and who has authorization to take prompt corrective measures to eliminate them."

Only a *qualified person* can conduct annual inspections of equipment; inspections of modified, repaired, and adjusted equipment; and inspections after equipment has been assembled. A *competent person* may conduct the work shift and monthly equipment inspections, as long as that person has been trained in the required elements of a shift inspection. The crane or derrick operator can be the inspector, but only if the operator meets the respective requirements for a qualified or competent person.

Crane safety is a function of crane inspection and maintenance, selection, rigging, loading, and use. While each crane will have its own specific requirements and limitations, some safe practices apply to all crane operations. Such safe practices include the following.

- The crane should be operated only by qualified and trained personnel.
- A "competent person" must inspect the crane and controls before it is used.
- The crane must be on a firm/stable surface and be level.
- During assembly and disassembly, pins should not be unlocked or removed unless the sections are blocked, secure, and stable.
- Outriggers should be fully extended.
- Areas inside the crane's swing radius should be barricaded.
- A clearance of at least 10 ft from overhead electric power lines should be watched for and maintained.
- All rigging must be inspected prior to use.
- Rig properly. Do not wrap hoist lines around the load.
- The correct load chart for the crane's current configuration and setup, load weight, and lift path must be used.
- The load chart capacity must not be exceeded.
- Before delivering the load, raise the load a few inches, hold, verify capacity/balance, and test the brake system.
- Do not use the crane's lift angle indicator for loads exceeding 75% of the rated capacity.
- Do not move loads over workers.
- Remain in voice and visual contact with observers and follow their signals.
- Observe the manufacturer's instructions and limitations.

14. CRANE LOAD CHARTS

Every crane has a *load chart* that specifies the crane's features, dimensions, and how its lifting capacity varies with configuration. A load chart actually consists of a set of tables of data organized in different ways. The chart includes data for operation with the outriggers extended, transport weight, and steering dimensions. The transport weight information, in turn, determines the trailer that would be required to transport the crane to the job site, how to load the crane on a trailer, the route to take, and what highway permits would be required. The load chart also specifies what counterweight is required and what the outrigger extension dimensions are.

Charts should be read conservatively. For example, when an angle is used that is not listed on the load chart, use the next lower boom angle noted on the load chart for determining the capacity of the crane. When using a particular radius that is not listed on the load chart, use the next larger radius measurement in the load chart for determining the capacity of the crane.

The chart's *lift capacity chart* (*lift table*) consists of a table of maximum loads. Each horizontal row of numbers corresponds to a particular radius from the center pin. This is the distance from the center pin to the center of lifted load. Each vertical column corresponds to a boom extension length. Maximum capacity is always measured by the shortest lift, usually over the rear of the crane, and with the outriggers fully extended.

The chart's *lift range chart* (*range diagram* or *range table*) illustrates how much boom length is needed to pick up and lift a load at given distances and vertical lift heights. The range chart is generally a part of the load chart decaling, as well as being part of the operating manual.

With higher angles of lift, the maximum load capacity decreases. The chart's *lift angle chart* illustrates the maximum lift with luffing and fixed jibs. With a luffing jib, the angle can be automatically adjusted from the operators cab. With a fixed jib, the angle is fixed.

The crane's *crane in motion chart* defines the lift capacity for *pick and carry* operation. The chart lists total capacities able to be picked at a 360° angle while stationary on wheels, while slowly rolling with the load at a 0° angle (creep), and while moving at a steady speed, usually 2.5 mi/hr. Each row corresponds to a particular radius. The maximum boom length that each weight can be carried at is also given.

Depending on how the load chart is organized, reductions in the gross capacity of a crane (as read from the load chart) may be required for the weights of the jib, block, ball, rigging, and lengths of the main and auxiliary ropes.

Example

The load chart for a Series C13M crane with an 18 ft jib is shown in the given table (at the end of the chapter). Loadline equipment deductions are: downhaul weight, 90 lbf; one sheave block, 185 lbf; two sheave blocks, 355 lbf. The crane is configured with 60 lbf of rigging and one sheave block with two parts of line. What is most nearly the crane's net capacity at a radius of 35 ft when the main boom is extended to 45 ft?

(A) 2200 lbf

(B) 2300 lbf

(C) 2600 lbf

(D) 2800 lbf

Solution

Although the crane has an 18 ft jib, an extension of 45 ft does not require the jib's use. Therefore, use the main part of the load chart.

The boom is extended to 45 ft, which is between the 40 ft and 48 ft columns on the table. Use the larger column (48 ft). At a radius of 35.0 ft, the gross crane capacity is 2600 lbf. Deductions are 185 lbf for one sheave block and 60 lbf for the rigging. The net crane capacity is

$$\begin{aligned}\text{net capacity} &= \text{gross capacity} - \text{deductions} \\ &= 2600 \text{ lbf} - 185 \text{ lbf} - 60 \text{ lbf} \\ &= 2355 \text{ lbf} \quad (2300 \text{ lbf})\end{aligned}$$

The answer is (B).

Construction

load radius (ft)	loaded boom angle (deg)	22 ft boom (lbf)	loaded boom angle (deg)	32 ft boom (lbf)	loaded boom angle (deg)	40 ft boom (lbf)	loaded boom angle (deg)	48 ft boom (lbf)	loaded boom angle (deg)	56 ft boom (lbf)	load radius (ft)	loaded boom angle (deg)	18 ft jib (lbf)
5	77.5	20,000											
6	74.5	17,300											
8	70.0	12,800	76.5	10,950	79.5	10,400							
10	63.0	10,000	73.0	8850	77.0	8650							
12	56.5	8800	69.0	7650	74.0	7250	76.5	7150					
14	50.0	7800	65.0	6850	70.5	6350	75.0	6150	77.0	5900	14	80.0	2800
16	42.5	6900	61.0	6050	68.5	5650	72.5	5400	75.5	5150	16	78.5	2700
18	34.0	6100	57.5	5450	65.0	5150	69.5	4900	73.5	4600	18	77.0	2500
20	23.0	5400	53.0	5050	62.0	4750	67.5	4600	71.0	4200	20	75.5	2300
25			40.5	4050	53.5	3850	61.0	3700	65.5	3400	25	72.0	2050
30			22.5	3200	43.5	3150	53.5	3000	60.0	2800	30	68.0	1750
35					31.0	2600	45.5	2600	53.5	2400	35	63.5	1500
40							36.0	2100	46.5	2000	40	59.0	1300
45							22.5	1750	39.0	1650	45	54.5	1100
50									29.0	1400	50	49.5	1000
55									13.0	1050	55	44.0	850
60											60	38.0	750
65											65	31.0	550
	0	4050	0	2350	0	1650	0	1250	0	850			

Topic XIV: Computational Tools

Chapter

59 Computer Software

1. CHARACTER CODING

Alphanumeric data refers to characters that can be displayed or printed, including numerals and symbols ($, %, &, etc.) but excluding *control characters* (tab, carriage return, form feed, etc.). Since computers can handle binary numbers only, all symbolic data must be represented by binary codes. *Coding* refers to the manner in which alphanumeric data and control characters are represented by sequences of bits.

The *American Standard Code for Information Interchange*, ASCII, is a seven-bit code permitting 128 (2^7) different combinations. It is commonly used in desktop computers, although use of the high order (eighth) bit is not standardized. ASCII-coded magnetic tape and disk files are used to transfer data and documents between computers of all sizes that would otherwise be unable to share data structures.

The *Extended Binary Coded Decimal Interchange Code*, EBCDIC (pronounced eb'-sih-dik), is in widespread use in IBM mainframe computers. It uses eight bits (one byte) for each character, allowing a maximum of 256 (2^8) different characters.

Since strings of binary digits (bits) are difficult to read, the *hexadecimal* (or "packed") format is used to simplify working with EBCDIC data. Each byte is converted into two strings of four bits each. The two strings are then converted to hexadecimal. Since $(1111)_2 = (15)_{10} = (\mathrm{F})_{16}$, the largest possible EBCDIC character is coded FF in hexadecimal.

Example

The American Standard Code for Information Interchange (ASCII) control character "LF" stands for

(A) line feed

(B) large format

(C) limit function

(D) large font

Solution

Lines of text displayed on monitors and printers are often followed by invisible carriage return (CR) and line feed (LF) control characters that determine the location of the next character.

The answer is (A).

2. PROGRAM DESIGN

A *program* is a sequence of computer instructions that performs some function. The program is designed to implement an *algorithm*, which is a procedure consisting of a finite set of well-defined steps. Each step in the algorithm usually is implemented by one or more instructions (e.g., READ, GOTO, OPEN, etc.) entered by the programmer. These original "human-readable" instructions are known as *source code statements*.

Except in rare cases, a computer will not understand source code statements. Therefore, the source code is translated into machine-readable object code and absolute memory locations. Eventually, an executable program is produced.

Programs use variables to store values, which may be known or unknown. A *declaration* defines a variable, specifies the type of data a variable can contain (e.g., INTEGER, REAL, etc.), and reserves space for the variable in the program's memory. *Assignments* give values to variables (e.g., X = 2). *Commands* instruct the program to take a specific action, such as END, PRINT, or INPUT. *Functions* are specific operations (e.g., calculating the SUM of several values) that are grouped into a unit that can be called within the program.

If the executable program is kept on disk or tape, it is normally referred to as *software*. If the program is placed in ROM (read-only memory) or EPROM (erasable programmable read-only memory), it is referred to as *firmware*. The computer mechanism itself is known as the *hardware*.

3. FLOWCHARTS

A *flowchart* is a step-by-step drawing representing a specific procedure or algorithm. Figure 59.1 illustrates the most common flowcharting symbols. The *terminal symbol* begins and ends a flowchart. The *input/output symbol* defines an I/O operation, including those to and from keyboard, printer, memory, and permanent data storage. The *processing symbol* and *predefined process symbol* refer to calculations or data manipulation. The *decision symbol* indicates a point where a decision must be made or two items are compared. The *connector symbol* indicates that the flowchart continues elsewhere. The *off-page symbol* indicates that the flowchart continues on the following page. Comments can be added in an *annotation symbol*.

Figure 59.1 *Flowcharting Symbols*

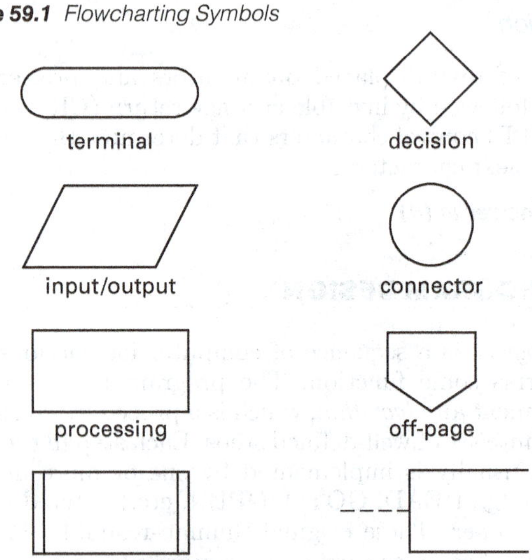

Example

The flowchart shown represents the summer training schedule for a college athlete.

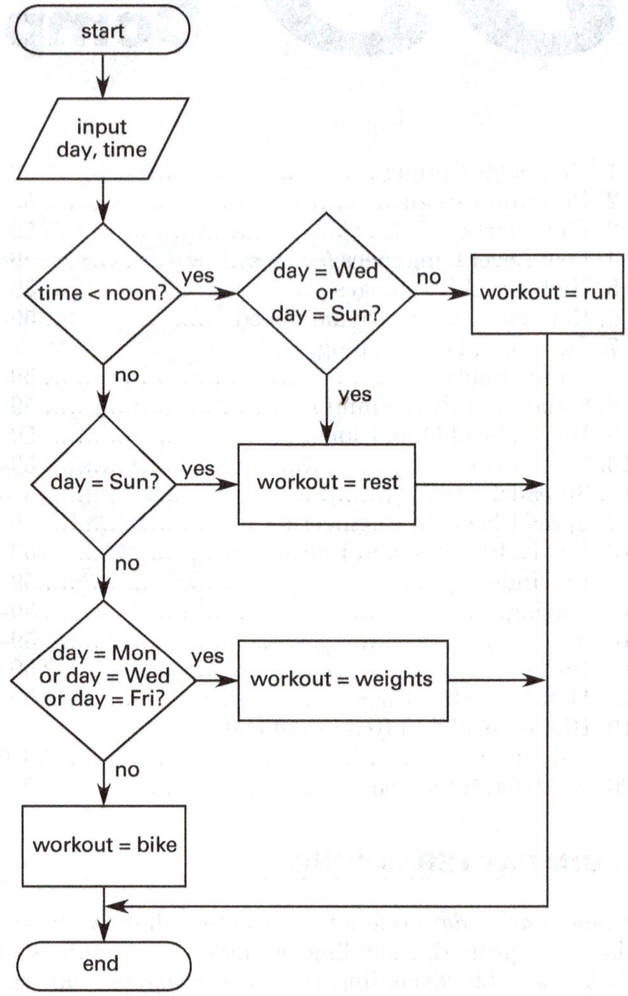

What is the regularly scheduled workout on Wednesday morning?

(A) bike

(B) rest

(C) run

(D) weights

Solution

The day and time being considered are Wednesday morning. At the first decision follow the "yes" branch, which leads to the question "Is the day Wednesday or Sunday?" Again the answer is "yes," so the scheduled workout is actually a resting day.

The answer is (B).

4. LOW-LEVEL LANGUAGES

Programs are written in specific languages, of which there are two general types: low-level and high-level. Low-level languages include machine language and assembly language.

Machine language instructions are intrinsically compatible with and understood by the computer's central processing unit (CPU). They are the CPU's native language. An instruction normally consists of two parts: the operation to be performed (*op-code*) and the operand expressed as a storage location. Each instruction ultimately must be expressed as a series of bits, a form known as *intrinsic machine code*. However, octal and hexadecimal coding are more convenient. In either case, coding a machine language program is tedious and seldom done by hand.

Assembly language is more sophisticated (i.e., is more symbolic) than machine language. Mnemonic codes are used to specify the operations. The operands are referred to by variable names rather than by the addresses. Blocks of code that are to be repeated verbatim at multiple locations in the program are known as *macros* (*macro instructions*). Macros are written only once and are referred to by a symbolic name in the source code.

Assembly language code is translated into machine language by an *assembler* (*macro-assembler* if macros are supported). After assembly, portions of other programs or function libraries may be combined by a *linker*. In order to run, the program must be placed in the computer's memory by a *loader*. Assembly language programs are preferred for highly efficient programs. However, the coding inconvenience outweighs this advantage for most applications.

5. HIGH-LEVEL LANGUAGES

High-level languages are easier to use than low-level languages because the instructions resemble English. High-level statements are translated into machine language by either an interpreter or a compiler. Table 59.1 shows a comparison of a typical ADD command in different computer languages. A *compiler* performs the checking and conversion functions on all instructions only when the compiler is invoked. A true stand-alone executable program is created. An *interpreter*, however, checks the instructions and converts them line by line into machine code during execution but produces no stand-alone program capable of being used without the interpreter. (Some interpreters check syntax as each statement is entered by the programmer. Some languages and implementations of other languages blur the distinction between interpreters and compilers. Terms such as *pseudo-compiler* and *incremental compiler* are used in these cases.)

Table 59.1 *Comparison of Typical ADD Commands*

language	instruction
intrinsic machine code	1111 0001
machine language	1A
assembly language	AR
high-level language	+

Example

A portion of computer code contains the instructions "LR," "SC," and "JL." Most likely, these are

(A) machine language

(B) assembly language

(C) object code

(D) interpreted code

Solution

Assembly language usually consists of short commands (e.g., "LR" for loading a value into the register, "SC" for storing a result in location C, and "JL" for jumping if the result is less than). Assembly language is translated into binary machine language (also known as "object code") by an interpreter or compiler program.

The answer is (B).

6. RELATIVE COMPUTATIONAL SPEED

Certain languages are more efficient (i.e., execute faster) than others. (Efficiency can also, but seldom does, refer to the size of the program.) While it is impossible to be specific, and exceptions abound, assembly language programs are fastest, followed in order of decreasing speed by compiled, pseudo-compiled, and interpreted programs.

Similarly, certain program structures are more efficient than others. For example, when performing a repetitive operation, the most efficient structure will be a single equation, followed in order of decreasing speed by a stand-alone loop and a loop within a subroutine. Incrementing the loop variables and managing the exit and entry points is known as *overhead* and takes time during execution.

7. STRUCTURE, DATA TYPING, AND PORTABILITY

A language is said to be *structured* if subroutines and other procedures each have one specific entry point and one specific return point. (Contrast this with BASIC, which permits (1) a GOSUB to a specific subroutine with a return from anywhere within the subroutine and (2) unlimited GOTO statements to anywhere in the main program.) A language has *strong data types* if integer and real numbers cannot be combined in arithmetic statements.

A *portable language* can be implemented on different machines. Most portable languages either are sufficiently rigidly defined (as in the cases of ADA and C) to eliminate variants and extensions, or (as in the case of Pascal) are compiled into an intermediate, machine-independent form. This so-called *pseudocode* (*p-code*) is neither source nor object code. The language is said to have been "ported to a new machine" when an interpreter is written that converts p-code to the appropriate machine code and supplies specific drivers for input, output, printers, and disk use. (Some companies have produced Pascal engines that run p-code directly.)

8. STRUCTURED PROGRAMMING

Structured programming (also known as *top-down programming*, *procedure-oriented programming*, and *GOTO-less programming*) divides a procedure or algorithm into parts known as subprograms, subroutines, modules, blocks, or procedures. (The format and readability of the source code—improved by indenting nested structures, for example—do not define structured programming.) Internal subprograms are written by the programmer; external subprograms are supplied in a library by another source. Ideally, the mainline program will consist entirely of a series of calls (references) to these subprograms. Liberal use is made of FOR/NEXT, DO/WHILE, and DO/UNTIL commands. Labels and GOTO commands are avoided as much as possible.

Very efficient programs can be constructed in languages that support *recursive calls* (i.e., permit a subprogram to call itself). Recursion requires less code, but recursive calls use more memory. Some languages permit recursion; others do not.

Variables whose values are accessible strictly within the subprogram are *local variables*. *Global variables* can be referred to by the main program and all other subprograms.

Calculations are performed in a specific order in an instruction, with the contents of parentheses done first. The symbols used for mathematical operations in programming are

+	add
−	subtract
*	multiply
/	divide

Raising one expression to the power of another expression depends on the language used. Examples of how X^B might be expressed are

X**B
X ^ B

Example

Two floating point numbers, A and B, are passed to a recursive subroutine. If number A is greater than or equal to 7, the subroutine returns the value of number A to the main program. If number A is less than the smaller of 7 and number B, number A is divided by 2. The subroutine then passes the quotient and number B to itself. Most likely, what condition might this subroutine encounter?

(A) overflow

(B) miscompare

(C) infinite loop

(D) type mismatch

Solution

Consider the case where A = 6 and B = 9. Since 6 is not greater than or equal to 7, control is not returned to the main program. Since 6 is less 7 (the smaller of 7 and 9), 6 is divided by 2, yielding a quotient of 3. The subroutine then passes A = 3 and B = 9 to itself. Since 3 (the new number A) is not greater than or equal to 7, control is not returned to the main program. Since 3 is less 7 (the smaller of 7 and 9), 3 is divided by 2, yielding a quotient of 1.5. The subroutine then passes A = 1.5 and B = 9 to itself. The process repeats indefinitely, since A will never be greater than 7. In some cases, the quotient might become small enough to terminate with an underflow error. (Nothing is said about the data type (variable class) of the number 7 or the comparison rules, so a type mismatch would require additional assumptions.)

The answer is (C).

Conditions and Statements

Following are brief descriptions of some commonly used structured programming functions.

IF THEN statements: In an IF <condition> THEN <action> statement, the condition must be satisfied, or the action is not executed and the program moves on to the next operation. Sometimes an IF THEN statement will include an ELSE statement in the format of IF <condition> THEN <action 1> ELSE <action 2>. If the condition is satisfied, then action 1 is executed. If the condition is not satisfied, action 2 is executed.

DO/WHILE loops: A set of instructions between the DO/WHILE <condition> and the ENDWHILE lines of code is repeated as long as the condition remains true.

The number of times the instructions are executed depends on when the condition is no longer true. The variable or variables that control the condition must eventually be changed by the operations, or the WHILE loop will continue forever.

DO/UNTIL loops: A set of instructions between the DO/UNTIL <condition> and the ENDUNTIL lines of code is repeated as long as the condition remains false. The number of times the instructions are executed depends on when the condition is no longer false. The variable or variables that control the condition must eventually be changed by the operations, or the UNTIL loop will continue forever.

FOR loops: A set of instructions between the FOR <counter range> and the NEXT <counter> lines of code is repeated for a fixed number of loops that depends on the counter range. The counter is a variable that can be used in operations in the loop, but the value of the counter is not changed by anything in the loop besides the NEXT <counter> statement.

GOTO: A GOTO operation moves the program to a number designator elsewhere on the program. The GOTO statement has fallen from favor and is avoided whenever possible in structured programming.

Example

A computer structured programming segment contains the following program segment.

```
Set G = 1 and X = 0
DO WHILE G ≤ 5
    X = G*X + 1
    G = X
ENDWHILE
```

What is the value of G after the segment is executed?

(A) 5

(B) 26

(C) 63

(D) The loop never ends.

Solution

The first execution of the WHILE loop results in

$$G = (1)(0) + 1 = 1$$
$$X = 1$$

The second execution of the WHILE loop results in

$$G = (1)(1) + 1 = 2$$
$$X = 2$$

The third execution of the WHILE loop results in

$$G = (2)(2) + 1 = 5$$
$$X = 5$$

The WHILE condition is still satisfied, so the instruction is executed a fourth time.

$$G = (5)(5) + 1 = 26$$
$$X = 26$$

The answer is (B).

9. HIERARCHY OF OPERATIONS

Operations in an arithmetic statement are performed in the order of exponentiation first, multiplication and division second, and addition and subtraction third. In the event that there are two consecutive operations with the same hierarchy (e.g., a multiplication followed by a division), the operations are performed in the order encountered, normally left to right (except for exponentiation, which is right to left).[1] Parentheses can modify this order; operations within parentheses are always evaluated before operations outside. If nested parentheses are present in an expression, the expression is evaluated outward starting from the innermost pair.

Example

Based on standard processing and hierarchy of operations, what would most nearly be the result of the following calculation?

$$459 + 306\ /\ (6*5\hat{}2 + 3)$$

(A) 3

(B) 5

(C) 459

(D) 461

Solution

The standard mathematical processing hierarchy is represented by the acronym "BODMAS": Brackets (parentheses), Order (exponentiation or powers), Division and

[1]In most implementations, a statement will be scanned from left to right. Once a left-to-right scan is complete, some implementations then scan from right to left; others return to the equals sign and start a second left-to-right scan. Parentheses should be used to define the intended order of operations.

Computational Tools

Multiplication (evaluated left to right), and Addition and Subtraction (evaluated left to right).

Evaluate the quantity in the parentheses first. Within the parentheses, evaluate the powered quantity first: 5^2 = 25. There are no more powered quantities within the parentheses. Evaluate division and multiplication within the parentheses from left to right: 6*25 = 150. Evaluate addition and subtraction within the parentheses: 150 + 3 = 153. This is the value of the parenthetical expression. The original expression is now 459 + 306/153. There are no brackets. There are no powers. Evaluate the division: 306/153 = 2. Evaluate the addition: 459 + 2 = 461.

The answer is (D).

10. SIMULATORS

A *simulator* is a computer program that replicates a real world system. A *digital model* is used in the simulation and incorporates variables representing the physical characteristics and behavioral properties of the system to be simulated (i.e., is a representation of the system itself). The simulator operates a model over a period of time to gain a detailed understanding of the system's characteristics and dynamics. Simulation makes it possible to test and examine multiple design scenarios before selecting or finalizing a design. Simulators can explore various design alternatives that may not be safe, feasible, or economically possible in real life.

Examples of simulators include the response of buildings to seismic forces, stormwater flowing through a drainage structure during a large storm event, automobiles merging onto a freeway during rush hour, and so on. In addition to testing various designs, simulators are used for training, education, and even entertainment.

11. SPREADSHEETS

Spreadsheet application programs (often referred to as *spreadsheets*) are computer programs that provide a table of values arranged in rows and columns and that permit each value to have a predefined relationship to the other values. If one value is changed, the program will also change other related values.

In a spreadsheet, the items are arranged in rows and columns. The rows are typically assigned with numbers (1, 2, 3, ...) along the vertical axis, and the columns are assigned with letters (A, B, C, ...), as is shown in Fig. 59.2.

A *cell* is a particular element of the table identified by an address that is dependent on the row and column assignments of the cell. For example, the address of the shaded cell in Fig. 59.2 is E3. A cell may contain a number, a formula relating its value to another cell or cells, or a label (usually descriptive text).

Figure 59.2 *Typical Spreadsheet Cell Assignments*

	A	B	C	D	E	F	G
1							
2							
3							
4							
5							
6							

When the contents of one cell are used for a calculation in another cell, the address of the cell being used must be referenced so the program knows what number to use.

When the value of a cell containing a formula is calculated, any cell addresses within the formula are calculated as the values of their respective cells. Cell addressing can be handled one of two ways: relative addressing or absolute addressing. When a cell is copied using *relative addressing*, the row and cell references will be changed automatically. *Absolute addressing* means that when a cell is copied, the row and cell references remain unchanged. Relative addressing is the default. Absolute addressing is indicated by a "$" symbol placed in front of the row or column reference (e.g., $C1 or C$1).

An *absolute cell reference* identifies a particular cell and will have a "$" before both the row and column designators. For example, A1 identifies the cell in the first column and first row, A3 identifies the cell in the first column and third row, and C1 identifies the cell in the third column and first row, regardless of the cell the reference is located in. If the absolute cell reference is copied and pasted into another cell, it continues referring to the exact same cell.

An *absolute column, relative row cell reference* has an absolute column reference (indicated with a "$") and a relative row reference. For example, the cell reference $A1 depends on what row it is entered in; if it is copied into a cell in the final row, it really refers to a cell that is in the first column in the final row.

If this reference is copied and pasted into a cell in the third row, the reference will become $A3. Similarly, a reference of $A3 in the second row refers to a cell in the first column one row below the current row. If this reference is copied and pasted into the third row, it becomes $A4.

A *relative column, absolute row cell reference* has a "$" on the row designator, and the column reference depends on the column it is entered in. For example, a cell reference of B$4 in the fourth column (column D) refers to a cell two columns to the left in the fourth row. If this reference is copied and pasted into the sixth column (column F), it becomes D$4.

A cell reference that does not include a "$" is entirely dependent on the cell in which is it located. For example, a cell reference to B4 in the cell C2 refers to a cell that is one column to the left and two rows below. If this reference is copied and pasted into cell D3, it becomes C5.

The syntax for calculations with rows, columns, or blocks of cells can differ from one brand of spreadsheet to another.

Cells can be called out in square or rectangular blocks, usually for a SUM function. The difference between the row and column designations in the call will define the block. For example, SUM(A1:A3) says to sum the cells A1, A2, and A3; SUM(D3:D5) says to sum the cells D3, D4, and D5; and SUM(B2:C4) says to sum cells B2, B3, B4, C2, C3, and C4.

Example

The cells in a spreadsheet are initialized as shown. The formula B1 + \$A\$1*A2 is entered into cell B2 and then copied into cells B3 and B4.

	A	B		
1	3	111		
2	4			
3	5			
4	6			
5				

What value will be displayed in cell B4?

(A) 123

(B) 147

(C) 156

(D) 173

Solution

When the formula is copied into cells B3 and B4, the relative references will be updated. The resulting spreadsheet (with formulas displayed) should look like

	A	B	
1	3	111	
2	4	B1 + A1*A2	
3	5	B2 + A1*A3	
4	6	B3 + A1*A4	
5			

$$\begin{aligned} B4 &= B3 + (A1)(A4) \\ &= B2 + (A1)(A3) + (A1)(A4) \\ &= B1 + (A1)(A2) + (A1)(A3) + (A1)(A4) \\ &= 111 + (3)(4) + (3)(5) + (3)(6) \\ &= 156 \end{aligned}$$

The answer is (C).

12. SPREADSHEETS IN ENGINEERING

Spreadsheets are powerful, widely used computational tools in various engineering disciplines. Their popularity is often attributed to widespread availability, ease of use, and robust functionality. A spreadsheet can be used to collect data, identify and analyze trends within a data set, perform statistical calculations, quickly execute a series of interconnected calculations, and determine the effect of one variable change on other related variables, among other uses.

Applications include calculating open channel flow, scheduling construction activities, determining distribution of moments on a beam, calculating design parameters for a batch reactor, calculating the efficiency of a pump or motor, graphing the relationship between voltages and currents in circuit analysis, and determining costs over time in cost estimation.

Some benefits of using spreadsheets for engineering calculations and applications include the ability to reproduce calculations faithfully, to save and later review input and results, to compare results based on different input values, and to plot results graphically. Drawbacks include the difficulty of programming complex calculations (though many spreadsheets are commercially available), the lack of transparency for how results are calculated, and the resulting challenge of debugging inaccurate results. Spreadsheet calculations can be used in iterative design calculations, but results may need to be validated by some other means.

13. FIELDS, RECORDS, AND FILE TYPES

A collection of *fields* is known as a *record*. For example, name, age, and address might be fields in a personnel record. Groups of records are stored in a *file*.

A *sequential file* structure (typical of data on magnetic tape) contains consecutive records and must be read starting at the beginning. An *indexed sequential file* is one for which a separate index file (see Sec. 59.14) is maintained to help locate records.

With a *random* (*direct access*) *file structure*, any record can be accessed without starting at the beginning of the file.

14. FILE INDEXING

It is usually inefficient to place the records of an entire file in order. (A good example is a mailing list with thousands of names. It is more efficient to keep the names in the order of entry than to sort the list each time names are added or deleted.) Indexing is a means of specifying the order of the records without actually changing the order of those records.

An *index* (*key* or *keyword*) file is analogous to the index at the end of this book. It is an ordered list of items with references to the complete record. One field in the data record is selected as the *key field* (*record index*). More than one field can be indexed. However, each field will require its own index file. The sorted keys are usually kept in a file separate from the data file. One of the standard search techniques is used to find a specific key.

15. SORTING

Sorting routines place data in ascending or descending numerical or alphabetical order.

With the method of *successive minima*, a list is searched sequentially until the smallest element is found and brought to the top of the list. That element is then skipped, and the remaining elements are searched for the smallest element, which, when found, is placed after the previous minimum, and so on. A total of $n(n-1)/2$ comparisons will be required. When n is large, $n^2/2$ is sometimes given as the number of comparisons.

In a *bubble sort*, each element in the list is compared with the element immediately following it. If the first element is larger, the positions of the two elements are reversed (swapped). In effect, the smaller element "bubbles" to the top of the list. The comparisons continue to be made until the bottom of the list is reached. If no swaps are made in a pass, the list is sorted. A total of approximately $n^2/2$ comparisons are needed, on the average, to sort a list in this manner. This is the same as for the successive minima approach. However, swapping occurs more frequently in the bubble sort, slowing it down.

In an *insertion sort*, the elements are ordered by rewriting them in the proper sequence. After the proper position of an element is found, all elements below that position are bumped down one place in the sequence. The resulting vacancy is filled by the inserted element. At worst, approximately $n^2/2$ comparisons will be required. On average, there will be approximately $n^2/4$ comparisons.

Disregarding the number of swaps, the number of comparisons required by the successive minima, bubble, and insertion sorts is on the order of n^2. When n is large, these methods are too slow. The *quicksort* is more complex but reduces the average number of comparisons (with random data) to approximately $n \times \log n/\log 2$, generally considered as being on the order of $n \log n$. (However, the quicksort falters, in speed, when the elements are in near-perfect order.) The maximum number of comparisons for a *heap sort* is $n \times \log_2 n = n \times \log n/\log 2$, but it is likely that even fewer comparisons will be needed.

16. SEARCHING

If a group of records (i.e., a list) is randomly organized, a particular element in the list can be found only by a *linear search* (*sequential search*). At best, only one comparison and, at worst, n comparisons will be required to find something (an event known as a *hit*) in a list of n elements. The average is $n/2$ comparisons, described as being on the order of n. (The term *probing* is synonymous with *searching*.)

If the records are in ascending or descending order, a binary search will be superior. (A binary search is unrelated to a binary tree. A binary tree structure (see Sec. 59.18) greatly reduces search time but does not use a sorted list.) The search begins by looking at the middle element in the list. If the middle element is the sought-for element, the search is over. If not, half the list can be disregarded in further searching since elements in that portion will be either too large or too small. The middle element in the remaining part of the list is investigated, and the procedure continues until a hit occurs or the list is exhausted. The maximum number of required comparisons in a list of n elements will be $\log_2 n = \log n/\log 2$ (i.e., on the order of $\log n$).

Example

A binary search is used to determine if a particular test number is present in a sequence of 101 integers derived from a Fibonacci sequence. The numbers are sorted in decreasing magnitude. The smallest number in the sequence is 144. What is the maximum number of comparisons needed to determine if the test number is present?

(A) 6

(B) 7

(C) 11

(D) 49

Solution

A binary search looks at half of an active list at a time. Comparing continues until the active list contains only a single value. The index of the middle integer in a sorted list of 101 integers is at index $(1+101)/2 = 51$. The first comparison is to integer 51.

The active list now contains 50 integers. The second comparison is to the middle integer at index $(1+50)/2 = 25$. The active list now contains a maximum of 25 items. The third comparison is to the integer at index $(26+50)/2 = 38$. The active list now contains 12 items. The fourth comparison is to the integer at index $(26+37)/2 = 31$. The active list now contains a maximum of 6 items. The fifth comparison is to the integer at index $(32+37)/2 = 35$. The active list now contains a maximum of 3 items. The sixth comparison is to the integer at index $(32+34)/2=33$. The active list now contains 1 item, which is used in the seventh and final comparison.

Since $\log_2 128 = 7$ (i.e., $2^7 = 128$), up to 128 integers can be searched with 7 comparisons.

The answer is (B).

17. HASHING

An index file is not needed if the record number (i.e., the storage location for a read or write operation) can be calculated directly from the key, a technique known as *hashing*. The procedure by which a numeric or nonnumeric key (e.g., a last name) is converted into a record number is called the *hashing function* or *hashing algorithm*. Most hashing algorithms use a remaindering modulus—the remainder after dividing the key by the number of records, n, in the list. Excellent results are obtained if n is a prime number; poor results occur if n is a power of 2. (Finding a record in this manner requires it to have been written in a location determined by the same hashing routine.)

Not all hashed record numbers will be valid. A *collision* occurs when an attempt is made to use a record number that is already in use. Chaining, linear probing, and double hashing are techniques used to resolve such collisions.

18. DATABASE STRUCTURES

Databases can be implemented as indexed files, linked lists, and tree structures; in all three cases, the records are written and remain in the order of entry.

An *indexed file* such as that shown in Fig. 59.3 keeps the data in one file and maintains separate index files (usually in sorted order) for each key field. The index file must be recreated each time records are added to the field. A *flat file* has only one key field by which records can be located. Searching techniques (see Sec. 59.16) are used to locate a particular record. In a *linked list* (*threaded list*), each record has an associated *pointer* (usually a record number or memory address) to the next record in key sequence. Only two pointers are changed when a record is added or deleted. Generally, a linear search following the links through the records is used. Figure 59.4(a) shows an example of a linked list structure.

Pointers are also used in *tree structures*. Each record has one or more pointers to other records that meet certain criteria. In a binary tree structure, each record has two pointers—usually to records that are lower and higher, respectively, in key sequence. In general, records in a tree structure are referred to as *nodes*. The first record in a file is called the *root node*. A particular node will have one node above it (the *parent* or *ancestor*) and one or more nodes below it (the *daughters* or *offspring*). Records are found in a tree by starting at the root node and moving sequentially according to the tree structure. The number of comparisons required to find a particular element is $1 + (\log n/\log 2)$, which is on the order of $\log n$. Figure 59.4(b) shows an example of a binary tree structure.

Figure 59.3 *Key and Data Files*

key file

key	record
ADAMS	3
JONES	2
SMITH	1
THOMAS	4

data file

record	last name	first name	age
1	SMITH	JOHN	27
2	JONES	WANDA	39
3	ADAMS	HENRY	58
4	THOMAS	SUSAN	18

Figure 59.4 *Database Structures*

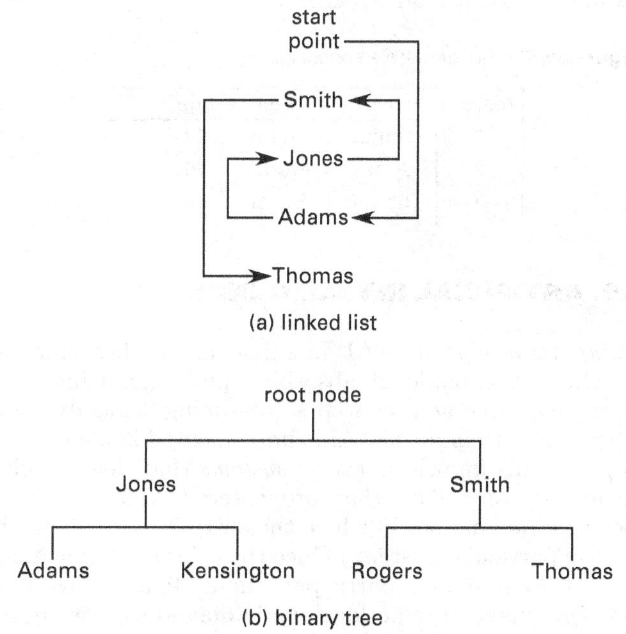

Computational Tools

19. HIERARCHICAL AND RELATIONAL DATA STRUCTURES

A *hierarchical database* contains records in an organized, structured format. Records are organized according to one or more of the indexing schemes. However, each field within a record is not normally accessible. Figure 59.5 shows an example of a hierarchical structure.

Figure 59.5 *A Hierarchical Personnel File*

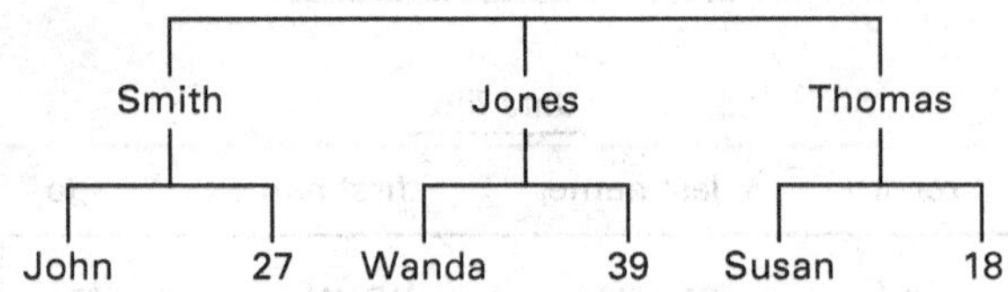

A *relational database* stores all information in the equivalent of a matrix. Nothing else (no index files, pointers, etc.) is needed to find, read, edit, or select information. Any information can be accessed directly by referring to the field name or field value. Figure 59.6 shows an example of relational structure.

Figure 59.6 *A Relational Personnel File*

rec. no.	last	first	age	
1	Smith	John	27	
2	Jones	Wanda	39	
3	Thomas	Susan	18	

20. ARTIFICIAL INTELLIGENCE

Artificial intelligence (AI) in a machine implies that the machine is capable of absorbing and organizing new data, learning new concepts, reasoning logically, and responding to inquiries. AI is implemented in a category of programs known as *expert systems* that "learn" rules from sets of events that are entered whenever they occur. (The manner in which the entry is made depends on the particular system.) Once the rules are learned, an expert system can participate in a dialogue to give advice, make predictions and diagnoses, or draw conclusions.

Topic XV: Engineering Economics

Chapter

60 Engineering Economics

Nomenclature

A	annual amount or annual value	\$
B	present worth of all benefits	\$
BV	book value	\$
C	initial cost, or present worth of all costs	\$
d	interest rate per period adjusted for inflation	decimal or %
D	depreciation	\$
f	inflation rate per period	decimal or %
F	future worth (future value)	\$
G	uniform gradient amount	\$
i	interest rate per period	decimal or %
j	number of compounding periods or years	–
m	number of compounding periods per year	–
n	total number of compounding periods or years	–
P	present worth (present value)	\$
r	nominal rate per year (rate per annum)	decimal or %
S	salvage value	\$
t	time	yr

Subscripts

0	initial
e	effective
j	jth year or period
n	final year or period

1. INTRODUCTION

In its simplest form, an *engineering economic analysis* is a study of the desirability of making an investment.[1] The decision-making principles in this chapter can be applied by individuals as well as by companies. The nature of the spending opportunity or industry is not important. Farming equipment, personal investments, and multimillion dollar factory improvements can all be evaluated using the same principles.

Similarly, the applicable principles are insensitive to the monetary units. Although *dollars* are used in this chapter, it is equally convenient to use pounds, yen, or euros.

Finally, this chapter may give the impression that investment alternatives must be evaluated on a year-by-year basis. Actually, the *effective period* can be defined as a day, month, century, or any other convenient period of time.

2. YEAR-END AND OTHER CONVENTIONS

Except in short-term transactions, it is simpler to assume that all receipts and disbursements (cash flows) take place at the end of the year in which they occur.[2] This is known as the *year-end convention*. The exceptions to the year-end convention are initial project cost (purchase cost), trade-in allowance, and other cash flows that are associated with the inception of the project at $t = 0$.

[1]This subject is also known as *engineering economics* and *engineering economy*. There is very little, if any, true economics in this subject.
[2]A *short-term transaction* typically has a lifetime of five years or less and has payments or compounding that are more frequent than once per year.

On the surface, such a convention appears grossly inappropriate since repair expenses, interest payments, corporate taxes, and so on seldom coincide with the end of a year. However, the convention greatly simplifies engineering economic analysis problems, and it is justifiable on the basis that the increased precision associated with a more rigorous analysis is not warranted (due to the numerous other simplifying assumptions and estimates initially made in the problem).

There are various established procedures, known as *rules* or *conventions*, imposed by the Internal Revenue Service on U.S. taxpayers. An example is the *half-year rule*, which permits only half of the first-year depreciation to be taken in the first year of an asset's life when certain methods of depreciation are used. These rules are subject to constantly changing legislation and are not covered in this book. The implementation of such rules is outside the scope of engineering practice and is best left to accounting professionals.

3. CASH FLOW

The sums of money recorded as receipts or disbursements in a project's financial records are called *cash flows*. Examples of cash flows are deposits to a bank, dividend interest payments, loan payments, operating and maintenance costs, and trade-in salvage on equipment. Whether the cash flow is considered to be a receipt or disbursement depends on the project under consideration. For example, interest paid on a sum in a bank account will be considered a disbursement to the bank and a receipt to the holder of the account.

Because of the time value of money, the timing of cash flows over the life of a project is an important factor. Although they are not always necessary in simple problems (and they are often unwieldy in very complex problems), *cash flow diagrams* can be drawn to help visualize and simplify problems that have diverse receipts and disbursements.

The following conventions are used to standardize cash flow diagrams.

- The horizontal (time) axis is marked off in equal increments, one per period, up to the duration of the project.
- *Receipts* are represented by arrows directed upward. *Disbursements* are represented by arrows directed downward. The arrow length is approximately proportional to the magnitude of the cash flow.
- Two or more transfers in the same period are placed end to end, and these may be combined.
- Expenses incurred before $t=0$ are called *sunk costs.* Sunk costs are not relevant to the problem unless they have tax consequences in an after-tax analysis.

For example, consider a mechanical device that will cost \$20,000 when purchased. Maintenance will cost \$1000 each year. The device will generate revenues of \$5000 each year for five years, after which the salvage value is expected to be \$7000. The cash flow diagram is shown in Fig. 60.1(a), and a simplified version is shown in Fig. 60.1(b).

Figure 60.1 *Cash Flow Diagrams*

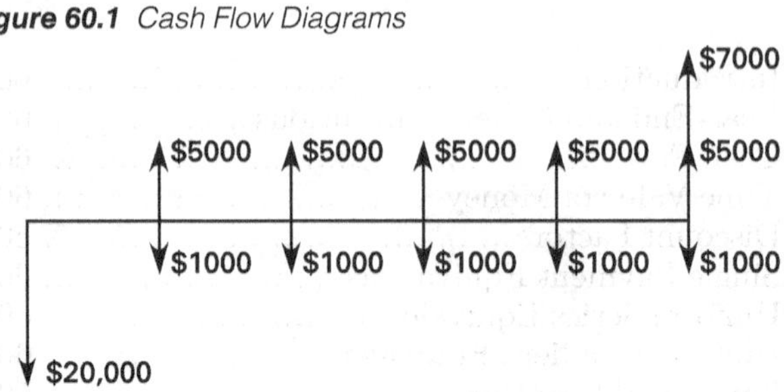

(a) cash flow diagram

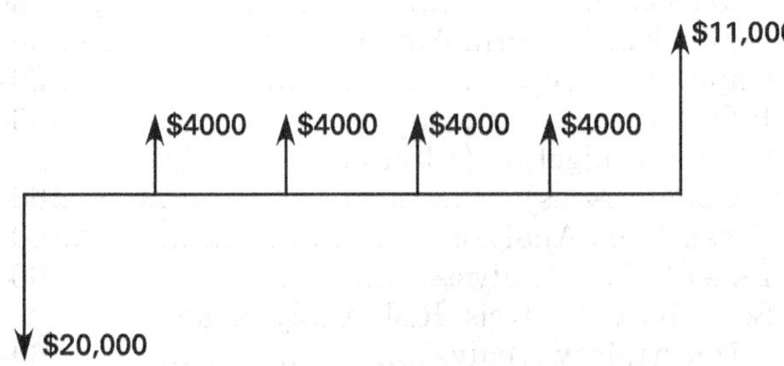

(b) simplified cash flow diagram

In order to evaluate a real-world project, it is necessary to present the project's cash flows in terms of standard cash flows that can be handled by engineering economic analysis techniques. The standard cash flows are single payment cash flow, uniform series cash flow, and gradient series cash flow.

A *single payment cash flow* can occur at the beginning of the time line (designated as $t=0$), at the end of the time line (designated as $t=n$), or at any time in between.

The *uniform series cash flow*, illustrated in Fig. 60.2, consists of a series of equal transactions starting at $t=1$ and ending at $t=n$. The symbol A (representing an *annual amount*) is typically given to the magnitude of each individual cash flow.

Figure 60.2 *Uniform Series*

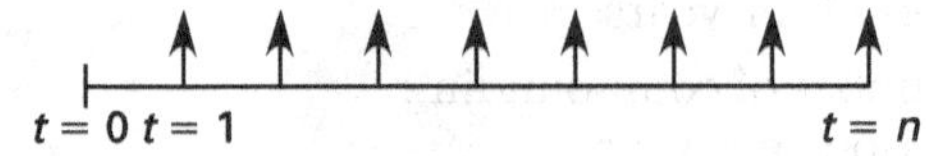

Notice that the cash flows do not begin at the beginning of a year (i.e., the year 1 cash flow is at $t=1$, not $t=0$). This convention has been established to accommodate the timing of annual maintenance and other cash flows for which the *year-end convention* is applicable. The year-end convention assumes that all receipts and disbursements take place at the end of the year in which they occur. The exceptions to the year-end convention are *initial project cost* (purchase cost), *trade-in allowance*, and other cash flows that are associated with the inception of the project at $t=0$.

The *gradient series cash flow*, illustrated in Fig. 60.3, starts with a cash flow (typically given the symbol G) at $t=2$ and increases by G each year until $t=n$, at which time the final cash flow is $(n-1)G$. The value of the gradient at $t=1$ is zero.

Figure 60.3 *Gradient Series*

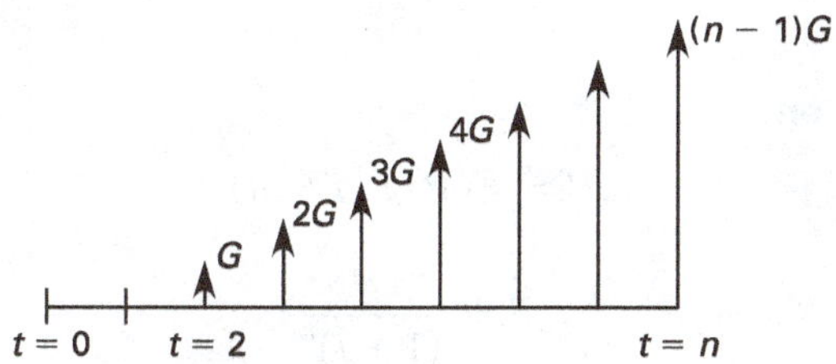

4. TIME VALUE OF MONEY

Consider \$100 placed in a bank account that pays 5% effective annual interest at the end of each year. After the first year, the account will have grown to \$105. After the second year, the account will have grown to \$110.25.

The fact that \$100 grows to \$105 in one year at 5% annual interest is an example of the *time value of money* principle. This principle states that funds placed in a secure investment will increase in value in a way that depends on the elapsed time and the interest rate.

The interest rate that is used in calculations is known as the *effective interest rate*. If compounding is once a year, it is known as the *effective annual interest rate*. However, effective quarterly, monthly, or daily interest rates are also used.

5. DISCOUNT FACTORS

Assume that there will be no need for money during the next two years, and any money received will immediately go into an account and earn a 5% effective annual interest rate. Which of the following options would be more desirable?

option a: receive \$100 now

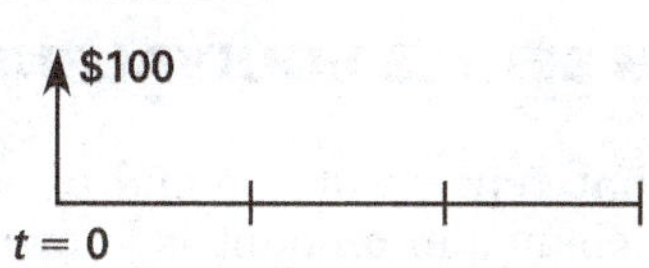

option b: receive \$105 in one year

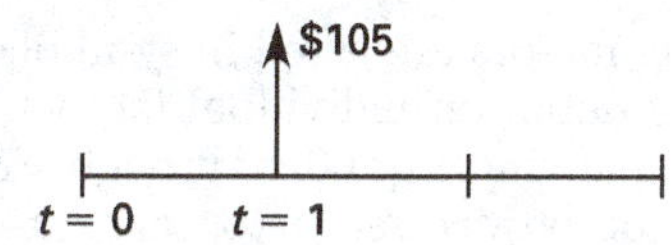

option c: receive \$110.25 in two years

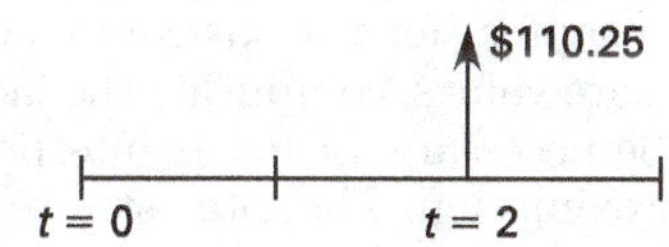

None of the options is superior under the assumptions given. If the first option is chosen, \$100 will be immediately placed into a 5% account, and in two years the account will have grown to \$110.25. In fact, the account will contain \$110.25 at the end of two years regardless of which option is chosen. Therefore, these alternatives are said to be *equivalent*.

The three options are equivalent only for money earning a 5% effective annual interest rate. If a higher interest rate can be obtained, then the first option will yield the most money after two years. So, equivalence depends on the interest rate, and an alternative that is acceptable to one decision maker may be unacceptable to another who invests at a higher rate. The procedure for determining the equivalent amount is known as *discounting*.

Table 60.1: Discount Factors

Table 60.1 *Discount Factors for Discrete Compounding*

factor name	converts	symbol	formula
single payment compound amount	P to F	$(F/P, i\%, n)$	$(1+i)^n$
single payment present worth	F to P	$(P/F, i\%, n)$	$(1+i)^{-n}$
uniform series sinking fund	F to A	$(A/F, i\%, n)$	$\frac{i}{(1+i)^n-1}$
capital recovery	P to A	$(A/P, i\%, n)$	$\frac{i(1+i)^n}{(1+i)^n-1}$
uniform series compound amount	A to F	$(F/A, i\%, n)$	$\frac{(1+i)^n-1}{i}$
uniform series present worth	A to P	$(P/A, i\%, n)$	$\frac{(1+i)^n-1}{i(1+i)^n}$
uniform gradient present worth	G to P	$(P/G, i\%, n)$	$\frac{(1+i)^n-1}{i^2(1+i)^n}-\frac{n}{i(1+i)^n}$
uniform gradient future worth*	G to F	$(F/G, i\%, n)$	$\frac{(1+i)^n-1}{i^2}-\frac{n}{i}$
uniform gradient uniform series	G to A	$(A/G, i\%, n)$	$\frac{1}{i}-\frac{n}{(1+i)^n-1}$

*See Eq. 60.9.

Description

The discounting factors are listed in symbolic and formula form. For more detail on individual factors, see the commentary accompanying Eq. 60.1 through Eq. 60.10. Normally, it will not be necessary to calculate factors from these formulas. Values of these cash flow (discounting) factors are tabulated in Table 60.4 through Table 60.13 (at the end of this chapter) for various combinations of i and n. For intermediate values, computing the factors from the formulas may be necessary, or linear interpolation can be used as an approximation. The interest rate used must be the effective rate per period for all discounting factor formulas. The basis of the rate (annually, monthly, etc.) must agree with the type of period used to count n. It would be incorrect to use an effective annual interest rate if n was the number of compounding periods in months.

6. SINGLE PAYMENT EQUIVALENCE

The equivalent future amount, F, at $t=n$, of any *present amount*, P, at $t=0$ is called the *future worth*. The equivalence of any future amount to any present amount is called the *present worth*. Compound amount factors may be used to convert from a known present amount to a future worth or vice versa.

Equation 60.1: Single Payment Future Worth

$$F = P(1+i)^n \qquad 60.1$$

Variation

$$F = P(F/P, i\%, n)$$

Description

The factor $(1+i)^n$ is known as the *single payment compound amount factor*. Rather than actually writing the formula for the compound amount factor (which converts a present amount to a future amount), it is common convention to substitute the standard functional notation of $(F/P, i\%, n)$, as shown in the variation equation. This notation is interpreted as, "Find F, given P, using an interest rate of $i\%$ over n periods."

Example

A 40-year-old consulting engineer wants to set up a retirement fund to be used starting at age 65. \$20,000 is invested now at 6% compounded annually. The amount of money that will be in the fund at retirement is most nearly

(A) \$84,000

(B) \$86,000

(C) \$88,000

(D) \$92,000

Solution

Determine the future worth of \$20,000 in 25 years. From Eq. 60.1,

$$\begin{aligned} F &= P(1+i)^n = (\$20{,}000)(1+0.06)^{25} \\ &= \$85{,}837 \quad (\$86{,}000) \end{aligned}$$

The answer is (B).

Equation 60.2: Single Payment Present Worth

$$P = F(1+i)^{-n} \qquad 60.2$$

Variations

$$P = F(P/F, i\%, n)$$

$$P = \frac{F}{(1+i)^n}$$

Description

The factor $(1+i)^{-n}$ is known as the *single payment present worth factor*.

Example

\$2000 will become available on January 1 in year 8. If interest is 5%, what is most nearly the present worth of this sum on January 1 in year 1?

(A) \$1330

(B) \$1350

(C) \$1400

(D) \$1420

Solution

From January 1 in year 1 to January 1 in year 8 is seven years. From Eq. 60.2, the present worth is

$$\begin{aligned} P &= F(1+i)^{-n} = (\$2000)(1+0.05)^{-7} \\ &= \$1421 \quad (\$1420) \end{aligned}$$

The answer is (D).

7. UNIFORM SERIES EQUIVALENCE

A cash flow that repeats at the end of each year for n years without change in amount is known as an *annual amount* and is given the symbol A. (This is shown in Fig. 60.2.)

Engineering Economics

Although the equivalent value for each of the n annual amounts could be calculated and then summed, it is more expedient to use one of the uniform series factors.

Equation 60.3: Uniform Series Future Worth

$$F = A\left(\frac{(1+i)^n - 1}{i}\right) \qquad 60.3$$

Variation

$$F = A(F/A, i\%, n)$$

Description

Use the *uniform series compound amount factor* to convert from an annual amount to a future amount.

Example

\$20,000 is deposited at the end of each year into a fund earning 6% interest. At the end of ten years, the amount accumulated is most nearly

(A) \$150,000

(B) \$180,000

(C) \$260,000

(D) \$280,000

Solution

The amount accumulated at the end of ten years is

$$\begin{aligned} F &= A\left(\frac{(1+i)^n - 1}{i}\right) \\ &= (\$20{,}000)\left(\frac{(1+0.06)^{10} - 1}{0.06}\right) \\ &= \$263{,}616 \quad (\$260{,}000) \end{aligned}$$

The answer is (C).

Equation 60.4: Uniform Series Annual Value of a Sinking Fund

$$A = F\left(\frac{i}{(1+i)^n - 1}\right) \qquad 60.4$$

Variation

$$A = F(A/F, i\%, n)$$

Description

A *sinking fund* is a fund or account into which annual deposits of A are made in order to accumulate F at $t = n$ in the future. Because the annual deposit is calculated as $A = F(A/F, i\%, n)$, the (A/F) factor is known as the *sinking fund factor*.

Example

At the end of each year, an investor deposits some money into a fund earning 7% interest. The same amount is deposited each year, and after six years the account contains \$1600. The amount deposited each time is most nearly

(A) \$190

(B) \$220

(C) \$240

(D) \$250

Solution

Use the sinking fund factor from Eq. 60.4 to find the annual value.

$$\begin{aligned} A &= F\left(\frac{i}{(1+i)^n - 1}\right) = (\$1600)\left(\frac{0.07}{(1+0.07)^6 - 1}\right) \\ &= \$224 \quad (\$220) \end{aligned}$$

The answer is (B).

Equation 60.5: Uniform Series Present Worth

$$P = A\left(\frac{(1+i)^n - 1}{i(1+i)^n}\right) \qquad 60.5$$

Variation

$$P = A(P/A, i\%, n)$$

Description

An *annuity* is a series of equal payments, A, made over a period of time. Usually, it is necessary to "buy into" an investment (a bond, an insurance policy, etc.) in order to fund the annuity. In the case of an annuity that starts at the end of the first year and continues for n years, the purchase price, P, is calculated using the *uniform series present worth factor*.

Example

A sum of money is deposited into a fund at 5% interest. \$400 is withdrawn at the end of each year for nine years, leaving nothing in the fund at the end. The amount originally deposited is most nearly

(A) $2600

(B) $2800

(C) $2900

(D) $3100

Solution

Find the present worth using the present worth factor.

$$P = A\left(\frac{(1+i)^n - 1}{i(1+i)^n}\right) = (\$400)\left(\frac{(1+0.05)^9 - 1}{(0.05)(1+0.05)^9}\right)$$
$$= \$2843 \quad (\$2800)$$

The answer is (B).

Equation 60.6: Uniform Series Annual Value Using the Capital Recovery Factor

$$A = P\left(\frac{i(1+i)^n}{(1+i)^n - 1}\right) \qquad 60.6$$

Variation

$$A = P(A/P, i\%, n)$$

Description

The *capital recovery factor* is often used when comparing alternatives with different lifespans. A comparison of two possible investments on the simple basis of their present values may be misleading if, for example, one alternative has a lifespan of 11 years and the other has a lifespan of 18 years. The capital recovery factor can be used to convert the present value of each alternative into its equivalent annual value, using the assumption that each alternative will be renewed repeatedly up to the duration of the longest-lived alternative.

8. UNIFORM GRADIENT EQUIVALENCE

A common situation involves a uniformly increasing cash flow. If the cash flow has the proper form (see Fig. 60.3), its present worth can be determined by using the *uniform gradient factor*, also called the *uniform gradient present worth factor*. The uniform gradient factor, $(P/G, i\%, n)$, finds the present worth of a uniformly increasing cash flow. By definition of a uniform gradient, the cash flow starts in year 2, not in year 1. Similar factors can be used to find the cash flow's future worth and annual worth.

There are three common difficulties associated with the form of the uniform gradient. The first difficulty is that the first cash flow starts at $t=1$. This convention recognizes that annual costs, if they increase uniformly, begin with some value at $t=1$ (due to the year-end convention), but do not begin to increase until $t=2$. The tabulated values of (P/G) have been calculated to find the present worth of only the increasing part of the annual expense. The present worth of the base expense incurred at $t=1$ must be found separately with the (P/A) factor.

The second difficulty is that, even though the $(P/G, i\%, n)$ factor is used, there are only $n-1$ actual cash flows. n must be interpreted as the *period number* in which the last gradient cash flow occurs, not the number of gradient cash flows.

Finally, the sign convention used with gradient cash flows can be confusing. If an expense increases each year, the gradient will be negative, since it is an expense. If a revenue increases each year, the gradient will be positive. In most cases, the sign of the gradient depends on whether the cash flow is an expense or a revenue.

Equation 60.7: Uniform Gradient Present Worth

$$P = G\left(\frac{(1+i)^n - 1}{i^2(1+i)^n} - \frac{n}{i(1+i)^n}\right) \qquad 60.7$$

Variation

$$P = G(P/G, i\%, n)$$

Description

Equation 60.7 is used to find the present worth, P, of a cash flow that is increasing by a uniform amount, G. This factor finds the value of only the increasing portion of the cash flow; if the value at $t=1$ is anything other than zero, its present worth must be found separately and added to get the total present worth.

Equation 60.8 and Eq. 60.9: Uniform Gradient Future Worth

$$F = G\left(\frac{(1+i)^n - 1}{i^2} - \frac{n}{i}\right) \qquad 60.8$$

$$F/G = (F/A - n)/i = (F/A) \times (A/G) \qquad 60.9$$

Variation

$$F = G(F/G, i\%, n)$$

Description

Equation 60.8 is used to find the future worth, F, of a cash flow that is increasing by a uniform amount, G. This factor finds the value of only the increasing portion of the cash flow; if the value at $t=1$ is anything other than zero, its future worth must be found separately and added to get the total future worth.

Equation 60.9 shows how the future worth factor, $(F/G, i\%, n)$, is closely related to (and, therefore, can be calculated quickly from) the factor $(F/A, i\%, n)$, or from the factors $(F/A, i\%, n)$ and $(A/G, i\%, n)$.

Equation 60.10: Uniform Gradient Uniform Series Factor

$$A = G\left(\frac{1}{i} - \frac{n}{(1+i)^n - 1}\right) \quad 60.10$$

Variation

$$A = G(A/G, i\%, n)$$

Description

Equation 60.10 is used to find the equivalent annual worth, A, of a cash flow that is increasing by a uniform amount, G. This factor finds the value of only the increasing portion of the cash flow; if the value at $t=1$ is anything other than zero, its annual worth must be found separately and added to get the total annual worth.

Example

The maintenance cost on a house is expected to be \$1000 the first year and to increase \$500 per year after that. Assuming an interest rate of 6% compounded annually, the maintenance cost over 10 years is most nearly equivalent to an annual maintenance cost of

(A) \$1900

(B) \$3000

(C) \$3500

(D) \$3800

Solution

Use the uniform gradient uniform series factor to determine the effective annual cost, remembering that the first year's cost, $A_1 = \$1000$, must be added separately.

The annual increase is $G = \$500$. The effective annual cost of the increasing portion of the costs alone is

$$\begin{aligned} A &= G\left(\frac{1}{i} - \frac{n}{(1+i)^n - 1}\right) \\ &= (\$500)\left(\frac{1}{0.06} - \frac{10}{(1+0.06)^{10} - 1}\right) \\ &= \$2011 \end{aligned}$$

The total effective annual cost is

$$\begin{aligned} A_{\text{total}} &= A_1 + A \\ &= \$1000 + \$2011 \\ &= \$3011 \quad (\$3000) \end{aligned}$$

The answer is (B).

9. FUNCTIONAL NOTATION

There are several ways of remembering what the functional notation means. One method of remembering which factor should be used is to think of the factors as *conditional probabilities.* The conditional probability of event A given that event B has occurred is written as $P\{A|B\}$, where the given event comes after the vertical bar. In the standard notational form of discounting factors, the given amount is similarly placed after the slash. The desired factor (i.e., A) comes before the slash. (F/P) would be a factor to find F given P.

Another method of remembering the notation is to interpret the factors algebraically. The (F/P) factor could be thought of as the fraction F/P. The numerical values of the discounting factors are consistent with this algebraic manipulation. The (F/A) factor could be calculated as $(F/P)(P/A)$. This consistent relationship can be used to calculate other factors that might be occasionally needed, such as (F/G) or (G/P).

10. NONANNUAL COMPOUNDING

If \$100 is invested at 5%, it will grow to \$105 in one year. If only the original principal accrues interest, the interest is known as *simple interest*, and the account will grow to \$110 in the second year, \$115 in the third year, and so on. Simple interest is rarely encountered in engineering economic analyses.

More often, both the principal and the interest earned accrue interest, and this is known as *compound interest.* If the account is compounded yearly, then during the second year, 5% interest continues to be accrued, but on \$105, not \$100, so the value at year end will be \$110.25. The value after the third year will be \$115.76, and so on.

The interest rate used in the discount factor formulas is the *interest rate per period*, i (called the *yield* by banks). If the interest period is one year (i.e., the interest is compounded yearly), then the interest rate per period, i, is equal to the *effective annual interest rate*, i_e. The effective annual interest rate is the rate that would yield the same accrued interest at the end of the year if the account were compounded yearly.

The term *nominal interest rate*, r (*rate per annum*), is encountered when compounding is more than once per year. The nominal rate does not include the effect of compounding and is not the same as the effective annual interest rate.

Equation 60.11: Effective Annual Interest Rate

$$i_e = \left(1 + \frac{r}{m}\right)^m - 1 \qquad 60.11$$

Description

The effective annual interest rate, i_e, can be calculated if the nominal rate, r, and the number of compounding periods per year, m, are known. If there are m compounding periods during the year (two for semiannual compounding, four for quarterly compounding, twelve for monthly compounding, etc.), the *effective interest rate per period*, i, is r/m. The effective annual interest rate, i_e, can be calculated from the effective interest rate per period by using Eq. 60.11.

Example

Money is invested at 5% per annum and compounded quarterly. The effective annual interest rate is most nearly

(A) 5.1%

(B) 5.2%

(C) 5.4%

(D) 5.5%

Solution

The rate per annum is the nominal interest rate. Use Eq. 60.11 to calculate the effective annual interest rate.

$$\begin{aligned} i_e &= \left(1 + \frac{r}{m}\right)^m - 1 = \left(1 + \frac{0.05}{4}\right)^4 - 1 \\ &= 0.05095 \quad (5.1\%) \end{aligned}$$

The answer is (A).

11. DEPRECIATION

Tax regulations do not generally allow the purchase price of an asset or other property to be treated as a single deductible expense in the year of purchase. Rather, the cost must be divided into portions, and these artificial expenses are spread out over a number of years. The portion of the cost that is allocated to a given year is called the *depreciation*, and the period of years over which these portions are spread out is called the *depreciation period* (also known as the *service life*).

When depreciation is included in an engineering economic analysis problem, it will increase the asset's after-tax present worth (profitability). The larger the depreciation is, the greater the profitability will be. For this reason, it is desirable to make the depreciation in each year as large as possible and to accelerate the process of depreciation as much as possible.

The *depreciation basis* of an asset is that part of the asset's purchase price that is spread over the depreciation period. The depreciation basis may or may not be equal to the purchase price.

A common depreciation basis is the difference between the purchase price and the expected salvage value at the end of the depreciation period (i.e., depreciation basis = $C - S_n$).

In the *sum-of-the-years' digits* (SOYD) method of depreciation, the digits from 1 to n inclusive are added together. An easy way to calculate this sum is the formula

$$\sum_{j=1}^{n} j = \frac{n(n+1)}{2}$$

The depreciation in year j is found from

$$D_j = \frac{n + 1 - j}{\sum_{j=1}^{n} j}(C - S_n)$$

Using this method, the depreciation from one year to the next decreases by a constant amount.

Equation 60.12: Straight Line Method

$$D_j = \frac{C - S_n}{n} \qquad 60.12$$

Description

With the *straight line method*, depreciation is the same each year. The depreciation basis $(C - S_n)$ is divided uniformly among all of the n years in the depreciation period.

Engineering Economics

Example

A computer will be purchased at \$3900. The expected salvage value at the end of its service life of 10 years is \$1800. Using the straight line method, the annual depreciation for this computer is most nearly

(A) \$210

(B) \$230

(C) \$260

(D) \$280

Solution

From Eq. 60.12, the annual depreciation using the straight line method is

$$D_j = \frac{C - S_n}{n} = \frac{\$3900 - \$1800}{10} = \$210$$

The answer is (A).

Equation 60.13: Modified Accelerated Cost Recovery System (MACRS)

$$D_j = (\text{factor})C \qquad 60.13$$

Values

Table 60.2 Representative MACRS Depreciation Factors

	recovery period (years)			
	3	5	7	10
year j	recovery rate (percent)			
1	33.33	20.00	14.29	10.00
2	44.45	32.00	24.49	18.00
3	14.81	19.20	17.49	14.40
4	7.41	11.52	12.49	11.52
5		11.52	8.93	9.22
6		5.76	8.92	7.37
7			8.93	6.55
8			4.46	6.55
9				6.56
10				6.55
11				3.28

Description

In the United States, property placed into service in 1981 and thereafter must use the *Accelerated Cost Recovery System* (ACRS), and property placed into service after 1986 must use the *Modified Accelerated Cost Recovery System* (MACRS) or another statutory method. Other methods, including the straight line method, cannot be used except in special cases.

Under ACRS and MACRS, the cost recovery amount in a particular year is calculated by multiplying the initial cost of the asset by a factor. (This initial cost is not reduced by the asset's salvage value.) The factor to be used varies depending on the year and on the total number of years in the asset's cost recovery period. These factors are subject to continuing legislation changes. Representative MACRS depreciation factors are shown in Table 60.2.

Example

A groundwater treatment system costs \$2,500,000. It is expected to operate for a total of 130,000 hours over a period of 10 years, and then have a \$250,000 salvage value. During the system's first year in service, it is operated for 6500 hours. Using the MACRS method, its depreciation in the third year is most nearly

(A) \$160,000

(B) \$250,000

(C) \$360,000

(D) \$830,000

Solution

MACRS depreciation depends only on the original cost, not on the salvage cost or hours of operation. From Table 60.2, the factor for the third year of a 10-year recovery period is 14.40%. From Eq. 60.13,

$$D_j = (\text{factor})C$$
$$D_3 = (0.1440)(\$2{,}500{,}000) = \$360{,}000$$

The answer is (C).

12. BOOK VALUE

The difference between the original purchase price and the accumulated depreciation is known as the *book value*, BV. The book value is initially equal to the purchase price, and at the end of each year it is reduced by that year's depreciation.

Figure 60.4 compares how the ratio of book value to initial cost changes over time under the straight line and the MACRS methods.

Equation 60.14: Book Value

$$BV = \text{initial cost} - \sum D_j \qquad 60.14$$

Description

In Eq. 60.14, BV is the book value at the end (not the beginning) of the jth year—that is, after j years of depreciation have been subtracted from the original purchase price.

Example

A machine initially costing \$25,000 will have a salvage value of \$6000 after five years. Using MACRS depreciation, its book value after the third year will be most nearly

(A) \$5500

(B) \$7200

(C) \$10,000

(D) \$14,000

Solution

Book value is the initial cost less the accumulated depreciation; the salvage value is disregarded. Use Eq. 60.14 and the MACRS factors for a five-year recovery period.

$$\begin{aligned}
\text{BV} &= \text{initial cost} - \sum D_j \\
&= \text{initial cost} - (D_1 + D_2 + D_3) \\
&= \text{initial cost} - \left(\begin{array}{r}(\text{factor}_1)(\text{initial cost}) \\ +(\text{factor}_2)(\text{initial cost}) \\ +(\text{factor}_3)(\text{initial cost})\end{array}\right) \\
&= (1 - \text{factor}_1 - \text{factor}_2 - \text{factor}_3)(\text{initial cost}) \\
&= (1 - 0.20 - 0.32 - 0.192)(\$25{,}000) \\
&= \$7200
\end{aligned}$$

The answer is (B).

13. EQUIVALENT UNIFORM ANNUAL COST

Alternatives with different lifespans will generally be compared by way of *equivalent uniform annual cost*, or EUAC. An EUAC is the annual amount that is equivalent to all of the cash flows in the alternative.

The EUAC differs in sign from all of the other cash flows. Costs and expenses expressed as EUACs, which would normally be considered negative, are considered positive. Conversely, benefits and returns are considered negative. The term *cost* in the designation EUAC serves to make clear the meaning of a positive number.

Figure 60.4 *Book Value with Straight Line and MACRS Methods*

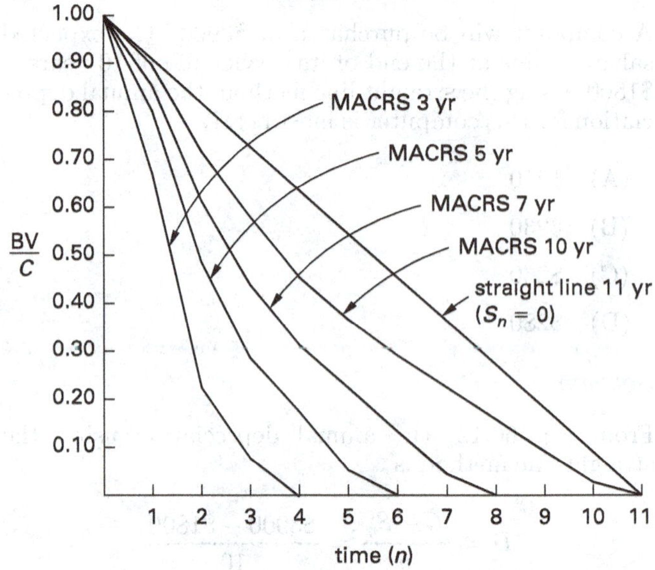

14. CAPITALIZED COST

The present worth of a project with an infinite life is known as the *capitalized cost*. Capitalized cost is the amount of money at $t = 0$ needed to perpetually support the project on the earned interest only. Capitalized cost is a positive number when expenses exceed income.

Normally, it would be difficult to work with an infinite stream of cash flows since most discount factor tables do not list factors for periods in excess of 100 years. However, the (A/P) discount factor approaches the interest rate as n becomes large. Since the (P/A) and (A/P) factors are reciprocals of each other, it is possible to divide an infinite series of annual cash flows by the interest rate in order to calculate the present worth of the infinite series.

Equation 60.15: Capitalized Costs for an Infinite Series

$$\text{capitalized costs} = P = \frac{A}{i} \qquad 60.15$$

Description

Equation 60.15 can be used when the annual costs are equal in every year. If the operating and maintenance costs occur irregularly instead of annually, or if the costs vary from year to year, it will be necessary to somehow determine a cash flow of equal annual amounts that is equivalent to the stream of original costs (i.e., to determine the EUAC).

Example

The construction of a volleyball court will cost \$1200, and annual maintenance cost is expected to be \$300.

Engineering Economics

At an effective annual interest rate of 5%, the project's capitalized cost is most nearly

(A) $2000

(B) $3000

(C) $7000

(D) $20,000

Solution

The cost of the project consists of two parts: the construction cost of $1200 and the annual maintenance cost of $300. The maintenance cost is an infinite series of annual amounts, so use Eq. 60.15 to find its present worth.

$$P_{\text{maintenance}} = \frac{A}{i} = \frac{\$300}{0.05} = \$6000$$

Add the present worth of the initial construction cost to get the total present worth (i.e., the capitalized cost) of the project.

$$P_{\text{total}} = P_{\text{construction}} + P_{\text{maintenance}} = \$1200 + \$6000 = \$7200 \quad (\$7000)$$

The answer is (C).

15. INFLATION

To be meaningful, economic studies must be performed in terms of constant-value dollars. Several common methods are used to allow for *inflation*. One alternative is to replace the effective annual interest rate, i, with a value adjusted for inflation, d.

Equation 60.16: Interest Rate Adjusted for Inflation

$$d = i + f + (i \times f) \qquad 60.16$$

Description

In Eq. 60.16, f is a constant *inflation rate* per year. The inflation-adjusted interest rate, d, can be used to compute present worth.

Example

An investment of $20,000 earns an effective annual interest of 10%. The value of the investment in five years, adjusted for an annual inflation rate of 6%, is most nearly

(A) $27,000

(B) $32,000

(C) $42,000

(D) $43,000

Solution

The interest rate adjusted for inflation is

$$d = i + f + (i \times f) = 0.10 + 0.06 + (0.10)(0.06) = 0.166$$

To determine the future worth of the investment, adjusted for inflation, use d instead of i in Eq. 60.1.

$$F = P(1+d)^n = (\$20{,}000)(1+0.166)^5 = \$43{,}105 \quad (\$43{,}000)$$

The answer is (D).

16. CAPITAL BUDGETING (ALTERNATIVE COMPARISONS)

In the real world, the majority of engineering economic analysis problems are alternative comparisons. In these problems, two or more mutually exclusive investments compete for limited funds. A variety of methods exists for selecting the superior alternative from a group of proposals. Each method has its own merits and applications.

Present Worth Analysis

When two or more alternatives are capable of performing the same functions, the economically superior alternative will have the largest present worth. The *present worth method* is restricted to evaluating alternatives that are mutually exclusive and that have the same lives. This method is suitable for ranking the desirability of alternatives.

Annual Cost Analysis

Alternatives that accomplish the same purpose but that have unequal lives must be compared by the *annual cost method.* The annual cost method assumes that each alternative will be replaced by an identical twin at the end of its useful life (i.e., infinite renewal). This method, which may also be used to rank alternatives according to their desirability, is also called the *annual return method* or *capital recovery method.*

The alternatives must be mutually exclusive and repeatedly renewed up to the duration of the longest-lived alternative. The calculated annual cost is known as the *equivalent uniform annual cost* (EUAC) or *equivalent annual cost* (EAC). Cost is a positive number when expenses exceed income.

Rate of Return Analysis

An intuitive definition of the *rate of return* (ROR) is the effective annual interest rate at which an investment accrues income. That is, the rate of return of an investment is the interest rate that would yield identical profits if all money was invested at that rate. Although this definition is correct, it does not provide a method of determining the rate of return.

The present worth of a \$100 investment invested at 5% is zero when $i = 5\%$ is used to determine equivalence. Therefore, a working definition of rate of return would be the effective annual interest rate that makes the present worth of the investment zero. Alternatively, rate of return could be defined as the effective annual interest rate that makes the benefits and costs equal.

A company may not know what effective interest rate, i, to use in engineering economic analysis. In such a case, the company can establish a minimum level of economic performance that it would like to realize on all investments. This criterion is known as the *minimum attractive rate of return*, or MARR.

Once a rate of return for an investment is known, it can be compared with the minimum attractive rate of return. If the rate of return is equal to or exceeds the minimum attractive rate of return, the investment is qualified (i.e., the alternative is viable). This is the basis for the rate of return method of alternative viability analysis.

If rate of return is used to select among two or more investments, an *incremental analysis* must be performed. An incremental analysis begins by ranking the alternatives in order of increasing initial investment. Then, the cash flows for the investment with the lower initial cost are subtracted from the cash flows for the higher-priced alternative on a year-by-year basis. This produces, in effect, a third alternative representing the costs and benefits of the added investment. The added expense of the higher-priced investment is not warranted unless the rate of return of this third alternative exceeds the minimum attractive rate of return as well. The alternative with the higher initial investment is superior if the incremental rate of return exceeds the minimum attractive rate of return.

Finding the rate of return can be a long, iterative process, requiring either interpolation or trial and error. Sometimes, the actual numerical value of rate of return is not needed; it is sufficient to know whether or not the rate of return exceeds the minimum attractive rate of return. This comparative analysis can be accomplished without calculating the rate of return simply by finding the present worth of the investment using the minimum attractive rate of return as the effective interest rate (i.e., $i = \text{MARR}$). If the present worth is zero or positive, the investment is qualified. If the present worth is negative, the rate of return is less than the minimum attractive rate of return and the additional investment is not warranted.

The present worth, annual cost, and rate of return methods of comparing alternatives yield equivalent results, but they are distinctly different approaches. The present worth and annual cost methods may use either effective interest rates or the minimum attractive rate of return to rank alternatives or compare them to the MARR. If the incremental rate of return of pairs of alternatives are compared with the MARR, the analysis is considered a rate of return analysis.

17. BREAK-EVEN ANALYSIS

Break-even analysis is a method of determining when the value of one alternative becomes equal to the value of another. It is commonly used to determine when costs exactly equal revenue. If the manufactured quantity is less than the *break-even quantity*, a loss is incurred. If the manufactured quantity is greater than the break-even quantity, a profit is made.

An alternative form of the break-even problem is to find the number of units per period for which two alternatives have the same total costs. Fixed costs are spread over a period longer than one year using the EUAC concept. One of the alternatives will have a lower cost if production is less than the break-even point. The other will have a lower cost if production is greater than the break-even point.

The *pay-back period*, PBP, is defined as the length of time, n, usually in years, for the cumulative net annual profit to equal the initial investment. It is tempting to introduce equivalence into pay-back period calculations, but the convention is not to.

18. BENEFIT-COST ANALYSIS

The *benefit-cost ratio method* is often used in municipal project evaluations where benefits and costs accrue to different segments of the community. With this method, the present worth of all benefits (irrespective of the beneficiaries), B, is divided by the present worth of all costs, C. If the benefit-cost ratio, B/C, is greater than or equal to 1.0, the project is acceptable. (Equivalent uniform annual costs can be used in place of present worths.)

When the benefit-cost ratio method is used, disbursements by the initiators or sponsors are *costs* and added to C. Disbursements by the users of the project are known as *disbenefits* and subtracted from B. It is often difficult to decide whether a cash flow should be regarded as a cost or a disbenefit. The placement of such cash flows can change the value of B/C, but cannot change whether B/C is greater than or equal to 1.0. For this reason, the benefit-cost ratio alone should not be used to rank competing projects.

If ranking is to be done by the benefit-cost ratio method, an incremental analysis is required, as it is for the rate-of-return method. The incremental analysis is accomplished by calculating the ratio of differences in benefits to differences in costs for each possible pair of alternatives. If the ratio exceeds 1.0, alternative 2 is superior to alternative 1. Otherwise, alternative 1 is superior.

Equation 60.17: Analysis Criterion

$$B - C \geq 0 \text{ or } B/C \geq 1 \qquad 60.17$$

Description

A project is acceptable if its benefit-cost ratio equals or exceeds 1 (i.e., $B/C \geq 1$). This will be true whenever $B - C \geq 0$.

Example

A large sewer system will cost $175,000 annually. There will be favorable consequences to the general public equivalent to $500,000 annually, and adverse consequences to a small segment of the public equivalent to $50,000 annually. The benefit-cost ratio is most nearly

(A) 2.2

(B) 2.4

(C) 2.6

(D) 2.9

Solution

The adverse consequences worth $50,000 affect the users of the project, not its initiators, so this is a disbenefit. The benefit-cost ratio is

$$B/C = \frac{\$500{,}000 - \$50{,}000}{\$175{,}000} = 2.57 \quad (2.6)$$

The answer is (C).

19. SENSITIVITY ANALYSIS, RISK ANALYSIS, AND UNCERTAINTY ANALYSIS

Data analysis and forecasts in economic studies require estimates of costs that will occur in the future. There are always uncertainties about these costs. However, these uncertainties are an insufficient reason not to make the best possible estimates of the costs. Nevertheless, a decision between alternatives often can be made more confidently if it is known whether or not the conclusion is sensitive to moderate changes in data forecasts. *Sensitivity analysis* provides this extra dimension to an economic analysis.

The sensitivity of a decision to various factors is determined by inserting a range of estimates for critical cash flows and other parameters. If radical changes can be made to a cash flow without changing the decision, the decision is said to be *insensitive* to uncertainties regarding that cash flow. However, if a small change in the estimate of a cash flow will alter the decision, that decision is said to be very *sensitive* to changes in the estimate. If the decision is sensitive only for a limited range of cash flow values, the term *variable sensitivity* is used. Figure 60.5 illustrates these terms.

An established semantic tradition distinguishes between risk analysis and uncertainty analysis. *Risk* is the possibility of an unexpected, unplanned, or undesirable event occurring (i.e., an occurrence not planned for or predicted in risk analysis). *Risk analysis* addresses variables that have a known or estimated probability distribution. In this regard, statistics and probability theory can be used to determine the probability of a cash flow varying between given limits. On the other hand, *uncertainty analysis* is concerned with situations in which there is not enough information to determine the probability or frequency distribution for the variables involved.

As a first step, sensitivity analysis should be performed one factor at a time to the dominant factors. Dominant cost factors are those that have the most significant impact on the present value of the alternative.[3] If warranted, additional investigation can be used to determine the sensitivity to several cash flows varying simultaneously. Significant judgment is needed, however, to successfully determine the proper combinations of cash flows to vary. It is common to plot the dependency of the present value on the cash flow being varied in a two-dimensional graph. Simple linear interpolation is used (within reason) to determine the critical value of the cash flow being varied.

20. ACCOUNTING PRINCIPLES

Basic Bookkeeping

An accounting or *bookkeeping system* is used to record historical financial transactions. The resultant records are used for product costing, satisfaction of statutory requirements, reporting of profit for income tax purposes, and general company management.

[3]In particular, engineering economic analysis problems are sensitive to the choice of effective interest rate, i, and to accuracy in cash flows at or near the beginning of the horizon. The problems will be less sensitive to accuracy in far-future cash flows, such as subsequent generation replacement costs.

Bookkeeping consists of two main steps: recording the transactions, followed by categorization of the transactions.[4] The transactions (receipts and disbursements) are recorded in a *journal* (*book of original entry*) to complete the first step. Such a journal is organized in a simple chronological and sequential manner. The transactions are then categorized (into interest income, advertising expense, etc.) and posted (i.e., entered or written) into the appropriate *ledger account.*[5]

Together, the ledger accounts constitute the *general ledger* or *ledger*. All ledger accounts can be classified into one of three types: *asset accounts*, *liability accounts*, and *owners' equity accounts*. Strictly speaking, income and expense accounts, kept in a separate journal, are included within the classification of owners' equity accounts.

Together, the journal and ledger are known simply as "the books" of the company, regardless of whether bound volumes of pages are actually involved.

Figure 60.5 *Types of Sensitivity*

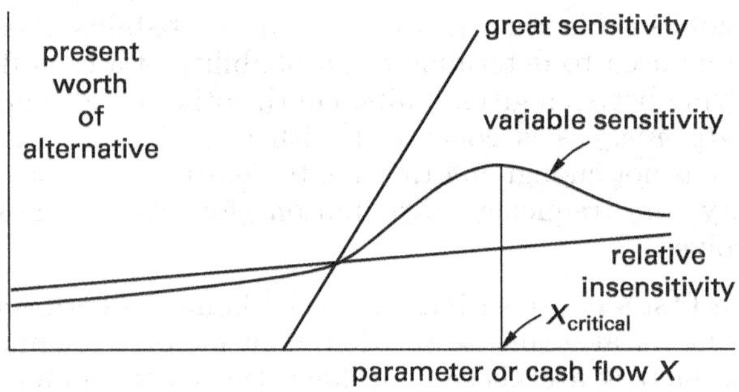

Balancing the Books

In a business environment, *balancing the books* means more than reconciling the checkbook and bank statements. All accounting entries must be posted in such a way as to maintain the equality of the *basic accounting equation*,

$$\text{assets} = \text{liability} + \text{owners' equity}$$

In a *double-entry bookkeeping system*, the equality is maintained within the ledger system by entering each transaction into two balancing ledger accounts. For example, paying a utility bill would decrease the cash account (an asset account) and decrease the utility expense account (a liability account) by the same amount.

Transactions are either *debits* or *credits*, depending on their sign. Increases in asset accounts are debits; decreases are credits. For liability and equity accounts, the opposite is true: Increases are credits, and decreases are debits.[6]

Cash and Accrual Systems[7]

The simplest form of bookkeeping is based on the *cash system*. The only transactions that are entered into the journal are those that represent cash receipts and disbursements. In effect, a checkbook register or bank deposit book could serve as the journal.

During a given period (e.g., month or quarter), expense liabilities may be incurred even though the payments for those expenses have not been made. For example, an invoice (bill) may have been received but not paid. Under the *accrual system*, the obligation is posted into the appropriate expense account before it is paid.[8] Analogous to expenses, under the accrual system, income will be claimed before payment is received. Specifically, a sales transaction can be recorded as income when the customer's order is received, when the outgoing invoice is generated, or when the merchandise is shipped.

Financial Statements

Each period, two types of corporate financial statements are typically generated: the *balance sheet* and *profit and loss* (P&L) *statement.*[9] The profit and loss statement, also known as a *statement of income and retained earnings*, is a summary of sources of *income* or *revenue* (interest, sales, fees charged, etc.) and *expenses* (utilities, advertising, repairs, etc.) for the period. The expenses are subtracted from the revenues to give a *net income* (generally, before taxes).[10] Figure 60.6 illustrates a simplified profit and loss statement.

[4]These two steps are not to be confused with the *double-entry bookkeeping method.*

[5]The two-step process is more typical of a *manual bookkeeping system* than a computerized *general ledger system*. However, even most computerized systems produce reports in journal entry order, as well as account summaries.

[6]There is a difference in sign between asset and liability accounts. An increase in an expense account is actually a decrease. The accounting profession, apparently, is comfortable with the common confusion that exists between debits and credits.

[7]There is also a distinction made between cash flows that are known and those that are expected. It is a *standard accounting principle* to record losses in full, at the time they are recognized, even before their occurrence. In the construction industry, for example, losses are recognized in full and projected to the end of a project as soon as they are foreseeable. Profits, on the other hand, are recognized only as they are realized (typically, as a percentage of project completion). The difference between cash and accrual systems is a matter of *bookkeeping*. The difference between loss and profit recognition is a matter of *accounting convention*. Engineers seldom need to be concerned with the accounting principles and conventions.

[8]The expense for an item or service might be accrued even *before* the invoice is received. It might be recorded when the purchase order for the item or service is generated, or when the item or service is received.

[9]Other types of financial statements (*statements of changes in financial position*, *cost of sales statements*, *inventory and asset reports*, etc.) also will be generated, depending on the needs of the company.

[10]Financial statements also can be prepared with percentages (of total assets and net revenue) instead of dollars, in which case they are known as *common size financial statements.*

Engineering Economics

The *balance sheet* presents the *basic accounting equation* in tabular form. The balance sheet lists the major categories of assets and outstanding liabilities. The difference between asset values and liabilities is the *equity*. This equity represents what would be left over after satisfying all debts by liquidating the company.

Figure 60.7 is a simplified balance sheet.

There are several terms that appear regularly on balance sheets.

- *current assets:* cash and other assets that can be converted quickly into cash, such as accounts receivable, notes receivable, and merchandise (inventory). Also known as *liquid assets.*
- *fixed assets:* relatively permanent assets used in the operation of the business and relatively difficult to convert into cash. Examples are land, buildings, and equipment. Also known as *nonliquid assets.*
- *current liabilities:* liabilities due within a short period of time (e.g., within one year) and typically paid out of current assets. Examples are accounts payable, notes payable, and other accrued liabilities.
- *long-term liabilities:* obligations that are not totally payable within a short period of time (e.g., within one year).

Analysis of Financial Statements

Financial statements are evaluated by management, lenders, stockholders, potential investors, and many other groups for the purpose of determining the *health of the company.* The health can be measured in terms of *liquidity* (ability to convert assets to cash quickly), *solvency* (ability to meet debts as they become due), and *relative risk* (of which one measure is *leverage*—the portion of total capital contributed by owners).

Figure 60.6 *Simplified Profit and Loss Statement*

revenue			
interest	2000		
sales	237,000		
returns	(23,000)		
net revenue		216,000	
expenses			
salaries	149,000		
utilities	6000		
advertising	28,000		
insurance	4000		
supplies	1000		
net expenses		188,000	
period net income			28,000
beginning retained earnings			63,000
net year-to-date earnings			91,000

Figure 60.7 *Simplified Balance Sheet*

ASSETS			
current assets			
cash	14,000		
accounts receivable	36,000		
notes receivable	20,000		
inventory	89,000		
prepaid expenses	3000		
total current assets		162,000	
plant, property, and equipment			
land and buildings	217,000		
motor vehicles	31,000		
equipment	94,000		
accumulated depreciation	(52,000)		
total fixed assets		290,000	
total assets			452,000
LIABILITIES AND OWNERS' EQUITY			
current liabilities			
accounts payable	66,000		
accrued income taxes	17,000		
accrued expenses	8000		
total current liabilities		91,000	
long-term debt			
notes payable	117,000		
mortgage	23,000		
total long-term debt		140,000	
owners' and stockholders' equity			
stock	130,000		
retained earnings	91,000		
total owners' equity		221,000	
total liabilities and owners' equity			452,000

The analysis of financial statements involves several common ratios, usually expressed as percentages. The following are some frequently encountered ratios.

- *current ratio:* an index of short-term paying ability.

$$\text{current ratio} = \frac{\text{current assets}}{\text{current liabilities}}$$

- *quick* (or *acid-test*) *ratio:* a more stringent measure of short-term debt-paying ability. The *quick assets* are defined to be current assets minus inventories and prepaid expenses.

$$\text{quick ratio} = \frac{\text{quick assets}}{\text{current liabilities}}$$

- *receivable turnover:* a measure of the average speed with which accounts receivable are collected.

$$\text{receivable turnover} = \frac{\text{net credit sales}}{\text{average net receivables}}$$

Engineering Economics

- *average age of receivables:* number of days, on the average, in which receivables are collected.

$$\text{average age of receivables} = \frac{365}{\text{receivable turnover}}$$

- *inventory turnover:* a measure of the speed, on the average, with which inventory is sold.

$$\text{inventory turnover} = \frac{\text{cost of goods sold}}{\text{average unit cost of inventory}}$$

- *days supply of inventory on hand:* number of days, on the average, that the current inventory would last.

$$\text{days supply of inventory on hand} = \frac{365}{\text{inventory turnover}}$$

- *book value per share of common stock:* number of dollars represented by the balance sheet owners' equity for each share of common stock outstanding.

$$\text{book value per share of common stock} = \frac{\text{common shareholders' equity}}{\text{number of outstanding shares}}$$

- *gross margin:* gross profit as a percentage of sales. (Gross profit is sales less cost of goods sold.)

$$\text{gross margin} = \frac{\text{gross profit}}{\text{net sales}}$$

- *profit margin ratio:* percentage of each dollar of sales that is net income.

$$\text{profit margin} = \frac{\text{net income before taxes}}{\text{net sales}}$$

- *return on investment ratio:* shows the percent return on owners' investment.

$$\text{return on investment} = \frac{\text{net income}}{\text{owners' equity}}$$

- *price-earnings ratio:* an indication of the relationship between earnings and market price per share of common stock; useful in comparisons between alternative investments.

$$\text{price–earnings} = \frac{\text{market price per share}}{\text{earnings per share}}$$

21. ACCOUNTING COSTS AND EXPENSE TERMS

The accounting profession has developed special terms for certain groups of costs. When annual costs are incurred due to the functioning of a piece of equipment, they are known as *operating and maintenance* (O&M) *costs.* The annual costs associated with operating a business (other than the costs directly attributable to production) are known as *general, selling, and administrative* (GS&A) *expenses.*

Direct labor costs are costs incurred in the factory, such as assembly, machining, and painting labor costs. *Direct material costs* are the costs of all materials that go into production.[11] Typically, both direct labor and direct material costs are given on a per-unit or per-item basis. The sum of the direct labor and direct material costs is known as the *prime cost.*

There are certain additional expenses incurred in the factory, such as the costs of factory supervision, stock-picking, quality control, factory utilities, and miscellaneous supplies (cleaning fluids, assembly lubricants, routing tags, etc.) that are not incorporated into the final product. Such costs are known as *indirect manufacturing expenses* (IME) or *indirect material and labor costs.*[12] The sum of the per-unit indirect manufacturing expense and prime cost is known as the *factory cost.*

Research and development (R&D) *costs* and *administrative expenses* are added to the factory cost to give the *manufacturing cost* of the product.

Additional costs are incurred in marketing the product. Such costs are known as *selling expenses* or *marketing expenses.* The sum of the selling expenses and manufacturing cost is the *total cost* of the product. Figure 60.8 illustrates these terms.[13] Typical classifications of expenses are listed in Table 60.3 (at the end of this chapter).

The distinctions among the various forms of cost (particularly with overhead costs) are not standardized. Each company must develop a classification system to deal with the various cost factors in a consistent manner. There are also other terms in use (e.g., *raw materials, operating supplies, general plant overhead*), but these terms must be interpreted within the framework of each company's classification system. Table 60.3 is typical of such classification systems.

22. COST ACCOUNTING

Cost accounting is the system that determines the cost of manufactured products. Cost accounting is called *job cost accounting* if costs are accumulated by part number

[11]There may be complications with pricing the material when it is purchased from an outside vendor and the stock on hand derives from several shipments purchased at different prices.

[12]The *indirect material and labor costs* usually exclude costs incurred in the office area.

[13]*Total cost* does not include income taxes.

or contract. It is called *process cost accounting* if costs are accumulated by departments or manufacturing processes.

Figure 60.8 Costs and Expenses Combined

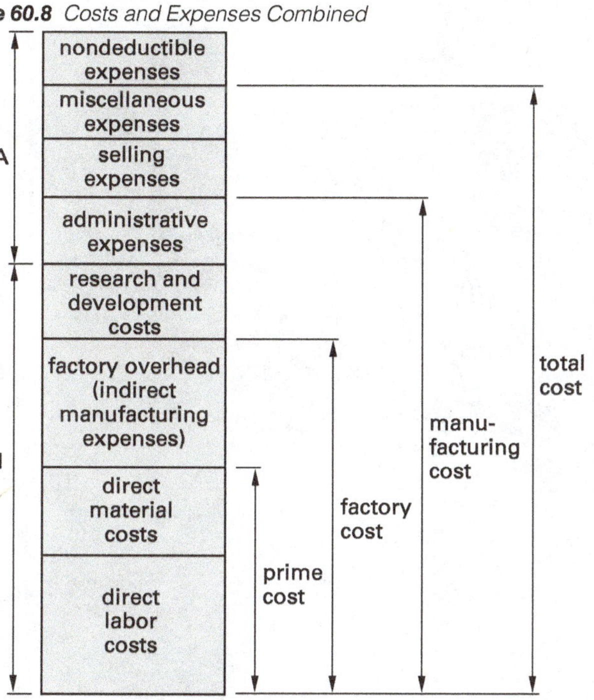

Cost accounting is dependent on historical and recorded data. The unit product cost is determined from actual expenses and numbers of units produced. Allowances (i.e., budgets) for future costs are based on these historical figures. Any deviation from historical figures is called a *variance*. Where adequate records are available, variances can be divided into *labor variance* and *material variance*.

When determining a unit product cost, the direct material and direct labor costs are generally clear-cut and easily determined. Furthermore, these costs are 100% variable costs. However, the indirect cost per unit of product is not as easily determined. Indirect costs (*burden*, *overhead*, etc.) can be fixed or semivariable costs. The amount of indirect cost allocated to a unit will depend on the unknown future overhead expense as well as the unknown future production (*vehicle size*).

A typical method of allocating indirect costs to a product is as follows.

step 1: Estimate the total expected indirect (and overhead) costs for the upcoming year.

step 2: Determine the most appropriate vehicle (basis) for allocating the overhead to production. Usually, this vehicle is either the number of units expected to be produced or the number of direct hours expected to be worked in the upcoming year.

step 3: Estimate the quantity or size of the overhead vehicle.

step 4: Divide expected overhead costs by the expected overhead vehicle to obtain the unit overhead.

step 5: Regardless of the true size of the overhead vehicle during the upcoming year, one unit of overhead cost is allocated per unit of overhead vehicle.

Once the prime cost has been determined and the indirect cost calculated based on projections, the two are combined into a *standard factory cost* or *standard cost*, which remains in effect until the next budgeting period (usually a year).

During the subsequent manufacturing year, the standard cost of a product is not generally changed merely because it is found that an error in projected indirect costs or production quantity (vehicle size) has been made. The allocation of indirect costs to a product is assumed to be independent of errors in forecasts. Rather, the difference between the expected and actual expenses, known as the *burden* (*overhead*) *variance*, experienced during the year is posted to one or more *variance accounts*.

Burden (overhead) variance is caused by errors in forecasting both the actual indirect expense for the upcoming year and the overhead vehicle size. In the former case, the variance is called *burden budget variance*; in the latter, it is called *burden capacity variance*.

Table 60.3 Typical Classification of Expenses

direct labor expenses
- machining and forming
- assembly
- finishing
- inspection
- testing

direct material expenses
- items purchased from other vendors
- manufactured assemblies

factory overhead expenses (indirect manufacturing expenses)
- supervision
- benefits
  - pension
  - medical insurance
  - vacations
- wages overhead
  - unemployment compensation taxes
  - social security taxes
  - disability taxes
- stock-picking
- quality control and inspection
- expediting
- rework
- maintenance
- miscellaneous supplies
  - routing tags
  - assembly lubricants
  - cleaning fluids
  - wiping cloths
  - janitorial supplies
- packaging (materials and labor)
- factory utilities
- laboratory
- depreciation on factory equipment

research and development expenses
- engineering (labor)
- patents
- testing
- prototypes (material and labor)
- drafting
- O&M of R&D facility

administrative expenses
- corporate officers
- accounting
- secretarial/clerical/reception
- security (protection)
- medical (nurse)
- employment (personnel)
- reproduction
- data processing
- production control
- depreciation on nonfactory equipment
- office supplies
- office utilities
- O&M of offices

selling expenses
- marketing (labor)
- advertising
- transportation (if not paid by customer)
- outside sales force (labor and expenses)
- demonstration units
- commissions
- technical service and support
- order processing
- branch office expenses

miscellaneous expenses
- insurance
- property taxes
- interest on loans

nondeductible expenses
- federal income taxes
- fines and penalties

Table 60.4 Factor Table i = 0.50%

n	P/F	P/A	P/G	F/P	F/A	A/P	A/F	A/G
1	0.9950	0.9950	0.0000	1.0050	1.0000	1.0050	1.0000	0.0000
2	0.9901	1.9851	0.9901	1.0100	2.0050	0.5038	0.4988	0.4988
3	0.9851	2.9702	2.9604	1.0151	3.0150	0.3367	0.3317	0.9967
4	0.9802	3.9505	5.9011	1.0202	4.0301	0.2531	0.2481	1.4938
5	0.9754	4.9259	9.8026	1.0253	5.0503	0.2030	0.1980	1.9900
6	0.9705	5.8964	14.6552	1.0304	6.0755	0.1696	0.1646	2.4855
7	0.9657	6.8621	20.4493	1.0355	7.1059	0.1457	0.1407	2.9801
8	0.9609	7.8230	27.1755	1.0407	8.1414	0.1278	0.1228	3.4738
9	0.9561	8.7791	34.8244	1.0459	9.1821	0.1139	0.1089	3.9668
10	0.9513	9.7304	43.3865	1.0511	10.2280	0.1028	0.0978	4.4589
11	0.9466	10.6670	52.8526	1.0564	11.2792	0.0937	0.0887	4.9501
12	0.9419	11.6189	63.2136	1.0617	12.3356	0.0861	0.0811	5.4406
13	0.9372	12.5562	74.4602	1.0670	13.3972	0.0796	0.0746	5.9302
14	0.9326	13.4887	86.5835	1.0723	14.4642	0.0741	0.0691	6.4190
15	0.9279	14.4166	99.5743	1.0777	15.5365	0.0694	0.0644	6.9069
16	0.9233	15.3399	113.4238	1.0831	16.6142	0.0652	0.0602	7.3940
17	0.9187	16.2586	128.1231	1.0885	17.6973	0.0615	0.0565	7.8803
18	0.9141	17.1728	143.6634	1.0939	18.7858	0.0582	0.0532	8.3658
19	0.9096	18.0824	160.0360	1.0994	19.8797	0.0553	0.0503	8.8504
20	0.9051	18.9874	177.2322	1.1049	20.9791	0.0527	0.0477	9.3342
21	0.9006	19.8880	195.2434	1.1104	22.0840	0.0503	0.0453	9.8172
22	0.8961	20.7841	214.0611	1.1160	23.1944	0.0481	0.0431	10.2993
23	0.8916	21.6757	233.6768	1.1216	24.3104	0.0461	0.0411	10.7806
24	0.8872	22.5629	254.0820	1.1272	25.4320	0.0443	0.0393	11.2611
25	0.8828	23.4456	275.2686	1.1328	26.5591	0.0427	0.0377	11.7407
30	0.8610	27.7941	392.6324	1.1614	32.2800	0.0360	0.0310	14.1265
40	0.8191	36.1722	681.3347	1.2208	44.1588	0.0276	0.0226	18.8359
50	0.7793	44.1428	1,035.6966	1.2832	56.6452	0.0227	0.0177	23.4624
60	0.7414	51.7256	1,448.6458	1.3489	69.7700	0.0193	0.0143	28.0064
100	0.6073	78.5426	3,562.7934	1.6467	129.3337	0.0127	0.0077	45.3613

Engineering Economics

***Table 60.5** Factor Table i = 1.00%*

n	P/F	P/A	P/G	F/P	F/A	A/P	A/F	A/G
1	0.9901	0.9901	0.0000	1.0100	1.0000	1.0100	1.0000	0.0000
2	0.9803	1.9704	0.9803	1.0201	2.0100	0.5075	0.4975	0.4975
3	0.9706	2.9410	2.9215	1.0303	3.0301	0.3400	0.3300	0.9934
4	0.9610	3.9020	5.8044	1.0406	4.0604	0.2563	0.2463	1.4876
5	0.9515	4.8534	9.6103	1.0510	5.1010	0.2060	0.1960	1.9801
6	0.9420	5.7955	14.3205	1.0615	6.1520	0.1725	0.1625	2.4710
7	0.9327	6.7282	19.9168	1.0721	7.2135	0.1486	0.1386	2.9602
8	0.9235	7.6517	26.3812	1.0829	8.2857	0.1307	0.1207	3.4478
9	0.9143	8.5650	33.6959	1.0937	9.3685	0.1167	0.1067	3.9337
10	0.9053	9.4713	41.8435	1.1046	10.4622	0.1056	0.0956	4.4179
11	0.8963	10.3676	50.8067	1.1157	11.5668	0.0965	0.0865	4.9005
12	0.8874	11.2551	60.5687	1.1268	12.6825	0.0888	0.0788	5.3815
13	0.8787	12.1337	71.1126	1.1381	13.8093	0.0824	0.0724	5.8607
14	0.8700	13.0037	82.4221	1.1495	14.9474	0.0769	0.0669	6.3384
15	0.8613	13.8651	94.4810	1.1610	16.0969	0.0721	0.0621	6.8143
16	0.8528	14.7179	107.2734	1.1726	17.2579	0.0679	0.0579	7.2886
17	0.8444	15.5623	129.7834	1.1843	18.4304	0.0643	0.0543	7.7613
18	0.8360	16.3983	134.9957	1.1961	19.6147	0.0610	0.0510	8.2323
19	0.8277	17.2260	149.8950	1.2081	20.8109	0.0581	0.0481	8.7017
20	0.8195	18.0456	165.4664	1.2202	22.0190	0.0554	0.0454	9.1694
21	0.8114	18.8570	181.6950	1.2324	23.2392	0.0530	0.0430	9.6354
22	0.8034	19.6604	198.5663	1.2447	24.4716	0.0509	0.0409	10.0998
23	0.7954	20.4558	216.0660	1.2572	25.7163	0.0489	0.0389	10.5626
24	0.7876	21.2434	234.1800	1.2697	26.9735	0.0471	0.0371	11.0237
25	0.7798	22.0232	252.8945	1.2824	28.2432	0.0454	0.0354	11.4831
30	0.7419	25.8077	355.0021	1.3478	34.7849	0.0387	0.0277	13.7557
40	0.6717	32.8347	596.8561	1.4889	48.8864	0.0305	0.0205	18.1776
50	0.6080	39.1961	879.4176	1.6446	64.4632	0.0255	0.0155	22.4363
60	0.5504	44.9550	1,192.8061	1.8167	81.6697	0.0222	0.0122	26.5333
100	0.3697	63.0289	2,605.7758	2.7048	170.4814	0.0159	0.0059	41.3426

Table 60.6 Factor Table $i = 1.50\%$

n	P/F	P/A	P/G	F/P	F/A	A/P	A/F	A/G
1	0.9852	0.9852	0.0000	1.0150	1.0000	1.0150	1.0000	0.0000
2	0.9707	1.9559	0.9707	1.0302	2.0150	0.5113	0.4963	0.4963
3	0.9563	2.9122	2.8833	1.0457	3.0452	0.3434	0.3284	0.9901
4	0.9422	3.8544	5.7098	1.0614	4.0909	0.2594	0.2444	1.4814
5	0.9283	4.7826	9.4229	1.0773	5.1523	0.2091	0.1941	1.9702
6	0.9145	5.6972	13.9956	1.0934	6.2296	0.1755	0.1605	2.4566
7	0.9010	6.5982	19.4018	1.1098	7.3230	0.1516	0.1366	2.9405
8	0.8877	7.4859	26.6157	1.1265	8.4328	0.1336	0.1186	3.4219
9	0.8746	8.3605	32.6125	1.1434	9.5593	0.1196	0.1046	3.9008
10	0.8617	9.2222	40.3675	1.1605	10.7027	0.1084	0.0934	4.3772
11	0.8489	10.0711	48.8568	1.1779	11.8633	0.0993	0.0843	4.8512
12	0.8364	10.9075	58.0571	1.1956	13.0412	0.0917	0.0767	5.3227
13	0.8240	11.7315	67.9454	1.2136	14.2368	0.0852	0.0702	5.7917
14	0.8118	12.5434	78.4994	1.2318	15.4504	0.0797	0.0647	6.2582
15	0.7999	13.3432	89.6974	1.2502	16.6821	0.0749	0.0599	6.7223
16	0.7880	14.1313	101.5178	1.2690	17.9324	0.0708	0.0558	7.1839
17	0.7764	14.9076	113.9400	1.2880	19.2014	0.0671	0.0521	7.6431
18	0.7649	15.6726	126.9435	1.3073	20.4894	0.0638	0.0488	8.0997
19	0.7536	16.4262	140.5084	1.3270	21.7967	0.0609	0.0459	8.5539
20	0.7425	17.1686	154.6154	1.3469	23.1237	0.0582	0.0432	9.0057
21	0.7315	17.9001	169.2453	1.3671	24.4705	0.0559	0.0409	9.4550
22	0.7207	18.6208	184.3798	1.3876	25.8376	0.0537	0.0387	9.9018
23	0.7100	19.3309	200.0006	1.4084	27.2251	0.0517	0.0367	10.3462
24	0.6995	20.0304	216.0901	1.4295	28.6335	0.0499	0.0349	10.7881
25	0.6892	20.7196	232.6310	1.4509	30.0630	0.0483	0.0333	11.2276
30	0.6398	24.0158	321.5310	1.5631	37.5387	0.0416	0.0266	13.3883
40	0.5513	29.9158	524.3568	1.8140	54.2679	0.0334	0.0184	17.5277
50	0.4750	34.9997	749.9636	2.1052	73.6828	0.0286	0.0136	21.4277
60	0.4093	39.3803	988.1674	2.4432	96.2147	0.0254	0.0104	25.0930
100	0.2256	51.6247	1,937.4506	4.4320	228.8030	0.0194	0.0044	37.5295

Table 60.7 *Factor Table i = 2.00%*

n	P/F	P/A	P/G	F/P	F/A	A/P	A/F	A/G
1	0.9804	0.9804	0.0000	1.0200	1.0000	1.0200	1.0000	0.0000
2	0.9612	1.9416	0.9612	1.0404	2.0200	0.5150	0.4950	0.4950
3	0.9423	2.8839	2.8458	1.0612	3.0604	0.3468	0.3268	0.9868
4	0.9238	3.8077	5.6173	1.0824	4.1216	0.2626	0.2426	1.4752
5	0.9057	4.7135	9.2403	1.1041	5.2040	0.2122	0.1922	1.9604
6	0.8880	5.6014	13.6801	1.1262	6.3081	0.1785	0.1585	2.4423
7	0.8706	6.4720	18.9035	1.1487	7.4343	0.1545	0.1345	2.9208
8	0.8535	7.3255	24.8779	1.1717	8.5830	0.1365	0.1165	3.3961
9	0.8368	8.1622	31.5720	1.1951	9.7546	0.1225	0.1025	3.8681
10	0.8203	8.9826	38.9551	1.2190	10.9497	0.1113	0.0913	4.3367
11	0.8043	9.7868	46.9977	1.2434	12.1687	0.1022	0.0822	4.8021
12	0.7885	10.5753	55.6712	1.2682	13.4121	0.0946	0.0746	5.2642
13	0.7730	11.3484	64.9475	1.2936	14.6803	0.0881	0.0681	5.7231
14	0.7579	12.1062	74.7999	1.3195	15.9739	0.0826	0.0626	6.1786
15	0.7430	12.8493	85.2021	1.3459	17.2934	0.0778	0.0578	6.6309
16	0.7284	13.5777	96.1288	1.3728	18.6393	0.0737	0.0537	7.0799
17	0.7142	14.2919	107.5554	1.4002	20.0121	0.0700	0.0500	7.5256
18	0.7002	14.9920	119.4581	1.4282	21.4123	0.0667	0.0467	7.9681
19	0.6864	15.6785	131.8139	1.4568	22.8406	0.0638	0.0438	8.4073
20	0.6730	16.3514	144.6003	1.4859	24.2974	0.0612	0.0412	8.8433
21	0.6598	17.0112	157.7959	1.5157	25.7833	0.0588	0.0388	9.2760
22	0.6468	17.6580	171.3795	1.5460	27.2990	0.0566	0.0366	9.7055
23	0.6342	18.2922	185.3309	1.5769	28.8450	0.0547	0.0347	10.1317
24	0.6217	18.9139	199.6305	1.6084	30.4219	0.0529	0.0329	10.5547
25	0.6095	19.5235	214.2592	1.6406	32.0303	0.0512	0.0312	10.9745
30	0.5521	22.3965	291.7164	1.8114	40.5681	0.0446	0.0246	13.0251
40	0.4529	27.3555	461.9931	2.2080	60.4020	0.0366	0.0166	16.8885
50	0.3715	31.4236	642.3606	2.6916	84.5794	0.0318	0.0118	20.4420
60	0.3048	34.7609	823.6975	3.2810	114.0515	0.0288	0.0088	23.6961
100	0.1380	43.0984	1,464.7527	7.2446	312.2323	0.0232	0.0032	33.9863

Table 60.8 Factor Table $i = 4.00\%$

n	P/F	P/A	P/G	F/P	F/A	A/P	A/F	A/G
1	0.9615	0.9615	0.0000	1.0400	1.0000	1.0400	1.0000	0.0000
2	0.9246	1.8861	0.9246	1.0816	2.0400	0.5302	0.4902	0.4902
3	0.8890	2.7751	2.7025	1.1249	3.1216	0.3603	0.3203	0.9739
4	0.8548	3.6299	5.2670	1.1699	4.2465	0.2755	0.2355	1.4510
5	0.8219	4.4518	8.5547	1.2167	5.4163	0.2246	0.1846	1.9216
6	0.7903	5.2421	12.5062	1.2653	6.6330	0.1908	0.1508	2.3857
7	0.7599	6.0021	17.0657	1.3159	7.8983	0.1666	0.1266	2.8433
8	0.7307	6.7327	22.1806	1.3686	9.2142	0.1485	0.1085	3.2944
9	0.7026	7.4353	27.8013	1.4233	10.5828	0.1345	0.0945	3.7391
10	0.6756	8.1109	33.8814	1.4802	12.0061	0.1233	0.0833	4.1773
11	0.6496	8.7605	40.3772	1.5395	13.4864	0.1141	0.0741	4.6090
12	0.6246	9.3851	47.2477	1.6010	15.0258	0.1066	0.0666	5.0343
13	0.6006	9.9856	54.4546	1.6651	16.6268	0.1001	0.0601	5.4533
14	0.5775	10.5631	61.9618	1.7317	18.2919	0.0947	0.0547	5.8659
15	0.5553	11.1184	69.7355	1.8009	20.0236	0.0899	0.0499	6.2721
16	0.5339	11.6523	77.7441	1.8730	21.8245	0.0858	0.0458	6.6720
17	0.5134	12.1657	85.9581	1.9479	23.6975	0.0822	0.0422	7.0656
18	0.4936	12.6593	94.3498	2.0258	25.6454	0.0790	0.0390	7.4530
19	0.4746	13.1339	102.8933	2.1068	27.6712	0.0761	0.0361	7.8342
20	0.4564	13.5903	111.5647	2.1911	29.7781	0.0736	0.0336	8.2091
21	0.4388	14.0292	120.3414	2.2788	31.9692	0.0713	0.0313	8.5779
22	0.4220	14.4511	129.2024	2.3699	34.2480	0.0692	0.0292	8.9407
23	0.4057	14.8568	138.1284	2.4647	36.6179	0.0673	0.0273	9.2973
24	0.3901	15.2470	147.1012	2.5633	39.0826	0.0656	0.0256	9.6479
25	0.3751	15.6221	156.1040	2.6658	41.6459	0.0640	0.0240	9.9925
30	0.3083	17.2920	201.0618	3.2434	56.0849	0.0578	0.0178	11.6274
40	0.2083	19.7928	286.5303	4.8010	95.0255	0.0505	0.0105	14.4765
50	0.1407	21.4822	361.1638	7.1067	152.6671	0.0466	0.0066	16.8122
60	0.0951	22.6235	422.9966	10.5196	237.9907	0.0442	0.0042	18.6972
100	0.0198	24.5050	563.1249	50.5049	1,237.6237	0.0408	0.0008	22.9800

Table 60.9 Factor Table $i = 6.00\%$

n	P/F	P/A	P/G	F/P	F/A	A/P	A/F	A/G
1	0.9434	0.9434	0.0000	1.0600	1.0000	1.0600	1.0000	0.0000
2	0.8900	1.8334	0.8900	1.1236	2.0600	0.5454	0.4854	0.4854
3	0.8396	2.6730	2.5692	1.1910	3.1836	0.3741	0.3141	0.9612
4	0.7921	3.4651	4.9455	1.2625	4.3746	0.2886	0.2286	1.4272
5	0.7473	4.2124	7.9345	1.3382	5.6371	0.2374	0.1774	1.8836
6	0.7050	4.9173	11.4594	1.4185	6.9753	0.2034	0.1434	2.3304
7	0.6651	5.5824	15.4497	1.5036	8.3938	0.1791	0.1191	2.7676
8	0.6274	6.2098	19.8416	1.5938	9.8975	0.1610	0.1010	3.1952
9	0.5919	6.8017	24.5768	1.6895	11.4913	0.1470	0.0870	3.6133
10	0.5584	7.3601	29.6023	1.7908	13.1808	0.1359	0.0759	4.0220
11	0.5268	7.8869	34.8702	1.8983	14.9716	0.1268	0.0668	4.4213
12	0.4970	8.3838	40.3369	2.0122	16.8699	0.1193	0.0593	4.8113
13	0.4688	8.8527	45.9629	2.1239	18.8821	0.1130	0.0530	5.1920
14	0.4423	9.2950	51.7128	2.2609	21.0151	0.1076	0.0476	5.5635
15	0.4173	9.7122	57.5546	2.3966	23.2760	0.1030	0.0430	5.9260
16	0.3936	10.1059	63.4592	2.5404	25.6725	0.0990	0.0390	6.2794
17	0.3714	10.4773	69.4011	2.6928	28.2129	0.0954	0.0354	6.6240
18	0.3505	10.8276	75.3569	2.8543	30.9057	0.0924	0.0324	6.9597
19	0.3305	11.1581	81.3062	3.0256	33.7600	0.0896	0.0296	7.2867
20	0.3118	11.4699	87.2304	3.2071	36.7856	0.0872	0.0272	7.6051
21	0.2942	11.7641	93.1136	3.3996	39.9927	0.0850	0.0250	7.9151
22	0.2775	12.0416	98.9412	3.6035	43.3923	0.0830	0.0230	8.2166
23	0.2618	12.3034	104.7007	3.8197	46.9958	0.0813	0.0213	8.5099
24	0.2470	12.5504	110.3812	4.0489	50.8156	0.0797	0.0197	8.7951
25	0.2330	12.7834	115.9732	4.2919	54.8645	0.0782	0.0182	9.0722
30	0.1741	13.7648	142.3588	5.7435	79.0582	0.0726	0.0126	10.3422
40	0.0972	15.0463	185.9568	10.2857	154.7620	0.0665	0.0065	12.3590
50	0.0543	15.7619	217.4574	18.4202	290.3359	0.0634	0.0034	13.7964
60	0.0303	16.1614	239.0428	32.9877	533.1282	0.0619	0.0019	14.7909
100	0.0029	16.6175	272.0471	339.3021	5638.3681	0.0602	0.0002	16.3711

Table 60.10 *Factor Table i = 8.00%*

n	P/F	P/A	P/G	F/P	F/A	A/P	A/F	A/G
1	0.9259	0.9259	0.0000	1.0800	1.0000	1.0800	1.0000	0.0000
2	0.8573	1.7833	0.8573	1.1664	2.0800	0.5608	0.4808	0.4808
3	0.7938	2.5771	2.4450	1.2597	3.2464	0.3880	0.3080	0.9487
4	0.7350	3.3121	4.6501	1.3605	4.5061	0.3019	0.2219	1.4040
5	0.6806	3.9927	7.3724	1.4693	5.8666	0.2505	0.1705	1.8465
6	0.6302	4.6229	10.5233	1.5869	7.3359	0.2163	0.1363	2.2763
7	0.5835	5.2064	14.0242	1.7138	8.9228	0.1921	0.1121	2.6937
8	0.5403	5.7466	17.8061	1.8509	10.6366	0.1740	0.0940	3.0985
9	0.5002	6.2469	21.8081	1.9990	12.4876	0.1601	0.0801	3.4910
10	0.4632	6.7101	25.9768	2.1589	14.4866	0.1490	0.0690	3.8713
11	0.4289	7.1390	30.2657	2.3316	16.6455	0.1401	0.0601	4.2395
12	0.3971	7.5361	34.6339	2.5182	18.9771	0.1327	0.0527	4.5957
13	0.3677	7.9038	39.0463	2.7196	21.4953	0.1265	0.0465	4.9402
14	0.3405	8.2442	43.4723	2.9372	24.2149	0.1213	0.0413	5.2731
15	0.3152	8.5595	47.8857	3.1722	27.1521	0.1168	0.0368	5.5945
16	0.2919	8.8514	52.2640	3.4259	30.3243	0.1130	0.0330	5.9046
17	0.2703	9.1216	56.5883	3.7000	33.7502	0.1096	0.0296	6.2037
18	0.2502	9.3719	60.8426	3.9960	37.4502	0.1067	0.0267	6.4920
19	0.2317	9.6036	65.0134	4.3157	41.4463	0.1041	0.0241	6.7697
20	0.2145	9.8181	69.0898	4.6610	45.7620	0.1019	0.0219	7.0369
21	0.1987	10.0168	73.0629	5.0338	50.4229	0.0998	0.0198	7.2940
22	0.1839	10.2007	76.9257	5.4365	55.4568	0.0980	0.0180	7.5412
23	0.1703	10.3711	80.6726	5.8715	60.8933	0.0964	0.0164	7.7786
24	0.1577	10.5288	84.2997	6.3412	66.7648	0.0950	0.0150	8.0066
25	0.1460	10.6748	87.8041	6.8485	73.1059	0.0937	0.0137	8.2254
30	0.0994	11.2578	103.4558	10.0627	113.2832	0.0888	0.0088	9.1897
40	0.0460	11.9246	126.0422	21.7245	259.0565	0.0839	0.0039	10.5699
50	0.0213	12.2335	139.5928	46.9016	573.7702	0.0817	0.0017	11.4107
60	0.0099	12.3766	147.3000	101.2571	1253.2133	0.0808	0.0008	11.9015
100	0.0005	12.4943	155.6107	2199.7613	27,484.5157	0.0800	–	12.4545

Engineering Economics

Table 60.11 Factor Table $i = 10.00\%$

n	P/F	P/A	P/G	F/P	F/A	A/P	A/F	A/G
1	0.9091	0.9091	0.0000	1.1000	1.0000	1.1000	1.0000	0.0000
2	0.8264	1.7355	0.8264	1.2100	2.1000	0.5762	0.4762	0.4762
3	0.7513	2.4869	2.3291	1.3310	3.3100	0.4021	0.3021	0.9366
4	0.6830	3.1699	4.3781	1.4641	4.6410	0.3155	0.2155	1.3812
5	0.6209	3.7908	6.8618	1.6105	6.1051	0.2638	0.1638	1.8101
6	0.5645	4.3553	9.6842	1.7716	7.7156	0.2296	0.1296	2.2236
7	0.5132	4.8684	12.7631	1.9487	9.4872	0.2054	0.1054	2.6216
8	0.4665	5.3349	16.0287	2.1436	11.4359	0.1874	0.0874	3.0045
9	0.4241	5.7590	19.4215	2.3579	13.5735	0.1736	0.0736	3.3724
10	0.3855	6.1446	22.8913	2.5937	15.9374	0.1627	0.0627	3.7255
11	0.3505	6.4951	26.3962	2.8531	18.5312	0.1540	0.0540	4.0641
12	0.3186	6.8137	29.9012	3.1384	21.3843	0.1468	0.0468	4.3884
13	0.2897	7.1034	33.3772	3.4523	24.5227	0.1408	0.0408	4.6988
14	0.2633	7.3667	36.8005	3.7975	27.9750	0.1357	0.0357	4.9955
15	0.2394	7.6061	40.1520	4.1772	31.7725	0.1315	0.0315	5.2789
16	0.2176	7.8237	43.4164	4.5950	35.9497	0.1278	0.0278	5.5493
17	0.1978	8.0216	46.5819	5.5045	40.5447	0.1247	0.0247	5.8071
18	0.1799	8.2014	49.6395	5.5599	45.5992	0.1219	0.0219	6.0526
19	0.1635	8.3649	52.5827	6.1159	51.1591	0.1195	0.0195	6.2861
20	0.1486	8.5136	55.4069	6.7275	57.2750	0.1175	0.0175	6.5081
21	0.1351	8.6487	58.1095	7.4002	64.0025	0.1156	0.0156	6.7189
22	0.1228	8.7715	60.6893	8.1403	71.4027	0.1140	0.0140	6.9189
23	0.1117	8.8832	63.1462	8.9543	79.5430	0.1126	0.0126	7.1085
24	0.1015	8.9847	65.4813	9.8497	88.4973	0.1113	0.0113	7.2881
25	0.0923	9.0770	67.6964	10.8347	98.3471	0.1102	0.0102	7.4580
30	0.0573	9.4269	77.0766	17.4494	164.4940	0.1061	0.0061	8.1762
40	0.0221	9.7791	88.9525	45.2593	442.5926	0.1023	0.0023	9.0962
50	0.0085	9.9148	94.8889	117.3909	1163.9085	0.1009	0.0009	9.5704
60	0.0033	9.9672	97.7010	304.4816	3,034.8164	0.1003	0.0003	9.8023
100	0.0001	9.9993	99.9202	13,780.6123	137,796.1234	0.1000	–	9.9927

Engineering Economics

Table 60.12 *Factor Table i = 12.00%*

n	P/F	P/A	P/G	F/P	F/A	A/P	A/F	A/G
1	0.8929	0.8929	0.0000	1.1200	1.0000	1.1200	1.0000	0.0000
2	0.7972	1.6901	0.7972	1.2544	2.1200	0.5917	0.4717	0.4717
3	0.7118	2.4018	2.2208	1.4049	3.3744	0.4163	0.2963	0.9246
4	0.6355	3.0373	4.1273	1.5735	4.7793	0.3292	0.2092	1.3589
5	0.5674	3.6048	6.3970	1.7623	6.3528	0.2774	0.1574	1.7746
6	0.5066	4.1114	8.9302	1.9738	8.1152	0.2432	0.1232	2.1720
7	0.4523	4.5638	11.6443	2.2107	10.0890	0.2191	0.0991	2.5515
8	0.4039	4.9676	14.4714	2.4760	12.2997	0.2013	0.0813	2.9131
9	0.3606	5.3282	17.3563	2.7731	14.7757	0.1877	0.0677	3.2574
10	0.3220	5.6502	20.2541	3.1058	17.5487	0.1770	0.0570	3.5847
11	0.2875	5.9377	23.1288	3.4785	20.6546	0.1684	0.0484	3.8953
12	0.2567	6.1944	25.9523	3.8960	24.1331	0.1614	0.0414	4.1897
13	0.2292	6.4235	28.7024	4.3635	28.0291	0.1557	0.0357	4.4683
14	0.2046	6.6282	31.3624	4.8871	32.3926	0.1509	0.0309	4.7317
15	0.1827	6.8109	33.9202	5.4736	37.2797	0.1468	0.0268	4.9803
16	0.1631	6.9740	36.3670	6.1304	42.7533	0.1434	0.0234	5.2147
17	0.1456	7.1196	38.6973	6.8660	48.8837	0.1405	0.0205	5.4353
18	0.1300	7.2497	40.9080	7.6900	55.7497	0.1379	0.0179	5.6427
19	0.1161	7.3658	42.9979	8.6128	63.4397	0.1358	0.0158	5.8375
20	0.1037	7.4694	44.9676	9.6463	72.0524	0.1339	0.0139	6.0202
21	0.0926	7.5620	46.8188	10.8038	81.6987	0.1322	0.0122	6.1913
22	0.0826	7.6446	48.5543	12.1003	92.5026	0.1308	0.0108	6.3514
23	0.0738	7.7184	50.1776	13.5523	104.6029	0.1296	0.0096	6.5010
24	0.0659	7.7843	51.6929	15.1786	118.1552	0.1285	0.0085	6.6406
25	0.0588	7.8431	53.1046	17.001	133.3339	0.1275	0.0075	6.7708
30	0.0334	8.0552	58.7821	29.9599	241.3327	0.1241	0.0041	7.2974
40	0.0107	8.2438	65.1159	93.0510	767.0914	0.1213	0.0013	7.8988
50	0.0035	8.3045	67.7624	289.0022	2,400.0182	0.1204	0.0004	8.1597
60	0.0011	8.3240	68.8100	897.5969	7,471.6411	0.1201	0.0001	8.2664
100	–	8.3332	69.4336	83,522.2657	696,010.5477	0.1200	–	8.3321

Engineering Economics

Table 60.13 Factor Table i = 18.00%

n	P/F	P/A	P/G	F/P	F/A	A/P	A/F	A/G
1	0.8475	0.8475	0.0000	1.1800	1.0000	1.1800	1.0000	0.0000
2	0.7182	1.5656	0.7182	1.3924	2.1800	0.6387	0.4587	0.4587
3	0.6086	2.1743	1.9354	1.6430	3.5724	0.4599	0.2799	0.8902
4	0.5158	2.6901	3.4828	1.9388	5.2154	0.3717	0.1917	1.2947
5	0.4371	3.1272	5.2312	2.2878	7.1542	0.3198	0.1398	1.6728
6	0.3704	3.4976	7.0834	2.6996	9.4423	0.2859	0.1059	2.0252
7	0.3139	3.8115	8.9670	3.1855	12.1415	0.2624	0.0824	2.3526
8	0.2660	4.0776	10.8292	3.7589	15.3270	0.2452	0.0652	2.6558
9	0.2255	4.3030	12.6329	4.4355	19.0859	0.2324	0.0524	2.9358
10	0.1911	4.4941	14.3525	5.2338	23.5213	0.2225	0.0425	3.1936
11	0.1619	4.6560	15.9716	6.1759	28.7551	0.2148	0.0348	3.4303
12	0.1372	4.7932	17.4811	7.2876	34.9311	0.2086	0.0286	3.6470
13	0.1163	4.9095	18.8765	8.5994	42.2187	0.2037	0.0237	3.8449
14	0.0985	5.0081	20.1576	10.1472	50.8180	0.1997	0.0197	4.0250
15	0.0835	5.0916	21.3269	11.9737	60.9653	0.1964	0.0164	4.1887
16	0.0708	5.1624	22.3885	14.1290	72.9390	0.1937	0.0137	4.3369
17	0.0600	5.2223	23.3482	16.6722	87.0680	0.1915	0.0115	4.4708
18	0.0508	5.2732	24.2123	19.6731	103.7403	0.1896	0.0096	4.5916
19	0.0431	5.3162	24.9877	23.2144	123.4135	0.1881	0.0081	4.7003
20	0.0365	5.3527	25.6813	27.3930	146.6280	0.1868	0.0068	4.7978
21	0.0309	5.3837	26.3000	32.3238	174.0210	0.1857	0.0057	4.8851
22	0.0262	5.4099	26.8506	38.1421	206.3448	0.1848	0.0048	4.9632
23	0.0222	5.4321	27.3394	45.0076	244.4868	0.1841	0.0041	5.0329
24	0.0188	5.4509	27.7725	53.1090	289.4944	0.1835	0.0035	5.0950
25	0.0159	5.4669	28.1555	62.6686	342.6035	0.1829	0.0029	5.1502
30	0.0070	5.5168	29.4864	143.3706	790.9480	0.1813	0.0013	5.3448
40	0.0013	5.5482	30.5269	750.3783	4,163.2130	0.1802	0.0002	5.5022
50	0.0003	5.5541	30.7856	3,927.3569	21,813.0937	0.1800	–	5.5428
60	0.0001	5.5553	30.8465	20,555.1400	114,189.6665	0.1800	–	5.5526
100	–	5.5556	30.8642	15,424,131.91	85,689,616.17	0.1800	–	5.5555

Engineering Economics

Topic XVI: Ethics and Professional Practice

61 Professional Practice

1. AGREEMENTS AND CONTRACTS

General Contracts

A *contract* is a legally binding agreement or promise to exchange goods or services.[1] A written contract is merely a documentation of the agreement. Some agreements must be in writing, but most agreements for engineering services can be verbal, particularly if the parties to the agreement know each other well.[2] Written contract documents do not need to contain intimidating legal language, but all agreements must satisfy three basic requirements to be enforceable (binding).

- There must be a clear, specific, and definite *offer* with no room for ambiguity or misunderstanding.
- There must be some form of conditional future *consideration* (i.e., payment).[3]
- There must be an *acceptance* of the offer.

There are other conditions that the agreement must meet to be enforceable. These conditions are not normally part of the explicit agreement but represent the conditions under which the agreement was made.

- The agreement must be *voluntary* for all parties.
- All parties must have *legal capacity* (i.e., be mentally competent, of legal age, not under coercion, and uninfluenced by drugs).
- The purpose of the agreement must be *legal*.

For small projects, a simple *letter of agreement* on one party's stationery may suffice. For larger, complex projects, a more formal document may be required. Some clients prefer to use a *purchase order*, which can function as a contract if all basic requirements are met.

Regardless of the format of the written document—letter of agreement, purchase order, or standard form—a contract should include the following features.[4]

- introduction, preamble, or preface indicating the purpose of the contract
- name, address, and business forms of both contracting parties
- signature date of the agreement
- effective date of the agreement (if different from the signature date)
- duties and obligations of both parties
- deadlines and required service dates
- fee amount
- fee schedule and payment terms
- agreement expiration date
- standard boilerplate clauses
- signatures of parties or their agents
- declaration of authority of the signatories to bind the contracting parties
- supporting documents

Example

Which feature is NOT a standard feature of a written construction contract?

(A) identification of both parties

(B) specific details of the obligations of both parties

(C) boilerplate clauses

(D) subcontracts

[1]Not all agreements are legally binding (i.e., enforceable). Two parties may agree on something, but unless the agreement meets all of the requirements and conditions of a contract, the parties cannot hold each other to the agreement.

[2]All states have a *statute of frauds* that, among other things, specifies what types of contracts must be in writing to be enforceable. These include contracts for the sale of land, contracts requiring more than one year for performance, contracts for the sale of goods over $500 in value, contracts to satisfy the debts of another, and marriage contracts. Contracts to provide engineering services do not fall under the statute of frauds.

[3]Actions taken or payments made prior to the agreement are irrelevant. Also, it does not matter to the courts whether the exchange is based on equal value or not.

[4]*Construction contracts* are unique unto themselves. Items that might also be included as part of the *contract documents* are the agreement form, the general conditions, drawings, specifications, and addenda.

Solution

A written contract should identify both parties, state the purpose of the contract and the obligations of the parties, give specific details of the obligations (including relevant dates and deadlines), specify the consideration, state the boilerplate clauses to clarify the contract terms, and leave places for signatures. Subcontracts are not required to be included, but may be added when a party to the contract engages a third party to perform the work in the original contract.

The answer is (D).

Agency

In some contracts, decision-making authority and right of action are transferred from one party (the owner, or *principal*) who would normally have that authority to another person (the *agent*). For example, in construction contracts, the engineer may be the agent of the owner for certain transactions. Agents are limited in what they can do by the scope of the agency agreement. Within that scope, however, an agent acts on behalf of the principal, and the principal is liable for the acts of the agent and is bound by contracts made in the principal's name by the agent.

Agents are required to execute their work with care, skill, and diligence. Specifically, agents have *fiduciary responsibility* toward their principal, meaning that the agent must be honest and loyal. Agents are liable for damages resulting from a lack of diligence, loyalty, and/or honesty. If the agents misrepresented their skills when obtaining the agency, they can be liable for breach of contract or fraud.

Standard Boilerplate Clauses

It is common for full-length contract documents to include important *boilerplate clauses.* These clauses have specific wordings that should not normally be changed, hence the name "boilerplate." Some of the most common boilerplate clauses are paraphrased here.

- Delays and inadequate performance due to war, strikes, and acts of God and nature are forgiven (*force majeure*).
- The contract document is the complete agreement, superseding all prior verbal and written agreements.
- The contract can be modified or canceled only in writing.
- Parts of the contract that are determined to be void or unenforceable will not affect the enforceability of the remainder of the contract (*severability*). Alternatively, parts of the contract that are determined to be void or unenforceable will be rewritten to accomplish their intended purpose without affecting the remainder of the contract.
- None (or one, or both) of the parties can (or cannot) assign its (or their) rights and responsibilities under the contract (*assignment*).
- All notices provided for in the agreement must be in writing and sent to the address in the agreement.
- Time is of the essence.[5]
- The subject headings of the agreement paragraphs are for convenience only and do not control the meaning of the paragraphs.
- The laws of the state in which the contract is signed must be used to interpret and govern the contract.
- Disagreements shall be arbitrated according to the rules of the American Arbitration Association.
- Any lawsuits related to the contract must be filed in the county and state in which the contract is signed.
- Obligations under the agreement are unique, and in the event of a breach, the defaulting party waives the defense that the loss can be adequately compensated by monetary damages (*specific performance*).
- In the event of a lawsuit, the prevailing party is entitled to an award of reasonable attorneys' and court fees.[6]
- Consequential damages are not recoverable in a lawsuit.

Example

An engineering consultant has signed a standard owner-engineer contract to build a scale model of a bridge in time for the owner to present the proposal to a financing committee. After the scale model has been built, it is destroyed in a building fire that consumes the consultant's building. The consultant is unable to rebuild the model in time. A breach of contract judgment against the consultant is most likely NOT obtainable due to which legal argument?

(A) caveat emptor

(B) privity of contract

(C) force majeure

(D) strict liability in tort

[5]Without this clause in writing, damages for delay cannot be claimed.

[6]Without this clause in writing, attorneys' fees and court costs are rarely recoverable.

Solution

Standard contracts have force majeure clauses that excuse nonperformance due to "acts of God" and other unforeseen events such as weather and acts of terrorism.

The answer is (C).

Subcontracts

When a party to a contract engages a third party to perform the work in the original contract, the contract with the third party is known as a *subcontract*. Whether or not responsibilities can be subcontracted under the original contract depends on the content of the *assignment clause* in the original contract.

Parties to a Construction Contract

A specific set of terms has developed for referring to parties in consulting and construction contracts. The *owner* of a construction project is the person, partnership, or corporation that actually owns the land, assumes the financial risk, and ends up with the completed project. The *developer* contracts with the architect and/or engineer for the design and with the contractors for the construction of the project. In some cases, the owner and developer are the same, in which case the term *owner-developer* can be used.

The *architect* designs the project according to established codes and guidelines but leaves most stress and capacity calculations to the *engineer*.[7] Depending on the construction contract, the engineer may work for the architect, or vice versa, or both may work for the developer.

Once there are approved plans, the developer hires *contractors* to do the construction. Usually, the entire construction project is awarded to a *general contractor*. Due to the nature of the construction industry, separate *subcontracts* are used for different tasks (electrical, plumbing, mechanical, framing, fire sprinkler installation, finishing, etc.). The general contractor who hires all of these different *subcontractors* is known as the *prime contractor* (or *prime*). (The subcontractors can also work directly for the owner-developer, although this is less common.) The prime contractor is responsible for all acts of the subcontractors and is liable for any damage suffered by the owner-developer due to those acts.

Construction is managed by an agent of the owner-developer known as the *construction manager*, who may be the engineer, the architect, or someone else.

Standard Contracts for Design Professionals

Several professional organizations have produced standard agreement forms and other standard documents for design professionals.[8] Among other standard forms, notices, and agreements, the following standard contracts are available.[9]

- standard contract between engineer and client
- standard contract between engineer and architect
- standard contract between engineer and contractor
- standard contract between owner and construction manager

Besides completeness, the major advantage of a standard contract is that the meanings of the clauses are well established, not only among the design professionals and their clients but also in the courts. The clauses in these contracts have already been litigated many times. Where a clause has been found to be unclear or ambiguous, it has been rewritten to accomplish its intended purpose.

Consulting Fee Structure

Compensation for consulting engineering services can incorporate one or more of the following concepts.

- *lump-sum fee:* This is a predetermined fee agreed upon by client and engineer. This payment can be used for small projects where the scope of work is clearly defined.
- *unit price:* Contract fees are based on estimated quantities and unit pricing. This payment method works best when required materials can be accurately identified and estimated before the contract is finalized. This payment method is often used in combination with a lump-sum fee.

[7]On simple small projects, such as wood-framed residential units, the design may be developed by a *building designer*. The legal capacities of building designers vary from state to state.

[8]There are two main sources of standardized construction and design agreements: the *Engineers Joint Contract Documents Committee* (EJCDC) and American Institute of Architects (AIA). Consensus documents, known as *ConsensusDOCS*, for every conceivable situation have been developed by EJCDC. EJCDC includes the American Society of Civil Engineers (ASCE), the American Council of Engineering Companies (ACEC), National Society of Professional Engineers' (NSPE's) Professional Engineers in Private Practice Division, Associated General Contractors of America (AGC), and more than fifteen other participating professional engineering design, construction, owner, legal, and risk management organizations, including the Associated Builders and Contractors; American Subcontractors Association; Construction Users Roundtable; National Roofing Contractors Association; Mechanical Contractors Association of America; and National Plumbing-Heating-Cooling Contractors Association. The AIA has developed its own standardized agreements in a less collaborative manner. Though popular with architects, AIA provisions are considered less favorable to engineers, contractors, and subcontractors who believe the AIA documents assign too much authority to architects, too much risk and liability to contractors, and too little flexibility in how construction disputes are addressed and resolved.

[9]The Construction Specifications Institute (CSI) has produced standard specifications for materials. The standards have been organized according to a UNIFORMAT structure consistent with ASTM Standard E1557.

- *cost plus fixed fee:* All costs (labor, material, travel, etc.) incurred by the engineer are paid by the client. The client also pays a predetermined fee as profit. This method has an advantage when the scope of services cannot be determined accurately in advance. Detailed records must be kept by the engineer in order to allocate costs among different clients.

- *per diem fee:* The engineer is paid a specific sum for each day spent on the job. Usually, certain direct expenses (e.g., travel and reproduction) are billed in addition to the per diem rate.

- *salary plus:* The client pays for the employees on an engineer's payroll (the salary) plus an additional percentage to cover indirect overhead and profit plus certain direct expenses.

- *retainer:* This is a minimum amount paid by the client, usually in total and in advance, for a normal amount of work expected during an agreed-upon period. Usually, none of the retainer is returned, regardless of how little work the engineer performs. The engineer can be paid for additional work beyond what is normal, however. Some direct costs, such as travel and reproduction expenses, may be billed directly to the client.

- *incentive:* This type of fee structure is based on established target costs and fees and lists minimum and maximum fees and an adjustment formula. The formula may be based on performance criteria such as budget, quality, and schedule. Once the project is complete, payment is calculated based on the formula.

- *percentage of construction cost:* This method, which is widely used in construction design contracts, pays the architect and/or the engineer a percentage of the final total cost of the project. Costs of land, financing, and legal fees are generally not included in the construction cost, and other costs (plan revisions, project management labor, value engineering, etc.) are billed separately.

Example

Which fee structure is a nonreturnable advance paid to a consultant?

(A) per diem fee

(B) retainer

(C) lump-sum fee

(D) cost plus fixed fee

Solution

A *retainer* is a (usually) nonreturnable advance paid by the client to the consultant. While the retainer may be intended to cover the consultant's initial expenses until the first big billing is sent out, there does not need to be any rational basis for the retainer. Often, a small retainer is used by the consultant to qualify the client (i.e., to make sure the client is not just shopping around and getting free initial consultations) and as a security deposit (to make sure the client does not change consultants after work begins).

The answer is (B).

Mechanic's Liens

For various reasons, providers of material, labor, and design services to construction sites may not be promptly paid or even paid at all. Such providers have, of course, the right to file a lawsuit demanding payment, but due to the nature of the construction industry, such relief may be insufficient or untimely. Therefore, such providers have the right to file a *mechanic's lien* (also known as a *construction lien, materialman's lien, supplier's lien,* or *laborer's lien*) against the property. Although there are strict requirements for deadlines, filing, and notices, the procedure for obtaining (and removing) such a lien is simple. The lien establishes the supplier's security interest in the property. Although the details depend on the state, essentially the property owner is prevented from transferring title of (i.e., selling) the property until the lien has been removed by the supplier. The act of filing a lawsuit to obtain payment is known as "perfecting the lien." Liens are perfected by forcing a judicial foreclosure sale. The court orders the property sold, and the proceeds are used to pay off any lienholders.

Discharge of a Contract

A contract is normally discharged when all parties have satisfied their obligations. However, a contract can also be terminated for the following reasons:

- mutual agreement of all parties to the contract
- impossibility of performance (e.g., death of a party to the contract)
- illegality of the contract
- material breach by one or more parties to the contract
- fraud on the part of one or more parties
- failure (i.e., loss or destruction) of consideration (e.g., the burning of a building one party expected to own or occupy upon satisfaction of the obligations)

Some contracts may be dissolved by actions of the court (e.g., bankruptcy), passage of new laws and public acts, or a declaration of war.

Extreme difficulty (including economic hardship) in satisfying the contract does not discharge it, even if it becomes more costly or less profitable than originally anticipated.

2. PROFESSIONAL LIABILITY

Breach of Contract, Negligence, Misrepresentation, and Fraud

A *breach of contract* occurs when one of the parties fails to satisfy all of its obligations under a contract. The breach can be *willful* (as in a contractor walking off a construction job) or *unintentional* (as in providing less than adequate quality work or materials). A *material breach* is defined as nonperformance that results in the injured party receiving something substantially less than or different from what the contract intended.

Normally, the only redress that an *injured party* has through the courts in the event of a breach of contract is to force the breaching party to provide *specific performance*—that is, to satisfy all remaining contract provisions and to pay for any damage caused. Normally, *punitive damages* (to punish the breaching party) are unavailable.

Negligence is an action, willful or unwillful, taken without proper care or consideration for safety, resulting in damages to property or injury to persons. "Proper care" is a subjective term, but in general it is the diligence that would be exercised by a reasonably prudent person.[10] Damages sustained by a negligent act are recoverable in a tort action. (See "Torts.") If the plaintiff is partially at fault (as in the case of *comparative negligence*), the defendant will be liable only for the portion of the damage caused by the defendant.

Punitive damages are available, however, if the breaching party was fraudulent in obtaining the contract. In addition, the injured party has the right to void (nullify) the contract entirely. A *fraudulent act* is basically a special case of *misrepresentation* (i.e., an intentionally false statement known to be false at the time it is made). Misrepresentation that does not result in a contract is a tort. When a contract is involved, misrepresentation can be a breach of that contract (i.e., *fraud*).

Unfortunately, it is extremely difficult to prove *compensatory fraud* (i.e., fraud for which damages are available). Proving fraud requires showing *beyond a reasonable doubt* (a) a reckless or intentional misstatement of a material fact, (b) an intention to deceive, (c) it resulted in misleading the innocent party to contract, and (d) it was to the innocent party's detriment.

For example, if an engineer claims to have experience in designing steel buildings but actually has none, the court might consider the misrepresentation a fraudulent action. If, however, the engineer has some experience, but an insufficient amount to do an adequate job, the engineer probably will not be considered to have acted fraudulently.

Example

The owner of a construction site is aware that the state driving license of one of its heavy machinery operators has been suspended for multiple driving under the influence (DUI) violations. In order to secure a desirable standing and contract with an abstinence-based commune, the owner misrepresents the non-drinking status of the construction crew. A serious injury occurs when the operator drives a loader over the leg of a member of the commune while intoxicated. Most likely, the commune will be able to obtain a judgment against the operator based on

(A) negligence

(B) breach of contract

(C) misrepresentation

(D) fraud

Solution

All elements necessary to obtain a judgement against the owner based on fraud are present in this scenario. The operator is, most likely, guilty only of negligence.

The answer is (A).

Torts

A *tort* is a civil wrong committed by one person causing damage to another person or person's property, emotional well-being, or reputation.[11] It is a breach of the rights of an individual to be secure in person or property. In order to correct the wrong, a civil lawsuit (*tort action* or *civil complaint*) is brought by the alleged injured party (the *plaintiff*) against the *defendant*. To be a valid *tort action* (i.e., lawsuit), there must have been injury (i.e., damage). Generally, there will be no contract between the two parties, so the tort action cannot claim a breach of contract.[12]

Tort law is concerned with compensation for the injury, not punishment. Therefore, tort awards usually consist of general, compensatory, and special damages and rarely include punitive and exemplary damages. (See "Damages" for definitions of these damages.)

[10]Negligence of a design professional (e.g., an engineer or architect) is the absence of a *standard of care* (i.e., customary and normal care and attention) that would have been provided by other engineers. It is highly subjective.

[11]The difference between a *civil tort* (*lawsuit*) and a *criminal lawsuit* is the alleged injured party. A *crime* is a wrong against society. A criminal lawsuit is brought by the state against a defendant.

[12]It is possible for an injury to be both a breach of contract and a tort. Suppose an owner has an agreement with a contractor to construct a building, and the contract requires the contractor to comply with all state and federal safety regulations. If the owner is subsequently injured on a stairway because there was no guardrail, the injury could be recoverable both as a tort and as a breach of contract. If a third party unrelated to the contract was injured, however, that party could recover only through a tort action.

Strict Liability in Tort

Strict liability in tort means that the injured party wins if the injury can be proven. It is not necessary to prove negligence, breach of explicit or implicit warranty, or the existence of a contract (*privity of contract*). Strict liability in tort is most commonly encountered in product liability cases. A defect in a product, regardless of how the defect got there, is sufficient to create strict liability in tort.

Case law surrounding defective products has developed and refined the following requirements for winning a strict liability in tort case. The following points must be proved.

- The product was defective in manufacture, design, labeling, and so on.
- The product was defective when used.
- The defect rendered the product unreasonably dangerous.
- The defect caused the injury.
- The specific use of the product that caused the damage was reasonably foreseeable.

Manufacturing and Design Liability

Case law makes a distinction between *design professionals* (architects, structural engineers, building designers, etc.) and manufacturers of consumer products. Design professionals are generally consultants whose primary product is a design service sold to sophisticated clients. Consumer product manufacturers produce specific product lines sold through wholesalers and retailers to the unsophisticated public.

The law treats design professionals favorably. Such professionals are expected to meet a *standard of care* and skill that can be measured by comparison with the conduct of other professionals. However, professionals are not expected to be infallible. In the absence of a contract provision to the contrary, design professionals are not held to be guarantors of their work in the strict sense of legal liability. Damages incurred due to design errors are recoverable through tort actions, but proving a breach of contract requires showing negligence (i.e., not meeting the standard of care).

On the other hand, the law is much stricter with consumer product manufacturers, and perfection is (essentially) expected of them. They are held to the standard of strict liability in tort without regard to negligence. A manufacturer is held liable for all phases of the design and manufacturing of a product being marketed to the public.[13]

Prior to 1916, the court's position toward product defects was exemplified by the expression *caveat emptor* ("let the buyer beware").[14] Subsequent court rulings have clarified that "... a manufacturer is strictly liable in tort when an article [it] places on the market, knowing that it will be used without inspection, proves to have a defect that causes injury to a human being."[15]

Although all defectively designed products can be traced back to a design engineer or team, only the manufacturing company is usually held liable for injury caused by the product. This is more a matter of economics than justice. The company has liability insurance; the product design engineer (who is merely an employee of the company) probably does not. Unless the product design or manufacturing process is intentionally defective, or unless the defect is known in advance and covered up, the product design engineer will rarely be punished by the courts.[16]

Example

An engineer designs and self-manufactures a revolutionary racing bicycle. The engineer uses finite element analysis (FEA) software to perfect the design, has the design checked by a reputable authority, and subjects the major components that he manufactures to nondestructive testing. After three years of heavier-than-anticipated usage, one of the engineer's bicycles disintegrates in a race, killing its rider. In his defense, the engineer may claim

(A) privity of contract

(B) standard of care

(C) statute of limitations

(D) contributory negligence

Solution

The scenario does not contain information about initial and ongoing testing, but the engineer has done everything described in a competent manner. While the engineer may still be held responsible in some manner, he

---

[13]The reason for this is that the public is not considered to be as sophisticated as a client who contracts with a design professional for building plans.

[14]1916, *MacPherson v. Buick*. MacPherson bought a Buick from a car dealer. The car had a defective wheel, and there was evidence that reasonable inspection would have uncovered the defect. MacPherson was injured when the wheel broke and the car collapsed, and he sued Buick. Buick defended itself under the ancient *prerequisite of privity* (i.e., the requirement of a face-to-face contractual relationship in order for liability to exist), since the dealer, not Buick, had sold the car to MacPherson, and no contract between Buick and MacPherson existed. The judge disagreed, thus establishing the concept of *third-party liability* (i.e., manufacturers are responsible to consumers even though consumers do not buy directly from manufacturers).

[15]1963, *Greenman v. Yuba Power Products*. Greenman purchased and was injured by an electric power tool.

[16]The engineer can expect to be discharged from the company. However, for strategic reasons, this discharge probably will not occur until after the company loses the case.

has met a normal standard of care in the design and manufacturing of his bicycles.

The answer is (B).

Damages

An injured party can sue for *damages* as well as for specific performance. Damages are the award made by the court for losses incurred by the injured party.

- *General* or *compensatory damages* are awarded to make up for the injury that was sustained.
- *Special damages* are awarded for the direct financial loss due to the breach of contract.
- *Nominal damages* are awarded when responsibility has been established but the injury is so slight as to be inconsequential.
- *Liquidated damages* are amounts that are specified in the contract document itself for nonperformance.
- *Punitive* or *exemplary damages* are awarded, usually in tort and fraud cases, to punish and make an example of the defendant (i.e., to deter others from doing the same thing).
- *Consequential damages* provide compensation for indirect losses incurred by the injured party but not directly related to the contract.

Insurance

Most design firms and many independent design professionals carry *errors and omissions insurance* to protect them from claims due to their mistakes. Such policies are costly, and for that reason, some professionals choose to "go bare."[17] Policies protect against inadvertent mistakes only, not against willful, knowing, or conscious efforts to defraud or deceive.

[17]Going bare appears foolish at first glance, but there is a perverted logic behind the strategy. One-person consulting firms (and perhaps, firms that are not profitable) are "judgment-proof." Without insurance or other assets, these firms would be unable to pay any large judgments against them. When damage victims (and their lawyers) find this out in advance, they know that judgments will be uncollectable. So the lawsuit often never makes its way to trial.

62 Ethics

1. CODES OF ETHICS

Creeds, Rules, Statutes, Canons, and Codes

It is generally conceded that an individual acting on his or her own cannot be counted on to always act in a proper and moral manner. Creeds, rules, statutes, canons, and codes all attempt to complete the guidance needed for an engineer to do "...the correct thing."

A *creed* is a statement or oath, often religious in nature, taken or assented to by an individual in ceremonies. For example, the *Engineers' Creed* adopted by the National Society of Professional Engineers (NSPE) is[1]

> As a Professional Engineer, I dedicate my professional knowledge and skill to the advancement and betterment of human welfare.
>
> I pledge...
>
> ... to give the utmost of performance;
>
> ... to participate in none but honest enterprise;
>
> ... to live and work according to the laws of man and the highest standards of professional conduct;
>
> ... to place service before profit, the honor and standing of the profession before personal advantage, and the public welfare above all other considerations.
>
> In humility and with need for Divine Guidance, I make this pledge.

A *rule* is a guide (principle, standard, or norm) for conduct and action in a certain situation, or a regulation governing procedure. A *statutory rule*, or statute, is enacted by the legislative branch of state or federal government and carries the weight of law. Some U.S. engineering registration boards have statutory *rules of professional conduct.*

A *canon* is an individual principle or body of principles, rules, standards, or norms. A *code* is a system of principles or rules. For example, the code of ethics of the American Society of Civil Engineers (ASCE) contains the following seven canons.

1. Engineers shall hold paramount the safety, health, and welfare of the public and shall strive to comply with the principles of sustainable development in the performance of their professional duties.
2. Engineers shall perform services only in areas of their competence.
3. Engineers shall issue public statements only in an objective and truthful manner.
4. Engineers shall act in professional matters for each employer or client as faithful agents or trustees and shall avoid conflicts of interest.
5. Engineers shall build their professional reputation on the merit of their services and shall not compete unfairly with others.
6. Engineers shall act in such a manner as to uphold and enhance the honor, integrity, and dignity of the engineering profession and shall act with zero tolerance for bribery, fraud, and corruption.
7. Engineers shall continue their professional development throughout their careers and shall provide opportunities for the professional development of those engineers under their supervision.

Example

Relative to the practice of engineering, which one of the following best defines "ethics"?

(A) application of United States laws

(B) rules of conduct

(C) personal values

(D) recognition of cultural differences

[1]The *Faith of an Engineer* adopted by the Accreditation Board for Engineering and Technology (ABET) is a similar but more detailed creed.

Solution

Ethics are the rules of conduct recognized in respect to a particular class of human actions or governing a particular group, culture, and so on.

The answer is (B).

Purpose of a Code of Ethics

Many different sets of *codes of ethics* (*canons of ethics*, *rules of professional conduct*, etc.) have been produced by various engineering societies, registration boards, and other organizations.[2] The purpose of these ethical guidelines is to guide the conduct and decision making of engineers. Most codes are primarily educational. Nevertheless, from time to time they have been used by the societies and regulatory agencies as the basis for disciplinary actions.

Fundamental to ethical codes is the requirement that engineers render faithful, honest, professional service. In providing such service, engineers must represent the interests of their employers or clients and, at the same time, protect public health, safety, and welfare.

There is an important distinction between what is legal and what is ethical. Many legal actions can be violations of codes of ethical or professional behavior. For example, an engineer's contract with a client may give the engineer the right to assign the engineer's responsibilities, but doing so without informing the client would be unethical.

Ethical guidelines can be categorized on the basis of who is affected by the engineer's actions—the client, vendors and suppliers, other engineers, or the public at large. (Some authorities also include ethical guidelines for dealing with the employees of an engineer. However, these guidelines are no different for an engineering employer than they are for a supermarket, automobile assembly line, or airline employer. Ethics is not a unique issue when it comes to employees.)

Example

Complete the sentence: "Guidelines of ethical behavior among engineers are needed because

(A) engineers are analytical and they don't always think in terms of right or wrong."

(B) all people, including engineers, are inherently unethical."

(C) rules of ethics are easily forgotten."

(D) it is easy for engineers to take advantage of clients."

Solution

Untrained members of society are at the mercy of the professionals (e.g., doctors, lawyers, engineers) they employ. Even a cab driver can take advantage of a new tourist who doesn't know the shortest route between two points. In many cases, the unsuspecting public needs protection from unscrupulous professionals, engineers included, who act in their own interest.

The answer is (D).

2. SUSTAINABILITY

Many professional societies' codes of ethics stress the importance of incorporating sustainability into engineering design and development. *Sustainability* (also known as *sustainable development* or *sustainable design*) encompasses a wide range of concepts and strategies. However, a general definition of sustainability is any design or development that seeks to minimize negative impacts on the environment so that the present generation's resource needs do not compromise the resource needs of a future generation. Examples of sustainable design principles include using renewable energy sources, conserving water, and using *sustainable materials* (materials sourced, manufactured, and transported with sustainability in mind).

3. NCEES MODEL LAW

Introduction[3]

Engineering is considered to be a "profession" rather than an occupation because of several important characteristics shared with other recognized learned professions, law, medicine, and theology: special knowledge, special privileges, and special responsibilities. Professions are based on a large knowledge base requiring extensive training. Professional skills are important to the well-being of society. Professions are self-regulating, in that they control the training and evaluation processes that admit new persons to the field. Professionals have autonomy in the workplace; they are expected to utilize their independent judgment in carrying out their professional responsibilities. Finally, professions are regulated by ethical standards.

The expertise possessed by engineers is vitally important to public welfare. In order to serve the public effectively, engineers must maintain a high level of technical competence. However, a high level of technical expertise without adherence to ethical guidelines is as much a

[2]All of the major engineering technical and professional societies in the United States (ASCE, IEEE, ASME, AIChE, NSPE, etc.) and throughout the world have adopted codes of ethics. Most U.S. societies have endorsed the *Code of Ethics of Engineers* developed by the Accreditation Board for Engineering and Technology (ABET), formerly the Engineers' Council for Professional Development (ECPD). The National Council of Examiners for Engineering and Surveying (NCEES) has developed its *Model Rules* as a guide for state registration boards in developing guidelines for the professional engineers in those states.

[3]Adapted from C. E. Harris, M. S. Pritchard, and M. J. Rabins, *Engineering Ethics: Concepts and Cases*, copyright © 1995 by Wadsworth Publishing Company, pg. 27–28.

threat to public welfare as is professional incompetence. Therefore, engineers must also be guided by ethical principles.

The ethical principles governing the engineering profession are embodied in codes of ethics. Such codes have been adopted by state boards of registration, professional engineering societies, and even by some private industries. An example of one such code is the NCEES Rules of Professional Conduct, found in Section 240 of *Model Rules* and presented here. As part of his/her responsibility to the public, an engineer is responsible for knowing and abiding by the code. Additional rules of conduct are also included in *Model Rules*.

The three major sections of Model Rules address (1) Licensee's Obligation to Society, (2) Licensee's Obligation to Employers and Clients, and (3) Licensee's Obligation to Other Licensees. The principles amplified in these sections are important guides to appropriate behavior of professional engineers.

Application of the code in many situations is not controversial. However, there may be situations in which applying the code may raise more difficult issues. In particular, there may be circumstances in which terminology in the code is not clearly defined, or in which two sections of the code may be in conflict. For example, what constitutes "valuable consideration" or "adequate" knowledge may be interpreted differently by qualified professionals. These types of questions are called *conceptual issues*, in which definitions of terms may be in dispute. In other situations, *factual issues* may also affect ethical dilemmas. Many decisions regarding engineering design may be based upon interpretation of disputed or incomplete information. In addition, *trade-offs* revolving around competing issues of risk vs. benefit, or safety vs. economics, may require judgments that are not fully addressed simply by application of the code.

No code can give immediate and mechanical answers to all ethical and professional problems that an engineer may face. Creative problem solving is often called for in ethics, just as it is in other areas of engineering.

Example

Which organizations typically do NOT enforce codes of ethics for engineers?

(A) technical societies (e.g., ASCE, ASME, IEEE)

(B) national professional societies (e.g., the National Society of Professional Engineers)

(C) state professional societies (e.g., the Michigan Society of Professional Engineers)

(D) companies that write, administer, and grade licensing exams

Solution

Companies that write, administer, and grade licensing exams typically do not enforce codes of ethics for engineers.

The answer is (D).

Licensee's Obligation to Society[4]

1. Licensees, in the performance of their services for clients, employers, and customers, shall be cognizant that their first and foremost responsibility is to the public welfare.

2. Licensees shall approve and seal only those design documents and surveys that conform to accepted engineering and surveying standards and safeguard the life, health, property, and welfare of the public.

3. Licensees shall notify their employer or client and such other authority as may be appropriate when their professional judgment is overruled under circumstances where the life, health, property, or welfare of the public is endangered.

4. Licensees shall be objective and truthful in professional reports, statements, or testimony. They shall include all relevant and pertinent information in such reports, statements, or testimony.

5. Licensees shall express a professional opinion publicly only when it is founded upon an adequate knowledge of the facts and a competent evaluation of the subject matter.

6. Licensees shall issue no statements, criticisms, or arguments on technical matters which are inspired or paid for by interested parties, unless they explicitly identify the interested parties on whose behalf they are speaking and reveal any interest they have in the matters.

7. Licensees shall not permit the use of their name or firm name by, nor associate in the business ventures with, any person or firm which is engaging in fraudulent or dishonest business of professional practices.

8. Licensees having knowledge of possible violations of any of these Rules of Professional Conduct shall provide the board with the information and assistance necessary to make the final determination of such violation.

Example

While working to revise the design of the suspension for a popular car, an engineer discovers a flaw in the design currently being produced. Based on a statistical

[4]Adapted from *NCEES FE Reference Handbook* (*NCEES Handbook*), 9th Ed., pg. 3–4, © 2013 by the National Council of Examiners for Engineering and Surveying® (ncees.org).

analysis, the company determines that although this mistake is likely to cause a small increase in the number of fatalities seen each year, it would be prohibitively expensive to do a recall to replace the part. Accordingly, the company decides not to issue a recall notice. What should the engineer do?

(A) The engineer should go along with the company's decision. The company has researched its options and chosen the most economic alternative.

(B) The engineer should send an anonymous tip to the media, suggesting that they alert the public and begin an investigation of the company's business practices.

(C) The engineer should notify the National Transportation Safety Board (NTSB), providing enough details for them to initiate a formal inquiry.

(D) The engineer should resign from the company. Because of standard nondisclosure agreements, it would be unethical as well as illegal to disclose any information about this situation. In addition, the engineer should not associate with a company that is engaging in such behavior.

Solution

The engineer's highest obligation is to the public's safety. In most instances, it would be unethical to take some public action on a matter without providing the company with the opportunity to resolve the situation internally. In this case, however, it appears as though the company's senior officers have already reviewed the case and made a decision. The engineer must alert the proper authorities, the NTSB, and provide them with any assistance necessary to investigate the case. To contact the media, although it might accomplish the same goal, would fail to fulfill the engineer's obligation to notify the authorities.

The answer is (C).

Licensee's Obligation to Employer and Clients

1. Licensees shall undertake assignments only when qualified by education or experience in the specific technical fields of engineering or surveying involved.

2. Licensees shall not affix their signatures or seals to any plans or documents dealing with subject matter in which they lack competence, nor to any such plan or document not prepared under their direct control and personal supervision.

3. Licensees may accept assignments for coordination of an entire project, provided that each design segment is signed and sealed by the licensee responsible for preparation of that design segment.

4. Licensees shall not reveal facts, data, or information obtained in a professional capacity without the prior consent of the client or employer except as authorized or required by law. Licensees shall not solicit or accept gratuities, directly or indirectly, from contractors, their agents, or other parties in connection with work for employers or clients.

5. Licensees shall make full prior disclosures to their employers or clients of potential conflicts of interest or other circumstances which could influence or appear to influence their judgment or the quality of their service.

6. Licensees shall not accept compensation, financial or otherwise, from more than one party for services pertaining to the same project, unless the circumstances are fully disclosed and agreed to by all interested parties.

7. Licensees shall not solicit or accept a professional contract from a governmental body on which a principal or officer of their organization serves as a member. Conversely, licensees serving as members, advisors, or employees of a government body or department, who are the principals or employees of a private concern, shall not participate in decisions with respect to professional services offered or provided by said concern to the governmental body which they serve.

Example

Plan stamping is best defined as the

(A) legal action of signing off on a project you didn't design but are taking full responsibility for

(B) legal action of signing off on a project you didn't design or check but didn't accept money for

(C) illegal action of signing off on a project you didn't design but did check

(D) illegal action of signing off on a project you didn't design or check

Solution

It is legal to stamp (i.e., sign off on) plans that you personally designed and/or checked. It is illegal to stamp plans that you didn't personally design or check, regardless of whether you got paid. It is legal to work as a "plan checker" consultant.

The answer is (D).

Ethics/ Prof. Prac.

Licensee's Obligation to Other Licensees

1. Licensees shall not falsify or permit misrepresentation of their, or their associates', academic or professional qualifications. They shall not misrepresent or exaggerate their degree of responsibility in prior assignments nor the complexity of said assignments. Presentations incident to the solicitation of employment or business shall not misrepresent pertinent facts concerning employers, employees, associates, joint ventures, or past accomplishments.
2. Licensees shall not offer, give, solicit, or receive, either directly or indirectly, any commission, or gift, or other valuable consideration in order to secure work, and shall not make any political contribution with the intent to influence the award of a contract by public authority.
3. Licensees shall not attempt to injure, maliciously or falsely, directly or indirectly, the professional reputation, prospects, practice, or employment of other licensees, nor indiscriminately criticize other licensees' work.

Example

Without your knowledge, an old classmate applies to the company you work for. Knowing that you recently graduated from the same school, the director of engineering shows you the application and resume your friend submitted and asks your opinion. It turns out that your friend has exaggerated his participation in campus organizations, even claiming to have been an officer in an engineering society that you are sure he was never in. On the other hand, you remember him as being a highly intelligent student and believe that he could really help the company. How should you handle the situation?

(A) You should remove yourself from the ethical dilemma by claiming that you don't remember enough about the applicant to make an informed decision.

(B) You should follow your instincts and recommend the applicant. Almost everyone stretches the truth a little in their resumes, and the thing you're really being asked to evaluate is his usefulness to the company. If you mention the resume padding, the company is liable to lose a good prospect.

(C) You should recommend the applicant, but qualify your recommendation by pointing out that you think he may have exaggerated some details on his resume.

(D) You should point out the inconsistencies in the applicant's resume and recommend against hiring him.

Solution

Engineers are ethically obligated to prevent the misrepresentation of their associates' qualifications. You must make your employer aware of the incorrect facts on the resume. On the other hand, if you really believe that the applicant would make a good employee, you should make that recommendation as well. Unless you are making the hiring decision, ethics requires only that you be truthful. If you believe the applicant has merit, you should state so. It is the company's decision to remove or not remove the applicant from consideration because of this transgression.

The answer is (C).

4. ETHICAL CONSIDERATIONS

Ethical Priorities

There are frequently conflicting demands on engineers. While it is impossible to use a single decision-making process to solve every ethical dilemma, it is clear that ethical considerations will force engineers to subjugate their own self-interests. Specifically, the ethics of engineers dealing with others need to be considered in the following order from highest to lowest priority.

- society and the public
- the law
- the engineering profession
- the engineer's client
- the engineer's firm
- other involved engineers
- the engineer personally

Example

To whom/what is a registered engineer's foremost responsibility?

(A) client

(B) employer

(C) state and federal laws

(D) public welfare

Solution

The purpose of engineering registration is to protect the public. This includes protection from harm due to conduct as well as competence. No individual or organization may legitimately direct a registered engineer to harm the public.

The answer is (D).

Ethics/ Prof. Prac.

Dealing with Clients and Employers

The most common ethical guidelines affecting engineers' interactions with their employer (the *client*) can be summarized as follows.[5]

- Engineers should not accept assignments for which they do not have the skill, knowledge, or time to complete.
- Engineers must recognize their own limitations. They should use associates and other experts when the design requirements exceed their abilities.
- The client's interests must be protected. The extent of this protection exceeds normal business relationships and transcends the legal requirements of the engineer-client contract.
- Engineers must not be bound by what the client wants in instances where such desires would be unsuccessful, dishonest, unethical, unhealthy, or unsafe.
- Confidential client information remains the property of the client and must be kept confidential.
- Engineers must avoid conflicts of interest and should inform the client of any business connections or interests that might influence their judgment. Engineers should also avoid the *appearance* of a conflict of interest when such an appearance would be detrimental to the profession, their client, or themselves.
- The engineers' sole source of income for a particular project should be the fee paid by their client. Engineers should not accept compensation in any form from more than one party for the same services.
- If the client rejects the engineer's recommendations, the engineer should fully explain the consequences to the client.
- Engineers must freely and openly admit to the client any errors made.

All courts of law have required an engineer to perform in a manner consistent with normal professional standards. This is not the same as saying an engineer's work must be error-free. If an engineer completes a design, has the design and calculations checked by another competent engineer, and an error is subsequently shown to have been made, the engineer may be held responsible, but will probably not be considered negligent.

Example

You are an engineer in charge of receiving bids for an upcoming project. One of the contractors bidding the job is your former employer. The former employer laid you off in a move to cut costs. Which of the following should you do?

I. say nothing

II. inform your present employer of the situation

III. remain objective when reviewing the bids

(A) II only

(B) I and II

(C) I and III

(D) II and III

Solution

Registrants should remain objective at all times and should notify their employers of conflicts of interest or situations that could influence the registrants' ability to make objective decisions.

The answer is (D).

Dealing with Suppliers

Engineers routinely deal with manufacturers, contractors, and vendors (*suppliers*). In this regard, engineers have great responsibility and influence. Such a relationship requires that engineers deal justly with both clients and suppliers.

An engineer will often have an interest in maintaining good relationships with suppliers since this often leads to future work. Nevertheless, relationships with suppliers must remain highly ethical. Suppliers should not be encouraged to feel that they have any special favors coming to them because of a long-standing relationship with the engineer.

The ethical responsibilities relating to suppliers are listed as follows.

- The engineer must not accept or solicit gifts or other valuable considerations from a supplier during, prior to, or after any job. An engineer should not accept discounts, allowances, commissions, or any other indirect compensation from suppliers, contractors, or other engineers in connection with any work or recommendations.
- The engineer must enforce the plans and specifications (i.e., the *contract documents*) but must also interpret the contract documents fairly.
- Plans and specifications developed by the engineer on behalf of the client must be complete, definite, and specific.

[5]These general guidelines contain references to contractors, plans, specifications, and contract documents. This language is common, though not unique, to the situation of an engineer supplying design services to an owner-developer or architect. However, most of the ethical guidelines are general enough to apply to engineers in the industry as well.

- Suppliers should not be required to spend time or furnish materials that are not called for in the plans and contract documents.
- The engineer should not unduly delay the performance of suppliers.

Example

In dealing with suppliers, an engineer may

(A) unduly delay vendor performance if the client agrees

(B) spend personal time outside of the contract to ensure adequate performance

(C) prepare plans containing ambiguous design-build references as cost-saving measures

(D) enforce plans and specifications to the letter, without regard to fairness

Solution

An engineer not only may, but is required to, ensure performance consistent with plans and specifications. If a job is intentionally or unintentionally underbid, the engineer will have to use personal time to complete the project.

The answer is (B).

Dealing with Other Engineers

Engineers should try to protect the engineering profession as a whole, to strengthen it, and to enhance its public stature. The following ethical guidelines apply.

- An engineer should not attempt to maliciously injure the professional reputation, business practice, or employment position of another engineer. However, if there is proof that another engineer has acted unethically or illegally, the engineer should advise the proper authority.
- An engineer should not review someone else's work while the other engineer is still employed unless the other engineer is made aware of the review.
- An engineer should not try to replace another engineer once the other engineer has received employment.
- An engineer should not use the advantages of a salaried position to compete unfairly (i.e., moonlight) with other engineers who have to charge more for the same consulting services.
- Subject to legal and proprietary restraints, an engineer should freely report, publish, and distribute information that would be useful to other engineers.

Dealing with (and Affecting) the Public

In regard to the social consequences of engineering, the relationship between an engineer and the public is essentially straightforward. Responsibilities to the public demand that the engineer place service to humankind above personal gain. Furthermore, proper ethical behavior requires that an engineer avoid association with projects that are contrary to public health and welfare or that are of questionable legal character.

- Engineers must consider the safety, health, and welfare of the public in all work performed.
- Engineers must uphold the honor and dignity of their profession by refraining from self-laudatory advertising, by explaining (when required) their work to the public, and by expressing opinions only in areas of their knowledge.
- When engineers issue a public statement, they must clearly indicate if the statement is being made on anyone's behalf (i.e., if anyone is benefitting from their position).
- Engineers must keep their skills at a state-of-the-art level.
- Engineers should develop public knowledge and appreciation of the engineering profession and its achievements.
- Engineers must notify the proper authorities when decisions adversely affecting public safety and welfare are made (a practice known as *whistle-blowing*).

Example

Whistle-blowing is best described as calling public attention to

(A) your own previous unethical behavior

(B) unethical behavior of employees under your control

(C) secret illegal behavior by your employer

(D) unethical or illegal behavior in a government agency you are monitoring as a private individual

Solution

"Whistle-blowing" is calling public attention to illegal actions taken in the past or being taken currently by your employer. Whistle-blowing jeopardizes your own good standing with your employer.

The answer is (C).

Competitive Bidding

The ethical guidelines for dealing with other engineers presented here and in more detailed codes of ethics no longer include a prohibition on *competitive bidding.* Until 1971, most codes of ethics for engineers considered competitive bidding detrimental to public welfare, since cost cutting normally results in a lower quality design. However, in a 1971 case against the National Society of Professional Engineers that went all the way to the U.S. Supreme Court, the prohibition against competitive bidding was determined to be a violation of the Sherman Antitrust Act (i.e., it was an unreasonable restraint of trade).

The opinion of the Supreme Court does not *require* competitive bidding—it merely forbids a prohibition against competitive bidding in NSPE's code of ethics. The following points must be considered.

- Engineers and design firms may individually continue to refuse to bid competitively on engineering services.
- Clients are not required to seek competitive bids for design services.
- Federal, state, and local statutes governing the procedures for procuring engineering design services, even those statutes that prohibit competitive bidding, are not affected.
- Any prohibitions against competitive bidding in individual state engineering registration laws remain unaffected.
- Engineers and their societies may actively and aggressively lobby for legislation that would prohibit competitive bidding for design services by public agencies.

Example

Complete the sentence: "The U.S. Department of Justice's successful action in the 1970s against engineering codes of ethics that formally prohibited competitive bidding was based on the premise that

(A) competitive bidding allowed minority firms to participate."

(B) competitive bidding was required by many government contracts."

(C) the prohibitions violated antitrust statutes."

(D) engineering societies did not have the authority to prohibit competitive bidding."

Solution

The U.S. Department of Justice's successful challenge was based on antitrust statutes. Prohibiting competitive bidding was judged to inhibit free competition among design firms.

The answer is (C).

63 Licensure

1. ABOUT LICENSING

Engineering licensing (also known as *engineering registration*) in the United States is an examination process by which a state's *board of engineering licensing* (typically referred to as the "engineers' board" or "board of registration") determines and certifies that an engineer has achieved a minimum level of competence.[1] This process is intended to protect the public by preventing unqualified individuals from offering engineering services.

Most engineers in the United States do not need to be licensed.[2] In particular, most engineers who work for companies that design and manufacture products are exempt from the licensing requirement. This is known as the *industrial exemption*, something that is built into the laws of most states.[3]

Nevertheless, there are many good reasons to become a licensed engineer. For example, you cannot offer consulting engineering services in any state unless you are licensed in that state. Even within a product-oriented corporation, you may find that employment, advancement, and managerial positions are limited to licensed engineers.

Once you have met the licensing requirements, you will be allowed to use the titles *Professional Engineer* (PE), *Structural Engineer* (SE), *Registered Engineer* (RE), and/or *Consulting Engineer* (CE) as permitted by your state.

Although the licensing process is similar in each of the 50 states, each has its own licensing law. Unless you offer consulting engineering services in more than one state, however, you will not need to be licensed in the other states.

2. THE U.S. LICENSING PROCEDURE

The licensing procedure is similar in all states. You will take two examinations. The full process requires you to complete two applications, one for each of the two examinations. The first examination is the *Fundamentals of Engineering* (FE) *examination*, formerly known (and still commonly referred to) as the *Engineer-In-Training* (EIT) *examination*.[4] This examination is designed for students who are close to finishing or have recently finished an undergraduate engineering degree. Seven versions of the exam are offered: chemical, civil, electrical and computer, environmental, industrial, mechanical, and other disciplines. Examinees are encouraged to take the module that best corresponds to their undergraduate degree. In addition to the discipline-specific topics, each exam covers subjects that are fundamental to the engineering profession, such as mathematics, probability and statistics, ethics, and professional practice.

The second examination is the *Professional Engineering* (PE) *examination*, also known as the *Principles and Practices* (P&P) *examination*. This examination tests your ability to practice competently in a particular engineering discipline. It is designed for engineers who have gained at least four years' post-college work experience in their chosen engineering discipline.

The actual details of licensing qualifications, experience requirements, minimum education levels, fees, and examination schedules vary from state to state. Contact your state's licensing board for more information. You will find contact information (websites, telephone numbers, email addresses, etc.) for all U.S. state and territorial boards of registration at **ppi2pass.com/stateboards**.

[1]Licensing of engineers is not unique to the United States. However, the practice of requiring a degreed engineer to take an examination is not common in other countries. Licensing in many countries requires a degree and may also require experience, references, and demonstrated knowledge of ethics and law, but no technical examination.
[2]Less than one-third of the degreed engineers in the United States are licensed.
[3]Only one or two states have abolished the industrial exemption. There has always been a lot of "talk" among engineers about abolishing it, but there has been little success in actually doing so. One of the reasons is that manufacturers' lobbies are very strong.
[4]The terms *engineering intern* (EI) and *intern engineer* (IE) have also been used in the past to designate the status of an engineer who has passed the first exam. These uses are rarer but may still be encountered in some states.

3. NATIONAL COUNCIL OF EXAMINERS FOR ENGINEERING AND SURVEYING

The *National Council of Examiners for Engineering and Surveying* (NCEES) in Seneca, South Carolina, writes, publishes, distributes, and scores the national FE and PE examinations.[5] The individual states administer the exams in a uniform, controlled environment as dictated by NCEES.

4. UNIFORM EXAMINATIONS

Although each state has its own licensing law and is, theoretically, free to administer its own exams, none does so for the major disciplines. All states have chosen to use the NCEES exams. The exams from all the states are graded by NCEES. Each state adopts the cut-off passing scores recommended by NCEES. These practices have led to the term *uniform examination*.

5. RECIPROCITY AMONG STATES

With minor exceptions, having a license from one state will not permit you to practice engineering in another state. You must have a professional engineering license from each state in which you work. Most engineers do not work across state lines or in multiple states, but some do. Luckily, it is not too difficult to get a license from every state you work in once you have a license from one of them.

All states use the NCEES examinations. If you take and pass the FE or PE examination in one state, your certificate or license will be honored by all of the other states. Upon proper application, payment of fees, and proof of your license, you will be issued a license by the new state. Although there may be other special requirements imposed by a state, it will not be necessary to retake the FE or PE examinations.[6] The issuance of an engineering license based on another state's licensing is known as *reciprocity* or *comity*.

[5]National Council of Examiners for Engineering and Surveying, 280 Seneca Creek Road, Seneca, SC 29678, (800) 250-3196, ncees.org.
[6]For example, California requires all civil engineering applicants to pass special examinations in seismic design and surveying in addition to their regular eight-hour PE exams. Licensed engineers from other states only have to pass these two special exams. They do not need to retake the PE exam.

Index

B

C

INDEX - C

D

E

INDEX - D

F

INDEX - G

INDEX - H

J

K

L

N

INDEX - M

O

P

INDEX - P

INDEX - P

Q

INDEX - R

INDEX - S

T

U

INDEX - T

Z